Gaither's Dictionary of Scientific Quotations

Volume II

Gaither's Dictionary of Scientific Quotations

A Collection of Quotations Pertaining to Archaeology, Architecture, Astronomy, Biology, Botany, Chemistry, Cosmology, Darwinism, Death, Engineering, Geology, Life, Mathematics, Medicine, Nature, Nursing, Paleontology, Philosophy, Physics, Probability, Science, Statistics, Technology, Theory, Universe, and Zoology

Carl C. Gaither

BA (Psychology), MA (Psychology), MA (Criminal Justice), MS (Mathematical Statistics)

and

Alma E. Cavazos-Gaither

BA (Spanish)

Volume II
Medicine and Art – Zoology

 Springer

Carl C. Gaither
502 Weiss Drive
Killeen, Texas 76542

Alma E. Cavazos-Gaither
502 Weiss Drive
Killeen, Texas 76542

ISBN: 978-0-387-49575-0   e-ISBN: 978-0-387-49577-4

Library of Congress Control Number: 2007938494

Printed on acid-free paper.

9 8 7 6 5 4 3 2 1

springer.com

This book is dedicated to
Timothy M. Donovan, Jr.
Aubree D. Moore-Woorley
Liliana Noemi Lopez
Annette Koth
James and Sharon Smith
Alice Pomeroy, USA (Retired)
Laura Rodriguez, USNR
Margaret Evans

And to the memory of
Rosa Cervantes (1952–1997)
Clifford C. Gaither, LTC USAF (Retired) (1917–2000)
Pearl Gaither, RN (1917–2002)
Maurice Moore II (1983–2004)
Magdalena Cavazos (1923–2004)
Pedro T. Villaneuva (1925–2006)

Preface

In putting before you, the reader, this collection of 18,000 quotations it seems fitting to discuss how a book such as this came about. In 1995 I told a librarian friend that I was deeply frustrated in my attempts to find quotations on statistics. I told her that although there were a few books with some quotations available, it was quite clear that each author had very different opinions of how to approach the subject. For example, both Alan Mackay and Isaac Asimov wrote books of science quotations that were organized thematically, but in them the quotations were often misstated and the documentation sparse or nonexistent. The books were, however, the state of the art for that time. Another example, Maurice B. Strauss's book *Familiar Medical Quotations*, provided good documentation but, understandably, the quantity of quotations pertaining to science as compared to medicine was limited. As I explained the failings of the extant literature, my friend looked up from her desk and quietly asked, "Well, why don't you compile one?"

I took this idea to my wife, who agreed to work with me on this task. Over a ten-year period we wrote a series of books that contained quotations from several fields of science. These books came to be known as the Speaking Series (Institute of Physics Publishing, Bristol, UK) and were written, like the current revised and greatly expanded compendium, for a broad audience of scientists as well as lay people like ourselves who do not claim expertise in the many scientific fields.

Science is a dynamic force in virtually every sphere of life. At this the beginning of the twenty-first century, few readers will need to be convinced of the enormous impact of science on art, politics, literature, commerce, education, communications, entertainment, judiciary matters, and—often intensely—on religion, and ethics. It is our opinion that the average reader of this book—whether engineer or technician, architect or artist, doctor or nurse, physicist or astronomer, poet or novelist, mathematician or statistician, teacher or student, atheist or believer—should find a great number of quotations pertaining to his or her individual interest. Furthermore, the juxtaposition of the many views may be thought-provoking.

A dictionary normally consists of an alphabetical arrangement of words and their meanings. In this dictionary instead of words we give an alphabetical arrangement of over 2,000 *thematically* organized catagories pertaining to science. Feedback from our previous books indicated that this format was preferred over an author-arranged selection of quotations. The presentation order of the quotations within each subject theme is alphabetical by author. Other quotations of a particular author can be found in the author index.

Our quotation choices were largely influenced by the availability of books, magazines, journals, and newspapers; in turn, to make it as simple as possible for the reader to obtain our sources, we provide our bibliographic references from what we hope are readily accessible sources. Also, for journal articles we strive to provide the actual page number where the quotation may be found, rather than just the first page of the article in which the quotation appears. Brief biographical information (birth/ death date and occupation) is given when at all possible. We were able to provide this contextual information because we were fortunate to have a publisher who did not deem the cost of including this information excessive.

The reader who needs to research a quotation in greater detail can use bibliographical information to find (1) other relevant data; (2) a fuller quote containing other interesting ideas; and (3) the context in which the quotation was used. Where we could not determine where a quotation was originally written we were obliged to use the quotation from a secondary source, and we list the reference where that has been done. Unfortunately, some very good quotations were bypassed and not included because we could not determine where they originated. As it is, about 100 of such quotations are included and have been credited, regrettably but by necessity, as "Author undetermined," "Source undetermined," or both. Despite unavoidable omissions, we hope that this book will provide a rich resource that allows you, the reader, to find relevant quotations or citations quickly, and will serve to inspire your search of the literature.

This dictionary, founded on the quotations from our nine previous books, contains over 7,000 additional quotations and provides by far the greatest number of scientific quotations that has appeared in any single published form to date. In addition, supporting information, such as source of the quotation and biographical information, are greatly expanded beyond any previously published effort.

Our three objectives in compiling this book were: First, to show the diversity and the richness of the various sciences from a variety of literary genres; second, to demonstrate that people from virtually every settled land and continent have given science a great deal of thought from 2000 BCE and earlier to the present time; and third, to provide a resource of thought-provoking ideas useful to anyone involved in just about any aspect of science or in any of the areas noted above, which are greatly influenced by the sciences.

In our attempt to fulfill these objectives we acted merely as collectors of quotations from many sources and from many areas of science. Here in this vast collection of quotations are the words of great philosophers and thought-influencers of science, past and present. Included are better known and lesser known thinkers of the classic Greek and Roman times, religious leaders, and philosophers from the Renaissance to the present. Many times an individual has spoken or written a statement pertaining to some aspect of science that was destined to live on and have meaning beyond the immediate context in which it was made. We hope you enjoy a pleasant and stimulating journey through the forest of ideas of scientists, laymen, politicians, novelists, playwrights, and poets about the human search for and attainment of scientific knowledge.

Max Delbrück, a physicist turned biologist, said in his Noble lecture "A Physicist's Renewed Look at Biology: Twenty Years Later" that "the books of the great scientists are gathering dust on the shelves of learned libraries." Somewhere else we read: "…often we rake in the litter of the printing press whilst a crown of gold and rubies is offered us in vain." Unfortunately, these "gems"— these ideas — are often lost to us before they have time to become established in the collective memory of readers. It has been our concern that much of this wit and wisdom is read once and returned to the library shelf to be heard of no more. It seemed that these ideas, hidden within obscure chapters of books, both fiction and nonfiction, or on pages between covers of long forgotten articles in journals, should once again see the light of day. Apart from the practical day-to-day use of doing so, it is valuable that a new generation see lost or forgotten quotable maxims, proverbs, aphorisms, epigrams, jokes, poetry, songs, and quotations so the young may appreciate their charm and interest.

We extend our thanks to the many publishers and authors for their kind permission to use copyrighted material from their works. For any inadvertent violation of copyright we beg forgiveness.

We would especially like to thank David Packer of Springer for his editorial guidance, and Kathleen McKenzie for her copyediting and her many valuable comments and suggestions, and for her help with fact checking.

We also wish to thank the following libraries for allowing us the use of their collections:

The Perry–Castañeda Library of the University of Texas, Austin, Texas.
The Physics-Math-Astronomy Library of the University of Texas, Austin, Texas.
The Life Science Library of the University of Texas, Austin, Texas.
The Engineering Library of the University of Texas, Austin, Texas.
The Geology Library of the University of Texas, Austin, Texas.
The Baylor University Library, Waco, Texas.
The Mary Hardin-Baylor Library, Belton, Texas.
The Central Texas College Library, Killeen, Texas.
The University of South West Texas Library, Georgetown, Texas.
The McNeese State University Library, Lake Charles, Louisiana.
The University of Richmond Library, Richmond, Virginia.
The Killeen Public Library, Killeen, Texas.

No claim for completeness is made, for completeness is impossible in a book of this type; nor has any attempt been made to provide balance in the quotations between the needs of the general reader and the specialist. It would have been impossible for us to document each person's favorite scientific quotation, and thus we know that this book will suffer the fate of other literary, artistic, or musical works that attempt a broad overview: Stern critics will find fault with the omission of what they perceive as an important quotation from their respective fields. We must ask these critics to remember that our aim in compiling this book has been to save both great and not-so-great words pertaining to science and to add unmistakable value to that which can be retrieved from the Internet, regardless of the time and effort expended by any who searches there.

Within these works we found surprising and often incredible quotations pertaining to science. Just as certain views about science represented in quotes of years long past are not necessarily those of the authors, certain opinions that are stated therein concerning women, and persons of various nationalities, creeds, and races, are clearly not reasonable in an age when belief in the equality of all people—including a person's inherent capacity to contribute to scientific thought—is a shared ideal. Steven Skiena stated in his book, The Algorithm Design Manual, "It is traditional for the author to magnanimously accept the blame for whatever deficiencies remain. I don't. Any errors, deficiencies, or problems in this book are somebody else's fault, but I would appreciate knowing about them so as to determine who is to blame." While we are in sympathy with this, we still believe that any errors are our responsibility and we would appreciate having them called to our attention. For our critics we are sure you will be able to suggest improvements.

Carl C. Gaither
Alma E. Cavazos-Gaither
Killeen, Texas
June 30. 2007

Contents

Contents (Volume I)

Contents

MEDICINE AND ART

Dubos, René Jules   1901–82
French-born American microbiologist and environmentalist

Directly or indirectly, the various forms of art reflect the strivings, the struggles, and the sufferings of mankind. The state of health and the ills of a society are recorded not only in the writings of its physicians and scholars but also in the themes and moods of its artists and poets.
Mirage of Health
Chapter VII (p. 215)
Harper & Brothers Publishers. New York, New York, USA. 1959

MEMORY

Osler, Sir William   1849–1919
Canadian physician and professor of medicine

Memory plays strange pranks with facts. The rocks and fissures and gullies of the mountain-side melt quickly into the smooth, blue outlines of the distant panorama. Viewed through the perspective of memory, an unrecorded observation, the vital details long since lost, easily changes its countenance and sinks obediently into the frame fashioned by the fancy of the moment.
Sir William Osler, Bart.
Osler, The Teacher (p. 51)
The Johns Hopkins Press. Baltimore, Maryland, USA. 1920

Thorne, Kip S.   1940–
American theoretical physicist

Memories are fallible; different people, experiencing the same events, may interpret and remember them in very different ways.
Black Holes and Time Warps: Einstein's Outrageous Legacy
Preface (p. 19)
W.W. Norton & Company, Inc. New York, New York, USA. 1994

MENSTRUATION

Butler, Brett
No biographical data available

I would like it if men had to partake in the same hormonal cycles to which we're subjected monthly. Maybe that's why men declare war — because they have a need to bleed on a regular basis.
In Roz Warren
Glibquips (p. 107)
Crossing Press, Freedom, California. USA. 1994

Crimmins, Cathy
Humorist

A period is just the beginning of a lifelong sentence.
In Roz Warren
Glibquips (p. 107)
Crossing Press. Freedom, California. USA. 1994

Harrison, R. J.
No biographical data available
Montagna, William
Dermatological researcher

The specific purpose of menstruation is obscure. Not least of its imponderables is why it should occur only in certain primates. It would seem to be a waste of tissue and essential substances such as iron. Several hundred milliliters of blood are lost each month and this is obviously a drain on a woman's reserves and a constant call on her blood-forming bone marrow. Apart from its social disadvantages and discomfort, it often leads to tiredness, bad temper and anemia. From a biological point of view, menstruation should not occur at all!
Man (p. 326)
Appleton-Century-Crofts. New York, New York, USA. 1973

Paglia, Camille   1947–
American social critic, intellect and writer

It is not menstrual blood per se which disturbs the imagination — unstaunchable as that red flood may be — but rather the albumen in the blood, the uterine sheds, placental jellyfish of the female sea. This is the chthonian matrix from which we rose. We have an evolutionary revulsion from slime, our site of biological origins. Every month, it is woman's fate to face the abyss of time and being, the abyss which is herself.
Sexual Personae. Art and Decadence from Nefertiti to Emily Dickinson
Yale University Press. New Haven, Connecticut, USA. 1990

Pliny (C. Plinius Secundus)   23–79
Roman savant and author

But nothing could easily be found that is more remarkable than the monthly flux or women. Contact with it turns new wine sour, crops touched by it become barren, grafts die, seeds in gardens are dried up, the fruit of trees falls off, the bright surface of mirrors in which it is merely reflected is dimmed, the edge of steel and the gleam of ivory are dulled, hives of bees die, even bronze and iron are at once seized by rust…
Natural History
Volume II, Book VII, sec 64
Harvard University Press. Cambridge, Massachusetts, USA. 1947

METAL

Baker, Russell   1925–
American writer and journalist

So there he is at last. Man on the moon. The poor magnificent bungler. He can't even get to the office without undergoing the agonies of the damned, but give him a little metal, a few chemicals, some wire and $20 or $30 billion dollars and, vroom!

Resource Recovery Act of 1969 (p. 1231)
Hearings, Ninety-first Congress
United States Congress. Senate. Committee on Public Works. Subcommittee on Air and Water Pollutions
US Government Printing Office. Washington, D.C. 1969

Byron, George Gordon, 6th Baron Byron   1788–1824
English Romantic poet and satirist

I think it may be of Corinthian brass
Which was a mixture of all metals, but
The brazen uppermost.
The Complete Poetical Works of Byron
Don Juan
Canto VI, 56–58
Houghton Mifflin Company. Boston, Massachusetts, USA. 1933

Flaubert, Gustave   1821–90
French novelist

Bronze. Metal of the classic centuries.
Dictionary of Accepted Ideas
M. Reinhardt. London, England. 1954

Henry, William
No biographical data available

The metals are not presented immediately to the hand of man, like the objects of the animal and vegetable kingdoms, but, they are, for the most part, buried in darkness, in the bowels of the earth, where they are so much disguised, by a combination of mixture with other substances, that they often appear entirely unlike themselves.
The Elements of Experimental Chemistry (Volume 2)
Notes (p. 389)
Thomas & Andres. Boston, Massachusetts, USA. 1814

Lavoisier, Antoine Laurent   1743–94
French chemist

One will not deny me, I trust, all the theory of oxidation and combustion; the analysis and decomposition of air by metals and combustible bodies …
Mémoires de Chimie Volume 2, (p. 87)
Dupont. Paris, France. 1803

Rawlings, Majorie Kinnan   1896–1953
American writer

Cast iron is so superior for cooking utensils to our modern aluminum that I not only cannot grieve for the pioneer hardship of cooking in iron over the hearth, but shall retire if necessary to the back yard with my two Dutch ovens, turning over all my aluminum cookers for airplanes with a secret delight.
Cross Creek
Charles Scribner's Sons. New York, New York, USA. 1942

METAPHOR

Calvin, William H.   1939–
Theoretical neurophysiologist

Kant said that our metaphors comprise the conceptual spectacles through which we view the world.… If we are to have meaningful, connected experiences; ones that we can comprehend and reason about; we must be able to discern patterns to our actions, perceptions, and conceptions. Underlying our vast network of interrelated literal meanings (all of those words about objects and actions) are those imaginative structures of understanding such as schema and metaphor, such as the mental imagery that allows us to extrapolate a path, or zoom in on one part of the whole, or zoom out until the trees merge into a forest.
The Cerebral Code (pp. 159–160)
The MIT Press. 1996

Capra, Fritjof   1939–
Austrian-born American physicist

Gradually, physicists began to realize that nature, at the atomic level, does not appear as a mechanical universe composed of fundamental building blocks, but rather as a network of relations, and that, ultimately, there are no parts at all in this interconnected web. Whatever we call a part is merely a pattern that has some stability and therefore captures our attention.
The Tao of Physics: An Exploration of the Parallels Between Modern Physics and Eastern Mysticism (p. 329)
Shambhala. Boston, Massachusetts, USA. 1991

Cole, K. C.   1942–
American science writer

So much of science consists of things we can never see: light "waves" and charged "particles"; magnetic "fields" and gravitational "forces"; quantum "jumps" and electron "orbits." In fact, none of these phenomena is literally what we say it is. Light waves do not undulate through empty space in the same way that water waves ripple over a still pond; a field is only a mathematical description of the strength and direction of a force; an atom does not literally jump from one quantum state to another, and electrons do not really travel around the atomic nucleus in orbits. The words we use are merely metaphors.
On Imagining the Unseeable
Discover Magazine, December 1982 (p. 70)

Davy, Sir Humphry   1778–1829
English chemist

…the tropes and metaphors of the speaker were like the brilliant wild flowers in a field of corn, very pretty, but which did very much hurt to the corn.
Consolations in Travel, or the Last Days of a Philosopher
Dialogue V (p. 253)
J. Murray. London, England. 1830

The works of scientific men are like the atoms of gold, of sapphire and diamonds, that exist in the mountain; they form no perceptible part of the mass of the mountain; they are neglected and unknown when it is entire; they are covered with vegetable mould, and by forests. But when time has sapped its foundation — when its fragments are scattered abroad by the elements, and its decayed materials carried down the rivers, then they glitter, and are found; then their immortality is known, and they are employed to ornament the diadems of emperors and the scepters of kings.

The Collected Works of Sir Humphry Davy (Volume 1)
Memories of the Life of Sir Humphry Davy
Chapter IV (p. 218)
London, England. 1839–1849

Fulford, Robert
No biographical data available

Metaphor, the life of language, can be the death of meaning. It should be used in moderation, like vodka. Writers drunk on metaphor can forget they are conveying information and ideas.

Globe & Mail (*Toronto*), December 4, 1996

Goleman, Daniel   1946–
American writer and psychologist

The logic of the emotional mind is associative; it takes elements that symbolize a reality, or trigger a memory of it, to be the same as that reality. That is why similes, metaphors and images speak directly to the emotional mind.… If the emotional mind follows this logic and its rules, with one element standing for another, things need not necessarily be defined by their objective identity: what matters is how they are perceived; things are as they seem.… Indeed, in emotional life, identities can be like a hologram in the sense that a single part evokes the whole.

Emotional Intelligence (p. 294)
Bloomsbury. London, England. 1996

Goodwin, Brian Carey   1931–
Biologist

The point…is not to conclude that there is something wrong with Darwin's theory because it is clearly linked to some very powerful cultural myths and metaphors. All theories have metaphorical dimensions which I regard as not only inevitable but also extremely important. For it is these dimensions that give depth and meaning to scientific ideas, that add to their persuasiveness, and colour the way we see reality.

How the Leopard Changed Its Spots: The Evolution of Complexity (p. 32)
Phoenix. London, England. 1994

Gould, Stephen Jay   1941–2002
American paleontologist, evolutionary biologist, and historian of science

If we must deal in metaphors [when discussing evolution], I prefer a very broad, low and uniform slope. Water drops randomly at the top and usually dries before flowing anywhere. Occasionally, it works its way downslope

and carves a valley to channel future flows. The myriad valleys could have arisen anywhere on the landscape. The current positions are quite accidental. If we could repeat the experiment, we might obtain no valleys at all, or a completely different system. Yet we now stand at the shore line contemplating the fine spacing of valleys and their even contact with the sea. How easy it is to be misled and to assume that no other landscape could possibly have arisen.

The Panda's Thumb: More Reflections in Natural History
Chapter 12 (p. 140)
W.W. Norton & Company, Inc. New York, New York, USA. 1980

Harré, Rom
No biographical data available

Metaphor and simile are the characteristic tropes of scientific thought, not formal validity of argument.

Varieties of Realism
Part I (p. 7)
Basil Blackwell. Oxford, England. 1986

Klarreich, E.
No biographical data available

Berry isn't speaking in metaphors. I've tried to play this music by putting a few thousand primes into my computer, he says but it's just a horrible cacophony. You'd actually need billions or trillions — someone with a more powerful machine should do it.

Prime Time
New Scientist, 11/11/00

McLuhan, Marshall   1911–80
Canadian educator, philosopher, and scholar
McLuhan, Eric
No biographical data available

…all words are metaphor…

The Laws of Media: The New Science
Chapter 3 (p. 120)
University of Toronto Press. Toronto, Ontario, Canada. 1988

Moore, James R.
No biographical data available

Clever metaphors die hard. Their tenacity of life approaches that of the hardiest micro-organisms. Living relics litter our language, their *raisons d'etre* forever past, ignored if not forgotten, and their present fascination seldom impaired by the confusions they may create.

The Post-Darwinian Controversies: A Study of the Protestant Struggle to Come to Terms with Darwin in Great Britain and America
Chapter I (p. 19)
Cambridge University Press. Cambridge, England. 1979

Poynting, John Henry   1852–1914
English physicist

To take an old but never-worn-out metaphor, the physicist is examining the garment of Nature, learning of how

many, or rather of how few different kinds of thread it is woven, finding how each separate thread enters into the pattern, and seeking from the pattern woven in the past to know the pattern yet to come.
Collected Scientific Papers
Presidential Address
The Mathematical and Physical Section
The British Association (Dover) 1899 (p. 603)
At The University Press. Cambridge, England. 1920

METAPHYSICS

Butler, Samuel   1612–80
English novelist, essayist, and critic

There is no drawing the line between physics and metaphysics. If you examine every day facts at all closely, you are a physicist; but if you press your physics at all home, you become a metaphysician; if you press your metaphysics at all home, you are in a fog.
The Note-Books of Samuel Butler (Volume 1)
1874–1883 (p. 259)
University Press of America, Inc. Lanham, Maryland, USA. 1984

Darwin, Charles Robert   1809–82
English naturalist

Origin of man now proved. — Metaphysics must flourish. — He who understands [the] baboon would do more towards metaphysics than Locke.
M Notebook
#84, 16 August 1838

Douglas, James   1753–1819
No biographical data available

Metaphysical truths can only be established by producing effects from corresponding causes; and though we may confront such demonstrative evidence with the immutable laws of mathematical decision, we must be sensible that there will still remain some pretense for doubt; thus the basis of that knowledge, which on these principles we have been long labouring to accomplish, will become an endless toil, an endless force for controversy: and having the passions and the prejudices of mankind to combat, which mathematical certainty can alone effectually suppress, we must content ourselves only with making converts of those who have minds sufficiently expansive without the shackles of Euclid, and the vanity of displaying their own learning and pedantry.
A Dissertation on the Antiquity of the Earth
Preface (pp. i-ii)
Printed at the Logographic Press. London, England. 1785

von Mises, Richard   1883–1953
Austrian-born American mathematician

There is no field that will always remain the special province of metaphysics and into which scientific research can never carry any light; there are no "eternally unexplorable" areas.
Positivism: A Study in Human Understanding
Chapter 21 (p. 273)
Harvard University Press. Cambridge, Massachusetts, USA. 1951

METEOR

Butler, Samuel   1612–80
English novelist, essayist, and critic

This hairy meteor did denounce
The fall of Scepters and of Crowns.
The Poetical Works of Samuel Butler (Volume 1)
First Part, Canto I, l. 245–246
Bell & Daldy. London, England. 1835

Caithness, James Balharrie
No biographical data available

Wonderful, shimmering trail of light,
Falling from whence on high!
Flooding the world in thy moment's flight
With the sense of a mystery!
Softly thy radiance works a spell,
Night is enhanced, as a note may swell
From a simple melody.
Pastime Poems
The Meteor
E. Macdonald. London, England. 1924

Darwin, Erasmus   1731–1802
English physician and poet

Ethereal Powers! you chase the shooting stars,
Or yoke the vollied lightnings to your cars.
The Botanic Garden
Part I, Canto I, II, l. 115
Jones & Company. London, England. 1825

Devaney, James   1890–1976
Australian poet, novelist, journalist and teacher

The coming of this lovely night
Lifted the world's great roof of blue
And bared the awful Infinite —
So grand an hour, so vast a view,
Abashed I stand each night anew:
When out of unimagined deeps
Spectacular you burst upon
The dark, and down the starry steeps
A trail of whitest fire you shone
One breathless moment — and were gone.
Where the Wind Goes
To a Falling Star
Angus & Robertson. Sydney, Australia. 1939

Dodd, Robert   1936–
No biographical data available

It is much too early to tell whether the idea of periodic impacts by extraterrestrial objects will blossom into a still grander view of the Earth's relation to the other members of the Sun's family or will wither before a fiery blast of new data, but it shows that the romance between geology and planetary astronomy that began with the manned space program is far from over.

Thunderstones and Shooting Stars
Chapter 11 (p. 186)
Harvard University Press. Cambridge, Massachusetts, USA. 1986

Dorman, Imogen
No biographical data available

Down thru the cold blue depths you've gone
A wanderer lone
Bearing a message written on
Metallic stone.
Cast from the planet of your birth
Thru cold you've flown,
Cold such as mortals on the earth
Have never known.
Heat you have found in wintry skies
Earth's atmosphere
Set you aflame, to many eyes
An omen drear.
Lucky the ones who, searching, read
Your message true,
Wanderer of the lightning speed
Down from the blue.

The Meteor
Popular Astronomy, Volume 38, Number 3, March 1930 (p. 133)

Dryden, John   1631–1700
English poet, dramatist, and literary critic

And oft, before tempestuous winds arise,
The seeming stars fall headlong from the skies,
And shooting through the darkness, gild the night
With sweeping flories and long trails of light.

The Poetical Works of Dryden
Virgil's Georgics, Book I, l. 501–504
The Riverside Press. Cambridge, Massachusetts, USA. 1949

Eddington, Sir Arthur Stanley   1882–1944
English astronomer, physicist, and mathematician

I would rather believe in ghosts than in hyperbolic meteors.

In David H. Levy
The Man Who Sold the Milky Way (p. 8)
University of Arizona Press. Tucson, Arizona. 1993

Frost, Robert   1874–1963
American poet

Have I not walked without an upward look
It was a risk I had to take — and took.

Complete Poems of Robert Frost
Bravado
Henry Holt & Company. New York, New York, USA. 1949

Did you stay up last night (the Magi did)
To see the star shower known as Leonid
That once a year by hand or apparatus

Is so mysteriously pelted at us?

Complete Poems of Robert Frost
A Loose Mountain (Telescopic)
Henry Holt & Company. New York, New York, USA. 1949

Hoffman, Jeffrey   1944–
American astronaut

Suddenly I saw a meteor go by underneath me. A moment later I found myself thinking, That can't be a meteor. Meteors burn up in the atmosphere above us; this was below us. Then, of course, the realization hit me [I was in space].

In Kevin W. Kelley
The Home Planet
With Plate 10
Addison-Wesley Publishing Company. Reading, Massachusetts, USA. 1988

Holmes, Charles N.
No biographical data available

Across the darkened dome of night
Where sun-kings reign till break of dawn,
A shooting star darts fast and bright,
Then like a spectral light is gone;
It fades from sight, and leaves behind
No more a trace than passing wind.

The Shooting Star
Source undetermined

Jeffers, Robinson   1887–1962
American poet

It was like the glittering night last October
When the earth swam through a comet's tail, and fiery serpents
Filled half of heaven.

In Tim Hunt (ed.)
The Collected Poetry of Robinson Jeffers (Volume 3)
The Double Axe: The Inhumanist (p. 283)
Stanford University Press. Stanford, California. USA. 1988

Jefferson, Thomas   1743–1826
3rd president of the United States

We certainly are not to deny whatever we cannot account for. A thousand phenomena present themselves daily which we cannot explain, but where facts are suggested, bearing no analogy with the laws of nature as yet known to us, their verity needs proof proportioned to their difficulty. A cautious mind will weigh the opposition of the phenomenon to everything hitherto observed, the strength of the testimony by which it is supported, and the error and misconceptions to which even our senses are liable. It may be very difficult to explain how the stone you possess came into the position in which it was found. But is it easier to explain how it got into the clouds from whence it is supposed to have fallen? The actual fact however is the thing to be established.

In Andrew A. Lipscomb (ed.)
The Writings of Thomas Jefferson

Volume 11 (p. 440)
Thomas Jefferson Memorial Association. Washington, D.C. 1905

I could more easily believe that two Yankee professors would lie than that stones would fall from the heaven.
In R.V. Jones
The Natural Philosophy of Flying Saucers
Physics Bulletin, Volume 19, 1968 (p. 225)

London, Jack   1876–16
American author

I would rather be a meteor, every atom of me in magnificent glow, than a sleepy and permanent planet.
StarDate, May/June 1955 (p. 3)

Martin, Florence Holcomb
No biographical data available

Slashed by the earth the comets orbit glares
With tiny meteors; each fiery tail
Now into incandescence sparks and flares
In earth's rare upper atmosphere.
The Riddle of the Skies
The Scientific Monthly, Volume LXXV, Number 2, August 1952 (p. 119)

Mitchell, Maria   1818–89
American astronomer and educator

…a meteor seems to come like a messenger from departed spirits.
In Eve Merrian
Growing Up Female in America
Maria Mitchell (p. 81)
Doubleday & Company, Inc. Garden City, New York, USA. 1971

Mooch
Fictional character

You mean that this little pebble's been out there hot-roddin' around the universe?
The Blob
Film (1958)

Plum, David
No biographical data available

Then bear us, O Earth, with our eyes upward gazing,
To the place where the Star-God his fireworks displays;
When countless as snowflakes are meteors blazing
With their red, green and orange and amber-like rays.
Meteors
New York Evening Post, November 20, 1866

Shakespeare, William   1564–1616
English poet, playwright, and actor

And certain stars shot madly from their spheres.
In *Great Books of the Western World* (Volume 26)
The Plays and Sonnets of William Shakespeare (Volume 1)
A Midsummer-Night's Dream
Act II, Scene i, l. 153
Encyclopædia Britannica, Inc. Chicago, Illinois, USA. 1952

Silliman, G. S.
No biographical data available

It seems not in accordance with ascertained science to ascribe mysterious appearances on the earth, or in its atmosphere, to causes preceding from the planets, or spheres, moving in space, independent of the earth and its system…. Is it not more in harmony with the integrity and perfection of His work that this phenomenon [meteorites] should originate in a meteorological process, than that the symmetry of the creation should be violated by a visit to the earth of a lone, foreign intruder from the depths of space?
On the Origin of Aerolites
W.C. Bryant. New York, New York, USA. 1859

Smythe, Daniel   1908–81
American poet

A curve of fire traces the dark
And warns us of a visitor.
It makes an unfamiliar mark
And then is seen no more.
The Meteor
Nature Magazine, Volume 50, Number 9, November 1957 (p. 493)

Teasdale, Sara   1884–1933
American writer and poet

I saw a star slide down the sky,
Blinding the north as it went by,
Too burning and too quick to hold,
Too lovely to be bought or sold,
Good only to make wishes on
And then forever to be gone.
The Collected Poems of Sara Teasdale
The Falling Star (p. 198)
Collier Books. New York, New York, USA. 1966

Tennyson, Alfred (Lord)   1809–92
English poet

Now slides the silent meteor on, and leaves
A shining furrow, as thy thoughts in me.
Alfred Tennyson's Poetical Works
The Princess, VII
Oxford University Press, Inc. London, England. 1953

The Bible

I saw a star that had fallen from heaven to the earth…
The Revised English Bible
Revelation 9:1–2
Oxford University Press, Inc. Oxford, England. 1989

Virgil   70 BCE– 19 BCE
Roman epic, didactic, and idyllic poet

As oft, from heaven unfixed, shoot flying stars,
And trail their locks behind them.
In *Great Books of the Western World* (Volume 13)
The Aeneid
Book V, l. 528–529 (pp. 200–201)
Encyclopædia Britannica, Inc. Chicago, Illinois, USA. 1952

Wells, H. G. (Herbert George)   1866–1946
English novelist, historian, and sociologist

When it struck our earth there was to be a magnificent spectacle, no doubt, for those who were on the right side of our planet to see; but beyond that nothing. It was doubtful whether we were on the right side. The meteor would loom larger and larger in the sky, but with the umbra of our earth eating its heart of brightness out, and at last it would be the whole sky, a sky of luminous green clouds, with a white brightness about the horizon west and east. Then a pause — a pause of not very exactly definite duration — and then, no doubt, a great blaze of shooting stars. They might be of some unwonted colour because of the unknown element that line in the green revealed. For a little while the zenith would spout shooting stars. Some, it was hoped, would reach the earth and be available for analysis.
Seven Famous Novels by H.G. Wells
In the Days of the Comet
Book I, Chapter 5 (p. 774)
Alfred A. Knopf. New York, New York, USA. 1934

METEORITE

Alexander, William
No biographical data available
Street, Arthur
No biographical data available

Meteorites usually consist of an alloy of iron with about 8 per cent of nickel, with a small amount of cobalt. No doubt primitive man, whose local culture was thus by accident raised from the level of the stone age to that of the iron age, thought metallic meteorites were valuable gifts from the gods. Nowadays, however, meteorites are hardly regarded as a useful source of iron. For one thing the delivery service is erratic and the unheralded arrival of a meteorite in one's back garden would be more embarrassing than profitable.
Metals In the Service of Man
Penguin Books Ltd. Harmondsworth, England. 1945

Chlandni, E. F. F.
No biographical data available

If the planets had a beginning, then either they must have formed from pieces of matter in an unconsolidated and chaotic state, which had been dispersed throughout a vast space before gravitational attraction gathered them into large masses; or else new planetary bodies were formed from the fragments of much larger ones that were broken to pieces, either by some external impact or by an internal explosion…. [I]t seems likely that many of these original pieces would not have joined the larger accumulating planets, because they were too far from them or traveling

at excessive velocities, but would have remained independent…continuing their journeys in space until each entered the sphere of attraction of some planet, whereupon it would fall, giving rise to the meteoritic phenomenon.
In John A. Wood
Meteorites and the Origin of Planets
Chapter 5 (p. 77)
McGraw-Hill Book Company, Inc. New York, New York, USA. 1968

Frost, Robert   1874–1963
American poet

Never tell me that not one star of all
That slipped from heaven at night and softly fall
Has been picked up with stones to build a wall.
Complete Poems of Robert Frost
A Star in a Stone-Boat
Henry Holt & Company. New York, New York, USA. 1949

Hamilton, W.
No biographical data available

The outside of every stone that has been found [in the Siennese territory], and has been ascertained to have fallen from the cloud near Sienna, is evidently freshly vitrified, and is black, having every sign of having passed through an extreme heat; when broken, the inside is of a light-gray color mixed with black spots, and some shining particles, which the learned here have decided to be pyrites….
In John A. Wood
Meteorites and the Origin of Planets
Chapter 2 (p. 13)
McGraw-Hill Book Company, Inc. New York, New York, USA. 1968

Kelvin, Lord William Thomson   1824–1907
Scottish engineer, mathematician, and physicist

Because we all confidently believe that there are at present, and have been from time immemorial, many worlds of life besides our own, we must regard it as probable in the highest degree that there are countless, seed-bearing meteoric stones moving about through space. If at the present instant no life existed upon this Earth, one such stone falling upon it might, by what we blindly call natural causes, lead to its becoming covered with vegetation…. The hypothesis that life originated on this Earth through moss-grown fragments from the ruins of another world may seem wild and visionary: all I maintain is that it is not unscientific.
Address of Sir William Thomson, Knt., L.L.D., F.R.S., President
Taylor & Francis. London, England. 1871

Meunier, M. S.
No biographical data available

[I have concluded that] the meteorites are pieces of debris from a disrupted planet. Now just as one can, from exhumed remains of extinct animals, reconstruct the beings of past epochs, so it should be possible by examining

meteorites to reconstruct the celestial body that supplies these fossil vestiges....
In John A. Wood
Meteorites and the Origin of Planets
Chapter 3 (p. 29)
McGraw-Hill Book Company, Inc. New York, New York, USA. 1968

Paneth, F. A.
No biographical data available

As is well known, the most exact way of determining the ages of rocks depends on the regularity of radioactive decay processes. Obviously the same method can be applied to meteorites.... In our present state of ignorance of how they were formed, we must admit the possibility that there may be meteorites substantially older than the oldest strata of the earth.
In John A. Wood
Meteorites and the Origin of Planets
Chapter 4 (p. 55)
McGraw-Hill Book Company, Inc. New York, New York, USA. 1968

Peattie, Donald Culrose   1896–1964
American botanist, naturalist, and author

Of all the astronomical events, the fall of a meteorite is the most unnerving and yet the most reassuring. Reassuring because it proves to us that the depths of space are inhabited by bodies made of the same elements we have here on earth, that, at rock bottom, a man and a star are built of the same stuff.
An Almanac for Moderns
August Eleventh (p. 155)
G.P. Putnam's Sons. New York, New York, USA. 1935

METHOD

Bernard, Claude   1813–78
French physiologist

...good methods can teach us to develop and use to better purpose the faculties with which nature has endowed us, while poor methods may prevent us from turning them to good account. Thus the genius of inventiveness, so precious in the sciences, may be diminished or even smothered by a poor method, while a good method may increase and develop it.
Translated by Henry Copley Greene
An Introduction to the Study of Experimental Medicine
Part One, Chapter II, Section ii (p. 35)
Henry Schuman, Inc. New York, New York, USA. 1927

Billroth, Theodor   1829–84
No biographical data available

The method of research, however, of positing the questions and solving the questions posited, is invariably the same, whether we have before us a blooming rose, a diseased grape-vine, a shining beetle, the spleen of a leopard, a

bird's feather, the intestines of a pig, the brain of a poet or philosopher, a sick poodle, or a hysterical princess.
The Medical Sciences in the German Universities (Part II)
The Descriptive Sciences (p. 53)
The Macmillan Company. New York, New York, USA. 1924

Camus, Albert   1913–60
Algerian-French novelist, essayist, and playwright

When one has no character one has to apply a method.
The Fall (p. 11)
Alfred A. Knopf. New York, New York, USA. 1958

Carroll, Lewis (Charles Dodgson)   1832–98
English writer and mathematician

The method employed I would gladly explain,
While I have it so clear in my head,
If I had but the time and you had but the brain —
But much yet remains to be said.
The Complete Works of Lewis Carroll
The Hunting of the Snark
Fit the Fifth (p. 771)
The Modern Library. New York, New York, USA. 1936

Chargaff, Erwin   1905–2002
Austrian biochemist

The availability of a large number of established methods serves in modern science often as a surrogate of thought.
Heraclitean Fire: Sketches from a Life Before Nature
Part III
Science or an Obsession (p. 170)
Rockefeller University Press. New York, New York, USA. 1978

Cohen, Morris Raphael   1880–1947
American philosopher

...the safety of science depends on there being men who care more for the justice of their methods than for any results obtained by their use.
An Introduction to Logic and Scientific Method
Chapter XX, Section 2 (p. 402)
Harcourt, Brace & Company. New York, New York, USA. 1934

Committee on the Conduct of Science

The fallibility of methods means that there is no cookbook approach to doing science, no formula that can be applied or machine that can be built to generate scientific knowledge.... The skillful application of methods to a challenging problem is one of the great pleasures of science.
On Being a Scientist
The Nature of Scientific Research (p. 6)
National Academy Press. Washington, D.C. 1989

Some methods, such as those governing the design of experiments or the statistical treatment of data, can be written down and studied. But many methods are learned only through personal experience and interactions with other scientists. Some are even harder to describe or

teach. Many of the intangible influences on scientific discovery — curiosity, intuition, creativity — largely defy rational analysis, yet they are often the tools that scientists bring to their work.
On Being a Scientist
The Nature of Scientific Research (p. 6)
National Academy Press. Washington, D.C. 1989

Descartes, René   1596–1650
French philosopher, scientist, and mathematician

Method consists entirely in the order and disposition of the objects towards which our mental vision must be directed if we would find out any truth.
In *Great Books of the Western World* (Volume 31)
Rules For the Direction of the Mind
Rule V (p. 7)
Encyclopædia Britannica, Inc. Chicago, Illinois, USA. 1952

Doyle, Sir Arthur Conan   1859–1930
Scottish writer

You know my method. It is founded upon the observation of trifles.
In William S. Baring-Gould (ed.)
The Annotated Sherlock Holmes (Volume 2)
The Bascombe Valley Mystery (p. 148)
Wings Books. New York, New York, USA. 1967

Pon my word Watson, you are coming along wonderfully. We have really done very well indeed. It is true that you have missed everything of importance, but you have hit upon the method…
In William S. Baring-Gould (ed.)
The Annotated Sherlock Holmes (Volume 1)
A Case of Identity (p. 411)
Wings Books. New York, New York, USA. 1967

Egler, Frank E.   1911–96
American botanist and ecologist

For all their value, the application of a method, alone, is not science, any more than a pile of bricks is architecture. I would sooner trust a good mind without a method than a good method without a mind.
The Way of Science
Methodology of Science (p. 36)
Hafner Publishing Company. New York, New York, USA. 1970

Concepts are games we play with our heads; methods are games we play with our hands, which at times are so handy they can be played without a head.
The Way of Science
Holism (p. 34)
Hafner Publishing Company. New York, New York, USA. 1970

Emerson, Ralph Waldo   1803–82
American lecturer, poet, and essayist

The method of nature: who could ever analyze it?
Ralph Waldo Emerson: Essays and Lectures
Nature: Addresses, and Lectures

The Method of Nature (p. 119)
The Library of America. New York, New York, USA. 1983

Gay-Lussac, Joseph Louis   1778–1850
French chemist and physicist

We are convinced that exactitude in experiments is less the outcome of faithful observation of the divisions of an instrument than of exactitude of method.
In Maurice Crosland
Gay-Lussac: Scientist and bourgeois
Chapter 3 (p. 70)
Cambridge University Press. Cambridge, England. 1978

Greenstein, George   1940–
American astronomer

The scientist immersed in research is more bound up by the methods he employs than by the object of his study.
Frozen Star
Chapter 1 (p. 6)
Freundlich Books. New York, New York, USA. 1983

Hertz, Heinrich   1857–94
German physicist

For the moment I am blundering without precise method. I repeat old experiments in this field and demonstrate others which pass through my head…I hope that, among the hundred remarkable phenomena which I come across, some light will shine from one or another.
In René Taton
Reason and Chance in Scientific Discovery
Chapter III (p. 4)
Philosophical Library. New York, New York, USA. 1957

Hilbert, David   1862–1943
German mathematician

…for he who seeks for methods without having a definite problem in mind seeks for the most part in vain.
Hilbert: Mathematical Problems
Bulletin of the American Mathematical Society, Volume 8, July 1902
(p. 444)

Hubble, Edwin Powell   1889–1953
American astronomer

The methods of science may be described as the discovery of laws, the explanation of laws by theories, and the testing of theories by new observations. A good analogy is that of the jigsaw puzzle, for which the laws are the individual pieces, the theories local patterns suggested by a few pieces, and the tests the completion of these patterns with pieces previously unconsidered.
The Nature of Science and Other Lectures
Part I, The Nature of Science (p. 11)
The Huntington Library, San Marino, California, USA. 1954

Lonergan, Bernard J. F.   1904–84
Canadian philosopher, theologian, and educator

It is in the measure that special methods acknowledge their common core in transcendental method, that norms common to all the sciences will be acknowledged, that a secure basis will be attained for tackling interdisciplinary problems, and that the sciences will be mobilized within a higher unity of vocabulary, thought and orientation, in which they will be able to make their quite significant contribution to the solution of fundamental problems.
Method in Theology
Chapter 1 (p. 23)
Herder & Herder. New York, New York, USA. 1972

Pavlov, Ivan Petrovich   1849–1936
Russian physiologist

With a good method even a rather untalented person can accomplish much.
In Daniel P. Todes
Pavlov's Physiology Factory: Experiment, Interpretation, Laboratory Enterprise
Chapter 3 (p. 101)
The Johns Hopkins University Press. Baltimore, Maryland, USA. 2002

Everything is in the method, in the chances of attaining a steadfast, lasting truth…
In Daniel P. Todes
Pavlov's Physiology Factory: Experiment, Interpretation, Laboratory Enterprise
Chapter 3 (p. 83)
The Johns Hopkins University Press. Baltimore, Maryland, USA. 2002

Pólya, George   1887–1985
Hungarian mathematician

What is the difference between a method and device? A method is a device which you use twice.
How to Solve It: A New Aspect of Mathematical Method
Part III, The Traditional Mathematics Professor (p. 208)
Princeton University Press. Princeton, New Jersey, USA. 1973

My method to overcome a difficulty is to go round it.
How to Solve It: A New Aspect of Mathematical Method
Part III, The traditional mathematics professor (p. 208)
Princeton University Press. Princeton, New Jersey, USA. 1973

Russell, Bertrand Arthur William   1872–1970
English philosopher, logician, and social reformer

In science the man of real genius is the man who invents a new method. The notable discoveries are often made by his successors, who can apply the method with fresh vigor, unimpaired by the previous labour of perfecting it; but the mental caliber of the thought required for their work, however brilliant, is not so great as that required by the first inventor of the method.
Mysticism and Logic and Other Essays
Chapter II, Section II (p. 41)
Longmans, Green & Company. London, England. 1925

Sagan, Carl   1934–96
American astronomer and author

The method of science, as stodgy and grumpy as it may seem, is far more important than the findings of science.
The Demon-Haunted World: Science as a Candle in the Dark
Chapter 1 (p. 22)
Random House, Inc. New York, New York, USA. 1995

Sir Joseph
Fictional character

Method is everything in archaeology, my boy. Why, we always deal with our finds in order.
The Mummy
Film (1940)

von Goethe, Johann Wolfgang   1749–1832
German poet, novelist, playwright, and natural philosopher

Content without method leads to fantasy; method without content to empty sophistry; matter without form to unwieldy erudition, form without matter to hollow speculation.
Scientific Studies (Volume 12)
Chapter VIII (p. 306)
Suhrkamp. New York, New York, USA. 1988

Walker, Kenneth   1882–1966
Physician

To understand the true function of science and to be able to evaluate its theories it will first be necessary to have a very clear idea of the method by which it works.
Meaning and Purpose
Chapter II (p. 16)
Jonathan Cape. London, England. 1944

Wilson, Edwin B.   1879–1964
American statistician

A method is a dangerous thing unless its underlying philosophy is understood, and none [is] more dangerous than the statistical. Our aim should be, with care, to avoid in the main erroneous conclusions. In a mathematical and strictly logical discipline the care is one of technique; but in the natural science and in statistics the care must extend not only over the technique but to the matter of judgment, as is necessarily the case in coming to conclusions upon any problem of real life where the complications are great. Over-attention to technique may actually blind one to the dangers that lurk about on every side — like the gambler who ruins himself with his system carefully elaborated to beat the game. In the long run it is only clear thinking, experienced methods, that win the strongholds of science.
The Statistical Significance of Experimental Data
Science, Volume 58, Number 1493, 10 August 1923 (p. 94)

METRICS

Lederman, Leon   1922–
American high-energy physicist

In the 1990s the United States, not to be left too far behind, is inching toward the metric system.
The God Particle: If the Universe Is the Answer, What Is the Question?
Chapter 4 (p. 108)
Houghton Mifflin Company. Boston, Massachusetts, USA. 1993

MICROBE

Donaldson, T. B.
No biographical data available

He, who fights Microbes Away
Will be an Immune, some fine Day.
An Apropos Alphabet with Immoral Conclusions by an Absent-Minded Beggar in Red & Blue
Letter K
W.S. Sterling & Company. New York, New York, USA. 1900

Dunne, Finley Peter   1867–1936
American journalist and humorist

…mickrobes is a vigitable, an' ivry man is like a conservatory full iv millyons iv these potted plants.
Mr. Dooley's Opinions
Christian Science (p. 5)
Harper. New York, New York, USA. 1906

Gillilan, Strickland   1869–1954
American poet-humorist

Adam Had 'em.
In Herbert V. Prochnow and Herbert V. Prochnow, Jr.
A Treasury of Humorous Quotations: For Speakers, Writers, and Home Reference
#79 (p. 5)
Harper & Row, Publishers. New York, New York, USA. 1969

Huxley, Aldous   1894–1963
English writer and critic

…think of the inexpugnable retreats for microbes prepared by Michelangelo in the curls of Moses' beard!
Time Must Have a Stop
Chapter III (p. 36)
The Sun Dial Press. Garden City, New York, USA. 1944

Muir, John   1838–1914
American naturalist

And surely all God's people, however serious and savage, great or small, like to play. Whales and elephants, dancing, humming gnats, and invisibly small mischievous microbes — all are warm with divine radium and must have lots of fun in them.
My Boyhood and Youth
Chapter V (pp. 149–150)
Houghton Mifflin Company. Boston, Massachusetts, USA. 1913

Nobel, Alfred   1833–96
Swedish chemist, engineer, inventor, and industrialist

The advance in scientific research and its ever widening sphere stirs the hope in us that the microbes, those of the soul as well as of the body, will gradually disappear, and that the only war humanity will wage in the future will be one against these microbes.
Quoted in Selman A. Waksman
Les Prix Nobel. The Nobel Prizes in 1952
Nobel banquet speech for award received in 1952
Nobel Foundation. Stockholm, Sweden. 1953

Osler, Sir William   1849–1919
Canadian physician and professor of medicine

In war the microbe kills more than the bullet.
In Harvey Cushing
The Life of Sir William Osler (Volume 2) (p. 427)
Clarendon Press. Oxford, England. 1925

Waksman, Selman A.   1888–1973
Ukrainian-born American biochemist

With the removal of the danger lurking in infectious diseases and epidemics, society can face a better future, can prepare for a time when other diseases not now subject to therapy will be brought under control. Let us hope that in contributing the antibiotics, the microbes will have done their part to make the world a better place to live in.
Les Prix Nobel. The Nobel Prizes in 1952
Nobel banquet speech for award received in 1952
Nobel Foundation. Stockholm, Sweden. 1953

Wolfe, Humbert   1885–1940
Poet and civil servant

The doctor lives by chicken pox,
by measles, and by mumps.
He keeps a microbe in a box
and cheers him when he jumps.
Cursory Rhymes
Poems Against Doctors II
E. Benn Limited. London, England. 1927

MICROBIOLOGY

Collard, Patrick
No biographical data available

Microbiology, like all the sciences, is founded upon the twin pillars of craft technique and philosophical speculation.
The Development of Microbiology
Chapter 1 (p. 1)
Cambridge University Press. London, England. 1976

MICROCOSM

Forbes, A.
No biographical data available

[A lake] is a little world within itself, a microcosm in which all the elemental forces are at work and the play of life goes on in full, but on a scale so small as to be easily grasped.

In F.E. Clements and V.E. Shelford
Bio-Ecology
Chapter I (p. 14)
John Wiley & Sons, Inc. New York, New York, USA. 1939

MICROPALEONTOLOGY

Lipps, Jere
No biographical data available

Micropaleontology is a strange subject. It is not easily defined, its history is fairly dull, and it seems to focus only on geologic topics. Biologists by and large ignore those organisms that when fossilized become microfossils. Evolutionary biologists disdain them. Paleobiologists snub them, and "micropaleontologists" seem not to know what to do with them as once living animals or plants. Most "micropaleontologists" are not trained in biology, and the literature of micropaleontology is an enormous edifice testifying to that fact. Although not a mere flunky of geology, micropaleontology is nevertheless largely a servant of geology, albeit an extremely powerful one.
What, If Anything, Is Micropaleontology?
Paleobiology, Volume 7, Number 2, 1981

MICROSCOPE

Bajer, Francis J.
No biographical data available

The curiosity of man remains undaunted since the dawn of civilization. It has manifested itself in many ways, not the least of which has been the burning desire to see but a little more or a little more clearly. Surely this is understandable when one realizes that almost all knowledge is first gleaned through visual inspection and observation.
Scanning a Tiny World's Wonders with the Magic of Electron Microscopy *Science Digest*, Volume 83, Number 3, March 1978 (p. 42)

Baker, Henry   1698–1774
English naturalist

When you employ the Microscope, shake off all Prejudice, nor harbor any favorite Opinions; for, if you do, 'tis not unlikely Fancy will betray you into Error, and make you see what you wish to see.
The Microscope Made Easy
Part I, Chapter XV, Cautions in Viewing Objects (p. 62)
Printed for R. Dodsley. London, England. 1743

Dickens, Charles   1812–70
English novelist

Yes, I have a pair of eyes, replied Sam, and that's just it. If they was a pair o' patent double million magnifyin' gas microscopes of hextra power, p'raps I might be able to see through a flight o' stairs and a deal door; but bein' only eyes, you see my wision's limited.
The Posthumous Papers of the Pickwick Club

Chapter XXXIV (p. 415)
Dodd, Mead & Company. New York, New York, USA. 1944

Dickinson, Emily   1830–86
American lyric poet

Faith is a fine invention
For gentlemen who see;
But microscopes are prudent
In an emergency.
The Complete Poems of Emily Dickinson
No. 185 (p. 87)
Little, Brown & Company. Boston, Massachusetts, USA. 1960

Eliot, George (Mary Ann Evans Cross)   1819–80
English novelist

...very close and diligent looking at living creatures, even through the best microscope, will leave room for new and contradictory discoveries.
Felix Holt, The Radical
Chapter XXII (p. 226)
William L. Allison Company. New York, New York, USA. No date

Holmes, Oliver Wendell   1809–94
American physician, poet, and humorist

I was sitting with my microscope,
upon my parlor rug,
With a very heavy quarto and a very lively bug;
The true bug had been organized
with only two antennae,
But the humbug in the copperplate would have them twice as many.
The Complete Poetical Works of Oliver Wendell Holmes
Nux Postcoenatica, Stanza 1
Houghton Mifflin Company. Boston, Massachusetts, USA. 1899

Hooke, Robert   1635–1703
English physicist

...me thinks it seems very probable, that nature has in these passages, as well as in those of Animal bodies, very many appropriated Instruments and contrivances, whereby to bring her designs and end to pass, which 'tis not improbable, but that some diligent Observer, if help'd with better Microscopes, may in time detect.
Micrographia
Observation, XVIII (p. 116)
Printed by Jo. Martyn & Ja. Allestry. London, England. 1665

Hugo, Victor   1802–85
French author, lyric poet, and dramatist

Where the telescope ends, the microscope begins, and which of the two has the grandest sight?
Les Miserables
Volume IV, Book III, Chapter 3 (p. 67)
The Heritage Press. New York, New York, USA. 1938

Lambert, Johann Heinrich   1728–77
Swiss-German mathematician and astronomer

In a grain of sand, in a drop of water, we discover worlds and inhabitants; besides, our best microscopes only shew us the whales and elephants of those worlds; they are still far from reaching the insects.
Translated by James Jacque
The System of the World
Part I, Chapter III (p. 12)
Printed for Vernor & Hood. London, England. 1800

Lichtenberg, Georg Christoph   1742–99
German physicist and satirical writer

Microscopes should be invented for every kind of investigation and, where that is impossible, experiments should be conducted on a large scale. This is the only direct road to new discovery.
In J.P. Stern
Lichtenberg: A Doctrine of Scattered Occasions
Further Excerpts from Lichtenberg's Notebooks (p. 294)
Indiana University Press. Bloomington, Indiana, USA. 1959

Mayo, William J.   1861–1939
American physician
Powers, Henry
No biographical data available

Of all the Inventions none there is Surpasses
the Noble Florentine's Dioptrick Glasses
For what a better, fitter guift Could bee
in this World's Aged Luciosity.
To help our Blindnesse so as to devize
a paire of new & Articicial eyes
By whose augmenting power wee now see more
than all the world Has ever doun Before.
In S. Bradbury
The Microscope Past and Present
In Commendation of ye Microscope (p. v)
Printed for R. Dodsley. London, England. 1743

Selye, Hans   1907–82
Austrian-American endocrinologist

The microscope can see things the naked eye cannot, but the reverse is equally true.
From Dream to Discovery: On Being a Scientist
McGraw-Hill Book Company, Inc. New York, New York, USA. 1950

Strindberg, Johann   1849–1912
Swedish dramatist and novelist

Then, is it reasonable to think that one can see, by looking in a microscope, what is going on in another planet?
Chief Contemporary Dramatists; Twenty Plays from the Recent Drama of England, Ireland, America, Germany, France, Belgium, Norway, Sweden, and Russia
Translated by N. Erichsen
The Father (p. 608)
Houghton Mifflin Company. Boston, Massachusetts, USA. 1915

van Leeuwenhoek, Antony   1632–1723
Dutch biology researcher and microscope developer

I have oft-times been besought, by divers gentlemen, to set down on paper what I have beheld through my newly invented Microscopia: but I have generally declined: first, because I have no style, or pen, wherewith to express my thoughts properly; secondly, because I have not been brought up to languages or arts, but only to business; and in the third place, because I do not gladly suffer contradiction or censure from others.
Observations Communicated to the Publisher in a Dutch Letter of the 9th of October 1676
Philosophical Transactions of the Royal Society of London,
Volume 12, 1677

Whewell, William   1794–1866
English philosopher and historian

…if the discoveries made by the Telescope should excite in any one's mind, difficulties respecting those doctrines of Natural Religion, — the adequacy of the Creator to the support and guardianship of all the animal life which may exist in the universe, — the discoveries of the Microscope may remove such difficulties: but…that train of thought which leads men to dwell upon such difficulties does not seem to be common.
Of the Plurality of Worlds
Chapter IV (p. 30)
John W. Parker & Son. London, England. 1853

Wood, John George   1827–1889
English writer on natural history

…even to those who aspire to no scientific eminence, the microscope is more than an amusing companion, revealing many of the hidden secrets of Nature, and unveiling endless beauties which were heretofore enveloped in the impenetrable obscurity of their own minuteness…a good observer will discover with a common pocket magnifier many a secret of nature which has escaped the notice of a whole array of dilettanti microscopists in spite of all their expensive and accurate instruments.
Common Objects of the Microscope
Chapter I (p. 2)
George Routledge & Sons. London, England. 1861

MIGRATION

O'Neill, Gerard K.   1927–92
American physicist

Every star around us is a favorable target for human migration. You don't have to wait for just those stars that happen to have earthlink planets; they may be very few and far between.
In Pamela Weintraub (ed.)
The Omni Interviews
Cosmic Colonies (p. 297)
Ticknor & Fields. New York, New York, USA. 1984

MILKY WAY

Alighieri, Dante   1265–1321
Italian poet and writer

…distinct with less and greater lights, the Galaxy so whitens between the poles of the world that it makes even the wise to question…
In *Great Books of the Western World* (Volume 21)
The Divine Comedy of Dante Alighieri
Paradise, Canto XIV, l. 97–100
Encyclopædia Britannica, Inc. Chicago, Illinois, USA. 1952

Chaucer, Geoffrey   1343–1400
English poet

See yonder, lo, the Galaxye
Which men clepeth the Milky Wey,
For hit is whyt.
The House of Fame
Book II
Chatto & Windus. London, England. 1908

de Fontenelle, Bernard le Bovier   1657–1757
French author

…you see that whiteness in the sky, which some call the milky-way; can you imagine what that is? Why, it is nothing but infinity of small stars, not to be seen by our eyes, because they are so very little; and they are sown so thick, one by another, that they seem to be one continued whiteness: I wish you had a glass to see this ant's nest of stars…
Conversations on the Plurality of Worlds
The Fifth Evening (pp. 159–160)
Printed for Peter Wilson. Dublin, Ireland. 1761

de Morgan, Augustus   1806–71
English mathematician and logician

I have often had the notion that all the nebula we see, including our own, which we call the Milky Way, may be particles of snuff in the box of a giant of a proportionately larger universe. Of course the minimum time — a million of years or whatever the geologists make it — which our little affair has lasted, is but a very small fraction of a second to the great creature in whose nose we shall all be in a few tens of thousands of millions of millions of millions of years.
A Budget of Paradoxes
Are Atoms Worlds (p. 377)
Longmans, Green & Company. London, England. 1872

Donne, John   1572–1631
English poet and divine

In that glistering circle in the firmament, which we call the Galaxie, the milkie way, there is not one starre of any of the six great magnitudes, which Astronomers proceed upon, belonging to that circle: it is a glorious circle, and possesseth a great part of heaven, and yet is all of so little starres, as have no name, no knowledge taken of them…
Donne's Sermons
Little Stars, Sermon 144 (p. 221)
Clarendon Press. Oxford, England. 1942

Hearn, Lafcadio   1850–1904
Greek-born American writer

In the silence of the transparent night, before the rising of the moon, the charm of the ancient tale sometimes descends upon me out of the scintillant sky, to make me forget the monstrous facts of science and the stupendous horror of space. Then I no longer behold the Milky Way, as that awful Ring of Cosmos, whose hundred million suns are powerless to lighten the abyss, but as the very Amanogwa itself — the river Celestial. I see the thrill of its shining stream, the mists that hover along the verge, and the watergrasses that bend in the winds of autumn. White Orihimé I see at her starry loom and the Ox that grazes on the farther shore — and I know that the falling dew is the spray of the Herdsman's oar.
The Writings of Lafcadio Hearn
Volume VIII, The Romance of the Milky Way (p. 257)
Houghton Mifflin Company. Boston, Massachusetts, USA. 1922

Herschel, Sir John Frederick William   1792–1871
English astronomer and chemist

As we are used to call the appearance of the heavens, where it is surrounded with a bright zone, the Milky Way, it may not be amiss to point out some other very remarkable Nebulae which cannot well be less, but are probably much larger than our own system; and, being also extended, the inhabitants of the planets that attend the stars which compose them must likewise perceive the same phenomena. For which reason they may also be called milky ways by way of distinction.
In Laurence A. Marschall
The Supernova Story
Chapter 2 (p. 34)
Plenum Press. New York, New York, USA. 1988

Joyce, James   1882–1941
Irish-born author

Bloom was pointing out all the stars and the comets in the heavens to Chris Callinan and the jarvey: the great bear and Hercules and the dragon, and the whole jingbang lot. But, by God, I was lost, so to speak, in the milky way.
Ulysses (p. 231)
Random House, Inc. New York, New York, USA. 1946

Kilmer, Joyce   1886–1918
American poet

God be thanked for the Milky Way that runs across the sky.

That's the path that my feet would tread whenever I have to die.

Some folks call it a Silver Sword, and some a Pearly Crown.

But the only thing I think it is, is Main Street, Heaventown.
Main Street and Other Poems
Main Street
George H. Doran Company. New York, New York, USA. 1917

Lambert, Johann Heinrich   1728–77
Swiss-German mathematician and astronomer

I am undecided whether or not the visible Milky Way is but one of countless others all of which form an entire system. Perhaps the light from these infinitely distant galaxies is so faint that we cannot see them.
In Eli Maor
To Infinity and Beyond: A Cultural History of the Infinite (p. 217)
Birkhäuser. Boston, Massachusetts, USA. 1987

Longfellow, Henry Wadsworth   1807–82
American poet

Showed the broad, white road in heaven,
Pathway of the ghosts, the shadows,
Running straight across the heavens,
Crowded with the ghosts, the shadows.
The Poetical Works of Henry Wadsworth Longfellow
The Song of Hiawatha, Hiawatha's Childhood
Houghton Mifflin Company. Boston, Massachusetts, USA. 1883

Milton, John   1608–74
English poet

A broad and ample road, whose dust is Gold,
And pavement Starss, as Starss to thee appear
Seen in the galaxy, that Milkie way…
In *Great Books of the Western World* (Volume 32)
Paradise Lost
Book VII, l. 577–579
Encyclopædia Britannica, Inc. Chicago, Illinois, USA. 1952

Ovid   43 BCE–17 AD
Roman poet

There is a high way, easily seen when the sky is clear. 'Tis called the Milky Way, famed for its shining whiteness.
Translated by Frank Justus Miller
Metamorphoses (Volume 1)
Book I, l. 168 (p. 15)
William Heinemann. London, England. 1916

Pasternak, Boris   1890–1960
Russian poet and novelist

With an awful, dreadful list
Towards other galaxies unknown
Ponderously turns the Milky Way…
Poems
Night
University of Michigan Press. Ann Arbor, Michigan, USA. 1959

And there, with frightful listing
Through emptiness, away
Through unknown solar systems

Revolves the Milky Way.
Fifty Poems
Night
George Allen & Unwin Ltd. London, England. 1963

Poincaré, Henri   1854–1912
French mathematician and theoretical astronomer

Consider now the Milky Way; there also we see an innumerable dust; only the grains of this dust are not atoms, they are stars; these grains move also with high velocities; they act at a distance one upon another, but this action is so slight at great distance that their trajectories are straight; and yet, from time to time, two of them may approach near enough to be deviated from their path, like a comet which had passed too near Jupiter. In a world, to the eyes of a giant for whom our suns would be as for us our atoms, the Milky Way would seem only a bubble of gas.
The Foundations of Science
Science and Method, Book IV
Chapter I (p. 524)
The Science Press. New York, New York, USA. 1913

Rich, Adrienne   1929–
American poet

Driving at night I feel the Milky Way
Streaming above me like the graph of a cry.
Leaflets, Poems 1965–1968
Ghazals 7/24/68: ii
W.W. Norton & Company, Inc. New York, New York, USA. 1969

Thoreau, Henry David   1817–62
American essayist, poet, and practical philosopher

This whole earth which we inhabit is but a point in space. How far apart, think you, dwell the two most distant inhabitants of yonder star, the breadth of whose disk cannot be appreciated by our instruments? Why should I feel lonely? is not our planet in the Milky Way?
The Writings of Henry David Thoreau (Volume 2)
Walden
Chapter V (p. 208)
Houghton Mifflin Company. Boston, Massachusetts, USA. 1893

Updike, John   1932–
American novelist, short story writer, and poet

The Milky Way, which used to be thought of as the path by which the souls of the dead traveled to Heaven, is an optical illusion; you could never reach it. Like fog, it would always thin out around you. It's a mist of stars we make by looking the long way through the galaxy…
The Centaur
Chapter I (p. 37)
Alfred A. Knopf. New York, New York, USA. 1995

Wright, Thomas   1711–86
English cosmologist

This is the great Order of Nature which I shall now endeavor to prove, and thereby solve the Phaenomena

of the Via Lactea; and in order thereto, I want nothing to be granted but what may easily be allowed, namely, that the Milky Way is formed of an infinite Number of small Stars.
An Original Theory or New Hypothesis of the Universe
Letter the Seventh (p. 62)
Printed for the Author. London, England. 1750

MIND

Adams, George   1750–95
English instrument maker

The human mind, like a mirror, must be smoothed and polished, freed from false imaginations and perverted notions, before it is fit to receive and reflect the light of truth, and just information.
Lectures on Natural and Experimental Philosophy (Volume 2)
Lecture XIV (p. 101)
Printed by R. Hindmarsh. London, England. 1794

Beveridge, William Ian Beardmore   1908–
Australian zoologist

Elaborate apparatus plays an important part in the science of today, but I sometimes wonder if we are not inclined to forget that the most important instrument in research must always be the mind of man.
The Art of Scientific Investigation
Preface (p. ix)
W.W. Norton & Company, Inc. New York, New York, USA. 1957

It is true that much time and effort is devoted to training and equipping the scientist's mind, but little attention is paid to the technique of making the best use of it.
The Art of Scientific Investigation
Preface (p. iv)
W.W. Norton & Company, Inc. New York, New York, USA. 1957

Blaise, Clarke

Our minds soar with instant connection, but our feet are stuck in temporal boots.
Time Lord
Chapter 1 (p. 19)
Weidenfeld & Nicolson. London, England. 2000

Boas, George   1891–1980
American philosopher

Though the solution of a problem may flash into the mind of the person without his knowing how it arose, nevertheless he has always done a good bit of thinking, puzzling, wondering about it before the flash occurs.
The Inquiring Mind
Chapter XV (p. 397)
Open Court Publishing Company. La Salle, Illinois, USA. 1959

Butler, Samuel   1612–80
English novelist, essayist, and critic

Chandrasekhar, Subrahmanyan   1910–95
Indian-born American astrophysicist

When a supremely great creative mind is kindled, it leaves a blazing trail that remains a beacon for centuries.
Newton and Michelangelo
Current Science, Volume 67, Number 7, 10 October 1994 (p. 499)

Darwin, Charles Robert   1809–82
English naturalist

My mind seems to have become a kind of machine for grinding general laws out of large collections of facts…
In Francis Darwin (ed.)
The Life and Letters of Charles Darwin (Volume 1)
Chapter II, Autobiography (p. 81)
D. Appleton & Company. New York, New York, USA. 1896

But then arises the doubt, can the mind of man, which has, as I full believe, been developed from a mind as low as that possessed by the lowest animals, be trusted when it draws such grand conclusions?
In Francis Darwin (ed.)
The Life and Letters of Charles Darwin (Volume 1)
Chapter VIII (p. 282)
D. Appleton & Company. New York, New York, USA. 1896

Doyle, Sir Arthur Conan   1859–1930
Scottish writer

To a great mind, nothing is little.
In William S. Baring-Gould (ed.)
The Annotated Sherlock Holmes (Volume 1)
A Study in Scarlet, Chapter 6 (p. 187)
Wings Books. New York, New York, USA. 1967

Dyson, Freeman J.   1923–
American physicist and educator

I do not make any clear distinction between mind and God. God is what mind becomes when it is passed beyond the scale of our comprehension. God may be considered to be either a world-soul or a collection of world-souls. We are the chief inlets of God on this planet at the present stage of his development.
Infinite in All Directions
Part One, Chapter Six (p. 119)
HarperCollins Publisher, Inc. New York, New York, USA. 1988

It appears to me that the tendency of mind to infiltrate and control matter is a law of nature. The infiltration of mind into the universe will not be permanently halted by any catastrophe or by any barrier that I can imagine. If our species does not choose to lead the way, others will do so, or may have already done so. If our species is extinguished, others will be wiser or luckier. Mind is patient. Mind has waited for 3 billion years on this planet before composing its first string quartet. It may have to wait for another 3 billion years before it spreads all over the galaxy. I do not expect that it will have to wait so long. But if necessary, it will wait. The universe is like a

fertile soil spread out all around us, ready for the seeds of mind to sprout and grow. Ultimately, late or soon, mind will come into its heritage.
Infinite in All Directions
Part One, Chapter Six (p. 118)
HarperCollins Publisher, Inc. New York, New York, USA. 1988

Eddington, Sir Arthur Stanley   1882–1944
English astronomer, physicist, and mathematician

Scientific theories have blundered no doubt in the past; they blunder no doubt today; yet we cannot doubt that along with the error there come gleams of a truth for which the human mind is impelled to strive.
Science and the Unseen World
Chapter II (p. 22)
The Macmillan Company. New York, New York, USA. 1929

If we are to discern controlling laws of Nature not dictated by the mind it would seem necessary to escape as far as possible from the cut-and-dried framework into which the mind is so ready to force everything that it experiences.
The Nature of the Physical World
Chapter X (p. 210)
The Macmillan Company. New York, New York, USA. 1930

Einstein, Albert   1879–1955
German-born physicist

The human mind is not capable of grasping the Universe. We are like a little child entering a huge library. The walls are covered to the ceiling with books in many different tongues. The child knows that something must have written these books. It does not know who or how. It does not understand the languages in which they are written. But the child notes a definite plan in the arrangement of the books — a mysterious order which it does not comprehend, but only dimly suspects.
In M. Taube
Evolution of Matter and Energy
Chapter 1 (p. 1)
Springer-Verlag. New York, New York, USA. 1985

Feynman, Richard P.   1918–88
American theoretical physicist

This law [the Law of Gravitation] has been called "the greatest generalization achieved by the human mind", and you can guess already from my introduction that I am interested not so much in the human mind as in the marvel of a nature which can obey such an elegant and simple law as this law of gravitation. Therefore our main concentration will not be on how clever we are to have found it all out, but on how clever nature is to pay attention to it.
The Character of Physical Law
Chapter 1 (p. 14)
BBC. London, England. 1965

In this chapter we shall discuss one of the most far-reaching generalizations of the human mind. While we are admiring the human mind, we should take some time off to stand in awe of a nature that could follow with such completeness and generality such an elegantly simple principle as the law of gravitation.
Six Easy Pieces: Essentials of Physics Explained by Its Most Brilliant Teacher
The Theory of Gravitation (p. 89)
Addison-Wesley Publishing Company. Reading, Massachusetts, USA. 1995

Fletcher, Colin   1922–2007
English backpacker and writer

In that first moment of shock, with my mind already exploding beyond old boundaries, I knew that something had happened to the way I looked at things.
The Man Who Walked Through Time
The Dream (p. 6)
Alfred A. Knopf. New York, New York, USA. 1967

Frazer, Sir James George   1854–1941
Scottish classicist and anthropologist

The mind of man refuses to acquiesce in the phenomena of sense. By an instinctive, an irresistible impulse it is driven to seek something beyond, something which it assumes to be more real and abiding than the shifting phantasmagoria of this sensible world.
The Worship of Nature (Volume 1)
Chapter I (p. 1)
Macmillan & Company Ltd. London, England. 1926

Gauss, Johann Carl Friedrich   1777–1855
German mathematician, physicist, and astronomer

Astronomy and Pure Mathematics are the magnetic poles toward which the compass of my mind ever turns.
In Franz Schmidt and Paul Stäckel (eds.)
Briefwechsel zwischen Carl Friedrich Gauss und Wolfgang Bolyai
Letter XXIII, Letter to Bolyai, June 30, 1803 (p. 55)
B.G. Teubner. Leipzig, Germany. 1899

Gould, Stephen Jay   1941–2002
American paleontologist and evolutionary biologist

I shall not, either in this forum or anywhere, resolve the age-old riddle of epistemology: How can we "know" the "realities" of nature? I will, rather, simply end by restating a point well recognized by philosophers and self-critical scientists, but all too often disregarded at our peril. Science does progress toward more adequate understanding of the empirical world, but no pristine, objective reality lies "out there" for us to capture as our technologies improve and our concepts mature. The human mind is both an amazing instrument and a fierce impediment — and the mind must be interposed between observation and understanding.
Dinosaur in a Haystack: Reflections in Natural History

Part Four, Chapter 16 (p. 214)
Random House, Inc. New York, New York, USA. 1995

Gregory, Sir Richard Arman   1864–1952
British science writer and journalist

Many men have been laughed at…for gazing heavenward when their minds might have been occupied with affairs of earth. There will always be the mind that strives to reach to the skies, and the scoffer who regards all such aspirations as folly.
Two men stood looking through the bars,
One saw mud, the other saw the stars.
Discovery; or, The Spirit and Service of Science
Chapter I (pp. 21–22)
Macmillan & Company Ltd. London, England. 1918

Herschel, Sir John Frederick William   1792–1871
English astronomer and chemist

A mind which has once imbibed a taste for scientific inquiry, and has learnt the habit of applying its principles readily to the cases which occur, has within itself an inexhaustible source of pure and exciting contemplations…
The Cabinet of Natural Philosophy
Part I, Chapter I, Section 11 (p. 14)
Longman, Rees, Orme, Brown & Green. London, England. 1831

Hey, Nigel S.   1936–
American science writer

Think of the myriad interconnected cells that we call a human being — living, dying and regenerating almost without our knowledge…. Think of the richness of their connectivity with the rest of creation. Think of their uncountable linkages with all the molecules — for example in air, in food, in water — that encounter us and enter into us almost without our knowledge, and leave us again, transformed, to visit some other resident, living or nonliving, of the world, the galaxy, the cosmos. This beautiful dance is what I see through the telescope of my mind.
Why People Need Space
Lecture, National Space Centre, October 2002

Holmes, Oliver Wendell   1809–94
American physician, poet, and humorist

Every now and then a man's mind is stretched by a new idea or sensation, and never shrinks back to its former dimensions.
The Autocrat of the Breakfast-Table
Chapter XI (p. 266)
Houghton Mifflin Company. Boston, Massachusetts, USA. 1891

Hugo, Victor   1802–85
French author, lyric poet, and dramatist

The mind, like Nature, abhors a vacuum.
The Novels and Poems of Victor Marie Hugo (p. 262)
Dumont, New York, New York, USA. 1896

Huxley, Thomas Henry   1825–95
English biologist

The mind is so constituted that it does not willingly rest in facts and immediate causes, but seeks always after a knowledge of the remoter links in the chain of causation.
Discourses Biological and Geological
On a Piece of Chalk
D. Appleton & Company. New York, New York, USA. 1897

Jevons, William Stanley   1835–82
English economist and logician

Summing up, then, it would seem as if the mind of the great discoverer must combine contradictory attributes. He must be fertile in theories and hypotheses, and yet full of facts and precise results of experience. He must entertain the feeblest analogies, and the merest guesses at truth, and yet he must hold them as worthless till they are verified in experiment. When there are any grounds of probability he must hold tenaciously to an old opinion, and yet he must be prepared at any moment to relinquish it when a clearly contradictory fact is encountered.
The Principles of Science: A Treatise on Logic and Scientific Method
Book IV, Chapter XXVI (p. 592)
Macmillan & Company Ltd. London, England. 1887

Kepler, Johannes   1571–1630
German astronomer

A mind accustomed to mathematical deduction, when confronted with the faulty foundations [of astrology] resists a long, long time, like an obstinate mule, until compelled by beating and curses to put its foot into that dirty puddle.
In Arthur Koestler
The Sleepwalkers
Part Four, Chapter I, Section 5 (p. 243)
The Macmillan Company. New York, New York, USA. 1966

Land, Edwin   1909–91
American scientist and inventor

Each stage of human civilization is defined by our mental structures: the concepts we create and then project upon the universe. They not only redescribe the universe but also in so doing modify it, both for our own time and for subsequent generations. This process — the revision of old cortical structures and the formulation of new cortical structures whereby the universe is defined — is carried on in science and art by the most creative and talented minds in each generation…
Remarks at Opening of New American Academy of Arts and Sciences
Cambridge, Massachusetts April 2, 1979

Lorentz, Hendrik Antoon   1853–1928
Dutch physicist

We wish to obtain a representation of phenomena and form an image of them in our minds. Till now, we have always attempted to form these images by means of the ordinary notions of time and space. These notions are perhaps innate; in any case they have been developed by our daily observations. For me, these notions are clear, and I confess that I am unable to gain any idea of physics without them.... For me, an electron is a corpuscle which at any given instant is situated at a determinate point of space, and if I believe that at the following instant this corpuscle is situated elsewhere, I attempt to imagine its path, which is a line in space. And if this electron meets an atom and penetrates into its interior and, after several adventures, leaves the atom, I attempt to construct a theory in which this electron has retained its individuality…I would like to retain this ideal of other days and describe everything that occurs in this world in terms of clear pictures.

In A. d'Abro
The Rise of the New Physics (Volume One)
Chapter XIII (p. 108)
Dover Publications, Inc. New York, New York, USA. 1951

Lovecraft, H. P. (Howard Phillips)   1890–1937
American writer of fantasy, horror, and science fiction

The most merciful thing in the world, I think, is the inability of the human mind to correlate all its contents. We live on a placid island of ignorance in the midst of black seas of infinity, and it was not meant that we should voyage far. The sciences, each straining in its own direction, have hitherto harmed us little; but some day the piecing together of dissociated knowledge will open such terrifying vistas of reality, and of our own frightful position therein, that we shall either go mad from the revelation or flee from the deadly light into the peace and safety of a new dark age.

The Call of Cthulhu
The Horror in Clay (p. 139)
Penguin Books. New York, New York, USA. 1999

Lucretius   ca. 99 BCE–55 BCE
Roman poet

Give your mind now to the true reasoning I have to unfold. A new fact is battling strenuously for access to your ears. A new aspect of the Universe is striving to reveal itself. But no fact is so simple that it is not harder to believe than to doubt at the first presentation.

On the Nature of the Universe
Book II, l. 1023
Penguin Books. New York, New York, USA. 1994

Mach, Ernst   1838–1916
Austrian physicist and philosopher

When the human mind, with its limited powers, attempts to mirror in itself the rich life of the world, of which it is itself only a small part, and which it can never hope to exhaust, it has every reason for proceeding economically.

Popular Scientific Lectures
The Economical Nature of Physical Inquiry (p. 186)
The Open Court Publishing Company. Chicago, Illinois, USA. 1898

Playfair, John   1748–1819
Scottish geologist, physicist, and mathematician

But to reason and to arrange are very different occupations of the mind; and a man may deserve praise as a mineralogist, who is but ill qualified for the researches of geology.

Illustrations of the Huttonian Theory of the Earth
Section 422 (p. 482)
Dover Publications, Inc. New York, New York, USA. 1964

Puiseux, P.
No biographical data available

It is an illogical peculiarity of the human mind that while it can not comprehend an infinite universe it readily refuses to limit it.

Annual Report of the Board of Regents of the Smithsonian Institution, 1912
The Year's Progress in Astronomy (p. 135)
Government Printing Office. Washington, D.C. 1913

Rosseland, Svein   1894–1985
Norwegian astronomer

Who has not experienced the mysterious thrill of springtime in a forest, with sunbeams flickering through the foliage, and the low humming of insect life? It is the feeling of unity with nature, which is the counterpart of the attitude of the scientist, analysing the sunbeams into light quanta and the soft rustling of the dragon-fly into condensations and rarefactions of the air. But what is lost in fleeting sentiment is more than regained in the feeling of intellectual security afforded by the scientific attitude, which may grow into a trusting devotion, challenging the peace of the religious mystic. For in the majestic growth of science, analytical in its experimental groping for detail, synthetic in its sweeping generalizations, we are watching at least one aspect of the human mind, which may be believed to have a future of dizzy heights and nearly unlimited perfectibility.

Theoretical Astrophysics: Atomic Theory and the Analysis of Stellar Atmospheres and Envelopes
Introduction (p. xi)
At The Clarendon Press. Oxford, England. 1936

Rowland, Henry Augustus   1848–1901
American physicist

I value in a scientific mind, most of all, that love of truth, that care in its pursuit, and that humility of mind which makes the possibility of error always present more than any other quality. This is the mind which has built up modern science to its present perfection, which has laid one stone upon the other with such care that it today offers to the world the most complete monument to human reason.

In Sir Richard Arman Gregory
Discovery; or, The Spirit and Service of Science
Chapter II (pp. 26–27)
Macmillan & Company Ltd. London, England. 1918

Smollett, Tobias George   1721–71
Scottish novelist

I find my spirits and my health affect each other recipro-
cally — that is to say, everything that decomposes my
mind produces a correspondent disorder in my body;
and my bodily complaints are remarkably mitigated by
those considerations that dissipate the clouds of mental
chagrin.
The Works of Tobias Smollett (Volume 6)
The Expedition of Humphry Clinker (p. 2340)
John D. Morris & Company. Philadelphia, Pennsylvania, USA. 1902

von Humboldt, Alexander   1769–1859
German naturalist and explorer

When the human mind first attempts to subject to its
control the world of physical phenomena, and strives by
meditative contemplation to penetrate the rich luxuri-
ance of living nature, and the mingled web of free and
restricted natural forces, man feels himself raised to a
height from whence, as he embraces the vast horizon,
individual things blend together in varied groups, and
appear as if shrouded in a vapory veil.
Cosmos: A Sketch of a Physical Description of the Universe (Volume 1)
Delineation of Nature. General Review of Natural Phenomena (p. 79)
Harper & Brothers. New York, New York, USA. 1869

That which, in the vagueness of our impressions, loses
all distinctness of form, like some distant mountain
shrouded from view by a veil of mist, is clearly revealed
by the light of mind, which, by its scrutiny into the causes
of phenomena, learns to resolve and analyze their different elements, assigning to each its individual character.
Cosmos: A Sketch of a Physical Description of the Universe (Volume 1)
Introduction (pp. 33–34)
Harper & Brothers, Publishers. New York, New York, USA. 1869

…besides the pleasure derived from acquired knowledge,
there lurks in the mind of man, and tinged with a shade
of sadness, an unsatisfied longing for something beyond
the present — a striving towards regions yet unknown
and unopened.
Cosmos: A Sketch of a Physical Description of the Universe (Volume 1)
Delineation of Nature. General Review of Natural Phenomena (p. 80)
Harper & Brothers, Publishers. New York, New York, USA. 1869

Waddington, Conrad Hal   1905–75
British biologist and paleontologist

We are part of nature, and our mind is the only instrument
we have, or can conceive of, for learning about nature or
about ourselves.
The Nature of Life
Chapter 5 (p. 124)
Harper & Row, Publishers. New York, New York, USA. 1960

Weidlein, Edward Ray
Chemical engineer

The endless frontiers of science now stretching to the
stars can provide rich opportunities for the best creative
minds.
Cooperation — A Responsibility of the Scientist
American Scientist, March 1962 (p. 35)

Weinberg, Steven   1933–
American nuclear physicist

…I do not believe that scientific progress is always best
advanced by keeping an altogether open mind. It is often
necessary to forget one's doubts and to follow the consequences of one's assumptions wherever they may lead
— the great thing is not to be free of theoretical prejudices, but to have the right theoretical prejudices. And
always, the test of any theoretical preconception is where
it leads.
The First Three Minutes
Chapter V (p. 119)
Basic Books, Inc., Publishers. New York, New York, USA. 1988

MINERAL

Smith, Godfrey

Human life would certainly have enjoyed more innocence and satisfaction, were it not for the riches and luster which nature dazzles their eyes with, and makes them
indefatigable searchers into the innermost recesses of the
earth, to her hidden treasures.
The Laboratory; or, School of Arts
Appendix
Of Mines and How to Discover Them
Printed by C. Whittingham. London, England. 1799

MINERAL: ALABASTER

Flaubert, Gustave   1821–90
French novelist

Alabaster. Its use is to describe the most beautiful parts
of a woman's body.
Dictionary of Accepted Ideas
M. Reinhardt. London, England. 1954

MINERAL: AMBER

Herrick, Robert   1591–1674
English poet

I saw a flie within a beade
Of amber cleanly buried.
In J. Max Patrick (ed.)
The Complete Poetry of Robert Herrick
The Amber Bead
W.W. Norton & Company, Inc. New York, New York, USA. 1968

Pope, Alexander   1688–1744
English poet

Pretty! in amber to observe the forms
Of hairs, or straws, or dirt, or grubs, or worms!
The things, we know, are neither rich nor rare,
But wonder how the devil they got there.
The Complete Poetical Works
Epistle to Arbuthnot, l. 169
Houghton Mifflin Company. New York, New York, USA. 1903

MINERAL: AMETHYST

The Bible

The foundations of the city were adorned with precious stones…the twelfth, an amethyst.
The Revised English Bible
Revelation 21:21
Oxford University Press, Inc. Oxford, England. 1989

MINERAL: CHALK

Huxley, Thomas Henry   1825–95
English biologist

A great chapter of the history of the world is written in the chalk. Few passages in the history of man can be supported by such an overwhelming mass of direct and indirect evidence as that which testifies to the truth of the fragment of the history of the globe…. Let me add, that few chapters of human history have a more profound significance for ourselves. I weigh my words well when I assert, that the man who should know the true history of the bit of chalk which every carpenter carries about in his breeches-pocket, though ignorant of all other history, is likely, if he will think his knowledge out to its ultimate results, to have a truer, and therefore a better, conception of this wonderful universe, and of man's relation to it, than the most learned student who is deep-read in the records of humanity and ignorant of those of Nature.
Collected Essays (Volume 8)
Discourses, Biological and Geological
On a Piece of Chalk (p. 4)
Macmillan & Company Ltd. London, England. 1904

The earth, from the time of the chalk to the present day, has been the theater of a series of changes as vast in their amount as they were slow in their progress. The area on which we stand has been first sea and then land for at least four alterations and has remained in each of these conditions for a period of great length.
Collected Essays (Volume 8)
Discourses, Biological and Geological
On a Piece of Chalk (p. 29)
Macmillan & Company Ltd. London, England. 1904

A small beginning has led us to a great ending. If I were to put the bit of chalk with which we started into the hot but obscure flame of burning hydrogen, it would presently shine like the sun. It seems to me that this physical metamorphosis is no false image of what has been the result of our subjecting it to a jet of fervent, though nowise brilliant, thought to-night. It has become luminous, and its clear rays, penetrating the abyss of the remote past, have brought within our ken some stages of the evolution of the earth. And in the shifting "without haste, but without rest" of the land and sea, as in the endless variation of the forms assumed by living beings, we have observed nothing but the natural product of the forces originally possessed by the substance of the universe.
Collected Essays (Volume 8)
Discourses, Biological and Geological
On a Piece of Chalk (p. 36)
Macmillan & Company Ltd. London, England. 1904

MINERAL: COAL

Huxley, Thomas Henry   1825–95
English biologist

The position of the beds which constitute the coal-measures is infinitely diverse. Sometimes they are tilted up vertically, sometimes they are horizontal, sometimes curved into great basins; sometimes they come to the surface, sometimes they are covered up by thousands of feet of rock. But, whatever their present position, there is abundant and conclusive evidence that every under-clay was once a surface soil. Not only do carbonized root-fibers frequently abound in these under-clays; but the stools of trees, the trunks of which are broken off and confounded with the bed of coal, have been repeatedly found passing into radiating roots, still embedded in the under-clay. On many parts of the coast of England, what are commonly known as "submarine forests" are…seen at low water. They consist, for the most part, of short stools of oak, beech, and fir-trees, still fixed by their long roots in the bed of blue clay in which they originally grew. If one of these submarine forest beds should be gradually depressed and covered up by new deposits, it would present just the same characters as an under-clay of the coal, if the Sigillaria and Lepidodendron of the ancient world were substituted for the oak, or the beech, of our own times.
Collected Essays (Volume 8)
Discourses, Biological and Geological
On the Formation of Coal (p. 147)
Macmillan & Company Ltd. London, England. 1904

MINERAL: CRYSTAL

Davidson, John   1857–1909
Scottish poet

"Who affirms that crystals are alive?"
I affirm it, let who will deny:

Crystals are engendered, wax and thrive,
Wane and wither; I have seen them die.
Fleet Street: And Other Poems
Snow
Grant Richards. London, England. 1909

Hauy, Abbé René Just 1743–1822
French mineralogist

A casual glance at crystals may lead to the idea that they were sports of nature, but this is simply an eloquent way of declaring our ignorance. With a thoughtful examination of them, we discover laws of arrangement.... How variable, and at the same time how precise and regular are these laws! How simple they are ordinarily, without losing anything of their significance.
Traité de Mineralogie (p. xiii)
Chez Louis. Paris, France. 1801

Hearn, Lafcadio 1850–1904
Writer, translator, and teacher

I feel like a white granular mass of amorphous crystals — my formula appears to be isomeric with Spasmotoxin. My aurochloride precipitates into beautiful prismatic needles. My Platinochloride develops octohedron crystals, with fine blue florescence. My physiological action is not indifferent. One millionth of a grain injected under the skin of a frog produced instantaneous death accompanied by an orange blossom odor.
In Elizabeth Bisland
The Life and Letters of Lafcadio Hearn (Volume 1)
Letter to George M. Gould, 1889 (p. 462)
Houghton Mifflin Company. Boston, Massachusetts, USA. 1900

Mann, Thomas 1875–1955
German-born American novelist

I shall never forget the sight. The vessel of crystallization was three-quarters full of slightly muddy water — that is, dilute waterglass — and from the sandy bottom there strove upwards a grotesque little landscape of variously colored growths: a confused vegetation of blue, green, and brown shoots which reminded one of algae, mushrooms, attached polyps, also moss, then mussels, fruit pods, little trees or twigs from trees, here, and there of limbs. It was the most remarkable sight I ever saw, and remarkable not so much for its profoundly melancholy nature."
Doktor Faustus
Chapter III (p. 19)
Alfred A. Knopf. New York, New York, USA. 1948

Marx, Carl M.
No biographical data available

In these crystalline structures, the formative forces of the earth seem to manifest themselves most directly, as if they were merely slumbering lightly beneath the rigid surface, resting from the first day of creation.
Norman E. Emerton

The Scientific Reinterpretation of Form
Chapter One (p. 19)
Cornell University Press. Ithaca, New York, USA. 1984

Shaw, George Bernard 1856–1950
Irish comic dramatist and literary critic

Tyndall declared that he saw in Matter the promise and potency of all forms of life, and with his Irish graphic lucidity made a picture of a world of magnetic atoms, each atom with a positive and a negative pole, arranging itself by attraction and repulsion in orderly crystalline structure. Such a picture is dangerously fascinating to thinkers oppressed by the bloody disorders of the living world. Craving for purer subjects of thought, they find in the contemplation of crystals and magnets a happiness more dramatic and less childish than the happiness found by mathematicians in abstract numbers, because they see in the crystals beauty and movement without the corrupting appetites of fleshly vitality.
Back to Methuselah
Preface (pp. lxii–lxiii)
Constable & Company Ltd. London, England. 1921

Thompson, Sir D'Arcy Wentworth 1860–1948
Scottish zoologist and classical scholar

Crystals lie outside the province of this book; yet snow-crystals, and all the rest besides, have much to teach us about the variety, the beauty and the very nature of form.
On Growth and Form (Volume 2)
Chapter IX (p. 696)
At The University Press. Cambridge, England. 1951

MINERAL: DIAMOND

Fleming, Ian 1935–
English novelist

It was domination by a beauty so pure that it held a kind of truth, a divine authority before which all other material things turned, like the bit of quartz, to clay. In these few minutes Bond understood the myth of diamonds, and he knew that he would never forget what he had suddenly seen inside the heart of this stone.
Diamonds Are Forever
Film (1971)

Gibran, Kahlil 1883–1931
Lebanese-American philosophical essayist

Perhaps time's definition of coal is the diamond.
Sand and Foam: A Book of Aphorisms (p. 55)
Alfred A. Knopf. New York, New York, USA. 1959

Meydendauer, A.
German mineralogist

The diamond can only be of cosmic origin, having fallen as a meteorite at later periods of the earth's formation.

The available localities of the diamond contain the residues of not very compact meteoric masses, which may, perhaps, have fallen in the prehistoric ages, and which have penetrated more or less deeply, according to the more or less resistant character of the surface where they fell. Their remains are crumbling away on exposure to the air and sun, and the rain has long ago washed away all prominent masses. The enclosed diamonds have remained scattered in the river-beds, while the fine, light matrix has been swept away.

In Frederick Houk Law
Science in Literature
The Romance of the Diamonds (p. 109)
Harper & Brothers. New York, New York, USA. 1929

Wells, H. G. (Herbert George)   1866–1946
English novelist, historian, and sociologist

Now, a year or so ago, I had occupied my leisure in taking a London science degree, so that I have a smattering of physics and mineralogy. The thing was not unlike an uncut diamond of the darker sort, though far too large, being almost as big as the top of my thumb. I took it, and saw it had the form of a regular octahedron, with the curved faces peculiar to the most precious of minerals. I took out my penknife and tried to scratch it — vainly. Leaning forward towards the gas-lamp, I tried the thing on my watch-glass, and scored a white line across that with the greatest ease.

Best Science Fiction Stories of H.G. Wells
The Diamond Maker
Dover Publications, Inc. New York, New York, USA. 1966

MINERAL: EMERALD

Salzberg, Hugh W.

Take white lead, one part, and of any glass you choose, two parts, fuse together in a crucible and then pour the mixture. To this crystal, add the urine of an ass and after forty days you will find emeralds.

From Caveman to Chemist
Chapter III (p. 36)
American Chemical Society. Washington, D.C. 1991

MINERAL: FLINT

Bierce, Ambrose   1842–1914
American newspaperman, wit, and satirist

FLINT, n. A substance much in use as a material for hearts. Its composition is silica, 98.00; oxide of iron, 0.25; alumina, 0.25; water, 1.50. When an editor's heart is made, the water is commonly left out; in a lawyer's more water is added — and frozen.

The Enlarged Devil's Dictionary (p. 96)
Doubleday & Company, Inc. Garden City, New York, USA. 1967

MINERAL: GRANITE

Cloos, Hans   1885–1951
German geologist

However silent, these large, simple shapes have a wordless language. They say that the granite has grown in secure and secluded depths, guided wholly by its own laws; and that no disturbing outside influence has intervened in the slow tranquil growth of the crystals.

Conversation with the Earth (p. 105)
Alfred A. Knopf. New York, New York, USA. 1953

MINERAL: JADE

Author undetermined

When I think of a wise man, he seems like jade. Wise men have seen in jade all different virtues. It is soft, smooth and shining like kindness. It is hard, fine and strong like intelligence. Its edges seem sharp but do not cut, like justice. It hangs down to the ground like humility. When struck, it gives a clear, ringing sound, like music. The stains in it, which are not hidden and which add to its beauty, are like thoughtfulness. Its brightness is like heaven while its firm substance, born of the mountain and the waters, is like the earth. That is why wise men love jade.

In Joan M. Hartman
Chinese Jade of Five Centuries
C.E. Tuttle Company. Rutland, Vermont, USA. 1969

MINERAL: LOADSTONE

Gilbert, William   1544–1603
English scientist and physician

….the more advanced one is in the science of the loadstone, the more trust he has in the hypotheses, and the greater the progress he makes; nor will one reach anything like certitude in the magnetic philosophy, unless all, or at all events most, of its principles are known to him.

In *Great Books of the Western World* (Volume 28)
On the Loadstone and Magnetic Bodies and on the Great Magnet the Earth
Preface (p. 2)
Encyclopædia Britannica, Inc. Chicago, Illinois, USA. 1952

MINERAL: MARBLE

Twain, Mark (Samuel Langhorne Clemens)   1835–1910
American author and humorist

Verily it is one thing to have cash and another to know how to spend it. The man ought to die a violent death that put it into people's heads to try to make cherished,

beloved, sacred homes out of such cold, ghostly, unfeeling stuff as marble — a material which God intended for only gravestones. You can build a house out of it, and put a door-plate on it, and call it a dwelling, but it isn't any use — it is bound to look like a mausoleum, after all. Stewart's house looks like a stately tomb, now, and after it is finished it will never look entirely natural without a hearse in front of it.
Letter to San Francisco *Alta California*
July 28, 1867

MINERAL: OPAL

Wilcox, Ella Wheeler   1850–1919
American poet and journalist

And lo! The beautiful Opal,
That rare and wondrous gem,
Where the moon and sun blend into one,
Is the child that was born to them.
How Salvator Won & Other Recitations
The Birth of the Opal
Edgar S. Werner. New York, New York, USA. 1891

MINERAL: PEARL

Gibran, Kahlil   1883–1931
Lebanese-American philosophical essayist

Perhaps the sea's definition of a shell is the pearl.
Sand and Foam: A Book of Aphorisms (p. 55)
Alfred A. Knopf. New York, New York, USA. 1959

A pearl is a temple built by pain around a grain of sand.
Sand and Foam: A Book of Aphorisms (p. 4)
Alfred A. Knopf. New York, New York, USA. 1959

MINERAL: SALT

Author undetermined

A Salt is a substance which has been naturalized.
Classroom Emanations
Journal of Chemical Education, Volume 2, Number 7, July 1925 (p. 611)

Boerhaave, Herman   1668–1738
Dutch chemist, physician, and botanist

Whomsoever is not acquainted with the taste of Salts will never arrive at the knowledge of our Arcana.
Elements of Chemistry (Volume 1)
Part III (p. 438)
Printed for J. & J. Pemberton. London, England. 1735

Darwin, Erasmus   1731–1802
English physician and poet

Hence with diffusive salt old Ocean steeps
His emerald shallows, and his sapphire deeps.
Oft in wide lakes, around their warmer brim

In hollow pyramids the crystals swim;
Or, fused by earth-born fires, in cubic blocks
Shoot their white forms, and harden into rocks.
The Botanic Garden
Part I, Canto II, V
Canto II, l. 120–125 (p. 29–30)
Jones & Company. London, England. 1825

Sagan, Carl   1934–96
American astronomer and author

But let us look a little more deeply at our microgram of salt. Salt happens to be a crystal in which, except for defects in the structure of the crystal lattice, the position of every sodium and chlorine atom is predetermined If we could shrink ourselves into this crystalline world, we would see rank upon rank of atoms in an ordered array, a regularly alternating structure — sodium, chlorine, sodium, chlorine, specifying the sheet of atoms we are standing on and all the sheets above us and below us. An absolutely pure crystal of salt could have the position of every atom specified by something like 10 bits of information. (Chlorine is a deadly poison gas employed in European battlefields in World War I. Sodium is a corrosive metal which burns upon contact with water. Together they make a placid and unpoisonous material, table salt. Why each of these substances has the properties it does is a subject called chemistry, which requires more than 10 bits of information to understand.) This would not strain the information-carrying capacity of the brain.
Broca's Brain: Reflections on the Romance of Science
Part I, Chapter 2 (p. 15)
Random House, Inc. New York, New York, USA. 1979

If the universe had natural laws that governed its behavior to the same degree of regularity that determines a crystal of salt, then, of course, the universe would be knowable. Even if there were many such laws, each of considerable complexity, human beings might have the capability to understand them all. Even if such knowledge exceeded the information-carrying capacity of the brain, we might store the additional information outside our bodies — in books, for example, or in computer memories — and still, in some sense, know the universe.
Broca's Brain: Reflections on the Romance of Science
Part I, Chapter 2 (p. 15)
Random House, Inc. New York, New York, USA. 1979

MINERAL: SANDSTONE

Abbey, Edward   1927–89
American environmentalist and nature writer

The sandstone walls rise higher than ever before, a thousand, two thousand feet above the water, rounding off on top as half-domes and capitols, golden and glowing in the sunlight, a deep radiant red in the shade.
Desert Solitaire

Down the River (p. 205)
Ballantine Books. New York, New York, USA. 1968

MINERAL: SAPPHIRE

Gübelin, Eduard
No biographical data available

In the close mesh of fiction and truth sapphire is more closely ensnared than any of its noble peers. Out of mankind's long acquaintance with it, towers of Babylonian dimensions have pressed heavily on its brazen back, built out of the tough ashlar of pagan and Christian magic, which sought to make use of supernatural powers through the stone of heavenly blue.
The Color Treasury of Gemstones
Sapphire: Lord Keeper of the Seals in the Gem Kingdom (p. 46)
Crowell. New York, New York, USA. 1975

MINERALOGIST

Landes, K. K.
No biographical data available

Let there be more geological mineralogists! The only requirements, outside of educational background, are a prodigious curiosity, a vivid imagination, and a thick skin.
Geological Mineralogy
American Mineralogist, Volume 31, Number 3 & 4, March–April 1946 (p. 134)

Russell, Bertrand Arthur William   1872–1970
English philosopher, logician, and social reformer

If you ask a mathematician, a mineralogist, a historian, or any other man of learning, what definite body of truths has been ascertained by his science, his answer will last as long as you are willing to listen.
The Problems of Philosophy
Chapter XV (p. 154)
Oxford University Press, Inc. London, England. 1959

MINERALOGY

Author undetermined

Mineralogy is the Alphabet of Geology.
In William Knight
Facts and Observations Towards Forming a New Theory of the Earth
Introduction (p. 3)

Verne, Jules   1828–1905
French novelist

I loved mineralogy, I loved geology. To me there was nothing like pebbles — and if my uncle had been in a little less of a fury, we should have been the happiest of families.
A Journey to the Center of the Earth

Chapter 1 (p. 6)
The Limited Editions Club. New York, New York, USA. 1966

von Goethe, Johann Wolfgang   1749–1832
German poet, novelist, playwright, and natural philosopher

…mineralogy is a science for the Understanding, for practical life; for its subjects are something dead which cannot rise again, and there is no room for synthesis.
In Johann Peter Eckermann
Conversations with Goethe
Friday, February 13, 1829 (p. 294)
J.M. Dent & Sons Ltd. London, England. 1970

MINING

Agricola, Georgius   1494–1555
German mineralogist

[I]f mining is a shameful and discreditable employment for a gentleman because slaves once worked mines, then agriculture also will not be a very creditable employment, because slaves once cultivated the fields, and even today do so among the Turks; nor will architecture be considered honest, because some slaves have been found skilful in that profession; nor medicine, because not a few doctors have been slaves; nor will any other worthy craft, because men captured by force of arms have practised it.
De Re Metallica
Book I (p. 23)
Dover Publications, Inc. New York, New York, USA. 1950

[I]nasmuch as the chief callings are those of the moneylender, the soldier, the merchant, the farmer, and the miner, I say, inasmuch as usury is odious, while the spoil cruelly captured from the possessions of the people innocent of wrong is wicked in the sight of God and man, and inasmuch as the calling of the miner excels in honour and dignity that of the merchant trading for lucre, while it is not less noble though far more profitable than agriculture, who can fail to realize that mining is a calling of peculiar dignity?
De Re Metallica
Book I (p. 24)
Dover Publications, Inc. New York, New York, USA. 1950

Muir, John   1838–1914
American naturalist

The drifts and tunnels in the rocks may perhaps be regarded as the prayers of the prospector, offered for the wealth he so earnestly craves; but like the prayers of any kind not in harmony with nature, they are unanswered.
Steep Trails
Chapter XVI (p. 203)
Norman S. Berg, Publisher. Dunwoody, Georgia, USA. 1970

Mining discoveries and progress, retrogression and decay, seem to have been crowded more closely against each other here than on any other portion of the globe.

Steep Trails
Chapter XVI (p. 198)
Norman S. Berg, Publisher. Dunwoody, Georgia, USA. 1970

MIRACLE

Bradbury, Ray   1920–
American writer

We live in miracles which cannot be explained. The scientist, the theologian, the artist — each attempts impossible explanations.
In Ray Bradbury, Arthur C. Clarke, Bruce Murray, Carl Sagan, and Walter Sullivan
Mars and the Mind of Man
Ray Bradbury (p. 139)
Harper & Row, Publishers. New York, New York, USA. 1973

Dawkins, Richard   1941–
English ethologist, evolutionary biologist, and popular science writer

Evolution is very possibly not, in actual fact, always gradual. But it must be gradual when it is being used to explain the coming into existence of complicated, apparently designed objects, like eyes. For if it is not gradual in these cases, it ceases to have any explanatory power at all. Without gradualness in these cases, we are back to miracle, which is simply a synonym for the total absence of explanation.
River Out of Eden: A Darwinian View of Life
Chapter 3 (p. 83)
Basic Books. New York, New York, USA. 1995

Dürrenmatt, Friedrich   1921–90
Swiss playwright and novelist

…in the realm of science there is nothing more repugnant than a miracle.
Translated by James Kirkup
The Physicists
Act One (p. 48)
Grove Press, Inc. New York, New York, USA. 1964

Einstein, Albert   1879–1955
German-born physicist

That this [analogy of the atom with the solar system] insecure and contradictory foundation was sufficient to enable a man of Bohr's unique instinct and tact to discover the major laws of the spectral lines and of the electron shells of the atoms together with their significance to chemistry appeared to me like a miracle — and appears to me as a miracle even today. This is the highest form of musicality in the sphere of thought.
In Paul Arthur Schlipp (ed.)
Albert Einstein: Philosopher-Scientist
Autobiographical Notes (p. 45, 47)
The Library of Living Philosophers, Inc. Evanston, Illinois, USA. 1949

Haldane, John Scott   1860–1936
Scottish physiologist

The more we discover as to physiological activity and inheritance, the more difficult does it become to imagine any physical or chemical description or explanation which could in any way cover the facts as to the persistent co-ordination. From the standpoint of the physical sciences the maintenance and reproduction of a living organism is nothing less than a standing miracle, and for that reason the co-ordinated maintenance of structure and activity is inconsistent with the physical conception of self-existent matter and energy.
The Philosophical Basis of Biology: Donnellan Lectures, University of Dublin, 1930
Lecture I, Mechanistic Biology (p. 11)
Doubleday, Doran & Company, Inc. Garden City, New York, USA. 1931

MIRROR

Durell, Clement V.   1882–1968
English mathematician

…a reception was held and the science departments were on view. A young lady, entering the physical laboratory and seeing an inverted image of herself in a large concave mirror, naively remarked to her companion: "They have hung that looking glass upside down."
Readable Relativity
Chapter II (p. 12)
Harper & Brothers. New York, New York, USA. 1960

MISERY

Darwin, Charles Robert   1809–82
English naturalist

…if the misery of our poor be caused not by the laws of nature, but by our institutions, great is our sin…
The Voyage of the Beagle
Chapter XXI (p. 500)
Heron Books. 1968

I am the most miserable, bemuddled, stupid dog in all England, and am ready to cry with vexation at my blindness and presumption.
In Francis Darwin (ed.)
The Life and Letters of Charles Darwin (Volume 1)
Letter to J.D. Hooker, July 14, 1857? (p. 461)
D. Appleton & Company. New York, New York, USA. 1896

MISTAKE

Braddon, Mary Elizabeth   1837–1915
English novelist

We spend the best part of our lives in making mistakes, and the poor remainder in reflecting how very easily we might have avoided them.
Aurora Floyd (p. 241)
Ward, Lock & Tyler. London, England. 1875

Gombrich, Ernst Hans   1909–2001
English art historian and scholar

In order to learn, we must make mistakes, and the most fruitful mistakes which nature could have implanted in us would be the assumption of even greater simplicities than we are likely to meet in this bewildering world of ours.… To probe a hole we first use a straight stick to see how far it takes us. To probe the visible world we use the assumption that things are simple until they prove to be otherwise.
In John Pottage
Geometrical Investigations (p. 15)
Addison-Wesley Publishing Company. Reading, Massachusetts, USA. 1983

Mayr, Ernst   1904–2005
German-born American biologist

In science one learns not only by one's own mistakes but by the history of the mistakes of others.
The Growth of Biological Thought: Diversity, Evolution and Inheritance
Chapter 1 (p. 20)
Harvard University Press. Cambridge, Massachusetts, USA. 1982

Miller, Henry George
No biographical data available

The more distinguished the doctor the more terrible the mistakes he has made — or will admit to.
Henry Millerisms
World Neurology, 9 April 1968 (p. 8)

Obruchev, Vladimir   1863–1956
Russian geologist and geographer

Be persistent and persevering, but never stubborn. Do not cling to your judgments. Remember that there are many clever people in the world liable to spot your mistakes. If they are right, be not reluctant to agree with them.
Compiled by V.V. Vorontsov
Words of the Wise: A Book of Russian Quotations
Translated by Vic Schneierson
Progress Publishers. Moscow, Russia. 1979

Siegel, Eli   1902–78
American philosopher, poet, critic, and founder of Aesthetic Realism

If a mistake is not a stepping stone, it is a mistake.
Damned Welcome
Aesthetic Realism, Maxims, Part One, #139 (p. 39)
Definition Press. New York, New York, USA. 1972

MITOCHONDRION

Gould, Stephen Jay   1941–2002
American paleontologist and evolutionary biologist

Surely, the mitochondrion that first entered another cell was not thinking about the future benefits of cooperation and integration; it was merely trying to make its own living in a tough Darwinian world.

Wonderful Life: The Burgess Shale and the Nature of History
Chapter V (p. 310)
W.W. Norton & Company, Inc. New York, New York, USA. 1989

MIXTURE

Huxley, Thomas Henry   1825–95
English biologist

Mix salt and sand, and it shall puzzle the wisest of men, with his mere natural appliances, to separate all the grains of sand from all the grains of salt; but a shower of rain will effect the same object in ten minutes.
Collected Essays (Volume 2)
Darwiniana
The Origin of Species (p. 76)
Macmillan & Company Ltd. London, England. 1904

MODEL

Ball, John
No biographical data available

To make progress in understanding all this, we probably need to begin with simplified (oversimplified?) models and ignore the critics' tirade that the real world is more complex. The real world is always more complex, which has the advantage that we shan't run out of work.
Memes as Replicators
Ethology and Sociobiology, Volume 5, Number 3, 1984 (p. 159)

Bianco, Margery Williams   1880–1944
English-American author

The Rabbit could not claim to be a model of anything, for he didn't know that real rabbits existed; he thought they were all stuffed with sawdust like himself, and he understood that sawdust was quite out-of-date and should never be mentioned in modern circles.
The Velveteen Rabbit: Or How Toys Become Real
Athenaeum Books for Young Readers. New York, New York, USA. 2002

Born, Max   1882–1970
German-born English physicist

All great discoveries in experimental physics have been due to the intuition of men who made free use of models, which were for them not products of the imagination, but representatives of real things.
Physical Reality
Philosophical Quarterly, Volume 3, Number 11, April 1953 (p. 140)

Box, George E. P.   1919–
English statistician

All models are wrong but some are useful.
Apocryphal

Bronowski, Jacob   1908–74
Polish-born English mathematician and polymath

The pre-eminence of astronomy rests on the peculiarity that it can be treated mathematically; and the progress of physics, and most recently biology, has hinged equally on finding formulations of their laws that can be displayed as mathematical models.
The Ascent of Man
Chapter 5 (p. 165)
Little, Brown & Company. Boston, Massachusetts, USA. 1973

Chargaff, Erwin   1905–2002
Austrian biochemist

Models — in contrast to those who sat for Renoir — improve with age.
Heraclitean Fire: Sketches from a Life before Nature
Part III
Science as a Profession (p. 171)
Rockefeller University Press. New York, New York, USA. 1978

Cheeseman, Peter
Australian computer scientist

The apparent simplicity of a model is due to a failure of imagination and limited data, unless the domain really is simple. If the world were really random, chemistry, cooking, and credit would not be possible, so our models cannot be figments of our imagination.
In J. Shrager and P. Langley (eds)
Computational Models of Scientific Discovery and Theory Formation
On Finding the Most Probable Model (p. 91)
Morgan Kaufmann Publishers. San Mateo, California, USA. 1990

Crick, Francis Harry Compton   1916–2004
English biochemist

...no good model ever accounted for all the facts, since some data was bound to be misleading if not plain wrong. A theory that did fit all the data would have been "carpentered" to do this and would thus be open to suspicion.
What Mad Pursuit: A Personal View of Scientific Discovery
Chapter 5 (p. 60)
Basic Books, Inc. New York, New York, USA. 1988

Davies, Paul Charles William   1946–
British-born physicist, writer, and broadcaster

When it comes to very highly organized systems, such as a living cell, the task of modeling by approximation to simple, continuous and smoothly varying quantities is hopeless. It is for this reason that attempts by sociologists and economists to imitate physicists and describe their subject matter by simple mathematical equations is rarely convincing.
The Cosmic Blueprint: New Discoveries in Nature's Creative Ability to Order the Universe
Chapter 3 (p. 22)
Simon & Schuster. New York, New York, USA. 1988

The incorporation of imaginary elements into physical theories is one of the most difficult practices for a professional physicist to justify to the layman. Of course,

if a particular feature, such as isotopic spin symmetry, renders the model a brilliant success, then the physicist can simply reply, "I put it in because it works!"
Superforce: The Search for a Grand Unified Theory of Nature
Chapter 4 (pp. 66–67)
Simon & Schuster. New York, New York, USA. 1984

A model of the universe does not require faith, but a telescope. If it is wrong, it is wrong.
Space and Time in the Modern Universe
Chapter 7 (p. 201)
Cambridge University Press. Cambridge, England. 1977

Deutsch, Karl W.   1912–92
Czech-born American international political scientist

Men think in terms of models.
Mechanism, Organism and Society
Philosophy of Science, Volume 18, Number 3, July 1951 (p. 230)

Eddington, Sir Arthur Stanley   1882–1944
English astronomer, physicist, and mathematician

Our model of Nature...should be like an engine with movable parts. We need not fix the position of any one lever; that is to be adjusted from time to time as the latest observations indicate. The aim of the theorist is to know the train of wheels which the lever sets in motion — that binding of the parts which is the soul of the engine.
The Internal Constitution of Stars
Nature, Volume 106, Number 2603, 2 September 1920 (p. 20)

Eigen, Manfred   1927–
German biophysicist

A theory has only the alternative of being right or wrong. A model has a third possibility: it may be right, but irrelevant.
In J. Mehra (ed.)
The Physicist's Conception of Nature: Symposium on the Development of the Physicist's Conception of Nature in the Twentieth Century
Chapter 30 (p. 618)
Reidel. Boston, Massachusetts, USA. 1973

Ferris, Timothy   1944–
American science writer

The model of the natural world we build in our minds by such a process will forever be inadequate, just a little cathedral in the mountains. Still it is better than no model at all.
The Red Limit: The Search for the Edge of the Universe
Preface (p. 8)
William Morrow & Company, Inc. New York, New York, USA. 1977

Feynman, Richard P.   1918–88
American theoretical physicist

...the more you see how strangely Nature behaves, the harder it is to make a model that explains how even the simplest phenomena actually work. So theoretical physics has given up on that.
QED: The Strange Theory of Light and Matter

Chapter 3 (p. 82)
Princeton University Press. Princeton, New Jersey, USA. 1985

Gould, Stephen Jay   1941–2002
American paleontologist and evolutionary biologist

I do not know that my view is more correct; I do not even think that "right" and "wrong" are good categories for assessing complex mental models of external reality — for models in science are judged [as] useful or detrimental, not as true or false.
Dinosaur in a Haystack: Reflections in Natural History
Part Three, Chapter 8 (p. 96)
Random House, Inc. New York, New York, USA. 1995

Greedman, D. A.
No biographical data available
Navidi, W. C.
No biographical data available

Models are often used to decide issues in situations marked by uncertainty. However statistical differences from data depend on assumptions about the process which generated these data. If the assumptions do not hold, the inferences may not be reliable either. This limitation is often ignored by applied workers who fail to identify crucial assumptions or subject them to any kind of empirical testing. In such circumstances, using statistical procedures may only compound the uncertainty.... Statistical modeling seems likely to increase the stock of things you think you know that ain't so.
Regression Models for Adjusting the 1980 Census
Statistical Science, Volume 1, Number 1, 1986 (p. 3)

Greenwood, H. J.
No biographical data available

Let us not grace loose thinking with the word "model."
On Models and Modeling
Canadian Mineralogist, Volume 27, 1989

Jeans, Sir James Hopwood   1877–1946
English physicist and mathematician

The making of models or pictures to explain mathematical formulae and the phenomena they describe is not a step towards, but a step away from reality; it is like making graven images of a spirit.... All the same, the mathematical physicist is still busily at work making graven images of the concepts of the wavemechanics.
The Mysterious Universe
Chapter V (p. 176, 177)
The Macmillan Company. New York, New York, USA. 1932

Kaplan, Abraham   1918–93
American philosopher of science, author, and educator

The words "model" and "mode" have, indeed, the same root; today, model building is science a la mode.
The Conduct of Inquiry: Methodology for Behavioral Science

Chapter VII, Section 30 (p. 258)
Chandler Publishing Company. San Francisco, California, USA. 1964

Karlin, Samuel
No biographical data available

The purpose of models is not to fit the data but to sharpen the questions.
11th R. A. Fisher Memorial Lecture
Royal Society 20 April 1983

Kelvin, Lord William Thomson   1824–1907
Scottish engineer, mathematician, and physicist

I never satisfy myself until I can make a mechanical model of a thing. If I can make a mechanical model, I understand it.
Baltimore Lectures on Molecular Dynamics, and the Wave Theory of Light (p. 270)
C.J. Clay & Sons. London, England. 1904

Kolb, Edward W. (Rocky)   1951–
American cosmologist

To try to ferret out the important and interesting objects from the multitude of things in the sky, every cosmologist looks at the universe through a filter of a model, for without the conceptual framework of a model the staggering number of things in the universe would overwhelm anyone.
Blind Watchers of the Sky
Chapter Eleven (p. 285)
Addison-Wesley Publishing Company. Reading, Massachusetts, USA. 1996

If you want to know whether you should take a model seriously or just regard it as a calculational tool, you can imagine that you are building a wall out of stones. For instance, if the stones represent models for various phenomena seen in the sky, and the wall represents all of astronomy, then choosing a model to explain a phenomenon is like choosing a stone to be incorporated into the wall. Sometimes the stone seems to fit naturally into a space in the wall; more often it has to be trimmed a bit to fit in. But it is impossible to judge whether it is a "beautiful" stone or the "correct" stone for that place in the wall when it is first inserted, because the true beauty or utility of the stone can't be judged in isolation from the rest of the wall. The two real criteria to judge the stone are whether it is one on which other stones can be placed and whether it exists harmoniously with the surrounding stones. If the stone not only fulfills the function of taking up spaces in the wall but also provides a platform on which to place other stones, it a beautiful stone.
Blind Watchers of the Sky
Chapter Ten (pp. 287–288)
Addison-Wesley Publishing Company. Reading, Massachusetts, USA. 1996

Lewis, C. S. (Clive Staples)   1898–1963
British author, scholar, and popular theologian

It is not impossible that our own Model will die a violent death, ruthlessly smashed by an unprovoked assault of new facts.... But I think it is more likely to change when, and because, far-reaching changes in the mental temper of our descendants demand that is should. The new Model will not be set up without evidence, but the evidence will turn up when the inner need for it becomes sufficiently great. It will be true evidence. But nature gives most of her evidence in answer to the questions we ask her. Here, as in the courts, the character of the evidence depends on the shape of the examination, and a good cross-examiner can do wonders.

The Discarded Image: An Introduction to Medieval and Renaissance Literature
Epilogue (pp. 222–223)
University Press. Cambridge, England. 1964

Lewis, Gilbert Newton   1875–1946
American chemist

As we continue the great adventure of scientific exploration our models must often be recast. New laws and postulates will be required, while those that we already have must be broadened, extended and generalized in ways that we are now hardly able to surmise.

The Anatomy of Science
Chapter VIII (p. 219)
Yale University Press. New Haven, Connecticut, USA. 1926

Lindley, David   1956–
English astrophysicist and author

There is no guarantee that any simple model will be able to explain everything.

The End of Physics: The Myth of a Unified Theory
Part I, Chapter 4 (p. 131)
Basic Books. New York, New York, USA. 1993

Lloyd, David
No biographical data available
Volkov, Evgenii I.
No biographical data available

One good experiment is worth a thousand models...; but one good model can make a thousand experiments unnecessary.

In C. Mosekilde and L. Moskilde (eds.)
Complexity, Chaos, and Biological Evolution
The Ultradian Clock: Timekeeping for Intracellular Dynamics (p. 51)
Plenum Press. New York, New York, USA. 1991

Maxwell, James Clerk   1831–79
Scottish physicist

As long as the training of a naturalist enables him to trace the action only of a particular material system, without giving him the power of dealing with the general properties of all such systems, he must proceed by the method so often described in histories of science

— he must imagine model after model of hypothetical apparatus, till he finds one which will do the required work. If this apparatus should afterwards be found capable of accounting for many of the known phenomena, and not demonstrably inconsistent with any of them, he is strongly tempted to conclude that his hypothesis is a fact, at least until an equally good rival hypothesis has been invented.

Tait's Thermodynamics
Nature, Volume XVII, Number 431, January 31, 1878 (p. 258)

Miall, Andrew
No biographical data available

There are those who try to generalize, synthesize, and build models, and there are those who believe nothing and constantly call for more data. The tension between these two groups is a healthy one; science develops mainly because of the model builders, yet they need the second group to keep them honest.

Principles of Sedimentary Basin Analysis
Chapter 8 (p. 363)
Springer-Verlag. New York, New York, USA. 1984

Milton, John   1608–74
English poet

Hereafter, when they come to model Heav'n
And calculate the Stars, how they will wield
The mightier frame, how build, unbind, contrive
To save appearances, how gird the Shear
With Centric and Eccentric scribbled ore,
Cycle and Epicycle, Orb in Orb.

In *Great Books of the Western World* (Volume 32)
Paradise Lost
Book VIII, l. 79–84
Encyclopædia Britannica, Inc. Chicago, Illinois, USA. 1952

Morrison, Foster
No biographical data available

Much of the technical literature is difficult to read, even for scientists and engineers. Even the best books tend to dwell on the mathematical models and don't give the slightest hint what to do if one is lucky enough to have some data.

The Art of Modeling Dynamic Systems: Forecasting for Chaos, Randomness & Determinism
Preface (p. vii)
John Wiley & Sons, Inc. New York, New York, USA. 1991

Oreskes, Naomi
No biographical data available
Belitz, K.
No biographical data available

A model, like a novel, may resonate with nature, but it is not a "real" thing. Like a novel, a model may be convincing — it may "ring true" if it is consistent with our experience of the natural world. But just as we may

wonder how much the characters in a novel are drawn from real life and how much is artifice, we might ask the same of a model: How much is based on observation and measurement of accessible phenomena, how much is convenience? Fundamentally, the reason for modeling is a lack of full access, either in time or space, to the phenomena of interest.
Science, Volume 263, 1944

Paulos, John Allen   1945–
American mathematician

The once-surprising existence of non-Euclidean models of Euclid's first four axioms can be seen as a sort of mathematical joke.
Once Upon a Number: The Hidden Mathematical Logic of Stories
Appendix: Humor and Computation (p. 132)
Basic Books. New York, New York, USA. 1998

Poincaré, Henri   1854–1912
French mathematician and theoretical astronomer

We should always aim toward the economy of thought. It is not enough to give models for imitation. It must be possible to pass beyond these models and, in place of repeating their reasoning at length each time, to sum this in a few words.
Annual Report of the Board of Regents of the Smithsonian Institution, 1909
The Future of Mathematics (p. 128)
Government Printing Office. Washington, D.C. 1910

Poynting, John Henry   1852–1914
English physicist

…while the building of Nature is growing spontaneously from within, the model of it, which we seek to construct in our descriptive science, can only be constructed by means of scaffolding from without, a scaffolding of hypotheses. While in the real building all is continuous, in our model there are detached parts which must be connected with the rest by temporary ladders and passages, or which must be supported till we can see how to fill in the understructure. To give the hypotheses equal validity with facts is to confuse the temporary scaffolding with the building itself.
Collected Scientific Papers
Part VII, Article 52 (p. 607)
At The University Press. Cambridge. 1920

Sciama, Dennis   1926–99
English physicist

Since we find it difficult to make a suitable model of a certain type, Nature must find it difficult too. This argument neglects the possibility that Nature may be cleverer than we are. It even neglects the possibility that we may be cleverer tomorrow than we are today.
In Neil de Grasse Tyson
Galactic Engines
Natural History, Volume 106, Number 4, May 1997 (p. 71)

Sophocles   496 BCE–406 BCE
Greek playwright

Nay, Knowledge must come through action; thou canst have no test which is not fanciful, save by trial.
In *Great Books of the Western World* (Volume 5)
The Plays of Sophocles
Trachiniae, l. 589
Encyclopædia Britannica, Inc. Chicago, Illinois, USA. 1952

Stewart, Ian   1945–
English mathematician and science writer

Construction of models, I said, was an art. On this occasion the art is conjuring: I can do no better than wave the magic wand and extract the rabbit from the hat.
Concepts of Modern Mathematics
Chapter 8 (p. 120)
Dover Publications, Inc. New York, New York, USA. 1995

Stocking, Martha
No biographical data available

Building statistical models is just like this. You take a real situation with real data, messy as this is, and build a model that works to explain the behavior of real data.
New York Times, February 10, 2000

von Neumann, John   1903–57
Hungarian-American mathematician

To begin, we must emphasize a statement which I am sure you have heard before, but which must be repeated again and again. It is that the sciences do not try to explain, they hardly even try to interpret, they mainly make models. By a model is meant a mathematical construct which, with the addition of certain verbal interpretations, describes observed phenomena. The justification of such a mathematical construct is solely and precisely that it is expected to work — that is, correctly to describe phenomena from a reasonably wide area. Furthermore, it must satisfy certain aesthetic criteria — that is, in relation to how much it describes, it must be rather simple.
The Neumann Compendium
Method in the Physical Sciences (p. 628)
World Scientific. Singapore. 1995

Walker, Marshall John
American physicist

Scientists have learned by humiliating experience that their model is not reality.
The Nature of Scientific Thought
Chapter XIV (p. 158)
Prentice-Hall, Inc., Englewood Cliffs, New Jersey, USA. 1963

Weisskopf, Victor Frederick   1908–2002
Austrian-American physicist

What is a model? A model is like an Austrian timetable. Austrian trains are always late. A Prussian visitor asks

the Austrian conductor why they bother to print time-tables. The conductor replies "If we did not, how would we know how late the trains are?
In H. Frauenfelder and E.M. Henley
Subatomic Physics
Part V (p. 351)
Prentice-Hall, Inc., Englewood Cliffs, New Jersey, USA. 1974

MOLAR SOLUTION

Author undetermined

A molar solution is one which contains one g.m.w. [gram molecular weight] per liter.
Classroom Emanations
Journal of Chemical Education, Volume 2, Number 7, July 1925 (p. 611)

MOLECULAR BIOLOGY

Chargaff, Erwin   1905–2002
Austrian biochemist

…molecular biology [is] the practice of biochemistry without a license.
Essays on Nucleic Acids
Amphisbaena
Elsevier Publishing Company. Amsterdam, Netherlands. 1963

Crick, Francis Harry Compton   1916–2004
English biochemist

…molecular biology can be defined as anything that interests molecular biologists.
Molecular Biology in the Year 2000
Nature, Volume 228, Number 5272, November 14, 1970 (p. 613 n)

Dobzhansky, Theodosius   1900–75
Russian-American scientist

Molecular biology is Cartesian in its inspiration.
The Biology of Ultimate Concern
Chapter 2 (p. 20)
The New American Library, Inc. New York, New York, USA. 1967

Kornberg, Arthur   1918–
American biochemist

Molecular biology falters when it ignores the chemistry of the DNA blueprint — the enzymes and proteins, and their products — the integrated machinery and framework of the cell.
The Two Cultures: Chemistry and Biology
Biochemistry, Volume 26, Number 22, November 3, 1987 (p. 6890)

Ludwig, Carl Friedrich Wilhelm   1816–95
German physiologist

Whenever the body of an animal is subdivided to its ultimate parts, one always finally arrives at a limited number of chemical atoms…. One draws the conclusion in harmony with this observation, that all forms of activity arising in the animal body must be a result of the simple attractions and repulsions which would be observed in the coming together of those elementary objects.
Quarterly Review, 2nd ed.
Winter, 1858

Luria, Salvador Edward   1912–91
Italian-American microbiologist

Molecular biology deals with questions of molecular structure, and therefore is biochemistry; but it is not the classical biochemistry that emerged earlier in the twentieth century out of the concerns of medical, agricultural, and industrial researchers. Molecular biology is genetics because it deals with genes, their functions, and their products; but, in contrast with classical genetics, it has dealt mainly with organisms such as bacteria and viruses rather than peas, maize or fruit flies, whose study had established the classical rules of genetics.
A Slot Machine, a Broken Test Tube: An Autobiography
The Science Path: II. The High Reaches (pp. 83–84)
Harper & Row, Publishers. New York, New York, USA. 1984

Maddox, John Royden   1925–
Welsh chemist and physicist

…coffee-breaks in molecular laboratories are as marked by speculation as in any other field, but the published literature gives the impression that its authors are more concerned with the correctness of their observations than with their significance. Those with the good fortune to have the time to think about the data accumulated in the literature would probably reap a rich harvest of understanding. The explanation of the unreflective state of molecular biology is easily accounted for: competitiveness.
The Dark Side of Molecular Biology
Nature, Volume 363, Number 6424, 6 May 1993 (p. 13)

Wolpert, Lewis   1929–
British embryologist

…the revolution in molecular biology changed the paradigm from metabolism to information.
The Unnatural Nature of Science
Chapter 5 (p. 93)
Harvard University Press. Cambridge, Massachusetts, USA. 1992

MOLECULAR HYPOTHESIS

Kelvin, Lord William Thomson   1824–1907
Scottish engineer, mathematician, and physicist

There must be something in this molecular hypothesis and that as a mechanical symbol it is certainly not a mere hypothesis, but a reality.
Baltimore Lectures on Molecular Dynamics and the Wave Theory of Light
Lecture I (p. 15)
At The University Press. Cambridge, England. 1905

MOLECULE

Aldersey-Williams, Hugh 1959–
English author and journalist

A molecule is a messy thing. It has a gangling skeleton whose bones are chemical bonds and whose joints are its component atoms.
The Most Beautiful Molecule
Chapter 1 (p. 11)
John Wiley & Sons, Inc. New York, New York, USA. 1995

Ball, Philip 1962–
English science writer

Once upon a time molecular scientists had to deduce all they knew about molecules from measurements made on many billions of them simultaneously. This can be a risky business, since we cannot always be sure how such measurements are related to the properties of individual molecules, just as the noise that emanates from a football stadium or theatre hall reveals nothing of the individual conversations people are having. But advances in experimental techniques that enable studies of single molecules…what they look like, how they interact, how they move…have over the past two decades opened up an entirely new realm of molecular studies. We are starting to get to know molecules in person.
Stories of the Invisible
Chapter 5 (p. 127)
Oxford University Press, Inc. Oxford, England. 2001

…molecules are the smallest units of meaning in chemistry. It is through molecules, not atoms, that one can tell stories in the sub-microscopic world. They are the words: atoms are just the letters…. And in molecules, as in words, the order in which the component parts are put together matters: "save" and "vase" do not mean the same thing.
Stories of the Invisible
Chapter 1 (p. 13)
Oxford University Press, Inc. Oxford, England. 2001

Barrow, Gordon M.
Chemist

The chemist must learn to live in, and to feel at home in, the world of molecules. It is not enough that he knows the chemical constitution and chemical reactions of the materials around him. To be really effective and successful, he must also develop an intimacy with the molecular world. He must fit himself into the molecular scale of things. He must put that first drummed-in chemical fact that molecules are small in the very back of his mind and replace it by a consciousness that molecules are real, intricate, structural arrangements of atoms in space.
The Structure of Molecules
Introduction (p. 1)
W.A. Benjamin, Inc. New York, New York, USA. 1964

Clifford, William Kingdon 1845–79
English philosopher and mathematician

…we look forward to the time when the structure and motions in the inside of a molecule will be so well known that some future Kant or Laplace will be able to make an hypothesis about the history and formation of matter.
In Leslie Stephen and Frederick Pollock (eds.)
Lectures and Essays (Volume 1)
Atoms (p. 190)
Macmillan & Company Ltd. London, England. 1879

Crick, Francis Harry Compton 1916–2004
English biochemist

Almost all aspects of life are engineered at the molecular level, and without understanding molecules we can only have a very sketchy understanding of life itself.
What Mad Pursuit: A Personal View of Scientific Discovery
Chapter 5 (p. 61)
Basic Books, Inc., Publishers. New York, New York, USA. 1988

Dawkins, Richard 1941–
English ethologist, evolutionary biologist, and popular science writer

[Molecules of living things] are put together in much more complicated patterns than the molecules of nonliving things, and this putting together is done following programs, sets of instructions for how to develop, which the organisms carry around inside themselves. Maybe they do vibrate and throb and pulsate with "irritability," and glow with "living" warmth, but these properties all emerge incidentally. What lies at the heart of every living thing is not a fire, not warm breath, not a "spark of life." It is information, words, instructions. If you want a metaphor, don't think of fire and sparks and breath. Think instead of a billion discrete, digital characters carved in tablets of crystal. If you want to understand life, don't think about vibrant, throbbing gels and oozes, think about information technology.
The Blind Watchmaker
Chapter 5 (p. 112)
W.W. Norton & Company, Inc. New York, New York, USA. 1986

Dennett, Daniel Clement 1942–
American philosopher

Any assortment of objects, especially "sticky" objects like molecules, randomly stirred for long enough will give rise to every conceivable possible combination.
Consciousness Explained (p. 11)
Little, Brown & Company. Boston, Massachusetts, USA. 1991

Frankel, Felice 1945–
Science photographer
Whitesides, George M.
American chemist

Molecules — like ants, lemmings, herring, people — are happiest when surrounded by their own kind.

On the Surface of Things: Images of the Extraordinary in Science
Introduction (p. 7)
Chronicle Books. San Francisco, California, USA. 1997

Harrison, George R.
No biographical data available

A farm is a factory where the energy of light is used to make cheap simple molecules into valuable complex molecules.
When Physics Goes Farming
The Atlantic Monthly, July 1937

Hoffmann, Roald   1937–
Polish-born American chemist

It's a wild dance floor there at the molecular level.
In Philip Ball
Designing the Molecular World: Chemistry at the Frontier (p. 83)
Princeton University Press. Princeton, New Jersey, USA. 1994

Men (and women) are not as different from molecules as they think.
The Metamict State
Men and Molecules (p. 43)
University of Central Florida Press. Orlando, Florida, USA. 1987

Jeans, Sir James Hopwood   1877–1946
English physicist and mathematician

If we assume that the last breath of, say, Julius Caesar has by now become thoroughly scattered through the atmosphere, then the chances are that each of us inhales one molecule of it with every breath we take.
An Introduction to the Kinetic Theory of Gases
Chapter II (p. 32)
At The University Press. Cambridge, England. 1940

Kropotkin, Peter Alekseyevich   1842–1921
Russian revolutionary and geographer

The molecule becomes a particle of the universe on a microscopic scale — a microcosmosos which lives the same life.
Recent Science
Nineteenth Century, Volume 34, 1893 (p. 252)

Latham, Peter Mere   1789–1875
English physician

The very existence of ultimate molecules, or atoms, with the qualities which we so confidently assign to them, is a matter of the purest conjecture; it is entirely a fiction of the mind.
An Essay on the Philosophy of Medical Science
Part I, Chapter 4
Lea & Blanchard. Philadelphia, Pennsylvania, USA. 1844

Macfie, Ronald Campbell   1867–1931
Poet and physician

Life in living tissue is like nothing perhaps so much as a candle-flame. In the candle-flame the molecules are composed and decomposed, yet the candle-flame keeps always the same shape; but let us change the environment of these dancing, partner-changing molecules — let us conduct away their heat by means of some copper wire, or let us deprive the flame of oxygen — and out goes the candle.

Protoplasm is only a slow flame, easily extinguished. It is easy to understand how a little thing — a needle, a few grains of poison — may destroy a large organism when we remember how intricately correlated it is in its minutest parts. Break but a single thread in the warp and woof of life and the whole wonderful web, with its pictures and patterns, all comes asunder. Take but a single brick out of the great house of life and it falls into ruin. The construction of the wonderful organisms of vegetables and animals is a miracle and mystery, but their death is merely a chemical or mechanical commonplace.
Science, Matter and Immortality
Chapter 21 (p. 268)
William & Norgate. London, England. 1909

…molecules of different elements throb in different ways, and thus produce different light waves, which, when analysed by a spectroscope…give definite characteristic colours…we can discover what any substance is by heating it so as to agitate its molecules, and then analysing, by a spectroscope, the ripples of light caused by the throbbing molecules. In this way we may be said to be able to tell any substance by feeling its pulse, or by listening to its heart, and the spectroscope may be compared to a stethoscope.
Science, Matter and Immortality
Chapter 4 (p. 55)
William & Norgate. London, England. 1909

Mann, Thomas   1875–1955
German-born American novelist

For the molecule was composed of atoms, and the atom was nowhere near large enough even to be spoken of as extraordinarily small. It was so small, such a tiny, early, transitional mass, a coagulation of the unsubstantial, of the not-yet-substantial and yet substance-like, of energy, that it was scarcely possible yet — or, if it had been, was now no longer possible — to think of it as material, but rather as mean and border-line between the material and immaterial.
Translated by H.T. Lowe-Poeter
The Magic Mountain
Chapter V (p. 283)
Alfred A. Knopf. New York, New York, USA. 1966

Maxwell, James Clerk   1831–79
Scottish physicist

As long as we have to deal with only two molecules, and have all the data given us, we can calculate the result of their encounter; but when we have to deal with millions

of molecules, each of which has millions of encounters in a second, the complexity of the problem seems to shut out all hope of a legitimate solution.
In W.D. Niven (ed.)
The Scientific Papers of James Clerk Maxwell (Volume 2)
Molecules (p. 373)
At The University Press. Cambridge, England. 1890

I come from empyrean fires
From microscopic spaces,
Where molecules with fierce desires,
Shiver in hot embraces
The atoms clash, the spectra flash,
Projected on the screen,
The double D, magnesian b,
And Thallium's living green.
The Life of James Clerk Maxwell
To the Chief Musician upon Nabla: A Tyndallic Ode (p. 634)
Macmillan & Company Ltd. London, England. 1882

…though in the course of ages catastrophes have occurred and may yet occur in the heavens, though ancient systems may be dissolved and new systems evolved out of their ruins, the molecules out of which these systems are built — the foundation-stones of the material universe — remain unbroken and unworn.
In W.D. Niven (ed.)
The Scientific Papers of James Clerk Maxwell (Volume 2)
Molecules (p. 377)
At The University Press. Cambridge, England. 1890

…in the heavens we discover by their light, and by their light alone, stars so distant from each other that no material thing can ever have passed from one to another; and yet this light, which is to us the sole evidence of the existence of these distant worlds, tells us also that each of them is built up of molecules of the same kinds as those which we find on earth. A molecule of hydrogen, for example, whether in Sirius or in Arcturus, executes its vibrations in precisely the same time.
In W.D. Niven (ed.)
The Scientific Papers of James Clerk Maxwell (Volume 2)
Molecules (pp. 375–376)
At The University Press. Cambridge, England. 1890

[Molecular science is] one of those branches of study which deal with things invisible and imperceptible by our senses, and which cannot be subjected to direct experiment.
In W.D. Niven (ed.)
The Scientific Papers of James Clerk Maxwell (Volume 2)
Molecules (p. 361)
At The University Press. Cambridge, England. 1890

Montague, James J.
No biographical data available

Though men may boast of brain or brawn
And maids of soft attractions,
Such qualities depend upon
Their chemical reactions.
When Daniel, placid and serene,

Defied a den of lions,
He owed his calm, unflinching mien
To molecules and ions.
What's the Use of Worrying?
Industrial and Engineering Chemistry: News Edition, Volume 10, Number 20, 20 October 1932 (p. 257)

Morrison, Jim 1943–71
American singer, song writer and poet

Love hides in molecular structures.
Absolutely Live
Love Hides
Sung by The Doors
July 1969

Newman, Joseph S. 1892–1960
American poet

There's none to say how carbon first
Conceived it ocy-hydric thirst —
How nitrogen, in right proportion,
And sulphur joined the strange consortion —
But close upon the tenuous verge
Where shadows end, does life emerge,
And from these elemental five
Sprang proteid molecules alive!
Poems for Penguins and Other Lyrical Lapses
Biochemistry
Greenburg. New York, New York, USA. 1941

Quite recently to be exact,
Within a billion years, in fact.
Some time before the glacial drift
Had given the planet's face a lift,
A group of shameless molecules
Broke all the inorganic rules
And (C.I.O. epitomized!)
Spontaneously organized.
Poems for Penguins and Other Lyrical Lapses
Biology
Greenburg. New York, New York, USA. 1941

O'Brien, Flann 1911–66
Irish novelist and political commentator

Did you ever study the Mollycule Theory when you were a lad? he asked. Mick said not, not in any detail.

That is a very serious defalcation and an abstruse exacerbation, he said severely, but I'll tell you the size of it. Everything is composed of small Mollycules of itself, and they are flying around in concentric circles and arcs and segments and innumerable various other routes too numerous to mention collectively, never standing still or resting but spinning away and darting hither and thither and back again, all the time on the go.

…

Mollycules is a very intricate theorem and can be worked out with algebra but you would want to take it by degrees

with rulers and cosines and familiar other instruments and then at the wind-up not believe what you had proved at all.
The Dalkey Archive
Chapter 9 (pp. 87–88)
Hart-Davis, MacGibbon. London, England. 1968

von Baeyer, Hans Christian   1938–
Physicist and author

Coiled serpents capture the essence of the molecule that is missing from mechanical models — the element of mystery. They remind us that just beneath the surface of the dazzling atomic landscape recorded by modern technology, the paradoxes of quantum mechanics lurk like venomous snakes.
Taming the Atom
Chapter 5 (p. 88)
Random House, Inc. New York, New York, USA. 1992

Wald, George   1906–97
American biologist and biochemist

I have lived much of my life among molecules. They are good company. I tell my students to try to know molecules, so well that when they have some question involving molecules, they can ask themselves, What would I do if I were that molecule? I tell them, Try to feel like a molecule; and if you work hard, who knows? Some day you may get to feel like a big molecule!
Les Prix Nobel. The Nobel Prizes in 1967
Nobel banquet speech for award received in 1967
Nobel Foundation. Stockholm, Sweden. 1968

Weinberg, R. A.
No biographical data available

Can — and should — life be described in terms of molecules? For many, such description seems to diminish the beauty of Nature. For others of us, the wonder and beauty of nature are nowhere more manifest than in the submicroscopic plan of life.
The Molecules of Life
Scientific American, Volume 253, Number 4, October 1985 (p. 57)

Weiss, Paul A.   1898–1985
Chemist

...there is no phenomenon in a living system that is not molecular, but there is none that is only molecular, either.
Within the Gates of Science and Beyond
The Living System: Determinism Stratified (p. 270)
Hafner Publishing Company. New York, New York, USA. 1971

MOMENTUM

Descartes, René   1596–1650
French philosopher, scientist, and mathematician

We must consider motion in its two causes, the primary and universal cause, to which is due all the motion that is in the world, and the particular cause to which it is due that various portions of matter acquire the movements which before they had not. As to the former, it is evident to me that it must be attributed to God Himself, who in the beginning created matter along with motion and rest, and ever since has preserved these in the same quantity. For, though motion is nothing but a mode in the thing which is moved, yet it is of a definite amount that remains constant for the whole universe, though it varies in regard to the several parts.
Principles of Philosophy
Part II, 36, 42
E. Mellen Press. Lewistown, New York, USA. 1988

Duhem, Pierre-Maurice-Marie   1861–1916
French physicist and mathematician

If one wishes to draw a line of separation between the realm of ancient and modern science, it must be drawn at the instant when Jean Buridan conceived his theory of momentum, when he ceased to think of stars as kept in motion by certain divine beings and proclaimed that motions, celestial and terrestrial, are each controlled by the same mechanical laws.
Études, sur Leonardo da Vinci
A. Hermann. Paris, France. 1906–09

MONKEYS AND TYPEWRITERS

Eddington, Sir Arthur Stanley   1882–1944
English astronomer, physicist, and mathematician

...If I let my fingers wander idly over the keys of a typewriter it might happen that my screed made an intelligible sentence. If an army of monkeys were strumming on typewriters they might write all the books in the British Museum. The chance of their doing so is decidedly more favourable than the chance of the molecules returning to one half of the vessel.
The Nature of the Physical World
Chapter IV (p. 72)
The Macmillan Company. New York, New York, USA. 1930

Jeans, Sir James Hopwood   1877–1946
English physicist and mathematician

It was, I think, Huxley, who said that six monkeys, set to strum unintelligently on typewriters for millions of millions of years, would be bound in time to write all the books in the British Museum. If we examined the last page which a particular monkey had typed, and found that it had chanced, in its blind strumming, to type a Shakespeare sonnet, we should rightly regard the occurrence as a remarkable accident, but if we looked through all the millions of pages the monkeys had turned off in untold millions of years, we might be sure of finding a Shakespeare sonnet somewhere amongst them, the

product of the blind play of chance. In the same way, millions of millions of stars wandering blindly through space for millions of millions of years are bound to meet with every sort of accident, and so are bound to produce a certain limited number of planetary systems in time. Yet the number of these must be very small in comparison with the total number of stars in the sky.
The Mysterious Universe
Chapter I (p. 4)
The Macmillan Company. New York, New York, USA. 1932

Koestler, Arthur   1905–83
Hungarian-born English writer

Neo-Darwinism does indeed carry the nineteenth-century brand of materialism to its extreme limits — to the proverbial monkey at the typewriter, hitting by pure chance on the proper keys to produce a Shakespeare sonnet.
The Case of the Midwife Toad (p. 30)
New York, New York, USA. 1972

Russell, Bertrand Arthur William   1872–1970
English philosopher, logician, and social reformer

There is a special department of hell for students of probability. In this department there are many typewriters and many monkeys. Every time that a monkey walks on a typewriter, it types by chance one of Shakespeare's sonnets.
Nightmares of Eminent Persons
The Metaphysician's Nightmare (p. 29)
The Bodley Head. London, England. 1954

Russo, Richard   1949–
American novelist

In a novel, two characters discuss the glitch in a computer which causes it to scroll an endless series of meaningless symbols: He sighs. "It casts serious doubt on the old theory that an infinite number of monkeys at an infinite number of typewriters would eventually write the Great American Novel, doesn't it?"
Straight Man (p. 129)
Random House, Inc. New York, New York, USA. 1996

Synge, John L.   1897–1995
Irish mathematician and physicist

"But not the sonnets?" asked the Orc, quizzically. "Yes, of course," retorted the Plumber, "The sonnets too. And the Bible. And the Koran. And that poem of mine which you have just recited.…"

"But suppose," said the Orc, "that our monkey became very fond of some particular word, perhaps some naughty little four-letter word, and went on typing that word over and over again and never any other. I cannot see, in that case, how he would type even one play." "That would be quite an exceptional case," answered the Plumber. "Eddington had in mind a haphazard performance. The

monkey types the keys at random, and the outcome is governed by pure chance. And by pure chance the plays of Shakespeare emerge, after a long time of course." "Doubtless, doubtless," muttered the Orc, reflectively. "Poor monkey! How bored he would get! For he would reproduce the plays of Shakespeare not once but many times, in fact an infinite number of times."
Kandelman's Krim
Chapter Eleven (p. 145)
Jonathan Cape. London, England. 1957

MONOGRAPH

Doyle, Sir Arthur Conan   1859–1930
Scottish writer

"Oh, didn't you know?" he cried, laughing. "Yes, I have been guilty of several monographs. They are all upon technical subjects. Here, for example, is one 'Upon the Distinction between the Ashes of the Various Tobaccos.' In it I enumerate a hundred and forty forms of cigar, cigarette, and pipe tobacco, with coloured plates illustrating the difference in the ash. It is a point which is continually turning up in criminal trials, and which is sometimes of supreme importance as a clue. If you can say definitely, for example, that some murder has been done by a man who was smoking an Indian lunkah, it obviously narrows your field of search. To the trained eye there is as much difference between the black ash of a Trichinopoly and the white fluff of bird's-eye as there is between a cabbage and a potato."
In William S. Baring-Gould (ed.)
The Annotated Sherlock Holmes (Volume 1)
The Sign of the Four, Chapter 1 (p. 612)
Wings Books. New York, New York, USA. 1967

MONOPOLE

Dirac, Paul Adrian Maurice   1902–84
English theoretical physicist

From the theoretical point of view one would think that monopoles [magnets with one pole] should exist, because of the prettiness of the mathematics. Many attempts to find them have been made, but all have been unsuccessful. One should conclude that pretty mathematics by itself is not an adequate reason for nature to have made use of a theory. We still have much to learn in seeking for the basic principles of nature.
In Heinz R. Pagels
Perfect Symmetry: The Search for the Beginning of Time
Part Three, Chapter 1 (p. 284)
Simon & Schuster. New York, New York, USA. 1985

Gamow, George   1904–68
Russian-born American physicist

Two Monopoles worshipped each other,
And all of their sentiments clicked.

Still, neither could get to his brother,
Dirac was so fearfully strict!
Thirty Years That Shook Physics
Second Part (p. 202)
Doubleday & Company, Inc. Garden City, New York, USA. 1966

MONSTER

Miller, Hugh   1802–56
Scottish geologist and theologian

…the night comes on, and the shadows of the woods and rocks deepen: there are uncouth sounds along the beach and in the forest; and new monsters of yet stranger shape are dimly discovered moving amid the uncertain gloom.
Sketch-Book of Popular Geology
Lecture Forth (p. 151)
William P. Nimmo & Company. Edinburgh, Scotland. 1880

MOON

Abbey, Edward   1927–89
American environmentalist and nature writer

The old moon, like a worn and ancient coin, is still hanging in the west when I awake.
Desert Solitaire
Terra Incognita: Into the Maze (p. 289)
Ballantine Books. New York, New York, USA. 1968

Addison, Joseph   1672–1719
English essayist, poet, and statesman

Soon as the evening shades prevail,
The moon takes up the wondrous tale,
And nightly to the listening earth
Repeats the story of her birth.
In John Matthews Manley (ed)
English Poetry
Hymn (p. 220)
Ginn & Company. Boston, Massachusetts, USA. 1907

Alger, William R.   1822–1905
Unitarian minister and author

The moon is a silver pin-head vast,
That holds the heaven's tent-hangings fast.
Poetry of the Orient
The Use of the Moon
Roberts Brothers. Boston, Massachusetts, USA. 1866

Blake, William   1757–1827
English poet, painter, and engraver

The moon like a flower
In heaven's high bower,
With silent delight
Sits and smiles on the night.
The Complete Poetry and Prose of William Blake
The Moon
University of California Press. Berkeley, California, USA. 1982

Borman, Frank   1928–
American astronaut

The moon is a different thing to each of us.
From Apollo VIII
December 24, 1968

Bronte, Charlotte   1816–55
English author

Where, indeed, does the moon not look well? What is the scene, confined or expansive, which her orb does not follow? Rosy or fiery, she mounted now above a not distant bank; even while we watched her flushed ascent, she cleared to gold, and in a very brief space, floated up stainless into a now calm sky.
Life and Works of the Sisters Brontë (Volume 3)
Villette La Terrasse (p. 214)
AMS Press Inc. New York, New York, USA. 1973

Burton, Robert   1577–1640
English clergyman and scholar

Doth the moon care for the barking of a dog?
The Anatomy of Melancholy (Volume 2)
Part II, Sect. III, Memb. VII (p. 231)
AMS Press, Inc. New York, New York, USA. 1973

Burton, Sir Richard Francis   1821–90
English explorer

That gentle Moon, the lesser light, the Lover's lamp, the Swain's delight,

A ruined world, a globe burnt out, a corpse upon the road of night.
The Kasidah of Haji Abdu El-Yezdi (p. 10)
McBride. New York, New York, USA. 1929

Butler, Samuel   1612–80
English novelist, essayist, and critic

The moon pull'd off her veil of light,
That hides her face by day from sight
(Mysterious veil, of brightness made,)
That's both her lustre and her shade),
And in the lantern of the night,
With shining horns hung out her light.
The Poetical Works of Samuel Butler (Volume 1)
Part II, Canto I, l. 905
Bell & Daldy. London, England. 1835

He made an instrument to know
If the moon shine at full or no;
That would, as soon as e'er she shone straight,
Whether 'twere day or night demonstrate;
Tell what her d'ameter to an inch is,
And prove that she's not made of green cheese.
The Poetical Works of Samuel Butler (Volume 1)
Part II, Canto III, l. 261
Bell & Daldy. London, England. 1835

Carroll, Lewis (Charles Dodgson)   1832–98
English writer and mathematician

The moon was shining sulkily,
Because she thought the sun
Had got no business to be there
After the day was done —
The Complete Works of Lewis Carroll
Through the Looking-Glass
Chapter IV (p. 183)
The Modern Library. New York, New York, USA. 1936

Cawein, Madison Julius   1865–14
American poet

Into the sunset's turquoise marge
The moon dips, like a pearly barge;
Enchantment sails through magic seas,
To fairland Hesperides,
Over the hills and away.
Poems
At Sunset, Stanza 1
The Macmillan Company. New York, New York, USA. 1911

Collins, Michael   1880–1922
Irish soldier and politician

It was a totally different moon than I had ever seen before.
The moon that I knew from old was a yellow flat disk, and
this was a huge three-dimensional sphere, almost a ghostly
blue-tinged sort of pale white. It didn't seem like a very
friendly place or welcoming place. It made one wonder
whether we should be invading its domain or not.
In Kevin W. Kelley
The Home Planet
With Plate 39
Addison-Wesley Publishing, Inc. Reading, Massachusetts, USA. 1988

Conrad, Joseph   1857–1924
Polish-born English novelist

There is something haunting in the light of the moon; it
has all the dispassionateness of a disembodied soul, and
something of its inconceivable mystery.
Lord Jim
Chapter XXIV (p. 213)
Rinehart & Company, Inc. New York, New York, USA. 1957

Croly, George   1780–1860
German chemist and physician

How like a queen comes forth the lonely Moon
From the slow opening curtains of the clouds
Walking in beauty to her midnight throne!
Gems, Principally from the Antique
Diana
Printed for Hurst, Robinson & Company. London, England. 1822

Darwin, Erasmus   1731–1802
English physician and poet

And hail their queen, fair regent of the night.

The Botanic Garden
Part I, Canto II, III, l. 90
Jones & Company. London, England. 1825

Empedocles of Acragas   ca. 490 BCE–430 BCE
Greek pre-Socratic philosopher

A borrowed light, circular in form, it revolves about the
earth, as if following the track of a chariot.
In Arthur Fairbanks
The First Philosophers of Greece
Book I
Fragment 154 (p. 177)
Charles Scribner's Sons. New York, New York,
USA. 1898

Flammarion, Camille   1842–1925
French astronomer and author

Orb of dream and mystery, pale sun of the night, solitary
globe wandering in the silent firmament, the moon has in
all times and among all nations peculiarly attracted attention and thought.
Popular Astronomy: A General Description of the Heavens
Book II, Chapter VI (p. 145)
Chatto & Windus. London, England. 1894

Frost, Robert   1874–1963
American poet

The Moon for all her light and grace
Has never learned to know her place.
The notedest astronomers
Have set the dark aside for hers.
Complete Poems of Robert Frost
Two Leading Lights
Henry Holt & Company. New York, New York, USA. 1949

Fry, Christopher   1907–2005
English playwright

…the moon is nothing
But a circumambulating aphrodisiac
Divinely subsidized to provoke the world
Into a rising birth-rate.
The Lady's Not for Burning
Act Three (p. 66)
Oxford University Press, Inc. New York, New York, USA. 1950

Haggard, H. Rider   1856–1925
English novelist

The sky aft was dark as pitch, but the moon still shone
brightly ahead of us and lit up the blackness. Beneath its
sheen a huge white-topped breaker, twenty feet high or
more, was rushing on to us. It was on the break — the
moon shone on its crest and tipped its foam with light.
On it rushed beneath the inky sky, driven by the awful
squall behind it.
The Favorite Novels of H. Rider Haggard
She (p. 195)
Blue Ribbon Books, Inc. New York, New York, USA. 1928

Homer (Smyrns of Chios)   fl. 750 BCE
Greek poet

As when the stars shine clear, and the moon is bright —
there is not a breath of air, not a peak nor glade nor jutting
headland, but it stands out in the ineffable radiance that
breaks from the serene of heaven…
In *Great Books of the Western World* (Volume 4)
The Iliad of Homer
Book VIII, l. 555 (p. 56)
Encyclopædia Britannica, Inc. Chicago, Illinois, USA. 1952

Hood, Thomas   1582–98
English poet and editor

Mother of light! how fairly dost thou go
Over those hoary crests, divinely led!
Art thou that huntress of the silver bow
Fabled of old? Or rather dost thou tread
Those cloudy summits thence to gaze below,
Like the wild chamois from her Alpine snow,
Where hunters never climbed — secure from dread?
The Complete Poetical Works of Thomas Hood
Ode to the Moon
Greenwood Press, Publishers. Westport, Connecticut, USA. 1980

Huxley, Julian   1887–1975
English biologist, philosopher, and author

By death the moon was gathered in
Long ago, ah long ago;
Yet still the silver corpse must spin
And with another's light must glow.
Her frozen mountains must forget
Their primal hot volcanic breath,
Doomed to revolve for ages yet,
Void amphitheatres of death.
The Captive Shrew and Other Poems of a Biologist
Cosmic Death
Harper & Brothers. New York, New York, USA. 1933

Ingelow, Jean   1820–97
English poet and novelist

Such a slender moon, going up and up,
Waxing so fast from night to night,
And swelling like an orange flower-bud, bright,
Fated, methought, to round as to a golden cup,
And hold to my two lips life's best of wine.
The Poetical Works of Jean Ingelow
Songs of the Night Watches
The First Watch, pt. II
John B. Alden. New York, New York, USA. 1883

Jastrow, Robert   1925–
American space scientist
Newell, Homer E.
No biographical data available

The moon is the Rosetta stone of the solar system, and
to the student of the origin of the earth and planets, this

lifeless body is even more important than Mars and
Venus.
Why Land on the Moon?
The Atlantic Monthly, Volume 211, Number 2; August, 1963 (p. 43)

Jonson, Ben   1573?–1637
English dramatist and poet

Queen and huntress, chaste and fair,
Now the sun is laid to sleep,
Seated in thy silver car,
State in wonted manner keep.
Hesperus entreats thy light
Goddess, excellently bright!
Hymn
To Cynthia

Keats, John   1795–1821
English Romantic lyric poet

The moon put forth a little diamond peak
No bigger than an unobserved star,
Or tiny point of fairy scimitar.
The Complete Poetical Works and Letters of John Keats
Endymion
Book IV, l. 499
Houghton Mifflin Company. Boston, Massachusetts, USA. 1890

Lear, Edward   1812–88
English humorist and artist

They dined on mince, and slices of quince,
Which they ate with a runcible spoon;
And hand in hand, on the edge of the sand,
They danced by the light of the moon,
The moon, the moon,
They danced by the light of the moon.
In Tony Palazzo
Edward Lear's Nonsense Book
The Owl and the Pussycat
Garden City Books. Garden City, New York, USA. 1956

Lightner, Alice
No biographical data available

Queen of Heaven, fair of face,
Undefiled by alien feet;
Where the sun's untrammeled heat
Meets the cold of outer space;
Soon no more the Queen of Night,
For your conquest is in sight.
To the Moon
Nature Magazine, April 1957 (p. 213)

Longfellow, Henry Wadsworth   1807–82
American poet

Saw the moon rise from the water,
Rippling, rounding from the water,
Saw the flecks and shadows on it,
Whispered, "What is that, Nokomis?"
And the good Nokomis answered,

"Once a warrior very angry,
Seized his grandmother and threw her
Up into the sky at midnight;
Right against the moon he threw her;
'Tis her body that you see there."
The Poetical Works of Henry Wadsworth Longfellow
Hiawatha, Hiawatha's Childhood
Houghton Mifflin Company. Boston, Massachusetts,
USA. 1883

Lovell, James A.   1928–
American astronaut

The moon is essentially gray, no color. It looks like plaster of Paris, like dirty beach sand with lots of footprints in it.
Washington Post
25 December 1968

Milton, John   1608–74
English poet

…now glow'd the firmament
With living sapphires; Hesperus, that led
The starry host, rode brightest, till the moon,
Rising in clouded majesty, at length
Apparent queen, unveil'd her peerless light,
And o'er the dark her silver mantle threw.
In *Great Books of the Western World* (Volume 32)
Paradise Lost
Book IV, l. 604–609
Encyclopædia Britannica, Inc. Chicago, Illinois, USA. 1952

Moore, Thomas   1779–1852
Irish poet

The moon looks
On many brooks,
The brook can see no moon but this.
The Poetical Works of Thomas Moore
While Gazing on the Moon's Light
Lee & Shepard. Boston, Massachusetts, USA. 1873

Muir, John   1838–1914
American naturalist

The moon is looking down into the canon, and how marvelously the great rocks kindle to her light! Every dome, and brow, and swelling boss touched by her white rays, glows as if lighted with snow.
Steep Trails
Chapter II (p. 23)
Norman S. Berg, Publisher. Dunwoody, Georgia, USA. 1970

Rankin, William H.
No biographical data available

Someday I would like to stand on the moon, look down through a quarter of a million miles of space and say, "There certainly is a beautiful earth out tonight."
The Man Who Rode the Thunder
Prentice-Hall, Inc., Englewood Cliffs, New Jersey, USA. 1960

Robbins, Tom   1936–
American writer

Our Moon has surrendered none of its soft charm to technology. The pitter-patter of little spaceboots has in no way diminished its mystery.
Even Cowgirls Get the Blues
Chapter 19 (p. 60)
Houghton Mifflin Company. Boston, Massachusetts, USA. 1976

Ross, Sir Ronald   1857–1932
English bacteriologist

O Moon! When I look at thy beautiful face,
Careening along through the boundaries of space
The thought has quite frequently come to my mind
If ever I'll gaze on thy glorious behind.
In Harriet Monroe (ed)
Poetry
O Moon
Modern Poetry Association. Chicago, Illinois, USA.

Sappho   630 BCE–570 BCE
Greek lyric poet

Stars near the lovely moon cover their own bright faces when she is roundest and lights up the earth with her silver.
Poems by Sappho (p. 4)
Charles Scribner's Sons. New York, New York, USA. 1924

Serviss, Garrett P.   1851–1929
American science fiction writer

The imagination of mankind has never resisted the fascination of the moon.
Astronomy with the Naked Eye
Chapter XVIII (p. 226)
Harper & Brothers. New York, New York, USA. 1908

Shakespeare, William   1564–1616
English poet, playwright, and actor

Therefore the moon, the governess of floods,
Pale in her anger, washes all the air,
That rheumatic diseases do abound.
And thorough this distemperature we see
The seasons alter: hoary-headed frosts
Fall in the fresh lap of the crimson rose,
And on old Hiems' thin and icy crown
An odorous chaplet of sweet summer buds
Is, as in mockery, set.
In *Great Books of the Western World* (Volume 26)
The Plays and Sonnets of William Shakespeare (Volume 1)
A Midsummer-Night's Dream
Act II, Scene I
Encyclopædia Britannica, Inc. Chicago, Illinois, USA. 1952

It is the very error of the moon;
She comes more nearer earth than she was wont,
And drives men mad.
In *Great Books of the Western World* (Volume 27)
The Plays and Sonnets of William Shakespeare (Volume Two)

Othello, The Moor of Venice
Act V, Scene ii, l. 107–111
Encyclopædia Britannica, Inc. Chicago, Illinois, USA. 1952

Shelley, Mary   1797–1851
English Romantic writer

…the moon gazed on my midnight labours, while, with unrelaxed and breathless eagerness, I pursued nature to her hiding-places.
Frankenstein
Chapter 4 (p. 43)
Running Press. Philadelphia, Pennsylvania, USA. 1990

Shelley, Percy Bysshe   1792–1822
English poet

The young moon has fed
Her exhausted horn
With the sunset's fire.
The Complete Poetical Works of Percy Bysshe Shelley
Hellas Semi-Chorus II
Houghton Mifflin Company. Boston, Massachusetts, USA. 1901

Tennyson, Alfred (Lord)   1809–92
English poet

All night, through archways of the bridged pearl
And portals of pure silver, walks the moon.
Alfred Tennyson's Poetical Works
Sonnet
Oxford University Press, Inc. London, England. 1953

Thurber, James   1894–1961
American writer and cartoonist

"The moon is 300,000 miles away," said the Royal Mathematician. "It is round and flat like a coin, only it is made of asbestos, and it is half the size of this kingdom. Furthermore, it is pasted on the sky."
Many Moons
Harcourt Brace & Company. San Diego, California, USA. 1971

Tolstoy, Alexei   1882–1945
Russian writer

"Which is more useful, the Sun or the Moon?" asks Kuzma Prutkov, the renowned Russian philosopher, and after some reflection he answers himself: "The Moon is the more useful, since it gives us its light during the night, when it is dark, whereas the Sun shines only in the daytime, when it is light anyway."
Quoted by George Gamow
The Birth and Death of the Sun
Chapter I (p. 1)
The Viking Press. New York, New York, USA. 1945

Verne, Jules   1828–1905
French novelist

There is no one among you, my brave colleagues, who has not seen the Moon, or at least, heard speak of it.
From the Earth to the Moon and Round the Moon

From Earth to the Moon, Chapter II (p. 12)
A.L. Burt Company. New York, New York, USA. 1890

Williams, Dafydd (Dave) Rhys

We'll go back to the moon, this time for a much longer period of time. We'll build lunar outposts. We'll send a crew to Mars. There are no ifs around it. It's going to happen.
Ground Control to Dr. Dave
The McGill Reporter, Volume 31, Number 3 (8 October 1998)

MOON LANDING

Armstrong, Neil A.   1930–
American astronaut

That's one small step for a man, one giant leap for mankind.
Men Walk on Moon
New York Times, L5, column 3, 21 July 1969

Beckett, Chris
No biographical data available

The Moon landings were not about gathering data or testing hypotheses; they were about theatre, about the enactment of many mythical themes that were fed and nurtured by such a spectacular event. It was about the power of humankind, the power of technology, our ability to overcome the apparently impossible and to conquer not just Earth but the whole Universe. And yet at the same time it was about the smallness of humankind, our vulnerability the fact that we inhabit a single small planet, surrounded by emptiness…
New Scientist, November 11, 1989

Crew of Apollo 11
Here Men from The Planet Earth
First Set Foot upon The Moon
July, 1969 AD
We Came in Peace for All Mankind.
Plaque left behind on the moon's surface

Hoffer, Eric   1902–83
American longshoreman and philosopher

Our passionate preoccupation with the sky, the stars, and a God somewhere in outer space is a homing impulse. We are drawn back to where we came from.
Reactions to Man's Landing on the Moon Show Broad Variations in Opinions
New York Times, A6, column 2, 21 July 1969

Koestler, Arthur   1905–83
Hungarian-born English writer

Prometheus is reaching out for the stars with an empty grin on his face.

Reactions to Man's Landing on the Moon Show Broad Variations in Opinions
New York Times, A6, column 6, 21 July 1969

Nabokov, Vladimir   1899–1977
Russian-American writer

Treading the soil of the moon, palpating its pebbles, tasting the panic and splendor of the event, feeling in the pit of one's stomach the separation from terra…these form the most romantic sensation an explorer has ever known…this is the only thing I can say about the matter. The utilitarian results do not interest me.
Reactions to Man's Landing on the Moon Show Broad Variations in Opinions
New York Times, A6, column 5, 21 July 1969

MORPHOLOGY

Thompson, Sir D'Arcy Wentworth   1860–1948
Scottish zoologist and classical scholar

The waves of the sea, the little ripples on the shore, the sweeping curve of the sandy bay between the headlands, the outline of the hills, the shape of the clouds, all these are so many riddles of form, so many problems of morphology.
On Growth and Form (Volume 1)
Chapter I (p. 10)
At The University Press. Cambridge, England. 1951

MOTION

Burton, Robert   1577–1640
English clergyman and scholar

The heavens themselves run continually round, the sun riseth and sets, the moon increaseth, stars and planets keep their constant motions, the air is tossed by the winds, the waters ebb and flow, to their conservation no doubt, to teach us that we should ever be in motion.
The Anatomy of Melancholy (Volume 2)
Part II, Sect. II, Memb. IV (p. 80)
AMS Press, Inc. New York, New York, USA. 1973

Butterfield, Herbert   1900–79
English historian and philosopher of history

Of all the intellectual hurdles which the human mind has confronted and has overcome in the last fifteen hundred years, the one which seems to me to have been the most amazing in character and the most stupendous in the scope of its consequences is the one relating to the problem of motion…
The Origins of Modern Science
Chapter One (p. 3)
The Macmillan Company. New York, New York, USA. 1961

Carroll, Lewis (Charles Dodgson)   1832–98
English writer and mathematician

"Well, in our country," said Alice, still panting a little, "you'd generally get to somewhere else — if you ran very fast for a long time, as we've been doing."

"A slow sort of country!" said the Queen. "Now here, you see, it takes all the running you can do, to keep in the same place. If you want to get somewhere else, you must run at least twice as fast as that."
The Complete Works of Lewis Carroll
Through the Looking-Glass
Chapter II (p. 166)
The Modern Library. New York, New York, USA. 1936

Dee, John   1527–1609
English mathematician and occultist

Whatever is in the universe is continuously moved by some species of motion.
Translated by Wayne Schumaker
John Dee on Astronomy
XVI (p. 129)
University of California Press. Berkeley, California, USA. 1978

Descartes, René   1596–1650
French philosopher, scientist, and mathematician

…God always preserves in the world just so much motion as He impressed on it at its first creation.
Principles of Philosophy
Part II, 36, 42
E. Mellen Press. Lewistown, New York, USA. 1988

Eddington, Sir Arthur Stanley   1882–1944
English astronomer, physicist, and mathematician

We not merely deduce [a] three-dimensional world; we see it. But we have no such aid in synthesising different motions. Perhaps if we had been endowed with two eyes moving with different velocities our brains would have developed the necessary faculty; we should have perceived a kind of relief in the fourth dimension so as to combine into one picture the aspect of things seen with different motions.
Space, Time and Gravitation: An Outline of the General Relativity Theory
Chapter II (p. 32)
At The University Press. Cambridge, England. 1921

It is curious that the philosophical denial of absolute motion is readily accepted, whilst the denial of absolute simultaneity appears to many people revolutionary.
Space, Time and Gravitation: An Outline of the General Relativity Theory
Chapter III (p. 51)
At The University Press. Cambridge, England. 1921

Galilei, Galileo   1564–1642
Italian physicist and astronomer

My purpose is to set forth a very new science dealing with a very ancient subject. There is, in nature, perhaps nothing older than motion, concerning which the books

written by philosophers are neither very few nor small; nevertheless, I have discovered by experiment some properties of it which are worth knowing and which have not hitherto been either observed or demonstrated.
In *Great Books of the Western World* (Volume 28)
Dialogues Concerning the Two New Sciences
Third Day, Change of Position (p. 197)
Encyclopædia Britannica, Inc. Chicago, Illinois, USA. 1952

…we have decided to consider the phenomena of bodies falling with an acceleration such as actually occurs in nature and to make this definition of accelerated motion exhibit the essential features of observed accelerated motions.
In *Great Books of the Western World* (Volume 28)
Dialogues Concerning the Two New Sciences
Third Day, Naturally Accelerated Motion (p. 200)
Encyclopædia Britannica, Inc. Chicago, Illinois, USA. 1952

Hutton, W.
No biographical data available

Motion is the soul of the universe…
The Book of Nature Laid Open
Chapter XV (p. 173)
Joseph Milligan, Georgetown. 1822

Huygens, Christiaan   1629–95
Dutch mathematician, astronomer, and physicist

It is inconceivable to doubt that light consists in the motion of some sort of matter. For when one considers its production, one sees that here upon the earth it is chiefly engendered by fire and flame which contain without doubt bodies that are in rapid motion, since they dissolve and melt many other bodies, even the most solid; or when one considers its effects, one sees that when light is collected, as by concave mirrors, it has the property of burning as a fire does, that is to say, it disunites the particles of bodies. This is assuredly the mark of motion, at least in the true philosophy, in which once conceives the cause of all natural effects in terms of mechanical motions. This, in my opinion, we must necessarily do, or else renounce all hopes of ever comprehending anything in physics.
In *Great Books of the Western World* (Volume 34)
Treatise on Light
Chapter One. On Rays Propagated in Straight Lines (p. 553)
Encyclopædia Britannica, Inc. Chicago, Illinois, USA. 1952

Jeans, Sir James Hopwood   1877–1946
English physicist and mathematician

…the laws which nature obeys are less suggestive of those which a machine obeys in its motion than of those which a musician obeys in writing a fugue, or a poet in composing a sonnet. The motions of electrons and atoms do not resemble those of the parts of a locomotive so much as those of the dancers in a cotillion. And if the "true essence of substances" is for ever unknowable, it does not matter whether the cotillion is danced at a ball

in real life, or on a cinematography screen, or in a story of Boccaccio.
The Mysterious Universe
Chapter V (p. 168)
The Macmillan Company. New York, New York, USA. 1932

Leacock, Stephen   1869–1944
Canadian humorist

It was Einstein who made the real trouble. He announced in 1905 that there was no such thing as absolute rest. After that there never was.
The Boy I Left Behind Me
Chapter VI (p. 171)
The Bodely Head. London, England. 1947

Locke, John   1632–1704
English philosopher and political theorist

The parts of pure space are immovable, which follows from their inseparability; motion being nothing but change of distance between any two things; but this cannot be between parts that are inseparable; which therefore must needs be at perpetual rest one amongst other.
In *Great Books of the Western World* (Volume 35)
An Essay Concerning Human Understanding
Book II, Chapter XIII, Section 14 (p. 151)
Encyclopædia Britannica, Inc. Chicago, Illinois, USA. 1952

Lucretius   ca. 99 BCE–55 BCE
Roman poet

For whenever bodies fall through water and thin air, they must quicken their descents in proportion to their weights, because the body of water and subtle nature of air cannot retard everything in equal degree, but more readily give way [when] overpowered by the heavier…
In *Great Books of the Western World* (Volume 12)
Lucretius: on the Nature of Things
Book Two, l. 230–234 (p. 18)
Encyclopædia Britannica, Inc. Chicago, Illinois, USA. 1952

Maxwell, James Clerk   1831–79
Scottish physicist

Absolute space is conceived as remaining always similar to itself and immovable. The arrangement of the parts of space can no more be altered than the order of the portions of time. To conceive them to move from their places is to conceive a place to move away from itself.
Matter and Motion
Chapter I, Section 18
Dover. New York, New York, USA. 1953

Meredith, George   1828–1909
English novelist and poet

So may we read, and little find them cold:
Not frosty lamps illuminating dead space,
Not distant aliens, not senseless Powers.
The fire is in them whereof we are born;

The music of their motion may be ours.
A Reading of Earth
Meditation under Stars (p. 120)
Macmillan & Company Ltd. London, England. 1888

Newton, Sir Isaac   1642–1727
English physicist and mathematician

The quantity of motion is the measure of the same, arising from the velocity and quantity of matter conjointly.
Mathematical Principles of Natural Philosophy
Definitions, Definition II
E.P. Dutton & Company, Inc. New York, New York, USA. 1922

Plato   428 BCE–347 BCE
Greek philosopher

Motion…has many forms, and not one only; two of them are obvious enough even to wits no better than ours; and there are others, as I imagine, which may be left to wiser persons.
In *Great Books of the Western World* (Volume 7)
The Republic
Book VII, Section 530 (p. 396)
Encyclopædia Britannica, Inc. Chicago, Illinois, USA. 1952

Raman, Chandrasekhar Venkata   1888–1970
Indian physicist

Once during mass, Galileo in church
Conducted a major scientific search.
He measured with his pulse how a lamp did swing
That was to the ceiling tied with a string.
A Fable for Physicists, The Pendulum Period
The Physics Teacher, Volume 18, Number 7, October 1990 (p. 488)

Regnault, Nöel   1702–62
Jesuit mathematician

Nothing seems more clear at first than the Idea of Motion, and yet nothing is more obscure when one comes to search thoroughly into it.
Philosophical Conversations (Volume 1)
Conversation VI (p. 58)
Printed for W. Innys, C. Davis & N. Prevost. London, England. 1731

Sarpi, Fra Paolo   1552–1623
Venetian patriot, scholar, and church reformer

To give us the science of motion God and Nature have joined hands and created the intellect of Galileo.
In Morris Kline
Mathematics and the Physical World
Chapter 12 (p. 181)
Dover Publications, Inc. New York, New York, USA. 1981

Shakespeare, William   1564–1616
English poet, playwright, and actor

Two stars keep not their motion in one sphere.
In *Great Books of the Western World* (Volume 26)
The Plays and Sonnets of William Shakespeare (Volume 1)
The First Part of King Henry the Fourth

Act V, Scene iv
Encyclopædia Britannica, Inc. Chicago, Illinois, USA. 1952

Steele, Joel Dorman   1836–86
American educator and textbook writer

Rest is nowhere. The winds that come and go, the ocean that uneasily throbs along the shore, the earth that revolves about the sun, the light that darts through space — all tell of a universal of Nature. The solidest body hides within it inconceivable velocities. Even the molecules of garnet and iron have their orbits as do the stars, and move as ceaselessly.
Popular Physics
Chapter II (p. 19)
American Book Company. New York, New York, USA. 1896

Thierry, Paul Henri, Baron d'Holbach   1723–89
German-born French philosopher

Every thing in the universe is in motion; the essence of nature is to act; and if we consider attentively its parts, we shall see that there is not a particle which enjoys absolute repose.
Translated by M. Mirabaud
System of Nature or, The Laws of the Moral and Physical World (Volume First)
Part First, Chapter II (p. 27)
Published by R. Benson. Philadelphia, Pennsylvania, USA. 1808

Thomson, Thomas
No biographical data available

Substances may either be examined in a state of rest, or as acting upon each other and producing changes on each other. The knowledge derived from the first of these views, is called Natural History; that which we can obtain by the second, is distinguished by the name Science. But bodies cannot act upon each other without producing motion, and the motions produced by such actions are of two kinds; either so great as to be visible to our senses, and capable of being measured by the space passed over; or so small as not to be distinguishable by our senses, except by the effects produced. The phenomena connected with the first of these kinds of motions constitute what is called Natural Philosophy or Mechanical Philosophy in this country, and on the Continent, Physics. The phenomena connected with the imperceptible motion belong to the science called Chemistry.
History of the Royal Society from Its Institution to the End of the Eighteenth Century
Book III (p. 311)
Printed for Robert Baldwin. London, England. 1812

Tyndall, John   1820–93
Irish-born English physicist

But is it in the human mind to imagine motion without at the same time imagining something moved? Certainly not. The very conception of motion includes that of a moving body.

Light and Electricity (pp. 123–124)
D. Appleton & Company. New York, New York, USA. 1873

Walters, Marcia C.
No biographical data available

The fact that the photon gets mass from its motion
Is a widely accepted Einsteinion notion,
This doesn't apply to we mortals, alas —
For the smaller our motion the greater our mass.
Filler
The Physics Teacher, Volume 5, Number 8, November 1967 (p. 384)

Wells, H. G. (Herbert George)   1866–1946
English novelist, historian, and sociologist

"And here," he said, and opened the hand that held the glass. Naturally I winced, expecting the glass to smash. But so far from smashing, it did not even seem to stir; it hung in mid-air — motionless. "Roughly speaking," said Gibberne, "an object in these latitudes falls 16 feet in the first second. This glass is falling 16 feet in a second now. Only, you see, it hasn't been falling yet for the hundredth part of a second. That gives you some idea of the pace of my Accelerator."
28 Science Fiction Stories of H.G. Wells
The New Accelerator (p. 863)
Dover Publications, Inc. New York, New York, USA.1952

Whitman, Walt   1819–92
American poet, journalist, and essayist

The universe is a procession with measured and beautiful motion.
Complete Poetry and Collected Prose
Leaves of Grass
The Library of America. New York, New York, USA. 1982

Wittgenstein, Ludwig Josef Johann   1889–1951
Austrian-born English philosopher

The fact that we can describe the motions of the world using Newtonian mechanics tells us nothing about the world. The fact that we do, does tell us something about the world.
In John D. Barrow
The World Within the World (p. 77)
Clarendon Press. Oxford, England. 1988

Young, Joshua
No biographical data available

Said the earth to a ball falling free,
"You're enjoying this falling, I see."
The ball widened its eyes
And remarked with surprise,
"But it's you who is falling, not me!"
Physics Poems
The Physics Teacher, Volume 20, Number 9, December 1982 (p. 587)

MOUNTAIN

Austen, Jane   1775–1817
English writer

What are men to rocks and mountains?
Pride and Prejudice
Chapter 27
G. Allen. London, England. 1894

Austin, Mary Hunter   1868–1934
American novelist and essayist

Who shall say what another will find most to his liking in the streets of the mountains. As for me, once set above the country of the silver firs, I must go on until I find white columbine.
The Land of Little Rain
The Streets of the Mountain (p. 194–195)
Houghton Mifflin Company. Boston, Massachusetts, USA. 1903

Avicenna   908–1037
Islamic physician

Mountains may be due to two different causes. Either they are effects of upheavals of the crust of the earth, such as might occur during a violent earthquake, or they are the effect of water, which, cutting for itself a new route, has denuded the valleys, the strata being of different kinds, some soft, some hard. The winds and waters disintegrate the one, but leave the other intact. Most of the eminences of the earth have had this latter origin. It would require a long period of time for all such changes to be accomplished, during which the mountains themselves might be somewhat diminished in size. But that water has been the main cause of these effects is proved by the existence of fossil remains of aquatic and other animals on many mountains.
In John William Draper
History of the Intellectual Development of Europe (Volume 1)
Chapter XIII (pp. 410–411)
Harper & Brothers. New York, New York, USA. 1876

Brewster, Edwin Tenney   1866–1960
Educator

[M]ountains are always the children of the sea.
This Puzzling Planet
Chapter XIII (p. 217)
The Bobbs-Merrill Company, Indianapolis, Indiana. 1928

Burnet, Thomas   1635–1715
English cleric and scientist

There is nothing in Nature more shapeless and ill-figur'd than an old Rock or a Mountain, and all that variety that is among them is but the various modes of irregularity; so as you cannot make a better character of them, in short, than to say they are of all forms and figures, except regular.
The Sacred Theory of the Earth (2nd edition)
Book I, Chapter XI (p. 112)
Printed by R. Norton. London. 1691

The greatest objects of Nature are, methinks, the most pleasing to behold; and next to the great Concave of the

Heavens, and those boundless Regions where the Stars inhabit, there is nothing that I look upon with more pleasure than the wide Sea and the Mountains of the Earth.... And yet these Mountains we are speaking of, to confess the truth, are nothing but great ruins; but such as show a certain magnificence in Nature; as from old Temples and broken Amphitheaters of the Romans we collect the greatness of that people. But the grandeur of a Nation is less sensible to those that never see the remains and monuments that they have left, and those that never see the mountainous parts of the Earth, scarce ever reflect upon the causes of them, or what power in Nature could be sufficient to produce them.
The Sacred Theory of the Earth (2nd edition)
Book I, Chapter XI, Concerning the Mountains of the Earth (p. 109)
Printed by R. Norton. London. 1691

Look upon those great ranges of Mountains in Europe or in Asia, whereof we have given a short survey, in what confusion do they lie? They have neither form nor beauty, nor shape, nor order, no more than the Clouds in the Air. Then how barren, how desolate, how naked are they? how they stand neglected by Nature? neither the Rains can soften them, nor the Dews from Heaven make them fruitful.
The Sacred Theory of the Earth (2nd edition)
Book I, Chapter XI (p. 111)
Printed by R. Norton. London. 1691

Byron, George Gordon, 6th Baron Byron   1788–1824
English Romantic poet and satirist

Above me are the Alps,
The palaces of Nature, whose vast walls
Have pinnacled in clouds their snowy scalps,
And throned Eternity in icy halls …
The Complete Poetical Works of Byron
Childe Harold
Canto III Stanza 62
Houghton Mifflin Company. Boston, Massachusetts, USA. 1933

Where rose the mountains, there to him were friends …
The Complete Poetical Works of Byron
Childe Harold
canto III Stanza 13
Houghton Mifflin Company. Boston, Massachusetts, USA. 1933

Carr, William H.
No biographical data available

…the mountains are but the brothers of the hills.
The Stir of Nature
Chapter One (p. 26)
Oxford University Press, Inc. New York, New York, USA. 1930

Moore, Dudley   1935–2002
English actor and pianist
Cook, Peter   1937–1995
English comedian

DUDLEY: And will this wind be so mighty as to lay low the mountains of the earth?

PETER: No. It will not be quite as mighty as that. That is why we have come up on the mountain, you stupid nit. Up here we shall be safe—safe as houses.
Beyond the Fringe
The End of the World
British stage comedy revue. 1961

Cyrano Jones
Fictional character

Twice nothing is still nothing.
Star Trek
The Trouble with Tribbles
Television program
Season 2, 1967

Muir, John   1838–1914
American naturalist

Thousands of God's wild blessings will search you and soak you as if you were a sponge, and the big days will go by uncounted.
Our National Parks
Chapter I (p. 17)
Houghton Mifflin Company. Boston, Massachusetts, USA. 1901

Fear not, therefore, to try the mountain-passes. They will kill care, save you from deadly apathy, set you free, and call forth every faculty into vigorous, enthusiastic action. Even the sick should try these so-called dangerous passes, because for every unfortunate they kill, they cure a thousand.
Mountains of California
Chapter V (p. 79)
The Century Company. New York, New York, USA. 1911

Climb the mountains and get their good tidings. Nature's peace will flow into you as sunshine flows into trees. The winds will blow their own freshness into you and the storms their energy, while care will drop off like autumn leaves.
Our National Parks
Chapter II (p. 56)
Houghton Mifflin Company. Boston, Massachusetts, USA. 1901

The time will not be taken from the sum of your life. Instead of shortening, it will indefinitely lengthen it and make you truly immortal. Nevermore will time seem short or long, and cares will never again fall heavily on you, but gently and kindly as gifts from heaven.
Our National Parks
Chapter I (p. 19)
Houghton Mifflin Company. Boston, Massachusetts, USA. 1901

The mountains are fountains not only of rivers and fertile soil, but of men.
Steep Trails
Chapter III (p. 47)
Norman S. Berg, Publisher. Dunwoody, Georgia, USA. 1970

Here I could stay tethered forever with just bread and water, nor would I be lonely; loved friends and neighbors, as love for everything increased, would seem all the nearer however many the miles and mountains between us.
My First Summer in the Sierra
June 6 (p. 29)
Houghton Mifflin Company. Boston, Massachusetts, USA. 1911

…perhaps more than all, I was animated by a mountaineer's eagerness to get my feet in the snow once more, and my head into the clear sky, after lying dormant all winter at the level of the sea.
Steep Trails
Chapter IX (p. 128)
Norman S. Berg, Publisher. Dunwoody, Georgia, USA. 1970

Ruskin, John   1819–1900
English writer, art critic, and social reformer

[Mountains] seem to have been built for the human race, as at once their schools and cathedrals; full of treasures of illuminated manuscript for the scholar, kindly in simple lessons to the worker, quiet in pale cloisters for the thinker, glorious in holiness for the worshipper. …[G]reat cathedrals of the earth, with their gates of rock, pavements of cloud, choirs of stream and stone, altars of snow, and vaults of purple traversed by the continual stars…
Selections from the Works of John Ruskin
The Mountain Glory (p. 16)
Houghton Mifflin Company. Boston, Massachusetts, USA. 1908

Sedgwick, Adam   1854–1913
English geologist

My present objective is to convey some notion of the structure of the great mountain masses, and to show how the several parts are fitted one to another. This can only be done after great labour. The cliffs where the rocks are laid bare by the sea, the clefts and fissures in the hills and valleys, the deep grooves through which the water flows — all must in turn be examined; and out of such seeming confusion order will at length appear. We must, in imagination, sweep off the drifted matter that clogs the surface of the ground; we must suppose all the covering of moss and heath and wood to be torn away from the sides of the mountains, and the green mantle that lies near their feet to be lifted up; we may see the muscular integuments and sinews and bones of our mother Earth, and so judge of the parts played by each of them during those old convulsive movements whereby her limbs were contorted and drawn up into their present positions.
In John Hudson
Complete Guide to the Lakes
Second Letter
Longman & Company. London, England. 1842

Thoreau, Henry David   1817–62
American essayist, poet, and practical philosopher

The tops of mountains are among the unfinished parts of the globe, whither it is a slight insult to the gods to climb and pry into their secrets, and try their effects upon our humanity.
The Writings of Henry David Thoreau (Volume 3)
The Maine Woods
Ktaadn (p. 85)
Houghton Mifflin Company. Boston, Massachusetts, USA. 1893

MUCOUS

Pain, Roger H.
No biographical data available

If it is love that makes the world go round, then it is surely mucus and slime which facilitate its translational motion.
In Steven Vogel
Life's Devices: The Physical World of Animals
Chapter 9 (p. 177)
Princeton University Press. Princeton, New Jersey, USA. 1988

MULTIPLICATION

Chekhov, Anton Pavlovich   1860–1904
Russian author and playwright

There is no national science, just as there is no national multiplication table; what is national is no longer science.
Note-Book of Anton Chekhov (p. 18)
B.W. Huebsch, Inc. New York, New York, USA. 1921

Melrose, A. R.
No biographical data available

Twy-stymes, noun: 1 arithmetic if it has to do with the number two {2} {which is always a good number to have when doing anything}. 2 multiplication table of the number two {2}.
The Pooh Dictionary
Dutton Children's Books. New York, New York, USA. 1995

Zamyatin, Yevgeny   1884–1937
Russian novelist, playwright, and satirist

There are no more fortunate and happy people than those who live according to the correct, eternal laws of the multiplication table.
Translated by Gregory Zilboorg
We
Record Twelve (p. 64)
E.P. Dutton & Company, Inc. New York, New York, USA. 1952

The multiplication table is more wise and more absolute than the ancient god, for the multiplication table never (do you understand — never) makes mistakes!
Translated by Gregory Zilboorg
We
Record Twelve (pp. 63–64)
E.P. Dutton & Company, Inc. New York, New York, USA. 1952

MUON

Penman, Sheldon
No biographical data available

For the time being, however, the muon itself qualifies as a "riddle wrapped in a mystery inside an enigma."
The Muon
Scientific American, Volume 205, Number 1, July 1961 (p. 55)

Rabi, Isidor Isaac   1898–1988
Austrian-born American physicist

Who ordered that?
Attributed, upon learning of the muon

MUSEUM

Belloc, Hilaire   1870–1953
French-born poet and historian

The Dodo used to walk around,
And take the sun and air,
The sun yet warms his native ground —
The Dodo is not there!
The voice which used to squawk and squeak
Is now for ever dumb —
You may you see his bones and beak
All in the Mu-se-um.
Complete Verse
The Dodo (p. 238)
G. Duckworth. London, England. 1970

Edwards, R. Y.
No biographical data available

The physical heart of a museum is its collection, in fact having a collection is what makes a museum a museum, and most activity in most museums is involved with the acquisition, care, understanding, and use of their collections.
Research: A Museum Cornerstone
Occasional Papers of the British Columbia Provisional Museum, Volume 25, 1985 (p. 1)

Flower, Sir William Henry   1831–99
English zoologist

A museum is like a living organism; it requires constant and tender care; it must grow or it will perish.
In Archie F. Key
Beyond Four Walls: The Origins and Development of Canadian Museums
Chapter 6 (p. 52)
McClelland & Stewart Ltd. Toronto, Ontario, Canada. 1973

Goode, George Brown
No biographical data available

A finished museum is a dead museum, and a dead museum is a useless museum.

In Museums Association
The Principles of Museum Administration
Report of Proceedings, Newcastle, 1895 (p. 78)

St. Clair, George
No biographical data available

The crust of the earth, with its embedded fossils, must not be looked at as a well-filled museum, but as a poor collection made at hazard and at rare intervals.
Darwinism and Design, or, Creation by Evolution
Chapter III (p. 52)
Hodder & Stoughton. London, England. 1873

MUTATION

Crow, J. F.   American geneticist
No biographical data available

…we could still be sure on theoretical grounds that mutants would usually be detrimental. For a mutation is a random change of a highly organized, reasonably smoothly functioning human body. A random change in the highly integrated system of chemical processes which constitute life is certain to impair — just as a random interchange of connections in a television set is not likely to improve the picture.
Genetic Effects of Radiation
Bulletin of the Atomic Scientists, Volume 14, Number 1, January 14, 1958 (pp. 19, 20)

Dobzhansky, Theodosius   1900–75
Russian-American scientist

…a majority of mutations, both those arising in laboratories and those stored in natural populations, produce deteriorations of…viability, hereditary disease and monstrosities. Such changes it would seem, can hardly serve as evolutionary building blocks.
Genetics and the Origin of Species
Chapter III (p. 73)
Columbia University Press. New York, New York, USA. 1951

…the mutation process alone, not corrected and guided by natural selection, would result in degeneration and extinction rather than improved adaptiveness.
On Methods of Evolutionary Biology and Anthropology
American Scientist, Volume 45, 1957 (p. 385)

Heinlein, Robert A.   1907–88
American science fiction writer

"Mutation" is never an explanation; it is simply a name for an observed fact.
Time Enough for Love
Chapter IX (p. 246)
G.P. Putnam's Sons. New York, New York, USA. 1973

Huxley, Julian   1887–1975
English biologist, philosopher, and author

One would expect that any interference with such a complicated piece of chemical machinery as the genetic constitution would result in damage. And, in fact, this is so: the great majority of mutant genes are harmful in their effects on the organism.
Evolution in Action
Chapter 2 (p. 39)
Harper & Brothers. New York, New York, USA. 1953

Muller, Hermann Joseph   1890–1967
American geneticist

It is entirely in line with the accidental nature of natural mutations that extensive tests have agreed in showing the vast majority of them detrimental to the organism in its job of surviving and reproducing, just as changes accidentally introduced into any artificial mechanism are predominantly harmful to its useful operation.
How Radiation Changes the Genetic Constitution
Bulletin of the Atomic Scientists, Volume 11, Number 9, November 1955
(p. 331)

Pauling, Linus   1901–94
American chemist

Every species of plant and animal is determined by a pool of germ plasm that has been most carefully selected over a period of hundreds of millions of years. We can understand now why it is that mutations in these carefully selected organisms almost invariably are detrimental. The situation can be suggested by a statement by Dr. J.B.S. Haldane: "My clock is not keeping perfect time. It is conceivable that it will run better if I shoot a bullet through it; but it is much more probable that it will stop altogether." Professor George Beadle, in this connection, has asked: "What is the chance that a typographical error would improve Hamlet?"
No More War!
Chapter 4 (p. 53)
Dodd, Mead & Company. New York, New York, USA. 1958

MUTUALISM

Pound, Roscoe   1870–1964
American jurist

It is not necessary…in order to establish mutualism to show that the organisms do no injury to each other. Mutualism of the kind we meet with in the vegetable kingdom involves sacrifice on the part of the host. The parasite is not there gratuitously. It is there to steal from its host the living it is hereditarily and constitutionally indisposed to make for itself. If the host gains any advantage from the relation, it can only do so by sacrificing — by giving the parasite the benefit of its labor that it may subsist.
Symbiosis and Mutualism
The American Naturalist, Volume XXVII, Number 318, June 1893
(p. 519)

MYRMECOLOGIST

Hölldobler, Bert   1936–
German myrmecologist
Wilson, Edward O.   1929–
American biologist and author

Like all myrmecologists…we are prone to view the Earth's surface idiosyncratically, as a network of ant colonies. We carry a global map of these relentless little insects in our heads. Everywhere we go their ubiquity and predictable natures makes us feel at home, for we have learned to read part of their language and we understand certain designs of their social organization better than anyone understands the behavior of our fellow humans.
Journey to the Ants: A Story of Scientific Exploration
The Dominance of Ants (p. 1)
Harvard University Press. Cambridge, Massachusetts, USA. 1994

MYSTERY

Asimov, Isaac   1920–92
American author and biochemist

The mysteries of the universe and the questions that scientists strive to answer never come to an end. For that we should be grateful. A universe in which their were no mysteries for curious men to ponder would be a very dull universe indeed.
The Search for the Elements
Chapter 16 (p. 152)
Basic Books. New York, New York, USA. 1962

Author undetermined

"Give me the facts," said My Lord Judge, "thy conclusions are but the guess-work of imagination; which puzzle the brain, and tend not to solve the mystery."
In Colin Mackenzie
One Thousand Experiments in Chemistry
On Cover page
Printed for Sir Richard Phillips & Company. London, England. 1821

Burroughs, John   1837–1921
American naturalist and writer

…after science has done its best the mystery is as great as ever, and the imagination and the emotions have just as free a field as before.
Indoor Studies Science and Literature
Science and Literature (pp. 51–52)
Houghton Mifflin Company. Boston, Massachusetts, USA. 1889

Chargaff, Erwin   1905–2002
Austrian biochemist

It is the sense of mystery that, in my opinion, drives the true scientist: the same force, blindly seeing, deafly hearing, unconsciously remembering, that drive the larvae

into the butterfly. If he has not experienced, at least a few times in his life, this cold shudder down his spine, this confrontation with an immense, invisible face whose breath moves him to tears, he is not a scientist. The blacker the night, the brighter the light.
Heraclitean Fire: Sketches from a Life Before Nature
Part II
In the Light of Darkness (p. 114)
Rockefeller University Press. New York, New York, USA. 1978

Dawkins, Richard   1941–
English ethologist, evolutionary biologist, and popular science writer

There is mystery in the universe, beguiling mystery, but it isn't capricious, whimsical, frivolous in its changeability. The universe is an orderly place and, at a deep level, regions of it behave like other regions, times behave like other times.
Science, Delusion and the Appetite for Wonder
Richard Dimbleby Lecture, BBC1 Television, November 12th, 1996

Newton, Keats agreed with Lamb, had destroyed all the poetry of the rainbow, by reducing it to the prismatic colours.... Newton's dissection of the rainbow into light of different wavelengths led on to Maxwell's theory of electromagnetism and thence to Einstein's theory of special relativity. If you think the rainbow has poetic mystery, you should try relativity.
Unweaving the Rainbow: Science, Delusion and the Appetite for Wonder
Chapter 3 (p. 39)
Houghton Mifflin Company. Boston, Massachusetts, USA. 1998

The real universe has mystery enough to need no help from obscurantist hucksters.
The Real Romance in the Stars
The Independent, 31 Dec 1995

Mysteries do not lose their poetry when solved. Quite the contrary; the solution often turns out more beautiful than the puzzle and, in any case, when you have solved one mystery you uncover others, perhaps to inspire greater poetry.
Unweaving the Rainbow: Science, Delusion and the Appetite for Wonder
Chapter 3 (p. 41)
Houghton Mifflin Company. Boston, Massachusetts, USA. 1998

Doyle, Sir Arthur Conan   1859–1930
Scottish writer

It is a mistake to confound strangeness with mystery.
In William S. Baring-Gould (ed.)
The Annotated Sherlock Holmes (Volume 1)
A Study in Scarlet, Chapter 7 (p. 194)
Wings Books. New York, New York, USA. 1967

As a rule, said Holmes, the more bizarre a thing is the less mysterious it proves to be.
In William S. Baring-Gould (ed.)
The Annotated Sherlock Holmes (Volume 1)
The Red Headed League (p. 429)
Wings Books. New York, New York, USA. 1967

Einstein, Albert   1879–1955
German-born physicist

The most beautiful experience we can have is the mysterious. It is the fundamental emotion which stands at the cradle of true art and true science. Whoever does not know it and can no longer wonder, no longer marvel, is as good as dead, and his eyes are dimmed.
Ideas and Opinions
The World As I See It (p. 11)
Crown Publishers, Inc. New York, New York, USA. 1954

I claim credit for nothing. Everything is determined, the beginning as well as the end, by forces over which we have no control. It is determined for the insect as well as for the star. Human beings, vegetables, or cosmic dust, we all dance to a mysterious tune, intoned in the distance by an invisible player.
What Life Means to Einstein: An Interview by George Sylvester Viereck
The Saturday Evening Post, October 26, 1929 (p. 117)

Feynman, Richard P.   1918–88
American theoretical physicist

We choose to examine a phenomenon which is impossible, absolutely impossible to explain in any classical way, and which has in it the heart of quantum mechanics. In reality, it contains the only mystery.
The Feynman Lectures on Physics (Volume 1)
Chapter 37–1 (p. 37–2)
Addison-Wesley Publishing Company. Reading, Massachusetts, USA. 1983

...the idea that it [science] takes away mystery or awe or wonder in nature is wrong. It's quite the opposite. It's much more wonderful to know what something's really like than to sit there and just simply, in ignorance, say, "Oooh, isn't it wonderful!"
In Christopher Sykes
No Ordinary Genius: The Illustrated Richard Feynman
Chapter Four (p. 108)
W.W. Norton & Company, Inc. New York, New York, USA. 1994

With more knowledge comes a deeper, more wonderful mystery, luring one on to penetrate deeper still. Never concerned that the answer may prove disappointing, with pleasure and confidence we turn over each new stone to find unimagined strangeness leading on to more wonderful questions and mysteries — certainly a grand adventure.
What Do You Care What Other People Think?
The Value of Science (p. 243)
W.W. Norton & Company, Inc. New York, New York, USA. 1988

[The law of gravity] is not exact; Einstein had to modify it, and we know it is not quite right yet, because we have still to put the quantum theory in. That is the same with all our other laws — they are not exact. There is always an edge of mystery, always a place where we have some fiddling around to do yet. This may or may not be a property of Nature, but it certainly is common to all the laws as we know them today. It may be only a lack of knowledge.

The Character of Physical Law
Chapter 1 (p. 33)
BBC. London, England. 1965

It is a great adventure to contemplate the universe, beyond man, to contemplate what it would be like without man, as it was in a great part of its long history and as it is in a great majority of places. When this objective view is finally attained, and the mystery and majesty of matter are fully appreciated, to then turn the objective eye back on man viewed as matter, to view life as part of this universal mystery of greatest depth, is to sense an experience which is very rare, and very exciting. It usually ends up in laughter and a delight in the futility of trying to understand what this atom in the universe is, this thing — atoms with curiosity — that looks at itself and wonders why it wonders.

The Meaning of It All: Thoughts of a Citizen Scientist
Chapter II (p. 39)
Perseus Books. Reading, Massachusetts, USA. 1998

[T]hese scientific views end in awe and mystery, lost at the edge in uncertainty, but they appear to be so deep and so impressive that the theory that it is all arranged as a stage for God to watch man's struggle for good and evil seems inadequate.

The Meaning of It All: Thoughts of a Citizen Scientist
Chapter II (p. 39)
Perseus Books. Reading, Massachusetts, USA. 1998

…I can live with doubt and uncertainty and not knowing. I think it's much more interesting to live not knowing than to have answers which might be wrong. I have approximate answers and possible beliefs and different degrees of certainty about different things, but I'm not absolutely sure of anything and there are many things I don't know anything about, such as whether it means anything to ask why we're here…I don't have to know an answer, I don't feel frightened by not knowing things, by being lost in a mysterious universe without any purpose, which it is the way it really is so far as I can tell. It doesn't frighten me.

In Jeffrey Robbins (ed.)
The Pleasure of Finding Things Out: The Best Short Works of Richard P. Feynman
Chapter 1 (pp. 24–25)
Perseus Books. Cambridge, Massachusetts, USA. 1999

Fosdick, Harry Emerson   1878–1969
American clergyman and educator

I would rather live in a world where my life is surrounded by mystery than live in a world so small that my mind could comprehend it.

Riverside Sermons
The Mystery of Life (p. 22)
Harper & Brothers Publishers. New York, New York, USA. 1958

Kelvin, Lord William Thomson   1824–1907
Scottish engineer, mathematician, and physicist

When you call a thing mysterious, all that means is that you don't understand it.

In Sir Richard Arman Gregory
Discovery; or, The Spirit and Service of Science
Chapter IV (p. 56)
Macmillan & Company Ltd. London, England. 1918

Lightman, Alan   1948–
Physicist, novelist, and essayist

A painter paints a sunset, and a scientist measures the scattering of light. The beauty of nature lies in its logic as well as appearance. And we delight in that logic: The square of the orbital period of each planet equals the cube of its distance from the sun; the shape of a raindrop is spherical, to minimize the area of its surface. Why it is that nature should be logical is the greatest mystery of science. But it is a wonderful mystery.

Great Ideas in Physics
Introduction (p. 1)
McGraw-Hill Book Company, Inc. New York, New York, USA. 1997

Lindbergh, Anne Morrow   1906–2001
American aviator and writer

Today's mystery is not the old veiling by superstition of the things man does not understand, but a new unblinking gaze at the mysteries of the universe that may never be unveiled.

Earth Shine
Back to Earth (pp. 42–43)
Harcourt, Brace & World, Inc. New York, New York, USA. 1969

Lowell, Percival   1855–1916
American astronomer

Discoveries in science have a fatal facility for lying lost in the technical publications which record them. Few persons attend to what is not alluring and columns of figures form but an uninviting protocol to the learning within. Yet these very people would take the keenest interest in scientific progress could its beauty and real simplicity be adequately set out for them. For the whole object of science is to explain and make comprehensible the universe around us. Science consists in solving mysteries not, as the layman might imagine, in making them.

In William Graves Hoyt
Lowell and Mars
Chapter 7 (p. 96)
University of Arizona Press. Tucson, Arizona, USA. 1976

Margenau, Henry   1901–97
American physicist

I recognize no subjects and no facts which are alleged to be forever closed to inquiry or understanding: a mystery is but a challenge.

Open Vistas
Chapter III Section 3 (p. 76)
Yale University Press. New Haven, Connecticut, USA. 1961

Penrose, Roger   1931–
English mathematical physicist

…once you have put more and more of your physical world into a mathematical structure, you realize how profound and mysterious this mathematical structure is. How you can get all these things out of it is very mysterious…
In Alan Lightman and Robert Brawer (eds.)
Origins: The Lives and Worlds of Modern Cosmologists
Roger Penrose (p. 433)
Harvard University Press. Cambridge, Massachusetts, USA. 1990

Peterson, Ivars
Mathematics writer

The voyage of discovery into our own solar system has taken us from clockwork precision into chaos and complexity. This still unfinished journey has not been easy, characterized as it is by twists, turns, and surprises that mirror the intricacies of the human mind at work on a profound puzzle. Much remains a mystery. We have found chaos, but what it means and what its relevance is to our place in the universe remains shrouded in a seemingly impenetrable cloak of mathematical uncertainty.
Newton's Clock: Chaos in the Solar System
Chapter 12 (p. 293)
W.H. Freeman & Company. New York, New York, USA. 1993

Richards, Theodore William   1868–1928
American chemist

No one can predict how far we shall be enabled by means of our limited intelligence to penetrate into the mysteries of a universe immeasurably vast and wonderful; nevertheless, each step in advance is certain to bring new blessings to humanity and new inspiration to greater endeavor.
Annual Report of the Board of Regents of the Smithsonian Institution, 1911 (Faraday Lecture) The Fundamental Properties of the Elements
(p. 215)
Government Printing Office. Washington, D.C. 1912

Smullyan, Raymond   1919–
American mathematician and logician

…I certainly believe that some things may in principle be mysteries, but of what use is a hypothesis for explaining a mystery when the very hypothesis raises another mystery just as baffling as the one it explains?
5000 B.C. and Other Philosophical Fantasies
Chapter 13 (p. 165)
St. Martin's Press. New York, New York, USA. 1983

Wilson, Edward O.   1929–
American biologist and author

The unsolved mysteries of the rain forest are formless and seductive. They are like unnamed islands hidden in the blank spaces of old maps, like dark shapes glimpsed descending the far wall of a reef into the abyss.
The Diversity of Life
Chapter One (p. 7)
W.W. Norton & Company, Inc. New York, New York, USA. 1992

Young John Zachary   1907–97
English zoologist

The scientist is in a better position than anyone else to see that we are set about with mysteries. It is his business to grapple with ghosts every day of his life and he must refuse to allow them to be laid by the process of labeling them with a primitive nomenclature. The mysteries of the universe are too great to be expressed by such simple comparisons as are implicit in either the words "spirit" or "matter."
Doubt and Certainty in Science: A Biologist's Reflections on the Brain
Comment on the First Lecture (p. 23)
Oxford University Press, Inc. Oxford, England. 1960

MYSTICISM

Capra, Fritjof   1939–
Austrian-born American physicist

Science does not need mysticism and mysticism does not need science, but man needs both. Mystical experience is necessary to understand the deepest nature of things, and science is essential for modern life. What we need, therefore, is not a synthesis, but a dynamic interplay between mystical intuition and scientific analysis.
The Tao of Physics: An Exploration of the Parallels Between Modern Physics and Eastern Mysticism
Epigraph (p. 297)
Shambhala. Berkeley, California, USA. 1975

MYTH

Bernal, John Desmond   1901–71
Irish-born physicist and x-ray crystallographer

Science, in one aspect, is ordered technique; in another, it is rationalized mythology.
Science in History
Preface (p. ix)
Watts. London, England. 1957

Clift, Wallace B.
No biographical data available

However, for most of us, science functions like myth in that we have no personal experience in the matter. We put our trust in the scientific view given us by our culture and enshrined in its myths. If asked why leaves are green, most of us would probably mutter something about "chlorophyll." But unless we were specialists, we would simply be repeating the story of someone else's experience.
Jung and Christianity (pp. 62–63)
Publisher undetermined

Hubbard, Ruth 1924–
American biologist

The mythology of science asserts that with many different scientists all asking their own questions and evaluating the answers independently, whatever personal bias creeps into their individual answers is canceled out when the large picture is put together. This might conceivably be so if scientists were women and men from all sorts of different cultural and social backgrounds who came to science with very different ideologies and interests. But since, in fact, they have been predominantly university-trained white males from privileged social backgrounds, the bias has been narrow and the product often reveals more about the investigator than about the subject being researched.
Women Look at Biology Looking at Women
Have Only Men Evolved? (p. 31)
Schenkman Publishing Company. Cambridge, Massachusetts, USA. 1979

Jacob, François 1920–
French biologist

…myths and science fulfill a similar function: they both provide human beings with a representation of the world and the forces that are supposed to govern it.
The Possible and the Actual
Myth and Science (p. 9)
Pantheon Books. New York, New York, USA. 1982

Mahadeva, M.
No biographical data available

Myths are errors that result [both from] scientists bringing societal preconceptions into science and…scientists feeding society ideas that masquerade as science.
From Misinterpretations to Myths
Science Teacher, Volume 56, Number 4, 1989

Popper, Karl R. 1902–94
Austrian/British philosopher of science

Thus science must begin with myths — and with the criticism of myths…
In C.A. Mace (ed.)
British Philosophy in the Mid-Century
Philosophy of Science: A Personal Report, VII (p. 177)

Science never starts from scratch; it can never be described as free from assumptions; for at every instant it presupposes a horizon of expectations — yesterday's horizon of expectations, as it were. Today's science is built upon yesterday's science (and so it is the result of yesterday's searchlight); and yesterday's science, in turn, is based on the science of the day before. And the oldest scientific theories are built on pre-scientific myths, and these, in their turn, on still older expectations.
Objective Knowledge: An Evolutionary Approach
Appendix (pp. 346–347)
Clarendon Press. Oxford, England. 1972

Whyte, Lancelot Law 1896–1972
Scottish physicist

In the ultimate analysis science is born of myth and religion, all three being expressions of the ordering spirit of the human mind.
The Unconscious Before Freud
Chapter V (pp. 82–83)
Julian Friedmann Publishers. London, England. 1978

N

NAME

Abbey, Edward   1927–89
American environmentalist and nature writer

...the itch for naming things is almost as bad as the itch for possessing things.
Desert Solitaire
Terra Incognita: Into the Maze (p. 288)
Ballantine Books. New York, New York, USA. 1968

Agassiz, Jean Louis Rodolphe   1807–73
Swiss-born American naturalist, geologist, and teacher

The first thing to be determined about a new specimen is not its name, but its most prominent character. Until you know an animal, care not for its name.
In James Orton
Comparative Zoology, Structural and Systematic
Preceding Chapter I (p. 18)
Harper & Brothers. New York, New York, USA. 1877

Every art and science has a language of technical terms peculiar to itself. With those terms every student must make himself familiarly acquainted at the outset; and first of all, he will desire to know the names of the objects about which he is to be engaged.
Principles of Zoology
Introduction (p. xiii)
Gould, Kendall & Lincoln. Boston, Massachusetts, USA. 1848

Nothing is more to be deprecated than an over-appreciation of technicalities, valuing the name more highly than the thing; but some knowledge of this scientific nomenclature is necessary to every student of Nature.
Methods of Study in Natural History
Chapter II (p. 18)
Ticknor & Fields. Boston, Massachusetts, USA. 1863

Borland, Hal   1900–78
American writer

There is folk poetry in the common names; but science, devoted to order and systematic knowledge, insists on classifying and defining. The poet's buttercup is the botanist's Ranunculus. If you would walk with scientist as well as poet, learn both languages.
Beyond Your Doorstep: A Handbook to the Country
Chapter 15 (p. 359)
Alfred A. Knopf. New York, New York, USA. 1962

Carroll, Lewis (Charles Dodgson)   1832–98
English writer and mathematician

"What's the use of their having names," the Gnat said, "if they won't answer to them?"

"No use to them," said Alice; "but its useful to the people that name them, I suppose."
The Complete Works of Lewis Carroll
Through the Looking-Glass
Chapter III (p. 173)
The Modern Library. New York, New York, USA. 1936

"My name is Alice...."

"It's a stupid name enough!" Humpty Dumpty interrupted impatiently. "What does it mean?"

"Must a name mean something?" Alice asked doubtfully.

"Of course it must," Humpty Dumpty said with a short laugh; "my name means the shape I am.... With a name like yours, you might be any shape, almost."
The Complete Works of Lewis Carroll
Through the Looking-Glass
Chapter VI (p. 209)
The Modern Library. New York, New York, USA. 1936

Darwin, Charles Robert   1809–82
English naturalist

...I have lately been trying to get up an agitation (but I shall not succeed, and indeed doubt whether I have time and strength to go on with it), against the practice of Naturalists appending for perpetuity the name of the first describer to species. I look at this as a direct premium to hasty work, to naming instead of describing. A species ought to have a name so well known that the addition of the author's name would be superfluous, and a [piece] of empty vanity....Why should Naturalists append their own names to new species, when Mineralogists and Chemists do not do so to new substances?
In Francis Darwin (ed.)
The Life and Letters of Charles Darwin (Volume 1)
Letter to J.D. Hooker, October 6, 1848 (pp. 332–333)
D. Appleton & Company. New York, New York, USA. 1896

I do not think more credit is due a man for defining a species, than to a carpenter for making a box. But I am foolish and rabid against species-mongers, or rather against their vanity; it is useful and necessary work which must be done; but they act as if they had actually made the species, and it was their own property.
In Francis Darwin (ed.)
The Life and Letters of Charles Darwin (Volume 1)
Letter to Hugh Strickland, February 4, 1849 (pp. 338–339)
D. Appleton & Company. New York, New York, USA. 1896

How dreadfully difficult it is to name plants.
In Francis Darwin (ed.)
The Life and Letters of Charles Darwin (Volume 1)
Letter to J.D. Hooker, June 5, 1855 (p. 418)
D. Appleton & Company. New York, New York, USA. 1896

Doyle, Sir Arthur Conan   1859–1930
Scottish writer

Man, it's Witchcraft! Where in the name of all that is wonderful did you get those names?
In William S. Baring-Gould (ed.)
The Annotated Sherlock Holmes (Volume 1)
The Valley of Fear
Part I, Chapter 1 (p. 476)
Wings Books. New York, New York, USA. 1967

Eliot, George (Mary Ann Evans Cross)   1819–80
English novelist

The mere fact of naming an object tends to give definiteness to our conception of it — we have then a sign that at once calls up in our minds the distinctive qualities which mark out for us that particular object from all others.
The George Eliot Letters (Volume 1I) (p. 251)
Yale University Press. New Haven, Connecticut, USA. 1954–78

Ellis, Havelock   1859–1939
English sexuality researcher

For even the most sober scientific investigator in science, the most thoroughgoing Positivist, cannot dispense with fiction; he must at least make use of categories, and they are already fictions, analogical fictions, or labels, which give us the same pleasure as children receive when they are told the "name" of a thing.
The Dance of Life
Chapter III, Section II (p. 94)
Houghton Mifflin Company. Boston, Massachusetts, USA. 1923

Faraday, Michael   1791–1867
English physicist and chemist

…I am fully aware that names are one thing and science another.
Experimental Researches in Electricity (Volume 1)
Seventh Series, 666 (p. 198)
Richard and John Edward Taylor. London, England. 1839–1855

Ferris, G. F.
No biographical data available

The proper aim is not to *name* species but to *know* them.
Stanford University Publications: Biological Studies (Volume 5)
The Principles of Systematic Entomology (p. 105)
Stanford University. Stanford, California, USA. 1920–53

Gahan, A. B.
No biographical data available

Objects without names cannot well be talked about or written about; without descriptions they cannot be identified and such knowledge as may have accumulated regarding them is sealed.
The Role of the Taxonomist in Present Day Entomology
Entomological Society of Washington Proceedings, Volume 25, 1923 (p. 73)

Holmes, Oliver Wendell   1809–94
American physician, poet, and humorist

A Pseudo-science consists of a nomenclature, with a self-adjusting arrangement, by which all positive evidence, or such as favors its doctrines, is admitted, and all negative evidence, or such as tells against it, is excluded. It is invariably connected with some lucrative practical application.
The Professor at the Breakfast Table
Chapter VIII (p. 249)
Ticknor & Fields. Boston, Massachusetts, USA. 1860

Isidorus
No biographical data available

If you know not the names, the knowledge of things is wasted.
In Carl von Linné
Critica Botanica
Generic Names (p. 1)
The Ray Society. London, England. 1938

Juster, Norton   1929–
American architect and writer

Words and numbers are of equal value, for, in the cloak of knowledge, one is warp and the other woof. It is no more important to count the sands than it is to name the stars.
The Phantom Tollbooth
Chapter 6 (p. 77)
Alfred A. Knopf. New York, New York, USA. 1989

Lavoisier, Antoine Laurent   1743–94
French chemist

The impossibility of separating the nomenclature of a science from the science itself, is owing to this, that every branch of physical science must consist of three things; the series of facts which are the objects of the science, the ideas which represent these facts, and the words by which these ideas are expressed.
Elements of Chemistry in a New Systematic Order
Preface of the Author (p. xiv)
Printed for William Creech. Edinburgh, Scotland. 1790

Le Févre, Nicholas   1615–69
No biographical data available

This Principle, as well as the others, hath received several names; for it is called Oil, Natural Fire, Light, Vital Fire, Balsom of Life and of Sulphur, and besides, many other appellations have been given by the Sons of Art, which we will not fill up this Section: According to our usual custom, we will content our selves, with examining the nature of the thing, leaving the niceties of Names to the over-curious.
A Complete Body of Chymistry
Chapter III, Section IV (p. 25)
Printed for O. Pullyn. London, England. 1640

Linnaeus, Carl (von Linné)   1707–78
Swedish botanist

The first step of science is to know one thing from another. This knowledge consists in their specific distinctions; but in order that it may be fixed and permanent distinct names must be given to different things, and those names must be recorded and remembered.

In Sir James Edward Smith
A Selection of the Correspondence of Linnaeus and Other Naturalists from the Original Manuscripts (Volume 2) (p. 460)
Longman, Hurst, Rees, Orme and Brown. London, England. 1821

The first step in wisdom is to know the things themselves; this notion consists in having the true idea of the object; objects are distinguished and known by their methodical classification and appropriate naming; therefore Classification and Naming will be the foundation of our Science.

In P. F. Stevens
The Development of Biological Systematics: Antoine-Laurent de Jussieu, Nature and the Natural System
Chapter Nine (p. 201)
Columbia University Press. New York, New York, USA. 1994

For, even though the knowledge of the true and genuine Tree of Life, which might have delayed the coming of old age, is lost, still herbs remain and renew their flowers, and with perennial gratitude will always breathe forth the sweet memory of your names, and make them more enduring than marble, to outlive the names of kings and heroes. For wealth disappears, the most magnificent houses fall into decay, the most numerous family at some time or other comes to an end: the greatest states and the most prosperous kingdoms can be overthrown: but the whole of Nature must be blotted out before the race of plants passes away, and he is forgotten who in Botany held up the torch.

Critica Botanica
Generic Names (p. 68)
The Ray Society. London, England. 1938

Name and plant are two ideas, which ought to be so closely united that they cannot possibly be separated: in order to secure this, the plant ought to lend a hand to the name, and the name in its turn to the plant, while the name in its turn rejoices in the sound principle on which it was given: since there is no connexion between botanist and plant, there is also no sound principle in naming it after him: and so the naming is bad.

Critica Botanica
Generic Names (p. 61)
The Ray Society. London, England. 1938

Lloyd, C. G.
No biographical data available

Botanists meet and pass rules for the naming of plants, but they cannot agree on any set of rules, and never will as long as the members are vitally interested in the particular rules that perpetuate their own names and the plant names that have been proposed by themselves.

Personal Names in Nomenclature
The American Botanist, Volume 4, Number 3, March 1903 (p. 48)

Melville, Herman   1819–91
American novelist

I wonder whether mankind could not get along without all those names which keep increasing every day, and hour, and moment; till at last the very air will be full of them; and even in a great plain men will be breathing each other's breath, owing to the vast multitude of words they use that consume all of the air.... But people seem to have a great love for names; for to know a great many names seems to look like knowing a good many things.

Redburn
Chapter XIII
A. & C. Boni. New York, New York, USA. 1924

Page, Jake
No biographical data available

To name something is, in a sense, to own it.... [It] has been said that it is only by its name that anything can enter into thought and discourse. Naming, in other words, is a serious business.

Pastorale: A Natural History of Sorts
What Is in a Name? (p. 119)
W.W. Norton & Company, Inc. New York, New York, USA. 1985

Savory, Theodore
No biographical data available

...words are in themselves among the most interesting objects of study, and the names of animals and plants are worthy of more consideration than Biologists are inclined to give them.

Naming the Living World: An Introduction to the Principles of Biological Nomenclature
Preface (p. vii)
English Universities Press. London, England. 1962

Sylvester, James Joseph   1814–97
English mathematician

Perhaps I may without immodesty lay claim to the appellation of Mathematical Adam, as I believe that I have given more names (passed into general circulation) of the creatures of the mathematical reason than all the other mathematicians of the age combined.

Notes on a Proposed Addition to the Vocabulary of Ordinary Arithmetic
Nature, Volume 37, Number 946, December 15, 1887 (p. 152, fn 1)

Turnbull, Charles D.
No biographical data available

According to Genesis, "Adam gave names to all cattle and to the fowl of the air and to every beast of the field," and we have therefore authority to declare that although Father Adam named both birds and beasts, he woefully neglected the flower, and left the task to the fancy and haphazard desires of his descendants.

Concerning Nomenclature
The American Botanist, Volume 4, Number 3, March 1903 (p. 45)

Twain, Mark (Samuel Langhorne Clemens)   1835–1910
American author and humorist

Names are not always what they seem. The common Welsh name Bzjxxlwep is pronounced Jackson.
Following the Equator (Volume 1)
Chapter XXXVI (p. 339)
Harper & Brothers. New York, New York, USA. 1899

Whitehead, Alfred North   1861–1947
English mathematician and philosopher

…in the garden of Eden God saw the animals before he named them: in the traditional system [of education] children named the animals before they saw them.
Science and the Modern World
Chapter XIII (p. 285)
The Macmillan Company. New York, New York, USA. 1929

NATURAL HISTORY

Agassiz, Jean Louis Rodolphe   1807–73
Swiss-born American naturalist, geologist, and teacher

…Natural History must in good time become the analysis of the thoughts of the Creator of the Universe as manifested in the animal and vegetable kingdoms, as well as in the inorganic world.
Essay on Classification
Chapter I, Section XXXII (p. 137)
Harvard University Press. Cambridge, Massachusetts, USA. 1962

A laboratory of Natural History is a sanctuary where nothing profane should be tolerated. I feel less agony at improprieties in churches than in a scientific laboratory.
In David Stair Jordan
Popular Science Monthly, Volume 40, 1891

Borlase, William   1696–1772
Cornish antiquary

Natural History is the handmaid to Providence, collects into a narrow space what is distributed through the Universe, arranging and disposing the several Fossils, Vegetables and Animals, so as the mind may more readily examine and distinguish their beauties, investigate their causes, combinations, and effects, and rightly know how to apply them to the calls of private and public life.
The Natural History of Cornwell
The Air, Climate, Waters, Rivers, Lakes, Seas and Tides, to the Nobility and Gentry of the County of Cornwell (p. iv)
Publisher undetermined

Carroll, Lewis (Charles Dodgson)   1832–98
English writer and mathematician

In one moment I've seen what has hitherto been

Enveloped in absolute mystery,
And without extra charge I will give you at large
A Lesson in Natural History.
The Complete Works of Lewis Carroll
The Hunting of the Snark
Fit the Fifth (p. 771)
The Modern Library. New York, New York, USA. 1936

Darwin, Charles Robert   1809–82
English naturalist

What a splendid pursuit Natural History would be if it was all observing and no writing!
In Francis Darwin (ed.)
The Life and Letters of Charles Darwin (Volume 1I)
Letter to J.D. Hooker, February 3, 1868 (p. 258)
D. Appleton & Company. New York, New York, USA. 1896

Nobody but a person fond of Natural History can imagine the pleasure of strolling under cocoa-nuts in a thicket of bananas and coffee-plants, and an endless number of wildflowers.
In Francis Darwin (ed.)
The Life and Letters of Charles Darwin (Volume 1)
Chapter VI (pp. 201–202)
D. Appleton & Company. New York, New York, USA. 1896

Fleming, Donald
No biographical data available

For the colonial investigator himself, natural history was the ideal refuge from the more perilous enterprise of embarking upon theoretical constructions by which he would be pitched into naked competition with the best scholars of all countries. To be a forager for Linnaeus or correspondent of the Hookers might be an identity in science purchased by bondage to the local and particular; but it was also a shelter against the more bracing winds that would promptly blow upon any man who tired to grapple with undifferentiated Nature in physics.
Proceedings of the 10th International Congress of the History of Science, Ithaca, 1962
Science in Australia, Canada, and the United States: Some Comparative Remarks (p. 182)
Cornell University Press. Ithaca, New York, USA. 1964

Huxley, Thomas Henry   1825–95
English biologist

To a person uninstructed in natural history, his country or sea-side stroll is a walk through a gallery filled with wonderful works of art, nine-tenths of which have their faces turned to the wall. Teach him something of natural history, and you will place in his hands a catalogue of those which are worth turning around.
Lay Sermons, Addresses and Reviews
On the Educational Value of the Natural History Sciences (p. 91)
D. Appleton & Company. New York, New York, USA. 1872

Smellie, William   1740–95
Scottish encyclopedist

Natural History is the most extensive, and perhaps the most instructive and entertaining of all the sciences. It is the chief source from which human knowledge is derived. To recommend the study of it from motives of utility were to affront the understanding of mankind. Its importance, accordingly, in the arts of life, and in storing the mind with just ideas of external objects, as well as of their relations to the human race, was early perceived by all nations in their progress from rudeness to refinement.

In Buffon, Comte de Georges, Louis Leclerc
Natural History, General and Particular (Volume 1)
Preface by the Translator (p. ix)
T. Caldwell and W. Davies. London, England. 1812

NATURAL LAW

Adams, George   1750–95
English instrument maker

The end of natural philosophy is to increase either the knowledge or power of man, and enable him to understand the ways and procedure of nature. By discovering the laws of nature, he acquires knowledge, and obtains power; for when these laws are discovered, he can use them as rules of practice, to equal, subdue, or even excel nature by art.

Lectures on Natural and Experimental Philosophy (Volume 2)
Lecture XIV (pp. 100–101)
Printed by R. Hindmarsh. London, England. 1794

Vogt, Carl   1817–95
German physician and naturalist

The natural laws are rude unbending powers, which have neither morals nor heart.

In Ludwig Buchner
Force and Matter
Chapter VI (p. 35)
Trubner & Company. London, England. 1864

Whewell, William   1794–1866
English philosopher and historian

When we speak of material nature as being governed by laws, it is sufficiently evident that we use the term in a manner somewhat metaphorical.

The Bridgewater Treatises on the Power, Wisdom, and Goodness of God as Manifested in the Creation (Treatise III)
Astronomy and General Physics Considered with Reference to Natural Theology
Chapter II (p. 17)
Carey, Lea & Blanchard. Philadelphia, Pennsylvania, USA. 1833

NATURAL SCIENCE

Eddington, Sir Arthur Stanley   1882–1944
English astronomer, physicist, and mathematician

We need scarcely add that the contemplation in natural science of a wider domain than the actual leads to a far better understanding of the actual.

The Nature of the Physical World
Chapter XXII (p. 266–267)
The Macmillan Company. New York, New York, USA. 1930

Faraday, Michael   1791–1867
English physicist and chemist

I do not think that the study of natural science is so glorious a school for the mind that, with the laws impressed on all created things by the Creator, and the wonderful unity and stability of matter and the forces of matter, there cannot be a better school for the education of the mind.

In Sir Richard Arman Gregory
Discovery; or, The Spirit and Service of Science
Chapter I (p. 13)
Macmillan & Company Ltd. London, England. 1918

Whitehead, Alfred North   1861–1947
English mathematician and philosopher

…science conceived as resting on mere sense-perception, with no other source of observation, is bankrupt, so far as concerns its claim to self-sufficiency. Science can find no individual enjoyment in nature: Science can find no aim in nature: Science can find no creativity in nature; it finds mere rules of succession. These negations are true of Natural Science. They are inherent in it methodology.

Modes of Thought
Chapter III, Lecture Eight (p. 211)
The Macmillan Company. New York, New York, USA. 1938

NATURAL SELECTION

Bateson, William   1861–1926
English biologist and geneticist

Natural Selection is stern, but she has her tolerant moods.

In A.C. Seward
Darwin and Modern Science
Heredity and Variation in Modern Lights (p. 100)
University Press. Cambridge, England. 1910

Crick, Francis Harry Compton   1916–2004
English biochemist

Once we have become adjusted to the idea that we are here because we have evolved from simple chemical compounds by a process of natural selection, it is remarkable how many of the problems of the modern world take on a completely new light.

Of Molecules and Men
The Prospect Before Us (p. 93)
University of Washington Press. Seattle, Washington, USA. 1966

Darwin, Charles Robert   1809–82
English naturalist

To suppose that the eye with all its inimitable contrivances for adjusting the focus to different distances, for admitting different amounts of light, and for the correction of spherical and chromatic aberration, could have been formed by natural selection, seems, I freely confess, absurd in the highest degree.

In *Great Books of the Western World* (Volume 49)
The Origin of Species by Means of Natural Selection
Chapter VI (p. 185)
Encyclopædia Britannica, Inc. Chicago, Illinois, USA. 1952

We can no longer argue that, for instance, the beautiful hinge of a bivalve must have been made by an intelligent being, like the hinge of a door by man. There seems to be no more design in the variability of organic beings, and in the action of natural selection, than in the course which the wind blows.

In Francis Darwin (ed.)
The Life and Letters of Charles Darwin (Volume 1)
Chapter VIII (p. 279)
D. Appleton & Company. New York, New York, USA. 1896

When we descend to details, we can prove that no one species has changed (*i.e.*, we cannot prove that a single species has changed): nor can we prove that the supposed changes are beneficial, which is the groundwork of the theory. Nor can we explain why some species have changed and others have not.

In Francis Darwin (ed.)
The Life and Letters of Charles Darwin (Volume 2)
Chapter IV (p. 210)
Letter to G. Bentham, May 22, 1863
D. Appleton & Company. New York, New York, USA. 1887

Slow though the process of selection may be, if feeble man can do much by artificial selection, I can see no limit to the amount of change, to the beauty and complexity of the coadaptations between all organic beings, one with another and with their physical conditions of life, which may have been effected in the long course of time through nature's power of selection...

In *Great Books of the Western World* (Volume 49)
The Origin of Species by Means of Natural Selection
Chapter IV (p. 52)
Encyclopædia Britannica, Inc. Chicago, Illinois, USA. 1952

It may metaphorically be said that natural selection is daily and hourly scrutinising, throughout the world, the slightest variations; rejecting those that are bad, preserving and adding up all that is good; silently and insensibly working, whenever and wherever opportunity offers, at the improvement of each organic being in relation to its organic and inorganic condition of life. We see nothing of these slow changes in progress, until the hand of time has marked the long lapse of ages...

In *Great Books of the Western World* (Volume 49)
The Origin of Species by Means of Natural Selection
Chapter IV (p. 42)
Encyclopædia Britannica, Inc. Chicago, Illinois, USA. 1952

But then arises the doubt, can the mind of man, which has, as I fully believe been developed from a mind as low as that possessed by the lowest animal, be trusted when it draws such grand conclusions?

In Francis Darwin (ed.)

The Life and Letters of Charles Darwin (Volume 1)
Chapter VIII (p. 282)
D. Appleton & Company. New York, New York, USA. 1896

As natural selection acts solely by accumulating slight, successive, favorable variations, it can produce no great or sudden modification; it can act only by very short and slow steps. Hence the canon of "*Natura non facit saltum*" [Nature does not make jumps], [to] which every fresh addition to our knowledge tends to conform, is on this theory simply intelligible.

In *Great Books of the Western World* (Volume 49)
The Origin of Species by Means of Natural Selection
Chapter XV (p. 235)
Encyclopædia Britannica, Inc. Chicago, Illinois, USA. 1952

By the theory of natural selection all living species have been connected with the parent-species of each genus by differences not greater than we see between the varieties of the same species at the present day; and these parent-species, now generally extinct, have in their turn been similarly connected with more ancient species; and so on backwards, always converging to the common ancestor of each great class. So that the number of intermediate and transitional links, between all living and extinct species, must have been inconceivably great. But assuredly, if this theory be true, such have lived upon this earth.

In *Great Books of the Western World* (Volume 49)
The Origin of Species by Means of Natural Selection
Chapter X (p. 153)
Encyclopædia Britannica, Inc. Chicago, Illinois, USA. 1952

But just in proportion as this process of extermination has acted on an enormous scale, so must the number of intermediate varieties, which have formerly existed, be truly enormous. Why then is not every geological formation and every stratum full of such intermediate links? Geology assuredly does not reveal any such finely graduated organic chain; and this, perhaps, is the most obvious and serious objection which can be urged against the theory. The explanation lies, as I believe, in the extreme imperfection of the geological record.

In *Great Books of the Western World* (Volume 49)
The Origin of Species by Means of Natural Selection
Chapter X (p. 152)
Encyclopædia Britannica, Inc. Chicago, Illinois, USA. 1952

He who rejects this view of the imperfection of the geological record, will rightly reject the whole theory. For he may ask in vain where are the numberless transitional links which must formerly have connected the closely allied or representative species found in the successive stages of the same great formation?

In *Great Books of the Western World* (Volume 49)
The Origin of Species by Means of Natural Selection
Chapter XI (p. 179)
Encyclopædia Britannica, Inc. Chicago, Illinois, USA. 1952

As buds give rise by growth to fresh buds, and these, if vigorous, branch out and overtop on all sides many a

feebler branch, so by generation I believe it has been with the great Tree of Life, which fills with its dead and broken branches the crust of the earth, and covers the surface with its ever branching and beautiful ramifications.
In *Great Books of the Western World* (Volume 49)
The Origin of Species by Means of Natural Selection
Chapter IV (p. 64)
Encyclopædia Britannica, Inc. Chicago, Illinois, USA. 1952

I could show fight on natural selection having done and doing more for the progress of civilization than you seem inclined to admit. Remember what risk the nations of Europe ran, not so many centuries ago of being overwhelmed by the Turks, and how ridiculous such an idea now is! The more civilised so-called Caucasian races have beaten the Turkish hollow in the struggle for existence. Looking to the world at no very distant date, what an endless number of the lower races will have been eliminated by the higher civilized races throughout the world.
In Francis Darwin (ed.)
The Life and Letters of Charles Darwin (Volume 1)
Religion (p. 285)
Letter To W. Graham, July 3rd, 1881
D. Appleton & Company. New York, New York, USA. 1887

If it could be demonstrated that any complex organ existed, which could not possibly have been formed by numerous, successive, slight modifications, my theory would absolutely break down.
In *Great Books of the Western World* (Volume 49)
The Origin of Species by Means of Natural Selection
Chapter VI (p. 87)
Encyclopædia Britannica, Inc. Chicago, Illinois, USA. 1952

…extinction and natural selection go hand in hand.
In *Great Books of the Western World* (Volume 49)
The Origin of Species by Means of Natural Selection
Chapter VI (p. 80)
Encyclopædia Britannica, Inc. Chicago, Illinois, USA. 1952

[Evolution by natural selection] absolutely depends on what we in our ignorance call spontaneous or accidental variability. Let an architect be compelled to build an edifice with uncut stones, fallen from a precipice. The shape of each fragment may be called accidental. Yet the shape of each has been determined…by events and circumstances, all of which depend on natural laws; but there is no relation between these laws and the purpose for which each fragment is used by the builder. In the same manner the variations of each creature are determined by fixed and immutable laws; but these bear no relation to the living structure which is slowly built up through the power of selection.
The Variation of Animals and Plants Under Domestication
Chapter XXI (p. 236)
D. Appleton & Company. New York, New York, USA. 1896

Dawkins, Richard   1941–
English ethologist, evolutionary biologist, and popular science writer

All appearances to the contrary, the only watchmaker in nature is the blind forces of physics, albeit deployed in a very special way. A true watchmaker has foresight: he designs his cogs and springs, and plans their interconnections, with a future purpose in his mind's eye. Natural selection, the blind, unconscious, automatic process which Darwin discovered, and which we now know is the explanation for the existence and apparently purposeful form of all life, has no purpose in mind. It has no mind and no mind's eye. It does not plan for the future. It has no vision, no foresight, no sight at all. If it can be said to play the role of the watchmaker in nature, it is the blind watchmaker.
The Blind Watchmaker
Chapter 1 (p. 5)
W.W. Norton & Company, Inc. New York, New York, USA. 1986

Darwin's achievement, like Einstein's, is universal and timeless.
A Devil's Chaplain: Reflections on Hope, Lies, Science, and Love (p. 79)
Houghton Mifflin Company. Boston, Massachusetts, USA. 2003

Maybe we are neo-Darwinists today, but let us spell neo with a very small n! Our neo-Darwinism is very much in the spirit of Darwin himself.
A Devil's Chaplain: Reflections on Hope, Lies, Science, and Love (p. 80)
Houghton Mifflin Company. Boston, Massachusetts, USA. 2003

Natural selection is the only workable explanation for the beautiful and compelling illusion of "design" that pervades every living body and every organ. Knowledge of evolution may not be strictly useful in everyday commerce. You can live some sort of life and die without ever hearing the name of Darwin. But if, before you die, you want to understand why you lived in the first place, Darwinism is the one subject that you must study.
In John Maynard Smith
The Theory of Evolution
Forward (p. xvi)
Penguin Books Ltd. Harmondsworth, England. 1958

As an academic scientist I am a passionate Darwinian, believing that natural selection is, if not the only driving force in evolution, certainly the only known force capable of producing the illusion of purpose which so strikes all who contemplate nature.
A Devil's Chaplain: Reflections on Hope, Lies, Science, and Love (p. 10)
Houghton Mifflin Company. Boston, Massachusetts, USA. 2003

It is forever true that DNA is a double helix, true that if you are a chimpanzee (or an octopus or a kangaroo) [and] trace your ancestors back far enough you will eventually hit a shared ancestor. To a pedant, these are still hypotheses which might be falsified tomorrow. But they never will be.
A Devil's Chaplain: Reflections on Hope, Lies, Science, and Love
(pp. 17–18)
Houghton Mifflin Company. Boston, Massachusetts, USA. 2003

Fisher, Sir Ronald Aylmer   1890–1962
English statistician and geneticist

We may consequently state the fundamental theorems of Natural Selection in the form:

The rate of increase in fitness of any organism at any time is equal to its genetic variance in fitness at that time.
The Genetical Theory of Natural Selection
Chapter II (p. 37)
Dover Publications, Inc. New York, New York, USA. 1958

The million, million, million...to one chance happens once in a million, million, million...times no matter how surprised we may be that it results in us.
In K. Mather
Heredity, Volume 30, 1973

Natural Selection is not Evolution. Yet, ever since the two words have been in common use, the theory of Natural Selection has been employed as a convenient abbreviation for the Theory of Evolution by means of Natural Selection.... This has had the unfortunate consequences that the theory of Natural Selection itself has scarcely ever, if ever, received separate consideration.
The Genetical Theory of Natural Selection
Preface (p. vii)
Dover Publications, Inc. New York, New York, USA. 1958

...it was Darwin's chief contribution, not only to Biology but to the whole of natural science, to have brought to light a process by which contingencies a priori improbable, are given, in the process of time, an increasing probability, until it is their non-occurrence rather than their occurrence which becomes highly improbable.... Let the reader... attempt to calculate the prior probability that a hundred generations of his ancestry in the direct male line should each have left at least one son. The odds against such a contingency as it would have appeared to his hundredth ancestor (about the time of King Solomon) would require for their expression forty-four figures of the decimal notation; yet this improbable event has certainly happened.
In J.S. Huxley, A.C. Hardy and E.B. Ford (eds.)
Evolution as a Process
Retrospect of Criticisms of the Theory of Natural Selection (p. 91)
George Allen & Unwin Ltd. London, England. 1954

Ford, E. B.   1901–88
English ecological geneticist

...organisms automatically generate their own cycles of abundance and rarity and...the changes in selection pressures with which these are associated may greatly increase the speed of evolution.
Ecological Genetics
Chapter 3 (p. 34)
Chapman & Hall Ltd. London, England. 1971

Gould, Stephen Jay   1941–2002
American paleontologist, evolutionary biologist, and historian of science

Natural selection can only produce adaptation to immediately surrounding (and changing) environments. No feature of such local adaptation should yield any expectation of general progress (however such a vague term be defined). Local adaptation may as well lead to anatomical simplification as to greater complexity.
Full House
Chapter 12 (p. 139)
Harmony Books. New York, New York, USA. 1996

The theory of natural selection would never have replaced the doctrine of divine creation if evident, admirable design pervaded all organisms. Charles Darwin understood this, and he focused on features that would be out of place in a world constructed by perfect wisdom.... Darwin even wrote an entire book on orchids to argue that the structures evolved to ensure fertilization by insects are jerry-built of available parts used by ancestors for other purposes. Orchids are Rube Goldberg machines; a perfect engineer would certainly have come up with something better.

This principle remains true today. The best illustrations of adaptation by evolution are the ones that strike our intuition as peculiar or bizarre.
Ever Since Darwin: Reflections in Natural History
Chapter 10. Organic Wisdom or Why Should a Fly Eat Its Mother from Inside (p. 91)
W.W. Norton & Company, Inc. New York, New York, USA. 1977

Hamilton, William D.   1936–
English ethologist, evolutionary biologist, and popular science writer

To express the matter more vividly, in the world of our model organisms, whose behavior is determined strictly by genotype, we expect to find that no one is prepared to sacrifice his life for any single person, but that everyone will sacrifice it when he can thereby save more than two brothers, or four half-brothers, or eight first-cousins...
The Genetical Evolution of Social Behavior
The Journal of Theoretical Biology, Volume 7, 1964 (p. 16)

With very few exceptions, the only parts of the theory of natural selection which have been supported by mathematical models admit to no possibility of the evolution of any characters which are on average to the disadvantage of the individuals possessing them. If natural selection followed the classical model exclusively, species would not show any behavior more positively social than the coming together of the sexes and parental care.
The Genetical Evolution of Social Behavior
The Journal of Theoretical Biology, Part I, Volume 7, 1964

Monod, Jacques   1910–76
French biochemist

Drawn out of the realm of pure chance, the accident enters into that of necessity, of the most implacable certainties. For natural selection operates at the macroscopic level, the level of organisms.... In effect natural selection operates upon the products of chance and can feed nowhere else; but it operates in a domain of very demanding conditions, and from this domain chance is barred. It is not to chance but to these conditions that evolution owes its generally progressive course, its successive conquests, and the impression it gives of a smooth and steady unfolding.

Chance and Necessity: An Essay on the Natural Philosophy of Modern Biology
Chapter VII (pp. 118–119)
Vintage Books. New York, New York, USA. 1972

Shaw, George Bernard   1856–1950
Irish comic dramatist and literary critic

For "Natural Selection" has no moral significance: it deals with that part of evolution which has no purpose, no intelligence, and might more appropriately be called accidental selection, or better still, Unnatural Selection, since nothing is more unnatural than an accident. If it could be proved that the whole universe had been produced by such Selection, only fools and rascals could bear to live.

Back to Methuselah
Preface (p. liv)
Constable & Company Ltd. London, England. 1921

Waddington, Conrad Hal   1905–75
British biologist and paleontologist

The meaning of natural selection can be epigrammatically summarized as "the survival of the fittest." Here "survival" does not, of course, mean the bodily advance of a single individual outliving Methuselah. It implies, in its present-day interpretation, perpetuation as a source for future generations. That individual "survives" best which leaves the most offspring. Again, to speak of an animal as "fittest" does not necessarily imply that it is stronger or most healthy, or would win a beauty competition. Essentially it denotes nothing more than leaving most offspring. The general principle of natural selection, in fact, merely amounts to the statement that the individual which leaves most offspring are those which leave most offspring. It is a tautology.

The Strategy of the Genes: A Discussion of Some Aspects of Theoretical Biology
Chapter 3 (pp. 64–65)
George Allen & Unwin Ltd. London, England. 1957

Wallace, Alfred Russel   1823–1913
English humanist, naturalist, and geographer

The proof that there is a selective agency at work is, I think, to be found in the general stability of species during the period of human observation, notwithstanding the large amount of variability that has been proved to exist. If there were no selection constantly going on, why should it happen that the kind of variations that occur so frequently under domestication never maintain themselves in a state of nature? Examples of this class are

white blackbirds or pigeons, black sheep, and unsymmetrically marked animals generally. These occur not unfrequently, as well as such sports as six-toed or stump-tailed cats, and they all persist and even increase under domestication, but never in a state of nature; and there seems no reason for this but that in the latter case they are quickly eliminated through the struggle for existence — that is, by natural selection.

Variation and Natural Selection
Nature, Volume 44, Number 1144, October 1, 1891 (p. 518)

Wallin, Ivan E.   1883–1969
American biologist

Natural Selection, by itself, is not sufficient to determine the direction of organic evolution…. Natural Selection can only deal with that which has been formed; it has no creative powers. Any directing influence that Natural Selection may have in organic evolution, must, in the nature of the process, be secondary to some other unknown factor.

Symbioticism and the Origin of Species
Chapter I (p. 5)
Williams & Wilkins Company. Baltimore, Maryland, USA. 1927

Wright, Robert
American journalist and author

…natural selection "wants" us to behave in certain ways. But, so long as we comply, it doesn't care whether we are made happy or sad in the process, whether we get physically mangled, even whether we die. The only thing natural selection ultimately "wants" to keep in good shape is the information in our genes, and it will countenance any suffering on our part that serves this purpose.

The Moral Animal: Why We Are the Way We Are
Chapter 7 (pp. 162–163)
Vintage Books. New York, New York, USA. 1994

NATURAL THEOLOGY

Butler, Joseph   1692–1752
English bishop and exponent of natural theology

…the only distinct meaning of the word "natural" is stated, fixed, or settled; since what is natural as much requires and presupposes an intelligent agent to render it so, *i.e.*, to effect it continually or at stated times, as what is supernatural or miraculous [requires] to effect it for once.

The Analogy of Religion, Natural and Revealed to the Constitution and Course of Nature
Part 2, Chapter I (p. 105)
Harper & Brothers. New York, New York, USA. 1880

NATURALISM

Dembski, William A.   1960–
Mathematician and philosopher

Naturalism is the view that the physical world is a self-contained system that works by blind, unbroken natural laws. Naturalism doesn't come right out and say there's nothing beyond nature. Rather, it says that nothing beyond nature could have any conceivable relevance to what happens in nature. Naturalism's answer to theism is not atheism but benign neglect. People are welcome to believe in God, though not a God who makes a difference in the natural order.
The Design Revolution: Answering the Toughest Questions About Intelligent Design
Preface (p. 21)
InterVarsity Press. Downers Grove, Illinois, USA. 2004

Johnson, Philip
Law professor

The assumption of naturalism is the realm of speculative philosophy, and the rule against negative argument is arbitrary. It is as if a judge were to be charged with the crime.
Evolution as Dogma: The Establishment of Naturalism
First Things, October 1990

Scientists committed to philosophical naturalism do not claim to have found the precise answer to every problem, but they characteristically insist that they have the important problems sufficiently well in hand that they can narrow the field of possibilities to a set of naturalistic alternatives.

Absent that insistence, they would have to concede that their commitment to naturalism is based upon faith rather than proof. Such a concession could be exploited by promoters of rival sources of knowledge, such as philosophy and religion, who would be quick to point out that faith in naturalism is no more "scientific" (i.e., empirically based) than any other kind of faith.
Evolution as Dogma: The Establishment of Naturalism
First Things, October 1990

The worldview of scientific naturalism preserves a place for beliefs: a place, that is, among things to be explained by science. The Christian religion thus enters the university with a status precisely that of other comparable religious systems — say, the Aztec system of human sacrifice. Any individual, even a person of eminence in science, can make a personal choice to be "religious." Such choices are made on the basis of "faith," meaning subjective preference. A problem arises only if the Aztecs or the Christians claim access to knowledge. If they do that they are claiming that their own beliefs are normative for unbelievers. Only scientists can claim that kind of authority, because what the scientific community endorses constitutes knowledge, not belief. That is why Darwinian evolution can be taught in schools as fact, however strongly parents or students object, whereas a simple prayer acknowledging God as our Creator is deemed unacceptable — because somebody might object.

How the Universities Were Lost
First Things, March 1995

NATURALIST

Abbey, Edward   1927–89
American environmentalist and nature writer

For I am not a naturalist.... If a label is required say that I am one who loves unfenced country. The open range. Call me a ranger...The only higher honor I've ever heard of is to be called a man.
The Journey Home: Some Words in Defense of the American West
Introduction (p. xiii)
E.P. Dutton. New York, New York, USA. 1977

I am — really am — an extremist, one who lives and loves by choice far out on the very verge of things, on the edge of the abyss, where this world falls off into the depths of another. That's the way I like it.
The Journey Home: Some Words In Defense of the American West
Introduction (p. xiv)
E.P. Dutton & Company. New York, New York, USA. 1977

Agassiz, Jean Louis Rodolphe   1807–73
Swiss-born American naturalist, geologist, and teacher

The education of a naturalist now consists chiefly in learning how to compare.
Methods of Study in Natural History
Chapter II (p. 23)
Ticknor & Fields. Boston, Massachusetts, USA. 1863

Allen, Grant   1848–99
Naturalist

There are two kinds of naturalists, you know.... The superior class live in London or Paris, examine everything minutely with a big microscope, tack on inches of Greek nomenclature to an insignificant mite or bit of moss, and split hairs against anybody with marvelous dexterity. That's science. It dwells in a museum. For my part I detest it. The inferior class live in Europe, Asia, Africa or America, as fate or fancy carries; and, instead of looking at everything in a dried specimen, go out into the woods with rifle on shoulder, or box in hand, and observe the birds, and beasts, and green things of the earth, as God made them, in their own natural and lovely surroundings. That's natural history, old-fashioned, simple, commonplace natural history; and I, for my part, am an old-fashioned naturalist.
The Tents of Shem
Chapter I (p. 8)
Chatto & Windus. London, England. 1890

Darwin, Charles Robert   1809–82
English naturalist

It is well to remember that Naturalists value observations far more than reasoning...

In Francis Darwin (ed.)
The Life and Letters of Charles Darwin (Volume 2)
Darwin to Farrar, November 26, 1868 (p. 453)
D. Appleton & Company. New York, New York, USA. 1896

A naturalist's life would be a happy one if he had only to observe, and never to write.
In Francis Darwin (ed.)
The Life and Letters of Charles Darwin (Volume 2)
C. Darwin to C. Lyell, June 1st [1867] (p. 248)
D. Appleton & Company. New York, New York, USA. 1896

Einstein, Albert   1879–1955
German-born physicist

In every naturalist there must be a kind of religious feeling; for he cannot imagine that the connections into which he sees have been thought of by him for the first time. He rather has the feeling of a child, over whom a grown-up person rules.
Cosmic Religion, with Other Opinions and Aphorisms
On Science (pp. 100–101)
Covici-Fiede. New York, New York, USA. 1931

Forbes, Edward   1815–54
English naturalist

The naturalists of yore esteemed the ocean to be a treasury of wonders, and sought therein for monstrosities and organisms contrary to the law of nature, such as they interpreted it. The naturalists of our own time hold equal faith in the wonders of the sea, but seek therein rather for the links of nature's chain than for apparent exceptions.
The Natural History of the European Seas
Chapter I (pp. 4–5)
John van Voorst. London, England. 1859

Linnaeus, Carl (von Linné)   1707–78
Swedish botanist and explorer

After he [the traveler] has commenced his journey and has become transplanted, so to speak, into a new world, he should consider it his duty to observe everything, not carelessly or at random, but so that nothing will escape his keen vision and alert attention. In describing objects he must endeavor to depict nature so faithfully that he who reads the description must needs believe he is beholding the very things himself.
In A.G. Nathorst
Annual Report of the Board of Regents of the Smithsonian Institution,
1908
Carl von Linné as a Geologist (p. 711)
Government Printing Office. Washington, D.C. 1909

Montagu, George
No biographical data available

As natural history has, within the last half century, occupied the attention and pens of the ablest philosophers of the more enlightened parts of the globe, there needs no apology for the following sheets; since the days of dark-
ness are now past, when the researches of the naturalist were considered as trivial and uninteresting.
Testacea Britannica
Introduction (p. I)
J. White. London, England. 1803

Muir, John   1838–1914
American naturalist

Like Thoreau they see forests in orchards and patches of huckleberry brush, and oceans in ponds and drops of dew.
Our National Parks
Chapter I (p. 2)
Houghton Mifflin Company. Boston, Massachusetts, USA. 1901

Pavlov, Ivan Petrovich   1849–1936
Russian physiologist

For the naturalist everything lies in the method, in the chance of obtaining an unshakeable, lasting truth; and solely from this point of view, which for him is obligatory, the soul, as a naturalistic principle, is not only unnecessary but even harmful to his work, in vain limiting his courage and the depth of his analysis.
Experimental Psychology and Other Essays
Experimental Psychology and Psychopathology in Animals (p. 168)
Philosophical Library. New York, New York, USA. 1957

Riley, James Whitcomb   1849–1916
American poet

In gentlest worship has he bowed
To Nature. Rescued from the crowd
And din of town and thoroughfare,
He turns him from all worldly care
Unto the sacred fastness of
The forest, and the peace and love
That beats there prayer-like in the breeze.
The Complete Works of James Whitcomb Riley
Volume 7, The Naturalist
P.F. Collier & Son, Company. New York, New York, USA. 1916

Thoreau, Henry David   1817–62
American essayist, poet, and practical philosopher

Man cannot afford to be a naturalist, to look at Nature directly, but only with the side of his eye. He must look through and beyond her. To look at her is as fatal as to look at the head of Medusa. It turns the man of science to stone.
The Journal of Henry David Thoreau
March 23, 1853 (p. 43)
Houghton Mifflin Company. Boston, Massachusetts, USA. 1906

Wilson, Edward O.   1929–
American biologist and writer

For the naturalist every entrance into a wild environment rekindles an excitement that is childlike in spontaneity, often tinged with apprehension — in short, the way life ought to be lived, all the time.

The Future of Life
Chapter 6 (p. 146)
Alfred A. Knopf. New York, New York, USA. 2002

NATURE

Abbey, Edward   1927–89
American environmentalist and nature writer

There are no vacant lots in nature.
Desert Solitaire
Down the River (p. 189)
Ballantine Books. New York, New York, USA. 1968

There are enough cathedrals and temples and altars here for a Hindu pantheon of divinities. Each time I look up one of the secretive little side canyons I half expect to see not only the cottonwood tree rising over its tiny spring — the leafy god, the desert's liquid eye — but also a rainbow-colored corona of blazing light, pure spirit, pure being, pure disembodied intelligence, about to speak my name.

If a man's imagination were not so weak, so easily tired, if his capacity for wonder not so limited, he would abandon forever such fantasies of the supernal. He would learn to perceive in water, leaves and silence more than sufficient of the absolute and marvelous, more than enough to console him for the loss of the ancient dreams.
Desert Solitaire
Down the River (p. 200)
Ballantine Books. New York, New York, USA. 1968

Nature, like Maimonides said, is mainly a good place to throw beer cans on Sunday afternoons.
A Voice Crying in the Wilderness: Notes from a Secret Journal
Chapter 9 (p. 83)
St. Martin's Press. New York, New York, USA. 1989

Floating down a portion of Rio Colorado in Utah a rare month in spring, twenty-two years ago, a friend and I found ourselves passing through a world so beautiful it seemed and had to be eternal. Such perfection of being, we thought — these glens of sandstone, these winding corridors of mystery, leading each to its solitary revelation — could not possibly be changed.
Down the River
Part IV, Chapter 19 (p. 231)
E.P. Dutton. New York, New York, USA. 1982

Nature is indifferent to our love, but never unfaithful.
A Voice Crying in the Wilderness: Notes from a Secret Journal
Chapter 9 (p. 86)
St. Martin's Press. New York, New York, USA. 1989

I am here not only to evade for a while the clamor and filth and confusion of the cultural apparatus but also to confront, immediately and directly, if it's possible, the bare bones of existence, the elemental and fundamental, the bedrock which sustains us. I want to be able to look at and into a juniper tree, a piece of quartz, a vulture, a spider,

and see it as it is in itself, devoid of all humanly ascribed qualities, anti-Kantian, even the categories of scientific description. To meet God or Medusa face to face, even if it means risking everything human in myself. I dream of a hard and brutal mysticism in which the naked self merges with a nonhuman world and yet somehow survives still intact, individual, separate. Paradox and bedrock.
Desert Solitaire
The First Morning (p. 6)
Ballantine Books. New York, New York, USA. 1968

Ackerman, Diane   1948–
American writer

Nature neither gives nor expects mercy.
The Moon by Whale Light, and Other Adventures Among Bats and Crocodilians, Penguins and Whales
Chapter 4 (pp. 239–240)
Random House, Inc. New York, New York, USA. 1991

Just because we have evolved minds that crave order doesn't mean that nature is orderly. Evolution is a sleeping watchdog. It is possible for us to disturb it, or it may wake on its own. Either way, expect commotion.
The Rarest of the Rare: Vanishing Animals, Timeless Worlds
Introduction (p. xii)
Vintage Books. New York, New York, USA. 1997

[N]ature is also great fun. To pretend that nature isn't fun is to miss much of the joy of being alive…
The Rarest of the Rare: Vanishing Animals, Timeless Worlds
Introduction (p. xx)
Vintage Books. New York, New York, USA. 1997

Ackoff, Russell Lincoln   1919–
American operations research and systems scientist

Nature is not organized in the same way that universities are.
Toward an Idealized University
Management Science, Volume 15, December 1970 (p. B–127)

Adams, Abby   1939–
American astronaut and solar physicist

Nature is what wins in the end.
The Gardener's Gripe Book
What Is a Garden Anyway? (p. 10)
Workman Publishing. New York, New York, USA. 1995

Adams, George   1750–95
English instrument maker

The study of nature is as much distinguished from other subjects by the importance of its matter, as by the variety of its topics.
Lectures on Natural and Experimental Philosophy (Volume 2)
Lecture XII (p. 1)
Printed by R. Hindmarsh. London, England. 1794

Man has before him all nature, the whole world with which he is surrounded for the object of his view, and

the subject of his consideration; but his capacity is so circumscribed, his knowledge so straightened, his powers so limited, that he can by no means conceive the mechanism of so vast and complicate a structure.
Lectures on Natural and Experimental Philosophy (Volume 3)
Chapter XXXV (p. 510)
Printed by R. Hindmarsh. London, England. 1794

As you advance in the knowledge of nature's varieties, your mind will be opened, and you will find fresh ornament in truth, fresh dignity in devotion, and fresh reason in religion.
Lectures on Natural and Experimental Philosophy (Volume 2)
Lecture XII (p. 2)
Printed by R. Hindmarsh. London, England. 1794

…our views of nature are like the map of an inland country, where you see rivers without any sources continually discharging their waters without a sea to receive them; roads that you know not whence they come, nor whither they go; mountains, forests, and plains, cut off in the middle by the marginal lines of the paper: but even of those things which we know well, there is much that surpasses the extent of our faculties.
Lectures on Natural and Experimental Philosophy (Volume 3)
Chapter XXXV (pp. 510–511)
Printed by R. Hindmarsh. London, England. 1794

Addison, Joseph   1672–1719
English essayist, poet, and statesman

If there's a power above us, (and that there is all nature cries aloud through all her works) he must delight in virtue.
Cato
Act V, Scene 1
J. Dicks. London, England. 1883

Agassiz, Jean Louis Rodolphe   1807–73
Swiss-born American naturalist, geologist, and teacher

The eye of the Trilobite tells us that the sun shone on the old beach where he lived; for there is nothing in nature without a purpose; and when so complicated an organ was made to receive the light, there must have been light to enter it.
Geological Sketches
Chapter II (pp. 31–32)
Houghton Mifflin Company. Boston, Massachusetts, USA. 1886

Lay aside all conceit. Learn to read the book of nature for yourself. Those who have succeeded best have followed for years some slim thread which has once in a while broadened out and disclosed some treasure worth a life-long search.
In David Stair Jordan
Popular Science Monthly
Volume 40, 1891

The study of nature is an intercourse with the highest mind. You should never trifle with nature. At the lowest her works are the works of the highest powers — the highest something, in whatever way we may look at it.
In David Stair Jordan
Popular Science Monthly, Volume 40, 1891

As long as men inquire, they will find opportunities to know more upon these topics than those who have gone before them, so inexhaustibly rich is nature in the innermost diversity of her treasures of beauty, order, and intelligence.
Essay on Classification
Chapter II, Section I (p. 141)
Harvard University Press. Cambridge, Massachusetts, USA. 1962

I may say that here, as in most cases where the operations of nature interfere with the designs of man, it is not by a direct intervention on our part that we may remedy the difficulties, but rather by a precise knowledge of [nature's] causes, which may enable us, if not to check, at least to avoid the evil consequences.
Annual Report of the Superintendent of the Coast Survey, Showing the Progress of that Work During the Year Ending November, 1851
Extracts from the report of Professor Agassiz to the Superintendent of the Coast Survey, on the examination of the Florida reefs, keys, and coast (p. 158)
Printed by Robert Armstrong, Washington. 1852

…it must be for truth's sake, and not even for the sake of its usefulness to humanity, that the scientific man studies Nature.
Methods of Study in Natural History
Chapter II (p. 24)
Ticknor & Fields. Boston, Massachusetts, USA. 1863

Aldrich Thomas Bailey   1836–1907
American writer and editor

Nature, who loves to do a gentle thing even in her most savage moods, had taken one of those empty water-courses and filled it from end to end with forget-me-nots.
Queen of Sheba
IX (p. 205)
J.R. Osgood & Company. Boston, Massachusetts, USA. 1877

Aristotle   384 BCE–322 BCE
Greek philosopher

Nature proceeds little by little from things lifeless to animal life in such a way that it is impossible to determine the exact line of demarcation, nor on which side thereof an intermediate form should lie.
In *Great Books of the Western World* (Volume 8)
History of Animals
Book VIII, Chapter 1
Encyclopædia Britannica, Inc. Chicago, Illinois, USA. 1952

No one finds fault with defects which are the result of nature.
In *Great Books of the Western World* (Volume 8)
Ethics
Book III, Chapter 5
Encyclopædia Britannica, Inc. Chicago, Illinois, USA. 1952

Arnold, Matthew   1822–88
English poet and critic

Nature's great law,
and law of all men's minds? —
To its own impulse every creature stirs;
Live by thy light, and earth will live by hers!
The Poetical Works of Matthew Arnold
Religious Isolation
Stanza 4
Oxford University Press, Inc. New York, New York, USA. 1950

Man must begin, know this, where Nature ends; Nature and man can never be fast friends.
The Poetical Works of Matthew Arnold
In Harmony with Nature, l. 12–13
Oxford University Press, Inc. New York, New York, USA. 1950

Know, man hath all which Nature hath, but more,
And in that more lie all his hopes of good.
Nature is cruel, man is sick of blood;
Nature is stubborn, man would fain adore.
The Poetical Works of Matthew Arnold
In Harmony with Nature
Oxford University Press, Inc. New York, New York, USA. 1950

Atherton, Gertrude   1857–1948
American novelist

Nature is a wicked old matchmaker.
Senator North
Book II, VII (p. 174)
John Lane: The Bodley Head. New York, New York, USA. 1900

Audubon, John James   1785–1851
West Indian-born American ornithologist and artist

From Nature! — How often are these words used, when at a glance he who has seen the perfect and beautiful forms of birds, quadrupeds or other objects, as they have come from the hand of Nature, discovers that the representation is not that of living Nature!
Ornithological Biography (Volume 1)
The Yellow-Billed Cuckoo (p. 18)
Adam Black. Edinburgh, Scotland. 1831

Bacon, Sir Francis   1561–1626
English lawyer, statesman, and essayist

Those who become practically versed in nature are, the mechanic, the mathematician, the physician, the alchemist, and the magician, but all (as matters now stand) with faint efforts and meager success.
In *Great Books of the Western World* (Volume 30)
Novum Organum
First Book, Aphorism 5 (p. 107)
Encyclopædia Britannica, Inc. Chicago, Illinois, USA. 1952

…we can command nature only by obeying her…
In *Great Books of the Western World* (Volume 30)
Novum Organum
First Book, Aphorism 129 (p. 135)
Encyclopædia Britannica, Inc. Chicago, Illinois, USA. 1952

Man, as the minister and interpreter of nature, dies and understands as much as his observations on the order of nature, either with regard to things or the mind permit him, and neither knows or is capable of more.
In *Great Books of the Western World* (Volume 30)
Novum Organum
First Book, Aphorism 1 (p. 107)
Encyclopædia Britannica, Inc. Chicago, Illinois, USA. 1952

In nature things move violently to their place, and then calmly in their place.
Bacons Essays
Of Great Places (p. 27)
The Macmillan Company. New York, New York, USA. 1930

Bailey, Philip James   1816–1902
English poet

The course of Nature seems a course of Death,
And nothingness the whole substantial thing.
Festus: A Poem
Scene III (p. 58)
George Routledge & Sons, Ltd. London, England. 1893

Nature means Necessity.
Festus: A Poem
Dedication
George Routledge & Sons, Ltd. London, England. 1893

Baker, Henry   1698–1774
English naturalist

That Man is certainly the happiest, who is able to find out the greatest Number of reasonable and useful Amusements, easily attainable and within his Power: and, if so, he that is delighted with the Works of Nature, and makes them his Study must undoubtedly be happy, since every Animal, Flower, Fruit, or Insect, nay, almost every Particle of Matter, affords him an Entertainment.
The Microscope Made Easy
The Introduction (pp. xii–xiv)
Printed for R. Dodsley. London, England. 1743

Baron von Frankenstein
Fictional character

Nothing in nature is terrifying when one understands it.
The Son of Frankenstein
Film (1939)

Beaumont, Francis   1584–1616
English playwright and dramatic poet
Fletcher, John   1579–1625
Jacobean playwright

Nature too unkind;
That made no medicine for a troubled mind!
Philaster
Act III, Scene 1
D.C. Heath. Boston, Massachusetts, USA. 1906

Beebe, William 1877–1962
American ornithologist

There was the ending still unfinished, the finale buried in the future — and in this we find the fascination of Nature and Science.
In William H. Carr
The Stir of Nature
Chapter Thirteen (p. 167)
Oxford University Press, Inc. New York, New York, USA. 1930

Beston, Henry 1888–1968
American writer

The three great elemental sounds in nature are the sound of rain, the sound of wind in a primeval wood, and the sound of the outer ocean on a beach. I have heard them all, and of the three elemental voices, that of ocean is the most awesome, beautiful, and varied.
The Outermost House
Chapter III (p. 43)
Rinehart & Company. New York, New York, USA. 1928

Nature is a part of our humanity, and without some awareness and experience of that divine mystery, man ceases to be man.
The Outermost House
Forward (p. ix)
Rinehart & Company. New York, New York, USA. 1928

As well expect Nature to answer to your human values as to come into your house and sit in a chair.
The Outermost House
Chapter X (p. 221)
Rinehart & Company. New York, New York, USA. 1928

A year indoors is a journey along a paper calendar; a year in outer nature is the accomplishment of a tremendous ritual.
The Outermost House
Chapter IV (p. 59)
Rinehart & Company. New York, New York, USA. 1928

Bishop, Elizabeth 1911–79
American poet and writer

Nature repeats herself, or almost does: repeat, repeat, repeat, revise, revise, revise.
North Haven, l. 19–20
Farrar, Straus & Giroux. New York, New York, USA. 1984

Bloomfield, Robert 1766–1823
English poet

Strange to the world, he wore a bashful look,
The fields his study, nature was his book.
In John Aikin
Selected Works of the British Poets: In a Chronological Series from Falconer to Sir Walter Scott
The Farmer's Boy, Spring, l. 31
Thomas Wardle. Philadelphia, Pennsylvania, USA. 1838

Bohm, David 1917–92
American physicist

In nature nothing remains constant. Everything is in a perpetual state of transformation, motion, and change.
Causality and Chance in Modern Physics
Chapter One (p. 1)
University of Pennsylvania Press. Philadelphia, Pennsylvania, USA. 1957

Bohr, Niels Henrik David 1886–1962
Danish physicist

In our description of nature the purpose is not to disclose the real essence of phenomena but only to track down as far as possible relations between the multifold aspects of our experiences.
Atomic Theory and the Description of Nature
Introductory Survey (p. 18)
Cambridge University Press. Cambridge, England. 1934

Borland, Hal 1900–78
American writer

There are some things, but not too many, toward which the countryman knows he must be properly respectful if he would avoid pain, sickness, and injury. Nature is neither punitive nor solicitous, but she has thorns and fangs as well as bowers and grassy banks.
Beyond Your Doorstep: A Handbook to the Country
Chapter 13 (p. 303)
Alfred A. Knopf. New York, New York, USA. 1962

Nothing in nature is as simple as it sometimes seems when reduced to words.
The Enduring Pattern
Life — Flesh and Blood: Reptiles (p. 189)
Simon & Schuster. New York, New York, USA. 1959

Nature seems to look after her own only up to a certain point; beyond that they are supposed to fend for themselves.
The Enduring Pattern
Life — Flesh and Blood: Amphibians (p. 185)
Simon & Schuster. New York, New York, USA. 1959

Nature is an infinitely complex series of facts; it is not an object lesson, and it is not a ready-made sermon on conduct or morality.
The Enduring Pattern
A Place to Live: Time (p. 20)
Simon & Schuster. New York, New York, USA. 1959

Boyle, Robert 1627–91
English natural philosopher and theological writer

Nature always looks out for the preservation of the universe.
In Edward B. Davis and Michael Hunter (eds.)
A Free Enquiry into the Vulgarly Received Notions of Nature
Section IV (p. 31)
Cambridge University Press. Cambridge, England. 1996

It is one thing to be able to help nature to produce things, and another thing to understand well the nature of the things produc'd.
The Sceptical Chymist

The Third Part (p. 95)
Dawsons of Pall Mall. London, England. 1965

Bradley, Jr., John Hodgdon   1898–1962
American geologist

Nature flips the coin with a "Heads I win — tails you lose." She offers her children stagnation and degeneracy on the one hand, or over-specialization and extinction on the other.
Parade of the Living
Part III, Chapter XX (p. 290)
Coward-McCann, Inc. New York, New York, USA. 1930

Bridgman, Helen Bartlett

Nature seems positively to enjoy playing pranks which turn all preconceived notions topsy-turvy.
Gems
How It Began (p. 5)
Brooklyn, New York, USA. 1916

Bridgman, Percy Williams   1882–1961
American physicist

…our conviction that nature is understandable and subject to law arose from the narrowness of our horizons, and that if we sufficiently extend our range we shall find that nature is intrinsically and in its elements neither understandable nor subject to law…
The New Vision of Science
Harper's Magazine, Volume 158, March 1929 (p. 444)

Bronowski, Jacob   1908–74
Polish-born English mathematician and polymath

Nature is a network of happenings that do not unroll like a red carpet into time, but are intertwined between every part of the world; and we are among those parts. In this nexus, we cannot reach certainty because it is not there to be reached; it goes with the wrong model, and the certain answers ironically are the wrong answers. Certainty is a demand that is made by philosophers who contemplate the world from outside; and scientific knowledge is knowledge for action, not contemplation. There is no God's eye view of nature, in relativity, or in any science: only a man's eye view.
The Identity of Man
The Machinery of Nature, Section 6 (p. 38)
Doubleday & Company, Inc. Garden City, New York, USA. 1972

A new conception was being made…that whatever fundamental units the world is put together from, they are more delicate, more fugitive, more startling than we catch in the Butterfly Net of our senses.
The Ascent of Man
Chapter 11 (p. 364)
Little, Brown & Company. Boston, Massachusetts, USA. 1973

Bronte, Charlotte   1816–55
English author

The universal mother, Nature.
Jane Eyre
Chapter XXVIII (p. 320)
Harcourt, Brace & World, Inc. New York, New York, USA. 1962

Browne, Sir Thomas   1605–82
English writer and physician

There are no grotesques in nature; not anything framed to fill up empty cantons, and unnecessary spaces.
Religio Medici
Part XV
Elliot Stock. London, England. 1883

Now nature is not at variance with art, nor art with nature; they being both the servants of his providence. Art is the perfection of nature. Were the world now as it was the sixth day, there were yet a chaos. Nature hath made one world, and art another. In brief, all things are artificial; for nature is the art of God.
Religio Medici
Section 16
Elliot Stock. London, England. 1883

All things are artificial, for nature is the art of God.
Religio Medici
Part I, Section xvi (p. 29)
Elliot Stock. London, England. 1883

Browning, Robert   1812–89
English poet

I trust in Nature for the stable laws
Of beauty and utility. Spring shall plant
And Autumn garner to the end of time.
I trust in God — the right shall be the right
And other than the wrong, while he endures;
I trust in my own soul, that can perceive
The outward and the inward, Nature's good
And God's.
The Poems and Plays of Robert Browning
A Soul's Tragedy, Act I (p. 458)
The Modern Library. New York, New York, USA. 1934

…what I call God.
…fools call Nature…
The Poems and Plays of Robert Browning
The Pope, l. 1073–1074
The Modern Library. New York, New York, USA. 1934

Bryan, J. Ingram
No biographical data available

Nature does not tolerate the whimsical and the inane; all her structures are on principles, and she allows no others.
The Interpretation of Nature in English Poetry
Chapter I (p. 6)
Kaitakusha. Tokyo, Japan. 1932

Nature is not static but dynamic; she is not now what she will be, for she moves toward a goal.
The Interpretation of Nature in English Poetry
Chapter I (p. 7)
Kaitakusha. Tokyo, Japan. 1932

Bryant, William Cullen 1794–1878
American poet

To him who in the love of Nature holds
Communion with her visible forms, she speaks
A various language.
Poems
Thanatopsis
D. Appleton & Company. New York, New York, USA. 1874

Go forth under the open sky, and list
To Nature's teachings.
Poems
Thanatopsis
D. Appleton & Company. New York, New York, USA. 1874

Meredith, Owen (Edward Robert Bulwer-Lytton, 1st Earl Lytton) 1831–91
English statesman and poet

Nature is the great agent of the external universe…
The Last Days of Pompeii
Book One, VIII
George Routledge & Sons, Ltd. London, England. 1900

Burnet, Thomas 1635–1715
English cleric and scientist

SINCE I was first inclin'd to the Contemplation of Nature, and took pleasure to trace out the Causes of Effects, and the dependence of one thing upon another in the visible Creation, I had always, methought, a particular curiosity to look back into the first Sources and ORIGINAL of Things; and to view in my mind, so far as I was able, the Beginning and Progress of a RISING WORLD.
The Sacred Theory of the Earth (2nd edition)
Book I, Chapter I (p. 23)
Printed by R. Norton. London. 1691

Burney, Fanny 1752–1840
English novelist and diarist

…the lifeless symmetry of architecture, however beautiful the design and proportion, no man would be so mad as to put in competition with the animated charms of nature.
Evelina
Letter XXIII (p. 100)
J.M. Dent & Sons Ltd. London, England. 1909

Burroughs, John 1837–1921
American naturalist and writer

The love of nature is different from the love of science, though the two may go together.
The Breath of Life
Preface (p. vii)
Houghton Mifflin Company. Boston, Massachusetts, USA. 1915

Nature works with such simple means! A little more or a little of this or that, and behold the difference!
The Breath of Life
Chapter III (p. 55)
Houghton Mifflin Company. Boston, Massachusetts, USA. 1915

Nature teaches more than she preaches. There are no sermons in stone. It is easier to get a spark out of a stone than a moral.
Time and Change
The Gospel of Nature (p. 247)
Houghton Mifflin Company. Boston, Massachusetts, USA. 1912

Nature exists to the mind not as an absolute realization, but as a condition, as something constantly becoming.… It is suggestive and prospective; a body in motion, and not an object at rest.
Expression *The Atlantic Monthly*, Volume 6, Number XXXVII, November 1860 (p. 572)

Nature is not benevolent; Nature is just, gives pound for pound, measure for measure, makes no exceptions, never tempers her decrees with mercy, or winks at any infringement of her laws.
Harvest of a Quiet Eye: The Natural World of John Burroughs
The Gospel of Nature, 5 (p. 149)
Tamarack Press. Madison, Wisconsin, USA. 1976

Burton, Robert 1577–1640
English clergyman and scholar

See one promontory, said Socrates of old, one mountain, one sea, one river, & see all.
The Anatomy of Melancholy (Volume 1)
Part I, Sect. II, Memb. IV, Subsec. 7 (p. 422)
AMS Press, Inc. New York, New York, USA. 1973

Butler, Samuel 1612–80
English novelist, essayist, and critic

That one great lie she told about the earth being flat, when it was round all the time — and again how she stuck to it that the sun went round us when it was we who are going round her — this double falsehood has irretrievably ruined my confidence in her. There is no lie which she will not tell and stick to like a Gladstonian. How plausibly she told her tale, and how many ages was it before she was so much as suspected, and then when things did begin to look bad for her, how she brazened it out and what a desperate business it was to bring all her shifts and prevarications to book.
The Note-Books of Samuel Butler (Volume 1)
1874–1883 (p. 74)
University Press of America, Inc. Lanham, Maryland, USA. 1984

The progress of Nature is effected by steps that are often imperceptible and blend into one another with the utmost gentleness…
The Note-Books of Samuel Butler (Volume 1)
1874–1883 (p. 158)
University Press of America, Inc. Lanham, Maryland, USA. 1984

Byron, George Gordon, 6th Baron Byron 1788–1824
English Romantic poet and satirist

There is a pleasure in the pathless woods,
There is a rapture on the lonely shore,

There is society, where none intrudes,
By the deep sea, and music in its roar:
I love not man the less, but nature more.
Childe Harold's Pilgrimage
Canto IV,. Clxxvii–clxxxiv
Cassell. London, England. 1886

Cable, George W.   1844–1925
American writer and reformer

The book of nature is a catechism. But, after it answers the first question with "God," nothing but questions follow.
Madame Delphine
Chapter V
Charles Scribner's Sons. New York, New York, USA. 1896

Shall we ever subdue Nature and make her always submissive and compliant? Who knows what man may do with her when once he has got self, the universal self, under perfect mastery?
Bonaventure
Book III, XVIII (p. 291)
Charles Scribner's Sons. New York, New York, USA. 1888

Campbell, Jeremy C.
No biographical data available

Evidently nature can no longer be seen as matter and energy alone. Nor can all her secrets be unlocked with the keys of chemistry and physics, brilliantly successful as these two branches of science have been in our century. A third component is needed for any explanation of the world that claims to be complete. To the powerful theories of chemistry and physics must be added a late arrival: a theory of information. Nature must be interpreted as matter, energy, and information.
Grammatical Man: Information, Entropy, Language and Life
Chapter 1 (p. 16)
Simon & Schuster. New York, New York, USA. 1982

Campbell, Thomas   1777–1844
Scottish poet

There shall be love, when genial morn appears,
Like pensive Beauty smiling in her tears,
To watch the brightening roses of the sky,
And muse on Nature with a poet's eye.
The Complete Poetical Works
The Pleasures of Hope, Part ii, l. 98–101
Chadwyck-Healey. Cambridge, England. 1992

Carlyle, Thomas   1795–1881
English historian and essayist

Nature, like the Sphinx, is of womanly celestial loveliness and tenderness; the face and bosom of a goddess, but ending in claws and the body of a lioness…. Nature, Universe, Destiny, Existence, howsoever we name this grand unnamable Fact in the midst of which we live and struggle, is as a heavenly bride and conquest to the wise and brave, to them who can discern her behests and do them; a destroying fiend to them who cannot.
Past and Present
Chapter II (p. 7)
Chapman & Hall. London, England. 1843

Nature admits no lie; most men profess to be aware of this, but few in any measure lay it to heart.
Latter-Day Pamphlets
No. 5 (p. 170)
Chapman & Hall. London, England. 1850

Carson, Rachel   1907–64
American marine biologist and writer

The "control of nature" is a phrase conceived in arrogance, born of the Neanderthal age of biology and philosophy, when it was supposed that nature exists for the convenience of man. The concepts and practices of applied entomology for the most part date from that Stone Age of science. It is our alarming misfortune that so primitive a science has armed itself with the most modern and terrible weapons, and that in turning them against the insects it has also turned them against the earth.
Silent Spring
Chapter 17 (p. 297)
Houghton Mifflin Company. Boston, Massachusetts, USA. 1961

Exploring nature with your child is largely a matter of becoming receptive to what lies around you. It is learning again to use your eyes, ears, nostrils, and finger tips, opening up the disused channels of sensory impression.
The Sense of Wonder (p. 52)
Harper & Row, Publishers, New York 1984

A rainy day is the perfect time for a walk in the woods.
The Sense of Wonder (p. 30)
Harper & Row, Publishers, New York 1984

Cawein, Madison Julius   1865–14
American poet

I am a part of all you see
In Nature: part of all you feel:
I am the impact of the bee
Upon the blossom; in the tree
I am the sap — that shall reveal
The leaf, the bloom — that flows and flutes
Up from the darkness through its roots.
Poems
Penetralia
The Macmillan Company. New York, New York, USA. 1911

Chaisson, Eric J.   1946–
American astrophysicist

Without a brainy seat of consciousness and its inherent awareness of self and environment, galaxies would twirl and stars would shine, but no one or thing could comprehend the majesty of the reality that is nature.
The Life Era: Cosmic Selection and Conscious Evolution

Chapter 1 (p. 43)
The Atlantic Monthly Press. New York, New York, USA. 1987

Chargaff, Erwin   1905–2002
Austrian biochemist

We manipulate nature as if we were stuffing an Alsatian goose. We create new forms of energy; we make new elements; we kill crops; we wash brains. I can hear them in the dark sharpening their lasers.
The Paradox of Biochemistry
Columbia Forum, Volume 12, Number 2, Summer 1969 (p. 18)

What makes the study of nature so magnificent is its very giveness; it is because it is; it is as it is; and tolle, lege! (pick up and read!) remains its eternal admonition.
Voices in the Labyrinth: Nature, Man and Science
Chapter 3 (pp. 18–19)
The Seabury Press. New York, New York, USA. 1977

Chaucer, Geoffrey   1343–1400
English poet

Nature, the vicaire of the almyghty Lord…
The Complete Works of Geoffrey Chaucer
The Parliament of Fowls, l. 379
Houghton Mifflin Company. Boston, Massachusetts, USA. 1933

Child, Lydia M.   1802–80
American writer and abolitionist

That man's best works should be such bungling imitations of Nature's infinite perfection, matters not much: but that he should make himself an imitation, this is the fact which Nature moans over, and deprecates beseechingly. Be spontaneous, be truthful, be free, and thus be individuals! is the song she sings through warbling birds, and whispering pines, and roaring waves, and screeching winds.
Letters from New York
Letter XXXVIII (p. 276)
C.S. Francis & Company. New York, New York, USA. 1945

Chiras, Daniel D.
No biographical data available

In nature, virtually nothing is wasted.
Lessons from Nature: Learning to Live Sustainably on the Earth
Chapter 2 (pp. 31–32)
Island Press. Washington, D.C. 1992

Churchill, Charles   1731–64
English poet and satirist

Not without art, but yet to Nature true.
The Rosicad and the Apology
The Rosciad, l. 699
Lawrence & Bullen. London, England. 1891

It can't be nature, for it is not sense.
The Poems of Charles Churchill (Volume 2)
The Farewell, l. 201
Eyre & Spottiswoode Ltd. London, England. 1933

Cicero (Marcus Tullius Cicero)   106 BCE–43 BCE
Roman orator, politician, and philosopher

…ab interitu naturam abhorrere.
…nature shrinks from destruction.
Translated by H. Rackham
Cicero: De Finibus Bonorum, Et Malorum
De Finibus
V, XI, 31 (p. 427)
William Heinemann. London, England. 1931

Clarke, Arthur C.   1917–
English science and science fiction writer

How inappropriate to call this planet Earth when it is clearly Ocean.
In James E. Lovelock
Hands Up for the Gaia Hypothesis
Nature Volume 344, Number 6262, 8 March 1990 (p. 102)

…Nature always balances her books…
2061: Odyssey Three
Chapter 37 (p. 177)
Ballantine Books. New York, New York, USA. 1987

Close, Frank
Writer and physicist
Marten, Michael
No biographical data available
Sutton, Christine
No biographical data available

Even at subatomic level nature presents images of itself that reflect our own imaginings.
The Particle Explosion
Chapter 1 (p. 15)
Oxford University Press, Inc. Oxford, England. 1987

Coleridge, Samuel Taylor   1772–1834
English lyrical poet, critic, and philosopher

And what if all of animated nature
Be but organic harps diversely fram'd,
That tremble into thought, as o'er them sweeps,
Plastic and vast, one intellectual breeze,
At once the soul of each, and God of all?
The Complete Poetical Works of Samuel Taylor Coleridge (Volume 1)
The Eolian Harp, Stanza 4
The Clarendon Press. Oxford, England. 1912

In nature there is nothing melancholy.
The Complete Poetical Works of Samuel Taylor Coleridge (Volume 1)
The Nightingale, Stanza I, l. 15
The Clarendon Press. Oxford, England. 1912

Collingwood, Robin George   1889–1943
English historian and philosopher

The only condition on which there could be a history of nature is that the events of nature are actions on the part of some thinking being or beings, and that by studying these actions we could discover what were the thoughts which they expressed and think these thoughts for ourselves. This is a condition which probably no one will claim is fulfilled. Consequently the processes of nature are not historical processes and our knowledge of nature, though it may resemble history in certain superficial ways, e.g., by being chronological, is not historical knowledge.

The Idea of History
Part V, Section 5 (p. 302)
At The Clarendon Press. Oxford, England. 1967

Colman, George (The Younger)   1762–1836
English playwright

All argument will vanish before one touch of nature.
The Poor Gentleman
Act V, 1
J. Dicks. London, England. 1883

Coman, Dale Rex   1906–
American research physician and wildlife writer

My beloved planet has reminded me of my lowly and inconsequential place in her affairs, and I have no answer fit to offer, no excuses to give, no apologies I know how to phrase.
The Endless Adventure
The Off-Shore Islands (p. 9)
Henry Regnery Company. Chicago, Illinois, USA. 1972

Nature is neither harsh nor cruel nor sentimentally sweet and kind.
The Endless Adventure
From Mushrooms to Bats (p. 19)
Henry Regnery Company. Chicago, Illinois, USA. 1972

Commoner, Barry   1917–
American biologist, ecologist, and educator

Nature knows best.
The Closing Circle: Nature, Man & Technology
Chapter 2 (p. 41)
Alfred A. Knopf. New York, New York, USA. 1971

Connell, Joseph
No biographical data available
Sousa, Wayne
No biographical data available

If a balance of nature exists, it has proved exceedingly hard to demonstrate.
On the Evidence Needed to Judge Ecological Stability or Persistence
The American Naturalist, Volume 121, Number 6, June 1983 (p. 808)

Conrad, Joseph   1857–1924
Polish-born English novelist

…Nature — the balance of colossal forces…. Nature — the great artist.
Lord Jim
Chapter XIX (p. 179)
Rinehart & Company, Inc. New York, New York, USA. 1957

Copernicus, Nicolaus   1473–1543
Polish astronomer

But we should rather follow the wisdom of nature, which, as it takes very great care not to have produced anything superfluous or useless, often prefers to endow one thing with many effects.
In *Great Books of the Western World* (Volume 16)
On the Revolutions of the Heavenly Spheres
Book One, Chapter 10 (p. 526)
Encyclopædia Britannica, Inc. Chicago, Illinois, USA. 1952

Corbett, Jim   1875–1955
Indian-born hunter and naturalist

…for the book of nature has no beginning as it has no end. Open the book where you will at any point of your life, and if you have the desire to acquire knowledge, you will find it of intense interest, and no matter how long or how intently you study the pages, your interest will not flag for in nature there is no finality.
Jungle Lore
Chapter IV (p. 33)
Oxford University Press, Inc. New York, New York, USA. 1953

Cousins, Norman   1912–90
American editor and writer

It is unscientific to say that within the many billions of galactic systems, ours is the only planet that supports life in advanced form. Nature shuns one of a kind as much as it abhors a vacuum. Given infinite time and space, anything that occurs at one place or time in the universe will occur elsewhere or "elsewhen."
Rendezvous with Infinity
Cosmic Search Magazine, Volume 1, Number 1, January 1, 1979 (p. 30)

Cowper, William   1731–1800
English poet

Nature, exerting an unwearied power,
Forms, opens, and gives scent to every flower;
Spreads the fresh verdure of the field, and leads
The dancing Naiads through the dewy meads.
The Poetical Works of William Cowper
Table Talk, l. 690
John W. Lovell Company. New York, New York, USA. No date

Nor rural sights alone, but rural sounds,
Exhilarate the spirit, and restore
The tone of languid Nature.
The Poetical Works of William Cowper
The Task
Book I, The Sofa, l. 181
John W. Lovell Company. New York, New York, USA. No date

Nature indeed looks prettily in rhyme.
The Poetical Works of William Cowper
Retirement, l. 576
John W. Lovell Company. New York, New York, USA. No date

Crick, Francis Harry Compton   1916–2004
English biochemist

The basic trouble is that nature is so complex that many quite different theories can go some way to explaining the results.... [W]hat constraints can be used as a guide through the jungle of possible theories? It seems to me that the only useful constraints are contained in the experimental evidence.
What Mad Pursuit: A Personal View of Scientific Discovery
Chapter 13 (p. 141)
Basic Books, Inc., Publishers. New York, New York, USA. 1988

Crookes, Sir William   1832–1919
English chemist and physicist

…Nature — the word that stands for the baffling mysteries of the Universe. Steadily, unflinchingly, we strive to pierce the inmost heart of Nature, from what she is to reconstruct what she has been, and to prophesy what she yet shall be. Veil after veil we have lifted, and her face grows more beautiful, august, and wonderful, with every barrier that is withdrawn.
In William Walker Atkinson
Practical Mind Reading
Lesson I (p. 9)
Address
British Association for the Advancement of Science, Bristol, England
Advanced Thought Publishing Company. Chicago, Illinois, USA. 1908

Curie, Marie Sklodowska   1867–1934
Polish-born French physical chemist

All my life through, the new sights of Nature made me rejoice like a child.
Pierre Curie
Autobiographical Notes
Chapter I (p. 162)
The Macmillan Company. New York, New York, USA. 1926

da Vinci, Leonardo   1452–1519
Italian High Renaissance painter and inventor

Whoever flatters himself that he can retain in his memory all the effects of Nature is deceived, for our memory is not so capacious: therefore consult Nature for everything.
A Treatise on Painting
#365 (p. 156)
J.B. Nichols and Son. London, England. 1835

Necessity is the theme and the inventress of nature, the curb and law and theme.
In Jean Paul Richter
The Literary Works of Leonardo da Vinci (Volume 2)
Philosophical Maxims, 1135 (p. 237)
University of California Press. Berkeley, California, USA. 1977

Nature is constrained by the order of her own law which lives and works within her.

…

Nature never breaks her own law.
Leonardo da Vinci's Note Books
Book I, Life (p. 55)
Duckworth & Company. London, England. 1906

In nature there is no effect without cause; once the cause is understood there is no need to test it by experience.
In Jean Paul Richter
The Literary Works of Leonardo da Vinci (Volume 2)
Philosophical Maxims, 1148B (p. 239)
University of California Press. Berkeley, California, USA. 1977

Darwin, Charles Robert   1809–82
English naturalist

Truly the schemes and wonders of Nature are illimitable.
In Francis Darwin (ed.)
The Life and Letters of Charles Darwin (Volume 1)
Letter to C. Lyell, September 14, 1849 (p. 345)
D. Appleton & Company. New York, New York, USA. 1896

What a book a devil's chaplain might write on the clumsy, wasteful, blundering, low, and horribly cruel works of nature.
More Letters of Charles Darwin
Letter To J.D. Hooker, 13 July 1856 (p. 94)
D. Appleton and Company. New York, New York, USA. 1903

Nature will tell you a direct lie if she can.
In W.I.B. Beveridge
The Art of Scientific Investigation
Chapter Two (p. 25)
W.W. Norton & Company, Inc. New York, New York, USA. 1957

Nature…cares nothing for appearances, except insofar as they are useful to any being. She can act on every internal organ, on every shade of constitutional difference, on the whole machinery of life.
In *Great Books of the Western World* (Volume 49)
The Origin of Species by Means of Natural Selection
Chapter IV (p. 41)
Encyclopædia Britannica, Inc. Chicago, Illinois, USA. 1952

Darwin, Erasmus   1731–1802
English physician and poet

Nature may seem to have been niggardly to mankind in bestowing upon them so few senses; since a sense to have perceived electricity, and another to have perceived magnetism might have been of great service to them, many ages before these fluids were discovered by accidental experiment, but it is possible an increased number of senses might have incommoded us by adding to the size of our bodies.
The Botanic Garden
Part I, Canto I (p. 19, fn l. 365)
Jones & Company. London, England. 1825

In earth, sea, air, around, below, above,
Life's subtle woof in Nature's loom is wove,
Points glued to points in living line extends,
Touch'd by some goad approach the bending ends.
The Botanic Garden
Production of Life, Canto I, IV, l. 251–4
Jones & Company. London, England. 1825

Davy, Sir Humphry   1778–1829
English chemist

Oh, most magnificent and noble Nature!
Have I not worshipped thee with such a love
As never a mortal man before displayed?
Adored thee in thy majesty of visible creation,
And searched into thy hidden and mysterious ways
As Poet, as Philosopher, as Sage?
In J. Davy
Fragmentary Remains
Chapter I (p. 14)
John Churchill. London, England. 1858

Dawson, Sir John William   1820–99
Canadian geologist and educator

Few words are used among us more loosely than "nature."
Sometimes it stands for the material universe as a whole.
Sometimes it is personified as a sort of goddess, working her own sweet will with material things. Sometimes
it expresses the forces which act on matter, and again it
stands for material things themselves. It is spoken of as
subject to law, but just as often natural law is referred to
in terms which imply that nature itself is the lawgiver.
Some Salient Points in the Science of the Earth
Chapter XVIII (p. 481)
Hodder & Stroughton. London, England. 1893

de Fontenelle, Bernard le Bovier   1657–1757
French writer

There is no need of fance…do but trust your eyes, and
you will easily perceive how nature diversifies her works
in these several worlds.
Conversations on the Plurality of Worlds
The Third Evening (p. 95)
Printed for Peter Wilson. Dublin, Ireland. 1761

de Montaigne, Michel Eyquem   1533–92
French Renaissance writer

Let us a little permit Nature to take her own way; she better understands her own affairs than we.
Translated by Charles Cotton
In *Great Books of the Western World* (Volume 25)
The Essays of Michel Eyquem de Montaigne
Essays III, Chapter 13 (p. 528)
Encyclopædia Britannica, Inc. Chicago, Illinois, USA. 1952

Delacroix, Eugene   1798–1863
French romantic painter

The true wisdom of the philosopher ought to consist in
enjoying everything. Yet we apply ourselves to dissecting
and destroying everything that is good in itself, that has
virtue, albeit the virtue there is in mere illusions. Nature
gives us this life like a toy to a weak child. We want to
see how it all works; we break everything. There remains
in our hands and before our eyes, stupid and opened too
late, the sterile wreckage, fragments that will not again
make a whole. The good is so simple.
Translated by Walter Pach

The Journal of Eugene Delacroix
Tuesday, June 1, 1824 (p. 92)
Covici. New York, New York, USA. 1937

Desaguliers, J. T.   1683–1744
French-born English natural philosopher

Nature compell'd, his piercing Mind obeys,
And gladly shows him all her secret Ways;
'Gainst Mathematicks she has no Defence,
And yields t' experimental Consequence.
In H.N. Fairchild
Religious Trends in English Poetry (Volume 1)
The Newtonian System of the World (p. 357)
Columbia University Press. New York, New York, USA. 1939

Dickens, Charles   1812–70
English novelist

It is not easy to walk alone in the country without musing
upon something.
Little Dorrit
Book the First, Chapter XVI (p. 178)
Bradbury & Evans. London, England. 1857

…nature gives to every time and season some beauties
of its own, and from morning to night, as from the cradle
to the grave, is but a succession of changes so gentle and
easy, that we can scarcely mark their progress.
Nicholas Nickleby
Chapter XXII (p. 234)
Dodd, Mead & Company. New York, New York, USA. 1944

…we are all children of one great mother, Nature.
Bleak House (Part II)
Chapter XLIII (p. 605)
P.F. Collier & Son. New York, New York, USA. 1911

Dickinson, Emily   1830–86
American lyric poet

Nature — the Gentlest Mother is,
Impatient of no Child —
The feeblest — or the waywardest —
Her Admonition mild —
The Complete Poems of Emily Dickinson
No. 790 (p. 385)
Little, Brown & Company. Boston, Massachusetts, USA. 1960

How Strange that Nature does not knock, and yet does
not intrude!
Letters of Emily Dickinson
Letter to Mrs. J.S. Cooper (p. 395)
Robert Brothers. Boston, Massachusetts, USA. 1894

Dickinson, G. Lowes   1862–1932
English historian and political activist

I'm not much impressed by the argument you attribute to
Nature, that if we don't agree with her we shall be knocked
on the head. I, for instance, happen to object strongly to
her whole procedure: I don't much believe in the harmony
of the final consummation…and I am sensibly aware of

the horrible discomfort of the intermediate stages, the pushing, kicking, trampling of the host, and the wounded and dead left behind on the march. Of all this I venture to disapprove; then comes Nature and says, "but you ought to approve!" I ask why, and she says, "Because the procedure is mine." I still demur, and she comes down on me with a threat — "Very good, approve or no, as you like; but if you don't approve you will be eliminated!" "By all means," I say, and cling to my old opinion with the more affection that I feel myself invested with something of the glory of a martyr.... In my humble opinion it's nature, not I, that cuts a poor figure!
The Meaning of Good
Good as the End of Nature (p. 46)
Brimley Johnson & Ince. London, England. 1906

Diderot, Denis   1713–84
French encyclopedist and materialist philosopher

Man is merely a common product, the monster an uncommon product; both equally natural, equally necessary, equally part of this universal and general order of things.... And what is astonishing about this?.... All creatures intermingle with each other, consequently all species...everything is in perpetual flux.... Every animal is more or less man; every mineral is more or less plant; every plant more or less animal. There is nothing precise in nature...
Translated by Jean Stewart and Jonathan Kemp
Diderot: Interpreter of Nature
D'Alembert's Dream (pp. 78–79)
International Publishers. New York, New York, USA. 1938

Dillard, Annie   1945–
American poet, essayist, novelist, and writing teacher

Unfortunately, nature is very much a now-you-see-it, now-you-don't affair. A fish flashes, then dissolves in the water before my eyes like so much salt. Deer apparently ascend bodily into heaven; the brightest oriole fades into leaves.
Pilgrim at Tinker Creek
Chapter 2 (p. 16)
Harper's Magazine Press. New York, New York, USA. 1974

The general rule in nature is that live things are soft within and rigid without. We vertebrates are living dangerously...like so many peeled trees.
Pilgrim at Tinker Creek
Chapter 6, II (p. 91)
Harper's Magazine Press. New York, New York, USA. 1974

The creator is no puritan.... There is something that profoundly fails to be exuberant about these crawling, translucent lice and white, fat-bodied grubs, but there is an almost manic exuberance about a creator who turns them out, creature after creature after creature, and sets them buzzing and lurking and flying and flying and swimming about.
Pilgrim at Tinker Creek

Chapter 13, II (p. 233)
Harper's Magazine Press. New York, New York, USA. 1974

Nature will try anything once. This is what the sign of the insects says. If you're dealing with organic compounds, then let them combine. If it works, if it quickens, set it clacking in the grass; there's always room for one more...
Pilgrim at Tinker Creek
Chapter 4, II (p. 65)
Harper's Magazine Press. New York, New York, USA. 1974

Nature is, above all, profligate. Don't believe them when they tell you how economical and thrifty nature is, whose leaves return to the soil. Wouldn't it be cheaper to leave them on the tree in the first place?
Pilgrim at Tinker Creek
Chapter 4, II (p. 65)
Harper's Magazine Press. New York, New York, USA. 1974

In nature, improbabilities are the one stock in trade.
Pilgrim at Tinker Creek
Chapter 8, II (p. 144)
Harper's Magazine Press. New York, New York, USA. 1974

Dirac, Paul Adrian Maurice   1902–84
English theoretical physicist

It has become increasingly evident in recent times, however, that nature works on a different plan. Her fundamental laws do not govern the world as it appears in our mental picture in any very direct way, but instead they control a substraturn of which we cannot form a mental picture without introducing irrelevancies.
The Principles of Quantum Mechanics
Preface to the First Edition (p. vi)
At the Clarendon Press. Oxford, England. 1935

Dobzhansky, Theodosius   1900–75
Russian-American scientist

One may detest nature and despise science, but it becomes more and more difficult to ignore them. Science in the modern world is not an entertainment for some devotees. It is on its way to becoming everybody's business.
The Biology of Ultimate Concern
Chapter 1 (p. 9)
The New American Library, Inc. New York, New York, USA. 1967

Douglas, Andrew Ellicott   1867–1962
American astronomer

Nature is a book of many pages and each page tells a fascinating story to him who learns her language. Our fertile valleys and craggy mountains recite an epic poem of geologic conflicts. The starry sky reveals gigantic suns and space and time without end.
Annual Report of the Board of Regents of the Smithsonian Institution, 1922
Some Aspects of the Use of the Annual Rings of Trees in Climatic Study (p. 223)
Government Printing Office. Washington, D.C. 1924

Doyle, Sir Arthur Conan   1859–1930
Scottish writer

How sweet the morning air is! See how that one little cloud floats like a pink feather from some giant flamingo. Now the red rim of the sun pushes itself over the London cloud-bank. It shines on a good many folk, but on none, I dare bet, who are on a stranger errand than you or I. How small we feel with our petty ambitions and strivings in the presence of the great elemental forces of Nature.
In William S. Baring-Gould (ed.)
The Annotated Sherlock Holmes (Volume 1)
The Sign of the Four, Chapter 7 (p. 648)
Wings Books. New York, New York, USA. 1967

Draper, John William   1811–82
American scientist, philosopher, and historian

As a cataract shows from year to year an invariable shape, though the water composing it is perpetually changing, so the aspect of Nature is nothing more than a flow of matter presenting an impermanent form. The universe considered as a whole is unchangeable. Nothing is eternal but space, atoms, force. The forms of Nature that we see are essentially transitory, they must all pass away.
History of the Conflict between Religion and Science
Chapter I (p. 24)
D. Appleton & Company. New York, New York, USA. 1898

Dryden, John   1631–1700
English poet, dramatist, and literary critic

For Art may err, but Nature cannot miss.
The Poetical Works of Dryden
The Cock and the Fox, l. 452
The Riverside Press. Cambridge, Massachusetts, USA. 1949

By viewing nature, nature's handmaid, art,
Makes mighty things from small beginnings grow;
That fishes first to shipping did impart,
Their tail the rudder, and their head the prow.
The Poetical Works of Dryden
Annus Mirabilis
Stanza 155
The Riverside Press. Cambridge, Massachusetts, USA. 1949

du Bartas, Guillaume de Salluste   1544–90
French poet

Out of the book of Nature's learned breast.
Divine Weeks and Works
Second week, fourth day, Book III. 566
Humfrey Lownes. London, England. 1611

de Spinoza, Baruch   1632–77
Dutch philosopher

…Nature has set no end before herself, and…all final causes are nothing but human fictions.
Translated by William Hale White
Ethic: Demonstrated in Geometrical Order and Divided into Five Parts
Section 8 (p. 41)
Trubner & Company. London, England. 1883

Dulbecco, Renato   1914–
Italian-born American virologist

Nature does not abide by hard and fast rules — it follows opportunity.
The Design of Life
Chapter 4 (p. 88)
Yale University Press. New Haven, Connecticut, USA. 1987

In biology, once a door is opened, the space behind it is quickly filled.
The Design of Life
Chapter 9 (p. 190)
Yale University Press. New Haven, Connecticut, USA. 1987

Dumas, Jean Baptiste-Andre   1800–84
French biochemist

In the study of Nature conjecture must be entirely put aside, and vague hypothesis carefully guarded against. The study of Nature begins with facts, ascends to laws, and raises itself, as far as the limits of man's intellect will permit, to the knowledge of causes, by the threefold means of observation, experiment and logical deduction.
In Faraday Lectures
Lectures Delivered Before the Chemical Society
The First Faraday Lecture (p. 2)
The Chemical Society. London, England. 1928

Dunbar, Paul Laurence   1872–06
African-American poet

There is no rebel like Nature. She is an iconoclast.
The Uncalled
VI (p. 57)
International Association of Newspapers and Authors. New York, New York, USA, 1901

Durell, Clement V.   1882–1968
English mathematician

The scientists, in playing their game with Nature, are meeting an opponent on her own ground, who has not only made the rules of the game to suit herself, but may have even queered the pitch or cast a spell over the visiting team.
Readable Relativity
Chapter II (p. 11)
Harper & Brothers. New York, New York, USA. 1960

Nature is a conjurer for supermen. Generations of scientists have attempted to penetrate her secrets. Bit by bit the disguise is being torn away, but each new discovery seems only to open out fresh avenues demanding further exploration. Nature is a true woman, who will have the last word.
Readable Relativity
Chapter I (p. 9)
Harper & Brothers. New York, New York, USA. 1960

Eckert, Allan W.   1931–
American historian, naturalist and author

…in nature's book, everything has its place and its time; there exists a persistent interdependency of its creatures one upon another.
And there is never waste.
Wild Season
Epilogue (p. 244)
Little, Brown & Company. Boston, Massachusetts, USA. 1967

Eddington, Sir Arthur Stanley   1882–1944
English astronomer, physicist, and mathematician

The future is not predetermined, and Nature has no need to protect herself from giving away plans which she has not yet made.
New Pathways in Science
Chapter V, Section II (p. 102)
At The University Press. Cambridge, England. 1947

Philosophically the notion of a beginning of Nature is repugnant to me.
The End of the World: From the Standpoint of Mathematical Physics
Nature, Supplement, Volume 127, Number 3203, March 21, 1931 (p. 447)

So far as broader characteristics are concerned we see in Nature what we look for or are equipped to look for.
The Nature of the Physical World
Chapter XV (p. 330)
The Macmillan Company. New York, New York, USA. 1930

Egerton, E. N.
No biographical data available

The balance of nature has been a background assumption in natural history since antiquity.
Changing Concepts of the Balance of Nature
Quarterly Review of Biology, No. 48, Number 2, 1973 (p. 322)

Einstein, Albert   1879–1955
German-born physicist

What I see in Nature is a magnificent structure that we can comprehend only imperfectly, and that must fill a thinking person with a feeling of "humility." This is a genuinely religious feeling that has nothing to do with mysticism.
In Helen Dukas and Banesh Hoffman
Albert Einstein: The Human Side: New Glimpses from His Archives (p. 39)
Princeton University Press. Princeton, New Jersey, USA. 1979

Nature is not an engineer or contractor…
In Helen Dukas and Banesh Hoffman
Albert Einstein: The Human Side: New Glimpses from His Archives
Letter dated 12 November, 1930 (p. 92)
Princeton University Press. Princeton, New Jersey, USA. 1979

Eliot, George (Mary Ann Evans Cross)   1819–80
English novelist

Nature has her language, and she is not unveracious; but we don't know all the intricacies of her syntax just yet, and in a hasty reading we may happen to extract the very opposite of her real meaning.
Adam Bede
Chapter XV (p. 142)
Dodd, Mead & Company. New York, New York, USA. 1947

Emerson, Ralph Waldo   1803–82
American lecturer, poet, and essayist

To the intelligent, nature converts itself into a vast promise, and will not be rashly explained. Her secret is untold. Many and many an Oedipus arrives: he has the whole mystery teeming in his brain.
Ralph Waldo Emerson: Essays and Lectures
Essays: Second Series
Nature (p. 554)
The Library of America. New York, New York, USA. 1983

The first steps in Agriculture, Astronomy, Zoology (those first steps which the farmer, the hunter, and the sailor take,) teach that nature's dice are always loaded; that in her heaps and rubbish are concealed sure and useful results.
Ralph Waldo Emerson: Essays and Lectures
Nature: Addresses, and Lectures
Discipline (p. 27)
The Library of America. New York, New York, USA. 1983

The great mother Nature will not quite tell her secret to the coach or the steamboat, but says, One to one, my dear, is my rule also, and I keep my enchantments and oracles for the religious soul coming alone, or as good as alone, in true-love.
In James Elliot Cabot
A Memoir of Ralph Waldo Emerson (Volume 2)
Letter to Mrs. Emerson, 20 May 1871 (p. 650)
Houghton Mifflin Company. Boston, Massachusetts, USA. 1888

Nature, like a cautious testator, fires up her estate so as not to bestow it all on one generation, but has a forelooking tenderness and equal regard to the next and the next, and the fourth and the fortieth age.
The Complete Works of Ralph Waldo Emerson (Volume 10)
Society and Solitude
Farming (p. 143)
Houghton Mifflin Company. Boston, Massachusetts, USA. 1911

Nothing bizarre, nothing whimsical will endure. Nature is ever interfering with Art. You cannot build your house or pagoda as you will but as you must. Gravity, Wind, sun, rain, the size of men & animals, & such other aliens have more to say than the architect. Beneath the almighty necessity therefore I regard what is artificial in man's life & works as petty & insignificant by the side of what is natural. Every violation, every suicide, every miracle, every willfulness however large it may show near us, melts quickly into the All, & at a distance is not seen. The outline is as smooth as the curve of the moon.… A writer must have l'abandon, he must be content to stand aside & let truth & beauty speak for him, or he cannot expect to be heard far.
Journals of Ralph Waldo Emerson
May 28, 1836 (p. 56)
Houghton Mifflin Company. Boston, Massachusetts, USA. 1910

Nature, as we know her, is no saint…. She comes eating, drinking and sinning…
Ralph Waldo Emerson: Essays and Lectures
Essays: Second Series
Experience (p. 481)
The Library of America. New York, New York, USA. 1983

Nature is full of a sublime family likeness throughout her works and delights in startling us with resemblances in the most unexpected quarters.
Ralph Waldo Emerson: Essays and Lectures
Essays: First Series
History (p. 243)
The Library of America. New York, New York, USA. 1983

Nature never hurries: atom by atom, little by little, she achieves her work.
The Complete Works of Ralph Waldo Emerson (Volume 10)
Society and Solitude
Farming (p. 139)
Houghton Mifflin Company. Boston, Massachusetts, USA. 1903

Nature is a rag-merchant, who works up every shred and ort and end into new creations; like a good chemist, whom I found, the other day, in his laboratory, converting his old shirts into pure white sugar.
Ralph Waldo Emerson: Essays and Lectures
The Conduct of Life
Considerations by the Way (p. 1088)
The Library of America. New York, New York, USA. 1983

Nature is an endless combination and repetition of a very few laws. She hums the old well-known air through innumerable variations.
Ralph Waldo Emerson: Essays and Lectures
Essays: First Series
History (p. 243)
The Library of America. New York, New York, USA. 1983

Nature is no spendthrift, but takes the shortest way to her ends.
Ralph Waldo Emerson: Essays and Lectures
The Conduct of Life
Fate (p. 961)
The Library of America. New York, New York, USA. 1983

By fate, not option, frugal Nature gave
One scent to hyson and to wall-flower,
One sound to pine-groves and to waterfalls,
One aspect to the desert and the lake.
It was her stern necessity.
The Complete Works of Ralph Waldo Emerson (Volume 9)
Xenophanes (p. 137)
Houghton Mifflin Company. Boston, Massachusetts, USA. 1904

Evans, Howard Ensign   1919–2002
Entomologist

One's appreciation of nature is never more acute than when a bit of nature is injected into one's flesh.
The Pleasures of Entomology: Portraits of Insects and the People Who Study them
Chapter 18 (p. 221)
Smithsonian Institution Press. Washington, D.C. 1985

Fermi, Enrico   1901–54
Italian-born American physicist

Whatever nature has in store for mankind, unpleasant as it may be, men must accept, for ignorance is never better than knowledge.
In Laura Fermi
Atoms in the Family
Part II, Chapter 23 (p. 244)
The University of Chicago Press. Chicago, Illinois, USA. 1954

Feuerbach, Ludwig   1804–72
German philosopher

Nature returns no answer to the questions and lamentations of man; inexorably it refers him to himself.
In Ludwig Buchner
Force and Matter
Chapter VI (p. 35)
Trubner & Company. London, England. 1864

Fevre, R. W.
No biographical data available

The real truth is that, not only has man failed to overcome nature in any sphere whatsoever but that at best he has merely succeeded in getting hold of and lifting a tiny corner of the enormous veil which she has spread over her eternal mysteries and secret. He never creates anything. All he can do is discover something. He does not master nature but has only come to be the master of those living things who have not gained the knowledge he has arrived at by penetrating into some of nature's laws and mysteries. Apart from all this, an idea can never subject to its own sway those conditions which are necessary for the existence and development of mankind; for the idea itself has come only from man. Without man there would be no human idea in this world. The idea as such is therefore always dependent on the existence of man and consequently is dependent on those laws which furnish the conditions of his existence.
The Demoralization of Western Culture: Social Theory and the Dilemmas of Modern Living (p. 28)
Continuum. London, England. 2000

Feyerabend, Paul K.   1924–94
Austrian-born American philosopher of science

Everywhere science is enriched by unscientific methods and unscientific results…the separation of science and non-science is not only artificial but also detrimental to the advancement of knowledge. If we want to understand nature, if we want to master our physical surroundings, then we must use all ideas, all methods, and not just a small selection of them.
Against Method: Outline of an Anarchistic Theory of Knowledge
Chapter 18 (pp. 305, 306)
Verso. London, England. 1978

Trying to understand the way nature works involves a most terrible test of human reasoning ability. It involves

subtle trickery, beautiful tightropes of logic on which one has to walk in order not to make a mistake in predicting what will happen.
The Meaning of It All: Thoughts of a Citizen Scientist
Chapter I (p. 15)
Perseus Books. Reading, Massachusetts, USA. 1998

There was a moment when I knew how nature worked. It had elegance and beauty. The goddam thing was gleaming.
In Lee Edson
Two Men in Search of a Quark
New York Times Magazine, October 8, 1967

People may come along and argue philosophically that they like one better than another; but we have learned from much experience that all philosophical intuitions about what nature is going to do fail.
The Character of Physical Law
Chapter 2 (p. 53)
BBC. London, England. 1965

But see that the imagination of nature is far, far greater than the imagination of man. No one who did not have some inkling of this through observations could ever have imagined such a marvel as nature is.
The Meaning of It All: Thoughts of a Citizen Scientist
Chapter I (p. 10)
Perseus Books. Reading, Massachusetts, USA. 1998

Fielding, Henry   1707–54
English novelist, playwright, and barrister

Nature seems to wear one universal grin.
The Tragedy of Tragedies, or, The Life and Death of Tom Thumb the Great
Act I, Scene 1
Printed by S. Powell. Dublin, Ireland. 1730

Flammarion, Camille   1842–1925
French astronomer and writer

Nature, O immense, fascinating, infinite Nature! Who can divine, who can hear, the sounds of thy celestial harmony! What can we include in these childish formulae of our young science? We lisp an alphabet while the eternal Bible is still closed to us. But it is thus when all reading begins, and these first words are surer than all the antique affirmations of ignorance and human vanity.
Popular Astronomy: A General Description of the Heavens
Book II, Chapter III (p. 112)
Chatto & Windus. London, England. 1894

Nature is immense in the little as in the great, or, to speak more correctly, for here there is neither little nor great.
Popular Astronomy: A General Description of the Heavens
Book III, Chapter II (p. 239)
Chatto & Windus. London, England. 1894

…it has been said that nature has implanted in our bosoms a craving after the discovery of truth, and assuredly that glorious instinct is never more irresistibly awakened than

when our notice is directed to what is going on in the heavens.
Popular Astronomy: A General Description of the Heavens
Book III, Chapter VII (p. 328)
Chatto & Windus. London, England. 1894

Florio, John   1553?–1625
English teacher, writer, and translator

Nature is the right law.
Florio's Firste Fruites (p. 88)
Taihoku Imperial University. Formosa, Japan. 1936

Ford, Kenneth W.   1926–
American physicist

One of the elementary rules of nature is that, in the absence of a law prohibiting an event or phenomenon, it is bound to occur with some degree of probability. To put it simply and crudely: Anything that can happen does happen.
Magnetic Monopoles
Scientific American, Volume 209, Number 6, December 1963 (p. 122)

Foster, Sir Michael   1836–1907
English physiologist and educator

Nature is ever making signs to us, she is ever whispering to us the beginnings of her secrets; the scientific man must be ever on the watch, ready at once to lay hold of Nature's hint, however small, to listen to her whisper, however low.
In J.A. Thomson
Introduction to Science
Chapter I (p. 16)
Williams & Norgate Ltd. London, England. 1916

Fraenkel, Aviezri S.
Applied mathematician

Nature might be somehow more powerful than a digital computer.
New York Times, March 25, 1997 (p. C5, col. 6)

Fuller, R. Buckminster   1895–1983
American engineer and architect

Nature is trying very hard to make us succeed, but nature does not depend on us. We are not the only experiment.
Minneapolis Tribune, 30 April 1978

Garth, Sir Samuel   1661–1719
English physician and poet

As distant prospects please us, but when near
We find but desert rocks and fleeting air.
The Dispensary
Canto III, l. 27
Printed by J. Lister, at St. John's Gate. London, England. 1768

Gay, John   1685–1732
English poet and dramatist

But he who studies nature's laws

From certain truth his maxims draws.
The Poetical Works of John Gay (Volume 3)
Introduction to the Fables, l. 76–77
Lawrence & Bullen. London, England. 1893

Gillispie, Charles Coulston   1918–
French writer and editor of philosophy and history of science

[T]he renewals of the subjective approach to nature make a pathetic theme. Its ruins lie strewn like good intentions all along the ground traversed by science, until it survives only in strange corners like Lysenkoism [doctrine centered on belief in acquired characteristics] and anthroposophy, where nature is socialized or moralized.
The Edge of Objectivity: An Essay in the History of Scientific Ideas
Chapter V (pp. 199–200)
Princeton University Press. Princeton, New Jersey, USA. 1960

Gould, Stephen Jay   1941–2002
American paleontologist and evolutionary biologist

The need to distinguish sturdy facts (pervasive pattern) from shaky factual claim (single cases with dubious documentation) has never been more evident to me than in the current debate between evolutionists and so-called "scientific creationists." The fact of evolution is as sturdy as any claim in science. Its sturdiness resides in a pervasive pattern detected by several disciplines — for examples, the age of the earth and life as affirmed by astronomy and geology, and the pattern of imperfections in organisms that record a history of physical descent.
Hen's Teeth and Horses Toes
Quaggas, Coiled Oysters, and Flimsy Facts (p. 384)
W.W. Norton & Company, Inc. New York, New York, USA. 1983

I do not believe that nature frustrates us by design, but I rejoice in her intransigence nonetheless.
Hen's Teeth and Horses Toes
What, if Anything, Is a Zebra? (p. 365)
W.W. Norton & Company, Inc. New York, New York, USA. 1983

Gray, George W.
Freelance science writer

Our knowledge of nature is limited by our ability to apprehend the materials and the forces which meet us — both those of the Earth, which we encounter in their hurrying to and fro, and those of the Universe outside, which beat upon us from the stars and the darkness beyond the stars.
The Advancing Front of Science
Chapter I, Section 5 (pp. 20–21)
Whittlesey House. New York, New York, USA. 1937

Gray, Thomas   1716–71
English poet

E'en from the tomb the voice of nature cries,
E'en in our ashes live their wonted fires.
The Complete Poetical Works of Gray, Beattie, Blair, Collins, Thomson, and Kirke White
Elegy in a Country Churchyard

Stanza 23
J. Blackwood. London, England. 1800

Greene, Brian   1963–
American physicist

You must allow Nature to dictate what is, and what is not, sensible.
The Elegant Universe: Superstrings, Hidden Dimensions, and the Quest for the Ultimate Theory
Part II, Chapter 5 (p. 111)
W.W. Norton & Company, Inc. New York, New York, USA. 2003

Gregory, Dick   1932
American comedian and social activist

Nature is not affected by finance. If someone offered you ten thousand dollars to let them touch you on your eyeball without your blinking, you would never collect the money. At the very last moment, Nature would force you to blink your eye. Nature will protect her own.
The Shadow that Scares Me
Chapter VIII (p. 175)
Doubleday & Company, Inc. Garden City, New York, USA. 1968

Gregory, Sir Richard Arman   1864–1952
British science writer and journalist

The study of Nature is elevating, and its material value is of the highest, yet it is deplorably neglected with the result that only very rarely is the simplest scientific subject referred to accurately in the works of literary men.
Discovery; or, The Spirit and Service of Science
Chapter I (p. 13)
Macmillan & Company Ltd. London, England. 1918

Nature, like the rich man of the parable, requires importunate pleading before she will bestow any of her riches upon a suppliant at her temple.
Discovery; or, The Spirit and Service of Science
Chapter I (p. 12)
Macmillan & Company Ltd. London, England. 1918

Nature must be loved for herself and not for her dowry.
Discovery; or, The Spirit and Service of Science
Chapter III (p. 54)
Macmillan & Company Ltd. London, England. 1918

Haeckel, Ernst   1834–1919
German biologist and philosopher

The anthropomorphic notion of a deliberate architect and ruler of the world has gone forever from this field; the "eternal iron laws of nature" have taken its place.
The Riddle of the Universe
Chapter XIV (p. 267)
Watts & Company. London, England. 1900

Hales, Stephen   1677–1761
English physiologist and clergyman

...our reasonings about the wonderful and intricate operations of Nature are so full of uncertainty, that, as the

wise-man truly observes, hardly do we guess aright at the things that are upon earth, and with labour do we find the things that are before us.
Vegetable Staticks
Chapter VII (p. 181)
The Scientific Book Guild. London, England. 1961

Harkness, William   1837–1903
Scottish-American astronomer and surgeon

All nature is one, but for convenience of classification we have divided our knowledge into a number of sciences which we usually regard as quite distinct from each other.
Annual Report of the Board of Regents of the Smithsonian Institution, 1896
On the Magnitude of the Solar System (p. 93)
Government Printing Office. Washington, D.C. 1896

Hartwell, Leland H.   1939–
American genome scientist

Sometimes nature rewards foolish optimism.
Lex Prix Nobel. The Nobel Prizes in 2001
Nobel banquet speech for award received in 2001
Nobel Foundation. Stockholm, Sweden. 2002

Harvey, William   1578–1657
English physician

Nature…is the best and most faithful interpreter of her own secrets; and what she presents either more briefly or obscurely in one department, that she explains more fully and clearly in another.
In *Great Books of the Western World* (Volume 28)
Anatomical Exercises on the Generation of Animals
Dedication (p. 329)
Encyclopædia Britannica, Inc. Chicago, Illinois, USA. 1952

Hazlitt, William Carew   1834–1913
English bibliographer

Nature is stronger than reason: for nature is, after all, the text, reason but the comment.
In W. Carew Hazlitt (ed.)
The Round Table; Northcotes Conversations; Characteristics
Characteristics, CXXXV (p. 476)
George Bell & Sons. London, England. 1884

Heine, Heinrich   1797–1856
German poet

Nature, like a true poet, abhors abrupt transitions.
The German Classics of the Nineteenth and Twentieth Centuries
(Volume 6)
Translated by Charles Godfrey Leland
The Journey to the Harz (p. 73)
The German Publication Society. New York, New York, USA.
1913–1914

Nature knows how to produce the greatest effects with the most limited means.
The German Classics of the Nineteenth and Twentieth Centuries
(Volume 6)
Translated by Charles Godfrey Leland

The Journey to the Harz (p. 73)
The German Publication Society. New York, New York, USA.
1913–1914

Heinlein, Robert A.   1907–88
American science fiction writer

He shut up, realizing that grim old Mother Nature, red of tooth and claw, invariably punished damn fools who tried to ignore Her or repeal Her ordinances.
Time Enough for Love
Chapter VI (p. 205)
G.P. Putnam's Sons. New York, New York, USA. 1973

Heisenberg, Werner Karl   1901–76
German physicist and philosopher

The laws of nature [that] we formulate mathematically in quantum theory deal no longer with the particles themselves but with our knowledge of the elementary particles.
Daedalus
Volume 87, 1958 (p. 99)

Natural science does not simply describe and explain nature; it is part of the interplay between nature and ourselves; it describes nature as exposed to our method of questioning.
Physics and Philosophy: The Revolution in Modern Science
Chapter V (p. 81)
Harper & Row, Publishers. New York, New York, USA. 1958

Henley, William Ernest   1849–1903
English poet

What Nature has writ with her lusty wit
Is worded so wisely and kindly
That whoever has dipped in her manuscript
Must up and follow her blindly.
Echoes of Life and Death
Number XXXIII
T.B. Mosher. Portland, Maine, USA. 1908

Heraclitus   540 BCE–480 BCE
Greek philosopher

The real constitution of things is accustomed to hide itself.
In G.S. Kirk and J.E. Raven
The Pre-Socratic Philosophers: A Critical History with a Selection of Texts
Fragment 211 (p. 193)
At The University Press. Cambridge, England. 1963

Nature loves to hide.
Fragments
Fragment x (p. 4)
Publisher undetermined

Herschel, Sir John Frederick William   1792–1871
English astronomer and chemist

From the least of nature's work he may learn the greatest lesson.

A Preliminary Discourse on the Study of Natural Philosophy
Part I, Chapter I, Section 10 (p. 14)
Printed for Longman, Rees, Orme, Brown & Green. London, England. 1831

…Nature builds up by her refined and invisible architecture, with a delicacy eluding our conception, yet with a symmetry and beauty which we are never weary of admiring.
The Cabinet of Natural Philosophy
Part III, Chapter II, Section 292 (p. 263)
Longman, Rees, Orme, Brown & Green. London, England. 1831

Herschel, Friedrich Wilhelm   1738–1822
English astronomer

The phenomena of nature, especially those that fall under the inspection of the astronomer, are to be viewed, not only with the usual attention to facts as they occur, but with the eye of reason and experience.
An Account of Three Volcanoes in the Moon
Philosophical Transactions of the Royal Society of London, Volume 67, 1787 (p. 229)

Holton, Gerald   1922–
Research professor of physics and science history

The study of nature is a study of the artifacts that appear during an engagement between the scientist and the world in which he finds himself.
The Roots of the Complementarity
Daedalus, Number 4, Fall 1970 (p. 1019)

Hooke, Robert   1635–1703
English physicist

…the footsteps of Nature are to be trac'd, not in her ordinary course, but when she seems to be put to her shifts, to make many doublings and turnings, and to use some kind of art in endeavoring to avoid our discovery.
Micrographia
Preface (third page)
Printed by Jo. Martyn and Ja. Allestry. London, England. 1665

Horace (Quintus Horatius Flaccus)   65 BCE–8 BCE
Roman philosopher and dramatic critic

You may turn nature out of doors with violence, but she will still return.
Satires, Epistles, and Ars Poetica
Epistles, Book I, Epistle 10, l. 24
W. Heinemann. London, England. 1929

Housman, Alfred Edward   1859–1936
English poet, scholar and satirist

For nature, heartless, witless nature,
Will neither care nor know
What stranger's feet may find the meadow
And trespass there and go.
Last Poems
Number XL (p. 76)
Henry Holt & Company. New York, New York, USA. 1922

Hudson, William Henry   1841–1922
Argentinian/English ornithologist, naturalist, and author

Here Nature is unapproachable with her green, airy canopy, a sun-impregnated cloud — cloud above cloud — and though the highest may be reached by the eye, the beams yet filter through, illuming the wide spaces beneath — chambers succeeded by chamber, each with its own special lights and shadows.
Green Mansions
Chapter 2 (p. 28)
Grosset & Dunlap. New York, New York, USA. 1931

Huggins, Sir William   1824–1910
English astronomer

Since the time of Newton our knowledge of the phenomena of nature has wonderfully increased, but man asks, perhaps more earnestly now than in his days, What is the ultimate reality behind the reality of the perceptions? Are they only the pebbles of the beach with which we have been playing? Does not the ocean of ultimate reality and truth lie beyond?
Annual Report of the Board of Regents of the Smithsonian Institution, 1891
Celestial Spectroscopy (p. 102)
Government Printing Office. Washington, D.C. 1893

Hugo, Victor   1802–85
French author, lyric poet, and dramatist

Nature eludes calculation. Number is a grim pullulation. Nature is the thing that cannot be numbered.
Translated by Isabel F. Hapgood
The Toilers of the Sea
Part II, Book Third, Chapter III (p. 416)
The Heritage Press. New York, New York, USA. 1961

Nature has no candor. She shows herself to man with her face turned away.
Translated by Isabel F. Hapgood
The Toilers of the Sea
Part II, Book Third, Chapter III (p. 415)
The Heritage Press. New York, New York, USA. 1961

Because of nature's unity it has been concluded that she is simple. An error.
Translated by Isabel F. Hapgood
The Toilers of the Sea
Part II, Book Third, Chapter III (p. 405)
The Heritage Press. New York, New York, USA. 1961

Hutton, W.
No biographical data available

Every page of the volume of Nature is fraught with instruction.
The Book of Nature Laid Open
Chapter I (p. 1)
Joseph Milligan. Georgetown, Virginia, USA. 1822

Huxley, Thomas Henry   1825–95
English biologist

There is not throughout Nature a law of wider application than this, that a body impelled by two forces takes the direction of their resultant.
Collected Essays (Volume 2)
Darwiniana
The Origin of Species (p. 32)
Macmillan & Company Ltd. London, England. 1904

To every one of us the world was once as fresh and new as to Adam. And then, long before we were susceptible of any other mode of instruction, Nature took us in hand, and every minute of waking life brought its educational influence, shaping our actions into rough accordance with Nature's laws, so that we might not be ended untimely by too gross disobedience. Nor should I speak of this process of education as past for any one, be he as old as he may. For every man the world is as fresh as it was at the first day, and as full of untold novelties for him who has the eyes to see them. And Nature is still continuing her patient education of us in that great university, the universe, of which we are all members — Nature having no Test-Acts.
Collected Essays (Volume 3)
Science and Education
A Liberal Education; and Where to Find It (p. 84)
Macmillan & Company Ltd. London, England. 1904

Those who take honours in Nature's university, who learn the laws which govern men and things and obey them, are the really great and successful men in this world. The great mass of mankind are the "Poll," who pick up just enough to get through without much discredit. Those who won't learn at all are plucked; and then you can't come up again. Nature's pluck means extermination.
Collected Essays (Volume 3)
Science and Education
A Liberal Education; and Where to Find It (p. 85)
Macmillan & Company Ltd. London, England. 1904

The investigation of Nature is an infinite pasture-ground, where all may graze, and where the more bite, the longer the grass grows, the sweeter is its flavor, and the more it nourishes.
Collected Essays (Volume 1)
Method and Result
Administrative Nihilism (p. 282)
Macmillan & Company Ltd. London, England. 1904

Education is the instruction of the intellect in the laws of Nature, under which name I include not merely things and their forces, but men and their ways; and the fashioning of the affections and of the will into an earnest and loving desire to move in harmony with those laws.
Collected Essays (Volume 3)
Science and Education
A Liberal Education; and Where to Find It (p. 83)
Macmillan & Company Ltd. London, England. 1904

The student of Nature wonders the more and is astonished the less, the more conversant he becomes with her operations; but of all the perennial miracles she offers to his inspection, perhaps the most worthy of admiration is the development of a plant or of an animal from its embryo.
Collected Essays (Volume 2)
Darwiniana
The Origin of Species (p. 29)
Macmillan & Company Ltd. London, England. 1904

Nature is never in a hurry, and seems to have had always before her eyes the adage, "keep a thing long enough and you will find a use for it.
Collected Essays (Volume 8)
Discourses, Biological and Geological
On the Formation of Coal (p. 159)
Macmillan & Company Ltd. London, England. 1904

Harmonious order governing eternally continuous progress — the web and woof of matter and force interweaving by slow degrees, without a broken thread, that veil which lies between us and the Infinite — that universe which alone we know or can know; such is the picture which science draws of the world, and in proportion as any part of that picture is in unison with the rest, so may we feel sure that it is rightly painted.
Collected Essays (Volume 2)
Darwiniana
The Origin of Species (p. 59)
Macmillan & Company Ltd. London, England. 1904

…the "Law of Nature" is not a command to do, or to refrain from doing, anything. It contains, in reality, nothing but a statement of that which a given being tends to do under the circumstances of its existence; and which, in the case of a living and sensitive being, it is necessitated to do, if it is to escape certain kinds of disability, pain, and ultimate dissolution.
Collected Essays (Volume 1)
Method and Result
Natural and Political Rights (p. 349)
Macmillan & Company Ltd. London, England. 1904

Huygens, Christiaan   1629–95
Dutch mathematician, astronomer, and physicist

Nature seems to court variety in her Works, and may have made them widely different from ours either in their matter or manner of Growth, in their outward Shape, or their inward Contexture; she may have made them such as neither our Understanding nor Imagination can conceive.
The Celestial Worlds Discover'd, or, Conjectures Concerning the Planetary Worlds, their Inhabitants and Productions
Book the First, Not to be imagin'd too unlike ours (p. 22)
Printed for T. Childe. London, England. 1698

…we may mount from this dull Earth, and viewing it from on high, consider whether Nature has laid out all her cost and finery upon this small speck of Dirt. So, like Travelers into other distant Countries, we shall be better able to judge of what's done at home, know how to make a true estimate of, and set its own value upon every thing.

The Celestial Worlds Discover'd, or, Conjectures Concerning the
Planetary Worlds, their Inhabitants and Productions
Book the First, These Studies useful to Religion (p. 10)
Printed for T. Childe. London, England. 1698

Irwin, Keith Gordon
No biographical data available

It is nature, of course, that is the great chemist. Every growing plant is a marvelous chemical factory, every living thing a brilliant shifter of atoms from one bewildering compound to another. And down in the depths of the earth enormous forces operate to create the minerals that someday may be close to the earth's surface.
The Romance of Chemistry
Forward (p. xi)
The Viking Press. New York, New York, USA. 1959

James, William   1842–1910
American philosopher and psychologist

Visible nature is all plasticity and indifference, — a moral multiverse…and not a moral universe. To such a harlot we owe no allegiance; with her as a whole we can establish no moral communion; and we are free in our dealing with her several parts to obey or to destroy, and to follow no law but that of prudence in coming to terms with such of her particular features as will help us to our private ends.
The Will to Believe and Other Essays in Popular Philosophy
Is Life Worth Living? (p. 43)
Dover Publications, Inc. New York, New York, USA. 1956

It seems a priori improbable that the truth should be so nicely adjusted to our needs and powers.… In the great boarding-house of nature, the cakes and the butter and the syrup seldom come out so even and leave the plates so clean.
The Will to Believe and Other Essays in Popular Philosophy
The Will to Believe
Section VIII (p. 27)
Dover Publications, Inc. New York, New York, USA. 1956

Juvenal (Decimus Junius Juvenal)
Roman poet

Nunquam aliud Natura aliud Sapientia dicit.
Nature never says one thing, Wisdom another.
Satires
Chapter XIV, 321
Indiana University Press. Bloomington, Indiana, USA. 1958

Kant, Immanuel   1724–1804
German philosopher

…in every study of nature there can be only so much genuine science as there is a priori knowledge, by the same token, natural philosophy will contain genuine science only to the extent in which mathematics can be applied to it.
Translated by James Ellington
Metaphysical Foundations of Natural Science
Preface (p. 7)
The Bobbs–Merrill Company, Inc. Indianapolis, Indiana, USA. 1970

Kepler, Johannes   1571–1630
German astronomer

…the closer I approach her [Nature], the more petulant her games become, and the more she again and again sneaks out of the seeker's grasp just when he is about to seize her through some circuitous route. Nevertheless, she never ceases to invite me to seize her, as though delighting in my mistakes.
Translated by William H. Donahue
New Astronomy
Part IV, 58 (p. 573)
At the University Press. Cambridge, England. 1992

Kingsley, Charles   1819–75
English clergyman and writer

Nature's deepest laws, her own true laws, are her invisible ones.
Alton Locke, Taylor and Poet
Chapter XXXVIII (p. 289)
Macmillan & Company Ltd. London, England. 1911

…there is no lie in Nature; no discords in the revelations of science, in the laws of the Universe.
Alton Locke, Taylor and Poet
Chapter XVIII (p. 141)
Macmillan & Company Ltd. London, England. 1911

Kolb, Edward W. (Rocky)   1951–
American cosmologist

Nature, as read by patient observation and experiment, is the ultimate philosopher.
Blind Watchers of the Sky
Chapter Three (p. 59)
Addison-Wesley Publishing Company. Reading, Massachusetts, USA. 1996

Krutch, Joseph Wood   1893–1970
American naturalist, conservationist, and writer

To those who study her, Nature reveals herself as extraordinarily fertile and ingenious in devising means, but she has no ends which the human mind has been able to discover or comprehend.
The Modern Temper
Chapter Two, Section iii (p. 27)
Harcourt, Brace & Company. New York, New York, USA. 1929

Lamarck, Jean-Baptiste Pierre Antoine   1744–1829
French biologist

Do we not therefore perceive that by the action of the laws of organization…nature has in favorable times, places, and climates multiplied her first germs of animality, given place to developments of their organizations… and increased and diversified their organs? Then…aided by much time and by a slow but constant diversity of circumstances, she has gradually brought about in this respect the state of things which we now observe. How grand is this consideration, and especially how remote is it from all that is generally thought on this subject!

In Alpheus Spring Packard
Lamark, the Founder of Evolution: His Life and Work
Chapter 16 (p. 259)
Longmans, Green & Company. London, England. 1901

Nature has produced all the species of animals in succession, beginning with the most imperfect or simplest, and ending her work with the most perfect, so as to create a gradually increasing complexity in their organisation; these animals have spread at large throughout all the habitable regions of the globe, and every species has derived from its environment the habits that we find in it and the structural modifications which observation shows us.

Translated by Hugh Elliot
Zoological Philosophy: An Exposition with Regard to the Natural History of Animals
Chapter VII (p. 126)
The University of Chicago Press. Chicago, Illinois, USA. 1984

Laplace, Pierre Simon   1749–1827
French mathematician, astronomer, and physicist

Nature is so various in her productions and phenomena, that it is extremely difficult to ascertain their causes, hence it is requisite for a great number of men to unite their intellect and exertions in order to comprehend and develop her laws.

System of the World (Volume 2)
Book V, Chapter IV (p. 286)
Longman, Rees, Orme, Brown & Green. Dublin, Ireland. 1830

…[if] the result of a long series of precise observations approximates a simple relation so closely that the remaining difference is undetectable by observation and may be attributed to the errors to which they are liable, then this relation is probably that of nature.

Pierre Simon Laplace 1749–1827: A Life in Exact Science
Chapter 16 (p. 130)
Princeton University Press. Princeton, New Jersey, USA. 1997

Lawrence, Louise de Kiriline   1894–1992
Canadian naturalist, author, and nurse

Nature is a deep reality and whether we understand it or not it is true and elemental.

The Lovely and the Wild
Chapter Three (p. 33)
McGraw-Hill Book Company, Inc. New York, New York, USA. 1968

Leclerc, Georges-Louis, Comte de Buffon   1707–88
French naturalist

Nature turns upon two steady pivots, unlimited fecundity which she has given to all species; and those innumerable causes of destruction which reduce the product of this fecundity…

Natural History, General and Particular (Volume 5) (p. 88)
T. Caldwell and W. Davies. London, England. 1812

Nature is that system of laws established by the Creator for regulating the existence of bodies, and the succession of beings. Nature is not a body; for this body would comprehend every thing. Either is it a being; for this being would necessarily be God. But nature may be considered as an immense living power, which animates the universe, and which, in subordination to the first and supreme Being, began to act by his command, and its action is still continued by his concurrence or consent.

Natural History, General and Particular (Volume 6)
Of Nature, First View (p. 249)
T. Caldwell and W. Davies. London, England. 1812

Lederman, Leon   1922–
American high-energy physicist

The laws of nature must have existed before even time began in order for the beginning to happen. We say this, we believe it, but can we prove it?

The God Particle: If the Universe Is the Answer, What Is the Question?
Chapter 9 (p. 401)
Houghton Mifflin Company. Boston, Massachusetts, USA. 1993

Lewis, C. S. (Clive Staples)   1898–1963
British author, scholar, and popular theologian

If ants had a language they would, no doubt, call their anthill an artifact and describe the brick wall in its neighborhood an a natural object. Nature in fact would be for them all that was not "ant-made." Just so, for us, nature is all that is not man-made; the natural state of anything is its state when not modified by man.

Studies In Words
Nature (pp. 45–46)
The University Press. Cambridge, England. 1960

Linnaeus, Carl (von Linné)   1707–78
Swedish botanist and explorer

It is the exclusive property of man, to contemplate and to reason on the great book of nature. She gradually unfolds herself to him, who with patience and perseverance, will search into her mysteries; and when the memory of the present and of past generations shall be obliterated, he shall enjoy the high privilege of living in the minds of his successors, as he has been advanced in the dignity of his nature, by the labours of those who went before him.

In Thomas Steele Hall
A Source Book in Animal Biology (p. 32)
McGraw-Hill Book Company, Inc. New York, New York, USA. 1951

Longfellow, Henry Wadsworth   1807–82
American poet

Nature with folded hand seemed there,
Kneeling at her evening prayer!

The Poetical Works of Henry Wadsworth Longfellow
Voices of The Night
Prelude, Stanza 11
Houghton Mifflin Company. Boston, Massachusetts, USA. 1883

No tears
Dim the sweet look that Nature wears.
The Poetical Works of Henry Wadsworth Longfellow
Sunrise on the Hills, l. 35
Houghton Mifflin Company. Boston, Massachusetts, USA. 1883

So Nature deals with us, and takes away
Our playthings one by one, and by the hand
Leads us to rest so gently, that we go,
Scarce knowing if we wish to go or stay,
Being too full of sleep to understand
How far the unknown transcends the what we know.
The Poetical Works of Henry Wadsworth Longfellow
Nature, l. 9
Houghton Mifflin Company. Boston, Massachusetts, USA. 1883

And Nature, the old nurse, took
The child upon her knee,
Saying: "Here is a story-book
Thy Father has written for thee."
"Come, wander with me," she said,
"Into regions yet untrod;
And read what is still unread
In the manuscripts of God."
Fiftieth Birthday of Agassiz
Sackett & Wilhelms Lithographing Corp. New York, New York, USA.1935

Lorenz, Konrad   1903–89
Austrian zoologist

Much of the beauty and wonder of nature is based on the fact that organic life is directed towards goals — towards survival, reproduction, and the attainment of higher perfection.
In Niko Tinbergen
The Herring Gull's World: A Study of the Social Behavior of Birds
Foreword (p. vi)
Basic Books, Inc. Publishers. New York, New York, USA. 1960

Lubbock, Sir John   1834–1913
English banker, writer, and scientist

…we are not the only tenants of our farms — that the fields and hedges, woods and waters, all around us, teem with a complex, rich, and interesting life. …[N]ature will speak only to those who listen with love and sympathy…
In Henry C. McCook
Tenants of an Old Farm; Leaves from the Note-Book of a Naturalist
Introduction (p. vi)
George W. Jacobs & Company. Philadelphia, Pennsylvania, USA. 1895

Lucretius   ca. 99 BCE–55 BCE
Roman poet

This terror then and darkness of mind must be dispelled not by the rays of the sun and glittering shafts of day, but by the aspects and the law of nature; the warp of whose design we shall begin with this first principle, nothing is ever gotten out of nothing by divine power.
In *Great Books of the Western World* (Volume 12)
Lucretius: on the Nature of Things
Book One, l. 146 (pp. 2–3)
Encyclopædia Britannica, Inc. Chicago, Illinois, USA. 1952

Luther Standing Bear   1868–1939
Oglala Lakota chief, 1905–1939

Only to the white man was nature a "wilderness" and only to him was the land "infested" with "wild" animals and "savage" people. To us it was tame. Earth was bountiful and we were surrounded with the blessings of the Great Mystery. Not until the hairy man from the east came and with brutal frenzy heaped injustices upon us and the families that we loved was it "wild" for us. When the very animals of the forest began fleeing from his approach, then it was that for us the "Wild West" began.
Land of the Spotted Eagle
Boyhood (p. 38)
University of Nebraska Press. Lincoln, Nebraska, USA. 1978

Lyell, Sir Charles   1797–1875
English geologist

So in Geology, if we could assume that it is part of the plan of nature to preserve, in every region of the globe, an unbroken series of monuments to commemorate the vicissitudes of the organic creation, we might infer the sudden extirpation of species, and the simultaneous introduction of others, as often as two formations in contact include dissimilar organic fossils. But we must shut our eyes to the whole economy of the existing causes, aqueous, igneous, and organic, if we fail to perceive that such is not the plan of Nature.
Principles of Geology (Volume 3)
Chapter III (p. 34)
John Murray. London, England. 1830

Macfie, Ronald Campbell   1867–1931
Poet and physician

Nature never errs in the long-run. She made man from a shred of the Milky Way, and she may be trusted to look after the creature she has made.
Science, Matter and Immortality
Chapter XVIII (p. 224)
William & Norgate. London, England. 1909

Mach, Ernst   1838–1916
Austrian physicist and philosopher

In the infinite variety of nature many ordinary events occur; while others appear uncommon, perplexing, astonishing, or even contradictory to the ordinary run of things. As long as this is the case we do not possess a well-settled and unitary conception of nature.
The Science of Mechanics (5th edition)
Introduction (p. 6)
The Open Court Publishing Company. La Salle, Illinois, USA. 1942

Maclaurin, Colin   1698–1746
Scottish mathematician and natural philosopher

The processes of nature lie so deep, that, after all the pains we can take, much, perhaps, will remain undiscovered beyond the reach of human art or skill. But this is no reason why we should give ourselves up to the belief of fictions, be they ever so ingenious, instead of hearkening to the unerring voice of nature…
An Account of Sir Isaac Newton's Philosophical Discoveries, in Four Books
Book I, Chapter I (p. 12)
Printed for the Author's Children. London, England. 1748

A strong curiosity has prompted men in all times to study nature; every useful art has some connexion with the science; and the unexhausted beauty and variety of things makes it ever agreeable, new and surprising.
An Account of Sir Isaac Newton's Philosophical Discoveries, in Four Books
Book I, Chapter I (p. 3)
Printed for the Author's Children. London, England. 1748

Manning, Richard
No biographical data available

Now I take this thought as necessary to humble our science, to understand that everything we know was taught us by nature, but nature gave us brains evolved to their niche, so they are limited in their understanding. All that we know we have learned from nature, but we do not know all that nature knows.
Grassland
Chapter 12 (p. 263)
The Viking Press. New York, New York, USA. 1995

Marsh, George Perkins   1801–82
American scholar, writer, and statesman

Nature, left undisturbed, so fashions her territory as to give it almost unchanging permanence of form, outline, and proportion, except when shattered by geologic convulsions; and in these comparatively rare cases of derangement, she sets herself at once to repair the superficial damage, and to restore, as nearly as practicable, the former aspect of her dominion.
The Earth as Modified by Human Action: A New Edition of Man and Nature
Chapter I (p. 26)
Scribner, Armstrong & Company. New York, New York, USA. 1874

Mason, Frances
No biographical data available

Nature shows nothing finished and perfect in the beginning; she shows orderly divergence and an advance from lower to higher levels of creation.
Creation by Evolution
Editors Preface (p. vii)
The Macmillan Company. New York, New York, USA. 1928

McKibben, Bill   1960–
Freelance writer

The end of nature sours all my material pleasures. The prospect of living in a genetically engineered world sickens me. And yet it is toward such a world that our belief in endless material advancement hurries us. As long as that desire drives us, there is no way to set limits.
The End of Nature
A Path of More Resistance (p. 173)
Random House, Inc. New York, New York, USA. 1989

McLennan, Evan
No biographical data available

There is a charm for man in the study of Nature. It elevates his soul to real greatness. It frees his mind from stormy life, and thrills him with the purest joy.
Cosmical Evolution: A New Theory of the Mechanism of Nature
Introduction (p. 23)
Donohue, Henneberry & Company. Chicago, Illinois, USA. 1890

Meldola, R.
No biographical data available

It is only the active worker — the original investigator — who, by personal appeal to Nature through artificially imposed considerations, *i.e.*, experiment, or through observation, *i.e.*, ready-made phenomena, has come to understand fully what a fact really means in the scientific sense; to realise how laborious is the process of wooing truth and ambiguous are the answers often given by Nature to his cross-examinations.
In Sir Richard Arman Gregory
Discovery; or, The Spirit and Service of Science
Chapter III (p. 40)
Macmillan & Company Ltd. London, England. 1918

Melville, Herman   1819–91
American novelist

…nature is an immaculate virgin, forever standing unrobed before us.
Typee, Omoo, Mardi
Mardi
Chapter 137 (p. 1094)
The Library of America. New York, New York, USA. 1982

Mencken, H. L. (Henry Louis)   1880–1956
American journalist and literary critic

Nature abhors a moron.
A Mencken Chrestomathy
Chapter XXX (p. 616)
Alfred A. Knopf. New York, New York, USA. 1949

Mill, John Stuart   1806–73
English political philosopher and economist

Nature means the sum of all phenomena, together with the causes which produce them; including not only all that happens, but all that is capable of happening…

Three Essays on Religion
Nature (p. 5)
Longmans, Green, Reader & Dyer. London, England. 1875

Milton, John   1608–74
English poet

Wherefore did Nature power her bounties forth
With such a full and unwithdrawing hand,
Covering the earth with odours, fruits, flocks,
Thronging the seas with spawn innumerable,
But all to please and sate the curious taste?
In *Great Books of the Western World* (Volume 32)
Comus, l. 710
Encyclopædia Britannica, Inc. Chicago, Illinois, USA. 1952

Accuse not Nature, she hath done her part;
Do thou but thine…
In *Great Books of the Western World* (Volume 32)
Paradise Lost
Book VIII, l. 561–562
Encyclopædia Britannica, Inc. Chicago, Illinois, USA. 1952

Moleschott, Jacob   1822–93
Dutch scientist, physiologist, and philosopher

The law of nature is a stringent expression of necessity.
In Ludwig Buchner
Force and Matter
Chapter VI (p. 33)
Trubner & Company. London, England. 1864

Montgomery, Robert
No biographical data available

…not from Nature up to Nature's God,
But down from Nature's God look Nature through.
Luther: Or, the Spirit of the Reformation
A Landscape of Domestic Life
Francis Baisler. London, England. 1843

Morley, John 1st Viscount Morley of Blackburn   1838–1923
English statesman and writer

Nature, in her most dazzling aspects or stupendous parts,
is but the background and theatre of the tragedy of man.
Critical Miscellanies
Byron (p. 140)
Macmillan & Company Ltd. London, England. 1886

Morrison, A. Cressy   1884–1951
American scientist

Although nature, the great chemist, has provided man with
the prototypes and methods by which he has attempted,
with considerable success, to conquer his environment,
her motives and objectives have seldom been man's.
The beautiful silks with which man bedecks himself and
his womankind…were created for far different purposes
than those to which man has put them.
Man in a Chemical World
Chapter 2 (p. 13)
Charles Scribner's Sons. New York, New York, USA. 1937

Motherwell, William   1797–1835
Scottish poet

And we, with Nature's heart in tune,
Concerted harmonies.
In Frederick Saunders and Minnie K. Davis (eds.)
Gems of Genius in Poetry and Art
Jeannie Morrison
Thompson & Thompson. Chicago, Illinois, USA. 1899

Muir, John   1838–1914
American naturalist

Then to think of the infinite numbers of smaller fellow
mortals, invisibly small, compared with which the smallest
ants are as mastodons.
My First Summer in the Sierra
June 13 (p. 62)
Houghton Mifflin Company. Boston, Massachusetts, USA. 1911

Thus review the eventful past, we see Nature working
with enthusiasm like a man, blowing her volcanic forges
like a blacksmith blowing his smithy fires, shoving glaciers over the landscapes like a carpenter shoving his
planes, clearing, ploughing, harrowing, irrigating, planting, and sowing broadcast like a farmer and gardener,
doing rough work and fine work, planting sequoias and
pines, rosebushes and daisies; working in gems, filling every crack and hollow with them; distilling fine
essences; painting plants and shells, clouds, mountains,
all the earth and heavens, like an artist, ever working
toward beauty higher and higher.
Our National Parks
Chapter II (p. 73)
Houghton Mifflin Company. Boston, Massachusetts, USA. 1901

When we are with Nature we are awake, and we discover
many interesting things and reach many a mark we are
not aiming at.
In Linnie Marsh Wolfe (ed.)
John of the Mountains
Chapter VII, Section I, June, 1890 (p. 300)
Houghton Mifflin Company. Boston, Massachusetts, USA. 1938

None may wholly escape the Good of Nature, however
imperfectly exposed to her blessings.
Steep Trails
Chapter III (p. 48)
Norman S. Berg, Publisher. Dunwoody, Georgia, USA. 1970

None of Nature's landscapes are ugly so long as they are
wild.
Our National Parks
Chapter I (p. 4)
Houghton Mifflin Company. Boston, Massachusetts, USA. 1901

One is constantly reminded of the infinite lavishness and
fertility of Nature — inexhaustible abundance amid what
seems enormous waste. And yet when we look into any
of her operations that lie within reach of our minds, we
learn that no particle of her material is wasted or worn

out. It is eternally flowing from use to use, beauty to yet higher beauty.
Gentle Wilderness (p. 139)
Ballantine Books. New York, New York, USA. 1968

Nature…leading us with work…yet cheers us like a mother with tender prattle words of love…
In Linnie Marsh Wolfe (ed.)
John of the Mountains
Chapter II, Section 2, Undated (pp. 66–67)
Houghton Mifflin Company. Boston, Massachusetts, USA. 1938

Nature has always something rare to show us…and the danger to life and limb is hardly greater than one would experience crouching deprecatingly beneath a roof.
Mountains of California
Chapter X (p. 249)
The Century Company. New York, New York, USA. 1911

[I]n every walk with Nature one receives far more than he seeks.
Steep Trails
Chapter IX (p. 128)
Norman S. Berg, Publisher. Dunwoody, Georgia, USA. 1970

How fiercely, devoutly wild is Nature in the midst of her beauty loving tenderness.
My First Summer in the Sierra
July 29 (p. 177)
Houghton Mifflin Company. Boston, Massachusetts, USA. 1911

Musser, George
No biographical data available

The basic rules of nature are simple, but their consummation may never lose its ability to surprise.
From the Editors
Scientific American, Volume 280, Number 1, January 1999 (p. 6)

Newton, Sir Isaac   1642–1727
English physicist and mathematician

For Nature is very consonant and conformable to herself.
In Eugene Hecht
Optics
Book III, Part 1, Question 31 (p. 531)
Encyclopædia Britannica, Inc. Chicago, Illinois, USA. 1952

I wish we could derive the rest of the phenomena of Nature by the same kind of reasoning from mechanical principles, for I am induced by many reasons to suspect that they may all depend upon certain forces by which the particles of bodies, by some causes hitherto unknown, are either mutually impelled toward one another, and cohere in regular rigors, or are repelled and recede from one another.
In *Great Books of the Western World* (Volume 34)
Mathematical Principles
Preface to the First Edition (p. 2)
Encyclopædia Britannica, Inc. Chicago, Illinois, USA. 1952

Oliver, Mary   1935–
American poet

Nature, the total of all of us, is the wheel that drives our world; those who ride it willingly might yet catch a glimpse of a dazzling, even a spiritual restfulness, while those who are unwilling simply to hang on, who insist that the world must be piloted by man for his own benefit, will be dragged around and around all the same, gathering dust but no joy.
Blue Pastures
A Few Words (p. 92)
Harcourt Brace & Company. New York, New York, USA. 1995

Oppenheimer, J. Robert   1904–67
American theoretical physicist

Despite all the richness of what men have learned about the world of nature, of matter and of space, of change and of life, we carry with us today an image of the giant machine as a sign of what the objective world is really like.
Science and the Common Understanding
Chapter 1 (pp. 14–15)
Simon & Schuster. New York, New York, USA. 1954

O'Rourke, P. J.   1947–
Political satirist

Worship of nature may be ancient, but seeing nature as cuddlesome, hug-a-bear and too cute for words is strictly a modern fashion.
Parliament of Whores: A Lone Humorist Attempts to Explain the Entire U.S. Government
Dirt of the Earth (p. 196)
Vintage Books. New York, New York, USA. 1992

Orton, James   1830–77
Explorer

The Kingdom of Nature is a literal Kingdom. Order and beauty, law and dependence, are seen everywhere. Amidst the great diversity of the forms of life, there is unity; and this suggests that there is one general plan, but carried out in a variety of ways.
Comparative Zoology, Structural and Systematic
Chapter XXI (p. 222)
Harper & Brothers. New York, New York, USA. 1877

Pagels, Heinz R.   1939–88
American physicist and science writer

Nature avoids infinities.
Perfect Symmetry: The Search for the Beginning of Time
Part Four, Chapter 1 (p. 354)
Simon & Schuster. New York, New York, USA. 1985

Nature has been generous to astronomers, offering an abundance of different stars and galaxies at all stages in their lives to look at.
Perfect Symmetry: The Search for the Beginning of Time
Part One, Chapter 1 (p. 28)
Simon & Schuster. New York, New York, USA. 1985

Peattie, Donald Culrose   1896–1964
American botanist, naturalist and writer

Futile for science to try to discover what the forces of Nature are; it can only discover how they operate.
An Almanac for Moderns
March Twenty-Eighth (p. 10)
G.P. Putnam's Sons. New York, New York, USA. 1935

It is Nature herself, as we grow in comprehension of her, who weans us from our early faith.
An Almanac for Moderns
March Thirtieth (p. 12)
G.P. Putnam's Sons. New York, New York, USA. 1935

Peters, Ted
No biographical data available

Nature as we daily experience it is ambiguous, fraught with benefits and liabilities.
Playing God?: Genetic Determinism and Human Freedom
Playing God with DNA (p. 20)
Routledge. New York, New York, USA. 1997

Petrarch (Francesco Petrarca)   1304–74
Italian poet and humanist

There are fools who seek to understand the secrets of nature.
In Richard Olson
Science Deified and Science Defied: The Historical Significance of Science in Western Culture (Volume 1)
Chapter 7 (p. 210)
University of California Press. Berkeley, California, USA. 1982

Planck, Max   1858–1947
German physicist

In all cases, the quantum hypothesis has given rise to the idea, that in Nature, changes occur which are not continuous, but of an explosive nature.
Translated by R. Jones and D.H. Williams
A Survey of Physics: A Collection of Lectures and Essays
New Paths of Physical Knowledge (p. 51)
Methuen & Company Ltd. London, England. 1925

If one wishes to obtain a definite answer from Nature one must attack the question from a more general and less selfish point of view.
Translated by R. Jones and D.H. Williams
A Survey of Physics: A Collection of Lectures and Essays
The Unity of the Physical Universe (p. 15)
Methuen & Company Ltd. London, England. 1925

Pliny (C. Plinius Secundus)   23–79
Roman savant and writer

Hail, Nature, mother of all creation, and mindful that I alone of the men of Rome have praised thee in all thy manifestations, be gracious to me.
Natural History
Volume 10, Book XXXVII, sec 205
Harvard University Press. Cambridge, Massachusetts, USA. 1947

Poincaré, Henri   1854–1912
French mathematician and theoretical astronomer

The scientist does not study nature because it is useful; he studies it because he delights in it, and he delights in it because it is beautiful. If nature were not beautiful, it would not be worth knowing, and if nature were not worth knowing, life would not be worth living.
The Foundations of Science
Science and Method, Book I
Chapter I (p. 366)
The Science Press. New York, New York, USA. 1913

Pope, Alexander   1688–1744
English poet

See plastic Nature working to this end,
The single atoms each to other tend,
Attract, attracted to, the next in place
Form'd and impell'd its neighbor to embrace.
The Complete Poetical Works (Volume 2)
An Essay on Man
Epistle III, l. 9
Houghton Mifflin Company. New York, New York, USA. 1903

Eye nature's walks, shoot folly as it flies,
And catch the manners, living as they rise;
Laugh where we must, be candid where we can,
But vindicate the ways of God to man.
The Complete Poetical Works (Volume 2)
An Essay on Man
Epistle I, l. 13
Houghton Mifflin Company. New York, New York, USA. 1903

All are but parts of one stupendous whole,
Whose body Nature is, and God the soul;
That chang'd thro' all, and yet in all same,
Great in the earth as in th' ethereal frame;
Warms in the sun, refreshes in the breeze,
Glows in the stars, and blossoms in the trees;
Lives thro' all life, extends thro' all extent,
Spreads undivided, operates unspent;
Breathes in our soul, informs our mortal part,
As full, as perfect, in a hair as heart.
The Complete Poetical Works (Volume 2)
An Essay on Man
Epistle I, l. 267
Houghton Mifflin Company. New York, New York, USA. 1903

All nature is but art, unknown to thee.
The Complete Poetical Works (Volume 2)
An Essay on Man, Epistle I, l. 289
Houghton Mifflin Company. New York, New York, USA. 1903

Poynting, John Henry   1852–1914
English physicist

While the investigation of Nature is ever increasing our knowledge, and while each new discovery is a positive addition never again to be lost, the range of the investigation and the nature of the knowledge gained form the theme of endless discussion.
Collected Scientific Papers
Presidential Address
The Mathematical and Physical Section

The British Association (Dover) 1899 (p. 599)
At The University Press. Cambridge. 1920

Priestley, Joseph   1733–1804
English theologian and scientist

I view with rapture the glorious face of nature, and I admire its wonderful constitution, the laws of which are daily unfolding themselves to our view.
In F.W. Gobbs
Joseph Priestley: Adventure in Science and Champion of Truth
Chapter 10 (p. 168)
Thomas Nelson & Sons Ltd. London, England. 1965

Quammen, David   1948–
American science writer and naturalist

Nature grants no monopolies in resourcefulness. She does not even seem to hold much with the notion of portioning it out hierarchically. Gold, she decrees, is where you find it.
Natural Acts: A Sidelong View of Science and Nature
A Better Idea (p. 3)
Shocken Books. New York, New York, USA. 1985

Raman, Chandrasekhar Venkata   1888–1970
Indian physicist

The face of Nature as presented to us is infinitely varied, but to those who love her it is ever beautiful and interesting.
The New Physics: Talks on Aspects of Science
Chapter V (p. 29)
Philosophical Library, New York. 1951

Rey, Hans Augusto   1898–1977
Author and illustrator of children's books

No matter what part of nature one studies — microbes or Milky Ways — there is a point where one begins, but never an end.
The Stars: A New Way to See Them
Part 4 (p. 108)
Houghton Mifflin Company. Boston, Massachusetts, USA. 1967

Richet, Charles   1850–1935
French physiologist

Nature guards her secrets jealously: it is necessary to lay violent siege to her for a long time to discover a single one of them, however small it be.
The Natural History of a Savant
Chapter XIII (p. 149)
J.M. Dent & Sons Ltd. London, England. 1927

Rolleston, George   1916–2001
English physician and physiologist

Let us hope that in the interludes of rhetoric the logic of facts may find a moment to make itself heard. It will teach men…to hold of Nature that her ways are not as our ways, nor her thoughts as our thoughts.
Scientific Papers and Addresses (Volume 1)

Chapter IV (p. 61)
At The Clarendon Press. Oxford, England. 1884

Russell, Henry Norris   1877–1957
American astronomer

In the grandeur of its sweep in space and time, and the beauty and simplicity of the relations which it discloses between the greatest and the smallest things of which we know, it reveals as perhaps nothing else does, the majesty of the order about us which we call nature, and, as I believe, of that Power behind the order, of which it is but a passing shadow.
Annual Report of the Board of Regents of the Smithsonian Institution, 1923
Constitution of the Stars (p. 158)
Government Printing Office. Washington, D.C. 1925

Sagan, Carl   1934–96
American astronomer and author

Scientists do not seek to impose their needs and wants on Nature, but instead humbly interrogate Nature and take seriously what they find.
The Demon-Haunted World: Science as a Candle in the Dark
Chapter 2 (p. 32)
Random House, Inc. New York, New York, USA. 1995

Saunders, W. E.
Naturalist

Lovers of nature feel so confidently that their hobby is an enormous asset in life that there is no feeling of hesitancy in advocating that every person should become acquainted with new species of birds, trees, insects, etc., just as often as opportunity offers. And the time to do so is always NOW!
In R.J. Rutter (ed.)
W.E. Saunders, Naturalist: A Memorial Volume
Saunderisms (p. 50)
Federation of Ontario Naturalists. Toronto, Ontario, Canada. 1949

Sayers, Dorothy L.   1893–1957
English novelist and essayist
Eustace, R.
No biographical data available

Nature never worked by rule and compass.
The Documents in the Case
Letter 16, Agatha Milsom to Olive Farebrother (p. 56)
Victor Gollancz LTD, London, England; 1978

Schrieber, Hermann   1920–
Austrian historian

If nature and earth, those kindliest of mother-goddesses, were so aroused as to annihilate cities, men could only conclude that the fault was theirs.
Translated by Richard and Clara Winston
Vanished Cities
Part One (p. 4)
Alfred A. Knopf. New York, New York, USA. 1962

Schrödinger, Erwin   1887–1961
Austrian theoretical physicist

As our mental eye penetrates into smaller and smaller distances and shorter and shorter times, we find nature behaving so entirely differently from what we observe in visible and palpable bodies of our surroundings that no model shaped after our large-scale experiences can ever be "true." A complete satisfactory model of this type is not only practically inaccessible, but not even thinkable. Or, to be precise, we can, of course, think of it, but however we think it, it is wrong; not perhaps quite as meaningless as a "triangular circle," but more so than a "winged lion."
Science and Humanism
The Nature of Our "Models" (p. 25)
At The University Press. Cambridge, England. 1952

Sears, Paul Bigelow   1891–1990
American plant ecologist and conservationist

Nature is not to be conquered save on her own terms. She is not conciliated by cleverness or industry in devising means to defeat the operation of one of her laws through the workings of another.
Deserts on the March
Chapter I (p. 3)
University of Oklahoma Press. Norman, Oklahoma, USA. 1935

Selye, Hans   1907–82
Austrian-American endocrinologist

To me nature created man, and nature is superior.
In Denis Brian
Genius Talk: Conversations with Nobel Scientists and Other Luminaries
Chapter 13 (p. 267)
Plenum Press. New York, New York, USA. 1995

Seneca (Lucius Annaeus Seneca)   4 BCE–65 AD
Roman playwright

Nature does not turn out her work according to a single pattern; she prides herself upon her power of variation…
Physical Science in the Time of Nero, Being a Translation of the Quaestiones Naturales of Seneca
Book VII, Chapter XXVII (p. 301)
Macmillan & Company Ltd. London, England. 1910

Nature does not reveal all her secrets at once. We imagine we are initiated in her mysteries: we are, as yet, but hanging around her outer courts.
Physical Science in the Time of Nero, Being a Translation of the Quaestiones Naturales of Seneca
Book VII, Chapter XXXI (p. 306)
Macmillan & Company Ltd. London, England. 1910

Shakespeare, William   1564–1616
English poet, playwright, and actor

Thou, nature, art my goddess; to thy laws
My services are bound.
In *Great Books of the Western World* (Volume 27)
The Plays and Sonnets of William Shakespeare (Volume 2)

King Lear
Act I, Scene ii, l. 1–2
Encyclopædia Britannica, Inc. Chicago, Illinois, USA. 1952

One touch of nature makes the whole world kin,
That all with one consent praise new-born gawds,
Though they are made and moulded of things past,
And give to dust that is a little gilt
More laud than gilt o'er-dusted.
In *Great Books of the Western World* (Volume 27)
The Plays and Sonnets of William Shakespeare (Volume 2)
Troilus and Cressida
Act III, Scene iii, l. 175–179
Encyclopædia Britannica, Inc. Chicago, Illinois, USA. 1952

…nature is made better by no mean
But nature makes that mean.
So, over that art
Which you say adds to nature, is an art
That nature makes.
In *Great Books of the Western World* (Volume 27)
The Plays and Sonnets of William Shakespeare (Volume 2)
The Winter's Tale
Act IV, Scene iv, l. 88–90
Encyclopædia Britannica, Inc. Chicago, Illinois, USA. 1952

How sometimes nature will betray its folly,
Its tenderness, and make itself a pastime
To harder bosoms!
In *Great Books of the Western World* (Volume 27)
The Plays and Sonnets of William Shakespeare (Volume 2)
The Winter's Tale
Act I, Scene ii, l. 151–153
Encyclopædia Britannica, Inc. Chicago, Illinois, USA. 1952

In Nature's infinite book of secrecy
A little I can read.
In *Great Books of the Western World* (Volume 27)
The Plays and Sonnets of William Shakespeare (Volume 2)
Anthony and Cleopatra
Act I, Scene ii, l. 9–10
Encyclopædia Britannica, Inc. Chicago, Illinois, USA. 1952

Shaw, George Bernard   1856–1950
Irish comic dramatist and literary critic

The notion that Nature does not proceed by jumps is only one of the budget of plausible lies that we call classical education. Nature always proceeds by jumps. She may spend twenty thousand years making up her mind to jump; but when she makes it up at last, the jump is big enough to take us into a new age.
Back to Methuselah
Part II, XXXIII (p. 81)
Constable & Company Ltd. London, England. 1921

Smyth, Nathan A.
No biographical data available

By the pull of pleasure and prod of pain nature keeps the individual in tune with her purposes.
Through Science to God
Chapter X (p. 146)
The Macmillan Company. New York, New York, USA. 1936

Spencer, Herbert   1820–1903
English social philosopher

Nature's rules…have no exceptions.
Social Statics
Introduction, Lemma II (p. 39)
John Chapman. London, England. 1851

Is it not, indeed, an absurd and almost sacrilegious belief that the more a man studies Nature the less he reveres it? Think you that a drop of water, which to the vulgar eye is but a drop of water, loses anything in the eye of the physicist who knows that its elements are held together by a force which, if suddenly liberated, would produce a flash of lightning?
Education: Intellectual, Moral and Physical
A.L. Fowle. New York, New York, USA. 1860

Spenser, Edmund   1552–99
English poet

Yet neither spinnes, nor cards, ne cares nor fretts,
But to her mother Nature all her care she letts.
The Complete Poetical Works of Edmund Spenser
The Faerie Queene
Book II, Canto VI
Houghton Mifflin Company. Boston, Massachusetts, USA. 1908

Steele, Joel Dorman   1836–86
American educator and textbook writer

In Nature all is common, and no use is base. She keeps no selected elements done up in gilt papers for sensitive people.
A Fourteen Weeks Course in Chemistry
Conclusion (p. 223)
A.S. Barnes & Company. New York, New York, USA. 1870

Stevenson, Adlai E.   1900–65
American political leader and diplomat

Nature is neutral. Man has wrested from nature the power to make the world a desert or to make the deserts bloom.
High Fidelity Record Annual 1955
Speech (p. 338)
30 April 1946, House of Commons
J. B. Lippincott & Company. Philadelphia, Pennsylvania, USA. 1956

Nature is indifferent to the survival of the human species, including Americans.
Adlai's Almanac: The Wit and Wisdom of Stevenson of Illinois (p. 27)
H. Schuman. New York, New York, USA. 1952

Swann, William Francis Gray   1884–1962
Anglo-American physicist

There are times…in the growth of human thought when nature, having led man to the hope that he may understand her glories, turns for a time capricious and mockingly challenges his powers to harmonize her mysteries by revealing new treasures.
In Bernard Jaffe
Crucibles: The Story of Chemistry

Chapter XVI (p. 322)
Dover Publications. New York, New York, USA. 1976

Swift, Jonathan   1667–1745
Irish-born English writer

He said that new systems of nature were but new fashions, which would vary in every age; and even those who pretend to demonstrate them from mathematical principles, would flourish but a short period of time, and be out of vogue when that [system of nature] was determined.
In *Great Books of the Western World* (Volume 36)
Gulliver's Travels
Part III, Chapter VIII (pp. 118–119)
Encyclopædia Britannica, Inc. Chicago, Illinois, USA. 1952

Teale, Edwin Way   1899–1980
American naturalist

Nature is shy and noncommittal in a crowd. To learn her secrets, visit her alone or with a single friend, at most.
Circle of the Seasons
May 4 (p. 85)
Dodd, Mead & Company. New York, New York, USA. 1953

Tennyson, Alfred (Lord)   1809–92
English poet

Who trusted
God was love indeed
And love Creation's final law —
Tho' Nature, red in tooth and claw
With ravine, shriek'd against his creed.
Alfred Tennyson's Poetical Works
In Memoriam A.H.H., LVI, Stanza IV
Oxford University Press, Inc. London, England. 1953

Nothing in Nature is unbeautiful.
Alfred Tennyson's Poetical Works
Lover's Tale, l. 348
Oxford University Press, Inc. London, England. 1953

A void was made in Nature;
all her bonds
Crack'd; and I saw the flaring atom-streams
And torrents of her myriad universe
Ruining along the illimitable inane,
Fly on to clash together again…
Alfred Tennyson's Poetical Works
Lucretius, l. 37–39
Oxford University Press, Inc. London, England. 1953

Thierry, Paul Henri, Baron d'Holbach   1723–89
German-born French man of leisure

Man is only unhappy because he is ignorant of nature.
Translated by M. Mirabaud
System of Nature or, The Laws of the Moral and Physical World (Volume 1)
Preface by the Author (p. vii)
Published by R. Benson. Philadelphia, Pennsylvania, USA. 1808

…man disdains the study of nature, to pursue phantoms, which resemble the Will with the Wisp, which at once

terrifies and dazzles the benighted traveler, and which make him quit the simple road to truth, without pursuing which, he can never arrive at happiness.
Translated by M. Mirabaud
System of Nature or, The Laws of the Moral and Physical World (Volume 1)
Preface by the Author (p. vii)
Published by R. Benson. Philadelphia, Pennsylvania, USA. 1808

Thomson, J. Arthur   1861–1933
Scottish biologist

When we are thrilled with the wonder of the world, the heights and depths of things; when our Nature-feeling is informed with knowledge; when our science leaves us with a conviction of the mysteriousness of Nature — the unfathomed universe; when our philosophical outlook leads us towards a realisation of a meaning behind the process; then there may be a total reaction on our part worthy of the name of Natural Religion.
The System of Animate Nature (Volume 1)
Lecture I (p. 42)
William & Norgate. London, England. 1920

In her manifold opportunities Nature has thus helped man to polish the mirror of [man's] mind, and the process continues. Nature still supplies us with abundance of brain-stretching theoretical puzzles and we eagerly tackle them; there are more worlds to conquer and we do not let the sword sleep in our hand; but how does it stand with feeling? Nature is beautiful, gladdening, awesome, mysterious, wonderful, as ever, but do we feel it as our forefathers did?
The System of Animate Nature (Volume 1)
Lecture I (p. 25)
William & Norgate. London, England. 1920

Thomson, James   1700–48
Scottish poet

O nature!...
Enrich me with the knowledge of thy works;
Snatch me to Heaven.
Seasons
Autumn, l. 1,352
Printed by John Mycall. Newburyport, Massachusetts, USA. 1790

I care not, Fortune, what you me deny;
You cannot rob me of free Nature's grace,
You cannot shut the windows of the sky,
Through which Aurora shows her brightening face;
You cannot bar my constant feet to trace
The woods and lawns, by living stream, at eve.
Castle of Indolence
Canto II, Stanza 3
William Smith. London, England. 1842

Thoreau, Henry David   1817–62
American essayist, poet, and practical philosopher

Nature will bear the closest inspection; she invites us to lay our eye level with the smallest leaf, and take an insect view of its plain. She has no interstices; every part is full of life.
The Writings of Henry David Thoreau (Volume 9)
Natural History of Massachusetts (p. 132)
Houghton Mifflin Company. Boston, Massachusetts, USA. 1893

Nature must be viewed humanly to be viewed at all; that is, her scenes must be associated with humane affections, such as are associated with one's native place. She is most significant to a lover. A lover of Nature is preeminently a lover of man. If I have no friend, what is Nature to me? She ceases to be morally significant...
The Journal of Henry David Thoreau (Volume 4)
June 30, 1852 (p. 163)
Houghton Mifflin Company. Boston, Massachusetts, USA. 1906

I love Nature partly *because* she is not man, but a retreat from him. None of his institutions control or pervade her. There a different kind of right prevails. In her midst I can be glad with an entire gladness. If this world were all man, I could not stretch myself, I should lose all hope. He is constraint, she is freedom to me. He makes me wish for another world. She makes me content with this.
The Journal of Henry David Thoreau (Volume 4)
January 3, 1853 (p. 440)
Houghton Mifflin Company. Boston, Massachusetts, USA. 1906

I have a room all to myself; it is nature.
The Journal of Henry David Thoreau (Volume 4)
January 3, 1853 (p. 446)
Houghton Mifflin Company. Boston, Massachusetts, USA. 1906

I sit in my boat on Walden, playing the flute this evening, and see the perch, which I seem to have charmed, hovering around me, and the moon traveling over the bottom, which is strewn with the wrecks of the forest, and feel that nothing but the wildest imagination can conceive of the manner of life we are living. Nature is a wizard. The Concord nights are stranger than the Arabian nights.... Heaven lies above, because the air is deep.
Journal (Volume 1: 1837–1844)
May 27, 1841 (p. 311)
Princeton University Press. Princeton, New Jersey, USA. 1981

If we knew all the laws of Nature, we should need only one fact, or the description of one actual phenomenon, to infer all the particular results at that point. Now we know only a few laws, and our result is vitiated, not, of course, by any confusion or irregularity in Nature, but by our ignorance of essential elements in the calculation. Our notions of law and harmony are commonly confined to those instances which we detect; but the harmony which results from a far greater number of seemingly conflicting, but really concurring, laws, which we have not detected, is still more wonderful. The particular laws are as our points of view, as to the traveler, a mountain outline varies with every step, and it has an infinite number of profiles, though absolutely but one form. Even when cleft or bored through it is not comprehended in its entireness.

The Writings of Henry David Thoreau (Volume 2)
Walden
Chapter XVI (p. 448)
Houghton Mifflin Company. Boston, Massachusetts, USA. 1893

Thurlow, Lord Edward, 1st Baron Thurlow 1731–1806
English jurist and statesman

Nature is always wise in every part.
Select Poems
The Harvest Moon
Chadwyck-Healey. Cambridge, England. 1992

Tomonaga, Sin-Itiro 1906–79
Japanese physicist

We are too powerless to make assumptions based only on reasoning. We must beg instruction from Nature herself.
T. Miyazima (ed.)
Scientific Papers (Volume 1) (p. 545)
Misuzu-Shobo, Tokyo, Japan. 1971

Turgenev, Ivan 1818–83
Russian novelist and dramatist

Nature is no temple, but a workshop, and man is the worker therein.
Translated by Bernard Guilbert Guerney
Fathers and Sons
Chapter 9 (p. 58)
The Modern Library. New York, New York, USA. 1961

However much you knock at nature's door, she will never answer you in comprehensible words because she is dumb. She will utter a musical sound, or a moan like a harp string, but you don't expect a song from her.
On the Eve
Chapter I (p. 10)
Charles Scribner's Sons. New York, New York, USA. 1903–04

Tuttle, Hudson 1836–1910
American medium

We find in the constant harmony of nature a sufficient proof in favor of the immutability of its laws. Every miracle would involve their infraction; a process to which nature would submit as little as to any other intervention in its empire; in which every thing, from the gnat which dances in the sunbeam up to the human mind which issues from the brain, is governed by fixed principles.
In Ludwig Buchner
Force and Matter
Chapter VI (p. 38)
Trubner & Company. London, England. 1864

Twain, Mark (Samuel Langhorne Clemens) 1835–1910
American author and humorist

Architects cannot teach nature anything.
The Complete Essays of Mark Twain

A Memorable Midnight Experience (p. 30)
De Capo Press. New York, New York, USA. 2000

How blind and unreasoning and arbitrary are some of the laws of nature — most of them, in fact!
The Man That Corrupted Hadleyburg, and Other Stories and Essays
A Double-Barreled Detective Story
Chapter III (p. 296)
Harper & Brothers. New York, New York, USA. 1917

…Nature's attitude toward all life is profoundly vicious, treacherous and malignant.
Mark Twain's Notebook
Chapter XXIII (pp. 255–256)
Harper & Brothers. New York, New York, USA. 1935

Voltaire (François-Marie Arouet) 1694–1778
French writer

What are you, Nature? Live in you? But I have been searching for you for fifty years, and have never been able to find you.
The Works of Voltaire (Volume 12)
Philosophical Dictionary (Volume 8)
Nature (p. 48)
The St. Hubert Guild. Akron, Ohio, USA. 1901

von Baeyer, Adolf 1835–1917
German research chemist

What makes a great scientist? He must not command but listen; he must adapt himself to what he hears and reshape himself accordingly…. The ancient empiricists already did this. They put their ear to Nature. The modern scientist does the same…. Coming nearer to Nature has a very special effect on people. They develop very differently from someone who confronts Nature with preconceived ideas. Someone who approaches Nature with set ideas will, so to speak, stand before it like a general. He will want to issue orders to Nature.
In Richard Willstätter
From My Life: The Memoirs of Richard Willstatter
Chapter 6 (p. 140)
W.A. Benjamin. New York, New York, USA. 1965

von Baeyer, Hans Christian 1938–
German-born physicist and author

Not the scientist, but nature has the last word.
Rainbows, Snowflakes, and Quarks: Physics and the World Around Us
The Measure of Things (p. 189)
McGraw-Hill Book Company, Inc. New York, New York, USA. 1984

von Goethe, Johann Wolfgang 1749–1832
German poet, novelist, playwright, and natural philosopher

When a man of lively intellect first responds to Nature's challenge to be understood, he feels irresistibly tempted to impose his will upon the natural objects he is studying. Before long, however, they close in upon him with such force as to make him realize that he in turn must now acknowledge their might and hold in respect the authority they exert over him.

Goethe's Botanical Writings
Formation and Transformation (p. 21)
University of Hawaii Press. Honolulu, Hawaii, USA. 1952

We live in her midst and know her not. She is incessantly speaking to us, but betrays not her secret. We constantly act upon her, and yet have no power over her.
Translated by Thomas Huxley
Nature: Aphorisms by Goethe
Nature, Volume 1, Thursday, November 4, 1869 (p. 9)

Whoever wishes to deny nature as an organ of the divine must begin by denying all revelation.
In D. Miller (ed.)
Scientific Studies (Volume 12)
Chapter VIII (p. 303)
Suhrkamp. New York, New York, USA. 1988

That which is most unnatural is still Nature; the stupidest philistinism has a touch of her genius. Whoso cannot see her everywhere, sees her nowhere rightly.
Translated by Thomas Huxley
Nature: Aphorisms by Goethe
Nature, Volume 1, Thursday, November 4, 1869 (p. 9)

She tosses her creatures out of nothingness, and tells them not whence they came, nor whither they go. It is their business to run, she knows the road.
Translated by Thomas Huxley
Nature: Aphorisms by Goethe
Nature, Volume 1, Thursday, November 4, 1869 (p. 9)

She loves herself, and her innumerable eyes and affections are fixed upon herself. She has divided herself that she may be her own delight. She causes an endless succession of new capacities for enjoyment to spring up, that her insatiable sympathy may be assuaged.
Translated by Thomas Huxley
Nature: Aphorisms by Goethe
Nature, Volume 1, Thursday, November 4, 1869 (p. 9)

She rejoices in illusion. Whoso destroys it in himself and others, him she punishes with the sternest tyranny. Whoso follows her in faith, him she takes as a child to her bosom.
Translated by Thomas Huxley
Nature: Aphorisms by Goethe
Nature, Volume 1, Thursday, November 4, 1869 (p. 9)

She creates needs because she loves action. Wondrous! that she produces all this action so easily. Every need is a benefit, swiftly satisfied, swiftly renewed. Every fresh want is a new source of pleasure, but she soon reaches an equilibrium.
Translated by Thomas Huxley
Nature: Aphorisms by Goethe
Nature, Volume 1, Thursday, November 4, 1869 (p. 10)

She has always thought and always thinks; though not as a man, but as Nature. She broods over an all-comprehending idea, which no searching can find out.
Translated by Thomas Huxley

Nature: Aphorisms by Goethe
Nature, Volume 1, Thursday, November 4, 1869 (p. 9)

Every instant she commences an immense journey, and every instant she has reached her goal.
Translated by Thomas Huxley
Nature: Aphorisms by Goethe
Nature, Volume 1, Thursday, November 4, 1869 (p. 10)

Each of her works has an essence of its own; each of her phenomena a special characterisation: and yet their diversity is in unity.
Translated by Thomas Huxley
Nature: Aphorisms by Goethe
Nature, Volume 1, Thursday, November 4, 1869 (p. 9)

The one thing she seems to aim at is Individuality; yet she cares nothing for individuals. She is always building up and destroying; but her workshop is inaccessible.
Translated by Thomas Huxley
Nature: Aphorisms by Goethe
Nature, Volume 1, Thursday, November 4, 1869 (p. 9)

Nature goes her own way, and all that to us seems an exception is really according to order.
In Johann Peter Eckermann
Conversations with Goethe
Thursday, December 9, 1824 (p. 75)
J.M. Dent & Sons Ltd. London, England. 1970

Her crown is love. Through love alone dare we come near her. She separates all existences, and all tend to intermingle. She has isolated all things in order that all may approach one another. She holds a couple of draughts from the cup of love to be fair payment for the pains of a lifetime.
Translated by Thomas Huxley
Nature: Aphorisms by Goethe
Nature, Volume 1, Thursday, November 4, 1869 (p. 10)

Her mechanism has few springs — but they never wear out, are always active and manifold.
Translated by Thomas Huxley
Nature: Aphorisms by Goethe
Nature, Volume 1, Thursday, November 4, 1869 (p. 9)

It is not easy for us to grasp the vast, the super colossal, in nature; we have lenses to magnify tiny objects but none to make things smaller. And even for the magnifying glass we need eyes like Carus and Nees to profit intellectually from its use. However, since nature is always the same, whether found in the vast or the small, and every piece of turbid glass produces the same blue as the whole of the atmosphere covering the globe, I think it right to seek out prototypal examples and assemble them before me. Here, then, the enormous is not reduced; it is present within the small, and remains as far beyond our grasp as it was when it dwelt in the infinite.
In D. Miller (ed.)
Scientific Studies (Volume 12)
Chapter VIII (p. 304)
Suhrkamp. New York, New York, USA. 1988

We shall never succeed in exhausting the immeasurable riches of nature; and no generation of men will ever have cause to boast of having comprehended the total aggregation of phenomena.
Cosmos: A Sketch of a Physical Description of the Universe (Volume 1)
Introduction (p. 73)
Harper & Brothers. New York, New York, USA. 1869

Mere communion with nature, mere contact with the free air, exercise a soothing yet strengthening influence on the wearied spirit, calm the storm of passion, and soften the heart when shaken by sorrow to its inmost depths.
Cosmos: A Sketch of a Physical Description of the Universe (Volume 1)
Introduction (p. 25)
Harper & Brothers. New York, New York, USA. 1869

In order to depict nature in its exalted sublimity, we must not dwell exclusively on its external manifestations, but we must trace its image, reflected in the mind of man, at one time filling the dreamy land of physical myths with forms of grace and beauty, and at another developing the noble germ of artistic creations.
Cosmos: A Sketch of a Physical Description of the Universe (Volume 2)
Description of Nature by the Ancients (p. 20)
Harper & Brothers. New York, New York, USA. 1869

von Schelling, Friedrich Wilhelm Joseph   1775–1854
German philosopher

What then is that secret bond which couples our mind to Nature, or that hidden organ through which Nature speaks to our mind or our mind to Nature? For what we want is not that Nature should coincide with the laws of our mind by chance (as if through some third intermediary), but that she herself, necessarily and originally, should not only express, but even realize, the laws of our mind, and that she is, and is called, Nature only insofar as she does so.

Nature should be Mind made visible, Mind the invisible Nature. Here then, in the absolute identity of Mind in us and Nature outside us, the problem of the possibility of a Nature external to us must be resolved.
Translated by Errol E. Harris and Peter Heath
Ideas for a Philosophy of Nature as Introduction to the Study of this Science
Introduction (pp. 41–42)
Cambridge University Press. Cambridge, England. 1988

The purest exercise of man's rightful dominion over dead matter, which was bestowed upon him together with reason and freedom, is that he spontaneously operates upon Nature, determines her according to purpose and intention, lets her act before his eyes, and as it were spies on her at work. But that the exercise of this dominion is possible, he owes yet again to Nature, whom he would strive in vain to dominate, if he would not put her in conflict with herself and set her own forces in motion against her.

Translated by Errol E. Harris and Peter Heath
Ideas for a Philosophy of Nature as Introduction to the Study of this Science
Book I (p. 57)
Cambridge University Press. Cambridge, England. 1988

von Siemens, Werner   1816–1892
German inventor and entrepreneur

The deeper the insight we obtain into the mysterious workings of nature's forces…the more we are convinced that we are still standing in the vestibule of science; that an unexplored world still lies before us; and however much we may discover, we know not whether mankind will ever arrive at a full knowledge of nature.
In Gardner G. Hubbard
Annual Report of the Board of Regents of the Smithsonian Institution, 1891
The Evolution of Commerce (p. 660)
Government Printing Office. Washington, D.C. 1893

Walcott, Charles D.   1850–1927
Geologist

Nature has a habit of placing some of her most attractive treasures in places where it is difficult to locate and obtain them.
Annual Report of the Board of Regents of the Smithsonian Institution, 1915
Evidences of Primitive Life (p. 246)
Government Printing Office. Washington, D.C. 1916

Walker, John   1731–1803
English minister and educator

Nature consults no philosophers.
Lectures on Geology: Including Hydrography, Mineralogy, and Meteorology with an Introduction to Biology
Biographical Introduction (p. xxxi)
The University of Chicago Press. Chicago, Illinois, USA. 1966

Ward, Lester Frank   1898–1970
American sociologist

An entirely new dispensation has been given to the world. All the materials and forces of nature have been thus placed completely under the control of one of the otherwise least powerful of the creatures inhabiting the earth…. Nature has thus been made the servant of man.
Glimpses of the Cosmos (Volume 3)
Mind as a Social Factor (p. 370)
G.P. Putnam's Sons. New York, New York, USA. 1913

Warner, Charles Dudley   1829–1900
American editor and author

Nature is, in fact, a suggester of uneasiness, a promoter of pilgrimages and of excursions of the fancy which never come to any satisfactory haven.
Backlog Studies
Ninth Study, Section II (p. 203)
Houghton Mifflin Company. Boston, Massachusetts, USA. 1892

Watts, Alan Wilson   1915–73
American philosopher

The form of Christianity differs from the form of nature because in the Church and in its spiritual atmosphere we are in a universe that has been made. Outside the Church we are in a universe that has grown.
Nature, Man, and Woman
Part I, Chapter 1 (p. 40)
Vintage Books. New York, New York, USA. 1970

For the notion that the interrelatedness of nature is complex and highly detailed is merely the result of translating it into the linear units of thought. Despite its rigor and despite its initial successes, this is an extremely clumsy mode of intelligence. Just as it is a highly complicated task to drink water with a fork instead of a glass, so the complexity of nature is not innate but a consequence of the instruments used to handle it. There is nothing complex about walking, breathing, and circulating one's blood. Living organisms have developed these functions without thinking about them at all.
Nature, Man, and Woman
Part I, Chapter 2 (p. 62)
Vintage Books. New York, New York, USA. 1970

The rush of waterfalls and the babbling of streams are not loved for their resemblance to speech; the irregularly scattered stars do not excite us because of the formal constellations which have been traced out between them; and it is for no symmetry or suggestion of pictures that we delight in the patterns of foam, of the veins in rock, or of the black branches of trees in wintertime.
Nature, Man, and Woman
Part I, Chapter 5 (p. 124)
Vintage Books. New York, New York, USA. 1970

Webber, Charles Wilkins   1819–56
American explorer and journalist

God's own presence is felt lingering yet, as if, in love with his own work, he stayed to touch it again — creating new charms in multiplied duration.
Old Hicks, the Guide: Or, Adventures in the Comanche Country in Search of a Gold Mine
VIII
Harper & Brothers. New York, New York, USA. 1848

Wells, H. G. (Herbert George)   1866–1946
English novelist, historian, and sociologist

Has anything arisen to show…that where the life and breeding of every individual of a species is about equally secure, a degenerative process must not inevitably supervene?…Natural Selection grips us more grimly than it ever did, because the doubts thrown upon the inheritance of acquired characteristics have deprived us of our trust in education as a means of redemption for decadent families. In our hearts we wish that the case were not so, we all hate

Death and his handiwork; but the business of science is not to keep up the courage of men, but to tell the truth.
Bio-Optimism
Nature, Volume 52, Number 1348, August 29, 1895 (p. 411)

Weyl, Hermann   1885–1955
German mathematician

Once and for all I wish to record my unbounded admiration for the work of the experimenter in his struggle to wrest interpretable facts from an unyielding Nature who knows so well how to meet our theories with a decisive No — or with an inaudible Yes.
Translated by H.P. Robertson
The Theory of Groups and Quantum Mechanics
Introduction (p. xx)
Methuen & Company Ltd. London, England. 1931

Wheeler, John Archibald   1911–
American physicist and educator

…nature at the quantum level is not a machine that goes its inexorable way. Instead what answer we get depends on the question we put, the experiment we arrange, the registering device we choose. We are inescapably involved in bringing about that which appears to be happening.
In John Archibald Wheeler and Wojciech Hubert Zurek (eds.)
Quantum Theory and Measurement (p. 185)
Princeton University Press. Princeton, New Jersey, USA. 1982

Whitehead, Alfred North   1861–1947
English mathematician and philosopher

The primary task of a philosophy of natural science is to elucidate the concept of nature, considered as one complex fact for knowledge, to exhibit the fundamental entities and the fundamental relations between entities in terms of which all laws of nature have to be stated, and to secure that the entities and relations thus exhibited are adequate for the expression of all the relations between entities which occur in nature.
The Concept of Nature
Chapter II (p. 46)
At The University Press. Cambridge, England. 1920

Thus we gain from the poets the doctrine that a philosophy of nature must concern itself with at least these five notions: change, value, eternal objects, endurance, organism, interfusion.
Science and the Modern World
Chapter V (p. 127)
The Macmillan Company. New York, New York, USA. 1929

You cannot talk vaguely about Nature in general.
Nature and Life
Part I (p. 1)
At the University Press. Cambridge, England. 1934

We have to remember that while nature is complex with timeless subtlety, human thought issues from the

simple-mindedness of beings whose active life is less than half a century.
An Enquiry Concerning the Principles of Natural Knowledge
Part I (p. 15)
At the University Press. Cambridge, England. 1919

Nature, even in the act of satisfying anticipation, often provides a surprise.
Adventures of Ideas
Chapter VIII (p. 161)
The Macmillan Company. New York, New York, USA. 1956

…nature gets credit which in truth should be reserved for ourselves: the rose for its scent: the nightingale for his song: and the sun for his radiance. The poets are entirely mistaken. They should address their lyrics to themselves, and should turn them into odes of self-congratulation on the excellence of the human mind. Nature is a dull affair, soundless, scentless, colorless; merely the hurrying of material, endlessly, meaninglessly.
Science and the Modern World
Chapter III (p. 80)
The Macmillan Company. New York, New York, USA. 1929

Whitman, Walt   1819–92
American poet, journalist, and essayist

The fields of Nature long prepared and fallow,
the silent, cyclic chemistry,
The slow and steady ages plodding, the unoccupied
surface ripening, the rich ores forming beneath.
Complete Poetry and Collected Prose
Song of the Redwood Tree
The Library of America. New York, New York, USA. 1982

Wilde, Oscar   1854–1900
Irish wit, poet, and dramatist

Nature is so uncomfortable. Grass is hard and lumpy and damp, and full of dreadful insects.
The Works of Oscar Wilde (Volume 10)
Intentions
The Decay of Lying
AMS Press. New York, New York, USA. 1909

It seems to me that we all look at nature too much, and live with her too little.
The Complete Writings of Oscar Wilde
De Profundis (p. 158)
Nottingham Society. New York, New York, USA. 1907

And then Nature is so indifferent, so unappreciative. Whenever I am walking in the park here, I always feel that I am no more to her than the cattle that browse on the slope, or the burdock that blooms in the ditch.
The Works of Oscar Wilde (Volume 10)
Intentions
The Decay of Lying
AMS Press. New York, New York, USA. 1909

Willstätter, Richard   1872–1942
German chemist

It is the scientist's lot, as it is the artist's, to be less important than his work. He who is chosen to lift the veil from Nature's secrets will be easily overshadowed by the creation he has revealed and which will make him immortal.
From My Life: The Memoirs of Richard Willstätter
Chapter 6 (p. 141)
W.A. Benjamin. New York, New York, USA. 1965

Wilson, David Scofield
No biographical data available

Nature is present to naturalists the way God is to saints or the past is to humanists — not simply as a matter of fact but as an insistent and live reality.
In the Presence of Nature
Chapter I (p. 1)
University of Massachusetts Press. Amherst. 1978

Winchell, Alexander   1824–91
American geologist

Any thing which is a plan has been thought out. The plans of Nature are the expressions of the mind.
Walks and Talks in the Geological Field
Chapter XXXII (p. 190)
Chautauqua Press. New York, New York, USA. 1890

Wöhler, Friedrich   1800–82
German chemist
von Liebig, Justus   1803–73
German organic chemist

When in the dark province of organic nature, we succeed in finding a light point, appearing to be one of those inlets whereby we may attain to the examination and investigation of this province, then we have reason to congratulate ourselves, although conscious that the object before us is unexhausted.
Translated by James C. Booth
Researches Respecting the Radical of Benzoic Acid
American Journal of Science and Arts, Volume 26, Number 2, July 1834
(p. 261)

Woodbridge, Frederick James Eugene   1867–1940
American philosopher

The incorporation of man into nature may well do something to man, but it must also do something to nature. It is impossible that the word "nature" can mean the same after this incorporation that it meant before.
Nature and Mind: Selected Essays of Frederick J.E. Woodbridge (p. 7)
Columbia University Press. New York, New York, USA. 1937

Wordsworth, William   1770–1850
English poet

To the solid ground
Of Nature trusts the Mind that builds for aye.
The Complete Poetical Works of William Wordsworth
A Violent Tribe of Bards on Earth
Crowell. New York, New York, USA. 1888

Come forth into the light of things;
Let Nature be your teacher.
The Complete Poetical Works of William Wordsworth
The Tables Turned, l. 15–16
Crowell. New York, New York, USA. 1888

Nature never did betray
The heart that loved her.
The Complete Poetical Works of William Wordsworth
Lines Composed a Few Miles Above Tintern Abbey
Crowell. New York, New York, USA. 1888

As in the eye of Nature he has lived,
So in the eye of Nature let him die!
The Old Cumberland Beggar (last lines)
Crowell. New York, New York, USA. 1888

I have learned
To look on nature, not as in the hour
Of thoughtless youth; but hearing oftentimes
The still, sad music of humanity.
The Complete Poetical Works of William Wordsworth
Lines Composed a Few Miles above Tintern Abbey, l. 88–91
Crowell. New York, New York, USA. 1888

Wright, Thomas   1711–86
English cosmologist

…three of the finest Sights in Nature, are a rising Sun
at Sea, a verdant Landskip with a Rainbow, and a clear
Star-light Evening…
An Original Theory or New Hypothesis of the Universe
Letter the Fifth (p. 37)
Printed for the Author. London, England. 1750

Yogananda, Paramahansa   1893–1952
Indian yogi

Because modern science tells us how to utilize the powers of Nature, we fail to comprehend the Great Life in
back of all names and forms. Familiarity with Nature has
bred a contempt for her ultimate secrets; our relation with
her is one of practical business. We tease her, so to speak,
to discover the ways in which she may be forced to serve
our purposes; we make use of her energies, whose Source
yet remains unknown. In science our relation with Nature
is like that between an arrogant man and his servant; or,
in a philosophical sense, Nature is like a captive in the
witness box. We cross-examine her, challenge her, and
minutely weigh her evidence in human scales that cannot
measure her hidden values.
Autobiography of a Yogi
Chapter 35 (pp. 337–338)
Self-Realization Fellowship. Los Angeles, California, USA. 1971

Young, Edward   1683–1765
English poet and dramatist

The course of nature governs all!
The course of nature is the heart of God.
The miracles thou call'st for, this attest;
For say, could nature nature's course control?

But miracles apart, who sees Him not?
Night Thoughts
Night IX, l. 1,280
Printed by R. Nobels for R. Edwards. London, England. 1797

Read Nature, Nature is a friend of truth…
Night Thoughts
Printed by R. Nobels for R. Edwards. London, England. 1797

NEANDERTHAL

Constable, George   1941–
No biographical data available

Place him in a landscape of tall, waving grass, with the
sun shining down and the bubbling music of summer in
the air. Who is this man? He is an evolutionary bridge,
just shy of fully modern status. He is a true human
— our ancestor. We should regard him with honour,
because almost everything that we are springs directly
from him.
The Neanderthals
Chapter Five (p. 134)
Time-Life Books. New York, New York, USA. 1973

Margulis, Lynn   1938–
American cell biologist and evolutionist
Sagan, Dorion   1959–
American science writer

Neanderthalers, whatever else they were, were people.
They were artists and poets and buriers of the dead.
Microcosmos
Chapter 12 (p. 225)
Summit Books. New York, New York, USA. 1986

Wells, H. G. (Herbert George)   1866–1946
English novelist, historian, and sociologist

Hairy or grisly, with a big face like a mask, great brow
ridges and no forehead, clutching an enormous flint, and
running like a baboon with his head forward, and not,
like a man, with his head up, he must have been a fear-
some creature for our forefathers to come upon.
The Grisly Man
Storyteller Magazine, April 1921

NEBULA

Flammarion, Camille   1842–1925
French astronomer and writer

These are lights which glimmer on the frontiers of creation; they are the beginnings which show us the birth
of other universes; they are the voices of the past which
speak to us from the depths of the vanished ages.
Popular Astronomy: A General Description of the Heavens
Book VI, Chapter X (p. 668)
Chatto & Windus. London, England. 1894

Herschel, Friedrich Wilhelm   1738–1822
English astronomer

On the 15th of February, 1786, I discovered that one of my planetary nebula, had a spot in the center, which was more luminous than the rest, and with long attention, a very bright, round, well defined center became visible. I remained not a single moment in doubt, but that the bright center was connected with the rest of the apparent disk.
On Nebulous Stars, Properly So Called
Philosophical Transactions of the Royal Society of London, Volume 81, 1791

Hubble, Edwin Powell   1889–1953
American astronomer

The term nebulae offers the values of tradition…the term galaxies, the glamour of romance.
In Timothy Ferris
The Red Limit: The Search for the Edge of the Universe
Chapter 1 (p. 41)
William Morrow & Company, Inc. New York, New York, USA. 1977

Huggins, Sir William   1824–1910
English astronomer

On the evening of August 29, 1864, I directed the spectroscope for the first time to a planetary nebula in Draco. I looked into the spectroscope. No spectrum such as I had expected. A single bright line only!…The riddle of the nebulae was solved. The answer, which had come to us in the light itself, read: Not an aggregation of stars, but a luminous gas.
In George E. Hale
Annual Report of the Board of Regents of the Smithsonian Institution, 1902
Stellar Evolution in the Light of Recent Research (p. 155)
Government Printing Office. Washington, D.C. 1903

Hugo, Victor   1802–85
French author, lyric poet, and dramatist

A nebula is, as it were, a universe in the cocoon.
Translated by Isabel F. Hapgood
The Toilers of the Sea
Part II, Book Third, Chapter III (p. 413)
The Heritage Press. New York, New York, USA. 1961

Jeans, Sir James Hopwood   1877–1946
English physicist and mathematician

…nebulae are the birthplaces of the stars, so that each nebula consists of stars born and stars not yet born.
The Universe Around Us
Chapter I (p. 67)
The Macmillan Company. New York, New York, USA. 1929

NECESSITY

Huxley, Thomas Henry   1825–95
English biologist

Fact I know; and Law I know; but what is this Necessity, save an empty shadow of my own mind's throwing?
Collected Essays (Volume 1)
Method and Result
On the Physical Basis of Life (p. 161)
Macmillan & Company Ltd. London, England. 1904

NEURONS

Sherrington, Sir Charles   1857–1952
English physiologist

More than one way for doing the same thing is provided by the natural constitution of the nervous system. This luxury of means of compassing a given combination seems to offer the means of restitution of an act after its impairment or loss in one of its several forms.
Nobel Lectures, Physiology or Medicine 1922–1941
Nobel lecture for award received in 1932
Inhibition as a Coordinative Factor (p. 289)
Elsevier Publishing Company. Amsterdam, Netherlands. 1965

NEUROPHYSIOLOGY

Hodgkin, Alan L.   1914–98
English physiologist and biophysicist

Research in neurophysiology is much more like paddling a small canoe on a mountain river. The river which is fed by many distant springs carries you along all right, though often in a peculiar direction. You have to paddle quite hard to keep afloat. And sooner or later some of your ideas are upset and are carried downstream like an upturned canoe.
Lex Prix Nobel. The Nobel Prizes in 1963
Nobel banquet speech for award received in 1963
Nobel Foundation. Stockholm, Sweden. 1964

NEUROSCIENCE

Young John Zachary   1907–97
English zoologist

What would be the use of a neuroscience that cannot tell us anything about love?
Programs of the Brain
Chapter 14 (p. 143)
Oxford University Press. Oxford, England. 1978

NEUTRINO

Adams, Douglas   1952–2001
English author, comic radio dramatist, and musician

The chances of a neutrino actually hitting something as it travels through all this howling emptiness are roughly comparable to that of dropping a ball bearing at random from a cruising 747 and hitting, say, an egg sandwich.

The Ultimate Hitchhiker's Guide to the Galaxy
Mostly Harmless
Chapter 3 (p. 656)
The Ballantine Book Company. New York, New York, USA. 2002

Author undetermined

The neutrino is about as close to intangibility as we can get in this world — the human soul, perhaps is the next stage.
Engineering and Science, February 1973 (p. 15)

Mister Jordan
Takes neutrinos
And from those he
Builds the light.
And in pairs they
Always travel
One neutrino's
Out of sight.
In Abraham Pais
Inward Bound
To the tune "Mac the Knife" (p. 419)
Clarendon Press. Oxford, England. 1986

Crane, H. Richard
No biographical data available

Not everyone would be willing to say that he believes in the existence of the neutrino, but it is safe to say there is hardly one of us who is not served by the neutrino hypothesis as an aid in thinking about beta-decay process.
The Energy and Momentum Relations in the Beta-Decay and the Search for the Neutrino
Review of Modern Physics, Volume 20, Number 2, March 1948 (p. 278)

Eddington, Sir Arthur Stanley   1882–1944
English astronomer, physicist, and mathematician

The neutrino is just barely a fact.
The Two-Neutrino Experiment
Scientific American, Volume 208, Number 3, March 1963 (p. 60)

In an ordinary way I might say that I do not believe in neutrinos. But I have to reflect that a physicist may be an artist, and you never know where you are with artists. My old-fashioned kind of disbelief in neutrinos is scarcely enough. Dare I say that experimental physicists will not have sufficient ingenuity to make neutrinos? Whatever I may think, I am not going to be lured into a wager against the skill of the experimenters under the impression that it is a wager against the truth of a theory. If they succeed in making neutrinos, perhaps even in developing industrial application of them, I suppose I shall have to believe — though I may feel they have not been playing quite fair.
The Philosophy of Physical Science
Chapter VII, Section II (p. 112)
The Macmillan Company. New York, New York, USA. 1939

Gamow, George   1904–68
Russian-born American physicist

My mass is zero,
My Charge is the same.
You are my hero,
Neutrino's my name.
Thirty Years That Shook Physics
First Part (p. 188)
Doubleday & Company, Inc. Garden City, New York, USA. 1966

...one of the students asked whether the "Chadwick neutron" was the same "neutron" proposed by Pauli for the phenomena of beta transformation. "No," answered Fermi, *"il neutrone di Pauli è mol to più piccolo, cio è un neutrino."* The name stuck.
The Reality of Neutrinos
Physics Today, Volume 1, Number 3, July 1948 (p. 5)

Haag, Joel
No biographical data available

The Poet, J. Alfred Neutrino
Who subsisted sublimely on vino,
With a spin of one-half
Wrote his own epitaph:
"No rest-ness, no charge, no bambino."
In R.L. Weber
A Random Walk in Science (p. 138)
Institute of Physics Publishing. Bristol, England. 1973

Harari, Haim
Theoretical physicist

Neutrino physics is largely an art of learning a great deal by observing nothing.
Proceedings 13th International Conference on Neutrino Physics and Astrophysics
Boston, June 5–11, 1099 (p. 574)

Pauli, Wolfgang   1900–58
Austrian–born physicist

I have committed the ultimate sin, I have predicted the existence of a particle that can never be observed.
In Frank Wilczek and Betsy Devine
Longing For the Harmonies
Ego and Survival (p. 65)
W. W. Norton & Company, Inc., Publishers. New York, New York, USA. 1988

Perry, Georgette
No biographical data available

To trap them is almost impossible.
You may wait for months in a deep mine
Inside an anti-coincidence shield.
No charge deflects them.
Desireless, they cruise through the world
As if it's nothing, not there.
Twigs
Neutrinos, 1977

Pontecorvo, Bruno   1913–93
Italian-born English physicist

It is difficult to find a case where the word "intuition" characterises a human achievement better than in the case of the neutrino invention by Pauli.
Journel de Physique
Supplement C8, Volume 48, 1982 (p. 221)

Reines, Frederick   1918–
American physicist
Cowan, C.
No biographical data available

We are happy to inform you that we have definitely detected neutrinos from fission fragments by observing beta-decay of protons.
Translated by R. Schlapp
In Wolfgang Pauli, Charles Paul Enz and Karl von Meyenn (eds)
Writings on Physics and Philosophy
Telegram to W. Pauli
14 June 1956
Springer-Verlag. Berlin, Germany. 1994

Ruderman, M. A.
No biographical data available
Rosenfeld, A. H.
No biographical data available

Every second, hundreds of billions of these neutrinos pass through each square inch of our bodies, coming from above during the day and from below at night, when the sun is shining on the other side of the earth!
An Elementary Statement on Elementary Particle Physics
American Scientist, Volume 48, Number 2, June 1960 (p. 214)

Stenger, Victor J.   1935–
Physicist

Neutrinos are neither rare nor anomalous — just hard to detect.
Physics and Psychics: The Search for a World Beyond the Senses
Physics and Psychics, Chapter 1 (p. 20)
Prometheus Books. Buffalo, New York, USA. 1990

Updike, John   1932–
American novelist, short story writer and poet

Neutrinos, they are very small.
They have no charge and have no mass
And do not interact at all.
Telephone Poles and Other Poems
Cosmic Gall (p. 4)
Alfred A. Knopf. New York, New York, USA. 1969

NEUTRON

Chadwick, James   1891–1974
English physicist

I think we shall have to make a real search for the neutron.
In Henry Abraham Boorse and Lloyd Motz
The World of the Atom

Letter to E. Rutherford (p. 1293)
Basic Books, Inc., Publishers. New York, New York, USA. 1966

It is to be expected that many of the effects of a neutron in passing through matter should resemble those of a quantum of high energy, and it is not easy to reach the final decision between the two hypotheses. Up to the present, all the evidence is in favor of the neutron, while the quantum hypothesis can only be upheld if the conservation of energy and momentum be relinquished at some point.
Possible Existence of a Neutron *Nature*, Volume 129, Number 3252, February 27, 1932 (p. 312)

Gamow, George   1904–68
Russian-born American physicist

The Neutron has come to be.
Loaded with Mass is he.
Of Charge, forever free.
Pauli, do you agree?
Thirty Years That Shook Physics
Finale (p. 213)
Doubleday & Company, Inc. Garden City, New York, USA. 1966

Heisenberg, Werner Karl   1901–76
German physicist and philosopher

The basic idea is: shove all fundamental difficulties on to the neutron and practice quantum mechanics inside the nucleus.
In Abraham Pais
Inward Bound
Letter to Niels Bohr, 20 June 1932 (p. 413)
Clarendon Press. Oxford, England. 1986

Pauli, Wolfgang   1900–58
Austrian-born physicist

Dear Radioactive Ladies and Gentlemen,

As the bearer of these lines, for whom I pray the favor of a hearing will explain in more detail, I have…hit upon a desperate remedy for rescuing the "alternation law"….
This is the possibility that there might exist in the nuclei electrically neutral particles, which I call neutrons…
In Charles P. Enz
No Time to Be Brief
Chapter 6 (p. 215)
Oxford University Press, Inc. Oxford, England. 2002

NEWTONIAN MECHANICS

Davies, Paul Charles William   1946–
British-born physicist, writer, and broadcaster

However much we may feel free, everything that we do is, according to Laplace, completely determined. Indeed the entire cosmos is reduced to a gigantic clockwork mechanism, with each component slavishly and unfailingly executing it preprogrammed instructions to

mathematical precision. Such is the sweeping implication of Newtonian mechanics.
The Cosmic Blueprint: New Discoveries in Nature's Creative Ability to Order the Universe
Chapter 2 (p. 11)
Simon & Schuster. New York, New York, USA. 1988

NIGHT

Ackerman, Diane   1948–
American writer

It is nighttime on the planet Earth. But that is only a whim of nature, a result of our planet rolling in space at 1,000 miles per minute. What we call "night" is the time we spend facing the secret reaches of space, where other solar systems and, perhaps, other planetarians dwell. Don't think of night as the absence of day; think of it as a kind of freedom. Turned away from our sun, we see the dawning of far-flung galaxies. We are no longer sun-blind to the star-coated universe we inhabit.
A Natural History of the Senses
Vision, How to Watch the Sky (p. 245)
Random House, Inc. New York, New York, USA. 1990

Amaldi, Ginestra Giovene

The night sky looks like a giant fistful of glittering diamonds flung carelessly upon a black carpet.
Our World and the Universe Around Us (Volume 1)
The Universe (p. 13)
Abradale Press. New York, New York, USA. 1966

Atwood, Margaret   1939–
Canadian poet, novelist, and critic

Night falls. Or has fallen. Why is it that night falls, instead of rising, like the dawn? Yet if you look east, at sunset, you can see night rising, not falling; darkness lifting into the sky, up from the horizon, like a black sun behind cloud cover. Like smoke from an unseen fire, a line of fire just below the horizon, brushfire or a burning city. Maybe night falls because it's heavy, a thick curtain pulled up over the eyes. Wool blanket. I wish I could see in the dark.
The Handmaid's Tale
Chapter 30 (p. 201)
Houghton Mifflin Company. Boston, Massachusetts, USA. 1986

Beston, Henry   1888–1968
American writer

With lights and ever more lights, we drive the holiness and beauty of night back to the forests and the sea; the villages, the crossroads even, will have none of it.
The Outermost House
Chapter VIII (p. 168)
Rinehart & Company. New York, New York, USA. 1928

For a moment of night we have a glimpse of ourselves and of our world islanded in its stream of stars — pilgrims

of mortality, voyaging between horizons across the eternal seas of space and time.
The Outermost House
Chapter VIII (p. 176)
Rinehart & Company. New York, New York, USA. 1928

It is dark to-night, and over the plains of ocean the autumnal sky rolls up the winter stars.
The Outermost House
Chapter I (p. 18)
Rinehart & Company. New York, New York, USA. 1928

…today's civilization is full of people who have never even seen night. Yet to live thus, to know only artificial night, is as absurd and evil as to know only artificial day.
The Outermost House
Chapter VIII (p. 169)
Rinehart & Company. New York, New York, USA. 1928

de Saint-Exupéry, Antoine   1900–44
French aviator and writer

Night, the beloved. Night, when words fade and things come alive. When the destructive analysis of day is done, and all that is truly important becomes whole and sound again. When man reassembles his fragmentary self and grows with the calm of a tree.
Translated by Bernard Lamotte
Flight to Arras
Chapter I (p. 23)
Reynal & Hitchcock. New York, New York, USA. 1942

Diamond, Neil   1941–
American pop/folk singer, composer, and musician

I thank the Lord for the night time
To forget the day…
Classics: The Early Years (1966–1967)
Thank the Lord for the Nighttime
Columbia Records #38792. 1983

Murdin, Paul
British astronomer

Astronomers, literally, and human beings in general, figuratively, need the interruption of the night.
In Derek McNally
The Vanishing Universe
The Aims of Astronomy in Science and the Humanities: Why Astronomy Must Be Protected (p. 19)
Cambridge University Press. Cambridge, England. 1994

Stevenson, Robert Louis   1850–94
Scottish essayist and poet

Night is a dead monotonous period under a roof; but in the open world it passes lightly, with its stars and dews and perfumes, and the hours are marked by changes in the face of Nature.
Travels with a Donkey in the Cevennes
A Night Among the Pines (p. 79)
C. Kegan Paul & Company. London, England. 1879

Tagore, Rabindranath   1861–1941
Indian poet and philosopher

The night is like a dark child just born of her mother day. Millions of stars crowding round its cradle watch it, standing still, afraid lest it should wake up.
Personality
The World of Personality (p. 57)
The Macmillan Company. New York, New York, USA. 1917

NOCTURNAL

Holland, W. J.
No biographical data available

There are whole armies of living things, which when we go to sleep, begin to awaken; and when we awaken, go to sleep.
The Moth Book: A Popular Guide to a Knowledge of the Moths of North America
The World of the Dark (p. 77)
Doubleday, Page & Company. New York, New York, USA. 1904

There are two worlds; the world of sunshine, and the world of the dark. Most of us are more or less familiarly acquainted with the first; very few of us are well acquainted with the latter. Our eyes are well adapted to serve us in the daylight, but they do not serve us as well in the dark, and we therefore fail to know, unless we patiently study them, what wonders this world of the dark holds within itself.
The Moth Book: A Popular Guide to a Knowledge of the Moths of North America
The World of the Dark (p. 77)
Doubleday, Page & Company. New York, New York, USA. 1904

NONSENSE

Eddington, Sir Arthur Stanley   1882–1944
English astronomer, physicist, and mathematician

That which in the physical world shadows the nonsense in the mind affords no ground for its condemnation. In a world of aether and electrons we might perhaps encounter nonsense; we could not encounter damned nonsense.
The Nature of the Physical World
Conclusion (p. 345)
The Macmillan Company. New York, New York, USA. 1930

NOTATION

Brough, J. C.
No biographical data available

Though Frankland's notation commands admiration,
As something exceedingly clever,
And Mr. Kay Shuttleworth praises its subtle worth,
I give it up sadly for ever:
Its brackets and braces, and dashes and spaces,

And letters decreased and augmented
Are grimly suggestive of Lunes to make restive
A chemical printer demented.
I've tried hard, but vainly, to realize plainly
Those bonds of atomic connexion
Which Crum Brown's clear vision discerns with precision
Projecting in every direction.
In C.A. Russell
The History of Valencey
Chapter V (p. 106)
Humanities Press. New York, New York, USA. 1971

Cajori, Florian   1859–1930
Swiss-born American educator and mathematician

The miraculous powers of modern calculation are due to three inventions: the Arabic Notation, Decimal Fractions, and Logarithms.
A History of Mathematics
Europe During the Sixteenth, Seventeenth and Eighteenth Centuries (p. 149)
The Macmillan Company. London, England. 1919

Defoe, Daniel   1660–1731
English pamphleteer, journalist, and novelist

I cut every day a notch with my knife, and every seventh notch was as long again as the rest, and every first day of the month, as long again as that long one; and thus I kept my calendar, or weekly, monthly, and yearly reckoning of time.
Robinson Crusoe (p. 46)
Dodd, Mead & Company. New York, New York, USA. 1946

Dieudonné, Jean   1906–92
French mathematician and educator

This difficulty lead very gradually to the recognition of the need for a shorthand to make the sequence of operations easily comprehensible: here we have the problem of notation, which crops up again after every introduction of new objects, and which will probably never cease to torment mathematicians.
Mathematics — The Music of Reason
Chapter III, Section 7 (p. 49)
Springer-Verlag. Berlin, Germany. 1992

Frayn, Michael   1933–
English dramatist

We look at the taciturn, inscrutable universe, and cry, "Speak to me!"
Constructions
Number 7
Wildwood House. London, England. 1974

Glaisher, James Whitbread Lee   1848–1928
English mathematician

I have great faith in the power of well-chosen notation to simplify complicated theories and to bring remote ones near and I think it is safe to predict that the increased

knowledge of principles and the resulting improvements in the symbolic language of mathematics will always enable us to grapple satisfactorily with the difficulties arising from the mere extent of the subject.
Presidential Address, British Association for the Advancement of Science *Nature*, Section A, Volume 42, Number 1089, September 11, 1890 (p. 466)

Holland, John   1929–
American computer scientist

Mathematical notation is for the scientist what musical notation is for the composer.
Emergence: From Chaos to Order
Chapter 1 (p. 15)
Addison-Wesley Publishing, Inc. Reading, Massachusetts, USA. 1998

Russell, Bertrand Arthur William   1872–1970
English philosopher, logician, and social reformer

…for a good notation has a subtlety and suggestiveness which make it seem, at times, like a live teacher. Notational irregularities are often the first sign of philosophical errors, and a perfect notation would be a substitute for thought.
In Ludwig Wittgenstein
Tractatus Logico-Philosophicus
Introduction (pp. 17–18)
Routledge & Kegan Paul Ltd. London, England. 1922

Whitehead, Alfred North   1861–1947
English mathematician and philosopher

Before the introduction of the Arabic notation, multiplication was difficult, and the division even of integers called into play the highest mathematical faculties. Probably nothing in the modern world could have more astonished a Greek mathematician than to learn that, under the influence of compulsory education, the whole population of Western Europe, from the highest to the lowest, could perform the operation of division for the largest numbers. This fact would have seemed to him a sheer impossibility.… Our modern power of easy reckoning with decimal fractions is the most miraculous result of a perfect notation.
An Introduction to Mathematics
Chapter 5 (p. 39)
Oxford University Press, Inc. New York, New York, USA. 1958

NOVAE

Gaposchkin, Sergei   1898–1984
Russian-born astronomer

When the greatest, the cosmic,
And the most fascinating explosion
Has been probed by the fabulous light
Of the human (but god-like) mind,
It will be remembered
That you shouldered the task
Of exploring the Novae

With intrepid boldness
And richness of thought.
In Arthur Beer
Vistas in Astronomy (Volume 2)
Novae Observed (p. 1506)
Pergamon Press. New York, New York, USA. n.d.

NUCLEUS

Cudmore, Lorraine Lee
American cell biologist

Despite its comparatively prosaic name, like the unassuming nanny who turns out to be the head of the espionage network, the nucleus is the structural and actual center of the cell.
The Center of Life: A Natural History of the Cell
Cellular Evolution (p. 50)
New York Times Book Company. New York, New York, USA. 1977

Rutherford, Ernest   1871–1937
English physicist

It is my personal conviction that if we knew more about the nucleus, we should find it much simpler than we suppose. I am always a believer in simplicity, being a simple fellow.
Guttingen Lecture
December 14, 1931

NULL HYPOTHESIS

Dunnette, Marvin D.

…most of us still remain content to build our theoretical castles on the quicksand of merely rejecting the null hypothesis.
Fads, Fashions, and Folderol in Psychology
American Psychologist, Volume 21, 1966 (p. 345)

Fisher, Sir Ronald Aylmer   1890–1962
English statistician and geneticist

In relation to any experiment we may speak of this hypothesis as the "null hypothesis," and it should be noted that the null hypothesis is never proved or established, but is possibly disproved, in the course of experimentation. Every experiment may be said to exist only in order to give the facts a chance of disproving the null hypothesis.
The Design of Experiments
II, 8 (p. 16)
Hafner Publishing Company. New York, New York, USA. 1971

Tukey, John W.   1915–2000
American statistician

The worst, i.e., most dangerous, feature of "accepting the null hypothesis" is the giving up of explicit

uncertainty.... Mathematics can sometimes be put in such black-and-white terms, but our knowledge or belief about the external world never can.
The Philosophy of Multiple Comparisons
Statistical Science, Volume 6, Number 1, February 1991 (p. 100–101)

NUMBER

Aeschylus   525 BCE–426 BCE
Greek playwright

...but utterly without knowledge
Moiled, until I the rising of the stars
Showed them, and when they set, Though much obscure.
Moreover, number, the most excellent
Of all inventions, I for them devised,
And gave them writing that retaineth all,
The serviceable mother of the Muse.
In *Great Books of the Western World* (Volume 5)
The Plays of Aeschylus
Prometheus Bound, 457
Encyclopædia Britannica, Inc. Chicago, Illinois, USA. 1952

Archimedes of Syracuse   287 BCE–212 BCE
Sicilian mathematician

There are some, King Gelon, who think that the number of the sand is infinite in multitude; and I mean by the sand not only that which exists about Syracuse and the rest of Sicily but also that which is found in every region inhabited or uninhabited. Again there are some who, without regarding it as infinite, yet think that no number has been named which is great enough to exceed its multitude.... But I will try to show you by means of geometrical proofs, which you will be able to follow, that, of the numbers named by me, and given in the work which I sent to Zeuxippus, some exceed not only the number of the mass of sand equal in magnitude to the earth, but also that of a mass equal in magnitude to the universe.
The Works of Archimedes
The Sand-Reckoner
At The University Press. Cambridge, England. 1897

Aristotle   384 BCE–322 BCE
Greek philosopher

...the attributes of numbers are present in a musical scale and in the heavens...
In *Great Books of the Western World* (Volume 8)
Metaphysics
Book XIV, Chapter 3 (p. 622)
Encyclopædia Britannica, Inc. Chicago, Illinois, USA. 1952

Asimov, Isaac   1920–92
American author and biochemist

Human beings are very conservative in some ways and virtually never change numerical conventions once they grow used to them. They even come to mistake them for laws of nature.
Foundation and Earth (p. 376)
Doubleday & Company, Inc. Garden City, New York, USA. 1986

Auster, Paul   1947–
American writer

I've dealt with numbers all my life, of course, and after a while you begin to feel that each number has a personality of its own. A twelve is very different from a thirteen, for example. Twelve is upright, conscientious, intelligent, whereas thirteen is a loner, a shady character who won't think twice about breaking the law to get what he wants. Eleven is tough, an outdoorsman who likes tramping through woods and scaling mountains; ten is rather simpleminded, a bland figure who always does what he's told; nine is deep and mystical, a Buddha of contemplation.... Numbers have souls, and you can't help but get involved with them in a personal way.
The Music of Chance
Chapter 4 (p. 73)
Viking Penguin. New York, New York, USA. 1990

Bacon, Sir Francis   1561–1626
English lawyer, statesman, and essayist

...insomuch as we see in the schools both of Democritus and of Pythagoras, that the one did ascribe figure to the first seeds of things, and the other did suppose numbers to be the principles and originals of things.
In *Great Books of the Western World* (Volume 30)
The Advancement of Learning
Second Book, Chapter VIII, Section 1 (p. 46)
Encyclopædia Britannica, Inc. Chicago, Illinois, USA. 1952

Barrett-Browning, Elizabeth   1806–61
English poet

How do I love thee?
Let me count the ways.
The Complete Poetical Works of Elizabeth Barrett Browning
Sonnets from the Portuguese, XLIII
Houghton Mifflin Company. Boston, Massachusetts, USA. 1900

Begley, Sharon   1956–
Science editor

...number theory...is a field of almost pristine irrelevance to everything except the wondrous demonstration that pure numbers, no more substantial than Plato's shadows, conceal magical laws and orders that the human mind can discover after all.
New Answer for an Old Question
Newsweek, 5 July, 1993 (p. 53)

Bell, E. T. (Eric Temple)   1883–1960
Scottish-American mathematician and educator

Roughly it amounts to this: mathematical analysis as it works today must make use of irrational numbers (such

as the square root of two); the sense if any in which such numbers exist is hazy. Their reputed mathematical existence implies the disputed theories of the infinite. The paradoxes remain. Without a satisfactory theory of irrational numbers, among other things, Achilles does not catch up with the tortoise, and the earth cannot turn on its axis. But as Galileo remarked, it does. It would seem to follow that something is wrong with our attempts to compass the infinite.
Debunking Science
University of Washington Book Store. Seattle, Washington, USA. 1930

The algebraic numbers are spotted over the plane like stars against a black sky; the dense blackness is the firmament of the transcendentals.
Men of Mathematics (p. 569)
Simon & Schuster. New York, New York, USA. 1937

The next fundamental assumption of the Pythagoreans lies much deeper, so deep in fact that civilized man can scarcely hope to fetch it up to the full light of reason. Odd numbers are male; even numbers, female. We can only ask why, expecting no answer except possibly a hesitant allusion to a vestigial phallicism or a forgotten Orphism.
The Magic of Numbers
Chapter 14 (p. 155)
McGraw-Hill Book Company, Inc. New York, New York, USA. 1946

The theory of numbers is the last great uncivilized continent of mathematics. It is split up into innumerable countries, fertile enough in themselves, but all the more or less indifferent to one another's welfare and without a vestige of a central, intelligent government. If any young Alexander is weeping for a new world to conquer, it lies before him.
The Queen of the Sciences
Chapter VII (p. 91)
The Williams & Wilkins Company. Baltimore, Maryland, USA. 1931

Berry, Daniel M.
No biographical data available
Yavne, Moshe
No biographical data available

In the beginning, everything was void, and J. H. W. H. Conway began to create numbers. Conway said, "Let there be two rules which bring forth all numbers large and small. This shall be the first rule: Every number corresponds to two sets of previously created numbers, such that no member of the left set is greater than or equal to any member of the right set.

And the second rule shall be this: One number is less than or equal to another number if and only if no member of the first number's left set is greater than or equal to the second number, and no member of the second number's right set is less than or equal to the first number." And Conway examined these two rules he had made, and behold! they were very good.

The Conway Stones: What the Original Hebrew May Have Been
Mathematics Magazine, Volume 49, Number 4, September 1976 (p. 208)

Boethius   ca. 475–524
Roman philosopher and statesman

…in the science of numbers ought to be preferred as an acquisition before all others, because of its necessity and because of the great secrets and other mysteries which there are in the properties of numbers. All sciences partake of it, and it has need of none.
In J Fauvel and J Gray
The History of Mathematics: A Reader
Chapter 7
Section 7.B2 (p. 247)
Macmillan Press Ltd. Houndmills, England. 1987

Borel, Félix Edouard Justin Emile   1871–1956
French mathematician

One grain of wheat does not constitute a pile, nor do two grains, nor three and so on. On the other hand, everyone will agree that a hundred million grains of wheat do form a pile. What, then, is the threshold number? Can we say that 325,647 grains of wheat do not form a pile, but that 325,648 grains do? If it is impossible to fix a threshold number, it will also be impossible to know what is meant by a pile of wheat; the words can have no meaning, although, in certain extreme cases, everybody will agree about them.
Translated by Douglas Scott
Probability and Certainty
Chapter 8 (p. 98)
Walker & Company. New York, New York, USA. 1963

All of mathematics can be deduced from the sole notion of an integer; here we have a fact universally acknowledged today.
A l'analyse arithmetique du continu
Oeuvres 3, 1439–1485
Publisher undetermined

Borges, Jorge Luis   1899–1986
Argentine writer

The man who has learned that three plus one are four doesn't have to go through a proof of that assertion with coins, or dice, or chess pieces, or pencils. He knows it, and that's that. He cannot conceive a different sum. There are mathematicians who say that three plus one is a tautology for four, a different way of saying "four"… if three plus one can be two, or fourteen, then reason is madness.
Translated by Andrew Hurley
Shakespeare's Memory
Blue Tigers
Penguin. 1999

Bridgman, Percy Williams   1882–1961
American physicist

Nature does not count nor do integers occur in nature. Man made them all, integers and all the rest, Kroneker to the contrary notwithstanding.
The Way Things Are
Chapter IV (p. 100)
Harvard University Press. Cambridge, Massachusetts, USA. 1959

Buchanan, Scott   1895–1968
American educator and philosopher

Numbers are not just counters; they are elements in a system.
Poetry and Mathematics
Chapter III
The University of Chicago Press. Chicago, Illinois, USA. 1975

The theory of number is the epipoem of mathematics.
Poetry and Mathematics
Chapter III
The University of Chicago Press. Chicago, Illinois, USA. 1975

Burke, Edmund   1729–97
English statesman and philosopher

The starry heaven, though it occurs so very frequently to our view, never fails to excite an idea of grandeur. This cannot be owing to the stars themselves, separately considered. The number is certainly the cause. The apparent disorder augments the grandeur.
On the Sublime and the Beautiful
Part II, Sec. XIII (p. 139)
Printed for F.C. and J. Rivington and others. London, England. 1812

Butterworth, Brian
Neuroscientist

Although the idea that we have no bananas is unlikely to be a new one, or one that is hard to grasp, the idea that no bananas, no sheep, no children, no prospects are really all the same, in that they have the same numerosity, is a very abstract one.
The Mathematical Brain
Macmillan & Company Ltd. London, England. 1999

Clawson, Calvin C.
No biographical data available

Numbers, in fact, are the atoms of the universe, combining with everything else.
Mathematical Mysteries: The Beauty and Magic of Numbers
Chapter One (p. 22)
Plenum Press. New York, New York, USA. 1996

Comte, Auguste   1798–1857
French philosopher

There is no inquiry which is not finally reducible to a question of Numbers; for there is none which may not be conceived of as consisting in the determination of quantities by each other, according to certain relations.
The Positive Philosophy of Auguste Comte (Volume 1)
Book I, Chapter I (pp. 42–43)
John Chapman. London, England. 1853

Dantzig, Tobias   1884–1956
Russian mathematician

To attempt to apply rational arithmetic to a problem in geometry resulted in the first crisis in the history of mathematics. The two relatively simple problems — the determination of the diagonal of a square and that of the circumference of a circle — revealed the existence of new mathematical beings for which no place could be found within the rational domain.
In Eli Maor
To Infinity and Beyond: A Cultural History of the Infinite (p. 44)
Birkhäuser. Boston, Massachusetts, USA. 1987

Davies, Paul Charles William   1946–
British-born physicist, writer, and broadcaster

At this stage it might well seem that the infinity of natural numbers really is so big that it cannot be made any bigger, but this is wrong. In a celebrated theorem, Georg Cantor proved the seemingly impossible — that there are infinite sets which are so big that their elements cannot be counted, even with the infinity of natural numbers at one's disposal.
The Edge of Infinity
Chapter 2 (p. 30)
Simon & Schuster. New York, New York, USA. 1981

Dawkins, Richard   1941–
British ethologist, evolutionary biologist, and popular science writer

You can be moved to tears by numbers — provided they are encoded and decoded fast enough.
River Out of Eden: A Darwinian View of Life
Chapter 1 (p. 14)
Basic Books. New York, New York, USA. 1995

Dedekind, Richard   1831–1916
German mathematician

I regard the whole of arithmetic as a necessary, or at least natural, consequence of the simplest arithmetical act, that of counting, and counting itself as nothing else than the successive creation of the infinite series of positive integers in which each individual is defined by the one immediately preceding…
Translated by Wooster Woodruff Beman
Essays on the Theory of Numbers
Chapter I (p. 4)
The Open Court Publishing Company. Chicago, Illinois, USA. 1901

…numbers are free creations of the human mind; they serve as a means of apprehending more easily and more sharply the difference of things.
Translated by Wooster Woodruff Beman
Essays on the Theory of Numbers
The Nature and Meaning of Numbers
Preface to the First Edition (p. 31)
The Open Court Publishing Company. Chicago, Illinois, USA. 1901

Eco, Umberto   1932–
Italian novelist, essayist, and scholar

With numbers you can do anything you like. Suppose I have the sacred number 9 and I want to get a number 1314, date of the execution of Jacques de Molay — a date dear to anyone who, like me, professes devotion to the Templar tradition of knighthood. What do I do? I multiply nine by one hundred and forty-six, the fateful day of destruction of Carthage. How do I arrive at this? I divided thirteen hundred and fourteen by two, by three, *et cetera*, until I found a satisfying date. I could also have divided thirteen hundred and fourteen by 6.28, the double of 3.14, and I would have got two hundred and nine. That is the year Attalus I, king of Pergamon, ascended the throne. You see?
Translated by William Weaver
Foucault's Pendulum
Chapter 48 (pp. 288–289)
Harcourt Brace Jovanovich. San Diego, California, USA. 1988

Enzensberger, Hans Magnus   1929–
German writer, poet, translator, and editor

Still in a daze the next morning, Robert said to his mother, "Do you know the year I was born? It was 6×1 and 8×10 and 9×100 and 1×1000." I don't know what's got into the boy lately," said Robert's mother, shaking her head. "Here," she added, handing him a cup of hot chocolate, "maybe this will help. You say the oddest things." Robert drank his hot chocolate in silence. There are some things you can't tell your mother, he thought.
The Number Devil
The Second Night (p. 46)
Henry Holt & Company. New York, New York, USA. 1998

Euler, Leonhard   1707–83
Swiss mathematician and physicist

…we should take great care not to accept as true such properties of the numbers which we have discovered by observation and which are supported by induction alone. Indeed, we should use such a discovery as an opportunity to investigate more exactly the properties discovered and to prove or disprove them; in both cases we may learn something useful.
In G. Polya
Induction and Analogy in Mathematics (Volume 1)
Chapter I (p. 3)
Princeton University Press. Princeton, New Jersey, USA. 1954

Euripides   ca. 480 BCE–406 BCE
Greek playwright

Numbers are a fearful thing, and joined to craft a desperate foe.
In *Great Books of the Western World* (Volume 5)
The Plays of Euripides
Hecuba, l. 884
Encyclopædia Britannica, Inc. Chicago, Illinois, USA. 1952

Fabilli, Mary
No biographical data available

What would I do
without numbers?
A 7 there and a 3 here,
days in a month
months in a year
AD and BC
and all such symbols.
In Ernest Robson and Jet Wimp
Against Infinity
Numbers (p. 22)
Primary Press. Parker Ford, Pennsylvania, USA. 1979

Ferguson, Kitty
Science writer

Letting numbers take us where we can't go in person — whether that's to the top of a windmill or to the origin and borders of the universe — has been and still is one of humankind's favorite intellectual adventures.
Measuring the Universe: Our Historic Quest to Chart the Horizons of Space and Time
Prologue (p. 3)
Walker & Company. New York, New York, USA. 1999

Feynman, Richard P.   1918–88
American theoretical physicist

You know…there are about a hundred billion stars in the galaxy — ten to the eleventh power. That used to be considered a huge number. Today it's less than the national debt. We ought to call them "economical numbers."
In David L. Goodstein and Judith R. Goodstein
Feynman's Lost Lecture: The Motion of Planets Around the Sun
Chapter 2 (p. 62)
W.W. Norton & Company, Inc. New York, New York, USA. 1996

Gass, Fredrick
No biographical data available

Adam, did you find a good system for naming ordinals?
A: Ordinals? I thought you said "animals."
Constructive Ordinal Notation Systems
Mathematics Magazine, Volume 57, Number 3, May 1984 (p. 131)

Gibran, Kahlil   1883–1931
Lebanese-American philosophical essayist

And he who is versed in the science of numbers can tell of the regions of weight and measures, but he cannot conduct you thither.

For the vision of one man lends not its wings to another man.
The Prophet
On Teaching (pp. 56–57)
Alfred A. Knopf. New York, New York, USA. 1969

Ginsey, Gurney
No biographical data available

There are numbers I don't trust…take
One — too proud, too pointed, much too

Sure that all begins with him — and
Two, that sits cross-legged on the
Path and sneers as though he knows all secrets…
Numbers
Mathematics Magazine, Volume 38, Number 3, May 1965 (p. 168)

Gould, Stephen Jay   1941–2002
American paleontologist and evolutionary biologist

Numbers have undoubted powers to beguile and benumb, but critics must probe behind numbers to the character of arguments and the biases that motivate them.
An Urchin in the Storm: Essays About Books and Ideas
Chapter 8 (p. 144)
W.W. Norton & Company, Inc. New York, New York, USA. 1987

Hales, Stephen   1677–1761
English physiologist and clergyman

…since we are assured that the all-wise Creator has observed the most exact proportions, of number, weight, and measure, in the make of all things; the most likely way therefore, to get any insight into the nature of these parts of the creation, which come within our observation, must in all reason be to number, weigh and measure.
Vegetable Staticks
The Introduction (p. xxxi)
The Scientific Book Guild. London, England. 1961

Hardy, G. H. (Godfrey Harold)   1877–1947
English pure mathematician

The elementary theory of numbers should be one of the very best subjects for early mathematical instruction. It demands very little previous knowledge; its subject matter is tangible and familiar; the processes of reasoning which it employs are simple, general and few; and it is unique among the mathematical sciences in its appeal to natural human curiosity. A month's intelligent instruction in the theory of numbers ought to be twice as instructive, twice as useful, and at least ten times as entertaining as the same amount of calculus for engineers.
An Introduction to the Theory of Numbers
Bulletin of the American Mathematical Society, Volume 35, 1929 (p. 818)

Henle, James M.
American mathematician

Belief in a large number is no more daring, I should think, than belief in Tolstoy's *War and Peace*.
The Happy Formalist
The Mathematical Intelligencer, Volume 13, Number 1, Winter 1991 (p. 14)

Hofstadter, Douglas   1945–
American cognitive scientist and author

People enjoy inventing slogans which violate basic arithmetic but which illustrate "deeper" truths, such as "1 and 1 make 1" (for lovers), or "1 plus 1 plus 1 equals 1" (the Trinity).… Two raindrops running down a window pane merge; does one plus one make one? A cloud breaks up into two clouds — more evidence for the same? …Numbers as realities misbehave. However, there is an ancient and innate sense in people that numbers ought not to misbehave. There is something clean and pure in the abstract notion of number…and there ought to be a way of talking about numbers without always having the silliness of reality come in and intrude. The hard-edged rules that govern "ideal" numbers constitute arithmetic, and their more advanced consequences constitute number theory.
Godel, Escher, Bach: An Eternal Golden Braid
Part I, Chapter II (p. 56)
Basic Books, Inc. New York, New York, USA. 1979

Holmes, Oliver Wendell   1809–94
American physician, poet, and humorist

Dr. Hooke, the famous English mathematician and philosopher, made a calculation of the number of separate ideas the mind is capable of entertaining, which he estimated as 3,155,760,000.
Pages from an Old Volume of Life
Chapter VIII (p. 274, fn a)
Houghton Mifflin Company. Boston, Massachusetts, USA. 1895

Hume, David   1711–76
Scottish philosopher and historian

…I observe that when we mention any great number, such as a thousand, the mind has generally no adequate idea of it, but only a power of producing such an idea by its adequate idea of the decimals, under which the number is comprehended.
A Treatise of Human Nature
Book I, Part I, Section VII (p. 70)
Penguin Books. Baltimore, Maryland, USA. 1969

Huxley, Aldous   1894–1963
English writer and critic

A million million spermatozoa,
All of them alive:
Out of their cataclysm but one poor Noah
Dare hope to survive.
And among that billion minus one
Might have chanced to be
Shakespeare, another Newton, a new Donne —
But the One was Me.
Stories, Essays, and Poems
Fifth Philosopher's Song (p. 410)
J.M. Dent & Sons Ltd. London, England. 1937

Jacobi, Karl Gustav Jacob   1804–51
German mathematician

The God that reigns in Olympus is Number Eternal.
In Tobias Dantzig
Number: The Language of Science (4th edition)
Chapter 10 (p. 179)
The Macmillan Company. New York, New York, USA. 1954

Jevons, William Stanley   1835–82
English economist and logician

Number is but another name for diversity.
The Principles of Science: A Treatise on Logic and Scientific Method
Book I, Chapter VIII (p. 156)
Macmillan & Company. London, England. 1887

Johnson, Samuel   1696–1772
English critic, biographer, and essayist

Round numbers, she said, are always false.
In Mrs. Piozzi, Richard Cumberland, Bishop Percy and other book
Johnsoniana
Apothegms, Sentiment, Opinions, &c
G. Bell. London, England. 1884

Juster, Norton   1929–
American architect and writer

"How terribly confusing," he cried. "Everything here is called exactly what it is. The triangles are called triangles, the circles are called circles, and even the same numbers have the same name. Why, can you imagine what would happen if we named all the twos Henry or George or Robert or John or lots of other things? You'd have to say Robert plus John equals four, and if the four's name were Albert, things would be hopeless."
The Phantom Tollbooth
Chapter 14 (pp. 173–174)
Alfred A. Knopf. New York, New York, USA. 1989

Kaminsky, Kenneth
American mathematics professor, writer, and editor

How poor were we? Why, we were so poor we only had imaginary numbers to play with.
Professor Fogelfroe
Mathematics Magazine, Volume 69, Number 4, October 1996 (p. 303)

Kelvin, Lord William Thomson   1824–1907
Scottish engineer, mathematician, and physicist

When you can measure what you are speaking about and express it in numbers you know something about it; but when you cannot express it in numbers, your knowledge is of a meager and unsatisfactory kind: it may be the beginning of knowledge, but you have scarcely, in your thoughts, advanced to the stage of science, whatever the matter may be.
Popular Lectures and Addresses (Volume 1)
Lecture, Institution of Civil Engineering
3 May 1883 (p. 73)
Macmillan & Company Ltd. London, England. 1894

Klein, William
No biographical data available

Numbers are friends, for me, more or less. It doesn't mean the same for you, does it — 3,844? For you it's just a three and an eight and a four and a four. But I say, "Hi! 62 squared."

In Oliver Sacks
The Man Who Mistook His Wife for a Hat and Other Clinical Tales
The Twins (pp. 198–199)
Summit Books. New York, New York, USA. 1985

Kline, Morris   1908–92
American mathematics professor and writer

The theory of infinite numbers is only one of the creations of the nineteenth-century critical thinkers. Almost bizarre in its contents it is nevertheless both logical and useful.
Mathematics in Western Culture
Chapter XXV (p. 409)
Oxford University Press, Inc. New York, New York, USA. 1953

Kronecker, Leopold   1823–91
German mathematician

Number theorists are like lotus-eaters — having once tasted of this food they can never give it up.
In Howard W. Eves
Mathematical Circles (Volume 2)
Mathematical Circles Squared
302 (p. 149)
The Mathematical Association of America, Inc. 2003

Die Ganzen Zahlen hat Gott gemacht, andere ist Menschenwerk.

God created the natural numbers; everything else is man's handiwork.
In Sir Arthur Stanley Eddington
The Nature of the Physical World (p. 246)
At The University Press. Cambridge, England. 1929

Leibniz, Gottfried Wilhelm   1646–1716
German philosopher and mathematician

There is an old saying that God created everything according to weight, measure and number.
Philosophical Papers and Letters (Volume 1)
On the General Characteristic (pp. 339–340)
The University of Chicago Press. Chicago, Illinois, USA. 1956

Some things cannot be weighed, as having no force and power; some things cannot be measured, by reason of having no parts; but there is nothing which cannot be numbered.
In G. Frege
The Foundations of Arithmetic (p. 31e)
Northwestern University Press. Evanston, Illinois, USA. 1968…a
miracle of analysis, a monster of the ideal world, almost an amphibian
between being and not being.
In Walter R. Fuchs
Mathematics for the Modern Mind
Chapter 7, Section 7.1 (p. 168)
Macmillan & Company Ltd. New York, New York, USA. 1967

Lichtenberg, Georg Christoph   1742–99
German physicist and satirical writer

People don't like to choose lot #1 in a lottery. "Choose it," Reason cries loudly. "It has as good a chance of win-

ning the 12,000 thalers as any other." "In Heaven's name don't choose it," *a je ne sais quoi* whispers. "There's no example of such little numbers being listed before great winnings." And actually no one takes it.
Lichtenberg: Aphorisms & Letters
Aphorisms (p. 46)
Jonathan Cape. London, England. 1969

Locke, John   1632–1704
English philosopher and political theorist

For number applies itself to men, angels, actions, thoughts; everything that either doth exist, or can be imagined.
In *Great Books of the Western World* (Volume 35)
An Essay Concerning Human Understanding
Book II, Chapter XVI, Section 1 (p. 165)
Encyclopædia Britannica, Inc. Chicago, Illinois, USA. 1952

Longfellow, Henry Wadsworth   1807–82
American poet

Tell me not, in mournful numbers,
Life is but an empty dream!
The Poetical Works of Henry Wadsworth Longfellow
A Psalm of Life, Stanza 1
Houghton Mifflin Company. Boston, Massachusetts, USA. 1883

Maxwell, James Clerk   1831–79
Scottish physicist

Thus numbers may be said to rule the whole world of quantity, and the four rules of arithmetic may be regarded as the complete equipment of the mathematician.
In E.T. Bell
Men of Mathematics (p. xv)
Simon & Schuster. New York, New York, USA. 1937

Mazur, Barry   1937–
American mathematician

In the history of the concept of number, number has been adjective (three cows, three monads) and noun (three, pure and simple), and now…, number seems to be more like a verb (to triple).
Imagining Numbers
Part II, Chapter 8, Section 37 (p. 138)
Farrar, Straus & Giroux. New York, New York, USA. 2003

Parker, F. W.
No biographical data available

Number was born in superstition and reared in mystery… numbers were once made the foundation of religion and philosophy, and the tricks of figures have had a marvelous effect on a credulous people.
Talks on Pedagogics
Chapter IV (p. 64)
A.S. Barnes & Company. New York, New York, USA. 1909

Paulos, John Allen   1945–
American mathematician

The mathematician G. H. Hardy was visiting his protégé, the Indian mathematician Ramanujan, in the hospital. To make small talk, he remarked that 1729, the number of the taxi which had brought him, was a rather dull number, to which Ramanujan replied immediately, "No, Hardy! It is a very interesting number. It is the smallest number expressible as the sum of two cubes in two different ways.
Innumeracy
Examples and Principles (p. 6)
Hill & Wang. New York, New York, USA. 1988

Peirce, Charles Sanders   1839–1914
American scientist, logician, and philosopher

The rudest numerical scales, such as that by which the mineralogists distinguish different degrees of hardness, are found useful. The mere counting of pistils and stamens sufficed to bring botany out of total chaos into some kind of form. It is not, however, so much from counting as a measuring, not so much from the conception of number as from that of continuous quantity, that the advantage of mathematical treatment comes. Number, after all, only serves to pin us down to a precision in our thoughts which, however beneficial, can seldom lead to lofty conceptions, and frequently descends to pettiness.
Chance, Love and Logic: Philosophical Essays
The Doctrine of Chances (pp. 61–62)
Harcourt, Brace & Company, Inc. New York, New York, USA. 1923

Philolaus   ca. 480 BCE
Greek philosopher

All things which can be known have number; for it is not possible that without number anything can be either conceived or known.
In Carl B. Boyer
A History of Mathematics (p. 60)
John Wiley & Sons, Inc. New York, New York, USA. 1968

Plato   428 BCE–347 BCE
Greek philosopher

SOC: And all arithmetic and calculations have to do with number?
GLAUCON: Yes.
SOC: And they appear to lead the mind towards truth?
GLAUCON: Yes, in a very remarkable manner.
In *Great Books of the Western World* (Volume 7)
The Republic
Book VII, Section 525 (p. 393)
Encyclopædia Britannica, Inc. Chicago, Illinois, USA. 1952

Pliny (C. Plinius Secundus)   23–79
Roman savant and writer

Why do we believe that in all matters the odd numbers are more powerful.
Natural History
Volume 8, Book XXVIII, sec 23
Harvard University Press. Cambridge, Massachusetts, USA. 1947

Plotinus   ca. 205–70
Egyptian-Roman philosopher

Objects of sense are not unlimited and therefore the Number applying to them cannot be so. Nor is an enumerator able to number to infinity; though we double, multiply over and over again, we still end with a finite number…
In *Great Books of the Western World* (Volume 17)
The Six Enneads
Sixth Ennead VI.2 (p. 311)
Encyclopædia Britannica, Inc. Chicago, Illinois, USA. 1952

Pope, Alexander   1688–1744
English poet

As yet a child, nor yet a fool to fame,
I lisp'd in numbers, for the numbers came.
The Complete Poetical Works
An Epistle to Dr. Arbuthnot, l. 127
Houghton Mifflin Company. New York, New York, USA. 1903

Proclus   411–485
Greek philosopher

Wherever there is number, there is beauty.
In Morris Kline
Mathematical Thought from Ancient to Modern Times (p. 131)
Oxford University Press, Inc. New York, New York, USA. 1972

Russell, Bertrand Arthur William   1872–1970
English philosopher, logician, and social reformer

We are the finite numbers.
We are the stuff of the world.
Whatever confusion cumbers
The earth is by us unfurled.
We revere our master Pythagoras
And deeply despise every hag or ass.
Not Endor's witch nor Balaam's mount
We recognize as wisdom's fount.
But round and round in endless baller
We move like comets seen by Halley.
And honored by the immortal Plato
We think no later mortal great-o.
We follow the laws
Without a pause,
For we are the finite numbers.
The Collected Stories of Bertrand Russell
Nightmares of Eminent Persons, The Mathematician's Nightmare (p. 43)
George Allen & Unwin Ltd. London, England. 1972

Sagan, Carl   1934–96
American astronomer and author

Hiding between all the ordinary numbers was an infinity of transcendental numbers whose presence you would never have guessed until you looked deeply into mathematics.
Contact: A Novel
Chapter 1 (p. 21)
Simon & Schuster. New York, New York, USA. 1985

Sandburg, Carl   1878–1967
American poet and biographer

He was born to wonder about numbers.
Complete Poems
Number Man
Harcourt, Brace. New York, New York, USA. 1950

Shakespeare, William   1564–1616
English poet, playwright, and actor

This is the third time; I hope good luck lies in odd numbers…. There is a divinity in odd numbers, either in nativity, chance or death.
In *Great Books of the Western World* (Volume 27)
The Plays and Sonnets of William Shakespeare (Volume 2)
The Merry Wives of Windsor
Act V, Scene i, l. 2–3
Encyclopædia Britannica, Inc. Chicago, Illinois, USA. 1952

…I am ill at these numbers.
In *Great Books of the Western World* (Volume 27)
The Plays and Sonnets of William Shakespeare (Volume 2)
Hamlet, Prince of Denmark
Act II, Scene ii, l. 120
Encyclopædia Britannica, Inc. Chicago, Illinois, USA. 1952

Smith, Adam (George J. W. Goodman)

…those who live by numbers can also perish by them and it is a terrifying thing to have an adding machine write an epitaph, either way.
The Money Game
Chapter 7 (p. 84)
Random House, Inc. New York, New York, USA. 1968

Stoney, George Johnstone   1826–1911
Irish physicist

When interpreting nature's work, we are obliged frequently to speak of high numbers and small fractions.
Annual Report of the Board of Regents of the Smithsonian Institution, 1899
Survey of That Part of the Range of Nature's Operations Which Man Is Competent to Study (p. 207)
Government Printing Office. Washington, D.C. 1901

Sukoff, Albert
No biographical data available

Huge numbers are commonplace in our culture, but oddly enough the larger the number the less meaningful it seems to be…. Anthropologists have reported on the primitive number systems of some aboriginal tribes. The Yancos in the Brazilian Amazon stop counting at three. Since their word for "three" is "*poettarrarorincoaroac*," this is understandable.
Lotsa Hamburgers
Saturday Review of the Society, March 1973 (p. 6)

Synge, John L.   1897–1995
Irish mathematician and physicist

The northern ocean is beautiful, said the Orc, and beautiful the delicate intricacy of the snowflake before it melts and perishes, but such beauties are as nothing to him who delights in numbers, spurning alike the wild irrationality of life and the baffling complexities of nature's laws.
Kandelman's Krim
Chapter Six (p. 101)
Jonathan Cape. London, England. 1957

The Bible

...Mene, mene, tekel and u-pharsin

[Numbered, Numbered, Weighed, Divided]
The Revised English Bible
Daniel 5:25
Oxford University Press, Inc. Oxford, England. 1989

Virgil   70 BCE– 19 BCE
Roman epic, didactic, and idyllic poet

Uneven numbers are the god's delight.
In *Great Books of the Western World* (Volume 13)
The Eclogues
VIII, l. 77
Encyclopædia Britannica, Inc. Chicago, Illinois, USA. 1952

Waismann, Friedrich   1896–1959
Austrian mathematician, physicist, and philosopher

Will anyone seriously assert that the existence of negative numbers is guaranteed by the fact that there exist in the world hot assets and cold, and debts? Shall we refer to these things in the structure of arithmetic? Who does not see that thereby an entirely foreign element enters into arithmetic, which endangers the pureness and clarity of its concepts?
Introduction to Mathematical Thinking: The Formation of Concepts in Modern Mathematics
Chapter 2 (p. 15)
Frederick Ungar Publishing, Company. New York, New York, USA. 1951

Williams, Charles
No biographical data available

Nought usually comes at the beginning, Ralph said.

Not necessarily, said Sibyl. It might come anywhere. Nought isn't a number at all. It's the opposite of number.

Nancy looked up from the cards. Got you, aunt, she said. What about ten? Nought's a number there — it's part of ten.

Well, if you say that any mathematical arrangement of one and nought really makes ten — Sibyl smiled. Can it possibly be more than a way of representing ten?
The Greater Trumps
Victor Gollancz. London, England. 1932

NUMBER, FIBONACCI

Baumel, Judith   1956–
American poet

Learn the particular strength
of the Fibonacci series,
a balanced spiraling
outward of shapes,
those golden numbers
which describe dimensions
of sea shells, rams' horns,
collections of petals
and generations of bees.
The Weight of Numbers
Fibonacci (p. 21)
Wesleyan University Press, Middletown, Connecticut, USA; 1988

Lindon, J. A.
No biographical data available

Each wife of Fibonacci,
Eating nothing that wasn't starchy
Weighed as much as the two before her.
His fifth was some signora!
In Martin Gardner
Mathematical Circus
Chapter 13 (p. 152)
Alfred A. Knopf. New York, New York, USA. 1979

NUMBER THEORY

Barnett, I. A.
No biographical data available

...to discover mathematical talent, there is no better course in elementary mathematics than number theory. Any student who can work the exercises in a modern text in number theory should be encouraged to pursue a mathematical career.
The Theory of Numbers as A Required Course in the College Curriculum for Majors
American Mathematical Monthly, Volume 73, November 1966 (pp. 1002–1003)

Hardy, G. H. (Godfrey Harold)   1877–1947
English pure mathematician

The theory of numbers has always occupied a peculiar position among the purely mathematical sciences. It has the reputation of great difficulty and mystery among many who should be competent to judge; I suppose that there is no mathematical theory of which so many well qualified mathematicians are so much afraid.
The Sixth Josiah Willard Gibbs Lecture
Bulletin of the American Mathematical Society, Volume 35, 1929

Hilbert, David   1862–1943
German mathematician

A problem in number theory is as timeless as a true work of art.
In Legh Wilber Reid
The Elements of the Theory of Algebraic Numbers
Introduction
The Macmillan Company. New York, New York, USA. 1910

Mazur, Barry   1937–
American mathematician

[Number theory] produces, without effort, innumerable problems which have a sweet, innocent air about them, tempting flowers; and yet…number theory swarms with bugs, waiting to bite the tempted flower-lovers who, once bitten, are inspired to excesses of effort!
Number Theory as Gadfly
The American Mathematical Monthly, Volume 98, 1991

NURSING

Anderson, Peggy
No biographical data available

Nurses do whatever doctors and janitors won't do.
Nurse
Chapter 2 (p. 31)
Berkley. New York, New York, USA. 1979

The nurse's job is to help the patients get well, or help them to die.
Nurse
Chapter 1 (p. 20)
Berkley. New York, New York, USA. 1979

…nurse's play the same role on a regular floor that the electrocardiograph plays in the intensive care unit. They're the monitor.
Nurse
Chapter 1 (pp. 20–21)
Berkley. New York, New York, USA. 1979

Barnes, Djuna   1892–1982
American writer

The only people who really know anything about medical science are the nurses, and they never tell; they'd get slapped if they did.
Nightwood
La Somnambule (p. 40)
Harcourt, Brace & Company. New York, New York, USA. 1937

Beckett, Samuel   1906–89
Irish playwright

The patients seeing so much of the nurses and so little of the doctor, it was natural that they should regard the former as their persecutors and the latter as their savior.
The Collected Works of Samuel Beckett
Murphy
Chapter 9 (p. 158)
Grover Press, Inc., New York, New York, USA. 1970

Burns, Olive Ann   1924–90
Professional writer, journalist, and columnist

They ain't no feelin' in the world like takin' on somebody wilted and near bout gone, and you do what you can, and then all a-sudden the pore thang starts to put out new growth and git well.
Cold Sassy Tree
Chapter 3 (p. 12)
Ticknor & Fields. New York, New York, USA. 1984

Chekhov, Anton Pavlovich   1860–1904
Russian author and playwright

A doctor is called in, but a nurse sent for.
Note-Book of Anton Chekhov (p. 122)
B.W. Huebsch, Inc. New York, New York, USA. 1921

Cowper, William   1731–1800
English poet

The nurse sleeps sweetly, hir'd to watch the sick,
Whom snoring she disturbs.
The Poetical Works of William Cowper
The Task
Book I, The Sofa, l. 89–90
John W. Lovell Company. New York, New York, USA. No date

Di Bacco, Babs Z.
No biographical data available

Why modern doctors
Have more leisure time
For golf and cards
And things maritime
Than ever before
In history
While nurses don't,
Is a mystery.
Leisure Gap
American Journal of Nursing, January 1969 (p. 212)

Euripides   ca. 480 BCE–406 BCE
Greek playwright

Better be sick than tend the sick; the first is but a single ill, the last unites mental grief with manual toil.
In *Great Books of the Western World* (Volume 5)
The Plays of Euripides
Hippolytus, l. 186
Encyclopædia Britannica, Inc. Chicago, Illinois, USA. 1952

Hanson, Elayne Clipper
No biographical data available

It seems to be a well known fact
That nurses fairly ooze with tact.
Their smiles so warm,
And full of charm,
Match voices, low
And movements slow.
Then why, if I may venture bold

Are nurses' hands so icy cold?
Paradox
American Journal of Nursing, March 1969 (p. 672)

Jewett, Sarah Orne   1849–1909
American novelist and short story writer

She had no equal in sickness, and knew how to brew every old-fashioned dose and to make every variety of herb-tea, and when her nursing was put to an end by her patient's death, she was commander-in-chief at the funeral.
Deephaven
My Lady Brandon and the Widow Jim (p. 55)
Houghton Mifflin Company. Boston, Massachusetts, USA. 1895

Kipling, Rudyard   1865–1936
British writer and poet

Let us now remember many honourable women,
Such as bade us turn again when we were like to die.
Collected Verse of Rudyard Kipling
Dirge of Dead Sisters
Doubleday, Page. Garden City, New York, USA. 1915

Lewis, Lucille
No biographical data available

The central focus of nursing is to help the person cope with his physiological, psychological, and spiritual reactions to health problems and maintain his integrity in these experiences. To be therapeutic, the nurse must contribute to the wholeness of man, to the interrelationships of the parts to the whole, to the person's here-and-now as well as to his future, to the health of all the parts so that the person may attain and maintain his highest potential. All of these are essential.
This I Believe
Nursing Outlook, Volume 15, Number 5, May 1968 (p. 27)

Lydston, George Frank   1858–1923
American urologist

How absurd the situation! It is demanded that a woman should slave for three years and go through what would be a fair medical course, were it conscientiously given, for the privilege of finally earning a salary which, were she constantly employed — which, by the way, she never is — would just equal that of a good stenographer.
The Training School Fake and Its Victims
New York Medical Journal, Volume 79, 1904

Manfreda, Margurite Lucy
No biographical data available

People turn to God in times of crisis, and illness is among those times when people feel the need for spiritual guidance. Nurses, therefore, are in a unique position to bring spiritual aid to their patients and to the patients' families.
In Sharon Fish and Judith Allen Shelly
Spiritual Care: The Nurse's Role

Chapter I (p. 17)
InterVarsity Press. Dover Grove, Illinois, USA. 1978

Matthews, Marian
No biographical data available

Where are the interns I recall?
Fountains of wisdom, one and all.
Men in white, mature and strong —
They were the Doctors, never wrong.
Something happened to them, or me;
They're not the giants they used to be.
With stethoscopes like shiny toys
They seem to me like little boys.
Excuse me for crying on your shoulder —
They're not younger — I am older.
Interns
American Journal of Nursing, November 1968 (p. 2492)

Mayo, Charles Horace   1865–1939
American physician

The trained nurse has given nursing the human, or shall we say, the divine touch, and made the hospital desirable for patients with serious ailments regardless of their home advantages.
The Trained Nurse
Collected Papers of the Mayo Clinic & Mayo Foundation,
Volume 13, 1921

Nightingale, Florence   1820–1910
English nursing pioneer and statistician

For us who Nurse, our Nursing is a thing, which, unless in it we are making progress every year, every month, every week, — take my word for it we are going back.
Address
"Nightingale Fund" School
St Thomas Hospital, 1872

Conceit and Nursing cannot exist in the same person any more than new patches on an old garment.
Address
"Nightingale Fund" School
St. Thomas Hospital, 1872

The most important practical lesson that can be given to nurses is to teach them what to observe — how to observe — what symptoms indicate improvement — what the reverse — which are of importance — which are of none — which are the evidence of neglect — and of what kind of neglect.
Notes on Nursing: What It Is and What It Is Not
Chapter XIII (p. 59)
Harrison. London, England. 1859

I use the word nursing for want of a better. It has been limited to signify little more than the administration of medicines and the application of poultices. It ought to signify the proper use of fresh air, light, warmth, cleanliness, quiet, and the proper selection and administration of diet — all at

the least expense of vital power to the patient. It has been said and written scores of times, that every woman makes a good nurse. I believe, on the contrary, that the very elements of nursing are all but unknown.
Notes on Nursing: What It Is and What It Is Not
Notes on Nursing: What It Is and What It Is Not (p. 6)
Harrison. London, England. 1859

In watching disease, both in private houses and in public hospitals, the thing which strikes the experienced observer most forcibly is this, that the symptoms or the sufferings generally considered to be inevitable and incident to the disease are very often not symptoms of the disease at all, but of something quite different — of the want of fresh air, or of light, or of warmth, or of quiet, or of cleanliness, or of punctuality and care in the administration of diet, of each or of all of these. And this quite as much in private as in hospital nursing.
Notes on Nursing: What It Is and What It Is Not
Notes on Nursing: What It Is and What It Is Not (p. 5)
Harrison. London, England. 1859

Never to allow a patient to be wakened, intentionally or accidentally, is a sine qua non of all good nursing.
Notes on Nursing: What It Is and What It Is Not
Noise (p. 25)
Harrison. London, England. 1859

It seems a commonly received idea among men and even among women themselves that it requires nothing but a disappointment in love, the want of an object, a general disgust, or incapacity for other things to turn a woman into a good nurse.
Notes on Nursing: What It Is and What It Is Not
Conclusion (p. 75)
Harrison. London, England. 1859

A nurse who rustles (I am speaking of nurses professional and unprofessional) is the horror of a patient, though perhaps he does not know why.
Notes on Nursing: What It Is and What It Is Not
Noise (p. 27)
Harrison. London, England. 1859

Osler, Sir William   1849–1919
Canadian physician and professor of medicine

…the trained nurse has become one of the great blessings of humanity, taking a place beside the physician and the priest, and not inferior to either in her mission.
Aequanimitas, with Other Addresses to Medical Students, Nurses, and Practitioners of Medicine
Nurse and Patient (p. 156)
The Blakiston Company. Philadelphia, Pennsylvania, USA. 1932

Ray, John   1627–1705
English naturalist

A nurse's tongue is privileged to talk.
A Complete Collection of English Proverbs (p. 17)
Printed for G. Cowie. London, England. 1813

Richardson, Samuel   1689–1761
English novelist

Male nurses are unnatural creatures!
Sir Charles Grandison
Part 2, Volume 3, Letter XI (p. 58)
G. Allen. London, England. 1895

Roosevelt, Franklin Delano   1882–1945
32nd president of the United States

…I urge that the Selective Service Act be amended to provide for the induction of nurses into the armed forces…
Annual Message to Congress
January 6, 1945

Schmitz, Jacqueline T.
No biographical data available

Compressing an hour into a half,
Busy, yet heedful
Rushes the nurse
Bringing earnest solace to the
Sick and the needful.
Who must come first? How can she know?
Inverted scope, focused by Death,
Grants new perspective as He robs breath.
Point of View
Journal of Nursing, November 1968 (p. 2492)

NUTRITION

Davis, Adelle   1904–74
Nutritionist

Nutrition is a young subject; it has long been kicked around like a puppy that cannot take care of itself. Food faddists and crackpots have kicked it pretty cruelly.… They seem to believe that unless food tastes like Socratic hemlock, it cannot build health. Frankly, I often wonder what such persons plan to do with good health in case they acquire it.
Let's Eat Right to Keep Fit
Chapter 1 (p. 3)
New American Library. New York, New York, USA. 1970

…eat breakfast like a king, lunch like a prince, and dinner like a pauper.
Let's Eat Right to Keep Fit
Chapter 2 (p. 19)
New American Library. New York, New York, USA. 1970

O

OBJECTIVITY

Gould, Stephen Jay   1941–2002
American paleontologist and evolutionary biologist

Objectivity is not an unobtainable emptying of mind, but a willingness to abandon a set of preferences — for or against some view, as Darwin said — when the world seems to work in a contrary way.
Dinosaur in a Haystack: Reflections in Natural History
Part Three, Chapter 11 (p. 136)
Random House, Inc. New York, New York, USA. 1995

Objectivity cannot be equated with mental blankness; rather, objectivity resides in recognizing your preferences and then subjecting them to especially harsh scrutiny — and also in a willingness to revise or abandon your theories when the tests fail (as they usually do).
The Lying Stones of Marrakech: Penultimate Reflections in Natural History
The Proof of Lavoisier's Plates (pp. 104–105)
Harmony Books. New York, New York, USA. 2000

Schiebinger, Londa   1952–
American professor and writer of science history

Objectivity in science cannot be proclaimed, it must be built.
Nature's Body: Gender in the Making of Modern Science
Chapter 3 (p. 114)
Beacon Press. Boston, Massachusetts, USA. 1993

OBSCURATIONISM

Dawkins, Richard   1941–
English ethologist, evolutionary biologist, and popular science writer

Dawkin's Law of the Conservation of Difficulty states that obscurationism in an academic subject expands to fill the vacuum of its intrinsic simplicity.
A Devil's Chaplain: Reflections on Hope, Lies, Science, and Love (p. 8)
Houghton Mifflin Company. Boston, Massachusetts, USA. 2003

OBSERVATION

Abbott, Donald Putnam   1920–86
American marine biologist and professor

Get the experience of looking at fresh things. If you watch live animals, you gain clearer insights in shorter time than you would watching dead animals for much longer.
In Galen Howard Hilgard (ed.)
Observing Marine Invertebrates: Drawings from the Laboratory
Author's Preface (p. xvi)
Stanford University Press. Stanford, California, USA. 1987

Adams, Douglas   1952–2001
English author, comic radio dramatist, and musician

…a scientist must also be absolutely like a child. If he sees a thing, he must say that he sees it, whether it was what he thought he was going to see or not. See first, think later, then test. But always see first. Otherwise you will only see what you were expecting.
The Ultimate Hitchhiker's Guide to The Galaxy
So Long and Thanks for All the Fish
Chapter 31 (p. 587)
The Ballantine Book Company. New York, New York, USA. 2002

Altmann, Jeanne
Biologist

The true situation may be the opposite of the apparent one.
Baboon Mothers and Infants
Introduction (p. 6)
The University of Chicago Press. Chicago, Illinois, USA. 1980

Anscombe, Francis John   1918–2001
English-born American statistician

No observations are absolutely trustworthy.
Rejection of Outliers
Technometrics, Volume 2, Number 2, May 1960 (p. 124)

Argelander, Friedrich Wilhelm August   1799–1875
German astronomer

Observations buried in a desk are no observations.
In Harlow Shapley and Helen E. Howarth
A Source Book in Astronomy
Argelander (p. 237)
McGraw-Hill Book Company, Inc. New York, New York, USA. 1929

Arp, Halton Christian   1927–
American astronomer

Of course, if one ignores contradictory observations, one can claim to have an "elegant" or "robust" theory. But it isn't science.
Letters
Science News, Volume 140, Number 4 Jul 27, 1991 (p. 51)

Aurelius Antoninus, Marcus   121–180
Roman emperor

Consider that everything which happens, happens justly, and if thou observest carefully, thou wilt find it to be so.
In *Great Books of the Western World* (Volume 12)
The Meditations of Marcus Aurelius
Book IV, # 10 (p. 264)
Encyclopædia Britannica, Inc. Chicago, Illinois, USA. 1952

Ayres, Clarence Edwin   1891–1972
No biographical data available

When Moses emerged from the cloudy obscurity of Mount Sinai and stood before the people with the stone tablets in his hand, he announced that his laws were based on direct observation. It is not recorded that any one doubted him.

Science: The False Messiah
Chapter III (p. 42)
The Bobbs-Merrill Company Publishers, Indianapolis, Indiana, USA. 1927

Baker, Henry 1698–1774
English naturalist

Beware of determining and declaring your Opinion suddenly on any Object; for Imagination often gets the Start of judgment, and makes People believe they see Things, which better Observations will convince them could not possibly be seen: Therefore assert nothing till after repeated Experiments and Examinations in all Lights and in all Positions.

The Microscope Made Easy
Part I, Chapter XV, Cautions in Viewing Objects (p. 62)
Printed for R. Dodsley. London, England. 1743

Bauer, Georg (Agricola or Georgius Agricola) 1494–1555
German scholar and scientist

I have omitted all those things which I have not myself seen, or have not read or heard of from persons upon whom I can rely. That which I have neither seen, nor carefully considered after reading or hearing of, I have not written about. The same rule must be understood with regard to all my instruction, whether I enjoin things which ought to be done, or describe things which are usual, or condemn things which are done.

De Re Metallica
Preface (pp. xxx–xxxi)
Dover Publications, Inc. New York, New York, USA. 1950

Bernard, Claude 1813–78
French physiologist

Only within very narrow boundaries can man observe the phenomena which surround him; most of them naturally escape his senses, and mere observation is not enough.

Translated by Henry Copley Greene
An Introduction to the Study of Experimental Medicine
Part One, Chapter I (p. 5)
Henry Schuman, Inc. New York, New York, USA. 1927

Observation, then, is what shows facts.; experiment is what teaches about facts and gives experience in relation to anything.

Translated by Henry Copley Greene
An Introduction to the Study of Experimental Medicine
Part One, Chapter I, Section ii (p. 11)
Henry Schuman, Inc. New York, New York, USA. 1927

Speaking concretely, when we say "making experiments or making observations," we mean that we devote ourselves to investigation and to research, that we make

attempts and trials in order to gain facts from which the mind, through reasoning, may draw knowledge or instruction.

Speaking in the abstract, when we say, "relying on observation and gaining experience," we mean that observation is the mind's support in reasoning, and experience the mind's support in deciding, or still better, the fruit of exact reasoning applied to the interpretation of facts.

Observation, then, is what shows facts; experiment is what teaches about facts and gives experience in relation to anything.

Translated by Henry Copley Greene
An Introduction to the Study of Experimental Medicine
Part One, Chapter I, Section ii (p. 11)
Henry Schuman, Inc. New York, New York, USA. 1927

Men sometimes seem to confuse experiment with observation.

Translated by Henry Copley Greene
An Introduction to the Study of Experimental Medicine
Part One, Chapter I (p. 6)
Henry Schuman, Inc. New York, New York, USA. 1927

Blake, William 1757–1827
English poet, painter, and engraver

A fool sees not the same tree that a wise man sees.

The Complete Poetry and Prose of William Blake
The Marriage of Heaven and Hell, Proverbs of Hell, l. 8
University of California Press. Berkeley, California, USA. 1982

Bolles, Edmund Blair
No biographical data available

…yet there is a difference between scientific and artistic observation. The scientist observes to turn away and generalize; the artist observes to seize and use reality in all its individuality and peculiarity.

A Second Way of Knowing: The Riddle of Human Perception
Chapter 11 (p. 150)
Press Hall Press. New York, New York, USA. 1991

Box, George E. P. 1919–
English statistician

To find out what happens to a system when you interfere with it you have to interfere with it (not just passively observe it).

Use and Abuse of Regression
Technometrics, Volume 8, Number 4, November 1966 (p. 629)

Brownlee, Donald
American astronomer

It is no secret that our lives are patterned by personal experience, and by our observations of the experience of others. So it is with the entire human family, staring outward from a place that we know to others that we do not. Our knowledge of home helps us to understand the ways of planets beyond our ken.

In Nigel S. Hey
Solar System
How Rare Is the Earth? (p. 156)
Weidenfeld & Nicolson. London, England. 2002

Burroughs, John   1837–1921
American naturalist and writer

Unadulterated, unsweetened observations are what the real nature-lover craves. No man can invent incidents and traits as interesting as the reality.
Ways of Nature
Chapter I (p. 15)
Book for Libraries Press. Freeport, New York, USA. 1971

Carlyle, Thomas   1795–1881
English historian and essayist

Shakespeare says, we are creatures that look before and after: the more surprising that we do not look round a little, and see what is passing under our very eyes.
Sartor Resartus
Book I, Chapter I (p. 3)
Ginn & Company. Boston, Massachusetts, USA. 1897

Cohen, Morris Raphael   1880–1947
American philosopher

Accidental discoveries of which popular histories of science make mention never happen except to those who have previously devoted a great deal of thought to the matter. Observation unilluminated by theoretic reason is sterile.... Wisdom does not come to those who gape at nature with an empty head. Fruitful observation depends not as Bacon thought upon the absence of bias or anticipatory ideas, but rather on a logical multiplication of them so that having many possibilities in mind we are better prepared to direct our attention to what others have never thought of as within the field of possibility.
Reason and Nature
Chapter One, Section III (p. 17)
The Free Press, Publishers, Glencoe, Illinois, USA. 1931

da Vinci, Leonardo   1452–1519
Italian High Renaissance painter and inventor

Science is the observation of things possible, whether present or past; prescience is the knowledge of things which may come to pass.
The Literary Works of Leonardo da Vinci (Volume 2)
1148 (p. 239)
University of California Press. Berkeley, California, USA. 1977

Darwin, Charles Robert   1809–82
English naturalist

Some of my critics have said, "Oh, he is a good observer, but he has no power of reasoning!" I do not think that this can be true, for the "Origin of Species" is one long argument from the beginning to the end, and it has convinced not a few able men. No one could have written it without having some power of reasoning.
In Francis Darwin (ed.)
The Life and Letters of Charles Darwin (Volume 1)
Chapter II (p. 82)
D. Appleton & Company. New York, New York, USA. 1896

I have an old belief that a good observer really means a good theorist.
In Francis Darwin (ed.)
More Letters of Charles Darwin (Volume 1)
Letter 118, Darwin to Bates, 22 November, 1860 (p. 195)
D. Appleton & Company. New York, New York, USA. 1903

How odd it is that anyone should not see that all observation must be for or against some view if it is to be of any service!
In Francis Darwin (ed.)
More Letters of Charles Darwin (Volume 1)
Letter 133, Darwin to Henry Fawcett, Sept. 18, 1861 (p. 195)
D. Appleton & Company. New York, New York, USA. 1903

I am a firm believer that without speculation there is no good and original observation.
In Francis Darwin (ed.)
The Life and Letters of Charles Darwin (Volume 1)
To Wallace, December 22, 1857 (p. 465)
D. Appleton & Company. New York, New York, USA. 1896

Davy, Sir Humphry   1778–1829
English chemist

The grandest as well as the most correct views are those that have been gained by minute observation, and by the application of all the more precise and accurate methods of science.
The Collected Works of Sir Humphry Davy (Volume 1)
Memories of the Life of Sir Humphry Davy
Chapter III (p. 153)
Smith, Elder & Company. London, England. 1839–1849

Dickens, Charles   1812–70
English novelist

The bearings of this observation lays in the application on it.
The Works of Charles Dickens
Dombey and Son (Part I)
Chapter XXIII (p. 348)
P.F. Collier & Son. New York, New York, USA. 1911

Doyle, Sir Arthur Conan   1859–1930
Scottish writer

You see, but you do not observe. The distinction is clear.
In William S. Baring-Gould (ed.)
The Annotated Sherlock Holmes (Volume 1)
A Scandal in Bohemia (p. 349)
Wings Books. New York, New York, USA. 1967

The world is full of obvious things which nobody by any chance ever observes.
In William S. Baring-Gould (ed.)

The Annotated Sherlock Holmes (Volume 2)
The Hound of the Baskervilles, Chapter 3 (p. 18)
Wings Books. New York, New York, USA. 1967

Never trust impressions, my boy, but concentrate yourself upon details.
In William S. Baring-Gould (ed.)
The Annotated Sherlock Holmes (Volume 1)
A Case of Identity (p. 411)
Wings Books. New York, New York, USA. 1967

…it is my business to know things. Perhaps I have trained myself to see what others overlook.
In William S. Baring-Gould (ed.)
The Annotated Sherlock Holmes (Volume 1)
A Case of Identity (p. 406)
Wings Books. New York, New York, USA. 1967

Drake, Daniel   1785–1852
American physician

If observation be the soil, reading is the manure of intellectual culture.
Physician to the West: Selected Writings of Daniel Drake on Science and Society (p. 307)
University Press of Kentucky. Lexington, Kentucky, USA. 1970

du Noüy, Pierre Lecomte   1883–1947
French scientist

…I said that an observed fact only becomes a scientific fact when all the observers are in unanimous agreement.
The Road to Reason
Chapter I (pp. 29–30)
Longmans, Green & Company. London, England. 1949

Dumas, Jean Baptiste-Andre   1800–84
French biochemist

The art of observation and that of experimentation are very distinct. In the first case, the fact may either proceed from logical reasons or be mere good fortune; it is sufficient to have some penetration and a sense of truth in order to profit by it. But the art of experimentation leads from the first to the last link of the chain, without hesitation and without a blank, making successive use of Reason, which suggests an alternative, and of Experience, which decides on it, until, starting from a faint glimmer, the full blaze of light is reached.
In Sir Richard Arman Gregory
Discovery; or, The Spirit and Service of Science
Chapter VI (p. 126)
Macmillan & Company Ltd. London, England. 1918

Durkheim, Emile   1858–1917
French sociologist

Even one well-made observation will be enough in many cases, just as one well-constructed experiment often suffices for the establishment of a law.
Translated by Sarah A. Solovay and John H. Mueller
The Rules of Sociological Method

Chapter IV (p. 80)
The Free Press. Glencoe, Illinois, USA. 1938

Eddington, Sir Arthur Stanley   1882–1944
English astronomer, physicist, and mathematician

We should be unwise to trust scientific inference very far when it becomes divorced from opportunity for observational test.
The Internal Constitution of the Stars
Chapter I (p. 1)
At The University Press. Cambridge, England. 1930

For the truth of the conclusions of physical science, observation is the supreme Court of Appeal. It does not follow that every item which we confidently accept as physical knowledge has actually been certified by the Court; our confidence is that it would be certified by the Court if it were submitted. But it does follow that every item of physical knowledge is of a form which might be submitted to the Court. It must be such that we can specify (although it may be impracticable to carry out) an observational procedure which would decide whether it is true or not. Clearly a statement cannot be tested by observation unless it is an assertion about the results of observation. Every item of physical knowledge must therefore be an assertion of what has been or would be the result of carrying out a specified observational procedure.
The Philosophy of Physical Science
Chapter I, Section IV (pp. 9–10)
The Macmillan Company. New York, New York, USA. 1939

Let us suppose that an ichthyologist is exploring the life of the ocean. He casts a net into the water and brings up a fishy assortment. Surveying his catch, he proceeds in the usual manner of a scientist to systematise what it reveals…In applying this analogy, the catch stands for the body of knowledge which constitutes physical science, and the net for the sensory and intellectual equipment which we use in obtaining it. The casting of the net corresponds to observation; for knowledge which has not been or could not be obtained by observation is not admitted into physical science.
The Philosophy of Physical Science
Chapter II, Section I (p. 16)
The Macmillan Company. New York, New York, USA. 1939

I hope I shall not shock the experimental physicists too much if I add that it is also a good rule not to put overmuch confidence in the observational results that are put forward until they have been confirmed by theory…
New Pathways In Science
Chapter X, Section II (p. 211)
The Macmillan Company. New York, New York, USA. 1935

Einstein, Albert   1879–1955
German-born physicist

It is the theory which decides what we can observe.
In Werner Heisenberg
Physics and Beyond: Encounters and Conversations
Chapter 6 (p. 77)
Harper & Row, Publishers. New York, New York, USA. 1971

A man should look for what is, and not for what he thinks should be…
In Peter Michelmore
Einstein, Profile of the Man (p. 20)
Dodd & Mead Publishers. New York, New York, USA. 1962

Euler, Leonhard   1707–83
Swiss mathematician and physicist

It will seem not a little paradoxical to ascribe a great importance to observations even in that part of the mathematical sciences which is usually called Pure Mathematics, since the current opinion is that observations are restricted to physical objects that make impressions on the senses.
In G. Polya
Induction and Analogy in Mathematics (Volume 1)
Chapter I (p. 3)
Princeton University Press. Princeton, New Jersey, USA. 1954

…in the theory of numbers, which is still very imperfect, we can place our highest hopes in observations; they will lead us continually to new properties which we shall endeavor to prove afterwards.
In G. Polya
Induction and Analogy in Mathematics (Volume 1)
Chapter I (p. 3)
Princeton University Press. Princeton, New Jersey, USA. 1954

Fischer, Martin H.   1879–1962
German-American physician

You must acquire the ability to describe your observations and your experience in such language that whoever observes or experiences similarly will be forced to the same conclusion.
In Howard Fabing and Ray Marr
Fischerisms (p. 8)
C.C. Thomas. Springfield, Illinois, USA. 1944

Faraday, Michael   1791–1867
English physicist and chemist

If in such strivings, we…see but imperfectly, still we should endeavor to see, for even an obscure and distorted vision is better than none.
On the Conservation of Force
Philosophical Magazine, Volume 13, Number 4, 1857 (p. 238)

Feynman, Richard P.   1918–88
American theoretical physicist

The principle that the observation is the judge imposes a severe limitation to the kind of questions that can be answered. They are limited to questions that you can put this way: "if I do this, what will happen?" There are ways to try and see. Questions like, "should I do this?" and "what is the value of this?" are not of the same kind.
The Meaning of It All: Thoughts of a Citizen Scientist
Chapter I (p. 16)
Perseus Books. Reading, Massachusetts, USA. 1998

Galilei, Galileo   1564–1642
Italian physicist and astronomer

It is not in ancient tomes, but in close observations and personal consecration that a grain of truth may be found. It is so very easy to seek the significance of things in the papers of this or that man rather than in the works of nature which, ever alive and active, are constantly before our eyes.
In Helen Wright
Palomar: The World's Largest Telescope
Origin of the Telescope (p. 9)
The Macmillan Company. New York, New York, USA. 1952

Greer, Scott
No biographical data available

…the link between observation and formulation is one of the most difficult and crucial in the scientific enterprise. It is the process of interpreting our theory or, as some say, of "operationalizing our concepts." Our creations in the world of possibility must be fitted in the world of probability; in Kant's epigram, "Concepts without precepts are empty." It is also the process of relating our observations to theory; to finish the epigram, "Precepts without concepts are blind."
The Logic of Social Inquiry
Part III, Chapter 14 (p. 160)
Aldine Publishing Company. Chicago, Illinois, USA. 1969

Gregg, Alan   1890–1957
American medical educator and philosopher

…most of the knowledge and much of the genius of the research worker lie behind selection of what is worth observing. It is a crucial choice, often determining the success or failure of months of work, often differentiating the brilliant discoverer from the…plodder.
The Furtherance of Medical Research
Chapter I (p. 8)
Yale University Press. New Haven, Connecticut, USA. 1941

Gregory, Sir Richard Arman   1864–1952
British science writer and journalist

An accurate observation remains unaltered throughout the ages. Its scientific value is determined by its truth to Nature; and the more complete the testimony, the less room is there for elaboration by investigators in succeeding generations.
Discovery; or, The Spirit and Service of Science
Chapter IV (p. 70)
Macmillan & Company Ltd. London, England. 1918

In the world of natural knowledge, no authority is great enough to support a theory when a crucial observation has shown it to be untenable.
Discovery; or, The Spirit and Service of Science
Chapter I (p. 12)
Macmillan & Company Ltd. London, England. 1918

Grew, Nehemiah   1641–1712
Scientific writer and journalist

If…an inquiry into the Nature of Vegetation may be of good Import; It will be requisite to see, first of all, What may offer it self to be enquired of; or to understand, what or Scope is: That so doing, we may take our aim the better in making, and having made, in applying our Observations thereunto.
The Anatomy of Plants
An Idea of a Philosophical History of Plants (p. 3)
Printed by W. Rawlins. London, England. 1682

Hales, Stephen   1677–1761
English physiologist and clergyman

…it is from long experience chiefly that we are to expect the most certain rules of practice, yet it is withal to be remembered, that observations, and to put us upon the most probable means of improving any art, is to get the best insight we can into the nature and properties of those things which we are desirous to cultivate and improve.
Vegetable Staticks
The Conclusion (p. 214)
The Scientific Book Guild. London, England. 1961

Hanson, Norwood Russell   1924–67
American philosopher of science

The observer may not know what he is seeing: he aims only to get his observations to cohere against the background of established knowledge. This seeing is the goal of observation.
Patterns of Discovery
Chapter I (p. 20)
At The University Press. Cambridge, England. 1958

…there is more to seeing than what meets the eyeball.
Patterns of Discovery
Chapter I (p. 7)
At The University Press. Cambridge, England. 1958

Harvey, William   1578–1657
English physician

…the dull and unintellectual are indisposed to see what lies before their eyes…
In *Great Books of the Western World* (Volume 28)
An Anatomical Disquisition on the Motion of the Heart and Blood in Animals
Dedication (p. 268)
Encyclopædia Britannica, Inc. Chicago, Illinois, USA. 1952

Heisenberg, Werner Karl   1901–76
German physicist and philosopher

This assumption is not permissible in atomic physics; the interaction between observer and object causes uncontrollable and large changes in the system being observed, because of the discontinuous changes characteristic of atomic processes.
The Physical Principles of the Quantum Theory
Introductory (p. 3)
The University of Chicago Press. Chicago, Illinois, USA. 1930

This again emphasizes a subjective element in the description of atomic events, since the measuring device has been constructed by the observer, and we have to remember that what we observe is not nature in itself but nature exposed to our method of questioning.
Physics and Philosophy: The Revolution in Modern Science
Chapter III (p. 58)
Harper & Row, Publishers. New York, New York, USA. 1958

The idea that we do observe something already indicates something irreversible. If we draw a pencil line on a paper, for instance, we have established something which cannot be undone, so to speak. Every observation is irreversible, because we have gained information that cannot be forgotten.
In Paul Buckley and F. David Peat
Glimpsing Reality: Ideas in Physics and the Link to Biology
Werner Heisenberg (p. 12)
University of Toronto Press. Toronto, Ontario, Canada. 1996

…what we observe is not nature itself, but nature exposed to our method of questioning.
Physics and Philosophy: The Revolution in Modern Science
Chapter III (p. 58)
Harper & Row, Publishers. New York, New York, USA. 1958

A real difficulty in the understanding of this interpretation arises, however, when one asks the famous question: But what happens "really" in an atomic event? It has been said before that the mechanism and the results of an observation can always be stated in terms of the classical concepts. But what one deduces from an observation is a probability function, a mathematical expression that combines statements about possibilities or tendencies with statements about our knowledge of facts. So we cannot completely objectify the result of an observation, we cannot describe what "happens" between this observation and the next.
In T. Ferris (ed.)
World Treasury of Physics, Astronomy, and Mathematics
The Copenhagen Interpretation of Quantum Mechanics (pp. 90–91)
Little, Brown & Company. Boston, Massachusetts, USA. 1991

Herschel, Sir John Frederick William   1792–1871
English astronomer and chemist

There is scarcely any well-informed person, who, if he has but the will, has not also the power to add something essential to the general stock of knowledge, if he will only observe regularly and methodically some particular class of facts which may most excite his attention…

A Preliminary Discourse on the Study of Natural Philosophy
Part II, Chapter IV, Section 128 (p. 133)
Printed for Longman, Rees, Orme, Brown & Green. London, England.
1831

Seeing is in some respects an art which must be learnt. To make a person see with such a power is nearly the same as if I were asked to make him play one of Handel's fugues upon the organ. Many a night I have been practicing to see, and it would be strange if one did not acquire a certain dexterity by such constant practice.
In William Hoyt
Planets X and Pluto
Chapter 1 (p. 12)
The University of Arizona Press. Tucson, Arizona, USA. 1981

Hinshelwood, Sir Cyril   1897–1967
English chemist

Again and again, the key to a great discovery has been an unexpected observation.
Science and Scientists
Nature, Volume 207, Number 5001, September 4, 1965 (p. 1057)

Holton, Gerald   1922–
Research professor of physics and science history
Roller, Duane H. D.   ?–1994
Science historian

All intelligent endeavor stands with one foot on observation and the other on contemplation.
Foundations of Modern Physical Science
Chapter 13 (p. 218)
Addison-Wesley Publishing Company. Reading, Massachusetts, USA. 1950

Hooke, Robert   1635–1703
English physicist

The truth is, the Science of Nature has been already too long made only a work of the Brain and the Fancy: It is now high time that it should return to the plainness and soundness of Observations on material and obvious things.
Micrographia
Preface (Fifth page)
Printed by Jo. Martyn and Ja. Allestry. London, England. 1665

Hoyle, Sir Fred   1915–2001
English mathematician and astronomer

You take a number of radioactive nuclei of a particular kind, the number being chosen so that there's an even chance of one of them going off in a certain period of time, say ten seconds. Then for ten seconds you surround them with counters, or any other detecting device you might like to use. At the end of the time the question is, has one of them decayed or not. To decide this you take a look at your counters. The conventional notion is that the state of the counters decides whether a nucleus has gone off or not...my problem now concerns an individual case.... It is perfectly possible to put your counters, or your bubble chamber, your camera, all your gobbledegook in fact, into your calculations — and we know quite definitely that any attempt to get a definite answer out of calculation will prove completely fruitless. The thing that gives the answer isn't the camera or the counter, it's the actual operation of looking yourself at your equipment. It seems that only when we ourselves take a subjective decision can we improve our description of the world, over and above the uncertainty of our theories. I'm talking about quantum theories now.
October the First Is Too Late
Chapter Five (pp. 52–53)
Harper & Row, Publishers. New York, New York, USA. 1966

Hubble, Edwin Powell   1889–1953
American astronomer

...observation and theory are woven together, and it is futile to attempt their complete separation. Observation always involve theory. Pure theory may be found in mathematics, but seldom in science. Mathematics, it has been said, deals with possible worlds — logically consistent systems. Science attempts to discover the actual world we inhabit. So in cosmology, theory presents an infinite array of possible universes, and observation is eliminating them, class by class, until now the different types among which our particular universe must be included have become increasingly comprehensible.
The Realm of the Nebulae
Chapter I (p. 35)
Dover Publications, Inc. New York, New York, USA. 1958

Huxley, Thomas Henry   1825–95
English biologist

There is no question in the mind of anyone acquainted with the facts that, so far as observation and experiment can take us, the structure and the functions of the nervous system are fundamentally the same in an ape, or in a dog, and in a man. And the suggestion that we must stop at the exact point at which direct proof fails us, and refuse to believe that the similarity which extends so far stretches yet further, is no better than a quibble. Robinson Crusoe did not feel bound to conclude, from the single human footprint which he saw in the sand that the maker of the impression had only one leg.
Hume, with Helps to the Study of Berkeley (p. 123)
D. Appleton & Company. New York, New York, USA. 1896

James, William   1842–1910
American philosopher and psychologist

Round about the accredited and orderly facts of every science there ever floats a sort of dust-cloud of exceptional observations, of occurrences minute and irregular

and seldom met with, which it always proves more easy to ignore than to attend to...
The Will to Believe and Other Essays in Popular Philosophy and Human Immortality
What Psychical Research Has Accomplished (p. 299)
Dover Publications, Inc. New York, New York, USA. 1956

Jeans, Sir James Hopwood   1877–1946
English physicist and mathematician

Each observation destroys the bit of the universe observed, and so supplies knowledge only of a universe which has already become past history...
The New Background of Science
Chapter I (p. 2)
The University of Michigan Press. Ann Arbor, Michigan, USA. 1959

We can only see nature blurred by the clouds of dust we ourselves make; we can still only see the rainbow, but a sun of some sort must exist to produce the light by which we see it.
The New Background of Science
Chapter I (p. 4)
The University of Michigan Press. Ann Arbor, Michigan, USA. 1959

Jeffreys, Sir Harold   1891–1989
English astronomer and geophysicist

An observation, strictly, is only a sensation. Nobody means that we should reject everything but sensations. But as soon as we go beyond sensations we are making inferences.
Theory of Probability
General Questions (p. 412)
Clarendon Press. Oxford, England. 1961

Jones, Steve
No biographical data available

Observation and experiment are what count, not opinion and introspection. Few working scientists have much respect for those who try to interpret nature in metaphysical terms. For most wearers of white coats, philosophy is to science as pornography is to sex: it is cheaper, easier, and some people seem, bafflingly, to prefer it. Outside of psychology it plays almost no part in the functions of the research machine.
Review of *How the Mind Works* by Steve Pinker
The New York Review of Books, November 6, 1997 (p. 13)

Jonson, Ben   1573?–1637
English dramatist and poet

...let me alone to observe, till I turne my selfe into nothing but observation.
The Poetaster
Act II, Scene I, l. 193
Henry Holt & Company. New York, New York, USA. 1905

Kolb, Edward W. (Rocky)   1951–
American cosmologist

The art of science is knowing which observations to ignore and which are the key to the puzzle.
Blind Watchers of the Sky
Chapter Seven (p. 189)
Addison-Wesley Publishing Company. Reading, Massachusetts, USA. 1996

Kuhn, Thomas S.   1922–96
American historian of science

Observation and experience can and must drastically restrict the range of admissible scientific belief, else there would be no science. But they cannot alone determine a particular body of such belief. An apparently arbitrary element, compounded of personal and historical accident, is always a formative ingredient of the beliefs espoused by a given scientific community at a give time.
The Structure of Scientific Revolutions
Chapter I (p. 4)
The University of Chicago Press. Chicago, Illinois, USA. 1970

Langer, Susanne Knauth   1895–1985
American philosopher

The faith of scientists in the power of mathematics is so implicit that their work has gradually become less and less observation, and more and more calculation. The promiscuous collection and tabulation of data have given way to a process of assigning possible meanings, merely supposed real entities, to mathematical terms, working out the logical results, and then staging certain crucial experiments to check the hypothesis against the actual, empirical results. But the facts...accepted by virtue of these tests are not actually observed at all.
Philosophy in a New Key
Chapter I (pp. 19–20)
Harvard University Press. Cambridge, Massachusetts, USA. 1957

Observation has become almost entirely indirect; and readings take the place of genuine witness.
Philosophy in a New Key
Chapter I (p. 20)
Harvard University Press. Cambridge, Massachusetts, USA. 1957

The men in the laboratory...cannot be said to observe the actual objects of their curiosity at all. ... They sense data on which the propositions of modern science rest are, for the most part, little photographic spots and blurs, or inky curved likes on paper.... What is directly observable is only a sign of the "physical fact"; it requires interpretation to yield scientific propositions.
Philosophy in a New Key
Chapter I (p. 20)
Harvard University Press. Cambridge, Massachusetts, USA. 1957

Lee, Oliver Justin   1881–1964
American astronomer

Every bit of knowledge we gain and every conclusion we draw about the universe or about any part or feature

of it depends finally upon some observation or measurement. Mankind has had again and again the humiliating experience of trusting to intuitive, apparently logical conclusions without observations, and has seen Nature sail by in her radiant chariot of gold in an entirely different direction.
Measuring Our Universe: From the Inner Atom to Outer Space
Chapter 3 (p. 33)
Ronald Press Company. New York, New York, USA. 1950

Lewis, Gilbert Newton   1875–1946
American chemist

I claim that my eye touches a star as truly as my finger touches this table.
In George W. Gray
New Eyes on the Universe
The Atlantic Monthly, Volume 155, Number 5, May 1935 (p. 608)

Longair, Malcolm   1941–
Scottish physicist

Although by now a large amount of observational material is available, the implications of these observations are far from clear.
Contemporary Physics
Quasi-Stellar Radio Sources, Volume 8, Number 4, 1967 (p. 357)

Lonsdale, Dame Kathleen   1903–71
Irish-born English crystallographer

…observation is not enough, and it seems to me that in science, as in the arts, there is very little worth having that does not require the exercise of intuition as well as of intelligence, the use of imagination as well as of information.
Facts About Crystals
American Scientist, Volume 39, Number 4, October 1951 (p. 576)

Louis, Pierre-Charles-Alexandre   1787–1872
French physician

It behooves those who devote themselves to observation to be impressed by this truth (*i.e.*, that many "facts" grow old) and to realize that the best work is only good in relation to its time and that it awaits another, more exact and more complete.
Recherches Anatomiques, Pathologiques et Therapeutiques sur la Maladie Connue Sous les Noms de Gastro-Enterite (p. vii)
Publisher undetermined

Lubbock, Sir John   1834–1913
English banker, author, and scientist

What we do see depends mainly on what we look for. When we turn our eyes to the sky, it is in most cases merely to see whether it is likely to rain. In the same field the farmer will notice the crop, geologists the fossils, botanists the flowers, artists the coloring, sportsmen the cover for game. Though we may all look at the same things, it does not at all follow that we should see them.
The Beauties of Nature and the Wonders of the World We Live In
Introduction (pp. 3–4)
The Macmillan Company. New York, New York, USA. 1893

Mach, Ernst   1838–1916
Austrian physicist and philosopher

Everything which we observe in nature imprints itself uncomprehended and unanalyzed in our percepts and ideas which then, in their turn, mimic the processes of nature in their most general and most striking features.
The Science of Mechanics (5th edition)
Chapter I, Part II, Section 2 (p. 36)
The Open Court Publishing Company. La Salle, Illinois, USA. 1942

Macy, Arthur
No biographical data available

But I keep no log of my daily grog,
For what's the use o' being bothered? I drink a little more when the wind's offshore,
And most when the wind's from the no'th'ard.
Poems
The Indifferent Mariner
W.B. Clarke Company. Boston, Massachusetts, USA. 1905

Marschall, Laurence A.
American astronomer

A first-time deep-sky observer usually sees little more than a fuzzy glow against the blackness of the night. Thus, to the nonastronomer, once you've seen one celestial object, you've pretty much seen them all.
The Supernova Story
Chapter 1 (p. 2)
Plenum Press. New York, New York, USA. 1988

Marsland, Douglas
American biologist

The primary basis of all scientific thinking is observation.
Principles of Modern Biology (p. 12)
Holt, Rinehart, Winston. New York, New York, USA. 1969

Medawar, Sir Peter Brian   1915–87
Brazilian-born English zoologist

Innocent, unbiased observation is a myth.
Induction and Intuition in Scientific Thought
Chapter II, Section 2 (p. 28)
American Philosophical Society. Philadelphia, Pennsylvania, USA. 1969

It is not methodologically an exaggeration to say that Fleming eventually found penicillin because he had been looking for it. A thousand people might have observed whatever it was that he did observe without making anything of it or building upon the observation in any way; but Fleming had the right slot in his mind, waiting for it. Good luck is almost always preceded by an expectation that it will gratify. Pasteur is well known to have said

that fortune favors the prepared mind, and Fontenelle observed, *"Ces hasards ne sont que pour ceux qui jouent bien!"* ("These strokes of good fortune are only for those who play well!").
Advice to a Young Scientist
Chapter 11 (p. 90)
Basic Books, Inc. New York, New York, USA. 1979

Meredith, George   1828–1909
English novelist and poet

Observation is the most enduring of the pleasures of life…
Diana of the Crossways
Chapter XI (p. 104)
Charles Scribner's Sons. New York, New York, USA. 1924

Minnaert, M.
No biographical data available

It is indeed wrong to think that the poetry of Nature's moods in all their infinite variety is lost on one who observes them scientifically, for the habit of observation refines our sense of beauty and adds a brighter hue to the richly coloured background against which each separate fact is outlined. The connection between events, the relation of cause and effect in different parts of a landscape, unite harmoniously what would otherwise be merely a series of detached sciences.
The Nature of Light and Colour in the Open Air
Preface (p. v)
Dover Publications. New York, New York, USA. 1954

Mitchell, Maria   1818–89
American astronomer and educator

Nothing comes out more clearly in astronomical observations than the immense activity of the universe.
In Phebe Mitchell Kendall
Maria Mitchell: Life, Letters, and Journals
Chapter XI (p. 237)
Lee & Shepard. Boston, Massachusetts, USA. 1896

Morris, Richard   1939–2003
American physicist and science writer

Simple observation generally gets us nowhere. It is the creative imagination that increases our understanding by finding connections between apparently unrelated phenomena, and forming logical, consistent theories to explain them. And if a theory turns out to be wrong, as many do, all is not lost. The struggle to create an imaginative, correct picture of reality frequently tells us where to go next, even when science has temporarily followed the wrong path.
The Universe, the Eleventh Dimension, and Everything: What We Know and How We Know It
Part 3, Chapter 2 (p. 190)
Four Walls Eight Windows. New York, New York, USA. 1999

Moulton, Lord   1844–1921
English mathematician

When we are reduced to observation Science crawls.
In Alan Gregg
The Furtherance of Medical Research
Chapter I (p. 7)
Yale University Press. New Haven, Connecticut, USA. 1941

Müller, Johannes   1801–58
German physiologist

Observation is simple, indefatigable, industrious, upright, without any preconceived opinion. Experiment is artificial, impatient, busy, digressive, passionate, unreliable.
In V.J.E. Kruta
Purkyne Physiologist: A Short Account of His Contributions to the Progress of Physiology (p. 20)
Publisher undetermined

Newton, Sir Isaac   1642–1727
English physicist and mathematician

As in mathematics, so in natural philosophy the investigation of difficult things by the method of analysis ought ever to precede the method of composition. This analysis consists of making experiments and observations, and in drawing general conclusions from them by induction…. By this way of analysis we may proceed from compounds to ingredients, and from motions to the forces producing them; and in general from effects to their causes, and from particular causes to more general ones till the argument end in the most general. This is the method of analysis: and the synthesis consists in assuming the causes discovered and established as principles, and by them explaining the phenomena preceding from them, and proving the explanations.
In *Great Books of the Western World* (Volume 34)
Optics
Book III: Part I, Query 31
Encyclopædia Britannica, Inc. Chicago, Illinois, USA. 1952

O'Neil, William Matthew
No biographical data available

It urges the scientist, in effect, not to take risks incurred in moving far from the facts. However, it may properly be asked whether science can be undertaken without taking the risk of skating on the possibly thin ice of supposition. The important thing to know is when one is on the more solid ground of observation and when one is on the ice.
Fact and Theory: An Aspect of the Philosophy of Science
Chapter 8 (p. 154)
Sydney University Press. Sydney, Australia. 1969

Orwell, George (Eric Arthur Blair)   1903–50
English novelist and essayist

To see what is in front of one's nose requires a constant struggle.

In Sonia Orwell and Ian Angus (eds.)
The Collected Essays, Journalism and Letters of George Orwell: In Front of Your Nose, 1945–1950
1946, 36 (p. 125)
Harcourt, Brace & World. New York, New York, USA. 1968

Osler, Sir William   1849–1919
Canadian physician and professor of medicine

Observation plus thinking has given us the bodies of living creatures in health and disease. There have been two inherent difficulties — to get men to see straight and to get men to think clearly; but in spite of the frailty of the instrument, the method has been one of the most powerful ever placed in the hands of man.
The Pathological Institute of a General Hospital
Glasgow Medical Journal, Volume 76, 1911

Note with accuracy and care everything that comes within your professional ken.… Let nothing slip by you; the ordinary hum-drum cases of the morning routine may have been accurately described and pictured, but study each one separately as though it were new — so it is so far as your special experience goes; and if the spirit of the student is in you the lesson will be there.
Aequanimitas, with Other Addresses to Medical Students, Nurses, and Practitioners of Medicine
The Army Surgeon (p. 104)
The Blakiston Company. Philadelphia, Pennsylvania, USA. 1932

The whole art of medicine is in observation, as the old motto goes, but to educate the eye to see, the ear to hear and the finger to feel takes time, and to make a beginning, to start a man on the right path, is all that we can do. We expect too much of the student and we try to teach him too much. Give him good methods and a proper point of view, and all other things will be added, as his experience grows.
Aequanimitas, with Other Addresses to Medical Students, Nurses, and Practitioners of Medicine
The Hospital as a College (pp. 315–316)
The Blakiston Company. Philadelphia, Pennsylvania, USA. 1932

Man can do a great deal by observation and thinking, but with them alone he cannot unravel the mysteries of Nature. Had it been possible the Greeks would have done it; and could Plato and Aristotle have grasped the value of experiment in the progress of human knowledge, the course of European history might have been very different.
Man's Redemption of Man
Address, University of Edinburgh, July 1910 (p. 22)

Owen, Ed   1896–1981
American geologist

I wandered far and saw many things over a long time. Most of the things which I saw I did not understand. I looked about me and did not see that any others understood the complex pattern either. But as I wandered I could not escape the feel of things and of places and of the people in them.
In Samuel P. Ellison, Jr., Joseph J. Jones and Mirva Owen (eds.)
The Flavor of Ed Owen — A Geologist Looks Back
Introduction (p. 1)
Geology Foundation, University of Texas at Austin. Austin, Texas, USA. 1987

Pavlov, Ivan Petrovich   1849–1936
Russian physiologist

Observation collects whatever nature offers, whereas experimentation takes from nature whatever it requires.
In Ivan Valiela
Doing Science: Design, Analysis, and Communication of Scientific Research
Chapter I (p. 11)
Oxford University Press, Inc. Oxford, England. 2001

Planck, Max   1858–1947
German physicist

As long as Natural Philosophy exists, its ultimate highest aim will always be the correlating of various physical observations into a unified system, and, where possible, into a single formula.
Translated by R. Jones and D.H. Williams
A Survey of Physics: A Collection of Lectures and Essays
The Unity of the Physical Universe (p. 1)
Methuen & Company Ltd. London, England. 1925

Poincaré, Henri   1854–1912
French mathematician and theoretical astronomer

…to observe is not enough. We must use our observations, and to do that we must generalize.
The Foundations of Science
Science and Hypothesis, Part IV
Chapter IX (p. 127)
The Science Press. New York, New York, USA. 1913

Pope, Alexander   1688–1744
English poet

To observations which ourselves we make,
We grow more partial for th' observer's sake.
The Complete Poetical Works
Moral Essays, Epis. I, l. 11–12
Houghton Mifflin Company. New York, New York, USA. 1903

Popper, Karl R.   1902–94
Austrian/British philosopher of science

Some scientists find, or so it seems, that they get their best ideas when smoking; others by drinking coffee or whiskey. Thus there is no reason why I should not admit that some may get their ideas by observing or by repeating observations.
Realism and the Aim of Science
Part I, Chapter I (p. 36)
Rowman & Littlefield. Totowa, New Jersey, USA. 1983

Saxe, John Godfrey   1816–87
American poet

It was six men of Indostan
To learning much inclined,
Who went to see the Elephant
(Though all of them were blind),
That each by observation
Might satisfy his mind.
The Poetical Works of John Godfrey Saxe
The Parable of the Blind Men and the Elephant
Houghton Mifflin Company. Boston, Massachusetts, USA. 1892

Selye, Hans   1907–82
Austrian-American endocrinologist

If a scientist makes no important observation he deserves
no credit. But if a significant fact comes his way and he still
does not see its importance, he can only blame himself.
From Dream to Discovery: On Being a Scientist
McGraw-Hill Book Company, Inc. New York, New York, USA. 1950

Shakespeare, William   1564–1616
English poet, playwright, and actor

ARMANDO: How hast thou purchased this experience?
MOTH: By my penny of observation.
In *Great Books of the Western World* (Volume 26)
The Plays and Sonnets of William Shakespeare (Volume 1)
Love's Labour's Lost
Act III, Scene i, l. 23
Encyclopædia Britannica, Inc. Chicago, Illinois, USA. 1952

Steinbeck, John   1902–68
American novelist

There are good things to see in the tide pools and there
are exciting and interesting thoughts to be generated from
the seeing. Every new eye applied to the peep hole which
looks out at the world may fish in some new beauty and
some new pattern, and the world of the human mind must
be enriched by such fishing.
In Edward F. Ricketts, Jack Calvin, and Joel W. Hedgpeth
Between Pacific Tides
Prefaces (p. xi)
Stanford University Press. Stanford, California, USA. 1968

…one can live in a prefabricated world, smugly and
without question, or one can indulge perhaps the greatest
human excitement: that of observation to speculation to
hypothesis. This is a creative process, probably the highest and most satisfactory we know.
In Edward F. Ricketts, Jack Calvin and Joel W. Hedgpeth
Between Pacific Tides
Prefaces (p. xi)
Stanford University Press. Stanford, California, USA. 1968

Sterne, Laurence   1713–68
English novelist and humorist

What a large volume of adventures may be grasped within this little span of life by him who interests his heart in
everything and who, having eyes to see what time and
chance are perpetually holding out to him as he jour-
neyeth on his way, misses nothing he can fairly lay his
hands on.
The Life and Opinions of Tristram Shandy, Gentleman and a Sentimen-
tal Journey Through France and Italy (Volume 2)
In the Street (p. 251)
Macmillan & Company Ltd. London, England. 1900

Stewart, Ian   1945–
English mathematician and science writer

It's amazing how long it can take to see the obvious. But
of course it's only obvious now.
Nature's Numbers: The Unreal Reality of Mathematical Imagination
Chapter 3 (pp. 32–33)
Basic Books, Inc. New York, New York, USA. 1995

Swift, Jonathan   1667–1745
Irish-born English writer

That was excellently observ'd, say I, when I read a Pas-
sage in an Author, where his Opinion agrees with mine.
When we differ, there I pronounce him to be mistaken.
Satires and Personal Writings
Thoughts on Various Subjects (p. 416)
Oxford University Press, Inc. New York, New York, USA. 1965

Sylvester, James Joseph   1814–97
English mathematician

Most, if not all, of the great ideas of modern mathematics
have had their origin in observation.
A Plea for the Mathematician
Nature, Volume 1, Thursday, December 30, 1869 (p. 238)

Teale, Edwin Way   1899–1980
American naturalist

For observing nature, the best pace is a snail's pace.
Circle of the Seasons
July 14 (p. 150)
Dodd, Mead & Company. New York, New York, USA. 1953

The Bible

You have seen much but perceived little…
The Revised English Bible
Isaiah 42:20
Oxford University Press, Inc. Oxford, England. 1989

Thiele, T. N.   1838–1910
Danish astronomer

An isolated sensation teaches us nothing, for it does
not amount to an observation. Observation is a putting
together of several results of sensation which are or are
supposed to be connected with each other according to
the law of causality, so that some represent causes and
others their effects.
Theory of Observations (p. 2)
Cahales & Edwin Layton. London, England. 1903

Thomas, Lewis   1913–93
American physician and biologist

The role played by the observer in biological research is complicated but not bizarre: he or she simply observes, describes, interprets, maybe once in a while emits a hoarse shout, but that is that; the act of observing does not alter fundamental aspects of the things observed, or anyway isn't supposed to.
The Medusa and the Snail: More Notes of a Biology Watcher
An Apology (p. 88)
The Viking Press. New York, New York, USA. 1979

Thompson, Silvanus P.   1851–1916
English physics professor and author

The seemingly useless or trivial observation made by one worker leads on to a useful observation by another; and so science advances, "creeping on from point to point."
In Sir Richard Arman Gregory
Discovery; or, The Spirit and Service of Science
Chapter XI (p. 292)
Macmillan & Company Ltd. London, England. 1918

Thompson, William Robin   1887–1972
Canadian entomomoligist

The mathematical machine works with unerring precision; but what we get out of it is nothing more than a rearrangement of what we put into it. In the last analysis observation — the actual contact with real events — is the only reliable way of securing the data of natural history.
Science and Common Sense
Chapter Six (pp. 114–115)
Yale University Press. New Haven, Connecticut, USA. 1951

Thoreau, Henry David   1817–62
American essayist, poet, and practical philosopher

The question is not what you look at — but how you look & whether you see.
Journal (Volume 3: 1848–1851)
August 5, 1851 (pp. 354–355)
Princeton University Press. Princeton, New Jersey, USA. 1981

von Goethe, Johann Wolfgang   1749–1832
German poet, novelist, playwright, and natural philosopher

An extremely odd demand is often set forth but never met, even by those who make it; *i.e.*, that empirical data should be presented without any theoretical context, leaving the reader, the student, to his own devices in judging it. This demand seems odd because it is useless simply to look at something. Every act of looking turns into observation, every act of observation into reflection, every act of reflection into the making of associations; thus it is evident that we theorize every time we look carefully at the world.
In Douglas Miller
Scientific Studies (Volume 12)
Theory of Color
Preface (p. 159)
Suhrkamp. New York, New York, USA. 1988

von Humboldt, Alexander   1769–1859
German naturalist and explorer

True cosmical views are the result of observation and ideal combination, and of a long-continued communion with the external world; nor are they a work of a single people, but the fruits yielded by reciprocal communication, and by a great, if not general, intercourse between different nations.
Cosmos: A Sketch of a Physical Description of the Universe (Volume 2)
Physical Contemplation of the Universe (p. 116)
Harper & Brothers. New York, New York, USA. 1869

von Liebig, Justus   1803–73
German organic chemist

However numerous our observations may be, yet, if they only bear on one side of a question, they will never enable us to penetrate the essence of a natural phenomenon in its full significance.
Animal Chemistry
Preface (p. xxxii)
Johnson Reprint Corporation. New York, New York, USA. 1964

Wells, H. G. (Herbert George)   1866–1946
English novelist, historian, and sociologist

…nothing destroys the powers of general observation quite so much as a life of experimental science.
Seven Famous Novels by H.G. Wells
The Food of the Gods
Chapter 2 (p. 540)
Alfred A. Knopf. New York, New York, USA. 1934

Westaway, Frederic William
Science writer

It is the essence of good observation that the eye shall not only see a thing itself, but what parts that thing is composed. And if an observer is to become a successful investigator in any department of Science, he must have an extreme acquaintance with what has already been done in that particular department. Only then will he be prepared to seize upon any one of those minute indications which often connect phenomena apparently quite remote from each other. His eyes will thus be struck with any occurrence which, according to received theories, ought not to happen; for these are the facts which serve as clues to new discoveries.
Scientific Method: Its Philosophy and Its Practice
Chapter XVI, Section 3 (p. 196)
Blackie & Sons Ltd. London, England. 1919

Wheeler, John Archibald   1911–
American physicist and educator

Only by the analysis and interpretation of observations as they are made, and the examination of the larger implications of the results, is one in a satisfactory position to pose new experimental and theoretical questions of the greatest significance.
Elementary Particle Physics
American Scientist, Spring, April 1947 (p. 189)

Wheeler, John Archibald   1911–
American physicist and educator
Thorne, Kip S.   1940–
American theoretical physicist

May the universe in some strange sense be "brought into being" by the participation of those who participate?.... [T]he vital act is the act of participation. "Participator" is the incontrovertible new concept given by quantum mechanics. It strikes down the term "observer" of classical theory, the man who stands safely behind the thick glass wall and watches what goes on without taking part. It can't be done, quantum mechanics says.
Gravitation (p. 1273)
W.H. Freeman and Company. New York, New York, USA. 1973

Whitehead, Alfred North   1861–1947
English mathematician and philosopher

We habitually observe by the method of difference. Sometimes we see an elephant, and sometimes we do not. The result is that an elephant, when present, is noticed.
Process and Reality: An Essay in Cosmology
Part I, Chapter I, Section II (p. 6)
The Macmillan Company. New York, New York, USA. 1929

Wilson, Jr., E. Bright   1908–92
American physical chemist

Observations are useless until they have been interpreted.
An Introduction to Scientific Research
Chapter 8 (p. 169)
McGraw-Hill Book Company, Inc. New York, New York USA. 1952

Wöhler, Friedrich   1800–82
German chemist

My imagination is pretty active, but in thinking I am very slow. No one is less made to be a critic than I. The organ for philosophical thought I lack completely, as you well know, as completely as that for mathematics. Only for observing, do I possess, or at least I believe I do, a passable arrangement in my brain. A kind of instinct that allows me to become aware of relations among data may well be connected with [this arrangement].
In O. Theodor Benfey
From Vital Force to Structural Formulas
Chapter 3 (p. 18)
Houghton Mifflin Company. Boston, Massachusetts, USA. 1964

Wright, R. D.
No biographical data available

Whatever happened to the terms probability and observation? Are statements of high probability now to be deified by calling them truths? Does a set of consistent observations become fact? When I teach biology to the college student, the nature of information mandates that the class and I preserve a healthy skepticism regarding both the broad generalizations and the specific statements of the discipline. Fact and truth are terms we almost never use. There is nothing shameful in describing what we know as having a certain probability, following from observations that have a degree of imprecision. That's the nature of science, including the science of evolution.
Letters
BioScience, Volume 31, Number 11, December 1981 (p. 788)

Zinsser, Hans   1878–1940
American bacteriologist

The scientist takes off from the manifold observations of predecessors, and shows his intelligence, if any, by his ability to discriminate between the important and the negligible, by selecting here and there the significant stepping-stones that will lead across the difficulties to new understanding. The one who places the last stone and steps across the terra firma of accomplished discovery gets all the credit. Only the initiated know and honor those whose patient integrity and devotion to exact observation have made the last step possible.
As I Remember Him: The Biography of R.S.
Chapter XX (p. 332)
Little, Brown & Company. Boston, Massachusetts, USA. 1940

OBSERVATORY

Cerf, Bennett   1898–1971
American publisher and editor

Some weeks later the Einsteins were taken to the Mt. Wilson Observatory in California. Mrs. Einstein was particularly impressed by the giant telescope.
Try and Stop Me: A Collection of Anecdotes and Stories, Mostly Humorous
On the Telescope (p. 163)
Simon & Schuster. New York, New York, USA. 1944

Emerson, Ralph Waldo   1803–82
American lecturer, poet, and essayist

What is so good in a college as an observatory? The sublime attaches to the door and to the first stair you ascend; — and this is the road to the stars...
Journals of Ralph Waldo Emerson 1864–1876
November 14, 1865 (p. 118)
Houghton Mifflin Company. Boston, Massachusetts, USA. 1911

Lowell, Percival   1855–1916
American astronomer

A steady atmosphere is essential to the study of planetary detail; size of instrument being a very secondary matter. A large instrument in poor air will not begin to show what a smaller one in good air will. When this is recognized, as it eventually will be, it will become the fashion to put up observatories where they can see rather than be seen.

Mars
Preface (p. v)
Houghton Mifflin Company. Boston, Massachusetts, USA. 1895

Mitchell, Maria   1818–89
American astronomer and educator

There is no observatory in this land, nor in any land, probably, of which the question is not asked, "Are they doing anything? Why don't we hear from them? They should make discoveries, they should publish."

In Phebe Mitchell Kendall
Maria Mitchell: Life, Letters, and Journals
Chapter XI (p. 223)
Lee & Shepard. Boston, Massachusetts, USA. 1896

Rosseland, Svein   1894–1985
Norwegian astronomer

…an astronomical observatory of today looks more like a factory plant than an abode for philosophers. The poetry of constellations has given way to the lure of plate libraries, and the angel of cosmogenic speculation has been caught in a cobweb of facts insistently clamoring for explanations.

Theoretical Astrophysics: Atomic Theory and the Analysis of Stellar Atmospheres and Envelopes
Introduction (p. xi)
At The Clarendon Press. Oxford, England. 1936

Russell, Henry Norris   1877–1957
American astronomer

The good spectroscopist — to parody the old jest — might perhaps be permitted to go, when he died, instruments and all, and set up an observatory on the moon.

Where Astronomers Go When They Die
Scientific American, Volume 149, Number 3, September 1933 (p. 112)

OBSERVER

Agassiz, Jean Louis Rodolphe   1807–73
Swiss-born American naturalist, geologist, and teacher

He is lost, as an observer, who believes that he can, with impunity, affirm that for which he can adduce no evidence.

In Burt G. Wilder
Louis Agassiz, Teacher
The Harvard Graduate's Magazine, June, 1907

Cuvier, Georges   1769–1832
French zoologist and statesman

The observer listens to nature; the experimenter questions and forces her to unveil herself.

In Claude Bernard
An Introduction to the Study of Experimental Medicine
Part One, Chapter I (p. 6)
Henry Schuman, Inc. New York, New York, USA. 1927

de Chambaud, J. J. Ménuret
No biographical data available

The name of observer has been given to the physicist who is content to examine the phenomena just as nature presents them to him; he differs from the experimental physicist who combines…and who sees only the result of his own combinations. This latter one never sees nature as it is in fact; he pretends by his labor to render nature more accessible to the senses, to raise the mask which conceals it from our eyes, but often he disfigures it and renders it unintelligible. Nature is always unveiled and bare for him who has eyes — or it is covered only by a slight gauze which the eye and reflection easily pierce — and the pretended mask exists only in the imagination, usually quite limited, of the manipulator of experiments.

In D. Diderot and J.L. d'Alembert (eds.)
Observateur
Encyclopédie, ou Dictionnaire Raisonné des Sciences, des Arts et des Métiers, Volume 23 (p. 287D)

Eddington, Sir Arthur Stanley   1882–1944
English astronomer, physicist, and mathematician

Let the observer place himself so that he is, to the best of his knowledge, at rest. If he is a normal human being, he will seat himself in an arm-chair; if he is an astronomer, he will place himself on the sun or at the centre of the stellar universe.

Space, Time and Gravitation: An Outline of the General Relativity Theory
Chapter II (p. 38)
At The University Press. Cambridge, England. 1921

OBSTETRICS

Hosmer, William
No biographical data available

The present practice of medicine, especially obstetrics, must be set down not only as having an immoral tendency, but as, in itself, a gross, abusive, and shameless immorality.

Young Lady's Book: Or, Principles of Female Education
Chapter V (p. 191)
Miller, Orton & Mulligan. Buffalo, New York, USA. 1854

OCCAM'S RAZOR

Chekhov, Anton Pavlovich   1860–1904
Russian author and playwright

It is unfortunate that we try to solve the simplest questions cleverly, and therefore make them unusually complicated. We should seek a simple solution.
Note-Book of Anton Chekhov (p. 20)
B.W. Huebsch, Inc. New York, New York, USA. 1921

Crick, Francis Harry Compton   1916–2004
English biochemist

While Occam's razor is a useful tool in the physical sciences, it can be a very dangerous implement in biology. It is thus very rash to use simplicity and elegance as a guide in biological research.
What Mad Pursuit: A Personal View of Scientific Discovery
Chapter 13 (p. 138)
Basic Books, Inc. New York, New York, USA. 1988

Dixon, Malcom
No biographical data available

God doesn't always shave with Occam's razor.
In David Hall
Letters, God's Razor
New Scientist, Volume 142, Number 1922, April 23, 1994 (p. 51)

Gettings, Fred
No biographical data available

Simples sigillum veri
Cut causes, be merry
Slash 'em and dock 'em
Said William of Ockham
Wiping his razor
On the sleeve of his blazer.
In Renee Haynes
Signs of Secrecy
Times Literary Supplement, June 18, 1981 (p. 688)

Jeans, Sir James Hopwood   1877–1946
English physicist and mathematician

When two hypotheses are possible, we provisionally choose that which our minds adjudge to be the simpler, on the supposition that this is more likely to lead in the direction of the truth. It includes as a special case the principle of Occam's razor — *Entia non multiplicanda praeter necessitatem.*
Physics and Philosophy
Chapter VII (p. 183)
Dover Publications, Inc. New York, New York, USA. 1981

Newton, Sir Isaac   1642–1727
English physicist and mathematician

We are to admit no more causes of natural things than such as are both true and sufficient to explain their appearances.
The Mathematical Principles of Natural Philosophy
Book Three, Rule I (p. 270)
Printed for H.D. Symonds. London, England. 1803

Oppenheimer, J. Robert   1904–67
American theoretical physicist

We cannot in any sense be both the observers and the actors in any specific instance, or we shall fail properly to be either one or the other; yet we know that our life is built of these two modes, is part free and part inevitable, is part creation and part discipline, is part acceptance and part effort.
Science and the Common Understanding
Chapter 6 (p. 88)
Simon & Schuster. New York, New York, USA. 1954

Stenger, Victor J.   1935–
American physicist

The use of Occam's razor, along with the related critical, skeptical view toward any speculations about the unknown, is perhaps the most misunderstood aspect of the scientific method. People confuse doubt with denial. Science doesn't deny anything, but it doubts everything not required by the data. Note, however, that doubt does not necessarily mean rejection, just an attitude of disbelief that can be changed when the facts require it.
Physics and Psychics: The Search for a World Beyond the Senses
Chapter 1 (p. 26)
Prometheus Books. Buffalo, New York, USA. 1990

OCEAN

Aeschylus   525 BCE–426 BCE
Greek playwright

Ye waves
That o'er th' interminable ocean wreathe
Your crisped smiles.
Prometheus Bound, l. 95
Heritage Press. New York, New York, USA. 1966

Beebe, William   1877–1962
American ornithologist

The eternal one, the one most worthy and which will not pass from mind, the only other place comparable to these marvelous regions, must surely be naked space itself, out far beyond atmosphere, between the stars, where sunlight has no grip upon the dust and rubbish of planetary air, where the blackness of space, the shining planets, comets, suns, and stars must really be closely akin to the world of life as it appears to the eyes of an awed human being, in the open ocean, one half mile down.
Half Mile Down
Chapter 11 (p. 225)
Harcourt, Brace & Company. New York, New York, USA. 1934

Beston, Henry   1888–1968
American writer

The seas are the heart's blood of the earth.
The Outermost House
Chapter III (p. 47)
Rinehart & Company. New York, New York, USA. 1928

Bierce, Ambrose   1842–1914
American newspaperman, wit, and satirist

OCEAN, n. A body of water occupying about two-thirds of a world made for man — who has no gills.
The Enlarged Devil's Dictionary (p. 207)
Doubleday. Garden City, New York, USA. 1967

Bishop Joseph Hall   1574–1656
English bishop and satirist

There is many a rich stone laid up in the bowels of the earth, many a fair pearle in the bosome of the sea, that never was seene nor never shall bee.
The Works of the Right Reverend Father in God, Joseph Hall (Volume 1)
Contemplations (p. 115)
Printed by C. Whittingham. London, England. 1808

Browning, Robert   1812–89
English poet

The sea heaves up, hangs loaded o'er the land,
Breaks there, and buries its tumultuous strength.
The Poems and Plays of Robert Browning
Luria
Act I
The Modern Library. New York, New York, USA. 1934

Bryant, William Cullen   1794–1878
American poet

That make the meadows green; and, poured round all,
Old Ocean's gray and melancholy waste, — Are but the solemn decorations all
Of the great tomb of man.
Poems
Thanatopsis
D. Appleton. New York, New York, USA. 1874

Byron, George Gordon, 6th Baron Byron   1788–1824
English Romantic poet and satirist

Time writes no wrinkle on thine azure brow,
Such as Creation's dawn beheld, thou rollest now.
The Complete Poetical Works of Byron
Childe Harold
Canto IV, Stanza 182
Houghton Mifflin Company. Boston, Massachusetts, USA. 1933

Roll on, thou deep and dark blue Ocean — roll!
Ten thousand fleets sweep over thee in vain;
Man marks the earth with ruin — his control
Stops with the shore.
The Complete Poetical Works of Byron
Childe Harold
canto IV, Stanza 179
Houghton Mifflin Company. Boston, Massachusetts, USA. 1933

Carson, Rachel   1907–64
American marine biologist and author

There is no drop of water in the ocean, not even in the deepest parts of the abyss, that does not know and respond to the mysterious forces that create the tide. No other force that affects the sea is so strong.
The Sea Around Us
Part II, Chapter 3 (p. 149)
Oxford University Press, Inc. New York, New York, USA. 1989

Unmarked and trackless though it may seem to us, the surface of the ocean is divided into definite zones, and the pattern of the surface water controls the distribution of its life.
The Sea Around Us
Part I, Chapter 2 (p. 20)
Oxford University Press, Inc. New York, New York, USA. 1989

The ocean is the earth's greatest storehouse of minerals.
The Sea Around Us
Part III, Chapter 2 (p. 185)
Oxford University Press, Inc. New York, New York, USA. 1989

The edge of the sea is a strange and beautiful place. All through the long history of Earth it has been an area of unrest where waves have broken heavily against the land, where the tides have pressed forward over the continents, receded, and then returned. For no two successive days is the shore line precisely the same. Not only do the tides advance and retreat in their eternal rhythms, but the level of the sea itself is never at rest. It rises or falls as the glaciers melt or grow, as the floor of the deep ocean basins shift under its increasing load of sediments, or as the earth's crust along the continental margins warps up or down in adjustment to strain and tension. Today a little more land may belong to the sea, tomorrow a little less. Always the edge of the sea remains an elusive and indefinable boundary.
The Edge of the Sea
Chapter I (p. 1)
Houghton Mifflin Company. Boston, Massachusetts, USA. 1955

The face of the sea is always changing. Crossed by colors, lights, and moving shadows, sparkling in the sun, mysterious in the twilight, its aspects and its moods vary hour by hour.
The Sea Around Us
Part I, Chapter 3 (p. 29)
Oxford University Press, Inc. New York, New York, USA. 1989

The continents themselves dissolve and pass to the sea, in grain after grain of eroded land…
The Sea Around Us
Part III, Chapter 14 (p. 212)
Oxford University Press, Inc. New York, New York, USA. 1989

Every living thing of the ocean, plant and animal alike, returns to the water at the end of its own life span the materials that had been temporarily assembled to form its body. So there descends into the depths a gentle never-ending rain of the disintegrating particles of what once were living creatures of the sunlit surface waters, or of those twilight regions beneath.
Undersea
Atlantic Monthly, September 1937

Cornwall, Barry (Bryan Waller Procter)   1787–1874
English author

The sea! the sea! the open sea!
The blue, the fresh, the ever free!
Without a mark, without a bound,
It runneth the earth's wide regions round;
It plays with the clouds; it mocks the skies;
Or like a cradled creature lies.
In Richard Green Parker and J. Madison Watson (eds)
The National Fourth Reader
The Sailor's Song (p. 156)
Barns & Burr. New York, New York, USA. 1864

Dubos, René Jules   1901–82
French-born American microbiologist and environmentalist

The seas perhaps hold the highest hopes for continued life. Yet, what does man do to the seas? Not only does he grab with greed the creatures of the sea, he turns nature's cradle for life into a receptacle for garbage and filth. This is man whose life blood contains sodium, potassium, and calcium in almost the same proportions as they still exist in the environment of mother sea which encouraged his birth.
In Maurice F. Strong (ed.)
Who Speaks for Earth?
Unity Through Diversity (p. 44)
W.W. Norton & Company, Inc. New York, New York, USA. 1973

Emerson, Ralph Waldo   1803–82
American lecturer, poet, and essayist

Behold the Sea,
The opaline, the plentiful and strong,
Yet beautiful as is the rose in June,
Fresh as the trickling rainbow of July;
Sea full of food, the nourisher of kinds,
Purger of earth, and medicine of men;
Creating a sweet climate by my breath,
Washing out harms and griefs from memory,
And, in my mathematic ebb and flow,
Giving a hint of that which changes not.
The Complete Works of Ralph Waldo Emerson (Volume 9)
Seashore (p. 242)
Houghton Mifflin Company. Boston, Massachusetts, USA. 1904

Forbes, Edward   1815–54
English naturalist

Moreover it is becoming the Britons, whether scientific or unscientific, who boast at all fitting occasions of their aptitude to rule the waves, should know something of the population of their saline empire, especially of those parts of it immediately in contact with their terrestrial domain, and the coasts of the Continent to which our United Kingdom appertains.
The Natural History of the European Seas
Chapter I (p. 3)
John Van Voorst. London, England. 1859

...beneath the waves there are many dominions yet to be visited, and kingdoms to be discovered; and he who venturously brings up from the abyss enough of their inhabitants to display the physiognomy of the country, will taste that cup of delight, the sweetness of whose draught those only who have made a discovery know.
The Natural History of the European Seas
Chapter I (p. 11)
John Van Voorst. London, England. 1859

Gould, Hannah Flagg   1789–1865
American poet

Alone I walked on the ocean strand,
A pearly shell was in my hand;
I stooped, and wrote upon the sand
My name, the year, the day.
As onward from the sport I passed,
One lingering look behind I cast,
A wave came rolling high and fast,
And washed my lines away.
Poems
A Name in the Sand
Hilliard, Gray & Company. Boston, Massachusetts, USA. 1839

Gray, Thomas   1716–71
English poet

Full many a gem of purest ray serene,
The dark unfathomed caves of ocean bear.
The Complete Poetical Works of Gray, Beattie, Blair, Collins, Thomson, and Kirke White
Elegy in a Country Churchyard
Stanza 14
J. Blackwood. London, England. 1800

Henderson, Lawrence   1878–1942
American biochemist

No philosopher's or poet's fancy, no myth of a primitive people has ever exaggerated the importance, the usefulness, and above all the marvelous beneficence of the ocean for the community of living things.
The Fitness of the Environment: An Inquiry into the Biological Significance of the Properties of Matter
Chapter V, Section III (p. 190)
The Macmillan Company. New York, New York, USA. 1913

The regulatory devices of our modern laboratories have not yet succeeded in rivaling the oceans. Singly, certain conditions, for example, temperature, alkalinity, and concentration, may be more accurately regulated by man, though on a small scale only; but the regulation of all such properties together is not yet possible. The only known improvement upon the ocean is the body of a higher warm-blooded animal. Here, however, the processes of organic evolution have begun with the ocean, and in several respects merely perfected existing arrangements.
The Fitness of the Environment: An Inquiry into the Biological Significance of the Properties of Matter

Chapter V, Section III (p. 186)
The Macmillan Company. New York, New York, USA. 1913

Hess, Harry
No biographical data available

The birth of the oceans is a matter of conjecture, the subsequent history is obscure, and the present structure is just beginning to be understood. Fascinating speculation on these subjects has been plentiful, but not much of it predating the last decade holds water.
In A.E.J. Engel, Harold L. James, and B.F. Leonard (eds.)
Petrologic Studies — A Volume in Honor of A.F. Buddington
History of the Ocean Basins (p. 599)
The Geological Society of America. 1962

Heyerdahl, Thor   1914–2002
Norwegian ethnographer and adventurer

…bear in mind that the ocean currents circulate with no regard for political borderlines, and that nations can divide the land, but the revolving ocean, indispensable and yet vulnerable, will forever remain a common human heritage.
In Maurice F. Strong (ed.)
Who Speaks for Earth?
How Vulnerable Is the Ocean? (p. 63)
W.W. Norton & Company, Inc. New York, New York, USA. 1973

Horsfield, Brenda
No biographical data available
Stone, Peter Bennet
No biographical data available

…there is on the other hand some encouragement in the reflection that Oceanography has usually only ruined the reputations of people who dared to speculate too little and thought on too small a scale. She has smiled most benignly on those who backed the most daring and outrageous possibility…
The Great Ocean Business
Chapter 7 (p. 150)
Coward, McCann & Geoghegan. New York, New York, USA. 1972

Ingelow, Jean   1820–97
English poet and novelist

Quoth the Ocean, "Dawn! O fairest, clearest,
Touch me with thy golden fingers bland;
For I have no smile till thou appearest
For the lovely land."
The Poetical Works of Jean Ingelow
Winstanley
The Apology
John B. Alden, Publisher. New York, New York, USA. 1883

Kennedy, John F.   1917–63
35th president of the United States

Knowledge of the oceans is more than a matter of curiosity. Our very survival may hinge upon it.

General Government Matters: Department of Commerce, and Related Agencies Appropriations for 1862
Letter to the President of the Senate on Increasing the National Effort, in Oceanography, March 29, 1961 (p. 549)
U.S. Government Printing Office. Washington, D.C. 1961

Landor, Walter Savage   1775–1864
English poet and essayist

Past are three summers since she first beheld
The ocean; all around the child await
Some exclamation of amazement here:
She coldly said, her long-lasht eyes abased,
Is this the mighty ocean? is this all?
Gebir
Book V, l. 133–137
Woodstock Books. Oxford, England. 1993

But I have sinuous shells of pearly hue;
….
Shake one, and it awakens; then apply
Its polished lips to your attentive ear,
And it remembers its august abodes,
And murmurs as the ocean murmurs there.
Gebir
Book I, l. 169, 173–176
Woodstock Books. Oxford, England. 1993

Larcom, Lucy   1824–93
American writer

The land is dearer for the sea,
The ocean for the shore.
The Poetical Works of Lucy Larcom
On the Beach
Stanza 11
Houghton Mifflin Company. Boston, Massachusetts, USA. 1884

Lee-Hamilton, Eugene J.   1845–1907
English poet

The hollow sea-shell, which for years hath stood
On dusty shelves, when held against the ear
Proclaims its stormy parent, and we hear
The faint, far murmur of the breaking flood.
We hear the sea. The Sea? It is the blood
In our own veins, impetuous and near.
Sea-Shell Murmurs
The Living Age, Volume CLVI, January, February, March 1883 (p. 322)

Longfellow, Henry Wadsworth   1807–82
American poet

Would'st thou, — so the helmsman answered,
Learn the secret of the sea?
Only those who brave its dangers
Comprehend its mystery!
The Seaside and the Fireside
The Secret of the Sea
Stanza 8
Ticknor, Reed & Fields. Boston, Massachusetts, USA. 1850

Melville, Herman   1819–91
American novelist

Not only is the sea such a foe to man who is an alien to it, but it is also a fiend to its own offspring; worse than the Persian host who murdered his own guests; sparing not the creatures which itself hath spawned. Like a savage tigress that tossing in the jungle overlays her own cubs, so the sea dashes even the mightiest whales against the rocks, and leaves them there side by side with the split wrecks of ships. No mercy, no power but its own controls it. Panting and snorting like a mad battle steed that has lost its rider, the masterless ocean overruns the globe.
In *Great Books of the Western World* (Volume 48)
Moby Dick
Chapter 58 (p. 204)
Encyclopædia Britannica, Inc. Chicago, Illinois, USA. 1952

…that same image [of Narcissus] we ourselves see in all rivers and oceans. It is the image of the ungraspable phantom of life: and this is the key to it all.
In *Great Books of the Western World* (Volume 48)
Moby Dick
Chapter 1 (pp. 2–3)
Encyclopædia Britannica, Inc. Chicago, Illinois, USA. 1952

Milton, John   1608–74
English poet

…a dark
Illimitable ocean without bound,
Without dimension, where length, breadth, and height
And time and place are lost…
In *Great Books of the Western World* (Volume 32)
Paradise Lost
Book II, l. 891–894
Encyclopædia Britannica, Inc. Chicago, Illinois, USA. 1952

Mishima, Yukio   1925–70
Japanese writer

Down beneath the spray, down beneath the whitecaps, that beat themselves to pieces against the prow, there were jet-black invisible waves, twisting and coiling their bodies. They kept repeating their patternless movements, concealing their incoherent and perilous whims.
The Sound of Waves
Chapter 14 (p. 125)
Berkeley Publishing Group. New York, New York, USA. 1961

Montgomery, Robert
No biographical data available

And Thou, vast Ocean! on whole awful face
Time's iron feet can print no ruin trace.
Notes and Queries
The Omnipresence of the Deity
Part I, Stanza 20
Oxford University Press. London, England. 1849

Rossetti, Christina Georgina   1830–94
English poet

Why does the sea moan evermore?
Shut out from heaven it makes its moan,
It frets against the boundary shore;
All earth's full rivers cannot fill
The sea, that drinking thirsteth still.
In William Michael Rossetti
The Poetical Works of Christina Georgina Rossetti
By the Sea
Stanza 1
Macmillan & Company Ltd. London, England. 1911

Shelley, Percy Bysshe   1792–1822
English poet

There the sea I found
Calm as a cradled child in dreamless slumber bound.
The Poems of Percy Bysshe Shelley
The Revolt of Islam, Canto I, Stanza 15
Houghton Mifflin Company. Boston, Massachusetts, USA. 1901

Stoddard, Richard Henry   1825–1903
American critic and poet

Thou wert before the Continents, before
The hollow heavens, which like another sea
Encircles them and thee, but whence thou wert,
And when thou wast created, is not known,
Antiquity was young when thou wast old.
In Anna Ward (ed.)
Surf and Wave: The Sea as Sung by the Poets
Hymn to the Sea, l. 104
Thomas Y. Crowell & Company. New York, New York, USA. 1883

Taylor, Bayard   1825–78
American journalist and author

We follow and race
In shifting chase,
Over the boundless ocean-space!
Who hath beheld when the race begun?
Who shall behold it run?
In Anna Ward (ed.)
Surf and Wave: The Sea as Sung by the Poets
The Waves
Thomas Y. Crowell & Company. New York, New York, USA. 1883

Tennyson, Alfred (Lord)   1809–92
English poet

Break, break, break,
On thy cold gray stones, O sea!
And I would that my tongue could utter
The thoughts that arise in me.
Alfred Tennyson's Poetical Works
"Break, Break, Break"
Oxford University Press, Inc. London, England. 1953

Thoreau, Henry David   1817–62
American essayist, poet, and practical philosopher

The ocean is a wilderness reaching round the globe, wilder than a Bengal jungle, and fuller of monsters,

washing the very wharves of our cities and the gardens of our seaside residences.
Cape Cod
Chapter IX (p. 148)
Princeton University Press. Princeton,, New Jersey, USA. 2004

von Goethe, Johann Wolfgang   1749–1832
German poet, novelist, playwright, and natural philosopher

The sea is flowing ever,
The land retains it never.
The Works of Johann Wolfgang von Goethe
Hikmet Nameh
Book of Proverbs (p. 395)
J.H. Moore. Philadelphia, Pennsylvania, USA. 1901

Webb, Charles Henry   1834–1905
American writer

I send thee a shell from the ocean-beach;
But listen thou well, for my shell hath speech.
Hold to thine ear
And plain thou'lt hear
Tales of ships.
Vagrom Verse
With a Nantucket Shell
Ticknor & Company. Boston, Massachusetts, USA. 1889

Weisz, Paul B.   1919–
German-born American chemical engineer and biomedical researcher

The Pacific. You don't comprehend it by looking at a globe, but when you're traveling at four miles a second and it still takes you twenty-five minutes to cross it, you know it's big.
In Kevin W. Kelley
The Home Planet
With Plate 64
Addison–Wesley. Reading, Massachusetts, USA. 1988

Whitman, Walt   1819–92
American poet, journalist, and essayist

To me the sea is a continual miracle,
The fishes that swim — the rocks — the motion of the waves — the ships with men in them,
What stranger miracles are there?
Complete Poetry and Collected Prose
Miracles
The Library of America. New York, New York, USA. 1982

Wordsworth, William   1770–1850
English poet

I have seen
A curious child, who dwelt upon a tract
Of inland ground, applying to his ear
The convolutions of a smooth-lipped shell;
To which, in silence hushed, his very soul
Listened intensely; and his countenance soon
Brightened with joy; for from within were heard
Murmurings, whereby the monitor expressed
Mysterious union with its native sea.
Poems By William Wordsworth
Excursions, Book IV
Ginn and Company. Boston, Massachusetts, USA. 1897

OCEANOGRAPHY

Spilhaus, Athelstan   1911–78
South-African born American geophysicist and oceanographer

The science of oceanography is not a discipline but an adventure wherein any discipline or combination of disciplines may be focused on understanding and using the sea and all that is in it.
Annual Report of the Board of Regents of the Smithsonian Institution, 1964
The Future of Oceanography (p. 361)
Government Printing Office. Washington, D.C. 1965

ODDS

Jeans, Sir James Hopwood   1877–1946
English physicist and mathematician

Choose a point in space at random and the odds against it being occupied by a star are enormous.
The Universe Around Us
Chapter I (p. 102)
The Macmillan Company. New York, New York, USA. 1929

Stoppard, Tom   1937–
Czech-born English playwright

Life is a gamble at terrible odds — if it was a bet you wouldn't take it.
Rosencrantz and Guildenstern Are Dead
Act Three (p. 115)
Grove Press, Inc. New York, New York, USA. 1967

OMEGA POINT

Barrow, John D.   1952–
English theoretical physicist
Tipler, Frank   1947–
American physicist

If life evolves in all of the many universes in a quantum cosmology, and if life continue to exist in all of these universes, then all of these universes, which include all possible histories among them, will approach the Omega Point. At the instant the Omega Point is reached, life will have gained control of all matter and forces not only in a single universe, but in all universes whose existence is logically possible; life will have spread into all spatial regions in all universes which could locally exist, and will have stored an infinite amount of information, including all bits of knowledge which it is logically possible to know.
The Anthropic Cosmological Principle

Chapter 10 (p. 676)
Clarendon Press. Oxford, England. 1986

OPINION

Adams, George   1750–95
English instrument maker

Mankind are always ready to adopt or reject what accords with pre-conceived opinions, to make reason subservient to prejudice, and to reject without examination, whatever is discordant with a received system; thus closing the door of science, and excluding themselves from the benefit of light.
Lectures on Natural and Experimental Philosophy (Volume 1)
Lecture II (p. 27)
Printed by R. Hindmarsh. London, England. 1794

Bernard, Claude   1813–78
French physiologist

…no man's opinion, formulated in a theory or otherwise, may be deemed to represent the whole truth in the sciences. It is a guide, a light, but not an absolute authority. The revolution which the experimental method has effected in the sciences is this: it has put a scientific criterion in the place of personal authority. The experimental method is characterized by being dependent only on itself, because it includes within itself its criterion — experience.
Translated by Henry Copley Greene
An Introduction to the Study of Experimental Medicine
Part One, Chapter II, Section IV (p. 40)
Henry Schuman, Inc. New York, New York, USA. 1927

Bourne, William   1535–82
English mathematician

…for that my opinion doth differ from some of the ancient writers in natural Phylosophy, it is possible that it may be ytterly dislyked of and condemned to be of no trueth.
The Treasure for Travelers
The Fyfth Booke, To The Reader (p. 3)
Publisher undetermined

Browne, Sir Thomas   1605–82
English author and physician

Where we desire to be informed, 'tis food to contest with men above ourselves; but, to confirm and establish our opinions, 'tis best to argue with judgments below our own, that the frequent spoils and victories over their reasons may settle in ourselves an esteem and confirmed opinion of our own.
In Charles Sayle (ed.)
The Works of Sir Thomas Browne (Volume 1)
Religio Medici
The First Part, Section 6 (p. 12)
John Grant. Edinburgh, Scotland. 1912

Crum, H. A.
No biographical data available

Let us also remember that plants vary and opinions vary. One man's fish is another man's poison. One man's moss is another man's mess.
The Bryologist
Traditional Make-Do Taxonomy, Volume 88, 1985 (p. 22)

Hardy, G. H. (Godfrey Harold)   1877–1947
English pure mathematician

It is never worth a first class man's time to express a majority opinion. By definition there are plenty of others to do that.
A Mathematician's Apology
Foreword (p. 46)
Cambridge University Press. Cambridge, England. 1967

Heinlein, Robert A.   1907–88
American science fiction writer

Oh, I have strong opinions, but a thousand reasoned opinions are never equal to one case of diving in and finding out.
Time Enough for Love
Prelude, Chapter I (p. 31)
G.P. Putnam's Sons. New York, New York, USA. 1973

Hering, Constantine   1800–80
Father of American homeopathy

Among men of deliberate and acute reflection, no difference of opinion can exist relative to the truth of a discovery which rests upon the basis of actual experiment.
In S. Hahnemann
Organon of Homoeopathic Medicine
Preface (p. xii)
W. Radde. New York, New York, USA. 1843

Joubert, Joseph   1754–1824
French moralist

Our opinions are clouds between us and the clear skies of truth.
Translated by H.P. Collins
Pensées and Letters of Joseph Joubert
Chapter X (p. 83)
Books for Libraries Press, Freeport, New York, USA. 1972

Lippmann, Walter   1889–1974
American journalist and author

True opinions can prevail only if the facts to which they refer are known; if they are not known, false ideas are just as effective as true ones, if not a little more effective.
Liberty and the News
Liberty and the News (pp. 64–65)
Transaction Publishers. New Brunswick, New Jersey, USA. 1995

Locke, John   1632–1704
English philosopher and political theorist

New opinions are always suspected, and usually opposed, without any other reason but because they are not already common.
In *Great Books of the Western World* (Volume 35)
An Essay Concerning Human Understanding
Dedicatory Epistle (p. 85)
Encyclopædia Britannica, Inc. Chicago, Illinois, USA. 1952

Milton, John   1608–74
English poet

…opinion in good men is but knowledge in the making.
In *Great Books of the Western World* (Volume 32)
Areopagitica (p. 406)
Encyclopædia Britannica, Inc. Chicago, Illinois, USA. 1952

Pearson, Karl   1857–1936
English mathematician

We never think of taking the opinion of the man in the street on the reasons why the moon does not keep her calculated times; we do not ask his opinion on the value of the opsonic index; we recognise that these are problems which require special training and analysis wholly beyond his grasp, but we still think he is quite capable of expressing an opinion on whether the employment of women is good for her infants or not, although he may be in possession of no data, and although, if he were, he would be quite incapable of interpreting them.
Eugenics Laboratory Lecture Series
The Academic Aspect of the Science of National Eugenics, 7, 1911 (p. 20)

Terence   190 BCE–158 BCE
Roman comic dramatist

…as many men, so many opinions…
In T. A. Blythe
A Literal Translation of the Phormio by Terence, (p. 22)
Simkin, Marshall & Company. London, England. 1880

Twain, Mark (Samuel Langhorne Clemens)   1835–1910
American author and humorist

Our opinions do not really blossom into fruition until we have expressed them to someone else.
In Opie Read
Mark Twain and I
Five Quarts of Moonlight Juice (p. 38)
Reilly & Lee. Chicago, Illinois, USA. 1940

Opinions based upon theory, superstition, and ignorance are not very precious.
In Albert Bigelow Paine
Mark Twain's Letters (Volume 2)
Letter to J. H. Twitchell, 1/27/1900 (p. 695)
Harper & Brothers. New York, New York, USA. 1917

Young, Thomas   1773–1829
English polymath

The object of the present dissertation is not so much to propose any opinions which are absolutely new, as to refer some theories, which have been already advanced, to their original inventors, to support them by additional evidence, and to apply them to a great number of diversified facts, which have hitherto been buried in obscurity. Nor is it absolutely necessary in this instance to produce a single new experiment; for of experiments there is already an ample store.
On the Theory of Light and Colours
Philosophical Transactions of the Royal Society of London, Volume 92, 1802 (p. 12)

OPIUM

De Quincey, Thomas   1785–1859
English essayist

Oh! just, subtle, and mighty opium! that to the hearts of poor and rich alike, for the wounds that will never heal, and for "the pangs that tempt the spirit to rebel," bringest an assuaging balm; eloquent opium!
The Collected Writings of Thomas De Quincey (Volume 3)
Confessions of an English Opium Eater
Part II (p. 396)
A. & C. Black. London, England. 1897

Melville, Herman   1819–91
American novelist

…whenever my hypos get such an upper hand of me, that it requires a strong moral principle to prevent me from deliberately stepping into the street, and methodically knocking people's hats off — then, I account it high time to get to sea as soon as I can.
In *Great Books of the Western World* (Volume 48)
Moby Dick
Chapter 1 (p. 2)
Encyclopædia Britannica, Inc. Chicago, Illinois, USA. 1952

OPTICS

Day, Roger E.
No biographical data available

I wish I were a crystal lens,
With aplanatic face,
And lived at Number Seven Ten,
Illumination Place,
City of Glass.
Fantasy of Glass
The Physics Teacher, Volume 3, Number 6, September 1965 (p. 288)

Digges, Leonard   ca. 1520–59
English mathematician

But marvelous are the conclusions that may be performed by glasses concave and convex of Circulare and parabolicall formes, using for multiplication of beames sometime

the aide of Glasses transparent, which by fraction should unite or dissipate the images or figures presented by the reflection of other. By these kinde of Glasses or rather frames of them, placed in due Angles, yee may not only set out of the proportion of an whole region, ye may represent before your eye the lively image of every Towne, Village, &c and that in as little or great space or place as ye will prescribe, but also augment and dilate any parcell thereof.
A Geometrical Practical Treatise Named Pantometria, Divided into Three Bookes, Longimetra, Planimetra, and Stereometria
Chapter 21
Publisher undetermined

Grosseteste, Robert   1175–1253
English statesman

This part of Perspectiva, when well understood, shows us how we may make things a very long distance off appear as if placed very close, and larger near things appear very small, and how we may make small things placed at a distance appear any size we want, so that it may be possible for us to read the smallest letters at incredible distances, or to count sand, or grains, or seeds, or any sort of minute objects…
In A.C. Crombie
Science, Optics, and Music in Medieval and Early Modern Thought
Chapter 9, Section II (p. 198)
The Hambledon Press. London, England. 1990

Huygens, Christiaan   1629–95
Dutch mathematician, astronomer, and physicist

As happens in all the sciences in which Geometry is applied to matter, the demonstrations concerning Optics are founded on truths drawn from experiences.
A Treatise on Light
Chapter I (p. 1)
Macmillan & Company Ltd. London, England. 1912

Joyce, James   1882–1941
Irish-born author

He faced about and, standing between the awnings, held out his right arm at arm's length toward the sun. Wanted to try that often. Yes; completely. The tip of his little finger blotted out the sun's disc. Must be the focus where the rays cross.
Ulysses (p. 164)
Random House, Inc. New York, New York, USA. 1946

Marton, Ladislaus
No biographical data available

"Electron optics I believe,"
He often gravely said,
"Concern a branch of knowledge
That is way above my head."
Alice in Electronland
American Scientist, Volume 31, Number 3, July 1943 (p. 251)

ORBIT

Kepler, Johannes   1571–1630
German astronomer

The testimony of the ages confirm that the motions of the planets are orbicular.
New Astronomy
Part I, 1 (p. 115)
At the University Press. Cambridge, England. 1992

ORDER

Anaxagoras   ca. 500 BCE–428 BCE
Greek philosopher of nature

Mind orders all things.
In Fabre
The Glow-Worm (p. 234)
Hodder & Stoughton Ltd. London, England. 1919

Bacon, Sir Francis   1561–1626
English lawyer, statesman, and essayist

The human understanding is of its own nature prone to abstractions, and gives a substance and reality to things which are fleeting.
In John M. Robinson (ed)
The Philosophical Woks of Francis Bacon
Novum Organon
LI (p. 267)
George Routledge & Sons, Ltd. London, England. 1905

Birkhoff, Garrett   1911–96
American mathematician

…there is hidden order in Nature, to be found only by patient search.
Hydrodynamics: A Study in Logic, Fact, and Simulation
Conclusion (p. 179)
Princeton University Press. Princeton, New Jersey, USA. 1950

Brown, Thomas
No biographical data available

Even the rudest wanderer in the fields…finds that the profusion of blossoms around him — in the greater number of which he is able himself to discover many striking resemblances — may be reduced to some order of arrangement.
Quoted in Hugh Miller
The Testimony of the Rocks; of, Geology in Its Bearings on the Two Theologies, Natural and Revealed
Lecture First (p. 37)
Gould & Lincoln. Boston, Massachusetts, USA. 1857

Browne, Sir Thomas   1605–82
English author and physician

All things begin in order, so shall they end, and so shall they begin again; according to the ordainer of order, and the mysticall mathematicks of the City of Heaven.

In John Carter (ed.)
Urne Buriall and The Garden of Cyrus
The Garden of Cyrus, Chapter V (p. 114)
Cassell. London, England. 1932

Darwin, Charles Robert   1809–82
English naturalist

An organic being is a microcosm — a little universe, formed of a host of self-propagating organisms, inconceivably minute and numerous as the stars of heaven.
The Variation of Animals and Plants Under Domestication (Volume 2)
Chapter XXVII (p. 399)
D. Appleton & Company. New York, New York, USA

Davies, Paul Charles William   1946–
British-born physicist, writer, and broadcaster

The universe contains vastly more order than Earth-life could ever demand. All those distant galaxies, irrelevant for our existence, seem as equally well ordered as our own.
In Eugene F. Mallove
The Quickening Universe: Cosmic Evolution and Human Destiny (p. 61)
St. Martin's Press. New York, New York, USA. 1987

Eddington, Sir Arthur Stanley   1882–1944
English astronomer, physicist, and mathematician

If you take a pack of cards as it comes from the maker and shuffle it for a few minutes, all trace of the original systematic order disappears. The order will never come back however long you shuffle. Something has been done which cannot be undone, namely, the introduction of a random element in place of the arrangement.
The Nature of the Physical World
Chapter IV (p. 63)
The Macmillan Company. New York, New York, USA. 1930

Frankel, Felice   1945–
Science photographer
Whitesides, George M.
American chemist

Order is repetition, regularity, symmetry, simplicity. It forms the spine of our efforts to measure, control, and understand.
On the Surface of Things: Images of the Extraordinary in Science
Order (p. 63)
Chronicle Books. San Francisco, California, USA. 1997

Huntington, Edward V.   1874–1952
Mathematician

The fundamental importance of the subject of order may be inferred from the fact that all the concepts required in geometry can be expressed in terms of the concept of order alone.
The Continuum, and Other Types of Serial Order
Introduction (p. 2)
Harvard University Press. Cambridge, Massachusetts, USA. 1917

Huxley, Thomas Henry   1825–95
English biologist

…the man of science knows that here, as everywhere, perfect order is manifested; that there is not a curve of the waves, not a note in the howling chorus, not a rainbow glint on a bubble which is other than a necessary consequence of the ascertained laws of nature; and that with sufficient knowledge of the conditions competent physico-mathematical skill could account for, and indeed predict, every one of those 'chance' events.
In Francis Darwin (ed.)
The Life and Letters of Charles Darwin (Volume 1)
Chapter XIV (p. 554)
D. Appleton & Company. New York, New York, USA. 1896

Kline, Morris   1908–92
American mathematics professor and writer

Is there a law and order in this universe or is its behavior merely the working of chance and caprice? Will the Earth and other planets continue their motions around the sun or will some unknown body, coming from great distances, rush through our planetary system and alter the course of every planet? Cannot the sun some day explode, as other suns are doing daily, and burn us all to a crisp? Was man deliberately planted on a planet especially prepared for his existence or is he merely an insignificant concomitant of accidental cosmic circumstances?
Mathematics in Western Culture
Chapter XXIV (p. 374)
Oxford University Press, Inc. New York, New York, USA. 1953

Lewis, C. S. (Clive Staples)   1898–1963
British author, scholar, and popular theologian

To the modern man it seems simply natural that an ordered cosmos should emerge from chaos, that life should come out of the inanimate, reason out of instinct, civilization out of savagery, virtue out of animalism. This idea is supported in his mind by a number of false analogies: the oak coming from the acorn, the man from the spermatozoon, the modern steamship from the primitive coracle. The supplementary truth that every acorn was dropped by an oak, every spermatozoon derived from a man, and the first boat by something so much more complex than itself as a man of genius, is simply ignored. The modern mind accepts as a formula for the universe in general the principle "almost nothing may be expected to turn into almost everything" without noticing that the parts of the universe under our direct observation tell a quite different story.
Present Concerns: Essays by C.S. Lewis
Modern Man and His Categories of Thought (p. 63)
Harcourt Brace Jovanovich. New York, New York, USA. 1986

Lucretius   ca. 99 BCE–55 BCE
Roman poet

For verily not by design did the first-beginnings of things station themselves each in its right place guided by keen intelligence, nor did they bargain sooth to say what motions each should assume, but because many in number and shifting about in many ways throughout the universe they are driven and tormented by blows during infinite time past, after trying motions and unions of every kind at length they fall into arrangements such as those out of which our sum of things has been formed...
In *Great Books of the Western World* (Volume 12)
Lucretius: On the Nature of Things
Book One, l. 1020 (p. 13)
Encyclopædia Britannica, Inc. Chicago, Illinois, USA. 1952

Mann, Thomas   1875–1955
German-born American novelist

...order and simplification are the first steps toward the mastery of a subject — the actual enemy is the unknown.
The Magic Mountain
Chapter V
Encyclopaedic (pp. 245–246)
Alfred A. Knopf. New York, New York, USA. 1966

Miller, Jr., G. Tyler
No biographical data available

Man continually engages in attempts to create order, but only at the expense of greater disorder in the surroundings.
Energetics, Kinetics, and Life: An Ecological Approach (p. 200)
Wadsworth Publishing Company. Belmont, California, USA. 1971

Moulton, Forest Ray   1872–1952
American astronomer

To an astronomer the most remarkable and interesting thing about that part of the physical universe with which he has become acquainted is not its vast extent in space, nor the number and great masses of its stars, nor the violent forces that operate in the stars, nor in the long periods of astronomical time, but that which holds him awestruck is the perfect orderliness of the universe and the majestic succession of the celestial phenomena.
In H.H. Newman (ed.)
The Nature of the World and of Man
Astronomy (p. 30)
The University of Chicago Press. Chicago, Illinois, USA. 1927

Now we find ourselves a part of a Universal Order of which we did not dream and whose alphabet we are just beginning to learn. Instead of shrinking it to our measure, we contemplate its infinite orderliness and set no limits to the goal our race may hope to attain.
Astronomy
Chapter XVI (p. 533)
The Macmillan Company. New York, New York, USA. 1931

The orderliness of the universe is the supreme discovery in science; it is that which gives us hope that we shall be able to understand not only the exterior would but also our own bodies and our own mind.
In H.H. Newman (ed.)
The Nature of the World and of Man
Astronomy (p. 30)
The University of Chicago Press. Chicago, Illinois, USA. 1927

Oppenheimer, J. Robert   1904–67
American theoretical physicist

We cannot make much progress without a faith that in this bewildering field of human experience, which is so new and so much more complicated than we thought even five years ago, there is a unique and necessary order: not an order that we can tell a priori, not an order that we can see without experience, but an order which means that the parts fit into a whole and that the whole requires the parts.
The Constitution of Matter (p. 37)
Oregon State System of Higher Education. Eugene, Oregon, USA. 1956

One may only hope that what is at the moment just a picture of chaos will ultimately reveal again that deep harmony and order which one has always found in the physical world when one has pushed hard, and which is very beautiful indeed.
In Lincoln Barnett
Writing on Life: Sixteen Close-Ups
Physicist Oppenheimer (p. 358)
William Sloane Associates, Publishers. New York, New York, USA. 1951

Picard, Charles Emile   1856–1941
French mathematician

We no longer pretend to be able to grasp reality in a physical theory; we see in it rather an analytic or geometric mold useful and fertile for a tentative representation of phenomena, no longer believing that the agreement of a theory with experience demonstrates that the theory expresses the reality of things. Such statements have sometimes seemed discouraging; we ought rather to marvel that, with representations of things more or less distant and discolored, the human spirit has been able to find its way through the chaos of so many phenomena and to derive from scientific knowledge the ideas of beauty and harmony. It is no paradox to say that science puts order, at least tentative order, into nature.
In Lucienne Felix
The Modern Aspect of Mathematics (p. 31)
Basic Books, Inc. New York, New York, USA. 1960

Poincaré, Henri   1854–1912
French mathematician and theoretical astronomer

To obtain a result of real value, it is not enough to grind out calculations or to have a machine to put things in

order; it is not order alone, it is unexpected order, which is worth while. The machine may gnaw on the crude fact; the soul of the fact will always escape it.
The Foundations of Science
Science and Method, Book I
Chapter II (pp. 373–374)
The Science Press. New York, New York, USA. 1913

Pope, Alexander   1688–1744
English poet

Where order in variety we see,
And where, though all things differ, all agree.
The Complete Poetical Works
Windsor Forest, l. 15–16
Houghton Mifflin Company. New York, New York, USA. 1903

Reichenbach, Hans   1891–1953
German philosopher of science

…whereas inorganic nature was seen to be controlled by the laws of cause and effect, organic nature appeared to be governed by the law of purpose and means.
The Rise of Scientific Philosophy
Chapter 12 (p. 192)
University of California Press. Berkeley, California, USA. 1951

Russell, Bertrand Arthur William   1872–1970
English philosopher, logician, and social reformer

Dimensions, in geometry, are a development of order. The conception of a limit, which underlies all higher mathematics, is a serial conception. There are parts of mathematics which do not depend upon the notion of order, but they are very few in comparison with the parts in which this notion is involved.
Introduction to Mathematical Philosophy
Chapter IV (p. 29)
Dover Publications, Inc. New York, New York, USA. 1993

The notion of continuity depends upon that of order, since continuity is merely a particular type of order.
Mysticism and Logic and Other Essays
Chapter V (p. 91)
Longmans, Green & Company. London, England. 1925

Sarton, May   1912–95
American poet and novelist

I see a certain order in the universe and math is one way of making it visible.
As We Are Now (p. 38)
W.W. Norton & Company, Inc. New York, New York, USA. 1973

Shakespeare, William   1564–1616
English poet, playwright, and actor

The heavens themselves, the planets, and this centre,
Observe degree, priority, and place,
Insisture, course, proportion, season, form,
Office, and custom, in all line of order.
In *Great Books of the Western World* (Volume 27)

The Plays and Sonnets of William Shakespeare (Volume 2)
Troilius and Cressida
Act I, Scene iii, l. 85–88
Encyclopædia Britannica, Inc. Chicago, Illinois, USA. 1952

Whitehead, Alfred North   1861–1947
English mathematician and philosopher

In the first place, there can be no living science unless there is a widespread instinctive conviction in the existence of an Order of Things, and, in particular, of an Order of Nature.
Science and the Modern World
Chapter I (p. 5)
The Macmillan Company. New York, New York, USA. 1929

Wöhler, Friedrich   1800–82
German chemist
von Liebig, Justice   1803–73
German organic chemist

When in the dark province of organic nature, we succeed in finding a light point, appearing to be one of those inlets whereby we may attain to the examination and investigation of this province, then we have reason to congratulate ourselves, although conscious that the object before us is unexhausted.
American Journal of Science and Arts, Volume 26, 1834 (p. 261)

Yang, Chen Ning   1922–
Chinese-born American theoretical physicist

Nature possesses an order that one may aspire to comprehend.
Nobel Lectures, Physics 1942–1962
Nobel lecture for award received in 1957
The Law of Parity Conservation and Other Symmetry Laws of Physics (p. 394)
Elsevier Publishing Company. Amsterdam, Netherlands. 1964

ORGAN TRANSPLANT

Carrel, Alexis   1873–1944
French surgeon and biologist

Thus, while the problem of the transplantation of organs has been solved from a surgical point of view, we see that this by no means suffices to render such operations of definite surgical practicability, and it will only be through a more fundamental study of the biological relationships existing between living tissues that the problems involved will come to be solved and thereby render possible the benefits to humanity which we hope to see accomplished in the future.
Nobel Lectures, Physiology or Medicine 1901–1921
Nobel lecture for award received in 1912
Suture of Blood-Vessels and Transplantation of Organs (p. 464)
Elsevier Publishing Company. Amsterdam, Netherlands. 1967

ORGANIC CHEMISTRY

Berzelius, Jöns Jacob   1779–1848
Swedish chemist

We have reached a point where we are beginning to see a theory of organic compounds; but if, instead of letting this develop as our experience grows, we want to base it on isolated facts, considered without regard for their relations with the general system of our knowledge, and by giving explanations which do not harmonise with the principles of the science, and if, moreover, we want to conclude that this lack of agreement must lead us to reject as erroneous principles which are already well established on other grounds, then we shall never succeed in finding the truth.
Annals de chemie et de physique, Volume 71, 1839

Cram, Donald J.   1919–2001
American chemist
Cram, Jane M.
No biographical data available

No other profession is endowed with such a rich landscape, draws inspiration from so many fields of science, exercises the hand and mind in so many different ways, offers such opportunities to employ creative instincts, and mixes ideas, theory, and experiment on a daily basis. Hurrah for the science of organic chemistry, and for the joy it brings those who play the research game.
Container Molecules and Their Guests
Preface (p. vi)
Royal Society of Chemistry. Cambridge, England. 1994

Holmes, Oliver Wendell   1809–94
American physician, poet, and humorist

Drops of deliquescence glistened on his forehead,
Whitened round his feet the dust of efflorescence,
'Till one Monday morning when the flow suspended,
There was no De Sauty.
Nothing but a cloud of elements organic
C.O.H.N. Ferrum, Chlor. Flu. Sil. Potassa,
Calc. Sod. Phosph. Mag. Sulphur, Mang.? Alumin.?
Caprum?
Such as man is made of.
The Professor at the Breakfast Table
Chapter I
De Sauty (p. 33)
Houghton Mifflin Company. Boston, Massachusetts, USA. 1890

Hopkins, Frederick Gowland   1844–89
English biochemist

A very distinguished organic chemist long since dead, said to me in the late eighties: "The chemistry of the living? That is the chemistry of protoplasm; that is superchemistry; seek, my young friend, for other ambitions."
In Joseph Needham and Ernest Baldwin (eds.)
Hopkins & Biochemistry
Report of the British Association
Some Chemical Aspects of Life
1933 (p. 245)

Kekulé, Friedrich August   1829–96
German chemist

We define organic chemistry as the chemistry of carbon compounds. In doing this, we see no opposition between organic and inorganic compounds. What has been known for a long time as organic chemistry and which more usefully may be called the chemistry of carbon compounds, is rather only a special section of pure chemistry which is dealt with separately because the large number and special importance of carbon compounds seems to make a special field of study necessary.... It must be emphasized that organic chemistry does not deal with the study of the chemical processes in the organs of plants and animals.
Lehrbuch der Organischen Chemie (Volume 1) (p. 10)
Publisher undetermined

Thompson, Sir D'Arcy Wentworth   1860–1948
Scottish zoologist and classical scholar

The mysteries of organic chemistry are great, and the differences between its processes or reactions as they are carried out in the organism and in the laboratory are many; the actions, catalytic and other, which go on in the living cell, are of extraordinary complexity. But the contention that they are different in kind from ordinary chemical operations...would seem to be no longer tenable.
On Growth and Form (Volume 2)
Chapter IX (p. 652)
At The University Press. Cambridge, England. 1951

Thudichum, J. L. W.   1829–1901
Chemist

Organic chemistry is the child of medicine, and however far it may go on its way, with its most important achievements, it always returns to its parent.
On the Discoveries and Philosophy of Leibig
Journal of the Royal Society of Arts, Volume 24, 1876 (p. 141)

Ure, Andrew   1778–1857
Scottish physician

All of the elementary principles of organic nature may be considered as deriving the peculiar delicacy of their chemical equilibrium, and the consequent facility with which it may be subverted and new modeled, to the multitude of atoms grouped together in a compound. On this view, none of them should be expected to consist of a single atom of each component.
On the Ultimate Analysis of Vegetable and Animal Substances
Philosophical Transactions of the Royal Society of London, Volume 112, 1822 (pp. 468–469)

Wöhler, Friedrich   1800–82
German chemist

Organic chemistry just now is enough to drive one mad. It gives one the impression of a primeval, tropical forest full of the most remarkable things, a monstrous and boundless thicket, with no way of escape, into which one may well dread to enter.
In Edward Franklin Degering
An Outline of Organic Nitrogen Compounds
Letter to Berzelius, 28 January 1835 (p. 5)
University Lithoprinters. Ypsilanti, Michigan, USA. 1945

ORGANISM

Bernard, Claude   1813–78
French physiologist

We may, of course strike a balance between what a living organism takes in as nourishment and what it gives out in excretions.... This would be like trying to tell what happens inside a house by watching what goes in by the door and what comes out by the chimney.
An Introduction to the Study of Experimental Medicine
Part Two, Chapter II, Section IX
The Macmillan Company. New York, New York, USA. 1927

Evans, Howard Ensign   1919–2002
Entomologist

It has been said that for every problem concerning living things there is an organism ideal for its solution. It is probable that there are still undiscovered species living that hold the answers to problems that face us now or will in the future.
Pioneer Naturalist: The Discovery and Naming of North American Plants and Animals
Naturalists, Then and Now (p. 267)
Henry Holt & Company. New York, New York, USA. 1993

Hess, Walter   1881–1973
Swiss physiologist

A recognized fact which goes back to the earliest times is that every living organism is not the sum of a multitude of unitary processes, but is, by virtue of interrelationships and of higher and lower levels of control, an unbroken unity.
Nobel Lectures, Physiology or Medicine 1942–1962
Nobel lecture for award received in 1949
The Central Control of the Activity of Internal Organs (p. 247)
Elsevier Publishing Company. Amsterdam, Netherlands. 1964

Jacob, François   1920–
French biologist

And one of the deepest, one of the most general functions of living organisms is to look ahead, to produce future as Paul Valéry put it.
The Possible and the Actual

Time and the Invention of the Future (p. 66)
Pantheon Books. New York, New York, USA. 1982

Jones, J. S.
No biographical data available
Ebert, D.
No biographical data available

No organism can do everything. Every creature is restricted by constraints of various kinds. Many of these arise from the facts of history and the nature of evolution, both of which can proceed only from where they left off.
In R.J. Berry, T.J. Crawford and G.M. Hewitt (eds.)
Genes in Ecology
Life History and Mechanical Constraints on Reproduction in Genes, Cells and Waterfleas (p. 393)
Blackwell Scientific Publications. Oxford, England. 1992

Price, P. W.
No biographical data available

Visually stimulating organisms, the large, the colorful, the active, the aggressive, command our attention, while the secretive and insidious remain largely ignored.
Evolutionary Biology of Parasites
Chapter Eight (p. 171)
Princeton University Press. Princeton, New Jersey, USA. 1980

Savage-Rumbaugh, Sue
American psychologist
Lewin, Roger Amos
Anthropologist

All organisms with complex nervous systems are faced with the moment-by-moment question that is posed by life: What shall I do next?
Kanzi: The Ape at the Brink of the Human Mind
Chapter 10 (p. 255)
John Wiley & Sons, Inc. New York, New York, USA. 1994

Thomson, J. Arthur   1861–1933
Scottish biologist

The hosts of living organisms are not random creatures, they can be classified in battalions and regiments. Neither are they isolated creatures, for every thread of life is inter-twined with others in a complex web.
The System of Animate Nature (Volume 1)
Lecture II (p. 58)
William & Norgate. London, England. 1920

von Goethe, Johann Wolfgang   1749–1832
German poet, novelist, playwright, and natural philosopher

Basic characteristics of an individual organism: to divide, to unite, to merge into the universal, to abide in the particular, to transform itself, to define itself, and as living things tend to appear under a thousand conditions,

to arise and vanish, to solidify and melt, to freeze and flow, to expand and contract. Since these effects occur together, any or all may occur at the same moment.
In D. Miller (ed.)
Scientific Studies (Volume 12)
Chapter VIII (pp. 303–304)
Suhrkamp. New York, New York, USA. 1988

ORGANIZATION

Eddington, Sir Arthur Stanley   1882–1944
English astronomer, physicist, and mathematician

There is no doubt that the scheme of physics as it has stood for the last three-quarters of a century postulates a data at which either the entities of the universe were created in a state of high organization, or pre-existing entities were endowed with that organization, which they have been squandering ever since. Moreover, this organization is admittedly the antithesis of chance. It is something which could not occur fortuitously.
The Nature of the Physical World
Chapter IV (pp. 84–85)
The Macmillan Company. New York, New York, USA. 1930

Eiseley, Loren C.   1907–77
American anthropologist, educator, and author

Men talk much of matter and energy, of the struggle for existence that molds the shape of life. These things exist, it is true; but more delicate, elusive, quicker than fins in water, is that mysterious principle known as "organization," which leaves all other mysteries concerned with life stale and insignificant by comparison. For that without organization life does not persist is obvious. Yet this organization itself is not strictly the product of life, nor of selection. Like some dark and passing shadow within matter, it cups out the eyes' small windows or spaces the notes of a meadow lark's song in the interior of a mottled egg.
The Immense Journey
The Flow of the River (p. 26)
Vintage Books. New York, New York, USA. 1957

Huxley, Thomas Henry   1825–95
English biologist

Not only are all animals existing in the present creation organized to one of these five plans; but paleontology tends to show that in the myriad of past ages of which the earth's crust contains the records, no other plan of animal life made its appearance on our planet. A marvelous fact and one which seems to present no small obstacle in the way of the notion of the possibility of fortuitous development of animal life.
In Michael Foster and E. Ray Lankester (eds)
Scientific Memoirs of Thomas Huxley
Volume 1, On Natural History as Knowledge, Discipline and Power (p. 306
Publisher undetermined. 1901

Needham, Joseph   1900–95
English biochemist and sinologist

Organization and Energy are the two fundamental problems which all science has to solve.
Time: The Refreshing River
The Naturalness of the Spiritual World (p. 33)
The Macmillan Company. New York, New York, USA. 1943

...organization is not something fundamentally mystical and unamenable to scientific attack, but rather the basic problem confronting the biologist.... It is for us to investigate the nature of this biological organization, not to abandon it to the metaphysicians because the rules of physics do not seem to apply to it.
Order and Life
Chapter I (pp. 7, 17–18)
Yale University Press. New Haven, Connecticut, USA. 1936

Simpson, George Gaylord   1902–84
American paleontologist

The point about explanation in biology that I would particularly like to stress is this: to understand organisms one must explain their organization. It is elementary that one must know what is organized and how it is organized, but that does not explain the fact or the nature of the organization itself. Such explanation requires knowledge of how an organism came to be organized and what function the organization serves. Ultimate explanation in biology is therefore necessarily evolutionary.
This View of Life: The World of an Evolutionist
Chapter Six (p. 113)
Harcourt, Brace & World, Inc. New York, New York, USA. 1964

Szent-Györgyi, Albert   1893–1986
Hungarian-born American biochemist

One of the most basic principles of biology is organization, which means that two things put together in a specific way form a new unit, a system, the properties of which are not additive and cannot be described in terms of the properties of the constituents. As points may be connected to letters, letters to words, words to sentences, etc., so atoms can join to molecules, molecules to organelles, organelles to cells, etc., every level of organization having a new meaning of its own and offering exciting vistas and possibilities.
Bioenergetics
Chapter 6 (p. 39)
Academic Press. New York, New York, USA. 1957

Woodger, Joseph Henry   1894–1981
English biologist

The failure to take organization seriously is perhaps but another consequence of the rapid development of physics and chemistry as compared to other sciences, and the consequent dazzling effect this had on biological vision.
Biological Principles: A Critical Study

Part II, Chapter VI, B, 5 (p. 291)
Kegan Paul, Trench, Trubner & Company Ltd. London, England. 1929

If the concept of organization is of such importance as it appears to be it is something of a scandal that we have no adequate conception of it. The first duty of the biologist would seem to be to try and make clear this important concept. Some biochemists and physiologists…express themselves as though they really believed that if they concocted a mixture with the same chemical composition as what they call "protoplasm" it would proceed to "come to life." This is the kind of nonsense which results from forgetting or being ignorant of organization.
Biological Principles: A Critical Study
Part II, Chapter VI, B, 5 (p. 291)
Kegan Paul, Trench, Trubner & Company Ltd. London, England. 1929

ORGANS

Wallace, Alfred Russel   1823–1913
English humanist, naturalist, and geographer

…We have also here an acting cause to account for that balance so often observed in nature, — a deficiency in one set of organs always being compensated by an increased development of some others — powerful wings accompanying weak feet, or great velocity making up for the absence of defensive weapons; for it has been shown that all varieties in which an unbalanced deficiency occurred could not long continue their existence. The action of this principle is exactly like that of the centrifugal governor of the steam engine, which checks and corrects any irregularities almost before they become evident; and in like manner no unbalanced deficiency in the animal kingdom can ever reach any conspicuous magnitude, because it would make itself felt at the very first step, by rendering existence difficult and extinction almost sure soon to follow.
Journal of the Proceedings of the Linnean Society, Zoology, Volume 3, 1858 (pp. 61–62)

ORIGIN OF LIFE

Goldanskii, Vitalii   1923–2001
Soviet physicist and chemist

Two properties of living systems that are unique from the standpoint of physics, namely, self-replication and homochirality, may serve as Ariadne's thread in the labyrinth of hypotheses concerning this [origin-of-life] problem.
In J. and K. Tran Thon Van, J. C. Mounolou, J. Schneider and C. Mckay (eds.)
Frontiers of Life
Chirality, Origin of Life, and Evolution
Publisher undetermined

Oparin, Alexander Ivanovich   1894–1980
Russian biochemist

…when I began to be interested in the problem of the origin of life, in the early 1920s, the whole topic was in a state of crisis. It appeared as if it was a forbidden subject in the world of science. The problem was generally felt to be insoluble in principle using objective scientific research methods. It was felt that it belonged more to the sphere of faith than knowledge, and that, for this reason, serious scientists should not waste their time and effort on hopeless attempts to solve the problem.
Jubilee for Heterogenesis Research
New Scientist, Volume 142, 1974

Sagan, Carl   1934–96
American astronomer and author

Every human community has somehow or other tried to understand…deep questions of origins. Origin of our group, whatever it is, origin of our species, origin of life, origin of Earth, origin of the universe. I think you have to be made out of wood not to be interested in these questions. And there's no way to understand even the questions, much less the answers, without understanding science.
Speech
National meeting of the American Astronomical Society (January 5, 1993)

Teilhard de Chardin, Pierre   1881–1955
French Jesuit, paleontologist, and biologist

To push anything back into the past is equivalent to reducing it to its simplest element. Traced as far as possible in the direction of their origins, the last fibers of the human aggregate are lost to view and are merged in our eyes with the very stuff of the universe.
The Phenomenon of Man
Book One, Chapter I (p. 39)
Harper & Brothers. New York, New York, USA. 1959

Wächtershäuser, Günter
International patent lawyer

The chemist strives to explain the inanimate world by reference to mechanistic laws. The historian strives to understand the world of human culture by reference to a fabric of plans and purposes.… Nowhere is this encounter in sharper focus than in the problem of the origin of life.
The Origin of Life and Its Methodological Challenge
Journal of Theoretical Biology, Volume 187, 1997

ORIGINALITY

von Goethe, Johann Wolfgang   1749–1832
German poet, novelist, playwright, and natural philosopher

Those theories to which we ascribe originality are not so easily grasped, not so quickly epitomized and systematized. An author tends toward this or that way of thinking; but it is modified by his individuality, indeed, often simply by his presentation, by the peculiarity of the idiom in which he speaks and writes, by the change in times, by various considerations.

In Karl J. Fink
Goethe's History of Science
Chapter 9 (p. 115)
Cambridge University Press. Cambridge, England. 1991

ORNITHOLOGY

Audubon, John James   1785–1851
West Indian-born American ornithologist and artist

To render more pleasant the task you have imposed upon yourself, of following an author through the mazes of descriptive ornithology, permit me, kind reader, to relieve the tedium which may be apt now and then to come upon you, by presenting you with occasional descriptions of the scenery and manners of the land which has furnished the objects that engage your attention.

Ornithological Biography (Volume 1)
The Ohio (p. 29)
Adam Black. Edinburgh, Scotland. 1831

Author undetermined

…the philosophy of science is just about as useful to scientists as ornithology is to birds.

In S. Weinberg
Newtonianism, Reductionism and the Art of Congressional Testimony
Nature, Volume 330, Number 6147, 3–9 December 1987 (p. 433)

Darwin, Charles Robert   1809–82
English naturalist

I took much pleasure in watching the habits of birds, and even made notes on the subject. In my simplicity I remember wondering why every gentleman did not become an ornithologist.

In Francis Darwin (ed.)
The Life and Letters of Charles Darwin (Volume 1)
Chapter II (p. 32)
D. Appleton & Company. New York, New York, USA. 1896

Vidal, Gore   1925–
American essayist, novelist, and social/political commentator

To a man, ornithologists are tall, slender, and bearded so that they can stand motionless for hours, imitating kindly trees, as they watch for birds.

Armageddon? Essays 1983–1987
Mongolia (p. 131)
Vintage Books. New York, New York, USA. 1990

White, Gilbert   1720–93
English naturalist and cleric

A good ornithologist should be able to distinguish birds by their air [manner] as well as by their colours and shape; on the ground as well as on the wing, and in the bush as well as in the hand.

The Natural History of Selborne
Letter XLII
To Dianes Barrington
August 7, 1778
Robert M. McBride & Company. New York, New York, USA. 1925

OSMOTIC PRESSURE

van't Hoff, Jacobus Henricus   1852–1911
Dutch physical and organic chemist

In an investigation, whose essential aim was a knowledge of the laws of chemical equilibrium in solutions, it gradually became apparent that there is a deep-seated analogy — indeed, almost an identity — between solutions and gases, so far as their physical relations are concerned; provided that with solutions we deal with the so-called osmotic pressure, where with gases we are concerned with the ordinary elastic pressure.

Zeitschrift fur physikalische Chemie
The Role of Osmotic Pressure in the Analogy between Solutions and Gasses, Volume 1, 1887

OSTEOPATH

Mencken, H. L. (Henry Louis)   1880–1956
American journalist and literary critic

Osteopath — One who argues that all human ills are caused by the pressure of hard bone on soft tissue. The proof of his theory is to be found in the heads of those who believe it.

A Mencken Chrestomathy
Chapter XXX (p. 625)
Alfred A. Knopf. New York, New York, USA. 1949

OTHER WORLDS

Abbey, Henry   1842–1911
Author

When from the vaulted wonder of the sky
The curtain of the light is drawn aside,
And I behold the stars in all their wide
Significance and glorious mystery,
Assured that those more distant orbs are suns
Round which innumerable worlds revolve,
My faith grows strong, my day-born doubts dissolve,
And death, that dread annulment which life shuns,
Or fain would shun, becomes to life the way,
The thoroughfare to greater worlds on high,
The bridge from star to star. Seek how we may,
There is no other road across the sky;
And, looking up, I hear star-voices say:
"You could not reach us if you did not die."

The Poems of Henry Abbey
Faith's Vista
Kingston. New York, New York, USA. 1895

Jackson, Helen Hunt   1830–85
American writer and poet

Who knows what myriad colonies there are
Of fairest fields, and rich, undreamed-of gains
Thick planted in the distant shining plains
Which we call sky because they lie so far?
Oh, write of me, not "Died in bitter pains,"
But "Emigrated to another star!"
Helen Jackson's Poems
Emigravit
Robert Brothers. Boston, Massachusetts, USA. 1888

Magnus, Albertus   1206–1280
Scientist, philosopher, and theologian

Do there exist many worlds, or is there but a single world?
This is one of the most noble and exalted questions in the
study of Nature.
In G. McColley
The Seventeenth-Century Doctrine of a Plurality of Worlds
Annals of Science, Volume 1, Number 4, October 15, 1936 (p. 385)

Oersted, Hans Christian   1777–1851
Danish physicist and chemist

Dost thou perceive naught but machinery
In laws which guide the course along heaven's paths?
Look with a larger view around; behold
The unity of living thoughts, displayed
In countless varying forms. The mighty sun
Is but a twinkling star amidst the space
Infinite filled with worlds, whose suns, heaven's lamps,
Shine in our night…. Look
Upon the spangled heav'ns, there to discover
Thousands of blazing suns, encircled by
Companions numerous…. A race of beings behold
Struggling for mental power, knowledge divine.
The Soul in Nature: With Supplementary Contributions
The Balloon
H.G. Bohn. London, England. 1852

Tennyson, Alfred (Lord)   1809–92
English poet

The Moon's white cities, and the opal width
Of her small glowing lakes, her silver heights
Unvisited with dew of vagrant cloud,
And the unsounded, undescended depth
Of her black hollows. The clear galaxy
Shorn of its hoary lustre, wonderful,
Distinct and vivid with sharp points of light,
Blaze within blaze, an unimagin'd depth
And harmony of planet-girded suns
And moon-encircled planets, wheel in wheel,
Arch'd the wan sapphire. Nay — the hum of men,
Or other things talking in unknown tongues
And notes of busy life in distant worlds
Beat like a far wave on my anxious ear.
Alfred Tennyson's Poetical Works

Timbuctoo
Oxford University Press, Inc. London, England. 1953

And the suns of the limitless universe sparkled and
shone in the sky,
Flashing with fires as of God, but we knew that their
light was a lie —
Bright as with deathless hone — but, however they
sparkled and shone,
The dark little worlds running round them were worlds
of woe like our own.
Alfred Tennyson's Poetical Works
Despair, Stanza III
Oxford University Press, Inc. London, England. 1953

Whitman, Walt   1819–92
American poet, journalist, and essayist

Let your soul stand cool and composed before a million
universes.
Complete Poems and Collected Prose
Song of Myself
Section 48
The Library of America. New York, New York, USA. 1982

I was thinking this globe enough, till there sprang out so
noiseless
around me myriads of other globes.
Now, while the great thoughts of space and eternity fill
me, I will
measure myself by them;
And now, touch'd with the lives of other globes, arrived
as far
along as those of the earth,
Or waiting to arrive, or pass'd on farther than those of
the earth,
I henceforth no more ignore them, than I ignore my own
life,
Or the lives of the earth arrived as far as mine, or
waiting to arrive.
Complete Poetry and Collected Prose
Night on the Prairies
The Library of America. New York, New York, USA. 1982

OUTER SPACE

Hey, Nigel S.   1936–
American science writer

Human minds are being pulled into outer space by a
thin, strong filament of neural energy called wonder.
And this, to me, is a very good thing. As more of us let
our sense of wonder expand into the cosmos — so that
we comprehend the delicate smallness of our planet in
the scheme of things — we are gaining a special kind
of wisdom. My dream is that, with the blessing of good
fortune, this wisdom will eventually enable us to transcend the dangerous confusion of civilizations that are
maintained by coercion and misbelief. All sane persons

will comprehend their innate unity with the supernovas of which we are made. There will be no need for Utopia, for then we will have become meta-humans, siblings to all things that exist with and among the planets and the teeming stars. And then, perhaps, there will be peace at last.

Why People Need Space
Lecture, National Space Center, October 2002

MacLeod, Ken   1954–
Scottish science fiction writer

Outer space is, fundamentally, familiar. It's only the night sky, without the earth beneath your feet.

The Engines of Light
Cosmonaut's Keep (p. 1)
Tom Doherty Associates, LLC. New York, New York, USA. 2002

United Nations Treaty on the Exploration and Use of Space

The exploration and use of outer space, including the moon and other celestial bodies, shall be carried out for the benefit and in the interests of all countries, irrespective of their degree of economic or scientific development, and shall be the province of all mankind.

January 27, 1967

Webb, Jimmy   1946–
American music composer

I'll fly a starship, across the universe divine,

And when I reach the other side
I'll find a place to rest my spirit if I can
Perhaps I may become a highwayman again
Or I may simply be a single drop of rain
But I will remain, and I'll be back again
And again, and again, and again.

Ten Easy Pieces
Highwayman
CM Angel. 1996

OUTLIER

Green, Celia   1935–
English philosopher and psychologist

The fact that something is far-fetched is no reason why it should not be true; it cannot be as far-fetched as the fact that something exists.

The Decline and Fall of Science
Aphorisms (p. 1)
Hamilton. London, England. 1976

Hoyle, Sir Fred   1915–2001
English mathematician and astronomer

I don't see the logic of rejecting data just because they seem incredible.

In D.O. Edge and M.J. Mulkay
Astronomy Transformed: The Emergence of Radio Astronomy in Britain
Notes: Chapter 3 (p. 432, fn j)
John Wiley & Sons, Inc. New York, New York, USA. 1976

P

PAIN

Bell, Sir Charles 1774–1842
Scottish anatomist and surgeon

Pain is the necessary contrast to pleasure; it ushers us into existence or consciousness: it alone is capable of exciting the organs into activity: it is the compassion and the guardian of human life.
The Hand, Its Mechanism and Vital Endowments as Evincing Design
Chapter 7 (p. 211)
John Murray. London, England. 1852

Burney, Fanny 1752–1840
English novelist and diarist

When the dreadful steel was plunged into the breast — cutting through veins — arteries — flesh — nerves — I needed no more injunctions not to restrain my cries. I began a scream that lasted unintermittingly during the whole time of the incision — & I almost marvel that it rings not in my Ears still! so excruciating was the agony.
In A. Dally
Women Under the Knife: A History of Surgery
Letter to Esther Burney, 1811
Hutchinson Radius. London, England. 1991

Coates, Florence Earle 1850–1927
American poet

Ah, me! the Prison House of Pain! —
what lessons there are bought! —
Lessons of a sublimer strain
Than any elsewhere taught.
Poems (Volume 2)
The House of Pain
Houghton Mifflin Company. Boston, Massachusetts, USA. 1916

Dickinson, Emily 1830–86
American lyric poet

Pain has an element of blank;
It cannot recollect
Where it began, or if there were
A day when it was not.
The Complete Poems of Emily Dickinson
No. 650 (p. 323)
Little, Brown & Company. Boston, Massachusetts, USA. 1960

Emerson, Ralph Waldo 1803–82
American lecturer, poet, and essayist

He has seen but half the universe who never has been shewn the House of Pain.
The Complete Works of Ralph Waldo Emerson (Volume 12)
Natural History of Intellect
The Tragic (p. 405)
Houghton Mifflin Company. Boston, Massachusetts, USA. 1904

Hilton, John 1804–78
English surgeon

Pain the monitor, and Rest the cure, are starting points for contemplation which should ever be present to the mind of the surgeon in reference to his treatment.
Rest and Pain: A Course of Lectures on the Influence of Mechanical and Physiological Rest In the Treatment of Accidents and Surgical Diseases, and the Diagnostic Value of Pain (p. 500)
George Bell & Sons. London, England. 1892

Every pain has its distinct and pregnant signification, if we will but carefully search for it.
Rest and Pain: A Course of Lectures on the Influence of Mechanical and Physiological Rest In the Treatment of Accidents and Surgical Diseases, and the Diagnostic Value of Pain (p. 499)
George Bell & Sons. London, England. 1892

Hood, Thomas 1582–98
English poet and editor

Of all our pains, since man was curst,
I mean of body, not the mental,
To name the worst, among the worst,
The dental sure is transcendental;
Some bit of masticating bone,
That ought to help to clear a shelf:
But lets its proper work alone,
And only seems to gnaw itself.
The Complete Poetical Works of Thomas Hood
A True Story
Greenwood Press, Publishers. Westport, Connecticut, USA. 1980

Johnson, Samuel 1696–1772
English critic, biographer, and essayist

…those who do not feel Pain, seldom think that it is felt…
The Rambler (Volume 1)
No. 48, September 1, 1750 (p. 335)
Edward Earle. Philadelphia, Pennsylvania, USA. 1812

Latham, Peter Mere 1789–1875
English physician

It would be a great thing to understand Pain in all its meanings.
In William B. Bean
Aphorisms from Latham (p. 71)
Prairie Press. Iowa City, Iowa, USA. 1962

Mather, Cotton 1663–1728
American minister and religious writer

WHAT is Pain? Tis a Sensation produced on the Tension of a Nerve.
The Angel of Bethesda
Capsila IX (p. 54)
American Antiquarian Society and Barre Publishers. Barre, Massachusetts, USA. 1972

Robinson, Victor 1886–1947
Physician

The first cry of pain through the primitive jungle was the first call for a physician.
The Story of Medicine
Chapter I (p. 1)
The New York Home Library. New York, New York, USA. 1943

Schweitzer, Albert   1875–1965
Alsatian-German theologian and philosopher

Whosoever is spared personal pain must feel himself called to help in diminishing the pain of others.
Recalled on his death
September 4, 1965
Source undetermined

Thompson, Francis   1859–1907
English writer

Nothing begins, and nothing ends,
That is not paid with a moan;
For we are born in other's pain,
And perish in our own.
Complete Poetical Works of Francis Thompson
Daisy, Stanza 15
Boni & Liveright, Inc., Publishers. New York, New York, USA. 1923

Watson, Sir William   1858–1935
English author of lyrical and political verse

Pain with the thousand teeth.
The Poems of William Watson
The Dream of Man (p. 127)
Macmillan & Company. New York, New York, USA. 1893

PALEONTOLOGIST

Bracker, Milton
No biographical data available

Consider the sages who pulverize boulders,
And burrow for elbows and shinbones and shoulders,
And shovel the loot from a hill or a dale of it,
And lovingly carry off pail after pail of it.
P Is for Paleontology
Journal of Geological Education, Volume 19, Number 4, September 1971 (p. 192)

Bradley, Jr., John Hodgdon   1898–1962
American geologist

The paleontologist watching the rise and fall of races sees, with only actors and setting changed, one drama repeated. Thus like the jaded critic he knows the end before the final curtain. He sees that death is the penalty for life.
Parade of the Living
Part III, Chapter XVII (pp. 237–238)
Coward-McCann, Inc. New York, New York, USA. 1930

Brett-Surman, Michael   1950–
American paleontologist

Being a paleontologist is like being a coroner except all the witnesses are dead and all the evidence has been left out in the rain for 65 million years.
In Louie Psihoyos
Hunting Dinosaurs (p. vii)
Random House, Inc. New York, New York, USA. 1994

Colbert, Edwin H.   1905–2001
American vertebrate paleontologist

Any paleontologist worth his or her salt takes a great deal of pleasure in thinking of the discoveries he has made in the field and laboratory, but true satisfaction is in the publications that describe and interpret the fossils.
Digging into the Past: An Autobiography
Chapter VII (p. 126)
Dembner Books. New York, New York, USA. 1989

Cousteau, Jacques-Yves   1910–77
French naval officer and ocean explorer

A paleontologist holds the thread of evolution in his hands by combining the biological and geological evidence in fossils.
The Ocean World of Jacques Cousteau: The Adventure of Life
Chapter I (p. 13)
The World Publishing Company. New York, New York, USA. 1973

Gaudry, Jean-Albert   1827–1908
French paleontologist

It is the proper function of paleontologists to supply some proofs to the doctrine of evolution; it does not fall to them to explain the process by which the author of the world has produced the modifications. That study of processes is what is called Darwinism.... Assuredly it is a subject quite worthy of the attention of those naturalists that study the causes of the modifications of beings; but it is up to the physiologists, who experiment on living creatures, to teach us how the changes are produced today, and must have been produced formerly.... On this subject a paleontologist can avow his ignorance. All that he can say is that the discovery of vestiges buried in the bowels of the earth teach us that a constant harmony has presided at the transformation of the organic world.
Les Enchaenements
Revue des Deux Mondes, 9th Series, Volume 23, 1877 (p. 183)

Matthew, William Diller   1871–1930
Canadian-American paleontologist

...evolution is only one aspect of the order of nature, of the relations of cause and effect, of continuity of space and time, which pervade the universe and enable us to comprehend its simplicity of plan, its complexity of detail. The paleontologist, engaged in adding year by year to the mass of documents which record the history of life, in deciphering their meaning and interpreting their significance, has no more occasion to doubt

its continuity and orderly development than the historian has to doubt the continuity and consecutive evolution of human history, or the student of current affairs to doubt that the events of tomorrow.
Natural History, Volume 25, Number 2, 1925

Morris, Simon Conway   1951–
English paleontologist

As well as being lumps of patterned stone, fossils are also historical documents. History per se has had a bit of bad press recently.... There is a tension between the documentation of history (famously referred to as "one bloody thing after another"), and the search for universal principles that are ahistoric and possibly timeless. After a period in the doldrums, the bearers of the historical tidings, the paleontologists, are making tentative movements toward the legendary High Table where, just visible through the clouds of incense (and rhetoric), the high priests of evolutionary theory smile benignly.
The Phylogeny of Life and the Accomplishments of Phylogenetic Biology
Symposium at the University of Arizona. Tucson, Arizona, USA
October 11–13, 1996
Early Metazoan Radiations: What the Fossil Record Can and Cannot Tell Us

Simpson, George Gaylord   1902–84
American paleontologist

Not long ago paleontologists felt that a geneticist was a person who shut himself in a room, pulled down the shades, watched small flies disporting themselves in milk bottles, and thought that he was studying nature. A pursuit so removed from the realities of life, they said, had no significance for the true biologist. On the other hand, the geneticists said that paleontology had no further contributions to make to biology, that its only point has been the completed demonstration of the truth of evolution, and that it was a subject too purely descriptive to merit the name "science." The paleontologist, they believed, is like a man who undertakes to study the principles of the internal combustion engine by standing on a street corner and watching the motor cars whizz by.
Tempo and Mode in Evolution
Introduction (p. xv)
Columbia University Press. New York, New York, USA. 1944

Turney, John
No biographical data available

Take a complete, illustrated catalogue of London's National Gallery. Shred it into tiny pieces and cast them into the wind from the gallery's steps above Trafalgar Square. Wait a few weeks, then scour the square for surviving scraps of paper. Now try to reconstruct the history of painting from your haul. If you manage to produce a coherent story — schools, styles, genres, named painters and all — you are probably a paleontologist.
Review of "In Search of Deep Time"
New Scientist, 25 March 2000

PALEONTOLOGY

Grassé, Pierre P.   1895–1985
French zoologist

Naturalists must remember that the process of evolution is revealed only through fossil forms. A knowledge of paleontology is, therefore, a prerequisite; only paleontology can provide them with the evidence of evolution and reveal its course or mechanisms. Neither the examination of present beings, nor imagination, nor theories can serve as a substitute for paleontological documents. If they ignore them, biologists, the philosophers of nature, indulge in numerous commentaries and can only come up with hypotheses. That is why we constantly have recourse to paleontology, the only true science of evolution. From it we learn how to interpret present occurrences cautiously; it reveals that certain hypotheses considered certainties by their authors are in fact questionable or even illegitimate.
Evolution of Living Organisms: Evidence for a New Theory of Transformation
An Introduction to the Study of Evolution (p. 4)
Academic Press. New York, New York, USA. 1977

Bakker, Robert T.   1945–
American paleontologist

Paleontology is a very visual inquiry.... All paleontologists scribble on napkins at coffee breaks, making sketches to explain their thinking.
Brushing Up On Dinosaurs
Science News, October 4, 1986

Crawford, Osbert Guy Stanhope   1886–1957
English archaeologist

If archaeology is humanity revealed by its works, paleontology is life revealed by its own remains.
Man and His Past
Chapter VI (p. 70)
Oxford University Press, Inc. London, England. 1921

Dunbar, Carl O.
No biographical data available

If they stink, the remains belong to zoology, but if not, to paleontology.
In Alan M. Cvancara
Sleuthing Fossils: The Art of Investigating Past Life
Chapter 1 (p. 1)
John Wiley & Sons, Inc. New York, New York, USA. 1990

Hillery, Herbert
No biographical data available

There is no end to paleontology, there is no end to geology; and when the morning of the resurrection shall come, some paleontologist will be searching for some previously undiscovered species of extinct beings, and some geologist will be pecking away at the rocks to find some characteristics which have never been before ascertained. There is no end to it.

Congressional Record, Volume 23, 1892 (p. 4626)

Herzen, Aleksandr   1812–70
Russian political author

The small buds of organic chemistry, geology, paleontology, comparative anatomy have grown in our century into huge branches and borne fruit exceeding our wildest hopes. The world of the past, obedient to the mighty voice of science, has left the tomb to bear witness to the upheavals which accompany the evolution of the surface of the globe; the soil on which we live, this tombstone of the past life, is growing transparent, as it were; the stone vaults have opened, the interior of the rocks could not retain their secrets. Not only do the half-decayed, half-petrified vestiges again assume flesh, paleontology also strives to discover the law of the relation between geologic epochs and their complete flora and fauna. Then everything that ever lived will be resurrected in the human mind, will be saved from the sad fate of utter oblivion, and those whose bones have been completely decayed, whose phenomenal existence has been utterly obliterated, will be restored in the bright sanctuary of science where the temporal finds its repose and is perpetuated.

Selected Philosophical Works
Letters on the Study of Nature, Letter One (pp. 99–100)
Foreign Languages Publishing House. Moscow, Russia. 1956

Howard, Robert West
No biographical data available

The three volumes of [Lyell's] Principles of Geology, published between 1829 and 1833, became the essential textbook of the profession. In its discussion of fossils and the vital role they had played in the development of geology as an exact science, Lyell urged adoption of a Greek-rooted word, meaning "the science of early beings," as the professional name for research of the types of plant and animal fossils embedded in the Earth's layered crust. His suggestion was adopted throughout Europe and the Americas. Thus, three centuries after the curiosities of Leonardo da Vinci, the dawnseekers' science was given the name of paleontology.

The Dawnseekers: The First History of American Paleontology
Chapter 9 (p. 126)
Harcourt Brace Jovanovich. New York, New York, USA. 1975

Huxley, Thomas Henry   1825–95
English biologist

That application of the sciences of biology and geology, which is commonly known as palæontology, took its origin in the mind of the first person who, finding something like a shell, or a bone, naturally imbedded in gravel or rock, indulged in speculations upon the nature of this thing which he had dug out — this "fossil" — and upon the causes which had brought it into such a position.

Collected Essays (Volume 4)
The Rise and Progress of Palæontology (p. 24)
Macmillan & Company Ltd. London, England. 1904

Kielan-Jaworowska, Zofia   1925–
Polish paleontologist

No scientist familiar with the intellectual adventure of studying animals from times long past will have any hesitation in affirming that to travel millions of years into the past, which is what paleontological study amounts to, is much more fascinating than the most exotic geographical travel we are able to undertake today. The study of animals that lived on Earth millions of years ago is not merely a study of their anatomy, but first and foremost a study of the course of evolution on earth and of the laws that govern it.

Translated by the Israel Translation Society
Hunting for Dinosaurs
Chapter 15 (p. 176)
The MIT Press. Cambridge, Massachusetts, USA. 1969

Kitts, David B.
Evolutionist and paleontologist

Despite the bright promise that paleontology provides a means of "seeing" evolution, it has presented some nasty difficulties for evolutionists, the most notorious of which is the presence of "gaps" in the fossil record. Evolution requires intermediate forms between species and paleontology does not provide them…

Evolution, Volume 28, September 1974 (p. 467)

Medawar, Sir Peter Brian   1915–87
Brazilian-born English zoologist

It may seem that palaeontology is a science of pure speculation or inquisitiveness, and the palaeontologist the most unreal and useless of researchers; a man dedicated to retrospection, plunged living into the past, where he spends his days collecting the debris of dead things.

The Future of Man
Chapter IV, Part I, Section 3 (p. 66)
Harper & Row, Publishers. New York, New York, USA. 1964

Miller, Hugh   1802–56
Scottish geologist and theologian

Paleontology, or the science of ancient organisms, deals, as its subject, with all the plants and animals of all the geologic periods. It bears nearly the same sort of relation to the physical history of the past that biography does to the civil and political history of the past.

The Testimony of the Rocks; of, Geology in Its Bearings on the Two Theologies, Natural and Revealed
Lecture First (p. 33)
Gould & Lincoln. Boston, Massachusetts, USA. 1857

Osborn, Henry Fairfield   1857–1935
American paleontologist and geologist

Paleontology is the zoology of the past.
The Age of Mammals in Europe, Asia and North America
Chapter I (p. 1)
The Macmillan Company. New York, New York, USA. 1910

The preservation of extinct animals and plants in the rocks is one of the fortunate accidents of time, but to mistake this position as indicative of affinity [with zoology and botany] is about as logical as it would be to bracket the Protozoa, which are principally aquatic organisms, under hydrology, or the Insecta, because of their aerial life, under meteorology. No, this is emphatically a misconception which is still working harm in some museums and institutions of learning. Paleontology is not geology, it is zoology ; it succeeds only so far as it is pursued in the zoological and biological spirit.
The Present Problems of Paleontology
Popular Science Monthly, 1905 (p. 226)

Rudwick, Martin J. S.
Science historian

As paleontology now prepares for a great leap forward into a computerised age there is perhaps a danger that it may lose sight of its historic origins in the "steam age" of science and before.
The Meaning of Fossils, Episodes in the History of Paleontology
Chapter Five, Section XII (p. 266)
Macdonald. London, England. 1972

Twain, Mark (Samuel Langhorne Clemens)   1835–1910
American author and humorist

What a noble science is paleontology! And what really startling sagacity its votaries exhibit!
Collected Tales, Sketches, Speeches & Essays 1852–1890 (Volume 1)
A Brace of Brief Lectures on Science (p. 528)
The Library of America. New York, New York, USA. 1992

van der Gracht, W. A.   1873–1943
Dutch petroleum geologist

There are few subjects where there exists greater diversity of opinions regarding practically everything than in paleontology.
In C.G. Simpson
Mammals and the Nature of Continents
American Journal of Science, Volume 241, 1943 (p. 1)

von Buch, L.
No biographical data available

…through knowledge of [paleontology] we obtain not only the history of the Earth but also the history of life.
In Rudolf Daber and Jochen Helms (eds.)
Fossils: The Oldest Treasures that Ever Lived
Only a Slab of Transitional Limestone (p. 40)
T.H.F. Publications, Inc. Neptune City, New Jersey, USA. 1985

von Humboldt, Alexander   1769–1859
German naturalist and explorer

Palaeontological studies have brought charm and variety to the science of the rigid structures of this Earth. Petrified strata show us, preserved in their graves, the flora and fauna of past millennia. We climb upwards in time when, noting the spatial stratification conditions, we penetrate downwards from one stratum to the next. Long-vanished plant and animal life emerges before our eyes.
In Jochen Helms
Fossils: The Oldest Treasures that Ever Lived
Knowledge and Museums (p. 9)
T.H.F. Publications, Inc. Neptune City, New Jersey, USA. 1985

PANSPERMIA

Kelvin, Lord William Thomson   1824–1907
Scottish engineer, mathematician, and physicist

Should the time come when this earth comes into collision with another body, comparable in dimensions to itself…many great and small fragments carrying seeds of living plants and animals would undoubtedly be scattered through space. Hence, and because we all confidently believe that there are at present, and have been from time immemorial, many worlds of life besides our own, we must regard it as probable in the highest degree that there are countless seed-bearing meteoric stones moving about through space. If at the present instance no life existed upon this earth, one such stone falling upon it might, by what we blindly call natural causes, lead to its becoming covered with vegetation.
Popular Lectures and Addresses (Volume 2)
Presidential Address to the British Association, Edinburgh, 1871
British Association for the Advancement of Science, Volume 4, Number 262, 1871 (p. 201)
Macmillan & Company Ltd. London, England. 1894

PARABOLA

Allen, Woody   1935–
American film director and actor

She wore a short skirt and a tight sweater and her figure described a set of parabolas that could cause cardiac arrest in a yak.
Getting Even
Mr. Big (p. 139)
Random House, Inc. New York, New York, USA. 1971

Frere, John Hookham   1769–1846
British diplomat and man of letters

And first, the fair PARABOLA behold,
Her timid arms, with virgin blush, unfold!
Though, on one focus fixed, her eyes betray
A heart that glows with love's resistless sway…
In Charles Edmonds

Poetry of the Anti-Jacobin
The Loves of the Triangle, Canto II, l. 107–108
Printed for J. Wright, by W. Bulmer & Company. London, England. 1801

Galilei, Galileo   1564–1642
Italian physicist and astronomer

It has been observed that missiles and projectiles describe a curved path of some sort; however no one has pointed out the fact that this path is a parabola. But this and other facts, not few in number or less worth knowing, I have succeeded in proving; and what I consider more important, there have been opened up to this vast and most excellent science, of which my work is merely the beginning, ways and means by which other minds more acute than mine will explore its remote corners.
In *Great Books of the Western World* (Volume 28)
Dialogues Concerning the Two New Sciences
Third Day (p. 197)
Encyclopædia Britannica, Inc. Chicago, Illinois, USA. 1952

Shaw, George Bernard   1856–1950
Irish comic dramatist and literary critic

KNELLER:...take a sugar loaf and cut it slantwise, and you will get hyperbolas and parabolas, ellipses and ovals...
The Complete Plays of Bernard Shaw
In Good King Charles's Golden Days, Act I (p. 1358)
Odham's Press. London, England. 1950

PARADIGM

Barnes, Barry
Sociologist

...paradigms, the core of the culture of science, are transmitted and sustained just as is culture generally: scientists accept them and become committed to them as a result of training and socialization, and the commitment is maintained by a developed system of social control.
In Quentin Skinner (ed.)
The Return of Grand Theory in the Human Sciences
Thomas Kuhn (p. 89)
Cambridge University Press. Cambridge, England. 1985

Kuhn, Thomas S.   1922–96
American historian of science

The operations and measurements that a scientist undertakes in the laboratory are not "the given" of experience but rather "the collected with difficulty." They are not what the scientist sees — at least not before his research is well advanced and his attention focused.... Science does not deal in all possible laboratory manipulations. Instead, it selects those relevant to the juxtaposition of a paradigm with the immediate experience that that paradigm has partially determined.

The Structure of Scientific Revolutions
Chapter X (p. 126)
The University of Chicago Press. Chicago, Illinois, USA. 1970

Normal science does not aim at novelties of fact or theory and, when successful, finds none. New and unsuspected phenomena are, however, repeatedly uncovered by scientific research, and radical new theories have again and again been invented by scientists. History even suggests that the scientific enterprise has developed a uniquely powerful technique for producing surprises of this sort. If this characteristic of science is to be reconciled with what has already been said, then research under a paradigm must be a particularly effective way of inducing a paradigm change. That is what fundamental novelties of fact and theory do. Produced inadvertently by a game played under one set of rules, their assimilation requires the elaboration of another set. After they have become parts of science, the enterprise, at least of those specialists in whose particular field the novelties lie, is never quite the same again.
The Structure of Scientific Revolutions
Chapter VI (p. 52)
The University of Chicago Press. Chicago, Illinois, USA. 1970

A paradigm is what members of the scientific community share, and, conversely a scientific community consists of men who share a paradigm.
The Structure of Scientific Revolutions
Postscript (p. 176)
The University of Chicago Press. Chicago, Illinois, USA. 1970

PARADISE

Abbey, Edward   1927–89
American environmentalist and nature writer

When I write "paradise" I mean not only apple trees and golden women but also scorpions and tarantulas and flies, rattlesnakes and Gila monsters, sandstorms, volcanoes and earthquakes, bacteria and bear, cactus, yucca, bladderweed, ocotillo and mesquite, flash floods and quicksand, and yes — disease and death and the rotting of the flesh.
Desert Solitaire
Down the River (p. 190)
Ballantine Books. New York, New York, USA. 1968

Hilbert, David   1862–1943
German mathematician

No one...will drive us our of this paradise that Cantor has created for us!
Hilbert — Courant
Hilbert
Chapter XX (p. 177)
Springer-Verlag. New York, New York, USA. 1986

PARADOX

Bohr, Niels Henrik David   1886–1962
Danish physicist

How wonderful that we have met with a paradox. Now we have some hope of making progress.
In L.I. Ponomarev
The Quantum Dice (p. 75)
Institute of Physics Publishing. Bristol, England. 1993

Bourbaki, Nicholas
Mathematical discussion group

There is no sharply drawn line between those contradictions which occur in the daily work of every mathematician, beginner or master of his craft, as a result of more or less easily detected mistakes, and the major paradoxes which provide food for logical thought for decades and sometimes centuries.
In Bryan H. Bunch
Mathematical Fallacies and Paradoxes
Chapter 2 (p. 38)
Van Nostrand Reinhold Company. New York, New York, USA. 1982

Cudmore, Lorraine Lee
American cell biologist

It is a bizarre paradox we are facing, for we find that experimental scientists (who are supposed to be fair) at times make the Spanish Inquisition a model of fair hearings and unbiased judgment.
The Center of Life: A Natural History of the Cell
Cellular Evolution (p. 55)
New York Times Book Company. New York, New York, USA. 1977

de Morgan, Augustus   1806–71
English mathematician and logician

If I had before me a fly and an elephant, having never seen more than one such magnitude of either kind; and if the fly were to endeavour to persuade me that he was larger than the elephant, I might possibly be placed in a difficulty. The apparently little creature might use such arguments about the effect of distance, and might appeal to such laws of sight and hearing as I, if unlearned in those things, might be unable wholly to reject. But there were a thousand flies, all buzzing, to appearance, about the great creature and, to a fly, declaring, each one for himself, that he was bigger than the quadruped; and all giving different and frequently contradictory reasons; and each one despising and opposing the reasons of the others — I should feel quite at my ease…[to] say, My little friends, the case of each one of you is destroyed by the rest.
A Budget of Paradoxes
Introduction
The Open Court Publishing Company. Chicago, Illinois, USA. 1915

Eliot, George (Mary Ann Evans Cross)   1819–80
English novelist

Play not with paradoxes. That caustic which you handle in order to scorch others may happen to sear your own fingers and make them dead to the quality of things.
Felix Holt, the Radical
Chapter XIII (p. 151)
Wm. L. Allison Company. New York, New York, USA. No date

Falletta, Nicholas
No biographical data available

A paradox is truth standing on its head to attract attention.
The Paradoxicon (p. xvii)
Doubleday & Company, Inc. New York, New York, USA. 1983

Gilbert, Sir William Schwenck   1836–1911
English playwright and poet
Sullivan, Arthur   1842–1900
English composer

RUTH: A paradox?
KING: A paradox!
A most ingenious paradox!
We've quips and quibbles heard in flocks,
But none to beat this paradox!
The Complete Plays of Gilbert and Sullivan
Pirates of Penzance
Act II (p. 142)
W.W. Norton & Company, Inc. New York, New York, USA. 1976

How quaint the ways of paradox
At common sense she gaily mocks.
The Complete Plays of Gilbert and Sullivan
Pirates of Penzance
Act II (p. 168)
Random House, Inc. New York, New York, USA. 1936

Humphries, W. J.   1862–1949
American meteorologist and atmosphere scientist

The scientific paradox is only an exception to some familiar but too inclusive generalization. It, therefore, has both the appeal of the riddle and the charm of surprise — the surprise, the instant the truth is seen, of a sudden and unexpected discovery….
A Bundle of Meteorological Paradoxes
Annual Report of the Board of Regents of the Smithsonian Institution, 1920 (p. 183)
Government Printing Office. Washington, D.C. 1922

Kasner, Edward   1878–1955
American mathematician
Newman, James Roy   1911–66
Mathematician and mathematical historian

Perhaps the greatest paradox of all is that there are paradoxes in mathematics…because mathematics builds on the old but does not discard it, because its theorems are deduced from postulates by the methods of logic, in spite of its having undergone revolutionary changes we do not suspect it of being a discipline capable of engendering paradoxes.

Mathematics and the Imagination
Paradox Lost and Paradox Regained (p. 193)
Simon & Schuster. New York, New York, USA. 1940

Kelvin, Lord William Thomson   1824–1907
Scottish engineer, mathematician, and physicist

Paradoxes have no place in science. Their removal is the substitution of true for false statements and thoughts.
Popular Lectures and Addresses (Volume 1)
On Sun's Heat
Lecture
Royal Institution of Great Britain
January 21, 1887 (pp. 372–373)
Macmillan & Company Ltd. London, England. 1894

Rapoport, Anatol   1911–2007
Russian-born mathematician and biologist

Paradoxes have played a dramatic part in intellectual history, often foreshadowing revolutionary developments in science, mathematics, and logic. Whenever, in any discipline, we discover a problem that cannot be solved within the conceptual framework that supposedly should apply, we experience an intellectual shock. The shock may compel us to discard the old framework and adopt a new one. It is to this process of intellectual molting that we owe the birth of many of the major ideas in mathematics and science.
Escape from Paradox
Scientific American, Volume 217, Number 1, July 1967 (p. 50)

Rogers, Jr., Hartley
American mathematician

It is a paradox in mathematics and physics that we have no good model for the teaching of models.
In Lynn Arthur Steen
Mathematics Tomorrow
Physics and Mathematics (p. 232)
Springer-Verlag. New York, New York, USA. 1981

Rota, Gian-Carlo   1932–99
Italian-born American mathematician

Books on paradoxes in statistics are similar to mystery books. They have a faithful readership, and they follow a rigorous sequence in their presentation, like Greek tragedies.
Indiscrete Thoughts
Chapter XX (p. 224)
Birkhäuser. Boston, Massachusetts, USA. 1997

Russell, Bertrand Arthur William   1872–1970
English philosopher, logician, and social reformer

Although this may seem a paradox; all exact science is dominated by the idea of approximation.
In Jefferson Hane Weaver
The World of Physics (Volume 2)
K.2 (p. 22)
Simon & Schuster. New York, New York, USA. 1987

Schild, Alfred   1921–77
Physicist

Consider a pair of twins. Immediately after birth they are separated. One of them, the first one, remains on earth, the second on is put in a rocket ship and flown to Alpha Centauri at a pretty high speed, 99% that of light. Alpha Centauri is the nearest star; it is about four light-years away from us. As soon as the second twin gets to Alpha Centauri, he turns around and flies back to earth at the same high speed. When the two twins meet again, the first one, the one who stayed behind on earth, will be eight years old…he will be able to talk quite well and read a little bit. He may have finished second grade and be about to enter third. The second twin, the one who took the journey, on his return will be approximately one year old…. He will still need diapers, he will be barely able to walk, and he won't be able to talk much.
The Clock Paradox in Relativity Theory
The American Mathematical Monthly, Volume 66, Number 1, January 1959 (p. 1)

Schrödinger, Erwin   1887–1961
Austrian theoretical physicist

Attention had recently (A. Einstein, B. Podolsky, and N. Rosen, *Phys. Rev.* 47 (1935) 777) been called to the obvious but very disentangling measurement to one system, the representative obtained for the other system is by no means independent of the particular choice of observations which we select for that purpose and which by the way are entirely arbitrary. It is rather discomforting that the theory should allow a system to be steered or piloted into one or the other type of state at the experimenter's mercy in spite of his having no access to it.
Proceedings of the Cambridge Philosophical Society, Volume 11, 1935 (p. 555)

Shaw, George Bernard   1856–1950
Irish comic dramatist and literary critic

Paradoxes are the only truths.
Misalliance (p. 142)
Samuel French, Inc. London, England. 1957

Shimony, Abner   1928–
American physicist and philosopher of science

I hope that the rigor and beauty of the argument of EPR [Einstein–Podolsky–Rosen paradox] is apparent. If one does not recognize how good an argument it is — proceeding rigorously from premises which are thoroughly reasonable — then one does not experience an adequate intellectual shock when one finds out that the experimental evidence contradicts their conclusions. This shock should be as great as the one experienced by Frege when he read Russell's theoretical paradox and said, "Alas, arithmetic totters!"

Quoted by Franco Seller
Quantum Mechanics Versus Local Realism: The Einstein–Podolsky–Rosen Paradox
Chapter 1, Section 2 (p. 19)
Plenum Press. New York, New York, USA. 1988

Smith, E. E.   1890–1965
No biographical data available

With sufficient knowledge, any possible so-called paradox can be resolved.
Masters of the Vortex
Chapter 11 (p. 109)
Pyramid. New York, New York, USA. 1968

Sylvester, James Joseph   1814–97
English mathematician

As lightning clears the air of impalpable vapours, so an incisive paradox frees the human intelligence from the lethargic influence of latent and unsuspected assumptions. Paradox is the slayer of Prejudice.
The Collected Mathematical Papers of James Joseph Sylvester (Volume 3)
A Lady's Fan on Parallel Motion, and on an Orthogonal Web of Jointed Rods (p. 36)
University Press. Cambridge, England. 1904–1912

Wilde, Oscar   1854–1900
Irish wit, poet, and dramatist

The way of paradoxes is the way of truth. To test Reality we must see it on the tight-rope. When the Verities become acrobats we can judge them.
The Picture of Dorian Gray
Chapter 3 (p. 44)
The Modern Library. New York, New York, USA. 1992

PARASITE

Bishop of Birmingham

…the loathsome parasite is a result of the integration of mutations: it is both an exquisite example of adaptation to environment and ethically revolting.
Heredity and Predestination
Nature, Volume 126, Number 3187, November 29, 1930 (p. 842)

Brooks, Daniel R.   1951–
American evolutionary biologist
McLennan, Deborah A.   1955–
Canadian evolutionary biologist

Parasites are an enigma. To some people they are an unpleasant but unavoidable fact of life. To others they are, like Victorian ankles, an embarrassing topic to be avoided in polite conversation.
Parascript: Parasites and the Language of Evolution
Chapter 1 (p. 1)
Smithsonian Institution Press. Washington, D.C. 1993

Elton, Charles S.   1900–91
English biologist

The difference between the methods of a carnivore and a parasite is simply the difference between living upon capital and upon income; between the habits of the beaver, which cuts down a whole tree a hundred years old, and the bark-beetle, which levies a daily toll from the tissues of the tree; between the burglar and the blackmailer.
Animal Ecology
Chapter VI (pp. 72–73)
Sidgwick & Jackson, Ltd. London, England. 1927

Emerson, Ralph Waldo   1803–82
American lecturer, poet, and essayist

…each man, like each plant, has his parasites.
The Complete Works of Ralph Waldo Emerson (Volume 6)
The Conduct of Life
Chapter I (p. 45)
Houghton Mifflin Company. Boston, Massachusetts, USA. 1904

Frost, Robert   1874–1963
American poet

Will the blight end the chestnut?
The farmers rather guess not.
It keeps smoldering at the roots
And sending up new shoots
Till another parasite
Shall come to end the blight.
Complete Poems of Robert Frost
Evil Tendencies Cancel
Henry Holt & Company. New York, New York, USA. 1949

Mayr, Ernst   1904–2005
German-born American biologist

Parasitologists have accumulated, during the past decades, an amount of information that is truly formidable. This information is not only valuable for the parasitologist, but is also a potential gold-mine for the evolutionist and general biologist. Yet, much of this information is hidden away in a widely scattered and highly technical literature.
In J. G. Baer (ed.)
Premier Symposium sur la spécificité parasitaire des parasites des Vertébrés
Evolutionary Aspects of Host Specificity Among Parasites of Vertebrates
Université de Neuchâtel. Neuchâtel, Switzerland, 1957

Mr. Spock
Fictional character

A truly successful parasite is commensal, living in amity with its host, or even giving it positive advantages.… A parasite that regularly and inevitably kills its host cannot survive long, in the evolutionary sense, unless it multiplies with tremendous rapidity.… It is not pro-survival.

Star Trek II: The Wrath of Khan
Film (1982)

Noble, Elmer R.   1909–2001
American protozoologist and parasitologist

Noble, Glenn A.   1909–?
American biologist

Parasites as a whole are worthy examples of the inexorable march of evolution into blind alleys.
Parasitology: The Biology of Animal Parasites (3rd edition)
Section X, Chapter 25 (p. 572)
Lea & Febiger. Philadelphia, Pennsylvania, USA. 1971

Shakespeare, William   1564–1616
English poet, playwright, and actor

Unbidden guests
Are often welcomest when they are gone.
In *Great Books of the Western World* (Volume 26)
The Plays and Sonnets of William Shakespeare (Volume 1)
The First Part of King Henry the Sixth
Act II, Scene ii, l. 55–56
Encyclopædia Britannica, Inc. Chicago, Illinois, USA. 1952

van Beneden, P. J.
No biographical data available

In the ancient as well as the new world, more than one animal resembles somewhat the sharper leading the life of a great nobleman; and it is not rare to find, by the side of the humble pickpocket, the audacious brigand of the high road, who lives solely on blood and carnage. A great proportion of these creatures always escape, either by cunning, by audacity, or by superior villainy, from social retribution.
Animal Parasites and Messmates (p. xvii)
Henry S. King. London, England. 1876

Wilson, Edward O.   1929–
American biologist and author

Leishmaniasis, schistosomiasis, malignant tertian malaria, filariasis, echinococcosis, onchocerciasis, yellow fever, amoebic dysentery, bleeding bot-fly cysts... evolution has devised a hundred ways to macerate livers and turn blood into a parasite's broth.
Biophilia
Bernhardsdorp (pp. 12–13)
Harvard University Press. Cambridge, Massachusetts. 1984

PARKINSON'S DISEASE

Parkinson, James   1755–1824
English physician and paleontologist

The disease, respecting which the present inquiry is made, is of a nature highly afflictive.... The unhappy sufferer has considered it as an evil, from the domination of which he had not prospect of escape.

An Essay on the Shaking Palsy
Medical Classics, Volume 2, Number 10, June 1938

So slight nearly imperceptible are the first inroads of this malady, and so extremely slow its progress, that it rarely happens, that the patient can form any recollection of the precise period of its commencement. The first symptoms perceived are, a slight sense of weakness, with a proneness to trembling in some particular part, sometimes in the head, but most commonly in one of the hands and arms.
An Essay on the Shaking Palsy
Medical Classics, Volume 2, Number 10, June 1938

PARTICLE

Davy, Sir Humphry   1778–1829
English chemist

...the different bodies in nature are composed of particles or minute parts, individually imperceptible to the senses. When the particles are similar, the bodies they constitute are denominated simple, and when they are dissimilar, compound. The chemical phenomena result from the different arrangements of the particles of bodies; and the powers that produce these arrangements are repulsion, or the agency of heat, and attraction.
Syllabus of a Course of Lectures at the Royal Institution (p. 2)
Publisher undetermined. London, England. 1802

Fermi, Enrico   1901–54
Italian-born American physicist

If I could remember the names of all these particles, I'd be a botanist.
In A. Zee
Fearful Symmetry
Chapter 11 (p. 168)
Macmillan Publishing Company. New York, New York, USA. 1986

Feynman, Richard P.   1918–88
American theoretical physicist

We seem gradually to be groping toward an understanding of the world of sub-atomic particles, but we really do not know how far we have yet to go in this task.
Six Easy Pieces: Essentials of Physics Explained by Its Most Brilliant Teacher
Basic Physics (p. 45)
Addison-Wesley Publishing Company. Reading, Massachusetts, USA. 1995

One of the consequences is that things which we used to consider as waves also behave like particles, and particles behave like waves; in fact everything behaves the same way. There is no distinction between a wave and a particle. So quantum mechanics unifies the idea of the field and its waves, and the particles, all into one.
Six Easy Pieces: Essentials of Physics Explained by Its Most Brilliant Teacher

Basic Physics (p. 36)
Addison-Wesley Publishing Company. Reading, Massachusetts, USA.
1995

Feynman, Richard P.   1918–88
American theoretical physicist
Leighton, Robert B.   1919–97
American physicist
Sands, Matthew L.   1919–
American physicist

Quantum mechanics has many aspects. In the first place, the idea that a particle has a definite location and a definite speed is no longer allowed; that is wrong.
The Feynman Lectures on Physics (Volume 1)
Chapter 2–3 (p. 2–6)
Addison-Wesley Publishing Company. Reading, Massachusetts, USA.
1983

Glashow, Sheldon L.   1932–
American physicist

Tapestries are made by many artisans working together. The contributions of separate workers cannot be discerned in the complete work, and the loose and false threads have been covered over. So it is in our picture of particle physics.
Nobel Lectures, Physics 1971–1980
Nobel lecture for award received in 1979
Towards a Unified Theory — Threads in a Tapestry (p. 494)
World Scientific Publishing Company. Singapore. 1992

Gleick, James   1954–
American author, journalist, and essayist

Quantum mechanics taught that a particle was not a particle but a smudge, a traveling cloud of possibilities...
Genius: The Life and Science of Richard Feynman
M. I. T. (p. 89)
Pantheon Books. New York, New York, USA. 1992

Hein, Piet   1905–96
Danish poet and scientist

Nature, it seems is the popular name
for milliards and milliards and milliards
of particles playing their infinite game
of billiards and billiards and billiards.
Grooks II
Atomyriades
Doubleday & Company, Inc. Garden City, New York, USA. 1969

Heisenberg, Werner Karl   1901–76
German physicist and philosopher

We can not longer speak of the behavior of the particle independently of the process of observation. As a final consequence, the natural laws formulated mathematically in quantum theory no longer deal with the elementary particles themselves but with our knowledge of them.

Nor is it any longer possible to ask whether or not these particles exist in space and time objectively...
The Physicist's Conception of Nature
Chapter I (p. 15)
Greenwood Press, Publishers. Westport, Connecticut, USA. 1958

The mathematically formulated laws of quantum theory show clearly that our ordinary intuitive concepts cannot be unambiguously applied to the smallest particles. All the words or concepts we use to describe ordinary physical objects, such as position, velocity, color, size, and so on, become indefinite and problematic if we try to [apply them to] elementary particles.
Across the Frontiers
Chapter IX (p. 114)
Harper & Row, Publishers. New York, New York, USA. 1974

In the light of quantum theory these elementary particles are no longer real in the same sense as objects of daily life, trees or stones, but appear as abstractions derived from the real material of observation in the true sense.
On Modern Physics
Philosophical Problems (p. 13)
C.N. Potter. New York, New York, USA. 1961

Johnson, George   1952–
American science writer

In science's great chain of being, the particle physicists place themselves with the angels, looking down from the heavenly spheres on the chemists, biologists, geologists, meteorologists — those who are applying, not discovering, nature's most fundamental laws. Everything, after all, is made from subatomic particles. Once you have a concise theory explaining how they work, the rest should just be filigree.
New Contenders for a Theory of Everything
The New York Times, F1, Column 1, Tuesday, December 4, 2001

Regnault, Nöel   1702–62
Jesuit mathematician

The Imagination is lost here. Rather than the Minds; for if you divide a Particle into the most inconceivably minute Parts, the Mind will always find therein something that regards the West, and something that regards the East; and what regards the West, is not that which regards the East.
Philosophical Conversations (Volume 1)
Conversation I (p. 9)
Printed for W. Innys, C. Davis, and N. Prevost. London, England. 1731

Stewart, Ian   1945–
English mathematician and science writer
Cohen, Jack
Reproductive biologist

They ask wavy questions to decide whether it's a wave, and particle questions to decide whether it's a particle.

The Collapse of Chaos: Discovering Simplicity in a Complex World
Chapter 8 (p. 276)
The Viking Press. New York, New York, USA. 1994

Weinberg, Steven   1933–
American nuclear physicist

As a scientist, you're probably not going to get rich. Your friends and relatives probably won't understand what you're doing. And if you work in a field like elementary particle physics, you won't even have the satisfaction of doing something that is immediately useful. But you can get great satisfaction by recognizing that your work in science is a part of history.
Scientist: Four Golden Lessons
Nature, Volume 426, 2003 (p. 389)

Whitman, Walt   1819–92
American poet, journalist, and essayist

Oh amazement of things — even the least particle!
Complete Poems and Collected Prose
Song at Sunset
The Library of America. New York, New York, USA. 1982

PAST

Author undetermined

"Hands off the Past!" he cried. "No man is fit
To see or touch it till I've sieved each bit.
The Past is mine!" Well, now he's part of it.
Epitaph on an Archaeologist
Punch, February 12, 1986 (p. 63)

Ayer, Alfred Jules   1910–89
English philosopher

In practice, speculations about the past, if they are not to be entirely idle, must relate to the traces which the past has left.
The Central Questions of Philosophy
Chapter II (p. 25)
Weidenfeld & Nicolson. London, England. 1973

Barrow, John D.   1952–
English theoretical physicist

Things are as they are because they were as they were.
The Origin of the Universe
Chapter 1 (p. 17)
Basic Books. New York, New York, USA. 1994

Colbert, Edwin H.   1905–2001
American vertebrate paleontologist

The past is mysterious, ever so the farther we look back from our vantage point in the twentieth-century world. As we follow the procession of the year back through time the earth and its inhabitants seem to us less real and less substantial the more distantly they are removed from this age in which we live.
The Age of Reptiles
Chapter 1 Time, Tetrapods and Fossils (p. 1)
W.W. Norton & Company, Inc. New York, New York, USA. 1965

Crawford, Osbert Guy Stanhope   1886–1957
English archaeologist

The archaeologist who tries to re-create the past is like a craftsman at work upon a great building. At first he sees but dimly the plan of the whole, but as he warms to his work it gradually unfolds itself before him.
Man and His Past
Chapter XIX (p. 225)
Oxford University Press, Inc. London, England. 1921

Elton, G. R.
No biographical data available

The future is dark, the present burdensome; only the past, dead and finished, bears contemplation.
The Beaver, Volume 72, Number 4, Aug./Sept. 1992 (p. 4)

Gorky, Maxim   1868–1938
Soviet/Russian writer

You can't drive anywhere in a carriage of the past!
The Lower Depths
Act Four (p. 80)
Brentano's Publishers. New York, New York, USA. 1923

Hartley, L. P.   1895–1972
English writer

The past is a foreign country; they do things differently there.
The Go-Between
Prologue (p. 3)
Alfred A. Knopf. New York, New York, USA. 1954

Hawking, Stephen William   1942–
English theoretical physicist

…the light that we see from distant galaxies left them millions of years ago, and in the case of the most distant object we have seen, the light left some eight billions years ago. Thus, when we look at the universe, we are seeing it as it was in the past.
A Brief History of Time: From the Big Bang to Black Holes
Chapter 2 (p. 28)
Bantam Books. Toronto, Ontario, Canada. 1988

Horn, Alfred Aloysius   1854–1927
Traveler, trader, and adventurer

There's places in Africa where you get visions of primeval force…in Africa the Past has hardly stopped breathing.
Trader Horn: Being the Life and Works of Alfred Aloysius Horn
Chapter XXIII (pp. 257, 258)
Simon & Schuster. New York, New York, USA. 1927

Inscription

What is past is prologue.
Entrance to National Archives, Washington, D.C.

Koestler, Arthur 1905–83
Hungarian-born English writer

…man cannot inherit the past; he has to recreate it.
The Act of Creation
Book One, Part Two, Chapter XI (p. 266)
The Macmillan Company. New York, New York, USA. 1964

Kohl, Philip L. 1946–
American anthropologist

A real past, although blurred, can be glimpsed through archaeological materials.
Symbolic Cognitive Archaeology
Dialectical Anthropology, Volume 9, 1985 (p. 115)

Kubler, George 1912–96
American art historian

Knowing the past is as astonishing a performance as knowing the stars.
The Shape of Time: Remarks on the History of Things
Chapter 1 (p. 19)
Yale University Press. New Haven, Connecticut, USA. 1962

Leakey, Mary 1913–96
English archaeologist

Man's early tools and any insights we can get into the lifestyles and activities in succeeding stages of human evolution have been the aspects of the past that I have found the most absorbing, more so than the anatomical features linking or separating one fossil hominoid from another…
Disclosing the Past: An Autobiography
Chapter 16 (p. 211)
Doubleday & Company, Inc. Garden City, New York, USA. 1984

Lipe, William D.
American archaeologist

…consideration of the past removes us from the immediate concerns of the here and now…[and] plunges us directly into the larger common world which exists in the stream of time and hence bridges the mortality of generations.
In H. Cleere (ed.)
Approaches to the Archaeological Heritage: A Comparative Study of World Cultural Resource
Management Systems Value and Meaning in Cultural Resources (p. 10)
Cambridge University Press. London, England. 1984

Longfellow, Henry Wadsworth 1807–82
American poet

Let me review the scene,
And summon from the shadowy Past
The forms that once have been.
The Poetical Works of Henry Wadsworth Longfellow
A Gleam of Sunshine
Houghton Mifflin Company. Boston, Massachusetts, USA. 1883

Mann, Thomas 1875–1955
German-born American novelist

Very deep is the well of the past. Should we not call it bottomless?
Translated by H. T. Lowe-Porter
Joseph and His Brothers
Prelude (p. 3)
Alfred A. Knopf. New York, New York, USA. 1939

Newton, Sir Charles Thomas 1816–94
British archaeologist

The record of the Human Past is not all contained in printed books. Man's history has been graven on the rocks of Egypt, stamped on the bricks of Assyria, enshrined in the marble of the Parthenon — it rises before us a majestic Presence in the piled-up arches of the Coliseum — it lurks an unsuspected treasure amid the oblivious dust of archives and monasteries — it is embodied in all the heirlooms of religions, of races, of families; in the relics which affection and gratitude, personal or national, pride of country or pride of lineage, have preserved for us…
Essays on Art and Archaeology
Chapter I (p. 1)
Macmillan & Company Ltd. London, England. 1880

Orwell, George (Eric Arthur Blair) 1903–50
English novelist and essayist

Who controls the past controls the future: who controls the present controls the past.
Nineteen Eighty-Four
Part Three, Chapter II (p. 251)
Buccaneer Books. Cutchogue, New York, USA. 1949

Sandburg, Carl 1878–1967
American poet and biographer

I tell you the past is a bucket of ashes.
Complete Poems
Prairie
Harcourt, Brace. New York, New York, USA. 1950

Sir Joseph Whemple
Fictional character

We didn't come to Egypt to dig for medals! Much more is to be learned from studying bits of broken pottery than from all the sensational finds. Our job is to increase the sum of human knowledge of the past, not to satisfy our own curiosity.
The Mummy
Film (1940)

Toulmin, Stephen 1922–
Anglo-American philosopher

Goodfield, June
Science writer and historian

...no transformation in men's attitude to Nature — in their "common sense" — has been more profound than the change in perspective brought about by the discovery of the past. Rather than take this discovery for granted, it is almost preferable to exaggerate its significance.
The Discovery of Time
Introduction (pp. 17–18)
Harper & Row, Publishers. New York, New York, USA. 1965

Weigall, Arthur Edward   1880–1934
English Egyptologist and author

Man is by nature a creature of the present. It is only by an effort that he can consider the future, and it is often quite impossible for him to give any heed at all to the past.
The Glory of the Pharaohs
Chapter II (p. 35)
G.P. Putnam's Sons. New York, New York, USA. 1923

Wells, H. G. (Herbert George)   1866–1946
English novelist, historian and sociologist

[The] restoration of the past is one of the most astonishing adventures of the human mind.
The Grisly Folk
Storyteller Magazine, April 1921

Could anything be more dead, more mute and inexpressive to the inexpert eye than the ochreous fragments of bone and the fractured lumps of flint that constitute the first traces of something human in the world?
The Grisly Folk
Storyteller Magazine, April 1921

Whitman, Walt   1819–92
American poet, journalist, and essayist

The past, the infinite greatness of the past!
For what is the present, after all, but a growth out of the past.
Complete Poetry and Collected Prose
Leaves of Grass
Passage to India
The Library of America. New York, New York, USA. 1982

PATENT

O'Malley, John R.
No biographical data available

Almost every engineer is affected by the patent system.
Patents and the Engineer
Engineering Facts from Gatorland, Volume 4, Number 5, December 1967

Proverb

A patent is merely a title to a lawsuit.

In Frank Lewis Dyer
Edison. His Life and Inventions (Volume 2)
Chapter XXVIII (p. 700)
Harper & Brothers. New York, New York, USA. 1929

Roosevelt, Franklin Delano   1882–1945
32nd president of the United States

Patents are the key to our technology; technology is the key to production.
In Robert A. Buckles
Ideas, Inventions, and Patents: How to Develop and Protect Them
Chapter 1 (p. 1)
John Wiley & Sons, Inc. New York, New York, USA. 1957

PATENT MEDICINE

Adams, Samuel Hopkins   1871–1958
American author

With a few honorable exceptions the press of the United States is at the beck and call of the patent medicine. Not only do the newspapers modify news possibly affecting these interests, but they sometimes become their agents.
The Great American Fraud
Collier's Weekly, Volume 36 October 7, 1905 (p. 14)

PATHOLOGY

Mencken, H. L. (Henry Louis)   1880–1956
American journalist and literary critic

Pathology would remain a lovely science, even if there were no therapeutics, just as seismology is a lovely science, though no one knows how to stop earthquakes.
A Mencken Chrestomathy
Chapter XXX (pp. 625–626)
Alfred A. Knopf. New York, New York, USA. 1949

Virchow, Rudolf Ludwig Karl   1821–1902
German pathologist and archaeologist

Pathology has been released from the anomalous and isolated position which it has occupied for thousands of years. Through the application of its doctrines not only to diseases of man, but also to those of even the smallest and lowest of animals, and to those of plants, it helps to deepen biological knowledge, and to light up still further that region of the unknown which still envelops the intimate structure of living matter. It is no longer merely applied physiology — it has become physiology itself.
Translated by Lelland J. Rather
Disease, Life, and Man, Selected Essays
The Place of Pathology Among the Biological Sciences (p. 169)
Stanford University Press. Stanford, California, USA. 1958

Pathology also has its place in the science of biology, certainly a very honorable one, for to pathology we owe the realization that the contrast between health and disease is not to be sought in a fundamental difference of two

kinds of life, nor in an alteration of essence, but only in an alteration of conditions.
Translated by Lelland J. Rather
Disease, Life, and Man, Selected Essays
The Place of Pathology Among the Biological Sciences (p. 169)
Stanford University Press. Stanford, California, USA. 1958

PATIENT

Abernethy, John   1680–1740
Irish Presbyterian minister, theologian, and dissenter

Private patients, if they do not like me, can go elsewhere, but the poor devils in the hospital I am bound to take care of.
Memoirs of John Abernethy
Chapter V (p. 37)
Harper & Brothers. New York, New York, USA. 1853

Armour, Richard   1906–89
American poet

The perfect patient let us praise:
He's never sick on Saturdays,
In fact this wondrous, welcome sight
Is also never sick at night.
In waiting rooms he does not burn
But gladly sits and waits his turn,
And even, I have heard it said,
Begs others, "Please go on ahead."
He takes advice, he does as told,
He has a heart of solid gold.
He pays his bills, without a fail,
In cash, or by the same day's mail.
He has but one small fault I'd list:
He doesn't (what a shame!) exist.
The Medical Muse
Ideal Patient
McGraw-Hill Book Company, Inc. New York, New York, USA. 1963

Cushing, Harvey   1869–1939
American neurosurgeon

Every patient, he said, provided two questions — firstly what can be learnt from him and secondly what can be done for him.
In Robert Coope
The Quiet Art (p. 103)
E.S. Livingstone Ltd. Edinburgh, Scotland. 1952

de Madariaga, Salvador   1886–1978
Spanish writer and statesman

There are no diseases, there are only patients.
Essays with a Purpose
On Medicine (p. 174)
Hollis & Carter. London, England. 1954

Drake, Daniel   1785–1852
American physician

[There is no era in the life of a physician] in which his self-complacency is so exalted, as the time which passes between receiving his diploma with its blue ribbon, and receiving crepe and gloves, to wear at the funeral of his first patient.
Western Journal of Medicine and Surgery, New Series, II:355, October 1844

Eliot, T. S. (Thomas Stearns)   1888–1965
American-born British poet and playwright

REILLY: Most of my patients begin, Miss Coplestone, by telling me exactly what is the matter with them. And what I am to do about it.
The Collected Poems and Plays 1909–1950
The Cocktail Party, Act Two (p. 359)
Harcourt, Brace & World, Inc. New York, New York, USA. 1952

Helmuth, William Tod   1833–1902
American physician

She sent for me in haste to come and see,
What her condition for a cure might be.
Dear me! a patient — what a happy tone,
To have a patient and one all my own —
To have a patient and myself be feed,
Raised expectations very high indeed —
I saw a practice growing from the seed.
Scratches of a Surgeon
My First Patient (p. 61)
W.A. Chatterton & Company. Chicago, Illinois, USA. 1879

Heschel, Abraham J.   1907–72
Jewish theologian

The patient must not be defined as a client who contracts a physician for service; he is a human being entrusted to the cure of a physician.
The Insecurity of Freedom
The Patient as a Person (p. 31)
Farrar, Straus & Giroux. New York, New York, USA. 1966

Holmes, Oliver Wendell   1809–94
American physician, poet, and humorist

What I call a good patient is one who, having found a good physician, sticks to him till he dies.
Medical Essays
The Young Practitioner (p. 390)
Houghton Mifflin Company. Boston, Massachusetts, USA. 1911

Once in a while you will have a patient of sense, born with the gift of observation, from whom you may learn something.
Medical Essays
The Young Practitioner (pp. 382–383)
Houghton Mifflin Company. Boston, Massachusetts, USA. 1911

If you are making choice of a physician, be sure to get one, if possible, with a cheerful and serene countenance.
The Professor at the Breakfast Table
Chapter VI (p. 180)
Ticknor & Fields. Boston, Massachusetts, USA. 1860

Hubbard, Kin 1868–1930
American Democratic newspaper editor

Be kind t' th' henn egg. When sickness enters th' home an' th' patient comes thru th' crisis twenty pounds lighter than a straw hat, an' is propped up with pillows in th' bay window t' watch th' speedin', an' loved ones try t' tempt him with round steak, an' pickles an' near beer, he wearily waves 'em away. But with his first returnin' strength he squirms an' turns his listerless eyes toward th' kitchen an' says, in a voice weak an' scarcely audible, "Maw, I believe I could worry down an egg…"
Abe Martin: Hoss Sense and Nonsense (p. 53)
The Bobbs-Merrill Company, Indianapolis, Indiana, USA. 1926

Mayo, William J. 1861–1939
American physician

…the highly scientific development of this mechanistic age had led perhaps to some loss in appreciation of the individuality of the patient and to trusting largely to the laboratories and outside agencies which tended to make the patient not the hub of the wheel, but a spoke.
Edward Martin, M.D., 1859–1938
Collected Papers of the Mayo Clinic & Mayo Foundation, Volume 30, 1938

Morris, Robert Tuttle 1857–1945
Abdominal surgeon

It is the patient rather than the case which requires treatment.
Doctors Versus Folks
Chapter 2
Doubleday, Page & Company, Inc. Garden City, New York, USA. 1915

Newman, Sir George 1870–1948
English public health physician

There are four questions which in some form or other every patient asks his doctor: (a) What is the matter with me? This is diagnosis. (b) Can you put me right? This is treatment and prognosis. (c) How did I get it? This is causation. (d) How can I avoid it in future? This is prevention.
Preventive Medicine for the Medical Student
The Lancet, Volume 221, November 21, 1931 (p. 113)

Osler, Sir William 1849–1919
Canadian physician and professor of medicine

To study the phenomena of disease without books is to sail an uncharted sea, while to study books without patients is not to go to sea at all.
In Harvey Cushing
The Life of Sir William Osler (Volume 1)
Chapter IV (p. 67)
Clarendon Press. Oxford, England. 1925

Parrot, Max
No biographical data available

It is often been said that the technical aspects of medicine are easy. The difficult part is dealing with the personality of the patient, the so-called psychological or human factor. This takes up a great deal of the time of the practicing physician. It is harder on the doctor's constitution than all of the technical aspects of medicine. It may even cause his or her demise, in the case of a physician with an autonomic nervous system that can't take the heat.
In Irving Oyle
The New American Medical Show: Discovering the Healing Connection (p. 25)
Unity Press. Santa Cruz, California, USA. 1979

Potter, Stephen 1900–69
No biographical data available

If Patient turns out to be really ill, it is always possible to look grave at the same time and say "You realize, I suppose, that 25 years ago you'd have been dead?"
One-Upmanship
Chapter II (p. 28)
Henry Holt & Company. New York, New York, USA. 1952

Rhazes 865–925
Persian physician

The patient who consults a great many physicians is likely to have a very confused state of mind.
In Samuel Evans Massengill
A Sketch of Medicine and Pharmacy and a View of Its Progress by the Massengill Family from the Fifteenth to the Twentieth Century (p. 45)
The S.E. Massengill Company. Bristol, Tennessee, USA. 1943

Sacks, Oliver W. 1933–
American neurologist and author

There is only one cardinal rule: one must always listen to the patient.
Listening to the Lost
Newsweek, August 20, 1984 (p. 70)

Shakespeare, William 1564–1616
English poet, playwright, and actor

"How does your patient," doctor?
"Not so sick, my lord,
As she is troubled with thick-coming fancies."
In *Great Books of the Western World* (Volume 27)
The Plays and Sonnets of William Shakespeare (Volume 2)
Macbeth
Act V, Scene iii, l. 37–39
Encyclopædia Britannica, Inc. Chicago, Illinois, USA. 1952

PATTERN

Burns, Marilyn
No biographical data available

Searching for patterns is a way of thinking that is essential for making generalizations, seeing relationships, and understanding the logic and order of mathematics.

Functions evolve from the investigation of patterns and unify the various aspects of mathematics.
About Teaching Mathematics: A K–8 Resource
Patterns and Functions (p. 112)
Math Solutions Publications, USA. 1992

Derry, Gregory N.   1952–
American professor of physics

In trying to understand nature, we rarely attempt to grasp completely every possible detail. If we did, we'd be overwhelmed by the mass of inconsequential information. As a result, we would miss the truly interesting patterns and relationships that give us scientific insight.
What Science Is and How It Works
Chapter 6 (p. 69)
Princeton University Press. Princeton, New Jersey, USA. 1999

Gardner, Martin   1914–
American writer and mathematics games editor

If the cosmos were suddenly frozen, and all movement ceased, a survey of its structure would not reveal a random distribution of parts. Simple geometrical patterns, for example, would be found in profusion — from the spirals of galaxies to the hexagonal shapes of snow crystals. Set the clockwork going, and its parts move rhythmically to laws that often can be expressed by equations of surprising simplicity. And there is no logical or a priori reason why these things should be so.
Order and Surprise
Chapter 4 (p. 57)
Prometheus Books. Buffalo, New York, USA. 1983

Hofstadter, Douglas   1945–
American cognitive scientist and author

Yes, I am a relentless quester after the chief patterns of the universe — central organizing principles, clean and powerful ways to categorize what is "out there."
Metamagical Themas: Questing for the Essence of Mind and Pattern
Introduction (p. xxv)
Basic Books, Inc. New York, New York, USA. 1985

Huxley, Aldous   1894–1963
English writer and critic

The difference between a piece of stone and an atom is that an atom is highly organised, whereas the stone is not. The atom is a pattern, and the molecule is a pattern, and the crystal is a pattern; but the stone, although it is made up of these patterns, is just a mere confusion. It's only when life appears that you begin to get organisation on a larger scale. Life takes the atoms and molecules and crystals; but, instead of making a mess of them like the stone, it combines them into new and more elaborate patterns of its own.
Time Must Have a Stop
Chapter XIV (p. 145)
The Sun Dial Press. Garden City, New York, USA. 1944

MacArthur, Robert H.   1930–72
American ecologist

To do science is to search for repeated patterns, not simply to accumulate facts, and to do the science of geographical ecology is to search for patterns of plants and animal life that can be put on a map.
Geographical Ecology
Introduction (p. 1)
Harper & Row, Publishers, New York, New York, USA. 1972

Peterson, Ivars
Mathematics writer

In their search for patterns and logical connections, mathematicians face a vast, mysterious ocean of possibilities. Over the centuries, they have discovered an extensive archipelago of truth and beauty. Much of that accumulated knowledge is passed on to succeeding generations. Even more wonders await future explorers of deep, mathematical waters.
Islands of Truth: A Mathematical Mystery Cruise
Chapter 8 (p. 292)
W.H. Freeman & Company. New York, New York, USA. 1990

Stevens, Peter S.
No biographical data available

It turns out that those patterns and forms are peculiarly restricted, that the immense variety that nature creates emerges from the working and reworking of only a few formal themes. These limitations on nature bring harmony and beauty to the natural world.
Patterns in Nature
Chapter 1 (p. 3)
Little, Brown & Company. Boston, Massachusetts, USA. 1974

PAULI PRINCIPLE

Gamow, George   1904–68
Russian-born American physicist

We do not know why they have the masses they do; we do not know why they transform into another the way they do; we do not know anything! The one concept that stands like the Rock of Gibraltar in our sea of confusion is the Pauli principle.
The Exclusion Principle
Scientific American, Volume 201, Number 1, July 1959 (p. 86)

PENDULUM

Eco, Umberto   1932–
Italian novelist, essayist, and scholar

That was when I saw the Pendulum.
The sphere, hanging from a long wire set into the ceiling of the choir, swayed back and forth with isochronal majesty.

I knew — but anyone could have sensed it in the magic of that serene breathing — that the period was governed by the square root of the length of the wire and by pi, that number which, however irrational to sublunar minds, though a higher rationality binds the circumference and diameter of all possible circles. The time it took the sphere to swing from end to end was determined by an arcane conspiracy between the most timeless of measures: the singularity of the point of suspension, the duality of the plane's dimensions, the triadic beginning of π, the secret quadratic nature of the root, and the unnumbered perfection of the circle itself.

Translated by William Weaver
Foucault's Pendulum
Chapter 1 (p. 3)
Harcourt Brace Jovanovich, Publishers. San Diego, California, USA. 1988

Graham, L. A.
No biographical data available

Rock-a-bye baby in the tree top,
As a compound pendulum, you are a flop.
Your center of percussion is safe and low,
As one may see when the wind doth blow.
Your frequency of vibration is pretty small,
Frankly, I don't think you'll fall at all.

Ingenious Mathematical Problems and Methods
Mathematical Nursery Rhyme Number 2
Dover Publications, Inc. New York, New York, USA. 1959

PENICILLIN

Fleming, Sir Alexander   1881–1955
Scottish bacteriologist

I have been frequently asked why I invented the name "Penicillin." I simply followed perfectly orthodox lines and coined a word which explained that the substance penicillin was derived from a plant of the genus Penicillium just as many years ago the word "Digitalin" was invented for a substance derived from the plant Digitalis.

Nobel Lectures, Physiology or Medicine 1942–1962
Nobel lecture for award received in 1945
Penicillin (p. 83)
Elsevier Publishing Company. Amsterdam, Netherlands. 1964

PERCENTAGE

Barnes, Michael R.
No biographical data available

There's a 50 percent chance of anything — either it happens or it doesn't.

In Paul Dickson
The Official Explanations (p. B–9)
Delacorte Press. New York, New York, USA. 1980

Bloch, Arthur   1948–
American humorist

90% of everything is crap.

Murphy's Law
Sturgeon's Law (p. 21)
Price/Stern/Sloan, Publishers. Los Angeles, California, USA. 1981

Crichton, Michael   1942–
American novelist

John. Trust us on this, we have the figures. We are telling you with ninety-five percent confidence intervals how the people feel.

Rising Sun
Second Day (p. 255)
Ballantine Books. New York, New York, USA. 1993

"I did," Gerhard said. "But I don't know any more. We've passed the confidence limits already. They were about plus or minus two minutes for ninety-nine percent."

The Terminal Man
Chapter 6 (p. 157)
Alfred A. Knopf. New York, New York, USA. 1972

Jeans, Sir James Hopwood   1877–1946
English physicist and mathematician

When half a million babies are born in England in a year, we may say that 20 percent of them are born in London, 2 percent in Manchester, 1 percent in Bristol, and so on. But when we think of one baby born in a single minute of time, we cannot say that 20 percent of it was born in London, 2 percent in Manchester, and so on. We can only say that there is a 20 percent probability of its being born in London, a 2 percent probability of its being born in Manchester, and so on.

Physics and Philosophy
Chapter V (p. 136)
Dover Publications, Inc. New York, New York, USA. 1981

Twain, Mark (Samuel Langhorne Clemens)   1835–1910
American author and humorist

…I do not remember just when, for I was not born then and cared nothing for such things. It was a long journey in those days and must have been a rough and tiresome one. The village contained a hundred people and I increased the population by 1 percent. It was more than many of the best men in history could have done for a town. It may not be modest in me to refer to this but it is true.

Mark Twain's Autobiography (Volume 1)
Chapter Begun in Vienna (pp. 94–95)
Harper & Brothers. New York, New York, USA. 1924

PERCEPTION

Adams, Douglas   1952–2001
English author, comic radio dramatist, and musician

Everything you see or hear or experience in any way at all is specific to you. You create a universe by perceiving it.
The Ultimate Hitchhiker's Guide to The Galaxy
Mostly Harmless
Chapter 9 (p. 703)
The Ballantine Book Company. New York, New York, USA. 2002

Blake, William   1757–1827
English poet, painter, and engraver

As to that false appearance which appears to the reasoner
As of a Globe rolling thro' Voidness, it is a delusion of Ulro.
The Microscope knows not of this nor the Telescope: they alter
The ratio of the Spectator's Organs, but leave Objects untouch'd.
The Complete Poetry and Prose of William Blake
The Building of Time
University of California Press. Berkeley, California, USA. 1982

Cousins, Norman   1912–90
American editor and author

As we enlarge our sense of the cosmos, we are enlarging our consciousness. As we extend the reach of the mind, we are learning more about our potentialities. As we move beyond the human habitat, we are gaining perspective on ourselves as custodians of the planet.
Rendezvous with Infinity
Cosmic Search Magazine, Volume 1, Number 1, January 1, 1979 (pp. 30–31)

Murphy, Michael
No biographical data available

To a frog with its simple eye, the world is a dim array of grays and blacks. Are we like frogs in our limited sensorium, apprehending just part of the universe we inhabit? Are we as a species now awakening to the reality of multidimensional worlds in which matter undergoes subtle reorganizations in some sort of hyperspace?
The Future of the Body
Part I, Chapter 8 (p. 216)
Penguin Putnam, Inc. New York, New York, USA. 1992

Russell, Bertrand Arthur William   1872–1970
English philosopher, logician, and social reformer

Physics and perception are like two people on opposite sides of a brook which slowly widens as they walk; at first it is easy to jump across, but imperceptibly it grows more difficult, and at last a vast labor is required to get from one side to the other.
Analysis of Matter
Chapter XIV (p. 137)
Dover Publications, Inc., New York, New York, USA, 1954

Whitehead, Alfred North   1861–1947
English mathematician and philosopher

Our problem is, in fact, to fit the world to our perceptions, and not our perceptions to the world.
The Organization of Thought
Chapter VIII (p. 228)
Greenwood Press Publishers. Westport, Connecticut, USA. 1974

PERCUSSION

Auenbrugger, Leopold   1722–1809
Viennese physician

I present the reader with a new sign, which I have discovered for detecting diseases of the chest, This consists in the percussion of the human thorax, whereby according to the character of the particular sounds thence elicited, an opinion is formed of the internal state of that cavity.
New Invention by Means of Percussing the Human Thorax for Detecting Signs of Obscure Disease of the Interior of the Chest
December 31, 1761

Whitehead, Alfred North   1861–1947
English mathematician and philosopher

Even perfection will not bear the tedium of indefinite repetition.
Atlantic, September 29, 1979 (p. 244)

PERIODIC TABLE

Atkins, Peter William   1940–
English physical chemist and writer

The periodic table is arguably the most important concept in chemistry, both in principle and in practice. It is the everyday support for students, it suggests new avenues of research to professionals, and it provides a succinct organization of the whole of chemistry. It is a remarkable demonstration of the fact that the chemical elements are not a random clutter of entities but instead display trends and lie together in families. Anyone who seeks to be familiar with a scientist's-eye view of the world must be aware of the general form of the periodic table, for it is a part of scientific culture.
The Periodic Kingdom: A Journey into the Land of the Chemical Elements
Preface (pp. vii–viii)
Basic Books, Inc. New York, New York, USA. 1995

Bolton, Henry Carrington   1843–1903
American chemist, bibliographer, and historian

The periodic law has given to chemistry that prophetic power long regarded as the peculiar dignity of its sister science, astronomy.
In Joseph William Mellor
Mellor's Modern Inorganic Chemistry
Chapter 9 (p. 122)
Longmans. London, England. 1967

Mendeleyev, Dmitry   1834–1907
Russian chemist

I shall endeavor to show, as briefly as possible, ...how far the periodic law contributes to enlarge our range of vision. Before the promulgation of this law the chemical elements were mere fragmentary, incidental facts in Nature; there was no special reason to expect the discovery of new elements, and the new ones which were discovered from time to time appeared to be possessed of quite novel properties. The law of periodicity first enabled us to perceive undiscovered elements at a distance which formerly was inaccessible to chemical vision...

The Periodic Law of the Chemical Elements
Journal of the Chemical Society, Volume 55, 1889 (p. 648)

There must be some bond of union between mass and the chemical elements; and as the mass of a substance is ultimately expressed...in the atom, a functional dependence should exist and be discoverable between the individual properties of the elements and their atomic weights. But nothing, from mushrooms to a scientific law, can be discovered without looking and trying. So I began to look about and write down the elements with their atomic weights and typical properties, analogous elements and like atomic weights on separate cards, and this soon convinced me that the properties of elements are in periodic dependence upon their atomic weights.

In Thomas H. Pope (ed.)
Translated by George Kamensky
The Principles of Chemistry (Volume 2)
Longmans, Green & Company. London, England. 1905

By ordering the elements according to increasing atomic weight in vertical rows so that the horizontal rows contain analogous elements, still ordered by increasing atomic weight, one obtains the following arrangement, from which a few general conclusions may be derived.

David M. Knight (ed.)
Classical Scientific Papers — Chemistry, Second Series
On the Relationship of the Properties of the Elements to Their Atomic Weights (1869)
American Elsevier Publishing Company. New York, New York, USA. 1968

An established system is limited by its order of known or discovered elements. With the periodic and atomic relations now shown to exist between all the atoms and the properties of their elements, we see the possibility not only of noting the absence of some of them but even of determining, and with great assurance and certainty, the properties of these as yet unknown elements; it is possible to predict their atomic weight, density in the free state or in the form of oxides, acidity or basicity, degree of oxidation, and ability to be reduced and to form double salts and to describe the properties of the metalloorganic compounds and chlorides of the given element; it is even possible...to describe the properties of some compounds of these unknown elements in still greater detail. ...[A]t

the present time it is not possible to say when one of these bodies...will be discovered, yet the opportunity exists for finally convincing myself and other chemists of the truth of those hypotheses which lie at the base of the system I have drawn up.

A Natural System of the Elements and Its Use in Predicting the Properties of Undiscovered Elements
Journal of the Russian Chemical Society, Volume 3, 1871 (p. 25)

"I shall not form any hypothesis, either here nor further on to explain the nature of the periodic law. For first of all, the law itself is too simple; and secondly, this new subject has been too little studied yet, in its diverse parts for us to form any hypothesis."

In B. Bensaude-Vincent
Mendeléev's Periodic System (Part I)
British Journal for the History of Science, Volume 19, Number 61, March 1986 (p. 7)

Roscoe, Henry E.   1833–1915
English chemist

We must then find that these numbers regularly increase by a definite amount, *i.e.*, by the average age of a generation, which will be approximately the same in all the four families. Comparing the ages of the chemists themselves, we shall observe certain differences, but these are small in comparison with the period which has elapsed since the birth of their ancestors. Now each individual in this series of family trees represents a chemical element; and just as each family is distinguished by certain idiosyncrasies, so each group of the elementary bodies thus arranged shows distinct signs of consanguinity.

Report of the British Association of the Advancement of Science
1887 (p. 10)
Publisher undetermined

Sanderson, R. T.
No biographical data available

Students may readily be bewildered by the apparently fundamental lack of agreement among various periodic tables, and some may even acquire reasonable doubt as to whether chemists actually know what they are doing.

One More Periodic Table
Journal of Chemical Education, Volume 31, 1954 (p. 481)

PERPETUAL MOTION

Burroughs, John   1837–1921
American naturalist and writer

Physics proves to us the impossibility of perpetual motion among visible, tangible bodies, at the same time that it reveals to us a world where perpetual motion is the rule — the world of molecules and atoms.

The Breath of Life
Chapter IX (p. 190)
Houghton Mifflin Company. Boston, Massachusetts, USA. 1915

Hugo, Victor   1802–85
French author, lyric poet, and dramatist

Scientists have searched for a perpetuum mobile; they have found it: it is science itself.
In Rolf Huisgen
The Adventure Playground of Mechanisms and Novel Reactions
Has Chemistry Reached the Postmechanistic Era? (p. 244)
American Chemical Society. Washington, D.C. 1994

PESSIMISM

Rashevsky, Nicolas   1899–1972
Mathematical biophysicist

Pessimism is not a healthy thing in science, but neither is unrealistic optimism.
Mathematical Biophysics: Physico-Mathematical Foundations of Biology (Volume 2)
Chapter XXVIII (p. 307)
Dover Publications, Inc. New York, New York, USA. 1960

PEST CONTROL

Müller, Paul   1899–1965
Swiss chemist

The field of pest control is immense, and many problems impatiently await a solution. A new territory has opened up for the synthetics chemist, a territory which is still unexplored and difficult, but which holds out the hope that in time further progress will be made.
Nobel Lectures, Physiology or Medicine 1942–1962
Nobel lecture for award received in 1948
Dichloro-Diphenyl-Trichloroethane and Newer Insecticides (p. 236)
Elsevier Publishing Company. Amsterdam, Netherlands. 1964

PESTILENCE

Camus, Albert   1913–1960
Algerian-French author and philosopher

A pestilence isn't a thing made to man's measure; therefore we tell ourselves that pestilence is a mere bogy of the mind, a bad dream that will pass away. But it doesn't pass away and, from one bad dream to another, it is men who pass away.
The Plague
Part I, Chapter 5 (p. 37)
Vintage Books. New York, New York, USA. 1991

Hugo, Victor   1802–85
French author, lyric poet and dramatist

…death has a way of its own of harassing victory, and it causes pestilence to follow glory. Typhus is an annex of triumph.
Les Miserables
Volume 2, Book I, Chapter 2 (p. 7)
The Heritage Press. New York, New York, USA. 1938

PETRIFICATION

Leclerc, Georges-Louis, Comte de Buffon   1707–88
French naturalist

Petrification is the great means of nature to keep the transitory creatures of all epochs.
In Jochen Helms
Fossils: The Oldest Treasures that Ever Lived
The Berlin Specimen of the Primitive Bird Archaeopteryx (p. 94)
T.H.F. Publications, Inc. Neptune City, New Jersey, USA. 1985

PETROLOGY

Wyllie, Peter J.
Geologist

The results of experimental petrology…help to distinguish between possible and impossible processes.
In M.P. Atherton and C.D. Gribble (eds.)
Migmatites, Melting and Metamorphism
Experimental Studies on Biotite- and Muscovite-Granites and Some Crustal Magmatic Sources (p. 13)
Shiva Geology Series. 1983

PH.D.

Chargaff, Erwin   1905–2002
Austrian biochemist

The Ph.D. is essentially a license to start unlearning.
Voices in the Labyrinth: Nature, Man and Science
Chapter 1 (p. 2)
The Seabury Press. New York, New York, USA. 1977

Dyson, Freeman J.   1923–
American physicist and educator

The average student emerges at the end of the Ph.D. program, already middle-aged, overspecialized, poorly prepared for the world outside, and almost unemployable except in a narrow area of specialization. Large numbers of students for whom the program is inappropriate are trapped in it, because the Ph.D. has become a union card required for entry into the scientific job market.
From Eros to Gaia
Chapter 16 (p. 195)
Pantheon Books. New York, New York, USA. 1992

PHARMACIST

Eisenschiml, Otto   1880–1963
Austrian-American chemist and historian

The surgeon who has performed scores of brilliant operations is less talked about than the one who has inadvertently killed a patient; the pharmacist who has carefully

filled prescriptions for a lifetime remains obscure, but will gain publicity by a single oversight.
The Art of Worldly Wisdom: Three Hundred Precepts for Success Based on the Original Work of Baltasar Gracian
Part Seven (p. 87)
Duell, Sloan & Pearce. New York, New York, USA. 1947

Flexner, Abraham   1866–1959
American educator

The physician thinks, decides, and orders; the pharmacist obeys — obeys, of course, with discretion, intelligence, and skill — yet, in the end, obeys and does not originate. Pharmacy therefore is an arm added to the medical profession, a specially and distinctly higher form of handicraft, not a profession…
Is Social Work a Profession?
School and Society, Volume 1, 1915 (p. 905)

PHARMACY

Ghalioungui, Paul

The word pharmakon, whence pharmacy is derived, meant in Greek not only medicament, poison, or magical procedure, but also that which is slain to expiate the crimes of a city, like the scapegoat of Biblical times.… In other words, it meant "what carries off disease."
Magic and Medical Science in Ancient Egypt
Chapter II (p. 35)
Barnes and Nobles. New York, New York, USA. 1965

PHENOMENA

du Noüy, Pierre Lecomte   1883–1947
French scientist

When we speak of a phenomenon, we speak only of an event, or of a succession of events, arbitrarily isolated from the universe whose evolution they share. By isolating a fact in order to study it, we give it a beginning and an end, which are artificial and relative. In relation to the evolution of the universe, birth is not a beginning, and death is not an end. There are no more isolated phenomena in nature than there are isolated notes in a melody.
The Road to Reason
Chapter 2 (p. 53)
Longmans, Green & Company. London, England. 1949

Griffin Jay
Fictional character

There are some things in science which should be brought to light. There are others, doctor, which should be left alone.
The Mummy
Film (1940)

Haas, W. H.
American microbiologist

As most of us are aware, the world is now divided into two sets of phenomena, scientific and non-scientific.
The Teaching of Geography as a Science
Journal of Geography, Volume 30, 1931 (p. 323)

Hugo, Victor   1802–85
French author, lyric poet, and dramatist

Phenomena may well be suspected of anything, are capable of anything. Hypothesis proclaims the infinite; that is what gives hypothesis its greatness. Beneath the surface fact it seeks the real fact. It asks creation for her thoughts, and then for her second thoughts. The great scientific discoverers are those who hold nature suspect.
Translated by Isabel F. Hapgood
The Toilers of the Sea
Part II, Book Third, Chapter III (p. 415)
The Heritage Press. New York, New York, USA. 1961

Jevons, William Stanley   1835–82
English economist and logician

…every strange phenomenon may be a secret spring which, if rightly touched, will open the door to new chambers in the palace of nature.
The Principles of Science: A Treatise on Logic and Scientific Method
Book V, Chapter XXIX (p. 671)
Macmillan & Company Ltd. London, England. 1887

Laplace, Pierre Simon   1749–1827
French mathematician, astronomer, and physicist

The phenomena of nature are most often enveloped by so many strange circumstances, and so great a number of disturbing causes mix their influence, that it is very difficult to recognize them.
A Philosophical Essay on Probabilities
Chapter IX (p. 73)
Dover Publications, Inc. New York, New York, USA. 1951

Lederer, Charles   1906–76
American film writer and director

There are no enemies in science, professor. Only phenomena to study.
The Thing From Another World
Film (1951)

Ramón y Cajal, Santiago   1852–1934
Spanish neuropathologist

The intellect is presented with phenomena marching in review before the sensory organs. It can be truly useful and productive only when limiting itself to the modest tasks of observation, description, and comparison, and of classification that is based on analogies and differences.
Advice for a Young Investigator
Chapter 1 (p. 2)
The MIT Press. Cambridge, Massachusetts, USA. 1999

Spencer, Herbert   1820–1903
English social philosopher

Sad, indeed, is it to see how men occupy themselves with trivialities, and are indifferent to the grandest phenomena…
Education: Intellectual, Moral, and Physical
Chapter I (p. 73)
A.L. Fowle. New York, New York, USA. 1860

Wilson, Edward O.   1929–
American biologist and author

…all tangible phenomena, from the birth of stars to the workings of social institutions, are based on material processes that are ultimately reducible, however long and tortuous the sequences, to the laws of physics.
Consilience: The Unity of Knowledge
Chapter 12 (p. 266)
Alfred A. Knopf. New York, New York, USA. 1998

PHILOSOPHER

Cicero (Marcus Tullius Cicero)   106 BCE–43 BCE
Roman orator, politician, and philosopher

Somehow or other no statement is too absurd for some philosophers to make.
Translated by William Armistead Falconer
Cicero: De Senectute, De Amicitia, De Divinatione
De Divinatione, II, LVIII (p. 505)
Harvard University Press. Cambridge, Massachusetts, USA. 1938

Darwin, Charles Robert   1809–82
English naturalist

Why should the souls [of philosophers] be deeply vexed? The majesty of Fact is on their side, and the elemental forces of Nature are working for them. Not a star comes to the meridian at its calculated time but testifies to the justice of their methods — their beliefs are "one with the falling rain and with the growing corn." By doubt they are established, and open inquiry is their bosom friend.
Collected Essays (Volume 2)
Darwiniana
The Origin of Species (p. 53)
Macmillan & Company Ltd. London, England. 1904

Faraday, Michael   1791–1867
English physicist and chemist

The philosopher should be a man willing to listen to every suggestion, but determined to judge for himself. He should not be biased by appearances; have no favorite hypothesis; be of no school; and in doctrine have no master. He should not be a respecter of persons, but of things. Truth should be his primary object. If to these qualities he added industry, he may indeed hope to walk within the veil of the temple of Nature.
The Life and Letters of Faraday (Volume 1) (p. 220)
Longmans, Green & Company. London, England. 1870

Herschel, Sir John Frederick William   1792–1871
English astronomer and chemist

…to the natural philosopher there is no natural object unimportant or trifling.
A Preliminary Discourse on The Study of Natural Philosophy
Part I, Chapter I, Section 10 (p. 14)
Printed for Longman, Rees, Orme, Brown, and Green. London, England. 1831

Lindley, David   1956–
English astrophysicist and author

As philosophers have frequently found, the real world seems to messy, too stubbornly arbitrary, to be found out by the power of thought alone, no matter how fine the guiding sense of aesthetics.
The End of Physics: The Myth of a Unified Theory
Part III, Chapter 8 (p. 231)
Basic Books, Inc. New York, New York, USA. 1993

PHILOSOPHER'S STONE

Agassiz, Jean Louis Rodolphe   1807–73
Swiss-born American naturalist, geologist, and teacher

The philosopher's stone is no more to be found in the organic than the inorganic world; and we shall seek as vainly to transform the lower animal types into the higher ones by any of our theories, as did the alchemists of old to change the baser metals into gold.
Methods of Study in Natural History
Chapter XVI (p. 319)
Ticknor & Fields. Boston, Massachusetts, USA. 1863

PHILOSOPHY

Astaire, Fred   1899–1955
American dancer, actor, and singer
I wish I had your confidence…without your viewpoint.
Holiday Inn
Film (1942)

Barthelme, Donald   1931–89
American author

But I think everyone should have a little philosophy, Thomas said. It helps, a little. It helps. It is good. It is about half as good as music.
The Dead Father (p. 76)
Pocket Books. New York, New York, USA. 1975

Bell, R. P.
English chemist

The exact verbal definition of qualitative concepts is more often the province of philosophy than of physical science.
The Proton in Chemistry
Chapter II (p. 7)
Cornell University Press. Ithaca, New York, USA. 1959

Bernard, Claude 1813–78
French physiologist

I can no more accept a philosophy, then, which tries to assign boundaries to science, than a science which claims to suppress philosophic truths that are at present outside its own domain.
Translated by Henry Copley Greene
An Introduction to the Study of Experimental Medicine
Part Three, Chapter III, Section iv (p. 223)
Henry Schuman, Inc. New York, New York, USA. 1927

Born, Max 1882–1970
German-born English physicist

A philosophy in which the notions of chance and freedom are fundamental seems to me preferable to the almost inhuman determinism of the previous epoch — but that is no scientific argument.
Les Prix Nobel. The Nobel Prizes in 1954
Nobel banquet speech for award received in 1954
Nobel Foundation. Stockholm, Sweden. 1955

Burnet, Thomas 1635–1715
English cleric and scientist

Orators and Philosophers treat Nature after a very different manner…with all her graces and ornaments, and if there be anything which is not capable of that, they dissemble it, or pass it over slightly. But Philosophers view Nature with a more impartial eye, and without favor or prejudice give a just and free account [of] how they find all the parts of the Universe, some more, some less perfect.
The Sacred Theory of the Earth (2nd edition)
Book I, Chapter IX (p. 90)
Printed by R. Norton. London, England. 1691

Clarke, Samuel 1675–1729
English philosopher

'Tis of singular use, rightly to understand, and carefully to distinguish from hypotheses or mere suppositions, the true and certain consequences of experimental and mathematical philosophy; which do, with wonderful strength and advantage, to all such as are capable of apprehending them, confirm, establish, and vindicate against all objections, those great and fundamental truths of natural religion, which the wisdom of providence has at the same time universally implanted, in some degree, in the minds of persons even of the meanest capacities, not qualified to examine demonstrative proofs.
In H.G. Alexander
The Leibniz–Clarke Correspondence
Dedication (p. 6)
Philosophical Library Inc. New York, New York, USA. 1956

Dawson, Sir John William 1820–99
Canadian geologist and educator

It is a wise and thoughtful philosophy which can distinguish what is fixed and unchangeable from that which is fluctuating and capable of development.
Some Salient Points in the Science of the Earth
Chapter XII (p. 342)
Hodder & Stoughton. London, England. 1893

de Botton, Alain 1969–
Swiss-born English writer and television producer

Seneca believed…arguments are like eels: however logical, they may slip from the mind's weak grasp unless fixed there by imagery and style.
The Consolations of Philosophy (p. 92)
Vintage Books. New York, New York, USA. 2000

Dewey, John 1859–1952
American philosopher and educator

What would happen to philosophy…if it ceased to deal with the problem of reality and knowledge at large?…. From this point of view, the problem of philosophy concerns the *interaction* of our judgments about ends to be sought with the knowledge of the means for achieving them.
Quest for Certainty: A Study of the Relation of Knowledge and Action
Chapter II (pp. 36–37)
Minton, Balch & Company. New York, New York, USA. 1929

[Philosophy] has tried to combine acceptance of the conclusions of scientific inquiry as to the natural world with the acceptance of doctrines about the nature of mind and knowledge which originated before there was such a thing as systematic experimental inquiry. Between the two there is an inherent incompatibility.
Quest for Certainty: A Study of the Relation of Knowledge and Action
Chapter III (p. 49)
Minton, Balch & Company. New York, New York, USA. 1929

Disraeli, Benjamin, 1st Earl of Beaconsfield 1804–81
English prime minister, founder of Conservative Party, and novelist

Philosophy becomes poetry, and science imagination, in the enthusiasm of genius.
Miscllanies of Literature, by the Author of 'Curiosities of Literature' (p. 426)
G. Routledge. London, England. 1886

Donne, John 1572–1631
English poet and divine

And new Philosophy calls all in doubt,
The Element of fire is quite put out;
The sun is lost, and th' earth, and no man's wit
Can well direct him where to look for it.
And freely men confess that this world's spent,
When in the Planets and the Firmament
They seek so many new, they see that this
Is crumbled out again to his Atomies.
An Anatomy of the World

The First Anniversary, II, 205–212
Presented for presentation to members of the Roxburghe Club. Cambridge, England. 1951

Durant, William James   1885–1981
American historian and essayist

Science gives us knowledge, but only philosophy can give us wisdom.
The Story of Philosophy
Introduction (p. 3)
Simon & Schuster. New York, New York, USA. 1953

Eddington, Sir Arthur Stanley   1882–1944
English astronomer, physicist, and mathematician

The recent tendencies of science do…take us to an eminence from which we can look down into the deep waters of philosophy; and if I rashly plunge into them, it is not because I have confidence in my powers of swimming, but to try to show that the water is really deep.
The Nature of the Physical World
Chapter XIII (p. 276)
The Macmillan Company. New York, New York, USA. 1930

Einstein, Albert   1879–1955
German-born physicist

The Heisenberg–Bohr tranquilizing philosophy — or religion? — is so delicately contrived that, for the time being, it provides a gentle pillow for the true believer from which he cannot very easily be aroused. So let him lie there.
Letters on Wave Mechanics
Letter to Schrödinger, 31 May, 1928 (p. 31)

I have never belonged wholeheartedly to a country, a state, nor to a circle of friends, nor even to my own family. When I was still a rather precocious young man, I already realized most vividly the futility of the hopes and aspirations that most men pursue throughout their lives. Well-being and happiness never appeared to me as an absolute aim. I am even inclined to compare such moral aims to the ambitions of a pig.
In C.P. Snow
Variety of Men (p. 77)
Penguin Books, Harmondsworth, U.K.1969

Foster, Hannah W.   1758–1840
English writer

You ask me, my friend, whether I am in pursuit of truth, or [of] a lady? I answer, both. I hope and trust they are united; and really expect to find truth and the virtues and graces besides in a fair form.
The Coquette: The History of Eliza Norton (A Novel) (p. 10)
Oxford University Press. 1986

Hoyle, Sir Fred   1915–2001
English mathematician and astronomer

I realized that only in music could I find the answer I was seeking to the questions of the previous evening. Argument I could follow, it weighed with me, yet I could decide nothing from it.
October the First Is Too Late
Chapter Fourteen (p. 187)
Harper & Row, Publishers. New York, New York, USA. 1966

Inge, William Ralph   1860–1954
English religious leader and author

…science and philosophy can not be kept in water-tight compartments.
God and the Astronomers
Preface (p. vii)
W.W. Norton & Company, Inc. New York, New York, USA. 1978

Lewis, Gilbert Newton   1875–1946
American chemist

The average scientist, unequipped with the powerful lenses of philosophy, is a nearsighted creature, and cheerfully attacks each difficulty in the hope that it may prove to be the last.
The Anatomy of Science
Chapter I (p. 1)
Yale University Press. New Haven, Connecticut, USA. 1926

Mach, Ernst   1838–1916
Austrian physicist and philosopher

…every philosopher has his own private view of science, and every scientist his private philosophy.
Knowledge and Error: Sketches on the Psychology of Enquiry
Chapter I (p. 3)
D. Reidel Publishing Company. Dordrecht, Germany. 1976

Maclaurin, Colin   1698–1746
Scottish mathematician and natural philosopher

Is it not therefore the business of philosophy, in our present situation in the universe, to attempt to take in at once, in one view, the whole scheme of nature; but to extend, with great care and circumspection, our knowledge, by just steps, from sensible things as far as our observations or reasonings from them will carry us in our enquiries concerning either the greater motions and operations of nature, or her more subtle and hidden works.
An Account of Sir Isaac Newton's Philosophical Discoveries
Book I, Chapter I (p. 19)
Printed for the Author's Children. London, England. 1748

Nielsen, Kai
American-born Canadian philosopher

We must be on guard against the irrational heart of rationalism and not set out on the quest for certainty.
Ethics Without God (Revised edition) (p. 47)
Prometheus Books. Amherst, New York, USA. 1990

Paine, Thomas   1737–1809
Anglo-American political theorist and writer

Natural philosophy, mathematics and astronomy, carry the mind from the country to the creation, and give it a fitness suited to the extent.
Address to the People of England
Philadelphia, March, 1780

Raether, H.
No biographical data available

There are more things between cathode and anode than are dreamt of in your philosophy.
Electron Avalanches and Breakdown in Gases
Introduction (p. 1)
Butterworths. London, England. 1964

Russell, Bertrand Arthur William   1872–1970
English philosopher, logician, and social reformer

Philosophy, from the earliest times, has made greater claims, and achieved fewer results, than any other branch of learning.
Our Knowledge of the External World
Lecture I (p. 3)
The Open Court Publishing Company. Chicago, Illinois. 1914

…science is what you more or less know and philosophy is what you do not know.
Logic and Knowledge
The Philosophy of Logical Atomism (p. 281)
George Allen & Unwin Ltd. London, England. 1926

Shakespeare, William   1564–1616
English poet, playwright, and actor

There are more things in heaven and earth, Horatio, than are dreamt of in your philosophy.
In *Great Books of the Western World* (Volume 27)
The Plays and Sonnets of William Shakespeare (Volume 2)
Hamlet, Prince of Denmark
Act I, Scene v, l. 167–168
Encyclopædia Britannica, Inc. Chicago, Illinois, USA. 1952

Updike, John   1932–
American novelist, short story writer, and poet

The mad things dreamt up in the sky
Discomfort our philosophy.
Collected Poems 1953–1993
Skyey Developments (p. 334)
Alfred A. Knopf. New York, New York, USA. 1993

Whitehead, Alfred North   1861–1947
English mathematician and philosopher

Philosophy begins in wonder. And, at the end, when philosophic thought has done its best, the wonder remains. There have been added, however, some grasp of the immensity of things, some purification of emotion by understanding.
Modes of Thought

Chapter III, Lecture VIII (p. 232)
The Macmillan Company. New York, New York, USA. 1938

Philosophy asks the simple question, What is it all about?
Whitehead's Philosophy
Philosophical Review, Volume 46, Number 2, March 1937 (p. 178)

PHILOSOPHY OF SCIENCE

American Institute of Biological Science 1963

What is science? Is it a body of factual information? Is it a set of theories? Is it an activity or set of procedures for finding facts and developing theories? Science is really a combination of all three of these.
Biological Science: Molecules to Man (p. 3)
Houghton Mifflin Company. Boston, Massachusetts, USA. 1963

Asimov, Isaac   1920–92
American author and biochemist

Science is a process. It is a way of thinking, a manner of approaching and of possibly resolving problems, a route by which one can produce order and sense out of disorganized and chaotic observations. Through it we achieve useful conclusions and results that are compelling and upon which there is a tendency to agree.
"X" Stands for Unknown
Introduction (p. 10)
Doubleday & Company, Inc. Garden City, New York, USA. 1984

Ayala, Francisco J.   1934–
Spanish-born American biologist

Science is systematic organisation of knowledge about the universe on the basis of explanatory hypotheses which are genuinely testable. Science advances by developing gradually more comprehensive theories; that is, by formulating theories of greater generality which can account for observational statements and hypotheses which appear as prima facie unrelated.
Studies in the Philosophy of Biology: Reduction and Related Problems
Introduction (p. ix)
Macmillan & Company Ltd. London, England. 1974

Barrow, John D.   1952–
English theoretical physicist

The goal of science is to make sense of the diversity of Nature.
Theories of Everything: The Quest for Ultimate Explanation
Chapter One (p. 10)
The Clarendon Press. Oxford, England. 1991

Bauer, Henry H.   1931–
American chemist

Science is uniquely distinguished from other human practices: it is the only activity in which the constraints

of reality have brought to the quest for deep answers an effective consensus across all the variations that in other respects divide the human species.
Scientific Literacy and the Myth of the Scientific Method
Chapter 7 (p. 143)
University of Illinois Press. Urbana, Illinois, USA. 1992

Bernard, Claude   1813–78
French physiologist

True science suppresses nothing, but goes on searching and is undisturbed in looking straight at things that it does not yet understand.
Translated by Henry Copley Greene
An Introduction to the Study of Experimental Medicine
Part Three, Chapter III, Section iv (p. 223)
Henry Schuman, Inc. New York, New York, USA. 1927

True science teaches us to doubt and, in ignorance, to refrain.
Translated by Henry Copley Greene
An Introduction to the Study of Experimental Medicine
Part Two, Chapter II, Section VIII (p. 55)
Henry Schuman, Inc. New York, New York, USA. 1927

Bohm, David   1917–92
American physicist
Peat, D.
No biographical data available

The essential activity of science consists of thought, which arises in creative perception and is expressed through play. This gives rise to a process in which thought unfolds into provisional knowledge which then moves outward into action and returns as fresh perception and knowledge. This process leads to a continuous adoption of knowledge which undergoes constant growth, transformation, and extension. Knowledge is therefore not something rigid and fixed that accumulates indefinitely in a steady way but is a continual process of change. Its growth is closer to that of an organism than a data bank. When serious contradictions in knowledge [are] encountered, it is necessary to return to creative perception and free play, which act to transform existing knowledge. Knowledge apart from this cycle of activity, has no meaning.
Science, Order, and Creativity
Chapter One (p. 56)
Bantam Books. New York, New York, USA. 1987

Bondi, Sir Hermann   1919–2005
English mathematician and cosmologist

Science is driven forward by unexpected and surprising results emerging from new experiments or by the appearance of contradictions between theories previously thought compatible. Solving such problems as they arise is of the essence of our work. Thus science is not something strange and odd but the most human of pursuits.

The Philosopher of Science
Nature, Volume 358, Number 6385, 30 July, 1992 (p. 363)

Boulding, Kenneth E.   1910–93
English economist and social scientist

Science might also be defined as the process of substituting unimportant questions which can be answered for important questions which cannot.
Image: Knowledge in Life and Society
Chapter XI (p. 154)
The University of Michigan Press. Ann Arbor, Michigan, USA. 1956

Bushnell, Horace   1802–76
American Congregational minister

What is science, anyhow, but the knowledge of species? And if species do not keep their places, but go a masking or really becoming one another, in strange transmutations, what is there to know, and where is the possibility of science? If there is no stability or fixity in species, then, for aught that appears, even science itself may be transmuted into successions of music, and moonshine, and auroral fires. If a single kind is all kinds, then all are one, and since that is the same as none, there is knowledge no longer. The theory may be true, but it never can be proved, for that reason if no other. And when it is proved, if that must be the fact, we may well enough agree to live without religion.
Science and Religion
Putnam's Magazine, Volume 1, 1868 (p. 271)

Bronowski, Jacob   1908–74
Polish-born British mathematician and polymath

Science is the creation of concepts and their exploration in the facts. It has no other test of the concept than its empirical truth to fact.
Science and Human Values
The Sense of Human Dignity (p. 60)
Harper & Row, Publishers. New York, New York, USA. 1965

Science is not a mechanism but a human progress, and not a set of findings but the search for them.
Science and Human Values
The Sense of Human Dignity (p. 63)
Harper & Row, Publishers. New York, New York, USA. 1965

Science is a great many things…but in the end they all return to this: science is the acceptance of what works and the rejection of what does not. That needs more courage than we might think. It need more courage than we have ever found when we have faced our worldly problems.
The Common Sense of Science
Chapter IX, Section 6 (p. 148)
Harvard University Press. Cambridge, Massachusetts, USA. 1953

Carnap, Rudolf   1891–1970
American philosopher

Science is a system of statements based on direct experience, and controlled by experimental verification. Verification in science is not, however, of single statements but of the entire system or a sub-system of such statements.
The Unity of Science
Physics as a Universal Science, Section 3 (p. 42)
Thommes Press. Bristol, England. 1995

Burhoe, R. W.
Founding editor of *Zygon*

…in the usual sense a science is a discipline possessed of an empirically validated theoretical structure, which can indeed explain or account for and not simply describe, categorize, and correlate, patterns of human experience/behavior.
The Source of Civilization in the Natural Selection of Coadapted Information in Genes and Culture
Zygon, Volume 11, Number 3, September 1976 (p. 264)

Campbell, Norman R.   1880–1949
English physicist and philosopher

There are two aspects of science. First, science is a body of useful and practical knowledge and a method of obtaining it. It is science of this form which played so large a part in the destruction of war, and, it is claimed, should play an equally large part in the beneficent restoration of peace.… In its second form or aspect, science has nothing to do with practical life, and cannot affect it, except in the most indirect manner, for good or for ill. Science of this form is a pure intellectual study.… [I]ts aim is to satisfy the needs of the mind and not those of the body; it appeals to nothing but the disinterested curiosity of mankind.
What Is Science?
Chapter I (p. 1)
Dover Publications. New York, New York, USA. 1952

Cassirer, Ernst   1874–1945
German philosopher

Are we to be disgusted with science because it has not fulfilled our hopes or redeemed its promises? And are we, for this reason, to announce the "bankruptcy" of science, as is so often and so flippantly done? But this is rash and foolish: for we can hardly blame science just because we have not asked the right questions.
In David Hackett Fischer
Historian's Fallacies: Toward a Logic of Historical Thought
Chapter I (p. 3)
Harper & Row, Publishers. New York, New York, USA. 1970

Chargaff, Erwin   1905–2002
Austrian biochemist

The sciences have started to swell. Their philosophical basis has never been very strong. Starting as modest probing operations to unravel the works of God in the world, to follow its traces in nature, they were driven gradually to ever more gigantic generalizations. Since the pieces of the giant puzzle never seemed to fit together perfectly, subsets of smaller, more homogeneous puzzles had to be constructed, in each of which the fit was better.
Voices in the Labyrinth
Perspectives in Biology and Medicine, VII, Volume 18, Spring 1975 (p. 323)

In science, there is always one more Gordian knot than there are Alexanders. One could almost say that science, as it is practiced today, is an arrangement through which each Gordian knot, once cut, gives rise to two new knots, and so on. Out of one problem considered as solved, a hundred new ones arise; and this has created the myth of the limitlessness of the natural sciences. Actually, many sciences now look as feeble and emaciated as do mothers who have undergone too many deliveries.
Heraclitean Fire: Sketches from a Life Before Nature
Part II
More Foolish and More Wise (p. 116)
Rockefeller University Press. New York, New York, USA. 1978

Chesterton, G. K. (Gilbert Keith)   1874–1936
English author

Science finds facts in Nature, but Science is not Nature; because Science has co-ordinated ideas, interpretations and analyses; and can say of Nature what Nature cannot say for itself.
The Resurrection of Rome
Chapter IV (p. 126)
Dodd, Mead & Company. New York, New York, USA. 1930

Clark, Gordon H.   1902–85
American philosopher

The theologians who reply to…attacks [on relgious faith] are under a disadvantage. When a scientist or a philosopher argues against religion, he does not need to know much about religion; but when a theologian discusses science, he must know quite a lot. The scientist can get by if he understands no more than that Christians believe God to be an incorporeal spirit; but the theologian is called upon to discuss space, time, motion, energy, electrodynamics, the solar system, quantum theory, relativity, and other assorted items. There is something else the theologian must know, and something more important. In addition to a selection of particular pieces of information, such as the details just mentioned, the theologian must have an overall view of science as a whole. He must have a philosophy of science; that is, he must know what science is. Obviously he cannot compare, contrast, or relate religion and science unless he knows them both.… The scientific method is said to be the best, indeed, the only method for solving any problem, so that in every debate it is science, not theology, that has the last word. Since every curious and intelligent person naturally wishes to understand his own times, he must be prepared to give science sustained attention.

The Philosophy of Science (pp. 8–9)
Craig Press. Nutley, New Jersey, USA. 1964

Cohen, Morris Raphael   1880–1947
American philosopher

The certainty which science aims to bring about is not a psychologic feeling about a given proposition but a logical ground on which its claim to truth can be founded.
Reason and Nature
Chapter Three, Section II (A) (p. 84)
The Free Press, Publishers. Glencoe, Illinois, USA. 1931

Collingwood, Robin George   1889–1943
English historian and philosopher

The aim of science is to apprehend this purely intelligible world as a thing in itself, an object which is what it is independently of all thinking, and thus antithetical to the sensible world.… The world of thought is the universal, the timeless and spaceless, the absolutely necessary, whereas the world of sense is the contingent, the changing and moving appearance which somehow indicates or symbolizes it.
Essays in the Philosophy of Art
Outlines of a Philosophy of Art
Chapter 6, Section 27 (p. 142)
Indiana University Press. Bloomington, Indiana, USA. 1964

Science in general…does not consist in collecting what we already know and arranging it in this or that kind of pattern. It consists in fastening upon something we do not know, and trying to discover it.
The Idea of History
Introduction, Section 2 (p. 9)
At The Clarendon Press. Oxford, England. 1967

Courant, Richard   1888–1972
German-born American mathematician
Robbins, Herbert   1915–2001
American mathematician

A serious threat to the very life of science is implied in the assertion that mathematics is nothing but a system of conclusions drawn from definitions and postulates that must be consistent but otherwise may be created by the free will of the mathematician. If this description were accurate, mathematics could not attract any intelligent person. It would be a game with definitions, rules and syllogisms, without motivation or goal.
What Is Mathematics? (p. xvii)
Oxford University Press, Inc. London, England. 1941

The notion that the intellect can create meaningful postulational systems at its whim is a deceptive half-truth. Only under the discipline of responsibility to the organic whole, only guided by intrinsic necessity, can the free mind achieve results of scientific value.
What Is Mathematics? (p. xvii)
Oxford University Press, Inc. London, England. 1941

Davy, Sir Humphry   1778–1829
English chemist

Natural science is founded on minute critical views of the general order of events taking place upon our globe, corrected, enlarged, or exalted by experiments, in which the agents concerned are placed under new circumstances, and their diversified properties separately examined. The body of natural science, then, consists of facts; is analogy, — the relation of resemblance of facts by which its different parts are connected, arranged, and employed, either for popular use, or for new speculative improvements.
In John Davy (ed.)
The Collected Works of Sir Humphry Davy (Volume 8)
Introductory Lecture to the Chemistry of Nature (pp. 167–168)
Smith, Elder & Company. London, England. 1839–1840

Dawson, Sir John William   1820–99
Canadian geologist and educator

It is of the nature of true science to take nothing on trust or on authority. Every fact must be established by accurate observation, experiment, or calculation. Every law and principle must rest on inductive argument. The apostolic motto, "Prove all things, hold fast that which is good," is thoroughly scientific. It is true that the mere reader of popular science must often be content to take that on testimony which he cannot personally verify; but it is desirable that even the most cursory reader should fully comprehend the modes in which facts are ascertained and the reasons on which the conclusions are based.
The Chain of Life in Geological Time
Chapter I (p. 1)
Religious Tract Society. London, England. 1888

de Unamuno, Miguel   1864–1936
Spanish philosopher and writer

Wisdom is to science what death is to life, or, if you prefer it, wisdom is to death what science is to life.
Essays and Soliloquies
Some Arbitrary Reflections Upon Europeanization (p. 55)
Alfred A. Knopf. New York, New York, USA. 1925

Dennett, Daniel Clement   1942–
American philosopher

Science does not answer all good questions. Neither does philosophy. But for that very reason the phenomena of consciousness…do not need to be protected from science — or from the sort of demystifying philosophical investigation we are embarking on.… Looking on the bright side, let us remind ourselves of what has happened in the wake of earlier demystifications. We find no diminution of wonder; on the contrary, we find deeper beauties and more dazzling visions of the complexity of the universe than the protectors of mystery ever conceived. The "magic" of earlier visions was, for the most part, a

cover-up for frank failures of imagination, a boring dodge enshrined in the concept of a *deus ex machina*.
Consciousness Explained (pp. 22, 25)
Little, Brown & Company. Boston, Massachusetts, USA. 1991

…there is no such thing as philosophy-free science; there is only science whose philosophical baggage is taken on board without examination.
Darwin's Dangerous Idea
Chapter One, Section 1 (p. 21)
Simon & Schuster. New York, New York, USA. 1995

Descartes, René   1596–1650
French philosopher, scientist, and mathematician

Science in its entirety is true and evident cognition. He is no more learned who has doubts on many matters than the man who has never thought of them; nay he appears to be less learned if he has formed wrong opinions on any particulars. Hence it were better not to study at all than to occupy one's self with objects of such difficulty, that, owing to our inability to distinguish true from false, we are forced to regard the doubtful as certain; for in those matters any hope of augmenting our knowledge is exceeded by the risk of diminishing it. Thus in accordance with the above maxim we reject all such merely probable knowledge and make it a rule to trust only what is completely known and incapable of being doubted.
In *Great Books of the Western World* (Volume 31)
Rules for the Direction of the Mind
Rule II (p. 2)
Encyclopædia Britannica, Inc. Chicago, Illinois, USA. 1952

Einstein, Albert   1879–1955
German-born physicist

The aim of science is…a comprehension, as complete as possible, of the connection between the sense experience in [its] totality, and…the accomplishment of this aim by the use of a minimum of primary concepts and relations.
Out of My Later Years
Physics and Reality, I (p. 63)
Thames & Hudson. London, England. 1950

Science is the attempt to make the chaotic diversity of our sense experience correspond to a logically uniform system of thought.
Considerations Concerning the Fundaments of Theoretical Physics
Science, Volume 91, Number 2369, May 24, 1940 (p. 487)

Although it is true that it is the goal of science to discover rules which permit the association and foretelling of facts, this is not its only aim. It also seeks to reduce the connections discovered to the smallest possible number of mutually independent conceptual elements. It is in this striving after the rational unification of the manifold that it encounters its greatest successes, even though it is precisely this attempt which causes it to run the greatest risk of falling a prey to illusion. But whoever has undergone the intense experience of successful advances made in this domain, is moved by profound reverence for the rationality made manifest in existence.
Ideas and Opinions
Science and Religion (p. 49)
Crown Publishers, Inc. New York, New York, USA. 1954

The belief in the external world independent of the perceiving subject is the basis of all natural science.
Translated by Alan Harris
Essays in Science
Clerk Maxwell's Influence on the Evolution of the Idea of Physical Reality (p. 40)
Philosophical Library. New York, New York, USA. 1934

The grand aim of all science…is to cover the greatest number of empirical facts by logical deductions from the smallest number of hypotheses or axioms.
In Lincoln Barnett
The Meaning of Einstein's New Theory
Life, January 9, 1950 (p. 22)

Feynman, Richard P.   1918–88
American theoretical physicist

What is the fundamental hypothesis of science, the fundamental philosophy? …[It is that] the sole test of the validity of any idea is experiment.… We will invent some way to summarize the results of the experiment, and we do not have to be told ahead of time what this way will look like. If we are told that the same experiment will always produce the same result, that is all very well, but if when we try it, it does not, then it does not. We just have to take what we see, and then formulate all the rest of our ideas in terms of our actual experience.
Six Easy Pieces: Essentials of Physics Explained by Its Most Brilliant Teacher
Basic Physics (p. 32)
Addison-Wesley Publishing Company. Reading, Massachusetts, USA. 1995

Jeans, Sir James Hopwood   1877–1946
English physicist and mathematician

Science came to recognize that its only proper objects of study were the sensations that the objects of the external universe produced on our senses. The dictum *esse est percipi* was adopted whole-heartedly from philosophy — not because scientists had any predilection for an idealist philosophy, but because the assumption that things existed which could not be perceived had led them into a whole morass of inconsistencies and impossibilities. Those who did not adopt it were simply left behind, and the torch of those who did.
The Mathematical Aspect of the Universe
Philosophy, Volume VII, Number 25, January 1932 (p. 11)

Joad, Cyril Edwin Mitchinson   1891–1953
English philosopher and broadcasting personality

…the philosophising of the physicists is noticeably inferior to their physics, and eminent men are at the moment engaged in making all the mistakes which the philosophers made for themselves some three hundred years ago and have been engaged in detecting and correcting ever since. In particular it is thought that modern physics lends support to Idealism, and suggests, if it does not actually require, a religious interpretation of the universe.

Guide to Modern Thought
Chapter I (pp. 15–16)
Faber & Faber Ltd. London, England. 1936

Meredith, Patrick
No biographical data available

Hence a true philosophy of science must be a philosophy of scientists and laboratories as well as one of waves, particles and symbols.

Instruments of Communication
Chapter 2, Section 5 (p. 40)
Pergamon Press. Oxford, England. 1966

Moreland, J. P.   1936–
American philosopher

For the question What is the proper definition of science? is itself a philosophical question about science that assumes a vantage point above science; it is not a question of science. One may need to reflect on specific episodes in the history of science to answer the question. But the question and the reflection required to answer it are philosophical in nature, a point not diminished merely because a scientist may try to define science. When she does so, she is doing philosophy.

Christianity and the Nature of Science: A Philosophical Investigation
(pp. 20–21)
Baker Book House. Garnd Rapids, Michigan, USA. 1989

Popper, Karl R.   1902–94
Austrian/British philosopher of science

…I shall certainly admit a system as empirical or scientific only if it is capable of being tested by experience. These considerations suggest that not the verifiability but the falsifiability of a system is to be taken as a criterion of demarcation. In other words: I shall not require of a scientific system that it shall be capable of being singled out, once and for all, in a positive sense: but I shall require that its logical form shall be such that it can be singled out, by means of empirical tests, in a negative sense: it must be possible for an empirical scientific system to be refuted by experience.

The Logic of Scientific Discovery
Part I, Chapter I, Section 6 (p. 40)
Basic Books, Inc. New York, New York, USA. 1959

We do not know: we can only guess.

The Logic of Scientific Discovery
Part II, Chapter X, Section 85 (p. 278)
Basic Books, Inc. New York, New York, USA. 1959

Shapere, Dudley
No biographical data available

…philosophy of science is immune to the vicissitudes of science — the coming and going of particular theories; for those changes have to do with content of science, whereas the philosopher is concerned with its structure — not with specific theories, but with the meaning of "theory" itself.

Philosophical Problems of Natural Science
Introduction, Section IV (p. 9)
The Macmillan Company. New York, New York, USA. 1965

Simpson, George Gaylord   1902–84
American paleontologist

It is inherent in any acceptable definition of science that statements that cannot be checked by observation are not really about anything…or at the very least, they are not science.

The Nonprevalence of Humanoids
Science, Volume 143, Number 3608, February 21, 1964 (p. 770)

Spencer, Herbert   1820–1903
English social philosopher

Every science begins by accumulating observations, and presently generalizes these empirically; but only when it reaches the stage at which its empirical generalizations are included in a rational generalization does it become developed science.

The Data of Ethics
Chapter IV, Section 22a (p. 51)
William & Norgate. London, England. 1907

Torrance, Thomas F.
No biographical data available

In natural science we are concerned ultimately, not with convenient arrangements of observational data which can be generalized into universal explanatory form, but with movements of thought, at once theoretical and empirical, which penetrate into the intrinsic structure of the universe in such a way that there becomes disclosed to us its basic design and we find ourselves at grips with reality…. We cannot pursue natural science scientifically without engaging at the same time in meta-scientific operations.

Divine and Contingent Order (p. 3)
Oxford University Press, Inc. Oxford. 1981

Toulmin, Stephen   1922–
Anglo-American philosopher

Certainly, every statement in a science should conceivably be capable of being called in question, and of being shown empirically to be unjustified; for only so can the science be saved from dogmatism.

The Philosophy of Science
Harper & Row, Publishers. New York, New York, USA. 1960

Toynbee, Arnold J.   1852–83
English historian

[T]here will be differences in the degree of approximation to scientific study, ...determined by the nature of the part or aspect of the Universe under consideration. Study will be most scientific when its object is the physical structure of the Universe.... The object of study that will be the least amenable to scientific treatment is the nonphysical facet of human nature. Students in this field had better avoid letting themselves be tempted by the present-day prestige of the word "science" into applying that label to their own work.
Occasional Paper, The Institute for the Study of Science in Human Affairs
Science in Human Affairs: An Historian's View

van Fraassen, Bas C.   1941–
Dutch-born philosopher

To develop an empiricist account of science is to depict it as involving a search for truth only about the empirical world, about what is actual and observable.... It must involve throughout a resolute rejection of the demand for an explanation of the regularities in the observable course of nature, by means of truths concerning a reality beyond what is actual and observable, as a demand which plays no role in the scientific enterprise.
The Scientific Image
Chapter 6 (p. 203)
Clarendon Press. Oxford, England. 1990

...certain issues in philosophy of science (having to do with observation and the definition of a theory's empirical import) had been misconstrued as issues in philosophy of logic and of language. With respect to modality, I hold the exact opposite: important philosophical problems concerning language have been misconstrued as relating to the content of science and the nature of the world. This is not at all new, but is the traditional nominalist line.
The Scientific Image
Chapter 6 (p. 196)
Clarendon Press. Oxford, England. 1980

Weisskopf, Victor Frederick   1908–2002
Austrian-American physicist

Science developed only when men refrained from asking general questions such as: What is matter made of? How was the universe created? What is the essence of life? Instead they asked limited questions such as: How does an object fall? How does water flow in a tube? Thus, in place of asking general questions and receiving limited answers, they asked limited questions and found general answers. It remains a great miracle, that this process succeeded, and that the answerable questions became gradually more and more universal.
The Significance of Science
Science, Volume 176, Number 4031, April 14, 1972 (p. 143)

Whitehead, Alfred North   1861–1947
English mathematician and philosopher

The aim of science is to seek the simplest explanation of complex facts.... Seek simplicity and distrust it.
The Concept of Nature
Chapter VII (p. 163)
At The University Press. Cambridge, England. 1920

Science is simply setting out on a fishing expedition to see whether it cannot find some procedure which it can call measurement of space and some procedure which it can call the measurement of time, and something which it can call a system of forces, and something which it can call masses....
The Concept of Nature
Chapter VI (p. 139)
At The University Press. Cambridge, England. 1920

Wright, Chauncey   1830–75
American philosopher of science

Science asks no questions about the ontological pedigree or a priori character of a theory, but is content to judge it by its performance; and it is thus that a knowledge of nature, having all the certainty which the senses are competent to inspire, has been attained — a knowledge which maintains a strict neutrality toward all philosophical systems and concerns itself not with the genesis or a priori grounds of ideas.
The Philosophical Writings of Chauncey Wright
The Philosophy of Herbert Spencer (p. 8)
The Liberal Arts Press. New York, New York, USA. 1958

PHOSPHORUS

Author undetermined

Red phosphorus is used for matches so that people who are in the habit of chewing matches will not suffer.
Class-Room Chemical Emanations
Journal of Chemical Education, Volume 3, Number 1, 1926

PHOTOELECTRIC

Glashow, Sheldon L.   1932–
American physicist

Einstein examined the photoelectric effect, which is now so well understood that it is used to open the doors of supermarkets and elevators when you step through a beam of light. In 1905 it was still a mystery.
Interactions: A Journey Through the Mind of a Particle Physicist and the Matter of This World
Chapter 3 (p. 52)
Warner Books. New York, New York, USA. 1988

PHOTOGRAPHY

Adams, Robert   1937–
American photographer

No place is boring, if you've had a good night's sleep and have a pocket full of unexposed film.
Darkroom and Creative Camera Techniques, May 1995

PHOTON

Einstein, Albert   1879–1955
German-born physicist

Every physicist thinks that he knows what a photon is…. I spent my life to find out what a photon is and I still don't know it.
In Eugene Hecht
Optics
Chapter 1 (p. 9)
Encyclopædia Britannica, Inc. Chicago, Illinois, USA. 1952

Jespersen, James
No biographical data available
Fitz-Randolph, Jane
No biographical data available

We can think of the photons as being like a shower of snowballs flying back and forth between the two electrons. And like the opponents in a snowball fight, the electrons retreat from each other under the assault of the photons.
From Quarks to Quasars: A Tour of the Universe
Chapter 11 (p. 125)
Athenaeum. New York, New York, USA. 1987

Roberts, Michael
No biographical data available

While I, maybe, precisely seize
The elusive photon's properties
In a's and b's, set in bronze-
bright vectors, grim quaternions.
Notes on q, f, and y
The New Statesman, March 23, 1935

Rucker, Rudy   1946–
Science and science fiction author

A photon is a wavy yet solid little package that can zip through empty space without the benefit of any invisible jelly vibrating underfoot.
The Fourth Dimension: Toward a Geometry of Higher Reality
Chapter 6 (p. 73)
Houghton Mifflin Company. Boston, Massachusetts, USA. 1984

PHOTOSYNTHESIS

Baum, Harold
No biographical data available

When sunlight bathes the chloroplast, and photons are absorbed
The energy's transduced so fast that food is quickly stored,
Photosynthetic greenery traps light the spectrum through
Then dark pathway machinery fixes the CO_2.
The Biochemists' Handbook
Photosynthesis (Tune: Auld Lang Syne)
Van Nostrand. Princeton, New Jersey, USA. 1961

Pallister, William Hales   1877–1946
Canadian physician

The sunlight gives the stimulus
Which makes a plant of you;
Your chemic process puzzles us,
We look and see you do
Your photo-synthesis, and thus
Grow and divide in two.
Poems of Science
The Nature of Things, *Euglena viridis* (p. 5)
Playford Press. New York, New York, USA. 1931

Rabinowitch, Eugene   1901–73
Russian-born American biophysicist

In photosynthesis we are like travelers in an unknown country around whom the early morning fog slowly begins to rise, vaguely revealing the outlines of the landscape. It will be thrilling to see it in bright daylight!
In A Scientific American Book
The Physics and Chemistry of Life
Photosynthesis (p. 47)
Simon & Schuster. New York, New York, USA. 1955

PHYLOGENESIS

Haeckel, Ernst   1834–1919
German biologist and philosopher

Phylogenesis is the mechanical cause of ontogenesis. The connection between them is not of an external or superficial, but of a profound, intrinsic, and causal nature.
Anthropogenie, oder, Entwickelungsgeschichte des Menschen gemeinversthandliche wissenschaftliche Vortrage uber die Grundzuge der mensch
W. Engleman, Leipzig, Germany. 1874

PHYLOGENY

Abbott, Donald Putnam   1920–86
American marine biologist and professor

Cultivate a suspicious attitude toward people who do phylogeny.
In Galen Howard Hilgard (ed.)
Observing Marine Invertebrates: Drawings from the Laboratory
Author's Preface (p. xvi)
Stanford University Press. Stanford, California, USA. 1987

PHYSIC

Boorde, Andrew   1490–1549
English traveler, physician, and writer

A good cook is half a physician. For the chief physic (the counsel of a physician excepted) doth come from the kitchen; wherefore the physician and the cook for sick men must consult together for the preparation of meat for sick men. For if the physician, without the cook, prepared any meat, except he be very expert, he will make a wearish dish of meat, the which the sick cannot take.
The Wisdom of Andrew Boorde (p. 49)
Edgar Backus. Leicester, England. 1936

Colton, Charles Caleb   1780–1832
English sportsman and writer

No men despise physic so much as physicians, because no men so thoroughly understand how little it can perform.
Lacon; or Many Things in a Few Words
1:179
William Gowans. New York, New York, USA. 1849

Heurnius
No biographical data available

Many of them to get a fee, will give physic to every one that comes, when there is no cause.
In William Tod Helmuth
Scratches of a Surgeon
Medical Pomposity (p. 9)
W.A. Chatterton & Company. Chicago, Illinois, USA. 1879

Holmes, Oliver Wendell   1809–94
American physician, poet, and humorist

Not to take authority when I can have facts; not to guess when I can know; not to think a man must take physic because he is sick.
In Robert Coope
The Quiet Art (p. 101)
E.S. Livingstone Ltd. Edinburgh, Scotland. 1952

Lettsom, J. C.
No biographical data available

When people is ill, they comes to I,
I physics, bleed, and sweats 'em;
Sometimes they live, sometimes they die.
What's that to I? I lets 'em.
In William Davenport Adams
English Epigrams
On Dr. Lettsom, by Himself (cclxxii)
G. Routledge. London, England. 1878

Milton, John   1608–74
English poet

…in Physic, things of melancholic hue and quality are us'd against melancholy, sour against sour, salt to remove salt humours.
Samson Agonistes
On that Sort of Dramatic Poem Which Is Call'd Tragedy (p. 79)
The Doves Press. London, England. 1905

Pope, Alexander   1688–1744
English poet

Learn from the beasts the physic of the field.
The Complete Poetical Works (Volume 3)
Essay on Man, Epis. Iii, l. 174
Houghton Mifflin Company. New York, New York, USA. 1903

Proverb

Warre and Physicke are governed by the eye.
In George Herbert
Outlandish Proverbs
#906
Printed by T. Maxey for T. Garthwait. London, England. 1651

Ray, John   1627–1705
English naturalist

If physic do not work, prepare for the kirk.
A Complete Collection of English Proverbs (p. 149)
Printed for G. Cowie. London, England. 1813

Shakespeare, William   1564–1616
English poet, playwright, and actor

Throw physic to the dogs; I'll none of it.
In *Great Books of the Western World* (Volume 27)
The Plays and Sonnets of William Shakespeare (Volume 2)
Macbeth
Act V, Scene iii, l. 47
Encyclopædia Britannica, Inc. Chicago, Illinois, USA. 1952

'Tis time to give 'em physic, their diseases
Are grown so catching.
In *Great Books of the Western World* (Volume 27)
The Plays and Sonnets of William Shakespeare (Volume 2)
The Famous History of the Life of King Henry the Eighth
Act I, Scene iii, l. 36–37
Encyclopædia Britannica, Inc. Chicago, Illinois, USA. 1952

Take physic, pomp;
Expose thyself to feel what wretches feel.
In *Great Books of the Western World* (Volume 27)
The Plays and Sonnets of William Shakespeare (Volume 2)
King Lear
Act III, Scene ii, l. 33–34
Encyclopædia Britannica, Inc. Chicago, Illinois, USA. 1952

In this point
All his tricks founder, and he brings his physic
After his patient's death.
In *Great Books of the Western World* (Volume 27)
The Plays and Sonnets of William Shakespeare (Volume 2)
The Famous History of the Life of King Henry the Eighth
Act III, Scene ii, l. 39–41
Encyclopædia Britannica, Inc. Chicago, Illinois, USA. 1952

PHYSICAL LAW

Feynman, Richard P.   1918–88
American theoretical physicist

...there is...a rhythm and pattern between the phenomena of nature which is not apparent to the eye, but only to the eye of analysis; and it is these rhythms and patterns which we call Physical Laws.
The Character of Physical Law
Chapter 1 (p. 13)
BBC. London, England. 1965

PHYSICAL SCIENCE

Huxley, Thomas Henry   1825–95
English biologist

When simple curiosity passes into the love of knowledge as such, and the gratification of the aesthetic sense of the beauty of completeness and accuracy seems more desirable than the easy indolence of ignorance; when the finding out of the causes of things becomes a source of joy, and he is counted happy who is successful in the search, common knowledge of Nature passes into what our forefathers called Natural History, from whence there is but a step to that which used to be termed Natural Philosophy, and now passes by the name of Physical Science.
The Crayfish
Chapter I (p. 3)
D. Appleton & Company. New York, New York, USA. 1880

Mach, Ernst   1838–1916
Austrian physicist and philosopher

Physical science began in the witch's kitchen. It now embraces the organic and inorganic worlds, and with the physiology of articulation and the theory of the senses, has even pushed its researches, at times impertinently, into the province of mental phenomena.
Popular Scientific Lectures
Why Has Man Two Eyes? (p. 87)
The Open Court Publishing Company. Chicago, Illinois, USA. 1898

Thompson, Sir D'Arcy Wentworth   1860–1948
Scottish zoologist and classical scholar

How far...then mathematics will suffice to describe, and physics to explain, the fabric of the body, no man can foresee. It may be that all the laws of energy, and all the properties of matter, and all the chemistry of all the colloids are as powerless to explain the body as they are impotent to comprehend the soul. For my part, I think it is not so. Of how it is that the soul informs the body, physical science teaches me nothing; and that living matter influences and is influenced by mind is a mystery without a clue. Consciousness is not explained to my comprehension by all the nerve-paths and neurons of the physiologist; nor do I [explain by] physics how goodness shines in one man's face, and evil betrays itself in another. But of the construction and growth and workings of the body, as of all else that is of the earth earthy, physical science is, is, in my opinion, our only teacher and guide.

On Growth and Form (Volume 1)
Chapter I (p. 13)
At The University Press. Cambridge, England. 1951

Whitehead, Alfred North   1861–1947
English mathematician and philosopher

There can be no true physical science which looks first to mathematics for the provision of a conceptual model. Such a procedure is to repeat the errors of the logicians of the middle ages.
Principles of Relativity (p. 39)
Cambridge University Press. Cambridge, England. 1922

PHYSICIAN

Alexander the Great   356 BCE–323 BCE
Macedonian emperor

I am dying with the help of too many physicians.
Attributed

Allman, David
American physician

The dedicated physician is constantly striving for a balance between personal, human values, scientific realities and the inevitabilities of God's will.
Address to National Conference of Christian and Jews
The Brotherhood of Healing, 1 February 1958

Arnaldus de Villa Nova   1235–1313
Alchemist, astrologer and physician

...the physician must be learned in diagnosing, careful and accurate in prescribing, circumspect and cautious in answering questions, ambiguous in making prognosis, just in making promises; and he should not promise health because in doing so he would assume a divine function and insult God.
In Henry E. Sigerist (trans.)
Bedside Manners in the Middle Ages: The Treatise De Cautelis Medicorum Attributed to Arnald of Villanova
Quarterly Bulletin of Northwestern University Medical School, Volume 20, 1946

Aurelius Antoninus, Marcus   121–180
Roman emperor

Think continually how many physicians are dead after often contracting their eyebrows over the sick...
In *Great Books of the Western World* (Volume 12)
The Meditations of Marcus Aurelius
Book IV, # 48 (p. 267)
Encyclopædia Britannica, Inc. Chicago, Illinois, USA. 1952

Author undetermined

...a new physician must have a new church-yard...
In Robert Burton
The Anatomy of Melancholy (Volume 2)

Part 2, Sect. IV, Memb. I, subsect. 1 (p. 230)
AMS Press, Inc. New York, New York, USA. 1973

Bacon, Sir Francis 1561–1626
English lawyer, statesman, and essayist

Physicians are some of them so pleasing and conformable to the humor of the patient, as they press not the true cure of the disease; and some other are so regular in proceeding according to art for the disease, as they respect not sufficiently the condition of the patient. Take one of a middle temper; or if it may not be found in one man, combine two of either sort; and forget not to call as well the best acquainted with your body, as the best reputed of for his faculty.
Essays, Advancement of Learning, New Atlantis, and Other Pieces
The Essays or Counsels, Civil and Moral: I. Of Regiment of Health (p. 94)
Odyssey Press. New York, New York, USA. 1937

The weakness of patients, and sweetness of life, and nature of hope, maketh men depend upon physicians with all their defects.
In *Great Books of the Western World* (Volume 30)
Advancement of Learning
Second Book, Chapter X, Section 2 (p. 51)
Encyclopædia Britannica, Inc. Chicago, Illinois, USA. 1952

…it is the office of a physician not only to restore health, but to mitigate pain and dolors; and not only when such mitigation may conduce to recovery, but when it may serve to make a fair and easy passage.
In *Great Books of the Western World* (Volume 30)
Advancement of Learning
Second Book, Chapter X, Section 7 (p. 52)
Encyclopædia Britannica, Inc. Chicago, Illinois, USA. 1952

Bailey, Percival
No biographical data available

The function of the physician is to cure a few, help many, and comfort all.
Perspectives in Biology and Medicine, Volume 4, Number 254, 1961

Baldwin, Joseph G. 1815–64
American writer

Nobody knew who or what they were, except as they claimed, or as a surface view of their characters indicated. Instead of taking to the highway and magnanimously calling upon the wayfarer to stand and deliver, or to the fashionable larceny of credit without prospect or design of paying, some unscrupulous horse doctor would set up his sign as "Physician and Surgeon" and draw his lancet on you, or fire at random a box of pills into your bowels, with a vague chance of hitting some disease unknown to him, but with a better prospect of killing the patient, whom or whose administrator he charged some ten dollars a trial for his marksmanship.
The Flush Times of Alabama and Mississippi: A Series of Sketches

How the Times Served the Virginians (p. 89)
Louisiana State University Press. Baton Rouge, Louisiana, USA. 1987

Bass, Murray H.
No biographical data available

The ideal physician should be a combination of three persons — a clergyman, a fireman and a scientist. He must know how to handle and console the patient and his family…he must be ready to answer an "alarm" day and night; he must know the science of medicine…using its present potentialities to the utmost of his ability.
Clinical Pediatrics, Volume 3, Number 50, 1964

Bernard, Claude 1813–78
French physiologist

Medical personality is placed above science by physicians themselves; they seek their authority in tradition, in doctrines or in medical tact. This state of affairs is the clearest of proofs that the experimental method has by no means come into its own in medicine.
Translated by Henry Copley Greene
An Introduction to the Study of Experimental Medicine
Part One, Chapter Two, Section IV (p. 43)
Henry Schuman, Inc. New York, New York, USA. 1927

…physicians, in their treatment, often have to take account of the so-called influence of the moral over the physical, and also of any number of family and social considerations which have nothing to do with science. Therefore, an accomplished practising physician should be not only learned in his science, but also upright and endowed with keenness, tact and good sense.
Translated by Henry Copley Greene
An Introduction to the Study of Experimental Medicine
Part Three, Chapter IC, Section III (p. 206)
Henry Schuman, Inc. New York, New York, USA. 1927

Blackwell, Elizabeth 1821–1910
First woman to practice medicine in the United States

The true physician must possess the essential qualities of maternity. The sick are as helpless in his hands as the infant. They depend absolutely upon the insight and judgment, the honesty and hopefulness, of the doctor.
The Influence of Women in the Profession of Medicine (p. 11)
G. Bell. Baltimore, Maryland, USA. 1890

Bonaparte, Napoleon 1769–1821
French soldier and emperor of France

You are a physician, doctor. You would promise life to a corpse if he could swallow pills…
In J. Christopher Herold (ed.)
The Mind of Napoleon
Science and the Arts (pp. 137–138)
Columbia University Press. New York, New York, USA. 1955

In my opinion physicians kill as many people as we generals.

In J. Christopher Herold (ed.)
The Mind of Napoleon
Science and the Arts (p. 137)
Columbia University Press. New York, New York, USA. 1955

Brackenridge, Hugh Henry   1748–1816
American author and jurist

Gravity is the most practical qualification of the physician.
Modern Chivalry
Part II, Volume 1, Chapter X (p. 378)
American Book Company. New York, New York, USA. 1937

Brown, Michael S.   1941–
American physician

To apply tools of science, physicians must learn to think like scientists. They must acquire technical ability, taste in evaluating experiments, and a sense of creative adventure.
Les Prix Nobel. The Nobel Prizes in 1985
Nobel banquet speech for award received in 1985
Nobel Foundation. Stockholm, Sweden. 1986

Buchan, William   1729– 1805
Physician

Physicians, like other people, must live by their employment.
Domestic Medicine
Introduction (p. xviii)
Publisher undetermined. New York, New York, USA. 1816

No two characters can be more different than that of the honest physician and the quack; yet they have generally been much confounded.
Domestic Medicine
Introduction (p. xvi)
Publisher undetermined. New York, New York, USA. 1816

Burgess, Anthony   1917–93
English novelist

Keep away from physicians. It is all probing and guessing and pretending with them. They leave it to Nature to cure in her own time, but they take the credit. As well as very fat fees.
Nothing Like the Sun: A Story of Shakespeare's Love-Life
Chapter VIII (p. 180)
W.W. Norton & Company, Inc. New York, New York, USA. 1964

Byron, George Gordon, 6ᵗʰ Baron Byron   1788–
1824
English Romantic poet and satirist

This is the way physicians mend or end us,
Secundum aartem: but although we sneer
In health — when ill, we call them to attend us,
Without the least propensity to jeer.
The Complete Poetical Works of Byron
Don Juan
Canto X, Stanza 42
Houghton Mifflin Company. Boston, Massachusetts, USA. 1933

Camden, William   1551–1623
English historian

Few physicians live well.
Remains Concerning Britain
Proverbs (p. 322)
J.R. Smith. London, England. 1870

Carlyle, Thomas   1795–1881
English historian and essayist

The healthy know not of their health, but only the sick: this is the physician's aphorism.
Characteristics, by Thomas Carlyle; Favorite Poems, by Percy Bysshe Shelley; The Eve of St. Agnes; and Other Poems, by John Keats
Paragraph 1 (p. 3)
Houghton Mifflin Company. Boston, Massachusetts, USA. 1882

Clowes, William   1540–1604
English physician

When a physician or a surgeon comes to a man that lies sick and is in danger of death, yet by his judgment and skill, promises with God's help to cure him of his griefs and maladies, then the sick patient greatly rejoices and presently compares him to a god. But afterwards, being somewhat recovered, and perceiving good amendment, he says he is but an angel and not a god. Again, after he begins to walk abroad and to fall to his meat, truly he is then accounted no better than a man. In the end, when he happily comes for his money for the curing of his grievous sickness, he now reports him to be a devil and shuts the door.
Selected Writings
A Tragical History (p. 63)
Harvey & Blythe. London, England. 1948

Collins, Joseph
No biographical data available

The longer I practice medicine, the more I am convinced every physician should cultivate lying as a fine art. There are lies which contribute enormously to the success of the physician's mission of mercy and salvation.
Reader's Digest, May 1933 (p. 16)

Colton, Charles Caleb   1780–1832
English sportsman and writer

Physicians must discover the weaknesses of the human mind, and even condescend to humor them, or they will never be called in to cure the infirmities of the body.
Lacon; or Many Things in a Few Words
1.482
William Gowans. New York, New York, USA. 1849

Croll, Oswald   1560–1609
German chemist and physician

…a Physitian therefore should have both the Theory and Practice, he must both know and prepare his medicines…

Philosophy Reformed and Improved in Four Profound Tractates (p. 152)
Printed by M.S. for Lodowick Lloyd. London, England. 1657

A Physition should be born out of the Light or Grace and Nature of the inward and invisible Man…
Philosophy Reformed and Improved in Four Profound Tractates (p. 22)
Printed by M.S. for Lodowick Lloyd. London, England. 1657

Cushing, Harvey   1869–1939
American neurosurgeon

A physician is obligated to consider more than a diseased organ, more even than the whole man — he must view the man in his world.
In René J. Dubos
Man Adapting
Chapter XII (p. 342)
Yale University Press. New Haven, Connecticut, USA. 1965

Davies, Robertson   1913–95
Canadian novelist

I delivered my body into the hands of Learned Physicians this morning confiding that they may discover why I have hay fever. As soon as they got me out of my clothes I ceased to be a man to them, and they began to talk about me as though I did not understand English.
The Table Talk of Samuel Marchbanks (p. 194)
Clarke, Irwin. Toronto, Ontario, Canada. 1949

Belleville, Nicholas   1753–1831
French-born American physician

If you get one good doctor, you get one good thing, but if you get one bad doctor, you get one bad thing. If you have a lawsuit, you get a bad lawyer, you lose your suit — you can appeal; but if you have one bad doctor, and he kills you, then there can be no appeal.
In Stephen Wickes
History of Medicine in New Jersey
Part 2 (p. 143)
Martin R. Dennis & Company. Newark, New Jersey, USA. 1879

de Montaigne, Michel Eyquem   1533–92
French Renaissance writer

If your physician does not think it good for you to sleep, to drink wine, or to eat such and such meats, never trouble yourself; I will find you another that shall not be of that opinion…
In *Great Books of the Western World* (Volume 25)
The Essays
Book III, 13 (p. 528)
Encyclopædia Britannica, Inc. Chicago, Illinois, USA. 1952

Dekker, Thomas   1570–1632
English dramatist

A good physician comes to thee in the shape of an angel, and therefore let him boldly take thee by the hand, for he has been in God's garden, gathering herbs and sovereign roots to cure thee. The good physician deals in simples and will be simply honest with thee in they preservation.
In Robert Coope
The Quiet Art (p. 192)
E.S. Livingstone Ltd. Edinburgh, Scotland. 1952

Donne, John   1572–1631
English poet and divine

Whilst my Physitians by their love are growne
Cosmographers, and I their Mapp, who lie
Flat on this bed, that by them may be showne
That this is my South-west discoverie
per-fretum febris, by these streights to die.
In A.J. Smith (ed.)
The Complete English Poems
Hymne to God My God, in My Sicknesse, l. 6–10
St. Martin's Press. New York, New York, USA. 1971

I observe the Physician with the same diligence as hee the disease.
Devotions Upon Emergent Occasions
Meditation, VI (p. 29)
McGill-Queen's University Press. Montreal, Canada. 1975

Doyle, Sir Arthur Conan   1859–1930
Scottish writer

…if a gentleman walks into my room smelling of iodoform, with a black mark of nitrate of silver upon his right forefinger, and a bulge on the right side of his top-hat to show where he has secreted his stethoscope, I must be dull, indeed, if I do not pronounce him to be an active member of the medical profession.
In William S. Baring-Gould (ed.)
The Annotated Sherlock Holmes (Volume 1)
A Scandal in Bohemia (p. 349)
Wings Books. New York, New York, USA. 1967

Drake, Daniel   1785–1852
American physician

The young physician is not aware how soon his elementary knowledge — much of which is historical and descriptive, rather than philosophical — will fade from his mind, when he ceases to study. That which he possesses can only be retained by new additions.
Practical Essays on Medical Education, and the Medical Profession
Essay IV (p. 61)
Roff & Young. Cincinnati, Ohio. 1832

Professional fame, is the capital of a physician, and he must not suffer it to be purloined, even should its defence involve him in quarrels.
Practical Essays on Medical Education, and the Medical Profession
Essay VII (p. 99)
Roff & Young. Cincinnati, Ohio. 1832

Dryden, John   1631–1700
English poet, dramatist, and literary critic

The first Physicians by Debauch were made:

Excess began, and Sloth sustains the Trade.
The Poems of John Dryden (Volume 4)
To John Dryden, of Chesterton, l. 73 (p. 1530)
Longman. London, England. 1995

Duffy, John C.
No biographical data available

Litin, Edward M.
No biographical data available

These are the duties of a physician: First…to heal his mind and to give help to himself before giving it to anyone else.
Psychiatric Morbidity of Physicians
Journal of the American Medical Association, Volume 189, 1964
(p. 989)

Dumas, Alexandre   1824–95
French dramatist and novelist

The physician has a sacred mission on earth; and to fulfill it he begins at the source of life, and goes down to the mysterious darkness of the tomb.
The Count of Monte Cristo
Chapter 80 (p. 1000)
Grosset & Dunlap Publishers. New York, New York, USA. 1946

Eisenschiml, Otto   1880–1963
Austrian-American chemist and historian

The wise physician…knows when not to prescribe…
The Art of Worldly Wisdom: Three Hundred Precepts for Success Based on the Original Work of Baltasar Gracian
Part Six (p. 71)
Duell, Sloan & Pearce. New York, New York, USA. 1947

The wise physician avoids the knife; if he prescribes a bitter draft, he prescribes it in small doses or sweetens it to disguise its taste.
The Art of Worldly Wisdom: Three Hundred Precepts for Success Based on the Original Work of Baltasar Gracian
Part Seven (p. 87)
Duell, Sloan & Pearce. New York, New York, USA. 1947

Emerson, Ralph Waldo   1803–82
American lecturer, poet, and essayist

The physician prescribes hesitatingly out of his few resources…. If the patient mends, he is glad and surprised.
Ralph Waldo Emerson: Essays and Lectures
The Conduct of Life
Considerations by the Way (p. 1079)
The Library of America. New York, New York, USA. 1983

Field, Eugene   1850–95
American poet and journalist

When one's all right, he's prone to spite
The doctor's peaceful mission;
But when he's sick, it's loud and quick
He bawls for a physician.
The Poems of Eugene Field
Doctors
Charles Scribner's Sons. New York, New York, USA. 1910

No matter what conditions
Dyspeptic come to feaze,
The best of all physicians
Is apple pie and cheese!
The Poems of Eugene Field
Apple-Pie and Cheese, Stanza 5
Charles Scribner's Sons. New York, New York, USA. 1910

Fielding, Henry   1707–54
English novelist, playwright, and barrister

…as a wise general never despises his enemy, however inferior that enemy's force may be, so neither doth a wise physician ever despise a distemper, however inconsiderable.
The History of Tom Jones: A Foundling (Volume 1)
Book V, Chapter VIII (p. 229)
P.F. Collier & Son. New York, New York, USA. 1917

…every physician almost hath his favorite disease…
The History of Tom Jones: A Foundling (Volume 1)
Book II, Chapter 9 (p. 85)
P.F. Collier & Son. New York, New York, USA. 1917

…the gentleman of the Aesculapin art are in the right in advising, that the moment the disease has entered at one door, the physician should be introduced at the other.
The History of Tom Jones: A Foundling (Volume 1)
Book V, Chapter VII (p. 219)
P.F. Collier & Son. New York, New York, USA. 1917

Florio, John   1553?–1625
English teacher, writer, and translator

Unto a deadly disease, neyther
Phisition nor phisick wil serve.
Firste Fruites
Proverbs, Chapter 19
Da Capo Press. New York, New York, USA. 1969

From the phisito & Attorney,
keepe not the truth hidden.
Firste Fruites
Proverbs, Chapter 19
Da Capo Press. New York, New York, USA. 1969

Ford, John   1586–?1640
English dramatist

Physicians are the bodies' cobblers, rather than the Botchers, of men's bodies; as the one patches our tattered clothes, so the other solders our diseased flesh.
The Lovers Melancholy
Act I, Scene I (p. 13)
Da Capo Press. New York, New York, USA. 1970

Freeman, R. Austin   1862–1943
British physician and mystery novelist

Take the case of an aurist. You think that he lives by dealing with obscure and difficult middle and internal ear cases. Nothing of the kind. He lives on wax. Wax is the founda-tion of his practice. Patient comes to him deaf as a post. He does all the proper jugglery — tuning fork, otoscope, speculum, and so on, for the moral effect. Then he hikes out a good old plug of cerumen, and the patient hears perfectly. Of course he is delighted. Thinks a miracle has been performed.
The D'Arblay Mystery (p. 61)
Dodd, Mead & Company New York, New York, USA. 1926

Fox, Sir Theodore
No biographical data available

The patient may well be safer with a physician who is naturally wise than with one who is artificially learned.
Purposes of Medicine
The Lancet, Volume 2, October 23, 1965 (p. 801)

Fuller, Thomas   1608–61
English clergyman and author

Every man is a fool or a physician at forty.
Gnomologia: Adages and Proverbs, Wise Sentences, and Witty Sayings. Ancient and Modern, Foreign and British
No. 1428
Printed for Thomas and Joseph Allman. London, England. 1816

Commonly Physicians like beer are best when they are old; & Lawyers like bread when they are young and new.
The Holy and Profane State
Book II, Chapter I, Maxim VI (p. 50)
Printed for Thomas Tegg. London, England. 1841

Gisbourne, Thomas   1758–1846
English Anglican priest

It is frequently of much importance, not to the comfort only, but to the recovery of the patient, that he should be enabled to look upon his Physician as his friend.
An Enquiry into the Duties of Men
The Duties of Physicians (p. 398)
Printed by J. Davis. London, England. 1794

Gracian, Baltasar   1601–58
Spanish philosopher

The wise physician, if he has failed to cure, looks out for someone who, under the name of consultation, may help him carry out the corpse.
In Herbert V. Prochnow and Herbert V. Prochnow, Jr.
A Treasury of Humorous Quotations: For Speakers, Writers, and Home Reference
#4587 (p. 257)
Harper & Row, Publishers. New York, New York, USA. 1969

Gregg, Alan   1890–1957
American medical educator and philosopher

The true physician cannot remain outside the manifold of the events he observes.
Humanism and Science
Bulletin of the New York Academy of Sciences, Volume 17, 1941

Gregory, John   1724–73
Scottish physician and philosopher

I come now to mention the moral qualities peculiarly required in the character of a physician. The chief of these is humanity; that sensibility of heart which makes us feel for the distresses of our fellow creatures, and which of consequence incites in us the most powerful manner to help them.
Lectures on the Duties and Qualifications of a Physician (p. 19)
W. Strahan. London, England. 1772

Gull, Sir William Withey   1816–90
English physician

There are many good general practitioners, there is only one good universal practitioner — "a warm bed."
A Collection of The Published Writings (Volume 2) (p. viii)
New Sydenham Society. London, England. 1894

Harrison, Tinsley R.   1900–78
American physician

No greater opportunity, responsibility, or obligation can fall to the lot of a human being than to become a physician. In the care of the suffering he needs technical skill, scientific knowledge, and human understanding. He who uses these with courage, with humility, and with wisdom will provide a unique service for his fellow man and will build an enduring edifice of character within himself. The physician should ask of his destiny no more than this; he should be content with no less.
Principles of Internal Medicine (p. 1)
Blakiston. Philadelphia, Pennsylvania, USA. 1950

The true physician has a Shakespearean breadth of interest in the wise and the foolish, the proud and the humble, the stoic hero and the whining rouge. He cares for people.
Principles of Internal Medicine (4th ed.) (p. 7)
McGraw-Hill Book Company, Inc. New York, New York, USA. 1962

Hazlitt, William Carew   1834–1913
English bibliographer

One said Physitians had the best of it; for, if they did well, the world did proclaime it; if ill, the earth did cover it.
Shakespeare Jest Books (Volume 3)
Conceit, Cliches, Flashes and Whimzies, Number 127
Willis & Sotheran. London, England. 1864

One said a Physitian was naturall brother to the wormes, because he was ingendered out of man's corruption.
Shakespeare Jest Books (Volume 3)
Conceit, Clichés, Flashes and Whimzies, Number 42
Willis & Sotheran. London, England. 1864

Heberden, William   1710–1801
Physician

Plutarch says that the life of a Vestal virgin was divided into three portions; in the first of which she learned the duties of her profession, in the second she practiced them, and in the third she taught them to others. This is no bad model for the life of a physician.
Commentaries on the History and Cure of Diseases
Preface (p. vii)
T. Payne, Mews-Gate. London, England. 1802

Herophilus   325 BCE–255 BCE
Greek physician

He is the best physician who knows how to distinguish the possible from the impossible.
In Samuel Evans Massengill
A Sketch of Medicine and Pharmacy and a View of Its Progress by the Massengill Family from the Fifteenth to the Twentieth Century (p. 28)
The S.E. Massengill Company. Bristol, Tennessee, USA. 1943

Heschel, Abraham J.   1907–72
Jewish theologian

What manner of man is the doctor? Life abounds in works of achievement, in areas of excellence and beauty, but the physician is a person who has chosen to go to the areas of distress, to pay attention to sickness and affliction, to injury and anguish.
The Insecurity of Freedom
The Patient as a Person (p. 28)
Farrar, Straus & Giroux. New York, New York, USA. 1966

Hippocrates   460 BCE–377 BCE
Greek physician

…physicians are many in title but very few in reality.
In *Great Books of the Western World* (Volume 10)
Hippocratic Writings
The Law, 1 (p. 144)
Encyclopædia Britannica, Inc. Chicago, Illinois, USA. 1952

It appears to me a most excellent thing for the physician to cultivate Prognosis; for by foreseeing and foretelling, in the presence of the sick, the present, the past, and the future, and explaining the omissions which patients have been guilty of, he will be the more readily believed to be acquainted with the circumstances of the sick; so that men will have confidence to entrust themselves to such a physician.
In *Great Books of the Western World* (Volume 10)
Hippocratic Writings
The Book of Prognostics, 1 (p. 19)
Encyclopædia Britannica, Inc. Chicago, Illinois, USA. 1952

Hoffmann, Friedrich   1660–1742
German physician

The perfect physician must have not only the knowledge of medical art but also prudence and wisdom.
Fundamenta Medicianae

Physiology, Chapter I, 10 (p. 6)
American Elsevier. New York, New York, USA. 1971

The physician is the servant of nature, not her master; the principles of nature and of art are the same and hence the physician must work and act with nature.
Fundamenta Medicianae
Physiology, (p. 5)
American Elsevier. New York, New York, USA. 1971

Holmes, Oliver Wendell   1809–94
American physician, poet, and humorist

The life of a physician becomes ignoble when he suffers himself to feed on petty jealousies and sours his temper in perpetual quarrels.
Medical Essays
The Young Practitioner (p. 392)
Houghton Mifflin Company. Boston, Massachusetts, USA. 1911

The face of a physician, like that of a diplomat, should be impenetrable.
Medical Essays
The Young Practitioner (p. 388)
Houghton Mifflin Company. Boston, Massachusetts, USA. 1911

The old age of a physician is one of the happiest periods of his life. He is loved and cherished for what he has been, and even in the decline of his faculties there are occasions when his experience is still appealed to, and his trembling hands are looked to with renewing hope and trust…The young man feels uneasy if he is not continually doing something to stir up his patient's internal arrangements. The old man takes things more quietly, and is much more willing to let well enough alone.
Medical Essays by Oliver Wendell Holmes
Address
Graduating Class of the Bellevue Hospital College, March 2, 1871 (pp. 377, 395)
Classics of Medicine Library. Birmingham, Alabama, USA. 1987

But the practising physician's office is to draw the healing waters, and while he gives his time to this labor he can hardly be expected to explore all the sources that spread themselves over the wide domain of science. The traveler who would not drink of the Nile until he had tracked it to its parent lakes would be like to die of thirst; and the medical practitioner who would not use the results of many laborers in other departments without sharing their special toils, would find life far too short and art immeasurably too long.
Medical Essays
Scholastic and Bedside Teaching (p. 274)
Houghton Mifflin Company. Boston, Massachusetts, USA. 1911

…a physician's business is to avert disease, to heal the sick, to prolong life, and to diminish suffering…
Medical Essays
Scholastic and Bedside Teaching (p. 274)
Houghton Mifflin Company. Boston, Massachusetts, USA. 1911

The specialist is much like other people engaged in lucrative business. He is apt to magnify his calling, to make much of any symptom which will bring a patient within range of his battery of remedies.
Over the Teacups
Chapter VI (p. 129)
Houghton Mifflin Company. Boston, Massachusetts, USA. 1892

A physician who talks about ceremony and gratitude, and services rendered, and the treatment he got, surely forgets himself…
Medical Essays
The Contagious of Puerperal Fever (p. 115)
Houghton Mifflin Company. Boston, Massachusetts, USA. 1911

A man of very moderate ability may be a good physician if he devotes himself faithfully to the work.
Medical Essays
Scholastic and Bedside Teaching (p. 300)
Houghton Mifflin Company. Boston, Massachusetts, USA. 1911

Howells, William Dean   1837–1920
American realist novelist

I do not know how it is that clergymen and physicians keep from telling their wives the secrets confided to them; perhaps they can trust their wives to find them out for themselves whenever they wish.
The Rise of Silas Lapham
Chapter XXVII (p. 511)
Houghton Mifflin Company. Boston, Massachusetts, USA. 1912

Hufeland, Christoph Wilhelm   1762–1836
German physician

The physician must generalize the disease, and individualize the patient.
In Oliver Wendell Holmes
Medical Essays
Scholastic and Bedside Teaching (p. 275)
Houghton Mifflin Company. Boston, Massachusetts, USA. 1911

Hutchison, Sir Robert Grieve   1871–1960
English radiologist

From inability to let well alone; from too much zeal for the new and contempt for what is old; from putting knowledge before wisdom, science before art, and cleverness before common sense, from treating patients as cases, and from making the cure of the disease more grievous than the endurance of the same, Good Lord, deliver us.
British Medical Journal, Volume 1, 1953 (p. 671)

Jackson, James
No biographical data available

I have often remarked that, though a physician is sometimes blamed very unjustly, it is quite as common for him to get more credit than he is fairly entitled to; so that he has not, on the whole, any right to complain.

Letters to a Young Physician Just Entering Upon Practice
Letter II (p. 41)
Phillips, Sampson & Company. Boston, Massachusetts, USA. 1855

Jekyll, Joseph
No biographical data available

See, one physician, like a sculler, plies,
The patient lingers and by inches dies.
But two physicians, like a pair of oars,
Waft him more swiftly to the Stygian shores.
In Herbert V. Prochnow and Herbert V. Prochnow, Jr.
A Treasury of Humorous Quotations: For Speakers, Writers, and Home Reference
#1661 (p. 94)
Harper & Row, Publishers. New York, New York, USA. 1969

Jesus Christ

It is not the healthy who need a doctor, but the sick.
The Revised English Bible
Matthew 9:12
Oxford University Press, Inc. Oxford, England. 1989

Physician, heal yourself.
The Revised English Bible
Luke 4:23
Oxford University Press, Inc. Oxford, England. 1989

John of Salisbury   ca. 1115–80
English author and diplomatist

The common people say, that physicians are the class of people who kill other men in the most polite and courteous manner.
Policraticus
Book II, Chapter 29
Cambridge University Press. Cambridge, England. 1990

Johnson, Samuel   1696–1772
English critic, biographer, and essayist

A physician in a great city seems to be the mere plaything of fortune; his degree of reputation is for the most part totally casual; they that employ him know not his excellence; they that reject him know not his deficiency.
In William Osler
Aequanimitas, with Other Addresses to Medical Students, Nurses, and Practitioners of Medicine
Preceding Chapter VIII (p. 132)
The Blakiston Company. Philadelphia, Pennsylvania, USA. 1932

Jonsen, Albert
No biographical data available

…the absolute asceticism of the residency recreates, for the young physician, the sacrificial ethic of monastic medicine. That ethic is service: immediate response to the emergency room, to the demands of reports, unmitigated responsibility for correct decisions made promptly and communicated clearly; flagellating denial of sleep, self-indulgence, and frivolity, even to the point of depression and deterioration of personal life, of friendship and love.

Watching the Doctor
New England Journal of Medicine, Volume 308, Number 25, June 23, 1983 (p. 1534)

King, William H.
No biographical data available

The first requisite for a physician is spiritual character and the next requisites are sympathy and a sense of humor.
In R. Kagan (ed.)
Leaders of Medicine
Chapter V (p. 52)
The Medico-Historical Press. Boston, Massachusetts, USA. 1941

La Bruyére, Jean   1645–96
French satiric moralist

As long as men are liable to die and desirous to live, a physician will be made fun of, but he will be well paid.
The Characters of Jean La Bruyére
Characters 14.65
George Routledge & Sons, Ltd. London, England. 1929

Lamb, William   1779–1848
British prime minister

English physicians kill you, the French let you die.
In Elizabeth Longford
Queen Victoria: Born to Succeed
Chapter 5 (p. 69)
Harper & Row, Publishers. New York, New York, USA. 1964

Latham, Peter Mere   1789–1875
English physician

There are always two parties of the management of the disease — the physician and the patient.
In William B. Bean
Aphorisms from Latham (p. 21)
Prairie Press. Iowa City, Iowa, USA. 1962

We physicians had need be a self-confronting and a self-reproving race; for we must be ready, without fear or favor, to call in question our own Experience and to judge it justly; to confirm it, to repeal it, to reverse it, to set up the new against the old, and again to reinstate the old and give it preponderance over the new.
In William B. Bean
Aphorisms from Latham (pp. 93–94)
Prairie Press. Iowa City, Iowa, USA. 1962

The end of all the thought and labour of physicians is to make experiments with men's lives.
In William B. Bean
Aphorisms from Latham (p. 91)
Prairie Press. Iowa City, Iowa, USA. 1962

The best physicians have begun by being the physician of the poor.
In William B. Bean
Aphorisms from Latham (p. 25)
Prairie Press. Iowa City, Iowa, USA. 1962

Physicians are in a manner often called upon to be wiser than they possibly can be. Disease or imperfection of a vital organ is a fearfully interesting thing to him who suffers it, and he presses to learn all that is known, and often much more than is known about it.
In William B. Bean
Aphorisms from Latham (p. 26)
Prairie Press. Iowa City, Iowa, USA. 1962

Physicians, who have worthily achieved great reputation, become the refuge of the hopeless, and earn for themselves the misfortune of being expected to cure incurable diseases.
In William B. Bean
Aphorisms from Latham (p. 25)
Prairie Press. Iowa City, Iowa, USA. 1962

But Nature, in all her powers and operations, allows herself to be led, directed, and controlled. And to lead, direct, or control for purposes of good, this is the business of the physician.
In William B. Bean
Aphorisms from Latham (p. 24)
Prairie Press. Iowa City, Iowa, USA. 1962

I am persuaded that when the physician is called upon to perform great things, even to arrest destructive disease, and to save life, his skill in wielding the implements of his art rests mainly upon the right understanding of simple and single indications, and of the remedies which have power to fulfill them.
In William B. Bean
Aphorisms from Latham (p. 19)
Prairie Press. Iowa City, Iowa, USA. 1962

Longfellow, Henry Wadsworth   1807–82
American poet

You behold in me
Only a traveling Physician;
One of the few who have a mission
To cure incurable diseases,
Or those that are called so.
The Works of Henry Wadsworth Longfellow (Volume 5)
Christus, The Golden Legend, Part I (p. 144)
Houghton Mifflin Company. Boston, Massachusetts, USA. 1904–1917

Ludmerer, Kenneth M.
Physician

The thinking physician…is the one who in the practice of medicine asks not "What is there to do?" but "should it be done?"
Learning to Heal: The Development of American Medical Education (p. 280)
Basic Books, Inc., Publishers. New York, New York, USA. 1985

Luther, Martin   1483–1546
Leader of the Protestant Reformation

Able, cautious, and experienced physicians, are gifts of God. They are the ministers of nature, to whom human

life is confided; but a moment's negligence may ruin everything. No physician should take a single step, but in humility and the fear of God; they who are without the fear of God are mere homicides.
Translated by W. Hazlitt
The Table-Talk of Martin Luther (p. 383)
The Lutheran Publication Society. Philadelphia, Pennsylvania, USA. 1868

MacPhail, Sir Andrew 1864–1938
Canadian physician

I am well aware that in these days, when a student must be converted into a physiologist, a physicist, a chemist, a biologist, a pharmacologist, and an electrician, there is no time to make a physician of him.
British Medical Journal, Volume 1, 1933

Massinger, Philip 1583–1640
English dramatic poet

1 October What art can do, we promise; physic's hand
As apt is to destroy as to preserve,
If Heaven make not the med'cine: all this while,
Our skill hath combat hell with his disease;
But 'tis so arm'd, and a deep melancholy,
To be such in part with death, we are in fear
The grave must mock our labours.
The Plays of Philip Massinger (Volume 1)
The Virgin-Martyr, Act IV, Scene I (p. 76)
G. & W. Nicol. London, England. 1805

Mather, Cotton 1663–1728
American minister and religious writer

Of a Distemper we commonly say, To know the Cause, is Half the Cure. But, alas, how little Progress is there yett made in that Knowledge! Physicians talk about the Causes of Diseases. But their Talk is very Conjectural, very Uncertain, very Ambiguous; and often times a meer Jargon; and in it, they are full of Contradiction to One another.
The Angel of Bethesda
Capsila VII (p. 43)
American Antiquarian Society and Barre Publishers. Barre, Massachusetts, USA. 1972

Mayo, Charles Horace 1865–1939
American physician

The definition of a specialist as one who "knows more and more about less and less" is good and true. Its truth makes essential that the specialist, to do efficient work, must have some association with others who, taken altogether, represent the whole of which the specialty is only a part.
Surgery's Problems as They Affect the Hospital
Modern Hospital, Volume 51, September 1938

The true physician will never be satisfied just to pass his therapeutic wares over a counter.
Problems in Medical Education
Collected Papers of the Mayo Clinic & Mayo Foundation, Volume 18, 1926

Mencken, H. L. (Henry Louis) 1880–1956
American journalist and literary critic

The true physician does not preach repentance; he offers absolution.
Prejudices: Third Series
Chapter XIV, Section 5 (p. 269)
Alfred A. Knopf. New York, New York, USA. 1922

Meyer, Adolf 1866–1950
American neurologist and psychiatrist

I wonder how soon we shall be far enough along to have the physician ask: How much and what, if anything, is structural? how much Functional, somatic or metabolic? How much constitutional, psychogenic and social?
The "Complaint" as the Center of Genetic-Dynamic and Nosological Teaching in Psychiatry
New England Journal of Medicine, August 23, 1928

Miller H.
No biographical data available

The worst mistakes…must be laid at the door of the specialist rather than the general practitioner, who, from his intimate contact with sick people in their natural surroundings, often has a lively understanding of the nervous patient, and is able to see him and his problems as a whole.
The Recognition of Neurotic Illness
Practitioner, Volume 159, 1947

Molière (Jean-Baptiste Poquelin) 1622–1673
French playwright and actor

What will you do, sir, with four physicians? Is not one enough to kill any one body?
In Logan Clendening
Sourcebook of Medical History
Love's the Best Doctor
Act II, Scene I (p. 222)
General Publishing Company Ltd. Toronto, Ontario, Canada. 1942

Moore, Merrill
No biographical data available

If the average man is a harp on whom Nature occasionally plays, the physician is an instrument on whom the emotions are played continuously during his waking hours and that is not too good for any man.
In Mary Lou McDonough
Poet Physician: An Anthology of Medical Poetry Written by Physicians
Afterthought (p. 198)
C.C. Thomas. Springfield, Illinois, USA. 1945

More, Hannah 1745–1833
English religious writer

I used to wonder why people should be so fond of the company of their physician, till I recollected that he is the only person with whom one dares to talk continually of oneself, without interruption, contradiction or censure; I suppose that delightful immunity doubles their fees.
In William Roberts
Memoirs of the Life and Correspondence of Mrs. Hannah Moore (Volume 1)
Letter to Horace Walpole, 27 July 1789 (p. 317)
Harper & Brothers. New York, New York, USA. 1837

Nuland, Sherwin B.   1930–
American surgeon and teacher of bioethics and medicine

The very success of his esoteric therapeutics too often leads the physician to believe he can do what is beyond his doing and save those who, left to their own unhindered judgment, would choose not to be subjected to his saving.
How We Die: Reflections on Life's Final Chapter (p. 221)
Alfred A. Knopf. New York, New York, USA. 1994

Just as physicians must constantly admonish one another to seek the most subtle beginnings of disease, they must also forgive themselves when timing or circumstances frustrate their best intentions.
The Uncertain Art: The Whole Law of Medicine
The American Scholar, Summer, Volume 67, Number 3, 1998

Osler, Sir William   1849–1919
Canadian physician and professor of medicine

'Tis no idle challenge which we physicians throw out to the world when we claim that our mission is of the highest and of the noblest kind, not alone in curing disease but in educating the people in the laws of health, and in preventing the spread of plagues and pestilences…
In Harvey Cushing
The Life of Sir William Osler (Volume 1)
Chapter XVI (p. 408)
Clarendon Press. Oxford, England. 1925

To wrest from nature the secrets which have perplexed philosophers of all ages, to track to their sources the causes of disease, to correlate the vast stores of knowledge, that they may be quickly available for the prevention and cure of disease — these are our ambitions.
Aequanimitas, with Other Addresses to Medical Students, Nurses, and Practitioners of Medicine
Chauvinism in Medicine (p. 267)
The Blakiston Company. Philadelphia, Pennsylvania, USA. 1932

To investigate the causes of death, to examine carefully the condition of organs, after such changes have gone on in them as to render existence impossible and to apply such Knowledge to the prevention and treatment of disease, is one of the highest objects of the Physician…
In Harvey Cushing
The Life of Sir William Osler (Volume 1)
Chapter IV (p. 85)
Clarendon Press. Oxford, England. 1925

To prevent disease, to relieve suffering and to heal the sick — this is our work.
Aequanimitas, with Other Addresses to Medical Students, Nurses, and Practitioners of Medicine
Chauvinism in Medicine (p. 267)
The Blakiston Company. Philadelphia, Pennsylvania, USA. 1932

Permanence of residence, good undoubtedly for the pocket, is not always best for wide mental vision in the physician.
Aequanimitas, with Other Addresses to Medical Students, Nurses, and Practitioners of Medicine
The Army Surgeon (p. 101)
The Blakiston Company. Philadelphia, Pennsylvania, USA. 1932

No class of men needs friction so much as physicians; no class gets less. The daily round of busy practitioners tends to develop an egoism of a most intense kind, to which there is no antidote. The few setbacks are forgotten, the mistakes are often buried, and ten years of successful work tend to make a man touchy, dogmatic, intolerant of correction, and abominably self-centered. To this mental attitude the medical society is the best corrective, and a man misses a good part of his education who does not get knocked about a bit by his colleagues in discussions and criticisms…
In Harvey Cushing
The Life of Sir William Osler (Volume 1)
Chapter VXII (p. 447)
Clarendon Press. Oxford, England. 1925

Few men live lives of more devoted self-sacrifice than the family physician but he may become so completely absorbed in work that leisure is unknown.… There is danger in this treadmill life lest he lose more than health and time and rest — his intellectual independence. More than most men he feels the tragedy of isolation — that inner isolation so well expressed in Matthew Arnold's line — "We mortal millions live alone." Even in populous districts the practice of medicine is a lonely road which winds up-hill all the way and a man may easily go astray and never reach the Delectable Mountains unless he early finds those shepherd guides of which Bunyan tells, Knowledge, Experience, Watchful and Sincere. The circumstances of life mould him into a masterful, self-confident, self-centered man, whose worst faults often partake of his best qualities.
In Harvey Cushing
The Life of Sir William Osler (Volume 1)
Chapter XXI (p. 588)
Clarendon Press. Oxford, England. 1925

The physician who shows in his face the slightest alteration, expressive of anxiety or fear, has not his medullary centres under the highest control, and is liable to disaster at any moment. I have spoken this to you on many occasions, and have urged you to educate your nerve centres so that not the slightest dilator or contractor influence shall pass to the vessels of your face under any professional trial.

In Christopher Lawrence and Steven Shapin
Science Incarnate: Historical Embodiments on Natural Knowledge
(p. 171)
The University of Chicago Press. Chicago, Illinois, USA. 1998

It may be well for a physician to have pursuits outside his profession, but it is dangerous to let them become too absorbing.
In Harvey Cushing
The Life of Sir William Osler (Volume 1)
Chapter III (p. 67)
Clarendon Press. Oxford, England. 1925

A physician who does not use books and journals, who does not need a library, who does not read one or two of the best weeklies and monthlies, soon sinks to the level of the cross-counter prescriber, and not alone in practice, but in those mercenary feelings and habits which characterize a trade…
In Harvey Cushing
The Life of Sir William Osler (Volume 1)
Chapter XVII (p. 448)
Clarendon Press. Oxford, England. 1925

Ovid   43 BCE–17 AD
Roman poet

'Tis not always in a physician's power to cure the sick…
In Arthur Leslie Wheeler
Ovid with an English Translation
Ex Ponto, Book I, iii (p. 281)
Harvard University Press. Cambridge, Massachusetts, USA. 1924

Owen, John   1616–83
English Puritan divine and theologian

Physicians take Gold, but seldom give:
They Physick give, take none; yet healthy live.
A Diet They prescribe; the Sick must for't
Give Gold; Each other Thus supply-support.
Latine Epigrams
Book I, Number 53
Louisiana State University Press. Baton Rouge, Louisiana, USA. 1997

Paracelsus (Theophrastus Phillippus Aureolus Bombastus von Hohenheim)   1493–1541
Alchemist and mystic

The book of Nature is that which the physician must read; and to do so he must walk over the leaves.
Encyclopædia Britannica (9th edition), Volume 18 (p. 234)

Parkinson, John
No biographical data available

The common duty required of a physician lies in the recognition and treatment of disease. If he enlarges his study to cover life as affected by disease, and masters the psychology of the individual sick in body, he will widen his usefulness and reach a fuller life himself as a physician.
Annals of Internal Medicine

The Patient and the Physician
Address
32nd Annual Session of the American College of Physicians, St Louis, Missouri, April 11, 1951

Percival, Thomas   1740–1804
English physician, philosopher, and writer

The relations in which a physician stands to his patients, to his brethren, and to the public, are complicated, and multifarious; involving much knowledge of human nature, and extensive moral duties.
Medical Ethics
To E.C. Percival (p. viii)
Printed by S. Russell. Manchester, England. 1803

Hospital physicians and surgeons should minister to the sick, with due impressions of the importance of their office; reflecting that the ease, the health, and the lives of those committed to their charge depend on their skill, attention, and fidelity.
Medical Ethics
Chapter I (p. 9)
Printed by S. Russell. Manchester, England. 1803

Piozzi, Hester Lynch   1741–1821
English writer

A physician can sometimes parry the scythe of death, but has no power over the sand in the hourglass.
Letter to Fanny Burney, 12 November 1781

Plato   428 BCE–347 BCE
Greek philosopher

…no physician, in so far as he is a physician, considers his own good in what he prescribes, but the good of his patient; for the true physician is also a ruler having the human body as a subject, and is not a mere money-maker.
In *Great Books of the Western World* (Volume 7)
The Republic
Book I, Section 342 (p. 303)
Encyclopædia Britannica, Inc. Chicago, Illinois, USA. 1952

…so to in the body the good and healthy elements are to be indulged and the elements of disease are not to be indulged, but discouraged. And this is what the physician has to do, and in this the art of medicine consists: for medicine may be regarded generally as the knowledge of the loves and desires of the body, and how to satisfy them or not; and the best physician is he who is able to separate fair love from foul, or to convert one into the other; and he who knows how to eradicate and how to implant love, whichever is required, and can reconcile the most hostile elements in the constitution and make them loving friends, is a skillful practitioner.
In *Great Books of the Western World* (Volume 7)
Symposium
Section 186 (p. 156)
Encyclopædia Britannica, Inc. Chicago, Illinois, USA. 1952

...the most skillful physicians are those who, from their youth upwards, have combined with the knowledge of their art the greatest experience of disease; they had better not be robust in health, and should have had all manner of diseases in their own persons.
In *Great Books of the Western World* (Volume 7)
The Republic
Book 3, Section 408 (p. 337)
Encyclopædia Britannica, Inc. Chicago, Illinois, USA. 1952

Plutarch   46–119
Greek biographer and author

...a skillful physician, who, in a complicated and chronic disease, as he sees occasion, at one while allows his patient the moderate use of such things as please him, at another while gives him keen pains and drugs to work the cure.
In *Great Books of the Western World* (Volume 7)
Pericles (p. 129)
Encyclopædia Britannica, Inc. Chicago, Illinois, USA. 1952

Poe, Edgar Allan   1809–49
American short story writer

Is there — is there balm in Gilead? — tell me — tell me, I implore!
The Raven and Other Poems
The Raven, Stanza 15
Columbia University Press. New York, New York, USA. 1942

Prior, Matthew   1664–1721
English poet and diplomat

I sent for Ratcliffe; was so ill
That other doctors gave me over:
He felt my pulse — prescrib'd his pill,
And I was likely to recover.
But when the wit began to wheeze,
And wine had warm'd the politician
Cur'd yesterday of my disease,
I died last night of my physician.
In Helen & Lewis Melville
An Anthology of Humorous Verse
The Remedy Worse than the Disease
Dodd, Mead & Company. New York, New York, USA. 1924

Proverb

Where there are three physicians, there are two atheists.
In Oliver Wendell Holmes
Medical Essays
The Medical Profession in Massachusetts (p. 364)
Houghton Mifflin Company. Boston, Massachusetts, USA. 1911

Go not for every grief to the Physician, nor for every quarrel to the Lawyer, nor for every thirst to the pot.
In George Herbert
Outlandish Proverbs
#290
Printed by T. Maxey for T. Garthwait. London, England. 1651

Deceive not thy Physitian, Confessor, nor Lawyer.
In George Herbert
Outlandish Proverbs
#105
Printed by T. Maxey for T. Garthwait. London, England. 1651

There are more Physitians in health than drunkards.
In George Herbert
Outlandish Proverbs
#903
Printed by T. Maxey for T. Garthwait. London, England. 1651

The Physitian owes all to the patient, but the patient owes nothing to him but a little money.
In George Herbert
Outlandish Proverbs
#921
Printed by T. Maxey for T. Garthwait. London, England. 1651

God heales, and the Physitian hath the thankes.
In George Herbert
Outlandish Proverbs
#169
Printed by T. Maxey for T. Garthwait. London, England. 1651

A disobedient patient makes an unfeeling physician.
In Robert Christy
Proverbs, Maxims and Phrases of All Ages (p. 255)
G.P. Putnam's Sons. New York, New York, USA. 1888

Proverb, Chinese

The physician can cure the sick, but he cannot cure the dead.
In Robert Christy
Proverbs, Maxims and Phrases of All Ages (p. 259)
G.P. Putnam's Sons. New York, New York, USA. 1888

Proverb, German

When you call a physician call the judge to make your will.
In Robert Christy
Proverbs, Maxims and Phrases of All Ages (p. 259)
G.P. Putnam's Sons. New York, New York, USA. 1888

Proverb, Italian

From your confessor, lawyer and physician,
Hide not your case on no condition.
In Sir John Harrington
Metamorphosis of Ajax
The Second Section (p. 154)
Columbia University Press. New York, New York, USA. 1962

Proverb, Italian

Dove non va il sole, va il medico: Where the sunlight enters not, there goes the physician.
In Robert Means Lawrence
Primitive Psycho-Therapy and Quackery
The Blue-Glass Mania (p. 95)
Houghton Mifflin Company. Boston, Massachusetts, USA. 1910

Quarles, Francis   1592–1644
English poet

Physicians of all men are most happy; what good success soever they have, the world proclaimeth, and what faults they commit, the earth coverth.
Hieroglyphikes of the Life of Man
Part iv, Nicocles (p. 17)
Printed for M. Fleshor. London, England. 1638

Rabelais, François   ca. 1490–1553
French writer and physician

Happy is the physician, whose coming is desired at the declension of a disease.
In *Great Books of the Western World* (Volume 24)
Gargantua and Pantagruel
Pantagruel
Book 3, Chapter 41 (p. 209)
Encyclopædia Britannica, Inc. Chicago, Illinois, USA. 1952

Ray, John   1627–1705
English naturalist

The best physicians are Dr. Diet, Dr. Quiet, and Dr. Merryman.
A Complete Collection of English Proverbs (p. 34)
Printed for G. Cowie. London, England. 1813

Piss clear, and defy the physician.
A Complete Collection of English Proverbs (p. 35)
Printed for G. Cowie. London, England. 1813

Saint Augustine of Hippo   354–430
Theologian and doctor of the Church

…as it happens usually to him that having had experience of a bad physician, is fearful afterwards to trust himself with a good [physician]…
St. Augustine's Confessions (Volume 1)
Book VI, IV (p. 281)
William Heinemann. London, England. 1912

Scott, Sir Walter   1771–1832
Scottish novelist and poet

The praise of the physician…is the recovery of the patient.
The Talisman
Chapter VIII (p. 112)
Grosset & Dunlap. New York, New York, USA. 1929

…the sick chamber of the patient is the kingdom of the physician.
The Talisman
Chapter VII (p. 99)
Grosset & Dunlap. New York, New York, USA. 1929

…a slight touch of the cynic in manner and habits, gives the physician, to the common eye, an air of authority which greatly tends to enlarge his reputation.
The Complete Works of Sir Walter Scott (Volume 5)
The Surgeon's Daughter

Chapter I (p. 23)
Conner& Cooke. New York, New York, USA. 1833

Seegal, David
No biographical data available

The young physician today is so generously provided with a kit of diagnostic and therapeutic tools, his attention might be wisely directed to the question of "what not to do" as well as "what to do."
Journal of Chronic Diseases, Volume 17, 299, 1964

Selden, John   1584–1654
English jurist

Preachers say, do as I say, not as I do. But if the physician had the same disease upon him that I have, and he should bid me to do one thing, and he do quite another, could I believe him?
Table Talk of John Selden
Preaching #13 (p. 145)
J.M. Dent. London, England. 1899

Seneca (Lucius Annaeus Seneca)   4 BCE–65 AD
Roman playwright

The physician cannot prescribe by letter…he must feel the pulse.
Translated by Richard M. Gummere
Ad Lucilium Epistulae Morales (Volume 1)
Epistle xxii, Section 1 (p. 149)
Harvard University Press. Cambridge, Massachusetts, USA. 1925

Shakespeare, William   1564–1616
English poet, playwright, and actor

Trust not the physician;
His antidotes are poison, and he slays
More than you rob.
In *Great Books of the Western World* (Volume 27)
The Plays and Sonnets of William Shakespeare (Volume 2)
Timon of Athens
Act IV, Scene iii, l. 434–436
Encyclopædia Britannica, Inc. Chicago, Illinois, USA. 1952

Kill thy physician, and the fee bestow
Upon thy foul disease.
In *Great Books of the Western World* (Volume 27)
The Plays and Sonnets of William Shakespeare (Volume 2)
King Lear
Act I, Scene i, l. 164–165
Encyclopædia Britannica, Inc. Chicago, Illinois, USA. 1952

Sheridan, Richard Brinsley   1751–1816
English dramatist and politician

The art of the physician consists, in a great measure, in exciting hope, and other friendly passions and feelings.
Laconic Manual and Brief Remarker Containing Over a Thousand Subjects Alphabetically and Systematically Arranged (p. 330)
Robert Dick. Toronto, Ontario, Canada. 1853

…I had rather follow you to your grave, than see you owe your life to any but a regular bred physician.

St. Patrick's Day
Act II, Scene Justice Hoofe (p. 24)
Publisher undetermined

Sissman, Louis Edward   1928–76
Poet

The doctors — eleven of them, all told — marshaled their forces for a truly impressive attack on the disease. Everything, I felt was meticulously planned in some War Room in the depths of the hospital: the battery of tests in just such a sequence; the alternative battle plans contingent on the outcome of the tests; the choice of weapons — radiation or chemotherapy — for the mopping-up afterward.
The Atlantic
A Little Night Music: A Tangential Line
February 1972

Smollett, Tobias George   1721–71
Scottish novelist

The character of a physician, therefore, not only presupposes natural sagacity, and acquired erudition, but it also implies every delicacy of sentiment, every tenderness of nature, and every virtue of humanity.
The Life and Adventures of Sir Launcelot Greaves
Chapter XXIV (p. 192)
Oxford University Press, Inc. London, England. 1973

Stanton, Elizabeth Cady   1815–1902
American reformer

Besides the obstinacy of the nurse, I had the ignorance of the physicians to contend with.
Eighty Years and More (1815–1897) Reminiscences of Elizabeth Cady Stanton
Motherhood (pp. 118–119)
T. Fisher Unwin. London, England. 1898

Stevenson, Robert Louis   1850–94
Scottish essayist and poet

There are men and classes of men that stand above the common herd; the soldier, the sailor, and the shepherd not unfrequently; the artist rarely; rarelier still, the clergyman; the physician almost as a rule. He is the flower (such as it is) of our civilization; and when that stage of man is done with, and only remembered to be marveled at in history, he will be thought to have shared as little as any in the defects of the period, and most notably exhibited the virtue of the race. Generosity he has, such as is possible to those who practice an art, never to those who drive a trade; discretion, tested by a hundred secrets; tact, tried in a thousand embarrassments; and, what are more important, Herculean cheerfulness and courage. So it is that he brings air and cheer into the sick-room, and often enough, though not so often as he wishes, brings healing.
Underwoods
Preface
Charles Scribner's Sons. New York, New York, USA. 1887

Swift, Jonathan   1667–1745
Irish-born English writer

Physicians ought not to give their Judgment of Religion, for the same Reason that Butchers are not admitted to be Jurors upon Life and Death.
Satires and Personal Writings
Thoughts on Various Subjects (p. 410)
Oxford University Press, Inc. New York, New York, USA. 1965

Taylor, Jeremy   1613–67
English clergyman

…to preserve a man alive in the midst of so many chances, and hostilities, is as great a miracle as to create him…
Holy Living and Holy Dying (Volume 2)
Chapter I, Section 1, l. 7–9
At The Clarendon Press. Oxford, England. 1989

The Bible

Is there no balm in Gilead, no physician there?
The Revised English Bible
Jeremiah 8:22
Oxford University Press, Inc. Oxford, England. 1989

…Asa became gravely affected with disease in his feet; he did not seek guidance of the Lord but resorted to physicians.
The Revised English Bible
II Chronicles 16:12–13
Oxford University Press, Inc. Oxford, England. 1989

Thoreau, Henry David   1817–62
American essayist, poet, and practical philosopher

Priests and physicians should never look one another in the face. They have no common ground, nor is there any to mediate between them. When the one comes, the other goes. They could not come together without laughter, or a significant silence, for the one's profession is a satire on the other's, and either's success would be the other's failure.
The Writings of Henry David Thoreau (Volume 1)
A Week on the Concord and Merrimac Rivers
Wednesday (p. 339)
Houghton Mifflin Company. Boston, Massachusetts, USA. 1893

It is wonderful that the physician should ever die, and that the priest should ever live. Why is it that the priest is never called to consult with the Physician? It is because men believe practically that matter is independent of spirit. But what quackery? It is commonly an attempt to cure the disease of a man by addressing his body alone. There is a need of a physician who shall minister to both soul and body at once, that is to man. Now he falls between two stools.
The Writings of Henry David Thoreau (Volume 1)
A Week on the Concord and Merrimac Rivers
Wednesday (p. 339)
Houghton Mifflin Company. Boston, Massachusetts, USA. 1893

Tupper, Kerr Boyce
No biographical data available

Let a physician believe with all his heart that God meant him to be a physician, only a physician, wholly a physician, always a physician, then will he be a physician indeed, uncorrupted by the love of money, untainted by infection for fame, unintimidated by danger.... Have appetite for your life calling, and you will have aptitude for all its duties.
The Ideal Physician (p. 37)
Lea Brothers/Philadelphia, Pennsylvania, USA. 1899

Virchow, Rudolf Ludwig Karl   1821–1902
German pathologist and archaeologist

Only those who regard healing as the ultimate goal of their efforts can, therefore, be designated as physicians.
Translated by Lelland J. Rather
Disease, Life, and Man, Selected Essays
Standpoints in Scientific Medicine (1847) (p. 26)
Stanford University Press. Stanford, California, USA. 1958

...there are circumstances in which the split between scientific and practical medicine is so great that the learned physician can do nothing, while the practical physician knows nothing. Lord Bacon has said, *scientia est potentia*. Knowledge which is unable to support action is not genuine, and how unsure is activity without understanding! This split between science and practice is rather new; our century and our country have brought it into being.
Translated by Lelland J. Rather
Disease, Life, and Man
Standpoints in Scientific Medicine (p. 27)
Stanford University Press. Stanford, California, USA. 1958

Voltaire (François-Marie Arouet)   1694–1778
French writer

The Devil should not try his tricks on a clever physician. Those familiar with nature are dangerous for the wonder-workers. I advise the Devil always to apply to the faculty of theology — not to the medical faculty.
In Pearch Bailey
Voltaire's Relation to Medicine
Annals of Medical History, Volume 1, 1917 (p. 58)

Let nature be your first physician. It is she who made all.
The Works of Voltaire (Volume 11)
Philosophical Dictionary (Volume 7)
Medicine (p. 169)
The St. Hubert Guild. Akron, Ohio, USA. 1901

But nothing is more estimable than a physician who, having studied nature from his youth, knows the properties of the human body, the diseases which assail it, the remedies which will benefit it, exercises his art with caution, and pays equal attention to the rich and the poor.
The Works of Voltaire (Volume 12)
Philosophical Dictionary (Volume 8)

Physicians (pp. 199–200)
The St. Hubert Guild. Akron, Ohio, USA. 1901

von Ebner-Eschenbach, Marie   1830–1916
Austrian writer

Physicians are hated either on principle or for financial reasons.
Translated by David Scrase and Wolfgang Mieder
Aphorisms (p. 50)
Aridne Press. Riverside, California, USA. 1994

Webster, John   1580?–1625?
English playwright

Physicians are like kings —
they brook no contradiction.
The Duchess of Malfi
Act V, Scene II, l. 69–70
Chatto & Windus. London, England. 1958

Wordsworth, William   1770–1850
English poet

Physician art thou? one, all eyes,
Philosopher! a fingering slave,
One that would peep and botanize
Upon his mother's grave.
The Complete Poetical Works of William Wordsworth
A Poet's Epitaph
Crowell. New York, New York, USA. 1888

Young, Arthur   1741–1820
English traveler

...there is a great difference between a good physician and a bad one; yet very little between a good one and [no physician] at all.
Travels in France
9 September 1787 (p. 66)
G. Bell. London, England. 19112

PHYSICIST

Up to the time of the foundation of the Institute of Physics, the physicists had hardly been recognized as a member of one of the professions.
The Institute of Physics: Objects of the Institute (p. 5)

Adams, Douglas   1952–2001
English author, comic radio dramatist, and musician

Very strange people, physicists...in my experience the ones who aren't actually dead are in some way very ill.
The Long Dark Tea-Time of the Soul
Chapter II (p. 140)
Simon & Schuster. New York, New York, USA. 1990

It startled him even more when just after he was awarded the Galactic Institute's Prize for Extreme Cleverness he got lynched by a rampaging mob of respectable physicists

who had finally realized that the one thing they really couldn't stand was a smart-ass.
The Ultimate Hitchhiker's Guide to the Galaxy
The Hitchhiker's Guide to the Galaxy
Chapter 10 (p. 60)
The Ballantine Book Company. New York, New York, USA. 2002

Adams, Henry Brooks   1838–1918
American man of letters

…the future of Thought and therefore of History lies in the hands of physicists, and therefore the future historian must seek his education in the world of mathematical physics.
The Degradation of the Democratic Dogma
The Rule of Phase Applied to History (p. 283)
Peter Smith. New York, New York, USA. 1949

Author undetermined

We seek, we study, and we stare
At particles that weren't quite there.
In H. Arthur Klein
The World of Measurements
Song for A High-Energy Physicist (p. 180)
Simon & Schuster. New York, New York, USA. 1974

Baker, Adolph
No biographical data available

Physics is engaged neither in the development of time machines nor in the fabrication of bombs. But it is the business of physicists to take flights of fancy which carry them far beyond the boundaries imposed by current technology.
Modern Physics and Antiphysics
Chapter 3 (p. 27)
Addison-Wesley Publishing Company. Reading, Massachusetts, USA. 1970

Barnett, Lincoln   1909–79
Science writer

The young physicists are beyond all doubt the noisiest, rowdiest, most active and most intellectually alert group we have here. For them the world changes every week and they are simply delighted by it. A few days ago I asked one of them as they came bursting out of a seminar, "How did it go?" "Wonderful" he said. "Everything we knew about physics last week isn't true!"
Writing on Life: Sixteen Close-Ups
Physicist Oppenheimer (p. 378)
William Sloane Associates, Publishers. New York, New York, USA. 1951

Bergmann, P.
No biographical data available

In many aspects, the theoretical physicist is merely a philosopher in a working suit.
In Jean-Pierre Luminet
Black Holes (p. 51)
Cambridge University Press. New York, New York, USA. 1992

Birkhoff, George David   1884–1944
American mathematician

It is to be hoped that in the future more and more theoretical physicists will command a deep knowledge of mathematical principles; and also that mathematicians will no longer limit themselves so exclusively to the aesthetic development of mathematical abstractions.
Mathematical Nature of Physical Theories
American Scientist, Volume 31, Number 4, October 1943 (p. 286)

Boltzmann, Ludwig Edward   1844–1906
Austrian Physicist

$S = k \log w$
Carved on Boltzmann's gravestone

Brecht, Bertolt   1898–1956
German writer

VIRGINIA: Father says theologians have their bells to ring: physicists have their laughter.
Translated by John Willett
Life of Galileo
Scene 9 (p. 74)
Arcade Publishing. New York, New York, USA. 1994

Brillouin, Léon   1889–1969
French physicist

It is impossible to study the properties of a single mathematical trajectory. The physicist knows only bundles of trajectories, corresponding to slightly different initial conditions.
In John D. Barrow
The World Within the World (p. 277)
Clarendon Press. Oxford, England. 1988

Burroughs, William S.   1914–97
American writer

No atomic physicist has to worry, people will always want to kill other people on a mass scale.
The Adding Machine: Selected Essays
A Word to the Wise Guy (p. 29)
Seaver Books. New York, New York, USA. 1986

Crick, Francis Harry Compton   1916–2004
English biochemist

Physicists are all too apt to look for the wrong sorts of generalizations, to concoct theoretical models that are too neat, too powerful, and too clean. Not surprisingly, these seldom fit well with data. To produce a really good biological theory, one must try to see through the clutter produced by evolution to the basic mechanisms. What seems to physicists to be a hopelessly complicated process may have been what nature found simplest, because nature could build on what was already there.
What Mad Pursuit?: A Personal View of Scientific Discovery (p. 139)
Basic Books, Inc. New York, New York, USA. 1988

Cvitanovic, Predrag
Physicist

Indicative of the depth of mathematics lurking behind physicists' conjectures is that fact that the properties that one would like to establish about the renormalization theory of critical circle maps might turn out to be related to number-theoretic abysses such as the Riemann conjecture....
In C. Itzykson, et al. (eds.)
From Number Theory and Physics
Circle Maps: Irrationally Winding
Springer-Verlag New York, Inc. New York, New York, USA. 1992

Davies, Paul Charles William   1946–
British-born physicist, writer, and broadcaster
Brown, Julian R.
No biographical data available

Physicists, like theologians, are wont to deny that any system is in principle beyond the scope of their subject.
Superstrings: A Theory of Everything
Introduction (p. 1)
Cambridge University Press. Cambridge, England. 1988

Dicke, R. H.
No biographical data available

It is well known that carbon is required to make physicists.
Dirac's Cosmology and Mach's Principle
Nature, Volume 192, Number 4801, November 4, 1961 (p. 440)

Dirac, Paul Adrian Maurice   1902–84
English theoretical physicist

Some physicists may be happy to have a set of working rules leading to results in agreement with observation. They may think that this is the goal of physics. But it is not enough. One wants to understand how Nature works.
Proceedings of the Conference Perturbative Quantum Chromodynamics
Volume 74, 1981 (pp. 129–130)

Duhem, Pierre-Maurice-Marie   1861–1916
French physicist and mathematician

The watchmaker to whom one gives a watch that does not run will take it all apart and will examine each of the pieces until he finds out which one is damaged. The physician to whom one presents a patient cannot dissect him to establish the diagnosis. He has to guess the seat of the illness by examining the effect on the whole body. The physicist resembles a doctor, not a watchmaker.
Quelques reflexions au sujet de la physique experimentale
Revue des questions scientifiques., Volume 36, 1897 (p. 55)

...if the aim of physical theories is to explain experimental laws, theoretical physics is not an autonomous science; it is subordinate to metaphysics.
The Aim and Structure of Physical Theory
Part I, Chapter I (p. 10)
Princeton University Press. Princeton, New Jersey, USA. 1954

Dürrenmatt, Friedrich   1921–90
Swiss playwright and novelist

Dear Mobius. You have visitors. Now leave your physicist's lair for a moment and come in here.
Translated by James Kirkup
The Physicists
Act One (p. 37)
Grove Press, Inc. New York, New York, USA. 1964

It's ludicrous. Here we have hordes of highly paid physicists in gigantic state-supported laboratories working for years and years and years vainly trying to make some progress in the realm of physics, while you do it quite casually at your desk in this madhouse.
Translated by James Kirkup
The Physicists
Act Two (p. 75)
Grove Press, Inc. New York, New York, USA. 1964

Dyson, Freeman J.   1923–
American physicist and educator

Theoretical physicists are accustomed to living in a world which is removed from tangible objects by two levels of abstraction. From tangible atoms we move by one level of abstraction to invisible fields and particles. A second level of abstraction takes us from fields and particles to the symmetry-groups by which fields and particles are related. The superstring theory takes us beyond symmetry-groups to two further levels of abstraction. The third level of abstraction is the interpretation of symmetry-groups in terms of states in ten-dimensional space-time. The fourth level is the world of the superstrings by whose dynamical behavior the states are defined.
Infinite in All Directions
Part One, Chapter Two (p. 18)
Harper Collins Publisher, Inc. New York, New York, USA. 1988

Eddington, Sir Arthur Stanley   1882–1944
English astronomer, physicist, and mathematician

To the pure geometer the radius of curvature is an incidental characteristic — like the grin of the Cheshire cat. To the physicist it is an indispensable characteristic. It would be going too far to say that to the physicist the cat is merely incidental to the grin. Physics is concerned with interrelatedness such as the interrelatedness of cats and grins. In this case the "cat without a grin" and the "grin without a cat" are equally set aside as purely mathematical phantasies.
The Expanding Universe
Chapter IV, Section III (pp. 103–104)
The University Press. Cambridge. 1933

Wheresoever the carcass is, there will the eagles be gathered together, and where the symbols of the mathematical physicists flock, there presumably is some prey for them to settle on, which the plain man at least will prefer to call by a name suggestive of something more than passive emptiness.

New Pathways in Science
Chapter II, Section IV (p. 39)
The Macmillan Company. New York, New York, USA. 1935

Life would be stunted and narrow if we could feel no significance in the world around us beyond that which can be weighed and measured with the tools of the physicist or described by the metrical symbols of the mathematician.

In Arthur Beiser
The World of Physics
Introduction
Simon & Schuster. New York, New York, USA. 1987

Einstein, Albert   1879–1955
German-born physicist

The supreme task of the physicist is to arrive at those universal elementary laws from which the cosmos can be built up by pure deduction.

The World As I See It (p. 22)
Philosophical Library. New York, New York, USA. 1949

Dear Schrödinger: You are the only contemporary physicist, besides Laue, who sees that one cannot get around the assumption of reality — if only one is honest. Most of them simply do not see what sort of risky game they are playing with reality — reality as something independent of what is experimentally established.

In A.P. French & P.J. Kennedy (eds.)
Niels Bohr: A Centenary Volume
Letter, Albert Einstein to Erwin Schrödinger, December 22, 1950
(p. 143)
Harvard University Press. Cambridge, Massachusetts, USA. 1985

How wretchedly inadequate is the theoretical physicist as he stands before Nature — and before his students!

In Helen Dukas and Banesh Hoffman
Albert Einstein: The Human Side: New Glimpses from His Archives
Letter dated 15 March 1922 (p. 24)
Princeton University Press. Princeton, New Jersey, USA. 1979

If you want to find out anything from the theoretical physicists about the methods they use, I advise you to stick closely to one principle: Don't listen to their words, fix your attention on their deeds.

Ideas and Opinions
On the Method of Theoretical Physics (p. 270)
Crown Publishers, Inc. New York, New York, USA. 1954

…the supreme task of the physicist is the discovery of the most general elementary laws from which the world-picture can be deduced logically.

In Max Planck
Where Is Science Going?
Prologue (p. 10)
W.W. Norton & Company, Inc. New York, New York, USA. 1932

Feynman, Richard P.   1918–88
American theoretical physicist

A professor of theoretical physics always has to be told what to look for. He just uses his knowledge to explain the observations of the experimenters!"

What Do You Care What Other People Think?
The Cold Facts (p. 140)
W.W. Norton & Company, Inc. New York, New York, USA. 1988

The limited imagination of physicists: When we see a new phenomenon we try to fit it into the framework we already have…. It's not because Nature is really similar; it's because the physicists have only been able to think of the same damn thing, over and over again.

QED: The Strange Theory of Light and Matter
Chapter 4 (p. 149)
Princeton University Press. Princeton, New Jersey, USA. 1985

Physicists sometimes feel so superior and smart that other people would like to catch them out once on something. I will give you something to get them on. They should be utterly ashamed of the way they take energy and measure it in a host of different ways, with different names. It is absurd that energy can be measured in calories, in ergs, in electron volts, in foot pound, in B.T.U.s, in horsepower hours, in kilowatt hours — all measuring exactly the same thing…. For those who want some proof that physicists are human, the proof is in the idiocy of all the different units which they use for measuring energy.

The Character of Physical Law
Chapter 3 (p. 74)
BBC. London, England. 1965

Foster, G. C.
No biographical data available

…from the very outset of his investigations the physicist has to rely constantly on the aid of the mathematician, for even in the simplest cases, the direct result of his measuring operations are entirely without meaning until they have been submitted to more or less [a] mathematical discussion.

Mathematical and Physical Opening Address,
Nature, Section A, Volume 16, Number 407, August 16, 1887 (p. 312)

Gamow, George   1904–68
Russian-born American physicist

Now, Physicists, take warning,
Observe this sober test…
When new fleas are a-borning
Make sure they're fully dressed!

Thirty Years That Shook Physics
First Part (p. 193)
Doubleday & Company, Inc. Garden City, New York, USA. 1966

Gibbs, J. Willard   1839–1903
American mathematician

A mathematician may say anything he pleases, but a physicist must be at least partially sane.
In R.B. Lindsay
On the Relation of Mathematics and Physics
The Scientific Monthly, December 1944 (p. 456)

Green, Celia   1935–
English philosopher and psychologist

If you say to a theoretical physicist that something is inconceivable, he will reply: "It only appears inconceivable because you are naively trying to conceive it. Stop thinking and all will be well."
The Decline and Fall of Science
Aphorisms (pp. 2–3)
Hamilton. London, England. 1976

Greene, Brian   1963–
American physicist

Physicists are more like avant-garde composers, willing to bend traditional rules.… Mathematicians are more like classical composers…"
The Elegant Universe
Chapter 11 (p. 271)
W.W. Norton & Company, Inc. New York, New York, USA. 2003

Gribbin, John
English science writer and astronomer
Rees, Martin John   1942–
15th Astronomer Royal of England

…the fate of the Universe, like its present appearance, was imprinted right at the beginning, in the hot, dense fireball era. And to understand that era, and the nature of the relics it could have left behind, we enter the realm of the particle physicist.
Cosmic Coincidences: Dark Matter, Mankind, and Anthropic Cosmology
Part One, Chapter Three (p. 99)
Bantam Books. New York, New York, USA. 1989

Hanson, Norwood Russell   1924–67
American philosopher of science

Physicists do not start from hypotheses; they start from data. By the time a law has been fixed into an H-D [hypothetico-deductive] system, really original physical thinking is over.
Patterns of Discovery
Chapter IV (p. 70)
At The University Press. Cambridge, England. 1958

Heisenberg, Werner Karl   1901–76
German physicist and philosopher

Some physicists would prefer to come back to the idea of an objective real world whose smallest parts exist objectively in the same sense as stones or trees exist independently of whether we observe them. That, however, is impossible.
Physics and Philosophy: The Revolution in Modern Science
July, 1992
Harper & Row, Publishers. New York, New York, USA. 1958

The physicist may be satisfied when he has the mathematical scheme and knows how to use it for the interpretation of the experiments. But he has to speak about his results also to non-physicists who will not be satisfied unless some explanation is given in plain language. Even for the physicist the description in plain language will be the criterion of the degree of understanding that has been reached.
Physics and Philosophy: The Revolution in Modern Science
Chapter X (p. 168)
Harper & Row, Publishers. New York, New York, USA. 1958

Hoffmann, Banesh   1906–86
Mathematician and educator

They could but make the best of it, and went around with woebegone faces sadly complaining that on Mondays, Wednesdays and Fridays they must look on light as a wave; on Tuesdays, Thursdays and Saturdays, as a particle. On Sundays they simply prayed.
The Strange Story of the Quantum
Chapter IV (p. 42)
Dover Publications, Inc. New York, New York, USA. 1959

Hoyle, Sir Fred   1915–2001
English mathematician and astronomer

The physicist is mainly interested in the detailed structure of the threads in our tapestry. He is less interested in the broad pattern on the tapestry itself. The broadest pattern of all, on the scale of stars and galaxies, is the business of the astronomer.
Ten Faces of the Universe
The Astrophysicist's Universe (p. 55)
W.H. Freeman & Company. San Francisco, California, USA. 1977

Jeans, Sir James Hopwood   1877–1946
English physicist and mathematician

The physicist who can discard his human spectacles, and can see clearly in the strange new light which then assails his eyes, finds himself living in an unfamiliar world, which even his immediate predecessors would probably fail to recognize.
The New Background of Science
Chapter I (pp. 5–6)
The University of Michigan Press. Ann Arbor, Michigan, USA. 1959

Johnson, George   1952–
American science writer

Trying to capture the physicists' precise mathematical description of the quantum world with our crude words

and mental images is like playing Chopin with a boxing glove on one hand and a catcher's mitt on the other.
On Skinning Schrödinger's Cat
The New York Times, Section 4, Sunday, 2 June 1996 (p. 16)

Joyce, James   1882–1941
Irish-born author

As a physicist he had learned that of the 70 years of complete human life at least 2/7, viz. 20 years are passed in sleep.
Ulysses (p. 704)
Random House, Inc. New York, New York, USA. 1946

Krauss, Lawrence M.   1954–
American theoretical physicist

For the most part, physicists follow the same guidelines that have helped keep Hollywood movie producers rich: If it works, exploit it. If it still works, copy it.
Fear of Physics: A Guide for the Perplexed
Chapter 1 (p. 4)
Basic Books, Inc. New York, New York, USA. 1993

Krutch, Joseph Wood   1893–1970
American naturalist, conservationist, and writer

Electronic calculators can solve problems which the man who made them cannot solve; but no government-subsidized commission of engineers and physicists could create a worm…
The Twelve Seasons
March (p. 184)
W. Sloane Associates. New York, New York, USA. 1949

Kuhn, Thomas S.   1922–96
American historian of science

Looking at a contour map, the student sees lines on paper, the cartographer a picture of a terrain. Looking at a bubble-chamber photograph, the student sees confused and broken lines, the physicist a record of familiar subnuclear events. Only after a number of such transformations of vision does the student become an inhabitant of the scientist's world.
The Structure of Scientific Revolutions
Chapter X (p. 111)
The University of Chicago Press. Chicago, Illinois, USA. 1970

Kusch, Polykarp   1911–93
German-American physicist

Our early predecessors observed Nature as she displayed herself to them. As knowledge of the world increased, however, it was not sufficient to observe only the most apparent aspects of Nature to discover her more subtle properties; rather, it was necessary to interrogate Nature and often to compel Nature, by various devices, to yield an answer as to her functioning. It is precisely the role of the experimental physicist to arrange devices and procedures that will compel Nature to make a quantitative statement of her properties and behavior.
Nobel Lectures, Physics 1942–1962
Nobel lecture for award received in 1955
The Magnetic Moment of the Electron (p. 298)
Elsevier Publishing Company. Amsterdam, Netherlands. 1964

Ladenburg, Rudolf   1882–1952
German physicist

There are two kinds of physicists in Berlin: on the one hand was Einstein, and on the other all the rest.
In A.P. French
Einstein: A Centenary Volume
Chapter 4 (p. 125)
Harvard University Press. Cambridge, Massachusetts, USA. 1979

Lederman, Leon   1922–
American high-energy physicist

Today we have two groups of physicists both with the common aim of understanding the universe but with a large difference in cultural outlook, skills, and work habits. Theorists tend to come in late to work, attend grueling symposiums on Greek islands or Swiss mountaintops, take real vacations, and are at home to take out the garbage much more frequently. They tend to worry about insomnia…. Experimenters don't come in late — they never went home. During an intense period of lab work, the outside world vanishes and the obsession is total. Sleep is when you can curl up on the accelerator floor for an hour.
The God Particle: If the Universe Is the Answer, What Is the Question?
Chapter 1 (p. 14)
Houghton Mifflin Company. Boston, Massachusetts, USA. 1993

Lichtenberg, Georg Christoph   1742–99
German physicist and satirical writer

The myths of the physicists.
Lichtenberg: Aphorisms & Letters
Aphorisms (p. 57)
Jonathan Cape. London, England. 1969

Lindley, David   1956–
English astrophysicist and author

But by tradition the physicist, having found one level of order in nature, invariably wants to know, like the archaeologist digging down into the remains of Troy, whether there is another, more primitive layer underneath.
The End of Physics: The Myth of a Unified Theory
Part I, Chapter 3 (p. 98)
Basic Books, Inc. New York, New York, USA. 1993

Lodge, Sir Oliver   1851–1940
English physicist

But, notwithstanding any temptation to idolatry, a physicist is bound in the long run to return to his right mind; he must cease to be influenced unduly by superficial appearances, impracticable measurements, geometrical devices, and weirdly ingenious modes of expression; and remember that his real aim and object is absolute truth, however difficult of attainment that may be, that his function is to discover rather than to create, and that beneath and above and around all Appearances there exists a universe of full-bodied, concrete, absolute, reality.

Geometcisation of Physics, and Its Supposed Bias on the Michelson–Morley Experiment
Nature, Volume 106, Number 2677, February 17, 1921 (p. 800)

Marcus, Adrianne
No biographical data available

Let others lie about the universe,
make visible worlds. I am the keeper
of particles, custodian of stray
atoms.

In Steve Rasnic Tem (ed.)
The Umbral Anthology of Science Fiction Poetry
The Physicist's Purpose, 1978
Umbral Press. Denver, Colorado, USA. 1982

Mermin, Norman David   1935–
Mathematician

…contemporary physicists come in two varieties. Type 1 physicists are bothered by EPR [electronparamagnetic resonance] and Bell's theorem. Type 2 (the majority) are not, but one has to distinguish two subvarieties. Type 2a physicists explain why they are not bothered. Their explanations tend either to miss the point entirely (like Born's to Einstein) or to contain physical assertions that can be shown to be false. Type 2b are not bothered and refuse to explain why.

Is the Moon There When Nobody Looks? Reality and the Quantum Theory *Physics Today*, Volume 38, Number 4, April 1985 (p. 41)

Michelson, Albert Abraham   1852–1931
German-American physicist

If a poet could at the same time be a physicist, he might convey to others the pleasure, the satisfaction, almost the reverence, which the subject inspires. The aesthetic side of the subject is, I confess, by no means the least attractive to me. Especially is its fascination felt in the branch which deals with light…

Light Waves and Their Uses
Lecture I (p. 1)
The University of Chicago Press. Chicago, Illinois, USA. 1903

Nietzsche, Friedrich   1844–1900
German philosopher

We must be physicists in order…to be creative since so far codes of values and ideals have been constructed in ignorance of physics or even in contradiction to physics.

The Gay Science
Aphorism 335
Cambridge University Press. Cambridge, England. 2001

Oppenheimer, J. Robert   1904–67
American theoretical physicist

In some sort of crude sense which no vulgarity, no humor, no overstatement can quite extinguish, the physicists have known sin, and this is a knowledge which they cannot lose.

Expiation
Time, Volume 51, Number 8, 23 February 1948 (p. 94)

Pagels, Heinz R.   1939–88
American physicist and science writer

I once heard a story that physicists when they die go to a heavenly academy where their purpose is to lay down the laws of nature. But there is a rule they must obey: Any new law they make cannot contradict ones already discovered and verified by their colleagues back on earth. The legend says that Pauli, one of the sharpest critics of physics, is there now setting intellectual traps and doing physics tricks to foul our best efforts.

The Cosmic Code: Quantum Physics as the Language of Nature
Part III, Chapter 1 (p. 339)
Simon & Schuster. New York, New York, USA. 1982

Petroski, Henry   1942–
Civil engineer

Embedded in a matrix of mistakes
And slips of sighs, his next equation lies
About its symmetry. Among the lines
Of exercise and bold heuristic thrusts
Of algebra and calculus, it takes
His magic mirror mind to recognize
A juxtaposition that unifies
His theory of another universe.
Extracting the law from the accidents,
He calls it Theorem and proceeds to prove
It logically follows from stronger laws.
He makes some definitions and extends
The theorem more and more and marvels at the rules
His universe follows, effect from cause.

The Mathematical Physicist
Southern Humanities Review, Volume 8, Number 2, 1972 (p. 184)

Poincaré, Henri   1854–1912
French mathematician and theoretical astronomer

Nothing but facts are of importance. John Lackland passed by here. Here is something that is admirable. Here is a reality for which I would give all the theories in the world." That is the language of the historian. The physicist would say rather: "John Lackland passed by here; that makes no difference to me, for he never will pass this way again."

The Foundations of Science
Science and Hypothesis, Part IV
Chapter IX (p. 128)
The Science Press. New York, New York, USA. 1913

Rabi, Isidor Isaac   1898–1988
Austrian-born American physicist

I think physicists are the Peter Pans of the human race. They never grow up, and they keep their curiosity.
In Jeremy Bernstein
Experiencing Science
Part 1. Two Faces of Physics. Chapter 2. Rabi: The Modern Age (p. 102)
Basic Books, Inc. New York, New York, USA. 1978

Rees, Martin John   1942–
15th Astronomer Royal of England

The physicist is like someone who's watching people playing chess and, after watching a few games, he may have worked out what the moves in the game are. But understanding the rules is just a trivial preliminary on the long route from being a novice to being a grand master. So even if we understand all the laws of physics, then exploring their consequences in the everyday world where complex structures can exist is a far more daunting task, and that's an inexhaustible one I'm sure.
In Lewis Wolpert and Alison Richards
A Passion for Science
Chapter 3 (p. 37)
Oxford University Press, Inc. Oxford, England. 1988

Robinson, Howard A.
No biographical data available

Because physicists are a small group, they often suffer in many ways from psychoses similar to those found in political minorities. In an effort to keep their own individuality they feel it necessary to resist pressure from the outside and the result is…that a group of physicists tend to behave like an amoebae.
The Challenge of Industrial Physics
Physics Today, June 1948 (p. 7)

Rogers, Eric
No biographical data available

The physicist who does not enjoy watching a dime and a quarter drop together has no heart.
Astronomy for the Inquiring Mind
Preliminary Introduction (p. 4)
Princeton University Press. Princeton, New Jersey, USA.1982

Rorty, Richard   1931–
American philosopher

Here is one way to look at physics: the physicists are men looking for new interpretations of the Book of Nature. After each pedestrian period of normal science, they dream up a new model, a new picture, a new vocabulary, and then they announce that the true meaning of the Book has been discovered. But, of course, it never is, any more than the true meaning of Coriolanus or the Dunciad or the Phenomenology of the Spirit or the Philosophical Investigations. What makes them physicists is that their writings are commentaries on the writings of earlier interpreters of Nature, not that they all are somehow "talking about the same thing"…
Philosophy as a Kind of Writing
New Literary History, Volume 10, Number 1, Autumn 1978 (p. 141)

Ruelle, David   1935–
Belgian-French mathematical physicist

What is the origin of the urge, the fascination that drives physicists, mathematicians, and presumably other scientists as well? Psychoanalysis suggests that it is sexual curiosity. You start by asking where little babies come from, one thing leads to another, and you find yourself preparing nitroglycerine or solving differential equations. This explanation is somewhat irritating, and therefore probably basically correct.
Chance and Chaos
Chapter 26 (p. 164)
Princeton University Press. Princeton, New Jersey, USA. 1991

Russell, Henry Norris   1877–1957
American astronomer

If a first-rate physicist, well versed in all the knowledge acquired in the laboratory during the last quarter century on the structure and properties of the atom, should have lived his life on a planet so enshrouded by clouds that neither he nor others had ever glimpsed the starry heavens, yet if he had the imagination to conceive that immense quantities of matter might lie beyond the clouds, he would be able to picture the heavens much as they are, tell the probable maximum masses of the stars, their minimum distances, the range of their diameters and temperatures, the differences of their spectra, and in short to duplicate by prediction, not only in general features but in many of the finest details the actual appearance of the universe forever hidden from him.
Quoted in C.G. Abbot
Annual Report of the Board of Regents of the Smithsonian Institution, 1922
The Architecture of Atoms and a Universe Built of Atoms (p. 157)
Government Printing Office. Washington, D.C. 1924

Sagan, Carl   1934–96
American astronomer and author

Physicists had to invent words and phrases for concepts far removed from everyday experience. It was their fashion to avoid pure neologisms and instead to evoke, even if feebly, some analogous commonplace. The alternative was to name discoveries and equations after one another. This they did also. But if you didn't know it was physics they were talking, you might very well worry about them.
Contact: A Novel
Chapter 19 (p. 331)
Simon & Schuster. New York, New York, USA. 1985

Singer, Kurt   1886–1962
German philosopher

…the true mathematician and physicist know very well that the realms of the small and the great often obey quite different rules.
Mirror, Sword and Jewel: A Study of Japanese Characteristics
Chapter 5 (p. 75)
Croom Helm. London, England. 1973

Standen, Anthony
Anglo-American science writer

Physicists, being in no way different from the rest of the population, have short memories for what is inconvenient.
Science Is a Sacred Cow
Chapter III (p. 68)
Dutton. New York, New York, USA. 1950

Strutt, John William (Lord Rayleigh)   1842–1919
English physicist

The different habits of mind of the two schools of physicists sometimes lead them to the adoption of antagonistic views on doubtful and difficult questions. The tendency of the purely experimental school is to rely almost exclusively upon direct evidence, even when it is obviously imperfect, and to disregard arguments which they stigmatize as theoretical. The tendency of the mathematician is to over-rate the solidity of his theoretical structures, and to forget the narrowness of the experimental foundation upon which many of them rest.
Life of John William Strutt: Third Baron Rayleigh (p. 132)
University of Wisconsin Press. Madison, Wisconsin, USA. 1968

Thomson, Sir Joseph John   1856–1940
English physicist

There is a school of mathematical physicists which objects to the introduction of ideas which do not relate to things which can actually be observed and measured.… I hold that if the introduction of a quantity promotes clearness of thought, then even if at the moment we have no means of determining it with precision, its introduction is not only legitimate but desirable. The immeasurable of today may be the measurable of tomorrow.
In John D. Barrow
The World Within the World (p. 97)
Clarendon Press. Oxford, England. 1988

Toulmin, Stephen   1922–
Anglo-American philosopher

Natural historians…look for regularities of given forms, but physicists seek the form of given regularities.
The Philosophy of Science: An Introduction
Chapter II, Section 2.8 (p. 53)
Harper & Row, Publishers. New York, New York, USA. 1960

von Goethe, Johann Wolfgang   1749–1832
German poet, novelist, playwright, and natural philosopher

…cement, patch-up, and glue together, as witchdoctors do, the Newtonian doctrine, so that it could, as an embalmed corpse, preside in the style of ancient Egyptians, at the drinking bouts of physicists.
In S.L. Jaki
Goethe and the Physicists
American Journal of Physics, Volume 37 (p. 198)

Wald, George   1906–97
American biologist and biochemist

It would be a poor thing to be an atom in a universe without physicists, and physicists are made of atoms. A physicist is an atom's way of knowing about atoms.
The Fitness of the Environment: An Inquiry into the Biological Significance of the Properties of Matter
Foreword
Beacon Press. Boston, Massachusetts, USA. 1958

Weisskopf, Victor Frederick   1908–2002
Austrian-American physicist

There are three kinds of physicists, as we know, namely the machine builders, the experimental physicists, and the theoretical physicists. If we compare those three classes, we find that the machine builders are the most important ones, because if they were not there, we could not get to this small-scale region. If we compare this with the discovery of America, then, I would say, the machine builders correspond to the captains and ship builders who really developed the techniques at that time. The experimentalists were those fellows on the ships that sailed to the other side of the world and then jumped upon the new islands and just wrote down what they saw. The theoretical physicists are those fellows who stayed back in Madrid and told Columbus that he was going to land in India.
In Heinz R. Pagels
The Cosmic Code: Quantum Physics as the Language of Nature
Part II, Chapter 1 (p. 198)
Simon & Schuster. New York, New York, USA. 1982

Self-confidence is an important ingredient that makes for a successful physicist.
In L.M. Brown and L. Hoddeson
The Birth of Particle Physics
Growing Up with Field Theory: The Development of Quantum Electrodynamics (p. 75)
Cambridge University Press. Cambridge, England. 1983

Wheeler, John Archibald   1911–
American physicist and educator

The physicist does not have the habit of giving up something unless he gets something better in return.
In Cecil M. DeWitt and John A. Wheeler
Battelle Recontres: 1967 Lectures in Mathematics and Physics (p. 261)
W.A. Benjamin, Inc. New York, New York, USA. 1968

White, Stephen
No biographical data available

[Physicists] are, as a general rule, highbrows. They think and talk in long, Latin words, and when they write anything down they usually include at least one partial differential and three Greek letters.
A Newsman Looks at Physicists
Physics Today, Volume 1, Number 1, May 1948 (p. 15)

Wiener, Norbert   1894–1964
American mathematician

Experience has pretty well convinced the working physicist that any idea of nature which is not only difficult to interpret but which actively resists interpretation has not been justified as far as his past work is concerned, and therefore, to be an effective scientist, he must be naive, and even deliberately naive, in making the assumption that he is dealing with an honest God, and must ask his questions of the world as an honest man.
The Human Use of Human Beings
Chapter XI (p. 189)
Da Capo Press. New York, New York, USA. 1988

Zolynas, Al   1945–
American poet

And so, the closer he looks at things, the farther away they seem. At dinner, after a hard day at the universe, he finds himself slipping through his food. His own hands wave at him from beyond a mountain of peas. Stars and planets dance with molecules on his fingertips. After a hard day with the universe, he tumbles through himself, flies through the dream galaxies of his own heart. In the very presence of his family he feels he is descending through an infinite series of Chinese boxes.
The New Physics: Poems
The New Physics (p. 55)
Wesleyan University Press. Middletown, Connecticut, USA. 1979

PHYSICS

Achard, Franz Karl   1753–1821
German chemist and experimental physicist

Everyone now agrees that a physics lacking all connection with mathematics…would only be an historical amusement, fitter for entertaining the idle than for occupying the mind of a philosopher.
In J.L. Heilbron
Electricity in the 17th and 18th Centuries: A Study of Early Modern Physics (p. 74)
University of California Press. Berkeley, California, USA. 1979

Alvarez, Luis Walter   1911–88
American experimental physicist

There is no democracy in physics. We can't say that some second rate guy has as much right to opinion as Fermi.
In Daniel S. Greenberg
The Politics of Pure Science
Book One, Chapter II (p. 42)
New American Library. New York, New York, USA. 1967

Author undetermined

The Euclidean foundation of geometry is to the Gaussian foundation of geometry as the Newton particle concept of physics is to the Faraday–Maxwell concept of physics.
In Howard W. Eves
Mathematical Circles (Volume 2)
Mathematical Circles Squared
73 (p. 56)
The Mathematical Association of America, Inc. 2003

If you think, you experience time.
If you feel, you experience energy.
If you intuit, you experience wavelength.
If you sense, you experience space.
In Fred Alan Wolf
Star Wave: Mind Consciousness of Quantum Physics (p. 16)
Macmillan Publishing Company. New York, New York, USA. 1984

Bacon, Sir Francis   1561–1626
English lawyer, statesman, and essayist

We have no sound notions either in logic or physics; substance, quality, action, passion, and existence are not clear notions; much less weight, levity, density, tenuity, moisture, dryness, generation, corruption, attraction, repulsion, element, matter, form, and the like. They are all fantastical and ill-defined.
In *Great Books of the Western World* (Volume 30)
Novum Organum
First Book, Aphorism 15 (p. 108)
Encyclopædia Britannica, Inc. Chicago, Illinois, USA. 1952

Physic…is situate in a middle term or distance between natural history and metaphysic. For natural history describeth the variety of things; physic the causes, but variable or respective causes; and metaphysic the fixed and constant causes.
In *Great Books of the Western World* (Volume 30)
Advancement of Learning
Second Book, Chapter VII, Section 4 (p. 43)
Encyclopædia Britannica, Inc. Chicago, Illinois, USA. 1952

Ball, Walter William Rouse   1850–1925
English mathematician

The advance in our knowledge of physics is largely due to the application to it of mathematics, and every year it becomes more difficult for an experimenter to make any mark in the subject unless he is also a mathematician.
A Short Account of the History of Mathematics (p. 503)
Macmillan & Company Ltd. London, England. 1908

Bell, E. T. (Eric Temple)   1883–1960
Scottish-American mathematician and educator

Daniel Bernoulli has been called the father of mathematical physics.
In James R. Newman (ed.)
The World of Mathematics (Volume 2)
Kinetic Theory of Gases (p. 774)
Simon & Schuster. New York, New York, USA. 1956

Bergson, Henri   1859–1941
French philosopher

…physics is but logic spoiled.
Translated by Arthur Mitchell
Creative Evolution
Chapter IV (p. 320)
The Modern Library. New York, New York, USA. 1944

Berry, M. V.
No biographical data available

…in one of those unexpected connections that make theoretical physics so delightful, the quantum chorology of spectra turns out to be deeply connected to the arithmetic of prime numbers, through the celebrated zeros of the Riemann zeta function: the zeros mimic quantum energy levels of a classically chaotic system. The connection is not only deep but also tantalizing, since its basis is still obscure — though it has been fruitful for both mathematics and physics.
In R.J. Russell, P. Clayton, K. Wegter-McNelly and J. Polkinghorne (eds.)
Quantum Mechanics: Scientific Perspectives on Divine Action
Chaos and the Semiclassical Limit of Quantum Mechanics (Is the Moon There When Somebody Looks?)
University of Notre Dame Press. Notre Dame, Indiana, USA. 2002

Birkhoff, George David   1884–1944
American mathematician

It will probably be the new mathematical discoveries suggested through physics that will always be the most important, for from the beginning Nature has led the way and established the pattern which mathematics, the language of Nature, must follow.
The Mathematical Nature of Physical Theories
American Scientist, Volume 31, Number 4, October 1943 (p. 310)

Blackett, Lord Patrick Maynard Stuart   1897–1974
English physicist

Thus was born the vast modern subject of nuclear physics, which now gives such fertile research problems to so many of the world's physicists and, incidentally, such headaches to so many of the world's statesmen.
In J.B. Birks
Rutherford at Manchester
Memories of Rutherford (p. 104)
W.A. Benjamin. New York, New York, USA. 1962

Bohr, Niels Henrik David   1886–1962
Danish physicist

My starting point was rather the stability of matter, a pure miracle when considered from the standpoint of classical physics. By "stability" I mean that the same substances always have the same properties…
In Werner Heisenberg
Physics and Beyond: Encounters and Conversations
Chapter 3 (p. 39)
Harper & Row, Publishers. New York, New York, USA. 1971

It is wrong to think that the task of physics is to find out how nature is. Physics concerns what we can say about nature.
In N. Herbert
Quantum Reality: Beyond the New Physics
Chapter 3 (p. 45)
Anchor Press/Doubleday. Garden City, New York, USA. 1985

In physics…our problem consists in the co-ordination of our experience of the external world…
Atomic Theory and the Description of Nature
Introductory Survey (p. 1)
Cambridge University Press. Cambridge, England. 1934

…the new situation in physics is that we are both onlookers and actors in the great drama of existence.
Atomic Theory and the Description of Nature
Chapter IV (p. 119)
Cambridge University Press. Cambridge, England. 1934

Born, Max   1882–1970
German-born English physicist

The problem of physics is how the actual phenomena, as observed with the help of our sense organs aided by instruments, can be reduced to simple notions which are suited for precise measurement and used of the formulation of quantitative laws.
Experiment and Theory in Physics (pp. 8–9)
Cambridge University Press. Cambridge, England. 1944

Hope is a word one is unlikely to find in the literature of physics.
My Life and My Views
Chapter Six (p. 190)
Charles Scribner's Sons. New York, New York, USA. 1968

It is natural that a man should consider the work of his hands or his brain to be useful and important. Therefore nobody will object to an ardent experimentalist boasting of his measurements and rather looking down on the "paper and ink" physics of his theoretical friend, who on his part is proud of his lofty ideas and despises the dirty fingers of the other.
Experiment and Theory in Physics (p. 1)
Cambridge University Press. Cambridge, England. 1944

Boyle, Robert   1627–91
English natural philosopher and theological writer

I confess, that after I began…to discern how useful mathematicks may be made to physicks, I have often wished that I had employed the speculative part of geometry, and the cultivation of the specious Algebra I had been taught very young, a good part of that time and industry, that I had spent about surveying and fortification (of which I remember I once wrote an entire treatise) and other parts of practick mathematicks.

The Work of the Honourable Robert Boyles (Volume 4)
The Usefulness of Mathematicks to Natural Philosophy, Volume 3 (p. 426)
Printed for A. Millar. Ondon, England. 1744

Bragg, Sir William Henry   1862–1942
English physicist

On Mondays, Wednesdays, and Fridays we teach the wave theory and on Tuesdays, Thursdays, and Saturdays the corpuscular theory.

Electrons and Ether Waves, 23rd Robert Boyle Lecture
Scientific Monthly, Volume 4, Issue 2, 1922 (p. 11)

Brennan, Richard P.
Science writer

Physics can be expected to continue because it is, by its nature, open-ended and exploratory and because, at its heart, science is simply people asking questions.

Heisenberg Probably Slept Here
Epilogue (p. 249)
John Wiley & Sons, Inc. New York, New York, USA. 1997

Bronowski, Jacob   1908–74
Polish-born English mathematician and polymath

Physics becomes in those years the greatest collective work of art of the twentieth century.

The Ascent of Man
Chapter 10 (p. 328)
Little, Brown & Company. Boston, Massachusetts, USA. 1973

One aim of the physical sciences has been to give an exact picture of the material world. One achievement of physics in the twentieth century has been to prove that that aim is unattainable.

The Ascent of Man
Chapter 11 (p. 353)
Little, Brown & Company. Boston, Massachusetts, USA. 1973

Burbridge, Geoffrey
American astronomer

We live in an era when it seems legitimate to try everything conceivable within the known laws of physics, particularly in the absence of data.

Focal Point
Sky and Telescope, Volume 78, Number 6, June 1990 (p. 580)

Carnap, Rudolf   1891–1970
American philosopher

Physics originally began as a descriptive macrophysics, containing an enormous number of empirical laws with no apparent connections. In the beginning of a science, scientists may be very proud to have discovered hundreds of laws. But, as the laws proliferate, they become unhappy with this state of affairs; they begin to search for underlying principles.

An Introduction to the Philosophy of Science. (p. 244)
Clarendon Press. Oxford, England. 1988

…the facts and objects of the various branches of Science are fundamentally the same kind. For all branches are part of the unified Science, of Physics.

The Unity of Science
Unified Science in Physical Language, Section 7 (p. 101)
Thommes Press. Bristol, England. 1995

Cartwright, Nancy   1943–
Philosopher of physics

…the fundamental laws of physics do not describe true facts about reality. Rendered as descriptions of facts, they are false; amended to be true, they lose their explanatory force.

How the Laws of Physics Lie
Essay 3 (p. 54)
Clarendon Press. Oxford, England. 1983

Although philosophers generally believe in laws and deny causes, explanatory practice in physics is just the reverse.

How the Laws of Physics Lie
Essay 4 (p. 86)
Clarendon Press. Oxford, England. 1983

CERN Courier

The main goal of physics is to describe a maximum of phenomena with a minimum of variables.

In John N Shive and Robert L. Weber
In *Similarities in Physics*
Chapter 16 (p. 213)
John Wiley & Sons, Inc. New York, New York, USA. 1982

Compton, Karl Taylor   1887–1954
American educator and physicist

In the last fifty years physics has exerted a more powerful beneficial influence on the intellectual, economic and social life of the world than has been exerted in a comparable time by any other agency in history. Its influence has far exceeded that of wars, political alignment or social theories.

Science (supplement), Volume 84, Number 10, 1936

Comte, Auguste   1798–1857
French philosopher

The domain of physics is no proper field for mathematical pastimes. The best security would be in giving a geometrical training to physicists, who need not then

have recourse to mathematicians, whose tendency is to despise experimental science.
The Positive Philosophy of Auguste Comte
Book III, Chapter I (p. 220)
John Chapman. London, England. 1853

...the education of physicists must be more complicated than that of astronomers.
The Positive Philosophy of Auguste Comte
Book III, Chapter I (p. 222)
John Chapman. London, England. 1853

Condon, Edward Uhler   1902–74
American physicist

I take it to be the object of physics so to organize past experience and so to direct the acquisition of new experience that ultimately it will be possible to predict the outcome of any proposed experiment which is capable of being carried out — and to make the prediction in less time than it would have taken actually to carry out the proposed experiment. When this shall have been done I will say that man has a complete understanding of his physical environment. Others may ask more; with this I am satisfied.
The Philosophical Concepts of Modern Physics, Mathematical Models in Modern Physics
Journal of the Franklin Institute, Volume 225, Number 3, March 1938 (p. 257)

"All is fair in love and war" and, I might add, in theoretical physics.
Selected Popular Writings of E.U. Condon
Mathematical Models in Modern Physics (p. 96)
Springer-Verlag. New York, New York, USA. 1991

Crease, Robert P.
Science historian
Mann, Charles C.
American journalist and science writer

On August 2, 1932, Anderson obtained a stunningly clear photograph that shocked both men. Despite Millikan's protestations, a particle had indeed shot up like a Roman candle from the floor of the chamber, slipped through the plate, and fallen off to the left. From the size of the track, the degree of the curvature, and the amount of momentum lost, the particle's mass was obviously near to that of an electron. But the track curved the wrong way. The particle was positive. Neither electron, proton, or neutron, the track came from something that that had never been discovered before. It was, in fact, a "hole," although Anderson did not realize it for a while.... Anderson called the new particle a "positive electron"; positron was the name that stuck. Positrons were the new type of matter — antimatter — Dirac had been forced to predict by his theory. (The equation, he said later, had been smarter than he was.)

In T. Ferris (ed.)
World Treasury of Physics, Astronomy, and Mathematics
Uncertainty and Complementarity (p. 78)
Little, Brown & Company. Boston, Massachusetts, USA. 1991

Cropper, William N.
No biographical data available

Physics builds from observations. No physical theory can succeed if it is not confirmed by observations, and a theory strongly supported by observations cannot be denied.
Great Physicists
I, Mechanics (p. 3)
Oxford University Press, Inc. New York, New York, USA. 2001

Darrow, Karl Kelchner   1891–1982
American physicist

...it does not take an idea so long to become "classical" in physics as it does in the arts.
Bell System Technical Journal
Some Contemporary Advances in Physics V, Electrical Solids, Volume 3, 1924 (p. 621)

Davies, Paul Charles William   1946–
British-born physicist, writer, and broadcaster

Physics is the most pretentious of all the sciences, for it purports to address all of physical reality. The physicist may confess ignorance about a particular system — a snowflake, a living organism, a weather pattern — but he will never concede that it lies outside the domain of physics in principle. The physicist believes that the laws of physics, plus knowledge of the relevant boundary conditions, are sufficient to explain, in principle, every phenomenon in the universe. Thus the entire universe, from the smallest fragment of matter to the largest assemblage of galaxies, becomes the physicist's domain — vast natural laboratory for the interplay of lawful forces.
In P.C.W. Davies (ed.)
The New Physics
The New Physics: A Synthesis (p. 1)
Cambridge University Press. Cambridge, England. 1989

It is clear that for nature to produce a cosmos even remotely resembling our own, many apparently unconnected branches of physics have to cooperate to a remarkable degree.
The Accidental Universe (p. 111)
Cambridge University Press. Cambridge, England. 1984

It is no exaggeration to say that quantum mechanics had dominated twentieth-century physics and is far and away the most successful scientific theory in existence. It is indispensable for understanding subatomic particles, atoms and nuclei, molecules and chemical bonding, the structure of solids, superconductors and superfluids, the electrical and thermal conductivity of metals and semiconductors, the structure of stars, and much else. It has practical applications ranging from the laser to the microchip. All this from a theory that at first sight — and second sight — looks absolutely crazy! Neils Bohr, one of the founders of quantum mechanics, once remarked that anybody who is not shocked by the theory hasn't understood it.

In Richard P. Feynman
Six Easy Pieces: Essentials of Physics Explained by Its Most Brilliant Teacher
Introduction (p. xv, xvi)
Addison-Wesley Publishing Company. Reading, Massachusetts, USA. 1995

Davies, Paul Charles William   1946–
British-born physicist, writer, and broadcaster
Brown, Julian R.
No biographical data available

No science is more pretentious than physics, for the physicist lays claim to the whole universe as his subject matter.
Superstrings: A Theory of Everything
Introduction (p. 1)
Cambridge University Press. Cambridge, England. 1988

de Morgan, Augustus   1806–71
English mathematician and logician

Among the mere talkers, so far as mathematics are concerned, are to be ranked three out of four of those who apply mathematics to physics, who, wanting a tool only, are very impatient of everything which is not of direct aid to the actual methods which are in their hands.
In Robert Graves
Life of Sir William Rowan Hamilton (Volume 3) (p. 348)
Hodges, Figgis & Company. Dublin, Ireland. 1882–89

Descartes, René   1596–1650
French philosopher, scientist, and mathematician

I accept no principles of physics which are not also accepted in mathematics…
Principles of Philosophy
Part III, 4
Reidel. Dordrecht, Netherlands. 1983

I should consider that I know nothing about physics if I were able to explain only how things might be, and were unable to demonstrate that they could not be otherwise. For, having reduced physics to mathematics, the demonstration is now possible, and I think that I can do it within the small compass of my knowledge.
In A.C. Crombie
Descartes
Scientific American, Volume 201, Number 4, October 1959 (p. 160)

Deutsch, David   1953
Physicist

Anything that seems incomprehensible is regarded by science merely as evidence that there is something we have not yet understood, be it a conjuring trick, advanced technology or a new law of physics.
The Fabric of Reality
Chapter 6 (p. 138)
Penguin Books Ltd. London, England. 1998

Dilorenzo, Kirk
No biographical data available
Physics is the interrelationship of everything.
The Physics Teacher, Volume 14, Number 5, May 1976 (p. 315)

Dirac, Paul Adrian Maurice   1902–84
English theoretical physicist

Only questions about the results of experiments have a real significance and it is only such questions that theoretical physics has to consider.
The Principles of Quantum Mechanics (2nd edition)
Chapter I, Section 2 (p. 5)
At The Clarendon Press. Oxford, England. 1935

The present stage of physical theory is merely a steppingstone towards the better stages that we will have in the future. One can be quite sure that there will be better stages simply because of the difficulties that occur in the physics of today.
The Evolution of the Physicist's Picture of Nature
Scientific American, Volume 208, Number 5, May 1963 (p. 48)

Duhem, Pierre-Maurice-Marie   1861–1916
French physicist and mathematician

The development of physics incites a continual struggle between "nature that does not tire of providing" and reason that does not wish "to tire of conceiving."
The Aim and Structure of Physical Theory
Part I, Chapter II (p. 23)
Princeton University Press. Princeton, New Jersey, USA. 1954

Physics is not a machine one can take apart; one cannot try each piece in isolation and wait, to adjust it, until its solidity has been minutely checked. Physical science is a system that must be taken as a whole. It is an organism no part of which can be made to function without the remotest parts coming into play, some more, some less, but all in some degree.
Essays in the History and Philosophy of Science (p. 284)
Hackett Publishing Company. Indianapolis, Indiana, USA. 1996

…physics makes progress because experiment constantly causes new disagreements to break out between laws and facts, and because physicists constantly touch up and modify laws in order that they may more faithfully represent the facts.
The Aim and Structure of Physical Theory
Part II, Chapter V (p. 177)
Princeton University Press. Princeton, New Jersey, USA. 1954

A "Crucial Experiment" is Impossible in Physics.

The Aim and Structure of Physical Theory
Part II, Chapter VI (p. 188)
Princeton University Press. Princeton, New Jersey, USA. 1954

Dyson, Freeman J. 1923–
American physicist and educator

Physics is littered with the corpses of dead unified field theories.
In John D. Barrow
The World Within the World (p. 184)
Clarendon Press. Oxford, England. 1988

I am acutely aware of the fact that the marriage between mathematics and physics, which was so enormously fruitful in past centuries, has recently ended in divorce.
Missed Opportunities
Bulletin of the American Mathematical Society, Volume 78, 1972

…we have seen particle physics emerge as the playground of group theory.
In Joseph A. Gallian
Contemporary Abstract Algebra
Chapter 3 (p. 55)
D.C. Heath and Company. Lexington, Massachusetts, USA. 1994

Eddington, Sir Arthur Stanley 1882–1944
English astronomer, physicist, and mathematician

The external world of physics has become a world of shadows.
The Nature of the Physical World
Introduction (p. xvi)
The Macmillan Company. New York, New York, USA. 1930

Distance and duration are the most fundamental terms in physics; velocity, acceleration, force, energy, and so on, all depend on them; and we can scarcely make any statement in physics without direct or indirect reference to them.
In Ronald W. Clark
Einstein: The Life and Times
Part Two, Chapter 4 (p. 93)
The World Publishing Company. New York, New York, USA. 1971

I have not suggested that religion and free will can be deduced from modern physics…
New Pathways in Science
Chapter XIII, Section VI (p. 306)
The Macmillan Company. New York, New York, USA. 1935

I am afraid the knockabout comedy of modern atomic physics is not very tender towards our aesthetic ideals. The stately drama of stellar evolution turns out to be more like the hair-breadth escapades in the films. The music of the spheres has a painful suggestion of — jazz.
Stars and Atoms
Lecture I (p. 27)
Yale University Press. London, England. 1927

In the world of physics we watch a shadowgraph performance of familiar life. The shadow of my elbow rests on the shadow table as the shadow ink flows over the shadow paper…. The frank realisation that physical science is concerned with a world of shadows is one of the most significant of recent advances.
The Nature of the Physical World
Introduction (p. xi)
The Macmillan Company. New York, New York, USA. 1930

It is impossible to trap modern physics into predicting anything with perfect determinism because it deals with probabilities from the outset.
In James R. Newman (ed.)
The World of Mathematics (Volume 2)
Causality and Wave Mechanics (p. 1056)
Simon & Schuster. New York, New York, USA. 1956

Edelstein, Ludwig 1902–65
German scholar and historian of medicine

Physics…in antiquity remained closely connected with philosophy, and was predominantly concerned with the philosophical category of the "why," rather than the scientific category of the "how."
In Philip P. Wiener and Aaron Noland
Roots of Scientific Thought
Recent Trends in the Interpretation of Ancient Science (pp. 94–95)
Basic Books, Inc. New York, New York, USA. 1957

Ehrenfest, Paul 1880–1933
Austrian physicist

Physics is simple, but subtle.
In Victor F. Weisskopf
Physics in the Twentieth Century: Selected Essays
My Life as a Physicist (p. 3)
The MIT Press. Cambridge, Massachusetts, USA. 1972

Einstein, Albert 1879–1955
German-born physicist

Today we know that no approach which is founded on classical mechanics and electrodynamics can yield a useful radiation formula.
In B.L. van der Waerden
Sources of Quantum Mechanics
On the Quantum Theory of Radiation (p. 63)
Dover Publications. New York, New York, USA. 1968

What would physics look like without gravitation?
In Jean-Pierre Luminet
Black Holes (p. 114)
Cambridge University Press. New York, New York, USA. 1992

Physics constitutes a logical system of thought which is in a state of evolution, whose basis cannot be distilled, as it were, from experience by an inductive method, but can only be arrived at by free invention.
Out of My Later Years
Physics and Reality, Summary
Thames & Hudson. London, England. 1950

Physics is the attempt at the conceptual construction of a model of the real world and its lawful structure.
In Gerald Holton

Thematic Origins of Scientific Thought: Kepler to Einstein
Letter of November 28, 1930 to M. Schlick (p. 243)
Harvard University Press. Cambridge, Massachusetts, USA. 1973

Physics too deals with mathematical concepts; however, these concepts attain physical content only by the clear determination of their relation to the objects of experience.
Out of My Later Years
The Theory of Relativity (p. 41)
Thames & Hudson. London, England. 1950

That this insecure and contradictory foundation was sufficient to enable a man of Bohr's unique instinct and sensitivity to discover the principal laws of the spectral lines and of the electron shell of the atoms, together with their significance for chemistry appeared to me as a miracle — and appears to me a miracle even today.
Translated by Paul Arthur Schlipp
Albert Einstein: Autobiographical Notes (p. 43)
Open Court. La Salle, Illinois, USA. 1979

Reality is the real business of physics.
In Nick Herbert
Quantum Reality: Beyond the New Physics
Chapter 1 (p. 4)
Anchor Press. Garden City, New York, USA. 1985

Experience, of course, remains the sole criterion for the serviceability of mathematical constructions for physics, but the truly creative principle resides in mathematics.
In Philipp Frank
Modern Science and Its Philosophy
Chapter 16 (p. 297)
Harvard University Press, Cambridge, England. 1952

But in physics I soon learned to scent out the paths that led to the depths, and to disregard everything else, all the many things that clutter up the mind, and divert it from the essential. The hitch in this was, of course, the fact that one had to cram all this stuff into one's mind for the examination, whether one liked it or not.
In Robert H. March
Physics for Poets
Chapter 9 (p. 101)
McGraw-Hill Book Company, Inc. New York, New York, USA. 1996

I still lose my temper dutifully about physics. But I no longer flap my wings — I only ruffle my feathers. The majority of fools remain invincible.
In Sachi Sri Kantha
An Einstein Dictionary (p. 96)
Greenwood Press, Publishers. Westport, Connecticut, USA. 1996

I have become an evil renegade who does not wish physics to be used on probabilities.
In Sachi Sri Kantha
An Einstein Dictionary (p. 96)
Greenwood Press, Publishers. Westport, Connecticut, USA. 1996

In speaking here of "comprehensibility," the expression is used in its most modest sense. It implies: the production of some sort of order among sense impressions, this order being produced by the creation of general concepts, relations between these concepts and sense experience. It is in this sense that the world of our sense experiences is comprehensible. The fact that it is comprehensible is a miracle.
Out of My Later Years
Physics and Reality, Section 1
Thames & Hudson. London, England. 1950

In the matter of physics, the first lesson should contain nothing but what is experimental and interesting to see. A pretty experiment is in itself often more valuable than twenty formulae extracted from our minds; it is particularly important that a young mind that has yet to find its way about in the world of phenomena should be spared from formulae altogether. In [this mind] physics they play exactly the same weird and fearful part as the figures of dates in Universal History.
In A.P. French
Einstein: A Centenary Volume
Chapter 11 (p. 220)
Harvard University Press. Cambridge, Massachusetts, USA. 1979

If the basis of theoretical physics cannot be an inference from experience, but must be free invention, have we any right to hope that we shall find the correct way? Still more — does this correct approach exist at all, save in our imagination? To this I answer with complete assurance, that in my opinion there is the correct path; moreover, that it is in our power to find it.
In Philipp Frank
Einstein's Philosophy of Science
Review of Modern Physics, Volume 21, Number 3, July 1949 (p. 354)

...the development of physics has shown that at any given moment, out of all conceivable constructions, a single one has always proved itself decidedly superior to all the rest. Nobody who has really gone deeply into the matter will deny that in practice the world of phenomena uniquely determines the theoretical system, in spite of the fact that there is no logical bridge between phenomena and their theoretical principles; this is what Leibnitz described so happily as a "pre-established harmony."
Ideas and Opinions
Principles of Research (p. 224)
Crown Publishers, Inc. New York, New York, USA. 1954

Emerson, Ralph Waldo   1803–82
American lecturer, poet, and essayist

The axioms of physics translate the laws of ethics.
Ralph Waldo Emerson: Essays and Lectures
Nature: Addresses, and Lectures
Language (p. 24)
The Library of America. New York, New York, USA. 1983

On the platform of physics we cannot resist the contracting influences of so-called science.
The Complete Works of Ralph Waldo Emerson (Volume 33)

Essays: Second Series
Chapter II (p. 52)
Houghton Mifflin Company. Boston, Massachusetts, USA. 1904

How calmly and genially the mind apprehends one after another the laws of Physics!
Ralph Waldo Emerson: Essays and Lectures
Nature: Addresses, and Lectures
Discipline (p. 27)
The Library of America. New York, New York, USA. 1983

Faust
Fictional character

I have — alas — learned Valence Chemistry,
Theory of Groups, of the Electric Field,
And Transformation Theory as revealed
By Sophus Lie in eighteen-ninety-three.
Yet here I stand, for all my lore,
No wiser than I was before.
BLEGDAMSVEJ FAUST
Part First, Copenhagen Spring Conference, 1932

Ferguson, Arthlyn
No biographical data available

Bouncing a ball, flying a kite, blowing up a balloon — to a child it's play; to a scientist it's physics.
What's Physics?
The Physics Teacher, Volume 14, Number 5, May 1976 (p. 315)

Feynman, Richard P.   1918–88
American theoretical physicist

What do we mean by "understanding something?" We can imagine that this complicated array of moving things which constitutes "the world" is something like a great chess game being played by the gods, and we are observers of the game. We do not know what the rules of the game are: all we are allowed to do is to watch the playing. Of course, if we watch long enough we may eventually catch on to a few of the rules. The rules of the game are what we mean by fundamental physics. Even if we knew every rule, however, we might not be able to understand why a particular move is made in the game, merely because it is too complicated…
In P.C.W. Davies and J. Brown
The Feynman Lectures on Physics (Volume 1)
Chapter 2–1 (p. 2–1)
Addison-Wesley Publishing Company. Reading, Massachusetts, USA. 1983

The fact that I beat a drum has nothing to do with the fact that I do theoretical physics. Theoretical physics is a human endeavor, one of the higher developments of human beings — and this perpetual desire to prove that people who do it are human by showing that they do other things that a few other humans do (like playing bongo drums) is insulting to me. I'm human enough to tell you to go to hell.
In James Gleick

Genius: The Life and Science of Richard Feynman
Caltec (p. 364)
Pantheon Books. New York, New York, USA. 1992

The electron does anything it likes. It goes in any direction at any speed, forward or backward in time, however it likes…
In James Gleick
Genius: The Life and Science of Richard Feynman
Cornell (p. 250)
Pantheon Books. New York, New York, USA. 1992

Physics is to mathematics what sex is to masturbation.
In Lawrence M. Krauss
Fear of Physics: A Guide for the Perplexed
Chapter 2 (p. 27)
Basic Books, Inc. New York, New York, USA. 1993

In order for physics to be useful to other sciences in a theoretical way, other than in the invention of instruments, the science in question must supply to the physicist a description of the object in a physicist's language.
Six Easy Pieces: Essentials of Physics Explained by Its Most Brilliant Teacher
The Relation of Physics to Other Sciences (p. 64)
Addison-Wesley Publishing Company. Reading, Massachusetts, USA. 1995

…the behavior of things on a small scale is so fantastic, so wonderfully and marvelously different than anything on a large scale! You can say, "Electrons behave like waves" — no, they don't, exactly; "they act like particles" — no, they don't exactly; "they act like a kind of fog around the nucleus" — no, they don't, exactly. Well, if you would like to get a clear, sharp picture of an atom, so that you can tell correctly how it's going to behave — have a good image of reality, in other words — I don't know how to do it, because that image has to be mathematical. Strange!

I don't know how it is that we can write mathematical expressions and calculate what the thing is going to do without actually being able to picture it. It would be something like having a computer where you put some numbers in, and the computer can do the arithmetic to figure out what time a car will arrive at different destinations but it cannot picture the car.
In Christopher Sykes (ed.)
No Ordinary Genius: The Illustrated Richard Feynman
Chapter Six (p. 149)
W.W. Norton & Company, Inc. New York, New York, USA. 1994

Feynman, Richard P.   1918–88
American theoretical physicist
Leighton, Robert B.   1919–97
American physicist

In its efforts to learn as much as possible about nature, modern physics has found that certain things can never be "known" with certainty. Much of our knowledge must

always remain uncertain. The most we can know is in terms of probabilities.
The Feynman Lectures on Physics (Volume 1)
Chapter 6–5 (pp. 6–11)
Addison-Wesley Publishing Company. Reading, Massachusetts, USA. 1983

...the existence of the positive charge, in some sense, distorts, or creates a "condition" in space, so that when we put the negative charge in, it feels a force. This potentiality for producing a force is called an electric field.
The Feynman Lectures on Physics (Volume 1)
Chapter 2–2 (p. 2–4)
Addison-Wesley Publishing Company. Reading, Massachusetts, USA. 1983

Franklin, W. S.
No biographical data available

Physics is the science of the ways of taking hold of things and pushing them.
In R.B. Lindsay
The Broad Point of View in Physics
The Scientific Monthly, February 1932 (p. 115)

Fraser, Julius Thomas   1923–
No biographical data available

The task of asking nonliving matter to speak and the responsibility for interpreting its reply is that of physics.
Time: The Familiar Stranger
From the Diaries of a Timesmith (p. 358)
The University of Massachusetts Press. Amherst, Massachusetts, USA. 1987

Gamow, George   1904–68
Russian-born American physicist

I remember that once, walking with him to the institute, I mentioned Pascual Jordan's idea of how a star can be created from nothing, since at the point zero its negative gravitational mass defect is numerically equal to its positive rest mass. Einstein stopped in his tracks, and, since we were crossing a street, several cars had to stop to avoid running us down.
My World Line: An Informal Autobiography
Afterword (p. 150)
The Viking Press. New York, New York, USA. 1979

Gardner, Martin   1914–
American writer and mathematics games editor

In physics and chemistry, like all other branches of science, there is never a sharp line separating pseudo-scientific speculation from the theories of competent men.
Fads and Fallacies in the Name of Science
Chapter 7 (p. 80)
Dover Publications, Inc., New York, New York, USA; 1957

Gay-Lussac, Joseph Louis   1778–1850
French chemist and physicist

In the study of physics, we see what are called individual facts but which are by no means isolated and which are not independent of each other; on the contrary they are related to each other by laws which the physicist devotes all his attention to discovering. It is this which is a measure of the true progress of the science.
In Maurice Crosland
Gay-Lussac: Scientist and Bourgeois
Chapter 3 (p. 70)
Cambridge University Press. Cambridge, England. 1978

Geordi
Fictional character

Suddenly it's like the laws of physics went right out the window.
Star Trek: The Next Generation
True Q
Television program
Season 6, 1992

Goeppert-Mayer, Maria   1906–72
German-American physicist

Mathematics began to seem too much like puzzle solving. Physics is puzzle solving, too, but of puzzles created by nature, not by the mind of man.
In J. Dash
A Life of One's Own
Maria Goeppert-Mayer (p. 252)
Harper & Row, Publishers. New York, New York, USA. 1973

Greene, Brian   1963–
American physicist

To open our eyes to the true nature of the universe has always been one of physics' primary purposes.
The Fabric of the Cosmos
Chapter 1 (p. 12)
Alfred A. Knopf. New York, New York, USA. 2004

The arrow of time, through the defining role it plays in everyday life and its intimate link with the origin of the universe, lies at a singular threshold between the reality we experience and the more refined reality cutting-edge science seeks to uncover.
The Fabric of the Cosmos
Chapter 1 (p. 20)
Alfred A. Knopf. New York, New York, USA. 2004

Physicists generally do not spend their working days contemplating flowers in a state of cosmic awe and reverie. Instead, we devote much of our time to grappling with complex mathematical equations scrawled across well-scored chalkboards. Progress can be slow. Promising ideas, more often than not, lead nowhere. That's the nature of scientific research.
The Fabric of the Cosmos
Chapter 1 (p. 21)
Alfred A. Knopf. New York, New York, USA. 2004

Physicists spend a large part of their lives in a state of confusion. It's an occupational hazard. To excel in physics is to embrace doubt while walking the road to clarity.
The Fabric of the Cosmos
Chapter 16 (p. 470)
Alfred A. Knopf. New York, New York, USA. 2004

Space and time capture the imagination like no other scientific subject. For good reason. They form the arena of reality, the very fabric of the cosmos.
The Fabric of the Cosmos
Preface (p. ix)
Alfred A. Knopf. New York, New York, USA. 2004

It took the brashness of a Newton to plant the flag of modern scientific inquiry and never turn back.
The Fabric of the Cosmos
Chapter 1 (p. 22)
Alfred A. Knopf. New York, New York, USA. 2004

Nature does weird things. It lives on the edge. But it is careful to bob and weave from the fatal punch of logical paradox.
The Fabric of the Cosmos
Chapter 7 (p. 185)
Alfred A. Knopf. New York, New York, USA. 2004

Black holes have the universe's most inscrutable poker faces.
The Fabric of the Cosmos
Chapter 16 (p. 477)
Alfred A. Knopf. New York, New York, USA. 2004

…because observations are all we have, we take them seriously. We choose hard data and the framework of mathematics as our guides, not unrestrained imagination or unrelenting skepticism, and seek the simplest yet most wide-reaching theories capable of explaining and predicting the outcome of today's and future experiments.
The Fabric of the Cosmos
Preface (p. ix)
Alfred A. Knopf. New York, New York, USA. 2004

Gross, David   1941
American particle physicist

Progress in physics depends on the ability to separate the analysis of a physical phenomenon into two parts. First, there are the initial conditions that are arbitrary, complicated, and unpredictable. Then there are the laws of nature that summarize the regularities that are independent of the initial conditions.
Proceedings of the National Academy of Science USA
The Role of Symmetry in Fundamental Physics, Volume 93, Number 25, December 10, 1996

Hanson, Norwood Russell   1924–67
American philosopher of science

Physics is not applied mathematics. It is a natural science in which mathematics can be applied.
Patterns of Discovery

Chapter IV (p. 72)
At The University Press. Cambridge, England. 1958

Hasselberg, K. B.
No biographical data available

…as for physics, it has developed remarkably as a precision science, in such a way that we can justifiably claim that the majority of all the greatest discoveries in physics are very largely based on the high degree of accuracy which can now be obtained in measurements made during the study of physical phenomena…. [Accuracy of measurement] is the very root, the essential condition, of our penetration deeper into the laws of physics — our only way to new discoveries.
Nobel Lectures, Physics 1901–1921
Presentation Speech to Michelson 1907 Nobel Award (p. 159)
Elsevier Publishing Company. Amsterdam, Netherlands. 1967

Heidegger, Martin   1889–1976
German philosopher

Modern physics is not experimental physics because it applies apparatus to the questioning of nature. Rather the reverse is true. Because physics, indeed already as pure theory, sets nature up to exhibit itself as a coherence of forces calculable in advance, it therefore orders its experiments precisely for the purpose of asking whether and how nature reports itself when set up in this way.
The Question Concerning Technology and Other Essays
Part I. The Question Concerning Technology (p. 21)
Harper & Row, Publishers. New York, New York, USA. 1977

Heinlein, Robert A.   1907–88
American science fiction writer

Physics doesn't have to have any use. It just is.
Time for the Stars (p. 138)
Charles Scribner's Sons. New York, New York, USA. 1956

Heisenberg, Werner Karl   1901–76
German physicist and philosopher

When I was a boy, my grandfather, who was a handicraftsman and knew how to do practical things, once met me when I put a cover on a wooden box…. He saw that I took the cover and I took a nail and I tried to hammer this one nail down to the bottom. "Oh", he said, "that is quite wrong what you do there, nobody can do it that way and it is a scandal to look at." I did not know what the scandal was, but then he said, "I will show you how you could do it." He took the cover and he took one nail, put it just a little bit through the cover into the box, and then the next nail a little bit, the third nail a little bit, and so on until all the nails were there. Only when everything was clear, when one could see, that all the nails would fit, then he would start to put the nails really into the box. So, I think this is a good description of how one should proceed in theoretical physics.

In International Centre for Theoretical Physics
From a Life of Physics. Evening Lectures at the International Centre For Theoretical Physics
Theory, Criticism and a Philosophy, My General Philosophy (p. 46)

Questions and answers, observations and determinations, are no longer directed at a general, metaphysical and theological understanding, but are delimited with modesty.... This modesty was largely lost during the nineteenth century. Physical knowledge was considered to make assertions about nature as a whole. Physicists wished to turn philosophers.... Today physics is undergoing a basic change, the most characteristic trait of which is a return to its original self-limitation.
The Physicist's Conception of Nature
Chapter 4 (p. 105)
Greenwood Press, Publishers. Westport, Connecticut, USA. 1958

Like all the other natural sciences, Physics advances by two distinct roads. On the one hand it operates empirically, and thus is enabled to discover and analyse a growing number of phenomena — in this instance, of physical facts; on the other hand it also operates by theory, which allows it to collect and assemble the known facts in one consistent system, and to predict new ones from the guidance of experimental research.
The Physicist's Conception of Nature
Chapter 6 (p. 158)
Greenwood Press, Publishers. Westport, Connecticut, USA. 1958

I remember discussions with Bohr which went through many hours till very late at night and ended almost in despair, and when at the end of the discussion I went alone for a walk in the neighboring park I repeated to myself again and again the question: "Can nature possibly be as absurd as it seemed to us in these atomic experiments?"
Physics and Philosophy: The Revolution in Modern Science
Chapter II (p. 42)
Harper & Row, Publishers. New York, New York, USA. 1958

Heyl, Paul R.
American Scientist

Physics is a state of mind.
In R.B. Lindsay
The Broad Point of View in Physics
The Scientific Monthly, February 1932 (p. 115)

Hilbert, David   1862–1943
German mathematician

Physics...is much too hard for physicists.
In Constance Reid
Hilbert — Courant
Hilbert (p. 127)
Springer-Verlag. New York, New York, USA. 1986

Hoyle, Sir Fred   1915–2001
English mathematician and astronomer

Hoyle, Geoffrey   1942–
English science fiction writer

"In physics," he said, "we plan. We plan months ahead, years ahead.... You astronomers don't plan, you rush around like a chicken without a head. Observe and observe and observe and all shall be revealed unto you."
The Inferno (p. 87)
Harper & Row, Publishers. New York, New York, USA. 1973

Huebner, Jay S.
No biographical data available

Physics is what a group of people who call themselves physicists do.
What's Physics?
The Physics Teacher, Volume 14, Number 5, May 1976 (p. 315)

Huxley, Thomas Henry   1825–95
English biologist

...nothing can be more incorrect than the assumption one sometimes meets with, that physics has one method, chemistry another, and biology a third.
Collected Essays (Volume 1)
Method and Result
The Progress of Science (p. 60)
Macmillan & Company Ltd. London, England. 1904

Icke, Vincent   1946–
No biographical data available

Physics is not difficult; it's just weird.... Physics is weird because intuition is false. To understand what an electron's world is like, you've got to be an electron, or jolly nearly. Intuition is forged in the hellish fires of the everyday world, which makes it so eminently useful in our daily struggle for survival. For anything else, it is hopeless.
The Force of Symmetry
Preface (p. xiii)
Cambridge University Press. Cambridge, England. 1995

Fiction writers worry about first as well as last sentences, but I don't have to do that: in the book of physics, there never is a final sentence.
The Force of Symmetry
Chapter 14 (p. 294)
Cambridge University Press. Cambridge, England. 1995

Jeans, Sir James Hopwood   1877–1946
English physicist and mathematician

The classical physics seemed to bolt and bar the door leading to any sort of freedom of the will; the new physics hardly does this; it almost seems to suggest that the door may be unlocked — if only we could find the handle. The old physics showed us a universe which looked more like a prison than a dwelling place. The new physics shows us a universe which looks as though it might conceivably form a suitable dwelling place for free men, and not a

mere shelter for brutes — a home in which it may at least be possible for us to mould events to our desires and live lives of endeavor and achievement.
Physics and Philosophy
Chapter VII (p. 216)
Dover Publications, Inc. New York, New York, USA. 1981

Kronecker is quoted as saying that in arithmetic God made the integers and man made the rest; in the same spirit we may perhaps say that in physics God made the mathematics and man made the rest.
Physics and Philosophy
Chapter I (p. 16)
Dover Publications, Inc. New York, New York, USA. 1981

...the tendency of modern physics is to resolve the whole material universe into waves, and nothing but waves. These waves are of two kinds: bottled-up waves, which we call matter, and unbottled waves, which we call radiation or light.
The Mysterious Universe
Chapter III (p. 77)
The Macmillan Company. New York, New York, USA. 1932

...physics tries to discover the pattern of events which controls the phenomena we observe. But we can never know what this pattern means or how it originates; and even if some superior intelligence were to tell us, we should find the explanation unintelligible.
Physics and Philosophy
Chapter I (p. 16)
Dover Publications, Inc. New York, New York, USA. 1981

Koyré, Alexandre   1892–1964
Russian-born French philosopher

Good physics is made a priori. Theory precedes fact. Experience is useless because, before any experience, we are already in possession of the knowledge we are seeking for. Fundamental laws of motion (and of rest), laws that determine the spatio-temporal behavior of material bodies, are laws of a mathematical nature. Of the same nature as those which govern relations and laws of figures and numbers. We find and discover them not in Nature, but in ourselves, in our mind, in our memory, as Plato long ago has taught us.
Galileo and the Scientific Revolution of the Seventeenth Century
The Philosophical Review, Volume 52, Number 3, July 1943 (p. 347)

Larrabee, Eric   1922–90
Historian

Some people think that physics was invented by Sir Francis Bacon, who was hit by an apple when he was sitting under a tree one day writing Shakespeare.
Humor from Harper's
Easy Road to Culture, Sort Of (p. 89)
Harper. New York, New York, USA. 1961

Lewis, C. S. (Clive Staples)   1898–1963
British author, scholar, and popular theologian

Without a parable modern physics speaks not to the multitudes.
In John D. Barrow
The World Within the World (p. 238)
Clarendon Press. Oxford, England. 1988

Lewis, Edwin Herbert   1866–1938
American rhetorician, novelist, and poet

To Marvin the advance of physics and chemistry was the most exciting thing on earth. The researchers were watching each other, checking each other, helping each other, bound to tell the exact truth no matter where it led. The two sciences were steadily becoming one science, and the great advance continued day by day as if one infinite reluctant mind were slowly revealing itself.
White Lightning
Chapter 56 (p. 242)
Covici-McGee. Chicago, Illinois, USA. 1933

Liebson, Morris
No biographical data available

"What will I learn here?" you might query.
You'll learn some math and Einstein's Theory.
Ask Teacher for an illustration.
He'll explain, "It's time dilation.
Length gets less. Mass gets more.
Time decreases. That's the law.
When things go so very, very fast.
Classical physics is of the past,
And to find what's really true,
We must seek the physics new.
Learning this is lots of fun
In our course called Physics 1.
Physics Inspires the Muses
The Physics Teacher, Volume 16, Number 9, December 1978 (p. 636)

Lindley, David   1956–
English astrophysicist and author

Physics may be complex, mathematical, and arcane, but it is not capricious. The inventors of strings and twenty-six dimensional spaces did not think up these things at random, simply to give themselves a new set of toys. There is a line of rational thinking that leads from the billiard-ball atoms of classical physics to the intangible mathematical entities of today. Physics is complicated because the world is complicated.
The End of Physics: The Myth of a Unified Theory
Prologue (p. 19)
Basic Books, Inc. New York, New York, USA. 1993

If particle physics is a mess, it is because that is the way the world appears to work.
The End of Physics: The Myth of a Unified Theory
Part I, Chapter 4 (p. 124)
Basic Books, Inc. New York, New York, USA. 1993

Lodge, Sir Oliver   1851–1940
English physicist

When a thing behaves as if it were alive, physics loses interest in it and hands it over to another section; for it is incompetent to deal with motions attributable to spontaneity and free will.
Contributions to a British Association Discussion on the Evolution of the Universe
Nature, Supplement, October 24, 1931 (p. 722)

Mach, Ernst   1838–1916
Austrian physicist and philosopher

Physics is experience, arranged in economical order.
In John N Shive and Robert L. Weber
Similarities in Physics
Preface (p. xi)
John Wiley & Sons, Inc. New York, New York, USA. 1982

I only seek to adopt in physics a point of view that need not be changed the moment our glance is carried over into the domain of another science; for ultimately, all must form one whole.
Translated by C.M. Williams
Analysis of Sensations and the Relation of the Physical to the Psychical
Chapter 1 (p. 30, fn 1)
Dover Publications, Inc. New York, New York, USA. 1959

Maimonides, Moses   1135–1204
Spanish-born philosopher, jurist, and physician

…he who wishes to attain to human perfection, must therefore first study Logic, next the various branches of Mathematics in their proper order, then Physics, and lastly Metaphysics.
The Guide for the Perplexed
Part I, Chapter XXXIV
E.P. Dutton & Company. New York, New York, USA. 1904

Maritain, Jacques   1882–1973
French philosopher

Few spectacles are as beautiful and moving for the mind as that of physics thus advancing toward its destiny like a huge throbbing ship.
Translated by Gerald B. Phelan
Distinguish to Unite or the Degrees of Knowledge
Chapter IV, section 12 (p. 165)
University of Notre Dame Press. Notre Dame, Indiana, USA. 1995

Mencken, H. L. (Henry Louis)   1880–1956
American journalist and literary critic

…it is now quite lawful for a Catholic woman to avoid pregnancy by a resort to mathematics, though she is still forbidden to resort to physics and chemistry.
Minority Report: H.L. Mencken's Notebooks
No. 62 (p. 52)
Alfred A. Knopf. New York, New York, USA. 1956

Mephisto

Can no one laugh?
Will no one drink?
I'll teach you Physics in a wink.…
BLEGDAMSVEJ FAUST
Part First, Copenhagen Spring Conference, 1932

Beware alone of Reason and of Science,
Man's highest powers, unholy in alliance.
You let yourself, through dazzling witchcraft, yield
To all temptations of the Quantum field.
Listen! As now the obstacles abate,
You'll know the fair Neutrino for your fate!
BLEGDAMSVEJ FAUST
Part First, Copenhagen Spring Conference, 1932

Millikan, Robert Andrews   1868–1953
American physicist

Physics has opened the eyes of mankind so that it can now see in a very new truth new worlds — a marvelous world of electrons, already quite well explored, which underlies our former world of atoms and molecules, a world of quanta, not yet well understood, which lies perhaps behind the ether.
Science and Life
Chapter IV (p. 67)
The Pilgrim Press. Boston, Massachusetts, USA. 1924

Mohapatra, Rabindra
Theoretical physicist

Most people who haven't been trained in physics probably think of what physicists do as a question of incredibly complicated calculations, but that's not really the essence of it. The essence of it is that physics is about concepts, wanting to understand the concepts, the principles by which the world works.
In Michio Kaku
Hyperspace: A Scientific Odyssey Through Parallel Universes, Time Warps, and the 10th Dimension
Chapter 7 (p. 152)
Oxford University Press, Inc. New York, New York, USA. 1995

If you want to do serious physics, sometime you just have to learn it.
As reported by Ernest Barreto, student
Quantum field theory class. 1994

Morgan, Thomas Hunt   1866–1945
American zoologist and geneticist

Physics has progressed because, in the first place, she accepted the uniformity of nature; because, in the next place, she early discovered the value of exact measurements; because, in the third place, she concentrated her attention on the regularities that underlie the complexities of phenomena as they appear to us; and lastly, and not the least significant, because she emphasized the importance of the experimental method of research. An ideal or crucial experiment is a study of an event, controlled so as to give a definite and measurable answer to a question — an answer in terms of specific theoretical ideas, or better still an answer in terms of better understood relations.

The Relation of Biology to Physics
Science, Volume 65, Number 1679, March 4, 1927 (p. 217)

Morrow, James 1947–
American author

Her eyes sprang fully opened, and she beheld Howard's rickety bookshelves. P-h-y-s-i-c-s. A coil of radiant energy shot from the word, flooding into her skull like a sunbeam passing through glass. She closed her eyes. Her dendrites danced. Her synapses sparkeld.

Only Begotten Daughter (p. 90)
Harcourt Incorporated. Orlando, Florida, USA. 1990

Newton, Sir Isaac 1642–1727
English physicist and mathematician

I do not define time, space, place, and motion, as being well known to all. Only I must observe, that the common people conceived those quantities under not other notions but from the relation they bear to sensible objects…. Absolute space, in its own nature, without relation to anything external remains always similar and immovable.

In *Great Books of the Western World* (Volume 34)
Mathematical Principles of Natural Philosophy
Definitions, Scholium (p. 8)
Encyclopædia Britannica, Inc. Chicago, Illinois, USA. 1952

It is indeed a matter of great difficulty to discover, and effectually to distinguish, the true motions of particular bodies from the apparent; because the parts of that immovable space, in which those motions are performed, do by no means come under the observation of our senses.

In *Great Books of the Western World* (Volume 34}
Mathematical Principles of Natural Philosophy
Definitions, Scholium (p. 12)
Encyclopædia Britannica, Inc. Chicago, Illinois, USA. 1952

Nietzsche, Friedrich 1844–1900
German philosopher

We…want to become…human beings who are new, unique, incomparable, who give themselves laws, who create themselves. To that end we must become the best learners and discoverers of everything that is lawful and necessary in the world: we must become physicists in order to be able to be creators in this sense — while hitherto all valuations and ideals have been based on ignorance of physics or were constructed so as to contradict it. Therefore: long live physics! And even more so that which compels us to turn to physics — our honesty!

The Gay Science

Fourth Book, Aphorism 335
Cambridge University Press. Cambridge, England. 2001

Noll, Ellis D.
No biographical data available

Physics is the science whose treehouse rests on the trunk of immutable physical law.

What's Physics?
The Physics Teacher, Volume 14, Number 5, May 1976 (p. 315)

Oman, John 1860–1939
English Presbyterian theologian

Beauty…is the goal of physics as it seeks to construe the order of the universe…

The Natural and the Supernatural
Value and Validity (p. 211)
The Macmillan Company. New York, New York, USA. 1931

Oppenheimer, J. Robert 1904–67
American theoretical physicist

Time and experience have clarified, refined and enriched our understanding of these notions. Physics has changed since then. It will change even more. But what we have learned so far, we have learned well. If it is radical and unfamiliar and a lesson that we are not likely to forget, we think that the future will be only more radical and not less, only more strange and not more familiar, and that it will have its own new insights for the inquiring human spirit.

In Lucienne Felix
The Modern Aspect of Mathematics (p. 31)
Basic Books, Inc. New York, New York, USA. 1960

The only thing that we can say about the properties of the ultimate particles is that we know nothing whatever about them.

In Cecilia Payne-Gaposchkin
Introduction to Astronomy
Chapter XIII, Section 5 (p. 339)
Prentice-Hall, Inc. Englewood Cliffs, New Jersey, USA. 1954

As you undoubtedly know, theoretical physics — what with the haunting ghosts of neutrinos, the Copenhagen conviction, against all evidence, that cosmic rays are protons, Born's absolutely unquantizable field theory, the divergence difficulties with the positron, and the utter impossibility of making a rigorous calculation of anything at all — is in a hell of a way.

In Alice Smith and Charles Weiner
Robert Oppenheimer, Letters and Reflections
Letter to F. Oppenheimer, 4 June 1934 (p. 181)
Harvard University Press. Cambridge, Massachusetts, USA. 1980

Pagels, Heinz R. 1939–88
American physicist and science writer

Bohr wondered how we could even talk about the atomic world — it was so far removed from human experience.

He struggled with this problem — how can we use ordinary language developed to cope with everyday events and objects to describe atomic events? Perhaps the logic inherent in our grammar was inadequate for the task.... The end of determinism meant not the end of physics but the beginning of a new vision of reality.
In T. Ferris (ed.)
World Treasury of Physics, Astronomy, and Mathematics
Uncertainty and Complementarity (p. 103, 110)
Little, Brown & Company. Boston, Massachusetts, USA. 1991

Pais, Abraham 1918–2000
Dutch-born physicist

It was a wonderful mess at that time. Wonderful! Just great! It was so confusing — physics at its best, when everything is confused and you know something important lies just around the corner.
In Robert Crease
The Second Creation: Makers of the Revolution in 20th Century Physics
Chapter 9 (p. 177)
The Macmillan Company. New York, New York, USA. 1986

...the state of particle physics...is...not unlike the one in a symphony hall before the start of a concert. On the podium one will see some but not all of the musicians. They are tuning up. Short brilliant passages are heard on some of the instruments; improvisations elsewhere; some wrong notes too. There is a sense of anticipation for the moment when the concert starts.
Particles
Physics Today, Volume 21, Number 2, May 1968 (p. 28)

Pines, David
No biographical data available

The central task of theoretical physics in our time is no longer to write down the ultimate equations but rather to catalog and understand emergent behavior in its many guises...
In George Johnson
Challenging Particle Physics as Path to Truth
The New York Times, F5, Columns 2 and 3, Tuesday, December 4, 2001

Planck, Max 1858–1947
German physicist

The chief law of physics, the pinnacle of the whole system is, in my opinion, the principle of least action.
Translated by R. Jones and D.H. Williams
A Survey of Physics: A Collection of Lectures and Essays
The Place of Modern Physics in the Mechanical View of Nature (p. 41)
Methuen & Company Ltd. London, England. 1925

Physics would occupy an exceptional position among all the other sciences if it did not recognize the rule that the most far-reaching and valuable results of investigation can only be obtained by following a road leading to a goal which is theoretically unobtainable. This goal is the apprehension of true reality.

The Universe in the Light of Modern Physics
Section 1 (p. 15)
Unwin Brothers Ltd. London, England. 1937

Physics is an exact Science and hence depends upon measurement, while all measurement itself requires sense-perception. Consequently all the ideas employed in Physics are derived from the world of sense-perception.
The Universe in the Light of Modern Physics
Section 1 (p. 7)
Unwin Brothers Ltd. London, England. 1937

Since Galileo's time, physics has achieved its greatest success by rejecting all teleological methods.
Translated by R. Jones and D.H. Williams
A Survey of Physics: A Collection of Lectures and Essays
The Principle of Least Action (p. 73)
Methuen & Company Ltd. London, England. 1925

Modern Physics impresses us particularly with the truth of the old doctrine which teaches that there are realities existing apart from our sense-perceptions, and that there are problems and conflicts where these realities are of greater value for us than the richest treasures of the world of experience.
The Universe in the Light of Modern Physics (p. 107)
George Allen & Unwin Ltd. London, England. 1931

In endeavoring to claim your attention for a short time, I would remark that our science, Physics, cannot attain its object by direct means, but only gradually along numerous and devious paths, and that therefore a wide scope is provided for the individuality of the worker. One works at one branch, another at another, so that the physical universe with which we are all concerned appears in different lights to different workers.
Translated by R. Jones and D.H. Williams
A Survey of Physics: A Collection of Lectures and Essays
The Unity of the Physical Universe (p. 1)
Methuen & Company Ltd. London, England. 1925

...the second law of thermodynamics appears solely as a law of probability, entropy as a measure of the probability, and the increase of entropy is equivalent to a statement that more probable events follow less probable ones.
Translated by R. Jones and D.H. Williams
A Survey of Physics: A Collection of Lectures and Essays
The Relation Between Physical Theories (p. 86)
Methuen & Company Ltd. London, England. 1925

Poincaré, Henri 1854–1912
French mathematician and theoretical astronomer

The science of physics does not only give us [mathematicians] an opportunity to solve problems, but helps us to discover the means of solving them, and it does this in two ways: it leads us to anticipate the solution and suggests suitable lines of argument.

The Foundations of Science
The Value of Science
The Science Press. New York, New York, USA. 1913

Quine, Willard Van Orman   1908–2000
American logician and philosopher

Physics investigates the essential nature of the world, and biology describes a local bump. Psychology, human psychology, describes a bump on the bump.
Theories and Things
Chapter 10 (p. 93)
Harvard University Press. Cambridge, Massachusetts, USA. 1981

Rabi, Isidor Isaac   1898–1988
Austrian-born American physicist

I think that physics should be the central study in all schools. I don't mean physics as it is usually taught — very badly, as a bunch of tricks — but, rather, an appreciation of what it means, and a feeling for it. I don't want to turn everybody into a scientist, but everybody has to be enough of a scientist to see the world in the light of science — to be able to see the world as something that is tremendously important beyond himself, to be able to appreciate the human spirit that could discover these things, that could make instruments to inquire and advance into its own nature. I rate this so highly [because] with this education people would find something above their religious affiliations, and find a basic unity in the spirit of man.
In Jeremy Bernstein
Experiencing Science
Part 1. Two Faces of Physics. Chapter 2. Rabi: The Modern Age (p. 126)
Basic Books, Inc. New York, New York, USA. 1978

I think physics is infinite. You don't have to try to exhaust it in your generation, or in your lifetime.
In Jeremy Bernstein
Experiencing Science
Part 1 Two Faces of Physics Chapter 2 Rabi: The Modern Age (p. 56)
Basic Books, Inc. New York, New York, USA. 1978

Raman, Chandrasekhar Venkata   1888–1970
Indian physicist

The purpose of scientific study and research is to obtain an ever deeper understanding of the workings of nature. To the physicist falls the task of discovering the ultimate units or entities that constitute the material universe and of ascertaining the principles which govern their behavior.
The New Physics: Talks on Aspects of Science
Chapter II (p. 9)
Philosophical Library, New York. 1951

Reichenbach, Hans   1891–1953
German philosopher of science

If one knows physics for a distance only, if he hears merely strange names and mathematical formulae in it, he will, indeed, come to believe that it is an affair of the learned alone — ingeniously and wisely constructed, but without significance for men of other interests and problems.
Atoms and Cosmos
Chapter 19 (p. 293)
The Macmillan Company. New York, New York 1933

Richardson, Owen Willans   1879–1959
English physicist

The trouble with Physics at the present time is that there are so many workers making discoveries so fast, and important discoveries too, that it is difficult for any one worker to keep a balanced view of the state of the subject.
Lex Prix Nobel. The Nobel Prizes in 1928
Nobel banquet speech for award received in 1928
Nobel Foundation. Stockholm, Sweden. 1929

Roberts, Michael
No biographical data available
Thomas, E. R.
No biographical data available

The most brilliant discoveries in theoretical physics are not discoveries of new laws, but of terms in which the law can be discovered.
Newton and the Origin of Colours
Chapter I (p. 6)
G. Bell & Sons Ltd. London, England. 1934

Röntgen, Wilhelm Conrad   1845–1923
German physicist

To my view there are two methods of research, the apparatus and the calculation. Whoever prefers the first method is an experimenter; otherwise, he is a mathematical physicist. Both of them set up theories and hypotheses…
In Otto Glasser
Dr. W.C. Röntgen
Chapter II (p. 24)
Charles C. Thomas, Publisher. Springfield, Illinois, USA. 1945

Physics is a science which must be proved with honest effort. One can, perhaps, present a subject in such a manner that an audience of laymen may be convinced erroneously that it has understood the lecture. This, however, means a furthering a superficial knowledge, which is worse and more dangerous than none at all.
In Otto Glasser
Dr. W.C. Röntgen
Chapter VIII (p. 119)
Charles C. Thomas, Publisher. Springfield, Illinois, USA. 1945

Russell, Bertrand Arthur William   1872–1970
English philosopher, logician, and social reformer

The aim of physics, consciously or unconsciously, has always been to discover what we may call the causal skeleton of the world.

The Analysis of Matter
Chapter XXXVII (p. 391)
Harcourt, Brace & Company, Inc. New York, New York, USA. 1927

Physics is mathematical not because we know so much about the physical world, but because we know so little: it is only its mathematical properties that we can discover.
In John D. Barrow
The World Within the World (p. 278)
Clarendon Press. Oxford, England. 1988

Physics must be interpreted in a way which tends toward idealism, and perception in a way which tends toward materialism.
The Analysis of Matter
Chapter I (p. 7)
Harcourt, Brace & Company, Inc. New York, New York, USA. 1927

Naive realism leads to physics, and physics, if true, shows that naive realism is false. Therefore naive realism, if true, is false; therefore it is false.
In John D. Barrow
The World Within the World (p. 144)
Clarendon Press. Oxford, England. 1988

Broadly speaking, traditional physics has collapsed into two portions, truisms and geography.
The ABC of Relativity
Chapter XV
George Allen & Unwin Ltd. London, England. 1958

I come now to the statistical part of physics, which is concerned with the study of large aggregates. Large aggregates behave almost exactly as they were supposed to do before quantum theory was invented, so that in regard to them the older physics is very nearly right. There is, however, one supremely important law which is only statistical; this is the second law of thermodynamics. It states, roughly speaking, that the world is growing continuously more disorderly.
Scientific Metaphysics
The Scientific Outlook (p. 92)
George Allen & Unwin Ltd. London, England. 1931

It is obvious that a man who can see, knows things that a blind man cannot know; but a blind man can know the whole of physics.
The Analysis of Matter
Chapter XXXVII (p. 389)
Harcourt, Brace & Company, Inc. New York, New York, USA. 1927

Sandage, Allan   1926–
American astronomer

It is such a strange conclusion…it cannot really be true.
In Robert Jastrow
God and the Astronomers
Chapter 6 (p. 113)
W.W. Norton & Company, Inc. New York, New York, USA. 1978

Schlegel, Friedrich   1772–1829
German poet

It is in fact wonderful how physics — as soon as it is concerned not with technical purposes but with general results — without knowing it gets into cosmogony, astrology, theosophy, or whatever you wish to call it, in short, into a mystic discipline of the whole.
Translated by Ernst Behler and Roman Struc
Dialogue on Poetry and Literary Aphorisms
Talk on Mythology (p. 90)
The Pennsylvania State University Press. University Park, Pennsylvania, USA. 1968

Schrödinger, Erwin   1887–1961
Austrian theoretical physicist

Schrödinger: "Surely you realize the whole idea of quantum jumps is bound to end in nonsense…if the jump is sudden, Einstein's idea of light quanta will admittedly lead us to the right wave number, but them we must ask ourselves how precisely the electron behaves during the jump. Why does it not emit a continuous spectrum, as electromagnetic theory demands? And what law governs its motion during the jump? In other words, the whole idea of quantum jumps is sheer fantasy."

Niels Bohr: "What you say is absolutely correct. But it does not prove that there are no quantum jumps. It only proves that we cannot describe them, that the representational concepts with which we describe events in daily life and experiments in classical physics are inadequate when it comes to describing quantum jumps.
In Werner Heisenberg
Physics and Beyond: Encounters and Conversations
Chapter 6 (pp. 73–74)
Harper & Row, Publishers. New York, New York, USA. 1971

Research in physics has shown beyond the shadow of a doubt that in the overwhelming majority of phenomena whose regularity and invariability have led to the formulation of the postulate of causality, the common element underlying the consistency observed is chance.
What Is Natural Law
Clarendon Press. Oxford, England. 1988

Scotty
Fictional character

[To Captain Kirk on innumerable occasions] But I canna change the laws of physics, Captain!
Star Trek
Television program

Smith, Henry J. S.   1826–83
Irish mathematician

So intimate is the union between mathematics and physics that probably by far the larger part of the accessions to our mathematical knowledge have been obtained by the efforts of mathematicians to solve the problems set to them by experiment, and to create "for each successive class of phenomena, a new calculus or a new geom-

etry, as the case might be, which might prove not wholly inadequate to the subtlety of nature." Sometimes, indeed, the mathematician has been before the physicists, and it has happened that when some great and new question has occurred to the experimentalist or the observer, he has found in the armory of the mathematician the weapons which he has needed ready made to his hand. But, much oftener, the questions proposed by the physicist have transcended the utmost powers of the mathematics of the time, and a fresh mathematical creation has been needed to supply the logical instrument requisite to interpret the new enigma.
Presidential Address British Association for the Advancement of Science *Nature*, Section A, Volume 8, Number 204, September 25, 1873 (p. 450)

Snow, Charles Percy   1905–80
English novelist and scientist

He then gave me an explanation which I could not understand, although I had heard plenty of the jargon of nuclear physics from him and Luke. "Fission." "Neutrons." "Chain reaction." I could not follow. But I could gather that at last the sources of nuclear energy were in principle open to be set loose; and that it might be possible to make an explosive such as no one had realistically imagined.
The New Men (p. 11)
Charles Scribner's Sons. New York, New York, USA. 1955

I now believe that if I had asked an even simpler question — such as, What do you mean by mass, or acceleration, which is the scientific equivalent of saying, Can you read? — not more than one in ten of the highly educated would have felt that I was speaking the same language. So the great edifice of modern physics goes up, and the majority of the cleverest people in the western world have about as much insight into it as their neolithin ancestors would have had.
The Two Cultures: And a Second Look
Chapter I (p. 15)
At The University Press. Cambridge, England. 1964

Standen, Anthony
Anglo-American science writer

"[T]he extraordinary degree of dullness that pervades the laboratory periods of physics courses…[is] so acute that for many people it is the bitterest experience of their education."
Science Is a Sacred Cow
Chapter III (p. 83)
Dutton. New York, New York, USA. 1950

Physics is not about the real world, it is about "abstractions" from the real world, and this is what makes it so scientific.
Science Is a Sacred Cow
Chapter III (p. 61)
Dutton. New York, New York, USA. 1950

Sullivan, John William Navin   1886–1937
Irish mathematician

The present tendency of physics is toward describing the universe in terms of mathematical relations between unimaginable entities.
The Bases of Modern Science (p. 226)
Doubleday, Doran & Company, Inc. Garden City, New York, USA. 1929

Teilhard de Chardin, Pierre   1881–1955
French Jesuit, paleontologist, and biologist

The time has come to realise that an interpretation of the universe — even a positive one — remains unsatisfying unless it covers the interior as well as the exterior of things; mind as well as matter. The true physics is that which will, one day, achieve the inclusion of man in his wholeness in a coherent picture of the world.
The Phenomenon of Man
Forward (pp. 35–36)
Harper & Brothers. New York, New York, USA. 1959

The X-Files

MULDER: [I]n most of my work, the laws of physics rarely seems to apply.
Pilot
Television program
Season 1, 1993

Truesdell, Clifford   1919–2000
American mathematician, natural philosopher, historian of mathematics

Pedantry and sectarianism aside, the aim of theoretical physics is to construct mathematical models such as to enable us, from the use of knowledge gathered in a few observations, to predict by logical processes the outcomes in many other circumstances. Any logically sound theory satisfying this condition is a good theory, whether or not it be derived from "ultimate" or "fundamental" truth. It is as ridiculous to deride continuum physics because it is not obtained from nuclear physics as it would be to reproach it with lack of foundation in the Bible.
In Clifford Truesdell and Walter Noll
The Non-Linear Field Theories of Mechanics (2nd edition) (pp. 2–3)
Springer-Verlag. Berlin, Germany. 1992

Ulam, Stanislaw   1909–84
Polish-born mathematician

I should add here for the benefit of the reader who is not a professional physicist that the last thirty years or so have been a period of kaleidoscopically changing explanations of the increasingly strange world of elementary particles and of fields of force. A number of extremely talented theorists vie with each other in learned and clever attempts to explain and order the constant flow of experimental results which, or so it seems to me, almost perversely cast doubts about the just completed theoretical formulations.

Adventures of a Mathematician
Chapter 13 (p. 261)
Charles Scribner's Sons. New York, New York, USA. 1976

van Sant, Gus   1952–
American film editor

Like a disc jockey from Paradise, Howard flips Marie over and plays her B side. Every now and then she reaches for Sissy to include her, but the laws of physics insist on being obeyed.
Even Cowgirls Get the Blues
Screenplay (p. 34)
Faber & Faber Ltd. London, England. 1993

von Baeyer, Hans Christian   1938–
German-born physicist and author

When [an electron] is passing through the slits, it is a wave, when it is caught, it is a particle.... An atom, according to Bohr, represents a different reality from that of the ordinary world of our sense perceptions, and it is unreasonable to insist on forcing the language of our familiar macroscopic surroundings onto that alien mode of existence.
Taming the Atom
Chapter 13 (p. 197)
Random House, Inc. New York, New York, USA. 1992

von Goethe, Johann Wolfgang   1749–1832
German poet, novelist, playwright, and natural philosopher

Physics must be sharply distinguished from mathematics. The former must stand in clear independence, penetrating into the sacred life of nature in common with all the forces of love, veneration and devotion. The latter, on the other hand, must declare its independence of all externality, go its own grand spiritual way, and develop itself more purely than is possible so long as it tries to deal with actuality and seeks to adapt itself to things as they really are.
Werke
Schriften zur Naturwissenschaft, XXXIX (p. 92)
Temple-Verlag. Berlin, Germany. 1963

von Weizsäcker, Carl Friedrich (Baron)   1912–2007
German theoretical physicist and philosopher

Physics begins by facing a mystery. It transforms the mystery into a puzzle. It solves the puzzle. And it finds itself facing a new mystery.
In Pekka Lahti and Peter Mittelstaedt
Symposium on the Foundations of Modern Physics: 50 Years of the Einstein–Podolsky–Rosen Gedankenexperiment
Quantum Theory and Space-Time (p. 237)

Weinberg, Steven   1933–
American nuclear physicist

Our job in physics is to see things simply, to understand a great many complicated phenomena, in terms of a few simple principles.

In Robert K. Adair
The Great Design (p. 325)
Oxford University Press, Inc. New York, New York, USA. 1987

I think that is one of the great things about physics, that it is sufficiently precise that it makes predictions which can be disproved by observation, and which occasionally are. And, when you have that experience, you know that there is something out there that is not all just coming out of your closed society of fellow physicists. It's, I think, one of the things that I love so much about physics, the dialogue with nature; and this dialogue is not one in which nature always agrees with the physicists.
Does Physics Describe Reality?
The Challenge of the Universe
From Hypermind CD-ROM

Physics is not a finished logical system. Rather, at any moment it spans a great confusion of ideas, some that survive like folk epics from the heroic periods of the past, and others that arise like utopian novels from our dim premonitions of a future grand synthesis.
Gravitation and Cosmology: Principles and Applications of the General Theory of Relativity
Part I, Chapter I (p. 3)
John Wiley & Sons, Inc. New York, New York, USA. 1972

Wells, H. G. (Herbert George)   1866–1946
English novelist, historian, and sociologist

The science of physics is even more tantalizing than it was half a century ago, and, above the level of an elementary introduction, optics, acoustics and the rest, even less teachable. The more brilliant investigators rocket off into mathematical pyrotechnics and return to common speech with statements that are, according to the legitimate meanings of words, nonsensical.
Experiment in Autobiography
Chapter 5, Section 2 (p. 176)
The Macmillan Company. New York, New York, USA. 1934

Wheeler, John Archibald   1911–
American physicist and educator

No point is more central than this, that empty space is not empty. It is the seat of the most violent physics.
In Heinz R. Pagels
The Cosmic Code: Quantum Physics as the Language of Nature
Part II, Chapter 8 (p. 274)
Simon & Schuster. New York, New York, USA. 1982

Whitehead, Alfred North   1861–1947
English mathematician and philosopher

Physics refers to ether, electrons, molecules, intrinsically incapable of direct observation.
The Principle of Relativity with Application to Physical Science (p. 62)
The University Press. Cambridge, England. 1922

...in the present-day reconstruction of physics, fragments of the Newtonian concepts are stubbornly retained. The

result is to reduce modern physics to a sort of mystic chant over an unintelligible universe.
Modes of Thought
Chapter III, Lecture VII (p. 185)
The Macmillan Company. New York, New York, USA. 1938

Whyte, A. Gowans
Scottish writer

The progress of human thought is through metaphysics to physics.
The Triumph of Physics
The Rationalist Annual, 1931 (p. 28)

Wiener, Norbert 1894–1964
American mathematician

Physics — or so it is generally supposed — takes no account of purpose…
God and Golem, Inc.: A Comment on Certain Points Where Cybernetics Impinges on Religion
Chapter I (p. 5)
The MIT Press. Cambridge, Massachusetts, USA. 1964

Wigner, Eugene Paul 1902–95
Hungarian-born American physicist

We have ceased to expect from physics an explanation of all events, even of the gross structure of the universe, and we aim only at the discovery of the laws of nature, that is the regularities, of the events.
Nobel Lectures, Physics 1963–1970
Nobel lecture for award received in 1963
Events, Laws of Nature, and Invariance Principles (p. 9)
Elsevier Publishing Company. Amsterdam, Netherlands. 1972

Physics does not endeavor to explain nature. In fact, the great success of physics is due to the restriction of its objectives: it endeavors to explain the regularities in the behavior of objects. The renunciation of the broader aim, and the specification of the domain for which an explanation can be sought, now appears to us as an obvious necessity. In fact, the specification of the explainable may have been the greatest discovery of physics so far.
Nobel Lectures, Physics 1963–1970
Nobel lecture for award received in 1963
Events, Laws of Nature, and Invariance Principles (p. 2)
Elsevier Publishing Company. Amsterdam, Netherlands. 1972

Wilczek, Frank 1951–
American theoretical physicist

In physics, you don't have to go around making trouble for yourself — nature does it for you.
Longing for the Harmonies
How Asymptotic Freedom Discovered Me (p. 208)
W.W. Norton & Company, Inc. New York, New York, USA. 1988

Ziman, John M. 1925–2005
English physicist

Physics defines itself as the science devoted to discovering, developing and refining those aspects of reality that are amenable to mathematical analysis.
Reliable Knowledge
Chapter 2 (p. 28)
Cambridge University Press. Cambridge, England. 1978

The most astonishing achievements of science, intellectually and practically, have been in physics, which many people take to be the ideal type of scientific knowledge. In fact, physics is a very special type of science, in which the subject matter is deliberately chosen so as to be amenable to quantitative analysis.
Reliable Knowledge
Chapter 1 (p. 9)
Cambridge University Press. Cambridge, England. 1978

In the education of a physicist, we recount the bold voyages of great explorers — Newton and Einstein, Faraday and Bohr — in search of new laws of nature. They found and charted the continents on which we have built our cities of the mind and of art. Does anyone really suppose that similar vast and fertile territories are still waiting to be discovered and colonized? The unaccustomed rules that govern black holes and quasars in the cosmic deeps affect our lives no more than the icy crags of the Himalayas or the conjunctions of the planets.
Physics Bulletin, Volume 25, 1974 (p. 280)

Think of physics simply as the "fundamental" science and it is oversubscribed almost to bankruptcy. But define it as the science whose aim is to describe natural phenomena in the most mathematical or numerical language, and you will understand its past and have confidence in its future. The task of the modern physicist is to determine the mathematically comprehensible characteristics of the natural world and of human artifacts…
Physics Bulletin, Volume 25, 1974 (p. 280)

Zukav, Gary
American spiritual teacher

Unfortunately, when most people think of "physics", they think of chalkboards covered with undecipherable symbols of an unknown mathematics. The fact is that physics is not mathematics. Physics, in essence, is simple wonder at the way things are and a divine (some call it compulsive) interest in how that is so. Mathematics is the tool of physics. Stripped of mathematics, physics becomes pure enchantment.
The Dancing Wu Li Masters: An Overview of the New Physics
Part One
Wu Li?
Chapter I (p. 31)
William Morrow. New York, New York, USA. 1979

PHYSIOGNOMY

Miller, Hugh 1802–56
Scottish geologist and theologian

Physiognomy is no idle or doubtful science in connection with geology. The physiognomy of a country indicates almost invariably its geological character.
The Old Red Sandstone
Chapter XI (p. 201)
J.M. Dent & Sons Ltd. London, England. 1922

PHYSIOLOGIST

Collingwood, Robin George 1889–1943
English historian and philosopher

A physiologist…can certainly offer a definition of life; but this will only be an interim report on the progress of physiology to date. For him, as for the beginner, it is the nature of physiology that is relatively certain; it is the nature of life that is relatively vague.
The New Leviathan: Or Man, Society, Civilization and Barbarism
Part I, Chapter I. Aphorism I.47 (p. 3)
At The Clarendon Press. Oxford, England. 1942

Mayo, Charles Horace 1865–1939
American physician

Disease at times creates experiments that physiology completely fails to duplicate, and the wise physiologist can obtain clues to the resolution of many problems by studying the sick.
La funcion del higado en relacion con la cirugia
Annals de Circulation, Volume 2, April 1930

Pirenne, M. H.
No biographical data available

The soul, the mind, consciousness, thought, sensation, being nonmaterial, are not observable in physiological investigation like, say, nerve excitation or muscle contraction. Physiology gives no direct experimental evidence for them. Yet like all men, physiologists no doubt believe they have minds. Hence a dilemma.
British Journal for the Philosophy of Science, Volume 1, 1950/1951

PHYSIOLOGY

Gee, Samuel 1839–1911
Physician

Physiology owes more to medicine than medicine to physiology. Nature in disease performs vivisections for us. The greater and better part of what we know concerning the functions of the many organs of the body is derived from pathological observation and not from physiological experiment.
Medical Lectures and Aphorisms (p. 227)
Smith, Elder. London, England. 1902

Huxley, Thomas Henry 1825–95
English biologist

There is no side of the human mind which physiological study leaves uncultivated. Connected by innumerable ties with abstract science, Physiology is yet in the most intimate relation with humanity, and by teaching us that law and order, and a definite scheme of development, regulate even the strangest and wildest manifestations of individual life, she prepares the student to look for a goal even amidst the erratic wanderings of mankind, and to believe that history offers something more than an entertaining chaos — a journal of a toilsome, tragi-comic march nowhither.
Collected Essays (Volume 33)
Science and Education
On the Educational Value of the Natural History of Science (p. 59)
Macmillan & Company Ltd. London, England. 1904

A thorough study of Human Physiology is, in itself, an education broader and more comprehensive than much that passes under that name. There is no side of the intellect which it does not call into play, no region of human knowledge into which either its roots, or its branches, do not extend; like the Atlantic between the Old and the New Worlds, its waves wash the shores of the two worlds of matter and of mind; its tributary streams flow from both; through its waters, as yet unfurrowed by the keel of any Columbus, lies the road, if such there be, from the one to the other; far away from that North-west Passage of mere speculation, in which so many brave souls have been hopelessly frozen up.
Collected Essays (Volume 33)
Science and Education
Universities: Actual and Ideal (p. 220)
Macmillan & Company Ltd. London, England. 1904

Pavlov, Ivan Petrovich 1849–1936
Russian physiologist

It is perfectly clear that the horizon of medical observation of life is immeasurably wider than the sphere of vital phenomena which the physiologists have before their eyes in their laboratories. Hence the permanent incongruity between that which medicine knows, sees and empirically applies, and that which physiology can reproduce and explain.
Experimental Psychology and Other Essays
Concerning Trophic Innervation (p. 74)
Philosophical Library. New York, New York, USA. 1957

The outer limit of physiological knowledge, its goal, is to express this infinitely complex interrelationship of the organism with the surrounding world in the form of an exact scientific formula.
In Daniel P. Todes

Pavlov's Physiology Factory: Experiment, Interpretation, Laboratory Enterprise
Chapter 5 (p. 153)
The Johns Hopkins University Press. Baltimore, Maryland, USA. 2002

Starling, Ernest Henry   1866–1927
English physiologist

In physiology, as in all other sciences, no discovery is useless, no curiosity misplaced or too ambitious, and we may be certain that every advance achieved in the quest of pure knowledge will sooner or later play the part in the service of man.
The Linacre Lecture on the Law of the Heart (p. 147)
Publisher undetermined

PI

Beckmann, Petr   1924–93
Physicist

The digits beyond the first few decimal places are of no practical or scientific value. Four decimal places are sufficient for the design of the finest engines; ten decimals are sufficient to obtain the circumference of the earth to within a fraction of an inch if the earth were a smooth sphere…
A History of Pi
Chapter 10 (p. 100)
St. Martin's Press. New York, New York, USA. 1974

Chudnovsky, David
Mathematician

Maybe in the eyes of God pi looks perfect.
In Richard Preston
The Mountains of Pi
The New Yorker March 2, 1992

de Morgan, Augustus   1806–71
English mathematician and logician

…mysterious 3.14159…comes in at every door and window, and down every chimney.
A Budget of Paradoxes
Cyclometry (p. 393)
Longmans, Green. London, England. 1872

Duffin, R. J.
No biographical data available

God created the world and the integers, all in seven days. He then ordered two of his biotechnicians, James and Francis, to construct a genetic code for the fractional numbers. Moreover, they were to give special prominence to His favorite number, pi.
The Patron Saint of Mathematics
The Mathematical Intelligencer, Volume 15, Number 1, 1993 (p. 52)

Graham, L. A.
No biographical data available

Fiddle de dum, fiddle de dee,
A ring round the moon is pi times D;
But if a hole you want repaired,
You use the formula pi r^2.
Ingenious Mathematical Problems and Methods
Mathematical Nursery Rhyme Number 1
Dover Publications, Inc. New York, New York, USA. 1959

Little Jack Horner sat in a corner,
Trying to evaluate pi.
He disdained rule of thumb,
Found an infinite sum,
And exclaimed "It's REAL, nary an I."
Ingenious Mathematical Problems and Methods
Mathematical Nursery Rhyme Number 9
Dover Publications, Inc. New York, New York, USA. 1959

Kac, Mark   1914–84
Polish mathematician

Steinhaus, with his predilection for metaphors, used to quote a Polish proverb, *"Forturny kolem sie tocza"* (Luck runs in circles), to explain why pi, so intimately connected with circles, keeps cropping up in probability theory and statistics, the two disciplines which deal with randomness and luck.
Enigmas of Chance: An Autobiography
The Search for the Meaning of Independence (p. 55)
Harper & Row, Publishers. New York, New York, USA. 1985

Morgan, Robert
No biographical data available

The secret relationship
of line and circle, progress
and return, is always known,
transcendental and yet
a commonplace. And though
the connection is written
it cannot be written out
in full, never perfect, but
is exact and constant, is
eternal and everyday
as orbits of electrons,
chemical rings, noted here
in one brief sign as gateway
to completed turns and
the distance inside circles,
both compact and infinite.
Poetry
Pv. clxi, Number 4 (January, 1993) (p. 204)

Preston, Richard
No biographical data available

The digits of pi march to infinity in a predestined yet unfathomable code: they do not repeat periodically, seeming to pop up by blind chance, lacking any perceivable order, rule, reason, or design — "random" integers, ad infinitum.

The Mountains of Pi
The New Yorker, March 2, 1992

...pi is not the solution to any equation built from a less than infinite series of whole numbers. If equations are trains threading the landscape of numbers, then no train stops at pi.
The Mountains of Pi
The New Yorker, March 2, 1992

PILL

Crichton-Browne, Sir James   1840–1938
English physician

If you want fame and fortune, invent a pill.
The Doctor's After Thoughts (p. 14)
E. Benn Ltd. London, England. 1932

Fuller, Thomas   1608–61
English clergyman and author

If the pills were pleasant, they would not want gilding.
Gnomologia: Adages and Proverbs, Wise Sentences, and Witty Sayings. Ancient and Modern, Foreign and British
No. 2711
Printed for Thomas and Joseph Allman. London, England. 1816

Herrick, Robert   1591–1674
English poet

When his potion and his pill
His, or none, or little skill
Meet for nothing, but to kill;
Sweet Spirit comfort me!!
In J. Max Patrick (ed.)
The Complete Poetry of Robert Herrick
His Litanie, to the Holy Spirit (p. 132)
W.W. Norton & Company, Inc. New York, New York, USA. 1968

Jerrold, Douglas William   1803–57
English playwright, journalist, and humorist

A pill that the present moment is daily bread to thousands.
The Catspaw: A Comedy in Five Acts
Act I, Scene I
Published at the Punch Office. London, England. 1850

Molière (Jean-Baptiste Poquelin)   1622–1673
French playwright and actor

My lord Jupiter knows how to gild the pill.
Amphitryon
Act III, Scene X, l. 24
Harcourt, Brace & Company. New York, New York, USA. 1995

Ray, John   1627–1705
English naturalist

Apothecaries would not give pills in sugar unless they were bitter.
A Complete Collection of English Proverbs (p. 2)
Printed for G. Cowie. London, England. 1813

Shakespeare, William   1564–1616
English poet, playwright, and actor

When I was sick, you gave me bitter pills.
In *Great Books of the Western World* (Volume 26)
The Plays and Sonnets of William Shakespeare (Volume 1)
The Two Gentlemen of Verona
Act II, Scene iv, l. 149
Encyclopædia Britannica, Inc. Chicago, Illinois, USA. 1952

Stumpf, LaNore
No biographical data available

How is it that a little pill
Without a pair of eyes to see
Can travel down, and round and round
And figure out what's wrong with me?
Needed: Remote Control
American Journal of Nursing, April 1969 (p. 902)

PLANET

Author undetermined

The discovery of a new planet in a new way, by first finding where a planet ought to be, has given a fresh impulse to the enthusiasm of astronomers. All are looking to see if the motions of heavenly bodies in some other direction does not indicate that there are more weights in the scale on that side than have yet been seen.
Scientific American, Volume 2, Issue 26, March 20, 1847 (p. 203)

A new planet, it is said, has lately been discovered. This is not correct. The planet is as "old as the hills."
Scientific American, Volume 2, Issue 9, November 21, 1846 (p. 68)

Banks, Sir Joseph   1743–1820
English explorer and naturalist

Some of our astronomers here incline to the opinion that it is a planet and not a comet; if you are of that opinion it should forthwith be provided with a name [or] our nimble neighbors, the French, will certainly save us the trouble of Baptizing it.
In Constance A. Lubbock
The Herschel Chronicle (p. 95)
The Macmillan Company. New York, New York, USA. 1933

Blackmore, Sir Richard   1650–1729
English physician and writer

All these Illustrious Worlds, and many more,
Which by the Tube Astronomers explore;
And Millions which the Glass can ne'er descry,
Lost in the Wilds of vast Immensity,
Are Suns, are Centers, whose Superior Sway
Planets of various Magnitude obey.
The Poetical Works of Sir R. Blackmore: Containing Creation: A Philosophical Poem, in Seven Books
Book II, l. 536–541
Printed for C. Cooke. London, England. 1797

In beauteous Order all the Orbs advance,
And in their mazy complicated Dance,
Not in one part of all the Pathless Sky
Did any ever halt, or step awry.
The Poetical Works of Sir R. Blackmore: Containing Creation: A Philosophical Poem, in Seven Books
Book II, l. 91–94
Printed for C. Cooke. London, England. 1797

Burroughs, John   1837–1921
American naturalist and writer

The earth is not alone, it is not like a single apple on a tree; there are many apples on the tree, and there are many trees in the orchard.
The Breath of Life
Chapter XII (p. 289)
Houghton Mifflin Company. Boston, Massachusetts, USA. 1915

Chapman, Clark R.
Astronomer and asteroid researcher

Planets are like living creatures. They are born, full of life and activity. They mature, consume energy, and settle into established ways. Finally, they run down, become dormant, and die. On a human time scale planetary lives are virtually eternal. We see only a snapshot of each planet and can only surmise its evolution.
The Inner Planets: New Light on the Rocky Worlds of Mercury, Venus, Earth, the Moon, Mars, and the Asteroids
Chapter 6 (pp. 88–89)
Charles Scribner's Sons. New York, New York, USA. 1977

Chaucer, Geoffrey   1343–1400
English poet

The seven bodies I'll describe anon:
Sol, gold is, Luna's silver, as we see,
Mars iron, and quicksilver's Mercury,
Saturn is lead, and Jupiter is tin,
And Venus copper, by my father's kin!
In *Great Books of the Western World* (Volume 22)
The Canterbury Tales
Canon Yeoman's Tale (p. 476)
Encyclopædia Britannica, Inc. Chicago, Illinois, USA. 1952

de Bergerac, Cyrano   1619–55
French dramatist

For my part, I…believe the Planets are Worlds about the Sun, and that the Fixed Stars are also Suns which have Planets about them, that's to say, Worlds which because of their smallness, and that their borrowed light can-not reach us, are not discernible by Men in this World…
In Roger A MacGowan and Frederick I. Ordway, III
Intelligence in the Universe
The Comical History of the States and Empires of the World and of the Sun (p. 1)
Prentice-Hall, Englewood Cliffs, New Jersey, USA. 1966

de Fontenelle, Bernard le Bovier   1657–1757
French author

…you must go a great way to prove that the Earth may be a Planet, the Planets so many Earths, and all the Stars Worlds.
In Roger A MacGowan and Frederick I. Ordway, III
Intelligence in the Universe (p. 75)
Prentice-Hall, Englewood Cliffs, New Jersey, USA. 1966

Doyle, Sir Arthur Conan   1859–1930
Scottish writer

"But the Solar System!" I protested.

"What the deuce is it to me?" [Sherlock Holmes] interrupted impatiently: "You say that we go round the sun. If we went round the moon it would not make a pennyworth of difference to me…"
In William S. Baring-Gould (ed.)
The Annotated Sherlock Holmes (Volume 1)
A Study in Scarlet, Chapter 2 (p. 154)
Wings Books. New York, New York, USA. 1967

Dudley Manlove
Fictional character

Do you still believe it impossible we exist? You didn't actually think you were the only inhabited planet in the universe. How can any race be so stupid?
Plan 9 from Outer Space
Film (1959)

Eddington, Sir Arthur Stanley   1882–1944
English astronomer, physicist, and mathematician

If the planets of the solar system should fail us, there remain some thousands of millions of stars which we have been accustomed to regard as suns ruling attendant systems of planets. It has seemed a presumption, bordering almost on impiety, to deny them life of the same order of creation as ourselves. It would indeed be rash to assume that nowhere else has Nature repeated the strange experiment which she has performed on the earth.
Man's Place in the Universe
Harper's Magazine, October 1928 (p. 573)

Eiseley, Loren C.   1907–77
American anthropologist, educator and author

Things get odder on this planet, not less so.
The Unexpected Universe
Chapter Ten (p. 232)
Harcourt, Brace & World, Inc. New York, New York, USA. 1969

Emerson, Ralph Waldo   1803–82
American lecturer, poet, and essayist

He who knows what sweets and virtues are in the ground, the waters — the planets, the heavens, and how to come at these enchantments, is the rich and royal man.
Ralph Waldo Emerson: Essays and Lectures
Essays: Second Series
Nature (p. 543)
The Library of America. New York, New York, USA. 1983

Feynman, Richard P.   1918–88
American theoretical physicist

…what makes planets go around the sun? At the time of Kepler some people answered this problem by saying that there were angels behind them beating their wings and pushing the planets around in orbit. As you will see, the answer is not very far from the truth. The only difference is that the angels sit in a different direction and their wings push inwards.
The Character of Physical Law
Chapter 1 (p. 18)
BBC. London, England. 1965

Hammond, Allen Lee
No biographical data available

With the beginning of direct exploration of the solar system, planetary science has revived to become not only respectable but one of the active, forefront areas of research. How active can be gauged by the assessment, widely agreed on, that the rate of new discoveries and the rate of obsolescence of old ideas have never been so rapid as at present. Investigators are now confronted with such an overwhelming array of new observations and theories that what amounts to a revolution in understanding the solar system is in progress.
Exploring the Solar System(s): An Emerging New Perspective
Science, Volume 186, Number 4165, 22 November 1974 (p. 720)

Herschel, Friedrich Wilhelm   1738–1822
English astronomer

It has generally been supposed that it was a lucky accident that brought this new star to my view; this is an evident mistake. In the regular manner I examined every star of the heavens, not only of that magnitude but many far inferior, it was that night its turn to be discovered.
In Constance A. Lubbock
The Herschel Chronicle (pp. 78–79)
The Macmillan Company. New York, New York, USA. 1933

Hey, Nigel S.   1936–
American science writer

Each of us is part of an endless drama that started billions of years ago, when a gargantuan cloud of cosmic gas and dust began to collect within the firmament. Then our galaxy, the Milky Way, came to be, and, within it, our solar system. The Sun is the great nucleus of this little cell, radiating incredible amounts of energy to its daughter planets, furnishing all the ingredients of life and all the analogues of life that manifest themselves as simple movement and change.
Solar System
Chapter 1 (p. 11)
Weidenfeld & Nicolson. London, England. 2002

I used to wonder, why bother? The other planets and moons are too inhospitable for us ever to visit, let alone colonize. But this is shallow thinking. Our destiny is in space. We will always want to explore new frontiers, even when separated physically by great gulfs of space and time. The need is in our genes. And who knows what wonders we may find out there.
Solar System
Introduction (p. 8)
Weidenfeld & Nicolson. London, England. 2002

Homer (Smyrns of Chios)   fl. 750 BCE
Greek poet

The thick tresses of gold with which Vulcan had crested the helmet floated round it, and as the evening star that shines brighter than all others through the stillness of night, even such was the gleam of the spear which Achilles poised in his right hand, fraught with the death of noble Hector.
In *Great Books of the Western World* (Volume 4)
The Iliad of Homer
Book XXII, l. 317 (p. 158)
Encyclopædia Britannica, Inc. Chicago, Illinois, USA. 1952

At length as the Morning Star was beginning to herald the light which saffron-mantled Dawn was soon to suffuse over the sea, the flames fell and the fire began to die.
In *Great Books of the Western World* (Volume 4)
The Iliad of Homer
Book XXIII, l. 226 (p. 163)
Encyclopædia Britannica, Inc. Chicago, Illinois, USA. 1952

Huygens, Christiaan   1629–95
Dutch mathematician, astronomer, and physicist

…the rest of the Planets have their Dress and Furniture, nay and their Inhabitants too as well as this Earth of ours.
The Celestial Worlds Discover'd, or, Conjectures Concerning the Planetary Worlds, Their Inhabitants and Productions
Book the First (p. 2)
Printed for T. Childe. London, England. 1698

Lockyer, Joseph Norman   1836–1920
English astronomer and physicist

The work of the true man of Science is a perpetual striving after a better and closer knowledge of the planet on which his lot is cast, and of the universe in the vastness of which that planet is lost.
Studies in Spectrum Analysis
Chapter I (p. 1)
D. Appleton & Company. New York, New York, USA. 1878

Marcy, Geoffrey
American astronomer

Look at how perfect this thing is. It's like a jewel. You've got circular orbits. They're all in the same plane. They're all going around in the same direction…. It's perfect, you know. It's gorgeous. It's almost uncanny.

Jumping Jupiter! Is Our Solar System a Rarity?
Washington Post, Monday, February 15, 1999 (p. A3)

Marlowe, Christopher   1564–93
English poet

…whose faculties can comprehend
The wondrous architecture of the world,
And measure every wand'ring planet's course…
Tamburlaine the Great
Scene VIII
Odyssey Press. Indianapolis, Indiana, USA. 1974

Miller, Hugh   1802–56
Scottish geologist and theologian

The planet which we inhabit is but one vessel in the midst of a fleet sailing on through the vast ocean of space, under convoy of the sun.
Geology Versus Astronomy: Or the Conditions and the Periods; Being a View of the Modifying Effects of Geologic Discovery on the Old Astronomic Inferences Respecting the Plurality of Inhabited Worlds
Chapter II (p. 14)
Glasgow, Scotland. 1857

Molière (Jean-Baptiste Poquelin)   1622–1673
French playwright and actor

We had a narrow escape, Madame, While asleep;
A neighboring planet did pass us close by,
Cutting a swathe right through our whirlpool;
Had its path led to a collision with mother earth,
She would have shattered in pieces like glass.
Les Femmes Savantes
Act IV, Scene iii
Oxford University Press. Oxford, England. 1974

Morrow, Jeff   1907–93
American actor

A lifeless planet. And yet, yet still serving a useful purpose, I hope. Yes, a sun. Warming the surface of some other world. Giving light to those who may need it.
This Island Earth
Film (1955)

Redfern, Martin
No biographical data available

We are no longer the victims of our planet, we are the custodians of it. Through our inconsiderate greed for land and our disregard for pollution, we bite the hand that feeds us. But we do so at our own peril. We still have our eggs in one basket, all our people in one planet. We need to care for that planet and take responsibility for it. But we also need to progress with the search for new homes and the technology to take us to the stars.
The Earth: A Very Short Introduction
Epilogue (p 132)
Oxford University Press, Inc. Oxford, England. 2003

Shapley, Harlow   1885–1972
American astronomer

Millions of planetary systems must exist, and billions is the better word. Whatever the methods of origin, and doubtless more than one type of genesis has operated, planets may be the common heritage of all stars except those so situated that planetary materials would be swallowed up by greater masses or cast off through gravitational action.
Of Stars and Men: Human Response to an Expanding Universe
Chapter 7 (p. 112)
Beacon Press. Boston, Massachusetts, USA. 1958

Siegel, Eli   1902–78
American philosopher, poet, critic and founder of Aesthetic Realism

The planets show grandeur and nicety in their operations; the question is, how did they learn this?
Damned Welcome
Aesthetic Realism, Maxims, Part One, #50 (p. 26)
Definition Press. New York, New York, USA. 1972

Standage, Tom
English journalist and author

A planet is, by definition, an unruly object.
The Neptune File
Chapter 2 (p. 19)
Walker & Company. New York, New York, USA. 2000

Swedenborg, Emanuel   1688–1772
Swedish scientist, theosophist, and mystic

[There are] many earths, inhabited by man…thousands, yea, ten thousands of earths, all full of inhabitants…not only in this solar system, but also beyond it, in the starry heaven.
The Earths in Our Solar System, Which Are Called Planets, and Earths in the Starry Heavens
New Church Board of Publication. New York, New York, USA. 1876

Tagore, Rabindranath   1861–1941
Indian poet and philosopher

Through millions and millions of years,
The stars shine,
Fiery whirlpools revolve and rise
In the dark ever-moving current of time.
In this current
The earth is a bubble of mud…
Translated by Indu Dutt
Our Universe (p. 43)
Jaico Publishing House. Bombay, India. 1969

Teilhard de Chardin, Pierre   1881–1955
French Jesuit, paleontologist, and biologist

Despite their vastness and splendor the stars cannot carry the evolution of matter much beyond the atomic series: it is only on the very humble planets, on them alone, that the mysterious ascent of the world into the sphere of higher complexity has a chance to take place. However inconsiderable they may be in the history of sidereal

bodies, however accidental their coming into existence, the planets are finally nothing less than the key-points of the Universe. It is through them that the axis of life now passes; it is upon them that the energies of an Evolution principally concerned with the building of large molecules is now concentrated.
The Future of Man
Chapter VI, Part I, Section I (p. 114)
Harper & Row, Publishers. New York, New York, USA. 1964

Tennyson, Alfred (Lord)   1809–92
English poet

This world was once a fluid haze of light,
Till toward the centre set the starry tides,
And eddied into suns, that wheeling cast
The planets.
Alfred Tennyson's Poetical Works
The Princess, Part Second, l. 101–103
Oxford University Press, Inc. London, England. 1953

Tombaugh, Clyde   1906–97
American astronomer

Behold the heavens and the great vastness thereof, for a planet could be anywhere therein.
Thou shalt dedicate thy whole being to the search project with infinite patience and perseverance.
Thou shalt set no other work before thee, for the search shall keep thee busy enough.
Thou shalt take the plates [photographs] at opposition time lest thou be deceived by asteroids near their stationary positions.
Thou shalt duplicate the plates of a pair at the same hour angle lest refraction distortions overtake thee.
Thou shalt give adequate overlap of adjacent plate regions lest the planet play hide and seek with thee.
Thou must not become ill at the dark of the moon lest thou fall behind the opposition point.
Thou shalt have no dates except at full moon when long-exposure plates cannot be taken at the telescope.
Many false planets shall appear before thee, hundreds of them, and thou shalt check every one with a third plate.
In David H. Levy
Clyde Tombaugh: Discoverer of Planet Pluto
Chapter 12, Ten Special Commandments for a Would-Be Planet Hunter (p. 180)
University of Arizona Press. Tucson, Arizona, USA. 1991

Voltaire (François-Marie Arouet)   1694–1778
French writer

It would be very singular that all Nature, all the planets, should obey eternal laws, and there should be a little animal, five feet high who, in contempt of these laws, could act as he pleased, solely according to his caprice.
In John D. Barrow
The World Within the World (p. 55)
Clarendon Press. Oxford, England. 1988

PLANET: EARTH

Abbey, Edward   1927–89
American environmentalist and nature writer

We know so very little about this strange planet we live on, this haunted world where all answers lead only to more mystery.
The Crooked Word
Audubon Magazine, Volume 77, Number 6, 1975 (p. 24)

We are obliged, therefore, to spread the news, painful and bitter though it may be for some to hear, that all living things on earth are kindred.
Desert Solitaire
The Serpents of Paradise (p. 24)
Ballantine Books. New York, New York, USA. 1968

Yes. Feet on earth. Knock on wood. Touch stone. Good luck to all.
Desert Solitaire
Bedrock and Paradox (p. 301)
Ballantine Books. New York, New York, USA. 1968

The earth is not a mechanism but an organism, a being with its own life and its own reasons, where the support and sustenance of the human animal is incidental.
The Journey Home: Some Words in Defense of the American West
Chapter 21 (p. 225)
E.P. Dutton. New York, New York, USA. 1977

The world is what it is, no less and no more, and therein lies its entire and sufficient meaning.
A Voice Crying in the Wilderness: Notes from a Secret Journal
Chapter 9 (p. 89)
St. Martin's Press. New York, New York, USA. 1989

I am not an atheist but an earthiest. Be true to the earth.
Desert Solitaire
Down the River (p. 208)
Ballantine Books. New York, New York, USA. 1968

Ackerman, Diane   1948–
American writer

Long ago, Earth bunched its granite to form the continents, ground molar Alps and Himalayas, rammed Africa and Italy into Europe, gnashing its teeth, till mountain ranges buckled and churned, and oceans (salty once rivers bled flavor from the seasoned earth) gouged their kelpy graves. And the rest is history…
The Planets: A Cosmic Pastoral
Earth, III (p. 36)
William Morrow & Company, Inc. New York, New York, USA. 1976

Agassiz, Jean Louis Rodolphe   1807–73
Swiss-born American naturalist, geologist, and teacher

…there was a time when our earth was in a state of igneous fusion, when no ocean bathed it and no atmosphere surrounded it, when no wind blew over it and no rain fell upon it, but an intense heat held all its materials in

solution. In those days the rocks which are now the very bones and sinews of our mother Earth — her granites, her porphyries, her basalts, her syenites — were melted into a liquid mass.
Geological Structures
Chapter I (p. 2)
Ticknor & Fields. Boston, Massachusetts, USA. 1866

Airy, George Biddell   1801–92
English astronomer

Since Astronomy first assumed the form of a Science, the inquiry into the Figure and dimensions of the Earth has always excited the interest of Philosophers. It can hardly be doubted that in the mind of a reflecting man there would always be a desire to know the nature of the Planet upon which he existed; but without Science of an exalted order, it would be impossible for him to gratify his curiosity.
Encyclopaedia Metropolitana, Volume 5, 1845

Arendt, Hannah   1906–75
Political philosopher

The earth is the very quintessence of the human condition…
The Human Condition
Prologue (p. 2)
The University of Chicago Press. Chicago, Illinois, USA. 1958

Aristotle   384 BCE–322 BCE
Greek philosopher

There is much change, I mean, in the stars which are overhead, and the stars seen are different, as one moves northward or southward. Indeed there are some stars seen in Egypt and in the neighborhood of Cyprus which are not seen in the northerly regions; and stars, which in the north are never beyond range of observation, in those regions rise and set. All of which goes to show not only that the earth is circular in shape, but also that it is a sphere of no great size: for otherwise the effect of so slight a change of place would not be so quickly apparent.
On the Heavens
Book II, Chapter 14
Encyclopædia Britannica, Inc. Chicago, Illinois, USA. 1952

Ball, Philip   1962–
English science writer

What is so special about the Earth, then, is not that it is a world of water, but that the water is marine blue — we have oceans, not just glassy sheets of bright ice. Perhaps, soon after the solar system was formed, blue worlds were commonplace, until one by one they turned pearly or ruddy, or became shrouded in bright acid. And then there we were, a lone blue dot, waiting for life to begin.
Life's Matrix: A Biography of Water
Part One, Chapter 4 (p. 111)
Farrar, Straus & Giroux. New York, New York, USA. 2000

Barrell, Joseph   1869–1919
American geologist

The scheme of the Universe is more profound and the unknown is a little nearer than it was recently though to be. But such has been the progress of knowledge since man, in the days before the advent of science naively regarded the earth, his home, as firmament, created a few thousand years previously especially for his benefit.
In J.H.F. Umbgrove
The Pulse of the Earth
Chapter I (p. 2)
Martinus Nijhoff. The Hague, Netherlands. 1947

Bellamy, David   1933–
Botanist, author and broadcaster

The earth has a mass of 5.97×10^{24} kilograms…a big number and one that really matters because that is all the matter we have got.
Forces of Life: The Botanic Man
Chapter 2 (p. 24)
Crown Publishers. New York, New York, USA. 1979

Beston, Henry   1888–1968
American writer

Touch the earth, love the earth, honour the earth, her plains, her valleys, her hills, and her seas; rest your spirit in her solitary places.
The Outermost House
Chapter X (p. 222)
Rinehart & Company. New York, New York, USA. 1928

Borland, Hal   1900–78
American writer

I am not quite sure what the earth's business is, but I know it is not the nurturing of *Homo sapiens*, or ay one species of animal or plant.
Borland Country
Foreword (p. 7)
J.B. Lippincott Company. Philadelphia, Pennsylvania, USA. 1971

Bradley, Jr., John Hodgdon   1898–1962
American geologist

The earth, however, never forgets. While men are sleeping she is awake, silently strengthening the cords of he influence. When men make boast of their conquests she is not concerned, for she knows that the limits of human attainment are the limits she chooses to set. When they strut through the kingdom they think they have conquered, she tightens the strings that hold them to her hand.
Autobiography of Earth
Chapter XII (p. 331)
Coward-McCann, Inc. New York, New York, USA. 1935

The Earth is a selfish mother who would keep her children forever at her breast.
Autobiography of Earth

Chapter XII (p. 331)
Coward-McCann, Inc. New York, New York, USA. 1935

Broad, William   1951–
Science writer
Wade, Nicholas
British-born scientific writer

The ultimate gatekeeper of science is neither peer reviews, nor referees, nor replication, nor the universalism implicit in all three mechanisms. It is time. In the end, bad theories don't work, fraudulent ideas don't explain the world so well as true ideas do. The ideal mechanisms by which science should work are applied to a large extent in retrospect.... Time and the invisible boot that kicks out all useless science are the true gatekeepers of science. But these inexora-ble mechanisms take years, sometimes more than a millennium, to operate. During the interval, fraud may flourish, particularly if it can find shelter under the mantle of immunity that scientific elitism confers.
Betrayers of the Truth (p. 106)
Simon & Schuster. New York, New York, USA. 1982

Burnet, Thomas   1635–1715
English cleric and scientist

We must therefore be impartial where the Truth requires it, and describe the Earth as it is really in it self; and though it be handsome and regular enough to the eye in certain parts of it, single tracts and single Regions; yet if we consider the whole surface of it, or the whole Exteriour Region, 'tis as a broken and confus'd heap of bodies, plac'd in no order to one another, nor with any correspondency or regularity of parts: And such a body as the Moon appears to us, when 'tis look'd upon with a good Glass, rude and ragged; as it is also represented in the modern Maps of the Moon; such a thing would the Earth appear if it was seen from the Moon. They are both in my judgment the image or picture of a great Ruine, and have the true aspect of a World lying in its rubbish.
The Sacred Theory of the Earth (2nd edition)
Book I, Chapter IX (p. 91)
Printed by R. Norton. London. 1691

Burroughs, William S.   1914–97
American writer

After one look at this planet any visitor from outer space would say 'I WANT TO SEE THE MANAGER.'
The Adding Machine: Selected Essays
Women: A Biological Mistake (p. 124)
Seaver Books. New York, New York, USA. 1986

Byron, George Gordon, 6th Baron Byron   1788–1824
English Romantic poet and satirist

He saw with his own eyes the moon was round,

Was also certain that the earth was square.
Because he had journey'd fifty miles, and found
No sign that it was circular anywhere.
The Complete Poetical Works of Byron
Don Juan
Canto V
Houghton Mifflin Company. Boston, Massachusetts, USA. 1933

Chamberlain, Rollin T.
American geologist

Just as the written life of some famous man properly commences with a portrayal of his family antecedents, so any real history of the earth should begin with the activities of the sun and the origin of its present family of planets.
In H.H. Newman (ed.)
The Nature of the World and of Man
The Origin and Early Stages of the Earth (p. 31)
The University of Chicago Press. Chicago, Illinois, USA. 1927

Chief Seattle   ca. 1784–1866
Chief of the Duwamish, Suquamish, and allied Indian tribes

You must teach your children that the ground beneath their feet is the ashes of your grandfathers. So that they will respect the land, tell your children that the earth is rich with the lives of our kin. Teach your children what we have taught our children, that the earth is our mother. Whatever befalls the earth befalls the sons of the earth. If men spit upon the ground, they spit upon themselves.
Catch the Whisper of the Wind: Collected Stories and Proverbs from Native Americans
Attributed to Chief Seattle (p. 41)
Health Communications, Inc. Deerfield Beach, Florida, USA. 1995

Cloos, Hans   1885–1951
German geologist

The earth gives us more knowledge of ourselves than all the books, because it resists us.
Conversations with the Earth
Chapter VII (p. 99)
Alfred A. Knopf. New York, New York, USA. 1953

The earth is large and old enough to teach modesty; and yet it is small enough to be comprehended and to be learned from, as our understanding of it increases.
Conversation with the Earth
Prologue (p. 8)
Alfred A. Knopf. New York, New York, USA. 1953

Earth: beautiful, round, colorful planet. You carry us safely through the emptiness and deadness of space. Graciously you cover the black abyss with air and water. You turn as towards the sun, that we may be warm and content, that we may wander, with open eyes, through your meadows, and look upon your splendor. And then you turn us away from the too fiercely burning sun, that we may rest in the coolness of the night from life's heat and the struggle of the day.
Conversation with the Earth

Prologue (p. 3)
Alfred A. Knopf. New York, New York, USA. 1953

Coleridge, Samuel Taylor 1772–1834
English lyrical poet, critic, and philosopher

Earth! Thou mother of numberless children, the nurse and the mother,
Sister thou of the stars, and beloved by the Sun, the rejoicer!
Guardian and friend of the moon, O Earth, whom the comets forget not,
Yea, in the measureless distance wheel round and again they behold thee!
The Complete Poetical Works of Samuel Taylor Coleridge (Volume 1)
Hymn to the Earth (p. 328)
The Clarendon Press. Oxford, England. 1912

Coman, Dale Rex 1906–
American research physician and wildlife writer

The planet Earth is a doomed ship being blown to inevitable disaster by the winds of universal dynamics, a miniscule sacrifice to the attainment of that inscrutable destiny held secret by the galaxies as they rush through the long darkness of space.
The Endless Adventure
Once There Was a Planet (p. 181)
Henry Regnery Company. Chicago, Illinois, USA. 1972

I peer into the endlessness of space. The lights from a host of suns leap across the millions of miles of darkness to reach this spinning bit of star stuff we call Earth. This outcropping of rock has unknowingly endureth the passage of millennia. The footpads of extinct species have pattered over it. An Indian in moccasins once stood upon it. Forests have appeared, disappeared and reappeared around it. It has been washed by floods, buried in ice and baked by the sun. Here it will lie through the ages yet to come, silently awaiting the next phase of a planet's destiny.
The Endless Adventure
Once There Was a Planet (p. 185)
Henry Regnery Company. Chicago, Illinois, USA. 1972

da Vinci, Leonardo 1452–1519
Italian High Renaissance painter and inventor

[We] may say that the earth has a spirit of growth; that its flesh is the soil, its bones are the successive strata of the rocks which form the mountains, its muscles are the tufa stone, its blood the springs of its waters. The lake of blood that lies about the heart is the ocean; its breathing is by the increase and decrease of the blood in its pulses, and even so in the earth is the flow and ebb of the sea. And the heat of the spirit of the world is the fire which is spread throughout the earth; and the dwelling-place of its creative spirit is in the fires, which in diverse parts of the earth are breathed out in baths and sulphur mines, and in volcanoes…

Leonardo da Vinci's Note Books (pp. 130–131)
Duckworth & Company. London, England. 1906

Dubos, René Jules 1901–82
French-born American microbiologist and environmentalist

The incredible beauty of the earth as seen from space results largely from the fact that our planet is covered with living things. What gives vibrant colors and exciting variety to the surface of the earth is the fact that it is literally a living organism.
Federal Highway Act of 1970 and Miscellaneous Bills
United States Congress. Senate. 1970

The spaceship Earth is the cage within and against which man has developed in his evolutionary past and continues to develop his biological and mental characteristics. As the terrestrial environment deteriorates so does humanness and the quality of human life.
Reason Awake
Chapter 5 (pp. 191–192)
Columbia University Press. New York, New York, USA. 1970

Dunlap, Ellen L.
No biographical data available

The building of the earth was dramatic beyond our imagination. The shaping of the continents and the ocean depths required titanic convulsions. The Supreme One did not shout the Earth into instant being any more than an architect would order the instant erection of a skyscraper. The ground work must be laid, and, little by little, the job progresses. It required several billions of years to make the Earth what it is today, and the job is not yet done.
From Aunt Nellie's Notebook
Nature Magazine, January 1958 (p. 17)

Eddington, Sir Arthur Stanley 1882–1944
English astronomer, physicist, and mathematician

My subject [astronomy] disperses the galaxies, but it unites the Earth.
In Arthur Beer (ed.)
Vistas in Astronomy (Volume 2)
Meeting of the International Astronomical Union, Cambridge, Massachusetts, USA., September, 1932 (p. i)

Ehrenreich, Barbara 1941–
American social critic and essayist

Some of us still get all weepy when we think about the Gaia Hypothesis, the idea that earth is a big furry goddess-creature who resembles everybody's mom in that she knows what's best for us. But if you look at the historical record — Krakatoa, Mt. Vesuvius, Hurricane Charley, poison ivy, and so forth down the ages — you have to ask yourself: Whose side is she on, anyway?
The Worst Years of Our Lives
The Great Syringe Tide (p. 55)
Pantheon Books. New York, New York, USA. 1981

Einstein, Albert   1879–1955
German-born physicist

There has been an earth for a little more than a billion years. As for the question of the end of it I advise: Wait and see.
In Eli Maor
To Infinity and Beyond: A Cultural History of the Infinite (p. 182)
Birkhäuser. Boston, Massachusetts, USA. 1987

Ficino, Marsilio   1433–1499
Early Italian humanist philosopher

We see the Earth give birth, thanks to varieties of seeds, to a multitude of trees and animals, nourish them, and make them grow; we see her cause even stones to grow as her teeth, vegetable life as hairs, as long as they remain connected to their roots, while if they are removed or unearthed they cease growing. Could we say that the breast of this female lacks life, she who spontaneously gives birth and nourishes so many off-spring, who sustains herself and whose back carries teeth and hair?
Theologica plantonica (Volume 1) (p. 144)
Harvard University Press. Cambridge, Massachusetts, USA. 1917

Galilei, Galileo   1564–1642
Italian physicist and astronomer

For my part, I consider the earth very noble and admira-ble precisely because of the diverse alterations, changes, generations, etc., that occur in it incessantly. If, not being subject to any change, it were a vast desert of sand or a mountain of jasper, or if at the time of the flood the waters which covered it had frozen, and it had remained an enormous globe of ice where nothing was ever born or ever altered or changed, I should deem it a useless lump in the universe, devoid of any activity and, in a word, superfluous and essentially nonexistent.
Dialogues Concerning the Two Chief World Systems
The First Day (p. 58)
University of California Press. Berkeley, California, USA. 1953

Gray, George W.
Freelance science writer

Perhaps the Earth is a clod, but if so it is a vibrant clod, responsive to an endless symphony — or cacophony — of cosmic influences.
The Advancing Front of Science
Chapter II (p. 25)
Whittlesey House. New York, New York, USA. 1937

Guiterman, Arthur   1871–1943
Poet

We dwell within the Milky Way,
Our Earth, a paltry little mommet,
Suspended in a grand array
Of constellation, moon and comet.

Gaily the Troubadour
Outline of the Universe (p. 70)
E.P. Dutton & Company, Inc. New York, New York, USA. 1936

Hageman, Samuel M.   1848–1905
American clergyman and poet

Earth is but the frozen echo of the silent voice of God.
Silence
Silence, Stanza XIX
Dodd, Mead and Company. New York, New York, USA. 1877

Hardy, Thomas   1840–1928
English poet and regional novelist

Let me enjoy the earth no less
Because the all-enacting Might
That fashioned forth its loveliness
Had other aims than my delight.
Collected Poems of Thomas Hardy
Let Me Enjoy the Earth
Macmillan & Company Ltd. London, England. 1920

Hillel, Daniel
No biographical data available

We have all come out of the earth, and are its children. The earth has always nurtured us, despite our scornful abuse, and we can no longer continue to behave as its ungrateful offspring. It is time for us, as *Homo sapiens curans*, to nurture the earth in return.
Out of the Earth: Civilization and the Life of the Soil
Chapter 30 (p. 283)
The Free Press. New York, New York, USA. 1991

Hoyle, Sir Fred   1915–2001
English mathematician and astronomer

…once a photograph of the Earth, taken from the outside, is available, a new idea as powerful as any in history will be let loose.
In Martin Redfern
The Earth: A Very Short Introduction
Chapter 1 (p. 1)
Oxford University Press, Inc. Oxford, England. 2003

Humphrey, Hubert H.   1911–78
38th vice-president of the United States

As we begin to comprehend that the earth itself is a kind of manned spaceship hurtling through the infinity of space — it will seem increasingly absurd that we have not better organized the life of the human family.
Speech, 26 September 1966

Hutton, James   1726–97
Scottish geologist, chemist, and naturalist

This globe of the earth is a habitable world; and on its fitness for this purpose, our sense of wisdom in its forma-tion must depend.
The Theory of the Earth (Volume 1)
Part I, Chapter I, Section I (p. 4)
Messrs. Cadwell, Junior, and Davies. London, England. 1795

When we trace the parts of which this terrestrial system is composed, and when we view the general connection of those several parts, the whole presents a machine of a peculiar construction by which it is adapted to a certain end. We perceive a fabric, erected in wisdom, to obtain a purpose worthy of the power that is apparent in the production of it.

The Theory of the Earth (Volume 1)
Part I, Chapter I, Section I (p. 3)
Messrs. Cadwell, Junior, and Davies. London, England. 1795

Huygens, Christiaan 1629–95
Dutch mathematician, astronomer, and physicist

…how vast those Orbs must be, and how inconsiderable this Earth, the Theatre upon which all our mighty Designs, all our Navigations, and all our Wars are transacted, is when compared to them. A very fit Consideration, and matter of Reflection, for those Kings and Princes who sacrifice the Lives of so many People, only to flatter their Ambition in being Masters of some pitiful corner of this small Spot.

The Celestial Worlds Discover'd, or, Conjectures Concerning the Planetary Worlds, Their Inhabitants and Productions
Book the Second. The Immense Distance Between the Sun and the Planets Illustrated (pp. 141–142)
Printed for T. Childe. London, England. 1698

Irwin, James 1930–91
American astronaut

The Earth reminded us of a Christmas tree ornament hanging in the blackness of space. As we got farther and farther away it diminished in size. Finally it shrank to the size of a marble, the most beautiful marble you can imagine. That beautiful, warm, living object looked so fragile, so delicate, that if you touched it with a finger it would crumble and fall apart. Seeing this has to change a man, has to make a man appreciate the creation of God and the love of God.

In Kevin W. Kelley
The Home Planet
With Plate 38
Addison-Wesley Publishing Company. Reading, Massachusetts, USA. 1988

Jeans, Sir James Hopwood 1877–1946
English physicist and mathematician

Standing on our microscopic fragment of a grain of sand, we attempt to discover the nature and purpose of the universe which surrounds our home in space and time.

The Mysterious Universe
Chapter I (p. 3)
The Macmillan Company. New York, New York, USA. 1932

So long as the earth was believed to be the center of the universe the question of life on the other worlds could hardly arise; there are no other worlds in the astronomical sense, although a heaven above and a hell beneath might form adjuncts to this world.

Is There Life on the Other Worlds
Annual Report of the Board of Regents of the Smithsonian Institution, 1942
(p. 145)
Government Printing Office. Washington, D.C. 1943

Jeffers, Robinson 1887–1962
American poet

It is only a little planet
But how beautiful it is.

The Beginning and the End and Other Poems
How Beautiful It Is (p. 29)
Random House, Inc. New York, New York, USA. 1963

Johnson, Lyndon B. 1908–73
36th president of the United States

Think of our world as it looks from that rocket that's heading toward Mars. It is like a child's globe, hanging in space, the continents stuck to its side like colored maps. We are all fellow passengers on a dot of earth. And each of us, in the span of time, has really only a moment among our companions.

Inaugural Address, January 20, 1965

Kahn, Fritz 1888–1958
German-born American writer and conceptual medical illustrator

This is the universe: infinity. Space without beginning, without end, dark, empty, cold. Through the silent darkness of this space more gleaming spheres, separated from each other by inconceivable distances. Around them again inconceivably far away, like bits of dust lost in immensity, circle smaller dark spheres, receiving light and life from their "mother suns." One of these little spheres in the light of one of the countless suns in endless space, is our earth. This is man's home in the universe.

Design of the Universe
Chapter One (p. 2)
Crown Publishers. New York, New York, USA. 1954

Kelvin, Lord William Thomson 1824–1907
Scottish engineer, mathematician, and physicist

Considering the almost certain truth that the earth was built up of meteorites falling together, we may follow in imagination the whole process of shrinking from gaseous nebula to liquid lava and metals, and solidification of liquid from central regions outward.

The Age of the Earth as An Abode Fitted for Life
Science, New Series, Volume 9, Number 229, May 19, 1898 (p. 706)

All these reckonings of the history of underground heat, the details of which I am sure you do not wish me to put before you at present, are founded on the very sure assumption that the material of our present solid earth all round its surface was at one time a white-hot liquid.

The Age of the Earth As An Abode Fitted for Life
Science, New Series, Volume 9, Number 229, May 19, 1898 (p. 672)

MacLeish, Archibald   1892–1982
American poet and Librarian of Congress

To see the earth as we now see it, small and beautiful in that eternal silence where it floats, is to see ourselves as riders on the earth together, brothers on that bright loveliness in the unending night — brothers who see now they are truly brothers.
Riders on the Earth
Bubble of Blue Air (p. xiv)
Houghton Mifflin Company. Boston, Massachusetts, USA. 1978

Marvin, Ursula
American geologist

In learning about the earth — a most fundamental preoccupation of man — geologists are limited to direct examination of its outermost surface and that of its nearest neighbor, the moon. All knowledge of the deep interior of our planet and of the nature of other planets in the solar system is gathered by remote sensing devices. Characteristically, the signals from these devices are interpreted differently by different scientists and every new advance tends to raise new problems, leaving us acutely aware of the limitations of our knowledge.
Continental Drift: Evolution of a Concept
Postscript (p. 207)
Smithsonian Institution Press. Washington, D.C. 1973

Masson, David
No biographical data available

Each orb has had it history. For ours,
It blazed and steamed, cooled and contracted, till,
Tired of mere vaporing within the grasp
Of ruthless condensation, it assumed
The present form, proportions, magnitude —
Our Tidy ball, axeled eight thousand miles.
In Alexander Winchell
World-Life or Comparative Geology
Chapter III (p. 338)
S.C. Griggs & Company. Chicago, Illinois, USA. 1883

Momaday, N. Scott   1934–
Native American writer

Once in his life a man ought to concentrate his mind upon the remembered earth, I believe. He ought to give himself up to a particular landscape in his experience, to look at it from as many angles as he can, to wonder about it, to dwell upon it. He ought to imagine that he touches it with his hands at every season and listens to the sounds that are made upon it. He ought to imagine the creatures there and all the faintest motions of the wind. he ought to recollect the glare of noon and all the colors of the dawn and dusk.
The Way to the Rainy Mountain
The Closing, XXIV (p. 83)
University of New Mexico Press. Albuquerque, New Mexico, USA. 1969

Montague, C. E.
No biographical data available

The earth with no history to it — what it would be if it had all been made only last night and were not a worn ancient face, seamed, stained, and engraved with endless cross-hatching of documentary wrinkles, its mountains the ruins of more wondrous height now all but erased.
In A.C. Seward
Plant Life Through the Ages: A Geological and Botanical Retrospect
Chapter 1 (p. 1)
Hafner Publishing Company. New York, New York, USA. 1959

Morton, Oliver
Science and technology editor

If the space age has opened new ways of seeing mere matter, though, it has also fostered a strange return to something reminiscent of the pre-Copernican universe. The life that Lowell and his like expected elsewhere has not appeared, and so the Earth has become unique again. The now-iconic image of a blue-white planet floating in space, or hanging over the deadly deserts of the moon, reinforces the Earth's isolation and specialness. And it is this exceptionalism that drives the current scientific thirst for finding life elsewhere, for finding a cosmic mainstream of animation, even civilization, in which the Earth can take its place. It is both wonderful and unsettling to live on a planet that is unique.
Mapping Mars: Science, Imagination and the Birth of a World
A Point of Warlike Light (p. 14)
Fourth Estate. London, England. 2002

Muir, John   1838–1914
American naturalist

…when we contemplate the whole globe as one great dewdrop, striped and dotted with continents and islands, flying through space with other stars all singing and shining together as one, the whole universe appears as an infinite storm of beauty.
Travels in Alaska
Chapter I (p. 5)
Houghton Mifflin Company. Boston, Massachusetts, USA. 1915

Newman, Joseph S.   1892–1960
American poet

This ball was once a glowing mass
Of mixed and superheated gas
Which cooled to liquid, shrank in girth,
Solidified and turned to earth.
Poems for Penguins and Other Lyrical Lapses
Geology
Greenburg. New York, New York, USA. 1941

Peattie, Donald Culrose   1896–1964
American botanist, naturalist, and author

Old earth is great with her children, the bulb and the grub, and the sleepy mammal and the seed.

An Almanac for Moderns
April Sixth (p. 19)
G.P. Putnam's Sons. New York, New York, USA. 1935

Platt, John R.
No biographical data available

The earth is finite, and when we have come to the ends of it, we have come to the ends of it.
In Robert M. Hutchins and Mortimer J. Adler (eds.)
The Great Ideas Today, 1968
The New Biology and the Shaping of the Future (p. 124)
Encyclopædia Britannica, Inc. Chicago, Illinois, USA. 1966

Pollard, William
No biographical data available

…the earth with its vistas of breathtaking beauty, its azure seas, beaches, mighty mountains, and soft blanket of forest and steppe is a veritable wonderland in the universe. It is a gem of rare and magic beauty hung in a trackless space filled with lethal radiations and accompanied in its journey by sister planets which are either viciously hot or dreadfully cold, arid, and lifeless chunks of raw rocks. Earth is choice, precious, and sacred beyond all comparison or measure.
In Michael Hamilton (ed.)
This Little Planet
God and His Creation (p. 59)
Charles Scribner's Sons. New York, New York, USA. 1970

Pupin, Michael
Physicist

Our terrestrial globe is a celestial casting, and he who like myself learned the language of the foundry in his early youth will ask the human question: What is the mission of this celestial casting, this old celestial wanderer through the mighty stream of chaotic solar radiation? Is it only to receive its final tempering from the solar furnace which gave it its birth, and to smooth out its jagged surface by the erosive action of the waters which solar radiation carries in ceaseless succession of cycles from the oceans to the higher continental elevations? The answer to this human question is obvious; it is this: The highest mission of this celestial casting, which we call affectionately "our mother earth," is to provide a congenial home for a new universe, "the universe of organic life."
The New Reformation: From Physical to Spiritual Realities
Chapter VII, Section IV (pp. 235–236)
Charles Scribner's Sons. New York, New York, USA. 1928

Robinson, Victor   1886–1947
Physician

Earth is her own historian, and in every age writes her story in forests and deserts, on rocks and in river-beds.
The Story of Medicine
Chapter I (p. 1)
The New York Home Library. New York, New York, USA. 1943

Sagan, Carl   1934–96
American astronomer and author

There are some hundred billion (10^{11}) galaxies, each with, on the average a hundred billion stars. In all the galaxies, there are perhaps as many planets as stars, $10^{11} \times 10^{11} = 10^{22}$, ten billion trillion. In the face of such overpowering numbers, what is the likelihood that only one ordinary star, the Sun, is accompanied by an inhabited planet? Why should we, tucked away in some forgotten corner of the Cosmos, be so fortunate? To me, it seems far more likely that the universe is brimming over with life. But we humans do not yet know. We are just beginning our explorations. The only planet we are sure is inhabited is a tiny speck of rock and metal, shining feebly by reflected sunlight, and at this distance utterly lost.
Cosmos
Chapter I (p. 5, 7)
Random House, Inc. New York, New York, USA. 1980

Our planet is a lonely speck in the great enveloping cosmic dark. In our obscurity, in all this vastness, there is no hint that help will come from elsewhere to save us from ourselves.
Pale Blue Dot: A Vision of the Human Future In Space
Chapter 1 (p. 9)
Random House, Inc. New York, New York, USA. 1994

If we are to understand the Earth, we must have a comprehensive knowledge of the other planets.
The Solar System
Scientific American, Volume 233, Number 3, 1975 (p. 27)

Schneider, Herman   1905–2003
Polish-born American educator and author
Schneider, Nina
No biographical data available

The story of the earth is in a leaf and in a stone; in a cloud and in the sea. The leaf was once a stone; the cloud was once the sea. The earth tells its story over and over again-the leaf will become a stone, the cloud will become the sea again.
Rocks, Rivers and the Changing Earth: A First Book About Geology
Part One. A Leaf and A Stone (p. 3)
William R. Scott, Inc. New York, New York, USA. 1952

Scrope, George Poulett   1797–1876
English geologist and political economist

Towards the end of the last century, men of science became convinced of the futility of those crude and fanciful speculations on the original state of the earth, in which cabinet geologists had for some time indulged; and justly perceived that the only sure road to the true history of our planet lies in a minute and practical study of those portions of its surface which are open to our examination, and in their comparison with the results of those changes

and operations which the ever-active hand of Nature is still carrying on upon that surface.
Memoir on the Geology of Central France
Preface (p. v)
Longman, Rees, Orme, Brown & Green. London, England. 1827

Sedgewick Seti   1854–1913
English geologist

The earth is our cradle,
The solar system our kindergarten,
The galaxy our middle-school and
The universe our university.
Filler material
Cosmic Search Magazine, Volume 1, Number 1, January 1979 (p. 24)

Sexton, Anne   1928–74
American poet

God owns heaven, but He craves the earth.
The Awful Rowing Toward God
The Earth
Houghton Mifflin Company. Boston, Massachusetts, USA. 1975

Shaler, Nathaniel Southgate   1841–1906
American geologist

…earth-lore is not a discrete science at all, but is that way of looking at the operations of energy in the physical, chemical and organic series which introduces the elements of space and time into the considerations and which furthermore endeavors to trace the combination of the various trends of action in the stages of the developments of the earth. It is in these peculiarities of geology that we find the basis of its value in education and in the general culture of society.
Relations of Geologic Science to Education
Bulletin of the American Geological Society, Volume 7, Number x, 1896 (p. 319)

Taine, Hippolyte   1828–93
French critic and historian

Amid this vast and overwhelming space and in these boundless solar archipelagoes, how small is our own sphere, and the earth, what a grain of sand!
Translated by John Durand
The Ancient Regime
Book Third, Chapter II (p. 175)
Henry Holt & Company. New York, New York, USA. 1881

Teilhard de Chardin, Pierre   1881–1955
French Jesuit, paleontologist, and biologist

Without being overawed by the improbable, let us now concentrate our attention on the planet we call Earth. Enveloped in the blue mist of oxygen which its life breathes, it floats at exactly the right distance from the sun to enable the higher chemisms to take place on its surface. We do well to look at it with emotion. Tiny and

isolated though it is, it bears clinging to its flanks the destiny and future of the Universe.
The Future of Man
Chapter VI, Part I, Section I (p. 114)
Harper & Row, Publishers. New York, New York, USA. 1964

Tennyson, Alfred (Lord)   1809–92
English poet

Many an aeon moulded earth before her highest, man, was born,
Many an aeon too may pass when earth is manless and forlorn,
Earth so huge and yet so bounded — pools of salt and plots of land — Shallow skin of green and azure — chains of mountains, grains of sand!
Alfred Tennyson's Poetical Works
Locksley Hall, Sixty Years After, Stanza 103
Oxford University Press, Inc. London, England. 1953

Thomas, Lewis   1913–93
American physician and biologist

Viewed from the distance of the moon, the astonishing thing about the earth, catching the breath, is that it is alive. The photographs show the dry, pounded surface of the moon in the foreground, dry as an old bone. Aloft, floating free beneath the moist, gleaming, membrane of bright blue sky, is the rising earth, the only exuberant thing in this part of the cosmos.
The Lives of a Cell: Notes of a Biology Watcher
The World's Biggest Membrane (p. 145)
The Viking Press. New York, New York, USA. 1974

The word for earth, at the beginning of the Indo-European language thousands of years ago (no one knows for sure how long ago) was *dhghem*. From this word, meaning simply earth, came our word humus, the handiwork of soil bacteria. Also, to teach us the lesson, [came the words] humble, human, and humane. There is the outline of a philological parable here.
In Lynn Margulis and Dorion Sagan
Microcosmos
Foreword (p. 12)
Summit Books. New York, New York, USA. 1986

The overwhelming astonishment, the queerest structure we know about so far in the whole universe, the greatest of all cosmological scientific puzzles, confounding all our efforts to comprehend it, is the earth.
Late Night Thoughts on Listening to Mahler's Ninth Symphony
The Corner of the Eye (p. 16)
Viking Press. New York, New York, USA. 1983

I have been trying to think of the earth as a kind of organism, but it is no go. I cannot think of it this way. It is too big, too complex, with too many working parts lacking visible connections. The other night, driving through a hilly, wooded part of southern New England, I wondered about this. If not like an organism, what is it like, what is

it most like? Then, satisfactorily for that moment, it came to me: it is most like a single cell.
The Lives of a Cell: Notes of a Biology Watcher
The Lives of a Cell (p. 5)
The Viking Press. New York, New York, USA. 1974

Thoreau, Henry David   1817–62
American essayist, poet, and practical philosopher

The earth is not a mere fragment of dead history, stratum upon stratum like the leaves of a book, to be studied by geologists and antiquaries chiefly, but living poetry like the leaves of a tree, which precede flowers and fruit — not a fossil earth, but a living earth; compared with whose great central life all animal and vegetable life is merely parasitic. Its throes will heave our exuviae from their graves…You may melt your metals and cast them into the most beautiful moulds you can; they will never excite me like the forms which this molten earth flows out into.
The Writings of Henry David Thoreau (Volume 2)
Walden
Chapter XVII (p. 476)
Houghton Mifflin Company. Boston, Massachusetts, USA. 1893

Vitousek, Peter Mooney
No biographical data available
Lubchenco, Harold A.
No biographical data available

…we are changing Earth more rapidly than we are understanding it.
Human Domination of Earth's Ecosystems
Science, Volume 277, Number 5325, July 25, 1997 (p. 498)

Vizinczey, Stephen   1933–
Hungarian author

Is it possible that I am not alone in believing that in the dispute between Galileo and the Church, the Church was right and the centre of man's universe is the earth?
Truth and Lies in Literature: Essays and Reviews
Rules of the Game (p. 269)
Atlantic Monthly Press. Boston, Massachusetts, USA. 1986

Voltaire (François-Marie Arouet)   1694–1778
French writer

"But then to what end?" asked Candide, "was the world formed?"

"To make us mad," said Martin.
The Best Known Works of Voltaire
Candide
Chapter XXI (p. 57)
Blue Ribbon Books. New York, New York, USA. 1940

Wells, H. G. (Herbert George)   1866–1946
English novelist, historian, and sociologist

Our earth…is a spinning globe. Vast though it seems to us, it is a mere speck of matter in the greater vastness of space.
The Outline of History (Volume 1)
Book I, Chapter I, Section 2 (p. 13)
Garden City Books. Garden City, New York, USA. 1961

Whipple, Fred L.   1906–2004
Pioneer in comet research

Our Earth seems so large, so substantial, and so much with us that we tend to forget the minor position it occupies in the solar family of planets. Only by a small margin is it the largest of the other terrestrial planets. True, it does possess a moderately thick atmosphere that overlies a thin patchy layer of water and it does have a noble satellite, about 1/4 its diameter. These qualifications of the Earth, however, are hardly sufficient to bolster our cosmic egotism. But, small as is the Earth astronomically, it is our best-known planet and therefore deserves and has received careful study.
Earth, Moon and Planets
The Earth (p. 60)
Grosset & Dunlap, Publishers. New York, New York, USA. 1958

Whitman, Walt   1819–92
American poet, journalist, and essayist

The earth never tires;
The earth is rude, silent, incomprehensible at first —
Nature is rude and incomprehensible at first —
Be not discouraged — keep on — there are divine things well envelop'd;
I swear to you there are divine things more beautiful than words can tell.
Complete Poetry and Collected Prose
Song of the Open Road
The Library of America. New York, New York, USA. 1982

In this broad earth of ours,
Amid the measureless grossness and the slag,
Enclosed and safe within its central heart
Nestles the seed perfection.
Complete Poetry and Collected Prose
Song of the Universal
The Library of America. New York, New York, USA. 1982

Winchell, Alexander   1824–91
American geologist

The stability of the solid earth is instability itself.
Walks and Talks in the Geological Field
Part I, Chapter XVIII (p. 102)
Chautauqua Press. New York, New York, USA. 1890

Young, Louise B.
Science writer

Time flows on…the planet continues to spin on its path through the unknown reaches of space. We cannot guess its destination or its destiny. The beautiful blue bubble of

matter holds many wonders still unrealized and a mysterious future waiting to unfold.
The Blue Planet
Chapter 14 (p. 266)
Little, Brown & Company. Boston, Massachusetts, USA. 1983

PLANET: JUPITER

Ackerman, Diane   1948–
American writer

Vibrant as an African trade-bead with bonechips in orbit round it, Jupiter floods the night's black scullery, all those whirlpools and burbling aerosols little changed since the solar-system began.
The Planets: A Cosmic Pastoral
Jupiter (p. 81)
William Morrow & Company, Inc. New York, New York, USA. 1976

Sizzi, Francisco
Astronomer

The satellites [Jupiter's moons] are invisible to the naked eye and therefore can have no influence on the earth, and therefore would be useless, and therefore do not exist.
In Oliver Lodge
Pioneers of Science and the Development of Their Scientific Theories
(p. 106)
Dover Publications, Inc. New York, New York, USA. 1926

PLANET: MARS

Barnard, Edward Emerson   1857–1923
American astronomer

To save my soul I can't believe in the canals as Schiaparelli draws them…. I verily believe…that the canals as depicted by Schiaparelli are a fallacy and that they will be so proved before many oppositions are past.
NASA Serial Publication
Letter to Simon Newcomb, September 11, 1894 (p. 6)
Scientific and Information Technical Office, NASA, Washington, D.C. 1962

Boynton, William
No biographical data available

The signal we have been getting loud and clear is there is a lot of ice on Mars.
Evidence of Plentiful Water on Mars
The Associated Press, 2 March 2002

Bradbury, Ray   1920–
American writer

We are all…children of this universe. Not just Earth, or Mars, or this System, but the whole grand fireworks. And if we are interested in Mars at all, it is only because we wonder over our past and worry terribly about our possible future.

In Ray Bradbury, Arthur C. Clarke, Bruce Murray, Carl Sagan and Walter Sullivan
Mars and the Mind of Man
Forward (p. x)
Harper & Row, Publishers. New York, New York, USA. 1973

Cosmo Kramer
Fictional character

I've never been to Mars but I imagine it's quite lovely.
Seinfield
TV series
The pilot (1) 1993

de Fontenelle, Bernard le Bovier   1657–1757
French author

Mars, who affords nothing curious that I know of; his day is rather more than half an hour longer than ours, but his year is twice as long, wanting about a month and near a half. He is about four times less than the earth, and the sun seems not altogether so large and so bright to him, as it appears to us. But let us leave Mars, he is not worthy our stay…
Conversations on the Plurality of Worlds
The Fourth Evening (pp. 118–119)
Printed for Peter Wilson. Dublin, Ireland. 1761

Galilei, Galileo   1564–1642
Italian physicist and astronomer

I dare not affirm that I am able to observe the phases of Mars; nonetheless, if I am not mistaken, I believe I have seen that it is not perfectly round.
Letter to Benedetto Castelli, December 30, 1610
Source undetermined

Huygens, Christiaan   1629–95
Dutch mathematician, astronomer, and physicist

I am apt to believe that the Land in Mars is of a blacker Colour than that of Jupiter or the Moon, which is the reason of his appearing of a Copper Colour, and his reflecting a weaker Light than is proportionable to his distance from the Sun…. His Light and Heat is twice, and sometimes three times less than ours, to which I suppose the Constitution of his Inhabitants is answerable.
The Celestial Worlds Discover'd, or, Conjectures Concerning the Planetary Worlds, Their Inhabitants and Productions
Section 11, Book 2 (p. 111)
Printed for T. Childe. London, England. 1698

Kepler, Johannes   1571–1630
German astronomer

…[the] motions [of Mars] provide the only possible access to the hidden secrets of astronomy, without which we would remain forever ignorant of those secrets.
New Astronomy
Part II, 7 (p. 185)
At The University Press. Cambridge, England. 1992

Kuiper, Gerard P.   1905–73
Dutch-born American astronomer

The hypothesis of plant life…appears still the most satisfactory explanation of the various kinds of dark markings and their complex seasonal and secular changes.
In Steven J. Dick
Life on Other Worlds: The 20ᵗʰ Century Extraterrestrial Life Debate
Chapter 2 (p. 25)
Cambridge University Press. Cambridge, England. 1998

Leovy, Conway B.
No biographical data available

Unlike the moon, whose story appears essentially to have ended one or two billion years ago, Mars is still evolving and changing. On Mars, as on the earth, the most pervasive agent of change is the planet's atmosphere, itself the product of the sorting of the planet's initial constituents that began soon after it condensed from the primordial cloud of dust and gas that gave rise to the solar system 4.6 billion years ago.
The Atmosphere of Mars
Scientific American, Volume 237, Number 1, July 1977 (p. 34)

Longfellow, Henry Wadsworth   1807–82
American poet

There is no light in earth or heaven
But the cold light of stars;
And the first watch of night is given
To the red planet Mars.
The Poetical Works of Henry Wadsworth Longfellow
The Light of Stars, Stanza 2
Houghton Mifflin Company. Boston, Massachusetts, USA. 1883

Lowell, Percival   1855–1916
American astronomer

To account for these phenomena, the explanation that at once suggests itself is, that a direct transference of water takes place over the face of the planet, and that the canals are so many waterways.
Mars
Canals (p. 164)
Houghton Mifflin Company. Boston, Massachusetts, USA. 1895

There are celestial sights more dazzling, spectacles that inspire more awe, but to the thoughtful observer who is privileged to see them well, there is nothing in the sky so profoundly impressive as the canals of Mars.
Mars as the Abode of Life
Part II, Notes (p. 228)
The Macmillan Company. New York, New York, USA. 1908

Thus, not only do the observations we have scanned lead to the conclusion that Mars at this moment is inhabited, but they land us at the further one that these denizens are of an order whose acquaintance was worth the making. Whether we ever shall come to converse with them in any more instant way is a question upon which science at present has no data to provide. More important to us is the fact that they exist, made all the more interesting by their precedence of us in the path of evolution.
Mars as the Abode of Life
Part I, Chapter VI (p. 215)
The Macmillan Company. New York, New York, USA. 1908

The struggle for existence in their planet's decrepitude and decay would tend to evolve intelligence to cope with circumstances growing momentarily more and more adverse. But, furthermore, the solidarity that the conditions prescribe would conduce to a breadth of understanding sufficient to utilize it. Intercommunication over the whole globe is made not only possible, but obligatory. This would lead to the easier spreading over it of some dominant creature, — especially were this being of an advanced order of intellect, — able to rise above its bodily limitations to amelioration of the conditions through the exercise of the mind.
Mars as the Abode of Life
Part I, Chapter IV (p. 143)
The Macmillan Company. New York, New York, USA. 1908

Malin, Michael
Science and technology editor

The Mars we are trying to explore does not exist
In William Sheehan and Stephen James O'Meara
Mars: The Lure of the Red Planet
Prometheus Books. Buffalo, New York, USA. 2001

Morton, Oliver
Science and technology editor

Yet if the Earth is a single isolated planet, the human world is less constrained. The breakdown of the equation between planets and worlds works both ways. If there can now be planets that are not worlds, then there can be worlds that spread beyond planets — and ours is doing so. Our spacecraft and our imaginations are expanding our world. This projection of our world beyond the Earth is for the most part a very tenuous sort of affair. It is mostly a matter of imagery and fantasy. Mars, though, might make it real — which is why Mars matters.
Mapping Mars: Science, Imagination and the Birth of a World
A Point of Warlike Light (p. 14)
Fourth Estate. London, England. 2002

Mars is not an independent world, held together by the memories and meanings of its own inhabitants. But nor is it no world at all. More than any other planet we have seen, Mars is like the Earth. It is not very like the Earth. Its gravity is weak, its atmosphere thin, its surface sealess, its soil poisonous, its sunlight deadly in its levels of ultraviolet, its climate beyond frigid. It would kill you in an instant. But it is earthlike enough that it is possible to imagine some of us going there and experiencing this new part of our human world in the way we've always experienced the old part-from the inside. The fact that

humans could feasibly become Martians is the strongest of the links between Mars and Earth.
Mapping Mars: Science, Imagination and the Birth of a World
A Point of Warlike Light (p. 14)
Fourth Estate. London, England. 2002

Murray, Bruce   1932–
American planetologist

The Mars we had found was just a big moon with a thin atmosphere and no life. There were no Martians, no canals, no water, no plants, no surface characteristics that even faintly resembled Earth's.
Journey into Space: The First Three Decades of Space Exploration
Chapter 1 (p. 43)
W.W. Norton & Company, Inc. New York, New York, USA. 1989

Extending out from the chaotic terrain…are some extraordinary channels, which are also found in a number of other localities on the planet. It is hard to look at these channels without considering the possibility that they were cut by flowing water.
Mars from Mariner 9
Scientific American, Volume 228, Number 1, January 1973 (p. 58)

Schiaparelli, G. V.   1835–1910
Italian astronomer

What strange confusion! What can all this mean? Evidently the planet has some fixed geographical details, similar to those of the Earth.… Comes a certain moment, all this disappears to be replaced by grotesque polygonations and germinations which, evidently, attach themselves to represent apparently the previous state, but it is a gross mask, and I say almost ridiculous.
Corrispondenza su Marte (Volume 2)
Schiaparelli to Terby, June 8, 1888
Letter to François Terby, June 8, 1888
No. 1, 1894
Domus Galilaeana. Pisa, Italy. 1965

Sheeham, William
No biographical data available
O'Meara, Stephen James
No biographical data available

Who is to say that [Mars] will not — like a hardy seed lying dormant beneath the snow of a long winter — come once more to life, and in so doing once more quicken our fondest hopes of life beyond Earth?
Mars: The Lure of the Red Planet
Epilogue (p. 323)
Prometheus Books. Amherst, New York, USA. 2001

Mars is but a tiny pinprick in the vast fabric of space-time, a mere mote in the solar beam.
Mars: The Lure of the Red Planet
Chapter 2 (p. 27)
Prometheus Books. Amherst, New York, USA. 2001

Swift, Jonathan   1667–1745
Irish-born English writer

They have likewise discovered two lesser stars, or satellites, which revolve around Mars, whereof the innermost is distant from the centre of the primary planet exactly three of his diameters, and the outermost five; the former revolves in the space of ten hours, and the latter in twenty-one and a half; so that the squares of their periodical times are very near in the same proportion with the cubes of their distances from the centre of Mars, which evidently shows them to be governed by the same law of gravitation that influences the other heavenly bodies.
In *Great Books of the Western World* (Volume 36)
Gulliver's Travels
Part III, Chapter III (p. 102)
Encyclopædia Britannica, Inc. Chicago, Illinois, USA. 1952

Wallace, Alfred Russel   1823–1913
English humanist, naturalist, and geographer

The conclusion…is therefore irresistible — that animal life, especially in its higher forms, cannot exist on the planet. Mars, therefore, is not only uninhabited by intelligent beings such as Mr. Lowell postulates, but is absolutely UNINHABITABLE.
Is Mars Habitable?
Chapter VIII (p. 110)
Macmillan & Company Ltd. London, England. 1907

Washburn, Mark
No biographical data available

The red fire of Mars burns as bright as ever in the night sky and in the hearts of men. Mars has always been much more than just the next planet out from the sun. Mars is the place where dreams and reality meet — and form new dreams for the curious and questing people of the earth to follow.
Mars At Last!
Chapter 14 (p. 277)
G.P. Putnam's Sons. New York, New York, USA. 1977

PLANET: MERCURY

Ackerman, Diane   1948–
American writer

A prowling holocaust keeling low in the sky heads westward for another milk run. The Sun never sets on the Mercurian empire: it only idles on each horizon and lurches back, broiling the same arc across the sky.
The Planets: A Cosmic Pastoral
Mercury (p. 15)
William Morrow & Company, Inc. New York, New York, USA. 1976

Blackmore, Sir Richard   1650–1729
English physician and writer

Mercurius nearest to the Central Sun,

Does in an Oval Orbit circling run:
But rarely is the Object of our Sight,
In Solar Glory sunk and more prevailing Light.
The Poetical Works of Sir R. Blackmore: Containing Creation: A Philosophical Poem, in Seven Books
Book II, l. 511–514
Printed for C. Cooke. London, England. 1797

de Fontenelle, Bernard le Bovier   1657–1757
French author

…Mercury is the bedlam of the universe…
Conversations on the Plurality of Worlds
The Fourth Evening (p. 105)
Printed for Peter Wilson. Dublin, Ireland. 1761

PLANET: NEPTUNE

Clerke, Agnes Mary   1842–1907
Irish astronomer

Forever invisible to the unaided eye of man, a sister-globe to our earth was shown to circulate, in frozen exile, at 30 times its distance from the sun. Nay, the possibility was made apparent that the limits of our system were not even thus reached, but that yet profounder abysses of space might shelter obedient, though little favoured members of the solar family, by future astronomers to be recognized through the sympathetic thrillings of Neptune, even as Neptune himself was recognized through the tell-tale deviations of Uranus.
A Popular History of Astronomy During the Nineteenth Century
Part I, Chapter IV (p. 82)
A. & C. Black. London, England. 1908

PLANET: SATURN

Huygens, Christiaan   1629–95
Dutch mathematician, astronomer, and physicist

Annulo cingitur, tenui, plano, nusquam cohaerente, ad eclipticam inclinato

[It is surrounded by a thin flat ring, inclined to the ecliptic, and nowhere touches the body of the planet]
De Saturni luna observatio nova
The Hague, Netherlands. 1656

Keill, John   1671–1721
Scottish mathematician and natural philosopher

…Saturn has an Ornament peculiar to himself, for he is dignified with a Ring which surrounds his middle, and does no where touch his Body; but by an exact Libration and Equiponderancy of all its Parts, sustains it self like an Arch, and being thus suspended by Geometry, it is kept from falling upon his Body.
An Introduction to the True Astronomy
Lecture III (p. 25)
Printed for Bernard Lintot. London, England. 1721

Melville, Herman   1819–91
American novelist

Seat thyself sultanically among the moons of Saturn.
In *Great Books of the Western World* (Volume 48)
Moby Dick
Chapter 107 (p. 343)
Encyclopædia Britannica, Inc. Chicago, Illinois, USA. 1952

Pallister, William Hales   1877–1946
Canadian physician

The planet Saturn, to the naked eye
Appears an oval star; in seeking why
The telescope shows us a startling sight
Which seems some lovely vision on a night
Of dreams. A giant, wide, sunlit, tilted ring,
More strange than any other heavenly thing.
Poems of Science
Other Worlds and Ours, Saturn (p. 205)
Playford Press. New York, New York, USA. 1931

Thayer, John H.

If you want to see a picture painted as only the hand of God can paint it, go with me to Saturn…
Saturn. The Wonder of the Worlds
Popular Astronomy, Volume 37, Number 263, March 1919 (p. 175)

PLANET: URANUS

Herschel, Friedrich Wilhelm   1738–1822
English astronomer

In the fabulous ages of ancient times the appellations of Mercury, Venus, Mars, Jupiter, and Saturn were given to the planets as being the names of their principal heroes and divinities. In the present more philosophical era, it would hardly be allowable to have recourse to the same method, and call on Juno, Pallas, Apollo, or Minerva for a name to our new heavenly body…. I cannot but wish to take this opportunity of expressing my sense of gratitude, by giving the name Georgium Sidus, to a star [Uranus], which (with respect to us) first began to shine under His auspicious reign.
In James Sime
William Herschel and His Work
Chapter V, Letter to Sir Joseph Banks (p. 74)
Charles Scribner's Sons. New York, New York, USA. 1900

PLANET: VENUS

Ball, Sir Robert S.   1840–1913
Astronomer

The lover of nature turns to admire the sunset, as every lover of nature will. In the golden glory of the west a beauteous gem is seen to glitter; it is the evening star — the planet Venus…. All the heavenly host — even

Sirius and Jupiter — must pale before the splendid lustre of Venus, the unrivalled queen of the firmament.
The Story of the Heavens
Venus (p. 140)
Cassell & Company Ltd. London, England. 1885

Blackmore, Sir Richard   1650–1729
English physician and writer

Venus the next, whose lovely Beams adorn
As well the Dewy Eve, as opening Morn,
Does her fair Orb in beauteous Order turn.
The Poetical Works of Sir R. Blackmore: Containing Creation: A Philosophical Poem, in Seven Books
Book II, l. 515–517
Printed for C. Cooke. London, England. 1797

Hunter, Robert   1941–
American lyricist and poet

Counting stars by candlelight, all are dim but one is bright;
The spiral light of Venus, rising first and shining best,
Oh, from the northwest corner, of a brand new crescent moon,

crickets and cicadas sing, a rare and different tune…
Terrapin Station
Terrapin Station
Arista Records. 1977

Tennyson, Alfred (Lord)   1809–92
English poet

For a breeze of morning moves,
And the planet of love is on high,
Beginning to faint in the light that she loves
On a bed of daffodil sky,
To faint in the light of the sun she loves,
To faint in his light, and to die.
Alfred Tennyson's Poetical Works
Maude, Part I, Section XXII, Stanza II
Oxford University Press, Inc. London, England. 1953

Twain, Mark (Samuel Langhorne Clemens)   1835–1910
American author and humorist

An occultation of Venus is not half so difficult as an eclipse of the Sun, but because it comes seldom the world thinks it's a grand thing.
Collected Tales, Sketches, Speeches, & Essays 1891–1910
More Maxims of Mark (p. 945)
The Library of America. New York, New York, USA. 1992

PLANKTON

Hyerdahl, Thor   1914–2002
Norwegian ethnographer and adventurer

Some looked like fringed, fluttering spooks cut out of cellophane paper, while others resembled red-beaked birds with hard shells instead of feathers. There was no end to Nature's extravagant inventions in the plankton world.
Translated by F.H. Lyon
Kon-Tiki
Chapter 5 (p. 139)
Rand McNally & Company. Chicago, Illinois, USA. 1950

PLANT

Bailey, William Whitman   1843–1914
American botanist

Beginners almost always collect their plants too young; they have a nervous fear that they will not last.
The Botanical Collector's Handbook
Naturalists' Handy Series, Number 3 (p. 29)
Publisher undetermined. Salem, Massachusetts, USA. 1881

No division of the vegetable kingdom has attracted more deserved attention than that of the sea-weeds or sea-mosses. Throughout the world they have found their earnest students and devoted admirers. It is not alone for their intrinsic beauty that they are loved. Their collection involves the visiting of romantic cliffs — of shores strewn with the ocean's debris, of caves, and hollows, and even of the deep sea itself. The pursuit is always fascinating, and sometimes even perilous. A spice of danger does not deter the heroic algologist. Like "one who gathers sapphire, fearful trade!" he hangs suspended from crags, or ventures at low tide upon the slippery rocks over which the spray is dashing. There need not, however, be danger in the study. Many ladies have been successful gatherers of sea-weeds, and in the albums of many a watering-place belle may be seen choice specimens, self-collected. The plants need not be studied at all, if one prefers the simple collection and preservation, but it is always pleasanter to know something of the habits, uses, and even names of the objects which one treasures.
The Botanical Collector's Handbook
Naturalists' Handy Series, Number 3 (pp. 46–47)
Publisher undetermined. Salem, Massachusetts, USA. 1881

Borland, Hal   1900–78
American writer

There are no idealists in the plant world and no compassion. The rose and the morning glory know mercy. Bindweed, the morning glory, will quickly choke its competitors to death, and the fencerow rose will just as quietly crowd out any other plant that tried to share its roothold. Idealism and mercy are human terms and human concepts.
Book of Days
22 July 1976 (pp. 188–189)
Alfred A. Knopf. New York, New York, USA. 1976

Burroughs, John   1837–1921
American naturalist and writer

I know of nothing in vegetable nature that seems so really to be born as the ferns. They emerge from the ground rolled up, with a rudimentary and "touch-me-not" look, and appear to need a maternal tongue to lick them into shape. The sun plays the wet-nurse to them, and very soon they are out of that uncanny covering in which they come swathed and take their places with other green things.
Signs and Seasons
A Spring Relish (p. 193)
Houghton Mifflin Company. Boston, Massachusetts, USA. 1886

Clute, Willard N.
American botanist

One of the redeeming features of rubbish heaps, ballast grounds and waste lands is that they furnish a lurking place for numerous wanderers and outcasts of the vegetable kingdom.
A Plant Immigrant
The American Botanist, Volume 1, Number 2, August 1901 (p. 18)

Emerson, Ralph Waldo 1803–82
American lecturer, poet, and essayist

To every plant there are two powers; one shoots down as rootlet, and one upward as tree.
The Complete Works of Ralph Waldo Emerson (Volume 8)
Letters and Social Aims
Chapter I (p. 71)
Houghton Mifflin Company. Boston, Massachusetts, USA. 1904

The root of the plant is not unsightly to science...
The Complete Works of Ralph Waldo Emerson (Volume 2)
Essays: First Series
Chapter VI (p. 196)
Houghton Mifflin Company. Boston, Massachusetts, USA. 1904

Plants are the young of the world, vessels of health and vigor; but they grope ever upwards towards consciousness; the trees are imperfect men, and seem to bemoan their imprisonment, rooted in the ground.
Ralph Waldo Emerson: Essays and Lectures
Essays: Second Series
Nature (p. 547)
The Library of America. New York, New York, USA. 1983

Gatty, M. S. 1809–73
English writer

It was once prettily said by a lady who cultivated flowers, that she had "buried many a care in her garden"; and the sea-weed collector can often say the same of his garden — at the shore; as many a loving disciple could testify, who, having taken up the pursuit originally as a resource against weariness, or a light possible occupation during hours of sickness, has ended by an enthusiastic love, which throws a charm over every sea-place on the coast, however dull and ugly to the world in general; makes every day spent there too short, and every visit

too quickly ended. Only let there be sea, and plenty of low, dark rocks stretching out, peninsular-like, into it; and only let the dinner-hour be fixed for high-water time, — and the loving disciple asks no more of fate.
British Sea-Weeds: Drawn from Professor Harvey's Phycologia Britannica (Volume 1)
Introduction (p. VII)
Bell & Daldy. London, England. 1872

Gerard, John 1545–1612
English botanist

Among the manifold creatures of God (right Honorable, and my singular good Lord) that have all in all ages diversely entertained many excellent wits, and drawne them to the contemplation of the divine wisdome, none have provoked mens' studies more, or satisfied their desires so much as Plants have done, and that upon Just and worthy causes: For if delight may provoke mens' labor, what greater delight is there than to behold the earth appareled with plants, as with a robe of embroidered worke, set with Orient pearles, and garnished with great diversitie of rare and costly jewels?
The Herball or Generall Historie of Plantes
The Epistle Dedicatorie
Bonham and I. Norton. London, England. 1597

Although my paines have not been spent (Courteous Reader) in the gracious discoverie of golden mines, nor in the tracing after silver veines, whereby my native country might be enriched with such merchandise as it hath most in request and admiration; yet hath my labour (I trust) been otherwise profitably employed, in descrying of such a harmlesse treasure of herbes, trees, and plants, as the earth frankely without violence offereth unto our most necessarie uses.
The Herball or Generall Historie of Plantes
To the Courteous and Well-Willing Reader
Bonham and I. Norton. London, England. 1597

Gleason, Henry Allan 1882–1975
American botanist

Every species of plant is a law unto itself.
The Individualistic Concepts of the Plant Association
Bulletin of the Torrey Botanical Club, Volume 53, 1926 (p. 26)

Haldane, John Burdon Sanderson 1892–1964
English biologist

The simplest plants, such as the green algae growing in stagnant water or on the bark of trees, are mere round cells. The higher plants increase their surface by putting out leaves and roots. Comparative anatomy is largely the story of the struggle to increase surface in proportion to volume.
In James R. Newman (ed.)
The World of Mathematics (Volume 2)
On Being the Right Size (p. 954)
Simon & Schuster. New York, New York, USA. 1956

Hemans, Felicia D.   1793–1835
English poet

Oh! Call us not *weeds*, but flowers of the sea,
For lovely, and gay, and bright-tinted are we!
Our Blush is as deep as the rose of thy bowers,
Then call us not *weeds*, we are Ocean's gay flowers.
Not nursed like the plants of the summer parterre
Whose gales are but sights of an evening air
Our exquisite, fragile and delicate forms,
Are the prey of the Ocean, when vexed with his storms.
The Poetical Works of Mrs. Felicia Hemans
Ocean Flowers and Their teachings
Crosby, Nichols, Lee & Company. Boston, Massachusetts, USA. 1860

Linnaeus, Carl (von Linné)   1707–78
Swedish botanist and explorer

For wealth disappears, the most magnificent houses
fall into decay, the most numerous family at some time
or another comes to an end: the greatest and the most
prosperous kingdoms can be overthrown: but the whole
of Nature must be blotted out before the race of plants
passes away, and he is forgotten who in Botany held up
the torch.
Critica Botanica
Generic Names (p. 68)
The Ray Society. London, England. 1938

Muir, John   1838–1914
American naturalist

Found a lovely lily (*Calochortus albus*) in a shady
adenostoma thicket near Coulterville, in company
with *Adiantum chilense*. It is white with a faint
purplish tinge inside at the base of the petals, a most
impressive plant, pure as snow crystal, one of the plant
saints…must love and be made so much the purer by
every time it is seen. It puts the roughest mountaineer
on his good behavior.
My First Summer in the Sierra
June 6 (p. 22)
Houghton Mifflin Company. Boston, Massachusetts, USA. 1911

Well, perhaps I may yet become a proper cultivated plant,
cease my wanderings and for it a so called pillar or something in society, but if so, I must, like a revived Meth-
odist, care to love what I hate and to hate what I most
intensely and devoutly love.
In Linnie Marsh Wolfe (ed.)
John of the Mountains
Chapter II, Section 5. Plants and Humans (p. 90)
Houghton Mifflin Company. Boston, Massachusetts, USA. 1938

The plants are as busy as the animals, every cell in a swirl
of enjoyment, humming like a hive, singing the old new
song of creation.
Our National Parks
Chapter II (p. 70)
Houghton Mifflin Company. Boston, Massachusetts, USA. 1901

It drapes all the branches from top to bottom, hanging
in long silver-gray skeins, reaching a length of not less
than eight or ten feet, and when slowly waving in the
wind they produce a solemn funereal effect singularly
impressive.
A Thousand Mile Walk to the Gulf
Chapter IV (p. 68)
Houghton Mifflin Company, Boston Massachusetts, USA. 1916

Nuttall, Thomas   1786–1859
English botanist

To acquire a knowledge of the vegetable world, so pleasing to all observers, it may not perhaps be amiss to anticipate the dry detail of technical phrases, which has but too
often deterred, at the very portal of Flora's temple, the
enquirer into the nature and character of this beautiful
and useful tribe of beings, and begin, at once, by examining plants as we naturally find them, in the manner
our predecessors must have done, from whom we have
received their history.
An Introduction to Systematic and Physiological Botany
Part I, Chapter I (p. 1)
Hillard & Brown. Cambridge, England. 1830

Shelley, Percy Bysshe   1792–1822
English poet

A Sensitive Plant in a garden grew,
And the young winds fed it with silver dew,
And it opened its fan-like leaves to the light,
And clothed them beneath the kisses of night.
The Complete Poetical Works of Percy Bysshe Shelley
The Sensitive Plant, Part I, Stanza 1
Houghton Mifflin Company. Boston, Massachusetts, USA. 1901

Turner, William
No biographical data available

Although (most mighty and Christian Prince) there be
many noble and excellent arts and sciences, which no
man doubteth, but that almighty God the author of all
goodness hath given unto us by the hands of the heathen,
as necessary unto the use of mankind, yet is there none
among them all which is so openly commended by the
verdict of any holy writer in the Bible, as is the knowledge of plants, herbs and trees…
In George T.L. Chapman and Marilyn N. Tweddle (eds.)
A New Herball
Part I (p. 213)
Cambridge University Press. Cambridge, England. 1995

Twain, Mark (Samuel Langhorne Clemens)   1835–1910
American author and humorist

Sage-brush is a very fair fuel, but as a vegetable it is a
distinguished failure. Nothing can abide the taste of it but
the jackass and his illegitimate child the mule.
Roughing It (Volume 1)

Chapter III (p. 32)
Harper & Brothers Publishers. New York, New York, USA. 1899

von Goethe, Johann Wolfgang   1749–1832
German poet, novelist, playwright, and natural philosopher

We will see the entire plant world, for example, as a vast sea which is as necessary to the existence of individual insects as the oceans and rivers are to the existence of individual fish, and we will observe that an enormous number of living creatures are born and nourished in this ocean of plants. Ultimately we will see the whole world of animals as a great element in which one species is created, or at least sustained, by and through another. We will no longer think of connections and relationships in terms of purpose or intention. This is the only road to progress in understanding how nature expresses itself from all quarters and in all directions as it goes about its work of creation.
In D. Miller (ed.)
Scientific Studies (Volume 12)
Chapter II (p. 55)
Suhrkamp. New York, New York, USA. 1988

The Primal Plant is going to be the strangest creature in the world, which Nature herself must envy me. With this model and the key to it, it will be possible to go on for ever inventing plants and know that their existence is logical; that is to say, if they do not actually exist, they could, for they are not the shadowy phantoms of a vain imagination, but possess an inner necessity and truth. The same law will be applicable to all other living organisms.
Translated by W.H. Auden and Elizabeth Mayer
Italian Journey
Letter to Herder
May 17, 1787 (p. 305)
Pantheon Books. New York, New York, USA. 1962

Anyone who pays a little attention to the growth of plants will readily observe that certain of their external members are sometimes transformed so that they assume — either wholly or in some lesser degree — the form of the members nearest in the series.

Thus, for example, the usual process by which a single flower becomes double, is that, instead of filaments and anthers, petals are developed; these either show a complete resemblance in form and color to the other leaves of the corolla, or they still carry some visible traces of the origin.

If we note that it is in this way possible for the plant to take a step backwards and thus to reverse the order of growth, we shall obtain so much the more insight into Nature's regular procedure; and we shall make the acquaintance of the laws of transmutation, according to which she produces one part from another, and sets before us the most varied forms through modification of a single organ.
An Attempt to Interpret the Metamorphosis of Plants, Introduction
Section 1 and Section 3
Chronica Botanica, Volume 10, Number 2, Summer 1946 (p. 91)

von Humboldt, Alexander   1769–1859
German naturalist and explorer

Even the child longs to pass the hills or the seas which enclose his narrow home; yet, when his eager steps have borne him beyond those limits, he pines, like the plant, for his native soil; and it is by this touching and beautiful attribute of man — this longing for that which is unknown, and this fond remembrance of that which is lost — that he is spared from an exclusive attachment to the present.
Cosmos: A Sketch of a Physical Description of the Universe (Volume 1)
Conclusion of the Subject (p. 358)
Harper & Brothers. New York, New York, USA. 1869

PLATE TECTONICS

Bailey, Edward Battersby   1881–1965
English geologist

Even those who have more sympathy with man's endeavor than with the affairs of Nature may take an interest in the Science of Tectonics. Knowledge, after all, is of human creation; and, as a rule, the knowledge of the structure of a mountain chain comes as the reward of glorious struggle, both physical and mental.
Tectonic Essays
Introduction (p. 1)
At The Clarendon Press. Oxford, England. 1935

King, B. C.
No biographical data available
King, G. C. P.
No biographical data available

They [plates] can't curl down; they must curl up
To form a kind of dish
To stop the oceans spilling out
And losing all the fish.
Letters to Nature
Nature, Volume 232, Number 5305, July 2, 1971 (p. 37)

Ovid   43 BCE–17 AD
Roman poet

I have myself seen what once was solid land changed into sea; and again I have seen land made from the sea.
Translated by Frank Justus Miller
Metamorphoses (Volume 2)
Book XV, l. 263 (p. 383)
William Heinemann. London, England. 1916

Ward, Peter D.
American paleontologist
Brownlee, Donald
American astronomer

Plate tectonics plays at least three crucial roles in maintaining animal life: It promotes biological productivity; it

promotes diversity (the hedge against mass extinction); and it helps maintain equable temperatures, a necessary requirement for animal life. It may be that plate tectonics is the central requirement for life on a planet and that it is necessary for keeping a world supplied with water.
Rare Earth: Why Complex Life Is Uncommon in the Universe
Most Crucial Element of the Rare Earth Hypothesis? (p. 220)
Springer-Verlag. New York, New York, USA. 2000

Wilson, John Tuzo   1908–93
Canadian geologist and geophysicist

Formerly, most scientists of the earth thought of as one rigid body with fixed continents and permanent ocean basins, rather scientists now consider the earth to be broken into six large plates and several smaller ones, which very slowly move and jostle one another like blocks of ice on a river that is breaking up in the spring thaw.... Each continent does not constitute one plate, but rather each is incorporated with the surrounding ocean floor into a plate that is larger than the continent, just as a raft of logs may be frozen into a sheet of ice.
In Scientific American
Readings from Scientific American
Continents Adrift and Continents Aground
Preface (p. v)
W.H. Freeman & Company. San Francisco, California, USA. 1976

PMS

Bates, Rhonda
No biographical data available

My doctor said "I've got good news and I got bad news. The good news is you don't have Premenstrual Syndrome. The bad news is — you're a bitch!"
In Roz Warren
Glibquips (p. 122)
Crossing Press, Freedom, California. USA. 1994

Hankla, Susan
Professional writer

God grant me the serenity to change the things about me and others I cannot stand
And to stand the things about me an others I cannot change
And the insight to know the difference
Between a PMS day and a normal day
So no one gets hurt.
In Roz Warren
Glibquips (p. 122)
Crossing Press, Freedom, California. USA. 1994

POINT

Warner, Sylvia Townsend   1893–1978
English novelist and poet

He took out his pocket knife and whittled the end of the stick. Then he tried again.
"What is this?"
"A smaller hole."
"Point," said Mr. Fortune suggestively.
"Yes, I mean a smaller point."

"No, not quite. It is a point. but it is not smaller. Holes may be of different sizes, but no point is larger or smaller than another point."
Mr. Fortune's Maggot
Mr. Fortune's Maggot (p. 108)
New York Review of Books. New York, New York, USA. 1927

...if a given point were not in a given place it would not be there at all.
Mr. Fortune's Maggot
Mr. Fortune's Maggot (p. 110)
New York Review of Books. New York, New York, USA. 1927

POINT OF VIEW

Mach, Ernst   1838–1916
Austrian physicist and philosopher

No point of view has absolute, permanent validity. Each has importance only for some given end.
The Analysis of Sensations and the Relation of the Physical to the Psychical
Chapter I (p. 37)
The Open Court Publishing Company. Chicago, Illinois, USA. 1914

He who knows only one view or one form of a view does not believe that another has ever stood in its place, or that another will ever succeed; he neither doubts nor tests.
History and Root of the Principle of the Conservation of Energy
Chapter I (p. 17)
The Open Court Publishing Company. Chicago, Illinois, USA. 1911

POLLUTION

Abbey, Edward   1927–89
American environmentalist and nature writer

Our world is so full of beautiful things: fruit and ideas and women and good men and banjo music and onions with purple skins. A virtual Paradise. But even Paradise can be damned, flooded, overrun, generally mucked up by fools in pursuit of paper profits and plastic happiness.
Down the River
Part II, Chapter 8 (p. 233)
E.P. Dutton. New York, New York, USA. 1982

Ames, Bruce   1928–
American biochemist

We are living in a sea of chemicals that have not been tested for mutagenicity or carcinogenicity.
In Roger Lewin
Cancer Hazards in the Environment
New Scientist, Volume 69, Number 984, January 22, 1976 (p. 168)

Carson, Rachel   1907–64
American marine biologist and author

These sprays, dusts, and aerosols are now applied almost universally to farms, gardens, forests, and homes — non-selective chemicals that have the power to kill every insect, the "good" and the "bad," to still the song of birds and the leaping of fish in the streams, to coat the leaves with a deadly film, and to linger on in soil — all this though the intended target may be only a few weeds or insects. Can anyone believe it is possible to lay down such a barrage of poisons on the surface of the earth without making it unfit for all life? They should not be called "insecticides," but "biocides."
Silent Spring
Chapter 2 (pp. 7–8)
Houghton Mifflin Company. Boston, Massachusetts, USA. 1961

For the first time in the history of the world, every human being is now subjected to contact with dangerous chemicals, from the moment of conception until death.
Silent Spring
Chapter 3 (p. 15)
Houghton Mifflin Company. Boston, Massachusetts, USA. 1961

As crude a weapon as a cave man's club, the chemical barrage has been hurled against the fabric of time.
Silent Spring
Chapter 17 (p. 297)
Houghton Mifflin Company. Boston, Massachusetts, USA. 1961

Eliot, T. S. (Thomas Stearns)   1888–1965
American expatriate poet and playwright

There are flood and drouth
Over the eyes and in the mouth,
Dead water and dead sand
Contending for the upper hand.
The parched eviscerate soil
Gapes at the vanity of toil,
Laughs without mirth.
This is the death of earth.
The Collected Poems and Plays 1909–1950
Little Gidding, Part II, stanza 2 (p 140)
Harcourt, Brace & World, Inc. New York, New York, USA. 1952

Lovelock, James Ephraim   1919–
English scientist

There is only one pollution…people.
Gaia: A New Look at Life on Earth
Chapter 7 (p. 114)
Oxford University Press, Inc. Oxford, England. 2000

Peacock, Thomas Love   1785–1866
English writer

They have poisoned the Thames and killed the fish in the river. A little further development of the same wisdom and science will complete the poisoning of the air, and kill the dwellers on the banks…I almost think it is the destiny of science to exterminate the human race.
Gryll Grange
Chapter 1 (p. 11)
Penguin Books. Harmondsworth, England. 1949

Shakespeare, William   1564–1616
English poet, playwright, and actor

…this most excellent canopy, the air, look you, this brave o'erhanging firmament, this majestical roof fretted with golden fire, why, it appears no other thing to me than a foul and pestilent congregation of vapours.
In *Great Books of the Western World* (Volume 27)
The Plays and Sonnets of William Shakespeare (Volume 2)
Hamlet, Prince of Denmark
Act II, Scene ii, l. 311–315
Encyclopædia Britannica, Inc. Chicago, Illinois, USA. 1952

Taylor, John
No biographical data available

Then by the Lords Commissioners, and also
By my good King (whom all true subjects call so),
I was commanded with the Water Baylie,
To see the rivers cleaned, both night and dayly.
Dead Hogges, Dogges, Cates and well flayed Carryon Horses,
Their Noysom Corpses soyled the Water Courses;
Both Swines' and Stable dynge, beasts' guts and garbage,
Street dirt, with Gardners' Weeds and Rotten Herbage.
And from those Waters' filthy putrifaction
Our Meat and Drinke were made, which bred Infection.
Myself and partner, with cost paines and Travell,
Saw all made clean, from Carryon, Mud and Gravell,
And now and then was punisht a Delinquent,
By which good meanes away the filth and stink went.
Unknown, An Echo from the Past
The American Biology Teacher, Volume 35, Number 4, April 1973
(p. 208)

Toffler, Alvin   1928–
American writer and futurist

…industrial vomit…fills our skies and seas. Pesticides and herbicides filter into our foods. Twisted automobile carcasses, aluminum cans, non-returnable glass bottles and synthetic plastics form immense kitchen middens in our midst as more and more of our detritus resists decay. We do not even begin to know what to do with our radioactive wastes — whether to pump them into the earth, shoot them into outer space, or pour them into the oceans. Our technological powers increase, but the side effects and potential hazards also escalate.
Future Shock
Chapter 19 (p. 380)
Random House, Inc. New York, New York, USA. 1979

POPULATION

Malthus, Thomas Robert   1776–1834
English economist and sociologist

Population, when unchecked, increases in a geometrical ratio. Subsistence increases only in an arithmetical ratio. A slight acquaintance with numbers will show the immensity of the first power in comparison of the second.
In E.A. Wrigley and David Souden (eds.)
The Works of Thomas Malthus (Volume 1)
An Essay on the Principle of Population (p. 9)
Houghton Mifflin Company. Boston, Massachusetts, USA, 1885–1886

POSITION

Ridley, B. K.
No biographical data available

Imagine a billiard ball as the only inhabitant of the universe. What position does it have? The question has no meaning, for position can only be defined with respect to another position, which we call an origin, and there is nothing to define where the origin is.
Time, Space and Things
Chapter 3 (p. 41)
Cambridge University Press. Cambridge, England. 1984

POSITRON

Eddington, Sir Arthur Stanley   1882–1944
English astronomer, physicist, and mathematician

A positron is a hole from which an electron has been removed; it is a bung-hole which would be evened up with its surroundings if an electron were inserted.... You will see that the physicist allows himself even greater liberty than the sculptor. The sculptor removes material to obtain the form he desires. The physicist goes further and adds material if necessary — an operation which he describes as removing negative. He fills up a bung-hole, saying he is removing a positron.
The Philosophy of Physical Science
Chapter VIII, Section II (pp. 120–121)
The Macmillan Company. New York, New York, USA. 1939

Hacking, Ian   1936–
Canadian-born philosopher of science

Now how does one alter the charge on the niobium ball? "Well at that stage," said my friend, "we spray it with positrons to increase the charge or with electrons to decrease the charge." From that day forth I've been a scientific realist. So far as I'm concerned, if you can spray them then they are real.
Representing and Intervening (p. 23)
Cambridge University Press. Cambridge, England. 1983

POSSIBILITY

Armstrong, David Malet   1926–
Australian philosopher

The Naturalist theory of possibility now to be advanced will be called a Combinatorial theory. It traces the very idea of possibility to the idea of the combinations — all the combinations — of given, actual elements.
A Combinatorial Theory of Possibility
Part II, Chapter 3, Section I (p. 37)
Cambridge University Press. Cambridge, England. 1989

Eliot, George (Mary Ann Evans Cross)   1819–80
English novelist

We know what a masquerade all development is, and what effective shapes may be disguised in helpless embryos. — In fact, the world is full of hopeful analogies and handsome dubious eggs called possibilities.
Middlemarch
Book I, Chapter X (p. 82)
Clarendon Press. Oxford, England. 1986

POSTULATE

Russell, Bertrand Arthur William   1872–1970
English philosopher, logician, and social reformer

The method of "postulating" what we want has many advantages; they are the same as the advantages of theft over honest toil. Let us leave them to others and proceed with our honest toil.
Introduction to Mathematical Philosophy
Chapter VII (p. 71)
Dover Publications, Inc. New York, New York, USA. 1993

POWER

Boulton, Matthew   1728–1809
English engineer

"Ha! Boulton," said the king. "It is long since we have seen you at court. Pray, what business are you now engaged in?"
"I am engaged, your Majesty, in the production of a commodity which is the desire of kings."
"And what is that? What is that?"
"POWER, your majesty!"
In Ralph Stein
The Great Inventions
The Steam Engine (p. 24)
Playboy Press. Chicago, Illinois, USA. 1976

Morison, George S.   1842–1903
Civil engineer

Fire, animal strength, and written language have in turn advanced men and nations; something like a new capacity was developed with the discovery of explosives and again in the invention of printing; but the capacity of man has always been limited to his own individual strength and that of the men and animals he could control. His capacity is no longer so limited; man has now learned to manufacture power, and with the manufacture of power a new epoch began.

The New Epoch as Developed by the Manufacture of Power
Chapter I (p. 4)
Houghton Mifflin Company. Boston, Massachusetts, USA. 1903

PRAYER

Ayres, Clarence Edwin   1891–1972
No biographical data available

I believe in atoms, molecules, and electrons, matter of heaven and earth, and electrical energy its only form. I believe in modern science, conceived by Copernicus and borne out by Newton, which suffered under the Inquisition, was persecuted and anathematized, but rose to be the right hand of civilization as a consequence of the fact it rules the quick and the dead. I believe in the National Research Council, the communion of scientists, the publication of discoveries, the control of nature, and progress everlasting. Amen.

Science: The False Messiah
Chapter X (p. 129)
The Bobbs-Merrill Company. Indianapolis, Indiana, USA. 1927

Conoley, Gillian   1955–
Poet

I had only prayer, prayer and science.

Beckon
American Poetry Review, Volume 25, Number 2, March-April 1996 (p. 9)

Fiedler, Edgar R.   1916–2003
American economist

Thank God for Compensating errors.

Across the Board
The Three R's of Economic Forecasting — Irrational, Irrelevant and Irreverent, June 1977

Hammond, Kenneth R.
No biographical data available
Adelman, Leonard
No biographical data available

Lord, Please find me a one-armed statistician…so I won't always hear "on the other hand…"

Paraphrasing Edmund Muskie
Science, Values, and Human Judgment
Science, Volume 194, Number 4263, 22 October 1976 (p. 390)

Howe, E. W.
No biographical data available

What is the thing we call Common Sense? It is prayer practically applied, assistance given hope.

Sinner Sermons: A Selection of the Best Paragraphs of E.W. Howe (p. 7)
Girard, Kansas, USA. 1926

Plato   428 BCE–347 BCE
Greek philosopher

…I call upon God, and beg him to be our savior out of a strange and unwanted enquiry, and to bring us to the heaven of probability.

In *Great Books of the Western World* (Volume 7)
Timaeus
Section 48 (p. 456)
Encyclopædia Britannica, Inc. Chicago, Illinois, USA. 1952

Southgate, Theresa
No biographical data available

I am like a chemical compound in Your Laboratory of Life, O Lord, a compound from which the element Perfection has not yet been isolated, a compound in which the properties of the element Perfection are disguised by combination with earthly vanities. Take me then, O Lord, analyze me according to my good and evil constituents and isolate the pure element Perfection, as a chemist analyzes and separates from a substance all foreign matter. Analyze me that I may learn to know myself and that I may emerge a pure element, worthy to be included in the group of elements already freed from the bonds of their earthy life.

First, grind me in the mortar of childish whims, that I may emerge a composite sample of Your Likeness. Weigh me on the balance of Your generosity and decide how great a sample I shall be in Your Laboratory of Life. Then, ignite me in the furnace of Your love, that the carbon dioxide of earthly vanities be driven off. Cool me with the balm of Your mercy. Dissolve me in Your grace and filter me through the fine mesh of earthly trials so that Imperfections may be banished. Precipitate my evil tendencies with the strong precipitate the gelatinous silicate of earthly attachments which draw me from You. Imprison me in Your love with the mordant of sacrifice. Digest me in the length of my life, that my good deeds will grow and that self-satisfaction shall not be occluded with them. Blast out any impurities that may be introduced and finally, seal me forever, a pure substance in the container of your Eternal Happiness.

A Chemist's Prayer
Journal of Chemical Education, Volume 23, Number 10, October 1946 (p. 507)

Tukey, John W.   1915–2000
American statistician

The physical sciences are used to "praying over" their data, examining the same data from a variety of points of view. This process has been very rewarding, and has led to many extremely valuable insights. Without this sort of flexibility, progress in physical science would have been much slower. Flexibility in analysis is often to be had honestly at the price of a willingness not to demand that what has already been observed shall establish, or prove, what analysis suggests. In physical science generally, the results of praying over the data are thought of as something to be put to further test in another experiment, as indications rather than conclusions.

The Future of Data Analysis
The Annals of Mathematical Statistics, Volume 33, Number 1, March 1962 (p. 46)

Wheelock, John   1754–1817
No biographical data available

Oh, Lord, we thank thee for the Oxygen Gas; we thank Thee for the Hydrogen Gas; and for all the gases. We thank Thee for the Cerebrum; we thank Thee for the Cerebellum; and for the Medulla Oblongata. Amen.

Apocryphal

PRECISION

Davy, Sir Humphry   1778–1829
English chemist

Simplicity and precision ought to be the characteristics of a scientific nomenclature: words should signify things, or the analogies of things, and not opinions.

Elements of Chemical Philosophy
Part I, Volume 1, Introduction (p. 46)
Printed for J. Johnson & Company. London, England. 1812

Herschel, Sir John Frederick William   1792–1871
English astronomer and chemist

[Precision] is the very soul of science; and its attainment afford the only criterion, or at least the best, of the truth of theories, and the correctness of experiments.

A Preliminary Discourse on the Study of Natural Philosophy
Part II, Chapter IV, Section 115 (p. 122)
Printed for Longman, Rees, Orme, Brown & Green. London, England. 1831

Queneau, Raymond   1903–76
French poet, novelist, and publisher

In a bus of the S-line, 10 meters long, 3 wide, 6 high, at 3 km 600 m from its starting point, loaded with 48 people, at 12.17 p.m., a person of the masculine sex aged 27 years 3 months and 8 days, 1 m 72 cm tall and weighing 65 kg and wearing a hat 35 cm in height round the crown of which a ribbon 60 cm long, interpolated a man aged 48 years 4 months and 3 days, 1 m 68 cm tall and weighing 77 kg, by means of 14 words whose enunciation lasted 5 seconds and which alluded to some involuntary displacements of from 15 to 20 mm. Then he went and sat down about 1 m 10 cm away. 57 minutes later he was 10 meters away from the suburban entrance to the gare Saint-Lazare and was walking up and down over a distance of 30 m with a friend aged 28, 1 m 70 cm tall and weighing 71 kg who advised him in 15 words to move by 5 cm in the direction of the zenith a button which was 3 cm in diameter.

Exercises in Style
Precision (pp. 37–38)
New Direction Publishing Corporation. New York, New York, USA. 1981

Thompson, Sir D'Arcy Wentworth   1860–1948
Scottish zoologist and classical scholar

Dream apart, numerical precision is the very soul of science.

On Growth and Form (Volume 1)
Chapter I (p. 2)
At The University Press. Cambridge, England. 1951

PREDICTION

Armstrong, Neil A.   1930–
American astronaut

Science has not yet mastered prophecy. We predict too much for the next year and yet far too little for the next ten.

Address to Joint Sessions of Congress, September 16, 1969

Asimov, Isaac   1920–92
American author and biochemist

It is one thing to be able to make predictions. It is another to listen to the predictions you have made and to act upon them.

The Road to Infinity
Chapter 1 (p. 3)
Avon Books. New York, New York, USA. 1979

Bohr, Niels Henrik David   1886–1962
Danish physicist

It is very difficult to make an accurate prediction, especially about the future.

In Timothy Ferris (ed.)
The Mind's Sky: Human Intelligence in a Cosmic Context
The Manichean Heresy (p. 181)
Bantam Books. New York, New York, USA. 1992

Comte, Auguste   1798–1857
French philosopher

The aim of every science is foresight (prevoyance). For the laws of established observation of phenomena are generally employed to foresee their succession. All men, however little advanced make true predictions, which are always based on the same principle, the knowledge of the future from the past.

In Bertrand de Jouvenel
The Art of Conjecture

Chapter 11 (p. 111)
Basic Books, Inc. New York, New York, USA. 1967

Darwin, Charles Robert   1809–82
English naturalist

Anyone who attempts to predict the history of the next ten years is a rash man, and if he attempts to make his forecast for a century he is very properly regarded as so foolhardy as not to be worth listening to at all.
The Next Million Years
Introduction (p. 13)
Doubleday & Company, Inc. Garden City, New York, USA. 1953

du Noüy, Pierre Lecomte   1883–1947
French scientist

The aim of science is not so much to search for truth, or even truths, as to classify our knowledge and to establish relations between observable phenomena in order to be able to predict the future in a certain measure and to explain the sequence of phenomena in relation to ourselves.
Between Knowing and Believing
The Road to Reason (p. 188)
McKay. New York, New York, USA. 1967

The aim of science is to foresee, and not, as has often been said, to understand. Science describes facts, objects and phenomena minutely, and tries to join them by what we call laws, so as to be able to predict events in the future.
Human Destiny
Chapter 2 (p. 13)
Longmans, Green & Company. London, England. 1947

Dyson, Freeman J.   1923–
American physicist and educator

In the long run, qualitative changes always outweigh quantitative ones. Quantitative predictions of economic and social trends are made obsolete by qualitative changes in the rules of the game. Quantitative predictions of technological progress are made obsolete by unpredictable new inventions. I am interested in the long run, the remote future, where quantitative predictions are meaningless. The only certainty in that remote future is that radically new things will be happening.
Disturbing the Universe
Chapter 17 (p. 192)
Basic Books, Inc. New York, New York, USA. 1979

Hacking, Ian   1936–
Canadian-born philosopher of science

Cutting up fowl to predict the future is, if done honestly and with as little interpretation as possible a kind of randomization. But chicken guts are hard to read and invite flights of fancy or corruption.
The Emergence of Probability
An Absent Family of Ideas (p. 3)
Cambridge University Press. Cambridge, England. 1975

Kaplan, Abraham   1918–93
American philosopher of science, author, and educator

…if we can predict successfully on the basis of a certain explanation, we have good reason, and perhaps the best of reason, for accepting the explanation.
The Conduct of Inquiry: Methodology for Behavioral Science
Chapter IX, Section 40 (p. 350)
Chandler Publishing Company. San Francisco, California, USA. 1964

Kendrew, John   1917–99
English biochemist

Scientists cannot predict the future any better than anyone else — even about their own field of research.
The Thread of Life
Chapter 10 (p. 110)
Harvard University Press. Cambridge, Massachusetts, USA. 1966

Kluckhohn, Clyde   1905–60
American anthropologist

…it is one thing to be able to make some useful predictions as to what is likely to happen.… It is quite another thing to interfere, willfully to introduce new complications into an already tortuous social maze.
Mirror for Man: The Relation of Anthropology to Modern Life
Chapter X (p. 263)
McGraw-Hill Book. New York, New York, USA. 1949

Mill, John Stuart   1806–73
English political philosopher and economist

Of all truths relating to phenomena, the most valuable to us are those which relate to the order of their succession. On a knowledge of these is founded every reasonable anticipation of future facts, and whatever power we possess of influencing those facts to our advantage. Even the laws of geometry are chiefly of practical importance to us as being a portion of the premises from which the order of the succession of phenomena may be inferred.
A System of Logic, Rationative and Inductive
Book III, Chapter 5, Section 1 (p. 212)
Longmans, Green, Reader & Dyer. London, England. 1906

Rowling, J. K.   1965–
English author

The consequences of our actions are always so complicated, so diverse, that predicting the future is a very difficult business indeed.
Harry Potter and The Prisoner of Azkaban
Chapter Twenty-Two (p. 426)
Scholastic Press. New York, New York, USA. 1999

Russell, Bertrand Arthur William   1872–1970
English philosopher, logician, and social reformer

Science is the attempt to discover, by means of observation, and reasoning based upon it, first, particular facts about the world, and then laws connecting facts with one

another and (in fortunate cases) making it possible to pre-dict future occurrences.
Religion and Science
Grounds of Conflict (p. 8)
Henry Holt & Company. New York, New York, USA. 1935

Samuelson, Paul A.
No biographical data available

Wall Street indexes predicted nine out of the last five recessions!
Science and Stocks
Newsweek, September 19, 1966 (p. 92)

Toulmin, Stephen   1922–
English philosopher

Prediction is all very well; but we must make sense of what we predict. The mainspring of science is the con-viction that by honest, imaginative enquiry we can build up a system of ideas about Nature which has some legiti-mate claim to 'reality'.
The Philosophy of Science: An Introduction
Chapter 6 (p. 115)
Indiana University Press. Bloomington, Indiana, USA. 1961

Wheeler, John Archibald   1911–
American physicist and educator
Thorne, Kip S.   1940–
American theoretical physicist

The universe starts with a big bang, expands to a maxi-mum dimension, then recontracts and collapses: no more awe-inspiring prediction was ever made. It is preposter-ous. Einstein himself could not believe his own predic-tion.
Gravitation
Part X, Chapter 44 (p. 1196)
W.H. Freeman & Company. San Francisco, California, USA. 1973

Young, Louise B.
Science writer

Our most imaginative projections will pale beside the reality that takes shape tomorrow.
The Unfinished Universe
Chapter 10 (p. 197)
Simon & Schuster. New York, New York, USA. 1986

PREHISTORIC MAN

Aeschylus   525 BCE–426 BCE
Greek playwright

How, first beholding, they beheld in vain,
And hearing, heard not, but, like shapes in dreams,
Mixed all things wildly down the tedious time,
Nor knew to build a house against the sun
With wickered sides, nor any woodcraft knew,
But lived, like silly ants, beneath the ground

In hollow caves unsunned.
There, came to them
No steadfast sign of winter, nor of spring
Flower-perfumed, nor of summer full of fruit,
But blindly and lawlessly they did all things…
In Elizabeth Barrett-Browning
Prometheus Bound and Other Poems
Prometheus Bound
Scene: At the Rocks

James, William   1842–1910
American philosopher and psychologist

Bone of our bone and flesh of our flesh are these half-brutish pre-historic brothers. Girdled about with the immense darkness of this mysterious universe even as we are, they were born and died, suffered and struggled. Given over to fearful crime and passion, plunged in the blackest ignorance, preyed upon by hideous and gro-tesque delusions, yet steadfastly serving the profoundest of ideals in their fixed faith that existence in any form is better than non-existence, they ever rescued trium-phantly from the jaws of ever-imminent destruction the torch of life, which, thanks to them, now lights the world for us.
The Will to Believe, and Other Essays in Popular Philosophy and Human Immortality
Human Immortality (p. 33)
Dover Publications, Inc. New York, New York, USA. 1956

Leakey, Richard Erskine   1944–
Kenyan paleoanthropologist and politician

Needless to say, language and consciousness, which are among the most prized features of *Homo Sapiens*, leave no trace in the prehistoric record.
The Origin of Humankind
Preface (p. xiv)
Basic Books, Inc. New York, New York, USA. 1994

PREHISTORY

Clark, Grahame   1907–95
English archaeologist

The study of prehistory stands in no more need of jus-tification than exploration of the physical nature and mathematical properties of the universe, the investi-gation of all the multifarious forms of life, or for that matter the practice of the arts or the cultivation of speculative philosophy. Each in its own way enlarges the range of human experience and enriches the quality of human life.
Aspects of Prehistory
Chapter 1 (p. 4)
University of California Press. Berkeley, California, USA. 1970

Dunnell, Robert C.   1942-
American archaeologist

Like its sister discipline, sociocultural anthropology, prehistory has a tendency to invent a term for its own sake and then argue about what it means for twenty years rather than defining the term in the first place.
Systematics In Prehistory
Introduction (p. 4)
The Free Press. New York, New York, USA. 1971

Peale, Rembrandt 1778–1860
American neoclassical painter

The revolutions which have happened on our earth, by which its original appearance has been successively changed, have, at all times, commanded the attention of the learned, and excited various speculations concerning the time, cause and manner; and although we may never learn much on a subject so extensive, so remote and so wonderful, yet as far as facts will authorize us, we may safely proceed....
An Historical Disquisition on the Mammoth
Introduction (p. 1)
Printed for E. Lawrence. London, England. 1803

Wilson, Sir Daniel 1816–92
English-born Canadian archaeologist

In the application of the term Prehistoric — introduced, if I mistake not, for the first time in this work, — it was employed originally in reference to races which I then assigned reasons for believing had preceded the oldest historical ones in Britain and Northern Europe. But since then the term has become identified with a comprehensive range of speculative and inductive research, in which the archaeologist labours hand in hand with the geologist and ethnologist, in solving some of the most deeply interesting problems of modern science.
Prehistoric Annals of Scotland (Volume 1)
Preface (p. xiv)
Macmillan & Company Ltd. London, England. 1863

PRESCRIPTION

Chekhov, Anton Pavlovich 1860–1904
Russian author and playwright

What did your uncle die of?
Instead of fifteen Butkin drops, as the doctor prescribed, he took sixteen.
Note-Book of Anton Chekhov (p. 37)
B.W. Huebsch, Inc. New York, New York, USA. 1921

Helmuth, William Tod 1833–1902
American physician

Term pain "neuralgia," or if the man be stout,
Cry out, "Dear Sir, you have rheumatic gout."
Tap on the chest — some awful sounds they hear,
Then satisfied, declare, "The case is clear,"
Draw forth a paper, seize the magic quill,
And write in mystic signs, "Cathartic pill."

Scratches of a Surgeon
Medical Pomposity (p. 11)
W.A. Chatterton & Company. Chicago, Illinois, USA. 1879

Holmes, Oliver Wendell 1809–94
American physician, poet, and humorist

Pliny says, in so many words, that the cerates and cataplasms, plasters, collyria, and antidotes, so abundant in his time, as in more recent days, were mere tricks to make money.
Currents and Counter-Currents in Medical Science
Address
Massachusetts Medical Society at the Annual Meeting, May 30, 1860
Ticknor & Fields. Boston, Massachusetts, USA. 1861

Part of the blame of over-medication must, I fear, rest with the profession, for yielding to the tendency to self-delusion, which seems inseparable from the practice of the art of healing.
Currents and Counter-Currents in Medical Science
Address
Massachusetts Medical Society at the Annual Meeting, May 30, 1860
Ticknor & Fields. Boston, Massachusetts, USA. 1861

Latham, Peter Mere 1789–1875
English physician

To bring many important remedies together, and unite them by a lucky combination, and compress them within a small compass, and so place them within the common reach, all this gives a facility of prescribing which is hurtful to the advance of medical experience. The facility of prescribing is a temptation to prescribe; and, under this temptation, there is a lavish expenditure continually going on of important remedies in the mass, of which the prescribers have made no sufficient experiment in detail.
In William B. Bean
Aphorisms From Latham (p. 60)
Prairie Press. Iowa City, Iowa, USA. 1962

Thoreau, Henry David 1817–62
American essayist, poet, and practical philosopher

There are sure to be two prescriptions diametrically opposite. Stuff a cold and starve a cold are but two ways.
The Writings of Henry David Thoreau (Volume 1)
A Week on the Concord and Merrimac Rivers
Wednesday (p. 338)
Houghton Mifflin Company. Boston, Massachusetts, USA. 1893

Twain, Mark (Samuel Langhorne Clemens) 1835–1910
American author and humorist

It would be a good thing for the world at large, however unprofessional it might be, if medical men were required by law to write out in full the ingredients named in their prescriptions. Let them adhere to the Latin, or Fejee, if they choose, but discard abbreviations, and form their

letters as if they had been to school one day in their lives, so as to avoid the possibility of mistakes on that account.
Damages Awarded
San Francisco Morning Call, 10/1/1864

Wynter, Dr.
No biographical data available

Tell me from whom, fat-headed Scot,
Thou didst thy system learn;
From Hippocrates thou hadst it not,
Nor Celsus, not Pitcairn.
Suppose that we own that milk is good,
And say the same of grass;
The one for babes is only food,
The other for an ass.
Doctor! our new prescription try
(A friend's advice forgive);
Eat grass, reduce thyself, and die;-
Thy patients then may live.
In William Davenport Adams
English Epigrams
On Doctor Cheyne, the Vegetarian, cclxxvi
G. Routledge. London, England. 1878

PRESENT

Dillard, Annie   1945–
American poet, essayist, novelist, and writing teacher

Catch it if you can. The present is an invisible electron; its lightning path traced faintly on a blackened screen is fleet, and fleeing, and gone.
Pilgrim at Tinker Creek
Chapter 6, I (p. 79)
Harper's Magazine Press. New York, New York, USA. 1974

General Motors

The present is but an instant between an infinite past and a hurrying future.
General Motors 1964 Futurama
Audio narration accompanying the ride

PRESERVATION

Linnaeus, Carl (von Linné)   1707–78
Swedish botanist

To perpetuate the established course of nature in a continued series, the divine wisdom has thought fit, that all living creatures should constantly be employed in producing individuals, that all natural things should contribute and lend a helping hand towards preserving every species, and lastly that the death and destruction of one thing should always be subservient to the restitution of another.
Translated by B. Stillingfleet

Miscellaneous Tracts Relating to Natural History, Husbandry, and Physick (p. 32)
R. and J. Dodsley. London, England. 1759

PRIME NUMBER

Auster, Paul   1947–
American writer

Prime numbers. It was all so neat and elegant. Numbers that refuse to cooperate, that don't change or divide, numbers that remain themselves for all eternity.
The Music of Chance
Chapter 4 (pp. 73–74)
Viking Penguin. New York, New York, USA. 1990

Bombieri, Enrico   1940–
Italian mathematician

To me, that the distribution of prime numbers can be so accurately represented in a harmonic analysis is absolutely amazing and incredibly beautiful. It tells of an arcane music and a secret harmony composed by the prime numbers.
The Sciences
Prime Territory: Exploring the Infinite Landscape at the Base of the Number System, Sept/Oct 1992

Crandall, Robert W.   1940–
Economist
Pomerance, Carl
Number theorist

Prime numbers belong to an exclusive world of intellectual conceptions. We speak of those marvelous notions that enjoy simple, elegant description, yet lead to extreme — one might say unthinkable — complexity in the details. The basic notion of primality can be accessible to a child, yet no human mind harbors anything like a complete picture. In modern times, while theoreticians continue to grapple with the profundity of the prime numbers, vast toil and resources have been directed toward the computational aspect, the task of finding, characterizing, and applying the primes in other domains.
Prime Numbers: A Computational Perspective
Chapter 1 (p. 1)
Springer-Verlag. New York, New York, USA. 2001

Davis, Philip J.   1923–
American mathematician
Hersh, Reuben   1927–
American mathematician

Some order begins to emerge from this chaos when the primes are considered not in their individuality but in the aggregate; one considers the social statistics of the primes and not the eccentricities of the individuals.
The Mathematical Experience

The Prime Number Theorem (p. 213)
Birkhäuser. Boston, Massachusetts, USA. 1981

Doxiadis, Apostolos   1953–
Writer

The seeming absence of any ascertained organizing principle in the distribution of the succession of the primes
had bedeviled mathematicians for centuries and given
Number Theory much of its fascination. Here was a great
mystery indeed, worthy of the most exalted intelligence:
since the primes are the building blocks of the integers
and the integers the basis of our logical understanding
of the cosmos, how is it possible that their form is not
determined by law? Why isn't "divine geometry" apparent in their case?
Uncle Petros and Goldbach's Conjecture (p. 84)
Faber & Faber Ltd. London, England. 2000

du Sautoy, Marcus
English mathematician and writer

...despite their apparent simplicity and fundamental
character, prime numbers remain the most mysterious
objects studied by mathematicians. In a subject dedicated
to finding patterns and order, the primes offer the ultimate challenge.
The Music of the Primes
Chapter 1 (p. 5)
HarperCollins Publisher, Inc. New York, New York, USA. 2003

The primes have been a constant companion in our exploration of the mathematical world yet they remain the most
enigmatic of all numbers. Despite the best efforts of the
greatest mathematical minds to explain the modulation
and transformation of this mystical music, the primes
remain an unanswered riddle.
The Music of the Primes
Chapter 12 (pp. 314–315)
HarperCollins Publisher, Inc. New York, New York, USA. 2003

The search for the secret source that fed the primes had
been going on for over two millennia. The yearning for
this elixir had made mathematicians all too susceptible to
Bombieri's [April Fools announcement of a proof of the
Riemann Hypothesis in 1997]. For years, many had sim-
ply been too frightened to go anywhere near this notori-
ously difficult problem.
The Music of the Primes
Chapter 1 (p. 13)
HarperCollins Publisher, Inc. New York, New York, USA. 2003

We have all this evidence that the Riemann zeros are vibra-
tions, but we don't know what's doing the vibrating.
The Music of the Primes
Chapter 11 (p. 280)
HarperCollins Publisher, Inc. New York, New York, USA. 2003

[The Riemann] zeros did not appear to be scattered at
random. Riemann's calculations indicated that they were

lining up as if along some mystical ley line running
through the landscape.
The Music of the Primes
Chapter 4 (p. 99)
HarperCollins Publisher, Inc. New York, New York, USA. 2003

Prime numbers present mathematicians with one of the
strangest tensions in their subject. On the one hand a
number is either prime or it isn't. No flip of a coin will
suddenly make a number divisible by some smaller number. Yet there is no denying that the list of primes looks
like a randomly chosen sequence of numbers. Physicists
have grown used to the idea that a quantum die decides
the fate of the universe, randomly choosing at each throw
where scientists will find matter. But it is something of an
embarrassment to have to admit that these fundamental
numbers on which mathematics is based appear to have
been laid out by Nature flipping a coin, deciding at each
toss the fate of each number. Randomness and chaos are
anathema to the mathematician. Despite their randomness, prime numbers — more than any other part of
our mathematical heritage — have a timeless, universal
character. Prime numbers would be there regardless of
whether we had evolved sufficiently to recognise them.
The Music of the Primes
Chapter 1 (p. 6)
HarperCollins Publisher, Inc. New York, New York, USA. 2003

Riemann's insight followed his discovery of a mathematical looking-glass through which he could gaze at
the primes. Alice's world was turned upside down when
she stepped through her looking-glass. In contrast, in the
strange mathematical world beyond Riemann's glass,
the chaos of the primes seemed to be transformed into
an ordered pattern as strong as any mathematician could
hope for. He conjectured that this order would be maintained however far one stared into the never-ending world
beyond the glass. His prediction of an inner harmony on
the far side of the mirror would explain why outwardly
the primes look so chaotic. The metamorphosis provided
by Riemann's mirror, where chaos turns to order, is one
which most mathematicians find almost miraculous. The
challenge that Riemann left the mathematical world was
to prove that the order he thought he could discern was
really there.
The Music of the Primes
Chapter 1 (p. 9)
HarperCollins Publisher, Inc. New York, New York, USA. 2003

Riemann had found a passageway from the familiar
world of numbers into a mathematics which would have
seemed utterly alien to the Greeks who had studied prime
numbers two thousand years before. He had innocently
mixed imaginary numbers with his zeta function and
discovered, like some mathematical alchemist, the mathematical treasure emerging from this admixture of elements that generations had been searching for. He had

crammed his ideas into a ten-page paper, but was fully aware that his ideas would open up radically new vistas on the primes.
The Music of the Primes
Chapter 2 (p. 58)
HarperCollins Publisher, Inc. New York, New York, USA. 2003

Armed with his prime number tables, Gauss began his quest. As he looked at the proportion of numbers that were prime, he found that when he counted higher and higher a pattern started to emerge. Despite the randomness of these numbers, a stunning regularity seemed to be looming out of the mist.
The Music of the Primes
Chapter 2 (p. 47)
HarperCollins Publisher, Inc. New York, New York, USA. 2003

Gauss had heard the first big theme in the music of the primes, but it was one of his students, Riemann, who would truly unleash the full force…of the hidden harmonies that lay behind the cacophony of the primes.
The Music of the Primes
Chapter 2 (p. 58)
HarperCollins Publisher, Inc. New York, New York, USA. 2003

For centuries, mathematicians had been listening to the primes and hearing only disorganised noise. These numbers were like random notes wildly dotted on a mathematical stave with no discernible tune. Now Riemann had found new ears with which to listen to these mysterious tones. The sine-like waves that Riemann had created from the zeros in his zeta landscape revealed some hidden harmonic structure.
The Music of the Primes
Chapter 4 (p. 93)
HarperCollins Publisher, Inc. New York, New York, USA. 2003

The revelation that the graph appears to climb so smoothly, even though the primes themselves are so unpredictable, is one of the most miraculous in mathematics and represents one of the high points in the story of the primes. On the back page of his book of logarithms, Gauss recorded the discovery of his formula for the number of primes up to N in terms of the logarithm function. Yet despite the importance of the discovery, Gauss told no one what he had found. The most the world heard of his revelation were the cryptic words, "You have no idea how much poetry there is in a table of logarithms."
The Music of the Primes
Chapter 2 (p. 50)
HarperCollins Publisher, Inc. New York, New York, USA. 2003

The primes are jewels studded throughout the vast expanse of the infinite universe of numbers that mathematicians have explored down the centuries. For mathematicians they instill a sense of wonder: 2, 3, 5, 7, 11, 13, 17, 19, 23… — timeless numbers that exist in the same world independent of our physical reality. They are Nature's gift to the mathematician.

The Music of the Primes
Chapter 1 (p. 5)
HarperCollins Publisher, Inc. New York, New York, USA. 2003

It seems paradoxical that the fundamental objects on which we build our order-filled world of mathematics should behave so wildly and unpredictably.
The Music of the Primes
Chapter 2 (p. 45)
HarperCollins Publisher, Inc. New York, New York, USA. 2003

Maybe we have become so hung up on looking at the primes from Gauss's and Riemann's perspective that what we are missing is simply a different way to understand these enigmatic numbers. Gauss gave an estimate for the number of primes, Riemann predicted that the guess is at worst the square root of N off its mark, Littlewood showed that you can't do better than this. Maybe there is an alternative viewpoint that no one has found because we have become so culturally attached to the house that Gauss built.
The Music of the Primes
Chapter 12 (p. 312)
HarperCollins Publisher, Inc. New York, New York, USA. 2003

Littlewood wrote to Hardy about [Ramanujan]: "it is not surprising that he would have been [misled], unsuspicious as he presumably is of the diabolical malice inherent in the primes."
The Music of the Primes
Chapter 6 (p. 139)
HarperCollins Publisher, Inc. New York, New York, USA. 2003

Littlewood's proof…revealed that prime numbers are masters of disguise. They hide their true colours in the deep recesses of the universe of numbers, so deep that witnessing their true nature may be beyond the computational power of humankind. Their true behavior can be seen only through the penetrating eyes of abstract mathematical proof.
The Music of the Primes
Chapter 5 (p. 130)
HarperCollins Publisher, Inc. New York, New York, USA. 2003

…Gauss liked to call [number theory] "the Queen of Mathematics." For Gauss, the jewels in the crown were the primes, numbers which had fascinated and teased generations of mathematicians.
The Music of the Primes
Chapter 2 (p. 22)
HarperCollins Publisher, Inc. New York, New York, USA. 2003

Erdös, Paul   1913–96
Hungarian mathematician

God may not play dice with the universe, but something strange is going on with the prime numbers.
In D. Mackenzie
Homage to an Itinerant Master
Science, Volume 275, Number 5301, 7 February, 1997 (p. 759)

Euler, Leonhard 1707–83
Swiss mathematician and physicist

Mathematicians have tried in vain to this day to discover some order in the sequence of prime numbers, and we have reason to believe that it is a mystery into which the human mind will never penetrate. To convince ourselves, we have only to cast a glance at tables of primes, which some have taken the trouble to compute beyond a hundred thousand, and we should perceive at once that there reigns neither order nor rule.
Collected Works
Serial 1, Volume 2 (p. 241)
Publisher undetermined

Gardner, Martin 1914–
American writer and mathematics games editor

The primes...[are] exasperating, unruly integers that refuse to be divided...by any integers except themselves and one.
In Eli Maor
To Infinity and Beyond: A Cultural History of The Infinite (p. 21)
Birkhäuser. Boston, Massachusetts, USA. 1987

Gauss, Johann Carl Friedrich 1777–1855
German mathematician, physicist, and astronomer

The problem of distinguishing prime numbers from composite numbers and of resolving the latter into their prime factors is known to be one of the most important and useful in arithmetic. It has engaged the industry and wisdom of ancient and modern geometers to such an extent that it would be superfluous to discuss the problem at length.... Further, the dignity of the science itself seems to require that every possible means be explored for the solution of a problem so elegant and so celebrated.
Disquisitiones Arithmeticae
Article 329 (p. 326)
Yale University Press. New Haven, Connecticut, USA. 1965

Gonek, S.
Mathematician

If there are lots of zeros off the line — and there might be — the whole picture is just horrible, horrible, very ugly. It's an Occam's razor sort of thing, you either have absolutely beautiful behavior of prime numbers, they behave just like you want them to behave, or else it's really bad.
In K. Sabbagh
The Riemann Hypothesis: The Greatest Unsolved Problem in Mathematics
Chapter 8 (p. 135)
Farrar, Straus & Giroux. New York, New York, USA. 2002

Gowers, Timothy 1963–
English mathematician

Although the prime numbers are rigidly determined, they somehow feel like experimental data.
Mathematics: A Very Short Introduction

Chapter 7 (p. 121)
Oxford University Press, Inc. Oxford, England. 2002

Hardy, G. H. (Godfrey Harold) 1877–1947
English pure mathematician

...317 is a prime, not because we think so, or because our minds are shaped in one way rather than another, but because it is so, because mathematical reality is built that way.
A Mathematician's Apology
Chapter 24 (p. 130)
Cambridge University Press. Cambridge, England. 1967

Jutila, M.
No biographical data available

I sometimes have the feeling that the number system is comparable with the universe that the astronomer is studying...The number system is something like a cosmos.
In K. Sabbagh
Beautiful Mathematics
Prospect, January 2002

Motohashi, Yoichi
No biographical data available

[Primes] are full of surprises and very mysterious...They are like things you can touch.... In mathematics most things are abstract, but I have some feeling that I can touch the primes, as if they are made of a really physical material. To me, the integers as a whole are like physical particles.
The Riemann Hypothesis: The Greatest Unsolved Problem in Mathematics
Chapter 1 (p. 22)
Farrar, Straus & Giroux. New York, New York, USA. 2002

Queneau, Raymond 1903–76
French poet, novelist, and publisher

When One made love to Zero
spheres embraced their arches
and prime numbers caught their breath...
Pounding the Pavement, Beating the Bush, and Other Paraphysical Poems
Sines
Unicorn Press. Greensboro, North Carolina, USA. 1985

Sagan, Carl 1934–96
American astronomer and author

Do we know what the sequence of numbers is? Okay, here, we can do it in our heads...fifty-nine, sixty-one, sixty-seven...seventy-one.... Aren't these all prime numbers? A little buzz of excitement circulated through the control room. Ellie's own face momentarily revealed a flutter of something deeply felt, but this was quickly replaced by a sobriety, a fear of being carried away, an apprehension about appearing foolish, unscientific.
Contact: A Novel
Chapter 4 (p. 78)
Simon & Schuster. New York, New York, USA. 1985

Stewart, Ian   1945–
English-mathematician and science writer

Who would have imagined that something as straight-forward as the natural numbers (1, 2, 3, 4…) could give birth to anything so baffling as the prime numbers (2, 3, 5, 7, 11…)?
Jumping Champions
Scientific American, Volume 283, Number 6, December 2000 (p. 106)

Sylvester, James Joseph   1814–97
English mathematician

[Tschebycheff] was the only man ever able to cope with the refractory character and erratic flow of prime numbers and to confine the stream of their progression with algebraic limits, building up, if I may so say, banks on either side which that stream, devious and irregular as are its windings, can never overflow.
In E. Kramer
The Nature and Growth of Mathematics
Chapter 21 (p. 503)
Hawthorn Books, Inc. New York, New York, USA. 1970

I have sometimes thought that the profound mystery which envelops our conceptions relative to prime numbers depends upon the limitations of our faculties in regard to time, which like space may be in essence poly-dimensional and that this and other such sort of truths would become self-evident to a being whose mode of perception is according to *superficially* as opposed to our own limitation to *linearly* extended time.
The Collected Mathematical Papers of James Joseph Sylvester (Volume 4)
On Certain Inequalities Relating to Prime Numbers (p. 600)
University Press. Cambridge, England. 1904–1912

Tenenbaum, G.
No biographical data available

Addition and multiplication equip the set of positive natural numbers {1, 2, 3…} with a double structure of Abelian semigroup. The first is associated with a total order relation, and is generated by the single number 1. The second, reflecting the partial order of divisibility has an infinite number of generators: the prime numbers. Defined since antiquity, this key concept has yet to deliver up all its secrets — and there are plenty of them.
Introduction to Analytic and Probabilistic Number Theory (p. 299)
Cambridge University Press. Cambridge, England. 1995

Tenenbaum, G.
No biographical data available
France, M. Mendés
No biographical data available

One of the remarkable aspects of the distribution of prime numbers is their tendency to exhibit global regularity and local irregularity. The prime numbers behave like the "ideal gases" which physicists are so fond of. Considered from an external point of view, the distribution is — in broad terms — deterministic, but as soon as we try to describe the situation at a given point, statistical fluctuations occur as in a game of chance where it is known that on average the heads will match the tail but where, at any one moment, the next throw cannot be predicted.
Translated by Philip G. Spain
The Prime Numbers and Their Distribution (p. 51)
American Mathematical Society. Providence, Rhode Island, USA. 2000

Prime numbers try to occupy all the room available (meaning that they behave as randomly as possible), given that they need to be compatible with the drastic constraint imposed on them, namely to generate the ultra-regular sequence of integers. This idea underpins the majority of conjectures concerning prime numbers: everything which is not trivially forbidden should actually happen.
Translated by Philip G. Spain
The Prime Numbers and Their Distribution (p. 51)
American Mathematical Society. Providence, Rhode Island, USA. 2000

As archetypes of our representation of the world, numbers form, in the strongest sense, part of ourselves, to such an extent that it can legitimately be asked whether the subject of study of arithmetic is not the human mind itself. From this a strange fascination arises: how can it be that these numbers, which lie so deeply within ourselves, also give rise to such formidable enigmas? Among all these mysteries, that of the prime numbers is undoubtedly the most ancient and most resistant.
Translated by Philip G. Spain
The Prime Numbers and Their Distribution (p. 1)
American Mathematical Society. Providence, Rhode Island, USA. 2000

Weyl, Hermann   1885–1955
German mathematician

The mystery that clings to numbers, the magic of numbers, may spring from this very fact, that the intellect, in the form of the number series, creates an infinite manifold of well-distinguished individuals. Even we enlightened scientists can still feel it, e.g., in the impenetrable law of the distribution of prime numbers.
Philosophy of Mathematics and Natural Science
Part I, Chapter I (p. 7)
Princeton University Press. Princeton, New Jersey, USA. 1949

Zagier D.
No biographical data available

I hope that…I have communicated a certain impression of the immense beauty of the prime numbers and the endless surprises which they have in store for us.
The First 50 Million Prime Numbers
The Mathematical Intelligencer, Volume 0 August 1977

…there is no apparent reason why one number is prime and another not. To the contrary, upon looking at these numbers one has the feeling of being in the presence of one of the inexplicable secrets of creation.

The First 50 Million Prime Numbers
The Mathematical Intelligencer, Volume 0 August 1977

PRIMORDIAL

de Maupassant, Guy   1850–93
French writer

Nothing is more impressive, nothing more disquieting, more terrifying occasionally, than a fen. Why should a vague terror hang over these low plains covered with water? Is it the low rustling of the rushes, the strange Will-o'-the-wisp light, the silence which prevails on calm nights, the still mists which hang over the surface like a shroud; or is it the almost inaudible splashing, so slight and so gentle, yet sometimes more terrifying than the cannons of men of the thunders of skies, which make these marshes resemble countries which none has dreamed of, terrible countries concealing an unknown and dangerous secret?

No, something else belongs to it — another mystery, profounder and graver, floats amid these thick mists, perhaps the mystery of the creation itself! For was it not in stagnant and muddy water, amid the heavy humidity of moist land under the heat of the sun, that the first germ of life pulsated and expanded to the day?

A Selection From the Writings of Guy de Maupassant (Volume 1)
Chapter 7, Love
President Publishing Company. New York, New York, USA. 1903

Newman, Joseph S.   1892–1960
American poet

A highly speculative void
Divides the germ and anthropoid
But we've discovered certain clues
In fossilized primordial ooze
Where ancient polyps lived and died
And countless myriads multiplied.

Poems for Penguins and Other Lyrical Lapses
Biology
Greenburg. New York, New York, USA. 1941

Shakespeare, William   1564–1616
English poet, playwright, and actor

In the cauldron boil and bake;
Eye of newt and toe of frog,
Wool of bat and tongue of dog,
Adder's fork and blind-worm's sting,
Lizard's leg and howlet's wing...

In *Great Books of the Western World* (Volume 27)
The Plays and Sonnets of William Shakespeare (Volume 2)
Macbeth
Act IV, Scene i, l. 13–17
Encyclopædia Britannica, Inc. Chicago, Illinois, USA. 1952

PRINCIPLE

Adams, John   1735–1826
2nd president of the United States

The reasoning of mathematicians is founded on certain and infallible principles. Every word they use conveys a determinate idea, and by accurate definitions they excite the same ideas in the mind of the reader that were in the mind of the writer. When they have defined the terms they intend to make use of, they premise a few axioms, or self-evident principles, that every one must assent to as soon as proposed. They then take for granted certain postulates, that no one can deny them, such as, that a right line may be drawn from any given point to another, and from these plain, simple principles they have raised most astonishing speculations, and proved the extent of the human mind to be more spacious and capacious than any other science.

Works (Volume 2)
Diary (p. 21)
Boston, Massachusetts, USA. 1850

Bilaniuk, Oleksa-Myron   1926–
Polish/Ukrainian-American physicist
Sudarshan, E. C.   1931–
Indian-American physicist

There is an unwritten precept in modern physics, often facetiously referred to as Gell-Mann's totalitarian principle, which states that in physics "anything which is not prohibited is compulsory." Guided by this sort of argument we have made a number of remarkable discoveries, from neutrinos to radio galaxies.

Particles Beyond the Light Barrier
Physics Today, Volume 22, Number 5, May 1969 (p. 44)

Pólya, George   1887–1985
Hungarian mathematician

This principle is so perfectly general that no particular application is possible.

How to Solve It: A New Aspect of Mathematical Method
Part III. The Traditional Mathematics Professor (p. 208)
Princeton University Press. Princeton, New Jersey, USA. 1973

PROBABILITY

Adams, Douglas   1952–2001
English author, comic radio dramatist, and musician

TRILLIAN: Five to one against and falling...four to one against and falling...three to one...two...one...Probability factor one to one...we have normality...I repeat we have normality...anything you still can't cope with is therefore your own problem.

The Original Hitchhiker Radio Script
Fit the Second (p. 42)
Harmony Books. New York, New York, USA. 1983

FORD: Arthur, This is fantastic, we've been picked up by a ship with the new Infinite Improbability Drive, this is really incredible, Arthur.... Arthur, what's happening?

ARTHUR: Ford, there's an infinite number of monkeys outside who want to talk to us about this script for Hamlet they've worked out.
The Original Hitchhiker Radio Script
Fit the Second (pp. 41–42)
Harmony Books. New York, New York, USA. 1983

Arbuthnot, John   1667–1735
Scottish mathematician and physician

The Reader may here observe the Force of Numbers, which can be successfully applied, even to those things, which one would imagine are subject to no Rules. There are very few things which we know, which are not capable of being reduc'd to a Mathematical Reasoning, and when they cannot, its a Sign our Knowledge of them is very small and confus'd; And where mathematical reasoning can be had, its as great folly to make use of any other, as to grope for a thing in the dark, when you have a Candle standing by you. I believe the Calculation of the Quantity of Probability might be improved to a very useful and pleasant Speculation, and applied to a great many Events which are accidental, besides those of Games…
Of the Laws of Chance
Preface
Benjamin Motte. London, England. 1692

Atkins, Russell   1926–
Poet, composer, editor, and teacher

…dogs are random.
Probability and Birds in the Yard
Poem

Austen, Jane   1775–1817
English writer

Are no probabilities to be accepted, merely because they are not certainties?
Sense and Sensibility (Volume 1)
Chapter 15 (p. 68)
Oxford University Press. Oxford, England. 1980

Bagehot, Walter   1826–77
English journalist

Life is a school of probability.
In Rudolf Flesch
The New Book of Unusual Quotations
Harper & Row, Publishers. New York, New York, USA. 1966

Barrow, John D.   1952–
English theoretical physicist
Tipler, Frank   1947–
American physicist

In a randomly infinite Universe, any event occurring here and now with finite probability must be occurring simultaneously at an infinite number of other sites in the Universe. It is hard to evaluate this idea any further, but one thing is certain: if it is true then it is certainly not original!
The Anthropic Cosmological Principle
Chapter 4.6 (p. 249)
Clarendon Press. Oxford, England. 1986

Barry, Frederick   1876–1943
Historian of science

In short, these fundamental elements of scientific knowledge assimilate and grow, coalesce and separate and recombine, shrink and wane, die and come to life again; and while they persist they are never more than probable.
The Scientific Habit of Thought: An Informal Discussion of the Source and Character of Dependable Knowledge (p. 139)
Columbia University Press. New York, New York, USA. 1927

Blake, William   1757–1827
English poet, painter, and engraver

…all is to them a dull round of probabilities and possibilities.
The Complete Poetry and Prose of William Blake
The Ancient Britons
University of California Press. Berkeley, California, USA. 1982

Bleckley, Logan E.   1827–1907
American lawyer

…it is always probable that something improbable will happen.
Warren v. Purtell, 63 *Georgia Reports* 428, 430 (1879)

Boole, George   1815–64
English mathematician

Probability is expectation founded upon partial knowledge. A perfect acquaintance with all the circumstances affecting the occurrence of an event would change expectation into certainty, and leave neither room nor demand for a theory of probabilities.
Collected Logical Works (Volume 2)
An Investigation of the Law of Thought, Chapter XVI (p. 258)
The Open Court Publishing Company. La Salle, Illinois, USA. 1952

Borel, Félix Edouard Justin Emile   1871–1956
French mathematician

Probabilities must be regarded as analogous to the measurement of physical magnitudes; that is to say, they can never be known exactly, but only within certain approximation.
Translated by Maurice Baudin
Probabilities and Life
Chapter Three, Section 6 (pp. 32–33)
Dover Publications. New York, New York, USA. 1962

The principles on which the calculus of probabilities is based are extremely simple and as intuitive as the reasonings which lead an accountant through his operations.
Translated by Maurice Baudin

Probabilities and Life
Introduction (p. 1)
Dover Publications. New York, New York, USA. 1962

...just as price is not the sole element of our decision when we make a purchase, probability alone must not dictate our decision in the matters of a risk.
Translated by Maurice Baudin
Probabilities and Life
Chapter 3, Section 6 (p. 32)
Dover Publications. New York, New York, USA. 1962

Born, Max   1882–1970
German-born English physicist

As far as I can see, the only foundation of the doctrine of probability, which (though not satisfactory for a mind devoted to the "absolute") seems at least not more mysterious than science as a whole, is the empirical attitude: The laws of probability are valid just as any other physical law in virtue of the agreement of their consequences with experience.
Experiment and Theory in Physics (pp. 26–27)
Cambridge University Press. Cambridge, England. 1944

If Gessler had ordered William Tell to shoot a hydrogen atom off his son's head by means of a particle and had given him the best laboratory instruments in the world instead of a cross-bow, Tell's skill would have availed him nothing. Hit or miss would have been a matter of chance.
In Sir Arthur Stanley Eddington
New Pathways in Science
Chapter IV, Section III (p. 82)
The Macmillan Company. New York, New York, USA. 1935

Bostwick, Arthur Elmore   1860–1942
American librarian

It is easier to make true misleading statements in the subject of probabilities than anywhere else.
The Theory of Probabilities
Science, Volume 3, Number 54, January 10, 1896 (p. 66)

Boswell, James   1740–95
Scottish biographer and diarist

JOHNSON: "If I am well acquainted with a man, I can judge with great probability how he will act in any case, without his being restrained by my judging. God may have this probability increased to certainty."
The Life of Samuel Johnson (Volume 2)
April 15, 1778 (pp. 209–210)
J.M. Dent & Sons Ltd. London, England. 1938

Meredith, Owen (Edward Robert Bulwer-Lytton, 1st Earl Lytton)   1831–91
English statesman and poet

...fate laughs at probabilities!
Eugene Aram

Book First, Chapter 10 (p. 72)
John W. Lovell Company. New York, New York, USA. n.d.

Burney, Fanny   1752–1840
English novelist and diarist

The play of imagination, in the romance of early youth, is rarely interrupted with scruples of probability.
In Edward A Bloom and Lillian D. Bloom (eds.)
Camilla, or, A Picture of Youth
Book II, Chapter V (p. 102)
Oxford University Press. Oxford, England. 1983

Bush, Vannevar   1890–1974
American electrical engineer and physicist

If scientific reasoning were limited to the logical processes of arithmetic, we should not get very far in our understanding of the physical world. One might as well attempt to grasp the game of poker entirely by the use of the mathematics of probability.
As We May Think
Atlantic Monthly, July 1945

Butler, Joseph   1692–1752
English bishop and exponent of natural theology

But to us, probability is the very guide to life.
The Analogy of Religion
Introduction (p. 76)
Henry G. Bohn. London, England. 1852

Cardozo, Benjamin N.   1870–1938
American jurist

...law, like other branches of social science, must be satisfied to test the validity of its conclusions by the logic of probabilities rather than the logic of certainty.
The Growth of the Law (p. 33)
Yale University Press. New Haven, Connecticut, USA. 1924

Coats, R. H.
No biographical data available

...the electron is just a "smear of probability."
Science and Society
Journal of the American Statistical Association, Volume 34, Number 205, March 1939 (p. 6)

Cohen, John
No biographical data available

Unlike almost all mathematics, I agree completely with your statement that every probability evaluation is a probability evaluation, that is, something to which it is meaningless to apply such attributes as right, wrong, rational, etc.
Chance, Skill, and Luck: The Psychology of Guessing and Gambling
Chapter 2, Part I (p. 28)
Penguin Books. Baltimore, Maryland, USA. 1960

Crichton, Michael   1942–
American novelist

Harry sighed irritably, pulled out a sheet of paper. It's a probability equation? He wrote: $p = f_p n_h f_l f_i f_c$
"What it means," Harry Adams said, "is that the probability, p, that intelligent life will evolve in any star system is a function of the probability that the star will have planets, the number of habitable planets, the probability that simple life will evolve on a habitable planet, the probability that intelligent life will evolve from simple life, and the probability that intelligent life will attempt interstellar communication within five billion years. That's all the equation says."
Sphere: A Novel
The Briefing (pp. 28–29)
Ballantine Books. New York, New York, USA. 1987

But the point is that we have no facts," Harry said. "We must guess at every single one of these probabilities."
Sphere: A Novel
The Briefing (p. 29)
Ballantine Books. New York, New York, USA. 1987

Crofton, M. W.
British mathematician

The mathematical theory of probability is a science which aims at reducing to calculation, where possible, the amount of credence due to propositions or statements, or to the occurrence of events, future or past, more especially as contingent or dependent upon other propositions or events the probability of which is known.
Encyclopædia Britannica (9th edition)
Probability

Dampier-Whetham, William   1867–1952
English scientific writer

Indeed the intellectual basis of all empirical knowledge may be said to be a matter of probability, expressible only in terms of a bet.
A History of Science
Chapter III (p. 155)
The Macmillan Company. New York, New York, USA. 1936

Darwin, Charles Robert   1809–82
English naturalist

As for a future life, every man must judge for himself between conflicting vague probabilities.
In Francis Darwin (ed.)
The Life and Letters of Charles Darwin (Volume 1)
Chapter VIII (p. 277)
D. Appleton & Company. New York, New York, USA. 1896

de Cervantes, Miguel   1547–1616
Spanish novelist, playwright, and poet

…I would reply that fiction is all the better the more it looks like truth, and gives the more pleasure the more probability and possibility there is about it.
In *Great Books of the Western World* (Volume 29)
The History of Don Quixote de la Mancha
Part I, Chapter 47 (p. 184)
Encyclopædia Britannica, Inc. Chicago, Illinois, USA. 1952

de Jouvenel, Bertrand   1903–87
French man of letters

We defined the art of conjecture, or stochastic art, as the art of evaluating as exactly as possible the probabilities of things, so that in our judgments and actions we can always base ourselves on what has been found to be the best, the most appropriate, the most certain, the best advised; this is the only object of the wisdom of the philosopher and the prudence of the statesman.
Translated by Nikita Lary
The Art of Conjecture
Introduction, 3 (p. 21, note 19)
Basic Books, Inc. New York, New York, USA. 1967

de Leeuw, A. L.
No biographical data available

The laws of chance tell us what is probable, but not what is certain to happen. They do not predict. They do not tell us what will, but what may happen.
Rambling Through Science
Gambling (p. 88)
Whittlesey House. London, England. 1932

de Moivre, Abraham   1667–1754
French-born mathematician

The Probability of an Event is greater or less, according to the number of chances by which it may happen, compared with the whole number of chances by which it may either happen or fail.
The Doctrine of Chances: or, A Method of Calculating the Probabilities of Events in Play (3rd edition)
Introduction (p. 1)
Printed for Millar. London, England. 1756

de Morgan, Augustus   1806–71
English mathematician and logician

No part of mathematics or mathematical physics involves considerations so strange or so difficult to handle correctly, and there is no subject upon which opinions have been more freely hazarded by the ignorant, or rational dissent more unambiguously expressed by the learned.
Encyclopædia Metropolitiana (Volume 2)
Theory of Probabilities (p. 393)

Deming, William Edwards   1900–93
American statistician, educator, and consultant

The statistician's report to management should not talk about probabilities. It will merely give outside margins of error for the results of chief importance.
Sample Design inin Business Research (p. 13)
John Wiley & Sons, Inc. New York, New York, USA. 1960

Descartes, René   1596–1650
French philosopher, scientist, and mathematician

...when it is not in our power to determine what is true, we ought to act accordingly to what is most probable...
Discourse on the Method of Rightly Conducting the Reason and Seeking For Truth in the Sciences
Part III (p. 67)
Simpkon, Marshal, and Company. London, England. 1850

Diaconis, Persi   1945–
American mathematician

Our brains are just not wired to do probability problems very well.
The Search for Randomness
Talk, March 29, 1989

Doyle, Sir Arthur Conan   1859–1930
Scottish writer

We are coming now rather into the region of guesswork, said Dr. Mortimer.
Say, rather, into the region where we balance probabilities and choose the most likely. It is the scientific use of the imagination, but we have always some material basis on which to start our speculation.
In William S. Baring-Gould (ed.)
The Annotated Sherlock Holmes (Volume 2)
The Hound of the Baskervilles, Chapter 4 (p. 24)
Wings Books. New York, New York, USA. 1967

Eddington, Sir Arthur Stanley   1882–1944
English astronomer, physicist, and mathematician

There can be no unique probability attached to any event or behavior: we can only speak of "probability in the light of certain given information", and the probability alters according to the extent of the information.
The Nature of the Physical World
Chapter XIV (pp. 314–315)
The Macmillan Company. New York, New York, USA. 1930

But it is necessary to insist more strongly than usual that what I am putting before you is a model — the Bohr model atom — because later I shall take you to a profounder level of representation in which the electron, instead of being confined to a particular locality, is distributed in a sort of probability haze all over the atom...
New Pathways in Science
Chapter II, Section III (p. 34)
The Macmillan Company. New York, New York, USA. 1935

In most modern theories of physics probability seems to have replaced aether as "the nominative of the verb 'to undulate'."
New Pathways in Science
Chapter VI, Section I (p. 110)
The Macmillan Company. New York, New York, USA. 1935

Edgeworth, Francis Ysidro   1845–1926
Irish economist and statistician

The Calculus of Probabilities is an instrument which requires the living hand to direct it.
Metretike, or The Method of Measuring Probability and Utility (p. 18)
Temple. London, England. 1887

Probability may be described, agreeably to general usage, as importing partial incomplete belief.
The Philosophy of Chance
Mind, Volume 9, 1884

I hope that you flourish in Probabilities.
Quoted in Stephen M. Stigler
The History of Statistics
Letter from Edgeworth to Pearson, 11 September 1893 (p. 326)

It is a useful discipline to walk in a world where, though the objects themselves are fixed, their images are ever vibrating through a large part of their own dimensions. The Calculus of Probabilities...conveys a lesson which is required for the study of social science, the power of contemplating general tendencies through the wavering medium of particulars.
On Methods of Ascertaining Variation in the Rate of Births, Deaths, and Marriages
Journal of the Royal Statistical Society, Volume 48, 1885 (p. 633)

Eliot, George (Mary Ann Evans Cross)   1819–80
English novelist

Secrets are rarely betrayed or discovered according to any program our fear has sketched out. Fear is almost always haunted by terrible dramatic scenes, which recur in spite of the best-argued probabilities against them...
The Mill on the Floss
Book V, V (p. 317)
J.M. Dent & Sons Ltd. London, England. 1908

Still there is a possibility — even a probability — the other way.
The George Eliot Letters (Volume 2) (p. 127)
Yale University Press. New Haven, Connecticut, USA. 1954–1978

But I see no probability of my being able to be with you before your other Midsummer visitors arrive.
The George Eliot Letters (Volume 2) (p. 160)
Yale University Press. New Haven, Connecticut, USA. 1954–1978

...ignorance gives one a large range of probabilities.
Daniel Deronda II
Book II, Chapter XIII (p. 136)
A.L. Burt Company. New York, New York, USA. 18??

Evanovich, Janet   1943-
American writer

I graduated from Douglass College without distinction. I was in the top 98% of my class and damn glad to be there. I slept in the library and daydreamed my way through history lecture. I failed math twice, never fully grasping probability theory. I mean, first off, who cares if you pick a black ball or a white ball out of the bag? And second, if

you're bent over about the color, don't leave it to chance. Look in the damn bag and pick the color you want.
Hard Eight (pp. 227–228)
St. Martin's Press. New York, New York, USA. 2002

Feller, William   1906–70
Yugoslavian-born American mathematician

Probability is a mathematical discipline with aims akin to those, for example, of geometry or analytical mechanics. In each field we must carefully distinguish three aspects of the theory: (a) the formal logical content, (b) the intuitive background, (c) the applications. The character, and the charm, of the whole structure cannot be appreciated without considering all three aspects in their proper relation.
An Introduction to Probability Theory and Its Applications (Volume 1)
Introduction (p. 1)
John Wiley & Sons, Inc. New York, New York, USA. 1957

All possible "definitions" of probability fall short of the actual practice.
An Introduction to Probability Theory and Its Applications (Volume 1)
Chapter I (p. 19)
John Wiley & Sons, Inc. New York, New York, USA. 1957

Feynman, Richard P.   1918–88
American theoretical physicist

Nature permits us to calculate only probabilities.
QED: The Strange Theory of Light and Matter
Chapter 1 (p. 19)
Princeton University Press. Princeton, New Jersey, USA. 1985

A philosopher once said "It is necessary for the very existence of science that the same conditions always produce the same results." Well, they do not.
The Character of Physical Law
Chapter 6 (p. 147)
BBC. London, England. 1965

Forbes, J. D.
No biographical data available

...the ratios or probabilities of which we have been speaking have no absolute signification with reference to an event which has occurred...They represent only the state of expectation of the mind of a person before the event has occurred, or having occurred before he is informed of the results.
On the Alleged Evidence for a Physical Connection Between Stars Forming Binary or Multiple Groups
The London, Edinburgh and Dublin Philosophical Magazine and Journal of Science, Third Series, December 1850 (p. 406)

Freeman, R. Austin   1862–1943
British physician and mystery novelist

It is a question of probabilities....
A Certain Dr. Thorndyke
Thorndyke Makes a Beginning (p. 209)
Dodd, Mead & Company. New York, New York, USA. 1928

Froude, James Anthony   1818–94
English historian and biographer

Philosophy goes no further than probabilities, and in every assertion keeps a doubt in reserve.
Short Studies on Great Subjects (Volume 2)
Calvinism (p. 51)
Charles Scribner's Sons. New York, New York, USA. 1890

Fry, Thornton C.
No biographical data available

But if probability measures the importance of our state of ignorance it must change its value whenever we add new knowledge. And so it does.
Probability and Its Engineering Uses (2nd edition)
Chapter VI (p. 145)
D. Van Nostrand Company. Princeton, New Jersey, USA. 1965

After all, without the experiment — either a real one or a mathematical model — there would be no reason for a theory of probability.
Probability and Its Engineering Uses (2nd edition)
Chapter II (p. 23)
D. Van Nostrand Company. Princeton, New Jersey, USA. 1965

Gay, John   1685–1732
English poet and dramatist

Let men suspect your tale untrue,
Keep probability in view.
John Gay: Poetry and Prose
Fables. The Painter Who Pleased Nobody and Everybody, l. 1
At The Clarendon Press. Oxford, England. 1974

Gibbon, Edward   1737–94
English historian

Such a fact is probable, but undoubtedly false.
In *Great Books of the Western World* (Volume 40)
The Decline and Fall of the Roman Empire
Notes: Chapter XXIV, 116 (p. 794)
Encyclopædia Britannica, Inc. Chicago, Illinois, USA. 1952

Gilbert, William   1544–1603
English scientist and physician

Men are deplorably ignorant with respect to natural things, an modern philosophers, as though dreaming in the darkness, must be aroused and taught the uses of things, the dealing with things; they must be made to quit the sort of learning that comes only from books, and that rests only on vain arguments from probability and upon conjecture.
In *Great Books of the Western World* (Volume 28)
On the Loadstone and Magnetic Bodies and on the Great Magnet the Earth
Book First, Chapter 10
Encyclopædia Britannica, Inc. Chicago, Illinois, USA. 1952

Gilbert, Sir William Schwenck   1836–1911
English playwright and poet

Sullivan, Arthur   1842–1900
English composer

Of what there is no manner of doubt,
No probable, possible, shadow of doubt,
No possible doubt whatever.
The Complete Plays of Gilbert and Sullivan
The Gondoliers
Act I (p. 466)
W.W. Norton & Company, Inc. New York, New York, USA. 1976

Gissing, George   1857–1903
English novelist

Of course, if your work is strong, and you can afford to
wait; the probability is that half a dozen people will at last
begin to shout that you have been monstrously neglected,
as you have.
New Grub Street
Interim (p. 411)
The Modern Library. New York, New York, USA. 1926

Good, I. J.
No biographical data available

Although there are at least five kinds of probability, we
can get along with just one kind.
Kinds of Probability…
Science, Volume 129, 1959

Gracian, Baltasar   1601–58
Spanish philosopher

Whereas wisdom favors the probabilities, folly favors
only the possibilities.
In Thomas G. Corvan
The Best of Gracian (p. 38)
Philosophical Library. New York, New York, USA. 1964

Wisdom does not trust to probabilities; it always marches
in the midday light of reason.
In Rudolf Flesch
The New Book of Unusual Quotations
Harper & Row, Publishers. New York, New York, USA. 1966

It is only by mature meditation on the possibilities and
probabilities of future events — that we can elude the
tortuous troubles of the tomorrows.
In Thomas G. Corvan
The Best of Gracian (p. 22)
Philosophical Library. New York, New York, USA. 1964

Gumperson, R. F.
Physicist

The outcome of a given desired probability will be
inverse to the degree of desirability.
Gumperson's Law
Changing Times, Volume 11, Number 11, November, 1957 (p. 46)

…the contradictory of a welcome probability will assert
itself whenever such an eventuality is likely to be most
frustrating.

Gumperson's Law
Changing Times, Volume 11, Number 11, November, 1957 (p. 46)

Hamming, Richard W.   1915–98
Mathematician

Probability is too important to be left to the experts.
The Art of Probability for Scientists and Engineers
Chapter 1 (p. 4)
Westview Press. Boulder, Colorado, USA. 1991

Hammond, Henry

The only seasonable inquiry is, Which is of probables the
most, or of improbables the least, such.
In Robert Sanderson and Izaak Walton
Works (Volume 5)
A Letter to Dr. Sanderson (p. 319)
At The University Press. Oxford, England. 1854

Harris, Errol E.
No biographical data available

Probability is truth in some degree…
Hypothesis and Perception: The Roots of Scientific Method
The Logic of Construction (p. 342)
George Allen & Unwin Ltd. London, England. 1970

Harrison, Edward Robert   1919–2007
English-born American cosmologist

We have a wraithlike quantum world of ghostly waves
where all is fully determined and predictable. Yet, when
we translate it into our observed world of sensible things
and their events, we are limited to the concept of chance
and the language of probability. What happens at the
interface of the quantum world and the observed world
may be this or may be that.
Masks of the Universe
Chapter 8 (p. 124)
Macmillan Publishing Company. New York, New York, USA. 1985

Herbert, Nick
American physicist

probability = (possibility)^2
Quantum Reality: Beyond the New Physics
Chapter 6 (p. 96)
Anchor Press. Garden City, New York, USA. 1985

Holmes, Oliver Wendell   1809–94
American physician, poet, and humorist

No priest or soothsayer that ever lived could hold his own
against Old Probabilities.
Pages from an Old Volume of Life
Chapter X (p. 327)
Houghton Mifflin Company. Boston, Massachusetts, USA. 1895

Hooker, Richard   1554–1600
English writer and theologian

As for probabilities, what thing was there ever set down so agreeable with sound reason but some probable show against it might be made.
In S. Austin Allibone
Prose Quotations from Socrates to Macaulay
Probability
J.B. Lippincott. Philadelphia, Pennsylvania, USA. 1903

Howe, E. W.
No biographical data available

A reasonable probability is the only certainty.
Sinner Sermons: A Selection of the Best Paragraphs of E.W. Howe (p. 23)
Girard. Kansas, USA. 1926

Hume, David   1711–76
Scottish philosopher and historian

…all knowledge resolves itself into probability…
A Treatise of Human Nature
Book I, Part IV, Section 1 (p. 232)
Penguin Books. Baltimore, Maryland, USA. 1969

Hunter, Evan   1926–2005
American writer

Now, your Honor; in much the same way that there are laws governing our society, there are also laws governing chance, and these are called the laws of probability, and it is against these that we must examine the use of an identical division number.
The Paper Dragon
Tuesday, Chapter 6
Delacorte Press. New York, New York, USA. 1966

Huxley, Aldous   1894–1963
English writer and critic

Magic and devils offend our sense of probabilities.
Proper Studies
Varieties of Intelligence (p. 7)
Chatto & Windus. London, England. 1957

Huxley, Thomas Henry   1825–95
English biologist

The scientific imagination always restrains itself within the limits of probability.
Collected Essays (Volume 5)
Science and Christian Traditions
Science and Pseudo-Science (p. 124)
Macmillan & Company Ltd. London, England. 1904

Huygens, Christiaan   1629–95
Dutch mathematician, astronomer, and physicist

We know nothing very certainly, but everything only probably, and the probability has degrees that are widely different.
Oeuvres
Complètes de Christiaan Huygens (p. 298)
Publisher undetermined

In such noble and sublime Studies as these, 'tis a Glory to arrive at Probability, and the search itself rewards the pains. But there are many degrees of Probable, some nearer Truth than others, in the determining of which lies the chief exercise of our Judgment.
The Celestial Worlds Discover'd, or, Conjectures Concerning the Planetary Worlds, Their inhabitants and Productions
Book the First. Conjectures Not Useless, Because Not Certain (p. 10)
Printed for T. Childe. London, England. 1698

James, P. D.   1920–
English writer

Juries hate scientific evidence.

They think they won't be able to understand it so naturally they can't understand it. As soon as you step into the box you see a curtain of obstinate incomprehension clanging down over their minds. What they want is certainty. Did this paint particle come from this car body? Answer yes or Number None of those nasty mathematical probabilities we're so fond of.
Death of an Expert Witness
Book II, Chapter III (p. 96)
Warner Books, Inc. New York, New York, USA. 1992

Jefferson, Thomas   1743–1826
3rd president of the United States

Perhaps an editor might begin a reformation in some way as this. Divide his paper into four chapters, heading the 1st, Truth. 2d, Probabilities. 3d, Possibilities. 4th, Lies.
The Writings of Thomas Jefferson (Volume 9)
Letter to John Norvell, June 11, 1807
G.P. Putnam's Sons. New York, New York, USA. 1898

Kac, Mark   1914–84
Polish mathematician

To the author the main charm of probability theory lies in the enormous variability of its applications. Few mathematical disciplines have contributed to as wide a spectrum of subjects, a spectrum ranging from number theory to physics, and even fewer have penetrated so decisively the whole of our scientific thinking.
Lectures in Applied Mathematics (Volume 1)
Probability and Related Topics in Physical Sciences, Preface (p. ix)
Interscience Publishers, Ltd. London, England. 1959

Kasner, Edward   1878–1955
American mathematician
Newman, James Roy   1911–66
Mathematician and mathematical historian

Equiprobability in the physical world is purely a hypothesis. We may exercise the greatest care and the most accurate of scientific instruments to determine whether or not a penny is symmetrical. Even if we are satisfied that it is, and that our evidence on that point is conclusive,

our knowledge, or rather our ignorance, about the vast number of other causes which affect the fall of the penny is so abysmal that the fact of the penny's symmetry is a mere detail. Thus, the statement "head and tail are equiprobable" is at best an assumption.
Mathematics and the Imagination
Chance and Chanceability (p. 251)
Simon & Schuster. New York, New York, USA. 1940

Keynes, John Maynard   1883–1946
British economist

Probability is, so far as measurement is concerned, closely analogous to similarity.
A Treatise on Probability
Chapter III (p. 28)
Harper & Row, Publishers. New York, New York, USA. 1962

It has been pointed out already that no knowledge of probabilities, less in degree than certainty, helps us to know what conclusions are true, and that there is no direct relation between the truth of a proposition and its probability. Probability begins and ends with probability.
A Treatise on Probability
Part V (p. 322)
Harper & Row, Publishers. New York, New York, USA. 1962

It is difficult to find an intelligible account of the meaning of "probability," or of how we are ever to determine the probability of any particular proposition; and yet treatises on the subject profess to arrive at complicated results of the greatest precision and the most profound practical importance.
A Treatise on Probability
Chapter IV (p. 51)
Harper & Row, Publishers. New York, New York, USA. 1962

…others have suggested seriously a "barometer of probability."
A Treatise on Probability
Chapter III (p. 20)
Harper & Row, Publishers. New York, New York, USA. 1962

Kolmogorov, Andrei N.   1903–87
Russian physicist and mathematician

The theory of probability as mathematical discipline can and should be developed from axioms in exactly the same way as Geometry and Algebra.
Foundations of the Theory of Probability
Chapter 1. Elementary Theory of Probability (p. 1)
Chelsea Publishing Company. New York, New York, USA. 1956

Kosko, Bart
American engineer

Which is easier to believe in, probability or God?…The ultimate fraud is the scientific atheist who believes in probability.
Fuzzy Thinking
Chapter 3 (p. 50)
Hyperion. New York, New York, USA. 1993

Probability has turned modern science into a truth casino.
Fuzzy Thinking
Chapter 1 (p. 12)
Hyperion. New York, New York, USA. 1993

Kyburg, Jr., II. E.
No biographical data available
Smokler, H. E.
No biographical data available

…there is no problem about probability: it is simply a nonnegative, additive set function, whose maximum value is unity.
Studies in Subjective Probability
Introduction (p. 3)
John Wiley & Sons, Inc. New York, New York, USA. 1964

Laplace, Pierre Simon   1749–1827
French mathematician, astronomer, and physicist

One of the great advantages of the calculus of probabilities is to teach us to distrust first opinions.
A Philosophical Essays on Probabilities
Chapter XVI (p. 164)
Dover Publications, Inc. New York, New York, USA. 1951

The probability of events serves to determine the hope or the fear of persons interested in their existence.
A Philosophical Essays on Probabilities
Chapter IV (p. 20)
Dover Publications, Inc. New York, New York, USA. 1951

The regularity which astronomy shows us in the movements of the comets doubtless exists in all phenomena.
The curve described by a simple molecule or air or vapour is regulated in a manner just as certain as the planetary orbits; the only difference between them is that which comes from our ignorance.
A Philosophical Essay on Probabilities
Chapter II (p. 6)
Dover Publications, Inc. New York, New York, USA. 1951

It is remarkable that a science, which commenced with the consideration of games of chance, should be elevated to the rank of the most important subjects of human knowledge.
A Philosophical Essay on Probabilities
Chapter XVII (p. 195)
Dover Publications, Inc. New York, New York, USA. 1951

Leibniz, Gottfried Wilhelm   1646–1716
German philosopher and mathematician

…the art of weighing probabilities is not yet even partly explained, though it would be of great importance in legal matters and even in the management business.
Philosophical Papers and Letters (Volume 1)
Letter to John Frederick, Duke of Brunswick Hanover (p. 399)
The University of Chicago Press. Chicago, Illinois, USA. 1956

Lewis, C. S. (Clive Staples)   1898–1963
British author, scholar, and popular theologian

We may not be able to get certainty, but we can get probability, and half a loaf is better than no bread.
Christian Reflections
Historicism (p. 111)
William B. Erdmanns Publishing. Grand Rapids, Michigan, USA. 1997

Lewis, Clarence Irving   1883–1964
American philosopher

There is no such thing as *the* probability of four aces in one hand, or *the* probability of anything else. Given all the relevant data which there are to be known, everything is either certainly true or certainly false.
Mind and the World-Order: Outline of a Theory of Knowledge
Chapter X (p. 330)
Charles Scribner's Sons. New York, New York, USA. 1929

…empirical knowledge is exclusively a knowledge of probabilities…
Mind and the World-Order: Outline of a Theory of Knowledge
Chapter XI (p. 345)
Charles Scribner's Sons. New York, New York, USA. 1929

A "poor evaluation" of the probability of anything may reflect ignorance of relevant data which "ought" to be known…
Mind and the World-Order: Outline of a Theory of Knowledge
Chapter X (p. 331)
Charles Scribner's Sons. New York, New York, USA. 1929

The only knowledge a priori is purely analytic; all empirical knowledge is probable only.
Mind and the World-Order: Outline of a Theory of Knowledge
Chapter X (p. 309)
Charles Scribner's Sons. New York, New York, USA. 1929

Lincoln, Abraham   1809–65
16th president of the United States

The probability that we may fall in the struggle ought not to deter us from the support of a cause we believe to be just; it shall not deter me.
The Sub-Treasury
Speech, Springfield, Illinois, December 26, 1839

Lindley, Dennis V.   1923–
American statistician

Are we probabilists, believers, or fuzzifiers?
Comment: A Tale of Two Wells
Statistical Science, Volume 2, Number 1, February 1987 (p. 38)

Locke, John   1632–1704
English philosopher and political theorist

Probability is the appearance of agreement upon fallible proofs.
In *Great Books of the Western World* (Volume 35)
An Essay Concerning Human Understanding
Book IV, Chapter XV, Section 1 (p. 365)
Encyclopædia Britannica, Inc. Chicago, Illinois, USA. 1952

Probability is likeliness to be true…
In *Great Books of the Western World* (Volume 35)
An Essay Concerning Human Understanding
Book IV, Chapter XV, Section 3 (p. 365)
Encyclopædia Britannica, Inc. Chicago, Illinois, USA. 1952

The mind ought to examine all the grounds of probability, and upon a due balancing the whole, reject or receive it proportionably to the preponderancy of probability on the one side or the other.
In S. Austin Allibone
Prose Quotations from Socrates to Macaulay
Probability
J.B. Lippincott. Philadelphia, Pennsylvania, USA. 1903

Ludlum, Robert   1927–2001
American author

It was a desperate strategy, based on probabilities, but it was all he had left.
The Bourne Supremacy
Chapter 24 (p. 365)
Random House, Inc. New York, New York, USA. 1986

It wasn't a probability anymore, it was a reality.
The Bourne Supremacy
Chapter 18 (p. 256)
Random House, Inc. New York, New York, USA. 1986

Masters, Dexter   1908–89
American writer

If absolutes had disappeared under the inquiries of science, and apparently they had, …then the only rational procedure, the only procedure consistent with man's development, was to follow where the probabilities led.
The Accident
Part I, Chapter 3 (p. 19)
Alfred A. Knopf. New York, New York, USA. 1955

Meyer, Agnes   1887–1970
American author and journalist

We can never achieve absolute truth but we can live hopefully by a system of calculated probabilities. The law of probability gives to natural and human sciences — to human experience as a whole — the unity of life we seek.
Education for a New Morality
Chapter 3 (p. 21)
Macmillan Publishing Company. New York, New York, USA. 1957

Moroney, M. J.
American statistician

There are certain notions which it is impossible to define adequately. Such notions are found to be those based on universal experience of nature. Probability is such a notion. The dictionary tells me that "probable" means "likely." Further reference gives the not very helpful information that "likely" means "probable."
Facts from Figures

The Laws of Chance (p. 4)
Penguin Books Ltd. Harmondsworth, England. 1951

Muggeridge, Malcolm   1903–90
English journalist and social critic

The probability is, I suppose that the Monarchy has become a kind of ersatz religion. Chesterton once remarked that when people [cease] to believe in God they do not believe in nothing, but in anything.
New Statesman, 1955

Pascal, Blaise   1623–62
French mathematician and physicist

Probability — Each one can employ it; no one can take it away.
In *Great Books of the Western World* (Volume 33)
Pensées
Section XIV, 913
Encyclopædia Britannica, Inc. Chicago, Illinois, USA. 1952

Take away probability, and you can no longer please the world; give probability, and you can no longer displease it.
In *Great Books of the Western World* (Volume 33)
Pensées
Section XIV, 918
Encyclopædia Britannica, Inc. Chicago, Illinois, USA. 1952

But is it probable that probability gives assurance?
In *Great Books of the Western World* (Volume 33)
Pensées
Section XIV, 908
Encyclopædia Britannica, Inc. Chicago, Illinois, USA. 1952

Pearl, Judea
Computer scientist and statistician

Probabilities are summaries of knowledge that is left behind when information is transferred to a higher level of abstraction.
Probabilistic Reasoning in Intelligent Systems: Network of Plausible Inference
Chapter 1 (p. 21)
Morgan Kaufmann Publishers, Inc. San Mateo, California, USA. 1988

Pearson, E. S.   1895–1980
English statistician

Hitherto the user has been accustomed to accept the function of probability theory laid down by the mathematicians; but it would be good if he could take a larger share in formulating himself what are the practical requirements that the theory should satisfy in applications.
The Choice of Statistical Test Illustrated on the Interpretation of Data Classed in a 2 × 2 Table
Biometrika, Volume 34, Number 35, 1948 (p. 142)

Peirce, Charles Sanders   1839–1914
American scientist, logician, and philosopher

This branch of mathematics [probability] is the only one, I believe, in which good writers frequently get results entirely erroneous.
Writings of Charles S. Peirce (Volume 3)
The Doctrine of Chances, II (p. 279)
Indiana University Press. Bloomington, Indiana, USA. 1986

The relative probability of this or that arrangement of Nature is something which we should have a right to talk about if universes were as plenty as blackberries, if we could put a quantity of them in a bag, shake them well up, draw out a sample, and examine them to see what proportion of them had one arrangement and what proportion another. But, even in that case, a higher universe would contain us, in regard to whose arrangements the conception of probability could have no applicability.
The Probability of Induction
Popular Science Monthly, Volume 12, April 1878 (p. 714)

…it may be doubtful if there is a single extensive treatise on probabilities in existence which does not contain solutions absolutely indefensible.
Writings of Charles Sanders Peirce (Volume 3)
The Doctrine of Chances, II (p. 279)
Indiana University Press. Bloomington, Indiana, USA. 1986

Planck, Max   1858–1947
German physicist

Nature prefers more probable to less probable states.… Heat flows from a body of high temperature to a body of lower temperature, because the state of equal temperature is more probable than a state of unequal distribution of temperature.
Translated by R. Jones and D.H. Williams
A Survey of Physics: A Collection of Lectures and Essays
The Unity of the Physical Universe (p. 15)
Methuen & Company Ltd. London, England. 1925

Plato   428 BCE–347 BCE
Greek philosopher

I know too well that these arguments from probabilities are impostors, and unless great caution is observed in the use of them, they are apt to be deceptive.
In *Great Books of the Western World* (Volume 7)
Phaedo
Section 92 (p. 238)
Encyclopædia Britannica, Inc. Chicago, Illinois, USA. 1952

Poincaré, Henri   1854–1912
French mathematician and theoretical astronomer

No matter how solidly founded a prediction may appear to us, we are never absolutely sure that experiment will not contradict it, if we undertake to verify it.… It is far better to foresee even without certainty than not to foresee at all.
The Foundations of Science
Science and Hypothesis, Part IV

Chapter IX (p. 129)
The Science Press. New York, New York, USA. 1913

The very name calculus of probabilities is a paradox. Probability opposed to certainty is what we do not know, and how can we calculate what we do not know?
The Foundations of Science
Science and Hypothesis, Part IV
Chapter XI (p. 155)
The Science Press. New York, New York, USA. 1913

Predicted facts…can only be probable.
The Foundations of Science
Science and Hypothesis, Part IV
Chapter XI (p. 155)
The Science Press. New York, New York, USA. 1913

Popper, Karl R.   1902–94
Austrian/British philosopher of science

I think that we shall have to get accustomed to the idea that we must not look upon science as a "body of knowledge" but rather as a system of hypotheses; that is to say, as a system of guesses or anticipations which in principle cannot be justified, but with which we work as long as they stand up to tests, and of which we are never justified in saying that we know that they are "true" or "more or less certain" or even "probable."
The Logic of Scientific Discovery
New Appendices, Two Notes on Induction and Demarcation 1933–1934 (p. 317)
Basic Books, Inc. New York, New York, USA. 1959

The most important application of the theory of probability is to what we may call "chance-like" or "random" events, or occurrences. These seem to be characterized by a peculiar kind of incalculability which makes one disposed to believe — after many unsuccessful attempts — that all known rational methods of prediction must fail in their case. We have, as it were, the feeling that not a scientist but only a prophet could predict them. And yet, it is just this incalculability that makes us conclude that the calculus of probability can be applied to these events.
The Logic of Scientific Discovery
Part II, Chapter VII, Section 49 (p. 150)
Basic Books, Inc. New York, New York, USA. 1959

Pratchett, Terry   1948–
English author

Understanding is the first step towards control. We now understand probability…
The Dark Side of the Sun (p. 37)
St. Martin's Press. New York, New York, USA. 1976

"You haven't heard of probability math? You, and tomorrow you become Chairman of the Board of Widdershinss and heir to riches untold? Then first we will talk, and then we will eat."
The Dark Side of the Sun (p. 13)
St. Martin's Press. New York, New York, USA. 1976

I can't pretend to understand probability math. But if the universe is so ordered, so — immutable — that the future can be told by a handful of numbers, then why need we go on living?
The Dark Side of the Sun (p. 22)
St. Martin's Press. New York, New York, USA. 1976

Prior, Matthew   1664–1721
English poet and diplomat

In this case probability must atone for want of Truth.
The Literary Works of Matthew Prior
Solomon, Preface (p. 309)
Clarendon Press. Oxford, England. 1959

Ramsey, Frank Plumpton   1903–30
English mathematician

I think I perceive or remember something but am not sure; this would seem to give me some ground for believing it, contrary to Mr. Keynes' theory, by which the degree of belief in it which it would be rational for me to have is that given by the probability relation between the proposition in question and the things I know for certain.
In R.B. Braithwaite (ed.)
The Foundation of Mathematics and Other Logical Essays
Truth and Probability (p. 189)
Kegan, Paul, Trench, Trubner & Company. London, England. 1931

Redfield, Roy A.
No biographical data available

Good and bad come mingled always. The long-time winner is the man who is not unreasonably discouraged by persistent streaks of ill fortune, [who is] not at other times made reckless with the thought that he is fortune's darling. He keeps a cool head and trusts in the mathematics of probability, or as often said, the law of averages.
Factors of Growth in a Law Practice (p. 168)
Callaghan & Company. Mundelein, Illinois, USA. 1962

Reichenbach, Hans   1891–1953
German philosopher of science

To say that observations of the past are certain, whereas predictions are merely probable, is not the ultimate answer to the question of induction; it is only a sort of intermediate answer, which is incomplete unless a theory of probability is developed that explains what we should mean by "probable" and on what ground we can assert probabilities.
The Rise of Scientific Philosophy
Chapter 5 (p. 93)
University of California Press. Berkeley, California, USA. 1951

The study of inductive inference belongs to the theory of probability, since observational facts can make a theory only probable but will never make it absolutely certain.
The Rise of Scientific Philosophy
Chapter 14 (p. 231)
University of California Press. Berkeley, California, USA. 1951

Rota, Gian-Carlo 1932–99
Italian-born American mathematician

The distance between probability and statistical mechanics is diminishing, and soon we won't be able to tell which is which. We will be rid of the handwaving arguments with which mathematically illiterate physicists have been pestering us.
Indiscrete Thoughts
Chapter XX (p. 228)
Birkhäuser. Boston, Massachusetts, USA. 1997

Conditional probability tends to be viewed as one technique for calculating probabilities. Actually, there is more to it than meets the eye: it has an interpretation which is ordinarily passed over in silence. Philosophers take notice.
Indiscrete Thoughts
Chapter XX (p. 227)
Birkhäuser. Boston, Massachusetts, USA. 1997

According to quantum mechanics, it cannot be known what an atom will do in given circumstances; there are a definite set of alternatives open to it, and it chooses sometimes one, sometimes another. We know in what proportion of cases one choice will be made, in what proportion a second, or a third, and so on. But we do not know any law determining the choice in an individual instance. We are in the same position as a booking-office clerk at Paddington, who can discover, if he chooses, what proportion of travelers from that station go to Birmingham, what proportion to Exeter, and so on, but knows nothing of the individual reasons which lead to one choice in one case and another in another.
Religion and Science
Determinism (p. 152)
Henry Holt & Company. New York, New York, USA. 1935

Sartre, Jean-Paul 1905–80
French existentialist philosopher and novelist

When we want something, we always have to reckon with probabilities.
The Philosophy of Existentialism
Part 1. The Humanism of Existentialism (p. 46)
Philosophical Library, New York, New York, USA; 1965

…all views are only probable, and a doctrine of probability which is not bound to a truth dissolves into thin air. In order to describe the probable, you must have a firm hold on the true. Therefore, before there can by any truth whatsoever, there must be absolute truth.
The Philosophy of Existentialism
Part 1 The Humanism of Existentialism (p. 51)
Philosophical Library, New York, New York, USA; 1965

Schiller, Ferdinand Canning Scott 1864–1937
English philosopher

…it is better to be satisfied with probabilities than to demand impossibilities and starve.

In Charles Singer (ed.)
Studies in the History and Method of Science (Volume 1)
Scientific Discovery and Logical Proof (p. 272)
At The Clarendon Press. Oxford, England. 1917

Shakespeare, William 1564–1616
English poet, playwright, and actor

'Tis pretty, sure, and very probable…
In *Great Books of the Western World* (Volume 26)
The Plays and Sonnets of William Shakespeare (Volume 1)
As You Like It
Act III, Scene v, l. 11
Encyclopædia Britannica, Inc. Chicago, Illinois, USA. 1952

Sherwood, Thomas
No biographical data available

It is a nice idea, of course, that numbers must prove something true: after all, one patient is a case-report, two are a series. But all that numbers can do is tell us what is probable, and probability can lead us into terrible mistakes — like Bertrand Russell's chicken. Day in, day out, throughout its life, the chicken got breakfast as the farmer arrived in the morning. Thus it had clearly discovered a highly probable natural law: farmer = food — until, that is, the morning when the farmer very naturally arrived to wring its neck instead.
Science in Radiology
Lancet, Volume 1, 1978 (p. 594)

South, Robert
No biographical data available

That is accounted probable which has better arguments producible for it than can be brought against it.
In S. Austin Alibone
Prose Quotations from Socrates to Macaulay
Probability
J.B. Lippincott. Philadelphia, Pennsylvania, USA. 1903

Stoppard, Tom 1937–
Czech-born English playwright

If we postulate…that within un-, sub- or supernatural forces the probability is that the law of probability will not operate as a factor, then we must accept that the probability of the first part will not operate as a factor within un-, sub- or supernatural forces. And since it obviously hasn't been doing so, we can take it that we are not held within un-, sub- or supernatural forces after all; in all probability, that is.
Rosencrantz and Guildenstern Are Dead
Act One (p. 17)
Grove Press, Inc. New York, New York, USA. 1967

The Bible

For our knowledge and our prophecy alike are partial…
The Revised English Bible
I Corinthians 13:9
Oxford University Press, Inc. Oxford, England. 1989

Tillotson, John   1630–94
Archbishop of Canterbury

Though moral certainty be sometimes taken for a high degree of probability, which can only produce a doubtful assent, yet it is also frequently used for a firm assent to a thing upon such grounds as fully satisfy a prudent man.
In S. Austin Allibone
Prose Quotations from Socrates to Macaulay
Probability
J.B. Lippincott. Philadelphia, Pennsylvania, USA. 1903

Toffler, Alvin   1928–
American writer and futurist

The management of changes is the effort to convert certain possibles into probables, in pursuit of agreed-on preferables.
Future Shock
Chapter 20 (p. 407)
Random House, Inc. New York, New York, USA. 1979

Voltaire (François-Marie Arouet)   1694–1778
French writer

From generation to generation skepticism increases; and probability diminishes; and soon probability is reduced to zero.
The Portable Voltaire
Philosophical Dictionary, Truth (p. 217)
The Viking Press. New York, New York, USA. 1959

He who has heard the thing told by twelve thousand eyewitnesses, has only twelve thousand probabilities, equal to one strong probability, which is not equal to certainty.
The Portable Voltaire
Philosophical Dictionary, Truth (p. 217)
The Viking Press. New York, New York, USA. 1959

von Clausewitz, Carl   1780–1831
Prussian soldier

In short, absolute, so-called mathematical factors never find a firm basis in military calculations. From the very start there is an interplay of possibilities, probabilities, good luck and bad that weaves its way throughout the length and breadth of the tapestry. In the whole range the human activities war most closely resembles a game of cards.
On War
Chapter 1, 21 (p. 86)
The Modern Library. New York, New York, USA. 1943

von Humboldt, Alexander   1769–1859
German naturalist and explorer

Whatever must be characterized as mere probability lies beyond the domain of physical description of the universe; science must not wander into the cloudland of cosmological dreams.
Cosmos: A Sketch of a Physical Description of the Universe (Volume 4)

Conclusion (p. 230)
Harper & Brothers. New York, New York, USA. 1869

von Mises, Richard   1883–1953
Austrian-born American mathematician

The theory of probability can never lead to a definite statement concerning a single event.
Probability, Statistics, and Truth
First Lecture (p. 33)
Dover Publications, Inc. New York, New York, USA. 1981

…if one talks of the probability that the two poems known as the Iliad and the Odyssey have the same author, no reference to a prolonged sequence of cases is possible and it hardly makes sense to assign a numerical value to such a conjecture.
Mathematical Theory of Probability and Statistics (pp. 13–14)
Academic Press. New York, New York, USA. 1964

Walker, Marshall John
American physicist

One can locate an octopus by giving the coordinates of his beak, but it would be unwise to forget that neighboring coordinates for two or three yards out in all directions have a considerable probability of being occupied by octopus at a given instant.
The Nature of Scientific Thought
Chapter V (p. 65)
Prentice-Hall, Inc. Englewood Cliffs, New Jersey, USA. 1963

Whitehead, Alfred North   1861–1947
English mathematician and philosopher

…Positivistic science is solely concerned with observed fact, and must hazard no conjecture as to the future. If observed fact be all we know, then there is no other knowledge. Probability is relative to knowledge. There is no probability as to the future within the doctrine of Positivism.
Adventures of Ideas
Chapter VIII (p. 160)
The Macmillan Company. New York, New York, USA. 1956

Whyte, Lancelot Law   1896–1972
Scottish physicist

Only a certain probability remains of a one-to-one association of any spatial feature now with a similar feature a moment later. It is sheer luck, in a sense, that any physical apparatus stays put, for the laws of quantum mechanics allow it a finite, though small, probability of dispersing while one is not looking, or even while one is.
Essay on Atomism: From Democritus to 1960
Chapter 2 (pp. 25–26)
Wesleyan University Press. Middletown, Connecticut, USA. 1961

If the universe is a mingling of probability clouds spread through a cosmic eternity of space-time, how is there as much order, persistence, and coherent transformation as there is?

Essay on Atomism: From Democritus to 1960
Chapter 2 (p. 27)
Wesleyan University Press. Middletown, Connecticut, USA. 1961

Wilde, Oscar   1854–1900
Irish wit, poet, and dramatist

GILBERT: No ignoble consideration of probability, that cowardly concession to the tedious repetitions of domestic or public life, effect it ever.
Complete Writings of Oscar Wilde (p. 143)
Nottingham Society. New York, New York, USA. 1907

Wilder, Thornton   1897–1975
American playwright and novelist

Ashley had no competitive sense and no need for money, but he took great interest in the play of numbers. He drew up charts analyzing the elements of probability in the various games. He had a memory for numbers and symbols.
The Eighth Day
II. Illinois to Chile (p. 123)
Harper & Row, Publishers, New York, New York, USA, 1967

Woodward, Robert Simpson   1849–1924
American scientist and teacher

The theory of probabilities and the theory of errors now constitute a formidable body of knowledge of great mathematical interest and of great practical importance. Though developed largely through the applications to the more precise sciences of astronomy, geodesy, and physics, their range of applicability extends to all the sciences; and they are plainly destined to play an increasingly important role in the development and in the applications of the sciences of the future. Hence their study is not only a commendable element in a liberal education, but some knowledge of them is essential to a correct understanding of daily events.
Probability and Theory of Errors
Author's Preface (p. 4)
John Wiley & Sons, Inc. New York, New York, USA. 1906

PROBABLE ERROR

Russell, Bertrand Arthur William   1872–1970
English philosopher, logician, and social reformer

Who ever heard a theologian preface his creed, or a politician conclude his speech with an estimate of the probable error of his opinion.
In Edwin Hubble
The Nature of Science and Other Lectures
Part I. The Nature of Science (p. 10)
The Huntington Library. San Marino, California, USA. 1954

Student (William Sealy Gossett)

An experiment may be regarded as forming an individual of a "population" of experiments which might be performed under the same conditions. A series of experiments is a sample drawn from this population. Now any series of experiments is only of value in so far as it enables us to form a judgment as to the statistical constants of the population to which the experiments belong. In a great number of cases the question finally turns on the value of a mean, either directly, or as the mean difference between the two quantities.
The Probable Error of a Mean
Biometrika, Volume 6, 1908

PROBLEM

Ackerman, Diane   1948–
American writer

Part of the irony of environmentalism is questing for solutions when you know you're part of the problem.
The Rarest of the Rare: Vanishing Animals, Timeless Worlds
Insect Love (p. 156)
Vintage Books. New York, New York, USA. 1997

Agnew, Ralph Palmer
American mathematician

Working a problem is like cutting down a tree and reading a problem is like looking at a tree. We spend part of our time swinging axes to develop our muscles, and we spend part of our time looking around to keep us from being dolts. The mathematical and scientific forests really are interesting, and we should all enjoy chopping and looking at the scenery.
Differential Equations
Chapter 1 (p. 6)
McGraw-Hill Book Company. New York, New York, USA. 1972

Alger, John R. M.
American engineer
Hays, Carl V.
No biographical data available

...a problem in the stage of being "recognized" is a highly emotional subject.
Creative Synthesis in Design (p. 13)
Prentice-Hall. Englewood Cliffs, New Jersey, USA. 1964

Anderson, Poul   1926–2001
American science fiction writer

I have yet to see any problem, however complicated, which, when you looked at it in the right way, did not become still more complicated.
In William Thorpe article
Reduction v. Organicism
New Scientist, Volume 43, Number 66, 25 September 1969 (p. 638)

Bahn, Paul
Archaeologist

It takes very special qualities to devote one's life to problems with no attainable solutions and to poking around in dead people's garbage: words like "masochistic", "nosy", and "completely batty" spring readily to mind.
Bluff Your Way in Archaeology (p. 7)
Ravette Books. West Sussex, England. 1989

Berkeley, Edmund C.   1909–88
American computer theoretician

Most problems have either many answers or no answer. Only a few problems have a single answer.
Computers and Automation
Right Answers — A Short Guide for Obtaining Them, Volume 18, Number 10, September 1969 (p. 20)

Beveridge, William Ian Beardmore   1908–
Australian zoologist

Too often I have been able to [do] little more than indicate the difficulties likely to be met — yet merely to be forewarned is often help.
The Art of Scientific Investigation
Preface (p. ix)
W.W. Norton & Company, Inc. New York, New York, USA. 1957

Bloch, Arthur   1948–
American humorist

Inside every large problem is a small problem struggling to get out.
Murphy's Law
Hoare's Law of Large Problems (p. 50)
Price/Stern/Sloan, Publishers. Los Angeles, California, USA. 1981

Bragg, Sir William Lawrence   1890–1971
Australian-born English physicist

I am sure that when the first circumnavigators of the world returned from their voyage they were told by friends that some Greek philosopher…had held that the world was round and that they might have spared their trouble. The world is either round or flat, and endless discussion might have been carried on for ages between opposing schools who held one view or the other. The real contribution to settling the problem was made by the circumnavigators.
The Physical Sciences
Science, Volume 79, Number 2046, March 16, 1934 (p. 240)

Chambers, Robert   1802–71
Science writer

Man is seen to be an enigma only as an individual; in mass, he is a mathematical problem.
Vestiges of the Natural History of Creation (p. 333)
W.R. Chambers. Edinburgh, Scotland. 1884

Chesterton, G. K. (Gilbert Keith)   1874–1936
English author

It isn't that they can't see the solution. It is that they can't see the problem.

The Scandal of Father Brown
The Point of the Pin (p. 949)
Dodd, Mead & Company. New York, New York, USA. 1935

Churchill, Winston Spencer   1882–1965
British prime minister, statesman, soldier, and author

In the course of a few bewildering years we have found ourselves the master or indeed the servants of gigantic powers which confront us with problems never known before.
In R. James (ed.)
Winston S. Churchill — His Complete Speeches 1897–1963
Volume 8 (p. 8563)
Chelsea House Publishers. New York, New York, USA. 1974

Clarke, Arthur C.   1917–
English science and science fiction writer

Sorry to interrupt the festivities, but we have a problem.
2001: A Space Odyssey
IV Abyss, Chapter 21 (p. 120)
New American Library, New York, New York, USA. 1968

Cleaver, Eldridge   1935–98
American civil rights leader and author

…you're either part of the solution or part of the problem.
Speech
San Francisco, 1968

Cohen, I. Bernard   1914–2003
American physicist and science historian

The fundamental postulate of the history of science is that the scientists of the past were just as intelligent as we are and that, therefore, the problems that baffled them would have baffled us too, had we been living then.
Franklin and Newton
Chapter Two (p. 39)
Harvard University Press. Cambridge, Massachusetts, USA. 1966

Collingwood, Robin George   1889–1943
English historian and philosopher

You can only solve a problem which you recognize to be a problem.
The New Leviathan; or Man, Society, Civilization and Barbarism
Part I, Chapter I, aphorism 2.66 (p. 13)
At The Clarendon Press. Oxford, England. 1942

Commoner, Barry   1917–
American biologist, ecologist, and educator

The freedom to choose his own problem is often the scientist's most precious possession.
Is Science Getting Out of Hand?
The Science Teacher, Volume 30, October 1963

Compton, Karl Taylor   1887–1954
American educator and physicist

Neither curiosity nor ingenuity is a modern impulse.... The distinctive feature of science and technology at the present time is the accelerated pace of their development. This is partly due to continually improved techniques and organization, and it is partly due to the great accumulation of knowledge and art, because the more information and tools we have at our disposal, the more powerful can be the attack on any new problem.

A Scientist Speaks: Excerpts from Addresses by Karl Taylor Compton During the Years 1930–1949 (pp. 1–2)
Undergraduate Association, Massachusetts Institute of Technology.
Cambridge, Massachusetts, USA. 1955

Condorcet, Marie Jean   1743–94
French philosopher and mathematician

If a scholar poses himself a new problem, he can attack it fortified by the pooled resources of all his predecessors.
In Maurice Daumas
Scientific Instruments of the 17th and 18th Centuries and Their Makers
Eulogy for J. de Vaucanson before the Academie of Sciences (p. 119)

Cross, Hardy   1885–1959
American professor of civil and structural engineering

In general the problems of civil engineers are given to them by God Almighty. They are the problems of nature. On the other hand mechanical and electrical work has problems which man, to a certain extent, has created for himself.
In Lenox H. Lohr
Centennial of Engineering: History and Proceedings of Symposia: 1852–1952
Professional Aspects of Mechanical Engineering (p. 150)
Centennial of Engineering. Chicago, Illinois. 1952

Douglas, A. Vibert   1894–1988
Canadian astronomer

It is a solemn thought that no man liveth unto himself. It is equally true that no star, no atom, no electron, no ripple of radiant energy, exists unto itself. All the problems of the physical universe are inextricably bound up with one another in the relations of space and time.
From Atoms to Stars
The Atlantic Monthly, Volume 144, August 1929 (p. 165)

Doyle, Sir Arthur Conan   1859–1930
Scottish writer

Every problem becomes very childish when once it is explained to you...
In William S. Baring-Gould (ed.)
The Annotated Sherlock Holmes (Volume 2)
The Adventure of the Dancing Men (p. 258)
Wings Books. New York, New York, USA. 1967

"My mind," he said, "rebels at stagnation. Give me problems, give me work, give me the most abstruse cryptogram, or the most intricate analysis, and I am in my own proper atmosphere."

In William S. Baring-Gould (ed.)
The Annotated Sherlock Holmes (Volume 1)
The Sign of the Four, Chapter 1 (p. 611)
Wings Books. New York, New York, USA. 1967

In solving a problem of this sort, the grand thing is to be able to reason backwards. That is a very useful accomplishment, and a very easy one, but people do not practice it much. In the every-day affairs of life it is more useful to reason forwards, and so the other comes to be neglected. There are fifty who can reason synthetically for one who can reason analytically.
In William S. Baring-Gould (ed.)
The Annotated Sherlock Holmes (Volume 1)
A Study in Scarlet, Chapter 14 (p. 231)
Wings Books. New York, New York, USA. 1967

It is quite a three pipe problem, and I beg that you won't speak to me for fifty minutes.
In William S. Baring-Gould (ed.)
The Annotated Sherlock Holmes (Volume 1)
The Red-Headed League (p. 428)
Wings Books. New York, New York, USA. 1967

du Preez, Peter
No biographical data available

The reason for the rapid advance of the problem-solving capacity of natural sciences is that scientists are trained to introduce theoretical variations, to test them empirically, and to preserve and propagate those innovations which survive whatever tests have been proposed.
A Science of Mind: The Quest for Psychological Reality
Part II, Chapter 7 (p. 123)
Academic Press Ltd. London, England. 1991

Dyson, Freeman J.   1923–
American physicist and educator

The difference between a text without problems and a text with problems is like the difference between learning to read a language and learning to speak it.
Disturbing the Universe (p. 13)
Harper & Row, Publishers. New York, New York, USA. 1979

Easton, Elmer C.
No biographical data available

All of the problems with which engineers are normally concerned have to do with the satisfying of some human want.
An Engineering Approach to Creative Thinking
Ceramic Age, September 1955 (p. 28)

Ehrenberg, A. S. C.
No biographical data available

Many problems arise year after year. The answers, if only we knew them, should therefore also be similar year after year.
Data Reduction

Chapter 4 (p. 56)
John Wiley & Sons Ltd. London, England. 1975

Einstein, Albert   1879–1955
German-born physicist

There are so many unsolved problems in physics. There is so much that we do not know; our theories are far from adequate.
In I. Bernard Cohen
An Interview with Einstein
Scientific American, Volume 193, Number 1, July 1955 (p. 69)

Einstein, Albert   1879–1955
German-American physicist
Infeld, Leopold   1898–1968
Polish physicist

The importance of a problem should not be judged by the number of pages devoted to it.
The Evolution of Physics
Preface (p. ix)
Simon & Schuster. New York, New York, USA. 1961

Emerson, Ralph Waldo   1803–82
American lecturer, poet, and essayist

"What we know, is a point to what we do not know." Open any recent journal of science, and weigh the problems suggested concerning Light, Heat, Electricity, Magnetism, Physiology, Geology, and judge whether the interest of natural science is likely to be soon exhausted.
The Complete Works of Ralph Waldo Emerson (Volume 1)
Nature: Addresses, and Lectures
Nature, Chapter V (p. 39)
Houghton Mifflin Company. Boston, Massachusetts, USA. 1904

Feynman, Richard P.   1918–88
American theoretical physicist

No problem can be solved without it dragging in its wake new problems to be solved.
Selected Papers of Richard Feynman with Commentary
The Present Status of Quantum Electrodynamics (p. 134)
World Scientific. Singapore. 2000

If we want to solve a problem that we have never solved before, we must leave the door to the unknown ajar.
What Do You Care What Other People Think?
The Value of Science (p. 247)
W. W. Norton & Company, Inc. New York, New York, USA. 1988

Fleming, J. A.
No biographical data available

Whilst we derive satisfaction from the thought that so much valuable discovery and invention has already rewarded the labors of workers in many lands, we have but to glance around us to see in all directions, in connection with it, unsolved problems, untrodden paths, wide fields of knowledge ripe for harvest in which the sickle of the reaper has never yet been moved.

Annual Report of the Board of Regents of the Smithsonian institution, 1907
Recent Contributions to Electric Wave Telegraphy (p. 193)
Government Printing Office. Washington, D.C. 1908

Flexner, Abraham   1866–1959
American educator

…science, in the very act of solving problems, creates more of them.
Universities: American, English, German
Chapter I, Section v (p. 19)
Oxford University Press, Inc. Oxford, England. 1930

Frazier, A. W.
No biographical data available

Often problems not solved earlier have not been posed earlier.
Hydrocarbon Processing
The Practical Side of Creativity, Volume 45, Number 1, January 1966

Fredrickson, A. G.   1932–
No biographical data available

To be aware that a problem exists is the prerequisite for any attempt to solve the problem.
The Dilemma of Innovating Societies
Chemical Engineering Education, Volume 4, Summer 1969 (p. 148)

Gould, Stephen Jay   1941–2002
American paleontologist and evolutionary biologist

…without a commitment to science and rationality in its proper domain, there can be no solution to the problems that engulf us. Still, the Yahoos never rest.
Ever Since Darwin: Reflections in Natural History
Chapter 17 (p. 146)
W.W. Norton & Company, Inc. New York, New York, USA. 1977

Halmos, Paul R.   1916–2006
Hungarian-born American mathematician

A teacher who is not always thinking about solving problems — ones he does not know the answer to — is psychologically simply not prepared to teach problem solving to his students.
I Want to Be a Mathematician
Chapter 14 (p. 322)
Springer-Verlag. New York, New York, USA. 1985

Hawkins, D.
No biographical data available

There are many things you can do with problems besides solving them. First you must define them, pose them. But then of course you can also refine them, depose them, or expose them or even dissolve them! A given problem may send you looking for analogies, and some of these may lead you astray, suggesting new and different problems, related or not to the original. Ends and means can get reversed. You had a goal, but the means you found

didn't lead to it, so you found a new goal they did lead to. It's called play. Creative mathematicians play a lot; around any problem really interesting they develop a whole cluster of analogies, of playthings.
In Necia Grant Cooper (ed.)
From Cardinals to Chaos
The Spirit of Play (p. 44)
Cambridge University Press. Cambridge, England. 1988

Heisenberg, Werner Karl   1901–76
German physicist and philosopher

...only those revolutions in science will prove fruitful and beneficial whose investigators try to change as little as possible and limit themselves to the solution of a particular and clearly defined problem. Any attempt to make a clean sweep of everything or to change things arbitrarily leads to utter confusion.
Physics and Beyond: Encounters and Conversations
Chapter 12 (p. 148)
Harper & Row, Publishers. New York, New York, USA. 1971

Heller, Joseph   1923–99
American writer

He was pinched perspiringly in the epistemological dilemma of the skeptic, unable to accept solutions to problems he was unable to dismiss as unsolvable. He was never without misery and never without hope.
Catch-22
Chapter 25 (p. 275)
Dell Publishing Company, Inc. New York, New York, USA. 1985

Herschel, Sir John Frederick William   1792–1871
English astronomer and chemist

The great problems which offer themselves on all hands for solution, problems which the wants of the age force upon us as practically interesting, and with which its intellect feels itself competent to deal, are far more complex in their conditions, and depend on data which to be of use must be accumulated in far greater masses, collected over an infinitely wider field, and worked upon with a greater and more systematized power than has sufficed for the necessities of astronomy. The collecting, arranging, and duly combining these data are operations which, to be carried out to the extent of the requirements of modern science, lie utterly beyond the reach of all private industry, mean, or enterprise. Our demands are not merely for a slight and casual sprinkling to refresh and invigorate an ornamental or luxurious product, but for a copious, steady, and well-directed stream, to call forth from a soil ready to yield it, an ample, healthful, and remunerating harvest.
Essays from the Edinburgh and Quarterly Reviews with Addresses and Other Pieces
Terrestrial Magnetism (pp. 110–111)
Longman, Brown, Green, Longmans & Roberts. London, England. 1857

Herstein, I. N.
No biographical data available

The value of a problem is not so much in coming up with the answer as in the ideas and attempted ideas it forces on the would-be solver.
Topics in Algebra
Preface (p. vi)
Xerox College Publishing. Waltham, Massachusetts, USA. 1964

Hilbert, David   1862–1943
German mathematician

As long as a branch of science offers an abundance of problems, so long is it alive; a lack of problems foreshadows extinction of the cessation of independent development. Just as every human undertaking pursues certain objects, so also mathematical research requires its problems. It is by the solution of problems that the investigator tests the temper of his steel; he finds new methods and new outlooks, and gains a wider and freer horizon.
Mathematical Problems
Bulletin of the American Mathematical Society, Volume 8, July 1902
(p. 438)

It is difficult and often impossible to judge the value of a problem correctly in advance; for the final award depends upon the gain which science obtains from the problem.
Mathematical Problems
Bulletin of the American Mathematical Society, Volume 8, July 1902
(p. 438)

Hodnett, Edward   1901–84
Illustration historian

You have to identify the real problem, and you have to identify the total problem.
The Art of Problem Solving
Part I, Chapter 2 (p. 12)
Harper & Brothers. New York, New York, USA. 1955

Problems often boil down to the simple form of a dilemma. A dilemma presents a choice of two solutions to a problem, both of which are unsatisfactory.
The Art of Problem Solving
Part II, Chapter 8 (p. 63)
Harper & Brothers. New York, New York, USA. 1955

An unstated problem cannot be solved. Many problems go unsolved for centuries for lack of adequate statement.
The Art of Problem Solving
Part I, Chapter 3 (p. 19)
Harper & Brothers. New York, New York, USA. 1955

...being able to predict which problems you are not likely to solve is good for your peace of mind.
The Art of Problem Solving
Part I, Chapter 1 (p. 6)
Harper & Brothers. New York, New York, USA. 1955

Hoyle, Sir Fred   1915–2001
English mathematician and astronomer

It is almost a matter of principle that in any difficult unsolved problem the right method of attack has not been found; failure to solve important problems is rarely due to the inadequacy in the handling of technical details.
Man in the Universe
Chapter 2 (p. 20)
Columbia University Press. New York, New York, USA. 1966

Huxley, Julian   1887–1975
English biologist, philosopher, and author

The time has gone by when the intelligent public needs to be reminded of the practical utility of science, or of the fact that the investigation of any problem, however apparently remote from everyday life, may be fraught with the most valuable consequences.
The Century Illustrated Monthly Magazine
Searching for the Elixir of Life, Volume 103, Number 4, February 1922 (p. 629)

Ingle, Dwight J.   1907–78
Biologist and endocrinologist

When you are trying to solve problems and are searching for new and creative ideas, let your mind be free-wheeling. Enjoy unbridled fancy. After you get an idea that seems important and plausible, it can be tested by evidence and reason.
Is It Really So?: A Guide to Clear Thinking
Chapter 18 (p. 129)
The Westminster Press. Philadelphia, Pennsylvania, USA. 1976

Kaplan, Abraham   1918–1993
American philosopher of science, author, and educator

Give a small boy a hammer and he will find that everything he encounters needs pounding. It comes as no particular surprise that a scientist formulates problems in a way which requires for their solution just those techniques in which he himself is skilled…
The Conduct of Inquiry: Methodology for Behavioral Science (p. 28)
Chandler Publishing Company. San Francisco, California, USA. 1964

Kettering, Charles Franklin   1876–1958
American engineer and inventor

A problem is not solved in a laboratory. It is solved in some fellow's head. All the apparatus is for is to get his head turned around so that he can see the thing right.
In T.A. Boyd
Professional Amateur
Part II Chapter XII (pp. 102–103)
E.P. Dutton & Company, Inc. New York, New York, USA. 1957

No one should pick a problem, or make a resolution, unless he realizes that the ultimate value of it will offset the inevitable discomfort and trouble that always goes along with the accomplishment of anything worth while. So let us not waste our time and effort on some trivial thing.
Short Stories of Science and Invention: A Collection of Radio Talks by C.F. Kettering
Patience (p. 59)
General Motors. Detroit, Michigan, USA. 1955

As long as we try and patiently do our best to solve the problem, although we may not get the answer we are looking for, we always get something — even if it is only the valuable experience.
Short Stories of Science and Invention: A Collection of Radio Talks by C.F. Kettering
Purple Dye, Sun Glasses and Malaria (p. 115)
General Motors. Detroit, Michigan, USA. 1955

But in picking that problem be sure to analyze it carefully to see that it is worth the effort. It takes just as much effort to solve a useless problem as a useful one.
Short Stories of Science and Invention: A Collection of Radio Talks by C.F. Kettering
Research Is a State of Mind (p. 11)
General Motors, Detroit, Michigan, USA. 1955

I often think we have so many facilities that we lose track of the problem. Problems, as you know, are solved in the mind of some intensely interested person.
Short Stories of Science and Invention: A Collection of Radio Talks by C.F. Kettering
Christmas Lecturer (p. 57)
General Motors, Detroit, Michigan, USA. 1955

Kiepenheuer, Karl
No biographical data available

For the astronomer, the inexhaustible store of problems in the world he has set out to conquer remains the real mainspring of all his arduous researches.
The Sun
Conclusion (p. 158)
The University of Michigan Press. Ann Arbor, Michigan, USA. 1959

Lewis, Gilbert Newton   1875–1946
American chemist

Indeed it seems hardly likely that much progress can be made in the solution of the difficult problems relating to chemical combination by assigning in advance definite laws of force between the positive and negative constituents of an atom, and then on the basis of these laws building up mechanical models of the atom.
The Atom and the Molecule
Journal of the American Chemical Society, Volume 38, Number 1, 1916 (p. 773)

Mach, Ernst   1838–1916
Austrian physicist and philosopher

There is no problem in all mathematics that cannot be solved by direct counting. But with the present implements of mathematics many operations can be performed in a few minutes which without mathematical methods would take a lifetime.
Popular Scientific Lectures

The Economical Nature of Physical Inquiry (p. 197)
The Open Court Publishing Company. Chicago, Illinois, USA. 1898

No man should dream of solving a great problem unless he is so thoroughly saturated with his subject that everything else sinks into comparative insignificance.
Popular Scientific Lectures
The Part Played by Accident in Invention and Discovery (pp. 273–274)
The Open Court Publishing Company. Chicago, Illinois, USA. 1898

Every real problem can and will be solved in due course without supernatural divination, entirely by accurate observation and close, searching thought.
Popular Scientific Lectures
On Sensations of Orientation (p. 308)
The Open Court Publishing Company. Chicago, Illinois, USA. 1898

Maddox, John Royden   1925–
Welsh chemist and physicist

The problems that remain unsolved are gargantuan. They will occupy our children and their children and on and on for centuries to come, perhaps even for the rest of time.
What Remains to Be Discovered
Conclusion (p. 378)
The Free Press. New York, New York, USA. 1998

Michener, James A.   1907?–97
American novelist

At NACA [fictional space agency] we solve everything eventually. That's our job, and now it's yours.… At NACA…there are no insoluble problems. Only time-consuming ones.
Space
Chapter III (p. 175)
Random House, Inc. New York, New York, USA. 1982

Nehru, Jawaharla   1889–1969
Former prime minister of India

It is science alone that can solve the problems of hunger and poverty, of insanitation and illiteracy, of superstition and deadening custom and tradition, of vast resources running to waste, of a rich country inhabited by starving people.
Speech
Proceedings of the National Institute of Science of India, Volume 27, 1960 (p. 564)

Nietzsche, Friedrich   1844–1900
German philosopher

There are dreadful people who, instead of solving a problem, complicate it for those who deal with it and make it harder to solve. Whoever does not know how to hit the nail on the head should [be] entreated not [to] hit it at all.
The Complete Works of Friedrich Nietzsche (Volume Seven)
Human, All Too Human
The Wanderer and His Shadow, Part Two, Number 326
Macmillan Publishing Company. New York, New York, USA. 1924

Oppenheimer, J. Robert   1904–67
American theoretical physicist

…we probably have no very good idea today of the range of problems that will be accessible to science.
The Flying Trapeze: Three Crises for Physicists
Space and Time (p. 2)
Oxford University Press, Inc. London, England. 1964

Pallister, William Hales   1877–1946
Canadian physician

Science solves life's problems, but she must solve them one-at-a-time. Her course and methods are evolutionary. She cannot solve insoluble problems; they must first become soluble. She grows, like every other plant, only from powdered fock, uses only the chemical constituents which are soluble.
Poems of Science
The Nature of Things (p. 14)
Playford Press. New York, New York, USA. 1931

Pearse, A. S.
No biographical data available

A scientist has his circulating medium in problems. He deals in and develops problems as a broker deals in stocks and bonds. When his problems are completed he "sells" them to the scientific world by publication, usually at his own expense.
Adventure, Romance and Science
Science, Volume 58, Number 1492, 3 August, 1923 (p. 78)

Pendry, John   1944–
English theoretical physicist

It has been said that tackling a new scientific problem is like going into a darkened room. First you fall over the furniture, then you collide with other people in the room; arguments might develop. With time things settle down, as you learn where most of the furniture is and don't fall over so often. Eventually someone finds the light switch and everything becomes obvious.
Positively Negative
Nature, Volume 423, Number 6935, May 1, 2003 (p. 22)

Poe, Edgar Allan   1809–49
American short story writer

The great problem is at length solved! The air, as well as the earth and the ocean, has been subdued by science, and will become a common and convenient highway for mankind.
In H. Beaver (ed.)
The Science Fiction of Edgar Allan Poe
The Balloon Hoax (p. 11)
Penguin Books. Hammondsworth, England. 1976

Poincaré, Henri   1854–1912
French mathematician and theoretical astronomer

To give a complete mechanical explanation of electrical phenomena, reducing the laws of physics to the fundamental principles of dynamics is a problem that has attracted many investigators.... If the problem admitted of only one solution, the possession of this solution, which would be the truth, could not be bought too dearly.

In Frederick Vreeland

Maxwell's Theory and Wireless Telegraphy. Part 1. Maxwell's Theory and Hertzian Oscillations. Part 2

Part One, Chapter I (p. 1)

McGraw Publishing Company. New York, New York, USA. 1904

Pólya, George 1887–1985
Hungarian mathematician

Solving problems is a practical art, like swimming, or skiing, or playing a piano; you can learn it only by imitation and practice...

Mathematical Discovery; or Understanding, Learning, and Teaching Problem Solving (Volume 1)

Preface (p. v)

John Wiley & Sons, Inc. New York, New York, USA. 1966

Popper, Karl R. 1902–94
Austrian/British philosopher of science

Science never pursues the illusory aim of making its answers final or even probable. Its advance is rather toward an infinite yet attainable aim: that of ever discovering new, deeper, and more general problems.

The Logic of Scientific Discovery

Part II, Chapter X, Section 85 (p. 281)

Basic Books, Inc. New York, New York, USA. 1959

...a young scientist who hopes to make discoveries is badly advised if his teacher tells him: "Go round and observe" and...well advised if his teacher tells him: "Try to learn what people are discussing nowadays in science. Find out where difficulties arise, and take an interest in disagreements. These are the questions which you should take up." In other words, you should study the problems of the day. This means that you pick up, and try to continue, a line of inquiry which has the whole background of the earlier development of science behind it.

Conjectures and Refutations: The Growth of Scientific Knowledge

Chapter 4 (p. 129)

Harper & Row, Publishers. New York, New York, USA. 1963

...science should be visualized as progressing from problem to problem — to problems of ever increasing depth.... Problems crop up especially when we are disappointed in our expectations, or when our theories involve us in difficulties, in contradictions; and these may arise either within a theory, or between two different theories, or as the result of a clash between our theories and our observations.... Thus science starts from problems, and not from observations; though observations may give rise to a problem, especially if they are unexpected; that is to say, if they clash with our expectations or theories.

Conjectures and Refutations: The Growth of Scientific Knowledge

Chapter 10, Section VI (p. 222)

Harper & Row, Publishers. New York, New York, USA. 1963

...I think there is only one way to science — or to philosophy, for that matter: to meet a problem, to see its beauty and fall in love with it; to get married to it, and to live with it happily, till death do ye part — unless you should meet another and even more fascinating problem, or unless, indeed, you should obtain a solution. But even if you do obtain a solution, you may then discover, to your delight, the existence of a whole family of enchanting though perhaps difficult problem children for whose welfare you may work, with a purpose, to the end of your days.

Realism and the Aim of Science

Preface, 1956 (p. 8)

Rowman & Littlefield. Totowa, New Jersey, USA.

Porter, George 1920–2002
English chemist

To solve a problem is to create new problems, new knowledge immediately reveals new areas of ignorance, and the need for new experiments.

Nobel Lectures, Chemistry 1963–1970

Nobel lecture for award received in 1967

Flash Photolysis and Some of Its Applications (p. 261)

Elsevier Publishing Company. Amsterdam, Netherlands. 1972

Rabinow, Jacob 1910–99
Inventor

...the creating of the problem is as big an invention as the solving of the problem — sometimes, a much greater invention.

In Daniel V. DeSimone

Education for innovation

The Process of Invention (p. 84)

Pergamon Press. New York, New York, USA. 1968

Ramón y Cajal, Santiago 1852–1934
Spanish neuropathologist

The forging of new truth almost always requires severe abstention and renunciation. During the so-called intellectual incubation period, the investigator should ignore everything unrelated to the problem of interest, like a somnambulist attending only to the voice of the hypnotist.

Advice for a Young Investigator

Chapter 3 (p. 35)

The MIT Press. Cambridge, Massachusetts, USA. 1999

I believe that excessive admiration for the work of great minds is one of the most unfortunate preoccupations of intellectual youth — along with a conviction that certain problems cannot be attacked, let alone solved, because of one's relatively limited abilities.

Advice for a Young Investigator

Chapter 2 (p. 9)
The MIT Press. Cambridge, Massachusetts, USA. 1999

Rapoport, Anatol 1911–
Russian-born mathematician and biologist

…the problems scientists are called on to solve are for the most part selected by the scientists themselves. For example, our Department of Defense did not one day decide that it wanted an atomic bomb and then order the scientists to make one. On the contrary, it was Albert Einstein, a scientist, who told Franklin D. Roosevelt, a decision maker, that such a bomb was possible.
Science, Conflict and Society: Readings from Scientific American
The Use and Misuse of Games Theory (p. 286)
W. H. Freeman & Company. San Francisco, California, USA. 1969

Raymond, Eric S.
No biographical data available

Often, the most striking and innovative solutions come from realizing that your concept of the problem was wrong.
The Cathedral and the Bazaar: Musings on Linux and Open Source by an Accidental Revolutionary
The Cathedral and the Bazaar (p. 40)
O'Riley. Beijing, China. 2001

Roszak, Theodore 1933–
American social critic

If a problem does not have a technical solution, it must not be a real problem. It is but an illusion…a figment born of some regressive cultural tendency.
The Making of a Counter Culture: Reflections on the Technocratic Society and Its Youthful Opposition
Chapter I (p. 10)
Doubleday & Company, Inc. Garden City, New York, USA; 1969

Russell, Bertrand Arthur William 1872–1970
English philosopher, logician, and social reformer

I am sorry that I have had to leave so many problems unsolved. I always have to make this apology, but the world really is rather puzzling and I cannot help it.
In John G. Slater (ed.)
The Collected Papers of Bertrand Russell (Volume 8)
The Philosophy of Logical Atomism
Lecture V (p. 211)
George Allen & Unwin Ltd. London, England. 1986

Russell, Henry Norris 1877–1957
American astronomer

The unsolved problems of Nature have a distinctive fascination, though they still far outnumber those which have even approximately been resolved.
The Solar System and Its Origin
Chapter I (p. 1)
The Macmillan Company. New York, New York, USA. 1935

Shaw, George Bernard 1856–1950
Irish comic dramatist and literary critic

Religion is always right. Religion solves every problem and thereby abolishes problems from the Universe. Religion gives us certainty, stability, peace and the absolute. It protects us against progress which we all dread. Science is the very opposite. Science is always wrong. It never solves a problem without raising ten more problems.
In B. Patch
Thirty Years with G.B.S.
Chapter Twelve (p. 235)
Dodd, Mead & Company. New York, New York, USA. 1951

My business tonight will be very largely to raise difficulties. That is all the use I am really in this world.
The New York Times
Shaw Expounds Socialism as World Panacea
December 12, 1926

…all problems are finally scientific problems.
The Doctor's Dilemma
Preface on Doctors
The Technical Problem (p. lxxxiii)
Brentano's. New York, New York, USA. 1920

Simon, H.
No biographical data available

Problem formulation in science is to be understood by looking at the continuity of the whole stream of scientific endeavor.
In Robert G. Colodny (ed.)
Mind and Cosmos
Scientific Discovery and the Psychology of Problem Solving (p. 37)
University of Pittsburgh Press. Pittsburgh, Pennsylvania, USA. 1966

Simon, Herbert Alexander 1916–2001
American social scientist

The capacity of the human mind for formulating and solving complex problems is very small compared with the size of problems whose solution is required for objectively rational behavior in the real world — or even for a reasonable approximation to such objective rationality.
Models of Man: Social and Rational; Mathematical Essays on Rational Human Behavior in a Social Setting
Part IV (p. 198)
John Wiley & Sons, Inc. New York, New York, USA. 1957

The more difficult and novel the problem, the greater is likely to be the amount of trial and error required to find a solution. At the same time, the trial and error is not completely random or blind; it is, in fact, rather highly selective. The new expressions that are obtained by transforming given ones are examined to see whether they represent progress toward the goal. Indications of progress spur further search in the same direction; lack of progress signals the abandonment of a line of search. Problem solving requires selective trial and error.
The Sciences of the Artificial
Chapter 4 (pp. 95–96)
The MIT Press. Cambridge, Massachusetts, USA. 1969

…human problem solving, from the most blundering to the most insightful, involves nothing more than varying mixtures of trial and error and selectivity.
The Sciences of the Artificial
Chapter 4 (p. 97)
The MIT Press. Cambridge, Massachusetts, USA. 1969

Simpson, N. F.   1919–
English playwright

And suppose we solve all the problems it presents? What happens? We end up with more problems than we started with. Because that's the way problems propagate their species. A problem left to itself dries up or goes rotten. But fertilize a problem with a solution — you'll hatch out dozens.
New English Dramatists 2
A Resounding Tinkle, Act I, Scene 1 (pp. 80–81)
Penguin Books. London, England. 1960

Szent-Györgyi, Albert   1893–1986
Hungarian-born American biochemist

Somehow, problems get into my blood and they don't give me peace, they torture me. I have to get them out of my system, and there is but one way to get them out — by solving them. A problem solved is no problem at all, it just disappears.
On Scientific Creativity
Perspectives in Biology and Medicine, Volume 5, Number 2, Winter 1962 (p. 176)

Tatum, Edward   1909–75
American biochemist

As in any scientific research, a problem clearly seen is already half solved.
Nobel Lectures, Physiology or Medicine 1942–1962
Nobel lecture for award received in 1958
A Case History in Biological Research (p. 610)
Elsevier Publishing Company. Amsterdam, Netherlands. 1964

Thurstone, Louis Leon   1887–1955
American pioneer of psychometrics and psychophysics

Every scientific problem can be stated most clearly if it is thought of as a search for the nature of the relation between two definitely stated variables. Very often a scientific problem is felt and stated in other terms, but it cannot be so clearly stated in any way as when it is thought of as a function by which one variable is shown to be dependent upon or related to some other variable.
The Fundamentals of Statistics (p. 187)
The Macmillan Company. New York, New York, USA. 1925

Weil, Simone   1909–43
French philosopher and mystic

Our science is like a store filled with the most subtle intellectual devices for solving the most complex problems, and yet we are almost incapable of applying the elementary principles of rational thought.
In George A. Panichas (Ed(ed.)
The Simone Weil Reader
The Power of Words (p. 271)
McKay. New York, New York, USA. 1977

Wiesner, Jerome Bert   1915–94
Educational administrator

Some problems are just too complicated for rational logical solutions. They admit of insights, not answers.
In D. Lang
Profiles: A Scientist's Advice, II
New Yorker, 26 January 1963

Wilson, Jr., E. Bright   1908–92
American physical chemist

Many scientists owe their greatness not to their skill in solving problems but to their wisdom in choosing them. It is therefore worth considering the points on which this choice can be based.
An Introduction to Scientific Research
Chapter 1 (p. 1)
McGraw-Hill Book Company, Inc. New York, New York USA. 1952

Yalow, Rosalyn   1921–
American medical physicist

We bequeath to you, the next generation, our knowledge but also our problems. While we still live, let us join hands, hearts and minds to work together for their solution so that your world will be better than ours and the world of your children even better.
Les Prix Nobel. The Nobel Prizes in 1977
Nobel banquet speech for award received in 1977
Nobel Foundation. Stockholm, Sweden. 1978

PROGRESS

Abbey, Edward   1927–89
American environmentalist and nature writer

[T]here is a cloud on my horizon. A small dark cloud no bigger than my hand.

Its name is progress.
Desert Solitaire
Industrial Tourism and the National Parks (p. 48)
Ballantine Books. New York, New York, USA. 1968

Belinsky, Vissarion Grigorievich   1811–48
Russian writer and literary critic

Without the striving for infinity there is no life, no development, and no progress.
Compiled by V.V. Vorontsov
Words of the Wise: A Book of Russian Quotations
Translated by Vic Schneierson
Progress Publishers. Moscow, Russia. 1979

Boorstin, Daniel J.   1914–2004
American historian

One of the great obstacles to progress is not ignorance, but the illusion of knowledge.
The Discoverers
Part One, Chapter II (p. 86)
Random House, Inc. New York, New York, USA. 1983

Chernyshevsky, Nikolai Gavrilovich   1828–89
Russian socialist reformer

To renounce progress is as silly as to renounce the Earth's force of gravitation.
Compiled by V.V. Vorontsov
Words of the Wise: A Book of Russian Quotations
Translated by Vic Schneierson
Progress Publishers. Moscow, Russia. 1979

Clerke, Agnes Mary   1842–1907
Irish astronomer

Progress is the result, not so much of sudden flights of genius, as of sustained, patient, often commonplace endeavor; and the true lesson of scientific history lies in the close connection which it discloses between the most brilliant developments of knowledge and the faithful accomplishment of his daily task by each individual thinker and worker.
A Popular History of Astronomy During the Nineteenth Century
Part I, Chapter VI (p. 108)
A. & C. Black. London, England. 1908

Curie, Marie Sklodowska   1867–1934
Polish-born French physical chemist

…I was taught that the way of progress is neither swift nor easy…
Pierre Curie
Autobiographical Notes
Chapter I (p. 167)
The Macmillan Company. New York, New York, USA. 1926

Eddington, Sir Arthur Stanley   1882–1944
English astronomer, physicist, and mathematician

The knowledge that progress will inevitably lead to a readjustment of ideas must instill a writer with caution; but I believe that excessive caution is not to be desired. There can be no harm in building hypotheses, and weaving explanations which seem best fitted to our present partial knowledge. These are not idle speculations if they help us, even temporarily, to grasp the relations of scattered facts, and to organise our knowledge.
Stellar Movements and the Structure of the Universe
Preface (p. v)
Macmillan & Company Ltd. London, England. 1914

Feyerabend, Paul K.   1924–94
Austrian-born American philosopher of science

The only principle that does not inhibit progress is: Anything goes.
Against Method: Outline of an Anarchistic Theory of Knowledge
Chapter 1 (p. 23)
Verso. London, England. 1978

Galton, Sir Francis   1822–1911
English anthropologist, explorer, and statistician

If we summon before our imagination in a single mighty host, the whole number of living things from the earliest date at which terrestrial life can be deemed to have probably existed, to the latest future at which we may think it can probably continue, and if we cease to dwell on the mis-carriages of individual lives or single generations, we shall plainly perceive that the actual tenantry of the world progresses in a direction that may in some sense be described as the greatest happiness of the greatest number.
Inquiries into Human Faculty and Its Development
The Observed Order of Events (pp. 194–195)
AMS Press. New York, New York, USA. 1973

Goddard, Robert H.   1882–1945
American physicist

How many more years I shall be able to work on the problem, I do not know; I hope, as long as I live. There can be no thought of finishing, for "aiming at the stars," both literally and figuratively, is a problem to occupy generations, so that no matter how much progress one makes, there is always the thrill of just beginning.
The Papers of Robert H. Goddard (Volume 2)
R.H. Goddard to H.G. Wells
April 20, 1932 (p. 823)
McGraw-Hill Book Company. New York, New York, USA. 1970

Jastrow, Joseph   1863–1944
Polish-born psychologist

Mind in the making follows no straightforward progression; its many wanderings in the quest for truth compose a cyclopedia of error and vain solutions far more than orderly annals of successful advance.
In Joseph Jastrow (ed.)
The Story of Human Error
Introduction (p. 2)
D. Appleton-Century Company, Inc. New York, New York, USA. 1936

Lavrov, Pyotr
No biographical data available

The physical, intellectual, and ethical development of the personality and the embodiment of truth and justice in social forms — this, it seems to me, is the brief formula that encompasses everything we can regard as progress.
Compiled by V.V. Vorontsov
Words of the Wise: A Book of Russian Quotations
Translated by Vic Schneierson
Progress Publishers. Moscow, Russia. 1979

Lawrence, Ernest   1901–58
American physicist

No individual is alone responsible for a single stepping stone along the path of progress, and where the path is smooth progress is most rapid.
Les Prix Nobel. The Nobel Prizes in 1939
Nobel banquet speech for award received in 1939
Nobel Foundation. Stockholm, Sweden. 1940

Lodge, Sir Oliver   1851–1940
English physicist

The present is an epoch of astounding activity in physical science. Progress is a thing of months and weeks, almost days. The long line of isolated ripples of past discovery seen blending into a might wave, on the crest of which one begins to discern some oncoming magnificent generalization. The suspense is becoming feverish, at times almost painful. One feels like a boy who has been long strumming on the silent keyboard of a deserted organ, into the chest of which an unseen power begins to blow a vivifying breath.
Modern View of Electricity
Lecture III
The Discharge of a Leyden Jar (pp. 382–382)
Macmillan & Company Ltd. London, England. 1889

Medawar, Sir Peter Brian   1915–87
Brazilian-born English zoologist

To deride the hope of progress is the ultimate fatuity, the last word in poverty of spirit and meanness of mind.
The Hope of Progress
Introduction (p. 1)
Anchor Books. Garden City, New York, USA. 1973

Nirenberg, Marshall W.   1927–
American biochemist and geneticist

One individual alone creates only a note or so that blends with those produced by others.
Les Prix Nobel. The Nobel Prizes in 1968
Nobel banquet speech for award received in 1968
Nobel Foundation. Stockholm, Sweden. 1969

Poincaré, Lucien   1862–1920
French physicist

There are no limits to progress, and the field of our investigations has no boundaries. Evolution will continue with invincible force. What we today call the unknowable, will retreat further and further before science, which will never stay her onward march. Thus physics will give greater and increasing satisfaction to the mind by furnishing new interpretations of phenomena; but it will accomplish, for the whole of society, more valuable work still, by rendering, by the improvements it suggests, life every day more easy and more agreeable, and by providing mankind with weapons against the hostile forces of Nature.
The New Physics and Its Evolution
Chapter XI (p. 328)
Kegan Paul, Trench, Trubner & Company Ltd. London, England. 1907

Ramsay, Sir William   1852–1916
English chemist

Progress is made by trial and failure; the failures are generally a hundred times more numerous than the successes; yet they are usually left unchronicled.
Essays Biographical and Chemical
Chemical Essays
Radium and Its Products (p. 179)
Archibald Constable & Company Ltd. London, England. 1908

Sarton, George   1884–1956
Belgian-born American scholar and writer

The history of science is the only history which can illustrate the progress of mankind. In fact, progress has no definite and unquestionable meaning in fields other than the fields of science.
The Study of The History of Science (p. 5)
Harvard University Press. Cambridge, Massachusetts, USA. 1936

The saints of today are not necessarily more saintly than those of a thousand years ago; our artists are not necessarily greater than those of early Greece; they are likely to be inferior; and, of course, our men of science are not necessarily more intelligent than those of old; yet one thing is certain, their knowledge is at once more extensive and more accurate. The acquisition and systemization of positive knowledge is the only human activity that is truly cumulative and progressive.
Introduction to the History of Science (Volume 1)
Introductory Chapter (p. 3)
The Williams & Wilkins Company. Baltimore, Maryland, USA. 1927

Schmidt, O. Y.
No biographical data available

There can be no progress in science and education in the absence of political progress.
Compiled by V.V. Vorontsov
Words of the Wise: A Book of Russian Quotations
Translated by Vic Schneierson
Progress Publishers. Moscow, Russia. 1979

Serres, Michel   1930–
French philosopher

But, irresistibly, I cannot help thinking that this idea is the equivalent of those ancient diagrams we laugh at today, which place the Earth at the center of everything, or our galaxy at the middle of the universe, to satisfy our narcissism. Just as in space we situate ourselves at the center, at the navel of things in the universe, so for time, through progress, we never cease to be at the summit, on the cutting edge, at the state-of-the-art of development. It follows that we are always right, for the simple, banal, and

naïve reason that we are living in the present moment. The curve traced by the idea of progress thus seems to me to sketch or project into time the vanity and fatuousness expressed spatially by that central position. Instead of inhabiting the heart or the middle of the world, we are sojourning at the summit, the height, the best of truth.
Conversations on Science, Culture, and Time
Second Conversation
Method (pp. 48–49)
The University of Michigan Press. Ann Arbor, Michigan, USA. 1995

Shaw, George Bernard 1856–1950
Irish comic dramatist and literary critic

Sir Patrick: Lord! Yes. Modern science is a wonderful thing. Look at your great discovery! Look at all the great discoveries! Where are they leading to? Why, right back to my poor dear old father's ideas and discoveries. He's been dead now over forth years. Oh, it's very interesting.

Ridgeton: Well, there's nothing like progress, is there?
The Doctor's Dilemma
Act I (p. 11)
Brentano's. New York, New York, USA. 1909

Stoppard, Tom 1937–
Czech-born English playwright

Don't confuse progress with perfectibility. A great poet is always timely. A great philosopher is an urgent need. There's no rush for Isaac Newton. We were quite happy with Aristotle's cosmos. Personally, I preferred it. Fifty-five crystal spheres geared to God's crankshaft is my idea of a satisfying universe.
Arcadia
Act II, Scene Five (p. 61)
Faber & Faber Ltd. London, England. 1993

von Liebig, Justus 1803–73
German organic chemist

To resolve an enigma, we must have a perfectly clear conception of the problem. there are many ways to the highest pinnacle of a mountain, but those only can hope to reach it who keep the summit constantly in view. All our labour and all our efforts, if we strive to attain it through a morass, only serve to cover it more completely with mud; our progress is impeded by difficulties of our own creation, and at last even the greatest strength must give way when so absolutely wasted.
Animal Chemistry
Part II
The Metamorphosis of Tissues (p. 125)
Johnson Reprint Corporation. New York, New York, USA. 1964

Walker, Kenneth 1882–1966
No biographical data available

We travel through life with so much mental luggage that it is advisable occasionally to pause and take stock of it

in order that we may get rid of those ideas which impede our progress.
Meaning and Purpose
Chapter I (p. 14)
Jonathan Cape. London, England. 1944

Whitehead, Alfred North 1861–1947
English mathematician and philosopher

Too many apples from the tree of systematized knowledge lead to the fall of progress.
Modes of Thought
Chapter I, Lecture Three (p. 79)
The Macmillan Company. New York, New York, USA. 1938

PROOF

Auster, Paul 1947–
American writer

I had made an empirical discovery and it carried all the weight of a mathematical proof.
The Book of Illusions (pp. 9–10)
Picador. New York, New York, USA. 2002

Bell, E. T. (Eric Temple) 1883–1960
Scottish-American mathematician and educator

There is a sharp disagreement among competent men as to what can be proved and what cannot be proved, as well as an irreconcilable divergence of opinions as to what is sense and what is nonsense.
Debunking Science (p. 18)
University of Washington Book Store. Seattle, Washington, USA. 1930

Bernal, John Desmond 1901–71
Irish-born physicist and x-ray crystallographer

…the cardinal rule in science is that a statement must be provable — but that does not mean that it has to be proved now.
In S.W. Fox (ed.)
The Origins of Prebiological Systems and of Their Molecular Matrices
The Folly of Probability, Discussion (pp. 53–55)
Academic Press. New York, New York, USA. 1965

Blake, William 1757–1827
English poet, painter, and engraver

What is now proved was once only imagined.
The Complete Poetry and Prose of William Blake
The Marriage of Heaven and Hell
University of California Press. Berkeley, California, USA. 1982

Buchanan, Scott 1895–1968
American educator and philosopher

The best proofs in mathematics are short and crisp like epigrams, and the longest have swings and rhythms that are like music.
Poetry and Mathematics

Chapter 1 (p. 36)
The University of Chicago Press. Chicago, Illinois, USA. 1975

Cabell, James Branch   1879–1958
American essayist and novelist

"But I can prove it by mathematics, quite irrefutably. I can prove anything you require of me by whatever means you may prefer," said Jurgen, modestly, "for the simple reason that I am a monstrous clever fellow."
Jurgen: A Comedy of Justice
Chapter 32 (p. 236)
Robert M. McBride & Company. New York, New York, USA. 1925

Davis, Philip J.   1923–
American mathematician
Hersh, Reuben   1927–
American mathematician

Proof serves many purposes simultaneously.... Proof is respectability. Proof is the seal of authority. Proof, in its best instance, increases understanding by revealing the heart of the matter. Proof suggests new mathematics.... Proof is mathematical power, the electric voltage of the subject which vitalizes the static assertions of the theorems.
The Mathematical Experience
Proof (p. 151)
Birkhäuser. Boston, Massachusetts, USA. 1981

Proof is for cosmetic purposes and also to reduce somewhat the edge of insecurity on which one always lives.
The Mathematical Experience
A Physical Look at Mathematics (p. 48)
Birkhäuser. Boston, Massachusetts, USA. 1981

He rests his faith on rigorous proof; he believes that the difference between a correct proof and an incorrect one is an unmistakable and decisive difference. He can think of no condemnation more damning than to say of a student, "He doesn't even know what a proof is."
The Mathematical Experience
The Ideal Mathematician (p. 34)
Birkhäuser. Boston, Massachusetts, USA. 1981

de Morgan, Augustus   1806–71
English mathematician and logician

Would Mathematicals — forsooth —
If true, have failed to prove truth?
Would not they — if they could — submit
Some overwhelming proofs of it?
A Budget of Paradoxes
The Moon's Rotation (p. 262)
Longmans, Green. London, England. 1872

Proof requires a person who can give and a person who can receive...
A Budget of Paradoxes
The Moon's Rotation (p. 262)
Longmans, Green. London, England. 1872

Dedekind, Richard   1831–1916
German mathematician

In science nothing capable of proof ought to be accepted without proof.
Translated by Wooster Woodruff Beman
Essays on the Theory of Numbers
The Nature and Meaning of Numbers
Preface to the First Edition (p. 31)
The Open Court Publishing Company. Chicago, Illinois, USA. 1901

Euclid of Alexandria   325 BCE–265 BCE
Greek mathematician

Quod erat demonstrandum (Q.E.D.)

Which was to be proved.
The Thirteen Books of Euclid's Elements
Element I, Proposition 5
At The University Press. Cambridge, England. 1906

Evans, Bergen   1904–78
Author

"You can't prove it isn't so!" is as good as Q.E.D. in folk logic.
The Natural History of Nonsense
Chapter 19 (p. 264)
Alfred A. Knopf. New York, New York, USA. 1947

Gleason, Andrew M.
Mathematician

...proofs really aren't there to convince you that something is true — they're there to show why it is true.
In D. Albers, G. Alexanderson and C. Reid (eds.)
More Mathematical People
Andrew M. Gleason (p. 86)
Harcourt Brace Jovanovich. New York, New York, USA. 1990

Hamming, Richard W.   1915–98
Mathematician

Some people believe that a theorem is proved when a logically correct proof is given; but some people believe a theorem is proved only when the student sees why it is inevitably true. The author tends to belong to this second school of thought.
Coding and Information Theory
Chapter 9 (p. 155)
Prentice-Hall, Inc. Englewood Cliffs, New Jersey, USA. 1980

Hilbert, David   1862–1943
German mathematician

...it is an error to believe that rigor in the proof is the enemy of simplicity...
Hilbert: Mathematical Problems
Bulletin of the American Mathematical Society, Volume 8, 2nd Series,
October 1901–July 1902 (p. 441)

Hoyle, Sir Fred   1915–2001
English mathematician and astronomer

What constitutes proof in one generation is not the same thing as proof in another.
Of Men and Galaxies
Motives and Aims of the Scientist (pp. 16–17)
University of Washington Press. Seattle, Washington, USA. 1964

Lenstra, Jr., H. W.

A math talk without a proof is like a movie without a love scene.
AMS-MAA 2002 annual meeting
San Diego, January 8, 2002

Lowell, Percival   1855–1916
American astronomer

Now, between the truths we take for granted because of their age, and those we question because of their youth, we are apt to forget that in both, proof is nothing but preponderance of probability.
Mars
Chapter I, 1 (p. 6)
Houghton Mifflin Company. Boston, Massachusetts, USA. 1895

Manin, Yu I.
No biographical data available

…a good proof is one that makes us wiser.
A Course in Mathematical Logic (p. 51)
Springer-Verlag. New York, New York, USA. 1977

Nicholas Bourbaki
Mathematical discussion group

Indeed every mathematician knows that a proof has not been "understood" if one has done nothing more than verify step by step the correctness of the deductions of which it is compose d and has not tried to gain a clear insight into the ideas which have led to the construction of this particular chain of deductions in preference to every other one.
Quoted in Douglas M. Campbell and John C. Higgins
Mathematics: People, Problems, Results (Volume 3)
In Richard A. De Millo, Richard J. Lipton and Alan J. Perlos
Social Processes and Proofs of Theorems and Programs (p. 25)
Wadsworth, Inc. Belmont, California, USA. 1984

Pearson, Karl   1857–1936
English mathematician

…we must remember that because a proposition has not yet been proved, we have no right to infer that its converse must be true.
The Grammar of Science
Chapter IV, Section 17 (p. 179)
Charles Scribner's Sons. London, England. 1892

Platt, John R.
No biographical data available

There is no such thing as proof in science — because some later alternative explanation may be as good or better — so that science advances only by disproofs. There is no point in making hypotheses that are not falsifiable, because such hypotheses do not say anything: it must be possible for an empirical scientific system to be refuted by experience.
Strong Inference
Science, Volume 146, Number 3641, 16 October 1964 (p. 350)

Shakespeare, William   1564–1616
English poet, playwright, and actor

Be sure of it: give me the ocular proof…
In *Great Books of the Western World* (Volume 27)
The Plays and Sonnets of William Shakespeare (Volume 2)
Othello, The Moor of Venice
Act III, Scene iii, l. 360
Encyclopædia Britannica, Inc. Chicago, Illinois, USA. 1952

And this may help to thicken other proofs
That do demonstrate thinly.
In *Great Books of the Western World* (Volume 27)
The Plays and Sonnets of William Shakespeare (Volume 2)
Othello, The Moor of Venice
Act III, Scene iii, l. 429–431
Encyclopædia Britannica, Inc. Chicago, Illinois, USA. 1952

Stewart, Ian   1945–
English mathematician and science writer

Proofs knit the fabric of mathematics together, and if a single thread is weak, the entire fabric may unravel.
Nature's Numbers: The Unreal Reality of Mathematical Imagination
Chapter 3 (p. 45)
Basic Books, Inc. New York, New York, USA. 1995

An intuitive proof allows you to understand why the theorem must be true; the logic merely provides firm grounds to show that it is true.
Concepts of Modern Mathematics
Chapter 1 (p. 5)
Dover Publications, Inc. New York, New York, USA. 1995

Sylvester, James Joseph   1814–97
English mathematician

Divide et impera: is as true in algebra as in statecraft; but no less true and even more fertile is the maxim *auge et impera*. The more to do or to prove, the easier the doing or the proof.
The Collected Mathematical Papers of James Joseph Sylvester (Volume 3)
Proof of the Fundamental Theorem of Invariants (1878) (p. 126)
At The University Press. Cambridge, England. 1904–1912

It always seems to me absurd to speak of a complete proof, or of a theorem being rigorously demonstrated. An incomplete proof is no proof, and a mathematical truth not rigorously demonstrated is not demonstrated at all.
The Collected Mathematical Papers of James Joseph Sylvester (Volume 2) (p. 200)
At The University Press. Cambridge, England. 1904–12

Truzzi, Marcello   1935–2003
Danish-born American sociology professor

And when such claims are extraordinary, that is, revolutionary in their implications for established scientific generalizations already accumulated and verified, we must demand extraordinary proof.
Editorial
Zetetic Scholar, Volume 1, Number 1, Fall/Winter 1976 (p. 4)

Tymoczko, Thomas   1943–86
Logician

A proof is a construction that can be looked over, reviewed, verified by a rational agent. We often say that a proof must be perspicuous or capable of being checked by hand. It is an exhibition, a derivation of the conclusion, and it needs nothing outside itself to be convincing. The mathematician surveys the proof in its entirety and thereby comes to know the conclusion.
The Four Color Problems
Journal of Philosophy, Volume 76, 1979

Ward, Peter D.
American paleontologist
Brownlee, Donald
No biographical data available

Proof is a rarity in science.
Rare Earth: Why Complex Life Is Uncommon in the Universe
Preface (p. ix)
Springer-Verlag New York, Inc. New York, New York, USA. 2000

White, Arthur
No biographical data available

A teacher once, having some fun,
In presenting that two equals one,
Remained quite aloof
From his rigorous proof;
But his class was convinced and undone.
Mathematical Magazine, Volume 64, Number 2, April 1991 (p. 91)

PROPHECY

Clarke, Arthur C.   1917–
English science and science fiction writer

With monotonous regularity, apparently competent men have laid down the law about what is technically possible or impossible — and have been proved utterly wrong, sometimes while the ink was scarcely dry from their pens.
Profiles of the Future: An Inquiry into the Limits of the Possible
Chapter 1 (p. 1)
Harper & Row, Publishers. New York, New York, USA. 1973

Before one attempts to set up in business as a prophet, it is instructive to see what success others have made of this

dangerous occupation — and it is even more instructive to see where they have failed.
Profiles of the Future: An Inquiry into the Limits of the Possible
Chapter 1 (p. 1)
Harper & Row, Publishers. New York, New York, USA. 1973

PROPOSITION

Keyser, Cassius Jackson   1862–1947
American mathematician

If he contend, as sometimes he will contend, that he has defined all his terms and proved all his propositions, then either he is a performer of logical miracles or he is an ass; and, as you know, logical miracles are impossible.
Mathematical Philosophy: A Study of Fate and Freedom

Russell, Bertrand Arthur William   1872–1970
English philosopher, logician, and social reformer

The belief or unconscious conviction that all propositions are of the subject-predicate form — in other words, that every fact consists in some thing having some quality — has rendered most philosophers incapable of giving any account of the world of science and daily life.
Our Knowledge of the External World
Lecture II (p. 45)
The Open Court Publishing Company. Chicago, Illinois. 1914

…I wish to propose for the reader's favourable consideration a doctrine which may, I fear, appear wildly paradoxical and subversive. The doctrine in question is this: that it is undesirable to believe a proposition when there is no ground whatever for supposing it true…
Skeptical Essays
Chapter I (p. 11)
W.W. Norton & Company, Inc. New York, New York, USA. 1928

Shaw, George Bernard   1856–1950
Irish comic dramatist and literary critic

My method for examining any proposition is to take its two extremes, both of them impracticable; make a scale between them; and try to determine at what point on the scale it can best be put into practice. A mother who has to determine the temperature of her baby's bath has two fixed limits to work between. The baby must not be boiled and must not be frozen.
Everybody's Political What's What?
Chapter 20 (p. 162)
Dodd, Mead & Company. New York, New York, USA. 1944

Whitehead, Alfred North   1861–1947
English mathematician and philosopher

It is more important that a proposition be interesting than that it be true. This statement is almost a tautology. For the energy of operation of a proposition in an occasion of experience is its interest and is its importance. But of

course a true proposition is more apt to be interesting than a false one.
Process and Reality: An Essay in Cosmology
Part III, Chapter IV, Section II (pp. 395–396)
The Macmillan Company. New York, New York, USA. 1929

PROTON

Ball, Philip   1962–
English science writer

As far as atoms are concerned, protons and electrons are like knives and forks at the dinner table; no matter how big the table, there are equal numbers of each.
Life's Matrix: A Biography of Water
Part One, Chapter 1 (p. 7)
Farrar, Straus & Giroux. New York, New York, USA. 2000

Dyson, Freeman J.   1923–
American physicist and educator

The most serious uncertainty affecting the ultimate fate of the universe is the question whether the proton is absolutely stable against decay into lighter particles. If the proton is unstable, all matter is transitory and must dissolve into radiation.
Time Without End: Physics and Biology in an Open Universe
Reviews of Modern Physics, Volume 51, Number 3, July 1979

PROTOPLASM

Bradley, Jr., John Hodgdon   1898–1962
American geologist

Life carries on. Through all the tortures inflicted by a changing earth, through glacial cold and desert thirst, through flood and famine, the protoplasm has climbed with unabating vigor toward the future.
Parade of the Living
Part III, Chapter XVIII (p. 248)
Coward-McCann, Inc. New York, New York, USA. 1930

PROVINCIAL REGION

Forbes, Edward   1815–54
English naturalist

Everyone knows that the same animals and plants are not found everywhere…but that they are distributed so as to be gathered together in distinct zoological and botanical provinces, of greater or less extent, according to their degree of limitation by physical conditions, whether features of the earth's outline or climate.
The Natural History of the European Seas
Chapter I (p. 1)
John Van Voorst. London, England. 1859

PSEUDOSCIENCE

Sagan, Carl   1934–96
American astronomer and author

Pseudoscience is embraced, it might be argued, in exact proportion as real science is misunderstood.
The Demon-Haunted World: Science as a Candle in the Dark
Chapter 1 (p. 15)
Random House, Inc. New York, New York, USA. 1995

PSYCHICAL CONSTITUTION

Pavlov, Ivan Petrovich   1849–1936
Russian physiologist

Essentially only one thing in life interests us: our psychical constitution, the mechanism of which was and is wrapped in darkness. All human resources, art, religion, literature, philosophy and historical sciences, all of them join in bringing light in this darkness.
Nobel Lectures, Physiology or Medicine 1901–1921
Nobel lecture for award received in 1904
Physiology of Digestion (pp. 154–155)
Elsevier Publishing Company. Amsterdam, Netherlands. 1967

PUBLIC SPEAKING

Sylvester, James Joseph   1814–97
English mathematician

When called upon to speak in public [the mathematician] feels as a man might…who has passed all his life in peering through a microscope, and is suddenly called upon to take charge of an astronomical observatory. He has to get out of himself, as it were, and change the habitual focus of his vision.
The Collected Mathematical Papers of James Joseph Sylvester
(Volume 3)
An Inquiry into Newton's Rule for the Discovery of Imaginary Roots
(p. 73)
University Press. Cambridge, England. 1904–1912

PURITY

Nilson, Lars Fredrik
No biographical data available

On the purity of substances depends the perfection of the whole.
In Mary Elvira Weeks
The Discovery of the Elements (p. 407)
Journal of Chemical Education. Easton Pennsylvania, USA. 1956

PURPOSE

Einstein, Albert   1879–1955
German-born physicist

How strange is the lot of us mortals! Each of us is here for a brief sojourn; for what purpose he knows not, though he sometimes thinks he senses it. But without deeper reflection one knows from daily life that one exists for other people — first of all for those upon whose smiles and well-being our own happiness is wholly dependent, and then for the many, unknown to us, to whose destinies we are bound by the ties of sympathy. A hundred times every day I remind myself that my inner and outer life are based on the labors of other men, living and dead, and that I must exert myself in order to give in the same measure as I have received and am still receiving.
Ideas and Opinions
The World as I See It (p. 8)
Crown Publishers, Inc. New York, New York, USA. 1954

Townson, Robert   1763–1827
Australian scholar and scientist

Plan and design are in all Nature's works, though universal discord and confusion seem to prevail, and though certain ruin awaits her fairest productions.
Philosophy of Mineralogy
Chapter III (p. 32)
Printed for the author. London, England. 1798

PYRAMID

Bonaparte, Napoleon   1765–1821
French general

From the top of those pyramids, forty centuries look down on you.
In Ralph Waldo Emerson
English Traits and Representative Men
Representative Men, Chapter VII (p. 324)
Oxford University Press, Inc. Oxford, England. 1934

Kipling, Rudyard   1865–1936
British writer and poet

Who shall doubt "the secret hid
Under Cheops' pyramid"

Was that the contractor did
Cheops out of several million?
Rudyard Kipling's Verse
A General Summary
Hodder & Stoughton. London, England. 1919

Shelley, Percy Bysshe   1792–1822
English poet

Nile shall pursue his changeless way:
Those Pyramids shall fall;
Yea! Not a stone shall stand to tell
The spot where on they stood.
Their very site shall be forgotten,
As is their builder's name.
The Complete Poetical Works of Percy Bysshe Shelley
Queen Mab
Houghton Mifflin Company. Boston, Massachusetts, USA. 1901

Thoreau, Henry David   1817–62
American essayist, poet, and practical philosopher

As for the pyramids, there is nothing to wonder in at them so much as the fact that so many men could be found degraded enough to spend their lives constructing a tomb for some ambitious booby, whom it would have been wiser and manlier to have drowned in the Nile, and then given the body to the dogs.
The Writings of Henry David Thoreau (Volume 2)
Walden
Chapter I (p. 93)
Houghton Mifflin Company. Boston, Massachusetts, USA. 1893

PYTHAGORAS

Browne, Sir Thomas   1605–82
English author and physician

I have often admired the mystical way of Pythagoras, and the secret magic of numbers.
Religio Medici
12
Elliot Stock. London, England. 1883

Q

QUACK

Aesop ca. 620 BCE–560 BCE
Greek fabulist and author

A frog once upon a time came forth from his home in the marsh and proclaimed to all the beasts that he was a learned physician, skilled in the use of drugs and able to heal all diseases. A Fox asked him, "How can you pretend to prescribe for others when you are unable to heal your own lame gait and wrinkled skin?"
The Quack Frog
SeaStar Books. New York, New York, USA. 2000

Ames, Nathaniel 1708–64
American almanac maker

Where silly quacks are most respected, there honest doctors are neglected. Petty Attorneys and Quack Doctors are like Wolves and scabbed Sheep among the Flock. One devours and the other breeds the rot.
An Astronomical Diary, or, an Almanack for ... 1734
Printed for the Booksellers and sold at their shops. Boston, Massachusetts, USA. 1734

Bernstein, Al
American writer and stage performer

You can usually tell a quack doctor by his bill.
Quote, the Weekly Digest, July 28, 1968 (p. 77)

Bishop, Samuel
No biographical data available

When quacks, as quacks may by good luck, to be sure,
Blunder out at haphazard a desperate cure,
In the prints of the day, with due pomp and parade,
Case, patient, and doctor are amply display'd.
And this is quite just — and no mortal can blame it;
If they save a man's life, they've a right to proclaim it
But there's reason to think they might save more lives still,
Did they publish a list of the numbers they kill!
In William Davenport Adams
English Epigrams
Audi Alteram Partem, cclxxxv
G. Routledge. London, England. 1878

Clowes, William 1540–1604
Surgeon and medical author

Yea, nowadays, it is too apparent to see how tinkers, tooth-drawers, peddlers, ostlers, carters, porters, horse-gelders, and horse-leeches, idiots, apple-squires, broom-men, bawds, witches, conjurers, soothsayers and sow-gelders, rogues, ratcatchers, runagates and proctors of Spittlehouses, with such other like rotten and stinking weeds do in town and country, without order, honesty or skill, daily abuse both Physic and Surgery, having no more perseverance, reason or knowledge in this art than has a goose, but only a certain blind practice, without wisdom or judgment, and most commonly use one remedy for all diseases and one way of curing to all persons, both old and young, men, and women and children, which is as possible to perform or to be true as for a shoemaker with one last to make a shoe fit for every man's foot, and this is one principal cause [why] so many perish.
Selected Writings
Of Blind Buzzards and Cracking Cumbatters (pp. 77–78)
Harvey & Blythe. London, England. 1948

Colton, Charles Caleb 1780–1832
English sportsman and writer

It is better to have recourse to a quack, if he can cure the disorder, although he cannot explain it, than to a physician, if he can explain our disease, but cannot cure it.
Lacon; or Many Things in a Few Words
1:170
William Gowans. New York, New York, USA. 1849

Crabbe, George 1754–1832
English poet

A potent quack, long versed in human ills,
Who first insults the victim whom he kills...
The Poetical Works of George Crabbe
The Village L. 282–283 (p. 15)
Oxford University Press, Inc. London, England. 1908

Graves, Richard
No biographical data available

A doctor, who, for want of skill,
Did sometimes cure — and sometimes kill;
Contriv'd at length, by many a puff,
And many a bottle fill'd with stuff,
To raise his fortune, and his pride;
And in a coach, forsooth! must ride.
His family coat long since worn out,
What arms to take, was all the doubt.
A friend, consulted on the case,
Thus answer'd with a sly grimace:
"Take some device in your own way,
Neither too solemn nor too gay;
Three ducks, suppose; white, grey, or black;
And let your motto be, Quack! quack!"
In William Davenport Adams
English Epigrams
A Doctor's Motto, cclxxxi
G. Routledge. London, England. 1878

Hood, Thomas 1799–1845
English poet and editor

Not one of these self-constituted saints,
Quacks — not physicians — in the cure of souls.

The Complete Poetical Works of Thomas Hood
Ode to Rae Wilson ESQ.
Greenwood Press, Publishers. Westport, Connecticut, USA. 1980

Jenner, Edward   1749–1823
English physician

I've dispatc'd, my dear madam, this scrap of a letter,
To say that Miss —
— is very much better.
A Regular Doctor no longer she lacks,
And therefore I've sent her a couple of Quacks.
In William Davenport Adams
English Epigrams
Sent to a Patient, with the Present of a Couple of Ducks, cclxxiii
G. Routledge. London, England. 1878

Lydston, George Frank   1858–1923
American urologist

The quack doesn't find out what the matter is but, to the patient's cost, he does find a lot of things that do not exist, and all because the reputable physician flouted as imaginary conditions which, to the patient's sensitive and morbid mind, are always terribly real.
Sexual Neurasthenia and the Prostate
Medical Record, Volume 81, 1912

Massinger, Philip   1583–1640
English dramatic poet

Out, you impostors!
Quacksalving, cheating mountebanks! your skill
Is to make sound men sick, and sick men kill.
The Plays of Philip Massinger (Volume 1)
The Virgin-Martyr, Act IV, Scene I (p. 78)
G. & W. Nicol. London, England. 1805

Shaw, George Bernard   1856–1950
Irish comic dramatist and literary critic

Did I hear from the fireside armchair the bow-wow of the old school defending its drugs? Ah, believe me, Paddy, the world would be healthier if every chemist's shop in England were demolished. Look at the papers! full of scandalous advertisements of patent medicines! a huge commercial system of quackery and poison. Well, whose fault is it? Ours. I say, ours. We set the example. We spread the superstition. We taught the people to believe in bottles of doctor's stuff; and now they buy it at the stores instead of consulting a medical man.
The Doctor's Dilemma
Act I (p. 27)
Brentano's. New York, New York, USA. 1920

Wycherley, William   1640–1760
English dramatist

A quack is as fit for a pimp as a midwife for a bawd: they are still but in their way, both helpers of nature.
The Country Wife

Act 1 (p. 5)
Random House, Inc. New York, New York, USA. 19—

QUALITIES

Darwin, Charles Robert   1809–82
English naturalist

I am inclined to agree with Francis Galton in believing that education and environment produce only a small effect on the mind of any one, and that most of our qualities are innate.
In Francis Darwin (ed.)
The Life and Letters of Charles Darwin (Volume 1)
Chapter I (p. 21)
D. Appleton & Company. New York, New York, USA. 1896

Gregory, Sir Richard Arman   1864–1952
British science writer and journalist

The only qualifications required for the study of Nature's story-book are devotion to truth, and sincerity of spirit; all the other qualities will come to the possessor of these, and a habit of mind will be developed that tries to face all facts squarely and honestly, despises shams and false conventions, and exposes superstition whenever it is encountered.
Discovery; or, The Spirit and Service of Science
Chapter III (p. 44)
Macmillan & Company Ltd. London, England. 1918

QUANTIFICATION

Platt, John R.
No biographical data available

Today we preach that science is not science unless it is quantitative. We substitute correlation for causal studies., and physical equations for organic reasoning. Measurements and equations are supposed to sharpen thinking, but…they more often tend to make the thinking non-causal and fuzzy.
Strong Inference
Science, Volume 146, Number 3641, 16 October 1964 (pp. 351–352)

Sagan, Carl   1934–96
American astronomer and author

Quantify. If whatever it is you're explaining has some measure, some numerical quantity attached to it, you'll be much better able to discriminate among competing hypotheses. What is vague and qualitative is open to many explanations.
The Demon-Haunted World: Science as a Candle in the Dark
Chapter 12 (p. 211)
Random House, Inc. New York, New York, USA. 1995

Whitehead, Alfred North   1861–1947
English mathematician and philosopher

Elegant intellects which despise the theory of quantity are but half developed.
The Aims of Education
Presidential Address
Mathematical Association of England, 1916

QUANTUM MECHANICS

Abbey, Edward   1927–89
American environmentalist and nature writer

Quantum mechanics provides us with an approximate, plausible, conjectural explanation of what actually is, or was, or may be taking place inside a cyclotron during a dark night in February.
A Voice Crying in the Wilderness: Notes from a Secret Journal
Chapter 10 (p. 93)
St. Martin's Press. New York, New York, USA. 1989

Barrow, John D.   1952–
English theoretical physicist

We live in the in-between world…betwixt the "devil" of the quantum world and the "deep blue sea" of curved space.
The World Within The World (p. 161)
Clarendon Press. Oxford, England. 1988

Belinfante, Frederik Jozef   1913–1991
Dutch-born American physicist

If I get the impression that Nature itself makes the decisive choice what possibility to realize, where quantum theory says that more than one outcome is possible, then I am ascribing personality to Nature, that is to something that is always everywhere. Omnipresent eternal personality which is omnipotent in taking the decisions that are left undetermined by physical law is exactly what in the language of religion is called God.
In John D. Barrow
The World Within the World (p. 157)
Clarendon Press. Oxford, England. 1988

Bohr, Niels Henrik David   1886–1962
Danish physicist

Anyone who is not shocked by quantum theory has not understood it.
In N.C. Panda
Maya in Physics (p. 73)
Motilal Banarsdass Publishers. Delhi, India. 1991

If anybody says he can think about quantum problems without getting giddy, that only shows he has not understood the first thing about them.
In Ruth Moore
Niels Bohr (p. 127)
MIT Press. Cambridge, Massachusetts, USA. 1985

There is no quantum world. There is only an abstract quantum physical description. It is wrong to think that the task of physics is to find out how nature is. Physics concerns what we can say about nature.
The Philosophy of Niels Bohr
Bulletin of the Atomic Scientists, Volume 19, Number 7, September 1963 (p. 12)

…the fundamental postulate of the indivisibility of the quantum is itself from the classical point of view, an irrational element which inevitably requires us to forgo a causal mode of description and which, because of the coupling between phenomena and their observation, forces us to adopt a new mode of description designated as complementary in the sense that any given application of classical concepts precludes the simultaneous use of other classical concepts which in a different connection are equally necessary for the elucidation of the phenomena.
Atomic Theory and the Description of Nature
Introductory Survey (p. 10)
Cambridge University Press. Cambridge, England. 1934

Born, Max   1882–1970
German-born English physicist

[In quantum mechanics] we have the paradoxical situation that observable events obey laws of chance, but that the probability for these events itself spreads according to laws which are in all essential features causal laws.
Natural Philosophy of Cause and Chance
Chapter IX (p. 103)
At The Clarendon Press. Oxford, England. 1949

Bridgman, Percy Williams   1882–1961
American physicist

The explanatory crisis which now confronts us in relativity and quantum phenomena is but a repetition of what has occurred many times in the past…. Every kitten is confronted with such a crisis at the end of nine days.
The Logic of Modern Physics
Chapter II (p. 42)
The Macmillan Company. New York, New York, USA. 1927

Calvin, William H.   1939–
Theoretical neurophysiologist

Quantum mechanics is probably essential to consciousness in about the same way as crystals were once essential to radios, or spark plugs are still essential to traffic jams. Necessary, but not sufficient.
How Brains Think: Evolving Intelligence, Then and Now
Chapter 3 (p. 36)
Basic Books, Inc. New York, New York, USA. 1996

Captain Janeway
Fictional character

Who wanted to muck around in the dirt when you could be studying quantum mechanics?
STAR TREK: Voyager
Resolutions

Television program
Season 2, 1996

Cole, K. C. 1946–
American science writer

The introduction of quantum theory in the early 1920s marked one of the greatest revolutions in all of physical science. It could not (cannot) adequately be described in metaphors borrowed from our previous view of reality, because many of those metaphors no longer apply. This inability to imagine quantum goings-on led to the popular perception that the realm of the inner atom is fuzzy, elusive, murky, and uncertain. On the contrary, most physicists would agree that what quantum theory has brought to science is exactly the opposite — concreteness and clarity.
First You Build a Cloud and Other Reflections on Physics as a Way of Life
Chapter Seven (p. 113)
Harcourt Brace & Company. New York, New York, USA. 1999

DeWitt, Bryce 1923–2004
American theoretical physicist
Graham, Neill
No biographical data available

No development of modern science has had a more profound impact on human thinking than the advent of quantum theory. Wrenched out of centuries-old thought patterns, physicists of generation ago found themselves compelled to embrace a new metaphysics. The distress which the reorientation caused continues to the present day. Basically physicists have suffered a severe loss: their hold on reality.
Resource IQM-1 on the Interpretation of Quantum Mechanics
American Journal of Physics, Volume 39, 1971

Dirac, Paul Adrian Maurice 1902–84
English theoretical physicist

…the main object of physical science is not the provision of pictures, but in the formulation of laws governing phenomena and the application of these laws to the discovery of new phenomena. If a picture exists, so much the better; but whether a picture exists or not is a matter of only secondary importance. In the case of atomic phenomena no picture can be expected to exist in the usual sense of the word "picture," by which is meant to model functioning essentially on classical lines. One may extend the meaning of the word "picture" to include any way of looking at the fundamental laws which make their self-consistency obvious. With this extension, one may acquire a picture of atomic phenomena by becoming familiar with the laws of quantum theory.
The Principles of Quantum Mechanics (2nd edition)
Chapter I, Section 4 (p. 10)
At The Clarendon Press. Oxford, England. 1935

Dyson, Freeman J. 1923–
American physicist and educator

…Dick Feynman told me about his "sum over histories" version of quantum mechanics. "The electron does anything it likes," he said. "It goes in any direction at any speed, forward or backward in time, however it likes, and then you add up the amplitudes and it gives you the wave function." I said to him, "You're crazy." But he wasn't.
In Harry Woolf (ed.)
Some Strangeness In the Proportion
Chapter 23 (p. 376)
Addison-Wesley Publishing Company, Inc. Reading, Massachusetts, USA. 1980

Eddington, Sir Arthur Stanley 1882–1944
English astronomer, physicist, and mathematician

Rather against my better judgment I will try to give a rough impression of the theory. It would probably be wiser to nail up over the door of the new quantum theory a notice, "Structural alterations in progress — No admittance except on business", and particularly to warn the doorkeeper to keep out prying philosophers.
The Nature of the Physical World
Chapter X (p. 211)
The Macmillan Company. New York, New York, USA. 1930

A very useful kind of operator is the selective operator. In my schooldays a foolish riddle was current — "How do you catch lions in the desert?" Answer: "In the desert you have a lot of sand and a few lions; so you take a sieve and sieve out the sand and the lions remain." I recall it because it describes one of the most usual methods used in quantum theory for obtaining anything that we wish to study.
New Pathways in Science
Chapter XII, Section III (p. 263)
The Macmillan Company. New York, New York, USA. 1935

Einstein, Albert 1879–1955
German-born physicist

This theory [quantum theory] reminds me a little of the system of delusions of an exceedingly intelligent paranoiac, concocted of incoherent elements of thoughts.
In Arthur Fine
The Shaky Game: Einstein, Realism, and the Quantum Theory
Letter of July 5, 1952 to D. Lipkin (p. 1)
University of Chicago Press. Chicago, Illinois, USA. 1986

Quantum mechanics is certainly imposing. But an inner voice tells me that it is not yet the real thing. The theory says a lot, but does not bring us any closer to the secret of the Old One. I, at any rate, am convinced that He does not throw dice.
In Ronald W. Clark
Einstein: The Life and Times
Letter to Max Born, 1926 (p. 340)
The World Publishing Company. New York, New York, USA. 1971

The quantum theory gives me a feeling very much like yours. One really ought to be ashamed of its success, because it has been obtained with the Jesuit maxim: "Let not thy left hand know what thy right hand doeth."
The Born–Einstein Letters: Correspondence Between Albert Einstein and Max and Hedwig Born from 1916 to 1955
Letter to Max Born, June 4, 1919 (p. 11)
Walker & Company. New York, New York, USA. 1971

I cannot seriously believe in [the quantum theory] because it cannot be reconciled with the idea that physics should represent a reality in time and space, free from spooky actions at a distance.
The Born–Einstein Letters: Correspondence Between Albert Einstein and Max and Hedwig Born from 1916 to 1955
Letter to Max Born, March 1948 (p. 158)
Walker & Company. New York, New York, USA. 1971

[Quantum theory] If this is correct, it signifies the end of physics as a science.
In L.I. Ponomarev
The Quantum Dice (p. 80)
Institute of Physics Publishing. Bristol, England. 1993

The more one chases after quanta, the better they hide themselves.
In Helen Dukas and Banesh Hoffman
Albert Einstein: The Human Side: New Glimpses from His Archives
Letter to Paul Ehrenfest, 12 July 1924 (p. 69)
Princeton University Press. Princeton, New Jersey, USA. 1979

Ekert, Artur   1961
Polish/British quantum physicist

Possibly the best way to agitate a group of jaded but philosophically inclined physicists is to buy them a bottle of wine and mention interpretations of quantum mechanics. It is like opening a Pandora's box. I have been amused to discover that the number of viewpoints often exceeds the number of participants.
Physics World
Pet Theories of Quantum Mechanics, December 1995

Ferris, Timothy   1944–
American science writer

Gertrude Stein said of modern art, "A picture may seem extraordinarily strange to you and after some time not only does it not seem strange but it is impossible to find what there was in it that was strange." Quantum physics isn't like that. The longer you look at it, the stranger it gets.
The Whole Shebang: A State-of-the Universe's Report
Quantum Weirdness (p. 265)
Simon & Schuster. New York, New York, USA. 1996

Feynman, Richard P.   1918–88
American theoretical physicist

The theory of quantum electrodynamics describes Nature as absurd from the point of view of common sense. And it agrees with experiment. So I hope you can accept Nature as She is — absurd.
QED: The Strange Theory of Light and Matter
Chapter 1 (p. 10)
Princeton University Press. Princeton, New Jersey, USA. 1985

There was a time when the newspapers said that only twelve men understood the theory of relativity. I do not believe that there ever was such a time. There might have been a time when only one man did, because he was the only guy who caught on, before he wrote his paper. But after people read the paper a lot of people understood the theory of relativity in some way or other, certainly more than twelve. On the other hand I think I can safely say that nobody understands quantum mechanics…. Do not keep saying to yourself, if you can possibly avoid it, "But how can it be like that?" because you will get "down the drain", into a blind alley from which nobody has yet escaped. Nobody knows how it can be like that.
The Character of Physical Law
Chapter 6 (p. 129)
BBC. London, England. 1965

Feynman, Richard P.   1918–88
American theoretical physicist
Leighton, Robert B.   1919–97
American physicist
Sands, Matthew L.   1919–
American physicist

It is possible in quantum mechanics to sneak quickly across a region which is illegal energetically.
The Feynman Lectures on Physics (Volume 3)
Chapter 8–6 (p. 8–12)
Addison-Wesley Publishing Company. Reading, Massachusetts, USA. 1983

…there are certain situations in which the peculiarities of quantum mechanics can come out in a special way on a large scale.
The Feynman Lectures on Physics (Volume 3)
Chapter 21–1 (p. 21–1)
Addison-Wesley Publishing Company. Reading, Massachusetts, USA. 1983

Gell-Mann, Murray   1929–
American physicist

All of modern physics is governed by that magnificent and thoroughly confusing discipline called quantum mechanics, invented more than fifty years ago. It has survived all tests and there is no reason to believe that there is any flaw in it. We suppose that it is exactly correct. Nobody understands it, but we all know how to use it and how to apply it to problems; and so we have learned to live with the fact that nobody can understand it.
In Frank Durham and Robert D. Purrington (eds.)
Some Truer Method: Reflections on the Heritage of Newton

Chapter 2 (p. 51)
Columbia University Press. New York, New York, USA. 1990

Harrison, Edward Robert   1919–2007
English-born American cosmologist

…in the impalpable and seemingly inconsequential entities of the quantum world, one finds the true music and magic of nature.
Masks of the Universe
Chapter 8 (p. 123)
Macmillan Publishing Company. New York, New York, USA. 1985

Hawking, Stephen William   1942–
English theoretical physicist

You would have to fly around the world four hundred million times to add one second to your life; but your life would be reduced by more than that by all those airline meals.
Black Holes and Baby Universes and Other Essays
Chapter Eight (p. 72)
Bantam Books. New York, New York, USA. 1987

Heisenberg, Werner Karl   1901–76
German physicist and philosopher

Quantum theory reminds us of the old wisdom that when searching for harmony in life we must forget that in the drama of existence we are ourselves both players and spectators.
In Denis Alexander
Beyond Science
Chapter Two (p. 48)
Lion Publishing. Berkhamsted, Hertz, England. 1972

Quantum theory thus provides us with a striking illustration of the fact that we can fully understand a connection though we can only speak of it in images and parables.
Physics and Beyond: Encounters and Conversations
Chapter 17 (p. 210)
Harper & Row, Publishers. New York, New York, USA. 1971

The problem of quantum theory centers on the fact that the particle picture and the wave picture are merely two different aspects of one and the same physical reality.
The Physical Principles of the Quantum Theory
Translated by Carl Ekhart and Frank C. Hoyt (p. 177)
The University of Chicago Press. Chicago, Illinois, USA. 1930

If anything like mechanics were true then one would never understand the existence of atoms. Evidently there exists another [type of mechanics —] "quantum mechanics."
In Keith Hannabuss
An Introduction to Quantum Theory
Letter to Wolfgang Pauli, June 21, 1925 (p. 21)
Oxford University Press, Inc. Oxford, England. 1997

Joyce, James   1882–1941
Irish-born author

I am working out a quantum theory about it for it is really most tantalizing state of affairs.
Finnegans Wake
Book I (p. 149)
The Viking Press. New York, New York, USA. 1939

Kaku, Michio
Theoretical physicist

…it is often stated that of all the theories proposed in this century, the silliest is quantum theory. In fact, some say that the only thing that quantum theory has going for it is that it is unquestionably correct.
Hyperspace : A Scientific Odyssey Through Parallel Universes, Time Warps, and the 10th Dimension
Chapter 12 (p. 262)
Oxford University Press, Inc. New York, New York, USA. 1995

Kramers, Hendrick Anthony   1894–1952
Physicist

The theory of quanta is similar to other victories in science; for some months you smile at it, and then for years you weep.
In L.I. Ponomarev
The Quantum Dice (p. 80)
Institute of Physics Publishing. Bristol, England. 1993

The theory of quanta can be likened to a medicine that cures the disease but kills the patient.
In L.I. Ponomarev
The Quantum Dice (p. 81)
Institute of Physics Publishing. Bristol, England. 1993

Krauss, Lawrence M.   1954–
American theoretical physicist

…what we really should be discussing is "the interpretation of classical mechanics" — that is, how can the classical world we see — which is only an approximation of the underlying reality, which in turn is quantum mechanical in nature — be understood in terms of the proper quantum mechanical variables? If we insist on interpreting quantum mechanical phenomena in terms of classical concepts, we will inevitably encounter phenomena that seem paradoxical, or impossible.
The Physics of Star Trek
Chapter Nine (pp. 150–151)
Harp Perennial Publishers. New York, New York, USA. 1995

Lawrence, D. H. (David Herbert)   1885–1930
English writer

I like relativity and quantum theories
because I don't understand them
and they make me feel as if space shifted
about like a swan that can't settle,
refusing to sit still and be measured;
and as if the atom were an impulsive thing
Always changing its mind.
In Vivian de Sola Pinto and Warren Roberts (eds.)

The Complete Poems of D.H. Lawrence
Relativity (p. 524)
Viking Press. New York, New York, USA. 1973

Lindley, David   1956–
English astrophysicist and author

[I]t is misleading to say that "measurement affects the thing measured" because that can seem to imply that a quantum object was in some definite but unknown state, but was then disturbed by an act of measurement and is now in some other state. Rather, measurement gives definition to quantities that were previously indefinite; there is no meaning that can be given to a quantity until it is measured.
Where Does the Weirdness Go? Why Quantum Mechanics Is Strange,
but Not as Strange as You Think
Which Way Did the Photon Go? (p. 60)
Basic Books, Inc. New York, New York, USA. 1996

Although quantum mechanics provides explanations of the results of experiments, those explanations tend not, in our minds, to add up to an understanding. But why should they? It's the job of science to provide theories and models that give us an accurate picture of the way the world works, but we are not free also to demand that these theories should conform to our prior expectations of the way we would like the world to work, or think it ought to work. If science sometimes provides explanations without giving us what we would regard as an understanding, the deficiency belongs to us, not to science.
Where Does the Weirdness Go? Why Quantum Mechanics Is Strange,
but Not as Strange as You Think
You Can Push It Around, But You Can't Get Rid of It (p. 125)
Basic Books, Inc. New York, New York, USA. 1996

The Moon really is there, after all, when no one's looking. In a general sense, Einstein's comment was correct: quantum mechanics demands that a measurement be made in order for the Moon really to exist at a particular spot. But the new insight afforded by the decoherence argument is that the rain of solar photons onto the Moon's surface is enough of a physical process to constitute a "measurement" — it's enough to get rid of superposed states, which is what we want a measurement to accomplish. No actual observation is required, and the whole process carries on efficiently and relentlessly without any intervention of human action, let alone human consciousness. The world works in its own way, and doesn't need us to look at it.
Where Does the Weirdness Go? Why Quantum Mechanics Is Strange,
but Not as Strange as You Think
In Which Einstein's Moon Is Restored (p. 204)
Basic Books. New York, New York, USA. 1996

The microworld is not a simple place, and physicists have therefore not been able to keep their theories of it simple.
The End of Physics: The Myth of a Unified Theory

Part I (p. 24)
Basic Books, Inc. New York, New York, USA. 1993

Moser, David
No biographical data available

Quantum Particles: the dreams that stuff is made of.
In Douglas Hofstadter
Metamagical Themas: Questing for the Essence of Mind and Pattern
Section IV, Chapter 20 (p. 473)
Basic Books, Inc. New York, New York, USA. 1985

Pagels, Heinz R.   1939–88
American physicist and science writer

Another way the old physics differs from the quantum physics is the way the determinism of a clock differs from the contingency of a pinball machine.
The Cosmic Code: Quantum Physics as the Language of Nature
Foreword (p. 13)
Simon & Schuster. New York, New York, USA. 1982

Pauli, Wolfgang   1900–58
Austrian-born physicist

Physics is a blind alley again. In any case, it has become too difficult for me, and I would prefer to be a comedian in the cinema, or something like that, and hear no more about physics.
In L.I. Ponomarev
The Quantum Dice (p. 81)
Institute of Physics Publishing. Bristol, England. 1993

I know a great deal. I know too much. I am a quantum ancient.
In Jeremy Bernstein
Experiencing Science
Part 1. Two Faces of Physics. Chapter 2. Rabi: The Modern Age (p. 102)
Basic Books, Inc. New York, New York, USA. 1978

Peat, F. David
Theoretical physicist

The choice before us is either to abandon any hope of knowing the nature of quantum reality or to accept a nonlocal universe.
Einstein's Moon (p. 124)
Contemporary Books. Chicago, Illinois, USA. 1990

Planck, Max   1858–1947
German physicist

My futile attempts to fit the elementary quantum of action somehow into the classical theory continued for a number of years and they cost me a great deal of effort. Many of my colleagues saw in this something bordering on a tragedy. But I feel differently about it, for the thorough enlightenment I thus received was all the more valuable. I now knew for a fact that the elementary quantum of action played a far more significant part in physics than I had originally been inclined to suspect, and this recogni-

tion made me see clearly the need for the introduction of totally new methods of analysis and reasoning in the treatment of atomic problems.
Scientific Autobiography and Other Papers
A Scientific Autobiography (pp. 44–45)
Philosophical Library. New York, New York, USA. 1949

Polkinghorne, John   1930–
British physicist, Episcopal priest, and writer

Quantum theory is both stupendously successful as an account of the small-scale structure of the world and it is also the subject of unresolved debate and dispute about its interpretation. That sounds rather like being shown an impressively beautiful palace and being told that no one is quite sure whether its foundations rest on bedrock or shifting sand.
The Quantum World
Chapter 1 (p. 1)
Princeton University Press. Princeton, New Jersey, USA. 1984

Robinson, Arthur L.
No biographical data available

In short, quantum mechanics, special relativity, and realism cannot all be true.
Quantum Mechanics Passes Another Test
Science, Volume 217, Number 4558, July 30, 1982 (p. 435)

Rothman, Tony   1953–
American cosmologist

Quantum mechanics — the theory that explains phenomena on the size of atoms — is right. It is also so conceptually weird that physicists to this day feel uncomfortable with it.
Instant Physics: From Aristotle to Einstein, and Beyond
Chapter 7 (p. 159)
Ballentine Books. New York, New York, USA. 1995

Sagan, Carl   1934–96
American astronomer and author

How can light simultaneously be a wave and a particle? It might be better to think of it as something else, neither a wave nor a particle, something with no ready counterpart in the everyday world of the palpable, that under some circumstances partakes of the properties of a wave, and, under others, of a particle. This wave-particle dualism is another reminder of a central humbling fact: Nature does not always conform to our predispositions and preferences, to what we deem comfortable and easy to understand.
Billions & Billions: Thoughts on Life and Death at the Brink of the Millennium
Chapter 4 (p. 37)
Random House, Inc. New York, New York, USA. 1997

Schrödinger, Erwin   1887–1961
Austrian theoretical physicist

I don't like it, and I'm sorry I ever had anything to do with it.
In John Gribbin
In Search of Schrödinger's Cat: Quantum Physics and Reality (p. v)
Bantam Books. New York, New York, USA. 1984

If all this damned quantum jumping were really here to stay, I should be sorry I ever got involved with quantum theory.
In Werner Heisenberg
Physics and Beyond: Encounters and Conversations
Chapter 6 (p. 75)
Harper & Row, Publishers. New York, New York, USA. 1971

Stapledon, Olaf   1886–1950
English author

…whenever a creature was faced with several possible courses of action, it took them all, thereby creating many distinct temporal dimensions and distinct histories of the cosmos. Since in every evolutionary sequence of the cosmos there were very many creatures, and each was constantly faced with many possible courses, and the combinations of all their courses were innumerable, an infinity of distinct universes exfoliated from every moment of every temporal sequence in this cosmos.
Last and First Man and Star Maker
Chapter XV, 2 (p. 426)
Dover Publications, Inc. New York, New York, USA. 1968

Stenger, Victor J.   1935–
American physicist

This type of schizophrenic behavior is not confined to photons alone. Electrons, neutrons, and other entities that normally appear as localized particles also can't seem to decide whether they are waves or particles. It all depends on what you try to measure. If you look for localized electrons, neutrons, or photons, you find them. If, on the other hand, you set up an experiment designed to measure wave properties, you find these too. We look at the world through colored glasses, and so it should not surprise us that the world appears a different color when we change to another pair.
Physics and Psychics: The Search for a World Beyond the Senses
Chapter 10 (p. 215)
Prometheus Books. Buffalo, New York, USA. 1990

Trefil, James   1938–
American physicist

But…we recognize that the wave-particle duality does not arise because of anything paradoxical about the behavior of elementary particles, but simply from the fact that we have asked the wrong question. If we had asked "How does an elementary particle behave?" instead of asking "Does it behave like a particle or a wave?", we would have been able to give a perfectly sensible answer. An elementary particle is not a particle in the sense that

a bullet is, and it is not a wave like the surf. It exhibits some properties that we normally associate with each of these kinds of things, but it is an entirely new kind of phenomenon.
From Atoms to Quarks: An Introduction to the Strange World of Particle Physics (Revised edition), 1994
Charles Scribner's Sons. New York, New York, USA. 1980

Wheeler, John Archibald   1911–
American theoretical physicist and educator

There may be no such thing as the "glittering central mechanism of the universe" to be seen behind a glass wall at the end of the trail. Not machinery but magic may be the better description of the treasure that is waiting.
In Nick Herbert
Quantum Reality: Beyond the New Physics
Chapter 2 (p. 29)
Anchor Press. Garden City, New York, USA. 1985

Nothing is more important about quantum physics than this: it has destroyed the concept of the world as "sitting out there." The universe afterwards will never be the same.
Quoted by Jefferson Hane Weaver
The World of Physics (Volume 2)
N.10 (p. 427)
Simon & Schuster. New York, New York, USA. 1987

So the quantum, fiery creative force of modern physics, has burst forth in eruption after eruption and for all we know the next may be the greatest of all.
In Franco Selleri
Quantum Mechanics Versus Local Realism: The Einstein–Podolsky–Rosen Paradox
Chapter 1, Section 3 (p. 47)
Plenum Press. New York, New York, USA. 1988

…if one really understood the central point and its necessity in the construction of the world, one ought to be able to state it in one clear, simple sentence. Until we see the quantum principle with this simplicity we can well believe that we do not know the first thing about the universe, about ourselves, and about our place in the universe.
In Francesco de Finis (ed.)
Relativity, Quanta and Cosmology in the Development of the Scientific Thought of Albert Einstein (Volume 2)
The Quantum and the Universe

Yang, Chen Ning   1922–
Chinese-born American theoretical physicist

To those of us who were educated after light and reason had struck in the final formulation of quantum mechanics, the subtle problems and the adventurous atmosphere of these pre-quantum mechanics days, at once full of promise and despair, seem to take on an almost eerie quality. We could only wonder what it was like when to reach correct conclusions through reasonings that were manifestly inconsistent constituted the art of the profession.

Elementary Particles: A Short History of Some Discoveries in Atomic Physics
Chapter 1 (p. 9)
Princeton University Press. Princeton, New Jersey, USA. 1962

von Baeyer, Hans Christian   1938–
German-born physicist and author

That, in a nutshell, is the mystery of the quantum: When an electron is observed, it is a particle, but between observations its map of potentiality spreads out like a wave. Compared to the electron, even a platypus is banal.
Taming the Atom
Chapter 3 (p. 51)
Random House, Inc. New York, New York, USA. 1992

Wolf, Fred Alan   1934–
American theoretical physicist, writer, and lecturer

The quantum is that embarrassing little piece of thread that always hangs from the sweater of space-time. Pull it and the whole thing unravels.
Star Wave: Mind Consciousness of Quantum Physics
The Macmillan Company. New York, New York, USA. 1984

Zee, Anthony
American physicist

Welcome to the strange world of the quantum, where one cannot determine how a particle gets from here to there. Physicists are reduced to bookies, posting odds on the various possibilities.
Fearful Symmetry
Chapter 10 (p. 141)
Macmillan Publishing Company. New York, New York, USA. 1986

QUARK

Author undetermined

O! O! you eight colourful guys
You won't let quarks materialize
You're tricky, but now we realize
You hold together our nucleus.
In Frank Wilczek and Betsy Devine
Longing for the Harmonies
Chapter 18 (p. 200)
W.W. Norton & Company, Inc. New York, New York, USA.1988

Heisenberg, Werner Karl   1901–76
German physicist and philosopher

Even if quarks should be found (and I do not believe that they will be), they will not be more elementary than other particles, since a quark could be considered as consisting of two quarks and one anti-quark, and so on. I think we have learned from experiments that by getting to smaller and smaller units, we do not come to fundamental units, or indivisible units, but we do come to a point where division has no meaning. This is a result of the experiments of the last twenty years, and I am afraid that some physicists simply ignore this experimental fact.

In Paul Buckley and F. David Peat
Glimpsing Reality: Ideas in Physics and the Link to Biology
Werner Heisenberg (p. 15)
University of Toronto Press. Toronto, Ontario, Canada. 1996

Joyce, James   1882–1941
Irish-born author

Three quarks for Muster Mark!
Sure he hasn't got much of a bark
And sure any he has it's all beside the mark.
Finnegans Wake
Book II (p. 383)
The Viking Press. New York, New York, USA. 1939

Melnechuk, Theodore
Neuroscientist

Poor Gell-Man seeks
But fails to find
The fractioned freaks
He bore in mind.
And yet a Quark,
Yea, better, three,
Exist in stark
Reality.
The Hunting of the Quark
The Physics Teacher, Volume 7, Number 7, October 1969 (p. 415)

Stenger, Victor J.   1935–
American physicist

Today's quarks and leptons can be viewed as metaphors of the underlying reality of nature, though metaphors that are objectively and rationally defined and are components of theories that have great predictive power. And that's the difference between the metaphors of science and those of myth: scientific metaphors work.... In the pragmatic view of truth of William James, science is true because it works. Science may not be the only path to the truth, but it is the best one we have yet been able to discover.
Physics and Psychics: The Search for a World Beyond the Senses
Chapter 4 (p. 79)
Prometheus Books. Buffalo, New York, USA. 1990

Taylor, Richard E.   1929–
Canadian-born American physicist

The quarks and the stars were here when you came, and they will be here when you go. They have no sense of humor so, if you want a world where more people smile, you will have to fix things yourselves.
Les Prix Nobel. The Nobel Prizes in 1990
Nobel banquet speech for award received in 1990
Nobel Foundation. Stockholm, Sweden. 1991

QUASAR

Mundell, Carole
English astronomer

...observing quasars is like observing the exhaust fumes of a car from a great distance and then trying to figure out what is going on under the hood.
A New Look at Quasars
Scientific American, Volume 278, Number 6, June 1998 (p. 57)

QUATERNION

Bell, E. T. (Eric Temple)   1883–1960
Scottish-American mathematician and educator

...Frenchmen, Germans, and Italians, urging their respective substitutes for quaternions, added to the din. By the second decade of the twentieth century there was a babble of conflicting vector algebras, each fluently spoken only by its inventor and his few chosen disciples. If, at any time in the brawling half-century after 1862, the bickering sects had stopped quarreling for half an hour to listen attentively to what Grassmann was doing his philosophical best to tell them, the noisy battle would have ended as abruptly as a thunderclap. Such, at any rate, seems to have been the opinion of Gibbs. In retrospect, the fifty-year war between quaternions and its rivals for scientific favor, appears as an interminable sequence of duels fought with stuffed clubs in a vacuum over nothing.
The Development of Mathematics (p. 208)
McGraw-Hill Book Company, Inc. New York, New York, USA. 1945

QUESTION

Adams, Douglas   1952–2001
English author, comic radio dramatist, and musician

"I checked it very thoroughly," said the computer, "and that quite definitely is the answer. I think the problem, to be quite honest with you, is that you've never actually known what the question is."
The Ultimate Hitchhiker's Guide to the Galaxy
The Hitchhiker's Guide to the Galaxy
Chapter 28 (p. 121)
The Ballantine Book Company. New York, New York, USA. 2002

Alvarez, Luis Walter   1911–88
American experimental physicist

Much of the work we do as scientists involves filling in the details about matters that are basically understood already, or applying standard techniques to new specific cases. But occasionally there is a question that offers an opportunity for a really major discovery.
T. Rex and the Crater of Doom
Chapter 2 (p. 42)
Princeton University Press. Princeton, New Jersey, USA. 1997

Beecher, Henry Ward   1813–87
American Congregational preacher and orator

Never ask a question if you can help it; and never let a thing go unknown for the lack of asking a question if you can't help it.
In James Orton
Comparative Zoology, Structural and Systematic
Preceding Chapter XXI (p. 222)
Harper & Brothers. New York, New York, USA. 1877

Bloor, David
English sociologist and philosopher of science

To ask questions of the sort which philosophers address to themselves is usually to paralyze the mind…
Knowledge and Social Imagery
Chapter Three (p. 52)
The University of Chicago Press. Chicago, Illinois, USA. 1991

Bohm, David   1917–92
American physicist

…it is frequently realised that half the battle is over when we know what are the right questions to ask.
On the Relationship Between Methodology in Scientific Research and the Content of Scientific
British Journal for the Philosophy of Science, Volume 12, Number 46, 12, August 1961 (p. 105)

Boltzmann, Ludwig Edward   1844–1906
Austrian physicist

If a general intends to conquer a hostile city, he will not consult his map for the shortest road leading there; rather he will be found to make the most various detours, and every hamlet, even if quite off the path, will become a valuable point of leverage for him, if only he can take it; impregnable places will be isolated. Likewise, the scientist asks not what are the currently most important questions, but, "Which are at present solvable?", or sometimes simply, "In which can we make some small but genuine advance?"
In Brian McGuinness (ed.)
Theoretical Physics and Philosophical Problems. Selected Writings
The Second Law of Thermodynamics (p. 13–14)
Reidel Publishing Company. Boston, Massachusetts, USA. 1974

…but all the more splendid is the success when, groping in the thicket of special questions, we suddenly find a small opening that allows a hitherto undreamt of outlook on the whole.
In Brian McGuinness (ed.)
Theoretical Physics and Philosophical Problems. Selected Writings
The Second Law of Thermodynamics (p. 14)
Reidel Publishing Company. Boston, Massachusetts, USA. 1974

Bombieri, Enrico   1940–
Italian mathematician

When things get too complicated, it sometimes makes sense to stop and wonder: Have I asked the right question?

Prime Territory: Exploring the Infinite Landscape at the Base of the Number System
The Sciences, Volume 32, Number 5, 1992

Chargaff, Erwin   1905–2002
Austrian biochemist

Science is wonderfully equipped to answer the question "How?" But it gets terribly confused when you ask it the question "Why?"
Voices in the Labyrinth: Nature, Man and Science
Chapter 1 (p. 8)
The Seabury Press. New York, New York, USA. 1977

Colby, Frank Moore   1865–1925
American educator and writer

Every man ought to be inquisitive through every hour of his great adventure down to the day when he shall no longer cast a shadow in the sun. For if he dies without a question in his heart, what excuse is there for his continuance?
In Hans Selye
From Dream to Discovery: On Being a Scientist
Why Should You Do Research (p. 10)
McGraw-Hill Book Company, Inc. New York, New York, USA. 1950

Dawkins, Richard   1941–
British ethologist, evolutionary biologist, and popular science writer

There may be some deep questions about the cosmos that are forever beyond science. The mistake is to think that they are therefore not beyond religion too.
A Devil's Chaplain: Reflections on Hope, Lies, Science, and Love
(p. 149)
Houghton Mifflin Company. Boston, Massachusetts, USA. 2003

Feynman, Richard P.   1918–88
American theoretical physicist

There are all kinds of interesting questions that come from a knowledge of science, which only adds to the excitement and mystery and awe of a flower. It only adds. I don't understand how it subtracts.
What Do You Care What Other People Think?
Further Adventures of a Curious Character, The Making of a Scientist (p. 11)
W.W. Norton & Company, Inc. New York, New York, USA.1988

So right away I found out something about biology: it was very easy to find a question that was very interesting, and that nobody knew the answer to. In physics you had to go a little deeper before you could find an interesting question that people didn't know.
Surely You're Joking, Mr. Feynman!: Adventures of a Curious Character
A Map of a Cat? (p. 71)
W.W. Norton & Company, Inc. New York, New York, USA.1985

Fischer, D. H.
No biographical data available

Questions are the engines of intellect, the cerebral machines which convert energy to motion, and curiosity to controlled inquiry.
Historian's Fallacies: Toward a Logic of Historical Thought
Chapter I (p. 3)
Harper & Row, Publishers. New York, New York, USA. 1970

Gore, George   1826–1909
English electrochemist

The area of scientific discovery enlarges rapidly as we advance; every scientific truth now known yields many questions yet to be answered. To some of these questions it is possible to obtain answers at the present time, others may only be decided when other parts of science are more developed.
The Art of Scientific Discovery
Part I, Chapter III (p. 27)
Longmans, Green & Company. London, England. 1878

Gould, Stephen Jay   1941–2002
American paleontologist and evolutionary biologist

Questions are not neutral; they presuppose a list of assumptions that may be long and complex.
Dinosaur in a Haystack: Reflections in Natural History
Part Three, Chapter 11 (p. 136)
Random House, Inc. New York, New York, USA. 1995

Supporters assume that the greatness and importance of a work correlates directly with its stated breadth of achievement: minor papers solve local issues, while great works claim to fathom the general and universal nature of things. But all practicing scientists know in their bones that successful studies require strict limitations. One must specify a particular problem with an accessible solution, and then find a sufficiently simple situation where attainable facts might point to a clear conclusion. Potential greatness then arises from cascading implications toward testable generalities. You don't reach the generality by direct assault without proper tools. One might as well dream about climbing Mount Everest wearing a T-shirt and tennis shoes and carrying a backpack containing only an apple and a bottle of water.
Writing in the Margins
Natural History, Volume 7, Number 9, 1998 (p. 19)

Greene, Brian   1963–
American physicist

Sometimes attaining the deepest familiarity with a question is our best substitute for actually having the answer.
The Elegant Universe
Chapter 14 (p. 365)
W.W. Norton & Company, Inc. New York, New York, USA. 2003

Heisenberg, Werner Karl   1901–76
German physicist and philosopher

We ask, "What does a proton consist of? Can an electron be divided or is it indivisible? Is a photon simple or compound?" But all these questions are wrongly put, because words such as "divide" or "consists of" have to a large extent lost their meaning. It must be our task to adapt our thinking and speaking — indeed our scientific philosophy — to the new situation created by the experimental evidence. Unfortunately this is very difficult. Wrong questions and wrong pictures creep automatically into particle physics and lead to developments that do not fit the real situation in nature.
The Nature of Elementary Particles
Physics Today, Volume 29, Number 3, March 1976 (p. 37)

Our scientific work in physics consists in asking questions about nature in the language that we possess and trying to get an answer from experiment by the means that are at our disposal. In this way quantum theory reminds us, as Bohr has put it, of the old wisdom that when searching for harmony in life one must never forget that in the drama of existence we are ourselves both players and spectators. It is understandable that in our scientific relation to nature our own activity becomes very important when we have to deal with parts of nature into which we can penetrate only by using the most elaborate tools.
Physics and Philosophy: The Revolution in Modern Science
Chapter III (p. 58)
Harper & Row, Publishers. New York, New York, USA. 1958

Hoffer, Eric   1902–83
American longshoreman and philosopher

To spell out the obvious is often to call it in question.
The Passionate State of Mind, and Other Aphorisms
No. 220
Harper & Brothers. New York, New York, USA. 1955

Horrobin, David F.   1939–2003
Medical researcher

One needs to be neither particularly observant nor particularly arrogant to realise that the majority of the human race is capable of understanding the nature of the universe in only the simplest and crudest terms. The truth about the universe is clearly beyond the comprehension of most men. Most human brains are incapable of framing appropriate questions, let alone of providing adequate answers.
Science Is God
Chapter 2 (p. 16)
Medical and Technical Publishing Company Ltd. Aylesbury, England. 1969

Hoyle, Sir Fred   1915–2001
English mathematician and astronomer

…in science answers are not important, it is the questions that are important.
In Philip Morrison

Nothing Is Too Wonderful to Be True
Less May Be More (p. 219)
The American Institute of Physics. Woodbury, New York, USA. 1995

Huxley, Elspeth   1907–97
English writer

The best way to find things out…is not to ask questions at all. If you fire off a question, it is like firing of a gun; bang it goes, and everything takes flight and runs for shelter. But if you sit quite still and pretend not to be looking, all the little facts will come and peck round your feet, situations will venture forth from thickets, and intentions will creep out and sun themselves on a stone; and if you are very patient you will see and understand a great deal more than a man with a gun.
The Flame Trees of Thika
Chapter Twenty-Eight (p. 272)
William Morrow & Company. New York, New York, USA. 1959

Kelvin, Lord William Thomson   1824–1907
Scottish engineer, mathematician, and physicist

Questions of personal priority…however interesting they may be to the persons concerned, sink into insignificance in the prospect of any gain of deeper insight into the secrets of nature.
In Silvanus P. Thompson
Annual Report of the Board of Regents of the Smithsonian Institution, 1908
The Kelvin Lecture: The Life and Work of Lord Kelvin (p. 752)
Government Printing Office. Washington, D.C. 1909

Kundera, Milan   1929–
Czech-born writer

…the only truly serious questions are the ones that even a child can formulate. Only the most naive of questions are truly serious. They are questions with no answers. A question with no answer is a barrier that cannot be breached. In other words, it is questions with no answers that set the limits of human possibilities, describe the boundaries of human existence.
Translated by Michael Henry Heim
The Unbearable Lightness of Being
Part Four, Section 6 (p. 139)
Harper & Row, Publishers. New York, New York, USA. 1984

Landau, Lev   1908–68
Russian physicist

Physicists have learned that certain questions cannot be asked, not because the level of our knowledge does not yet permit us to find the answer, but because such an answer simply isn't stored in nature.
In Alexandre Dorozynski
The Man They Wouldn't Let Die
Chapter 7 (p. 108)
Secker & Warburg. London, England. 1966

Leggett, A. J.
No biographical data available

In those exciting but frustrating fields of knowledge, or perhaps one should say ignorance, where physics tangles with philosophy, the difficulties usually lie less in finding answers to well-posed questions than in formulating the fruitful questions in the first place.
The Encyclopaedia of Ignorance: Everything You Ever Wanted to Know About the Unknown
The "Arrow of Time" and Quantum Mechanics (p. 102)
Pergamon Press. Oxford, England. 1977

Little, T. M.
No biographical data available

The purpose of an experiment is to answer questions. The truth of this seems so obvious, that it would not be worth emphasizing were it not for the fact that the results of many experiments are interpreted and presented with little or no reference to the questions that were asked in the first place.
Interpretation and Presentation of Results
Hortscience, Volume 16, 1981 (pp. 637–640)

MacRobert, Alan
Editor

Valid physical questions face us for which our physics is utterly inadequate. This can only be a sign that we stand at a great frontier of science, one that will form a cutting edge of inquiry for generations to come, with results we cannot guess.
Beyond the Big Bang
Sky & Telescope, Volumes 65–66, March 1983 (p. 213)

Maxwell, James Clerk   1831–79
Scottish physicist

There are some questions in Astronomy, to which we are attracted rather on account of their peculiarity, as the possible illustration of some unknown principle, than from any direct advantage which their solution would afford to mankind.
On the Stability of the Motion of Saturn's Rings
Macmillan & Company Ltd. London, England. 1859

Medawar, Sir Peter Brian   1915–87
Brazilian-born English zoologist

I do not believe that there is any intrinsic limitation upon our ability to answer the questions that belong to the domain of natural knowledge and fall therefore within the agenda of scientific enquiry.
The Strange Case of the Spotted Mice and Other Classic Essays on Science
On "The Effecting of All Things Possible"
Oxford University Press, Inc. New York, New York, USA. 1996

Midgley, Mary   1919–
English moral philosopher

The astonishing successes of western science have not been gained by answering every kind of question, but precisely by refusing to. Science has deliberately set narrow limits to the kinds of questions that belong to it, and further limits to the questions peculiar to each branch. It has practiced an austere modesty, a rejection of claims to universal authority.
Can Science Save Its Soul?
New Scientist, 1 August 1992 (p. 25)

Morris, Desmond   1928–
Zoologist and ethnologist

We never stop investigating. We are never satisfied that we know enough to get by. Every question we answer leads on to another question. This has become the greatest survival trick of our species.
The Naked Ape
Chapter Four (p. 130)
McGraw-Hill Book Company, Inc. New York, New York, USA. 1967

Payne-Gaposchkin, Celia   1900–79
British-American astronomer

Whenever we look in nature we can see spiral forms in the uncurling fern, the snail, the nautilus shell, the hurricane, the stirred cup of coffee, the water that swirls out of a wash bowl. Perhaps we shouldn't be surprised to see spirals in the great star systems whirling in space. Yet they remain a great, intriguing question.
Why Do Galaxies Have a Spiral Form?
Scientific American, Volume 189, Number 3, September 1953 (p. 89)

Peirce, Charles Sanders   1839–1914
American scientist, logician, and philosopher

Who would have said, a few years ago, that we could ever know of what substances stars are made of whose light may have been longer in reaching us than the human race existed? Who can be sure of what we shall now know in a few hundred years? Who can guess what would be the result of continuing the pursuit of science for ten thousand years, with the activity of the last hundred? And if it were to go on for a million, or a billion, or any number of years you please, how is it possible to say that there is any question which might not ultimately be solved?
Values in a Universe of Chance
How to Make Ideas Clear (p. 134)
Stanford University Press. Stanford, California, USA. 1958

…all the followers of science are fully persuaded that the processes of investigation, if only pushed far enough, will give one certain solution to each question to which they can be applied.… This great law is embodied in the conception of truth and reality. The opinion which is fated to be ultimately agreed to by all who investigate is what we mean by the truth, and the object represented in this opinion is the real.
In H.S. Thayer (ed.)

Pragmatism: The Classic Writings
How to Make Ideas Clear (p. 97)
New American Library. New York, New York, USA. 1970

Popper, Karl R.   1902–94
Austrian/British philosopher of science

…a young scientist who hopes to make discoveries is badly advised if his teacher tells him, "Go round and observe," and he is well advised if his teacher tells him: "Try to learn what people are discussing nowadays in science. Find out where difficulties arise, and take an interest in disagreements. These are the questions which you should take up."
Conjectures and Refutations: The Growth of Scientific Knowledge
Chapter 4 (p. 129)
Harper & Row, Publishers. New York, New York, USA. 1963

Ramsay, Sir William   1852–1916
English chemist

Whosoever asks shall receive, but he must ask sensible questions in definite order, so that the answer to the first suggests a second, and the reply to the second suggests a third, and so on.
Essays Biographical and Chemical
Chemical Essays
How Discoveries Are Made (p. 128)
Archibald Constable & Company Ltd. London, England. 1908

Russell, Bertrand Arthur William   1872–1970
English philosopher, logician, and social reformer

Clearly our first problem must be to define the issue, since nothing is more prolific of fruitless controversy than an ambiguous question.
Determinism and Physics
Proceeding of the University of Durham Philosophical Society, 1936

Sagan, Carl   1934–96
American astronomer and author

There are no forbidden questions in science, no matters too sensitive or delicate to be probed, no sacred truths.
The Demon-Haunted World: Science as a Candle in the Dark
Chapter 2 (p. 31)
Random House, Inc. New York, New York, USA. 1995

Seignobos, Charles   1854–1942
French historian

It is useful to ask oneself questions, *but very dangerous to answer them.*
In Marc Bloch
The Historian's Craft
Introduction (p. 14)
Manchester University Press. Manchester, England. 2004

Shaw, George Bernard   1856–1950
Irish comic dramatist and literary critic

No question is so difficult to answer as that to which the answer is obvious.
Saturday Review, January 26, 1895

Silver, Brian L.
Israeli professor of physical chemistry

The Big Questions may be beyond the capabilities of the human computer, just as dogs will never understand jokes. We understand a very great deal about what forces do but are far from finalizing the discussion about what forces are. Maybe we never will. Newton very specifically refused to commit himself as to what gravitational force was, but he nevertheless accounted for the movements of Earth and Moon and deduced the masses of the Earth and the Sun by knowing only what gravity does. We have discovered forces that Newton never knew, but basically we still only define force by what it does.
The Ascent of Science
Part II, Chapter 3 (p. 30)
Solomon Press Book. New York, New York, USA. 1998

Steinbeck, John   1902–68
American novelist

The literature of science is filled with answers found when the question propounded had an entirely different direction and end.
Sea of Cortez
Chapter 17 (pp. 179–180)
Paul P. Appel, Publisher. Mount Vernon, New York, USA. 1982

Stewart, Ian   1945–
English mathematician and science writer
Cohen, Jack
Reproductive biologist

The history of science, broadly speaking, is the tale of a lengthy battle to dig out the secret simplicities of a complicated world. It is an astonishing story of insignificant humanity's triumph over huge mysteries.
The Collapse of Chaos: Discovering Simplicity in a Complex World
Chapter 1 (p. 28)
Viking Press. New York, New York, USA. 1994

Trefil, James   1938–
American physicist

Great questions in science — questions like the ones Herschel raised about the structure of the universe — are seldom answered by ivory-tower types engaging in pure thought. They are answered by people who are willing to get down into the trenches and grapple with nature. If that means casting your own telescope mirrors, as Herschel did, so be it.
Reading the Mind of God: In Search of the Principle of Universality
Charles Scribner's Sons. New York, New York, USA. 1989

Virchow, Rudolf Ludwig Karl   1821–1902
German pathologist and archaeologist

If only people would finally stop finding points of disagreement in the personal characteristics and external circumstances of investigators! It does not matter at all whether someone is a professor of clinical medicine or of theoretical pathology, whether he is a practitioner or a hospital physician, if only he possesses material for observation. In addition, it is not of decisive significance whether he confronts an overwhelming or a modest amount of material, if only he understands how to exploit it. And to do this he must know what he wants: in other words, he must be in a position to put the right questions and to find the right methods for answering them.
Translated by Lelland J. Rather
Disease, Life, and Man
Cellular Pathology (p. 77)
Stanford University Press. Stanford, California, USA. 1958

Walker, Kenneth   1882–1966
Physician

We must accept the fact that the scientist can answer only a few of the questions we ask him and never the question of "why?."
Meaning and Purpose
Chapter VIII (p. 80)
Jonathan Cape. London, England. 1944

Weisskopf, Victor Frederick   1908–2002
Austrian-American physicist

It was absolutely marvelous working for Pauli. You could ask him anything. There was no worry that he would think a particular question was stupid, since he thought all questions were stupid.
Working for Pauli
American Journal of Physics, Volume 45, Number 5, May 1977 (p. 422)

Wilde, Oscar   1854–1900
Irish wit, poet, and dramatist

Questions are never indiscreet. Answers sometimes are.
The Plays of Oscar Wilde
An Ideal Husband
Act I (p. 10)
The Modern Library. New York, New York, USA. No date

Wittgenstein, Ludwig Josef Johann   1889–1951
Austrian-born English philosopher

As long as I continue to come across questions in more remote regions which I can't answer, it is understandable that I should still not be able to find my way around regions that are less remote. For how do I know that what stands in the way of an answer here is not precisely what is preventing me from clearing away the fog over there?
Translated by Peter Winch
Culture and Value (p. 66e)
The University of Chicago Press. Chicago, Illinois, USA. 1980

I may find scientific questions interesting, but they never really grip me. Only conceptual and aesthetic questions

do that. At bottom I am indifferent to the solution of scientific problems; but not the other sort.
Translated by Peter Winch
Culture and Value (p. 79e)
The University of Chicago Press. Chicago, Illinois, USA. 1980

Zinkernagel, Rolf M. 1944–
Swiss immunologist and pathologist

To ask questions, to search for answers, to do research — I mean re-search in nature what is already there, but has not been revealed, so far is the most fascinating and the most exciting thing we can dream of doing and what we would like to continue doing.
Les Prix Nobel. The Nobel Prizes in 1996
Nobel banquet speech for award received in 1996
Nobel Foundation. Stockholm, Sweden. 1997

QUESTIONNAIRE

Hauge, Bernt K.
No biographical data available

[One] feature of questionnaires is that they give [an] excellent opportunity to gather useless information in such a way that it can be handled by data machines, a handling that can give a mysterious authority of exactness to the most incredible nonsense.
Etcetera
The Physics Teacher, Volume 15, Number 9, December 1977 (p. 575)

QUOTATION

Dubos, René Jules 1901–82
French-born American microbiologist and environmentalist

History is replete with anecdotes and *bons mots* relating to statesmen, soldiers, artists, philosophers, and most other types of notables; but even a well-informed man finds it difficult to enliven talk with quotations from scientists.
The Dreams of Reason
Chapter 3 (p. 40)
Columbia University Press. New York, New York, USA. 1961

R

RACISM

Huntington, Ellsworth 1876–1947
American geographer

The climate of many countries seems to be one of the great reasons why idleness, dishonesty, immorality, stupidity, and weakness of will prevail. If we can conquer climate, the whole world will become stronger and nobler.
Civilization and Climate (p. 294)
University Press of the Pacific. Honolulu, Hawaii, USA. 2001

Lewin, Roger Amos
Anthropologist

Racism, as we would characterize it today, was explicit in the writings of virtually all the major anthropologists of the first decades of this century, simply because it was the generally accepted world view. The language of the epic tale so often employed by Arthur Keith, Grafton Elliot Smith, Henry Fairfield Osborn, and their contemporaries fitted perfectly an imperialistic view of the world, in which Caucasians were the most revered product of a grand evolutionary march to nobility.
Bones of Contention
Chapter 13, Man's Place in Nature (p. 307)
Simon & Schuster Inc. New York, New York, USA. 1987

Mandela, Nelson 1918–
First president of South Africa

The doctors and nurses treated me in a natural way as though they had been dealing with blacks on a basis of equality all their lives. It reaffirmed my long-held belief that education was the enemy of prejudice. These were men and women of science, and science had no room for racism.
The Long Walk to Freedom: The Autobiography of Nelson Mandela
(p. 492)
Little, Brown, Boston & Company. Boston, Massachusetts, USA. 1994

RADIATION

Bryson, Bill 1951–
American-born travel author

Incidentally, disturbance from cosmic background radiation is something we have all experienced. Tune your television to any channel it doesn't receive, and about 1 percent of the dancing static you see is accounted for by this ancient remnant of the Big Bang. The next time you complain that there is nothing on, remember that you can always watch the birth of the universe.
A Short History of Nearly Everything
Chapter 1 (p. 12)
Broadway Books. New York, New York, USA. 2003

Eddington, Sir Arthur Stanley 1882–1944
English astronomer, physicist, and mathematician

It has been widely supposed that the ultimate fate of protons and electrons is to annihilate one another, and release the energy of their constitution in the form of radiation. If so it would seem that the universe will finally become a ball of radiation, becoming more and more rarefied and passing into longer and longer wave-lengths. The longest waves of radiation are Hertzian waves of the kind used in broadcasting. About every 1500 million years this ball of radio waves will double its diameter; and it will go on expanding in geometrical progression for ever. Perhaps then I may describe the end of the physical world as — one stupendous broadcast.
New Pathways in Science
Chapter III, Section VI (p. 71)
The Macmillan Company. New York, New York, USA. 1935

Gamow, George 1904–68
Russian-born American physicist

Radiation is like butter, which can be bought or returned to the grocery store only in quarter-pound packages, although the butter as such can exist in any desired amount (not less, though, than one molecule!).
Thirty Years That Shook Physics
Chapter 1 (pp. 22–23)
Doubleday & Company, Inc. Garden City, New York, USA. 1966

Planck, Max 1858–1947
German physicist

Either the quantum of action was a fictional quantity, then the whole deduction of the radiation law was in the main illusionary and represented nothing more than an empty nonsignificant play on formulae, or the derivation of the radiation law was based on sound physical conception.
In Jefferson Hane Weaver
The World of Physics (Volume 2)
N.1 (p. 284)
Simon & Schuster. New York, New York, USA. 1987

RADICAL

von Liebig, Justus 1803–73
German organic chemist
Dumas, Jean Baptiste-Andre 1800–84
French biochemist

...in inorganic chemistry the radicals are simple; in organic chemistry they are compounds — that is the sole difference.
In William H. Brock

Justus von Liebig
Chapter 3 (p. 81)
Cambridge University Press. Cambridge, England. 1997

Mark, Herman F.   1898–1992
Polymer chemist

The concept of "free radicals" was not known in 1920 — well perhaps in politics, but not in chemistry.
From Small Organic Molecules to Large: A Century of Progress
Chemistry Study in Vienna (p. 15)
American Chemical Society, Washington, D.C. 1993

RADIO ASTRONOMY

Christiansen, Chris
No biographical data available

Radio astronomy was not born with a silver spoon in its mouth. Its parents were workers. One parent was the radio-telescope, the other was radar.
Daily Telegraph (*Sydney*), August 25, 1952

Commentary

It has been demonstrated that a receiving set of great delicacy in New Jersey will get a new kind of static from the Milky Way. This is believed to be the longest distance anybody ever went to look for trouble.
The New Yorker Magazine, 17 June 1933

Davies, Paul Charles William   1946–
British-born physicist, writer, and broadcaster

It is a striking thought that ten years of radio astronomy have taught humanity more about the creation and organization of the universe than thousands of years of religion and philosophy.
Space and Time in the Modern Universe (p. 211)
Cambridge University Press. Cambridge, England. 1977

Gingerich, Owen   1930–
American astronomer

But even if radio astronomy has not so much destroyed our older astronomical viewpoint, it has enormously enlarged and enriched it. It is like that magical moment in the old Cinerama, when the curtains suddenly opened still further, unveiling the grandeur of the wide screen. Optical astronomy in the 1950s, on that narrow, central screen, offered a quiescent view of a slowly burning universe, the visible radiations from thermal disorder. But then the curtains abruptly parted, adding a grand and breathtaking vista, a panorama of swift and orderly motions that revealed themselves through the synchrotron radiation they generated — the so-called violent universe.
In W.T. Sullivan III
The Early Years of Radio Astronomy: Reflections Fifty Years After Jansky's Discovery
Radio Astronomy and the Nature of Science (p. 404)
Cambridge University Press. Cambridge, England. 1984

Kraus, John   1910–2004
Radio astronomer

The radio sky is no carbon copy of the visible sky; it is a new and different firmament, one where the edge of the universe stands in full view and one which bears the telltale marks of a violent past.
Big Ear
Chapter 21 (p. 166)
Cygnus-Quasar Books. Powell, Ohio, USA. 1976

Mitton, Simon
No biographical data available

During the last 20 years radio astronomers have led a revolution in our knowledge of the Universe that is paralleled only by the historic contributions of Galileo and Copernicus. In particular, the poetic picture of a serene Cosmos populated by beautiful wheeling galaxies has been replaced by a catalogue of events of astonishing violence: a primeval fireball, black holes, neutron stars, variable quasars and exploding galaxies.
Newest Probe of the Radio Universe
New Scientist, Volume 56, Number 816, 19 October 1972 (p. 138)

Unsold, Albrecht   1905–95
Astrophysicist

The old dream of wireless communication through space has now been realized in an entirely different manner than many had expected. The cosmos' short waves bring us neither the stock market nor jazz from distant worlds. With soft noises they rather tell the physicist of the endless love play between electrons and protons.
In W.T. Sullivan, III
Classics in Radio Astronomy
Preface (p. xiii)
R. Reidel Publishing Company. Dordrecht, Netherlands. 1982

RAIN FOREST

Fuertes, Louis Agassiz   1874–1920
American ornithologist

The principal sensation one gets in the tropical forests is the mystery of the unknown voices. Many of these remain forever mysteries unless one stays long and seeks diligently.
Annual Report of the Board of Regents of the Smithsonian Institution, 1915
Impressions of the Voices of Tropical Birds (p. 313)
Government Printing Office. Washington, D.C. 1916

RAINBOW

Twain, Mark (Samuel Langhorne Clemens)   1835–1910
American author and humorist

We have not the reverent feeling for the rainbow that the savage has, because we know how it is made. We have lost as much as we gained by prying into that matter.
A Tramp Abroad
Chapter XLIII (p. 318)
Penguin Books. New York, New York, USA. 1997

One can enjoy a rainbow without necessarily forgetting the forces that made it.
Europe and Elsewhere
Queen Victoria's Jubilee (p. 210)
Harper & Brothers. New York, New York, USA. 1923

RAMIFICATION

Sylvester, James Joseph   1814–97
English mathematician

The theory of ramification is one of pure colligation, for it takes no account of magnitude or position; geometrical lines are used, but these have no more real bearing on the matter than those employed in genealogical tables have in explaining the laws of procreation.
The Collected Mathematical Papers of James Joseph Sylvester (Volume 3)
On Recent Discoveries in Mechanical Conversion of Motion (p. 23)
University Press. Cambridge, England. 1904–1912

RANDOMNESS

Cohen, John
No biographical data available

…nothing is so alien to the human mind as the idea of randomness.
Chance, Skill, and Luck
Chapter 2, Part IV (p. 42)
Penguin Books. Baltimore, Maryland, USA. 1960

Fisher, Sir Ronald Aylmer   1890–1962
English statistician and geneticist

The postulate of randomness thus resolves itself into the question, "of what population is this a random sample?" which must frequently be asked by every practical statistician.
On the Mathematical Foundation of Theoretical Statistics
Philosophical Transactions of the Royal Society of London, Volume A222, 1922 (p. 313)

James, William   1842–1910
American philosopher and psychologist

If I should throw down a thousand beans at random upon a table, I could doubtless, by eliminating a sufficient number of them, leave the rest in almost any geometrical pattern you might propose to me, and you might then say that that pattern was the thing prefigured beforehand, and that the other beans were mere irrelevance and packing material. Our dealings with Nature are just like this.

The Varieties of Religious Experience
Lecture XVIII (p. 429)
The Modern Library. New York, New York, USA. 196?

Leucippus   5th century BCE
Greek philosopher of Atomism

Nothing occurs at random, but everything for a reason and by necessity.
In G.S. Kirk, J.E. Raven and M. Schofield (eds.)
The Pre-Socratic Philosophers: A Critical History with a Selection of Texts
Aetius I.25.4 (p. 420)
At The University Press. Cambridge, England. 1963

Peterson, Ivars
Mathematics and physics writer

We are surrounded by jungles of randomness. With our mathematical and statistical machetes, we can hack out extensive networks of trails and clearings that provide for most of our day-to-day needs and make sense of some fraction of human experience. The vast jungle, however, remains close at hand, never to be taken for granted, never to reveal all its secrets — and always teasing the inquiring mind.
The Jungle of Randomness: A Mathematical Safari
Chapter 10 (p. 203)
John Wiley & Sons, Inc. New York, New York, USA. 1998

Sophocles   496 BCE–406 BCE
Greek playwright

IOCLASTA: Nay, what should mortal[s] fear, for whom the degree of fortune are supreme, and who hath clear foresight of nothing? 'Tis best to live at random, as one may.
In *Great Books of the Western World* (Volume 5)
The Plays of Sophocles
Oedipus the King, l. 997
Encyclopædia Britannica, Inc. Chicago, Illinois, USA. 1952

Szent-Györgyi, Albert   1893–1986
Hungarian-born American biochemist

The usual answer to this question is that there was plenty of time to try everything. I could never accept this answer. Random shuttling of bricks will never build a castle or a Greek temple, however long the available time. A random process can build meaningful structures only if there is some kind of selection between meaningful and nonsense mutations.
Molecular Evolution: Prebiological and Biological
The Evolutionary Paradox and Biological Stability (p. 111)
Plenum Press. New York, New York, USA. 1972

RANDOM DIGITS

Dickens, Charles   1812–70
English novelist

Anyone who considers arithmetical methods of producing random digits is, of course, in a state of sin.
Oliver Twist
Chapter LI
P.F. Collier & Son, Company. New York, New York, USA. 1912

RANDOM NUMBER

Coveyou, R. R.
No biographical data available

The generation of random numbers is too important to be left to chance.
Random Number Generation Is Too Important to Be Left to Chance
Studies in Applied Mathematics, Volume 3, 1970

RATIOCINATION

Keyser, Cassius Jackson   1862–1947
American mathematician

When the greatest of American logicians, speaking of the powers that constitute the born geometrician, had named Conception, Imagination, and Generalization, he paused. Thereupon from one of the audience there came the challenge, "What of reason?" The instant response, not less just than brilliant, was: "Ratiocination — that is but the smooth pavement on which the chariot rolls."
In Columbia University
Lectures on Science, Philosophy and Art 1907–1908 (p. 31)
New York, New York, USA. 1908

REACTION

Baudrimont, A. E.
No biographical data available

A chemical reaction cannot take place without a movement of the atoms. Consequently a reaction…cannot and will never be able to indicate the arrangement of the atoms in a combination…. For a reaction, by establishing a molecular movement, destroys the preceding arrangements of the atoms. Therefore, being able to extract a compound substance from a combination does not mean that this compound already existed in this combination.
In S.C. Kapoor
The Origins of Laurent's Organic Classification
Isis, Volume 60, 1960 (p. 493)

Hoffmann, Roald   1937–
Polish-born American chemist

there was no question that the reaction
but transient colors were seen
in the slurry of sodium methoxide in dichloromethane
and we got a whole lot of products
for which we can't sort out the kinetics
the next slide will show

the most important part
very rapidly
within two minutes
and I forgot to say on further warming
we get in fact the keytone…
The Metamict State
Next Slide Please (p. 51)
University of Central Florida Press. Orlando, Florida, USA. 1987

Lippmann, Walter   1889–1974
American journalist and author

The reaction of one chemical element to another chemical element is always correct, is never misled by misinformation, by untruth, and by illusion.
Essays in the Public Philosophy
Chapter VIII (p. 92)
Little, Brown & Company. Boston, Massachusetts, USA. 1955

READING

Bacon, Sir Francis   1561–1626
English lawyer, statesman, and essayist

Read not to contradict and confute, nor to believe and take for granted, nor to find talk and discourse, but to weigh and consider.
Bacon's Essays
Of Studies (p. 210)
Donohue, Henneberry & Company. Chicago, Illinois, USA. 1883

Fisher, Sir Ronald Aylmer   1890–1962
English statistician and geneticist

Fairly large print is a real antidote to stiff reading.
In J.H. Bennett (ed.)
Natural Selection, Heredity, and Eugenics
Letter to K. Sisam, 31 May 1929 (p. 20)
Clarendon Press. Oxford, England. 1983

Huxley, Thomas Henry   1825–95
English biologist

I MUST adopt a fixed plan of studies, for unless this is done I find time slips away without knowing it — and let me remember this — that it is better to read a little and thoroughly, than cram a crude undigested mass into my head, though it be great in quantity.
The Life and Letters of Thomas Henry Huxley (Volume 1)
Chapter 1.1
Macmillan & Company Ltd. London, England. 1903

REAL (BEING)

Bianco, Margery Williams   1880–1944
Author

"What is REAL?" asked the Rabbit one day, when they were lying side by side near the nursery fender, before Nana came to tidy up the room. "Does it mean having

things that buzz inside you and a stick-out handle?"
The Velveteen Rabbit: Or How Toys Become Real
Athenaeum Books for Young Readers. New York, New York, USA.
2002

You become. It takes a long time. That's why it doesn't often happen to people who break easily, or have sharp edges, or who have to be carefully kept. Generally, by the time you are Real, most of your hair has been loved off, and your eyes drop out, and you get loose in the joints and very shabby. But these things don't matter at all, because once you are Real you can't be ugly, except to people who don't understand.
The Velveteen Rabbit: Or How Toys Become Real
Athenaeum Books for Young Readers. New York, New York, USA.
2002

Einstein, Albert   1879–1955
German-born physicist

The important point for us to observe is that all these constructions and the laws connecting them can be arrived at by the principle of looking for the mathematically simplest concepts and the link between them. In the limited number of mathematically existent simple field types, and the simple equations possible between them, lies the theorist's hope of grasping the real in all its depth.
Ideas and Opinions
On the Methods of Theoretical Physics (p. 275)
Crown Publishers, Inc. New York, New York, USA. 1954

REALITY

Bohr, Niels Henrik David   1886–1962
Danish physicist

The conception of the objective reality of the elementary particles has evaporated in a curious way, not into the fog of some new, obscure reality concept, but into the transparent clarity of a mathematics that represents no longer the behavior of the elementary particles but rather our knowledge of this behavior.
The Representation of Reality in Contemporary Physics
Daedalus, 87(3), 1958

…an independent reality in the ordinary physical sense can neither be ascribed to the phenomena nor to the agencies of observation.
Atomic Theory and the Description of Nature
Chapter II (p. 54)
Cambridge University Press. Cambridge, England. 1934

Born, Max   1882–1970
German-born English physicist

The simple and unscientific man's belief in reality is fundamentally the same as that of the scientist.
Physics in My Generation
On the Meaning of Physical Theories (p. 16)
Springer-Verlag New York, Inc. New York, New York, USA. 1969

Bronowski, Jacob   1908–74
Polish-born English mathematician and polymath

Reality is not an exhibit for man's inspection, labeled "Do not touch." There are no appearances to be photographed, no experiences to be copied, in which we do not take part. Science, like art, is not a copy of nature but a re-creation of her.
Science and Human Values
The Creative Mind (p. 20)
Harper & Row, Publishers. New York, New York, USA. 1965

Brooks, Harvey
No biographical data available

A more problematic example is the parallel between the increasingly abstract and insubstantial picture of the physical universe which modern physics has given us and the popularity of abstract and non-representational forms of art and poetry. In each case the representation of reality is increasingly removed from the picture which is immediately presented to us by our senses.
Scientific Concepts and Cultural Change
Daedalus, Winter 1965

Burtt, E. A.
No biographical data available

Man begins to appear for the first time in the history of thought as an irrelevant spectator and insignificant effect of the great mathematical system which is the substance of reality.
The Metaphysical Foundations of Modern Physical Science (p. 80)
Doubleday, Doran & Company, Inc. Garden City, New York, USA. 1954

Cerf, Bennett   1898–1971
American publisher and editor

The best of them was the conversation between Ginsberg, who demanded to know what reality was, and Garfinkle, who brazenly attempted to explain it to him.
Try and Stop Me: A Collection of Anecdotes and Stories, Mostly Humorous
Jokes About Relativity (p. 163)
Simon & Schuster. New York, New York, USA. 1944

Cromer, Alan   1935–
American physicist and educator

Reality has far more wonders than all the tales of Arabia, giving us in return for our lost feeling of omnipotence some knowledge of the external world, some control over and responsibility for our lives, and even a touch of humility.
Uncommon Sense: The Heretical Nature of Science
Chapter 10 (p. 207)
Oxford University Press, Inc. New York, New York, USA. 1993

Dampier-Whetham, William   1867–1952
English scientific writer

The physicist analyzes matter into particles, and finds that their forces and motions can be described in mathematical terms. The materialist pushes this scientific result into philosophy, and says that there is no other reality.
In Lloyd William Taylor
Physics: The Pioneer Science (Volume 1)
Chapter 14 (p. 181)
Houghton Mifflin Company. Boston, Massachusetts, USA. 1941

Drees, Willem B.
Dutch philosopher of science and religion

That natural reality is assumed rather than explained is not proof for the existence of a creator. Introducing god as an explanatory notion only shifts the locus of the question: why would such a god exist? And, it is possible that the universe just happens to exist, without explanation.
In Victor J. Stenger
Has Science Found God?: The Latest Results in the Search for Purpose in the Universe
Chapter Seven (p. 163)
Prometheus Books. Amherst, New York, USA. 2003

Dürrenmatt, Friedrich   1921–90
Swiss playwright and novelist

Our researches are perilous, our discoveries are lethal. For us physicists there is nothing left but to surrender to reality.
Translated by James Kirkup
The Physicists
Act Two (p. 81)
Grove Press, Inc. New York, New York, USA. 1964

Eddington, Sir Arthur Stanley   1882–1944
English astronomer, physicist, and mathematician

It is by looking into our own nature that we first discover the failure of the physical universe to be co-extensive with our experience of reality. The "something to which truth matters" must surely have a place in reality what ever definition of reality we may adopt.
New Pathways in Science
Chapter XIV, Section II (p. 317)
The Macmillan Company. New York, New York, USA. 1935

Egler, Frank E.   1911–96
American botanist and ecologist

Reality is not what is; it is what the layman wishes it to be.
The Way of Science
Science Concepts (p. 22)
Hafner Publishing Company. New York, New York, USA. 1970

Einstein, Albert   1879–1955
German-born physicist

Pure logical thinking cannot yield us any knowledge of the empirical world; all knowledge of reality starts from experience and ends in it. Propositions arrived at by purely logical means are completely empty of reality.
In Paul Arthur Schlipp (ed.)
Albert Einstein: Philosopher-Scientist
Einstein's Conception of Science (p. 391)
The Library of Living Philosophers, Inc. Evanston, Illinois, USA. 1949

Physics is an attempt conceptually to grasp reality as it is thought independently of its being observed. In this sense one speaks of "physical reality."
In Paul Arthur Schlipp (ed.)
Albert Einstein: Philosopher-Scientist
Autobiographical Notes (p. 81)
The Library of Living Philosophers, Inc. Evanston, Illinois, USA. 1949

Space has devoured ether and time; it seems to be on the point of swallowing up also the field and the corpuscles, so that it alone remains as the vehicle of reality.
In R. Thiel
And There Was Light (p. 345)
New American Library. New York, New York, USA. 1960

All our science, measured against reality, is primitive and childlike — good yet it is the most precious thing we have.
The Physics Teacher, April 1970 (p. 200)

Einstein, Albert   1879–1955
German-born physicist
Infeld, Leopold   1898–1968
Polish physicist

In our endeavor to understand reality we are somewhat like a man trying to understand the mechanism of a closed watch. He sees the face and moving hands, and even hears the ticking, but he as no way of opening the case. If he is ingenious enough he may form some picture of a mechanism which could be responsible for all the things he observes, but he may never be quite sure his picture is the only one which could explain his observations.
The Evolution of Physics
On Clew Remains (p. 31)
Simon & Schuster. New York, New York, USA. 1961

Frankel, Felice   1945–
Science photographer
Whitesides, George M.
American chemist

Our reality is illusion: We don't know for sure what's out there.
On the Surface of Things: Images of the Extraordinary in Science
Illusion (p. 121)
Chronicle Books. San Francisco, California, USA. 1997

Frost, Robert   1874–1963
American poet

You're searching, Joe
For things that don't exist.
I mean beginnings
Ends and beginnings

Ends and beginnings — there are no such things
There are only middles.
Complete Poems of Robert Frost
Mountain Interval
Henry Holt & Company. New York, New York, USA. 1949

Gribbin, John
English science writer and astronomer

Don't look here for any "eastern mysticism", spoon bending or ESP. Do look here for the true story of quantum mechanics, a truth far stranger than any fiction.... The question this book addresses is "What is reality?" The answer(s) may surprise you; you may not believe them.
In Search of Schrödinger's Cat: Quantum Physics and Reality
Introduction (p. xvi)
Bantam Books. New York, New York, USA. 1984

Heisenberg, Werner Karl   1901–76
German physicist and philosopher

[The probability wave] meant a tendency for something. It was a quantitative version of the old concept of "Potentia" in Aristotelian philosophy. It introduced something standing in the middle between the idea of an event and the actual event, a strange kind of physical reality just in the middle between possibility and reality.
Physics and Philosophy: The Revolution in Modern Science
Chapter II (p. 41)
Harper & Row, Publishers. New York, New York, USA. 1958

Hilbert, David   1862–1943
German mathematician

It has become perfectly clear that physics does not deal with the material world or with the contents of reality, but rather, what it perceives is merely the formal constitutions of reality.
In Walter R. Fuchs
Mathematics for the Modern Mind
Chapter 10, Section 10.1 (p. 240)
The Macmillan Company. New York, New York, USA. 1967

Hubble, Edwin Powell   1889–1953
American astronomer

...sometimes, through the strangely compelling experience of mystical insight, a man knows beyond the shadow of a doubt, that he has been in touch with a reality that lies behind mere phenomena. He himself is completely convinced, but he cannot communicate the certainty. It is a private revelation. He may be right, but unless we share his ecstasy we cannot know.
The Nature of Science and Other Lectures
Part I, The Nature of Science (p. 19)
The Huntington Library, San Marino, California, USA. 1954

Jacobi, Abraham   1830–1919
Pioneer of pediatrics

It is one thing to build an educational tower in the air at your library table, and another to face its actual appearance under the existing circumstances.
In R. Kagan (ed.)
Leaders of Medicine
Chapter IV (p. 41)
The Medico-Historical Press. Boston, Massachusetts, USA. 1941

Jeans, Sir James Hopwood   1877–1946
English physicist and mathematician

...the most outstanding achievement of twentieth-century physics is not the theory of relativity with its welding together of space and time, or the theory of quanta with its present apparent negation of the laws of causation, or the dissection of the atom with the resultant discovery that things are not what they seem; it is the general recognition that we are not yet in contact with ultimate reality.
The Mysterious Universe
Chapter V (pp. 150–151)
The Macmillan Company. New York, New York, USA. 1932

Kaufmann, William J., III   1942–94
American astronomer

From the moment of birth, our daily experiences strongly enforce the notion that reality is comprehensible. The fact that a rock released from your hand always falls down or that the moon goes through its phases every 29 1/2 days implies order rather than chaos to the rational human mind. To discover this order, to understand the basic and underlying qualities of all physical objects, to comprehend the fundamental principles that dictate the behavior of reality: this is the business of science.
Introduction to Particles and Fields (p. 1)
Scientific American, Inc., 1953

Mead, George H.   1863–1931
American philosopher, sociologist and psychologist

...the ultimate touchstone of reality is a piece of experience found in an unanalyzed world. The approach to the crucial experiment may be a piece of torturing analysis, in which things are physically and mentally torn to shreds, so that we seem to be viewing the dissected tissues of objects in ghostly dance before us, but the actual objects in the experimental experience are the common things of which we say that seeing is believing, and of whose reality we convince ourselves by handling. We extravagantly advertise the photograph of the path of an electron, but in fact we could never have given as much reality to the electrical particle as does now inhabit it if the photograph had been of naught else than glistening water vapour.
The Philosophy of the Act
Chapter II (p. 32)
The University of Chicago, Chicago, Illinois, USA; 1938

Olson, Sigurd F.   1899–1982
American conservationist

Flashes of insight or reality are sunbursts of the mind.
Reflections from the North Country
Flashes of Insight (p. 131)
Alfred A. Knopf. New York, New York, USA. 1976

Pagels, Heinz R.   1939–88
American physicist and science writer

We may begin to see reality differently simply because the computer…provides a different angle on reality.
The Dreams of Reason: The Computer and the Rise of the Sciences of Complexity
Preface (p. 13)
Simon & Schuster. New York, New York, USA. 1988

Palmieri, M.
No biographical data available

Since the dawn of human intelligence man has tried to form for himself a conception of the outside world which would correspond to the truest reality. But to determine what this reality is has proved to be a task of no mean import, and we still stand bewildered and wondering at the door of what has been and remains for mankind the greatest of all mysteries: the nature of ultimate reality.
Relativity: An Interpretation of Einstein's Theory
Introduction
Forbush Publishing Company. Los Angeles. 1931

Raymo, Chet   1936–
American physicist and science writer

Science is a map of reality.
The Virgin and the Mousetrap: Essays in Search of the Soul of Science
Chapter 16 (p. 147)
The Viking Press. New York, New York, USA. 1991

Reeve, F. D.
No biographical data available

Because all things balance — as on a wheel — and we cannot see nine-tenths of what is real, our claims of self-reliance are pieced together by unpanned gold. The whole system is a game: the planets are the shells; our earth, the pea. May there be no moaning of the bar. Like ships at sunset in a reverie, We are shadows of what we are.
Coasting
The American Poetry Review, Volume 24, Number 4, July–August 1995 (p. 38)

Riordan, Michael
Physicist

Subatomic reality is a lot like that of a rainbow, whose position is defined only relative to an observer. This is not an objective property of the rainbow-in-itself but involves such subjective elements as the observer's own position. Like the rainbow, a subatomic particle becomes fully "real" only through the process of measurement.
The Hunting of the Quark
Chapter 1 (p. 39)
Touchstone Books/Simon & Schuster. New York, New York, USA. 1987

Smith, David
No biographical data available

Everything imagined is reality. The mind cannot conceive unreal things.
The Private Thoughts of David Smith
Vogue, November 15, 1968 (p. 198)

Trilling, Lionel   1905–75
American critic, author, and teacher

In the American metaphysic, reality is always material reality, hard, resistant, unformed, impenetrable, and unpleasant.
The Liberal Imagination
Reality in America, ii (p. 13)
Charles Scribner's Sons. New York, New York, USA. 1950

Wald, George   1906–97
American biologist and biochemist

A scientist lives with all reality. There is nothing better. To know reality is to accept it, and eventually to love it.
Lex Prix Nobel. The Nobel Prizes in 1967
Nobel banquet speech for award received in 1967
Nobel Foundation. Stockholm, Sweden. 1968

Walgate, Robert
No biographical data available

…what the scientist must now admit is that in many problems of great consequence to people reality may not be accessible, in practice, through entirely manipulative and analytical methods.
Breaking Through the Disenchantment
New Scientist, September 18, 1975 (p. 667)

Weinberg, Steven   1933–
American nuclear physicist

When we say that a thing is real we are simply expressing a sort of respect.
Dreams of a Final Theory: The Scientist's Search for the Ultimate Laws of Nature
Chapter II (p. 46)
Pantheon Books. New York, New York, USA. 1992

Physical reality remains so mysterious even to physicists because of the extreme improbability that it was constructed to be understood by the human mind.
The Form of Nature
Bulletin of the American Academy of Arts and Sciences, Volume 29, Number 4, 1976

Weyl, Hermann   1885–1955
German mathematician

A picture of reality drawn in a few sharp lines cannot be expected to be adequate to the variety of all its shades. Yet even so the draftsman must have the courage to draw the lines firm.
Philosophy of Mathematics and Natural Science

Appendix D (p. 274)
Princeton University Press. Princeton, New Jersey, USA. 1949

Wheeler, John Archibald   1911–
American physicist and educator

What we call reality consists…of a few iron posts of observation between which we fill an elaborate papier-mâché of imagination and theory.
In Harry Woolf (ed.)
Some Strangeness in the Proportion
Chapter 22
Fig. 22.10 (p. 358)
Addison-Wesley Publishing Company. Reading, Massachusetts, USA. 1980

Whitehead, Alfred North   1861–1947
English mathematician and philosopher

Progress in truth — truth of science and truth of religion — is mainly a progress in the framing of concepts, in discarding artificial abstractions or partial metaphors, and in evolving notions which strike more deeply into the root of reality.
Religion in the Making
Truth and Criticism (p. 127)
New American Library. New York, New York, USA. 1960

Yeats, William Butler   1865–1939
Irish poet and playwright

IILLE: The rhetorician would deceive his neighbors,
The sentimentalist himself; while art
Is but a vision of reality.
The Collected Poems of W.B. Yeats
Ego Dominus Tuus (p. 159)
The Macmillan Company. New York, New York, USA. 1956

REASONING

Abbey, Edward   1927–89
American environmentalist and nature writer

Reason is the newest and rarest thing in human life, the most delicate child of human history.
A Voice Crying in the Wilderness: Notes from a Secret Journal
Chapter 10 (p. 91)
St. Martin's Press. New York, New York, USA. 1989

Arnauld, Antoine   1612–94
French philosopher, lawyer, and mathematician

We use Reason for improving the Sciences; whereas we ought to use the Sciences for improving our Reason.
The Port-Royal Logic
Preface
Printed for T.B. and J. Taylor. London, England. 1696

Barnett, P. A.
No biographical data available

…the reasoning of mathematics is a type of perfect reasoning.
Common Sense in Education and Teaching
Chapter IX (p. 222)
Longmans, Green & Company. London, England. 1899

Beaumarchais, Pierre-Augustin Caron de   1732–99
French dramatist

It is not necessary to believe things in order to reason about them.
The Barber of Seville
Act V, Scene 4
Pioneer Classics. Long Beach, California, USA. 1994

Beck, Lewis White   1913–97
American scholar in German philosophy

In the logic of science there is a principle as important as that of parsimony: it is that of sufficient reason.
The "Natural Science Ideal" in the Social Sciences
The Scientific Monthly, Volume LXVIII, June 1949 (p. 393)

Bernard, Claude   1813–78
French physiologist

Pile up facts or observations as we may, we shall be none the wiser. To learn, we must necessarily reason about what we have observed, compare the facts and judge them by other facts used as controls.
Translated by Henry Copley Greene
An Introduction to the Study of Experimental Medicine
Part One, Chapter I, Section iv (p. 16)
Henry Schuman, Inc. New York, New York, USA. 1927

Reasoning will always be correct when applied to accurate notions and precise facts; but it can lead only to error when the notions or facts on which it rests were originally tainted with error or inaccuracy.
Translated by Henry Copley Greene
An Introduction to the Study of Experimental Medicine
Introduction (p. 2)
Henry Schuman, Inc. New York, New York, USA. 1927

Beveridge, William Ian Beardmore   1908–
Australian zoologist

Every experience and history teach us that in the biological and medical sciences reason seldom can progress far from the facts without going astray.
The Art of Scientific Investigation
Chapter Seven (p. 81)
W.W. Norton & Company, Inc. New York, New York, USA.1957

The role of reason in research is not so much in exploring the frontiers of knowledge as in developing the findings of the explorers.
The Art of Scientific Investigation
Chapter Seven (p. 91)
W.W. Norton & Company, Inc. New York, New York, USA.1957

A useful habit for scientists to develop is that of not trusting ideas based on reason only.... Practically all reasoning is influenced by feelings, prejudice and past experience, albeit often subconsciously.
The Art of Scientific Investigation
Chapter Seven (p. 87)
W.W. Norton & Company, Inc. New York, New York, USA.1957

Brophy, Brigid   1929–95
English novelist

Reason is necessarily the language of moral, political, and scientific argument: not because reason is holy or on some elevated plane, but because it isn't; because it is accessible to all humans; because, as well as working, it can be seen to work.
In Stanley and Rosiland Godlovitch and John Harris (eds.)
Animals, Men and Morals: An Enquiry into the Maltreatment of Non-Humans
In Pursuit of Fantasy (p. 126)
Taplinger Publishing Company. New York, New York, USA. 1972

Browne, Sir Thomas   1605–82
English author and physician

Every man's own reason is his best Oedipus.
Religio Medici
Part I, Section 6
Elliot Stock. London, England. 1883

Bryant, William Cullen   1794–1878
American poet

I would make
Reason my guide.
Poems
Conjunction of Jupiter and Venus
D. Appleton & Company. New York, New York, USA. 1874

Burton, Sir Richard Francis   1821–90
English explorer

Reason is Life's sole arbiter, the magic Laby'rinth's single clue...
The Kasidah of Haji Abdu El-Yezdi
Part vii, Stanza xxi
Citadel Press. New York, New York, USA. 1965

Chrysostom, John   349–c.407
Archbishop of Constantinople and preacher

...there is nothing that has been created without some reason, even if human nature is incapable of knowing precisely the reason for them all.
Homilies on Genesis
7.14
Catholic University of American Press. Washington, D.C. 1986

Congreve, William   1670–1729
English dramatist

...error lives
Ere reason can be born.
The Mourning Bride
Act III, Scene I
J. Dicks. London, England. 1883

Darwin, Charles Robert   1809–82
English naturalist

It is a fatal fault to reason whilst observing, though so necessary beforehand and so useful afterwards.
The Autobiography of Charles Darwin, 1809–1882: With Original Omissions Restored
Appendix, Quotations (p. 159)
Harcourt, Brace. New York, New York, USA. 1959

Descartes, René   1596–1650
French philosopher, scientist, and mathematician

Those long chains of reasoning, simple and easy as they are, of which geometricians make use in order to arrive at the most difficult demonstrations, had caused me to imagine that all those things which fall under the cognizance of man might very likely be mutually related in the same fashion...
In *Great Books of the Western World* (Volume 31)
Discourse on the Method of Rightly Conducting the Reason
Part II (p. 47)
Encyclopædia Britannica, Inc. Chicago, Illinois, USA. 1952

I believe that all those to whom God has given the use of reason are bound to use it mainly to know Him and to know themselves. This is where I endeavored to begin my own research, and I can say that I would have been unable to find the foundation of physics had I not sought after them in this way
In William R. Shea
The Magic of Numbers and Motion: The Scientific Career of René Descartes
Chapter Eight (p. 166)
Science History Publications Canton, Massachusetts, USA. 1991

Dewey, John   1859–1952
American philosopher and educator

Reason is experimental intelligence, conceived after the pattern of science, and used in the creation of social arts; it has something to do. It liberates man from the bondage of the past, due to ignorance and accident hardened into custom. It projects a better future and assists man in its realization. And its operation is always subject to test in experience...The principles which man projects as guides...are not dogmas. They are hypotheses to be worked out in practice, and to be rejected, corrected and expanded as they fail or succeed in giving our present experience the guidance it requires. We may call them programmes of action, but since they are to be used in making our future acts less blind, more directed, they are flexible. Intelligence is not something possessed once for all. It is in constant process of forming, and its retention requires constant alertness in observing consequences, an open-minded will to learn and courage in re-adjustment.

Reconstruction in Philosophy
Chapter IV (p. 96)
Beacon Press. Boston, Massachusetts, USA. 1920

Doyle, Sir Arthur Conan   1859–1930
Scottish writer

You reasoned it out beautifully, I exclaimed in unfeigned admiration. It is so long a chain, and yet every link rings true.
In William S. Baring-Gould (ed.)
The Annotated Sherlock Holmes (Volume 1)
The Red Headed League (p. 438)
Wings Books. New York, New York, USA. 1967

Like all Holmes's reasoning the thing seemed simplicity itself when it was once explained.
In William S. Baring-Gould (ed.)
The Annotated Sherlock Holmes (Volume 2)
Stockbroker's Clerk (p. 154)
Wings Books. New York, New York, USA. 1967

I can see nothing, said I, handing it back to my friend.

On the contrary, Watson, you can see everything. You fail, however, to reason from what you see. You are too timid in drawing your inferences.
In William S. Baring-Gould (ed.)
The Annotated Sherlock Holmes (Volume 2)
The Adventure of the Blue Carbuncle (p. 453)
Wings Books. New York, New York, USA. 1967

I feel that there is reason lurking in you somewhere, so we will patiently grope round for it.
The Lost World
Chapter IV (p. 52).
The Colonial Press. Clinton, Massachusetts, USA. 1959

Ah! my dear Watson, there we come into those realms of conjecture where the most logical mind may be at fault.
In William S. Baring-Gould (ed.)
The Annotated Sherlock Holmes (Volume 2)
The Adventure of the Empty House (p. 348)
Wings Books. New York, New York, USA. 1967

Drummond, William, Sir
No biographical data available

…he who will not reason is a bigot; he who cannot is a fool; and he who dares not is a slave.
Academical Questions
Preface (p. xv)
Scholar's Facsimiles & Reprints. Delmar, New York, USA. 1984

Dubos, René Jules   1901–82
French-born American microbiologist and environmentalist

Reason dreams of an empire of knowledge, a mansion of the mind. Yet sometimes we end up living in a hotel by its side.
The Dreams of Reason: The Computer and the Rise of the Sciences of Complexity
Chapter 14 (p. 333)
Simon & Schuster. New York, New York, USA. 1988

Edgeworth, Francis Ysidro   1845–1926
Irish economist and statistician

…our reasoning appears to become more accurate as our ignorance becomes more complete; that when we have embarked upon chaos we seem to drop down into a cosmos.
The Philosophy of Chance
Mind, Volume 9, 1884 (p. 229)

Einstein, Albert   1879–1955
German-born physicist

We have thus assigned to pure reason and experience their places in a theoretical system of physics. The structure of the system is the work of reason: the empirical contents and their mutual relations must find their representation in the conclusions of the theory. In the possibility of such a representation lie the sole value and justification of the whole system, and especially of the concepts and fundamental principles which underlie it. Apart from that, these latter are free inventions of human intellect, which cannot be justified either by the nature of that intellect or in any other fashion a priori.
Ideas and Opinions
On the Methods of Theoretical Physics (p. 272)
Crown Publishers, Inc. New York, New York, USA. 1954

Eldridge, Paul   1888–1982
American educator

Reason is the shepherd trying to corral life's vast flock of wild irrationalities.
Maxims for a Modern Man
2194
T. Yoseloff. New York, New York, USA. 1965

Epictetus   ca. 55–135
Greek philosopher

Since it is Reason which shapes and regulates all other things, it not ought itself to be left in disorder.
Discourses
Chapter XVII
G. Bell & Sons. London, England. 1908

Fersman, A. E.   1883–1945
Geochemist and mineralogist

There are no bounds to fantasy, no limits to the penetration of reason, and none to the technical powers that conquer nature.
Compiled by V.V. Vorontsov
Words of the Wise: A Book of Russian Quotations
Translated by Vic Schneierson
Progress Publishers. Moscow, Russia. 1979

Feyerabend, Paul K.   1924–94
Austrian-born American philosopher of science

Copernicanism and other essential ingredients of modern science survived only because reason was frequently overruled in their past.
Against Method: Outline of an Anarchistic Theory of Knowledge
Analytical Index (p. 13)
Verso. London, England. 1978

Galilei, Galileo   1564–1642
Italian physicist and astronomer

SALVIATI: Now, since I wish to convince you by demonstrative reasoning rather than to persuade you by mere probabilities, I shall suppose that you are familiar with present-day mechanics…
In *Great Books of the Western World* (Volume 28)
Dialogues Concerning the Two New Sciences
First Day (p. 133)
Encyclopædia Britannica, Inc. Chicago, Illinois, USA. 1952

In science the authority embodied in the opinion of thousands is not worth a spark of reason in one man.
In Pedro Redondi
Galileo: Heretic
Chapter 2 (p. 37)
Princeton University Press. Princeton, New Jersey, USA.1987

Gay-Lussac, Joseph Louis   1778–1850
French chemist and physicist

The scientific glory of a country may be considered in some measure, as an indication of its innate strength. The exaltation of Reason must necessarily be connected with the exaltation of the other faculties of the mind; and there is one spirit of enterprise, vigor and conquest in science, arts, and arms.
In Maurice Crosland
Gay-Lussac: Scientist and Bourgeois
Chapter 4 (p. 80)
Cambridge University Press. Cambridge, England. 1978

Gore, George   1826–1909
English electrochemist

That which is beyond reason at present may not be so in the future; but it has now no place in science for want of a basis of verified truth.
The Art of Scientific Discovery
Chapter III (p. 25)
Longmans, Green & Company. London, England. 1878

Grew, Nehemiah   1641–1712
Scientific writer and journalist

He that speaketh Reason may be rather satisfied in being understood, than believed.
The Anatomy of Vegetables Begun
Preface
Printed for Spencer Hickman. London, England. 1672

Hubbard, Elbert   1856–1915
American editor, publisher, and author

REASON: The arithmetic of the emotions.
The Roycroft Dictionary Concocted by Ali Baba and the Bunch on Rainy Days (p. 126)
The Roycrofters. East Aurora, New York, USA. 1914

Hutton, James   1726–97
Scottish geologist, chemist, and naturalist

…to reason without data is nothing but delusion.
The Theory of the Earth (Volume 1)
Part I, Chapter III (p. 281)
Messrs. Cadwell, Junior & Davies. London, England. 1795

John of Salisbury   ca. 1115–80
English author and diplomatist

Reason, therefore, is a mirror in which all things are seen…
In John van Laarhoven (ed.)
Entheticus Maior and Minor (Volume 1)
Part II, Section I, Notes from Epicuris, l. 657
E.J. Brill. Leidn, Netherlands. 1987

Johnson, Samuel   1696–1772
English critic, biographer, and essayist

We may take fancy for a companion, but must follow Reason as our guide.
The Life of Samuel Johnson (Volume 1)
Letter to Boswell 1774 (p. 474)
Sir Isaac Pitman & Sons, Ltd. London, England. 1907

Memory is the purveyor of reason, the power which places those images before the mind upon which the judgment is to be exercised, and which treasures up the determinations that are once passed, as the rules of future action, or grounds of subsequent conclusions.
The Rambler (Volume 1)
No. 41, August 7, 1750 (p. 296)
Edward Earle. Philadelphia, Pennsylvania, USA. 1812

Joos, Georg   1894–1959
German physicist

As soon as we inquire into the reasons for the phenomena, we enter the domain of theory, which connects the observed phenomena and traces them back to a single "pure" phenomena, thus bringing about a logical arrangement of an enormous amount of observational material.
Theoretical Physics
Introduction (p. 1)
Blackie & Son Ltd. London, England. 1968

Kant, Immanuel   1724–1804
German philosopher

Mathematics and physics are the two theoretical sciences of reason, which have to determine their objects a priori.
In *Great Books of the Western World* (Volume 42)
Critique of Pure Reason
Preface to the Second Edition (p. 5)
Encyclopædia Britannica, Inc. Chicago, Illinois, USA. 1952

Human reason has this peculiar fate that in one species of its knowledge it is burdened by questions which, as prescribed by the very nature of reason itself, it is not able to ignore, but which, as transcending all its powers, it is also not able to answer.
In *Great Books of the Western World* (Volume 42)
Critique of Pure Reason
Preface to First Edition (p. 1)
Encyclopædia Britannica, Inc. Chicago, Illinois, USA. 1952

Kasner, Edward   1878–1955
American mathematician
Newman, James Roy   1911–66
Mathematician and mathematical historian

One of the difficulties arising out of the subjective view of probability results from the principle of insufficient reasons. This principle…holds that if we are wholly ignorant of the different ways an event can occur and therefore have no reasonable ground for preference, it is as likely to occur one way as another.
Mathematics and the Imagination
Chance and Chanceability (p. 229)
Simon & Schuster. New York, New York, USA. 1940

Lamarck, Jean-Baptiste Pierre Antoine
1744–1829
French biologist

Reason is not a faculty; still less is it a torch or entity of any kind; but it is a special condition of the individual's intellectual faculties; a condition that is altered by experience, gradually improves and controls the judgments, according as the individual exercises his intellect.
Translated by Hugh Elliot
Zoological Philosophy: An Exposition with Regard to the Natural History of Animals
Chapter VIII (p. 401)
The University of Chicago Press. Chicago, Illinois, USA. 1984

Lambert, Johann Heinrich   1728–77
Swiss-German mathematician and astronomer

Reason hates the false décor by which advocates of error try to make her more impressive. She recognizes that one should doubt where reasons are not sufficient, that one should present each reason stripped of all that makes it apparent, and then she will assist if proper tools are on hand, and reserve for herself to pronounce the sentence, or if she suspends it, to indicate what is required to proceed to the conclusion.
Translated by Stanley Jaki
Cosmological Letters on the Arrangement of the World-Edifice
Twentieth Letter (p. 186)
Science History Publications. New York, New York, USA. 1976

Medawar, Sir Peter Brian   1915–87
Brazilian-born English zoologist

…scientific reasoning is a kind of dialogue between the possible and the actual, what might be and what is in fact the case…
Induction and Intuition in Scientific Thought
Chapter III, Section 1 (p. 48)
American Philosophical Society. Philadelphia, Pennsylvania, USA. 1969

Miller, Hugh   1802–56
Scottish geologist and theologian

In the geologic, as in other departments,
What can we reason but from what we know?
The Old Red Sandstone
Geological Evidences in Favour of Revealed Religion (p. 280)
J.M. Dent & Sons Ltd. London, England. 1922

Minnick, Wayne C.   1915–2006
Professor of communications

This kind of reasoning has weaknesses, of course, as do all forms of reasoning. If the correspondence between two things compared is not complete, that is, if significant differences can be shown to exist, then the argument collapses.
The Art of Persuasion (p. 16)
Houghton Mifflin Company. New York, New York, USA. 1968

Moulton, Forest Ray   1872–1952
American astronomer

…reason and the laws of nature…have become a sort of intellectual telescope, as it were, with which modern science looks back across the geological ages and discerns, at least in outline, the chief steps of the evolution of the inanimate and of the organic world; and, similarly, penetrates the future to a time when this earth will cease to be suited for the abode of life.
In H.H. Newman (ed.)
The Nature of the World and of Man
Astronomy (p. 2)
The University of Chicago Press. Chicago, Illinois, USA. 1927

Newton, Sir Isaac   1642–1727
English physicist and mathematician

A Vulgar Mechanick can practice what he has been taught or seen done, but it he is an error he knows not how to find it out and correct it, and if you put him out of his road, he is at a stand; Whereas he that is able to reason nimbly and judiciously about figure, force and motion is never at rest till he gets over every rub.
In Richard S. Westfall
Never at Rest: A Biography of Isaac Newton
Letter to Nathaniel Hawes, 25 May 1694 (p. ii)
Cambridge University Press. Cambridge, England. 1980

Osler, Sir William   1849–1919
Canadian physician and professor of medicine

With reason, science never parts company, but with feeling, emotion, passion, what has she to do? They are

not of her; they owe her no allegiance. She may study, analyze, and define, she can never control them, and by no possibility can their ways be justified to her.
Aequanimitas, with Other Addresses to Medical Students, Nurses, and Practitioners of Medicine
The Leaven of Science (p. 93)
The Blakiston Company. Philadelphia, Pennsylvania, USA. 1932

Parkington, J. E.
No biographical data available

The ability to sort out stone implements into "types" as one would playing-cards into suits is of minor importance compared to the reasons underlying the tendency for implements to cluster into "ideal forms." It is not the group "handaxes" which is important but, as Plato might have said, "handaxeness."
Stone Implements as Information
The Interpretation of Archaeological Evidence, Goodwin Series, Number 1, June 1972 (p. 12)

Pascal, Blaise   1623–62
French mathematician and physicist

The last proceeding of reason is to recognise that there is an infinity of things which are beyond it.
In *Great Books of the Western World* (Volume 33)
Pensées
Section IV, 267
Encyclopædia Britannica, Inc. Chicago, Illinois, USA. 1952

Peirce, Charles Sanders   1839–1914
American scientist, logician, and philosopher

Every work of science great enough to be remembered for a few generations affords some exemplification of the defective state of the art of reasoning of the time when it was written; and each chief step in science has been a lesson in logic.
Inquiry and Belief
The Popular Science Monthly, Volume 12, 1877–1878

Pope, Alexander   1688–1744
English poet

What can we reason but from what we know?
The Complete Poetical Works (Volume 2)
An Essay on Man
Epistle I, l. 18
Houghton Mifflin Company. New York, New York, USA. 1903

Price, Bartholomew   1818–98
English mathematician and educator

…the reasoning process [employed in mathematics] is not different from that of any other branch of knowledge…but there is required, and in a great degree, that attention of mind which is in some part necessary for the acquisition of all knowledge, and in this branch [it] is indispensably necessary. This must be given in its fullest intensity.… [T]he other elements especially characteristic

of a mathematical mind are quickness in perceiving logical sequence, love of order, methodical arrangement and harmony, distinctness of conception.
Treatise on Infinitesimal Calculus (Volume 3) (p. 6)
At The Clarendon Press. Oxford, England. 1868

Recorde, Robert   1510?–58
English mathematician and writer

You are to farre deceived, and therefore I interrupt your woordes, for all things are to bee governed by reason.
The Castle of Knowledge
The Fourth Treatise (p. 243)
Imprinted by R. Wolfe. London, England. 1556

If reasons reache transcende the skye,
Why shoulde it then to earthe be bounde?
The witte is wronged and leadde awrye,
If mynde be maried to the grounde.
The Castle of Knowledge
The Preface
Imprinted by R. Wolfe. London, England. 1556

…who so ever will travail in the sciences with profit, must lean rather to reason, than to authority, else he may be deceaved.
The Castle of Knowledge
The Fourth Treatise (p. 182)
Imprinted by R. Wolfe. London, England. 1556

Romanoff, Alexis Lawrence   1892–1980
Russian soldier and scientist

Reasoning goes beyond the analysis of facts.
Encyclopedia of Thoughts
Aphorisms 1973
Ithaca Heritage Books. Ithaca, New York, USA. 1975

Russell, Bertrand Arthur William   1872–1970
English philosopher, logician, and social reformer

Reason is a harmonizing, controlling force rather than a creative one.
Our Knowledge of the External World
Lecture I (p. 21)
The Open Court Publishing Company. Chicago, Illinois. 1914

Supposing you got a crate of oranges that you opened, and you found all the top layer of oranges bad, you would not argue, "The underneath ones must be good, so as to redress the balance"; You would say, "Probably the whole lot is a bad consignment"; and that is really what a scientific person would say about the universe.
Why I Am Not a Christian: And Other Essays on Religion and Related Subjects
Why I Am Not A Christian (p. 13)
Watts. London, England. 1927

Shakespeare, William   1564–1616
English poet, playwright, and actor

FOOL: "The reason why the seven stars are no more than seven is a pretty reason."

LEAR: "Because they are not eight?"
FOOL: "Yes, indeed. Thou wouldst make a good fool."
In *Great Books of the Western World* (Volume 27)
The Plays and Sonnets of William Shakespeare (Volume 2)
King Lear
Act I, Scene v, l. 38–40
Encyclopædia Britannica, Inc. Chicago, Illinois, USA. 1952

Good reason must, of force, give place to better.
In *Great Books of the Western World* (Volume 26)
The Plays and Sonnets of William Shakespeare (Volume 1)
Julius Caesar
Act IV, Scene iii, l. s03
Encyclopædia Britannica, Inc. Chicago, Illinois, USA. 1952

His reasons are as two grains of wheat hid in two bushels
of chaff: you shall seek all day ere you find them, and
when you have them, they are not worth the search.
In *Great Books of the Western World* (Volume 26)
The Plays and Sonnets of William Shakespeare (Volume 1)
The Merchant of Venice
Act I, Scene i, l. 115
Encyclopædia Britannica, Inc. Chicago, Illinois, USA. 1952

Spencer-Brown, George   1923–
English mathematician and polymath

The concept of randomness arises partly from games of
chance. The word "chance" derives from the Latin *cadentia*
signifying the fall of a die. The word "random" itself comes
from the French *randir* meaning to run fast or gallop.
Probability and Scientific Inference
Chapter VII (p. 35)
Longmans, Green & Company. London, England. 1957

von Goethe, Johann Wolfgang   1749–1832
German poet, novelist, playwright, and natural philosopher

Reason is applied to what is developing, practical understanding to what is developed. The former does not ask,
What is the Purpose? and the latter does not ask, What is
the source? Reason takes pleasure in development; practical understanding tries to hold things fast so that it can
use them.
Scientific Studies (Volume 12)
Chapter VIII (p. 308)
Suhrkamp. New York, New York, USA. 1988

The texture of this world is made up of necessity and
chance. Human reason holds the balance between them,
treating necessity as the basis of existence, but manipu-
lating and directing chance, and using it.
In Eric A. Blackall (ed.)
Wilhelm Meister's Apprenticeship
Book One, Chapter Seventeen (p. 38)
Princeton University Press. Princeton, New Jersey, USA. 1995

von Helmholtz, Hermann   1821–94
German scientist and philosopher

The iron labor of conscious logical reasoning demands
great perseverance and great caution; it moves on but

slowly, and is rarely illuminated by brilliant flashes of
genius. It knows little of that facility with which the most
varied instances come thronging into the memory of the
philologist or historian. Rather is it an essential condition of the methodical progress of mathematical reasoning that the mind should remain concentrated on a single
point, undisturbed alike by collateral ideas on the one
hand, and by wishes and hopes on the other, and moving
on steadily in the direction it has deliberately chosen.
Vorträge and Reden
Ueber das Verhaltniss der Naturwissenschaften zur Gesammtheit der
Wissenschaft, Bd. 1, 1896 (p. 178)
English novelist, historian, and sociologist
Friedrich Viewig & Sohn. Brunswick, Germany. 1896

Wells, H. G. (Herbert George)   1866–1946
English novelist, historian, and sociologist

"It's against reason?" said Filby.

"What reason?" said the Time Traveler.
In Robert M. Hutchins and Mortimer J. Adler (eds.)
The Great Ideas Today, 1971
The Time Machine
Chapter One (p. 451)
Encyclopædia Britannica, Inc. Chicago, Illinois, USA. 1971

Whitehead, Alfred North   1861–1947
English mathematician and philosopher

The art of reasoning consists in getting hold of the subject at the right end, of seizing on the few general ideas
that illuminate the whole, and of persistently organizing
all subsidiary facts round them. Nobody can be a good
reasoner unless by constant practice he has realized the
importance of getting hold of the big ideas and hanging
on to them like grim death.
In W.W. Sawyer
Prelude to Mathematics
Presidential Address to the London Branch of the Mathematical Association, 1914 (p. 183)
Penguin Books Limited. London, England. 1960

Wigner, Eugene Paul   1902–95
Hungarian-born American physicist

The great mathematician fully, almost ruthlessly, exploits
the domain of permissible reasoning and skirts the imper-
missible. That his recklessness does not lead him into a
morass of contradiction is a miracle in itself. Certainly it
is hard to believe that our reasoning power was brought,
by Darwin's process of natural selection, to the perfection
which it seems to possess.
The Unreasonable Effectiveness of Mathematics in Natural Science
Communications on Pure and Applied Mathematics, Volume 13, Number 1, 1960 (p. 3)

Wright, Frances   1795–1852
Scottish-born American reformer

The best road to correct reasoning is by physical science; the way to trace effects to causes is through physical science; the only corrective, therefore, of superstition is physical science.
Course of Popular Lectures
Lecture 3
Published by the author. Philadelphia, Pennsylvania, USA. 1836

RECOGNITION

Oppenheimer, J. Robert   1904–67
American theoretical physicist

Whatever the individual motivation and belief of the scientist, without the recognition from his fellow men of the value of his work, in the long term science will perish.
Atomic Weapons
Proceeding of the American Philosophical Society, Volume 90, Number 1, 1946

RECORD

Coues, E.
No biographical data available

Don't trust your memory; it will trip you up; what is clear now will grow obscure; what is found will be lost. Write down everything while it is fresh in your mind; write it out in full. Time so spent will be time saved in the end, when you offer your researches to the discriminating public.
Field Ornithology (pp. 44–45)
Naturalist's Agency. Salem, Massachusetts, USA. 1874

Dickens, Charles   1812–70
English novelist

When found, make a note of.
The Works of Charles Dickens
Dombey and Son (Part I)
Chapter XV (p. 217)
P.F. Collier & Son. New York, New York, USA. 1911

Grove, Sir William   1811–96
English chemist

It would be vain to attempt specifically to predict what may be the effect of Photography on future generations. A Process by which the most transient actions are rendered permanent, by which facts write their own annals in a language that can never be obsolete, forming documents which prove themselves, — must interweave itself not only with science but with history and legislature.
Lecture
Progress of Physical Science since the opening of the London Institution, (19 January 1842)

RECOVERY

Byron, George Gordon, 6th Baron Byron   1788–1824
English Romantic poet and satirist

Despair of all recovery spoils longevity,
And makes men's miseries of alarming brevity.
The Complete Poetical Works of Byron
Don Juan
Canto II, Stanza LXIV
Houghton Mifflin Company. Boston, Massachusetts, USA. 1933

Massinger, Philip   1583–1640
English dramatic poet

O my Doctor,
I never shall recover.
The Bondman
Act I, Scene I
Printed for A. Bettesworth. London, England. 1719

Petrarch (Francesco Petrarca)   1304–74
Italian poet and humanist

I once heard a physician of high standing in his profession say: …If a hundred men, or a thousand of the same age and general constitution and accustomed to the same diet, should all fall victim to a disease at the same time, and if half of them should follow the prescriptions of our contemporary doctors, and if the other half should be guided by their natural instinct and common sense, with no doctors at all, I have no doubt that the latter group would do better.
In M. Bishop
Letters from Petrarch
Book V, 3 (p. 250)
Indiana University Press. Bloomington, Indiana, USA. 1966

RECTANGLE

Frere, John Hookham   1769–1846
British diplomat and man of letters

Alas! that partial Science should approve
The sly RECTANGLE'S too licentious love!
In Charles Edmonds
Poetry of the Anti-Jacobin
The Loves of the Triangle, Canto II, l. 75–76
Printed for J. Wright, by W. Bulmer & Company. London, England. 1801

RECURSION

Papert, Seymour   1928–
South African mathematician

Of all ideas I have introduced to children, recursion stands out as the one idea that is particularly able to evoke an excited response.
Mindstorms: Children, Computers and Powerful Ideas

Chapter 3 (p. 71)
Basic Books, Inc. New York, New York, USA. 1980

Young, Louise B.
Science writer

Whatever can be done once can always be repeated.
The Mystery of Matter
Introduction (p. 15)
Oxford University Press, Inc. New York, New York, USA. 1965

RED SHIFT

Boas, Jr., Ralph P.   1912–92

Consider the Pitiful Plight
Of a runner who wasn't too bright,
But sprinted so fast
He vanished at last
By red-shifting himself out of sight.
Reprinted in Ralph P. Boas, Jr.
Lion Hunting & Other Mathematical Pursuits (p. 103)
Mathematical Association of America. Washington, D.C. 1995

Gamow, George   1904–68
Russian-born American physicist

The discovery of the red shift in the spectra of distant stellar galaxies revealed the important fact that our universe is in the state of uniform expansion, and raised an interesting question as to whether the present features of the universe could be understood as the result of its evolutionary development.... We conclude first of all that the relative abundance of various atomic species (which were found to be essentially the same all over the observed region of the universe) must represent the most ancient archaeological document pertaining to the history of the universe.
The Evolution of the Universe
Nature, Volume 162, Number 4122, October 1948 (p. 680)

Gray, George W.
Free lance science writer

...just as the shifting of bookkeeping accounts into the red measures disintegrating, scattering, dissipating financial resources, so the shifting of starlight into the red indicates disintegrating, scattering, dissipating physical resources. It says that the universe is running down.... To entertain this preposterous idea of all these massive star systems racing outward was to accept a radically new picture of the cosmos — a universe in expansion, a vast bubble blowing, distending, scattering, thinning out into gossamer, losing itself. The snug, tight, stable world of Einstein had room for no such flights.
Universe in the Red
The Atlantic Monthly, Volume 1151, Number 2, February 1933
(p. 233, 236)

Jeans, Sir James Hopwood   1877–1946
English physicist and mathematician

Another possibility...is that the universe retains its size, while we and all material bodies shrink uniformly. The red shift we observe in the spectra of the nebulae is then due to the fact that the atoms which emitted the light millions of years ago were larger than the present-day atoms with which we measured the light — the shift is, of course, proportional to distance.
Contributions to a British Association Discussion on the Evolution of the Universe
Supplement to Nature, Volume 128, Number 3234 November 1931 (pp. 703–704)

Stapledon, Olaf   1886–1950
English author

I noticed that the sun and all the stars in his neighborhood were ruddy. Those at the opposite pole of the heaven were of an icy blue. The explanation of this strange phenomenon flashed upon me. I was still traveling, and traveling so fast that light itself was not wholly indifferent to my passage. The overtaking undulations took long to catch me. They therefore affected me as slower pulsations than they normally were, and I saw them therefore as red. Those that met me on my headlong flight were congested and shortened, and were seen as blue.
Last and First Men and Star Maker
Star Maker, Chapter III (p. 262)
Dover Publications, Inc. New York, New York, USA. 1968

REDUCTIONISM

Commoner, Barry   1917–
American biologist, ecologist, and educator

There is, indeed, a specific fault in our system of science, and in the resultant understanding of the natural world.... This fault is reductionism, the view that effective understanding of a complex system can be achieved by investigating the properties of its isolated parts. The reductionist methodology, which is so characteristic of much of modern research, is not an effective means of analyzing the vast natural systems that are threatened by degradation.
The Closing Circle: Nature, Man and Technology
Chapter 10 (p. 189)
Alfred A. Knopf. New York, New York, USA. 1971

d'Abro, Abraham
No biographical data available

...in spite of its achievements, thermodynamics suffers from the limitations common to all phenomenological theories. Because it restricts its attention to the macroscopic properties of bodies, it fails to anticipate many phenomena which find their interpretation in the interplay

of underlying microscopic processes, and which have since been clarified by the more speculative theories of the hidden-occurrence type.
The Rise of the New Physics (Volume 1)
Chapter XXI (pp. 371–372)
Dover Publications, Inc. New York, New York, USA. 1951

Dyson, Freeman J.   1923–
American physicist and educator

My message is that science is a human activity, and the best way to understand it is to understand the individual human beings who practice it. Science is an art form and not a philosophical method. The great advances in science usually result from new tools rather than from new doctrines. If we try to squeeze science into a single philosophical viewpoint such as reductionism, we are like Procrustes chopping off the feet of his guests when they do not fit onto his bed.
The Scientist as Rebel
New York Times Book Review, May 25, 1995

Eiseley, Loren C.   1907–77 American anthropologist, educator, and author

In the end, science as we know it has two basic types of practitioners. One is the educated man who still has a controlled sense of wonder before the universal mystery, whether it hides in a snail's eye or within the light that impinges on that delicate organ. The second kind of observer is the extreme reductionist who is so busy stripping things apart that the tremendous mystery has been reduced to a trifle, to intangibles not worth troubling one's head about.
The Star Thrower
Science and the Sense of the Holy (p. 190)
Times Books. New York, New York, USA. 1978

REFEREE

Broad, William   1951–
Science writer
Wade, Nicholas
British-born scientific writer

The ultimate gatekeeper of science is neither peer reviews, nor referees, nor replication, nor the universalism implicit in all three mechanisms. It is time. In the end, bad theories don't work, fraudulent ideas don't explain the world so well as true ideas do. The ideal mechanisms by which science should work are applied to a large extent in retrospect .... Time and the invisible boot that kicks out all useless science are the true gatekeepers of science. But these inexora-ble mechanisms take years, sometimes more than a millennium, to operate. During the interval, fraud may flourish, particularly if it can find shelter under the mantle of immunity that scientific elitism confers.
Betrayers of the Truth (p. 106)
Simon & Schuster. New York, New York, USA. 1982

Magueijo, Joao   1967–
Theoretical physicist and cosmologist

Peer review is an unpaid and usually anonymous activity. Perhaps for this reason the average referee report is sloppy and sleazy. Reports usually reveal that the referee has not read the paper. Acceptance or rejection often reflects the personal relationship between authors and referee. Publishers have always been reluctant to open their files to historians of science and sociologists. Clearly they are embarrassed to reveal how little science, and how much sociology, there is in their files.
Electronic Archives and the Death of Journals
http://theory.ic.ac.uk/Ðmagueijo/com.pdf

Zoman, John M.
No biographical data available

The referee is the lynchpin about which the whole business of Science is pivoted.
Public Knowledge: An Essay Concerning the Social Dimension of Science
Chapter 6 (p. 111)
At The University Press. Cambridge, England. 1968

REFORM

Hilbert, David   1862–1943
German mathematician

We have reformed mathematics, the next thing is to reform physics, and then we'll go on to chemistry.
Hilbert — Courant
Hilbert
Chapter XVI (p. 129)
Springer-Verlag. New York, New York, USA. 1986

REGRESSION

Cardozo, Benjamin N.   1870–1938
American jurist

Where the line is to be drawn the important and the trivial cannot be settled by a formula.
Jacob & Youngs v. Kent, 230
New York Reports 239, 243, 1921

Fiedler, Edgar R.   1916–2003
American economist

Most economists think of God as working great multiple regressions in the sky.
The Three R's of Economic Forecasting — Irrational, Irrelevant and Irreverent
Across the Board, June 1977

Juster, Norton   1929–
American architect and author

Once upon a time, there was a sensible straight line who was hopelessly in love with a dot.
The Dot and the Line: A Romance in Lower Mathematics
The Film (1965)

RELATION

Buchanan, Scott   1895–1968
American educator and philosopher

Science is an allegory that asserts that the relations between the parts of reality are similar to the relations between terms of discourse.
Poetry and Mathematics
Chapter 5 (pp. 96–97)
The University of Chicago Press. Chicago, Illinois, USA. 1975

Dingle, Herbert   1890–1978
English astrophysicist

…if, as we must surely do, we wish to characterize science by the elements in it that persist and grow, and not by that which continually changes, we must recognize…the progressive discovery of relations between the various constituents of our experience…. Amid all the changes of theories and pictures and conceptions, the relations remain and steadily accumulate. Franklin found that lightning was a manifestation of the electric ether revealed in laboratory experiments. The electric ether has disappeared, and other theories of electricity have in turn succeeded it and disappeared also, but the relation between lightning and laboratory sparks remains. Maxwell established a relation between light and electromagnetic oscillations. His ether also has gone, but the relation stays. All permanent advances in science are discoveries of relations between phenomena, and the factor in science that shows a steady uninterrupted growth is the extent of the field of related observations.
The Scientific Adventure: Essays in the Hisory and Philosophy of Science
Chapter One (p. 40)
Pitman. London, England. 1952

Durant, William James   1885–1981
American historian and essayist

Science tells us how to heal and how to kill; it reduces the death rate in retail and then kills us wholesale in war; but only wisdom — desire coordinated in the light of all experience — can tell us when to heal and when to kill. For a fact is nothing except in relation to a purpose and a whole.
The Story of Philosophy
Introduction (p. 2)
Simon & Schuster. New York, New York, USA. 1953

Fiske, John   1842–1901
American philosopher and historian

The ability to imagine relations is one of the most indispensable conditions of all precise thinking. No subject can be named in the investigation of which it is not imperatively needed; but it can be nowhere else so thoroughly acquired as in the study of mathematics.
Darwinism and Other Essays (p. 296)
Houghton Mifflin Company. Boston, Massachusetts, USA. 1993

Keyser, Cassius Jackson   1862–1947
American mathematician

To be is to be related.
Mole Philosophy and Other Essays
Chapter XVII (p. 94)
E.P. Dutton & Company, Inc. New York, New York, USA. 1927

Schukarev, A. N.
No biographical data available

At present one can consider it universally acknowledged that among the phenomena of inanimate nature there is no arbitrary will; here the unshakable connection between phenomena rule with complete authority — relations which we call laws. In the invariance of these relations we are even inclined to see the characteristic sign which differentiates the inanimate from the living.
In Michael D. Gordin
A Well-Ordered Thing: Dmitrii Mendeleev and the Shadow of the Periodic Table
Chapter 2 (p. 15)
Basic Books, Inc. New York, New York, USA. 2004

RELATIVITY

Asimov, Isaac   1920–92
American author and biochemist

No physicist who is even marginally sane doubts the validity of special relativity.
In Timothy Ferris (ed.)
The World Treasury of Physics, Astronomy and Mathematics
The Two Masses (p. 186)
Little Brown & Company. Boston, Massachusetts, USA. 1991

Special relativity is so much a part not only of physics but of everyday life, that it is no longer appropriate to view it as the special "theory" of relativity. It is a fact…
Was Einstein Right? (p. 246)
Basic Books, Inc. New York, New York, USA.

Davies, Paul Charles William   1946–
British-born physicist, writer, and broadcaster
Gribbin, John   1946–
English science writer and astronomer

All the implications of special relativity…have been confirmed by direct experiments. There are still people who believe it is "just a theory." But they are wrong.
The Matter Myth: Dramatic Discoveries That Challenge Our

Understanding of Physical Reality (p. 85)
Simon & Shuster. New York, New York, USA. 1992

Durell, Clement V.   1882–1968
English mathematician

Relativity without mathematics may be compared with "Painless Dentistry," or "Skiing without Falling," or "Reading without Tears."
Readable Relativity
Preface (p. ix)
Harper & Brothers. New York, New York, USA. 1960

Eddington, Sir Arthur Stanley   1882–1944
English astronomer, physicist, and mathematician

Results of measurements are the subject-matter of physics; and the moral of the theory of relativity is that we can only comprehend what the physical quantities stand for if we first comprehend what they are.
The Mathematical Theory of Relativity
Conclusion (p. 240)
At The University Press. Cambridge, England. 1930

We walk the stage of life, performers of a drama for the benefit of the cosmic spectator. As the scene proceeds, he notices that the actors are growing smaller and the action quicker. When the last act opens the curtain rises on midgets rushing through their parts at frantic speed.
The Expanding Universe
Chapter III, Section VI (p. 91–92)
The University Press. Cambridge. 1933

Einstein, Albert   1879–1955
German-born physicist

There is something attractive in presenting the evolution of a sequence of ideas in as brief a form as possible, and yet with a completeness sufficient to preserve throughout the continuity of development. We shall endeavor to do this for the Theory of Relativity, and to show that the whole ascent is composed of small, almost self-evident steps of thought.
A Brief Outline of the Development of the Theory of Relativity
Nature, Volume 106, Number 2677, 17 February 1921 (p. 782)

When you are courting a nice girl an hour seems like a second. When you sit on a red-hot cinder a second seems like an hour. That's relativity.
News Chronicle, 14 March 1949

Ought we to smile at the man and say that he errs in his conclusion? I do not believe we ought to if we wish to remain consistent; we must rather admit that his mode of grasping the situation violates neither reason nor known mechanical laws. Even though it is being accelerated with respect to the "Galilean space" first considered, we can nevertheless regard the chest as being at rest. We have thus good grounds for extending the principle of relativity to include bodies of reference which are accelerated with respect to each other, and as a result we have gained a powerful argument for a generalised postulate of relativity.
Translated by Robert W. Lawson
Relativity: The Special and General Theory
Part II, Chapter 20 (p. 88)
Pi Press. New York, New York, USA. 2005

I sometimes ask myself how it came about that I was the one to develop the theory of relativity. The reason, I think, is that a normal adult never stops to think about problems of space and time. These are things which he has thought of as a child. But my intellectual development was retarded, as a result of which I began to wonder about space and time only when I had already grown up.
In John D. Barrow
Theories of Everything: The Quest for Ultimate Explanation
Chapter Three (p. 68)
The Clarendon Press. Oxford. London. 1991

The meaning of relativity…has been widely misunderstood. Philosophers play with the word, like a child with a doll. Relativity, as I see it, merely denotes that certain physical and mechanical facts, which have been regarded as positive and permanent, are relative with regard to certain other facts in the sphere of physics and mechanics. It does not mean that everything in life is mischievously topsy-turvy.
What Life Means to Einstein: An Interview by George Sylvester Viereck
The Saturday Evening Post, October 26, 1929 (p. 17)

Greene, Brian   1963–
American physicist

…general relativity and quantum mechanics, when combined, begin to shake, rattle, and gush with steam like a red-lined automobile.
The Elegant Universe
Chapter 1 (p. 4)
W.W. Norton & Company, Inc. New York, New York, USA.2003

Haldane, R. B.   1856–1928
British liberal and labor politician

It is only a world embodying the principle of relativity, in the form which the doctrine entails, that can be said to exhibit the character of mind, with its exclusion of disconnected fragments and relations.
The Reign of Relativity (p. 138)
Yale University Press. New Haven, Connecticut, USA. 1921

Harrison, B.
No biographical data available
Thorne, Kip S.   1940–
American theoretical physicist

If one intends to abandon Relativity, here is the place [black holes] to do so. Otherwise one is on the way to a new world of physics, both classical and quantum. Here we go!

In Jean-Pierre Luminet
Black Holes (p. 117)
Cambridge University Press. New York, New York, USA. 1992

Holton, Gerald 1922–
Research professor of physics and science history

The cliché became, erroneously, "everything is relative"; whereas the point is that out of the vast flux one can distill the very opposite: "some things are invariant."
Einstein, History, and Other Passions: The Rebellion Against Science at the End of the Twentieth Century
Part 1, Chapter 6 (p. 131)
Addison-Wesley Publishing Company. Reading, Massachusetts, USA. 1996

Relativity theory, of course, does not find that truth depends on the point of view of the observer but, on the contrary, reformulates the laws of physics so that they hold good for all observers, no matter how they move or where they stand. Its central meaning is that the most valued truths in science are independent of the point of view.... Einstein did not prove the work of Newton wrong; he provided a larger setting within which some limitations, contradictions, and asymmetries in the earlier physics disappeared.
Einstein, History, and Other Passions: The Rebellion Against Science at the End of the Twentieth Century
Part 1, Chapter 2 (p. 48)
Addison-Wesley Publishing Company. Reading, Massachusetts, USA. 1996

Krauss, Lawrence M. 1954–
American theoretical physicist

Einstein was thus faced with the following apparent problem. Either give up the principle of relativity, which appears to make physics possible by saying that the laws of physics are independent of where you measure them, as long as you are in a state of uniform motion; or give up Maxwell's beautiful theory of electromagnetism and electromagnetic waves. In a truly revolutionary move, he chose to give up neither.... It is a testimony to his boldness and creativity not that he chose to throw out existing laws that clearly worked, but rather that he found a creative way to live within their framework. So creative, in fact, that it sounds nuts.
Fear of Physics: A Guide for the Perplexed
Chapter 3 (p. 78)
Basic Books, Inc. New York, New York, USA. 1993

Indeed, long before the Star Trek writers conjured up warp fields, Einstein warped spacetime, and, like the Star Trek writers, he was armed with nothing other than his imagination. Instead of imagining twenty-second-century starship technology, however, Einstein imagined an elevator. He was undoubtedly a great physicist, but he probably never would have sold a screenplay.
The Physics of Star Trek

Chapter Three (p. 31)
Harp Perennial Publishers. New York, New York, USA. 1995

L. L. Cool J. 1968–
American hip hop artist and actor

Grab hold of a hot pan and a second can seem like an hour. Put your hands on a hot woman and an hour can seem like a second.
Deep Blue Sea
Film (1999)

Lindley, David 1956–
English astrophysicist and author

Relativity removes from physics the authoritarian rule of classical physics, with its absolute space and time, and replaces it not with anarchy, in which all participants have their own rules, but with perfect democracy, in which the same rules govern all.... It may seem unsatisfactory to respond that there can be only one correct theory of space and time, and that Einstein's happens to be it, but for the physicist such an answer has to suffice. Relativity, like other physical theories, is a set of rules based on a number of crucial assumptions, and experiment and observation bear it out. That is all we ever ask of physical theories, and to ask for some further statement of why the special theory of relativity supplanted absolute Newtonian spacetime is to search for a truth beyond the domain of science.
The End of Physics: The Myth of a Unified Theory
Part I, Chapter 2 (p. 60, 61)
Basic Books, Inc. New York, New York, USA. 1993

Lindon, J. A.
English writer of comic verse

When they questioned her, answered Miss Bright,
"I was there when I got home that night;
So I slept with myself,
Like two shoes on a shelf,
Put-up relatives shouldn't be tight!"
In Martin Gardner
Time Travel and Other Mathematical Bewilderments
Chapter One (p. 9)
W.H. Freeman & Company. New York, New York, USA. 1988

Mach, Ernst 1838–1916
Austrian physicist and philosopher

I can accept the theory of relativity as little as I can accept the existence of atoms and other such dogmas.
In Stephen Pile
The Book of Heroic Failures
Routledge & Kegan Paul. London, England. 1979

Nabokov, Vladimir 1899–1977
Russian-American writer

At this point, I suspect, I should say something about my attitude to "Relativity." It is not sympathetic. What many

cosmogonists tend to accept as an objective truth is really the flaw inherent in mathematics which parades as truth.
Ada or Ardor: A Family Chronicle
Part Four (p. 543)
McGraw-Hill Book Company, Inc. New York, New York, USA. 1969

Oppenheimer, J. Robert   1904–67
American theoretical physicist

General relativity has very few connections with any other part of physics and, as I said, is something that we might just now be beginning to discover…
The Flying Trapeze: Three Crises for Physics
Space and Time (p. 23)
Oxford University Press, Inc. London, England. 1964

Page, Leigh   1884–1952
American physicist

The rotating armatures of every generator and motor in this age of electricity are steadily proclaiming the truth of the relativity theory to all who have ears to hear.
Filler material
American Journal of Physics, Volume 43, Number 4, April 1975 (p. 330)

Rindler, Wolfgang   1952–
German writer

Relativity has taught us to be wary of time.
Essential Relativity (p. 203)
Van Nostrand Company, Inc. New York, New York, USA. 1969

Rogers, Eric
No biographical data available

Since Relativity is a piece of mathematics, popular accounts that try to explain it without mathematics are almost certain to fail.
Physics for the Inquiring Mind
Chapter 31 (p. 472)
Princeton University Press. Princeton, New Jersey, USA. 1960

Rothman, Tony   1953–
American cosmologist

Relativity does not mean everything is relative. And the brilliance of Einstein's discoveries is so great that no amount of journalistic overkill has managed to dim it. Einstein and Bach are the only two people who deserve their reputations.
Instant Physics: From Aristotle to Einstein, and Beyond
Chapter 5 (p. 115)
Ballantine Books. New York, New York, USA. 1995

Russell, Bertrand Arthur William   1872–1970
English philosopher, logician, and social reformer

Einstein's theory of relativity is probably the greatest synthetic achievement of the human intellect up to the present time.
N.Y. Times, April 19, 1955

Sciama, Dennis   1926–99
English physicist

General relativity contains within itself the seeds of its own destruction.
In John D. Barrow
The World Within the World (p. 306)
Clarendon Press. Oxford, England. 1988

Shaw, George Bernard   1856–1950
Irish comic dramatist and literary critic

The orbit of the electron observes no law: it chooses one path and rejects another: it is as capricious as the planet Mercury, who wanders from his road to warm his hands at the sun. All is caprice: the calculable world has become incalculable.
Bernard Shaw's Plays
Too True to Be Good: A Political Extravaganza
Act III
W.W. Norton & Company, Inc. New York, New York, USA. 1970

Thomson, Sir Joseph John   1856–1940
English physicist

It [relativity] was not a discovery of an outlying island, but of a whole continent of new scientific ideas of the greatest importance to some of the most fundamental questions connected with physics.
Eclipse Showed Gravity Variation: Hailed as Epochmaking
The New York Times, November 9, 1919 (p. 6)

Weyl, Hermann   1885–1955
German mathematician

It is as if a wall which separated us from Truth has collapsed. Wider expanses and greater depths are now exposed to the searching eye of knowledge, regions of which we had not even a presentiment. It has brought us much nearer to grasping the plan that underlies all physical happening.
Translated by Henry L. Brose
Space — Time — Matter
Preface to the First Edition (p. ix)
Dover Publications, Inc. New York, New York, USA. 1922

Williams, W.
No biographical data available

You hold that time is badly warped,
That even light is bent;
I think I get the idea there,
If this is what you meant;
The mail the postman brings me today,
Tomorrow will be sent.
In Ronald W. Clark
Einstein: The Life and Times
Part Four, Chapter 12 (p. 330)
The World Publishing Company. New York, New York, USA. 1971

RELIABILITY

Adams, George   1750–95
English instrument maker

The mind of man admits with reluctance the truth of every testimony concerning matters of fact, which happen to be repugnant to the uniform experience of his senses; hence the general backwardness to believe the miracles in the Bible: and hence the Dutchman, who informed the king of Siam that water in his country would sometimes in cold weather be found so hard, that men walked upon it, and that it would bear an elephant, was esteemed a person unworthy of credit. Hitherto, says the king, I have believed the strange things you told me, because I looked upon you as a sober man, but now I am sure you lie.
Lectures on Natural and Experimental Philosophy (Volume 1)
Lecture VI (pp. 279–280)
Printed by R. Hindmarsh. London, England. 1794

RELIGION

Barrow, John D.   1952–
English theoretical physicist

If a "religion" is defined to be a system of ideas that contains unprovable statements, then Gödel has taught us that, not only is mathematics a religion, it is the only religion that can prove itself to be one.
Between Inner Space and Outer Space (p. 88)
Oxford University Press, Inc. New York, New York, USA. 1999

Einstein, Albert   1879–1955
German-born physicist

It is quite clear to me that the religious paradise of youth, which [I] lost, was a first attempt to free myself from the chains of the "merely personal," from an existence which is dominated by wishes, hopes, and primitive feelings.
In Gerald Holton
Einstein, History, and Other Passions: The Rebellion Against Science at the End of the Twentieth Century
Part Two, Chapter 8 (p. 172)
Addison-Wesley Publishing Company. Reading, Massachusetts, USA. 1996

[My] deep religiosity…found an abrupt ending at the age of twelve, through the reading of popular scientific books.
In Gerald Holton
Einstein, History, and Other Passions: The Rebellion Against Science at the End of the Twentieth Century
Part Two, Chapter 8 (p. 172)
Addison-Wesley Publishing Company. Reading, Massachusetts, USA. 1996

Ferrer, Francisco   1849–1909
Spanish anarchist

When the masses become better informed about science, they will feel less need for help from supernatural Higher Powers. The need for religion will end when man becomes sensible enough to govern himself.
In James A. Haught (ed.)
2000 Years of Disbelief: Famous People with the Courage to Doubt
Part Six: The Early Twentieth Century Chapter 52 (p. 224)
Prometheus Books. Amherst, New York, USA. 1996

Gould, Stephen Jay   1941–2002
American paleontologist, evolutionary biologist, and historian of science

I do get discouraged when some of my colleagues tout their private atheism (their right, of course, and in many ways my own suspicion as well) as a panacea for human progress against a caricature of "religion," erected as a straw man for rhetorical purposes.… If these colleagues wish to fight superstition, irrationalism, philistinism, ignorance, dogma, and a host of other insults to the human intellect, then God bless them — but don't call this enemy "religion."
Rocks of Ages
The Two False Paths of Irenics (pp. 209–210)
The Ballantine Publishing Group. New York, New York, USA. 1999

RENORMALIZATION

Berry, Sir Michael
No biographical data available

In The Renormalization Group method you take a structure you don't understand and convert it to another structure you don't understand. You keep doing it until you finally understand.
2002 Gibbs Lecture, San Diego, California, January 6, 2002

REPAIR

Adams, Douglas   1952–2001
English author, comic radio dramatist, and musician

The major difference between a thing that might go wrong and a thing that cannot possibly go wrong is that when a thing that cannot possibly go wrong goes wrong, it usually turns out to be impossible to get at and repair.
The Ultimate Hitchhiker's Guide to the Galaxy
Mostly Harmless
Chapter 12 (p. 720)
Ballantine Books. New York, New York, USA. 2002

REPLICA

Whitehead, Alfred North   1861–1947
English mathematician and philosopher

For, whereas you can make a replica of an ancient statue, there is no possible replica of an ancient state of mind. There can be no nearer approximation [of one] than that which a masquerade bears to real life.

Science and the Modern World
Chapter IX (p. 200)
The Macmillan Company. New York, New York, USA. 1929

REPORT

Kettering, Charles Franklin   1876–1958
American engineer and inventor

Some technical reports are so dry and dusty…that if you put a pile of them in a hydraulic press and apply millions of pounds of pressure to it, not a drop of juice will run out.
Professional Amateur
Part III, Chapter XXI (p. 215)
E.P. Dutton & Company, Inc. New York, New York, USA. 1957

REPRODUCTION

Bateson, William   1861–1926
English biologist and geneticist

I know nothing which to a man well trained in scientific knowledge and method brings so vivid a realisation of our ignorance of the nature of life as the mystery of cell-division…. It is this power of spontaneous division which most sharply distinguishes the living from the non-living…. The greatest advance I can conceive in biology would be the discovery of the instability which leads to the continued division of the cell. When I look at a dividing cell I feel as an astronomer might do if he beheld the formation of a double star: that an original act of creation is taking place before me.
In Louise B. Young (ed.)
The Mystery of Matter
Aspects of Immortality, Death, and Reproduction (p. 403)
Oxford University Press, Inc. New York, New York, USA. 1965

Fisher, Sir Ronald Aylmer   1890–1962
English statistician and geneticist

We can set no limit to human potentialities; all that is best in man can be bettered…The ordinary social reformer sets out with a belief that no environment can be too good for humanity; it is without contradicting this that the eugenist may add that man can never be too good for his environment.
Some Hopes of a Eugenist
Eugenics Review, 5:309, 1914

Fletcher, Joseph   1905–91
Anglican theologian and founder of the theory of situational ethics

Our basic ethical choice as we consider man's new control over himself, over his body and his mind as well as over his society and environment, is still what it was when primitive men holed up in caves and made fires. Chance versus control. Should we leave the fruits of human reproduction to take shape at random, keeping our children dependent on accidents of romance and genetic endowment, of sexual lottery or what one physician calls "the meiotic roulette of his parents' chromosomes?" Or should we be responsible about it, that is, exercise our rational and human choice, no longer submissively trusting to the blind worship of raw nature?
The Ethics of Genetic Control: Ending Reproductive Roulette
Chapter I. Trying to be Natural (p. 36)
Prometheus Books. Buffalo, New York, USA. 1988

Huxley, Julian   1887–1975
English biologist, philosopher, and author

The pioneers of Eugenic Insemination by Donor…will be accused of mortal sin, of theological impropriety, of immoral and unnatural practices. But they can take heart from what has happened in the field of birth control, and can be confident that the rational control of reproduction aimed at the prevention of human suffering and frustration and the promotion of human well-being and fulfillment will in the not too distant future come to be recognized as a moral imperative.
Eugenics Review, Volume 54, 1963 (p. 123)

Pearson, Karl   1857–1936
English mathematician

A majority of the community would probably also admit today that the physical characters of man are inherited with practically the same intensity as the like characters in cattle and horses. But few, however…, apply the results which flow from such acceptance to their own conduct in life.
On the Laws of Inheritance in Man
Biometrika, Volume 3, 1904 (p. 131)

Russell, Bertrand Arthur William   1872–1970
English philosopher, logician, and social reformer

…impregnation will be regarded in an entirely different manner, more in the light of a surgical operation, so that it will be thought not ladylike to have it performed in the natural manner.
The Scientific Outlook
Chapter XVI (p. 262)
George Allen & Unwin Ltd. London, England. 1931

The Bible

Generations come and generations go…
The Revised English Bible
Ecclesiastes 1:2
Oxford University Press, Inc. Oxford, England. 1989

Zihlman, Adrienne
American paleoanthropologist

As with most things in life, the debate centers on two themes: food and sex; or to give it a proper academic tone: diet and reproduction.

Sex, Sexes and Sexism in Human Origins
Yearbook of Physical Anthropology, Volume 30, 12 April 1985 (p. 11)

RESEARCH

Abbe, Cleveland   1838–1916
American meteorologist

The ultimate goal of scientific research is not the collection of facts furnished by explorations and surveys,
not even the exact data furnished by the most laborious measurements as in astronomy, geodesy, chemistry,
and physics. Neither is it the framing of a few generalizations and inductions, such as the general idea of
evolution; nor is it the establishment of some isolated
fundamental laws, such as the attraction of gravitation,
the conservation of energy, the mechanical equivalent
of heat, the atomic weights and their periodic law.
Research aims to go deeper than all this and show how
these laws and phenomena result necessarily from a
few simple premises — not premises in the sense of
assumption, but axioms that are just as truly the basis of
the physical universe as Euclid's axioms are the basis
of geometry.
*Annual Report of the Board of Regents of the Smithsonian Institution,
1907*
The Progress of Science as Illustrated by the Development of
Meteorology (p. 287)
Government Printing Office. Washington, D.C. 1908

Adams, George   1750–95
English instrument maker

The best directed and most successful researches only
inform us how little is known, and give us no cause to be
satisfied with the discoveries they have made.
Lectures on Natural and Experimental Philosophy (Volume 3)
Chapter XXXV (p. 512)
Printed by R. Hindmarsh. London, England. 1794

The more diligent our search, the more accurate our
scrutiny, the more we are convinced that our labours
can never finish, and that subjects inexhaustible remain
behind still unexplored.
Lectures on Natural and Experimental Philosophy (Volume 1)
Lecture X (p. 421)
Printed by R. Hindmarsh. London, England. 1794

Agnew, Neil McK.
No biographical data available
Pyke, Sandra W.
No biographical data available

Research is like a love affair. The ingredients include:
(1) your image of the girl; (2) the real girl as she would
appear to you if you…had access to all information about
her; and (3) the bits, pieces, or samples of information
you have, some of it clear, some of it vague, some of
it twisted by memory or biased senses…. Changing a

once-loved picture is a very painful process, and we know
the degrees to which a lover will go to ignore, twist, and
blink away negative data…
The Science Game
Leaping to Conclusions (p. 128)
Prentice-Hall, Inc. Englewood Cliffs, New Jersey, USA. 1969

Asimov, Isaac   1920–92
American author and biochemist

…One can appreciate and take pleasure in the achievements of science even though he does not himself have
a bent for creative work in science…. Initiation into the
magnificent world of science brings great aesthetic satisfaction, inspiration to youth, fulfillment of the desire to
know, and a deeper appreciation of the wonderful potentialities and achievements of the human mind.
Asimov's New Guide to Science
What Is Science? (p. 15)
Basic Books, Inc. New York, New York, USA. 1984

Author undetermined

Great cabinets may be unlocked by little keys.
Astronomical Observations
Nature, Volume 4, May 11, 1871 (p. 31)

Bachrach, Arthur J.
No biographical data available

…people don't usually do research the way people who
write books about research say that people do research.
Psychological Research: An Introduction
Introduction (pp. 19–20)
Random House, Inc. New York, New York, USA. 1965

Ball, Sir Robert S.   1840–1913
Astronomer

Just as the astronomer staggers our powers of conception by the description of appalling distances and stupendous periods of time, and relies with confidence on
the evidence which convinces him of the reality of his
statements, so the physicist avails himself of a like potent
method of research to study distances so minute and time
so brief that the imagination utterly fails to realize them.
*Annual Report of the Board of Regents of the Smithsonian Institution,
1893*
Atoms and Sunbeams (p. 127)
Government Printing Office. Washington, D.C. 1894

Barrie, Sir James M.   1860–1937
Scottish journalist, writer, and dramatist

…those hateful persons…Original Researchers…
My Lady Nicotine
Chapter XIII (p. 85)
Charles Scribner's Son's. New York, New York, USA. 1921

Bates, Marston   1906–74
American zoologist

Research is the process of going up alleys to see if they are blind.
In Jefferson Hane Weaver
The World of Physics (Volume 2)
K.6 (p. 63)
Simon & Schuster. New York, New York, USA. 1987

Baum, L. Frank   1856–1919
American author

"But, dear me, in that case you will never find your lost brother!" exclaimed the girl.

"Maybe not; but it's my duty to try," answered Shaggy. "I've wandered so far without finding him, but that only proves he is not where I've been looking."
Tik-Tok of Oz
Chapter Six
The Reilly & Lee Company. Chicago, Illinois, USA. 1914

Belloc, Hilaire   1870–1953
French-born poet and historian

…anyone of common mental and physical health can practice scientific research…. Anyone can try by patient experiment what happens if this or that substance be mixed in this or that proportion with some other under this or that condition. Anyone can vary the experiment in any number of ways. He that hits in this fashion on something novel and of use will have fame…. The fame will be the product of luck, and industry. It will not be the product of special talent.
Essays of a Catholic
Science as the Enemy of Truth (pp. 226–227)
The Macmillan Company. New York, New York, USA. 1931

Beveridge, William Ian Beardmore   1908–
Australian zoologist

People in most other walks of life can allow themselves the indulgence of fixed ideas and prejudices which make thinking so much easier…but the research worker must try to keep his mind malleable and avoid holding set ideas in science. We have to strive to keep our mind receptive and to examine suggestions made by others fairly and on their own merits, seeking arguments for as well as against them. We must be critical, certainly, but beware lest ideas be rejected because an automatic reaction causes us to see only the arguments against them. We tend especially to resist ideas competing with our own.
The Art of Scientific Investigation
Chapter Seven (p. 86)
W.W. Norton & Company, Inc. New York, New York, USA.1957

Research is one of those highly complex and subtle activities that usually remain quite unformulated in the minds of those who practice them. This is probably why most scientists think that it is not possible to give any formal instruction in how to do research.
The Art of Scientific Investigation

Preface (pp. ix–x)
W.W. Norton & Company, Inc. New York, New York, USA.1957

The research worker remains a student all his life.
The Art of Scientific Investigation
Chapter One (p. 1)
W.W. Norton & Company, Inc. New York, New York, USA.1957

Anyone with an alertness of mind will encounter in the course of an investigation numerous interesting side issues that might be pursued. It is a physical impossibility to follow up all of these. The majority are not worth following, a few will reward investigation and the occasional one provides the opportunity of a lifetime. How to distinguish the promising clues is the very essence of the art of research.
The Art of Scientific Investigation
Chapter Three (p. 35)
W.W. Norton & Company, Inc. New York, New York, USA.1957

Birch, Arthur J.   1915–1995
Australian chemist

The details of a research career are dictated partly by accident, but largely by inclinations and temperament.
To See the Obvious
Prelude, and Evolution of a Chemist (p. 7)
American Chemical Society. Washington, D.C. 1995

Bradley, A. C.   1851–1935
English literary scholar

Research, though toilsome, is easy; imaginative vision, though delightful, is difficult.
Oxford Lectures on Poetry
Shakespeare's Theatre and Audience (p. 362)
Macmillan & Company Ltd. London, England. 1909

Brown, J. Howard
No biographical data available

A man may do research for the fun of doing it but he can not expect to be supported for the fun of doing it.
The Biological Approach to Bacteriology
Journal of Bacteriology, Volume 18, Number 1, January 1932 (p. 9)

Browning, Robert   1812–89
English poet

…as is your sort of mind,
So is your sort of search: you'll find
What you desire.
The Poems and Plays of Robert Browning
Easter Day, Part vii, l. 3 (p. 501)
The Modern Library. New York, New York, USA. 1934

Bunge, Mario   1919–
Argentine philosopher and physicist

Most scientists are prepared to grant that the chief theoretical (that is, nonpragmatic) aim of scientific research is to answer, in an intelligible, exact, and testable way,

five kinds of questions, namely those beginning with what (or how), where, when, whence, and why.... [T]he Five W's of Science. (Only radical empiricists deny that science has an explanatory function, and restrict the task of scientific research to the description and prediction of observable phenomena.) Also, most scientists would agree that all five W's are gradually (and painfully) being answered through the establishment of scientific laws, that is, general hypotheses about the patterns of being and becoming.
Causality: The Place of the Causal Principle in Modern Science
Chapter 10 (p. 248)
Harvard University Press. Cambridge, Massachusetts, USA. 1959

Bush, Vannevar   1890–1974
American electrical engineer and physicist

Basic research leads to new knowledge. It provides scientific capital. It creates the fund from which the practical applications of knowledge must be drawn. New products and new processes do not appear full-grown. They are founded on new principles and new conceptions, which in turn are painstakingly developed by research in the purest realms of science.
Endless Horizons
Chapter 5 (pp. 52–53)
Public Affairs Press. Washington, D.C. 1946

In these circumstances it is not at all strange that the workers sometimes proceed in erratic ways. There are those who are quite content, given a few tools, to dig away, unearthing odd blocks, piling them up in the view of fellow workers and apparently not caring whether they fit anywhere or not.... Some groups do not dig at all, but spend all their time arguing as to the exact arrangement of a cornice or an abutment. Some spend all their days trying to pull down a block or two that a rival has put in place. Some, indeed, neither dig nor argue, but go along with the crowd, scratch here and there, and enjoy the scenery. Some sit by and give advice, and some just sit.
Endless Horizons
Chapter 17 (p. 180)
Public Affairs Press. Washington, D.C. 1946

Capra, Fritjof   1939–
Austrian-born American physicist

Scientists, therefore, are responsible for their research not only intellectually but also morally...the results of quantum mechanics and relativity theory have opened up two very different paths for physics to pursue. They may lead us — to put it in extreme terms — to the Buddha or to the bomb, and it is up to each of us to decide which path to take.
The Turning Point
Chapter II (p. 87)
A Bantam Book. New York, New York, USA. 1983

Carrel, Alexis   1873–1944
French surgeon and biologist

In researches dealing with physics and chemistry, and also with physiology, one always attempts to isolate relatively simple systems, and to determine their exact conditions.
Man the Unknown
Chapter 2, Section 5 (pp. 50–51)
Harper & Brothers. New York, New York, USA. 1939

Caullery, Maurice   1868–1958
French biologist

The double danger of research into this type of phenomenon lies, on the one hand, in bringing...preconceived ideas of too subjective a nature, bordering on an illusory anthropomorphism, and on the other hand, trying to reduce complex facts to simple elementary reactions.
Translated by Averil M. Lysaght
Parasitism and Symbiosis
Chapter I (p. 2)
Sidgwick & Jackson Limited. London, England. 1952

Chesterton, G. K. (Gilbert Keith)   1874–1936
English author

Research is the search of people who don't know what they want.
The G.K. Chesterton Calendar
May 25
Cecil Palmer & Hayward. London, England. 1916

Crick, Francis Harry Compton   1916–2004
English biochemist

In research the front line is almost always in a fog.
What Mad Pursuit: A Personal View of Scientific Discovery
Chapter 3 (p. 35)
Basic Books, Inc. New York, New York, USA. 1988

Cussler, Clive
American author
Dirgo, Craig
No biographical data available

Research is the key. You can never do enough research. This is so vital I'll repeat it. You can never to enough research.... Research can either lower the odds or tell you it's hopeless.
The Sea Hunters
Introduction (p. 28)
Simon & Schuster. New York, New York, USA. 1996

da Vinci, Leonardo   1452–1519
Italian High Renaissance painter and inventor

Nothing is written as the result of new researches.
Leonardo da Vinci's Note Books (p. 53)
Duckworth & Company. London, England. 1906

Dawkins, Richard   1941–
English ethologist, evolutionary biologist, and popular science writer

Existing science must be overthrown not by casual anecdotes but by the most rigorous research, repeated, dissected, and repeated again.
Putting Away Childish Things
Skeptical Inquirer, Jan/Feb 1995 (p. 31)

Day, R. A.
No biographical data available

The goal of scientific research is publication. Scientists, starting as graduate students, are measured primarily not by their innate knowledge of either broad or narrow scientific subjects, and certainly not by their wit or charm; they are measured, and become known (or remain unknown), by their publications.
How to Write and Publish a Scientific Paper (3rd edition) (p. vii)
Oryx Press. Phoenix, Arizona, USA. 1988

Dessauer John   1792–1871
English astronomer and chemist

All the efforts of the researcher to find other models, conceptions, different mathematical forms, better linguistic modes of expression, to do justice to newly discovered layers of being mean self-transformation. The researcher in his place is the human being in self-transformation to more profound insight into what is given.
Universitas: A Quarterly German Review of the Arts and Sciences,
Volume 26, Number 4, April 6, 1984 (p. 316)

Dr. Gil
No biographical data available

There! Little Research, don't you cry —
You'll be a paper by and by.
Three observations — half a page of notes,
Will bring prostration to seven other blokes.
The Professor Sings to His Brain Child
Industrial and Engineering Chemistry: News Edition, Volume 11, Number 9, 19 May 1933 (p. 149)

Einstein, Albert   1879–1955
German-born physicist

When a man after long years of searching chances upon a thought which discloses something of the beauty of this mysterious universe, he should not therefore be personally celebrated. He is already sufficiently paid by his experience of seeking and finding.
New York Times, 128:18, Section 4, November 10, 1978

Freeman, R. Austin   1862–1943
British physician and mystery novelist

…in scientific research there is no…division of function. The investigator is at once judge, jury, and witness. His knowledge is first-hand, and hence he knows the exact value of his evidence. He can hold a suspended judgment. He can form alternative opinions and act upon both alternatives. He can construct hypotheses and try them out. He is hampered by no rules but those of his own making. Above all, he is able to interrogate things as well as persons.
A Certain Dr. Thorndyke
Thorndyke Connects the Links (pp. 277–278)
Dodd, Mead & Company. New York, New York, USA. 1928

George, William H.
No biographical data available

Scientific research is not itself a science: it is still an art or craft.
The Scientist in Action: A Scientific Study of His Methods
Four Qualities of Scientific Research (p. 29)
Williams & Norgate Ltd. London, England. 1936

Gibbs, J. Willard   1839–1903
American mathematician

One of the principal objects of theoretical research in any department of knowledge is to find the point of view from which the subject appears in its greatest simplicity.
In G. K. Batchelor
Preoccupations of a Journal Editor
Journal of Fluid Mechanics, Volume 106, 1981

Green, Celia   1935–
English philosopher and psychologist

Research is a way of taking calculated risks to bring about incalculable consequences.
The Decline and Fall of Science
Aphorisms (p. 1)
Hamilton. London, England. 1976

The way to do research is to attack the facts at the point of greatest astonishment.
The Decline and Fall of Science
Aphorisms (p. 1)
Hamilton. London, England. 1976

Gregg, Alan   1890–1957
American medical educator and philosopher

Research has been defined as a guerrilla warfare on the unknown. In the rigorous uncertainties of such campaigns the investigator must be prepared to swap horses in mid-stream and to discard some very dear items of accumulated baggage of belief or personal pride, whenever intellectual honesty calls for such sacrifices.
The Furtherance of Medical Research
Chapter III (pp. 87–88)
Yale University Press. New Haven, Connecticut, USA. 1941

Heisenberg, Werner Karl   1901–76
German physicist and philosopher

If I were asked what was Christopher Columbus' greatest achievement in discovering America, my answer would not be that he took advantage of the spherical shape of the earth to get to India by the western route — this idea had occurred to others before him — or that he prepared his expeditions meticulously and rigged his ships most expertly — that, too, others could have done equally well. His most remarkable feat was the decision to leave the known regions of the world and to sail westward, far beyond the point from which provisions could have gotten him back home again.

Physics and Beyond: Encounters and Conversations
Chapter 6 (p. 70)
Harper & Row, Publishers. New York, New York, USA. 1971

…the subject matter of research is no longer nature in itself, but nature subjected to human questioning…

In Aldous Huxley
Literature and Science
Chapter 25 (p. 76)
Harper & Row, Publishers. New York, New York, USA. 1963

Heschel, Abraham J.   1907–72
Jewish theologian

Scientific research is an entry into the endless, not a blind alley; Solving one problem, a greater one enters our sight. One answer breeds a multitude of new questions; explanations are merely indications of greater puzzles. Everything hints at something that transcends it; the detail indicates the whole, the whole, its idea, the idea, its mysterious root. What appears to be a center is but a point on the periphery of another center. The totality of a thing is actual infinity.

Analog Science Fiction/Science Fact Magazine, Volume CIV, Number 12, December 1984 (p. 63)

Hubble, Edwin Powell   1889–1953
American astronomer

Research men attempt to satisfy their curiosity, and are accustomed to use any reasonable means that may assist them toward the receding goal. One of the few universal characteristics is a healthy skepticism toward unverified speculations. These are regarded as topics for conversation until tests cant be devised. Only then do to they attain the dignity of subjects for investigation.

The Realm of the Nebulae
Introduction (p. 6)
Dover Publications, Inc. New York, New York, USA. 1958

Hurston, Zora Neale   1891–1960
American author and anthropologist

Research is formalized curiosity. It is poking and prying with a purpose.

Dust Tracks on a Road
Chapter X (p. 174)
University of Illinois Press. Urbana, Illinois, USA. 1984

Jaffe, Bernard   1896–1968
American science writer

There is no last word or ultimate solution in the adventure of scientific research.

Michelson and the Speed of Light
Chapter XII (p. 171)
Doubleday & Company, Inc. Garden City, New York, USA. 1960

Jevons, William Stanley   1835–82
English economist and logician

So-called original research is now regarded as a profession, adopted by hundreds or men, and communicated by a system of training.

The Principles of Science: A Treatise on Logic and Scientific Method
Book IV, Chapter XXVI (p. 574)
Macmillan & Company Ltd. London, England. 1887

Johnson, Harry G.   1923–79
American economist

To an important extent, indeed, scientific research has become the secular religion of materialistic society; and it is somewhat paradoxical that a country whose constitution enforces the strict separation of church and state should have contributed so much public money to the establishment and propagation of scientific pessimism.

In National Academy of Sciences
Basic Research and National Goals: A Report to the Committee on Science and Astronautics Federal Support of Basic Research: Some Economic Issues
Note 4 (p. 141)
U.S. Government Printing Office. Washington, D.C. 1965

Kettering, Charles Franklin   1876–1958
American engineer and inventor

We find that in research a certain amount of intelligent ignorance is essential to progress; for if you know too much, you won't try the thing.

In T.A. Boyd
Professional Amateur
Part II (p. 106)
E.P. Dutton & Company, Inc. New York, New York, USA. 1957

Kettering, Charles Franklin   1876–1958
American engineer and inventor
Smith, Beverly
No biographical data available

[Research] may use a laboratory or it may not. It is purely a principle, and everybody can apply it in his own life. It is simply a way of trying to find new knowledge and ways of improving things which you are not satisfied with.

Ten Paths to Fame and Fortune
The American Magazine, December 1937 (p. 14)

Kline, Morris   1908–92
American mathematics professor and writer

Mathematical research is also becoming highly professionalized in the worst sense of that term. Research

performed voluntarily and sincerely by devoted souls, research as a relish of knowledge, is to be welcomed even if the results are minor. But hothouse-grown research, which crowds the journals and promotes only promotion, is a drag on science.
Why the Professor Can't Teach: Mathematics and the Dilemma of University Education
Chapter 3 (p. 67)
St. Martin's Press. New York, New York, USA. 1977

Kropotkin, Peter Alekseyevich 1842–1921
Russian revolutionary and geographer

…all great researches, all discoveries revolutionizing science, have been made outside academies and universities, wither by men rich enough to remain independent, like Darwin and Lyell, or by men who undermined their health by working in poverty, and often in great straits, losing endless time for want of a laboratory, and unable to procure the instruments or books necessary to continue their researches, but persevering against hope, and often dying before they had reached the end in view. Their name is legion.
The Conquest of Bread
Chapter IX, Section IV (p. 103)
Vanguard Press. New York, New York, USA. 1926

Lasker, Albert D. 1901–70

"Research," he said, "is something that tells you that a jackass has two ears."
In John Gunther
Taken at the Flood: The Story of Albert D. Lasker (p. 96)
Harper & Brothers. New York, New York, USA. 1960

Mach, Ernst 1838–1916
Austrian physicist and philosopher

The aim of research is the discovery of the equations which subsist between the elements of phenomena.
Popular Scientific Lectures
The Economical Nature of Physical Inquiry (p. 205)
The Open Court Publishing Company. Chicago, Illinois, USA. 1898

…scientific research is somewhat like unraveling complicated tangles of strings, in which luck is almost as vital as skill and accurate observation.
Knowledge and Error: Sketches on the Psychology of Enquiry
Chapter I (p. 10)
D. Reidel Publishing Company. Dordrecht, Netherlands. 1976

Mallove, Eugene F. 1947–2004
Editor

The image of a searching man held by Einstein and by Isaac Newton two hundred years before him is of a being wading in the shallows of the ocean of physical reality — trying to fathom the entirety by sampling only a part.
The Quickening Universe: Cosmic Evolution and Human Destiny
Prologue (p. xvi)
St. Martin's Press. New York, New York, USA. 1987

Human beings individually have only a brief time in this world to form an image of the cosmos. Their minds are like film in a camera of awareness. Birth and death are the opening and closing of the shutters. Yet generations of striving to understand have led to a picture of the universe far more complete than any of us alone could have hoped to develop.
The Quickening Universe: Cosmic Evolution and Human Destiny
Prologue (pp. xviii–xix)
St. Martin's Press. New York, New York, USA. 1987

Mayr, Ernst 1904–2005
German-born American biologist

…research not only brings us abundant joy but it also gives us a deep sense of humility.
In Walter Shropshire, Jr. (ed.)
The Joys of Research
Evolutionary Biology (p. 157)
Smithsonian Institution Press. Washington, D.C. 1981

Medawar, Sir Peter Brian 1915–87
Brazilian-born English zoologist

The scientist values research by the size of its contribution to that huge, logically articulated structure of ideas which is already, though not yet half built, the most glorious accomplishment of mankind.
Two Conceptions of Science
Encounter, Volume 143, August 1965

If politics is the art of the possible, research is surely the art of the soluble. Both are immensely practical-minded affairs.
The Act of Creation
New Statesman, Volume 19, June 1964 (p. 950)

Mizner, Wilson 1876–1933
American playwright

If you steal from one author, it's plagiarism; if you steal from many, it's research.
In Alva Johnson
The Legendary Mizners
Chapter 4, The Sport (p. 66)
Farrar, Straus & Young. New York, New York, USA. 1953

Oppenheimer, J. Robert 1904–67
American theoretical physicist

Research is action; and the question I want to leave in a very raw and uncomfortable form with you is how to communicate this sense of action to our fellow men who are not destined to devote their lives to the professional pursuit of new knowledge.
The Open Mind
Chapter VII (p. 129)
Simon & Schuster. New York, New York, USA. 1955

We have done the devil's work. Now we have come back to our real job, which is to devote ourselves exclusively to research.

In Karl Jasper
La Bombe Atomique et l' Avenir de L'Homme (p. 360)
Buchet-Chassel. Paris, France. 1963

Peabody, A. P.
No biographical data available

No man becomes proficient in any science who does not transcend system, and gather up new truth for himself in the boundless field of research.
In James Orton
Comparative Zoology, Structural and Systematic
Preceding Chapter XXI (p. 222)
Harper & Brothers. New York, New York, USA. 1877

Perutz, Max F.   1914–2002
Austrian-born English biochemist

…research consists of formulation of imaginative hypotheses that are open to falsification by experiment.
Is Science Necessary?
How to Become a Scientist (p. 199)
E.P. Dutton & Company, Inc. New York, New York, USA. 1989

Platt, Sir Robert   1900–78
English physician

The conventional picture of the research worker is that of a rather austere man in a white coat with a background of complicated glassware. My idea of a research worker, on the other hand, is a man who brushes his teeth on the left side of his mouth only so as to use the other side as a control and see if tooth-brushing has any effect on the incidence of caries.
British Medical Journal, Volume 1, 1953 (p. 577)

Recorde, Robert   1510?–58
English mathematician and writer

The time seemeth longe (bee it never so shorte indeed) to hym that desirously looketh for any thing: for as the obtaining of it bringeth great pleasure, namelye the thinge itselfe being profitable, so the wante thereof causeth displeasure and cotinuall grief tyll the desire be eyther fully satisfied, other partly (at the least) accomplished.
The Castle of Knowledge
The First Treatise (p. 1)
Imprinted by R. Wolfe. London, England. 1556

Reichenbach, Hans   1891–1953
German philosopher of science

The reliance on the concrete is the basis of both the charm and the power of physical research.
Atom and Cosmos
Chapter 4 (p. 75)
The Macmillan Company. New York, New York, USA. 1933

Richet, Charles   1850–1935
French physiologist

Understand this clearly; that the right method, even for obtaining a useful practical result, is not to worry about the practice, but to concentrate intensely on pure investigation, without being hampered by any parasitic considerations other than whatever conduces to greater facility for research.
The Natural History of a Savant
Chapter XII (p. 134)
J.M. Dent & Sons Ltd. London, England. 1927

The gift for investigation appears at an early age: the demon of research speaks to men whilst they are still young.
The Natural History of a Savant
Chapter VI (pp. 38–39)
J.M. Dent & Sons Ltd. London, England. 1927

Robinson, James Harvey   1863–1936
American historian

Research is mainly looking for things that are not there and attempting processes that will not occur.
The Humanizing of Knowledge
Chapter II (p. 32)
George H. Doran Company. New York, New York, USA. 1923

Romanoff, Alexis Lawrence   1892–1980
Russian soldier and scientist

Scientific research is based chiefly on creative thinking.
Encyclopedia of Thoughts
Aphorisms 219
Ithaca Heritage Books. Ithaca, New York, USA. 1975

Scientific research provides the shortest route to useful practice.
Encyclopedia of Thoughts
Aphorisms 112
Ithaca Heritage Books. Ithaca, New York, USA. 1975

Sarewitz, Daniel
No biographical data available

A lone scientist, frizzy-haired and bespectacled — the absent-minded, benevolent genius lost in thought; or perhaps the dedicated experimentalist clad in a white coat and laboring madly among the condensers, Van de Graaf generators, computers, and even electrode-covered cadavers: These are typical public images of scientific research.
Frontiers of Illusion
Chapter 3 (p. 31)
Temple University Press. Philadelphia, Pennsylvania, USA. 1996

Sarnoff, David   1891–1971
Russian-born American broadcasting executive

The wonderful thing about research is that the more of it you do, the more of it there is left to do.
Research and Industry: Partners in Progress
Address to the Board of Directors of the Stanford Research Institute
November 14, 1951 (p. 13)

Scalera, Mario
No biographical data available

There is no practical purpose here. There is simply man's insatiable curiosity, his abhorrence of the unknown — the desire to see, in the confusing phenomena of nature, the law, the order, that underlies them. This kind of urge has its own reward...the reward that comes to a man who suddenly sees order shaping out of chaos — this is what we call fundamental research.
An Industrial Research Director Views Fundamental Research
Chemical and Engineering News, April 21, 1958 (p. 85)

Schild, Alfred   1921–77
Physicist

If one can tell ahead of time what one's research is going to be, the research problem cannot be very deep and may be said to be almost nonexistent.
On the Matter of Freedom: The University and the Physical Sciences
Bulletin
Canadian Association of University Teachers, Volume 11, Number 4, 1963

Schön, Donald A.   1930–97
American philosopher of practice and learning theory

He emphasizes the key issue of the starting point of research. In real-world practice, problems do not present themselves to the practitioner as givens. They must be constructed from the materials of problematic situations which are puzzling, troubling and uncertain. In order to convert a problematic situation to a problem, a practitioner must do a certain kind of work.
The Reflective Practitioner: How Professionals Think in Action (p. 40)
Basic Books, Inc. New York, New York, USA. 1983

Smith, Homer W.
Renal physiologist

On every scientist's desk there is a drawer labeled UNKNOWN in which he files what are at the moment unsolved questions, lest through guess-work or impatient speculation he come upon incorrect answers that will do him more harm than good. Man's worst fault is opening the drawer too soon. His task is not to discover final answers but to win the best partial answers that he can, from which others may move confidently against the unknown, to win better ones.
From Fish to Philosopher
Chapter XIII (p. 210)
Little, Brown & Company. Boston, Massachusetts, USA. 1953

Smith, Theobald   1859–1934
American pathologist

The joy of research must be found in doing, since every other harvest is uncertain.
Letter from Dr. Theobald Smith,
Journal of Bacteriology Volume 27, Number 1, January 1934 (p. 20)

...it is the care we bestow on apparently trifling, unattractive and very troublesome minutiae which determines the results.
New York Medical Journal, Volume lii, 1890 (p. 485)

Stewart, Ian   1945–
English mathematician and science writer

The really important breakthroughs are always unpredictable. It is their very unpredictability that makes them important: they change our world in ways we didn't see coming.... There is nothing wrong with goal-oriented research as a way of achieving specific feasible goals. But the dreamers and the mavericks must be allowed some free rein, too. Our world is not static: new problems constantly arise, and old answers often stop working. Like Lewis Carroll's Red Queen, we must run very fast in order to stand still.
Nature's Numbers: The Unreal Reality of Mathematical Imagination
Chapter 2 (p. 29)
Basic Books, Inc. New York, New York, USA. 1995

Sutherland, Jr., Earl W.   1915–74
American pharmacologist and physiologist

I am fully convinced that medical research can offer one a happy and productive life. And if one has a little Viking spirit he can explore the world and people as no one else can do. The whole medical research area is wide open for exploration.
Les Prix Nobel. The Nobel Prizes in 1971
Nobel banquet speech for award received in 1971
Nobel Foundation. Stockholm, Sweden. 1972

Szent-Györgyi, Albert   1893–1986
Hungarian-born American biochemist

Research means going out into the unknown with the hope of finding something new to bring home.... The unknown is the unknown because one does not know what is there. If one knows what one will do and find in it, then it is not research any more and is not worth doing.
Research Grants
Perspectives in Biology and Medicine, Volume 18, Number 1, Autumn 1974 (p. 41)

Terence   190 BCE–158 BCE
Roman comic dramatist

Nothing is so difficult but that it may be found out by seeking.
Translated by Alexander Harvey
Heauton Timorumenos
Act iv, Scene 2, l. 675
Haldeman-Julius Company. Girard, Kansas, USA. 1925

The Bible

...seek, and you will find; knock, and the door will be opened to you.
The Revised English Bible

Matthew 7:7
Oxford University Press, Inc. Oxford, England. 1989

Thomas, Lewis   1913–93
American physician and biologist

In science in general, one characteristic feature is the awareness of error in the selection and pursuit of a problem. This is the most commonplace of criteria: if a scientist is going to engage in research of any kind, he has to have it on his mind, from the outset, that he may be on to a dud. You can tell a world-class scientist from the run-of-the-mill investigator by the speed with which he recognizes that he is heading into a blind alley. Blind alleys and garden paths leading nowhere are the principal hazards in research.
Late Night Thoughts on Listening to Mahler's Ninth Symphony
Viking Press. New York, New York, USA. 1983

Thompson, Elihu   1853–1937
American electrical engineer

Physical research by experimental methods is both a broadening and a narrowing field. There are many gaps yet to be filled, data to be accumulated, measurements to be made with great precision, but the limits within which we must work are becoming, at the same time, more and more defined.
Annual Report of the Board of Regents of the Smithsonian Institution, 1899
The Field of Experimental Research (p. 119)
Government Printing Office. Washington, D.C. 1901

Thorne, Kip S.   1940–
American theoretical physicist

In scientific research, as in life, many themes are pursued simultaneously by many different people; and the insights of one decade may spring from ideas that are several decades old but were ignored in the intervening years.
Black Holes and Time Warps: Einstein's Outrageous Legacy
Preface (p. 18)
W.W. Norton & Company, Inc. New York, New York, USA.1994

Veblen, Thorstein   1857–1929
Economist, social critic, and author

...the outcome of any serious research can only be to make two questions grow where only one grew before.
The Place of Science in Modern Civilization and Other Essays
The Evolution of the Scientific Point of View (p. 33)
The Viking Press, Inc. New York, New York, USA. 1942

von Braun, Wernher   1912–77
German-American rocket scientist

Basic research is when I'm doing what I don't know I'm doing.
In Jefferson Hane Weaver
The World of Physics (Volume 2)

K.6 (p. 63)
Simon & Schuster. New York, New York, USA. 1987

von Liebig, Justus   1803–73
German organic chemist

We were the first pioneers in unknown regions, and the difficulties in the way of keeping on the right path were sometimes insuperable. Now, when the paths of research are beaten roads, it is a much easier matter; but all the wonderful discoveries which recent times have brought forth were then our own dreams, whose realization we surely and without doubt anticipated.
Annual Report of the Board of Regents of the Smithsonian Institution, 1891
Autobiography (p. 267)
Government Printing Office. Washington, D.C. 1893

Weber, Robert L.
No biographical data avaialble

Much of the misunderstanding of scientists and how they work is due to the standard format of articles in scientific journals. With their terse accounts of successful experiments and well-supported conclusions they show little of the untidy nature of research at the frontiers of knowledge.
A Random Walk in Science
Introduction (p. xv)
Institute of Physics Publishing. Bristol, England. 1973

Weisskopf, Victor Frederick   1908–2002
Austrian-American physicist

It is difficult to distinguish clearly between fundamental and applied science, and any considerations of this kind can lead to dangerous oversimplifications. The success of basic research derives to a large extent from the close cooperation of basic and applied science. This close relation — often within the same scientist — provided tools of high quality, without which many fundamental discoveries could not have been made.
Physics in the Twentieth Century: Selected Essays
The Significance of Science (pp. 354–355)
The MIT Press. Cambridge, Massachusetts, USA. 1972

Wells, Carolyn   1862–1942
American writer

I think, for the rest of my life, I shall refrain from looking up things. It is the most ravenous time-snatcher I know. You pull one book from the shelf, which carries a hint or a reference that sends you posthaste to another book, and that to successive others. It is incredible, the number of books you hopefully open and disappointedly close, only to take down another with the same results.
The Rest of My Life
Chapter 8
J.B. Lippincott. Philadelphia, Pennsylvania, USA. 1937

Wells, H. G. (Herbert George)   1866–1946
English novelist, historian, and sociologist

The whole difference of modern scientific research from that of the Middle Ages, the secret of its immense success, lies in its collective character, in the fact that every fruitful experiment is published, every new discovery of relationships explained.
New Worlds For Old
Chapter II (p. 22)
The Macmillan Company. New York, New York, USA. 1918

In a sense scientific research is a triumph over natural instinct, over that mean instinct that makes men secretive, that makes a man keep knowledge to himself and use it slyly to his own advantage.
New Worlds for Old
Chapter II (pp. 22–23)
The Macmillan Company. New York, New York, USA. 1918

Wheeler, John Archibald   1911–
American physicist and educator

There is an age-old longing to understand the inner mystery of this strange and beautiful world of ours and our own little place in the scheme of things. Whoever knows a little and can give a little to the search wants to know more and give more.
At Home in the Universe
Be the Best to Give the Most (p. 80)
The American Institute of Physics. Woodbury, New York, USA. 1994

Whitney, Willis Rodney   1868–1958
American chemical and electrical engineer

The valuable attributes of [researchers] are conscious ignorance and active curiosity.
The Stimulation of Research in Pure Science Which Has Resulted from the Needs of Engineers and of Industry
Science, Volume LXV, Number 1862, March 25, 1927 (p. 289)

Wilson, Jr., E. Bright   1908–92
American physical chemist

Though the road may seem long and arduous, with many stretches of pure drudgery, when the end of a particular stage is reached, where the bits of evidence all fall together into a clear and unexpected pattern, there are few other human activities which can provide as much satisfaction; especially if, as is so often the case, the results turn out later to have applications in all sorts of unanticipated directions and help to give a clearer picture of the universe we live in and to make life in that universe more worth while.
An Introduction to Scientific Research
Conclusion (p. 364)
McGraw-Hill Book Company, Inc. New York, New York USA. 1952

Wittig, Georg   1897–1987
German chemist

Chemical research and mountaineering have much in common. If the goal or the summit is to be reached, both initiative and determination as well as perseverance are required. But after the hard work it is a great joy to be at the goal or the peak with its splendid panorama.
Nobel Lectures, Chemistry 1971–1980
Nobel lecture for award received in 1979
From Diyls to Ylides to My Idyll (p. 368)
World Scientific Publishing Company. Singapore. 1993

Wordsworth, William   1770–1850
English poet

Lost in the gloom of uninspired research.
The Complete Poetical Works of William Wordsworth
The Excursion, Despondency Corrected, l. 626
Crowell. New York, New York, USA. 1888

Yang, Chen Ning   1922–
Chinese-born American theoretical physicist

The necessary tendency toward bigness is unfortunate, as it hinders free and individual initiative. It makes research less intimate, less inspiring, and less controllable. However, it must be accepted as a fact of life. Let us take courage then in the knowledge that despite their physical bigness, the machines, the detectors, and indeed the experiments themselves are still based on ideas that have the same simplicity, the same intimacy and controllability that have always made research so exciting and inspiring.
Elementary Particles: A Short History of Some Discoveries in Atomic Physics
Chapter 2 (p. 40)
Princeton University Press. Princeton, New Jersey, USA. 1962

Yeats, William Butler   1865–1939
Irish poet and playwright

I had discovered, early in my researches, that their doctrine was no mere chemical fantasy, but a philosophy they applied to the world, to the elements, and to man himself.
Stories of Red Hanrahan, the Secret Rose, Rosa Alchemica
Rosa Alchemica (p. 192)
The Macmillan Company. New York, New York, USA. 1914

RESEARCH PLAN

van Noordwijk, A. J.   1949
Dutch-Canadian genetical ecologist

…however excellent multiannual planning, research-project management, and time recording may be, the scientist should always have some opportunity to test the idea that he got that morning while shaving.
The Bioassyist
Perspectives in Biology and Medicine, Volume 29, Number 2, Winter 1986 (p. 307)

Richter, Curt P.   1894–1988
American psychobiologist

...good researchers use research plans merely as starters and are ready to scrap them at once in the light of actual findings...
Free Research versus Design Research
Science, Volume 118, Number 3056, July 24, 1953 (p. 92)

Waksman, Selman A.   1888–1973
Ukrainian-born American biochemist

Good scientists use research plans merely as outlines to begin their investigations and are ready to give them up once they are not justified by actual findings. Experimental designs tend to give rise to "team research", which serves a purpose in developing and applying ideas; it rarely produces new ideas.
Searchers and Researchers
Perspectives in Biology and Medicine, Volume 7, Number 3, Spring 1964 (p. 312)

...a new problem has arisen — namely "planned research" versus the "individual investigator." There is a place for planned research. It can take a defined body of knowledge and lay out a set of experiments which will exploit this knowledge to its foreseeable limits. It can take a set of postulates and drive them home to their logical conclusions. It can do this with exhaustive thoroughness, economy, and speed. Within its limitations, it is efficient, expeditious, and authoritative. But there is a place also and a more important place for the random investigator. The role of planned research is to consolidate ground already won; the role of the random investigator is to seek out new worlds to conquer.
Searchers and Researchers
Perspectives in Biology and Medicine, Volume 7, Number 3, Spring 1964 (p. 311)

RESIDUAL

Herschel, Sir John Frederick William   1792–1871
English astronomer and chemist

Almost all the greatest discoveries in astronomy have resulted from the consideration of what we have elsewhere termed RESIDUAL PHENOMENA, of a quantitative or numerical kind, that is to say, of such portions of the numerical or quantitative results of observations as remain outstanding and unaccounted for after subducting and allowing for all that would result from the strict application of known principles.
Outlines of Astronomy
Part III, Chapter XVI (856) (p. 584)
Longman, Brown, Green & Longmans. London, England. 1849

RESPIRATION

Lavoisier, Antoine Laurent   1743–94
French chemist

Of all the phenomena of animal economy, none are more striking, nor more worthy of attention from physicists and physiologists than those accompanying respiration. If, on the other hand, we know little of the object of this singular function, we know, on the other hand, that it is so essential to life that it cannot be suspended for any time without exposing the animal to danger of immediate death.
Experiments on the Respiration of Animals and the Changes Which Happen to Air in Its Passage Through Their Lungs
Read to the Académie des Sciences
3 May, 1777

RESPONSIBILITY

Teller, Edward   1908–2003
Hungarian-born American nuclear physicist

Beyond the scientific responsibility to search the horizon of human knowledge, the responsibilities of scientists cannot be any greater than those of any other citizen in our democratic society. The consequences of scientific discoveries are the responsibility of the people.
Better a Shield than a Sword: Perspectives in Defense and Technology
Chapter 9 (p. 85)
The Free Press. New York, New York, USA. 1987

REST

Born, Max   1882–1970
German-born English physicist

It is odd to think that there is a word for something which, strictly speaking, does not exist, namely, "rest."
The Restless Universe
Chapter I (p. 1)
Dover Publications, Inc. New York, New York, USA. 1951

RESULT

Dante, Alighieri   1265–1321
Italian poet

Great flame follows a little spark.
In *Great Books of the Western World* (Volume 21)
The Divine Comedy of Dante Alighieri
Paradise
Canto I, l. 34
Encyclopædia Britannica, Inc. Chicago, Illinois, USA. 1952

Gauss, Johann Carl Friedrich   1777–1855
German mathematician, physicist and astronomer

I have had my results for a long; but I do not yet know how I am to arrive at them.
In L. Nelson
Socratic Method and Critical Philosophy
Attributed
Chapter IV (p. 89)
Yale University Press. New Haven, Connecticut, USA. 1949

Gay-Lussac, Joseph Louis   1778–1850
French chemist and physicist

I only present these conclusions with the greatest reserve, knowing myself how I have still to vary my experiments and how easy it is to err in the interpretation of results.
In Maurice Crosland
Gay-Lussac: Scientist and Bourgeois
Chapter 4 (p. 87)
Cambridge University Press. Cambridge, England. 1978

Maxwell, James Clerk   1831–79
Scottish physicist

The first process therefore in the effectual study of science must be one of simplification and reduction of results of previous investigation to a form in which the mind can grasp them. The results of this simplification may take the form of a purely mathematical formula or of a physical hypothesis.
On Faraday's Lines of Force
Transactions of the Cambridge Philosophical Society, 1856

Pauli, Wolfgang   1900–58
Austrian-born physicist

Never work too closely with experimenters. Allow the results to settle.
In Silvan S. Schweber
QED and the Men Who Made It: Dyson, Feynman, Schwinger, and Tomonaga
Chapter 10 (p. 594)
Princeton University Press. Princeton, New Jersey, USA. 1994

Poincaré, Henri   1854–1912
French mathematician and theoretical astronomer

To obtain a result of real worth it will not suffice to grind it out or to have a machine for putting our facts in order. It is not alone order but the unexpected order which is of real worth. The machine may grind upon the mere fact, but the soul of the fact will always escape it.
Annual Report of the Board of Regents of the Smithsonian Institution, 1909
The Future of Mathematics (p. 127)
Government Printing Office. Washington, D.C. 1910

Sagan, Carl   1934–96
American astronomer and author

Every scientist feels an affection for his or her ideas and scientific results. You feel protective of them. But you don't reply to critics: "Wait a minute, wait a minute; this is a really good idea. I'm very fond of it. It's done you no harm. Please don't attack it." That's not the way it goes. The hard but just rule is that if the ideas don't work, you must throw them away. Don't waste any neurons on what doesn't work. Devote those neurons to new ideas that better explain the data. Valid criticism is doing you a favor.
Wonder and Skepticism
Skeptical Inquirer, January/February 1995 (p. 24)

Wilson, Jr., E. Bright   1908–92
American physical chemist

One of the most difficult decisions which an experimenter has to make is whether or not to reject a result which seems unreasonably discordant.... The best procedure to use depends on what is known about the frequency of occurrence of wild values, on the cost of additional observations, and on the penalties for the various types of errors.... There is often a desire to disregard negative results on the grounds that conditions were not right or that the operator was not in the right mood. This is undoubtedly responsible for much pseudo science, psychic phenomena, and similar material.
An Introduction to Scientific Research
Chapter 9 (p. 256, 257, 257)
McGraw-Hill Book Company, Inc. New York, New York USA. 1952

RETROGRADE MOTION

Shakespeare, William   1564–1616
English poet, playwright, and actor

HELENA: Monsieur Parolles, you were born under a charitable star.
PAROLLES: Under Mars, I.
HELENA: I especially think, under Mars.
PAROLLES: Why under Mars?
HELENA: The wars have so kept you under that you must needs be born under Mars.
PAROLLES: When he was predominant.
HELENA: When he was retrograde, I think, rather.
PAROLLES: Why think you so?
HELENA: You go so much backward when you fight.
In *Great Books of the Western World* (Volume 27)
The Plays and Sonnets of William Shakespeare (Volume 2)
All's Well That Ends Well
Act I, Scene I
Encyclopædia Britannica, Inc. Chicago, Illinois, USA. 1952

REVOLUTION

Krauss, Lawrence M.   1954–
American theoretical physicist

Physics progresses not by revolutions, which do away with all that went before, but rather by evolutions, which exploit the best about what is already understood. Newton's laws will continue to be as true a million years from now as they are today, no matter what we discover at the frontiers of science.

The Physics of Star Trek
Chapter One (p. 8)
Harp Perennial Publishers. New York, New York, USA. 1995

RIDDLE

Einstein, Albert   1879–1955
German-born physicist

Out yonder there was this huge world, which exists independently of us human beings and which stands before us like a great, eternal riddle…

Translated by Paul Arthur Schlipp
Albert Einstein: Autobiographical Notes (p. 5)
Open Court. La Salle, Illinois, USA. 1979

RIEMANN HYPOTHESIS

Berry, M. V.
No biographical data available
Keating, J. P.
No biographical data available

If the Riemann Hypothesis is true…the function f(u) constructed from the primes has discrete spectrum; that is, the support of its Fourier transform is discrete. If the Riemann Hypothesis is false this is not the case. The frequencies then are reminiscent of the decomposition of a musical sound into its constituent harmonics. Therefore there is a sense in which we can give a one-line non technical statement of the Riemann hypothesis: "The primes have music in them."

The Riemann Zeros and Eigenvalue Asymptotics
SIAM Review, 41, Number 2 (1999) (p. 238)

Bombieri, Enrico   1940–
Italian mathematician

The failure of the Riemann hypothesis would create havoc in the distribution of prime numbers. This fact alone singles out the Riemann hypothesis as the main open question of prime number theory.

Prime Territory: Exploring the Infinite Landscape at the Base of the Number System
The Sciences, Sept/Oct 1992

Conrey, J. Brian

The Riemann Hypothesis (RH) has been around for more than 140 years, and yet now is arguably the most exciting time in its history to be working on RH. Recent years have seen an explosion of research stemming from the confluence of several areas of mathematics and physics.

The Riemann Hypothesis
Notices of the AMS (March 2003)

du Sautoy, Marcus   1965–
English mathematician and writer

[The Riemann] zeros did not appear to be scattered at random. Riemann's calculations indicated that they were lining up as if along some mystical ley line running through the landscape.

The Music of the Primes
Chapter 4 (p. 99)
HarperCollins Publisher, Inc. New York, New York, USA. 2003

As we shall see, Riemann's Hypothesis can be interpreted as an example of a general philosophy among mathematicians that, given a choice between an ugly world and an aesthetic one, Nature always chooses the latter.

The Music of the Primes
Chapter 2 (p. 55)
HarperCollins Publisher, Inc. New York, New York, USA. 2003

We have all this evidence that the Riemann zeros are vibrations, but we don't know what's doing the vibrating.

The Music of the Primes
Chapter 11 (p. 280)
HarperCollins Publisher, Inc. New York, New York, USA. 2003

As mathematicians navigate their way across the mathematical terrain, it as though all paths will necessarily lead at some point to the same awesome vista of the Riemann Hypothesis.

The Music of the Primes
Chapter 1 (p. 10)
HarperCollins Publisher, Inc. New York, New York, USA. 2003

In an interview, Hilbert explained that he believed the Riemann Hypothesis to be the most important problem "not only in mathematics but absolutely the most important."

The Music of the Primes
Chapter 5 (p. 114)
HarperCollins Publisher, Inc. New York, New York, USA. 2003

A solution to the Riemann Hypothesis offers the prospect of charting the misty waters of the vast ocean of numbers. It represents just a beginning in our understanding of Nature's numbers. If we can only find the secret of how to navigate the primes, who knows what else lies out there, waiting for us to discover?

The Music of the Primes
Chapter 1 (p. 18)
HarperCollins Publisher, Inc. New York, New York, USA. 2003

Until [the RH is proved], we shall listen enthralled by this unpredictable mathematical music, unable to master its twists and turns. The primes have been a constant companion in our exploration of the mathematical world yet they remain the most enigmatic of all numbers. Despite the best efforts of the greatest mathematical minds to explain the modulation and transformation of this mystical music, the primes remain an unanswered riddle. We still await the person whose name will live for ever as the mathematician who made the primes sing.

The Music of the Primes
Chapter 12 (p. 312)
HarperCollins Publisher, Inc. New York, New York, USA. 2003

Erdös, Paul   1913–96
Hungarian mathematician

To conclude, a somewhat daunting quote about the prime numbers from someone who was as familiar with them as anyone has ever been: "It will be millions of years before we'll have any understanding, and even then it won't be a complete understanding, because we're up against the infinite."
Interview with P. Hoffman
Atlantic Monthly, November 1987 (p. 74)

Heath-Brown, R.
Mathematician

[The Riemann Hypothesis has] no longer just analytic number theorists involved, but all mathematicians know about the problem, and many realize that they may have useful insights to offer. As far as I can see, a solution is as likely to come from a probabilist, geometer or mathematical physicist, as from a number theorist.
In K. Sabbagh
The Riemann Hypothesis: The Greatest Unsolved Problem in Mathematics
Chapter 17 (pp. 267–268)
Farrar, Straus & Giroux. New York, New York, USA. 2002

Ivic, A.
No biographical data available

…I don't believe or disbelieve the Riemann Hypothesis. I have a certain amount of data and a certain amount of facts. These facts tell me definitely that the thing has not been settled. Until it's been settled it's a hypothesis, that's all. I would like the Riemann Hypothesis to be true, like any decent mathematician, because it's a thing of beauty, a thing of elegance, a thing that would simplify many proofs and so forth, but that's all.
In K. Sabbagh
The Riemann Hypothesis: The Greatest Unsolved Problem in Mathematics
Chapter 17 (p. 269)
Farrar, Straus & Giroux. New York, New York, USA. 2002

Klarreich, E.
No biographical data available

Proving the Riemann hypothesis won't end the story. It will prompt a sequence of even harder, more penetrating questions. Why do the primes achieve such a delicate balance between randomness and order? And if their patterns do encode the behavior of quantum chaotic systems, what other jewels will we uncover when we dig deeper? Those who believe mathematics holds the key to the Universe might do well to ponder a question that goes back to the ancients: What secrets are locked within the primes?

Prime Time
New Scientist, November 11, 2000

Montgomery, H.
No biographical data available

So if you could be the Devil and offer a mathematician to sell his soul for the proof of one theorem — what theorem would most mathematicians ask for? I think it would be the Riemann Hypothesis.
In K. Sabbagh
The Riemann Hypothesis: The Greatest Unsolved Problem in Mathematics
Chapter 2 (p. 36)
Farrar, Straus & Giroux. New York, New York, USA. 2002

Sometimes I think that we essentially have a complete proof of the Riemann Hypothesis except for a gap. The problem is, the gap occurs right at the beginning, and so it's hard to fill that gap because you don't see what's on the other side of it.
In K. Sabbagh
The Riemann Hypothesis: The Greatest Unsolved Problem in Mathematics
Chapter 17 (p. 267)
Farrar, Straus & Giroux. New York, New York, USA. 2002

Motohashi, Yoichi
No biographical data available

…the Riemann Hypothesis will be settled without any fundamental changes in our mathematical thoughts, namely, all tools are ready to attack it but just a penetrating idea is missing.
In K. Sabbagh
The Riemann Hypothesis: The Greatest Unsolved Problem in Mathematics
Chapter 17 (p. 268)
Farrar, Straus & Giroux. New York, New York, USA. 2002

Sabbagh, K.
No biographical data available

For many mathematicians working on it, $1m is less important than the satisfaction that would come from finding a proof. Throughout my researches among the mathematicians' tribe (I have interviewed 30 in the past year), Riemann's Hypothesis was often described to me in awed terms. Hugh Montgomery of the University of Michigan said this was the proof for which a mathematician might sell his soul. Henryk Iwaniec, a Polish-American mathematician, sounded as if he were already discussing terms with Lucifer.

"I would trade everything I know in mathematics for the proof of the Riemann Hypothesis. It's gorgeous stuff. I'm only worried that I'll be unable to understand it. That would be the worst…"
Beautiful Mathematics
Prospect, January 2002

Sarnak, P.   1953–
South African-born American mathematician

Right now, when we tackle problems without knowing the truth of the Riemann hypothesis, it's as if we have a screwdriver. But when we have it, it'll be more like a bulldozer.
In E. Klarreich
New Scientist
Prime Time, November 11, 2000

The Riemann Hypothesis is the central problem and it implies many, many things. One thing that makes it rather unusual in mathematics today is that there must be over five hundred papers — somebody should go and count — which start Assume the Riemann Hypothesis, and the conclusion is fantastic. And those [conclusions] would then become theorems…With this one solution you would have proven five hundred theorems or more at once.
In K. Sabbagh
The Riemann Hypothesis: The Greatest Unsolved Problem in Mathematics
Chapter 14 (p. 222)
Farrar, Straus & Giroux. New York, New York, USA. 2002

If [the Riemann Hypothesis is] not true, then the world is a very different place. The whole structure of integers and prime numbers would be very different to what we could imagine. In a way, it would be more interesting if it were false, but it would be a disaster because we've built so much round assuming its truth.
In K. Sabbagh
The Riemann Hypothesis: The Greatest Unsolved Problem in Mathematics
Chapter 2 (p. 37)
Farrar, Straus & Giroux. New York, New York, USA. 2002

Stewart, Ian   1945–
English mathematician and science writer

One of the biggest problems of mathematics is to explain to everyone else what it is all about.
The Problems of Mathematics
Chapter 1 (p. 5)
Oxford University Press, Inc. Oxford, England. 1987

RIGHTS OF ANIMALS

Twain, Mark (Samuel Langhorne Clemens)   1835–1910
American author and humorist

I believe I am not interested to know whether Vivisection produces results that are profitable to the human race or doesn't. To know that the results are profitable to the race would not remove my hostility to it. The pains which it inflicts upon unconsenting animals is the basis of my enmity towards it, and it is to me sufficient justification of the enmity without looking further. It is so distinctly a matter of feeling with me, and is so strong and so deeply-rooted in my make and constitution, that I am sure I could not even see a vivisector vivisected with anything more than a sort of qualified satisfaction. I do not say I should not go and look on; I only mean that I should almost surely fail to get out of it the degree of contentment which it ought, of course, to be expected to furnish.
Letter, London Anti-Vivisection Society, May 26, 1899

RIGOR

Poincaré, Henri   1854–1912
French mathematician and theoretical astronomer

In mathematics rigor is not everything, but without it there would be nothing; a demonstration which is not rigorous is void.
Annual Report of the Board of Regents of the Smithsonian Institution, 1909
The Future of Mathematics (p. 127)
Government Printing Office. Washington, D.C. 1910

RISK

Fisher, Irving   1867–1947
American economist

Risk varies inversely with knowledge.
The Theory of Interest
Chapter IX (p. 221)
Porcupine Press. Philadelphia, Pennsylvania, USA. 1977

Florman, Samuel C.   1925–
Author and professional engineer

Good intentions and high moral standards do not help an engineer establish the limits of acceptable risk.
Blaming Technology
Moral Blueprints (p. 173)
St. Martin's Press. New York, New York, USA. 1981

RIVER

Abbey, Edward   1927–89
American environmentalist and nature writer

For the rest of the afternoon, keeping to the shady side, we drift down the splendid river, deeper and deeper and deeper into the fantastic.
Desert Solitaire
Down the River (p. 205)
Ballantine Books. New York, New York, USA. 1968

Burroughs, John   1837–1921
American naturalist and writer

The river idealizes the landscape. It multiplies and heightens the beauty of the day and season. A fair day it

makes more fair, and a wild, tempestuous day it makes more wild. The face of winter makes it doubly rigid and corpse-like, and to the face of spring it adds new youth and sparkle.
The Heart of Burrough's Journal (p. 94)
Houghton Mifflin Company. Boston, Massachusetts, USA.

Coleridge, Samuel Taylor   1772–1834
English lyrical poet, critic, and philosopher

From dark and icy caverns called you forth,
Down those precipitous, black, jagged rocks,
Forever shattered, and the same for ever?
Who gave you your invulnerable life,
Your strength, your speed, your fury and your joy,
Unceasing thunder and eternal foam?
A Library of Poetry
Hymn, Before Sunrise In the Vale of Chamouni, l. 41–46
J.B. Ford & Company. New York, New York, USA. 1874

Confucius   551 BCE–479 BCE
Chinese philosopher and reformer

Men of practical knowledge find their gratification among rivers.
In Lionel Giles (ed.)
The Analects of Confucius
Chinese University Press. Hong Kong. 1983

Dyer, John   1700?–58
Welsh clergyman and poet

And see the rivers how they run
Through woods and meads, in shade and sun,
Sometimes swift, sometimes slow, —
Wave succeeding wave, they go
A various journey to the deep,
Like human life to endless sleep!
In Thomas Campbell
Specimans of the British Poets
Granger Hill, l. 93
John Murray. London, England. 1841

Eliot, T. S. (Thomas Stearns)   1888–1965
American expatriate poet and playwright

I do not know much about gods; but I think that the river
Is a strong brown god — sullen, untamed and intractable
Patient to some degree, at first recognized as a frontier;
Useful, untrustworthy as a conveyor of commerce;
Then only a problem confronting the builder of bridges.
The problem once solved, the brown god is almost forgotten
By the dwellers in cities — ever, however, implacable,
Keeping his seasons and rages, destroyer, reminder
Of what men choose to forget.
Unhonoured, unpropitiated
By worshippers of the machine.
The Collected Poems and Plays 1909–1950

The Dry Salvages (p. 130)
Harcourt, Brace & World, Inc. New York, New York, USA. 1952

Emerson, Ralph Waldo   1803–82
American lecturer, poet, and essayist

The river knows the way to the sea:
Without a pilot it runs and falls,
Blessing all lands with its charity.
The Complete Works of Ralph Waldo Emerson (Volume 9)
Woodnotes
Part ii, Line 272 (p. 57)
Houghton Mifflin Company. Boston, Massachusetts, USA. 1904

Lubbock, Sir John   1834–1913
English banker, author, and scientist

It is as difficult for a river as for a man to get out of a groove.
The Beauties of Nature and the Wonders of the World We Live In
Chapter VIII (p. 279)
The Macmillan Company. New York, New York, USA. 1893

Maury, Matthew Fontaine   1806–73
American hydrographer and naval officer

There is a river in the ocean. In the severest droughts it never fails, and in the mightiest floods it never overflows. Its banks and its bottoms are of cold water, while its current is of warm. The Gulf of Mexico is its fountain, and its mouth is in the Arctic Seas. It is the Gulf Stream. There is in the world no other such majestic flow of waters. Its current is more rapid than the Mississippi or the Amazon.
The Physical Geography of the Sea
Chapter I (p. 25)
Harper & Brothers. New York, New York, USA. 1855

Muir, John   1838–1914
American naturalist

Tracing rivers to their fountains makes the most charming of travels. As the life blood of the landscapes, the best of the wilderness comes to their banks, and not one dull passage is found in all their eventful histories.
Steep Trails
Chapter V (p. 101)
Norman S. Berg, Publisher. Dunwoody, Georgia, USA. 1970

Palmer, Tim   1948–
No biographical data available

Rivers are exquisite in their abilities to nurture life, sublime in functioning detail, impressive in contributions of global significance.
Lifelines: The Case for River Conservation
Chapter One (p. 10)
Island Press. Washington, D.C. 1994

…rivers are magnets for the imagination, for conscious pondering and subconscious dreams, thrills, fears. People stare into the moving water, captivated, as they are when

gazing into a fire. What is it that draws and holds us? The rivers' reflections of our lives and experiences are endless.
Lifelines: The Case for River Conservation
Chapter One (p. 8)
Island Press. Washington, D.C. 1994

Pascal, Blaise   1623–62
French mathematician and physicist

Rivers are roads which move, and which carry us whither we desire to go.
In *Great Books of the Western World* (Volume 33)
Pensées
Section I, 17
Encyclopædia Britannica, Inc. Chicago, Illinois, USA. 1952

Playfair, John   1748–1819
Scottish geologist, physicist, and mathematician

A river, of which the course is both serpentine and deeply excavated in the rock, is among the phenomena by which the slow waste of the land, and also the cause of that waste, are most directly pointed out.
Illustrations of the Huttonian Theory of the Earth
Section 101 (p. 104)
Dover Publications, Inc. New York, New York, USA. 1964

Twain, Mark (Samuel Langhorne Clemens)   1835–1910
American author and humorist

The face of the water, in time, became a wonderful book…which told its mind to me without reserve, delivering its most cherished secrets as clearly as if it uttered them with a voice. And it was not a book to be read once and thrown aside, for it had a new story to tell every day.
Life on the Mississippi
Chapter IX (p. 77)
Harper & Row, Publishers. New York, New York, USA. 1951

Wallace, Alfred Russel   1823–1913
English humanist, naturalist, and geographer

These various-coloured waters may, we believe, readily be accounted for by the nature of the country the stream flows through. The fact that the most purely black-water rivers flow through districts of dense forest, and have granite beds, seems to show that it is the percolation of the water through decaying vegetable matter which gives it its peculiar colour. Should the stream, however, flow through any extent of alluvial country, or through any districts where it can gather much light-coloured sedimentary matter, it will change its aspect, and we shall have the phenomenon of alternating white and black water rivers. The Rio Branco and most of its tributaries rise in an open, rocky country, and the water there is pure and uncoloured; it must, therefore, be in the lower part of its course that it obtains the sediment that gives it so remarkably light a colour; and it is worthy of note, that all the other white-water tributaries of the Rio Negro run parallel to the Rio Branco, and, therefore, probably obtain their sediment from a continuation of the same deposits; only as they flow entirely through a forest district producing brown water, the result is not such a strikingly light tint as in the case of that river.
Journal of the Royal Geographical Society, Volume 23, 1853 (p. 213)

ROBOT

Hey, Nigel S.   1936–
American science writer

The craft is a space robot that is invested with the equivalents of eyes, ears, voice, and muscle. Each spacecraft takes with it the hopes and dreams of thousands of scientists and engineers and, most importantly, the special sense of wonder and imagination that is so great a part of human nature.
Solar System
Chapter 5 (p. 120)
Weidenfield & Nicolson. London, England. 2002

ROCK

Abbey, Edward   1927–89
American environmentalist and nature writer

Rocks, like louseworts and snail darters and pupfish and 3rd -world black, lesbian, feminist, militant poets, have rights, too. Especially the right to exist.
A Voice Crying in the Wilderness: Notes from a Secret Journal
Chapter 9 (p. 86)
St. Martin's Press. New York, New York, USA. 1989

Alvarez, Luis Walter   1911–88
American experimental physicist

Rocks are the key to Earth history, because solids remember but liquids and gasses forget. Retrieving these long-lost memories is the business of geologists ad paleontologists, of people who have chosen to be the historians of the Earth.
T. Rex and the Crater of Doom
Chapter I (p. 17)
Princeton University Press. Princeton, New Jersey, USA. 1997

Author undetermined

There are no books like a rock,
And nothing looks like a rock;
There are no meals like a rock,
And nothing feels like a rock;
Nothing stays like a rock,
Or decays like a rock.
There is nothing wrong with any man here

That can't be cured by putting him near
A lithic, igneous, metamorphic, sedimentary rock.
Roquiescat (Sung to "There Is Nothing Like a Dame")
The Pick and Hammer Club, Washington, D.C., May 2, 1952

Bradley, Jr., John Hodgdon   1898–1962
American geologist

Rocks are the graveyards of the past, and the student of fossil shells and bones sees the grim phenomenon in every guise.
Parade of the Living
Part III, Chapter XVII (p. 238)
Coward-McCann, Inc. New York, New York, USA. 1930

Burroughs, John   1837–1921
American naturalist and writer

The rocks have a history; gray and weather-worn, they are veterans of many battles; they have most of them marched in the ranks of vast stone brigades during the ice age; they have been torn from the hills, recruited from the mountain-tops, and marshaled on the plains and in the valleys; and now the elemental war is over, there they lie waging a gentle but incessant warfare with time, and slowly, oh, so slowly, yielding to its attacks!
Under the Apple-Trees
Chapter II (p. 42)
Houghton Mifflin Company. Boston, Massachusetts, USA. 1916

Daly, Reginald Aldworth   1871–1957
Canadian-American geologist

A final philosophy of the Earth's crust must be largely founded upon the unshakable facts known about igneous rocks.
Igneous Rocks and the Depths of the Earth: Containing Some Revised Chapters of "Igneous Rocks and their Origin"
Chapter I (p. 1)
McGraw-Hill Book Company, Inc. New York, New York, USA. 1933

Darwin, Charles Robert   1809–82
English naturalist

On first examining a new district nothing can appear more hopeless than the chaos of rocks; but by recording the stratification and nature of the rocks and fossils at many points, always reasoning and predicting what will be found elsewhere, light soon begins to dawn on the district, and the structure of the whole becomes more or less intelligible.
In Francis Darwin (ed.)
The Life and Letters of Charles Darwin (Volume 1)
Chapter II (p. 52)
D. Appleton & Company. New York, New York, USA. 1896

de Saint-Exupéry, Antoine   1900–44
French aviator and writer

A rock pile ceases to be a rock pile the moment a single man contemplates it, bearing within him the image of a cathedral.
Translated by Bernard Lamotte
Flight to Arras
Chapter XXII (p. 219)
Reynal & Hitchcock. New York, New York, USA. 1942

Fort, Charles   1874–1932
American writer

"I shall be scientific about it." Said Sir Isaac Newton — or virtually said he — "If there is no change in the direction of a moving body, the direction of a moving body is not changed." "But," continued he, "if something be changed, it is changed as much as it is changed." How do geologists determine the age of rocks? By the fossils in them. And how do they determine the age of fossils? By the rocks they're in. Having started with the logic of Euclid, I go on to the wisdom of Newton.
The Books of Charles Fort
Lo! (pp. 547–548)
Henry Holt & Company. New York, New York, USA. 1941

Le Guin, Ursula K.   1929–
American writer of science fiction and fantasy

The first thing about rocks is, they're old.… Rocks are in time in a different way than living things are, even the ancient trees. But then, the other thing about rocks is that they are place. Rocks are what a place is made of to start with and after all.… The stone is at the center.
Buffalo Gals and Other Animal Presences
Capra Press. Santa Barbara, California, USA. 1987

LeConte, John   1818–91
American physician and physicist

Here, then, we have the oldest known rocks. Are they, then, absolutely the oldest — the primitive rocks, as some imagine? By no means. They are stratified rocks, and therefore consolidated sediments, and therefore, also, the debris of still older rocks, of which we know nothing. Thus, we seek in vain for the absolutely oldest, the primitive crust.
A Compend of Geology
Part III, Chapter II (pp. 263–264)
American Book Company. New York, New York, USA. 1884

Levenson, Thomas
No biographical data available

Rock is the ultimate historian — what it is, and what remnants it contain are the only records of what the earth was like through virtually its entire lifetime.
Ice Time: Climate, Science, and Life on Earth
Chapter 1 (p. 3)
Harper & Row, Publishers. New York, New York, USA. 1989

Linnaeus, Carl (von Linné)   1707–78
Swedish botanist

The stony rocks are not primeval, but daughters of Time.
Systema Naturae
Ed. 5, Stockholm,1748 (p. 219)

Muir, John   1838–1914
American naturalist

Patient observation and constant brooding above the rocks, lying upon them for years as the ice did, is the way to arrive at the truths which are graven so lavishly upon them.
In William Frederic Badé
The Life and Letters of John Muir (Volume 1)
Letter to Mrs. Ezra S. Carr, October 1871 (p. 300)
Houghton Mifflin Company. Boston, Massachusetts, USA. 1924

…all the rocks seemed talkative, and more telling and lovable than ever. They are dear friends, and seemed to have warm blood gushing through their granite flesh; and I love them with a love intensified by long and close companionship.
Steep Trails
Chapter II (p. 19)
Norman S. Berg, Publisher. Dunwoody, Georgia, USA. 1970

O'Keefe, J. A.
No biographical data available

Liquids and gases forget, but rocks remember.
In G. Brent Dalrymple
The Age of the Earth
Chapter 7 (p. 305)
Stanford University Press. Stanford, California, USA. 1991

Read, Herbert Harold   1899–1970
English geologist
Watson, Janet
No biographical data available

…the best geologist is the one who has seen the most rocks.
Beginning Geology
Preface
Macmillan & Company Ltd. London, England. 1966

Seward, A. C.   1863–41
No biographical data available

Rocks are the source-books of geological history…
Plant Life Through the Ages
Chapter II (p. 5)
Hafner Publishing Company. New York, New York, USA. 1959

von Bubnoff, S.
No biographical data available

The materials from which the geologist draws his conclusions are rocks.
Fundamentals of Geology
Chapter II (p. 12)
Oliver & Boyd. Edinburgh, Scotland. 1963

Wells, H. G. (Herbert George)   1866–1946
English novelist, historian, and sociologist

The Record of the Rocks is like a great book that has been carelessly misused. All its pages are torn, worn, and defaced, and many are altogether missing.
The Outline of History (Volume 1)
Book I, Chapter III, Section 3 (p. 29)
Garden City Books. Garden City, New York, USA. 1961

White, Bailey   1950–
American writer

My Aunt Belle loves rocks. Her whole house used to be filled with rocks. Every flat surface was covered with slabs of amethyst crystal, piles of rainbow-colored labradorite, bowls full of fossilized sharks' teeth as big as a child's hand, and agate geodes lined with quartz crystals.… Every afternoon my Aunt Belle takes a bagful of rocks down to Shoney's Restaurant where she spreads them out on the Formica tabletop and says incantations over them while she drinks iced tea.
Sleeping at the Starlite Motel and Other Adventures on the Way Back Home
Rocks (p. 63)
Addison-Wesley Publishing Company. Reading, Massachusetts, USA. 1995

ROCKET

Goddard, Robert H.   1882–1945
American physicist

…I still seem to be alone in my enthusiasm for liquid-fueled rockets, but have a hunch that the time is coming when a good many will want to get aboard the bandwagon…
The Papers of Robert H. Goddard (Volume 3)
R.H. Goddard to T.E. Thompson
March 7, 1941 (p. 1386)
McGraw-Hill Book Company, Inc. New York, New York, USA. 1970

Woolley, Richard   1906–1986
Astronomer Royale of England

The whole procedure [of shooting rockets into space]… presents difficulties of so fundamental a nature, that we are forced to dismiss the notion as essentially impracticable, in spite of the author's insistent appeal to put aside prejudice and to recollect the supposed impossibility of heavier-than-air flight before it was actually accomplished.
Reviewing P.E. Cleator's "Rockets in Space"
Nature, March 14, 1936

ROCKFALL

Muir, John   1838–1914
American naturalist

The sound was inconceivably deep and broad and earnest, as if the whole earth, like a living creature, had at last found a voice and were calling to her sister planets.
Our National Parks
Chapter VIII (p. 263)
Houghton Mifflin Company. Boston, Massachusetts, USA. 1901

ROTATION OF EARTH

Archimedes of Syracuse   287 BCE–212 BCE
Sicilian mathematician

But Aristarchus of Samos brought out a book consisting of certain hypotheses, in which the premises lead to the conclusion that the universe is many times greater than that now so called. His hypotheses are that the fixed stars and the sun remain motion less, that the earth revolves about the sun in the circumference of a circle, the sun lying in the middle of the orbit, and that the sphere of the fixed stars, situated about the same center as the sun, is so great that the circle in which he supposes the earth to revolve bears such a proportion to the distance of the fixed stars as the center of the sphere bears to its surface.
In *Great Books of the Western World* (Volume 11)
The Sand-Recokoner (p. 520)
Encyclopædia Britannica, Inc. Chicago, Illinois, USA. 1952

Bacon, Sir Francis   1561–1626
English lawyer, statesman, and essayist

Nevertheless, in the system of Copernicus there are found many and great inconveniences; for both the loading of the earth with a triple motion is very incommodious, and the separation of the sun from the company of the planets, with which it has so many passions in common, is likewise a difficulty, and the introduction of so much immobility in nature, by representing the sun and stars as immoveable, especially being of all bodies the highest and most radiant, and making the moon revolve about the earth in an epicycle, and some other assumptions of his, are the speculations of one who cares not what fictions he introduces into nature, provided his calculations answer.
Descriptio Globi Intellectualis
Source undetermined

Blundeville, Thomas   fl. 1561
English author

Some also deny that the earth is in the middest of the world, and some affirme that it is mouable, as also Copernicus by way of supposition, and not for that he thought so in deede: who affirmeth that the earth turneth about, and that the sunne standeth still in the midst of the heauens, by help of which false supposition he hath made truer demonstrations of the motions and reuolutions of the celestiall Spheares, than euer were made before…
M. Blundeville His Exercises
Source undetermined

Brahe, Tycho   1546–1601
Danish astronomer

If Nicolaus Copernicus, the distinguished and incomparable master, in this work had not been deprived of exquisite and faultless instruments, he would have left us this science far more well-established. For he, if anybody, was outstanding and had the most perfect understanding of the geometrical and arithmetical requisites for building up this discipline. Nor was he in any respect inferior to Ptolemy; on the contrary, he surpassed him greatly in certain fields, particularly as far as the device of fit ness and compendious harmony in hypotheses is concerned. And his apparently absurd opinion that the Earth revolves does not obstruct this estimate, because a circular motion designed to go on uniformly about another point than the very center of the circle, as actually found in the Ptolemaic hypotheses of all the planets except that of the Sun, offends against the very basic principles of our discipline in a far more absurd and intolerable way than does the attributing to the Earth one motion or another which, being a natural motion, turns out to be imperceptible. There does not at all arise from this assumption so many unsuitable consequences as most people think.
Letter to Christopher Rothman, January 20, 1587
Source undetermined

Plutarch   46–119
Greek biographer and author

Some think that the earth remains at rest. But Philolaus the Pythagorean believes that, like the sun and moon, it revolves around the fire in an oblique circle. Heraclides of Pontus and Ecphantus the Pythagorean make the earth move, not in a progressive motion, but like a wheel in rotation from west to east around its own center.
In Nicholas Copernicus
On the Revolutions of the Heavenly Spheres
Preface (p. 508)
Encyclopædia Britannica, Inc. Chicago, Illinois, USA. 1952

RUIN

Macaulay, Rose   1881–1958
English writer

Of all ruins, possibly the most moving are those of long-deserted cities, fallen century by century into deeper decay, their forsaken streets grown over by forest and shrubs, their decadent buildings quarried and plundered down the years, gaping ruinous, the haunts of lizards and owls…the marble and gold of palaces, the laurel and jasmine of gardens, are now brambles and lagoons; the house built for Caesar is now dwelt in by lizards.…
Pleasure of Ruins
Chapter III (p. 255)
Walker & Company. New York, New York, USA. 1953

RULE

Arnheim, Rudolf   1904–
German-born author, film theorist, and psychologist

An orgy of self-expression is no more productive than blind obedience to rules.
Art and Visual Perception
Introduction (p. vii)
University of California Press. Berkeley, California, USA. 1957

Burton, Robert   1577–1640
English clergyman and scholar

No rule is so general, which admits not some exception…
The Anatomy of Melancholy (Volume 1)
Part I, Sect. II, Memb. II, Subsec. 3 (p. 264)
AMS Press, Inc. New York, New York, USA. 1973

de Cervantes, Miguel   1547–1616
Spanish novelist, playwright, and poet

There is no rule without an exception.
In *Great Books of the Western World* (Volume 29)
The History of Don Quixote de la Mancha
Part II, Chapter 18 (p. 258)
Encyclopædia Britannica, Inc. Chicago, Illinois, USA. 1952

Feyerabend, Paul K.   1924–94
Austrian-born American philosopher of science

…given any rule, however "fundamental" or "necessary" for science, there are always circumstances when it is advisable not only to ignore the rule, but to adopt its opposite.
Against Method: Outline of an Anarchistic Theory of Knowledge
Chapter 1 (p. 23)
Verso. London, England. 1978

Feynman, Richard P.   1918–88
American theoretical physicist

…the fact that there are rules at all to be checked is a kind of miracle; that it is possible to find a rule, like the inverse square law of gravitation, is some sort of miracle. It is not understood at all, but it leads to the possibility of prediction — that means it tells you what you would expect to happen in an experiment you have not yet done.
The Meaning of It All: Thoughts of a Citizen Scientist
Chapter I (p. 23)
Perseus Books. Reading, Massachusetts, USA. 1998

Galsworthy, John   1867–1933
English novelist and dramatist

KEITH: …I don't see the use in drawin' hard and fast rules. You only have to break 'em.
Eldest Son
Act I, Scene 2 (p. 13)
Charles Scribner's Sons. New York, New York, USA. 1913

Gardner, Martin   1914–
American writer and mathematics games editor

I shall add only the fantasy that God or Nature may be playing thousands, perhaps a countless number, of simultaneous Eleusis games [card games with secret rules] with intelligences on planets in the universe…Prophets and False Prophets come and go, and who knows when one round will end and another begin? Searching for any kind of truth is an exhilarating game. It is worth remembering that there would be no game at all unless the rules were hidden.
Mathematical Games
Scientific American, Volume 237, Number 4, October 1977 (p. 25)

Norton, Robert   1875–1932
No biographical data available

…every Art hath certain Rules and Principles…without the knowledge of which no man can attain unto a necessary perfection for practice thereof.…
The Gunner
The Preface to the Courteous Readers (second page)
J. Long. London, England. 1928

Wilson, John   1626–96
No biographical data available

…the Exception proves the Rule.
The Cheats
Appendix, The Author to the Reader, l. 27
W. Patterson. Edinburgh, Scotland. 1874

RUST

Chaucer, Geoffrey   1343–1400
English poet

If gold ruste, what shal iren do?
In *Great Books of the Western World* (Volume 22)
The Canterbury Tales
Prologue
The Parson, l. 50
Encyclopædia Britannica, Inc. Chicago, Illinois, USA. 1952

Kreutzberg, E. C.
No biographical data available

Rust and corrosion mean an enormous loss to Americans, greater than that caused by fire and flood combined, a loss of at least one billion dollars a year. Rust is a skin disease. Corrosion is an infectious internal disease like tuberculosis.
Nickel-Chromium Steels More Widely Used
Iron Trade Review, Volume 86, Number 16, 17 April 1930

Tennyson, Alfred (Lord)   1809–92
English poet

How dull it is to pause, to make an end,
To rust unburnish'd, not to shine in use.
Alfred Tennyson's Poetical Works
Ulysses, l. 22–23
Oxford University Press, Inc. London, England. 1953

S

SAGACITY

Diderot, Denis   1713–84
French encyclopedist and philosopher of materialism

I picture the vast realm of the sciences as an immense landscape scattered with patches of dark and light. The goal towards which we must work is either to extend the boundaries of the patches of light, or to increase their number. One of these tasks falls to the creative genius; the other requires a sort of sagacity combined with perfectionism.
In D. Adams (ed.)
Thoughts on the Interpretation of Nature and Other Philosophical Works
Section XIV (p. 42)
Clinamen Press. Manchester, England. 1999

Locke, John   1632–1704
English philosopher and political theorist

Those intervening ideas, which serve to show the agreement of any two others, are called proofs; and where the agreement or disagreement is by this means plainly and clearly perceived, it is called demonstration; it being shown to the understanding, and the mind made to see that it is so. A quickness in the mind to find out these intermediate ideas, (that shall discover the agreement or disagreement of any other) and to apply them right, is, I suppose, that which is called sagacity.
In *Great Books of the Western World* (Volume 35)
An Essay Concerning Human Understanding
Book VI, Chapter II, Section 3 (p. 310)
Encyclopædia Britannica, Inc. Chicago, Illinois, USA. 1952

Whewell, William   1794–1866
English philosopher and historian

The Conceptions by which Facts are bound together, are suggested by the sagacity of discoverers. This sagacity cannot be taught. It commonly succeeds by guessing; and this success seems to consist in framing several tentative hypotheses and selecting the right one. But a supply of appropriate hypotheses cannot be constructed by rules, nor without inventive talent.
The Philosophy of the Inductive Sciences Founded Upon Their History (Volume 2)
Aphorisms, Aphorisms Concerning Science, VIII (pp. 467–468)
John W. Parker. London, England. 1847

SAMPLE

Bloch, Arthur   1948–
American humorist

After painstaking and careful analysis of a sample, you are always told that it is the wrong sample and doesn't apply to the problem.

Murphy's Law
Fourth Law of Revision (p. 48)
Price/Stern/Sloan, Publishers. Los Angeles, California, USA. 1981

Cochran, William G.   1909–80
Scottish-born American statistician

Our knowledge, our attitudes, and our actions are based to a very large extent on samples.
Sampling Techniques (p. 1)
John Wiley & Sons. New York, New York, USA. 1953

A person's opinion of an institution that conducts thousands of transactions every day is often determined by the one or two encounters which he has had with the institution in the course of several years.
Sampling Techniques (p. 1)
John Wiley & Sons. New York, New York, USA. 1953

Cochran, William G.   1909–80
Scottish-born American statistician
Mosteller, Frederick   1916–2006
American statistician

In 1905, a physicist measuring the thermal conductivity of copper would have faced, unknowingly, a very small systematic error due to the heating of his equipment and sample by the absorption of cosmic rays, then unknown to physics. In early 1946, an opinion poller, studying Japanese opinion as to who won the war, would have faced a very small systematic error due to the neglect of the 17 Japanese holdouts, who were discovered later north of Saipan. These cases are entirely parallel. Social, biological and physical scientists all need to remember that they have the same problem, the main difference being the decimal place in which they appear.
Principles of Sampling
Journal of the American Statistical Association, Volume 49, 1954 (p. 31)

Deming, William Edwards   1900–93
American statistician, educator, and consultant

Sampling is the science and art of controlling and measuring the reliability of useful statistical information through the theory of probability.
Some Theory of Sampling (p. 3)
John Wiley & Sons. New York, New York, USA. 1950

If the cost of classifying a sampling unit were zero, one could always safely recommend fantastic plans of stratified sampling, with no worry about costs. The fact is, though, that there is always a price to pay…
Sample Design in Business Research (p. 320)
John Wiley & Sons. New York, New York, USA. 1960

A good sample-design is lost if it is not carried out according to plans.
Some Theory of Sampling (p. 241)
John Wiley & Sons. New York, New York, USA. 1950

Diconis, Persi 1945–
American mathematician
Mosteller, Frederick 1916–2006
American statistician

The law of truly large numbers states: With a large enough sample, any outrageous thing is likely to happen.
Methods for Studying Coincidences
Journal of the American Statistical Association, Volume 84, 1989 (p. 859)

Gilbert, Sir William Schwenck 1836–1911
English playwright and poet
Sullivan, Arthur 1842–1900
English composer

I've got a little list — I've got a little list...
I've got him on the list...
They never would be missed — they never would be missed!
The Complete Plays of Gilbert and Sullivan
The Mikado
Act I
The Modern Library. New York, New York, USA. 1936

Gissing, George 1857–1903
English novelist

He pointed to a heap of five or six hundred letters, and laughed consumedly.

"Impossible to read them all, you know. It seemed to me that the fairest thing would to be to shake them together, stick my hand in, and take our one by chance. If it didn't seem very promising, I would try a second time."
New Grub Street
The Way Hither (p. 62)
The Modern Library. New York, New York, USA. 1926

McNemar, Quinn 1900–86
American statistician

One does not have to read much of the current research literature in psychology, particularly in individual and social psychology, to realize that there exists a great deal of confusion in the minds of investigators as to the necessity of obtaining a truly representative sample, describing carefully how the sample was secured, and restricting generalizations to the universe, often ill-defined, from which the sample was drawn.
Psychological Bulletin
Sampling in Psychological Research
Journal of the American Statistical Association, Volume 37, Number 6, June 1940 (p. 33)

Mosteller, Frederick 1916–2006
American statistician

...weighing a sample appropriately is no more fudging the data than is correcting a gas volume for barometric pressure.
Principles of Sampling
Journal of the American Statistical Association, Volume 49, Number 265, 1964 (p. 33)

Slonim, Morris James
No biographical data available

Everyone who has poured a highball into the nearest potted plant after taking one sip has had some experience in sampling.
Sampling in a Nutshell (p. 1)
Simon & Schuster. New York, New York, USA. 1960

Sampling is only one component, but undoubtedly the most important one, of that broad based field of scientific method knows as statistics.
Sampling in a Nutshell (p. 7)
Simon & Schuster. New York, New York, USA. 1960

SAND

Carson, Rachel 1907–64
American marine biologist and author

Sand is a substance that is beautiful, mysterious, and infinitely variable; each grain on a beach is the result of processes that go back into the shadowy beginnings of life, or of the earth itself.
The Edge of the Sea
Chapter IV (p. 125)
Houghton Mifflin Company. Boston, Massachusetts, USA. 1955

Charlie Chan
Fictional character

Earthquake may shatter the rock, but sand upon which rock stood, in same old place.
Dark Alibi
Film (1946)

McCord, David 1897–1997
American poet

A handful of sand is an anthology of the universe...
Once and For All
Once and For All (p. 1)
Coward-McCann. New York, New York, USA. 1929

Simon, Anne W.
No biographical data available

The end product, rock's irreducible minimum, is sand. Hold it in your hands and you are in touch with the planet's essence. Each grain has been part of the Earth's solid crust at one time or another, eventually to be freed from rock to exist as a grain again, its particular structure intact.
The Thin Edge: Coast and Man in Crisis
Chapter 2 (pp. 17–18)
Harper & Row, Publishers. New York, New York, USA. 1978

SAVANT

Richet, Charles   1850–1935
French physiologist

For the savant, Science must be a religion. Everything that is discovered, be it great or small, has its origin in this faith.
Translated by Sir Oliver Lodge
The Natural History of a Savant
Chapter VI (p. 47)
J.M. Dent & Sons Ltd. London, England. 1927

SCATTERING

Rutherford, Ernest   1871–1937
English physicist

It was quite the most incredible event that has ever happened to me in my life. It was almost as incredible as if you fired a 15-inch shell at a piece of tissue paper and it came back and hit you. On consideration, I realized that this scattering backward must be the result of a single collision, and when I made calculations I saw that it was impossible to get anything of that order of magnitude unless you took a system in which the greater part of the mass of the atom was concentrated in a minute nucleus. It was then that I had the idea of an atom with a minute massive center carrying a charge.
In Joseph Needham and W. Pagel (ed.)
Background to Modern Science
From Aristotle to Galileo
The Development of the Theory of Atomic Structure (p. 68)
The Macmillan Company. New York, New York, USA. 1938

SCAVENGER

Austin, Mary Hunter   1868–1934
American novelist and essayist

Once at Red Rock, in a year of green pasture, which is a bad time for the scavengers, we saw two buzzards, five ravens, and a coyote feeding on the same carrion, and only the coyote seemed ashamed of the company.
The Land of Little Rain
The Scavengers (pp. 53–54)
Houghton Mifflin Company. Boston, Massachusetts, USA. 1903

SCENERY

Muir, John   1838–1914
American naturalist

The scenery is mostly of a comfortable, assuring kind, grand and inspiring without too much of that dreadful overpowering sublimity and exuberance which tend to discourage effort and cast people into inaction and superstition.

Steep Trails
Chapter XXI (p. 272)
Norman S. Berg, Publisher. Dunwoody, Georgia, USA. 1970

SCIATICA

Shakespeare, William   1564–1616
English poet, playwright, and actor

Thou cold sciatica,
Cripple our senators, that their limbs may halt
As lamely as their manners!
In *Great Books of the Western World* (Volume 27)
The Plays and Sonnets of William Shakespeare (Volume 2)
Timon of Athens
Act IV, Scene i, l. 23–25
Encyclopædia Britannica, Inc. Chicago, Illinois, USA. 1952

Which of your hips has the most profound sciatica?
In *Great Books of the Western World* (Volume 27)
The Plays and Sonnets of William Shakespeare (Volume 2)
Measure for Measure
Act I, Scene ii, l. 58
Encyclopædia Britannica, Inc. Chicago, Illinois, USA. 1952

SCIENCE

Abbey, Edward   1927–89
American environmentalist and nature writer

That which today calls itself science gives us more and more information, an indigestible glut of information, and less and less understanding.
Down the River
Part I
Down the River with Henry Thoreau
7 November 1980 (p. 29)
E.P. Dutton & Company. New York, New York, USA. 1982

Science is the whore of industry and the handmaiden of war.
A Voice Crying in the Wilderness: Notes from a Secret Journal
Chapter 10 (p. 93)
St. Martin's Press. New York, New York, USA. 1989

Adams, Henry Brooks   1838–1918
American man of letters

No sand-blast of science had yet skimmed off the epidermis of history, thought, and feeling.
In Ernest Samuels (ed.)
The Education of Henry Adams
Chapter VI (p. 90)
Houghton Mifflin Company. Boston, Massachusetts, USA. 1974

Man has mounted science, and is now run away with. I firmly believe that before many centuries more, science will be the master of men.
In J.C. Levenson, E. Samuels, C. Vandersee and V. Hopkins (eds.)
The Letters of Henry Adams: 1858–1868 (Volume 1)
Letter to Charles Francis Adams, Jr.
April 11, 1862 (p. 290)
Houghton Mifflin Company. Boston, Massachusetts, USA. 1938

My belief is that science is to wreck us, and that we are like monkeys monkeying with a loaded shell; we don't in the least know or care where our practically infinite energies come from or will bring us to.
Letters of Henry Adams (Volume 2)
Letter, August 10, 1902, to Brooks Adams (p. 392)
Houghton Mifflin Company. Boston, Massachusetts, USA. 1938

Akenside, Mark   1721–70
English poet and physician

Speak, ye, the pure delight, whose favour'd steps
The lamp of science, through the jealous maze
Of nature guides, when haply you reveal
Her secret honours.
The Poetical Works of Mark Akenside
The Pleasures of Imagination, Part II
Associated University Presses. Cranbury, New Jersey, USA. 1996

Alighieri, Dante   1265–1321
Italian poet and writer

…no Science demonstrates its own subject, but presupposes it.
The Convivo of Dante Alighieri
The Second Treatise, Chapter XIV (p. 114)
J.M. Dent & Sons Ltd. London, England. 1912

Allport, Susan   1950–
Naturalist and science writer

Science advances, it seems, less through scientific consensus than by means of a scientific melee, a free-for-all in which every scientist pushes his or her piece of the truth, knowing that only time will tell which piece best fits reality.
Explorers of the Black Box: The Search for the Cellular Basis of Memory
Chapter Ten (p. 263)
W.W. Norton & Company, Inc. New York, New York, USA.1986

Alves, Reuben
No biographical data available

Science is what it is, not what scientists think they do.
New York Times, July 13, 1979, A8 (p. 128)

Amiel, Henri-Frédéric   1821–81
Swiss philosopher, poet, and critic

Society lives by faith, develops by science.
Translated by Mrs. Humphrey Ward
Amiel's Journal
May 7, 1870 (p. 216)
A.L. Burt Company, Publishers. New York, New York, USA. 189?

Appleyard, Bryan   1951–
English author and journalist

Science is not a neutral or innocent commodity which can be employed as a convenience.… Rather it is spiritually corrosive, burning away at ancient authorities and traditions. It has shown itself unable to coexist with anything.

Understanding the Present: Science and the Soul of Modern Man
Chapter 1 (p. 9)
Doubleday. New York, New York, USA. 1992

Arabic Proverb

Science is a plant whose roots indeed are at Mecca, but its fruit ripens at Herat.
In Robert Christy
Proverbs, Maxims and Phrases of All Ages (p. 236)
G.P. Putnam's Sons. New York, New York, USA. 1888

Asimov, Isaac   1920–92
American author and biochemist

Science does not promise absolute truth, nor does it consider that such a thing necessarily exists. Science does not even promise that everything in the Universe is amenable to the scientific process.
"X" Stands for Unknown
Introduction (p. 10)
Doubleday & Company, Inc. Garden City, New York, USA. 1984

The process of science…involves a slow forward movement through the reachable portions of the Universe — a gradual unfolding of parts of the mystery.
"X" Stands for Unknown
Introduction (p. 10)
Doubleday & Company, Inc. Garden City, New York, USA. 1984

Astbury, William Thomas   1898–1961
English crystallographer and molecular biologist

…science is truly one of the highest expressions of human culture — dignified and intellectually honest, and withal a never-ending adventure. Personally, I feel much the same with regard to the more ecstatic moments in science as I do with regard to music. I see little difference between the thrill of scientific discovery and what one experiences when listening to the opening bars of the Ninth Symphony.
Science in Relation to the Community
School Science Review, Number 109, 1948 (p. 279)

Author undetermined

Science has no fear for dissent or for heresy. She collects facts eagerly, steadily, from generation to generation, the labors of one investigator being added to those of another, the speculations of one coalescing with those of another, in virtue of a necessary and admirable solidarity. Then, from these facts patiently observed, brought together, coordinated, classified, science deduces a law, a positive law, which is the expression of reality, of truth itself.
Scientific Miscellany
The Galaxy, Volume 17, January 1874 (p. 130)

The wonders of the heavens seem inexhaustible; each new adventure of science tasks the imagination and almost staggers the reason.

Scientific Miscellany
The Galaxy, Volume 11, February 1871 (p. 297)

...science has given us a new reading of nature, has opened the higher questions of life and human relations, has furnished a new method to the mind, and is fast becoming a new power in literature.
Scientific Miscellany
The Galaxy, Volume 11, January 1871 (p. 135)

Bacon, Sir Francis   1561–1626
English lawyer, statesman, and essayist

Those who have treated of the sciences have been either empirics or dogmatical. The former like ants only heap up and use their store, the latter like spiders spin out their own webs. The bee, a mean between both, extracts matter from the flower of the garden and the field, but works and fashions it by its own efforts.
In *Great Books of the Western World* (Volume 30)
Novum Organum
First Book, Aphorism 95 (p. 126)
Encyclopædia Britannica, Inc. Chicago, Illinois, USA. 1952

The divisions of the sciences are not like different lines that meet in one angle, but rather like the branches of trees that join in one trunk.
In J.A. Thomson
Introduction to Science
Chapter IV (p. 92)
Williams & Norgate Ltd. London, England. 1916

Science, being the wonder of the ignorant and unskillful, may be not absurdly called a monster. In figure and aspect it is represented as many-shaped, in allusion to the immense variety of matter with which it deals. It is said to have the face and voice of a woman, in respect of its beauty and facility of utterance. Wings are added because the sciences and the discoveries of science appeared and fly-aboard in an instant; the communication of knowledge being like that of one candle with another, which lights up at once. Claws, sharp and hooked, are ascribed to it with great elegance, because the axioms and arguments of science penetrate and hold fast the mind, so that it has no means of evasion or escape.
In Hugh Dick (Ed(ed.)
Selected Writings of Francis Bacon
Sphinx on Science (pp. 418–419)
Random House, Inc. New York, New York, USA. 1955

Even the effects already discovered are due to chance and experiment, rather than to the sciences; for our present sciences are nothing more than peculiar arrangements of matters already discovered, and not methods for discovery or plans for new operations.
In *Great Books of the Western World* (Volume 30)
Novum Organum
First Book, Aphorism 8 (p. 107)
Encyclopædia Britannica, Inc. Chicago, Illinois, USA. 1952

...the real and legitimate goal of the sciences is the endowment of human life with new inventions and riches.
In *Great Books of the Western World* (Volume 30)
Novum Organum
First Book, Aphorism 81 (p. 120)
Encyclopædia Britannica, Inc. Chicago, Illinois, USA. 1952

Balard, Antoine-Jérôme   1802–76
French chemist

Science appears to have as its mission not merely the satisfaction of man's need of learning and understanding everything, which characterizes the noblest of our faculties; it has another aim, doubtless less brilliant but perhaps more moral, I would almost say more sacred, which consists in coordinating the forces of nature to increase production and make men more nearly equal by the universality of comfort.
In Mary Elvira Weeks
The Discovery of the Elements (p. 438)
Journal of Chemical Education. Easton Pennsylvania, USA. 1956

Barnett, Lincoln   1909–79
American science writer

The quick harvest of applied science is the useable process, the medicine, the machine. The shy fruit of pure science is Understanding.
Life, January 9, 1950

Baruch, Bernard M.   1870–1965
American presidential advisor

Science has taught us how to put the atom to work. But to make it work for good instead of evil lies in the domain dealing with the principles of human duty. We are now facing a problem more of ethics than physics.
The Baruch Plan for Banning the Atom Bomb
Life, 24 June, 1946 (p. 35)

Barzun, Jacques   1907–
French-born American educator, historian, and educator

It is not clear to anyone, least of all the practitioners, how science and technology in their headlong course do or should influence ethics and law, education and government, art and social philosophy, religion and life of the affections. Yet science is an all-pervasive energy, for it is at once a mode of thought, a source of strong emotion, and a faith as fanatical as any in history.
Science: The Glorious Entertainment
To the Reader (p. 3)
Harper & Row, Publishers. New York, New York, USA. 1964

Baskerville, Charles
No biographical data available

I like to fancy scientific endeavor as the sea — calm and serene, supporting and mirroring that which is below it, bearing that which is upon it, reaching to and reflecting

that which is above it, moving all the while; yet, torn and rent at times by conflict from without and contest within, it runs; it beats against the shores of the unknown, making rapid progress here, meeting stubborn resistance there, compassing it, to destroy but to rebuild elsewhere; and the existence of those within it!
The Elements: Verified and Unverified
Science, New Series, Volume 19, Number 472, 15 January 1904 (p. 100)

Bass, William M.
American forensic anthropologist

Always think of the consequences of your actions both in the field and in the laboratory.
Human Osteology: A Laboratory and Field Manual of the Human Skeleton (3rd edition)
Missouri Archaeological Society. Columbia, Missouri, USA. 1987

Bates, Marston   1906–74
American zoologist

Science has put man in his place; one among the millions of kinds of living things crawling around on the surface of a minor planet circling a trivial star.
The Forest and the Sea: A Look at the Economy of Nature and the Ecology of Man
Chapter 1 (p. 5)
Random House, Inc. New York, New York, USA. 1960

Baudrillard, Jean   1929–
French cultural theorist

We can no longer say things appear unintelligible because science does not know enough about them. It seems that the more we know about them, the more unintelligible they become.
Translated by Chris Turner
Cool Memories
October 1983 (p. 144)
Verso. London, England. 1990

Bauer, Henry H.   1931–
American chemist

Quite in general, it is not the case that, because science has changed its mind in the past, therefore it might change its mind again in any direction and by any amount.
Scientific Literacy and the Myth of the Scientific Method
Chapter 4 (p. 66)
University of Illinois Press. Urbana, Illinois, USA. 1992

That science is inescapably a human activity does not mean that it is only or just a human activity, essentially similar to all other human activities.
Scientific Literacy and the Myth of the Scientific Method
Chapter 7 (p. 141)
University of Illinois Press. Urbana, Illinois, USA. 1992

That science is not everything should not blind us to the fact that it is the very best of what we do have. Just as those who benefit from individual therapy can take pride

from the persistent acts of will they exerted along the way, so humankind can take collective pride from the persistent determination to submit to reality therapy that has produced not only the science we now know but also an understanding of how to go about learning more.
Scientific Literacy and the Myth of the Scientific Method
Chapter 7 (p. 150)
University of Illinois Press. Urbana, Illinois, USA. 1992

Beard, Charles A.   1874–1948
American historian

A revolution in thought is at hand, a revolution as significant as the Renaissance: the subjection of science to ethical and esthetic purpose. Hence the next great survey undertaken in the name of the social sciences may begin boldly with a statement of values agreed upon, and then utilize science to discover the conditions, limitations, and methods involved in realization.
Limitations to the Application of Social Science Implied in Recent Social Trends
Social Forces, Volume 11, Number 4, May 1933 (p. 510)

Beattie, James   1735–1803
Scottish poet and essayist

'Twas thus by the glare of false science betray'd,
That leads, to bewilder; and dazzles, to blind…
The Complete Poetical Works of Gray, Beattie, Blair, Collins, Thomson, and Kirke White
The Hermit, Stanza 5
J. Blackwood. London, England. 1800

Beebe, William   1877–1962
American ornithologist

Thus was the ending still unfinished, the finale buried in the future — and in this we find the fascination of Nature and of Science. Who can be bored for a moment in the short existence vouchsafed us here; with dramatic beginnings barely hidden in the dust, with the excitement of every moment of the present, and with all of cosmic possibility lying just concealed in the future, whether of Betelgueze, of Amoeba or — of ourselves?
Edge of the Jungle
Chapter XII (p. 294)
Garden City Publishing Company, Inc. Garden City, New York, USA. 1925

Bennett, William Cox   1820–95
American poet

To what new realms of marvel, say,
Will conquering science war its way?
Poems
To a Boy
Stanza 1
Chapman & Hall. London, England. 1850

Bernal, John Desmond   1901–71
Irish-born physicist and x-ray crystallographer

Science is one of the most absorbing and satisfying pastimes, and as such it appeals in different ways to different types of personality. To some it is a game against the unknown where one wins and no one loses; to others, more humanly minded, it is a race between different investigators as to who should first wrest the prize from nature. It has all the qualities which make millions of people addicts of the crossword puzzle or the detective story, the only difference being that the problem has been set by nature or chance and not by man, that the answer cannot be got with certainty, and when they are found often raise far more questions than the original problem.
The Social Function of Science (p. 97)
The Macmillan Company. New York, New York, USA. 1939

…it is not possible in any published book to speak freely and precisely about the way science is run. The law of liable, reasons of State, and still more the unwritten code of the scientific fraternity itself forbid particular examples being held up alike for praise or blame.
The Social Function of Science (p. xv)
The Macmillan Company. New York, New York, USA. 1939

Bernard, Claude   1813–78
French physiologist

…my idea of the science of life…it is a superb and dramatically lighted hall which may be reached only by passing through a long and ghastly kitchen.
Translated by Henry Copley Greene
An Introduction to the Study of Experimental Medicine
Part One, Chapter III, Section iii (p. 15)
Henry Schuman, Inc. New York, New York, USA. 1927

Berthelot, Marcellin   1827–1907
French chemist

Science is essentially a collective work, pursued during the course of time by the efforts of a multitude of workers of every age and of every nation, succeeding themselves and associating by virtue of a tacit understanding for the search for pure truth and for the applications of that truth to the continuous betterment of the condition of all mankind.
In Camille Matignon
Annual Report of the Board of Regents of the Smithsonian Institution, 1907
Marcellin Berthelot (p. 684)
Government Printing Office. Washington, D.C. 1908

Birch, Arthur J.   1915–95
Australian chemist

Some people derive satisfaction from accumulating data, whereas others are content to dream and leave experiments to colleagues. Still others flit from flower to flower rather than learning more and more about one situation. The difference in approach is a matter of temperament, and we all must understand our own strengths. All workers ultimately contribute to the matrix of facts, ideas,

understandings, techniques, and visions that we know as science.
To See the Obvious
Random Conversations with the Editor (p. 195)
American Chemical Society. Washington, D.C. 1995

Black, Hugh
No biographical data available

The limits of science are not limits of its methods, but limits of its spheres.
Our Made-Over World
Everybody's Magazine, November 1914 (p. 710)

Black, Joseph   1728–99
Scottish chemist and physician

…if science be the discovery of the laws of nature, the knowledge of these laws will enable us to foresee what will be the result of any process, and must point out to us, in all cases, the means, and the best means, for producing any desired chemical effect: and here does our science repay, with a liberality unparalled in any other science, all her former obligations to the arts of life. From them did she borrow the many facts which excited her to speculate; and her occupation has at last enabled her to repay her debts with large interest, while she has grown rich in knowledge almost beyond hope.
Lectures on the Elements of Chemistry (Volume 1)
Lectures on Chemistry
Definitions (p. 20)
Printed for Mathew Carey. Philadelphia, Pennsylvania, USA. 1807

Blake, William   1757–1827
English poet, painter, and engraver

Art is the Tree of Life; Science is the Tree of Death
The Complete Poetry and Prose of William Blake
The Laocoön
University of California Press. Berkeley, California, USA. 1982

Blavatsky, Elena Petrovna   1831–91
Russian-born American theosophist

If there were such a thing as a void, a vacuum in Nature, one ought to find it produced, according to a physical law, in the minds of helpless admirers of the "lights" of Science, who pass their time in mutually destroying their teachings.
The Secret Doctrine
Section 17
Theosophy Company. Los Angeles, California, USA. 1925

Bloom, Allan   1930–92
American philosopher

Science, in freeing men, destroys the natural condition that makes them human. Hence, for the first time in history, there is the possibility of a tyranny grounded not on ignorance, but on science.
The Closing of the American Mind: How Higher Education Has Failed Democracy and Impoverished the Souls of Today's Students

Part Three, Swift's Doubts (p. 295)
Simon & Schuster. New York, New York, USA. 1987

Boas, George 1891–1980
American philosophy professor

Science is the art of understanding nature.
In Laurence M. Gould
Science and the Culture of Our Times
UNESCO Courier, February 1968 (p. 6)

Bohm, David 1917–92
American physicist
Peat, D.
No biographical data available

Science is essentially a public and social activity.
Science, Order, and Creativity
Chapter Two (p. 67)
Bantam Books. New York, New York, USA. 1987

Science is an attempt to understand the universe and humanity's relationship to nature.
Science, Order, and Creativity
Chapter One (p. 16)
Bantam Books. New York, New York, USA. 1987

Science is, however, at least in principle, dedicated to seeing any fact as it is, and to being open to free communication with regard not only to the fact itself, but also to the point of view from which it is interpreted.
Science, Order, and Creativity
Chapter Six (pp. 241–242)
Bantam Books. New York, New York, USA. 1987

Although science literally means "knowledge," the scientific attitude is concerned much more with rational perception through the mind and with testing such perceptions against actual fact, in the form of experiments and observations.
Science, Order, and Creativity
Chapter Six (p. 260)
Bantam Books. New York, New York, USA. 1987

Bohr, Niels Henrik David 1886–1962
Danish physicist

The importance of physical science for the development of general philosophical thinking rests not only on its contributions to our steadily increasing knowledge of that nature of which we ourselves are part, but also on the opportunities which time and again it has offered for examination and refinement of our conceptual tools.
Atomic Physics and Human Knowledge
Introduction (p. 1)
John Wiley & Sons, Inc. New York, New York, USA. 1958

The task of science is both to extend the range of our experience and reduce it to order.
Atomic Theory and the Description of Nature
Introductory Survey (p. 1)
Cambridge University Press. Cambridge, England. 1934

It is, indeed, perhaps the greatest prospect of humanistic studies to contribute through an increasing knowledge of the history of cultural development to that gradual removal of prejudices which is the common aim of all science.
Atomic Physics and Human Knowledge
Natural Philosophy and Human Cultures (p. 31)
John Wiley & Sons, Inc. New York, New York, USA. 1958

Bolton, Henry Carrington 1843–1903
American chemist, bibliographer, and historian

So rapid are the strides made by science in this progressive age and so boundless is its range, that those who view its career from without find great difficulty in following its diverse and intricate pathways, while those who have secured a footing within the same road are often quite unable to keep pace with its fleet movements and would fain retire from the unequal contest. It is not surprising, then, that those actually contributing to the advancement of science, pressing eagerly upward and onward, should neglect to look back upon the labors of those who precede them and should sometimes lose sight of the obligations which science owes to forgotten generations.
Notes on the Early Literature of Chemistry
Reprinted from *The American Chemist*, November 1875

Bonaparte, Napoleon 1769–1821
French soldier and emperor of France

The sciences, which have revealed so many secrets and destroyed so many prejudices, are destined to render us yet greater service. New truths, new discoveries will unveil secrets still more essential to the happiness of men — but only if we give our esteem to the scientists and our protection to the sciences.
In J. Christopher Herold (ed.)
The Mind of Napoleon
Science and the Arts (p. 135)
Columbia University Press. New York, New York, USA. 1955

Bondi, Sir Hermann 1919–2005
English mathematician and cosmologist

Throughout science there is a constant alternation between periods when a particular subject is in a state of order, with all known data falling neatly into their places, and a state of puzzlement and confusion, when new observations throw all neatly arranged ideas into disarray.
In Robert M. Hutchins and Mortimer J. Adler (eds.)
The Great Ideas Today, 1966
Astronomy and the Physical Sciences (p. 245)
Encyclopædia Britannica, Inc. Chicago, Illinois, USA. 1966

Born, Max 1882–1970
German-born English physicist

The satisfaction of the noble curiosity of the scholar is only one aspect of research. Science is also — and many say predominantly — a collective effort to obtain power over the forces of nature in the interest of human life.

The Restless Universe
Postscript (p. 297)
Dover Publications, Inc. New York, New York, USA. 1951

I now regard my former belief in the superiority of science over other forms of human thought and behavior as a self-deception due to youthful enthusiasm over the clarity of scientific thinking as compared with the vagueness of metaphysical systems.

Still, change of fundamental concepts and the failure to improve the moral standards of human society are no demonstration of the uselessness of science in the search for truth and for a better life.
Physics in My Generation
Preface (p. v)
Springer-Verlag New York, Inc. New York, New York, USA. 1969

Bosler, Jean
No biographical data available

As science advances, new questions appear before indeed the older ones, often badly put, are solved. But the latter often lose their interest, and as we proceed many untenable hypotheses which darkened our path are destroyed. And so, little by little, the knowledge we have of things progresses with a tidal motion which will doubtless end only with humanity.
Annual Report of the Board of Regents of the Smithsonian Institution, 1914
Modern Theories of the Sun (p. 160)
Government Printing Office. Washington, D.C. 1915

Bradbury, Ray   1920–
American writer

At base, science is no more than an investigation of a miracle we can never explain, and art is an interpretation of that miracle.
Ray Bradbury: 100 of His Most Celebrated Tales
June 2001: And the Moon Still Be as Bright (p. 421)
HarperCollins Publishers, Inc. New York, New York, USA. 2003

Bradley, Jr., John Hodgdon   1898–1962
American geologist

Thus along many battle fronts, the army of science advances on the strongholds of ignorance in the heart of earth. Speculations bristle like bayonets and collapse like papier-mâché. But though mistakes retard and darkness confuses, the army presses on.
Autobiography of Earth
Chapter VIII (p. 247)
Coward-McCann, Inc. New York, New York, USA. 1935

Brecht, Bertolt   1898–1956
German writer

GALILEO: One of the main reasons why the sciences are so poor is that they imagine they are so rich. It isn't their job to throw open the door to infinite wisdom but to put a limit to infinite error.
Translated by John Willett
Life of Galileo
Scene 9 (p. 74)
Arcade Publishing. New York, New York, USA. 1994

ANDREA: Science makes only one demand: contribute to science.
Translated by John Willett
Life of Galileo
Scene 14 (p. 107)
Arcade Publishing. New York, New York, USA. 1994

Bremer, J.
No biographical data available

What, then, is science according to common opinion? Science is what scientists do. Science is knowledge, a body of information about the external world. Science is the ability to predict. Science is power, it is engineering. Science explains, or gives causes and reasons.
What Is Science? Notes on the Nature of Science (pp. 37–38)
Harcourt, Brace & World, Inc. New York, New York, USA. 1962

Brewster, Edwin Tenney   1866–1960
Educator

Science, in fact, begins only as men confine themselves to accounting for the unknown by the known — not by the unknown by something about which they know still less.
This Puzzling Planet
Chapter V (p. 93)
The Bobbs-Merrill Company. Indianapolis, Indiana. 1928

Bridgman, Percy Williams   1882–1961
American physicist

Science is intelligence in action with no holds barred.
In Theodore Schick, Jr., and Lewis Vaughn
How to Think About Weird Things
Chapter 7 (p. 164)
The McGraw-Hill Companies. New York, New York, USA. 2002

Broad, William   1951–
Science writer
Wade, Nicholas
British-born scientific writer

In the acquisition of new knowledge, scientists are not guided by logic and objectivity alone, but also by such nonrational factors as rhetoric, propaganda, and personal prejudice. Scientists do not depend solely on rational thought, and have no monopoly on it. Science should not be considered the guardian of rationality in society, but merely one major form of its cultural expression.
Betrayers of the Truth (p. 9)
Simon & Schuster. New York, New York, USA. 1982

Science is not an abstract body of knowledge, but man's understanding of nature. It is not an idealized interrogation of nature by dedicated servants of truth, but a human

process governed by the ordinary human passions of ambition, pride, and greed, as well as by all the well-hymned virtues attributed to men of science.
Betrayers of the Truth (p. 223)
Simon & Schuster. New York, New York, USA. 1982

Bronowski, Jacob   1908–74
Polish-born British mathematician and polymath

Science has nothing to be ashamed of even in the ruins of Nagasaki. The shame is theirs who appeal to other values than the human imaginative value which science has evolved.
Science and Human Values
The Sense of Human Dignity (p. 73)
Harper & Row, Publishers. New York, New York, USA. 1965

The world today is made, it is powered by science; and for any man to abdicate an interest in science is to walk with open eyes towards slavery.
Science and Human Values
The Creative Mind (p. 6)
Harper & Row, Publishers. New York, New York, USA. 1965

The discoveries of Science, the works of art are explorations — more, are explosions, of a hidden likeness.
Science and Human Values
The Creative Mind (p. 19)
Harper & Row, Publishers. New York, New York, USA. 1965

The values of science derive neither from the virtues of its members, nor from the finger-wagging codes of conduct by which every profession reminds itself to be good. They have grown out of the practice of science, because they are the inescapable conditions for its practice.
Science and Human Values
The Sense of Human Dignity (p. 60)
Harper & Row, Publishers. New York, New York, USA. 1965

Like the voyages of the Spaniards into the fabulous West, Science even at its boldest does the will of history, and in turn helps to determine its movement.
The Common Sense of Science
Chapter VII, Section 1 (p. 97)
Harvard University Press. Cambridge, Massachusetts, USA. 1953

[Science] does not watch the world, it tackles it.
The Common Sense of Science
Chapter VII, Section 4 (p. 104)
Harvard University Press. Cambridge, Massachusetts, USA. 1953

All science is the search for unity in hidden likenesses.
Science and Human Values
The Creative Mind (p. 13)
Harper & Row, Publishers. New York, New York, USA. 1965

[This] is the essence of science: ask an impertinent question, and you are on the way to a pertinent answer.
The Ascent of Man
Chapter 4 (p. 153)
Little, Brown & Company. Boston, Massachusetts, USA. 1973

Bryan, William Jennings   1860–1925
American lawyer, orator, and politician

Evolution seems to close the heart to some of the plainest spiritual truths while it opens the mind to the wildest guesses advanced in the name of science.
The New York Times, Letter, 22 February 1922

Christians desire that their children shall be taught all the sciences, but they do not want them to lose sight of the Rock of Ages while they study the age of rocks…
Speech
Prepared for the Scopes Trial, 1925

Bryson, Bill   1951–
American author

The remarkable position in which we find ourselves is that we don't actually know what we actually know.
A Short History of Nearly Everything
Chapter 23 (p. 362)
Broadway Books. New York, New York, USA. 2003

Buchner, Ludwig   1824–99
German physician and philosopher

Science has gradually taken all the positions of the childish belief of the peoples; it has snatched thunder and lightning from the hands of the gods; the eclipse of the stars, and the stupendous powers of the Titans of the olden time have been grasped by the fingers of man.
Force and Matter
Chapter VI (p. 34)
Trubner & Company. London, England. 1864

Buckham, John Wright
No biographical data available

…the mind that has been trained simply or predominately in Science is an unconsciously meager and ill-furnished mind. The range of its interests is mainly technical and specialized. To look into a mind of this type is like looking into a laboratory. It is excellent as a workshop, but there are no pictures on the walls, no books, no flowers…. What are the resources of such a mind, its points of contact with human-kind?
The Passing of the Scientific Era
The Century Illustrated Monthly Magazine, August 1929 (p. 435)

Meredith, Owen (Edward Robert Bulwer-Lytton, 1st Earl Lytton)   1831–91
English statesman and poet

…science is not a club, it is an ocean; it is open to the cockboat as the frigate. One man carries across it a freightage of ingots, another may fish there for herrings. Who can exhaust the sea? who say to intellect, "the deeps of philosophy are preoccupied?"
The Caxtons
Book IV, III
G. Routledge & Company. London, England. 1848

Bunge, Mario   1919–
Argentine philosopher and physicist

The motto of science is not just *Pauca* but rather *Plurima ex paucissimis* — the most out of the least.
The Myth of Simplicity: Problems of Scientific Philosophy
Chapter 5, Section 4 (p. 82)
Prentice-Hall. Englewood Cliffs, New Jersey, USA. 1963

Bunting, Basil   1900–85
English modernist poet

I hate Science. It denies a man's responsibility for his own deeds, abolishes the brotherhood that springs from God's fatherhood. It is a hectoring, dictating expertise, which makes the least lovable of the Church Fathers seem liberal by contrast. It is far easier for a Hitler or a Stalin to find a mock-scientific excuse for persecution than it was for Dominic to find a mock-Christian one.
In Victoria Forde
The Poetry of Basil Bunting
Chapter 6, Letter of 1 January 1947 to Louis Zukofsky (p. 156)
Bloodaxe Books, Newcastle upon Tyne, England. 1991

Burroughs, John   1837–1921
American naturalist and writer

Science has made or is making the world over for us. It has builded us a new house — builded it over our heads while we were yet living in the old, and the confusion and disruption and the wiping-out of the old features and the old associations, have been, and still are, a sore trial — a much finer, more spacious and commodious house…but new, new, all bright and hard and unfamiliar. …
In the Noon of Science
The Atlantic Monthly, Volume 110, Number 3, September 1912 (p. 327)

Science is a capital or fund perpetually reinvested; it accumulates, rolls up, is carried forward by every new man. Every man of science has all the science before him to go upon, to set himself up in business with. What an enormous sum Darwin availed himself of and reinvested! Not so in literature; to every poet, to every artist, it is still the first day of creation, so far as the essentials of his task are concerned. Literature is not so much a fund to be reinvested as it is a crop to be ever new-grown.
The Writings of John Burroughs (Volume 17)
The Summit of the Years
In the Noon of Science (p. 64)
Houghton Mifflin Company. New York, New York, USA. 1913

Science enables us to understand our own ignorance and limitations, and so puts us at our ease amid the splendors and mysteries of creation.
The Writings of John Burroughs (Volume 17)
The Summit of the Years
In the Noon of Science (pp. 65–66)
Houghton Mifflin Company. New York, New York, USA. 1913

Science puts great weapons in men's hands for good or for evil, for war or for peace, for beauty or for ugliness, for life or for death, and how these weapons are used depends upon the motives that actuate us.
The Writings of John Burroughs (Volume 17)
The Summit of the Years
In the Noon of Science (p. 67)
Houghton Mifflin Company. New York, New York, USA. 1913

Bury, John Bagnell   1861–1927
English historian and classical scholar

Science has been advancing without interruption during the last three or four hundred years; every new discovery has led to new problems and new methods of solution, and opened up new fields for exploration. Hitherto men of science have not been compelled to halt, they have always found means to advance further. But what assurance have we that they will not come up against impassable barriers?
The Idea of Progress: An Inquiry into Its Origin and Growth
Introduction (p. 3)
Dover Publications. New York, New York, USA. 1955

Bush, Vannevar   1890–1974
American electrical engineer and physicist

Science has a simple faith, which transcends utility. Nearly all men of science, all men of learning for that matter, and men of simple ways too, have it in some form and in some degree. It is the faith that it is the privilege of man to learn to understand, and that this is his mission. If we abandon that mission under stress we shall abandon it forever, for stress will not cease. Knowledge for the sake of understanding, not merely to prevail, that is the essence of our being. None can define its limits, or set its ultimate boundaries.
Science Is Not Enough
Chapter X (p. 191)
William Morrow & Company, Inc. New York, New York, USA. 1967

Science does not exclude faith. … Science does not teach a harsh materialism. It does not teach anything beyond its boundaries, and those boundaries have been severely limited by science itself.
Modern Arms and Free Men
Threat and Bulwark (p. 183)
The MIT Press, Cambridge, Massachusetts, USA. 1968

Butler, Samuel   1612–80
English novelist, essayist, and critic

Science is being daily more and more personified and anthropomorphized into a god. By and by they will say that science took our nature upon him, and sent down his only begotten son, Charles Darwin, or Huxley, into the world so that those who believe in him, etc.; and they will burn people for saying that science, after all, is only an expression for our ignorance of our own ignorance.
In Geoffrey Keynes and Brian Hill (eds.)
Samuel Butler's Notebooks

Science (p. 233)
Jonathan Cape. London, England. 1951

If [science] tends to thicken the crust of ice on which, as it were, we are skating, it is all right. If it tries to find, or professes to have found, the solid ground at the bottom of the water it is all wrong.
In Geoffrey Keynes and Brian Hill (eds.)
Samuel Butler's Notebooks
Science (p. 110)
Jonathan Cape. London, England. 1951

Buttimer, Anne
Geographer

Strange indeed sounds the language of poets and philosophers; stranger still the refusal of science to read and hear its message.
Grasping the Dynamism of Lifeworld
Annals of the Association of American Geographers, Volume 66, 1976 (p. 277)

Buzzati-Traverso, Adriano   1913–83
Italian genetic scientist

Science is a game: it can be exhilarating, it can be useful, it can be frightfully dangerous. It is a play prompted by man's irrepressible curiosity to discover the universe and himself, and to increase his awareness of the world in which he lives and operates.
The Scientific Enterprise, Today and Tomorrow
Part I, Chapter 1 (p. 3)
UNES Company. Paris, France. 1977

Calder, Ritchie   1906–82
Scottish journalist

Science is a river which the explorer may encounter at any point along its course. He can follow it either to its source or to its delta or both.
Profile of Science
Introduction (p. 13)
George Allen & Unwin Ltd. London, England. 1951

Calvin, Melvin   1911–97
American biochemist

There is not a "pure" science. By this I mean that physics impinges on astronomy, on the one hand, and chemistry and biology on the other. And not only does each support neighbors, but derives sustenance from them. The same can be said of chemistry. Biology is, perhaps, the example par excellence today of an "impure" Science.
In Shirley Thomas
Men of Space. Profiles of the Scientists Who Probe for Life in Space
(Volume 6)
Melvin Calvin (p. 35)
Chilton Books. Philadelphia, Pennsylvania, USA. 1963

Campbell, Norman R.   1880–1949
English physicist and philosopher

...science [is] the study of those judgments concerning which universal agreement can be obtained.
What Is Science?
Chapter II (p. 32)
Dover Publications. New York, New York, USA. 1952

An audience of children of all ages gapes amazedly while the lecturer discourses glibly of times reckoned in millions of years and distances in thousands of millions of miles. But science has something better to offer than sensational journalism; nothing could be less characteristic of its spirit. The mere fact that the interest of the uninitiated can thus be easily stimulated with serious training suggests doubts of the value of the stimulus; nothing worth having in this world is to be had without effort.
Physics: The Elements
Chapter VIII (p. 226)
At The University Press. Cambridge, England. 1920

Campbell, Thomas   1777–1844
Scottish poet

When Science from Creation's face
Enchantment's veil withdraws,
What lovely visions yield their place
To cold material laws!
The Complete Poetical Works
To the Rainbow, l. 13–16
Chadwyck-Healey. Cambridge, England. 1992

Oh! star-eyed Science, hast thou wandered there,
To waft us home the message of dispair?
The Complete Poetical Works
Pleasures of Hope, Part II, l. 325
Chadwyck-Healey. Cambridge, England. 1992

Camus, Albert   1913–60
Algerian-French novelist, author, essayist, and philosopher

At the final stage you teach me that this wondrous and multicolored universe can be reduced to the atom and that the atom itself can be reduced to the electron. All this is good and I wait for you to continue. But you tell me of an invisible planetary system in which electrons gravitate around a nucleus. You explain this world to me with an image. I realize then that you have been reduced to poetry: I shall never know. Have I the time to become indignant? You have already changed theories. So that science that was to teach me everything ends up in a hypothesis, that lucidity founders in metaphor, that uncertainty is resolved in a work of art.
Translated by Justin O'Brien
The Myth of Sisyphus and Other Essays
An Absurd Reasoning (pp. 19–20)
Alfred A. Knopf. New York, New York, USA. 1961

Carlyle, Thomas   1795–1881
English historian and essayist

This world, after all our science and sciences, is still a miracle; wonderful, inscrutable, magical and more, to whosoever will think of it.
On Heroes and Hero Worship
Lecture I (p. 12)
John B. Alden, Publisher. New York, New York, USA. 1887

Carrel, Alexis   1873–1944
French surgeon and biologist

There is a strange disparity between the sciences of inert matter and those of life. Astronomy, mechanics, and physics are based on concepts which can be expressed, tersely and elegantly, in mathematical language. They have built up a universe as harmonious as the monuments of ancient Greece. They weave about it a magnificent texture of calculations and hypotheses. They search for reality beyond the realm of common thought up to unutterable abstractions consisting only of equations of symbols.
Man the Unknown
Chapter 1, Section 1 (p. 1)
Harper & Brothers. New York, New York, USA. 1939

Carson, Rachel   1907–64
American marine biologist and author

There is one quality that characterizes all of us who deal with the science of the earth and its life — we are never bored.
In Paul Brooks
The House of Life: Rachel Carson at Work
The Closing Journey (p. 324)
Houghton Mifflin Company. Boston, Massachusetts, USA. 1972

We live in a scientific age, yet we assume that knowledge of science is the prerogative of only a small number of human beings, isolated and priestlike in their laboratories. This is not true. The materials of science are the materials of life itself. Science is part of the reality of living; It is the what, the how and the why of everything in our experience.
In Paul Brooks
The House of Life: Rachel Carson at Work
Fame (p. 128)
Houghton Mifflin. Boston, Massachusetts, USA. 1972

The winds, the sea, and the moving tides are what they are. If there is wonder and beauty and majesty in them, science will discover these qualities. If they are not there, science cannot create them.
Acceptance Speech
1952 National Book Award

Chandrasekhar, Subrahmanyan   1910–95
Indian-born American astrophysicist

The pursuit of science has often been compared to the scaling of mountains, high and not so high. But who amongst us can hope, even in imagination, to scale the Everest and reach its summit when the sky is blue and the air is still, and in the stillness of the air survey the entire Himalayan range in the dazzling white of the snow stretching to infinity? None of us can hope for a comparable vision of nature and of the universe around us. But there is nothing mean or lowly in standing in the valley below and awaiting the sun to rise over Kinchinjunga.
Truth and Beauty: Aesthetics and Motivation in Science
Chapter 2, Section X (pp. 26–27)
The University of Chicago Press. Chicago, Illinois, USA. 1987

I am convinced that one's knowledge of the Physical Sciences is incomplete without a study of the Principia in the same way that one's knowledge of Literature is incomplete without a study of Shakespeare.
On Reading Newton's Principia at Age Apast Eighty
Current Science, Volume 67, Number 7, 10 October 1994 (p. 499)

Chargaff, Erwin   1905–2002
Austrian biochemist

To be a pioneer in science has lost much of its attraction: significant scientific facts and, even more, fruitful scientific concepts pale into oblivion long before their potential value has been utilized. New facts, new concepts keep crowding in and are in turn, within a year or two, displaced by even newer ones.… Now, however, in our miserable scientific mass society, nearly all discoveries are born dead; papers are tokens in a power game, evanescent reflections on the screen of a spectator sport, news items that do not outlive the day on which they appeared.
Heraclitean Fire: Sketches from a Life Before Nature
Part II
A Bouquet of Mortelles (p. 78, 81)
Rockefeller University Press. New York, New York, USA. 1978

To the scientist nature is like a mirror that breaks every thirty years; and who cares about the broken glass of past times?
Voices in the Labyrinth: Nature, Man and Science
Chapter 3 (p. 24)
The Seabury Press. New York, New York, USA. 1977

What counts, however, in science is to be not so much the first as the last.
Preface to a Grammar of Biology
Science, Volume 172, Number 3984, May 1971 (p. 639)

The sciences are extremely pedigree-conscious, and the road to the top of Mount Olympus is paved with letters of recommendation, friendly whispers at meetings, telephone calls at night.
Heraclitean Fire: Sketches from a Life Before Nature
Part I
No Hercules, No Crossroads (p. 32)
Rockefeller University Press. New York, New York, USA. 1978

In science you don't ask why, you ask how much.
Voices in the Labyrinth: Nature, Man and Science
Ouroboros (p. 128)
The Seabury Press. New York, New York, USA. 1977

Science cannot be a mass occupation, any more than the composing of music or the painting of pictures.
In Praise of Smallness — How Can We Return to Small Science
Perspectives in Biology and Medicine, Volume 23, Number 3, Spring 1980 (p. 373)

Science has become an eye without a head, a desperate attempt to fill holes with gaps. It came up to a lock, so it looked for the key; but it was a lock without a keyhole. The priests of truth are soiled with blood; their discoveries have become inventions, their pledges far from eternal. In a science in which one can say: "this is no longer true," nothing is true.
Voices in the Labyrinth: Nature, Man and Science
Chimaera (p. 151)
The Seabury Press. New York, New York, USA. 1977

The so-called exact sciences often are not as exact as is commonly believed. How often they infer the existence of a hat from the emergence of a rabbit!
Voices in the Labyrinth: Nature, Man and Science
Chapter 3 (p. 20)
The Seabury Press. New York, New York, USA. 1977

In science we always know much less than we believe we do.
Uncertainties Great, Is the Gain Worth the Risk?
Chemical and Engineering News, May 30, 1977

...in most sciences the question Why? is forbidden and the answer is actually to the question, How? Science is much better in explaining than in understanding, but it likes to mistake one for the other.
Voices in the Labyrinth
Perspectives in Biology and Medicine, Volume 18, Spring 1975 (p. 322)

...never before has science become so alienated from the common man, and he, in turn, so suspicious of science.
Heraclitean Fire: Sketches from a Life Before Nature
Part III
The Great Dilemma of the Life Sciences (p. 158)
Rockefeller University Press. New York, New York, USA. 1978

Chekhov, Anton Pavlovich   1860–1904
Russian author and playwright

Science is the most important, the most magnificent, and the most necessary element of life.
Compiled by V.V. Vorontsov
Words of the Wise: A Book of Russian Quotations
Translated by Vic Schneierson
Progress Publishers. Moscow, Russia. 1979

Chernin, Kim
No biographical data available

Science is not neutral in its judgments, not dispassionate, nor detached....
The Obsession: Reflections on the Tyranny of Slenderness
Chapter 3 (p. 37)
Harper & Row, Publishers. New York, New York, USA. 1981

Chernyshevsky, Nikolai Gavrilovich   1828–89
Russian socialist reformer

Science is the repository of the experience and thinking of the human race. It is mainly through science that the ideas, and then the morals and life of people, are improved.
Compiled by V.V. Vorontsov
Words of the Wise: A Book of Russian Quotations
Translated by Vic Schneierson
Progress Publishers. Moscow, Russia. 1979

Chesterton, G. K. (Gilbert Keith)   1874–1936
English author

When once one believes in a creed, one is proud of its complexity, as scientists are proud in the complexity of science. It shows how right it is in its discoveries. If it is right at all, it is a compliment to say that it is elaborately right. A stick might fit a hole or a stone a hollow by accident. But a key and a lock are both complex. And if a key fits a lock, you know it is the right key.
Orthodoxy
Chapter VI (p. 152)
John Lane Company. New York, New York, USA. 1918

Science, that nameless being, declared that the weakest must go to the wall; especially in Wall Street.
The Well and the Shallows
The Return to Religion (p. 74)
Sheed & Ward, Inc., New York, New York, USA. 1935

Science in the modern world has many uses; its chief use, however, is to provide long words to cover the errors of the rich.
Heretics
Cells and Celtophiles (p. 171)
Books for Libraries Press. Freeport, New York, USA. 1970

...physical science is like simple addition: it is either infallible or it is false.
All Things Considered
Science and Religion (p. 187)
John Lane Company. New York, New York, USA. 1908

...modern science cares far less for pure logic than a dancing Dervish.
Orthodoxy
Chapter II (p. 36)
John Lane Company. New York, New York, USA. 1918

[Modern science moves] toward the supernatural with the rapidity of a railway train.
Orthodoxy
Chapter IX (p. 277)
John Lane Company. New York, New York, USA. 1918

Churchill, Winston Spencer   1882–1965
British prime minister, statesman, soldier, and author

The latest refinements of science are linked with the cruelties of the Stone Age.

Speech
26 March 1942

Science bestowed immense new powers on man and at the same time created conditions which were largely beyond his comprehension and still more beyond his control.
Speech
March 31, 1949

Science has given to this generation the means of unlimited disaster or of unlimited progress. There will remain the greater task of directing knowledge lastingly towards the purpose of peace and human good.
Speech
New Delhi, January 3, 1944

My experience — and it is somewhat considerable — is that in these matters when the need is clearly explained by military and political authorities, science is always able to provide something. "Seek and ye shall find" has been borne out.
In F.B. Czarnomski
The Wisdom of Winston Churchill
Speech, Commons, June 7, 1935 (p. 327)
George Allen & Unwin Ltd. London, England. 1956

I have seldom seen a precise demand made upon science by the military which has not been met.
In F.B. Czarnomski
The Wisdom of Winston Churchill
Speech, Commons, March 16, 1950 (p. 327)
George Allen & Unwin Ltd. London, England. 1956

In the fires of science, burning with increasing heat every year, all the most dearly loved conventions are being melted down; and this is a process which is going continually to spread. In view of the inventions and discoveries which are being made for us, one might almost say every month, a unified direction of the war efforts of the three services would be highly beneficial.
In F.B. Czarnomski
The Wisdom of Winston Churchill
Speech, Commons, March 21, 1934 (p. 327)
George Allen & Unwin Ltd. London, England. 1956

It is arguable whether the human race have been gainers by the march of science beyond the steam engine. Electricity opens a field of infinite conveniences to ever greater numbers, but they may well have to pay dearly for them. But anyhow in my thought I stop short of the internal combustion engine which has made the world so much smaller. Still more must we fear the consequences of entrusting to a human race so little different from their predecessors of the so-called barbarous ages such awful agencies as the atomic bomb. Give me the horse.
In F.B. Czarnomski
The Wisdom of Winston Churchill
Speech, Royal College of Physicians, July 10, 1951 (p. 327)
George Allen & Unwin Ltd. London, England. 1956

Cobbe, Frances P.   1822–1904
English author

Then the Sorcerer Science entered, and where e'er he waved his wand

Fresh wonders and fresh mysteries rose on every hand.
The Pageant of Time, Stanza 1
Source undetermined

Coggan, Donald   1909–2000
101st archbishop of Canterbury

My ignorance of science is such that if anyone mentioned copper nitrate I should think he was talking about policeman's overtime.
New York Journal–American, September 20, 1961

Cohen, I. Bernard   1914–2003
American physicist and science historian

…all revolutionary advances in science may consist less of sudden and dramatic revelations than a series of transformations, of which the revolutionary significance may not be seen (except afterwards, by historians) until the last great step. In many cases the full potentiality and force of a most radical step in such a sequence of transformations may not even be manifest to its author.
The Newtonian Revolution: With Illustrations of the Transformation of Scientific Ideas
Chapter 4 (p. 162)
Cambridge University Press. Cambridge, England. 1980

Colbert, Edwin H.   1905–2001
American vertebrate paleontologist

It is too easy to think of science as something large and impersonal, as something outside the understanding of most of us, as something rather distant, and removed from the affairs of the average person. Even in this day, when the lives of all of us are touched every hour, and almost every minute, by the products of technology, the handmaiden of science, we are still inclined to accept the impersonal view of science. Science is so manifestly complex, so compartmentalized by specialization, and to all but those who are initiated into the priesthoods of these specializations, so largely incomprehensible, that we can hardly think of it in other than impersonal terms.
Men and Dinosaurs
Preface (p. v)
E.P. Dutton & Company, Inc. New York, New York, USA. 1968

Coles, Abraham   1813–91
American physician, hymnist, and poet

I value science — none can prize it more —
It gives ten thousand motives to adore.
Be it religious, as it ought to be,
The heart it humbles, and it bows the knee.

The Microcosm: And Other Poems
Christian Science
D. Appleton & Company. New York, New York, USA. 1880

Commoner, Barry   1917–
American biologist, ecologist, and educator

Science is triumphant with far-ranging success, but its triumph is somehow clouded by growing difficulties in providing for the simple necessities of human life on the earth.
Science and Survival
Chapter 2 (p. 9)
The Viking Press. New York, New York, USA. 1966

We seem to be entering a new world of technology, but the vehicle which is carrying us — science — shows dangerous signs of inadequacy for the voyage ahead.
Science and Survival
Chapter 4 (p. 63)
The Viking Press. New York, New York, USA. 1966

Compton, Karl Taylor   1887–1954
American educator and physicist

Fundamentally, science means simply knowledge of our environment. Combined with ingenuity, science becomes power.
A Scientist Speaks: Excerpts from Addresses by Karl Taylor Compton During the Years 1930–1949 (p. 2)
Undergraduate Association, MIT. Cambridge, Massachusetts, USA. 1955

I believe that the advent of modern science is the most important social event in all history.
A Scientist Speaks: Excerpts from Addresses by Karl Taylor Compton During the Years 1930–1949 (p. 2)
Undergraduate Association, MIT. Cambridge, Massachusetts, USA. 1955

Conant, James Bryant   1893–1978
American educator and scientist

There is only one proved method of assisting the advancement of pure science — that of picking men of genius, backing them heavily, and leaving them to direct themselves.
Letter
New York Times, August 13, 1945

Science is an interconnected series of concepts and schemes that have developed as a result of experimentation and observation and are fruitful of further experimentation and observation.
Science and Common Sense
Chapter Two (p. 25)
Yale University Press. New Haven, Connecticut, USA. 1951

Science is a dynamic undertaking directed to lowering the degree of the empiricism involved in solving problems.
Modern Science and Modern Man

Science and Human Conduct (p. 62)
Columbia University Press. New York, New York, USA. 1952

[Science is] the activity of people who work in laboratories and whose discoveries have made possible modern industry and medicine.
Science and Common Sense
Chapter Two (p. 23)
Yale University Press. New Haven, Connecticut, USA. 1951

Condon, Edward Uhler   1902–74
American physicist

Society is at this moment at the threshold of an undreamed-of mastery of our material environment, for science, which provides that mastery, is in its Golden Age.
Selected Popular Writings of E.U. Condon
Science and the National Welfare (p. 145)
Springer-Verlag. New York, New York, USA. 1991

The sciences, like those other truth-seeking activities of man, require a free environment, an environment, above all, free from fear, petty arbitrariness, and tyranny.
Selected Popular Writings of E.U. Condon
Science and the National Welfare (p. 155)
Springer-Verlag. New York, New York, USA. 1991

Condorcet, Marie Jean   1749–1827
French mathematician, astronomer, and physicist

This adventure of the physical sciences…could not be observed without enlightened men seeking to follow it up in the other sciences; at each step it held out to them the model to be followed.
In K.M. Baker
Condorcet: From Natural Philosophy to Social Mathematics
Chapter 2 (p. 85)
The University of Chicago Press. Chicago, Illinois, USA. 1975

In every century Princes have been found to love the sciences and even to cultivate them, to attract Savants to their palaces and to reward by their favors and their amity men who afforded them a sure and constant refuge from world-weariness, a sort of disease to which supreme power seems particularly prone.
Eloge des académiciens de l'Académie royal des sciences
Forward, I
Publisher undetermined

Constitution of the United States

The Congress shall have the Power…to promote the Progress of Science and useful Arts…
United States Constitution
Article I, Section 8

Cooper, Bernard   1951–
Physicist

At the rate science proceeds, rockets and missiles will one day seem like buffalo — slow, endangered grazers in the black pasture of outer space.
Harper's, January 1990

Cooper, Leon   1930–
American physicist

To say that science is logical is like saying that a painting is paint....
In George Johnson
In the Palaces of Memory: How We Build the Worlds Inside Our Heads
The Memory Machine (p. 194)
Alfred A. Knopf. New York, New York, USA. 1991

Cossons, Sir Neil   1939–
Chairman of English Heritage

Science's function is to describe how things work, not what they mean. That is a role for philosophers, artists, and writers.
Lancet, Volume 339, 1992

Crichton, Michael   1942–
American novelist

Finally, I would remind you to notice where the claim of consensus is invoked. Consensus is invoked only in situations where the science is not solid enough. Nobody says the consensus of scientists agrees that E = mc². Nobody says the consensus is that the sun is 93 million miles away. It would never occur to anyone to speak that way.
Lecture
Aliens Cause Global Warming, California Institute of Technology, January 17, 2003

I expected science to be, in Carl Sagan's memorable phrase, "a candle in a demon haunted world." And here, I am not so pleased with the impact of science. Rather than serving as a cleansing force, science has in some instances been seduced by the more ancient lures of politics and publicity. Some of the demons that haunt our world in recent years are invented by scientists.
Lecture
Aliens Cause Global Warming, California Institute of Technology, January 17, 2003

In recent years, much has been said about the post-modernist claims about science to the effect that science is just another form of raw power, tricked out in special claims for truth-seeking and objectivity that really have no basis in fact. Science, we are told, is no better than any other undertaking. These ideas anger many scientists, and they anger me. But recent events have made me wonder if they are correct.
Lecture
Aliens Cause Global Warming, California Institute of Technology, January 17, 2003

Crick, Francis Harry Compton   1916–2004
English biochemist

One of the striking characteristics of modern science is that it often moves so fast that a research worker can see rather clearly whether his earlier ideas, or those of his contemporaries, were correct or incorrect.
What Mad Pursuit: A Personal View of Scientific Discovery
Introduction (p. 3)
Basic Books, Inc. New York, New York, USA. 1988

Cromer, Alan   1935–
American physicist and educator

Science is the search for a consensus of rational opinion among all competent researchers.
Uncommon Sense: The Heretical Nature of Science
Chapter 8 (pp. 143–144)
Oxford University Press, Inc. New York, New York, USA. 1993

Science is the heretical belief that the truth about the real nature of things is to be found by studying the things themselves.
Uncommon Sense: The Heretical Nature of Science
Chapter 1 (p. 18)
Oxford University Press, Inc. New York, New York, USA. 1993

Science is overwhelmingly cumulative, not revolutionary, in its structure. This means that most of its established results — even those established recently — will be around forever. A particular result may be found to be an instance of a more general result, but its factualness, as far as it goes, will never change.
Uncommon Sense: The Heretical Nature of Science
Chapter 1 (p. 6)
Oxford University Press, Inc. New York, New York, USA. 1993

The notion that science and objective thinking are unnatural human activities seems quite radical at first. But when you think about it, monogamy, honesty, and democratic government are unnatural human behaviors as well. We are truly a species that has invented itself out of rather unpromising material. Our only claim to greatness is that we have at times gone against the grain of our own ego-centrism to forge a higher vision of the world.
Uncommon Sense: The Heretical Nature of Science
Preface (p. ix)
Oxford University Press, Inc. New York, New York, USA. 1993

Cromie, William J.   1930–
American journalist and writer

Science is not a mere "lump" of knowledge. It is disciplined thought, it is curiosity, it is creativity, it is the scientists themselves and the methods they use. It is the hope — the religion — that there is order in the universe; that man shall find that order; that someday he will be able to control the environment to which he is now little more than a slave. All this and more make up the dynamic, ever-changing whole called "science."
Exploring the Secrets of the Sea
Conclusion (p. 280)
Prentice-Hall, Inc. Englewood Cliffs, New Jersey, USA. 1962

Crothers, Samuel McChord   1857–1927
American clergyman and writer

On the coasts of the Dark Continent of Ignorance the several sciences have gained a foothold. In each case there is a well-defined country carefully surveyed and guarded. Within its frontiers the laws are obeyed, and all affairs are carried on in an orderly fashion. Beyond it is a vague "sphere of influence," a Hinter-land over which ambitious claims of suzerainty [foreign authority] are made; but the native tribes have not yet been exterminated, and life goes on very much as in the olden time.
The Gentle Reader
The Hinter-Land of Science (p. 231)
Houghton Mifflin Company. Boston, Massachusetts, USA. 1903

Science will not tolerate half knowledge nor pleasant imaginings, nor sympathetic appreciations; it must have definite demonstrations. The knowledge of the best that has been said and thought may be consoling, but it implies an unscientific principle of selection. It can be proved by statistics that the best things are exceptional.
The Gentle Reader
The Hinter-Land of Science (p. 228)
Houghton Mifflin Company. Boston, Massachusetts, USA. 1903

Crowley, Aleister   1875–1947
Poet and author

…science is always discovering odd scraps of magical wisdom and making a tremendous fuss about its cleverness!
The Confessions of Aleister Crowley: An Autobiography
Part Four, Chapter 64 (p. 593)
Arkana. 1989

Cudmore, Lorraine Lee
American cell biologist

Almost anyone can do science; almost no one can do good science.
The Center of Life: A Natural History of the Cell
Biochemical Evolution (p. 35)
New York Times Book Company. New York, New York, USA. 1977

…good science is almost always so very simple. After it has been done by someone else, of course.
The Center of Life: A Natural History of the Cell
Biochemical Evolution (p. 36)
New York Times Book Company. New York, New York, USA. 1977

Curie, Eve   1904–
French concert pianist and journalist

What does it matter to Science if her passionate servants are rich or poor, happy or unhappy, healthy or ill? She knows that they have been created to seek and to discover, and that they will seek and find until their strength dries up at its source. It is not in a scientist's power to struggle against his vocation: even on his days of disgust or rebellion his steps lead him inevitably back to his laboratory apparatus.
Madame Curie
Chapter XV (p. 193)
The Literary Guild of America, Inc. New York, New York, USA. 1937

Curie, Marie Sklodowska   1867–1934
Polish-born French physicist and chemist

In science we must be interested in things, not in persons.
In Eve Curie
Madame Curie
Chapter XVI (p. 222)
The Literary Guild of America, Inc. New York, New York, USA. 1937

After all, science is essentially international, and it is only through lack of the historical sense that national qualities have been attributed to it.
Memorandum
Intellectual Cooperation, June 16, 1926

da Vinci, Leonardo   1452–1519
Italian High Renaissance painter and inventor

Science is the captain, practice the soldiers.
Leonardo da Vinci's Note Books (p. 54)
Duckworth & Company. London, England. 1906

All science which end in words are dead the moment they come to life, except for their manual part, that is to say, the writing, which is the mechanical part.
The Literary Works of Leonardo da Vinci (Volume 1)
7a1148 (p. 35)
University of California Press. Berkeley, California, USA. 1977

d'Abro, Abraham
No biographical data available

Practically the whole of physical science is thus one mass of inference based ultimately, but not immediately, on direct knowledge.
The Rise of the New Physics (Volume 1)
Chapter II (p. 15)
Dover Publications, Inc. New York, New York, USA. 1951

Darwin, Charles Robert   1809–82
English naturalist

As for myself, I believe that I have acted rightly in steadily following, and devoting my life to Science. I feel no remorse from having committed any great sin, but have often and often regretted that I have not done more direct good to my fellow creatures.
In Francis Darwin (ed.)
The Life and Letters of Charles Darwin (Volume 2)
Chapter XVI (p. 530)
D. Appleton & Company. New York, New York, USA. 1896

Darwin, Sir Francis   1848–1925
English botanist

Forgive me for suggesting one caution; as Demosthenes said, "Action, action, action," was the soul of eloquence, so is caution almost the soul of science.
In Francis Darwin (ed.)

More Letters of Charles Darwin (Volume 2)
To Dohrn, January 4, 1870 (p. 444)
D. Appleton & Company. New York, New York, USA. 1903

But in science the credit goes to the man who convinces the world, not to the man to whom the idea first occurs.
First Galton Lecture Before the Eugenics Society
Eugenics Review, Volume 6, Number 1, 1914

How grand is the onward rush of science; it is enough to console us for the many errors which we have committed, and for our efforts being overlaid and forgotten in the mass of new facts and new views which are daily turning up.
In Francis Darwin (ed.)
The Life and Letters of Charles Darwin (Volume 2)
Darwin to Wallace, August 28, 1872 (p. 348)
D. Appleton & Company. New York, New York, USA. 1896

Data
Fictional character

Captain, the most elementary and valuable statement in science: "The beginning of wisdom is 'I do not know.'"
Star Trek: The Next Generation
Where Silence Has Lease
Television program
Season 2, 1988

Davies, Paul Charles William   1946–
British-born physicist, writer, and broadcaster

There is a popular misconception that science is an impersonal, dispassionate, and thoroughly objective enterprise. Whereas most other human activities are dominated by fashions, fads, and personalities, science is supposed to be constrained by agreed rules of procedure and rigorous tests. It is the results that count, not the people who produce them.

This is, of course, manifest nonsense.
In Richard Feynman
Six Easy Pieces: Essentials of Physics Explained by Its Most Brilliant Teacher
Preface (p. ix)
Addison-Wesley Publishing Company. Reading, Massachusetts, USA. 1995

Science may explain all the processes whereby the universe evolves its own destiny, but that still leaves room for there to be a meaning behind it all.
The Cosmic Blueprint: New Discoveries in Nature's Creative Ability to Order the Universe
Chapter 14 (p. 203)
Simon & Schuster. New York, New York, USA. 1988

Science remains a sort of witchcraft, its practitioners regarded with a mixture of awe and suspicion…
God and the New Physics
Chapter 1 (p. 3)
Simon & Schuster. New York, New York, USA. 1983

Conventional science attempts to explain things exactly, in terms of general principles. Any sort of explanation for the shape of a snowflake or a coastline could not be of this sort.
The Cosmic Blueprint: New Discoveries in Nature's Creative Ability to Order the Universe
Chapter 3 (p. 22)
Simon & Schuster. New York, New York, USA. 1988

…it is the job of science to solve mysteries without recourse to divine intervention.
The Fifth Miracle: The Search for the Origin of Life
Chapter 1 (p. 31)
Simon & Schuster. New York, New York, USA. 1996

Davies, Robertson   1913–95
Canadian novelist

Science, during the past hundred and fifty years, has gained formidable new authority, and it is to Science that we owe the increased longevity of the race, and the control of many of the terrible ills that afflict mankind. Science may cure disease, but can it confer health? Like all powerful gods, Science seeks to be the One True God, and as it writhes about the staff of Hermes it seeks to diminish and perhaps drive out the other god, the god of Humanism.
The Merry Heart
Chapter 5 (p. 98)
McClelland & Stewart. Toronto, Ontario, Canada. 1996

Davis, Kenneth S.   1912–99
American historian

In our time it has become all too easy to regard science as a vast impersonal force — a kind of Frankenstein's monster that, escaping human control, has forcibly seized us and carries us at terrifying speeds in directions we have not chosen towards ends unknown.
The Cautionary Scientists: Priestley, Lavoisier, and the Founding of Modern Chemistry
Introduction (p. 7)
Putnam. New York, New York, USA. 1966

Davis, Watson   1896–1967
No biographical data available

Science is a grand procession through the ages. Blaring trumpets, waving flags, and pomp are not its accompaniment. It travels the quieter roads of the intellect.
The Advance of Science
Chapter 32 (p. 375)
Doubleday, Doran & Company, Inc. Garden City, New York, USA. 1934

Davis, William Morris   1850–1934
American geomorphologist

Science is therefore not final any more than it is infallible.
In H. Shapley, H. Wright, and S. Rapport (eds.)
Readings in the Physical Sciences
The Reasonableness of Science (p. 25)
Appleton-Century-Crofts. New York, New York, USA. 1948

Davy, Sir Humphry   1778–1829
English chemist

There is now before us a boundless prospect of novelty in science; a country unexplored, but noble and fertile in aspect; a land of promise in philosophy.
The Collected Works of Sir Humphry Davy (Volume 1)
Memories of the Life of Sir Humphry Davy
Chapter III (p. 117)
Smith, Elder & Company. London, England. 1839–1840

There are very few persons who pursue science with true dignity.
Consolations in Travel, or the Last Days of a Philosopher
Dialogue V (p. 226)
J. Murray. London, England. 1830

Science has done much for man, but it is capable of doing still more; its sources of improvement are not not yet exhausted; the benefits that it has conferred ought to excite our hopes of its capability of conferring new benefits; and, in considering the progressiveness of our nature, we may reasonably look forwards to a state of greater cultivation and happiness than that which we at present enjoy.
A Discourse, Introductory to a Course of Lectures on Chemistry (p. 17)
Press of the Royal Institution of Great Britain. London. 1802

Dawkins, Richard   1941–
British ethologist, evolutionary biologist, and popular science writer

Science, like proper literary studies, can be hard and challenging but science is — also like proper literary studies — wonderful.
Unweaving the Rainbow: Science, Delusion and the Appetite for Wonder
Chapter 2 (p. 25)
Houghton Mifflin Company. Boston, Massachusetts, USA. 1998

Far from science not being useful, my worry is that it is so useful as to overshadow and distract from its inspirational and cultural value. Usually even its sternest critics concede the usefulness of science, while completely missing the wonder. Science is often said to undermine our humanity, or destroy the mystery on which poetry is thought to thrive.
Science, Delusion and the Appetite for Wonder
Richard Dimbleby Lecture, BBC1 Television, November 12th, 1996

de Balzac, Honoré   1799–1850
French novelist

Science is the language of the temporal world; love is that of the spiritual world. Man, indeed, describes more than he explains; while the angelic spirit sees and understands. Science saddens man; love enraptures the angel; science is still seeking, love has found. Man judges of nature in relation to itself; the angelic spirit judges of it in relation to heaven. In short to the spirits everything speaks.
The Works of Honoré Balzac (Volume 2)
Seraphita
Part II (p. 58)
Nottingham Society. New York, New York, USA. 1901

de Bono, Edward   1933–
Maltese psychologist and writer

When you have got somewhere interesting, that is the time to look back and pick out the surest way of getting there again. Sometimes it is very much easier to see the surest route to a place only after you have arrived. You may have to be at the top of a mountain to find the easiest way up.
New Think: The Use of Lateral Thinking in the Generation of New Ideas (p. 132)
Avon Books. New York, New York, USA. 1971

de Gourmont, Rémy   1858–1915
French critic and novelist

Science is the only truth and it is the great lie. It knows nothing, and people think it knows everything. It is misrepresented. People think that science is electricity, automobilism, and dirigible balloons. It is something very different. It is life devouring itself. It is the sensibility transformed into intelligence. It is the need to know stifling the need to live. It is the genius of knowledge vivisecting the vital genius.
Translated by Glenn S. Burne
Selected Writings
Art and Science (p. 172)
The University of Michigan Press, Ann Arbor, Michigan, USA. 1966

Science is the food of the intelligence.
Translated by Glenn S. Burne
Selected Writings
Art and Science (p. 171)
The University of Michigan Press, Ann Arbor, Michigan, USA. 1966

de Unamuno, Miguel   1864–1936
Spanish philosopher and writer

True science teaches, above all, to doubt and be ignorant.
Translated by J.E. Crawford Flitch
The Tragic Sense of Life in Men and in Peoples
Chapter V (p. 93)
Macmillan & Company Ltd. London, England. 1921

Science is a cemetery of dead ideas, even though life may issue from them.
Translated by J.E. Crawford Flitch
The Tragic Sense of Life in Men and in Peoples
Chapter V (p. 90)
Macmillan & Company Ltd. London, England. 1921

Science exists only in personal consciousness and thanks to it; astronomy, mathematics, have no other reality than that which they possess as knowledge in the minds of those who study and cultivate them.
Translated by J.E. Crawford Flitch
The Tragic Sense of Life in Men and in Peoples
The Starting-Point (pp. 30–31)
Macmillan & Company Ltd. London, England. 1921

del Rio, A. M.
No biographical data available

It is impossible that he who has once imbibed a taste for science can ever abandon it.

Analysis of an Alloy of Gold and Rhodium from the Parting House at Mexico
Annals of Philosophy, Volume 10, Number 2, October 1825

Delbrück, Max   1906–81
German-born American biologist

With science we can transcend our intuitions, just as with electronics we can transcend our eyes and ears. To the question of how such transcendence can have arisen in the course of biological evolution I have no satisfactory answer.
Mind from Matter
Twenty (p. 280)
Blackwell Scientific Publications, Inc. Palo Alto, California, USA. 1986

Descartes, René   1596–1650
French philosopher, scientist, and mathematician

Science is like a woman: if she stays faithful to her husband she is respected; if she becomes common property she grows to be despised.
In William R. Shea
The Magic of Numbers and Motion: The Scientific Career of René Descartes
Chapter Five (p. 114)
Science History Publications. Canton, Massachusetts, USA. 1991

The sciences are now masked, but when the masks are lifted, they will be seen in their beauty. Upon inspecting the chain of the sciences, it will not appear more difficult to remember them than a series of numbers.
In William R. Shea
The Magic of Numbers and Motion: The Scientific Career of René Descartes
Chapter Five (p. 101)
Science History Publications Canton, Massachusetts, USA. 1991

Dewar, James   1842–1923
English physicist and chemist

To serve in the scientific army, to have shown some initiative, and to be rewarded by the consciousness that in the eyes of his comrades he bears the accredited accolade of successful endeavor, is enough to satisfy the legitimate ambition of every earnest student of nature.
Annual Report of the Board of Regents of the Smithsonian Institution, 1902
History of Cold and the Absolute Zero (p. 240)
Government Printing Office. Washington, D.C. 1903

Dickinson, Emily   1830–86
American lyric poet

I climb the "Hill of Science,"
I view the landscape o'er;
such transcendental prospects,
I ne'er beheld before!
The Complete Poems of Emily Dickinson
No. 3 (p. 5)
Little, Brown & Company. Boston, Massachusetts, USA. 1960

Dickinson, G. Lowes   1862–1932
English historian and political activist

Science hangs in a void of nescience, a planet turning in the dark.
A Modern Symposium (p. 159)
Doubleday, Page & Company. Garden City, New York, USA. 1920

Disraeli, Benjamin, 1st Earl of Beaconsfield   1804–81
English prime minister, founder of Conservative Party, and novelist

What Art was to the ancient world, Science is to the modern…
Coningsby
Book IV, Chapter I (p. 126)
J.M. Dent & Sons Ltd. London, England. 1911

…the pursuit of science leads only to the insoluble.
Lothair
Chapter XVII (p. 70)
Longmans, Green & Company London, England. 1920

Dobie, J. Frank   1888–1964
American folklorist

Putting on the spectacles of science in expectation of finding the answer to everything looked at signifies inner blindness.
The Voice of the Coyote
Introduction (p. xvi)
Little, Brown & Company. Boston, Massachusetts, USA. 1949

Dobzhansky, Theodosius   1900–75
Russian-American scientist

Science has been called "the endless frontier." The more we know, the better we realize that our knowledge is a little island in the midst of an ocean of ignorance.
In Robert M. Hutchins and Mortimer J. Adler (eds.)
The Great Ideas Today, 1974
Advancement and Obsolescence in Science (p. 61)
Encyclopædia Britannica, Inc. Chicago, Illinois, USA. 1974

Science does more than collect facts; it makes sense of them. Great scientists are virtuosi of the art of discovering the meaning of what otherwise might seem barren observations.
In Robert M. Hutchins and Mortimer J. Adler (eds.)
The Great Ideas Today, 1974
Advancement and Obsolescence in Science (p. 56)
Encyclopædia Britannica, Inc. Chicago, Illinois, USA. 1974

Science is cumulative knowledge. Each generation of scientists works to add to the treasury assembled by its predecessors. A discovery made today may not be significant or even comprehensible by itself, but it will make sense in conjunction with what was known before. Indeed this will usually have been necessary to its achievement.
In Robert M. Hutchins and Mortimer J. Adler (eds.)
The Great Ideas Today, 1974
Advancement and Obsolescence in Science (p. 52)
Encyclopædia Britannica, Inc. Chicago, Illinois, USA. 1974

Dott, Jr., Robert H.
No biographical data available

Batten, Henry L.
No biographical data available

Science consists simply of the formulation and testing of hypotheses based on observational evidence; experiments are important where applicable, but their function is merely to simplify observation by imposing controlled conditions.
Evolution of the Earth (2nd edition)
Chapter 3 (p. 40)
McGraw-Hill Book Company, Inc. New York, New York, USA. 1976

Douglas, Mary   1921–
No biographical data available
Wildavsky, A.
No biographical data available

In our modern world people are supposed to live and die subject to known, measurable natural forces, not subject to mysterious moral agencies. That mode of reasoning, indeed, is what makes modern man modern. Science wrought this change between us and nonmoderns. It is hardly true, however, that their universe is more unknown than ours. For anyone disposed to worry about the unknown, science has actually expanded the universe about which we cannot speak with confidence.… This is the double-edge thrust of science, generating new ignorance with new knowledge. The same ability to detect causes and connections or parts per trillion can leave more unexplained than was left by cruder measuring instruments.
Risk and Culture: An Essay on the Selection of Technical and Environmental Dangers
Chapter III (p. 49)
University of California Press. Berkeley, California, USA. 1982

Doyle, Sir Arthur Conan   1859–1930
Scottish writer

Science seeks knowledge. Let the knowledge lead us where it will, we still must seek it. To know once for all what we are, why we are, where we are, is that not in itself the greatest of all human aspirations?
The Land of Mist; The Maracot Deep; and Other Stories
When the World Screamed (p. 430)
Doubleday, Doran & Company, Inc. Garden City, New York, USA. 1930

Dryden, John   1631–1700
English poet, dramatist, and literary critic

Science distinguishes a Man of Honour from one of those Athletick Brutes whom undeservedly we call Heroes.
Fables Ancient and Modern
Dedication
Printed for Jacob Tonson. London, England. 1700

du Noüy, Pierre Lecomte   1883–1947
French scientist

Science is very young when compared to the moral, spiritual, and religious ideas of humanity. It enjoys the prestige of a new toy. But we must not be misled. In spite of its youth and imperfections, it constitutes the best means of convincing us of the immensity and harmonious beauty of the universe, revealed by the infinite complexity of the apparently most simple phenomena. It is an orderly and confounding complexity which is a thousand times better qualified than ignorance to make us feel the omnipotence of the Creator.
Between Knowing and Believing
The Future of Spirit 1941 (p. 216)
McKay. New York, New York, USA. 1967

Either we have absolute confidence in our science and in the mathematical and other reasonings which enable us to give a satisfactory explanation of the phenomena surrounding us — in which case we are forced to recognize that certain fundamental problems escape us and that their explanation amounts to admitting a miracle — or else we doubt the universality of our science and the possibility of explaining all natural phenomena by chance alone; and we fall back on a miracle or a hyperscientific intervention.
Human Destiny
Chapter 3 (p. 36)
Longmans, Green & Company. London, England. 1947

Dubos, René Jules   1901–82
French-born American microbiologist and environmentalist

Science is not the product of lofty meditations and genteel behavior, it is fertilized by heartbreaking toil and long vigils — even if, only too often, those who harvest the fruit are but the laborers of the eleventh hour.
Louis Pasteur: Free Lance of Science
Chapter XIV (p. 389)
Little, Brown & Company. Boston, Massachusetts, USA. 1950

Science shows us what exists but not what to do about it.
The Dreams of Reason: The Computer and the Rise of the Sciences of Complexity
Chapter 14 (p. 325)
Simon & Schuster. New York, New York, USA. 1988

Science is still the versatile, unpredictable hero of the play, creating endless new situations, opening romantic vistas and challenging accepted concepts.
Louis Pasteur: Free Lance of Science
Chapter I (p. 15)
Little, Brown & Company. Boston, Massachusetts, USA. 1950

Science knows no country, because knowledge belongs to humanity, and is the torch which illuminates the world. Science is the highest personification of the nation because that nation will remain the first which carries the furthest the works of thought and intelligence.
Louis Pasteur: Free Lance of Science
Chapter III (p. 85)
Little, Brown & Company. Boston, Massachusetts, USA. 1950

Dyson, Freeman J.   1923–
American physicist and educator

Science is even more unpredictable than history. Every important discovery in science is by definition unpredictable. If it were predictable, it would not be an important discovery. The purpose of science is to create opportunities for unpredictable things to happen. When nature does something unexpected, we learn something about how nature works.

From Eros to Gaia
Chapter 6 (p. 68)
Pantheon Books. New York, New York, USA. 1992

It used to be said, before the recent era of revolutionary discoveries, that science was organized common sense. In the modern era it would be more accurate to define science as organized unpredictability.

From Eros to Gaia
Chapter 6 (p. 68)
Pantheon Books. New York, New York, USA. 1992

Eakin, Richard M.
American zoologist

If I had any advice to [give] you it is just this: love science but do not worship it. Put science in its proper place, ranking it along with philosophy and history, music and religion, literature and art. If I had my life to live again (Darwin says), I would make it a rule to read some poetry and listen to some music at least once every week.

Great Scientists Speak Again
Chapter 6 (p. 107)
University of California Press, Berkeley, California, USA, 1975

Eben, Aubrey
No biographical data available

Science is not a sacred cow. Science is a horse. Don't worship it. Feed it.

Reader's Digest, March 1963 (p. 67)

Eddington, Sir Arthur Stanley   1882–1944
English astronomer, physicist, and mathematician

To imagine that Newton's great scientific reputation is tossing up and down in these latter-day revolutions is to confuse science with omniscience.

The Nature of the Physical World
Chapter X (p. 202)
The Macmillan Company. New York, New York, USA. 1930

Science has its showrooms and its workshops. The public today, I think rightly, is not content to wander round the showrooms where the tested products are exhibited; they demand to see what is going on in the workshops. You are welcome to enter; but do not judge what you see by the standards of the showroom. We have been going round a workshop in the basement of the building of science. The light is dim, and we stumble sometimes. About us is confusion and mess which there hasn't been time to sweep away. The workers and their machines are enveloped in murkiness. But I think that something is being shaped here — perhaps

something rather big. I do not quite know what twill be when it is completed and polished for the showroom.

The Expanding Universe
Chapter IV, Section VII (p. 126)
The University Press. Cambridge, England. 1933

...unless science is to degenerate into idle guessing, the test of value of any theory must be whether it expresses with as little redundancy as possible the facts which it intended to cover.

Space, Time and Gravitation: An Outline of the General Relativity Theory
Chapter I (p. 29)
At The University Press. Cambridge, England. 1921

Edelman, Gerald M.   1929–
American biochemist and neuroscientist

Science is imagination in the service of the verifiable truth and that service is indeed communal. It cannot be rigidly planned. Rather, it requires freedom and courage and the plural contributions of many different kinds of people who must maintain their individuality while giving to the group.

Les Prix Nobel. The Nobel Prizes in 1972
Nobel banquet speech for award received in 1972
Nobel Foundation. Stockholm, Sweden. 1973

Egler, Frank E.   1911–96
American botanist and ecologist

Science is a product of man, of his mind; and science creates the real world in its own image.

The Way of Science
Science Concepts (p. 22)
Hafner Publishing Company. New York, New York, USA. 1970

...science...ever [reflects] a faith in the intelligibility of nature.

The Way of Science
The Nature of Science (p. 2)
Hafner Publishing Company. New York, New York, USA. 1970

Einstein, Albert   1879–1955
German-born physicist

One thing I have learned in a long life: that all our science, measured against reality, is primitive and childlike — and yet it is the most precious thing we have.

In Banesh Hoffman
Albert Einstein: Creator and Rebel
Preface (p. v)
The Viking Press. New York, New York, USA. 1972

Science as something existing and complete is the most objective thing known to man. But science in the making, science as an end to be pursued, is as subjective and psychologically conditioned as any other branch of human endeavor — so much so, that the question "what is the purpose and meaning of science?" receives quite different answers at different times and from different sorts of people.

The World as I See It
Address at Columbia University, New York (p. 137)
Philosophical Library. New York, New York, USA. 1949

Strange that science, which in the old days seemed harmless, should have evolved into a nightmare that causes everyone to tremble.
In G.J. Whitrow
Einstein: The Man and His Achievement
Chapter III (p. 89)
BBC. London, England. 1967

Science will stagnate if it is made to serve practical goals.
In Otto Nathan and Heinz Norden
Einstein on Peace
Chapter Thirteen (p. 402)
Simon & Schuster. New York, New York, USA. 1960

The whole of science is nothing more than a refinement of everyday thinking.
Out of My Later Years
Physics and Reality, I (p. 59)
Thames & Hudson. London, England. 1950

…science can only ascertain what is, but not what should be, and outside of its domain value judgments of all kinds remain necessary.
Out of My Later Years
Science and Religion, II (p. 25)
Thames & Hudson. London, England. 1950

…the fact that in science we have to be content with an incomplete picture of the physical universe is not due to the nature of the universe itself but rather to us.
In Max Planck
Where Is Science Going?
Prologue (p. 10)
W.W. Norton & Company, Inc. New York, New York, USA.1932

Science is a wonderful thing if one does not have to earn one's living at it.
In Helen Dukas and Banesh Hoffman
Albert Einstein: The Human Side: New Glimpses from His Archives
Letter to a California student March 24, 1951 (p. 57)
Princeton University Press. Princeton, New Jersey, USA. 1979

Einstein, Albert   1879–1955
German-born physicist
Infeld, Leopold   1898–1968
Polish physicist

Science is not just a collection of laws, a catalogue of unrelated facts. It is a creation of the human mind, with its freely invented ideas and concepts.
The Evolution of Physics
Physics and Reality (p. 294)
Simon & Schuster. New York, New York, USA. 1961

Eisenschiml, Otto   1880–1963
Austrian-American chemist and historian

Science seeks to build, not to destroy; to aid, not to hinder.

The Art of Worldly Wisdom: Three Hundred Precepts for Success Based on the Original Work of Baltasar Gracian
Part One (p. 8)
Duell, Sloan & Pearce. New York, New York, USA. 1947

Emelyanov, A. S.
No biographical data available

Science provides mankind with a great tool of cognition which makes it possible to reach unprecedented heights of abundance and equality. This determines the most important and most fruitful aspect of the social role of science, and as a result the social responsibility of scientists is growing.
In E.H.S. BurhopIn Maurice Goldsmith and Alan Mackay (eds.)
Society and Science
Scientist and Public Affairs (p. 31)
Simon & Schuster. New York, New York, USA. 1965

Emerson, Ralph Waldo   1803–82
American lecturer, poet, and essayist

All science has one aim, namely, to find a theory of nature.
Ralph Waldo Emerson: Essays and Lectures
Nature: Addresses, and Lectures
Introduction (p. 7)
The Library of America. New York, New York, USA. 1983

You must have eyes of science to see in the seed its nodes….
The Complete Works of Ralph Waldo Emerson (Volume 8)
Letters and Social Aims
Chapter I (p. 71)
Houghton Mifflin Company. Boston, Massachusetts, USA. 1904

When science is learned in love, and its powers are wielded by love, they will appear the supplements and continuations of the material creation.
The Complete Works of Ralph Waldo Emerson (Volume 2)
Essays: First Series
Art (p. 369)
Houghton Mifflin Company. Boston, Massachusetts, USA. 1904

What drops of all the sea of our science are baled up! and by what accident it is that these are exposed, when so many secrets sleep in nature!
Ralph Waldo Emerson: Essays and Lectures
Essays: Second Series
The Poet (p. 466)
The Library of America. New York, New York, USA. 1983

Science in England, in America, is jealous of theory, hates the name of love and moral purpose. There's revenge for this humanity. What manner of man does science make? The boy is not attracted. He says, I do not wish to be such a kind of man as my professor is.
The Complete Works of Ralph Waldo Emerson (Volume 6)
The Conduct of Life
Beauty (p. 284)
Houghton Mifflin Company. Boston, Massachusetts, USA. 1904

Something is wanting to science until it has been humanized.
The Complete Works of Ralph Waldo Emerson (Volume 4)

Representative Men
Chapter I (p. 10)
Houghton Mifflin Company. Boston, Massachusetts, USA. 1904

Scraps of science, of thought, of poetry are in the coarsest sheet, so that in every house we hesitate to burn a newspaper until we have looked it through.
The Complete Works of Ralph Waldo Emerson (Volume 7)
Society and Solitude
Chapter II (p. 24)
Houghton Mifflin Company. Boston, Massachusetts, USA. 1904

Science has shown the great circles in which Nature works....
The Complete Works of Ralph Waldo Emerson (Volume 7)
Society and Solitude
Chapter VI (p. 143)
Houghton Mifflin Company. Boston, Massachusetts, USA. 1904

Science is cold.
The Complete Works of Ralph Waldo Emerson (Volume 1)
Nature: Addresses and Lectures
An Address (p. 143)
Houghton Mifflin Company. Boston, Massachusetts, USA. 1904

Science is a search after identity, and the scientific whim is lurking in all corners.
Ralph Waldo Emerson: Essays and Lectures
The Conduct of Life
Beauty (p. 1108)
The Library of America. New York, New York, USA. 1983

Science surpasses the old miracles of mythology. ...
The Complete Works of Ralph Waldo Emerson (Volume 8)
Letters and Social Aims
Chapter VII (p. 207)
Houghton Mifflin Company. Boston, Massachusetts, USA. 1904

Empirical science is apt to cloud the sight, and, by the very knowledge of functions and processes, to bereave the student of the manly contemplation of the whole. The savant becomes unpoetic.
Ralph Waldo Emerson: Essays and Lectures
Nature: Addresses, and Lectures
Prospects (p. 43)
The Library of America. New York, New York, USA. 1983

The motive of science was the extension of man, on all sides, into Nature, till his hands should touch the stars, his eyes see through the earth, his ears understand the language of beast and bird, and the sense of the wind; and, through his sympathy, heaven and earth should talk with him. But that is not our science.
Ralph Waldo Emerson: Essays and Lectures
The Conduct of Life
Beauty (p. 1100)
The Library of America. New York, New York, USA. 1983

Intellectual science has been observed to beget invariably a doubt of the existence of matter.
The Complete Works of Ralph Waldo Emerson (Volume 1)
Nature: Addresses and Lectures
Nature (p. 56)
Houghton Mifflin Company. Boston, Massachusetts, USA. 1904

It is the last lesson of modern science that the highest simplicity of structure is produced, not by few elements, but by the highest complexity.
The Complete Works of Ralph Waldo Emerson (Volume 4)
Representative Men
Goethe; or, The Writer (p. 290)
Houghton Mifflin Company. Boston, Massachusetts, USA. 1904

...the path of science and of letters is not the way into nature. The idiot, the Indian, the child, and unschooled farmer's boy, stand nearer to the light by which nature is to be read, than the dissector or the antiquary.
Ralph Waldo Emerson: Essays and Lectures
Essays: First Series
History (p. 256)
The Library of America. New York, New York, USA. 1983

...as the power or genius of nature is ecstatic, so must its science or the description of it be.
The Complete Works of Ralph Waldo Emerson (Volume 1)
Nature: Addresses and Lectures
The Method of Nature (p. 213)
Houghton Mifflin Company. Boston, Massachusetts, USA. 1904

It is this domineering temper of the sensual world that creates the extreme need of the priests of science....
The Complete Works of Ralph Waldo Emerson (Volume 1)
Nature: Addresses and Lectures
Literary Ethics (p. 186)
Houghton Mifflin Company. Boston, Massachusetts, USA. 1904

Emmeche, Claus   1956–
Danish theoretical biologist

It has been said that science demystifies the world. It is closer to the truth to say that science, when it is at its best, opens the world up for us, bringing daily realities under a kind of magic spell and providing the means to see the limits of what we think we know, and the scope of what we do not at all understand.
Translated by Steven Sampson
The Garden in the Machine: The Emerging Science of Artificial Life
Chapter One (p. 13)
Princeton University Press. Princeton, New Jersey, USA. 1994

Everett, Edward   1794–1865
American statesman, educator, and orator

It usually happens in scientific progress, that when a great fact is at length discovered, it approves itself at once to all competent judges. It furnishes a solution to so many problems, and harmonizes with so many other facts, — that all the other data as it were crystallize at once about it.
The Uses of Astronomy
An Oration Delivered at Albany on the 28th of July, 1856 (p. 30)
Ross & Tousey. New York, New York, USA. 1856

Feigl, H.
No biographical data available

...science, properly interpreted, is not dependent on any sort of metaphysics. It merely attempts to cover a maximum of facts by a minimum of laws.
Naturalism and Humanism
American Quarterly, Volume 1, Number 2, Summer 1949 (p. 148)

Ferré, Nels F. S.   1908–71
Swedish-American theologian

Science is supposed by many to have banished every realm of the sacred; and behold, science becomes the sacred cow.
Faith and Reason
Chapter II (p. 43)
Harper & Brothers. New York, New York, USA. 1946

It is a sad experience to hear someone denounce science as the cause of modern chaos and destruction. Our technological advance may be abused and make of what could be a near heaven a near hell, but that is surely not the fault of science as such. Science has not failed man, but man has failed science.
Faith and Reason
Chapter II (p. 38)
Harper & Brothers. New York, New York, USA. 1946

Ferris, Timothy   1944–
American science writer

Far too many students accept the easy belief that they need not bother learning much science, since a revolution will soon disprove all that is currently accepted anyway. In such a climate it may be worth affirming that science really is progressive and cumulative, and that well-established theories, though they may turn out to be subsets of larger and farther-reaching ones — as happened when Newtonian mechanics was incorporated by Einstein into general relativity — are seldom proved wrong....
The Whole Shebang: A State-of-the Universe's Report
Preface (p. 13)
Simon & Schuster. New York, New York, USA. 1996

Science is not perfect, but neither is it just one more sounding board for human folly.
The Whole Shebang: A State-of-the Universe's Report
Preface (p. 13)
Simon & Schuster. New York, New York, USA. 1996

Feyerabend, Paul K.   1924–94
Austrian-born American philosopher of science

Science is neither a single tradition, nor the best tradition there is, except for people who have become accustomed to its presence, its benefits and its disadvantages. In a democracy it should be separated from the state just as churches are now separated from the state.
Against Method: Outline of an Anarchistic Theory of Knowledge
(p. 238)
Verso. London, England. 1978

Science is an essentially anarchistic enterprise....
Against Method: Outline of an Anarchistic Theory of Knowledge
Analytical Index (p. 10)
Verso. London, England. 1978

Science is not a closed book that is understood only after years of training. It is an intellectual discipline that can be examined and criticised by anyone who is interested, and that looks difficult and profound only because of a systematic campaign of obfuscation carried out by many scientists...
In E.D. Klemke, Robert Hollinger and A. David Kline
Introductory Reading in the Philosophy of Science
How to Defend Society Against Science (p. 62)
Prometheus Books. Buffalo, New York, USA. 1980

Feynman, Richard P.   1918–88
American theoretical physicist

We must, incidentally, make it clear from the beginning that if a thing is not a science, it is not necessarily bad. For example, love is not a science. So, if something is said not to be a science, it does not mean that there is something wrong with it; it just means that it is not a science.
Six Easy Pieces: Essentials of Physics Explained by Its Most Brilliant Teacher
The Relation of Physics to Other Sciences (p. 47)
Addison-Wesley Publishing Company. Reading, Massachusetts, USA. 1995

All science is intelligent inference; excessive literalism is a delusion, not a humble bowing to evidence.
Dinosaur in a Haystack: Reflections in Natural History
Part Three, Chapter 12 (p. 156)
Random House, Inc. New York, New York, USA. 1995

[An] aspect of science is its contents, the things that have been found out. This is the yield. This is the gold. This is the excitement, the pay you get for all the disciplined thinking and hard work. The work is not done for the sake of an application. It is done for the excitement of what is found out. ...it is almost impossible for me to convey in a lecture this important aspect, this exciting part, the real reason for science. And without understanding this you miss the whole point.
The Meaning of It All: Thoughts of a Citizen Scientist
Chapter I (p. 9)
Perseus Books. Reading, Massachusetts, USA. 1998

You cannot understand science and its relation to anything else unless you understand and appreciate [it as] the great adventure of our time. You do not live in your time unless you understand that this is...a wild and exciting thing.
The Meaning of It All: Thoughts of a Citizen Scientist
Chapter I (p. 9)
Perseus Books. Reading, Massachusetts, USA. 1998

Science can be defined as a method for, and a body of information obtained by, trying to answer only questions which can be put into the form: If I do this, what will happen?

Engineering and Science, Volume 19, 1956 (p. 23)

Science is a way to teach how something gets to be known, what is not known, to what extent things are known (for nothing is known absolutely), how to handle doubt and uncertainty, what the rules of evidence are, how to think about things so that judgments can be made, how to distinguish truth from fraud, and from show.
The Problem of Teaching Physics in Latin America
Engineering and Science, November 1963

Science alone of all the subjects contains within itself the lesson of the danger of belief in the infallibility of the greatest teachers in the preceding generation.... As a matter of fact, I can also define science another way: Science is the belief in the ignorance of experts.
In Jeffrey Robbins (ed.)
The Pleasure of Finding Things Out: The Best Short Works of Richard P. Feynman
Chapter 8 (p. 188)
Perseus Books. Cambridge, Massachusetts, USA. 1999

Although it is uncertain, it is necessary to make science useful. Science is only useful if it tells you about some experiment that has not been done; it is not good if it only tells you what just went on.
The Character of Physical Law
Chapter 7 (p. 164)
BBC. London, England. 1965

And so it is with Science. In a way it is key to the gates of heaven, and the same key opens the gates of hell, and we do not have any instructions as to which is which gate. Shall we throw away the key and never have a way to enter the gates of heaven? Or shall we struggle with the problem of which is the best way to use the key? That is, of course, a very serious question, but I think that we cannot deny the value of the key to the gates of heaven.
The Meaning of It All: Thoughts of a Citizen Scientist
Chapter I (p. 6–7)
Perseus Books. Reading, Massachusetts, USA. 1998

Is science of any value?
I think a power to do something is of value. Whether the result is a good thing or a bad thing depends on how it is used, but the power is a value.
Once in Hawaii I was asked to see a Buddhist temple. In the temple a man said, "I am going to tell you something that you will never forget." And then he said, "To every man is given the key to the gates of heaven. The same key opens the gates of hell."
The Meaning of It All: Thoughts of a Citizen Scientist
Chapter I (p. 6)
Perseus Books. Reading, Massachusetts, USA. 1998

It is necessary to teach both to accept and to reject the past with a kind of balance that takes considerable skill. Science alone of all the subjects contains within itself the lesson of the danger of belief in the infallibility of the greatest teachers of the preceding generation.
In Jeffrey Robbins (ed.)
The Pleasure of Finding Things Out: The Best Short Works of Richard P. Feynman
Chapter 8 (p. 188)
Perseus Books. Cambridge, Massachusetts, USA. 1999

If you thought before that science was certain — well, that is just an error on your part.
The Character of Physical Law
Chapter 3 (p. 77)
BBC. London, England. 1965

...it is imperative in science to doubt; it is absolutely necessary, for progress of science, to have uncertainty as a fundamental part of your inner nature.
Engineering and Science, Volume 19, 1956 (p. 21)

...science is of value because it can produce something.
What Do You Care What Other People Think?
The Value of Science (p. 241)
W.W. Norton & Company, Inc. New York, New York, USA.1988

Fischer, Emil   1852–1919
German chemist

The sciences are not abstract constructions, but rather the result of human endeavor; they are closely connected with the personalities and the fates of the dedicated researchers who developed them.
In Rolf Huisgen
Adolf von Baeyer's Scientific Achievements — A Legacy
Angewandet Chemie International Edition in English, Volume 25, Number 4, April 1986 (p. 297)

...science is and remains international.
In Albert Einstein
The World as I See It
The International Science (p. 50)
Philosophical Library. New York, New York, USA. 1949

Fiske, John   1842–1901
American philosopher and historian

...all human science is but the increment of the power of the eye....
The Destiny of Man Viewed in the Light of His Origin
Chapter VII (p. 60)
Houghton Mifflin Company. Boston, Massachusetts, USA. 1912

...there are moments when one passionately feels that this cannot be all. On warm June mornings in green country lanes, with sweet pine-odours, wafted in the breeze which sighs through the branches, and cloud-shadows flitting over far-off blue mountains, while little birds sing their love-songs, and golden-haired children weave garlands of wild roses; or when in the solemn twilight we listen to wondrous harmonies of Beethoven and Chopin that stir the heart like voices from an unseen world; at such times one feels that the profoundest answer which

science can give to our questionings is but a superficial answer after all.
The Unseen World, and Other Essays
I. The Unseen World, Part II (p. 56)
Houghton Mifflin Company. New York, New York, USA. 1876

Flaubert, Gustave   1821–90
French novelist

My kingdom is as wide as the world, and my desire has no limit. I go forward always, freeing spirits and weighing worlds, without fear, without compassion, without love, and without God. Men call me science.
The Temptation of Saint Anthony (p. 161)
The Modern Library. New York, New York, USA. 2001

Forbes, Edward   1815–54
English naturalist

People without independence have no business to meddle with science. It should never be linked with lucre.
In George Wilson and Archibald Geikie
Memoir of Edward Forbes, F.R.S.
Chapter XII (p. 392)
Macmillan & Company Ltd. Cambridge, England. 1861

Fort, Charles   1874–1932
American writer

Every science is a mutilated octopus. If its tentacles were not clipped to stumps, it would feel its way into disturbing contacts.
In Damon Knight
Charles Fort: Prophet of the Unexplained
A Charles Fort Sampler (p. vi)
Gollancz. London, England. 1971

Fox, Robin   1934–
English anthropologist, poet, and essayist

The conduct of science can lead to boring triviality. Even great results can be used to evil ends.
In Paul R. Gross, Norman Levitt, and Martin W. Lewis (eds.)
The Flight from Science and Reason
State of the Art/Science in Anthropology (p. 329)
New York Academy of Sciences. New York, New York, USA. 1996

Since science has no value agenda of its own it is always subject to hijacking by fanaticism and idealism.
In Paul R. Gross, Norman Levitt, and Martin W. Lewis (eds.)
The Flight from Science and Reason
State of the Art/Science in Anthropology (p. 329)
New York Academy of Sciences. New York, New York, USA. 1996

The point that science can be used for evil purposes is beside the point. Art and music can be used for evil purposes, but no one proposes abandoning either. Anything can be used for evil purposes. I am not going to stop listening to Wagner just because Hitler liked him.
In Paul R. Gross, Norman Levitt, and Martin W. Lewis (eds.)
The Flight from Science and Reason
State of the Art/Science in Anthropology (p. 332)
New York Academy of Sciences. New York, New York, USA. 1996

Franck, Georg   1946–
No biographical data available

Success in science is rewarded with attention. You gain full membership in the scientific community only by receiving the attention of your fellow scientists. Earning this attention "income" is a prime motive for becoming a scientist and for practicing science. In order to maximize this income, you have to employ your own attention in the most productive way. It does not pay to find things out anew that have been discovered already. Nor is reinvention rewarding in terms of the attention paid. It pays no pay attention to the work done by others.
Scientific Communication — A Vanity Fair?
Science, Volume 286, Number 5437, 1 October 1999 (p. 53)

Franklin, Benjamin   1706–90
American printer, scientist, and diplomat

The rapid progress true Science now makes occasions my regretting sometimes that I was born so soon. It is impossible to imagine the heights to which may be carried, in a thousand years, the power of man over matter. O that moral Science were in as fair a way of improvement, that men would cease to be wolves to one another, and that human beings would at length learn what they now improperly call humanity.
In Linus Pauling
College Chemistry
Letter to Joseph Priestley, 8 February 1780
Chapter 1 (p. 3)
W.H. Freeman & Company. San Francisco, California, USA. 1964

Freud, Sigmund   1856–1939
Austrian neurologist and co-founder of psychoanalysis

No, our science is no illusion. But an illusion it would be to suppose that what science cannot give us we can get elsewhere.
The Future of an Illusion
Chapter X (p. 56)
W.W. Norton & Company, Inc. New York, New York, USA.1961

Science in her perpetual incompleteness and insufficiency is driven to hope for her salvation in new discoveries and new ways of regarding things. She does well, in order not to be deceived, to arm herself with skepticism and to accept nothing new unless it has withstood the strictest examination.
Collected Papers
The Resistance to Psycho-Analysis (p. 4121)
The Hogarth Press. London, England. 1953

Freund, Ida   1863–1914
Austrian-born chemist

The object of all the Natural Sciences is the acquisition of knowledge concerning the natural objects surrounding us, as we apprehend them by our senses; of the changes occurring in these objects, together with the laws governing these changes; and of the more proximate or more ultimate causes to the operation of which are due the individual phenomena and the general laws comprising these.

The Study of Chemical Composition
Introduction (p. 1)
At The University Press. Cambridge, England. 1904

Friend, Julius W.
European historian
Feibleman, James K.   1904–87
American philosopher

The modern world, which has lost faith in so many causes, still accepts science nearly unchallenged. Science today occupies the position held by the Roman Church in the Middle Ages: as the single great authority in a world divided on almost every object of loyalty.

What Science Really Means
Chapter I (p. 11)
George Allen & Unwin Ltd. London, England. 1937

French Apothegm

Le scepticisme est le vrai flambeau de la science.
Doubt is the true torch of science.

In John Epps
The Life of Dr. Walker
Chapter IV (p. 101)
Whittaker, Treacher. London, England. 1831

Fromm, Erich   1900–80
German psychoanalyst

The pace of science forces the pace of the technique. Theoretical physics forces atomic energy on us; the successful production of the fission bomb forces upon us the manufacture of the hydrogen bomb. We do not choose our problems, we do not choose our products; we are pushed, we are forced — by what? By a system which has no purpose and goal transcending it, and which makes man its appendix.

The Sane Society
Chapter Five, Nineteenth-Century Capitalism (p. 83)
Fawcett Publications. Greenwich, Connecticut, USA. 1955

Frost, Robert   1874–1963
American poet

And how much longer a story has science
Before she must put out the light on the children
And tell them the rest of the story is dreaming?

Complete Poems of Robert Frost
Too Anxious for Rivers
Henry Holt & Company. New York, New York, USA. 1949

Where have those flowers and butterflies all gone
That science may have staked the future on?
He seems to say the reason why so much

Should come to nothing must be fairly faced....

Complete Poems of Robert Frost
Pod of the Milkweed
Henry Holt & Company. New York, New York, USA. 1949

Sarcastic Science, she would like to know,
In her complacent ministry of fear,
How we propose to get away from here
When she has made things so we have to go
Or be wiped out.

Complete Poems of Robert Frost
Why Wait for Science
Henry Holt & Company. New York, New York, USA. 1949

Fulbright, James William   1905–95
American politician

What a curious picture it is to find man, *Homo sapiens*, of divine origin, we are told, seriously considering going underground to escape the consequences of his own folly. With a little wisdom and foresight, surely it is not yet necessary to forsake life in the fresh air and in the warmth of the sunlight. What a paradox if our own cleverness in science should force us to live underground with the moles.

The Effect of the Atomic Bomb on American Foreign Policy
Congressional Record, November 2, 1945, Volume 91, Appendix (p. A4654)

Science has radically changed the conditions of human life on earth. It has expanded our knowledge and our power but not our capacity to use them with wisdom.

Old Myths and New Realities
Conclusion (p. 142)
Random House, Inc. New York, New York, USA. 1964

Gábor, Dennis   1900–79
Hungarian-English physicist

Science has never quite given man what he desired, not even in applied science. Man dreamt of wings; science gave him an easy chair which flies through the air.

Inventing the Future
The Future of the Uncommon Man (p. 162)
Secker & Warburg. London, England. 1963

Galbraith, John Kenneth   1908–2006
Canadian-American economist

The real accomplishment of modern science and technology consists in taking ordinary men, informing them narrowly and deeply, and then, through appropriate organization, arranging to have their knowledge combined with that of other specialized but equally ordinary men. This dispenses with the need for genius. The resulting performance, though less inspiring, is far more predictable.

The New Industrial State
Chapter VI, Section 2 (p. 62)
Houghton Mifflin Company. Boston, Massachusetts, USA. 1967

Gardner, Martin 1914–
American writer and mathematics games editor

...modern science should indeed arouse in all of us a humility before the immensity of the explored and a tolerance for crazy hypotheses.
Science: Good, Bad and Bogus
Chapter 22 (p. 246)
Prometheus Books. Buffalo, New York, USA. 1981

Garrod, Archibald 1857–1936
English physician

Science is not, as so many seem to think, something apart, which has to do with telescopes, retorts, and test tubes, and especially with nasty smells, but it is a way of searching out by observation, trial, and classification; whether the phenomenon investigated be the outcome of human activities, or of the more direct workings of nature's laws. Its methods admit of nothing untidy or slipshod, its keynote is accuracy and its goal is truth.
In Alexander G. Bearn
Archibald Garrod and the Individuality of Man
Chapter 8 (p. 97)
Clarendon Press. Oxford, England. 1993

Gerould, Katherine Fullerton 1879–1944
American writer

The great danger of the scientific obsession is not the destruction of all things that are not science, but the slow infection of those things.
Modes and Morals
The Extirpation of Culture (p. 87)
Charles Scribner's Sons. New York, New York, USA. 1920

Science has done great things for us; it has also pushed us hopelessly back. For, not content with filling its own place, it has tried to supersede everything else. It has challenged the super-eminence of religion; it has turned all philosophy out of doors except that which clings to its skirts; it has thrown contempt on all learning that does not depend on it; and it has bribed the skeptics by giving us immense material comforts.
Modes and Morals
The Extirpation of Culture (p. 85)
Charles Scribner's Sons. New York, New York, USA. 1920

The insidiousness of science lies in its claim to be not a subject, but a method. You could ignore a subject; no subject is all-inclusive. But a method can plausibly be applied to anything within the field of consciousness.
Modes and Morals
The Extirpation of Culture (p. 86)
Charles Scribner's Sons. New York, New York, USA. 1920

Gideonse, H. D.
No biographical data available

Science, as usually taught to liberal arts students, emphasizes results rather than method, and tries to teach technique rather than to give insight into and understanding of the scientific habit of thought. What is needed, however, is not a dose of metaphysics but a truly humanistic teaching of science.
In Lloyd William Taylor
Physics: The Pioneer Science (Volume 2)
Chapter 36 (p. 529)
Houghton Mifflin Company. Boston, Massachusetts, USA. 1941

Gill, Eric 1882–1940
English sculptor

Science is analytical, descriptive, informative. Man does not live by bread alone, but by science he attempts to do so. Hence the deadliness of all that is purely scientific.
It All Goes Together: Selected Essays (p. 117)
The Devin-Adair Company. New York, New York, USA. 1944

Ginger, Ray 1924–75
American historian

Science has explained everything it could explain, and it will continue to do so. Every effort to bar science from some areas on the ground that they were not susceptible to empirical investigation has had the effect of inhibiting science in other areas also. Man has progressed by exercising a humble confidence in the might of his own mind, not by throwing up his hands and shrugging his shoulders.
Six Days or Forever: Tennessee v. John Thomas Scopes
Chapter 11, Section III (p. 231)
Quadrangle Books. Chicago, Illinois, USA. 1958

Glass, H. Bentley 1906–2005
American geneticist

Science is not only to know, it is to do, and in the doing it has found its soul.
Science and Ethical Values
Chapter 3 (p. 101)
University of North Carolina Press. Chapel Hill, North Carolina, USA. 1965

...the general citizen of his country, the man in the street, must learn what science is, not just what it can bring about. Surely this is our primary task. If we fail in this, then within a brief period of years we may expect either nuclear devastation or worldwide tyranny. It is not safe for apes to play with atoms. Neither can men who have relinquished their birthright of scientific knowledge expect to rule themselves.
In Hilary J. Deason
A Guide to Science Reading
Revolution in Biology (pp. 25–26)
The New American Library. New York, New York, USA. 1966

Gluckman, Max 1911–75
English anthropologist

Science is cumulative. The apprentice in this generation can outdo his master of the last.
Politics, Law & Ritual
Chapter VII (p. 303)
Aldine Publishing Company. Chicago, Illinois, USA. 1965

Goddard, Robert H.   1882–1945
American physicist

Each must remember that no one can predict to what heights of wealth, fame, or usefulness he may rise until he has honestly endeavored, and he should derive courage from the fact that all sciences have been, at some time, in the same condition as he, and that it has often proved true that the dream of yesterday is the hope of today and the reality of tomorrow.
The Papers of Robert H. Goddard (Volume 1)
On Taking Things for Granted
Graduation oration (p. 66)
McGraw-Hill Book Company, Inc. New York, New York, USA. 1970

Gorky, Maxim   1868–1938
Russian writer

Science is becoming the nervous system of our time.
Compiled by V.V. Vorontsov
Words of the Wise: A Book of Russian Quotations
Translated by Vic Schneierson
Progress Publishers. Moscow, Russia. 1979

Science is humanity's superior reason, the sun which man has created of his own flesh and blood and has lit to illuminate the darkness of his hard life and to show the way to freedom, justice, and beauty.
Compiled by V.V. Vorontsov
Words of the Wise: A Book of Russian Quotations
Translated by Vic Schneierson
Progress Publishers. Moscow, Russia. 1979

For us natural science is the Archimedes' screw that alone can turn the world to face the sun of reason.
Compiled by V.V. Vorontsov
Words of the Wise: A Book of Russian Quotations
Translated by Vic Schneierson
Progress Publishers. Moscow, Russia. 1979

Humanity has no force more powerful and victorious than science.
Compiled by V.V. Vorontsov
Words of the Wise: A Book of Russian Quotations
Translated by Vic Schneierson
Progress Publishers. Moscow, Russia. 1979

Gortner, Ross Aiken
No biographical data available

Science will stagnate only when all will agree that only one interpretation can be drawn from a given series of data.
Selected Topics in Colloid Chemistry with Especial Reference to Biochemical Problems
Preface (p. vii)
Cornell University Press. Ithaca, New York, USA. 1937

Gould, Laurence M.   1896–1995
American polar explorer and geologist

Today, there is no other influence comparable with science in changing the foundations, indeed the very character of our lives. Science and its products determine our economy, dominate our industry, affect our health and welfare, alter our relations to all other nations, and determine the conditions of war and peace. Everyone who breathes is affected, and cannot remain impervious to them.
Science and the Culture of Our Times
UNESCO Courier, February 1968 (p. 4)

Gould, Stephen Jay   1941–2002
American paleontologist and evolutionary biologist

No factual discovery of science (statements about how nature "is") can, in principle, lead us to ethical conclusions (how we "ought" to behave) or to convictions about intrinsic meaning (the "purpose" of our lives).
Dorothy, It's Really Oz
Time, August 23, 1999 (p. 59)

Science is an integral part of culture. It's not this foreign thing, done by an arcane priesthood. It's one of the glories of the human intellectual tradition.
Independent (London), January 24, 1990

Science does progress toward more adequate understanding of the empirical world, but no pristine, objective reality lies "out there" for us to capture as our technologies improve and our concepts mature.
Dinosaur in a Haystack: Reflections in Natural History
Part Four, Chapter 16 (p. 214)
Random House, Inc. New York, New York, USA. 1995

Creative science is always a mixture of facts and ideas. Great thinkers are not those who can free their minds from cultural baggage and think or observe objectively (for such a thing is impossible), but people who use their milieu creatively rather than as a constraint.
An Urchin in the Storm: Essays About Books and Ideas
Chapter 6 (p. 103)
W.W. Norton & Company, Inc. New York, New York, USA. 1987

The net of science covers the empirical universe: What is it made of (fact) and why does it work this way (theory).
Non Overlapping Magisteria
Natural History, Volume 106, Number 2, March 1997 (p. 61)

Humanity has in course of time had to endure from the hand of science two great outrages upon its naive self-love. The first was when it realized that our earth was not the center of the universe, but only a speck in a world-system of a magnitude hardly conceivable…. The second was when biological research robbed man of his particular privilege of having been specially created and relegated him to a descent from the animal world.

Time's Arrow, Time's Cycle: Myth and Metaphor in the Discovery of Geological Time
Chapter 1 (p. 1)
Harvard University Press. Cambridge, Massachusetts, USA. 1987

Grassé, Pierre P.   1895–1985
French zoologist

There is no law against day dreaming, but science must not indulge in it.
Evolution of Living Organisms: Evidence for a New Theory of Transformation
Chapter IV (p. 104)
Academic Press. New York, New York, USA. 1977

Gray, Thomas   1716–71
English poet

Here rests his head upon the lap of Earth,
A youth to fortune and to fame unknown.
Fair Science frown's not on his humble birth,
And Melancholy mark's him for her own.
The Complete Poetical Works of Gray, Beattie, Blair, Collins, Thomson, and Kirke White
Elegy Written in a Country Churchyard, The Epitaph, Stanza 1
J. Blackwood. London, England. 1800

Gregory, Sir Richard Arman   1864–1952
British science writer and journalist

Science advances by opening completely new fields of knowledge upon which the literary man or investigator may exercise their intellectual activities, and the directions in which these domains are to be found are rarely indicated with success in romantic or in scientific literature.
Discovery; or, The Spirit and Service of Science
Chapter VI (p. 164)
Macmillan & Company Ltd. London, England. 1918

...I hope that my children, at least, if not I myself, will see the day, when ignorance of the primary laws and facts of science will be looked on as a defect, only second to ignorance of the primary laws of religion and morality.
Discovery; or, The Spirit and Service of Science
Chapter V (p. 92)
Macmillan & Company Ltd. London, England. 1918

...science is not to be measured by practical service alone, though it may contribute to material prosperity: it is an intellectual outlook, a standard of truth and a gospel of light.
Discovery; or, The Spirit and Service of Science
Preface (p. vi)
Macmillan & Company Ltd. London, England. 1918

Grove, Sir William   1811–96
English chemist

For my part I must say that science to me generally ceases to be interesting as it becomes useful.
In H.B.G. Casimir
Haphazard Reality: Half a Century of Science
Chapter 8 (p. 226)
Harper & Row, Publishers. New York, New York, USA. 1983

It would be vain to attempt specifically to predict what may be the effect of Photography on future generations. A Process by which the most transient actions are rendered permanent, by which facts write their own annals in a language that can never be obsolete, forming documents which prove themselves, — must interweave itself not only with science but with history and legislature.
Lecture
Progress of Physical Science Since the Opening of the London Institution, (19 January 1842)

Gruber, Howard E.   1922–2005
American psychology scholar and professor

The power and the beauty of science do not rest upon infallibility, which it has not, but on corrigibility, without which it is nothing.
The Origin of "The Origin of Species"
The New York Times Book Review, 22 July, 1979 (p. 7)

Gruenberg, Benjamin C.
No biographical data available

To vast numbers of men and women science appears as something altogether too remote from their interests or capacities to justify even a glance or a hope of grasping. It is something for the "highbrows" or wizards.
Science and the Public Mind
Chapter XIV (p. 152)
McGraw-Hill Book Company, Inc. New York, New York, USA. 1935

Guth, Alan   1947–
American physicist

...science is not merely a collection of facts, but is instead an ongoing detective story, in which scientists passionately search for clues in the hope of unraveling the mysteries of the universe.
The Inflationary Universe; the Quest for a New Theory of Cosmic Origins
Chapter 3 (p. 34)
Addison-Wesley Publishing Company. Reading, Massachusetts, USA. 1997

Guye, Charles Eugene
Swiss scientist

It is because we do not possess "science" that we have "sciences."
Physico-Chemical Evolution (p. 8)
Methuen & Company Ltd. London, England. 1925

Haggard, Howard W.
Physician

There is another class of explorers whose exploits are rarely heralded by the waving of flags, to whom few monuments are erected, and whose names find small place in world history. They are the explorers in science. They change no maps, but they change our ways of living.
In Bernard Jaffe

New World of Chemistry
Chapter 10 (p. 113)
Silver, Burdett & Company. New York, New York, USA. 1935

Hall, A. D.
No biographical data available

The true aim of science is the enrichment of life.
Nature, Volume 138, 1936 (p. 576)

Hall, Alfred Rupert   1920–
English historian of science
Hall, Marie Boas   1919–
English historian of science

It is hardly too much to say that the Middle Ages studied science as though it were theology and Aristotle's Physics as though it were the Bible.
A Brief History of Science
Chapter 6 (p. 78)
Iowa State University Press. Ames, Iowa, USA. 1988

Handler, Philip   1917–81
No biographical data available

My own belief is that science remains the most powerful tool we have yet generated to apply leverage for our future. It is the instrument which is most useful for guiding our own destinies, for assuring the condition of man in the years to come. I have much to hope that we will not abandon that tool, leaving us to our own brute devices.
Hearings
1971 National Science Foundation Authorization, Subcommittee on Science, Research and Development, House Committee on Science and Astronautics, 91st Congress, 2nd Session 1970 (p. 16)

Harari, Josué V.
No biographical data available
Bell, David F.
No biographical data available

Science is the totality of the world's legends. The world is the space of their inscription. To read and to journey are one and the same act.
In Michel Serres
Hermes: Literature, Science, Philosophy
Introduction (p. xxi)
The Johns Hopkins University Press. Baltimore, Maryland, USA. 1982

Hardy, Thomas   1840–1928
English poet and regional novelist

Well: what we gain by science is, after all, sadness, as the Preacher saith. The more we know of the laws & nature of the Universe the more ghastly a business we perceive it all to be — & the non-necessity of it.
In Richard Little Purdy and Michael Millgate (eds.)
The Collected Letters of Thomas Hardy (Volume 3)
Letter, February 27, 1902
Clarendon Press. Oxford, England. 1978

Harman, Willis
No biographical data available

Science is all about cause. That's why you have science; you're trying to find the explanation, the causes, for the phenomena. Now, if really everything is connected to everything, if there really is only a oneness, everything then affects everything, and everything is the cause of everything in a certain sense, so that the whole idea of causality has to be revised.
Thinking Allowed: Conversations on the Leading Edge of Knowledge and Discovery
Metaphysics and Modern Science, Part I: Consciousness and Science
Thinking Allowed Productions. Berkeley, California, USA.

Harrington, John W.
American geologist

Science is the progressive discovery of the nature of nature.
Dance of the Continents
The Lure of the Hunt (p. 30)
J.P. Tarcher. Los Angeles, California, USA. 1983

Harris, Errol E.
No biographical data available

Accordingly there are two main types of science, exact science...and empirical science...seeking laws which are generalizations from particular experiences and are verifiable (or, more strictly, "probabilities") only by observation and experiment.
Hypothesis and Perception: The Roots of Scientific Method
Prevalent Views of Science (p. 25)
George Allen & Unwin Ltd. London, England. 1970

Harrison, Jane   1850–1928
English classical scholar

Science has given us back something strangely like a World-Soul...
Ancient Art and Ritual
Chapter VII (p. 238)
Henry H. Holt. New York, New York, USA. 2002

Harvey, William   1578–1657
English physician

Although there is but one road to science, that, to wit, in which we proceed from things more known to things less known, from matters more manifest to matters more obscure; and universals are principally known to us, science bringing by reasoning from universals to particulars; still the comprehension of universals by the understanding is based upon the perception of individual things by the senses.
In *Great Books of the Western World* (Volume 28)
Anatomical Exercises on the Generation of Animals
Of the Manner and Order of Acquiring Knowledge (p. 332)
Encyclopædia Britannica, Inc. Chicago, Illinois, USA. 1952

Hauge, Philip   1913–2004
American scientist, editor, and administrator

Part of the strength of science is that it has tended to attract individuals who love knowledge and the creation of it. Just as important to the integrity of science have been the unwritten rules of the game. These provide recognition and approbation for work which is imaginative and accurate, and apathy or criticism for the trivial or inaccurate. ... Thus, it is the communication process which is at the core of the vitality and integrity of science. ... The system of rewards and punishments tends to make honest, vigorous, conscientious, hardworking scholars out of people who have human tendencies of slothfulness and no more rectitude than the law requires.
The Roots of Scientific Integrity
Science, Volume 139, 1963 (p. 3561)

Havel, Václav   1936–
Czech dramatist and essayist

Modern science abolishes as mere fiction the innermost foundations of our natural world: it kills God and takes his place on the vacant throne so henceforth it would be science that would hold the order of being in its hand as its sole legitimate guardian and so be the legitimate arbiter of all relevant truth. People thought they could explain and conquer nature — yet the outcome is that they destroyed it and disinherited themselves from it.
In L. Wolpert
The Unnatural Nature of Science
Introduction (p. ix)
Harvard University Press. Cambridge, Massachusetts, USA. 1992

Hawking, Stephen William   1942–
English theoretical physicist

The whole history of science has been the gradual realization that events do not happen in an arbitrary manner, but that they reflect a certain underlying order, which may or may not be divinely inspired.
A Brief History of Time: From the Big Bang to Black Holes
Chapter 8 (p. 122)
Bantam Books. Toronto, Ontario, Canada. 1988

In effect, we have redefined the task of science to be the discovery of laws that will enable us to predict events up to the limits set by the uncertainty principle.
A Brief History of Time: From the Big Bang to Black Holes
Chapter 11 (p. 173)
Bantam Books. Toronto, Ontario, Canada. 1988

Hayward, Jeremy
American physicist

Like Christianity, modern science teaches that these things of the world of senses are not really real, but that there is a more real reality, in Nature, behind these appearances, a permanent, unchanging reality in comparison to which the world of appearance is ever changing and is an accidental product of our sense organs. Unlike the "other world" of Christianity, which is world of spirit or mind, altogether without body, this "other world" of science is a world of matter, altogether without spirit, life, or mind. This ultimately real world is the world of particles (little bits of dead stuff), of space and time and of forces (gravitational, electromagnetic, and more recently the strong and weak nuclear forces).
Shifting Worlds, Changing Minds: Where the Sciences and Buddhism Meet
Chapter 1 (p. 14)
New Science Library Shambhala Publications, Inc. Boston, Massachusetts, USA. 1987

Hazlitt, William Carew   1834–1913
English bibliographer

The origin of all science is in the desire to know causes; and the origin of all false science and imposture is in the desire to accept false causes rather than none; or, which is the same thing, in the unwillingness to acknowledge our own ignorance.
The Atlas
February 15, 1829
Burke and the Edinburgh Phrenologists
This article is unsigned in the atlas but appears in P.P. Howe's *New Writings* by William Hazlitt, 1925

Heidel, W. A.
No biographical data available

It is an unwarranted assumption that ancient science differed in principle at any point from that of today.
In Julius W. Friend and James Feibleman
What Science Really Means
Chapter II (p. 26)
George Allen & Unwin Ltd. London, England. 1937

Heinlein, Robert A.   1907–88
American science fiction writer

The difference between science and the fuzzy subjects is that science requires reasoning, while those other subjects merely require scholarship.
Time Enough for Love
Second Intermission (p. 366)
G.P. Putnam's Sons. New York, New York, USA. 1973

If it can't be expressed in figures, it is not science; it is opinion.
Time Enough for Love
Intermission (p. 257)
G.P. Putnam's Sons. New York, New York, USA. 1973

Heisenberg, Werner Karl   1901–76
German physicist and philosopher

The assumption [is] that in the end it will always be possible to understand nature, even in every new field of experience, but that we may make no a priori assumptions about the meaning of the word "understand."
In Heinrich O. Proskauer
The Rediscovery of Color: Goethe versus Newton Today
Preface (p. ix)
Anthroposophic Press. Spring Valley, New York, USA. 1986

Science no longer confronts nature as an objective observer, but sees itself as an actor in this interplay between man and nature. The scientific method of analysing, explaining, and classifying has become conscious of its limitations.... Method and object can no longer be separated.
The Physicist's Conception of Nature
Chapter I (p. 29)
Greenwood Press, Publishers. Westport, Connecticut, USA. 1958

Almost every progress in science has been paid by a sacrifice, for almost every new intellectual achievement previous positions and conceptions had to be given up. Thus, in a way, the increase of knowledge and insight diminishes continually the scientist's claim to "understand" nature.
In A. Sarlemijn and M.J. Sparnaay (eds.)
Physics in the Making: Essays on Developments in 20th Century Physics: In Honour of H.B.G. Casimir on the Occasion of His 80th Birthday
Chapter I (p. 9)
North-Holland Publishing Company. Amsterdam, Netherlands. 1989

In science...it is impossible to open up new territory unless one is prepared to leave the safe anchorage of established doctrine and run the risk of a hazardous leap forward. With his relativity theory, Einstein had abandoned the concept of simultaneity, which was part of the solid ground of traditional physics, and, in so doing, outraged many leading physicists and philosophers and turned them into bitter opponents. In general, scientific progress calls for no more than the absorption and elaboration of new ideas — and this is a call most scientists are happy to heed.
Physics and Beyond: Encounters and Conversations
Chapter 6 (p. 70)
Harper & Row, Publishers. New York, New York, USA. 1971

Henderson, Lawrence   1878–1942
American biochemist

Science has finally put the old teleology to death. Its dismembered spirit, freed from vitalism and all material ties, immortal, alone lives on, and from such a ghost, science has nothing to fear.
The Fitness of the Environment: An Inquiry into the Biological Significance of the Properties of Matter
Chapter VIII, Section III, B (p. 311)
The Macmillan Company. New York, New York, USA. 1913

Henry, Joseph   1797–1878
Scottish-born American scientist

...science is the pursuit above all which impresses us with the capacity of man for intellectual and moral progress and awakens the human intellect to aspiration for a higher condition of humanity.
Inscription on the National Museum of American History, Washington, D.C.

...narrow minds think nothing of importance but their own favorite pursuit, but liberal views exclude no branch of science or literature...

Inscription on the National Museum of American History, Washington, D.C.

Herschel, Sir John Frederick William   1792–1871
English astronomer and chemist

Science is the knowledge of many, orderly and methodically digested and arranged, so as to become attainable to one.
The Cabinet of Natural Philosophy
Part I, Chapter II, Section 13 (p. 18)
Longman, Rees, Orme, Brown & Green. London, England. 1831

Science is of no party. Under the government, whether of Whig or Tory, she has often had to complain of the difficulty of making herself heard in recommendation of her objects; but those objects once recognized by a British government, are taken up in a spirit and with a liberality which ensures success, if success be possible.
Essays from the Edinburgh and Quarterly Reviews with Addresses and Other Pieces
Terrestrial Magnetism (pp. 112–113)
Longman, Brown, Green, Longmans & Roberts. London, England. 1857

Every student who enters upon a scientific pursuit, especially if at a somewhat advanced period of life, will find not only that he has much to learn, but much also to unlearn.
Outlines of Astronomy (2nd edition)
Part I, Introduction (p. 1)
Longman, Brown, Green & Longmans. London, England. 1849

...if science may be vilified by representing it as opposed to religion, or trammeled by mistaken notions of the danger of free enquiry, there is yet another mode by which it may be degraded from its native dignity, and that is by placing it in the light of a mere appendage to and caterer for our pampered appetites.
A Preliminary Discourse on the Study of Natural Philosophy
Part I, Chapter I, Section 7 (p. 10)
Printed for Longman, Rees, Orme, Brown & Green. London, England. 1831

Hertz, Heinrich   1857–94
German physicist

The rigor of science requires that we distinguish well the undraped figure of nature itself from the gay-coloured vesture with which we clothe it at our pleasure.
Letters to the Editor
In Ludwig Boltzmann
On Certain Question of the Theory of Gases
Nature, Volume 51, Number 1322, 28 February 1895 (p. 413)

Herzen, Aleksandr   1812–70
Russian political author

Superficial dilettantism and the narrow specialization of the scientists ex officio are the two banks of science which prevent the fertilizing waters of this Nile from overflowing.

Selected Philosophical Works
Dilettantism in Science (p. 52)
Foreign Languages Publishing House. Moscow, Russia. 1956

Science is a table abundantly laid for every man whose hunger is great enough, whose craving for spiritual nourishment has grown sufficiently insistent.
Selected Philosophical Works
Dilettantism in Science (p. 58)
Foreign Languages Publishing House. Moscow, Russia. 1956

Science, in the best sense of the word, shall come to be accessible to the people, and when it does it shall claim a voice in all practical matters.
Selected Philosophical Works
Dilettantism in Science (p. 69)
Foreign Languages Publishing House. Moscow, Russia. 1956

Science is strength; it shows the relations of things, their laws and interactions.
Compiled by V.V. Vorontsov
Words of the Wise: A Book of Russian Quotations
Translated by Vic Schneierson
Progress Publishers. Moscow, Russia. 1979

Hilts, Philip
No biographical data available

In all human activities, it is not ideas of machines that dominate; it is people. I have heard people speak of "the effect of personality on science." But this is a backward thought. Rather, we should talk about the effect of science on personalities. Science is not the dispassionate analysis of impartial data. It is the human, and thus passionate, exercise of skill and sense on such data.
Scientific Temperaments: Three Lives in Contemporary Science
Preface (pp. 11–12)
Simon & Schuster. New York, New York, USA. 1982

Science is not an exercise in which objectivity is prized.
Scientific Temperaments: Three Lives in Contemporary Science
Preface (pp. 11–12)
Simon & Schuster. New York, New York, USA. 1982

Hinshelwood, Sir Cyril   1897–1967
English chemist

Science is not the mere collection of facts, which are infinitely numerous and mostly uninteresting, but the attempt by the human mind to order these facts into satisfying patterns.
On the Structure of Physical Chemistry
Clarendon Press. Oxford, England. 1951

Hippocrates   460 BCE–377 BCE
Greek physician

There are, indeed, two things, knowledge and opinion, of which the one makes its possessor really to know, the other to be ignorant.
In *Great Books of the Western World* (Volume 10)
Hippocratic Writings

The Law, 4 (p. 144)
Encyclopædia Britannica, Inc. Chicago, Illinois, USA. 1952

Hitler, Adolf   1889–1945
Chancellor of Germany

Science is a social phenomenon, and like every other social phenomenon is limited by the injury or benefit it confers on the community.... The idea of free and unfettered science…is absurd.
In Hermann Rauschning
Hitler Speaks: A Series of Political Conversations with Adolf Hitler on His Real Aims (pp. 220–221)
Butterworth. London, England. 1939

Hoagland, Hudson   1899–1982
American physiologist

The fictions — that is, the hypotheses and theories — of science are not sacrosanct.
Science and the New Humanism
Science, Volume 143, Number 3062, 10 January 1964 (p. 112)

Hocking, R.
No biographical data available

It is an oversimplification to compare the impersonal aspect of science with the impersonal aspects of industrial society, and to deplore both in one breath. The former is an achievement of self-forgetful concentration upon truths about nature. The latter are deplorable to the extent that they exhibit crude power of men over men. By contrast, the selflessness of the scientific calling does silent honor to personal existence.
In T.J.J. Altizer, William A. Beardslee, and J. Harvey Young (eds.)
Truth, Myth, and Symbol
The Problem of Truth (p. 5)
Prentice-Hall, Inc. Englewood Cliffs, New Jersey, USA. 1962

Hodgson, Leonard   1889–1969
English theologian

…science can only deal with what is, and can say nothing about what ought to be, which is the concern of ethics; science can tell us about means to ends, but not about what the ends should be.
Theology in an Age of Science
An Inaugural Lecture, November 3, 1944 (p. 9)

Holmes, Oliver Wendell   1809–94
American physician, poet, and humorist

Science is the topography of ignorance.
The Writings of Oliver Wendell Holmes
Volume IX, Border Lines in Medical Science (p. 211)
Houghton Mifflin Company. Boston, Massachusetts, USA. 1891–1906

Go on, fair science; soon to thee
Shall nature yield he idle boast;
He vulgar fingers formed a tree,
But thou hast trained it to a post.
The Complete Poetical Works of Oliver Wendell Holmes

The Meeting of the Dryads (p. 412)
Houghton Mifflin Company. Boston, Massachusetts, USA. 1899

Holton, Gerald   1922–
Research professor of physics and science history
Roller, Duane H. D.   ?–1994
Science historian

Science is an ever-unfinished quest to discover facts and establish relationships between them.
Foundations of Modern Physical Science
Chapter 13 (p. 214)
Addison-Wesley Publishing Company. Reading, Massachusetts, USA. 1950

…science has grown almost more by what it has learned to ignore than by what it has had to take into account.
Foundations of Modern Physical Science
Chapter 2 (p. 25)
Addison-Wesley Publishing Company. Reading, Massachusetts, USA. 1950

Horgan, J.
No biographical data available

…to pursue science in a speculative, postempirical mode: that I call ironic science. Ironic science resembles literary criticism in that it offers points of view, opinions, which are, at best, interesting, which provoke further comment. But it does not converge on the truth. It cannot achieve empirically verifiable surprises that force scientists to make substantial revisions in their basic description of reality.
The End of Science: Facing the Limits of Knowledge in the Twilight of the Scientific Age
Introduction (p. 7)
Addison-Wesley Publishing Company. Reading, Massachusetts, USA. 1996

Hoyle, Sir Fred   1915–2001
English mathematician and astronomer

Science progresses by extending the territory over which its theories hold good…
Ten Faces of the Universe
The Origin of the Universe (p. 105)
W.H. Freeman & Company. San Francisco, California, USA. 1977

Hoyle, Sir Fred   1915–2001
English mathematician and astronomer
Hoyle, Geoffrey   1942–
English science fiction writer

The fragmentation of science is a source of difficulty to all teachers and to all students — the connection of one research area to another is not always apparent. This is because science is rather like a vast and subtle jig-saw puzzle, and the usual way to attack a jig-saw puzzle is to work simultaneously on several parts of it. Only at the end do we seek to fit the different parts of it together into a coherent whole.

In Eugene H. Kone and Helene J. Jordan (eds.)
The Greatest Adventure: Basic Research that Shapes Our Lives
Cosmology and Its Relation to the Earth (p. 22)
Rockefeller University Press. New York, New York, USA. 1974

Hubbard, Elbert   1856–1915
American editor, publisher, and author

SCIENCE: 1. The knowledge of the common people classified and carried one step further. 2. Accurate organized knowledge grounded on fact. 3. Classified superstition.
The Roycroft Dictionary Concocted by Ali Baba and the Bunch on Rainy Days (p. 134)
The Roycrofters. East Aurora, New York, USA. 1914

Hubbard, Gardiner G.   1822–97
American lawyer and educator

That which was unknown, science hath revealed.
Annual Report of the Board of Regents of the Smithsonian Institution, 1893
Relations of Air and Water to Temperature and Life (p. 265)
Government Printing Office. Washington, D.C. 1894

Hubbard, Ruth   1924–
American biologist

To overturn orthodoxy is no easier in science than in philosophy, religion, economics, or any of the other disciplines through which we try to comprehend the world and the society in which we live.
Women Look at Biology Looking at Women
Have Only Men Evolved? (p. 10)
Schenkman Publishing Company. Cambridge, USA. 1979

Hubble, Edwin Powell   1889–1953
American astronomer

There is a unity in science, connecting all its various fields. Men attempt to understand the universe, and they will follow clues which excite their curiosity wherever the clues may lead.
The Nature of Science and Other Lectures
Part I. The Nature of Science (p. 6)
The Huntington Library. San Marino, California, USA. 1954

Hugo, Victor   1802–85
French author, lyric poet, and dramatist

Science says the first word on everything, and the last word on nothing.
Victor Hugo's Intellectual Autobiography
Things of the Infinite
Funk & Wagnalls. New York, New York, USA. 1907

Huizinga, Johan   1872–1945
Dutch historian

Science, unguided by a higher abstract principle, freely hands over its secrets to a vastly developed and commercially inspired technology, and the latter, even less restrained by a supreme culture saving principle, with

the means of science creates all the instruments of power demanded from it by the organization of Might.
Translated by J.H. Huizinga
In the Shadow of Tomorrow
Chapter 9 (p. 93)
W.W. Norton & Company, Inc. New York, New York, USA. 1936

...that all science is merely a game can be easily discarded as a piece of wisdom too easily come by. But it is legitimate to enquire whether science is not liable to indulge in play within the closed precincts of its own method. Thus, for instance, the scientist's continuous penchant for systems tends in the direction of play.
Homo Ludens
Chapter XI (p. 203)
Roy Publishers. New York, New York, USA. 1950

Hume, David   1711–76
Scottish philosopher and historian

The sweetest and most inoffensive path of life leads through the avenues of science and learning; and whoever can either remove any obstructions in this way, or open up any new prospect, ought so far to be esteemed a benefactor to mankind.
In *Great Books of the Western World* (Volume 35)
An Enquiry Concerning Human Understanding
Enquiries Concerning the Human Understanding and Concerning the Principles of Morals
Section 1
Of the Different Species of Philosophy (p. 453)
Encyclopædia Britannica, Inc. Chicago, Illinois, USA. 1952

Huxley, Aldous   1894–1963
English writer and critic

To the twentieth century man of letters science offers a treasure of newly discovered facts and tentative hypotheses. If he accepts this gift and if, above all, he is sufficiently talented and resourceful to be able to transform the new materials into works of literary art, the twentieth century man of letters will be able to treat the age-old, and perennially relevant, theme of human destiny with a depth of understanding, a width of reference of which, before the rise of science, his predecessors (through no fault of their own, no defect of genius) were incapable.
Literature and Science
Chapter 29 (p. 87)
Harper & Row, Publishers. New York, New York, USA. 1963

We are living now, not in the delicious intoxication induced by the early successes of science, but in a rather grisly morning-after, when it has become apparent that what triumphant science has done hitherto is to improve the means for achieving unimproved or actually deteriorated ends.
Ends and Means
Chapter XIV (p. 268)
Chatto & Windus. London, England. 1938

Science is dangerous; we have to keep it most carefully chained and muzzled.
Brave New World
Chapter Sixteen (p. 270)
Harper & Brothers. New York, New York, USA. 1950

Science is a matter of disinterested observation, patient ratiocination within some system of logically correlated concepts. In real-life conflicts between reason and passion the issue is uncertain. Passion and prejudice are always able to mobilize their forces more rapidly and press the attack with greater fury; but in the long run (and often, of course, too late) enlightened self-interest may rouse itself, launch a counterattack and win the day for reason.
Literature and Science
Chapter 23 (p. 68)
Harper & Row, Publishers. New York, New York, USA. 1963

Science sometimes builds new bridges between universes of discourse and experience hitherto regarded as separate and heterogeneous. But science also breaks down old bridges and opens gulfs between universes that, traditionally, had been connected.
Literature and Science
Chapter 37 (p. 111)
Harper & Row, Publishers. New York, New York, USA. 1963

For Science in its totality, the ultimate goal is the creation of a monistic system in which — on the symbolic level and in terms of the inferred components of invisibility and intangibly fine structure — the world's enormous multiplicity is reduced to something like unity, and the endless successions of unique events of a great many different kinds get tidied and simplified into a single rational order. Whether this goal will ever be reached remains to be seen. Meanwhile we have the various sciences, each with its own system coordinating concepts, its own criterion of explanation.
Literature and Science
Chapter 3 (p. 9)
Harper & Row, Publishers. New York, New York, USA. 1963

...science has "explained" nothing; the more we know the more fantastic the world becomes and the profounder the surrounding darkness.
Along the Road
Part II. Views of Holland (p. 108)
Nan'-do. Tokyo, Japan. 1954

All science is based upon an act of faith — faith in the validity of the mind's logical processes, faith in the ultimate explicability of the world, faith that the laws of thought are laws of things.
Ends and Means
Chapter XIV (p. 258)
Chatto & Windus. London, England. 1938

Huxley, Julian   1887–1975
English biologist, philosopher, and author

Science…has not only turned her face outwards from man, but stripped him of all the robes of his divinity, turned him out of the palace that he had so laboriously built in the center of the world, and left him in rags, pitiably insignificant and suddenly transported to an outlying corner of the cosmos.
Harper's Monthly Magazine
Will Science Destroy Religion, April 1926 (p. 535)

Science, like Empires, have their rise and their time of flourishing, though not their decay.
What Dare I Think?: The Challenge of Modern Science to Human Action and Belief, Including the Henry La Barre Jayne Foundation Lectures
Chapter I (p. 1)
Harper & Brothers. New York, New York, USA. 1931

Huxley, Thomas Henry   1825–95
English biologist

Science in England does everything — but pay. You may earn praise but not pudding.
In Leonard Huxley
Life and Letters of Thomas Henry Huxley (Volume 1)
Chapter VII
Letter to his sister, April 17, 1852 (p. 108)
D. Appleton & Company. New York, New York, USA. 1901

Whatever happens, science may bide her time in patience and in confidence.
Collected Essays (Volume 5)
Science and Christian Traditions
An Episcopal Trilogy (p. 143)
Macmillan & Company Ltd. London, England. 1904

What men of science want is only a fair day's wages for more than a fair day's work…
Collected Essays (Volume 1)
Method and Result
Administrative Nihilism (p. 287)
Macmillan & Company Ltd. London, England. 1904

You have no idea of the intrigues that go on in this blessed world of science. Science is, I fear, no purer than any other region of human activity; though it should be. Merit alone is very little good; it must be backed by tact and knowledge of the world to do very much.
In Leonard Huxley
Life and Letters of Thomas Henry Huxley (Volume 1)
Chapter VII
Letter to his sister, March 5, 1852 (p. 105)
D. Appleton & Company. New York, New York, USA. 1901

The generalizations of science sweep on in ever-widening circles, and more aspiring flights, through limitless creation.
Letter
London Times, December 26, 1859

Nothing great in science has ever been done by men, whatever their powers, in whom the divine afflatus [inspiration] of the truth-seeker was wanting.

Collected Essays (Volume 1)
Method and Result
The Progress of Science (p. 56)
Macmillan & Company Ltd. London, England. 1904

Of the affliction caused by persons who think that what they have picked up from popular exposition qualifies them for discussing the great problems of science, it may be said, as the Radical toast said of the power of the Crown in bygone days, that it "has increased, is increasing, and ought to be diminished." The oddities of "English as she is spoke" might be abundantly paralleled by those of "science as she is misunderstood" in the sermon, the novel, and the leading article; and a collection of the grotesque travesties of scientific conceptions in the shape of essays on such trifles as "the Nature of Life" and the "Origin of All Things," which reach me, from time to time, might well be bound up with them.
Collected Essays (Volume 8)
Discourses, Biological and Geological
Preface (p. viii)
Macmillan & Company Ltd. London, England. 1904

No delusion is greater than the notion that method and industry can make up for lack of motherwit, either in science or in practical life.
Collected Essays (Volume 1)
Method and Result
The Progress of Science (p. 46)
Macmillan & Company Ltd. London, England. 1904

Science reckons many prophets, but there is not even a promise of a Messiah.
In Leonard Huxley
Life and Letters of Thomas Henry Huxley (Volume 2)
Letter dated March 1894 (p. 396)
D. Appleton and Company. New York, New York, USA. 1901

Posterity will cry shame on us if we do not remedy this deplorable state of things. Nay, if we live twenty years longer, our own consciences will cry shame on us.
It is my firm conviction that the only way to remedy it is to make the elements of physical science an integral part of primary education. I have endeavored to show you how that may be done for that branch of science which it is my business to pursue; and I can but add, that I should look upon the day when every schoolmaster throughout this land was a centre of genuine, however rudimentary, scientific knowledge as an epoch in the history of the country.
But let me entreat you to remember my last words. Addressing myself to you, as teachers, I would say, mere book learning in physical science is a sham and a delusion-what you teach, unless you wish to be impostors, that you must first know; and real knowledge in science means personal acquaintance with the facts, be they few or many.
Collected Essays (Volume 8)
Discourses, Biological and Geological

A Lobster; or, The Study of Zoology (p. 227)
Macmillan & Company Ltd. London, England. 1904

Science commits suicide when it adopts a creed.
Collected Essays (Volume 2)
Darwiniana
The Darwin Memorial (p. 252)
Macmillan & Company Ltd. London, England. 1904

So, the vast results obtained by Science are won by no mystical faculties, by no mental processes, other than those which are practised by every one of us, in the humblest and meanest affairs of life. A detective policeman discovers a burglar from the marks made by his shoe, by a mental process identical with that by which Cuvier restored the extinct animals of Montmartre from fragments of their bones.
Collected Essays (Volume 3)
Science and Education
On the Educational Value of the Natural History Sciences (p. 46)
Macmillan & Company Ltd. London, England. 1904

Extinguished theologians lie about the cradle of every science as the strangled snakes beside that of Hercules...
Collected Essays (Volume 2)
Darwiniana
The Origin of Species (p. 52)
Macmillan & Company Ltd. London, England. 1904

The whole of modern thought is steeped in science; it has made its way into the works of our best poets, and even the mere man of letters, who affects to ignore and despise science, is unconsciously impregnated with her spirit, and indebted for his best products to her methods.
Collected Essays (Volume 8)
Discourses, Biological and Geological
A Lobster; or, The Study of Zoology (p. 226)
Macmillan & Company Ltd. London, England. 1904

I believe that the greatest intellectual revolution mankind has yet seen is now slowly taking place by her agency. She is teaching the world that the ultimate court of appeal is observation and experiment, and not authority; she is teaching it to estimate the value of evidence; she is creating a firm and living faith in the existence of immutable moral and physical laws, perfect obedience to which is the highest possible aim of an intelligent being.
Collected Essays (Volume 8)
Discourses, Biological and Geological
A Lobster; or, The Study of Zoology (p. 226)
Macmillan & Company Ltd. London, England. 1904

Physical science, its methods, its problems, and its difficulties, will meet the poorest boy at every turn, and yet we educate him in such a manner that he shall enter the world as ignorant of the existence of the methods and facts of science as the day he was born. The modern world is full of artillery; and we turn out our children to do battle in it, equipped with the shield and sword of an ancient gladiator.
Collected Essays (Volume 8)

Discourses, Biological and Geological
A Lobster; or, The Study of Zoology (p. 226)
Macmillan & Company Ltd. London, England. 1904

The vast results obtained by Science are won by no mystical faculties, by no mental processes, other than those which are practiced by every one of us, in the humblest and meanest affairs of life. A detective policeman discovers a burglar from the marks made by his shoe, by a mental process identical with that by which Cuvier restored the extinct animals of Montmartre from fragments of their bones.
Collected Essays (Volume 3)
Science and Education
On the Educational Value of the National History Sciences (p. 45)
Macmillan & Company Ltd. London, England. 1904

Books are the money of Literature, but only the counters of Science.
Collected Essays (Volume 3)
Science and Education
Universities: Actual and Ideal (p. 213)
Macmillan & Company Ltd. London, England. 1904

Anybody who knows his business in science can make anything subservient to that purpose. You know it was said of Dean Swift that he could write an admirable poem upon a broomstick, and the man who has a real knowledge of science can make the commonest object in the world subservient to an introduction to the principles and greater truths of natural knowledge.
Collected Essays (Volume 3)
Science and Education
Address on Behalf of the National Association for the Promotion of Technical Education (p. 432)
Macmillan & Company Ltd. London, England. 1904

In science, as in art, and, as I believe, in every other sphere of human activity, there may be wisdom in a multitude of counselors, but it is only [obvious] in one or two of them.
Collected Essays (Volume 1)
Method and Result
The Progress of Science (p. 57)
Macmillan & Company Ltd. London, England. 1904

...the man of science, who, forgetting the limits of philosophical inquiry, slides from these formulae and symbols into what is commonly understood by materialism, seems to me to place himself on a level with the mathematician, who should mistake the x's and y's with which he works his problems for real entities — and with this further disadvantage, as compared with the mathematician, that the blunders of the latter are of no practical consequence, while the errors of systematic materialism may paralyse the energies and destroy the beauty of a life.
Collected Essays (Volume 1)
Method and Result
On the Physical Basis of Life (p. 165)
Macmillan & Company Ltd. London, England. 1904

…whatever evil voices may rage, Science, secure among the powers that are eternal, will do her work and be blessed.
Collected Essays (Volume 1)
Method and Result
Descartes' Discourse on Method (p. 198)
Macmillan & Company Ltd. London, England. 1904

Addressing myself to you, as teachers, I would say, mere book learning in physical science is a sham and a delusion — what you teach, unless you wish to be impostors, that you must first know; and real knowledge in science means personal acquaintance with the facts, be they few or many.
Collected Essays (Volume 8)
Discourses, Biological and Geological
A Lobster; or, The Study of Zoology (p. 227)
Macmillan & Company Ltd. London, England. 1904

Ingersoll, Robert G.   1833–99
American orator and lawyer

Reason, Observation, and Experience — the Holy Trinity of Science.
On the Gods and Other Essays
The Gods (p. 54)
Prometheus Books. Buffalo, New York, USA. 1990

Jacks, L. P.   1860–1955
English educator, philosopher, and Unitarian minister

Science is never static, never stagnant, never content with the boundary it has reached. It is always dynamic, always breaking bounds.... Science...abhors a limitation....
Is There a Foolproof Science?
The Atlantic Monthly, February 1924 (p. 231)

Science is the pursuer, life is the pursued....
Is There a Foolproof Science?
The Atlantic Monthly, February 1924 (p. 238)

Jacob, François   1920–
French biologist

Science advances metaphorically. It does not proceed in an orthogenic fashion moving inexorably forward in a straight line. It does not radiate along many branches like a growing tree. It moves from one view to another by a large leap, preceded by a radical shift in the scientist's mode of thought.
Translated by Betty E. Spillmann
The Logic of Life: A History of Heredity
Pantheon Books. New York, New York, USA. 1974

For science, there are many possible worlds; but the interesting one is the world that exists and has already shown itself to be at work for a long time. Science attempts to confront the possible with the actual.
The Possible and the Actual
Myth and Science (p. 12)
Pantheon Books. New York, New York, USA. 1982

Jacobi, Karl Gustav Jacob   1804–51
German mathematician

…Monsieur Fourier was of the opinion that the principal aim of Mathematics is to serve mankind and to explain natural phenomena; but a philosopher such as he ought to have known that the sole aim of science is the fulfillment of the human spirit, and that, accordingly, a question about numbers has as much significance as a question about the workings of the world.
Gesammelte Werke (Volume 1)
Letter to Legendre
July 2, 1830 (p. 454)
Publisher undetermined

James, William   1842–1910
American philosopher and psychologist

The aim of "science" is to attain conceptions so adequate and exact that we shall never need to change them.
The Principles of Psychology
The Perception of Things (p. 109)
Dover Publications, Inc. New York, New York, USA. 1950

Science herself consults her heart when she lays it down that the infinite ascertainment of fact and correction of false belief are the supreme goods for man.
The Will to Believe and Other Essays in Popular Philosophy
The Will to Believe
Section IX (p. 22)
Dover Publications, Inc. New York, New York, USA. 1956

Science as such assuredly has no authority, for she can only say what is, not what is not.
The Will to Believe and Other Essays in Popular Philosophy
Is Life Worth Living? (p. 56)
Dover Publications, Inc. New York, New York, USA. 1956

Science like life feeds on its own decay. New facts burst old rules; then newly developed concepts bind old and new together into a reconciling law.
The Will to Believe and Other Essays in Popular Philosophy
Psychical Research (p. 320)
Dover Publications, Inc. New York, New York, USA. 1956

Jastrow, Joseph   1863–1944
Polish-born psychologist

Theories rise and fall as better, truer theories replace them; yet it is unwarranted to conclude that science is truth for a day.
In Joseph Jastrow (ed.)
The Story of Human Error
Introduction (p. 34)
D. Appleton-Century Company, Inc. New York, New York, USA. 1936

Science, unlike the Bible, has no explanation for the occurrence of that extraordinary event. The universe, and everything that has happened in it since the beginning of time, are a grand effect without a known cause.
Until the Sun Dies
Chapter 2 (p. 21)
W.W. Norton & Company, Inc. New York, New York, USA. 1977

Jeans, Sir James Hopwood   1877–1946
English physicist and mathematician

To this present-day science adds that, at the farthest point she has so far reached, much, and possibly all, that was not mental has disappeared, and nothing new has come in that is not mental. Yet who shall say what we may find awaiting us round the next corner?
The New Background of Science
Chapter VIII (p. 307)
The University of Michigan Press. Ann Arbor, Michigan, USA. 1959

The infinitely great is never very far from the infinitely small in science…
Annual Report of the Board of Regents of the Smithsonian Institution, 1928
The Wider Aspect of Cosmogony (p. 170)
Government Printing Office. Washington, D.C. 1929

Jeffers, Robinson   1887–1962
American poet

Science is not to serve but to know. Science is for itself its own value, it is not for man…
In Tim Hunt (ed.)
The Collected Poetry of Robinson Jeffers (Volume 3)
The Double Axe: The Inhumanist (p. 291)
Stanford University Press. Stanford, California. USA. 1988

Science is an adoration; a kind of worship.
In Tim Hunt (ed.)
The Collected Poetry of Robinson Jeffers (Volume 3)
The Double Axe: The Inhumanist (p. 292)
Stanford University Press. Stanford, California. USA. 1988

Jefferson, Thomas   1743–1826
3rd president of the United States

The main objects of all science, the freedom and happiness of man…[are] the sole objects of all legitimate government.
In Andrew A. Lipscomb (ed.)
The Writings of Thomas Jefferson (Volume 12) (p. 369)
G. Putnam's Sons. New York, New York, USA. 1892–99

Jevons, William Stanley   1835–82
English economist and logician

Science arises from the discovery of Identity amidst Diversity.
The Principles of Science: A Treatise on Logic and Scientific Method
Book I, Chapter I (p. 1)
Macmillan & Company. London, England. 1887

Joad, Cyril Edwin Mitchinson   1891–1953
English philosopher and broadcasting personality

Man by the light of science can see his hands, and can catch a glimpse of himself, his past, and the patch upon which he stands; but around him in place of that known comfort and beauty he had anticipated, and in the first few moments falsely thought that he saw, is darkness still.
Philosophical Aspects of Modern Science
Chapter XI (p. 342)
George Allen & Unwin Ltd. London, England. 1939

Joffe, A. F.
No biographical data available

Science with its strict analysis of the facts, its persevering search for new, more consummate truths, and its relentless struggle against discovered mistakes and prejudices — science must saturate all or technics, our culture, and everyday life.
Compiled by V.V. Vorontsov
Words of the Wise: A Book of Russian Quotations
Translated by Vic Schneierson
Progress Publishers. Moscow, Russia. 1979

Johnson, Samuel   1696–1772
English critic, biographer, and essayist

The Sciences having long seen their votaries labouring for the benefit of mankind without reward, put up their petitions to Jupiter for a more equitable distribution of riches and honor…. A synod of the celestials was therefore convened, in which it was resolved that Patronage should descend to the assistance of the Sciences.
The Rambler (Volume 2)
No. 91, January 29, 1751 (p. 231)
Edward Earle. Philadelphia, Pennsylvania, USA. 1812

In science, which, being fixed and limited, admits of no other variety than such as arises from new methods of distribution, or new arts of illustration, the necessity of following the traces of our predecessors is indisputably evident; but there appears no reason why imagination should be subject to the same restraint…. The roads of science are narrow, so that they who travel them, must either follow or meet one another; but in the boundless regions of possibility, which fiction claims for her dominion, there are surely a thousand recesses unexplored, a thousand flowers unplucked, a thousand fountains unexhausted, combinations of imagery yet unobserved, and races of ideal inhabitants not hitherto described.
The Rambler (Volume 3)
No. 121, May 14, 1751 (pp. 89–90)
Edward Earle. Philadelphia, Pennsylvania, USA. 1812

Jones, Rufus M.   1863–1948
American writer and journal editor

Science has not closed, and will never close the soul's east window of divine surprise.
A Preface to Christian Faith in a New Age
Chapter II, Section IV (pp. 55–56)
The Macmillan Company. New York, New York, USA. 1932

Jones, Steve
No biographical data available

This is the essence of science. Even though I do not understand quantum mechanics or the nerve cell membrane, I trust those who do. Most scientists are quite ignorant about most sciences but all use a shared grammar that allows them to recognize their craft when they see it. The motto of the Royal Society of London is "*Nullius in verba*": trust not in words.
Review of *How the Mind Works* by Steve Pinker
The New York Review of Books, November 6, 1997 (p. 13)

Jordan, David Starr   1851–1931
American scientist and educator

Science must stop where the facts stop, or thereabout, the limit of "thereabout" covering all legitimate diversions and excursions of philosophy.
In Frances Mason
Creation by Evolution
Evolution — Its Meaning (p. 4)
The Macmillan Company. New York, New York, USA. 1928

Jung, Carl G.   1875–1961
Swiss psychiatrist and founder of analytical psychology

Science is the tool of the Western mind.… It is part and parcel of our understanding, and it obscures our insight only when it claims that the understanding it conveys is the only kind there is.
Translated by R.F.C. Hull
Alchemical Studies
Difficulties Encountered by a European in Trying to Understand the East (pp. 6–7)
Princeton University Press. Princeton, New Jersey, USA. 1967

Science is not indeed a perfect instrument, but it is a superb and invaluable tool that works harm only when it is taken as an end in itself.
Translated by R.F.C. Hull
Alchemical Studies
Difficulties Encountered by a European in Trying to Understand the East (p. 6)
Princeton University Press. Princeton, New Jersey, USA. 1967

Kaczynski, Theodore   1942–
American anarchist

Science marches on blindly…without regard to the real welfare of the human race or to any other standard, obedient only to the psychological needs of the scientists and of the government officials and corporate executives who provide the funds for research.
In Anne Eisenberg
The Unabomber and the Bland Decade
Scientific American, Volume 274, Number 4, April, 1998 (p. 35)

Kafka, Franz   1883–1924
German-language novelist

All science is methodology with regard to the Absolute. Therefore, there need be no fear of the unequivocally methodological. It is a husk, but not more than everything except the One.
Dearest Father: Stories and Other Writings
The Blue Octavo Notebooks
The Third Notebook
October 18, 1917

Kapitza, Pyetr Leonidovich   1894–1984
Russian physicist

The year that Rutherford died there disappeared for ever the happy days of free scientific work which gave us such delight in our youth. Science has lost her freedom. Science has become a productive force. She has become rich but she has become enslaved and part of her is veiled in secrecy.
Address to the Royal Society in Honour of Lord Rutherford
Nature, Volume 210, Number 5038, 17 May 1966 (p. 783)

Keller, Evelyn Fox   1936–
American scientist

To know the history of science is to recognize the mortality of any claim to universal truth.
Reflections on Gender and Science
Epilogue (pp. 178–179)
Yale University Press. New Haven, Connecticut, USA. 1985

A healthy science is one that allows for the productive survival of diverse conceptions of mind and nature and correspondingly diverse strategies.
Reflections on Gender and Science
Epilogue (p. 178)
Yale University Press. New Haven, Connecticut, USA. 1985

Keller, Helen   1880–1968
American author and lecturer

Science may have found a cure for most evils; but it has found no remedy for the worst of them all — the apathy of human beings.
My Religion
Part 1, Chapter 6
Swedenborg & Foundation, Inc. New York, New York, USA. 1927

Kellogg, Vernon L.   1867–1937
American zoologist

Science does not assume that it knows — despite the great deal that it does know — more than a very small part of the order of nature.
Some Things Science Doesn't Know
The World's Work, March 1926 (p. 528)

Kelvin, Lord William Thomson   1824–1907
Scottish engineer, mathematician, and physicist

Science is bound by the everlasting law of honour, to face fearlessly every problem which can fairly be presented to it. If a probable solution, consistent with the ordinary course of nature, can be found, we must not involve the abnormal act of Creative Power.
Popular Lectures and Addresses (Volume 2)

Presidential Address to the British Association, Edinburgh, 1871
(pp. 199–200)
Macmillan & Company Ltd. London, England. 1894

Kennedy, John F.   1917–63
35th president of the United States

Science contributes to our culture in many ways, as a creative intellectual activity in its own right, as the light which has served to illuminate man's place in the universe, and as the source of understanding of man's own nature.
Address to the National Academy of Sciences
Washington, D.C., 22 October, 1963

Let both sides seek to invoke the wonders of science instead of its terrors. Together let us explore the stars, conquer the deserts, eradicate disease, tap the ocean depths and encourage the arts and commerce.
Inaugural Address, January 20, 1961

In the years since man unlocked the power stored up within the atom, the world has made progress, halting but effective, toward bringing that power under human control. The challenge may be our salvation. As we begin to master the destructive potentialities of modern science, we move toward a new era in which science can fulfill its creative promise and help bring into existence the happiest society the world has ever known.
Address
National Academy of Sciences
Washington, D.C., October 22, 1963

Kettering, Charles Franklin   1876–1958
American engineer and inventor

So that we might kill one another more expertly, science found wonderful ways to live more comfortably, richly, to communicate more rapidly. So that we might exterminate one another more successfully, science showed us how we might all live longer and stronger…
In Paul de Kruif
America Comes Through a Crisis
Saturday Evening Post, 13 May 1933 (p. 3)

Keyser, Cassius Jackson   1862–1947
American mathematician

Science is destined to appear as the child and the parent of freedom blessing the earth without design. Not in the ground of need, not in bent and painful toil, but in the deep-centered play-instinct of the world, in the joyous mood of the eternal Being, her spirit, which is always young, Science has her origin and root; and her spirit, which is the spirit of genius in moments of elevation, is but a sublimated form of play, the austere and lofty analogue of the kitten playing with the entangled skein…
Mathematics (p. 44)
Columbia University Press. New York, New York, USA. 1907

King, Jr., Martin Luther   1929–68
American civil rights leader and clergyman

We have genuflected before the god of science only to find that it has given us the atomic bomb, producing fears and anxieties that science can never mitigate.
Strength to Love
Chapter XIII (p. 106)
Harper & Row, Publishers. New York, New York, USA. 1963

Kingsley, Charles   1819–75
English clergyman and author

For science is, I verily believe, like virtue, its own exceeding great reward.
Health and Education
Science (p. 289)
W. Isbister & Company. London, England. 1874

For from blind fear of the unknown, science does certainly deliver man. She does by man as he does by an unbroken colt. The colt sees by the road side some quite new object — a cast–away boot, an old kettle, or what not. What a fearful monster! What unknown terrific powers may it not posses! And the colt shies across the road, runs up the bank, rears on end; putting itself thereby, as many a man does, in real danger. What cure is there? But one, experience. So science takes us, as we should take the colt, gently by the halter; and makes us simply smell at the new monster; till after a few trembling sniffs, we discover, like the colt, that it is not a monster, but a kettle.
Health and Education
Science (p. 284)
W. Isbister & Company London, England. 1874

…it is the childlike, simple, patient, reverent heart, which science at once demands and cultivates. To prejudice or haste, to self-conceit or ambition, she proudly shuts her treasuries — to open them to men of humble heart, whom this world thinks simple dreamers — her Newtons, and Owens, and Faradays.
Alton Locke, Taylor and Poet
Chapter XVIII (p. 141)
Macmillan & Company Ltd. London, England. 1911

…Science was the child of Courage, and Courage the child of Knowledge.
Health and Education
Science (p. 259)
W. Isbister & Company London, England. 1874

Kipling, Rudyard   1865–1936
British writer and poet

There are times when Science does not satisfy.
With the Night Mail (p. 24)
Doubleday, Page & Company. New York, New York, USA. 1909

Kirby, William   1759–1850
Clergyman and entomologist

Mankind in general, not excepting even philosophers, are prone to magnify, often beyond its just merit, the science or pursuit to which they have addicted themselves, and to depreciate any that seems to stand in competition with their favorite: like the redoubted champions of romance, each thinks himself bound to take the field against every one that will not subscribe to the peerless beauty and accomplishments of his own Dulcinea.
An Introduction to Entomology; or, Elements of the Natural History of Insects
Letter I (p. 1)
Longman, Green, Longman & Roberts. London, England. 1860

Kirkpatrick, Clifford 1898–1970
American sociologist

Science recognizes no personal powers in the universe responsive to the prayers and needs of men.
Religion in Human Affairs
Chapter XVI (p. 470)
John Wiley & Sons, Inc. New York, New York, USA. 1929

Kline, Morris 1908–92
American mathematics professor and writer

Theoretical Science is a game of mathematical make-believe.
Mathematics: The Loss of Certainty
Chapter XIV (p. 325)
Oxford University Press, Inc. New York, New York, USA. 1980

Kliuchevsky, V. O. 1841–1911
Russian historian

Science is often identified with knowledge. This is a gross misunderstanding. Science is not merely knowledge, but also consciousness, that is, the skill of properly using knowledge.
Compiled by V.V. Vorontsov
Words of the Wise: A Book of Russian Quotations
Translated by Vic Schneierson
Progress Publishers. Moscow, Russia. 1979

Knight, David
No biographical data available

Most science is a very ordinary human activity not so very far removed from painting by numbers.
Ideas in Chemistry: A History of the Science
Introduction (p. 6)
Athlone. London, England. 1992

Knuth, Donald E. 1938–
Creator of TeX

Science is what we understand well enough to explain to a computer, Art is all the rest.
In Marko Petkovsek, Herbert S. Wilf and Doron Zeilberger
A=B
Foreword
A.K. Peters. Wellesley, Massachusetts, USA. 1996

Kofahl, R. E.
No biographical data available

Science is human experience systematically extended (by intent, methodology and instrumentation) for the purpose of learning more about the natural world and for the critical empirical testing and possible falsification of all ideas about the natural world. Scientific hypotheses may incorporate only elements of the natural empirical world, and thus may contain no element of the supernatural.
Correctly Redefining Distorted Science; A Most Essential Task
Creation Research Society Quarterly, Volume 23, 1986 (p. 112)

Köhler, Wolfgang 1887–1967
American psychologist

It would be interesting to inquire how many times essential advances in science have first been made possible by the fact that the boundaries of special disciplines are not respected. …at the present time it is of course quite customary for physicists to trespass on chemical ground, for mathematicians to do excellent work in physics, and for physicists to develop new mathematical procedures…trespassing is one of the most successful techniques in science.
Dynamics in Psychology
Retention and Recall (pp. 115–116)
Liveright Publishing Corporation. New York, New York, USA. 1940

Kolb, Edward W. (Rocky) 1951–
American cosmologist

Science does not proceed like a cookbook recipe in the making of a hypothesis, comparing its prediction with observations and either accepting or rejecting the hypothesis. There is always confusion at the leading edge of research, and there are always a few discrepant and contradictory pieces of information that can't be explained.
Blind Watchers of the Sky
Chapter Seven (pp. 193–194)
Addison-Wesley Publishing Company. Reading, Massachusetts, USA. 1996

The process of science involves heroic ideas as well as its share of stupidity.
Blind Watchers of the Sky
Preface (p. x)
Addison-Wesley Publishing Company. Reading, Massachusetts, USA. 1996

More often than not, the way science goes from point A to point B is by a random lurch through points X, Y, and Z. Even when great leaps of progress do occur, they only rarely come "out of the blue." Advances are nearly always preceded by years, decades, or even centuries of patient accumulation of facts and data and ideas.
Blind Watchers of the Sky
Chapter Two (p. 25)
Addison-Wesley Publishing Company. Reading, Massachusetts, USA. 1996

Koshland, Jr., Daniel E.   1920–
American biochemist

Science is not impressed with a conglomeration of data. It likes carefully constructed analysis of each problem.
Editorial
Science, Volume 263, Number 5144, 14 January 1994 (p. 155)

Krauss, Lawrence M.   1954–
American theoretical physicist

There are times, such as when the state school board in Kansas in 1999 removed evolution from its science curriculum, when I am reminded of Lavoisier, and shudder at the damage that can be done by ignorance combined with power. Even the magnificent modern edifice called science, built up over half a millennium of small increments toward the truth, is not safe from the vicissitudes of the political world. If, as Carl Sagan claimed, science is a "candle in the dark," banishing demons that haunted the benighted eras of mankind, it burns tenuously at best. One generation of ignorance, steeped in myth and mysticism, is all that may be needed to snuff it out.
Atom: An Odyssey from the Big Bang to Life on Earth...and Beyond
Chapter 13 (p. 172)
Little, Brown & Company. Boston, Massachusetts, USA. 2001

Science is based on limits: It proceeds by progressively finding out what is not possible, through experiment and theory, in order to determine how the universe might really function. It is worth recalling Sherlock Holmes's adage that when you have eliminated all other possibilities, whatever remains, no matter how improbable, is the truth. Because of this, the universe is a pretty remarkable place even without all the extras. The greatest gift science has bestowed upon humanity, in my opinion, is the knowledge that whether we like it or not, the universe is the way it is.
Beyond Star Trek: Physics from Alien Invasions to the End of Time
Epilogue (p. 173)
Basic Books, Inc. New York, New York, USA. 1997

Kroeber, Alfred Louis   1876–1960
American anthropologist

...it appears that the total work of science must be done on a series of levels which the experience of science gradually discovers.
The Nature of Culture (p. 121)
The University of Chicago Press. Chicago, Illinois, USA. 1952

Science has always promised two things not necessarily related — an increase first in our powers, second in our happiness or wisdom, and we have come to realize that it is the first and less important of the two promises which it has kept most abundantly.
The Modern Temper
Chapter Three (p. 43)
Harcourt, Brace & Company. New York, New York, USA. 1929

...the most important part of our lives — our sensations, emotions, desires, and aspirations — takes place in a universe of illusions which science can attenuate or destroy, but which it is powerless to enrich.
The Modern Temper
Chapter Three (p. 50)
Harcourt, Brace & Company. New York, New York, USA. 1929

Kubie, L. S.
No biographical data available

The primary achievement of science is the humility and honesty with which it constantly corrects its own errors. It is this that makes science the greatest of the humanities.
The Fostering of Creative Scientific Productivity
Daedalus, Volume 91, 1962 (p. 305)

Kuhn, Thomas S.   1922–96
American historian of science

To understand why science develops as it does, one need not unravel the details of biography and personality that lead each individual to a particular choice, though that topic has vast fascination. What one must understand, however, is the manner in which a particular set of shared values interacts with the particular experiences shared by a community of specialists to ensure that most members of the group will ultimately find one set of arguments rather than another decisive.
The Structure of Scientific Revolutions
Postscript–1969 (p. 200)
The University of Chicago Press. Chicago, Illinois, USA. 1970

Normal science...is predicated on the assumption that the scientific community knows what the world is like. Much of the success of the enterprise derives from the community's willingness to defend that assumption, if necessary at considerable cost.
The Structure of Scientific Revolutions
Chapter I (p. 5)
The University of Chicago Press. Chicago, Illinois, USA. 1970

The practice of normal science depends on the ability, acquired from exemplars, to group objects and situations into similarity sets which are primitive in the sense that the grouping is done without an answer to the question, "similar with respect to what?"
The Structure of Scientific Revolutions
Postscript–1969 (p. 200)
The University of Chicago Press. Chicago, Illinois, USA. 1970

...science...often suppresses fundamental novelties because they are necessarily subversive of its basic commitments.
The Structure of Scientific Revolutions
Chapter I (p. 5)
The University of Chicago Press. Chicago, Illinois, USA. 1970

Kusch, Polykarp   1911–93
German-American physicist

Science is the greatest creative impulse of our time. It dominates the intellectual scene and forms our lives, not only in the material things which it has given us, but also in that it guides our spirit. Science shows us truth and beauty and fills each day with a fresh wonder of the exquisite order which governs our world.
Address to University Students
December 10, 1955

Lamb, Charles   1775–1834
English essayist and critic

Science has succeeded to poetry, no less in the little walks of children than with men. Is there no possibility of averting this sore evil?
In Thoas Noon Talfourd
The Works of Charles Lamb: To which are Prefixed His Letters, and a Sketch of His Life (Volume 1)
Letter to S.T. Coleridge, October 23, 1802 (p. 118)
Harper & Brothers, Publishers. New York, New York, USA. 1838

Can we unlearn the arts that pretend to civilize, and then burn the world? There is a march of science; but who shall beat the drums for its retreat?
In Thoas Noon Talfourd
The Works of Charles Lamb: To which are Prefixed His Letters, and a Sketch of His Life (Volume 1)
Letter to George Dyer, December 20, 1830 (p. 292)
Harper & Brothers, Publishers. New York, New York, USA. 1838

In everything that relates to science, I am a whole Encyclopaedia behind the rest of the world.
Essays of Elia
The Old and the New Schoolmaster (p. 88)
Little, Brown & Company. Boston, Massachusetts. USA. 1896

Landsberg, Peter Theodore   1922–
No biographical data available

Everybody who takes up science has the ambition to become a successful scientist and make some discoveries. Most of us are disappointed because we do not make the really big and interesting discoveries. Or, if we do make them, we do not realize they are interesting, because other discoveries seem more important.
Mathematics Today, October 1902 (p. 135)

Lang, Andrew   1844–1912
Scottish scholar and man of letters

But science, like the spear of Achilles, can cure the wounds which herself inflicts.
The Disentanglers
Adventure of the Canadian Heiress (p. 399)
Longmans, Green Publishers. New York, New York, USA. 1902

Lapp, Ralph E.   1917–2004
American nuclear physicist

No one — not even the most brilliant scientist alive today — really knows where science is taking us. We are aboard a train which is gathering speed, racing down a track on which are an unknown number of switches leading to unknown destinations. No single scientist is in the cab, and there may be demons at the switch. Most of society is in the caboose looking backward. Some passengers, fearful that they have boarded an express train to hell, want to jump off before it is too late
The New Priesthood: The Scientific Elite and the Uses of Power
Chapter 2 (p. 29)
Harper & Row, Publishers. New York, New York, USA. 1965

Larrabee, Eric   1922–90
Historian

Science is a — what? a method, a faith, a body of facts, a structure of theories, an institution, a way of life, a finite number of duly qualified individuals, an infinity of relevance and possibility. For a large number of scientists, science is indescribable, but indisputably a thing: it is knowable, palpable, reliable, usable. They live with it and by it; it is simply and unequivocally there.
Commentary
Science and the Common Reader, June 1966 (p. 43)

Laudan, Larry   1945–
American philosopher of science

The aim of science is merely to secure theories with a high problem-solving effectiveness.
New Scientist, 1 August 1892 (p. 26)

Lavoisier, Antoine Laurent   1743–94
French chemist

When we begin the study of any science, we are in a situation, respecting that science, similar to that of children; and the course by which we have to advance is precisely the same which nature follow in the formation of their ideas.
In *Great Books of the Western World* (Volume 45)
Elements of Chemistry
Preface (p. 1)
Encyclopædia Britannica, Inc. Chicago, Illinois, USA. 1952

Science still has many chasms, which interrupt the series of facts and often render it expremely difficult to reconcile them with each other…
In *Great Books of the Western World* (Volume 45)
Elements of Chemistry
Preface (p. 2)
Encyclopædia Britannica, Inc. Chicago, Illinois, USA. 1952

Le Guin, Ursula K.   1929–
American writer of science fiction and fantasy

…it is only when science asks why, instead of simply describing how, that it becomes more than technology. When it asks why, it discovers Relativity. When it only shows how, it invents the atomic bomb and then puts its hands over its eyes and says, My God what have I done?…

Language of the Night
The Stalin in the Soul (p. 219)
Putnam. New York, New York, USA. 1979

Leary, Timothy   1920–96
American psychologist and educator

Science is all metaphor.
Contemporary Authors, Volume 107

Lebowitz, Fran   1951–
American comedian

Science is not a pretty thing. It is unpleasantly proportioned, outlandishly attired and often over-eager. What then is the appeal of science? What accounts for its popularity? And who gives it its start?
Metropolitan Life
Science (p. 104)
Fawcett Crest. New York, New York, USA. 1978

…modern science was largely conceived of as an answer to the servant problem and…is generally practiced by those who lack the flair for conversation.
Metropolitan Life
Science (p. 104)
Fawcett Crest. New York, New York, USA. 1978

Leclerc, Georges-Louis, Comte de Buffon   1707–88
French naturalist

The only good science is the knowledge of facts, and mathematical truths are only truths of definition, and completely arbitrary, quite unlike physical truths.
In L. Ducros
Les Encyclopedistes (p. 326)
Publisher undetermined

Lerner, Max   1902–92
American educator and author

It is not science that has destroyed the world, despite all the gloomy forebodings of the earlier prophets. It is man who has destroyed man.
Actions and Passions: Notes on the Multiple Revolution of Our Time
The Human Heart and Human Will (p. 3)
Simon & Schuster. New York, New York, USA. 1949

Lewis, Gilbert Newton   1875–1946
American chemist

The strength of science lies in its naiveté.
The Anatomy of Science
Chapter I (p. 1)
Yale University Press. New Haven, Connecticut, USA. 1926

Lewis, Wyndham   1882–1957
English author and painter

When we say "science" we can either mean any manipulation of the inventive and organizing power of the human intellect: or we can mean such an extremely different thing as the religion of science, the vulgarized derivative from this pure activity manipulated by a sort of priestcraft into a great religious and political weapon.
The Art of Being Ruled
Revolution and Progress, Chapter 1 (pp. 3–4)
Chatto & Windus. London, England. 1926

The puritanic potentialities of science have never been forecast. If it evolves a body of organized rites, and is established as a religion, hierarchically organized, things more than anything else will be done in the name of "decency." The coarse fumes of tobacco and liquors, the consequent tainting of the breath and staining of white fingers and teeth, which is so offensive to many women, will be the first things attended to.
The Art of Being Ruled
Chapter 7 (p. 210)
Chatto & Windus. London, England. 1926

Lichtenberg, Georg Christoph   1742–99
German physicist and satirical writer

There is no greater impediment to progress in the sciences than the desire to see it take place too quickly.
Translated by R.J. Hollingdale
Aphorisms
Notebook K, Aphorism 72
Penguin Classics. New York, New York, USA. 1990

The most heated defenders of a science, who can not endure the slightest sneer at it, are commonly those who have not made very much progress in it and are secretly aware of this defect.
Translated by R.J. Hollingdale
Aphorisms
Notebook F, aphorism 8
Penguin Classics. New York, New York, USA. 1990

Lindley, David   1956–
English astrophysicist and author

When scientists begin to wonder…how science is possible at all, which is ultimately what their questioning of the mathematical basis of science is about, they are searching for reassurance, for some proof that there really is a fundamental theory out there in the dark waiting to be hunted down.
The End of Physics: The Myth of a Unified Theory
Prologue (p. 4)
Basic Books, Inc. New York, New York, USA. 1993

Locke, John   1632–1704
English philosopher and political theorist

Nobody is under an obligation to know every thing. Knowledge and science in general is the business only of those who are at ease and leisure.
An Essay Concerning Human Understanding and a Treatise on the Conduct of the Understanding
A Treatise on the Conduct of the Understanding

Section 7 (p. 494)
James Kay, June & Company. Philadelphia, Pennsylvania, USA. ca.1850

Lorenz, Konrad   1903–89
Austrian zoologist

Truth, in science, can be defined as the working hypothesis best fitted to open the way to the next better one.
Translated by Marjorie Kerr Wilson
On Aggression
Chapter Fourteen (p. 288)
Harcourt, Brace & World, Inc. New York, New York, USA. 1963

Lowell, Percival   1855–1916
American astronomer

Now in science there exists two classes of workers. There are men who spend their days in amassing material, in gathering facts. They are the collectors of specimens in natural history, the industrious takers of routine measurements in physics and astronomy or the mechanical accumulators of photographic plates. Very valuable such collections are. They may not require much brains to get, but they enable other brains to get a great deal out of them later.... The rawer they are the better. For the less mind enters into them the more they are worth. When destitute altogether of informing intelligence, they become priceless, as they then convey nature's meaning unmeddled of man.... The second class of scientists are the architects of the profession. They are the men to whom the building up of science is due. In their hands, the acquired facts are put together to that synthesizing of knowledge from which new conceptions spring.... Though the gathering of material is good, without the informing mind to combine the facts they had forever remained barren of fruit.
In William Graves Hoyt
Lowell and Mars
Chapter 2 (p. 22)
University of Arizona Press. Tucson, Arizona, USA. 1976

Lubbock, Sir John   1834–1913
English banker, author, and scientist

Science, our Fairy Godmother, will, unless we perversely reject her help, and refuse her gifts, so richly endow us, that fewer hours of labour will serve to supply us with the material necessaries of life, leaving us more time to ourselves, more leisure to enjoy all that makes life best worth living.
The Beauties of Nature and the Wonders of the World We Live In
Introduction (p. 37)
Macmillan & Company New York, New York, USA. 1893

Lundberg, G. A.
No biographical data available

...no science tells us what to do with the knowledge that constitutes the science. Science only provides a car and a chauffeur for us. It does not directly, as science, tell us where to drive. The car and the chauffeur will take us into the ditch, over the precipice, against a stone wall, or into the highlands of age-long human aspirations with equal efficiency. If we agree as to where we want to go and tell the driver our goal, he should be able to take us there by one of a number of possible routes the costs and conditions of each of which the scientist should be able to explain to us.
Can Science Save Us?
Social Problems (p. 31)
Longmans, Green & Company New York, New York, USA. 1947

Lynch, Gary
No biographical data available

What you really need to do the best science is a tremendous tolerance for ambiguity. You have to be able to tolerate ambiguity. Because we as creatures are set up for some reason to see cause-and-effect. And what you really wind up doing is tolerating the fact that you have all these assumptions and all these uncertainties, and living with them. And when you really go into a novel area, what do you have to guide you? The more novel it is, the fewer the constraints. For a human being that is a very uncomfortable feeling.
In George Johnson
In the Palaces of Memory: How We Build the Worlds Inside Our Heads
Mucking Around in the Wetware (pp. 91–92)
Alfred A. Knopf. New York, New York, USA. 1991

Lysaght, Sidney R.   1860–1941
Irish writer

Science is the lamp which man has himself kindled. It has built him lighthouses on the dark shores of the unknown; but his dreams, his quests for truth, lead him beyond the waters which his little lamp of knowledge illuminates.
A Reading of Life
Chapter II (p. 54)
Macmillan & Company Ltd. London, England. 1936

Lyttleton, R. A.
English astronomer

...many very serious-minded, solid, and knowledgeable people work hard in science all their lives and produce nothing of the smallest importance, while others, few by comparison and perhaps seemingly carefree and not highly erudite, exhibit a serendipity of mind that enables them to have valuable ideas in any subject they may choose to take up.
In R. Duncan and M. Weston-Smith (eds.)
The Encyclopaedia of Ignorance: Everything You Ever Wanted to Know About the Unknown
The Nature of Knowledge (p. 10)
Pergamon Press. Oxford, England. 1977

MacArthur, Robert H.   1930–72
American ecologist

[N]ot all naturalists want to do science; many take refuge in nature's complexity as a justification to oppose any

search for patterns…. Doing science is not such a barrier to feeling or such a dehumanizing influence as is often made out. It does not take the beauty from nature.
Geographical Ecology
Introduction (p. 1)
Harper & Row, Publishers, New York, New York, USA. 1972

Macfie, Ronald Campbell   1867–1931
Poet and physician

The God of Science speaks in the thunder and smiles in the sunshine. He is so great that the stars eddy round his feet not ankle-high, yet so loving that He makes roses and sunsets for the human heart.
Science, Matter and Immortality
Chapter XVII (p. 207)
William & Norgate. London, England. 1909

Conceived aright, science must always lead to belief in the unseen and to hope of immortality; but Science must learn to recognize her own limitations — must learn to recognize that her logic is not conclusive when her postulates are dubious — and that she can only become a ruler of men's souls and a brightener of men's lives if she takes Poetry and Philosophy by the hand, and dwells with them in the temple of Beauty and Reverence.
Science, Matter and Immortality
Chapter XXIII (p. 300)
William & Norgate. London, England. 1909

It is time that men knew that Science does not write with the cold finger of a starfish; it is time that men realized that true science is not a mere compilation of dead facts; it is time that men understood that Science is flamboyant and alive.
Science, Matter and Immortality
Chapter XXIII (p. 297)
William & Norgate. London, England. 1909

Mach, Ernst   1838–1916
Austrian physicist and philosopher

All science has its origin in the needs of life.
The Science of Mechanics (5th edition)
Chapter V, Part II, Section 1 (p. 610)
The Open Court Publishing Company. La Salle, Illinois, USA. 1942

Every one who busies himself with science recognizes how unsettled and indefinite the notions are which he has brought with him from common life, and how, on a minute examination of things, old differences are effaced and new ones introduced.
Popular Scientific Lectures
The Forms of Liquids (pp. 1–2)
The Open Court Publishing Company. Chicago, Illinois, USA. 1898

…the entire course of the development of science will, as a matter of course, judge more freely and more correctly of the significance of any present scientific movement than they who, limited in their views to the age in which their own lives have been spent, contemplate merely the momentary trend that the course of intellectual events takes at the present moment.
The Science of Mechanics (5th edition)
Introduction (p. 8)
The Open Court Publishing Company. La Salle, Illinois, USA. 1942

The great results achieved by physical science in modern times — results not restricted to its own sphere but embracing that of other sciences which employ its help — have brought it about that physical ways of thinking and physical modes of procedure enjoy on all hands unwonted prominence, and that the greatest expectations are associated with their application.
The Analysis of Sensations and the Relation of the Physical to the Psychical
Chapter I (p. 1)
The Open Court Publishing Company. Chicago, Illinois, USA. 1914

The function of science, as we take it, is to replace experience. Thus, on the one hand, science must remain in the province of experience, but, on the other, must hasten beyond it, constantly expecting confirmation, constantly expecting the reverse. Where neither confirmation nor refutation is possible, science is not concerned…
The Science of Mechanics (5th edition)
Chapter IV, Part IV, Section 7 (p. 587)
The Open Court Publishing Company. La Salle, Illinois, USA. 1942

Physical science does not pretend to be a complete view of the world; it simply claims that it is working toward such a complete view in the future.
The Science of Mechanics (5th edition)
Chapter IV, Part II, Section 9 (p. 560)
The Open Court Publishing Company. La Salle, Illinois, USA. 1942

Science always has its origin in the adaptation of thought to some definite field of experience.
The Analysis of Sensations and the Relation of the Physical to the Psychical
Chapter I (p. 31)
The Open Court Publishing Company. Chicago, Illinois, USA. 1914

Science throws her treasures, not like a capricious fairy into the laps of a favored few, but into the laps of all humanity, with a lavish extravagance that no legend ever dreamt of!
Popular Scientific Lectures
The Economical Nature of Physical Inquiry (p. 189)
The Open Court Publishing Company. Chicago, Illinois, USA. 1898

Economy of communication and of apprehension is of the very essence of science. Herein lies its pacificatory, its enlightening, its refining element.
The Science of Mechanics (5th edition)
Introduction (p. 7)
The Open Court Publishing Company. La Salle, Illinois, USA. 1942

Is science itself anything more than — a business? Is not its task to acquire with the least possible work, in the least possible time, with the least possible thoughts, the greatest possible part of eternal truth?

Popular Scientific Lectures
The Forms of Liquids (p. 16)
The Open Court Publishing Company. Chicago, Illinois, USA. 1898

Maffei, Paolo 1926–
Italian astronomer

We are now moving beyond those concepts and the knowledge familiar to us in the first half of this century, and we are entering a world in which science and fantasy intertwine…
Translated by D.J.K. O'Connell
Beyond the Moon
Chapter 10 (p. 301)
The MIT Press. Cambridge, Massachusetts, USA. 1978

Magendie, Francois 1783–1855
French physiologist

I am a mere street scavenger of science. With hook in hand and basket on my back, I go about the streets of science collecting whatever I find.
In Rene Dubos
Louis Pasteur: Free Lance of Science
Chapter XIII (p. 363)
Little, Brown & Company. Boston, Massachusetts, USA. 1950

Mandelbrot, Benoit 1924–
French mathematician

I started looking in the trash cans of science…because I suspected that what I was observing was…perhaps very widespread. I attended lectures and looked in unfashionable periodicals…once in a while finding some interesting things. In a way it was a naturalist's approach, not a theoretician's approach. But my gamble paid off.
In James Gleick
Chaos: Making a New Science
A Geometry of Nature (p. 110)
The Viking Press. New York, New York, USA. 1987

Mara Corday
Fictional character

Science is science, but a girl must get her hair done.
Tarantula
Film (1955)

March, Robert H. 1937–
American professor of physics

Science is more than a mere attempt to describe nature as accurately as possible. Frequently the real message is well hidden, and a law that gives a poor approximation to nature has more significance than one which works fairly well but is poisoned at the root.
Physics for Poets
Chapter I (p. 17)
McGraw-Hill Book Company, Inc. New York, New York, USA. 1996

Margenau, Henry 1901–97
American physicist

It is in fact obvious that science should be pressed to say all it can about any problem which is at all susceptible to scientific treatment.
The Nature of Physical Reality: A Philosophy of Modern Physics
Chapter 2 (p. 12)
McGraw-Hill Book Company, Inc. New York, New York, USA. 1950

Margulis, Lynn 1938–
American cell biologist and evolutionist
Sagan, Dorion 1959–
American science writer

Science has become a social method of inquiring into natural phenomena, making intuitive and systematic explorations of laws which are formulated by observing nature, and then rigorously testing their accuracy in the form of predictions. The results are then stored as written or mathematical records which are copied and disseminated to others, both within and beyond any given generation. As a sort of synergetic, rigorously regulated group perception, the collective enterprise of science far transcends the activity within an individual brain.
Microcosmos
Chapter 12 (p. 233)
Summit Books. New York, New York, USA. 1986

Maritain, Jacques 1882–1973
French philosopher

Since science's competence extends to observable and measurable phenomena, not to the inner being of things, and to the means, not to the ends of human life, it would be nonsense to expect that the progress of science will provide men with a new type of metaphysics, ethics, or religion.
Science and Ontology
Bulletin of the Atomic Scientists, 1944, Volume 5 (p. 200)

Marshall, Alfred 1842–1924
English economist

…the mathematico-physical group of sciences…have this point in common, that their subject-matter is constant and unchanged in all countries and in all ages. …if the subject-matter of a science passes through different stages of development, the laws which apply to one stage will seldom apply without modification to others; the laws of science must have a development corresponding to that of the things of which they treat.
In A.C. Pigou (ed.)
Memorials of Alfred Marshall
Chapter VI (p. 154)
Macmillan & Company Ltd. London, England. 1925

Mason, James 1909–84
English actor

Don't you see what's at stake here? The ultimate aim of all science — to penetrate the unknown. Do you realize we know less about the earth we live on than about

the stars and the galaxies of outer space? The greatest mystery is right here, right under our feet.
A Journey to the Center of the Earth
Film (1959)

Matsen, F. Albert
No biographical data available

Science is defined as a set of observations and theories about observations.
The Role of Theory in Chemistry
Journal of Chemical Education, Volume 62, Number 5, May 1985
 (p. 365)

Maxwell, James Clerk   1831–79
Scottish physicist

It was a great step in science when men became convinced that, in order to understand the nature of things, they must begin by asking, not whether a thing is good or bad, noxious or beneficial, but of what kind it is? and how much is there of it? Quality and Quantity were then first recognized as the primary features to be observed in scientific inquiry.
In William H. George
The Scientist in Action: A Scientific Study of His Methods
British Association Address, 1870 (p. 15)
Williams & Norgate Ltd. London, England. 1936

McCarthy, Mary   1912–89
American writer

Modern neurosis began with the discoveries of Copernicus. Science made man feel small by showing him that the earth was not the center of the universe.
On the Contrary
Tyranny of the Orgasm (p. 168)
Farrar, Straus & Cudhay, New York, New York, USA, 1961

McLuhan, Marshall   1911–80
Canadian educator, philosopher, and scholar

Current illusion is that science has abolished all natural laws.
In Matie Molinaro, Corinne McLuhan, and William Toye (eds.)
Letters of Marshall McLuhan
Letter to Ezra Pound
January 1951
Oxford University Press, Inc. New York, New York, USA. 1987

Mead, Margaret   1901–78
American anthropologist

…the negative cautions of science are never popular. If the experimentalist would not commit himself, the social philosopher, the preacher and the pedagogue tried the harder to give a short-cut answer.
Coming of Age in Samoa
Chapter 1 (p. 3)
The Modern Library. New York, New York, USA. 1953

Medawar, Sir Peter Brian   1915–87
Brazilian-born English zoologist

Science is no more a classified inventory of factual information than history a chronology of dates. The equation of science with facts and of the humane arts with ideas is one of the shabby genteelisms that bolster up the humanist's self-esteem.
Two Conceptions of Science
Encounter, 143, August 1965

Science can only proceed on a basis of confidence, so that scientists do not suspect each other of dishonesty or sharp practice, and believe each other unless there is very good reason to do otherwise.
The Limits of Science
An Essay on Scians [Science] (p. 6)
Harper & Row, Publishers. New York, New York, USA. 1984

Science will persevere just as long as we retain a faculty we show no signs of losing: the ability to conceive — in no matter how imperfect or rudimentary a form — what the truth might be and retain also the inclination to ascertain whether our imaginings correspond to real life or not.
The Limits of Science
Chapter 4 (pp. 86–87)
Harper & Row, Publishers. New York, New York, USA. 1984

It is a layman's illusion that in science we caper from pinnacle to pinnacle of achievement and that we exercise a Method which preserves us from error. Indeed we do not; our way of going about things takes it for granted that we guess less often right than wrong, but at the same time ensures that we need not persist in error if we earnestly and honestly endeavor not to do so.
The Limits of Science
Notes, 3 (p. 101)
Harper & Row, Publishers. New York, New York, USA. 1984

…science is a great and glorious enterprise — the most successful, I argue, that human beings have ever engaged in. To reproach it for its inability to answer all the questions we should like to put to it is no more sensible than to reproach a railway locomotive for not flying or, in general, not performing any other operation for which it was not designed.
The Limits of Science
Preface (p. xiii)
Harper & Row, Publishers. New York, New York, USA. 1984

If we accept, as I fear we must, that science cannot answer questions about first and last things or about purposes, there is yet no known or conceivable limit to its power to answer questions of the kind science can answer…. Science will dry up only if scientists lose or fail to exercise the power or incentive to imagine what the truth might be.
Advice to a Young Scientist
Chapter 11 (p. 90)
Basic Books, Inc. New York, New York, USA. 1979

…the factual burden of a science varies inversely with its degree of maturity. As a science advances, particular facts are comprehended within, and therefore in a sense annihilated by, general statements of steadily increasing explanatory powers and compass. In all sciences we are being progressively relieved of the burden of singular instances, the tyranny of the particular. We need no longer record the fall of every apple.

The Art of the Soluble
Two Conceptions of Science (p. 114)
Methuen & Company Ltd. London, England. 1967

…nowadays we all give too much thought to the material blessings or evils that science has brought with it, and too little to its power to liberate us from the confinements of ignorance and superstition. The greatest liberation of thought achieved by the scientific revolution was to have given human beings a sense of future in this world.

The Art of the Soluble
Introduction (p. 15)
Methuen & Company Ltd. London, England. 1967

One can envisage an end of science no more readily than one can envisage an end of imaginative literature or the fine arts.

Advice to a Young Scientist
Chapter 11 (p. 90)
Basic Books, Inc. New York, New York, USA. 1979

Melville, Herman   1819–91
American novelist

…however baby man may brag of his science and skill, and however much, in a flattering future, that science and skill may augment; yet for ever and for ever, to the crack of doom, the sea will insult and murder him, and pulverize the stateliest, stiffest frigate he can make; nevertheless, by the continual repetition of these very impressions, man has lost that sense of the full awfulness of the sea which aboriginally belongs to it.

In *Great Books of the Western World* (Volume 48)
Moby Dick
Chapter 58 (p. 204)
Encyclopædia Britannica, Inc. Chicago, Illinois, USA. 1952

After science comes sentiment.

Typee, Omoo, Mardi
Mardi
Chapter 38 (p. 785)
The Library of America. New York, New York, USA. 1982

Mencken, H. L. (Henry Louis)   1880–1956
American journalist and literary critic

There is, in fact, no reason to believe that any given natural phenomenon, however marvelous it may seem today, will remain forever inexplicable. Soon or late the laws governing the production of life itself will be discovered in the laboratory, and man may set up business as a creator on his own account. The thing, indeed, is not only conceivable; it is even highly probable.

Treatise on the Gods

Chapter 5 (p. 241)
Vintage Books. New York, New York, USA. 1963

The notion that science does not concern itself with first causes — that it leaves the field to theology or metaphysics, and confines itself to mere effects — this notion has no support in the plain facts. If it could, science would explain the origin of life on earth at once — and there is every reason to believe that it will do so on some not too remote tomorrow. To argue that gaps in knowledge which will confront the seeker must be filled, not by patient inquiry, but by intuition or revelation, is simply to give ignorance a gratuitous and preposterous dignity.

Treatise on the Gods
Chapter 5 (p. 239)
Vintage Books. New York, New York, USA. 1963

Mendeleyev, Dmitry   1834–1907
Russian chemist

What has been sown for the field of science will grow up for the people's welfare.

Translated by George Kamensky
Principles of Chemistry (Volume 1)
Introduction
Longmans, Green & Company. London, England. 1891

While science is pursuing a steady onward movement, it is convenient from time to time to cast a glance back on the route already traversed, and especially to consider the new conceptions which aim at discovering the general meaning of the stock of facts accumulated from day to day in our laboratories.

The Periodic Law of the Chemical Elements
Journal of the Chemical Society, Volume 55, 1889 (p. 634)

The edifice of science not only requires material but also a plan, and necessitates the work of preparing the materials, putting them together, working out the plans and the symmetrical proportions of the various parts. To conceive, understand, and grasp the whole symmetry of the scientific edifice, including its unfinished portions, is equivalent to tasting that enjoyment only conveyed by the highest forms of beauty and truth.

Principles of Chemistry (Volume 1)
Preface (p. ix, fn 1)
Longmans, Green & Company. London, England. 1891

Science plays an auxiliary part in our lives, for it is merely a means to the attainment of wellbeing.

Compiled by V.V. Vorontsov
Words of the Wise: A Book of Russian Quotations
Translated by Vic Schneierson
Progress Publishers. Moscow, Russia. 1979

Knowing how contented, free and joyful is life in the realms of science, one fervently wishes that many would enter their portals.

Principles of Chemistry (Volume 1)
Preface (p. ix, fn 1)
Longmans, Green & Company. London, England. 1891

Meredith, George   1828–1909
English novelist and poet

Science is notoriously of slow movement.
The Ordeal of Richard Feverel
Chapter XLIV (p. 518)
The Modern Library. New York, New York, USA. 1927

Meyer, Agnes   1887–1970
American author and journalist

From the nineteenth century view of science as a god, the twentieth century has begun to see it as a devil. It behooves us now to understand that science is neither the one nor the other.
Education for a New Morality
Chapter 2 (p. 11)
The Macmillan Company. New York, New York, USA. 1957

Mill, John Stuart   1806–73
English political philosopher and economist

It is a common notion, or at least it is implied in many common modes of speech, that thoughts, feelings, and actions of sentient beings are not a subject of science.... This notion seems to involve some confusion of ideas, which it is necessary to begin by clearing up. Any facts are fitted, in themselves, to be a subject of science, which follows one another according to constant laws; although those laws may not have been discovered, nor even to be discoverable by our existing resources.
A System of Logic, Rationative and Inductive (Volume 2)
Book VI, Chapter 3, Section 1 (p. 426)
Longmans, Green, Reader & Dyer. London, England. 1868

Millikan, Robert Andrews   1868–1953
American physicist

We need science in education, and much more of it than we now have, not primarily to train technicians for the industries, which demand them, though that may be important, but much more to give everybody a little glimpse of the scientific mode of approach to life's problems, to give everyone some familiarity with at least one field in which the distinction between right and wrong is not always blurred and uncertain, to let him see that it is not true that "one opinion is as good as another"...
The Relationship of Science to Industry
Science, Volume 69, Number 1776, January 11, 1929 (p. 30)

The distinguishing feature of modern scientific thought lies in the fact that it begins by discarding all a priori conceptions about the nature of reality — or about the ultimate nature of the universe — such as had characterized practically all Greek philosophy and all medieval thinking as well, and takes instead, as its starting point, well-authenticated, carefully tested experimental facts, no matter whether these facts seen at the moment to fit into any general philosophical scheme or not — that is, no matter whether they seem at the moment to be reasonable or not.

Professor Einstein at the California Institute of Technology
Science, Volume 73, Number 1893, April 10, 1931 (p. 376)

It is to lighten man's understanding, to illuminate his path through life, and not merely to make it easy, that science exists.
In Frederick Houk Law
Science in Literature
Modern Physics (p. 318)
Harper & Brothers. New York, New York, USA. 1929

...Science walks forward on two feet, namely theory and experiment...Sometimes it is one foot which is put forward first, sometimes the other, but continuous progress is only made by the use of both — by theorizing and then testing, or by finding new relations in the process of experimenting and then bringing the theoretical foot up and pushing it beyond, and so on in unending alternation.
Nobel Lectures, Physics 1922–1941
Nobel lecture for award received in 1923
The Electron and the Light-Quant from the Experimental Point of View (p. 55)
Elsevier Publishing Company. Amsterdam, Netherlands. 1965

Milne, Edward Arthur   1896–1950
English astrophysicist and cosmologist

The Christmas message — which is also the Christian message — is *"Gloria in excelsis Deo"*...Glory to God in the highest and on earth peach among men of goodwill.... This is not a bad definition of the aim of all true science: the aim of rejoicing in the splendid mysteries of the world and universe we live in, and of attempting so to understand those mysteries that we can improve our command over nature, improve our conditions of life and so ensure peace...
Modern Cosmology and the Christian Idea of God
Chapter I (p. 1)
At The Clarendon Press. Oxford, England. 1952

Mitchell, Maria   1818–89
American astronomer and educator

The phrase "popular science" has in itself a touch of absurdity. That knowledge which is popular is not scientific.
In Phebe Mitchell Kendall
Maria Mitchell: Life, Letters, and Journals
Chapter VII (p. 138)
Lee & Shepard. Boston, Massachusetts, USA. 1896

Monod, Jacques   1910–76
French biochemist

In science, self-satisfaction is death. Personal self-satisfaction is the death of the scientist. Collective self-satisfaction is the death of the research. It is restlessness, anxiety, dissatisfaction, agony of mind that nourish science.
Obituary
News Science, Volume 109, June 5, 1976 (p. 359)

Montagu, Ashley   1905–99
English-born American anthropologist

As the god of contemporary man's idolatry, science is a two-handed engine, and as such science is too important a human activity to leave to the scientist.
Advertisement of Jacques Barzun's "Science: The Glorious Entertainment"
New York Times Book Review, April 26, 1964

More, Louis Trenchard   1870–1944
English physicist and biographer of Isaac Newton

Science has so many dazzling achievements to its credit; we have done so many things which seemed to be impossible, that the popular mind is apt to conclude that, if an explanation is given in the name of science, it must be true whether it be understood or not.
The Dogma of Evolution
Chapter Seven (p. 241)
Princeton University Press. Princeton, New Jersey, USA. 1925

Morgan, Lloyd   1852–1936
English psychologist

Science…deals exclusively with changes of configuration, and traces the accelerations which are observed to occur, leaving to metaphysics to deal with the underlying agency, if it exists.
The Interpretation of Nature
Chapter V (p. 62)
The Knickerbocher Press. New York, New York, USA. 1906

Morrow, James   1947–
American novelist

Everybody thinks he's being oh-so-deep when he says science doesn't have all the answers…. Science *does* have all the answers…. The problem is that we don't have all the science.
Only Begotten Daughter (p. 90)
Harcourt Incorporated. Orlando, Florida, USA. 1990

Moscovici, S.   1925–
Romanian-born French psychologist

Science has become involved in this adventure, our adventure, in order to renew everything it touches and warm all that it penetrates — the earth on which we live and the truths which enable us to live. At each turn it is not the echo of a demise, a bell tolling for a passing away that is heard, but the voice of rebirth and beginning, ever afresh, of mankind and materiality, fixed for an instant in their ephemeral permanence. That is why the great discoveries are not revealed on a deathbed like that of Copernicus, but offered like Kepler's on the road of dreams and passion.
Social Influence and Social Change (pp. 297–298)
Academic Press. London, England. 1980

Motto

Science Finds — Industry Applies — Man Conforms
Chicago World's Fair, 1933

Muller, Herbert J.   1905–80
American historian and educator

Although science is no doubt the Jehovah of the modern world, there is considerable doubt about the glory of its handiwork.
Science and Criticism: The Humanistic Tradition in Contemporary Thought
Chapter III (p. 59)
G. Braziller. New York, New York, USA. 1943

…men of science, men given to "realism," are likely to make a clean sweep of old interests and sentiments as so much rubbish. They regard religion as superstition, metaphysics as moonshine, art as primitive pastime, and all ritual as monkey-business.
Science and Criticism: The Humanistic Tradition in Contemporary Thought
Chapter I (p. 6)
G. Braziller. New York, New York, USA. 1943

Mumford, Lewis   1895–1990
American social philosopher

…however far modern science and technics have fallen short of their inherent possibilities, they have taught mankind at least one lesson: Nothing is impossible.
Technics and Civilization
Chapter VIII, Section 13 (p. 435)
Routledge & Kegan Paul Ltd. London, England, 1934

Munger, Theodore   1830–1910
American clergyman

Science cannot determine origin, and so cannot determine destiny. As it presents only a sectional view of creation, it gives only a sectional view of everything in creation.
In Jefferson Hane Weaver
The World of Physics (Volume 3)
U.1 (p. 212)
Simon & Schuster. New York, New York, USA. 1987

Needham, Joseph   1900–95
English biochemist and sinologist

…our proper conclusion seems to me to be that the conceptual framework of Chinese associative or coordinative thinking was essentially something different from that of European causal and "legal" or nomothetic thinking. That it did not give rise to 17th -century theoretical science is no justification for calling it primitive.
Science and Civilisation in China (Volume 2) (p. 286)
At The University Press. Cambridge, England. 1954

Nekrasov, Nikolai   1821–78
Russian poet

There is no science for the sake of science, no art for the sake of art — they exist for the sake of society, for the ennoblement and exaltation of man, to enrich his knowledge and provide his material comforts.
Compiled by V.V. Vorontsov
Words of the Wise: A Book of Russian Quotations

Translated by Vic Schneierson
Progress Publishers. Moscow, Russia. 1979

Newell, A.
No biographical data available

Scientific fields emerge as the concerns of scientists congeal around various phenomena. Sciences are not defined, they are recognized.
In R.C. Shank and K.M. Colby (eds.)
Computer Models of Thought and Language
Artificial Intelligence and the Concept of Mind (p. 1)
W.H. Freeman. San Francisco, California, USA. 1973

Newton, Roger G.
Physics professor and author

Science is not holy scripture, nor do its practitioners consider themselves priests protecting a glittering grail, forever unchanging and pure. What drives scientists on is the thirst to understand more and to use nature, to build rather than to exploit a comprehensible universe.
What Makes Nature Tick?
Epilogue (p. 234)
Harvard University Press. Cambridge, Massachusetts, USA. 1993

Science is, in fact, an intricate edifice erected from complex, imaginative designs in which esthetics is a more powerful incentive than utility. Beauty, finally, comprises its greatest intellectual appeal.
What Makes Nature Tick?
Epilogue (p. 236)
Harvard University Press. Cambridge, Massachusetts, USA. 1993

Nietzsche, Friedrich   1844–1900
German philosopher

To the man who works and searches in it, science gives much pleasure; to the man who learns its results, very little.
Translated by Marion Faber
Human, All Too Human: A Book for Free Spirits
Section Five, Number 251
Aphorism 205 (p. 153)
University of Nebraska Press. Lincoln, Nebraska, USA. 1984

Oh, how much is today hidden by science! Oh, how much it is expected to hide!
Translated by William A. Haussmann
The Genealogy of Morals
What Do Ascetic Ideals Mean?
Aphorism 23
Macmillan Publishing Company. New York, New York, USA. 1907

Science offends the modesty of all real women. It makes them feel as though it were an attempt to peek under their skin — or, worse yet, under their dress and ornamentation!
Beyond Good and Evil
Chapter IV, 127 (p. 83)
The Modern Library. New York, New York, USA. 1917

Science rushes headlong, without selectivity, without "taste," at whatever is knowable, in the blind desire to know all at any cost.

Translated by Marianne Cowan
Philosophy in the Tragic Age of the Greeks
Section 3 (p. 43)
A Gateway Edition. Chicago, Illinois, USA. 1962

The old God was seized by mortal terror. Man himself had been his greatest blunder; he had created a rival to himself; science makes men godlike — it is all up with priests and gods when man becomes scientific — Moral: science is the forbidden per se; it alone is forbidden. Science is the first of sins, the germ of all sins, the original sin. This is all there is of morality. — "Thou shalt not know": the rest follows from that.
Translated by H.L. Mencken
The Anti-Christ
Aphorism 48
Macmillan Publishing Company. New York, New York, USA. 1911

Nobel Prize Medal

Inventas vitam iuvat excoluisse per artes.
Let us improve life through science and art.
Inscribed on Nobel Prize Medal

Oberth, Hermann   1894–1989
German mathematician and physicist

The present state of science and of technological knowledge permits the building of machines that can rise beyond the limits of the atmosphere of the earth. After further development these machines will be capable of attaining such velocities that they — left undisturbed in the void of ether space — will not fall back to earth; furthermore, they will even be able to leave the zone of terrestrial attraction.
The Rocket to the Interplanetary Spaces
Publisher undetermined

O'Neill, Eugene   1888–1953
American playwright

DARRELL: Happiness hates the timid! So does Science!
Strange Interlude
Act Four (p. 152)
Boni & Liveright. New York, New York, USA. 1928

Oppenheimer, J. Robert   1904–67
American theoretical physicist

We live today in a world in which poets and historians and men of affairs are proud that they wouldn't even begin to consider thinking about learning anything of science, regarding it as the far end of a tunnel too long for any wise man to put his head into.
The Open Mind
Chapter VII (p. 128)
Simon & Schuster. New York, New York, USA. 1955

A subject is much harder to understand when no one understands it. The world is really an open place, but we start with such crude and limited experience, and our minds

are so determined by that experience, that when science carries us into new domains we are not always prepared for what we encounter, and we are floored by it.
In Edward Lueders
Writing on Life: Sixteen Close-Ups
Physicist Oppenheimer (p. 358)
William Sloane Associates, Publishers. New York, New York, USA. 1951

O'Rourke, P. J.   1947–
American political satirist

…to mistrust science and deny the validity of the scientific method is to resign your job as a human. You'd better go look for work as a plant or wild animal.
Parliament of Whores: A Lone Humorist Attempts to Explain the Entire U.S. Government
Dirt of the Earth (p. 1 97)
Vintage Books. New York, New York, USA. 1992

Orr, Louis
American Medical Association president

Science will never be able to reduce the value of a sunset to arithmetic. Nor can it reduce friendship or statesmanship to a formula. Laughter and love, pain and loneliness, the challenge of accomplishment in living, and the depth of insight into beauty and truth; these will always surpass the scientific mastery of nature.
Commencement Address, Emory University, June 6, 1960

Osler, Sir William   1849–1919
Canadian physician and professor of medicine

To the physician particularly, a scientific discipline is an incalculable gift, which leavens his whole life, giving exactness to habits of thought and tempering the mind with that judicious faculty of distrust which can alone, amid the uncertainties of practice, make him wise unto salvation. For perdition inevitably awaits the mind of the practitioner who has never had the full inoculation with the leaven, who has never grasped clearly the relations of science to his art, and who knows nothing and perhaps cares less, for the limitations of either.
Aequanimitas, with Other Addresses to Medical Students, Nurses, and Practitioners of Medicine
The Leaven of Science (p. 92)
The Blakiston Company. Philadelphia, Pennsylvania, USA. 1932

The future belongs to science. More and more she will control the destinies of the nations. Already she has them in her crucible and on her balances.
In Harvey Cushing
The Life of Sir William Osler (Volume 2) (p. 262)
Clarendon Press. Oxford, England. 1925

Ostwald, Friedrich Wilhelm   1853–1932
Latvian-born German chemist

The more perfect the theoretical evolution of the sciences becomes, the greater will be the scope of their explanations and at the same time the greater their practical importance.
On Chemical Energy
The Journal of the American Chemical Society, Volume 15, Number 8, August 1893 (p. 430)

Pagels, Heinz R.   1939–88
American physicist and science writer

This sense of the unfathomable beautiful ocean of existence drew me into science. I am awed by the universe, puzzled by it and sometimes angry at a natural order that brings such pain and suffering, Yet an emotion or feeling I have toward the cosmos seems to be reciprocated by neither benevolence nor hostility but just by silence. The universe appears to be a perfectly neutral screen unto which I can project any passion or attitude, and it supports them all.
Perfect Symmetry: The Search for the Beginning of Time
Part Four, Chapter 2 (p. 370)
Simon & Schuster. New York, New York, USA. 1985

Science is not the enemy of humanity but one of the deepest expressions of the human desire to realize that vision of infinite knowledge.
The Cosmic Code: Quantum Physics as the Language of Nature
Part III, Chapter 2 (p. 348)
Simon & Schuster. New York, New York, USA. 1982

Paglia, Camille   1947–
American social critic, intellectual, and writer

Modern bodybuilding is ritual, religion, sport, art, and science, awash in Western chemistry and mathematics. Defying nature, it surpasses it.
Sex, Art, and American Culture
Alice in Muscle Land (p. 82)
Vintage Books. New York, New York, USA. 1992

Pallister, William Hales   1877–1946
Canadian physician

You are the sum of what we know,
You are our might and main;
You are the whole of what is so,
The little we retain:
Our fond beliefs all come and go,
And you alone remain.
Poems of Science
Science (p. 39)
Playford Press. New York, New York, USA. 1931

Science works by the slow method of the classification of data, arranging the detail patiently in a periodic system into groups of facts, in series like the strata of the rocks. For each series there must be a vocabulary of special words which do not always make good sense when used in another series. But the laws of periodicity seem to hold throughout, among the elements and in every sphere of thought, and we must learn to co-ordinate the whole through our new conception of the reign of relativity.
Poems of Science

Men and the Stars (p. 88)
Playford Press. New York, New York, USA. 1931

Panunzio, Constantine   1884–1964
Italian sociologist

Science…involves active, purposeful search; it discovers, accumulates, sifts, orders, and tests data; it is a slow, painstaking, laborious activity; it is a search after bodies of knowledge sufficiently comprehensive to lead to the discovery of uniformities, sequential orders or so-called "laws"; it may be carried on by an individual, but it gains relevance only as it produces data which can be added to and tested by the findings of others.
Major Social Institutions
Chapter 20 (p. 322)
The Macmillan Company. New York, New York, USA. 1945

If science is to subserve human needs, it will continue to discover and catalogue "all the islands of the universe 300,000,000 or more light years distant," but it will not fiddle while Rome burns…
Major Social Institutions
Chapter 21 (p. 338)
The Macmillan Company. New York, New York, USA. 1945

Parin, V. V.
No biographical data available

Science breathes but one air — the oxygen of facts. New methods of research are the trees that clear its atmosphere of the carbon dioxide of inaccurate conclusions and saturate it with the oxygen of first discovered, seen and apprehended phenomena.
Compiled by V.V. Vorontsov
Words of the Wise: A Book of Russian Quotations
Translated by Vic Schneierson
Progress Publishers. Moscow, Russia. 1979

Pasteur, Louis   1822–95
French chemist

You bring me the deepest joy that can be felt by a man whose invincible belief is that Science and Peace will triumph over Ignorance and War, that nations will unite, not to destroy, but to build, and that the future will belong to those who will have done most for suffering humanity.
In Rene Vallery-Radot
The Life of Pasteur
Chapter XIV (pp. 450–451)
Doubleday, Doran & Company, Inc. Garden City, New York, USA. 1928

I am imbued with two deep impressions; the first, that science knows no country; the second, which seems to contradict the first, although it is in reality a direct consequence of it, that science is the highest personification of the nation. Science knows no country, because knowledge belongs to humanity, and is the torch which illuminates the world. Science is the highest personification of the nation

because that nation will remain the first which carries the furthest the works of thought and intelligence.
In Rene Dubos
Pasteur and Modern Science
Chapter 15. A Dedicated Life (p. 146)
Science Tech Publishers. Madison, Wisconsin, USA. 1988

I could never work for money, but I would always work for science.
In Sir Richard Arman Gregory
Discovery; or, The Spirit and Service of Science
Chapter I (p. 6)
Macmillan & Company Ltd. London, England. 1918

Pavlov, Ivan Petrovich   1849–1936
Russian physiologist

Only science, exact science about human nature itself, and the most sincere approach to it by the aid of the omnipotent scientific method, will deliver man from his present gloom, and will purge him from his contemporary shame in the sphere of interhuman relations.
Translated by Stephen G. Brush
Lectures on Conditioned Reflexes
Preface to the First Russian Edition (p. 41)
University of California Press. Berkeley, California, USA. 1964

Science moves in fits and starts, depending on the progress in methods of research. Every step forward in method takes us a step higher, affording a broader view of the horizon and of objects that were invisible before.
Compiled by V.V. Vorontsov
Words of the Wise: A Book of Russian Quotations
Translated by Vic Schneierson
Progress Publishers. Moscow, Russia. 1979

Remember that science demands from a man all his life. If you had two lives that would be not enough for you. Be passionate in your work and your searchings.
Bequest of Pavlov to the Academic Youth of his Country
Science, Volume 83, Number 2155, April 17, 1936 (p. 369)

Learn the ABC of science before you try to ascend to its summit. Never begin the subsequent without mastering the preceding. Never attempt to screen an insufficiency of knowledge even by the most audacious surmise and hypothesis.
Bequest of Pavlov to the Academic Youth of His Country
Science, Volume 83, Number 2155, April 17, 1936 (p. 369)

Here I now simply uphold and assert the absolute and incontestable right of natural science to operate wherever and whenever it is able to display its power. And who knows the limits to this!
Experimental Psychology and Other Essays
Natural Science and the Brain (p. 218)
Philosophical Library. New York, New York, USA. 1957

Peacock, Thomas Love   1785–1866
English writer

Science is one thing and wisdom is another. Science is an edged tool with which men play like children and cut

their own fingers. If you look at the results which science has brought in its train, you will find them to consist almost wholly in elements of mischief.... The day would fail if I should attempt to enumerate the evils which science has inflicted on mankind. *Gryll Grange*
Chapter 19 (p. 127)
Penguin Books. Harmondsworth, England. 1949

I almost think it is the ultimate destiny of science to exterminate the human race.
Gryll Grange
Chapter 19 (p. 127)
Penguin Books. Harmondsworth, England. 1949

Pearson, Karl   1857–1936
English mathematician

When every fact, every present or past phenomenon of that universe, every phase or present or past life therein, has been examined, classified, and co-ordinated with the rest, then the mission of science will be completed. What is this but saying that the task of science can never end till man ceases to be, till history is no longer made, and development itself ceases?
The Grammar of Science
Introductory, Section 5 (p. 15)
Charles Scribner's Sons. London, England. 1892

Science for the past is a description, for the future a belief...
The Grammar of Science
Chapter IV, Section 1 (p. 136)
Charles Scribner's Sons. London, England. 1892

Every great advance of science opens our eyes to facts which we have failed before to observe, and makes new demands on our powers of interpretation. This extension of the material of science into regions where our great-grandfathers could see nothing at all, or where they would have declared human knowledge impossible, is one of the most remarkable features of modern progress. Where they interpreted the motion of the planets of our own system, we discuss the chemical constitution of stars, many of which did not exist for them, for the telescopes could not reach them. Where they discovered the circulation of the blood, we see the physical conflict of living poisons within the blood, whose battles would have been absurdities for them.
The Grammar of Science
Introductory, Section 5 (p. 17)
Charles Scribner's Sons. London, England. 1892

Does science leave no mystery? On the contrary it proclaims mystery where others profess knowledge. There is mystery enough in the universe of sensation and in its capacity for containing those little corners of consciousness which project their own products, or order and law and reason, into an unknown and unknowable world. There is mystery enough here, only let us clearly distinguish it from ignorance within the field of possible knowledge. The one is impenetrable, the other we are daily subduing.
The Grammar of Science

Chapter III, Conclusion (p. 134)
Charles Scribner's Sons. London, England. 1892

Modern Science, as training the mind to an exact and impartial analysis of facts, is an education specifically fitted to promote sound citizenship.
The Grammar of Science
Introductory, Section 3 (p. 11)
Charles Scribner's Sons. London, England. 1892

Peattie, Donald Culrose   1896–1964
American botanist, naturalist, and author

Of our windows on the universe, science is set with the clearest pane; it is not warped or waved to make the images appear to support any dogma; the glass is not rose-tinted, neither is it leaded with a picture that shuts out the sun and, coming between the light of day and you, enforces the credence of the past upon the young present.
Flowering Earth
Chapter 18 (p. 244)
G.P. Putnam's Sons. New York, New York, USA. 1939

Peirce, Charles Sanders   1839–1914
American scientist, logician, and philosopher

Science, when it comes to understand itself, regards facts as merely the vehicle of eternal truth, while for Practice they remain the obstacles which it has to turn, the enemy of which it is determined to get the better.
The Essential Peirce: Selected Philosophical Writings (Volume 2)
The First Rule of Logic (p. 55)
Indiana University Press. Bloomington, Indiana, USA. 1998

It is a common observation that a science first begins to be exact when it is quantitatively treated. What are called the exact sciences are no other than the mathematical ones.
Chance, Love and Logic: Philosophical Essays
The Doctrine of Chances (p. 61)
Harcourt, Brace & Company, Inc. New York, New York, USA. 1923

Perl, Martin   1927–
American physicist

I was following an old idea in science: "If you can't understand a phenomenon, look for more examples of that phenomenon..."
Electron, Muon, and Tau Heavy Lepton — Are These the Truly Elementary Particles?
The Science Teacher, Volume 47, Number 9, December 1980 (pp. 18–19)

Perutz, Max F.   1914–2002
Austrian-born English biochemist

It seems to me that, just as the Church did in former times, science offers a safe niche where you can spend a quiet life classifying spiders, away from what E.M. Forster called the world of telegrams and anger.
Is Science Necessary?
How to Become a Scientist (p. 193)
E.P. Dutton & Company. New York, New York, USA. 1989

Pirsig, Robert M.   1928–
American writer

Science values static patterns.
Lila: An Inquiry Into Morals
Chapter 11 (p. 142)
Bantam Books. New York, New York, USA. 1991

Planck, Max   1858–1947
German physicist

That we do no construct the external world to suit our own ends in the pursuit of science, but that vice versa the external world forces itself upon our recognition with its own elemental power, is a point which ought to be categorically asserted again and again in these positivistic times. From the fact that in studying the happenings of nature we strive to eliminate the contingent and accidental and to come finally to what is essential and necessary, it is clear that we always look for the basic thing behind the dependent thing, for what is absolute behind what is relative, for the reality behind the appearance and for what abides behind what is transitory. In my opinion, this is characteristic not only of physical science but all of science.
Where Is Science Going?
Chapter VI (pp. 198–199)
W.W. Norton & Company, Inc. New York, New York, USA.1932

Science…means unresting endeavor and continually progressing development toward an aim which the poetic intuition may apprehend but which the intellect can never fully grasp.
Continuum, Volume 20, Number 5, February 1980 (p. 42)

Science cannot solve the ultimate mystery of nature. And that is because, in the last analysis, we ourselves are part of nature and therefore part of the mystery that we are trying to solve.
Where Is Science Going?
Epilogue (p. 217)
W.W. Norton & Company, Inc. New York, New York, USA.1932

Science does not mean an idle resting upon a body of certain knowledge; it means unresting endeavor and continually progressing development towards an aim, which the poetic intuition may apprehend, but which the intellect can never fully grasp.
Translated by W. H. Johnston
The Philosophy of Physics
Chapter II (p. 83)
W.W. Norton & Company, Inc. New York, New York, USA.1936

Exact science — what wealth of connotation these two words have! They conjure up a vision of a lofty structure, of imperishable slabs of stone firmly joined together, treasure-house of all wisdom, symbol and promise of the coveted goal for a human race thirsting for knowledge, longing for the final revelation of truth.
Scientific Autobiography and Other Papers

The Meaning and Limits of Exact Science (p. 80)
Philosophical Library. New York, New York, USA. 1949

The roots of exact science feed in the soil of human life.
Scientific Autobiography and Other Papers
The Meaning and Limits of Exact Science, Part IV (p. 112)
Philosophical Library. New York, New York, USA. 1949

…I had always looked upon the search for the absolute as the noblest and most worth while task of science.
Scientific Autobiography and Other Papers
A Scientific Autobiography (p. 46)
Philosophical Library. New York, New York, USA. 1949

…science is not contemplative repose amidst knowledge already gained, but is indefatigable work and an ever progressive development.
Scientific Autobiography and Other Papers
The Concept of Causality in Physics (p. 150)
Philosophical Library. New York, New York, USA. 1949

Plato   428 BCE–347 BCE
Greek philosopher

As being is to becoming, so is pure intellect to opinion. And as intellect is to opinion, so is science to belief…
In *Great Books of the Western World* (Volume 7)
The Republic
Book VII, Section 534 (p. 398)
Encyclopædia Britannica, Inc. Chicago, Illinois, USA. 1952

Podolsky, Boris   1896–1966
American physicist

In recent years the power of Science has received such popular recognition that the adjective scientific attached to merchandise or statement is known to give to such merchandise or to a statement prestige having definite advertising value. As a consequence the words science and scientific are frequently abused by those who find it profitable to borrow reputation instead of earning it.
What Is Science?
The Physics Teacher, Volume 3, Number 2, February 1965 (p. 71)

Poe, Edgar Allan   1809–49
American short story writer

Science! true daughter of old Time thou art
Who alterest all things with thy peering eyes!
Why prey'st thou thus upon the poet's heart,
Vulture! whose wings are dull realities!
The Raven and Other Poems
Sonnet — To Science
Columbia University Press. New York, New York, USA. 1942

Poincaré, Henri   1854–1912
French mathematician and theoretical astronomer

The advance of science is not comparable to the changes of a city, where old edifices are pitilessly torn down to give place to new, but to the continuous evolution of zoologic types which develop ceaselessly and end by becoming

unrecognizable to the common sight, but where an expert eye finds always traces of the prior work of the past centuries.
The Foundations of Science
The Value of Science, Introduction (p. 208)
The Science Press. New York, New York, USA. 1913

Man, then, can not be happy through science, but to-day he can be much less be happy without it.
The Foundations of Science
The Value of Science, Introduction (p. 206
The Science Press. New York, New York, USA. 1913

There is no science other than disinterested science.
In Stefan Amsterdamski
Between History and Method
Chapter V Crisis of the Modern Ideal (p. 94)
Kluwier Academic Publishers. Dordrecht, Netherlands. 1992

As science progress, it becomes more and more difficult to fit in the new facts when they will not fit in spontaneously. The older theories depend upon the coincidences of so many numerical results which can not be attributed to chance. We should not separate what has been joined together.
Annual Report of the Board of Regents of the Smithsonian Institution, 1912
The Ether and Matter (pp. 209–210)
Government Printing Office. Washington, D.C. 1913

…science is a rule of action which is successful…
The Foundations of Science
The Value of Science, Part III, Chapter X (p. 324)
The Science Press. New York, New York, USA. 1913

Polanyi, Michael   1891–1976
Hungarian-born English scientist, philosopher, and social scientist

This coherence of valuation throughout the whole range of science underlies the unity of science. It means that any statement recognized as valid in one part of science can, in general, be considered as underwritten by all scientists. It also results in a general homogeneity of and a mutual respect between all kinds of scientists, by virtue of which science forms an organic unity.
Science, Faith and Society
Authority and Conscience (p. 49)
The University of Chicago Press. Chicago, Illinois. 1964

The morsels of science which [the young scientist] picks up — even though often dry or else speciously varnished — instill in him the intimation of intellectual treasures and creative joys far beyond his ken. His intuitive realization of a great system of valid thought and of an endless path of discovery sustain him in laboriously accumulating knowledge and urge him on to penetrate into intricate brain-racking theories. Sometimes he will also find a master whose work he admires and whose manner and outlook he accepts for his guidance. Thus his mind will become assimilated to the premise of science. The scientific institution of reality henceforth shapes his perception. He learns the methods of scientific investigation and accepts the standards of scientific value.

Science, Faith and Society
Authority and Conscience (p. 44)
The University of Chicago Press. Chicago, Illinois. 1964

Pope, Alexander   1688–1744
English poet

Trace Science then, with Modesty thy guide;
First strip off all her equipage of Pride.
The Complete Poetical Works (Volume 2)
An Essay on Man
Epistle II, l. 43–44
Houghton Mifflin Company. New York, New York, USA. 1903

One science only will one genius fit,
So vast is art, so narrow human wit…
The Complete Poetical Works (Volume 2)
Essay on Criticism, Part I, l. 60–61
Houghton Mifflin Company. New York, New York, USA. 1903

Far eastward cast thine eye, from whence the Sun
And orient Science their brite course begun.
The Complete Poetical Works (Volume 4)
Duncaid, Book III, l. 73–74
Houghton Mifflin Company. New York, New York, USA. 1903

How Index-learning turns no student pale,
Yet holds the eel of science by the tail…
The Complete Poetical Works (Volume 4)
Duncaid, Book I, l. 279–80
Houghton Mifflin Company. New York, New York, USA. 1903

Popper, Karl R.   1902–94
Austrian/British philosopher of science

The empirical basis of objective science has thus nothing "absolute" about it. Science does not rest upon solid bedrock. The bold structure of its theories rises, as it were, above a swamp. It is like a building erected on piles. The piles are driven down from above into the swamp, but not down to any natural or "given" base; and when we cease our attempts to drive our piles into a deeper layer, it is not because we have reached firm ground. We simply stop when we are satisfied that they are firm enough to carry the structure, at least for the time being.
The Logic of Scientific Discovery
Part II, Chapter V, Section 30 (p. 111)
Basic Books, Inc. New York, New York, USA. 1959

Science is not a system of certain, or well-established statements, nor is it a system which steadily advances towards a state of finality.… Like Bacon we might describe our own contemporary science…as consisting of "anticipations, rash and premature," and as "prejudices."
The Logic of Scientific Discovery
Part II, Chapter X, Section 85 (p. 278)
Basic Books, Inc. New York, New York, USA. 1959

Science may be described as the art of systematic oversimplification.
The Observer, London, 1 August 1982

Science does not aim, primarily, at high probabilities. It aims at a high informative content, well backed by

experience. But a hypothesis may be very probable simply because it tells us nothing, or very little.
The Logic of Scientific Discovery
Appendix ix (p. 399)
Basic Books, Inc. New York, New York, USA. 1959

…science is most significant as one of the greatest spiritual adventures that man has yet known…
The Poverty of Historicism
Chapter III, Section 19 (p. 56)
The Beacon Press. Boston, Massachusetts, USA. 1957

…it is the aim of science to find satisfactory explanations, of whatever strikes us as being in need of explanation.
Objective Knowledge: An Evolutionary Approach
Chapter 5 (p. 191)
Clarendon Press. Oxford, England. 1972

Porterfield, Austin L.
No biographical data available

Science, in the broadest sense, is the entire body of the most accurately tested, critically established, systematized knowledge available about that part of the universe which has come under human observation. For the most part this knowledge concerns the forces impinging upon human beings in the serious business of living and thus affecting man's adjustment to and of the physical and the social world.… Pure science is more interested in understanding, and applied science is more interested in control…
Creative Factors in Scientific Research
Chapter II (p. 11)
Duke University Press. Durham, North Carolina, USA. 1941

Poteat, William Louis   1856–1938
American educator

Science confers power, not purpose. It is a blessing, therefore, if the purpose which it serves is good; it is a curse, if the purpose is bad.
Can a Man Be a Christian Today?
Part I, Section 2 (p. 27)
The University of North Carolina Press. Chapel Hill, North Carolina, USA. 1925

Powers, Richard   1957–
American novelist

Science is not about control. It is about cultivating a perpetual condition of wonder in the face of something that forever grows one step richer and subtler than our latest theory about it. It is about reverence, not mastery.
The Gold Bug Variations
Dialog
William Morrow & Company, Inc. New York, New York, USA. 1991

Praed, Winthrop   1802–39
English poet

Of science and logic he chatters,
As fine and as fast as he can;

Though I am no judge of such matters,
I'm sure he's a talented man.
The Poems of Winthrop Mackworth Praed
The Talented Man
Houghton Mifflin Company. Boston, Massachusetts, USA. 1909

Pratt, C. C.
No biographical data available

Science is a vast and impressive tautology.
The Logic of Modern Psychology
Chapter VI (p. 154)
The Macmillan Company. New York, New York, USA. 1939

Prescott, William Hickling   1796–1859
American historian

It is the characteristic of true science, to discern the impassable, but not very obvious, limits which divide the province of reason from that of speculation. Such knowledge comes tardily. How many ages have rolled away in which powers, that, rightly directed, might have revealed the great laws of nature, have been wasted in brilliant, but barren reveries on alchemy and astrology.
History of the Conquest of Mexico (Volume 1)
Book I, Chapter IV (p. 102)
J.B. Lippincott Company. Philadelphia, Pennsylvania, USA. 1891

Pribram, Karl   1919–
Austrian neurosurgeon

For the first time in three hundred years science is admitting spiritual values into its explorations.
In Pamela Weintraub (ed.)
The Omni Interviews
Holographic Brain (p. 133)
Ticknor & Fields. New York, New York, USA. 1984

Priestley, Joseph   1733–1804
English theologian and scientist

A successful pursuit of science makes a man the benefactor of all mankind and of every age.
Experiments and Observations on Different Kinds of Air (Volume 1)
The Preface (p. xxvii)
Thomas Pearson. Birmingham, England. 1790

Prigogine, Ilya   1917–2003
Russian-born Belgian physical chemist
Stengers, I.   1949–
Belgian philosopher

Science is part of the Darwinian struggle for life. It helps us to organize our experience. It leads us to economy of thought. Mathematical laws are nothing more than conventions useful for summarizing the results of possible experiments.
Order Out of Chaos
Chapter III 5. Ignoramus, Ignoramibus (p. 97)
Bantam Books. New York, New York, USA. 1984

Prior, Matthew 1664–1721
English poet and diplomat

Forc'd by reflective Reason I confess,
That human Science is uncertain guess.
In John Aikin
Select Works of the British Poets
Solomon, Book 1, l. 740
Longman, Hurst, Reese, Orme & Brown. London, England. 1820

Pritchett, V. S. 1900–97
English writer

A touch of science, even bogus science, gives an edge to the superstitious tale.
The Living Novel and Later Appreciations
An Irish Ghost (p. 123)
Random House, Inc. New York, New York, USA. 1964

Prout, Curtis
No biographical data available

The study of science suggests the need for humility.
Demand and Get the Best Health Care for You: An Eminent Doctor's Practical Advice (p. 148)
Faber & Faber. Boston, Massachusetts, USA. 1997

Quetelet, Adolphe 1794–1874
Belgian mathematician, astronomer, and statistician

The more progress physical sciences makes, the more they tend to enter the domain of mathematics, which is a kind of center to which they all converge. We may even judge of the degree of perfection to which a science has arrived by the facility with which it may be submitted to calculation.
In E. Mailly
Annual Report of the Board of Regents of the Smithsonian Institution, 1874
Eulogy of Quetelet (p. 173)
Government Printing Office. Washington, D.C. 1875

Quine, Willard Van Orman 1908–2000
American logician and philosopher

Science is like a boat, which we rebuild plank by plank while staying afloat in it. The philosopher and the scientist are in the same boat.
In George Johnson
In the Palaces of Memory: How We Build the World Inside Our Heads
The End of Philosophy (p. 222)
Alfred A. Knopf. New York, New York, USA. 1991

Quinet, Edgar 1803–75
French historian

Science is Christian, not when it condemns itself to the letter of things, but when, in the infinitely little, it discovers as many mysteries and as much depth and power as in the infinitely great.
Ultramontanism, or the Roman Church and Modern Society
The Roman Church and Science-Galileo
Lecture, May 7, 1844

Rabi, Isidor Isaac 1898–1988
Austrian-born American physicist

Science is a great game. It is inspiring and refreshing. The playing field is the universe itself.
New York Times, October 28, 1964 (p. 38)

Ramsay, Sir William 1852–1916
English chemist

No process is so perfect that there is not plenty of room for improvement. There is no finality in science. And that which today is a scientific toy may be to-morrow the essential part of an important industry.
Essays Biological and Chemical
The Great London Chemists
Section I (p. 19)
Archibald Constable & Company Ltd. London, England. 1908

Randall, J. H.
No biographical data available

[Science] swept man out of his proud position as the central figure and end of the universe, and made him a tiny speck on a third-rate planet revolving about a tenth-rate sun drifting in an endless cosmic ocean.
The Making of the Modern Mind: A Survey of the Intellectual Background of the Present Age
Chapter X (p. 226)
The Riverside Press. Cambridge, Massachusetts, USA. 1940

Randi, James 1928–
Canadian magician and scientific skeptic

I believe that science is best defined as a careful, disciplined, logical search for knowledge about any and all aspects of the universe, obtained by examination of the best available evidence and always subject to correction and improvement upon the discovery of better evidence.

What's left is magic, and it doesn't work.
The Mask of Nostradamus
Chapter Five (p. 66)
Prometheus Books. Buffalo, New York, USA. 1993

Ravetz, J. R.
No biographical data available

The obsolescence of the conception of science as the pursuit of truth results from several changes in the social activity of science. First, the heavy warfare with "theology and metaphysics" is over. Although a few sharp skirmishes still occur, the attacks on the freedom of science from this quarter are no longer significant. This is not so much because of the undoubted victory of science over its ancient contenders as for the deeper reason that the conclusions of natural science are no longer ideologically sensitive. What people, either the masses or the educated, believe about the inanimate universe or the biological

aspects of humanity is not relevant to the stability of society as it was once thought to be.
Scientific Knowledge and Its Social Problems
Chapter I (pp. 200–201)
Clarendon Press. Oxford, England. 1971

Raymo, Chet   1936–
American physicist and science writer

Science cannot be a repository of ultimate faith: It is a fulcrum upon which we can hope to balance the treasure of our knowledge against the claims of ignorance.
The Virgin and the Mousetrap: Essays in Search of the Soul of Science
Chapter 21 (p. 199)
The Viking Press. New York, New York, USA. 1991

…science is a spider's web. Confidence in any one strand of the web is maintained by the tension and resiliency of the entire web.
The Virgin and the Mousetrap: Essays in Search of the Soul of Science
Chapter 16 (p. 144)
The Viking Press. New York, New York, USA. 1991

Science, like the play of children, satisfies a deep-seated need for escape from the boredom of fixity and the trauma of chaos.
Focal Point
Sky and Telescope, Volume 81, Number 5, May 1991 (p. 460)

Renan, Ernest   1823–92
French philosopher and Orientalist

The lofty serenity of science becomes possible only on the condition of impartial criticism, which without regard for the beliefs of a certain portion of humanity, handles its imperturbable instrument with the inflexibility of the geometrician, without anger and without pity. The critic never insults.
The Future of Science
Chapter XV (p. 257)
Roberts Brothers. Boston, Massachusetts, USA. 1893

…science is a religion, science alone will henceforth make the creeds, science alone can solve for men the eternal problems, the solutions of which his nature imperatively demands.
The Future of Science
Chapter V (p. 97)
Roberts Brothers. Boston, Massachusetts, USA. 1893

…science must pursue its road without minding with whom it comes in collision. Let the others get out of the way. If it appears to raise objections against received dogmas, it is not for science but the received dogmas to be on the defensive and to reply to the objections. Science should behave as if the world were free from preconceived opinions, and not heed the difficulties it starts.
The Future of Science
Chapter V (p. 83)
Roberts Brothers. Boston, Massachusetts, USA. 1893

A little true science is better than a great deal of bad science. One is less liable to error by confessing one's ignorance than by fancying that one knows a great many things one does not.
The Future of Science
Preface (p. xix)
Roberts Brothers. Boston, Massachusetts, USA. 1893

Reynolds, William C.   1933–2004
American mechanical engineer
Perkins, Harry C.
No biographical data available

Concepts form the basis for any science. These are ideas, usually somewhat vague (especially when first encountered), which often defy really adequate definition. The meaning of a new concept can seldom be grasped from reading a one-paragraph discussion. There must be time to become accustomed to the concept, to investigate it with prior knowledge, and to associate it with personal experience. Inability to work with details of a new subject can often be traced to inadequate understanding of its basic concepts.
Engineering Thermodynamics
Chapter 1 (p. 4)
McGraw-Hill Book Company, Inc. New York, New York, USA. 1977

Richards, Ivor Armstrong   1893–1979
English literary critic

For science, which is simply our most elaborate way of pointing to things systematically, tells us and can tell us nothing about the nature of things in any ultimate sense. It can never answer any question of the form: What is so and so? [I]t can only tell us how so and so behaves. And it does not attempt to do more than this.
Science and Poetry
Chapter V (pp. 52–53)
Kegan Paul, Trench, Trubner & Company Ltd. London, England. 1926

Richards, Theodore William   1868–1928
American chemist

Every student of Science, even if he cannot start his journey where his predecessors left off, can at least travel their beaten track more quickly than they could while they were clearing the way: and so before his race is run he comes to virgin forest and becomes himself a pioneer.
Nobel Lectures, Chemistry 1901–1921
Nobel lecture for award received in 1914
Atomic Weights (p. 280)
Elsevier Publishing Company. Amsterdam, Netherlands. 1966

Richardson, Samuel   1689–1761
English novelist

Vast is the field of Science. The more a man knows, the more he will find he has to know.
Sir Charles Grandison (Volume 1)

Letter 11
Oxford University Press, Inc. Oxford, England. 1972

Richet, Charles   1850–1935
French physiologist

To neglect science is to exclude fair hope, to condemn ourselves to live in a uniform monotonous existence.
The Natural History of a Savant
Chapter XIII (p. 147)
J.M. Dent & Sons Ltd. London, England. 1927

The future and the happiness of humanity depend on science.
The Natural History of a Savant
Chapter XIII (p. 155)
J.M. Dent & Sons Ltd. London, England. 1927

No one has the right to encumber science with premature assertions.
The Natural History of a Savant
Chapter X (p. 122)
J.M. Dent & Sons Ltd. London, England. 1927

All…believe in the sovereignty of science; which like the grammar of Martine, rules even over kings, and imperiously subjects them to its laws.
The Natural History of a Savant
Chapter II (p. 13)
J.M. Dent & Sons Ltd. London, England. 1927

Ridley, Matt   1958–
English science writer

The fuel on which science runs is ignorance. Science is like a hungry furnace that must be fed logs from the forests of ignorance that surround us. In the process, the clearing we call knowledge expands, but the more it expands, the longer its perimeter and the more ignorance comes into view…. A true scientist is bored by knowledge; it is the assault on ignorance that motivates him — the mysteries that previous discoveries have revealed. The forest is more interesting than the clearing.
Genome: The Autobiography of a Species in 23 Chapters
Chapter 20 (p. 271)
HarperCollins Publishers. New York, New York, USA. 2000

Robinson, Sir Robert   1886–1975
English chemist

Science cannot be based on dogma or authority of any kind, nor on any institution or revelation, unless indeed it be of the Book of Nature that lies open before our eyes. We need not dwell on the processes of acquiring knowledge by observation, experiment, and inductive and deductive reasoning. The study of scientific method both in theory and practice is of great importance. It is inherent in the philosophy that the record may be imperfect and the conceptions erroneous; the potential fallibility of our science is not only acknowledged but also insisted upon.

Science and the Scientist
Nature, Volume 176, Number 4479, September 3, 1955 (p. 434)

Romanoff, Alexis Lawrence   1892–1980
Russian soldier and scientist

Pure science has no part in politics.
Encyclopedia of Thoughts
Aphorisms 1281
Ithaca Heritage Books. Ithaca, New York, USA. 1975

Religions offer unbounded faith; science, logical preciseness; and the arts, creative imagination.
Encyclopedia of Thoughts
Aphorisms 1471
Ithaca Heritage Books. Ithaca, New York, USA. 1975

Science and technology may lead to self-destruction; humanities to sensible social recovery.
Encyclopedia of Thoughts
Aphorisms 2937
Ithaca Heritage Books. Ithaca, New York, USA. 1975

Science is advanced by husbands, but wives are often behind them.
Encyclopedia of Thoughts
Aphorisms 91
Ithaca Heritage Books. Ithaca, New York, USA. 1975

Dedication to pure science or fine arts often is incompatible with a desire for economic gain.
Encyclopedia of Thoughts
Aphorism 1457
Ithaca Heritage Books. Ithaca, New York, USA. 1975

Into the life of a cultured man enter science, art, and poetic philosophy.
Encyclopedia of Thoughts
Aphorisms 1220
Ithaca Heritage Books. Ithaca, New York, USA. 1975

Ross, Sir Ronald   1857–1932
English bacteriologist

We must not accept any speculations merely because they now appear pleasant, flattering, or ennobling to us. We must be content to creep upwards step by step, planting each foot on the firmest finding of the moment, using the compass and such other instruments as we many have, observing without either despair or contempt the clouds and precipices above and beneath us. Especially our duty at present is to better our present foothold; to investigate; to comprehend the forces of nature, to set our State rationally in order; to stamp down disease in body, mind, and government; to lighten the monstrous misery of our fellows, not by windy dogmas, but by calm science.
In Sir Richard Arman Gregory
Discovery; or, The Spirit and Service of Science
Chapter VIII (p. 233)
Macmillan & Company Ltd. London, England. 1918

Roszak, Theodore   1933–
American social critic

Science *uses* the senses but does not *enjoy* them; finally buries them under theory, abstraction, mathematical generalization
Where the Wasteland Ends
Chapter 9 (p. 280)
Doubleday & Company. Garden City, New York, USA. 1973

Rothman, Tony   1953–
American cosmologist

Principle of Literary Oversight: Textbooks may be straightforward and succinct, but the path of science is crooked and tortuous.
Instant Physics: From Aristotle to Einstein, and Beyond
Introduction (p. xi)
Ballentine Books. New York, New York, USA. 1995

Roux, Joseph   1725–93
French hydrographer

Science is for those who learn; poetry, for those who know.
Meditations of a Parish Priest
Part I, Number 71 (p. 43)
Thomas Y. Crowell & Company New York, New York, USA. 1886

Rubin, Harry
No biographical data available

...one of the great pitfalls of science is the fallacy of misplaced concreteness. Scientists seem to prefer questionable explanations to no explanation at all.
Does Somatic Mutation Cause Most Cancers?
Journal of the National Cancer Institute, Volume 64, Number 5. May 1980 (p. 999)

Ruse, Michael   1940–
English historian and philosopher of science

Science, like most human cultural phenomena, has evolved. What was allowable in the early nineteenth century is not necessarily allowable in the late twentieth century. Specifically, science today does not break with law. And this is what counts for us. We want criteria of science for today, not for yesterday.
Response to the Commentary: Pro Judice
Science, Technology and Human Values, Volume 7, Number 41, Fall 1982 (p. 21)

Ruskin, John   1819–1900
English writer, art critic, and social reformer

Science does its duty, not in telling us the causes of spots in the sun, but in explaining to us the laws of our own life, and the consequences of their violation.
In Henry Attwell
Thoughts from Ruskin
33 (p. 29)
Longmans, Green & Company, New York, New York, USA; 1901

Russell, Bertrand Arthur William   1872–1970
English philosopher, logician, and social reformer

Science, ever since the time of the Arabs, has had two functions: (1) to enable us to know things, and (2) to enable us to do things.
The Impact of Science on Society
Chapter II (p. 18)
Simon & Schuster. New York, New York, USA. 1938

Science has always prided itself on being empirical and believing only what could be verified.
In Robert E. Egener and Lester E. Denonn (eds.)
The Basic Writings of Bertrand Russell
Limitations on the Scientific Method (p. 623–24)
Simon & Schuster. New York, New York, USA. 1961

Even if the open windows of science at first make us shiver after the cozy indoor warmth of traditional humanizing myths, in the end the fresh air brings vigor, and the great spaces have a splendor of their own.
What I Believe
Chapter I (p. 14)
E.P. Dutton & Company. New York, New York, USA. 1925

In science men have discovered an activity of the very highest value in which they are no longer, as in art, dependent for progress upon the appearance of continually greater genius, for in science the successors stands upon the shoulders of their predecessors; where one man of supreme genius has invented a method, a thousand lesser men can apply it.
A Free Man's Worship and Other Essays
Chapter 3 (first published as "The Free Man's Worship" in December 1903)
George Allen & Unwin Ltd. London, England. 1917

A life devoted to science is therefore a happy life, and its happiness is derived from the very best sources that are open to dwellers on this troubled and passionate planet.
Mysticism and Logic and Other Essays
Chapter II, Section II (p. 45)
Longmans, Green & Company London, England. 1925

Sagan, Carl   1934–96
American astronomer and author

There is no other species on Earth that does science. It is, so far, entirely a human invention, evolved by natural selection in the cerebral cortex for one simple reason: it works. It is not perfect. It can be misused. It is only a tool. But it is by far the best tool we have, self-correcting, ongoing, applicable to everything.
Cosmos
Chapter XIII (p. 333)
Random House, Inc. New York, New York, USA. 1980

One of the great commandments of science is, "Mistrust arguments from authority."
The Demon-Haunted World: Science as a Candle in the Dark
Chapter 2 (p. 28)
Random House, Inc. New York, New York, USA. 1995

Science is an attempt, largely successful, to understand the world, to get a grip on things, to get hold of ourselves,

to steer a safe course. Microbiology and meteorology now explain what only a few centuries ago was considered sufficient cause to burn women to death.
The Demon-Haunted World: Science as a Candle in the Dark
Chapter 2 (p. 26)
Random House, Inc. New York, New York, USA. 1995

Science involves a seemingly self-contradictory mix of attitudes: On the one hand, it requires an almost complete openness to all ideas, no matter how bizarre and weird they sound, a propensity to wonder.... But at the same time, science requires the most vigorous and uncompromising skepticism, because the vast majority of ideas are simply wrong, and the only way you can distinguish the right from the wrong, the wheat from the chaff, is by critical experiment and analysis.
Wonder and Skepticism
Skeptical Inquirer, Jan/Feb 1995 (p. 24)

[Science is not popular] is...the fault of the educational system. We do not teach how to think. This is a very serious failure that may even, in a world rigged with 60,000 nuclear weapons, compromise the human future.
The Burden of Skepticism
Skeptical Inquirer, Fall 1987

Science demands a tolerance for ambiguity. Where we are ignorant, we withhold belief. Whatever annoyance the uncertainty engenders serves a higher purpose: It drives us to accumulate better data. This attitude is the difference between science and so much else. Science offers little in the way of cheap thrills. The standards of evidence are strict. But when followed they allow us to see far, illuminating even a great darkness.
Pale Blue Dot: A Vision of the Human Future in Space
Chapter 20 (p. 365)
Random House, Inc. New York, New York, USA. 1994

Reasoned disputation is the lifeblood of science — as is, sadly, infrequently the case in the intellectually more anemic arena of politics.
The Cosmic Connection: An Extraterrestrial Perspective
Preface (p. ix)
Dell Publishing, Inc. New York, New York, USA. 1975

Science — pure science, science not for any practical application but for its own sake — is a deeply emotional matter for those who practice it, as well as for those non-scientists who every now and then dip in to see what's been discovered lately.
The Demon-Haunted World: Science as a Candle in the Dark
Chapter 19 (p. 330)
Random House, Inc. New York, New York, USA. 1995

Cutting off fundamental, curiosity-driven science is like eating the seed corn. We may have a little more to eat next winter, but what will we plant so we and our children will have enough to get through the winters to come?
The Demon-Haunted World: Science as a Candle in the Dark
Chapter 23 (p. 400)

Random House, Inc. New York, New York, USA. 1995

It is a supreme challenge for the popularizer of science to make clear the actual, tortuous history of its great discoveries and the misapprehensions and occasional stubborn refusal by its practitioners to change course.
The Demon-Haunted World: Science as a Candle in the Dark
Chapter 1 (p. 22)
Random House, Inc. New York, New York, USA. 1995

Science is far from a perfect instrument of knowledge. It's just the best we have.
The Demon-Haunted World: Science as a Candle in the Dark
Chapter 2 (p. 27)
Random House, Inc. New York, New York, USA. 1995

Sandage, Allan   1926–
American astronomer

Science is the only self-correcting human institution, but it also is a process that progresses only by showing itself to be wrong.
In Alan Lightman and Roberta Brawer
Origins: The Lives and Worlds of Modern Cosmologists
Allan Sandage (p. 82)
Harvard University Press. Cambridge, Massachusetts, USA. 1990

Santayana, George (Jorge Augustín Nicolás Ruiz de Santillana)   1863–1952
Spanish-born American philosopher

Science is a half-way house between private sensation and universal vision.
The Life of Reason; or The Phases of Human Progress
Part V, Chapter I (p. 385)
Charles Scribner's Sons. New York, New York, USA. 1953

Sarnoff, David   1891–1971
Russian-born American broadcasting executive

At their best, at their most creative, science and engineering are attributes of liberty — noble expressions of man's God-given right to investigate and explore the universe without fear of social or political or religious reprisals.
In Emily Davie (ed.)
Profile of America: An Autobiography of the U.S.A
Electronics — Today and Tomorrow
Crowell. New York, New York, USA. 1954

Scatchard, George   1892–1973
American physical chemist

Since science progresses by building block upon block, it is important to examine the structure from time to time to make sure that there are no badly fitted blocks, none which are being made to carry more than their proper capacity and none which might be made more useful.
Equilibrium Thermodynamics and Biological Chemistry
Science, Volume 95, Number 2454, January 9, 1942 (p. 27)

Schiebinger, Londa   1952–
Science historian

Only recently have we begun to appreciate that who does science affects the kind of science that gets done. How, then, has our knowledge of nature been influenced by struggles determining who is included in science and who is excluded, which projects are pursued and which ignored, whose experiences are validated and whose are not, and who stands to gain in terms of wealth or well-being and who does not?

Nature's Body: Gender in the Making of Modern Science
Introduction (p. 3)
Beacon Press. Boston, Massachusetts, USA. 1993

Schiller, Ferdinand Canning Scott   1864–1937
English philosopher

To Archimedes once came a youth, who for knowledge was thirsting,
Saying, "Initiate me into the science divine,
Which for my country has borne forth fruit of such wonderful value,
And which the walls of the town 'gainst the Sambuca protects.
"Calls't thou the science divine? It is so," the wise man responded;
"But it was so, my son, ere it avail'd for the town.
Wouldst thou have fruit from her only, e'en mortals wit that can provide thee;
Wouldst thou the goddess obtain, seek not the woman in Her!"

In Edgar Alfred Bower
The Poems of Schiller
Archimedes and the Student (p. 262)
John W. Parker & Son. London, England. 1851

Science: To one, she is the exalted and heavenly Goddess; to another she is a capable cow which keeps him supplied with butter.

In Folke Dovring
Knowledge and Ignorance: Essays on Lights and Shadows
Chapter Ten (p. 141)
Praeger. Westport, Connecticut, USA. 1998

Schneer, Cecil J.   1923–
American science historian and mineralogist

The primary importance of science and the characteristic that distinguished it from other philosophies and arts is its usefulness. The remarkable thing about science is the extent to which nature and the world appear to adhere to the rules and constructions of science.

The Evolution of Physical Science
Grove Press, Inc. New York, New York, USA. 1960

Schrödinger, Erwin   1887–1961
Austrian theoretical physicist

…there is a tendency to forget that all science is bound up with human culture in general, and that scientific findings, even those which at the moment appear the most advanced and esoteric and difficult to grasp, are meaningless outside their cultural context.

Are There Quantum Jumps?
The British Journal for the Philosophy of Science, Volume 3, 1952 (p. 109)

A theoretical science unaware that those of its constructs considered relevant and momentous are destined eventually to be framed in concepts and words that have a grip on the educated community and become part and parcel of the general world picture — a theoretical science, I say, where this is forgotten, and where the initiated continue musing to each other in terms that are, at best, understood by a small group of close fellow travelers, will necessarily be cut off from the rest of cultural mankind; in the long run it is bound to atrophy and ossify however virulently esoteric chat may continue within its joyfully isolated groups of experts.

Are There Quantum Jumps?
The British Journal for the Philosophy of Science, Volume 3, 1952 (p. 110)

…the scientific picture of the real world around me is very deficient. It gives a lot of factual information, puts all our experience in a magnificently consistent order, but it is ghastly silent about all and sundry that is really near to our heart, that really matters to us. It cannot tell us a word about red and blue, bitter and sweet, physical pain and physical delight; it knows nothing of beautiful and ugly, good or bad, God and eternity. Science sometimes pretends to answer questions in these domains, but the answers are very often so silly that we are not inclined to take them seriously.

Nature and the Greeks
Chapter VII (p. 93)
At The University Press. Cambridge, England. 1954

Schwartz, John   1941–
American theoretical physicist

Science is a long movie, and the news media generally take mere snapshots.

If You Seek the Truth, Don't Trash the Science
Washington Post, 21 February, 1999 (p. B–1)

Seifriz, William
No biographical data available

Let me give full credit to the young and enthusiastic research workers, full of high-energy phosphate bonds. What I deplore is their attitude of mind. Science has become tough and students learn to accept it that way.

A New University
Science, Volume 120, Number 3107, 16 July 1954 (p. 89)

Shapiro, Harry L.   1902–90
American physical anthropologist

Science, like organic life, has ramified by expanding into unoccupied areas and then adapting itself to the special requirements encountered there.

Symposium on the History of Anthropology, the History and Development of Physical Anthropology
American Anthropologist, Volume 61, Number 3, 1959 (p. 371)

…as the diversified forms of animals, plants, and insects make evident by their morphology and their function the characteristics of ecological niches whose very existence might otherwise escape notice, so the diversity of techniques and concepts of scientific specialties by their very formulation reveal aspects of nature we would not have suspected. Anthropology, like other branches of science, has also embodied in its structure whole new worlds rich in insights into the development and nature of man.
Symposium on the History of Anthropology, the History and Development of Physical Anthropology
American Anthropologist, Volume 61, Number 3, 1959 (p. 371)

Shapiro, Robert   1935–
American DNA researcher and author

Science is not a given set of answers but a system for obtaining answers. The method by which the search is conducted is more important than the nature of the solution. Questions need not be answered at all, or answers may be provided and then changed. It does not matter how often or how profoundly our view of the universe alters, as long as these changes take place in a way appropriate to science. For the practice of science, like the game of baseball, is covered by definite rules.
Origins: A Skeptic's Guide to the Creation of Life on Earth
Chapter One (p. 33)
Summit Books. New York, New York, USA. 1986

Science is not the place for those who want certainty, who wish the truths they learned in childhood to reassure them in their old age. Surprises occur, and alter our perception of reality — for example, the discovery of radioactivity or the genetic role of DNA…. When we treat each new observation and theory with skepticism, retaining our doubt until it has passed the test of experience, and then place it alongside our other acquisitions with the care of a collector who has acquired a valued object after a long search, then we can experience the joy of science. It is this joy, rather than an insistence on an immediate answer, that is likely to be our reward as we continue to search for the origin of life. But even in this conclusion, let us exercise some caution. We may be closer to the answer than we think.
Origins: A Skeptic's Guide to the Creation of Life on Earth
Chapter Thirteen (p. 312)
Summit Books. New York, New York, USA. 1986

Shaw, George Bernard   1856–1950
Irish comic dramatist and literary critic

Science becomes dangerous only when it imagines that it has reached its goal.
The Doctor's Dilemma
Preface on Doctors
The Latest Theories (p. xc)
Brentano's. New York, New York, USA. 1920

Science is always simple and always profound. It is only the half-truths that are dangerous.
The Doctor's Dilemma
Act I (p. 24)
Brentano's. New York, New York, USA. 1920

Shelley, Mary   1797–1851
English Romantic writer

You seek for knowledge and wisdom, as I once did; and I ardently hope that the gratification of your wishes may not be a serpent to sting you, as mine has been.
Frankenstein
Letter 4 (p. 27)
Running Press. Philadelphia, Pennsylvania, USA. 1990

The ambition of the enquirer seemed to limit itself to the annihilation of those visions on which my interest in science was chiefly founded. I was required to exchange chimeras of boundless grandeur for realities of little worth.
Frankenstein
Chapter 3 (p. 38)
Running Press. Philadelphia, Pennsylvania, USA. 1990

Life and death appeared to me ideal bounds, which I should first break through, and pour a torrent of light into our dark world.
Frankenstein
Chapter 4 (p. 43)
Running Press. Philadelphia, Pennsylvania, USA. 1990

"Man," I cried, "how ignorant art thou in thy pride of wisdom!"
Frankenstein
Chapter 23 (p. 141)
Running Press. Philadelphia, Pennsylvania, USA. 1990

Shelley, Percy Bysshe   1792–1822
English poet

The cultivation of those sciences which have enlarged the limits of the empire of man over the external world, has, for want of the poetical faculty, proportionally circumscribed those of the internal world; and man, having enslaved the elements, remains himself a slave.
In Fanny Delisle
A Study of Shelley's "A Defence of Poetry" (Volume 1)
Line 1223 (p. 138)
Institute fur Englische Sprache und Literatur. Salzburg, Austria. 1974

Shermer, Michael   1954–
American science writer

What separates science from all other human activities (and morality has never been successfully placed on a scientific basis) is its commitment to the tentative nature of all its conclusions. There are no final answers in science, only varying degrees of probability. Even scientific "facts" are just conclusions confirmed to such an extent that it would be reasonable to offer temporary agreement, but that assent is never final.
Why People Believe Weird Things: Pseudoscience, Superstition, and Other Confusions of Our Time

Part 2, Chapter 8 (p. 124)
Henry H. Holt. New York, New York, USA. 2002

Science is not the affirmation of a set of beliefs but a process of inquiry aimed at building a testable body of knowledge constantly open to rejection or confirmation. In science, knowledge is fluid and certainty fleeting. That is at the heart of its limitations. It is also its greatest strength.

Why People Believe Weird Things: Pseudoscience, Superstition, and Other Confusions of Our Time
Part 2, Chapter 8 (p. 124)
Henry H. Holt. New York, New York, USA. 2002

Shu, Frank H.
American theoretical astrophysicist

Science has a beauty and uplifting spirit which rivals any of the other cultural attainments of humanity. This aesthetic response arose in a recent congressional hearing. When asked how particle physics contributes to the defense of our country, Robert Wilson replied that it makes the country worth defending.

The Physical Universe: An Introduction to Astronomy (p. 101)
University Science Books. Mill Valley, California, USA. 1982

Siegel, Eli   1902–78
American philosopher, poet, critic, and founder of Aesthetic Realism

Science comes from the knowing that you want to know.
Damned Welcome
Aesthetic Realism, Maxims, Part One, #298 (p. 68)
Definition Press. New York, New York, USA. 1972

Silver, Brian L.
Israeli professor of physical chemistry

One point is incontestable: the "truth" of science must always remain open to critical scrutiny and will sometimes have the status of a beauty queen: looks good today, but next year she'll be dethroned. That is because the real test of a scientific theory is not whether it is "true." The real test is whether it works.

The Ascent of Science
Part I, Chapter 2 (p. 24)
Solomon Press Book. New York, New York, USA. 1998

Science is not a harmless intellectual pastime. In the last two centuries we have moved from being simply observers of nature to being, in a modest but growing way, its controller. Concomitantly, we have occasionally disturbed the balance of nature in ways that we did not always understand. Science has to be watched. The layman can no longer afford to stand to one side, ignorant of the meaning of advances that will determine the kind of world that his children will inhabit–and the kind of children that he will have. Science has become part of the human race's way of conceiving of and manipulating its future. The manipulation of the future is not a question to be left to philosophers. The answers can affect the

national budget, the health of your next child, and the long-term prospects for life on this planet.
The Ascent of Science
Preface (p. xiv)
Solomon Press Book. New York, New York, USA. 1998

Science, man's greatest intellectual adventure, has rocked his faith and engendered dreams of a material Utopia. At its most abstract, science shades into philosophy; at its most practical, it cures disease. It has eased our lives and threatened our existence. It aspires, but in some very basic ways fails, to understand the ant and the Creation, the infinitesimal atom and the mind-bludgeoning immensity of the cosmos. It has laid its hand on the shoulders of poets and politicians, philosophers and charlatans. Its beauty is often apparent only to the initiated, its perils are generally misunderstood, its importance has been both over- and underestimated, and its fallibility, and that of those who create it, is often glossed over or malevolently exaggerated.
The Ascent of Science
Preface (p. xiii)
Solomon Press Book. New York, New York, USA. 1998

…the history of science may be a trail littered with broken theories and discarded concepts, science is also a triumph of reason, luck, and above all imagination. There are few more successful, exciting, or strange journeys.
The Ascent of Science
Preface (p. xiv)
Solomon Press Book. New York, New York, USA. 1998

Simon, Herbert Alexander   1916–2001
American social scientist

The central task of a natural science is to make the wonderful commonplace: to show that complexity, correctly viewed, is only a mask for simplicity; to find pattern hidden in apparent chaos.
The Sciences of the Artificial
Chapter I (p. 1)
The MIT Press. Cambridge, Massachusetts, USA. 1969

Simpson, George Gaylord   1902–84
American paleontologist

The important distinction between science and those other systematizations [*i.e.*, art, philosophy, and theology] is that science is self-testing and self-correcting. Here the essential point of science is respect for objective fact. What is correctly observed must be believed…the competent scientist does quite the opposite of the popular stereotype of setting out to prove a theory; he seeks to disprove it.
Notes on the Nature of Science
Notes on the Nature of Science by a Biologist (p. 9)
Harcourt, Brace & World, Inc. New York, New York, USA. 1962

As for the scope of science, it includes everything known to exist or to happen in the material universe. Since the

arts, philosophy, and theology do exist in the material universe, they too are within the scope of science and can properly be studied as psychological, anthropological, and biological phenomena.
Notes on the Nature of Science
Notes on the Nature of Science by a Biologist (pp. 11–12)
Harcourt, Brace & World, Inc. New York, New York, USA. 1962

Singer, Charles   1876–1960
Historian of science and medicine

To succeed in science it is necessary to receive the tradition of those who have gone before us. In science, more perhaps than in any other study, the dead and the living are one.
In Lloyd William Taylor
Physics: the Pioneer Science (Volume 1)
Chapter 15 (p. 182)
Houghton Mifflin Company. Boston, Massachusetts, USA. 1941

Slosson, Edwin E.   1865–1929
American chemist and journalist

Science consists in learning from nature how to surpass nature.
Spun Logs
The Scientific Monthly, December 1925

Most people think of science as a serious and solemn thing, a strain upon the strongest intellect.
So it is for the pioneers of scientific progress, but not for those who merely follow along behind.
Chats on Science
Introduction (p. 1)
G. Bell & Sons Ltd. London, England. 1924

Smith, Adam   1723–1790
Scottish moral philosopher and founder of modern economic theory

Science is the great antidote to the poison of enthusiasm and superstition.
An Inquiry into the Nature and Causes of the Wealth of Nations
Book V, Chapter I, part III, Section III (p. 748)
The Modern Library. New York, New York, USA. 1937

Smith, Henry Preserved   1847–1927
American Biblical scholar

More and more, science has become not only increasingly necessary as a foundation for professional skill, but has come to be regarded as the most valuable instrument of culture.
In Lloyd William Taylor
Physics: The Pioneer Science (Volume 1)
Chapter 28 (p. 395)
Houghton Mifflin Company. Boston, Massachusetts, USA. 1941

…whether as the new salvation or the new superstition, science has modeled the whole life of the modern world.… All modern production of wealth, all contemporary life, depend on the knowledge of nature acquired by science. But more than that, religion, politics,

philosophy, art and literature have capitulated to science, or at least receded before her. There is no department of human activity today untouched with the spirit of experiment and of mathematics.
A History of Modern Culture (Volume 1)
Epilogue (p. 606)
Peter Smith. Gloucester, Massachusetts, USA. 1957

Smith, Sydney   1771–1845
English clergyman, writer, and wit

Science is his forte, and omniscience his foible.
In Isaac Todhunter
William Whewell (Volume 1)
Conclusion (p. 410)
Macmillan & Company Ltd. London, England. 1876

Smolin, Lee   1955–
American theoretical physicist

Science is, above everything else, a search for an understanding of our relationship with the rest of the universe.
The Life of the Cosmos
Part One, Chapter One (p. 23)
Oxford University Press, Inc. New York, New York, USA. 1997

Smyth, Nathan A.
No biographical data available

To recognize with science that beyond our horizon lies impenetrable mystery will serve but to increase our reverence for the glory of the whole.
Through Science to God
Chapter I (p. 5)
The Macmillan Company. New York, New York, USA. 1936

Snow, Charles Percy   1905–80
English novelist and scientist

But after the idyllic years of science, we passed into a tempest of history; and by an unfortunate coincidence, we passed into a technological tempest, too.
In Paul C. Obler and Herman A. Estrin (eds.)
The New Scientist: Essays on the Methods and Value of Modern Science
The Moral Un-Neutrality of Science (p. 135)

Soddy, Frederick   1877–1956
English chemist

As science advances and most of the more accessible fields of knowledge have been gleaned of their harvest, the need for more and more powerful and elaborate appliances and more and more costly materials ever grows.
Science and Life
Science and the State (p. 60)
E.P. Dutton & Company, Inc. New York, New York, USA. 1920

Somerville, Mary   1780–1872
English mathematician

Science, regarded as the pursuit of truth, which can only be attained by patient and unprejudiced investigation, wherein nothing is too great to be attempted, nothing so minute as to be justly disregarded, must ever afford occupation of consummate interest and subject of elevated meditation.

On the Connexion of the Physical Sciences
Section I (p. 2)
John Murray. London, England. 1834

Sorokin, Pitirim A.   1889–1968
Russian-born American sociologist

Any science, at any moment of its historical existence, contains not only truth but also much that is half-truth, sham-truth, and plain error.

Fads and Foibles in Modern Sociology
Preface (p. v)
Henry Regnery. Chicago, Illinois, USA. 1956

Spark, Muriel   1918–2006
Scottish novelist

Art and religion first; then philosophy; lastly science. That is the order of the great subjects of life, that's their order of importance.

The Prime of Miss Jean Brodie
Chapter 2 (p. 39)
Macmillan & Company Ltd. London, England. 1961

Spencer, Herbert   1820–1903
English social philosopher

What knowledge is of most worth? — the uniform reply is — Science. This is the verdict on all counts.

Education: Intellectual, Moral, and Physical
Chapter I (p. 84)
A.L. Fowle. New York, New York, USA. 1860

For direct self-preservation, or the maintenance of life and health, the all-important knowledge is — Science. For the indirect self-preservation which we call gaining a livelihood, the knowledge of greatest value is — Science. For that interpretation of national life, past and present, without which the citizen cannot rightly regulate his conduct, the indispensable key is — Science. Alike for the most perfect production and highest enjoyment of art in all its forms, the needful preparation is still — Science. And for the purposes of discipline — intellectual, moral, religious — the most efficient study is, once more — Science.

Education: Intellectual, Moral, and Physical
Chapter I (pp. 84–85)
A.L. Fowle. New York, New York, USA. 1860

Only when Genius is married to Science, can the highest results be produced.

Education: Intellectual, Moral, and Physical
Chapter I (p. 70)
A.L. Fowle. New York, New York, USA. 1860

Devotion to science is a tacit worship — a tacit recognition of worth in the things one studies; and by implication

in their cause. It is not a mere lip-homage, but a homage expressed in actions — not a mere professed respect, but a respect proved by the sacrifice of time, thought, and labour.

In Sir Richard Arman Gregory
Discovery; or, The Spirit and Service of Science
Chapter III (p. 41)
Macmillan & Company Ltd. London, England. 1918

Spencer-Brown, George   1923–
English mathematician and polymath

Left to itself, the world of science slowly diminishes as each result classed as scientific has to be reclassed as anecdotal or historical…Science is a continuous living process; it is made up of activities rather than records; and if the activities cease it dies.

Probability and Scientific Inference
Chapter XV (p. 107)
Longmans, Green & Company. London, England. 1957

Stanislaus, Leszczynski (Stanislaus I)   1677–1766
King of Poland

Science when well digested is nothing but good sense and reason.

Maxims
No. 43
Publisher undetermined

Stansfield, William D.   1930–
American biologist

Most scientific theories, however, are ephemeral. Exceptions will likely be found that invalidate a theory in one or more of its tenets. These can then stimulate a new round of research leading either to a more comprehensive theory or perhaps to a more restrictive (*i.e.*, more precisely defined) theory. Nothing is ever completely finished in science; the search for better theories is endless. The interpretation of a scientific experiment should not be extended beyond the limits of the available data. In the building of theories, however, scientists propose general principles by extrapolation beyond available data. When former theories have been shown to be inadequate, scientists should be prepared to relinquish the old and embrace the new in their never-ending search for better solutions. It is unscientific, therefore, to claim to have "proof of the truth" when all that scientific methodology can provide is evidence in support of a theory.

The Science of Evolution
Introduction (p. 8)
Macmillan Publishing Company. New York, New York, USA. 1977

Stenger, Victor J.   1935–
American physicist

No one ever said science was easy, and nobody, scientist or not, should be expected to fall over and play dead when a challenge to existing knowledge is made. If a new

idea has sufficient merit, it should ultimately overcome any resistance, no matter how strong.... Resistance to new ideas is part of the process of science. A worthy new idea must overcome barriers of doubt and skepticism, and even occasional irrational objections. But if an idea has merit, it will eventually climb over these barriers.
Physics and Psychics: The Search for a World Beyond the Senses
Chapter 3 (p. 65)
Prometheus Books. Buffalo, New York, USA. 1990

Sterne, Laurence   1713–68
English novelist and humorist

Sciences may be learned by rote, but Wisdom not.
The Life and Opinions of Tristram Shandy, Gentleman and A Sentimental Journey Through France and Italy (Volume 1)
Book V, Chapter XXXII (p. 356)
Macmillan & Company Ltd. London, England. 1900

Stevenson, Robert Louis   1850–94
Scottish essayist and poet

Science writes of the world as if with the cold finger of a starfish; it is all true; but what is it when compared to the reality of which it discourses, where hearts beat high in April, and death strikes, and hills totter in the earthquake, and there is a glamour over all the objects of sight, and a thrill in all noises for the ear, and Romance herself has made her dwelling among men? So we come back to the old myth, and hear the goat-footed piper making the music which is itself the charm and terror of things; and when a glen invites our visiting footsteps, fancy that Pan leads us thither with a gracious tremolo; or when our hearts quail at the thunder of the cataract, tell ourselves that he has stamped his hoof in the nigh thicket.
Virginibus Puerisque and Familiar Studies of Men and Books
Pan's Pipe (p. 108)
J.M. Dent & Sons Ltd. London, England. No date

Steward, J. H.
No biographical data available

It is the unhappy lot of science that it must clear the ground of flimsy and fanciful structures built upon false premises and errors of fact before it can build anew.
Annual Report of the Board of Regents of the Smithsonian Institution, 1936
Petroglyphs of the United States (p. 407)
Government Printing Office. Washington, D.C. 1937

Stewart, Ian   1945–
English mathematician and science writer

Science is a collective activity, and the actions of each individual resonate and interact with those of the others in a pattern so gigantic that we can no more comprehend the whole than a blood cell can comprehend how its host feels when it mashes a finger in a car door.
The Problems of Mathematics
Chapter 20 (p. 234)
Oxford University Press, Inc. Oxford, England. 1987

Stoker, Bram   1847–1912
English writer

There are mysteries which men can only guess at, which age by age they may solve only in part.
Dracula
Chapter XV (p. 217)
Ameron House. Mattituck, New York, USA. No date

Strutt, John William (Lord Rayleigh)   1842–1919
English physicist

There are some great men of science whose charm consists in having said the first word on a subject, in having introduced some new idea which has proved fruitful; there are others whose charm consists perhaps in having said the last word on the subject, and who have reduced the subject to logical consistency and clearness.
Life of John William Strutt: Third Baron Rayleigh
Chapter XVII (p. 310)
University of Wisconsin Press. Madison, Wisconsin, USA. 1968

Sullivan, John William Navin   1886–1937
Irish mathematician

Science, like everything else that man has created, exists, of course, to gratify certain human needs and desires. The fact that it has been steadily pursued for so many centuries, that it has attracted an ever-wider extent of attention, and that it is now the dominant intellectual interest of mankind, shows that it appeals to a very powerful and persistent group of appetites.
The Limitations of Science
Introduction (p. 7)
New American Library. New York, New York, USA. 1956

...science deals with but a partial aspect of reality, and there is no faintest reason for supposing that everything science ignores is less real than what it accepts...Why is it that science forms a closed system? Why is it that the elements of reality it ignores never come in to disturb it? The reason is that all the terms of physics are defined in terms of one another. The abstractions with which physics begins are all it ever has to do with.
The Limitations of Science
Chapter 6, Section IV (p. 147)
New American Library. New York, New York, USA. 1956

Swann, William Francis Gray   1884–1962
Anglo-American physicist

The forerunners in the march of science do not often come heralded by much ceremony suggestive of the power that lies behind them. Often in apparent trivialities do they reveal themselves — trivialities so void of spectacular content that but few can be found who deem it worth while to listen to their story.
Annual Report of the Board of Regents of the Smithsonian Institution, 1928
Three Centuries of Natural Philosophy (p. 237)
Government Printing Office. Washington, D.C. 1929

Swenson, Jr., Lloyd S.
No biographical data available

The interplay of thought and action, theory and experiment, individuals and institutions in science is both comic and tragic, despite the actors' common belief that their lines are delivered as if for a triumphal pageant rather than a tragiocomic play.
Genesis of Relativity: Einstein in Context
Preface (p. xiii)
Burt Franklin & Company, Inc. New York, New York, USA. 1979

Tatishchev, Vasilii Nikitich   1686–1750
Russian historian and geographer

Freedom is not an essential and basic condition for the growth of science; the care and diligence of government authorities are the most important conditions for this development.
USA
OMNI Magazine, Volume 3, Number 1, October 1980 (p. 41)

Teall, J. J. Harris   1849–1924
British geologist

The chief glory of science is, not that it produces an amelioration of the conditions under which we live, but that it continually enlarges our view, introduces new ideas, new ways of looking at things, and thus contributes in no small degree to the intellectual development of the human race.
Annual Report of the Board of Regents of the Smithsonian Institution, 1902
The Evolution of Petrological Ideas (p. 288)
Government Printing Office. Washington, D.C. 1903

Teller, Edward   1908–2003
Hungarian-born American nuclear physicist

Science introduces consistency and simplicity into a world that without them appears confused and random.
Better a Shield than a Sword: Perspectives in Defense and Technology
Chapter 29 (p. 215)
The Free Press. New York, New York, USA. 1987

Science, like music or art, is not something that can or should be practiced by everybody. But we want all children to be able to enjoy music, to be able to tell good music from poor music, so we teach them to appreciate music in a discriminating manner. That should be the aim in science education for the nonscientist.
Better a Shield than a Sword: Perspectives in Defense and Technology
Chapter 28 (p. 208)
The Free Press. New York, New York, USA. 1987

Teller, Edward   1908–2003
Hungarian-born American nuclear physicist
Teller, Wendy
No biographical data available

If there ever was a misnomer, it is "exact science." Science has always been full of mistakes. The present day is no exception. And our mistakes are good mistakes; they require a genius to correct them. Of course, we do not see our own mistakes.
Conversations on the Dark Secrets of Physics
Chapter 3 (p. 37)
Plenum Press. New York, New York, USA. 1991

Temple, Frederick   1821–1902
Anglican prelate and archbishop of Canterbury

The regularity of nature is the first postulate of Science; but it requires the very slightest observation to show us that, along with this regularity, there exists a vast irregularity which Science can only deal with by exclusion from its province.
The Relations Between Religion and Science (p. 99)
Macmillan & Company. New York, New York, USA. 1884

Temple, G.
No biographical data available

…any serious examination of the basic concepts of any science is far more difficult than the elaboration of their ultimate consequences.
Turning Points in Physics: A Series of Lectures Given at Oxford University in Trinity Term, 1958 (p. 68)
Interscience Publishers. New York, New York, USA. 1959

Tennyson, Alfred (Lord)   1809–92
English poet

Science moves, but slowly, slowly, creeping on from point to point.
Alfred Tennyson's Poetical Works
Locksey Hall, Stanza 60
Oxford University Press, Inc. London, England. 1953

…nourishing a youth sublime
With the fairy tales of science, and the long result of time.
Alfred Tennyson's Poetical Works
Locksley Hall, Stanza 6
Oxford University Press, Inc. London, England. 1953

Thom, René   1923–2002
French mathematician

If we admit a priori that science is just acquisition of knowledge, that is, building an inventory of all observable phenomena in a given disciplinary domain — then, obviously, any science is empirical.
In J. Casti and A. Karlvist (eds.)
Newton to Aristotle: Toward a Theory of Models for Living Systems
Causality and Finality in Theoretical Biology

Thomas, Lewis   1913–93
American physician and biologist

You either have science or you don't, and if you have it you are obliged to accept the surprising and disturbing pieces of information, even the overwhelming and upheaving ones, along with the neat and promptly useful bits. It is like that.

The Medusa and the Snail: More Notes of a Biology Watcher
The Hazard of Science (p. 73)
The Viking Press. New York, New York, USA. 1979

The central task of science is to arrive, stage by stage, at a clearer comprehension of nature, but this does not mean, as it is sometimes claimed to mean, a search for mastery over nature.
Late Night Thoughts on Listening to Mahler's Ninth Symphony
Humanities and Science (p. 153)
Viking Press. New York, New York, USA. 1983

The essential wildness of science as a manifestation of human behavior is not generally perceived. As we extract new things of value from it, we also keep discovering parts of the activity that seem in need of better control, more efficiency, less unpredictability.
The Lives of a Cell: Notes of a Biology Watcher
Natural Science (p. 100)
The Viking Press. New York, New York, USA. 1974

Science began by fumbling. It works because the people involved in it work, and work together. They become excited and exasperated, they exchange their bits of information at a full shout, and, the most wonderful thing of all, they keep at one another.
Late Night Thoughts on Listening to Mahler's Ninth Symphony
Alchemy
The Viking Press. New York, New York, USA. 1983

Thompson, A. R.
No biographical data available

...science has not only helped to destroy popular traditions that might have nourished a modern spirit of admiration, but has fostered a wintry skepticism, making man appear not an imperfect angel, but a super-educated monkey.
In R. Foerster (ed.)
Humanism and America: Essays on the Outlook of Modern Civilization
The Dilemma of Modern Tragedy (p. 129)
Farrar & Rinehart Inc. New York, New York, USA. 1930

Thomson, J. Arthur   1861–1933
Scottish biologist

When science makes minor mysteries disappear, greater mysteries stand confessed. For one object of delight whose emotional value science has inevitably lessened — as Newton damaged the rainbow for Keats — science gives back double.
The Outline of Science (Volume 4)
Chapter XXXVIII (p. 1176)
G.P. Putnam's Sons. New York, New York, USA. 1937

To the grand primary impression of the world power, the immensities, the pervading order, and the universal flux, with which the man of feeling has been nurtured from the old, modern science has added thrilling impressions of manifoldedness, intricacy, uniformity, inter-relatedness, and evolution. Science widens and clears the emotional

window. There are great vistas to which science alone can lead, and they make for elevation of mind.
The Outline of Science (Volume 4)
Chapter XXXVIII (pp. 1176–1177)
G.P. Putnam's Sons. New York, New York, USA. 1937

The opposition between science and feeling is largely a misunderstanding. As one of our philosophers has remarked, Science is in a true sense "one of the humanities."
The Outline of Science (Volume 4)
Chapter XXXVIII (p. 1177)
G.P. Putnam's Sons. New York, New York, USA. 1937

Science expresses a quite specific endeavor to get phenomena under intellectual control, so that we can think of them economically and clearly in relation to the rest of our science, and so that we can use them as a basis for secure prediction and effective action.
The System of Animate Nature (Volume 1)
Lecture I (p. 8)
William & Norgate. London, England. 1920

Science as science never asks the question Why? That is to say, it never inquires into the meaning, or significance, or purpose of this manifold Being, Becoming, and Having Been.... Thus science does not pretend to be a bedrock of truth.
In Bertrand Russell
Religion and Science
Mysticism (p. 175)
Henry Holt & Company. New York, New York, USA. 1935

Science makes so many permanent discoveries, which are never contradicted though often transcended, that she acquires an assured confidence which has only been equaled by that of Theology.
The System of Animate Nature (Volume 1)
Lecture I (p. 13)
William & Norgate. London, England. 1920

Science is one of the pathways toward the truth, but there are other pathways.
The New World of Science
The Atlantic Monthly, June 1930

Science is not wrapped up with any particular body of facts; it is characterized as an intellectual attitude. It is not tied down to any peculiar methods of inquiry; it is simply sincere critical thought, which admits conclusions only when these are based on evidence.
Introduction to Science
Chapter I (p. 27)
Williams & Norgate Ltd. London, England. 1916

Science is always setting forth on Columbus voyages, discovering new worlds and conquering them by understanding.
The Outline of Science (Volume 1)
Introduction (p. 3)
G.P. Putnam's Sons. New York, New York, USA. 1937

Science is frankly empirical in method and aim; it seeks to discover the laws of concrete being and becoming, and to formulate these in the simplest terms, which are either immediate data of experience or verifiably derived therefrom.

The System of Animate Nature (Volume 1)
Lecture I (p. 39)
William & Norgate. London, England. 1920

Science is a particular way of looking at the world, but it is not the only way.

The New World of Science
The Atlantic Monthly, June 1930

Great stores of wealth are awaiting the scientific "Open Sesame"; a great heightening of the standard of health will be attainable in a few generation if men of good-will take science as their torch.

The Outline of Science (Volume 4)
Chapter XXXVIII (p. 1180)
G.P. Putnam's Sons. New York, New York, USA. 1937

Is it science that satisfies man's soul, or is it the attendant feeling and imagining which the study of Nature evokes?

The System of Animate Nature (Volume 1)
Lecture I (p. 27)
William & Norgate. London, England. 1920

...science gives Man from time to time a greatly increased mastery over Nature; science, with its analytical triumphs, ever tends to diminish, in the shallowminded, the saving grace of wonder; and science is ever dispelling the darkness that oppresses the mind.

The System of Animate Nature (Volume 1)
Lecture I (pp. 41–42)
William & Norgate. London, England. 1920

...science aims at description in terms of the lowest common denominators available; while religion and philosophy aim at interpretation in terms of the greatest common measure.

The New World of Science
The Atlantic Monthly, June 1930

Thomson, Sir George   1892–1975
English physicist

The influence of science on men's lives comes in two rather different ways — one through the ideas themselves, and the other through their material consequences.

The New Industrial Revolution
Bulletin of the Atomic Scientists, Volume 13, Number 1, January 1957 (p. 9)

[The method of science is] a collection of pieces of advice, some general, some rather special, which may help to guide the explorer in his passage through the jungle of apparently arbitrary facts.... In fact, the sciences differ so greatly that it is not easy to find any sort of rule which applies to all without exception.

The Inspiration of Science
Chapter II (p. 7)
Oxford University Press, Inc. London, England. 1961

Thoreau, Henry David   1817–62
American essayist, poet, and practical philosopher

There is a chasm between knowledge and ignorance which the arches of science can never span.

The Writings of Henry David Thoreau (Volume 1)
A Week on the Concord and Merrimack Rivers
Sunday (p. 125)
Houghton Mifflin Company. Boston, Massachusetts, USA. 1893

What an admirable training is science for the more active warfare of life! Indeed, the unchallenged bravery which these studies imply, is far more impressive than the trumpeted valor of the warrior.

The Writings of Henry David Thoreau (Volume 9)
Natural History of Massachusetts (p. 131)
Houghton Mifflin Company. Boston, Massachusetts, USA. 1893

Science with its retorts would have put me to sleep; it was the opportunity to be ignorant that I improved. It suggested to me that there was something to be seen if one had eyes. It made a believer of me more than before. I believed that the woods were not tenantless, but choke-full of honest spirits as good as myself any day, — not an empty chamber, in which chemistry was left to work alone, but an inhabited house, — and for a few moments I enjoyed fellowship with them.

The Writings of Henry David Thoreau (Volume 3)
The Maine Woods, the Allegash and East Branch (pp. 247–248)
Houghton Mifflin Company. Boston, Massachusetts, USA. 1893

Let us consider under what disadvantages Science has hitherto labored before we pronounce thus confidently on her progress.

The Writings of Henry David Thoreau (Volume 4)
Paradise (to Be) Regained (p. 301)
Houghton Mifflin Company. Boston, Massachusetts, USA. 1893

Already nature is serving all those uses which science slowly derives on a much higher and grander scale to him that will be served by her. When the sunshine falls on the path of the poet, he enjoys all those pure benefits and pleasures which the arts slowly and partially realize from age to age. The winds which fan his cheek waft him the sum of that profit and happiness which their lagging inventions supply.

The Writings of Henry David Thoreau (Volume 4)
Paradise (to Be) Regained (p. 302)
Houghton Mifflin Company. Boston, Massachusetts, USA. 1893

Thorne, Kip S.   1940–
American theoretical physicist

Science is a community enterprise. The insights that shape our view of the Universe come not from a single person or a small handful, but from the combined efforts of many.

Black Holes and Time Warps: Einstein's Outrageous Legacy
Preface (p. 18)
W.W. Norton & Company, Inc. New York, New York, USA.1994

Thurber, James 1894–1961
American writer and cartoonist

Science has zipped the atom open in a dozen places, it can read the scrawling on the Rosetta stone as glibly as a literary critic explains Hart Crane, but it doesn't know anything about playwrights.
Collecting Himself: James Thurber on Writing and Writers, Humor and Himself
Roaming in the Gloaming (p. 194)
Harper & Row, Publishers. New York, New York, USA. 1989

Tolstoy, Leo 1828–1910
Russian writer

What is called science today consists of a haphazard heap of information, united by nothing, often utterly unnecessary, and not only failing to present one unquestionable truth, but as often as not containing the grossest errors, today put forward as truths, and tomorrow overthrown.
What Is Religion?
Chapter I (p. 3)
T.Y. Crowell. New York, New York, USA. 1899

The highest wisdom has but one science — the science of the whole — the science explaining the whole creation and man's place in it.
In *Great Books of the Western World* (Volume 51)
War and Peace
Book Five, Chapter II (p. 197)
Encyclopædia Britannica, Inc. Chicago, Illinois, USA. 1952

Toynbee, Arnold J. 1852–83
English historian

There have been many definitions of the word "science." Perhaps the most generally accepted one is that science is a form of study in which there can be an exact knowledge of the present and the past and, through this, an infallible prediction of the future. If this is what science means, then no study made by a human mind can be completely scientific.
Occasional Paper, The Institute for the Study of Science in Human Affairs
Science in Human Affairs: An Historian's View

Tucker, Wilson 1914–2006
American mystery and science fiction writer

…Science tends to frighten those who are infrequently exposed to it, while the practitioners of science are often the most misunderstood people in the world.…
The Year of the Quiet Sun (p. 78)
Ace. New York, New York, USA. 1970

Tudge, Colin 1943–
English science writer

The true role of science is not to change the universe but more fully to appreciate it.
The Engineer in the Garden: Genes and Genetics (p. 361)
Hill & Wang. New York, New York, USA. 1993

Twain, Mark (Samuel Langhorne Clemens) 1835–1910
American author and humorist

Science is as sorry as you are that this year's science is no more like last year's science than last year's was like the science of twenty years gone by. But science cannot help it. Science is full of change. Science is progressive and eternal. The scientists of twenty years ago laughed at the ignorant men who had groped in the intellectual darkness of twenty years before. We derive pleasure from laughing at them.
Collected Tales, Sketches, Speeches, and Essays 1852–1890 (Volume 1)
A Brace of Brief Lectures on Science (p. 538)
The Library of America. New York, New York, USA. 1992

[W]hat we most admire is the vast capacity of that intellect which, without effort, takes in at once all the domains of science — all the past, the present and the future, all the errors of two thousand years, all the encouraging signs of the passing times, all the bright hopes of the coming age.
Is Shakespeare Dead?
Chapter X (p. 124)
Oxford University Press, Inc. London, England. 1996

Union Carbide and Carbon

More Jobs Through Science
Advertising slogan

Urey, Harold Clayton 1893–1981
American chemist

To those of us who spend our lives working on scientific problems, science is a great intellectual adventure of such interest that nothing else we ever do can compare with it. We are attempting to understand the order of a physical universe, vast in extent in space and time, and most complicated and beautiful in its details.
In Shirley Thomas
Men of Space. Profiles of the Scientists Who Probe for Life in Space (Volume 6)
Harold C. Urey (p. 212)
Chilton Books. Philadelphia, Pennsylvania, USA. 1963

Valéry, Paul 1871–1945
French poet and critic

Science is feasible when the variables are few and can be enumerated; when their combinations are distinct and clear. We are tending toward the condition of science and aspiring to do it. The artist works out his own formulas; the interests of science lies in the art of making science.
In Jackson Mathews (ed.)
The Collected Works of Paul Vallery (Volume 14)
Analects (p. 191)
Princeton University Press. Princeton, New Jersey, USA. 1971

Science means simply the aggregate of all the recipes that are always successful. All the rest is literature.
In J. Matthews (ed.)

Collected Works (Volume 14)
Analects
Princeton University Press. Princeton, New Jersey, USA. 1971

Vash
Fictional character

Well, when it comes to choosing between science and profit, I'll choose profit every time.
Star Trek: Deep Space Nine
Q-Less
Television program
Season 1, 1993

Vernadskii, Vladimir Ivanovich   1863–1945
Russian mineralogist

Science is alone and the routes to its achievement are alone. They are independent from the ideas of man, from his aspirations and wishes, from the social tenor of his life, from his philosophical, social, and religious theories. They are independent from his will and from his world outlook — they are primordial.
In Loren R. Graham
The Soviet Academy of Sciences and the Communist Party, 1927–1932
Chapter III (p. 80)
Princeton University Press. Princeton, New Jersey, USA. 1967

Verne, Jules   1828–1905
French novelist

[I]n the cause of science men are expected to suffer.
A Journey to the Center of the Earth
Chapter 6 (p. 33)
The Limited Editions Club. New York, New York, USA. 1966

When science has sent forth her fiat — it is only to hear and obey.
A Journey to the Center of the Earth
Chapter 11 (p. 73)
The Limited Editions Club. New York, New York, USA. 1966

Science, great, mighty and in the end unerring…science has fallen into many errors — errors which have been fortunate and useful rather than otherwise, for they have been the stepping stones to truth.
A Journey to the Center of the Earth
Chapter 28 (p. 182)
The Limited Editions Club. New York, New York, USA. 1966

We may brave human laws, but we cannot resist natural ones.
Translated by Mercier Lewis
Twenty Thousand Leagues Under the Sea
Part Two, Chapter 15 (p. 249)
Nelson Doubleday, Inc. Garden City, New York, USA. 1900

Virchow, Rudolf Ludwig Karl   1821–1902
German pathologist and archaeologist

Science in itself is nothing, for it exists only in the human beings who are its bearers.
Translated by Lelland J. Rather

Disease, Life, and Man, Selected Essays
Standpoints in Scientific Medicine (1847) (pp. 29–30)
Stanford University Press. Stanford, California, USA. 1958

Has not science the noble privilege of carrying on its controversies without personal quarrels.
In F.H. Garrison
Bulletin of the New York Academy of Medicine, Volume 4, 1928 (p. 995)

Voltaire (François-Marie Arouet)   1694–1778
French writer

True science necessarily carries tolerance with it.
Letter to Madame d'Epinay
Correspondance de Voltaire, 1881 edition, Volume 12, July 6, 1766 (p. 329)

von Baer, Carl Ernst   1792–1876
Prussian-Estonian biologist

Science…is, in its source, eternal; in its operation, not limited by time and space; in its scope, immeasurable; in its problem, endless; in its goal, unattainable.
In Sir Richard Arman Gregory
Discovery; or, The Spirit and Service of Science
Chapter III (p. 53)
Macmillan & Company Ltd. London, England. 1918

von Frisch, Karl   1886–1982
Austrian ethnologist

Science advances but slowly, with halting steps. But does not therein lie her eternal fascination? And would we not soon tire of her if she were to reveal her ultimate truths too easily?
A Biologist Remembers
To Munich for the Fifth Time (p. 178)
Pergamon Press. Oxford, England. 1967

von Goethe, Johann Wolfgang   1749–1832
German poet, novelist, playwright, and natural philosopher

Sciences destroy themselves in two ways: by the breadth they reach and by the depth they plumb.
Scientific Studies (Volume 12)
Chapter VIII (p. 305)
Suhrkamp. New York, New York, USA. 1988

Four epochs of science:
childlike,
poetic, superstitious;
empirical,
searching, curious;
dogmatic,
didactic, pedantic;
ideal,
methodical, mystical.
Scientific Studies (Volume 12)
Chapter VIII (pp. 304–305)
Suhrkamp. New York, New York, USA. 1988

Germans — and they are not alone in this — have a knack of making the sciences unapproachable.
Scientific Studies (Volume 12)

Chapter VIII (p. 306)
Suhrkamp. New York, New York, USA. 1988

In general the sciences put some distance between themselves and life, and make their way back to it only by a roundabout path.
Scientific Studies (Volume 12)
Chapter VIII (p. 306)
Suhrkamp. New York, New York, USA. 1988

A crisis must necessarily arise when a field of knowledge matures enough to become a science, for those who focus on details and treat them as separate will be set against those who have their eye on the universal and try to fit the particular into it.
Scientific Studies (Volume 12)
Chapter VIII (p. 305)
Suhrkamp. New York, New York, USA. 1988

von Humboldt, Alexander 1769–1859
German naturalist and explorer

Science is the labor of mind applied to nature…
Cosmos: A Sketch of a Physical Description of the Universe (Volume 1)
Introduction (p. 76)
Harper & Brothers. New York, New York, USA. 1869

Waddington, Conrad Hal 1905–75
British biologist and paleontologist

Science is the organized attempt of mankind to discover how things work as causal systems.
The Scientific Attitude
Forward (p. 9)
Penguin Books. Middlesex, England. 1941

Science as a whole certainly cannot allow its judgment about facts to be distorted by ideas of what ought to be true, or what one may hope to be true.
The Scientific Attitude
Science Is Not Neutral (p. 25)
Penguin Books, Middlesex, England. 1941

…science, if given its head, is not just cold efficiency; its attitude is tolerant, friendly and humane. It has already become the dominant inspiration of human culture, so that modern poetry, painting and architecture derive their most constructive ideas from scientific thought.
The Scientific Attitude
The Scientific Attitude (p. 1)
Penguin Books, Middlesex, England. 1941

Wald, George 1906–97
American biologist and biochemist

Science goes from question to question; big questions, and little, tentative answers. The questions as they age grow ever broader, the answers are seen to be more limited.
Les Prix Nobel. the Nobel Prizes in 1967
Nobel banquet speech for award received in 1967
Nobel Foundation. Stockholm, Sweden. 1968

The trouble with most of the things that people want is that they get them. No scientist needs to worry on that score. For him there is always the further horizon. Science goes from question to question; big questions, and little, tentative answers. The questions as they age grow ever broader, the answers are seen to be more limited.
Les Prix Nobel. the Nobel Prizes in 1967
Nobel banquet speech for award received in 1967
Nobel Foundation. Stockholm, Sweden. 1968

Waterman, Alan T. 1892–1967
American physicist

Science, in its pure form, is not concerned with where discoveries may lead; its disciples are interested only in discovering the truth.
Imagination of Science and Society
American Behavioral Scientist, Volume VI, Number 4, December 1962
(p. 3)

Watson, David Lindsay 1901–73
No biographical data available

The main vehicle of science is not the published formulations of laws and experiments in books and periodicals. This vehicle is, first and foremost, men who are worthy of them, who can understand and use the laws. But more than this: the vehicle is also the pattern of the society that can produce such men.
Scientists Are Human
Chapter I (p. 3)
Watts. London, England. 1938

Science sprawls over all the horizons of the modern mind like some vast cloudbank. The outlook and method of science penetrate relentlessly the strata of daily custom into the caverns of the unconscious mind itself. Science is by far the most powerful intellectual phenomenon of modern times, inexorably laying down the law in regions far from the laboratory, and subtly governing, by its techniques and devices, our modes of life and ways of thinking.
Scientists Are Human
Chapter I (p. 1)
Watts. London, England. 1938

Watson, James D. 1928–
American geneticist and biophysicist

…good science as a way of life is sometimes difficult. It often is hard to have confidence that you really know where the future lies. We must thus believe strongly in our ideas, often to point where they may seem tiresome and bothersome and even arrogant to our colleagues.
Les Prix Nobel. The Nobel Prizes in 1962
Nobel banquet speech for award received in 1962
Nobel Foundation. Stockholm, Sweden. 1963

Weaver, Warren 1894–1978
American mathematician

The desirable adjuncts of modern living, although in many instances made possible by science, certainly do not constitute science.
Science and Imagination: Selected Papers
Chapter 1
Basic Books, Inc. New York, New York, USA. 1967

Science is not technology, it is not gadgetry, it is not some mysterious cult, it is not a great mechanical monster! Science is an adventure of the human spirit. It is essentially an artistic enterprise, stimulated largely by the universe, served largely by disciplined imagination, and based largely on faith in the reasonableness, order, and beauty of the universe of which man is part.
In Walter Orr Robek
Science, a Well Spring of our Discontent
American Scientist, Volume 55, Number 1, March 1957 (p. 3)

It is hardly necessary to argue, these days, that science is essential to the public. It is becoming equally true, as the support of science moves more and more to state and national sources, that the public is essential to science. The lack of general comprehension of science is thus dangerous both to science and the public, these being interlocked aspects of the common danger that scientists will not be given the freedom, the understanding, and the support that are necessary for vigorous and imaginative development.
In Hilary J. Deason
A Guide to Science Reading
Science and People (p. 38)
The New American Library. New York, New York, USA. 1966

Weber, Max   1864–1920
German founder of modern sociology and economic thinker

Science today is a "vocation" organized in special disciplines in the service of self-clarification and knowledge of interrelated facts. It is not the gift of grace of seers and prophets dispensing sacred values and revelations, nor does it partake of the contemplation of sages and philosophers about the meaning of the Universe.
In H.H. Gerth and C. Wright Mills (eds.)
From Max Weber: Essays in Sociology
Science as a Vocation (p. 152)
Oxford University Press, Inc. New York, New York, USA. 1970

Weil, Simone   1909–43
French philosopher and mystic

To us, men of the West, a very strange thing happened at the turn of the century; without noticing it, we lost science, or at least the thing that had been called by that name for the last four centuries. What we now have in place of it is something different, radically different, and we don't know what it is. Nobody knows what it is.
Translated by Richard Rees
On Science, Necessity, and the Love of God
Classical Science and After, Chapter I (p. 3)
Oxford University Press, Inc. London, England. 1968

Science is voiceless; it is the scientist who talks.
On Science, Necessity, and the Love of God
Reflections on Quantum Theory (p. 57)
Oxford University Press, Inc. London, England. 1968

Science today will either have to seek a source of inspiration higher than itself or perish.
Gravity and Grace
Intelligence and Grace (p. 119)
Routledge & Kegan Paul. London, England. 1952

Science only offers three kinds of interest: 1. Technical applications. 2. A game of chess. 3. A road to God. (Attractions are added to the game of chess in the shape of competitions, prizes, and medals.)
Gravity and Grace
Intelligence and Grace (p. 119)
Routledge & Kegan Paul. London, England. 1952

Science is today regarded by some as a mere catalogue of technical recipes, and others as a body of pure intellectual speculations which are sufficient unto themselves; the former set too little value on the intellect, the latter on the world.
Translated by Arthur Wills and John Petrie
Oppression and Liberty
Theoretical Picture of a Free Society (pp. 104–105)
Routledge & Kegan Paul Ltd. London, England. 1958

Weinberg, Steven   1933–
American nuclear physicist

…there is an essential element in science that is cold, objective, and nonhuman…the laws of nature are as impersonal and free of human values as the rules of arithmetic…. Nowhere do we see human value or human meaning.
Reflections of a Working Scientist
Daedalus, Volume 103, 1974 (p. 3)

Weiss, Paul A.   1898–1985
Chemist

Science, to some, is Lady Bountiful, to others is the Villain of the Century. Some years ago, a book called it our "Sacred Cow," and certainly to many it has at least the glitter of the "Golden Calf." Glorification at one extreme, vituperation at the other…
Within the Gates of Science and Beyond
Science Looks at Itself (p. 25)
Hafner Publishing Company. New York, New York, USA. 1971

Weisskopf, Victor Frederick   1908–2002
Austrian-American physicist

Science cannot develop unless it is pursued for the sake of pure knowledge and insight. It will not survive unless it is used intensely and wisely for the betterment of humanity and not as an instrument of domination by one group over another.
Physics in the Twentieth Century: Selected Essays

The Significance of Science (p. 364)
The MIT Press. Cambridge, Massachusetts, USA. 1972

Science is curiosity, discovering things and asking why.
The Privilege of Being a Physicist
Chapter 4 (p. 31)
W.H. Freeman & Company. New York, New York, USA. 1989

Wells, H. G. (Herbert George)   1866–1946
English novelist, historian, and sociologist

The science hangs like a gathering fog in a valley, a fog which begins nowhere and goes nowhere, an incidental, unmeaning inconvenience to passers-by.
The Works of H.G. Wells (Volume 9)
A Modern Utopia
Chapter 3, Section 3
Charles Scribner's Sons. London, England. 1924–27

Science is a match that man has just got alight. He thought he was in a room — in moments of devotion, a temple — and that his light would be reflected from and display walls inscribed with wonderful secrets and pillars carved with philosophical systems wrought into harmony. It is a curious sensation, now that the preliminary splutter is over and the flame burns up clear, to see his hands and just a glimpse of himself and the patch he stands on visible, and around him, in place of all that human comfort and beauty he anticipated — darkness still.
The Rediscovery of the Unique
The Fortnightly Review, New Series 50, July 1891

Weyl, Hermann   1885–1955
German mathematician

Modern science, insofar as I am familiar with it through my own scientific work, mathematics and physics make the world appear more and more an open one.... Science finds itself compelled, at once by the epistemological, the physical and the constructive-mathematical aspect of its own methods and results, to recognize this situation. It remains to be added that science can do no more than show us this open horizon; we must not by including the transcendental sphere attempt to establish anew a closed (though more comprehensive) world.
The Open World: Three Lectures in the Metaphysical Implications of Science
Preface (p. v)
Yale University Press. New Haven, Connecticut, USA. 1932

We must await the further development of science, perhaps for centuries, before we can design a true and detailed picture of the interwoven texture of Matter, Life and Soul. But the old classical determinism of Hobbes, and Laplace need not oppress us any longer.
The Open World: Three Lectures in the Metaphysical Implications of Science
Lecture II (p. 55)
Yale University Press. New Haven, Connecticut, USA. 1932

Modern science in so far as I am familiar with it through my own scientific work, mathematics and physics, make the world appear more and more as an open one, as a world not closed but pointing beyond itself...science finds itself compelled, at once by the epistemological, the physical and the constructive-mathematical aspect of its own methods and results, to recognize this situation. It remains to be added that science can do no more than show us this open horizon; we must not by including the transcendental sphere attempt to establish anew a closed (though more comprehensive) world.
In A.S. Eddington
New Pathways in Science
Chapter XIV (p. 309)

...science would perish without the continuous interplay between its facts and constructions on the one hand and the imagery of ideas on the other.
Philosophy of Mathematics and Natural Science
Preface (p. vi)
Princeton University Press. Princeton, New Jersey, USA. 1949

Wheeler, John Archibald   1911–
American physicist and educator

...the pursuit of science is more than the pursuit of understanding. It is driven by the creative urge, the urge to construct a vision, a map, a picture of the world that gives the world a little more beauty and coherence than it had before.
Geons, Black Holes, and Quantum Foam: A Life in Physics
Chapter 3 (p. 84)
W.W. Norton & Company, Inc. New York, New York, USA.1998

...the human activity that we call science is not science unless it is the uncovering or discovery of something new.
At Home in the Universe
Be the Best to Give the Most (p. 76)
The American Institute of Physics. Woodbury, New York, USA. 1994

Whetham, Sir William Cecil Dampier   1867–1952
English scientific writer

But beyond the bright searchlights of science,
Out of sight of the windows of sense,
Old riddles still bid us defiance,
Old questions of Why and of Whence.
Recent Development of Physical Science (p. 10)
John Murray. London, England. 1927

Whewell, William   1794–1866
English philosopher and historian

Science begins with common observation of facts; but even at this stage, requires that the observations be precise. Hence the sciences which depend upon space and number were the earliest formed. After common observation, comes Scientific Observation and Experiment.
The Philosophy of the Inductive Sciences Founded upon Their History
(Volume 2)

Aphorisms, Aphorisms Concerning Science, VII (p. 467)
John W. Parker. London, England. 1847

The tendency of the sciences has long been an increasing proclivity of separation and dismemberment. The mathematician turns away from the chemist; the chemist from the naturalist; between the mathematician and the chemist is to be interpolated a "physician" (we have no English name for him), who studies heat, moisture and the like.
Quarterly Review, Volume 51, 1834 (p. 59)

The principles which constituted the triumph of preceding stages of science may appear to be subverted and ejected by later discoveries, but in fact they are (so far as they are true) taken up into the subsequent doctrines and included in them. They continue to be an essential part of the science. The earlier truths are not expelled but absorbed, not contradicted but extended; and the history of each science which may thus appear like a succession of revolutions is, in reality, a series of developments.
History of the Inductive Sciences, from the Earliest to the Present Time
(Volume the First)
Introduction (p. 10)
John W. Parker. London, England. 1837

…two things are requisite to science — facts and ideas…
History of the Inductive Sciences, from the Earliest to the Present Time
(Volume 1)
Book I, Chapter III, Section 2 (p. 79)
John W. Parker. London, England. 1837

Man is the interpreter of Nature, Science is the right interpretation.
The Philosophy of the Inductive Sciences Founded upon Their History
(Volume 2)
Aphorisms Concerning Ideas, Aphorism I (p. 443)
John W. Parker. London, England. 1847

White, Leslie Alvin   1900–75
American anthropologist

Science is sciencing.
Science Is Sciencing
Philosophy of Science, Volume 5, 1938

Whitehead, Alfred North   1861–1947
English mathematician and philosopher

Science is the organisation of thought.
The Organisation of Thought
Chapter VI (p. 106)
Greenwood Press Publishers. Westport, Connecticut, USA. 1974

Science is in the minds of men, but men sleep and forget, and at their best in any one moment of insight entertain but scanty thoughts. Science therefore is nothing but a confident expectation that relevant thoughts will occasionally occur.
An Enquiry Concerning the Principles of Natural Knowledge
Part I, Chapter I (p. 10)
At The University Press. Cambridge, England. 1919

Science is either an important statement of systematic theory correlating observations of a common world or is the daydream of a solitary intelligence with a taste for the daydream of publication.
Process and Reality: An Essay in Cosmology
Part IV, Chapter V, Section IV (p. 502)
The Macmillan Company. New York, New York, USA. 1929

Science has always suffered from the vice of overstatement. In this way conclusions true within strict limitations have been generalized dogmatically into a fallacious universality.
The Function of Reason
Chapter I (p. 22)
Beacon Press. Boston, Massachusetts, USA. 1929

Science is even more changeable than theology.
Science and the Modern World
Chapter XII (p. 183)
The Macmillan Company. New York, New York, USA. 1929

Aristotle discovered all the half-truths which were necessary to the creation of science.
In Lucien Price
Dialogues of Alfred North Whitehead
Dialogue XLII September 11, 1945 (p. 344)
Little Brown. Boston, Massachusetts, USA. 1954

Science is a river with two sources, the practical source and the theoretical source. The practical source is the desire to direct our actions to achieve predetermined ends.… The theoretical source is the desire to understand.
The Organisation of Thought
Chapter VI (p. 106)
Greenwood Press Publishers. Westport, Connecticut, USA. 1974

A science which hesitates to forget its founders is lost.
The Organisation of Thought
Chapter VI (p. 115)
Greenwood Press Publishers. Westport, Connecticut, USA. 1974

Whyte, Lancelot Law   1896–1972
Scottish physicist

Science starts with an assumption which is always present, though it may be unconscious, may be forgotten, and may sometimes even be denied.
Accent on Form: An Anticipation of the Science of Tomorrow
Chapter IV (p. 59)
Harper & Brothers. New York, New York, USA. 1954

Wigner, Eugene Paul   1902–95
Hungarian-born American physicist

There is no natural phenomenon that is comparable with the sudden and apparently accidentally timed development of science, except perhaps the condensation of a super-saturated gas or the explosion of some unpredictable explosives. Will the fate of science show some similarity to one of these phenomena?
Proceedings of the American Philosophical Society, Volume 94, Number 5, 1950

Wilde, Oscar   1854–1900
Irish wit, poet, and dramatist

The advantage of the emotions is that they lead us astray, and the advantage of Science is that it is not emotional.
The Picture of Dorian Gray
Chapter 3 (p. 45)
The Modern Library. New York, New York, USA. 1992

Science can never grapple with the irrational. That is why there is no future before it, in this world.
The Plays of Oscar Wilde
An Ideal Husband
Act I (p. 9)
The Modern Library. New York, New York, USA. No date

Science is out of the reach of morals, for her eyes are fixed upon eternal truths.
The Works of Oscar Wilde (Volume 10)
Intentions
The Critic as Artist, Part 2 (p. 394)
AMS Press. New York, New York, USA. 1909

Wilson, Edward O.   1929–
American biologist and author

To a considerable degree science consists in originating the maximum amount of information with the minimum expenditure of energy. Beauty is the cleanness of line in such formulations along with symmetry, surprise, and congruence with other prevailing beliefs.
Biophilia
The Poetic Species (p. 60)
Harvard University Press. Cambridge, Massachusetts. 1984

Important science is not just any similarity glimpsed for the first time. It offers analogues that map the gateways to unexplored terrain.
Biophilia
The Poetic Species (p. 67)
Harvard University Press. Cambridge, Massachusetts. 1984

Wittgenstein, Ludwig Josef Johann   1889–1951
Austrian-born English philosopher

Man has to awaken to wonder — and so perhaps do peoples. Science is a way of sending him to sleep again.
Translated by Peter Winch
Culture and Value (p. 5e)
The University of Chicago Press. Chicago, Illinois, USA. 1980

Wolpert, Lewis   1929–
British embryologist

When we come to face the problems before us — poverty, pollution, overpopulation, illness — it is to science that we must turn, not to gurus. The arrogance of scientists is not nearly as dangerous as the arrogance that comes from ignorance.
In Mary Midgley
Can Science Save Its Soul?
New Scientist, Volume 135, Number 1832,1 August 1992 (p. 24)

Wordsworth, William   1770–1850
English poet

Science appears but what in truth she is,
Not as our glory and our absolute boast,
But as a succedaneum and a prop
To our infirmity. No officious slave
Art thou of that false secondary power
By which we multiply distinctions, then
Deem that our puny boundaries are things
That we perceive, and not that we have made.
The Complete Poetical Works of William Wordsworth
The Prelude, Book II, l. 212–219
Crowell. New York, New York, USA. 1888

Wright, Chauncey   1830–75
American philosopher of science

The accidental causes of science are only "accidents" relatively to the intelligence of a man.
The Philosophical Writings of Chauncey Wright
The Genesis of Species (p. 37)
The Liberal Arts Press. New York, New York, USA. 1958

Wright, Frank Lloyd   1869–1959
American architect

I have seemed to belittle the nature of our time and the great achievements of science, but I have intended to do neither because I believe human nature still sound, and recognize that science has done a grand job as well; but well I know that Science cannot save us.
An Organice Architecture, Speech
London, England, May 1939

Yates, Frances   1899–1981
English historian

Is not all science a gnosis, an insight into the nature of the All, which proceeds by successive revelations?
Giordano Bruno and the Hermetic Tradition
Chapter XXII (p. 452)
The University of Chicago Press. Chicago, Illinois, USA. 1964

Ziman, John M.   1925–2005
British physicist

Penicillin is not Science, any more than a cathedral is Religion or a witness box is Law.
In E.D. Klemke, Robert Hollinger and A. David Kline
Introductory Readings in the Philosophy of Science
What Is Science? (p. 36)
Prometheus Books. Buffalo, New York, USA. 1980

In science, to echo Beethoven's dictum about music, "Everything should be both surprising and expected."
Reliable Knowledge
Chapter 3 (fn 17, p. 71)
Cambridge University Press. Cambridge, England. 1978

Zinkernagel, Rolf M.   1944–
Swiss immunologist and pathologist

...in science there are collectors, classifiers, compulsory tidiers-up and permanent contesters, detectives, some artists and many artisans, there are poet-scientists and philosophers and even a few mystics.
Les Prix Nobel. The Nobel Prizes in 1996
Nobel banquet speech for award received in 1996
Nobel Foundation. Stockholm, Sweden. 1997

Zinsser, Hans   1878–1940
American bacteriologist

Science is but a method. Whatever its material, an observation accurately made and free of compromise to bias and desire, and undeterred by consequence, is science.
Untheological Reflections
The Atlantic Monthly, July 1929 (p. 91)

SCIENCE, AGE OF

Compton, Karl Taylor   1887–1954
American educator and physicist

We live in an age of science. I do not say "an age of technology" for every age has been an age of technology. We recognize this when we describe past civilizations as the Stone Age, the Bronze Age, and the Age of Steam or of Steel, thus implicitly admitting that the stage of civilization is determined by the tools at man's disposal — in other words, by his technology....
A Scientist Speaks: Excerpts from Addresses by Karl Taylor Compton During the Years 1930–1949 (p. 1)
Undergraduate Association, MIT. Cambridge, Massachusetts, USA. 1955

Kronenberger, Louis   1904–80
American author and critic

Nominally a great age of scientific inquiry, ours has actually become an age of superstition about the infallibility of science; of almost mystical faith in its nonmystical methods; above all...of external verities; of traffic-cop morality and rabbit-test truth.
Company Manners: A Cultural Inquiry into American Life
Chapter 4 (p. 94)
The Bobbs-Merrill Company, Inc. Indianapolis, Indiana, USA. 1954

Russell, Bertrand Arthur William   1872–1970
English philosopher, logician, and social reformer

To say that we live in an age of science is a common place, but like most common places, it is only partially true. From the point of view of our predecessors, if they could view our society, we should, no doubt, appear to be very scientific, but from the point of view of our successors, it is probable that the exactly opposite would seem to be the cause.
The Scientific Outlook
Introduction (p. 9)
George Allen & Unwin Ltd. London, England. 1931

[T]heoretical science...is an attempt to understand the world. Practical science, which is an attempt to change the world, has been important from the first, and has continually increased in importance, until it has almost ousted theoretical science from men's thoughts....
A History of Western Philosophy
Book Three, Part I, Chapter I (p. 492–493, 493)
Simon & Schuster. New York, New York, USA. 1945

The triumph of science has been mainly due to its practical utility, and there has been an attempt to divorce this aspect from that of theory, thus making science more and more a technique, and less and less a doctrine as to the nature of the world. The penetration of this point of view to philosophers is very recent.
A History of Western Philosophy
Book Three, Part I, Chapter I (p. 492–493, 493)
Simon & Schuster. New York, New York, USA. 1945

SCIENCE, APPLIED

Bacon, Sir Francis   1561–1626
English lawyer, statesman, and essayist

Even when men build any science and theory upon experiment, yet they almost always turn with premature and hasty zeal to practise, not merely on account of the advantage and benefit to be derived from it, but in order to seize upon some security in a new undertaking of their not employing the remainder of their labor unprofitably, and by making themselves conspicuous, to acquire a greater name for their pursuit.
In *Great Books of the Western World* (Volume 30)
Novum Organum
First Book, Aphorism 70 (p. 116)
Encyclopædia Britannica, Inc. Chicago, Illinois, USA. 1952

Compton, Karl Taylor   1887–1954
American educator and physicist

Applied science is not an end in itself, but it is the most powerful means ever discovered for supplying the opportunity to secure the finest things of life.
A Scientist Speaks: Excerpts from Addresses by Karl Taylor Compton During the Years 1930–1949 (p. 9)
Undergraduate Association, MIT. Cambridge, Massachusetts, USA. 1955

Einstein, Albert   1879–1955
German-born physicist

Why does this magnificent applied science which saves work and makes life easier bring us so little happiness? The simple answer runs: Because we have not yet learned to make sensible use of it.
Einstein Seeks Lack in Applying Science
The New York Times, February 17, 1931 (p. 6)

It is not enough that you should understand about applied science in order that your work may increase man's

blessings. Concern for the man himself and his fate must always form the chief interest of all technical endeavors; concern for the great unsolved problems of the organization of labor and the distribution of our mind shall be a blessing and not a curse to mankind. Never forget this in the midst of your diagrams and equations.
Einstein Seeks Lack in Applying Science
The New York Times, February 17, 1931 (p. 6)

Huxley, Aldous   1894–1963
English writer and critic

Applied Science is a conjurer, whose bottomless hat yields impartially the softest of Angora rabbits and the most petrifying of Medusas.
Tomorrow and Tomorrow and Tomorrow and Other Essays
The Desert (p. 82)
Harper & Brothers. New York, New York, USA. 1956

Huxley, Thomas Henry   1825–95
English biologist

I often wish that this phrase "applied science," had never been invented. For it suggests that there is a sort of scientific knowledge of direct practical use, which can be studied apart from another sort of scientific knowledge, which is of no practical utility, and which is termed "pure science." But there is no more complete fallacy than this.
Collected Essays (Volume 3)
Science and Education
Science and Culture (p. 137)
Macmillan & Company Ltd. London, England. 1904

Pasteur, Louis   1822–95
French chemist

There does not exist a category of science to which one can give the name applied science. There are science and the applications of science, bound together as the fruit of the tree which bears it.
Revue Scientifique
Pourquoi la France n'a pas trouvé hommes supérieurs au montent du péril (1871)

Porter, George   1920–2002
English chemist

To feed applied science by starving basic science is like economising on the foundations of a building so that it may be built higher.
Lest the Edifice of Science Crumble
New Scientist, Volume 111, Number 1524, September 1986 (p. 16)

Wheeler, Edgar C.
No biographical data available

…[researchers] cannot remain indefinitely in any field of pure research. For every time they come upon a new bit of knowledge, almost instantly they discover some practical application. Thus the dividing line between pure science and applied science becomes thin.

Makers of Lightning
The World's Work, January 1927 (p. 271)

SCIENCE, COMMUNICATION OF

Casimir, Hendrik B. G.   1909–2000
Dutch physicist

There exists today a universal language that is spoken and understood almost everywhere: it is Broken English. [It] is used by the waiters in Hawaii, prostitutes in Paris and ambassadors in Washington, by businessmen from Buenos Aires, by scientists at international meetings and by dirty-postcard peddlers in Greece — in short, by honorable people like myself all over the world.
Haphazard Reality: Half a Century of Science
Chapter 4 (p. 122)
Harper & Row, Publishers. New York, New York, USA. 1983

Chargaff, Erwin   1905–2002
Austrian biochemist

There is no real popularization [of science] possible, only vulgarization that in most instances distorts the discoveries beyond recognition.
Bitter Fruits from the Tree of Knowledge
Perspectives in Biology and Medicine, Volume 16, Summer, 1973 (p. 491)

…Scientists, like little fishes, swim in schools. When we open one of our scientific journals these days, we find a very uneven distribution of topics. Some important fields are almost entirely neglected, others seem to explode into bursts of unbelievable mediocrity. Really valuable contributions in the fields most in vogue at present probably are just as scarce as those dealing with the stepchildren of present-day biochemistry. But not all disciplines make it so easy to call each mush a "homogenate", each soup a "partially purified extract", and so to speak — when you have nothing whatever — of a "system." There is a real danger that our science may suffocate in its own excrements.
Essays on Nucleic Acids
Chapter 10 (p. 162)
Elsevier Publishing Company. Amsterdam, Netherlands. 1963

Chesterton, G. K. (Gilbert Keith)   1874–1936
English author

Scientific phrases are used like scientific wheels and piston-rods to make swifter and smoother yet the path of the comfortable.
Orthodoxy
Chapter VIII (p. 230)
John Lane Company. New York, New York, USA. 1918

Dancoff, S. M.
American physicist

When you set out in a new field and choose a terminology you have the choice of using old, familiar words in new meanings or else you can make up new words for

the new meanings. If you use old words you make the theory look homey and inviting, but you run the risk of confusing the issue every time the old word is used. If you use new words you make the thing look excessively highbrow and frighten off any who might be interested.
Does the Neutrino Really Exist?
Bulletin of the Atomic Scientists, Volume 8, Number 5, June 1952 (p. 139)

Dornan, Christopher   1957–
American journalism professor

Science is seen as an avenue of access to assured findings, and scientists — in the dissemination of these findings — as the initial sources. The members of the laity are understood purely as recipients of this information. Journalists and public relations personnel are viewed as intermediaries through which the scientific findings filter. The task for science communication is to transmit as much information as possible with maximum fidelity.
Some Problems of Conceptualizing the Issues of "Science and the Media" *Critical Studies in Mass Communication*, Volume 7, Number 1, March 1990 (p. 51)

Feynman, Richard P.   1918–88
American theoretical physicist

We have a habit in writing articles published in scientific journals to make the work as finished as possible, to cover all the tracks, to not worry about the blind alleys or describe how you had the wrong idea first, and so on. So there isn't any place to publish, in a dignified manner what you actually did in order to do the work....
Nobel Lectures, Physics 1963–1970
Nobel lecture for award received in 1965
The Development of the Space-Time View of Quantum Electrodynamics (p. 155)
Elsevier Publishing Company. Amsterdam, Netherlands. 1972

Fischer, Martin H.   1879–1962
German-American physician

You must learn to talk clearly. The jargon of scientific terminology which rolls off your tongues is mental garbage.
In Howard Fabing and Ray Marr
Fischerisms
C.C. Thomas. Springfield, Illinois, USA. 1944

Gibbs, J. Willard   1839–1903
American mathematician

...science is, above all, communication.
In H.N. Parton
Science Is Human
Science and the Liberal Arts (p. 11)
University of Otago Press. Dunedin, New Zealand. 1972

Huxley, Thomas Henry   1825–95
English biologist

...there is assuredly no more effectual method of clearing up one's own mind on any subject than by talking it over

so to speak, with men of real power and grasp, who have considered it from a totally different point of view.
Collected Essays (Volume 1)
Method and Result
Animal Automatism (p. 202)
Macmillan & Company Ltd. London, England. 1904

Large, E. C.
American author

[I]t was contended that...compartments labeled Chemistry, Mycology, Bacteriology...were never really fishtanks for myopic specialists to swim about in, but merely convenient departments in one splendid and sunlit edifice of science, separated at the most by glass walls, decorated with the flags of all nations, and provided with innumerable intercommunicating doors. If so many stacks of old scientific papers got piled up on each side of the glass partitions that in the end no one could see through them, that was certainly regrettable; and if some of the doors were locked for periods ranging from a decade to a century, well, that also was a pity — but who wanted to work in a draught?
The Advance of the Fungi
Chapter XXIII (p. 317)
Henry Holt & Company. New York, New York, USA. 1940

Lemke, J.
No biographical data available

True Dialogue occurs when teachers ask questions to which they do not presume to already know the "correct answer."
Talking Science: Language, Learning and Values
Chapter 3 (p. 55)
Ablex Publishing Corporation. Norwood, New Jersey, USA. 1990

Loomis, Frederic Brewster   1873–1937
American geologist

Everyone, who is alert as he wanders about this world, wants to know what he is seeing and what it is all about. Here and there with the aid of capable guides a few have been introduced into the sphere that wide and fascinating knowledge of Nature which has been so rapidly accumulated during this and the latter part of the last century. It is a full treasure house constantly being enriched, but unfortunately the few who have been initiated have soon acquired technical language and habit, so that their knowledge and new acquisitions are communicated to but a few.
Field Book of Common Rocks and Minerals
Preface (p. vii)

Macdonald, Sharon
No biographical data available

...science communication involves selection and definition, not just of which "facts" are presented to the public, but of what is to count as science and of what kind of entity or enterprise science is to be. That is, science communicators act as authors of science for the public. They may also, however, by dint of their own institutional status, give implicit stamps of approval or disapproval to particular visions or versions of science. That is, they may act as authors with special authority on science — as authorisers of science.

In Alan Irwin and Brian Wynne (eds.)
Misunderstanding Science? The Public Reconstruction of Science and Technology
Authorizing Science: Public Understanding of Science in Museums (p. 152)
Cambridge University Press. Cambridge, England. 1996

Mach, Ernst   1838–1916
Austrian physicist and philosopher

Science is communicated by instruction, in order that one man may profit by the experience of another and be spared the trouble of accumulating it for himself; and thus, to spare posterity, the experiences of whole generations are stored up in libraries.

The Science of Mechanics (5th edition)
Chapter IV, Part IV, Section 1 (p. 578)
The Open Court Publishing Company. La Salle, Illinois, USA. 1942

Michelson, Albert Abraham   1852–1931
German-American physicist

Science, when it has to communicate the results of its labor, is under the disadvantage that its language is but little understood. Hence it is that circumlocution is inevitable and repetitions are difficult to avoid.

Light Waves and Their Uses
Lecture I (p. 1)
The University of Chicago Press. Chicago, Illinois, USA. 1903

Moore, John A.
American writer and professor of genetics and biology

…recall some of the lectures you may have heard recently. Did you always know why the research had been done? Was it clear what problem was being illuminated by the data presented?

Science as a Way of Knowing
American Zoologist, Volume 24, Number 2, 1984 (p. 471)

Moravcsik, M. J.
No biographical data available

New theories, when first proposed, may appear on the first page of the New York Times, but their demise, a few years later, never makes even page 68.

Research Policy, Volume 17, 1988 (p. 293)

Neal, Patricia   1926–
American actress

Gort, *Klaatu berada nikto*!
The Day the Earth Stood Still
Film (1951)

Oppenheimer, J. Robert   1904–67
American theoretical physicist

The true responsibility of a scientist, as we all know, is to the integrity and vigor of his science. And because most scientists, like all men of learning, tend in part also to be teachers, they have a responsibility for the communication of the truths they have found. This is at least a collective, if not an individual responsibility. That we should see in this any insurance that the fruits of science will be used for man's benefit, or denied to man when they make for his distress or destruction, would be a tragic naiveté.

The Open Mind
Chapter V (p. 91)
Simon & Schuster. New York, New York, USA. 1955

Often the very fact that the words of science are the same as those of our common life and tongues can be more misleading than enlightening, more frustrating to understanding than recognizably technical jargon.

Science and the Common Understanding
Chapter 1 (p. 5)
Simon & Schuster. New York, New York, USA. 1954

It is proper to the role of the scientist that he not merely find new truth and communicate it to his fellows, but that he teach, that he try to bring the most honest and intelligible account of new knowledge to all who will try to learn.

The Open Mind: Lectures
Prospects in the Arts and Sciences (p. 138)
Simon & Schuster. New York, New York, USA. 1955

Parton, H. N.
No biographical data available

Scientists have the duty to communicate, firstly with each other, that is with those who are interested in the same or allied problems, and secondly with laymen: by layman I mean anyone not familiar with their special science, for specialization has raised the level of scientific achievement so much, that chemists, for example, are usually laymen in say, biology; we may hope, intelligent laymen.

Science Is Human
Science and the Liberal Arts (p. 12)
University of Otago Press. Dunedin. 1972

Aldous Huxley, in a lecture on his grandfather, said that all communication is literature, and even in scientific writing there is wide room for the exercise of art.

Science Is Human
Science and the Liberal Arts (p. 14)
University of Otago Press. Dunedin. 1972

Pool, Ithiel de Sola   1917–84
No biographical data available

Computing and communication are becoming one…
Technologies Without Boundaries: On Telecommunications in a Global Age
Part I, Chapter I (p. 8)
Harvard University Press. Cambridge, Massachusetts, USA. 1990

Priestley, Joseph 1733–1804
English theologian and scientist

When for the sake of a little more reputation, men can keep brooding over a new fact, in the discovery of which they might, possibly, have very little real merit, till they think they can astonish the world with a system as complete as it is new, and give mankind a high idea of their judgment and penetration; they are justly punished for their ingratitude to the fountain of all knowledge, and for their want of a genuine love of science and of mankind, in finding their boasted discoveries anticipated, and the field of honest fame pre-occupied, by men, who, from a natural ardour of mind engage in philosophical pursuits, with an ingenious simplicity immediately communicate to others whatever occurs to them in their inquiries.
Experiment and Observations on Different Kinds of Air (Volume 1)
The Preface (pp. xvii–cviii)
Printed by Thomas Pearson. Birmingham, England. 1740

Roe, Anne 1904–1991
American psychologist

Nothing in science has any value to society if it is not communicated. . . .
The Making of a Scientist
Chapter I (p. 17)
Greenwood Press, Publishers. Westport, Connecticut, USA. 1973

Schrödinger, Erwin 1887–1961
Austrian theoretical physicist

Bohr's…approach to atomic problems…is really remarkable. He is completely convinced that any understanding in the usual sense of the word is impossible. Therefore the conversation is almost immediately driven into philosophical questions, and soon you no longer know whether you really take the position he is attacking, or whether you really must attack the position that he is defending.
In W. Moore
Schrodinger: Life and Thoughts
Chapter 6, Letter to W. Wein 1926 (p. 228)
Cambridge University Press. Cambridge, England. 1989

If you cannot — in the long run — tell everyone what you have been doing, your doing has been worthless.
Science and Humanism: Physics in Our Time
The Spiritual Bearing of Science on Life (pp. 8–9)
At The University Press. Cambridge, England. 1952

von Goethe, Johann Wolfgang 1749–1832
German poet, novelist, playwright, and natural philosopher

The present age has a bad habit of being abstruse in the sciences. We remove ourselves from common sense without opening up a higher one; we become transcendent, fantastic, fearful of intuitive perception in the real world, and when we wish to enter the practical realm, or need to, we suddenly turn atomistic and mechanical.
Scientific Studies (Volume 12)

Chapter VIII (pp. 308–309)
Suhrkamp. New York, New York, USA. 1988

Whitehead, Alfred North 1861–1947
English mathematician and philosopher

Nobody has a right to speak more clearly than he thinks.
Washingtonian, Volume 15, Number 143, November 1979

Wiener, Norbert 1894–1964
American mathematician

[T]he more probable the message, the less information it gives. Clichés, for example, are less illuminating than great poems.
The Human Use of Human Beings
Chapter I (p. 21)
Da Capo Press. New York, New York, USA. 1988

Ziman, John M. 1925–2005
British physicist

The cliché of scientific prose betrays itself "Hence we arrive at the conclusion that.…" The audience to which scientific publications are addressed is not passive; by its cheering or booing, its bouquets or brickbats, it actively controls the substance of the communications that it receives.
Public Knowledge: An Essay Concerning the Social Dimension of Science
Chapter 1 (p. 9)
Cambridge University Press. Cambridge, England. 1968

It is not enough to observe, experiment, theorize, calculate and communicate; we must also argue, criticize, debate, expound, summarize, and otherwise transform the information that we have obtained individually into reliable, well established, public knowledge.
Information, Communication, Knowledge
Nature, Volume 224, Number 5217, October 25, 1969 (p. 324)

SCIENCE, HISTORY OF

Appleton, Sir Edward 1892–1965
English physicist

…the history of science has proved that fundamental research is the lifeblood of individual progress and that the ideas which lead to spectacular advances spring from it.
In J. Edwin Holmstrom
Records and Research in Engineering and Industrial Science
Chapter One (p. 7)
Chapman & Hall. London, England. 1956

Asimov, Isaac 1920–92
American author and biochemist

A number of years ago, when I was a freshly-appointed instructor, I met, for the first time, a certain eminent historian of science. At the time I could only regard him with tolerant condescension. I was sorry for a man who, it seemed to me, was forced to hover about the edges of

science…. In a lifetime of being wrong at many a point, I was never more wrong. It was I, not he, who was wandering in the periphery. It was he, not I, who lived in the blaze. I had fallen victim to the fallacy of the "growing edge"; the belief that only the very frontier of scientific advance counted; that everything that had been left behind by that advance was faded and dead.
Adding a Dimension
Introduction (p. 7)
Lancer Books. New York, New York, USA. 1969

Bernal, John Desmond   1901–71
Irish-born physicist and x-ray crystallographer

The whole history of modern science, has been that of a struggle between ideas derived from observation and practice, and pre-conceptions derived from religious training. It was not…that Science had to fight an external enemy, the Church; it was that the Church itself — its dogmas, its whole way of conceiving the universe — was within the scientists themselves…. After Newton, God ruled the visible world by means of Immutable Laws of Nature, set in action by one creative impulse, but He ruled the moral world by means of absolute intimations of moral sanctions, implanted in each individual soul, reinforced and illuminated by Revelation and the Church….
In W.H. Waddington
Science and Ethics
A Marxist Critique (pp. 115–116)
George Allen & Unwin Ltd. London, England. 1942

The role of God in the material world has been reduced stage by stage with the advance of Science, so much so that He only survives in the vaguest mathematical form in the minds of older physicists and biologists.
In W.H. Waddington
Science and Ethics
A Marxist Critique (pp. 115–116)
George Allen & Unwin Ltd. London, England. 1942

Butterfield, Herbert   1900–79
English historian and philosopher of history

The greatest obstacle to the understanding of the history of science is our inability to unload our minds of modern views about the nature of the universe.
The History of Science, Origins and Results of the Scientific Revolution: A Symposium
Dante's View of the Universe (p. 15)
Free Press. Glencoe, Illinois, USA. 1953

One of the safest speculations that we could make…[is] that very soon the history of science is going to acquire an importance…incommensurate with anything that it has hitherto possessed. It…is no longer merely an account of one of many human activities like the history of music or…of cricket…. Because it deals with one of the main constituents of the modern world and the modern mind, we cannot construct a respectable history of Europe or a tolerable survey of western civilization without it. It is going to be as important for us for the understanding of ourselves as Graeco-Roman antiquity was for Europe during a period of over a thousand years.
The History of Science and the Study of History
Harvard Library Bulletin, Volume 13, 1959 (pp. 330–331)
Free Press. Glencoe, Illinois, USA. 1953

[The scientific revolution] outshines everything since the rise of Christianity and reduces the Renaissance and Reformation to the ranks of mere episodes, mere internal displacements, within the system of medieval Christendom…it looms so large as the real origin of the modern world and the modern mentality that our customary periodization of European history has become an anachronism and an encumbrance.
The Origins of Modern Science
Introduction (pp. vii–viii)
The Macmillan Company. New York, New York, USA. 1961

Chamberlain, Owen   1920–2006
American physicist

The most that any scientist can ask is that he help to lay a few stones of a partially-built edifice that we call scientific knowledge. To him this edifice is a beautiful structure, although it will never be finished.
Les Prix Nobel. the Nobel Prizes in 1959
Nobel banquet speech for award received in 1959
Nobel Foundation. Stockholm, Sweden. 1960

Cohen, I. Bernard   1914–2003
American physicist and science historian

History with the history of science, to alter slightly an apothegm of Lord Bacon, resembles a statue of Polyphemus without his eye — that very feature being left out which most marks the spirit and life of the person. My own thesis is complementary: science taught…without a sense of history is robbed of those very qualities that make it worth teaching to the student of the humanities and the social sciences.
In I. Bernard Cohen and Fletcher G. Watson (eds.)
General Education in Science (p. 71)
The History of Science and the Teaching of Science (p. 71)
Harvard University Press. Cambridge, Massachusetts, USA. 1952

Conant, James Bryant   1893–1978
American educator and scientist

We can put it down as one of the principles learned from the history of science that a theory is only overthrown by a better theory, never merely by contradictory facts.
On Understanding Science
Chapter II (p. 36)
Yale University Press. New Haven, Connecticut, USA. 1947

The history of science demonstrates beyond doubt that the really revolutionary and significant advances come not from empiricism but from new theories.

Modern Science and Modern Man
Science and Technology (p. 30)
Columbia University Press. New York, New York, USA. 1952

Darwin, Charles Robert   1809–82
English naturalist

Great is the power of steady misrepresentation — but the history of science shows how, fortunately, this power does not long endure.
In *Great Books of the Western World* (Volume 49)
The Origin of Species by Means of Natural Selection
Chapter XV (p. 239)
Encyclopædia Britannica, Inc. Chicago, Illinois, USA. 1952

Draper, John William   1811–82
American scientist, philosopher, and historian

The history of science is not a mere record of isolated discoveries; it is a narrative of the conflict of two contending powers, the expansive force of the human intellect on one side, and the compression arising from traditional faith and human interest on the other.
History of the Conflict Between Religion and Science
Preface (p. vi)
D. Appleton and Company. New York, New York, USA. 1898

Duhem, Pierre-Maurice-Marie   1861–1916
French physicist and mathematician

…the history of science alone can keep the physicist from the mad ambitions of dogmatism as well as the despair of Pyrrhonnian skepticism.
The Aim and Structure of Physical Theory
Part II, Chapter VII (p. 270)
Princeton University Press. Princeton, New Jersey, USA. 1954

Feyerabend, Paul K.   1924–94
Austrian-born American philosopher of science

The history of science, after all, does not just consist of facts and conclusions drawn from facts. It also contains ideas, interpretations of facts, problems created by conflicting interpretations, mistakes, and so on. On closer analysis we even find that science knows no "bare facts" at all but that the "facts" that enter our knowledge are already viewed in a certain way and are, therefore, essentially ideational.
Against Method: Outline of an Anarchistic Theory of Knowledge
Introduction (p. 19)
Verso. London, England. 1978

…the history of science will be as complex, chaotic, full of mistakes, and entertaining as the ideas it contains, and these ideas in turn will be as complex, chaotic, full of mistakes, and entertaining as are the minds of those who invented them.
Against Method: Outline of an Anarchistic Theory of Knowledge
Introduction (p. 19)
Verso. London, England. 1978

Fisher, Sir Ronald Aylmer   1890–1962
English statistician and geneticist

More attention to the History of Science is needed, as much by scientists as by historians, and especially by biologists, and this should mean a deliberate attempt to understand the thoughts of the great masters of the past, to see in what circumstances or intellectual milieu their ideas were formed, where they took the wrong turning or stopped short on the right track.
Natural Selection from the Genetical Standpoint
Australian Journal of Science, Volume 22, 1959

Foster, Sir Michael   1836–1907
English physiologist and educator

When we look into the past of science and trace our the first buddings of what afterwards grow to be umbrageous branches, it sometimes seems as if every time, and almost every year, marked an epoch.
Annual Report of the Board of Regents of the Smithsonian Institution, 1898
Recent Advances in Science, and Their Bearing on Medicine and Surgery (p. 345)
Government Printing Office. Washington, D.C. 1899

It is one of the lessons of the history of science that each age steps on the shoulders of the ages which have gone before.
In Lloyd William Taylor
Physics: The Pioneer Science (Volume 1)
Chapter 5 (p. 51)
Houghton Mifflin Company. Boston, Massachusetts, USA. 1941

Geikie, Sir Archibald   1835–1924
English geologist

In science, as in all other departments of inquiry, no thorough grasp of a subject can be gained unless the history of its development is clearly appreciated.
The Founders of Geology
Lecture I (p. 1)
Macmillan & Company Ltd. London, England. 1897

Hall, Alfred Rupert   1920–
English historian of science

The difficulty [in understanding science history] is the greater because the history of science is not, and cannot be, a tight unity. The different branches of science are themselves unlike in complexity, in techniques, and in their philosophy. They are not all affected equally, or at the same time, by the same historical factors, whether internal or external. It is not even possible to trace the development of a single scientific method, some formulation of principles and rules of operating which might be imagined as applicable to every scientific inquiry, for there is no such thing.
The Scientific Revolution, 1500–1800
Introduction (p. xiv)
Longmans, Green & Company. London, England. 1954

Hall, Marie Boas   1919–
English historian of science

For the…student whose chief interest does lie in science, for whom history as a course of study so often seems to deal solely with subjects remote from his intellectual turn of mind, the history of science provides a valid and useful point of contact with history, through which he may learn to develop wider humanistic interests. For the non-scientist, bored and baffled by the technical problems of science, the history of science may provide some insight into the scientific point of view and prevent the feeling of isolation which too often makes the scientist and the humanist appear to move in separate worlds.
History of Science (p. 1)
American Historical Association. Washington, D.C. 1958

Holton, Gerald   1922–
Research professor of physics and science history

And yet, on looking into the history of science, one is over-whelmed by evidence that all too often there is no regular procedure, no logical system of discovery, no simple, continuous development. The process of discovery has been as varied as the temperament of the scientist.
Thematic Origins of Scientific Thought: Kepler to Einstein
Chapter 11 (pp. 384–385)
Harvard University Press. Cambridge, Massachusetts, USA. 1973

Huxley, Thomas Henry   1825–95
English biologist

…any one acquainted with the history of science will admit that its progress has meant, in all ages and now more than ever, the extension of the province of matter and causation, and the gradual banishment from human thought of what we call spirit and spontaneity.
Collected Essays (Volume 1)
Method and Result
On the Physical Basis of Life (p. 159)
Macmillan & Company Ltd. London, England. 1904

Knickerbocker, William Skinkle   1892–1972
American professor of English and author

…the history of science is as inspiring in its human values as are the legends of the saints. Contemplate the heroism of a Galileo, the patience of a Darwin, the humility of a Pasteur; a modern eleventh chapter of Hebrews might be written listing the names of all those men of faith who by quiet work, unremitting in their zeal, one by one discovered facts which made man's lot easier and happier in what was otherwise to him a hostile and unhappy universe.
Classics of Modern Science
Preface
Alfred A Knopf. New York, New York, USA. 1927
Houghton Mifflin Company. Boston, Massachusetts, USA. 1904

Kuhn, Thomas S.   1922–96
American historian of science

Though the gap seems small, there is no chasm that more needs bridging than that between the historian of ideas and the historian of science.

International Encyclopedia of the Social Sciences
Volume 14, History of Science (p. 78)
The Macmillan Company. New York, New York, USA. 1968

Lakatos, Imre   1922–74
Hungarian-born philosopher

Philosophy of science without history of science is empty; history of science without philosophy of science is blind.
In R. Buck and R. Cohen (eds.)
Boston Studies in the Philosophy of Science (Volume 8)
History of Science and Rational Reconstructions (p. 91)
D. Reidel Publishing Company. Dordrecht, Netherlands.

Lavoisier, Antoine Laurent   1743–94
French chemist

…if I had allowed myself to enter into long dissertations on the history of the science, and the works of those who have studied it, I must have lost sight of the true object I had in view, and produced a work, the reading of which must have been extremely tiresome to beginners. It is not to the history of the science, or of the human mind, that we are to attend in an elementary treatise: Our only aim ought to be ease and perspicuity, and with the utmost care to keep every thing out of view which might draw aside the attention of the student; it is a road which we should be continually rendering more smooth, and from which we should endeavor to remove every obstacle which can occasion delay.
Elements of Chemistry in a New Systematic Order
Preface of the Author (pp. xxxii–xxxiii)
Printed for William Creech. Edinburgh, Scotland. 1790

Lévi-Strauss, Claude   1908–
French social anthropologist and structuralist

…scientific knowledge advances haltingly and is stimulated by contention and doubt.
Translated by John and Doreen Weightman
The Raw and the Cooked
Overture (p. 7)
Harper & Row, Publishers. New York, New York, USA. 1975

Libby, Walter   1867–1955
American science historian

The history of science has something to offer to the humblest intelligence. It is a means of imparting a knowledge of scientific facts and principles to unschooled minds.
An Introduction to the History of Science
Preface (p. v)
Houghton Mifflin Company. Boston, Massachusetts, USA. 1917

The history of science is an aid in scientific research. It places the student in the current of scientific thought, and gives him a clue to the purpose and necessity of the theories he is required to master. It presents science as the constant pursuit of truth rather than the formulation of truth long since revealed; it shows science as progressive

rather than fixed; dynamic rather than static, a growth to which each may contribute.
An Introduction to the History of Science
Preface (p. v)
Houghton Mifflin Company. Boston, Massachusetts, USA. 1917

The history of science is hostile to the spirit of caste. It shows the sciences rising from daily needs and occupations, formulated by philosophy, enriching philosophy, giving rise to new industries, which react in turn upon the sciences.
An Introduction to the History of Science
Preface (p. vi)
Houghton Mifflin Company. Boston, Massachusetts, USA. 1917

The history of science studies the past for the sake of the future. It is a story of continuous progress. It is rich in biographical material. It shows the sciences in their interrelations, and saves the student from narrowness and premature specialization.
An Introduction to the History of Science
Preface (p. vi)
Houghton Mifflin Company. Boston, Massachusetts, USA. 1917

Mach, Ernst   1838–1916
Austrian physicist and philosopher

[N]ot only a knowledge of the ideas that have been accepted and cultivated by subsequent teachers is necessary for the historical understanding of a science, but also that the rejected and transient thoughts of the inquirers, nay even apparently erroneous notions, may be very important and very instructive. The historical investigation of the development of a science is most needful, lest the principles treasured up in it become a system of half-understood prescripts, or worse, a system of prejudices.
The Science of Mechanics (5th edition)
Chapter II, Part VIII, Section 7 (p. 316)
The Open Court Publishing Company. La Salle, Illinois, USA. 1942

Historical investigation not only promotes the understanding of that which now is, but also brings new possibilities before us, by showing that which exists to be in great measure conventional and accidental. From the higher point of view at which different paths of thought converge we may look about us with freer vision and discover routes before unknown.
The Science of Mechanics (5th edition)
Chapter II, Part VIII, Section 7 (p. 316)
The Open Court Publishing Company. La Salle, Illinois, USA. 1942

The knowledge of the development of a science rests on the study of writings in their historical sequence and in their historical connection.
The Science of Mechanics (5th edition)
Chapter I, Part V, Section 9 (p. 97)
The Open Court Publishing Company. La Salle, Illinois, USA. 1942

Macquer, Pierre Joseph   1718–84
French chemist

As the History of any Science ought to relate the labours, the discoveries, and the errors of the cultivators of that Science; and to shew the obstacles which they have been obliged to surmount, and the mistaken paths into which they have sometimes been misled; it cannot therefore fail of being very useful to persons engaged in the same pursuits.
A Dictionary of Chemistry (Volume 1)
A Preliminary Discourse Concerning the Origin and Progress of Chemistry (p. 1)
Printed for T. Caldwell & R.F. Elmsly. London, England. 1771

Popper, Karl R.   1902–94
Austrian/British philosopher of science

The history of science, like the history of all human ideas, is a history of irresponsible dreams, of obstinacy, and of error.
Conjectures and Refutations: The Growth of Scientific Knowledge
Chapter 10, Section I (p. 216)
Harper & Row, Publishers. New York, New York, USA. 1963

Richet, Charles   1850–1935
French physiologist

In the history of science, nobody has left his mark on the world unless he has been, in this sense, an innovator.
The Natural History of a Savant
Chapter VI (p. 38)
J.M. Dent & Sons Ltd. London, England. 1927

Rostand, Jean   1894–1972
French biologist and philosopher

By showing us the extreme diversity of the factors involved in scientific creativity, the history of science teaches us that we should open the doors of our laboratories more widely. If we put that lesson into practice, our reflection on the past will have had a beneficial effect on the future.
Humanly Possible: A Biologist's Note on the Future of Mankind
On the History of Science (p. 182)
Saturday Review Press. New York, New York, USA. 1970

If there is one notion that clearly emerges from the history of science, and from which we can learn something, it is, I believe, the extreme diversity of the personal qualities and abilities that have contributed to the advancement of our knowledge.
Humanly Possible: A Biologist's Note on the Future of Mankind
On the History of Science (p. 180)
Saturday Review Press. New York, New York, USA. 1970

Sarton, George   1884–1956
Belgian-born American scholar and writer

From the point of view of the history of science, transmission is as essential as discovery.
Introduction to the History of Science (Volume 2)
Introductory Chapter (p. 15)
The Williams & Wilkins Company. Baltimore, Maryland, USA. 1927

The study of history, and especially of the history of science, may thus be regarded, not only as a source of wisdom and humanism, but also as a regulator for our consciences: it helps us not to be complacent, arrogant, too sanguine of success, and yet remain grateful and hopeful, and never to cease working quietly for the accomplishment of our own task.

The History of Science and the New Humanism
Chapter IV (p. 191)
Indiana University Press. Bloomington, Indiana, USA. 1962

Schweizer, Karl W.
American professor of history

One of the obstructions to a genuine appreciation of history is the existence of a vague unformulated assumption that historical research merely seeks to disinter a fossilized past — merely digs into the memory to recover things which the human race once knew before. On the basis of such an assumption it is possible for people to have the feeling that history can never produce anything which is fundamentally novel, but merely fills our minds with the lumber of bygone ages.

Herbert Butterfield: Essays on the History of Science
Chapter II (p. 19)
Edwin Mellon Press. Lewiston, New York, USA. 1998

Silver, Brian L.
Israeli professor of physical chemistry

The essence of scientific history has been conflict.

The Ascent of Science
Preface (p. xiii)
Solomon Press Book. New York, New York, USA. 1998

Tannery, Paul   1843–1904
French mathematician and historian of science

The scientist in so far as he is a scientist is only drawn to the history of the particular science that he studies himself; he will demand that this history be written with every possible technical detail, for it is only thus that it can supply him with materials of any possible utility. But what he will particularly require is the study of the thread of ideas and the linking together of discoveries. His chief object is to rediscover in its original form the expression of his predecessors' actual thoughts, in order to compare them with his own; and to unravel the methods that served in the construction of current theories, in order to discover at what point and towards what goal an effort towards innovation may be made.

In A. Rupert Hall
Can the History of Science Be History?
The British Journal for the History of Science, Volume 4, Part III, Number 15, June 1969 (p. 212)

Turner, H. H.
No biographical data available

It is a familiar fact that there are epochs in the history of a science when it acquires new vigor; when new branches are put forth and old branches bud afresh or blossom more plenteously. The vivifying cause is generally to be found either in the majestic form of the discovery of a new law of nature or in the humbler guise of the invention of a new instrument of research.

Annual Report of the Board of Regents of the Smithsonian Institution, 1904
Some Reflections Suggested by the Application of Photography to
Astronomical Research (p. 171)
Government Printing Office. Washington, D.C. 1905

Weisskopf, Victor Frederick   1908–2002
Austrian-American physicist

It was a heroic period [about 1922–1930] without any parallel in the history of science, the most fruitful and interesting one of modern physics.... In this great period of physics, Bohr and his associates touched the nerve of the universe. The intellectual eye of man was opened to the inner workings of nature.

In A.P. French and P.J. Kennedy (eds.)
Niels Bohr: A Centenary Volume
Niels Bohr, the Quantum, and the World (p. 22)
Harvard University Press. Cambridge, Massachusetts, USA. 1985

Wells, H. G. (Herbert George)   1866–1946
English novelist, historian, and sociologist

History is no exception amongst the sciences; as the gaps fill in, the outline simplifies; as the outlook broadens, the clustering multitude of details dissolve into general laws.

The Outline of History
Introduction (p. vi)
The Macmillan Company. New York, New York, USA. 1921

Whewell, William   1794–1866
English philosopher and historian

We may best hope to understand the nature and conditions of real knowledge by studying the nature and conditions of the most certain and stable portions of knowledge which we already possess: and we are most likely to learn the best methods of discovering truth by examining how truths, now universally recognized, have really been discovered. *The Philosophy of the Inductive Sciences Founded upon Their History* (Volume 1)
Book I, Chapter I (p. 1)
John W. Parker. London, England. 1847

[T]here do exist among us doctrines of solid and acknowledged certainty, and truths of which the discovery has been received with universal applause. These constitute what we commonly term Sciences; and of these bodies of exact and enduring knowledge, we have within our reach so large and varied a collection, that we may examine them, and the history of their formation, with good prospect of deriving from the study such instruction as we seek.

The Philosophy of the Inductive Sciences Founded upon Their History
(Volume 1)
Book I, Chapter I (p. 2)
John W. Parker. London, England. 1847

It will be universally expected that a history of Inductive Science should…afford us some indication of the most promising mode of directing our future efforts to add to its extent and completeness.
History of the Inductive Sciences, from the Earliest to the Present Time
(Volume the First)
Introduction (p. 5)
John W. Parker. London, England. 1837

[T]he progress of knowledge is the main action of our drama; and all the events which do not bear upon this, though they may relate to the cultivation and the cultivators of philosophy, are not a necessary part of our theme.
History of the Inductive Sciences, from the Earliest to the Present Time
(Volume the First)
Introduction (pp. 9, 12)
John W. Parker. London, England. 1837

Whitehead, Alfred North   1861–1947
English mathematician and philosopher

Science is concerned with the facts of bygone transition. History relates the aim at ideals. And between Science and History, lies the operation of the Deistic impulse of energy. It is the religious impulse in the world which transforms the dead facts of Science into the living drama of History. For this reason Science can never foretell the perpetual novelty of History.
Modes of Thought
Chapter II, Lecture Five (p. 142)
The Macmillan Company. New York, New York, USA. 1938

Williams, L. Pearce
American historian of science

…the history of science is a professional and rigorous discipline demanding the same level of skills and scholarship as any other scholarly field. It is time for the scientists to realize that he studies nature and others study him. He is no more nor no less competent to comment on his own activities and the activities of his fellow scientist than is the politician. Critical political history is rarely written by the politician and the same is true of the history of science.
Letter to the Editor
Scientific American, Volume 214, Number 6, June 1966 (p. 8)

Willstätter, Richard   1872–1942
German chemist

I consider the teaching and study of the historical development of science as indispensable.… Our textbooks fail in this respect.
In Rolf Huisgen
Adolf von Baeyer's Scientific Achievements — A Legacy
Angewandet Chemie International Edition in English, Volume 25, Number 4, April 1986 (p. 297)

SCIENCE, MAN OF

Barrie, Sir James M.   1860–1937
Scottish journalist, writer, and dramatist

The man of science appears to be the only man who has something to say, just now — and the only man who does not know how to say it.
Applied Physics, Volume 2, 1963

Berkeley, George   1685–1753
Irish prelate and metaphysical philosopher

Query: Whether the difference between a mere computer and a man of science be not, that the one computes on principles clearly conceived, and by rules evidently demonstrated, whereas the other doth not?
In A. Luce and T. Jessop (eds.)
The Works of George Berkeley, Bishop of Cloyne (Volume 4)
The Analyst
Nelson. London, England. 1948

Bernard, Claude   1813–78
French physiologist

Men of science, then, do not seek for the pleasure of seeking; they seek the truth to possess it, and they possess it already within the limits expressed in the present state of science.
Translated by Henry Copley Greene
An Introduction to the Study of Experimental Medicine
Part Three, Chapter IV, Section IV (p. 222)
Henry Schuman, Inc. New York, New York, USA. 1927

…men of science must not halt on the road; they must climb ever higher and strive toward perfection; they must always seek as long as they see anything to be found.
Translated by Henry Copley Greene
An Introduction to the Study of Experimental Medicine
Part Three, Chapter IV, Section IV (p. 222)
Henry Schuman, Inc. New York, New York, USA. 1927

Bradley, Omar   1893–1981
American Army general

With the monstrous weapons man already has, humanity is in danger of being trapped in this world by its moral adolescents. Our knowledge of science has already outstripped our capacity to control it. We have many men of science, too few men of God.
Address
Boston, November 10, 1948

Butler, Samuel   1612–80
English novelist, essayist, and critic

I do not know whether my distrust of men of science is congenital or acquired, but I think I should have transmitted it to descendants.
In Geoffrey Keynes and Brian Hill (eds.)
Samuel Butler's Notebooks
Myself and Distrust of Men of Science (p. 32)
Jonathan Cape. London, England. 1951

If [men of science] are worthy of the name, [they] are indeed about God's path and about his bed and spy out all his ways.

In Geoffrey Keynes and Brian Hill (eds.)
Samuel Butler's Notebooks
Men of Science (p. 204)
Jonathan Cape. London, England. 1951

Chesterton, G. K. (Gilbert Keith)   1874–1936
English author

Far away in some strange constellation in skies infinitely remote, there is a small star, which astronomers may some day discover. At least, I could never observe in the faces or demeanor of most astronomers or men of science any evidence that they had discovered it; though as a matter of fact they were walking about on it all the time. It is a star that brings forth out of itself very strange plants and very strange animals; and none stranger than the men of science.
The Everlasting Man
Chapter I (p. 1)
Dodd, Mead & Company. New York, New York, USA. 1925

…the ordinary scientific man is strictly a sentimentalist. He is a sentimentalist in this essential sense, that he is soaked and swept away by mere associations.
Orthodoxy
Chapter IV (pp. 94–95)
John Lane Company. New York, New York, USA. 1918

Clifford, William Kingdon   1845–79
English philosopher and mathematician

A man of science…explains as much as ever he can, and then he says, "This is all I can do; for the rest you must ask the next man."
In Leslie Stephen and Frederick Pollock (eds.)
Lectures and Essays (Volume 2)
Body and Mind (p. 32)
Macmillan & Company. London, England. 1879

Darwin, Charles Robert   1809–82
English naturalist

Children are one's greatest happiness, but often and often a still greater misery. A man of science ought to have none — perhaps not a wife; for then there would be nothing in this wide world worth caring for, and a man might (whether he could is another question) work away like a Trojan.
In Francis Darwin (ed.)
More Letters of Charles Darwin (Volume 1)
Letter 139, Darwin Asa Gray, July 11, 1862 (p. 202)
D. Appleton & Company. New York, New York, USA. 1903

…my success as a man of science, whatever this may have amounted to, has been determined, as far as I can judge, by complex and diversified mental qualities and conditions. Of these, the most important have been — the love of science, unbounded patience in long reflecting over any subject, industry in observing and collecting facts, and a fair share of invention as well as of common-sense. With such moderate abilities as I possess,

it is truly surprising that I should have influenced to a considerable extent the belief of scientific men on some important points.
In Francis Darwin (ed.)
The Life and Letters of Charles Darwin (Volume 1)
Chapter II (p. 85)
D. Appleton & Company. New York, New York, USA. 1896

Dumas, Jean Baptiste-Andre   1800–84
French biochemist

The recollection of an already long life has permitted me to become acquainted with a great variety of personages. And if I call on memory to picture to me how the type of true happiness is realized on earth I do not see it under the form of the powerful man clothed in high authority, nor under that of the rich man to whom the splendors of luxury and the delicacies of well-being are granted, but under that of the man of science, who consecrates his life to penetrating the secrets of Nature and to the discovery of new truths.
In Sir Richard Arman Gregory
Discovery; or, The Spirit and Service of Science
Chapter I (p. 18)
Macmillan & Company Ltd. London, England. 1918

Eddington, Sir Arthur Stanley   1882–1944
English astronomer, physicist, and mathematician

Verily, it is easier for a camel to pass through the eye of a needle than for a scientific man to pass through a door. And whether the door be barn door or church door it might be wiser that he should consent to be an ordinary man and walk in rather that wait till all the difficulties involved in a really scientific ingress are resolved.
The Nature of the Physical World
Chapter XV (p. 342)
The Macmillan Company. New York, New York, USA. 1930

Einstein, Albert   1879–1955
German-born physicist

It has often been said, and certainly not without justification, that the man of science is a poor philosopher. Why then should it not be the right thing for the physicist to let the philosopher do the philosophizing? …At a time like the present, when experience forces us to seek a newer and more solid foundation, the physicist cannot simply surrender to the philosopher the critical contemplation of the theoretical foundations; for, he himself knows best, and feels more surely where the shoe pinches. In looking for a new foundation, he must try to make clear in his own mind just how far the concepts which he uses are justified, and are necessities.
Physik and Realität
Journal of the Franklin Institute, Volume 221, 1936

Emerson, Ralph Waldo   1803–82
American lecturer, poet, and essayist

We hearken to the man of science, because we anticipate the sequence in natural phenomena which he uncovers.
The Complete Works of Ralph Waldo Emerson (Volume 4)
Representative Men
Chapter IV (p. 170)
Houghton Mifflin Company. Boston, Massachusetts, USA. 1904

Froude, James Anthony   1818–94
English historian and biographer

The secrets of nature have been opened out to us on a thousand lines; and men of science of all creeds can pursue side by side their common investigations.
Short Studies on Great Subjects (Volume 1)
Times of Erasmus, Desderius and Luther, Lecture I (p. 41)
Longmans, Green & Company. London, England. 1879

Gregory, Sir Richard Arman   1864–1952
British science writer and journalist

The scientific man has to work for truth so far as her ways can be comprehended by him, but he is never more than a trustee for posterity, and has no authority to define the functions or limit the freedom of those who follow him.
Discovery; or The Spirit and Service of Science
Chapter II (p. 30)
Macmillan & Company Ltd. London, England. 1918

The man of science, by virtue of his training, is alone capable of realising the difficulties — often enormous — of obtaining accurate data upon which just judgment may be based.
Discovery; or, The Spirit and Service of Science
Chapter III (p. 40)
Macmillan & Company Ltd. London, England. 1918

To the popular mind, a man of science is a callous necromancer who has cut himself off from communion with his fellows, and has thereby lost the throbbing and compassionate heart of a full life: he is a Faust who has not yet made a bargain with Mephistopheles, and is therefore without human interest.
Discovery; or, The Spirit and Service of Science
Preface (p. v)
Macmillan & Company Ltd. London, England. 1918

Hall, Asaph   1829–1907
American astronomer

When men are striving for the discovery of truth in its various manifestations, they learn that it is by correcting the mistakes of preceding investigators that progress is made, and they have charity for criticism. Hence persecution for difference of opinion becomes an absurdity. The labours of scientific men are forming a great body of doctrine that can be appealed to with confidence in all countries. Such labours bring people together, and tend to break down national barriers and restrictions.
In Sir Richard Arman Gregory
Discovery; or, The Spirit and Service of Science

Chapter II (pp. 30–31)
Macmillan & Company Ltd. London, England. 1918

Huxley, Thomas Henry   1825–95
English biologist

The man of science has learned to believe in justification, not by faith, but by verification.
Collected Essays (Volume 1)
Method and Result
On Improving Natural Knowledge (p. 41)
Macmillan & Company Ltd. London, England. 1904

Laplace, Pierre Simon   1749–1827
French mathematician, astronomer, and physicist

The isolated man of science can dedicate himself without fear to dogmatism; he hears only from afar contradictions of his ideas. But in a scientific society the impact of dogmatic ideas soon results in their destruction, and the desire to win one another over to their point of view establishes necessarily among members the convention of admitting only the results of observations and calculation.
In Maurice Crosland
Gay-Lussac: Scientist and Bourgeois
Chapter 2 (p. 34)
Cambridge University Press. Cambridge, England. 1978

Mather, Kirtley F.   1888–1978
American geologist

To the man of science every event in the history of the universe is a miracle. It is both awe-inspiring and significant, a "sign and wonder."
In Edward H. Cotton
Has Science Discovered God?
Sermons from Stones (p. 3)
Thomas Y. Crowell Company, Publishers. New York, New York, USA. 1931

Melville, Herman   1819–91
American novelist

…a man of true science…uses but few hard words, and those only when none other will answer his purpose; whereas the smatterer in science…thinks, that by mouthing hard words, he proves that he understands hard things.
White Jacket
Chapter LXIII (p. 277)
Northwestern University Press. Evanston, Illinois, USA. 1970

Pearson, Karl   1857–1936
English mathematician

The scientific man has above all things to strive at self-elimination [elimination of self] in his judgments.…
The Grammar of Science
Introductory, Section 2 (p. 7)
Charles Scribner's Sons. London, England. 1892

Poincaré, Henri   1854–1912
French mathematician and theoretical astronomer

The true man of science has no such expression in his vocabulary as useful science…if there can be no science for science's sake there can be no science.
In James Kip Finch
Engineering and Science
Technology and Culture, Fall 1961 (p. 330)

Popper, Karl R.   1902–94
Austrian/British philosopher of science

…it is not his possession of knowledge, of irrefutable truth, that makes the man of science, but his persistent and recklessly critical quest for truth.
The Logic of Scientific Discovery
Part II, Chapter X, Section 85 (p. 281)
Basic Books, Inc. New York, New York, USA. 1959

Renan, Ernest   1823–92
French philosopher and Orientalist

With the saints, the heroes, the great men of all ages we may fearlessly compare our men of scientific minds, given solely to the research of truth, indifferent to fortune, often proud of their poverty, smiling at the honors they are offered, as careless of flattery as of obloquy, sure of the worth of that they are doing, and happy because they possess truth.
The Nobility of Science
Scientific American, Volume 40, Number 20, New Series, 17 May 1879 (p. 310)

Robinson, James Harvey   1863–1936
American historian
Beard, Charles A.   1874–1948
American historian

It may well be that men of science, not kings, or warriors, or even statesmen are to be the heroes of the future.
The Development of Modern Europe: An Introduction to the Study of Current History (Volume 2)
Chapter XXXI (p. 421)
Ginn & Company. Boston, Massachusetts, USA. 1908

Ross, Sir Ronald   1857–1932
English bacteriologist

A witty friend of mine once remarked that the world thinks of the man of science as one who pulls out his watch and exclaims, "Ha! half an hour to spare before dinner: I will just step down to my laboratory and make a discovery." Who but men of science themselves are to blame for such a misconception? Out of the many memoirs…[o]ur books of science are records of results rather than of that sacred passion for discovery which leads to them. Yet many discoveries have really been the climax of an intense drama… in which the protagonists are man and nature, and the issue a decision for all the ages.
Memoirs
Preface (pp. v–vi)
Publisher undetermined

Russell, Bertrand Arthur William   1872–1970
English philosopher, logician, and social reformer

The man of science looks for facts that are significant, in the sense of leading to general laws; and such facts are frequently quite devoid of intrinsic interest.
The Scientific Outlook
Chapter I (p. 49)
George Allen & Unwin Ltd. London, England. 1931

All the conditions of happiness are realized in the life of the man of science.
The Conquest of Happiness
Chapter X (p. 146)
Liverwright Publishing Corporation. New York, New York, USA. 1930

Spencer, Herbert   1820–1903
English social philosopher

Only the sincere man of science (and by this title we do not mean the mere calculator of distances, or analyser of compounds, or labeler of species; but him who through lower truths seeks higher, and eventually the highest) — only the genuine man of science, we say, can truly know how utterly beyond, not only human knowledge, but human conception, is the Universal Power of which Nature, and Life, and Thought are manifestations.
Education: Intellectual, Moral, and Physical
Chapter I (p. 84)
A.L. Fowle. New York, New York, USA. 1860

Suits, C. G.
American physicist

I've never met that "coldly calculating man of science" whom the novelists extol.… I doubt that he exists; and if he did exist I greatly fear that he would never make a startling discovery or invention.
In Frederic Brownell
Heed that Hunch
The American Magazine, December 1945 (p. 142)

Sullivan, John William Navin   1886–1937
Irish mathematician

…outside their views on purely scientific matters there is nothing characteristic of men of science.
Aspects of Science
Scientific Citizen (p. 120)
J. Cape & H. Smith. New York, New York, USA. 1927

Thoreau, Henry David   1817–62
American essayist, poet, and practical philosopher

He is not a true man of science who does not bring some sympathy to his studies, and expect to learn something by behavior as well as by application. It is childish to rest in the discovery of mere coincidences, or of partial and extraneous laws. The study of geometry is a petty and idle exercise of the mind if it is applied to no larger system than the starry one.

The Writings of Henry David Thoreau (Volume 1)
A Week on the Concord and Merrimack Rivers
Friday (p. 477)
Houghton Mifflin Company. Boston, Massachusetts, USA. 1893

The true man of science will know nature better by his finer organization; he will smell, taste, see, hear, feel, better than other men. His will be a deeper and finer experience. We do not learn by inference and deduction and the application of mathematics to philosophy, but by direct intercourse and sympathy. It is with science as with ethics, — we cannot know truth by contrivance and method; the Baconian is as false as any other, and with all the helps of machinery and the arts, the most scientific will still be the healthiest and friendliest man, and possess a more perfect Indian wisdom.
The Writings of Henry David Thoreau (Volume 9)
Natural History of Massachusetts (pp. 161–162)
Houghton Mifflin Company. Boston, Massachusetts, USA. 1893

Twain, Mark (Samuel Langhorne Clemens)   1835–1910
American author and humorist

The surest way for a nation's scientific men to prove that they were proud and ignorant was to claim to have found out something fresh in the course of a thousand years or so. Evidently the peoples of this book's day regarded themselves as children, and their remote ancestors as the only grown-up people that had existed. Consider the contrast:… our own scientific men may and do regard themselves as grown people and their grandfathers as children. The change…is probably the most sweeping that has ever come over mankind in the history of the race. It is the utter reversal, in a couple of generations, of an attitude which had been maintained without challenge or interruption from the earliest antiquity.… The change from reptile to bird was not more tremendous, and it took longer.
The Complete Humorous Sketches and Tales of Mark Twain
A Majestic Literary Fossil (p. 534)
Hanover House. Garden City, New York, USA. 1961

von Goethe, Johann Wolfgang   1749–1832
German poet, novelist, playwright, and natural philosopher

Scientific man is supposed to limit himself to his immediate surroundings. However if he should occasionally want to step forth as a poet, he certainly should not be prevented from doing so.
In Karl J. Fink
Goethe's History of Science
Chapter 9 (p. 125)
Cambridge University Press. Cambridge, England. 1991

von Helmholtz, Hermann   1821–94
German scientist and philosopher

In fact, men of science form, as it were, an organised army, labouring on behalf of the whole nation, and generally under its direction and at its expense, to augment the stock of such knowledge as may serve to promote industrial enterprises, to increase wealth, to adorn life, to improve political and social relations, and to further the moral development of individual citizens.
Popular Lectures on Scientific Subjects
Lecture I
Volume 2, 1846 (p. 28)
D. Appleton & Company. New York, New York, USA. 1885

Wells, H. G. (Herbert George)   1866–1946
English novelist, historian, and sociologist

The training of a scientific man is a training in what an illiterate lout would despise as a weakness, it is a training in blabbing, in blurting things out, in telling just as plainly as possible and as soon as possible what it is he has found.
New Worlds for Old
Chapter II (p. 23)
The Macmillan Company. New York, New York, USA. 1918

Whitehead, Alfred North   1861–1947
English mathematician and philosopher

No man of science wants merely to know. He acquires knowledge to appease his passion for discovery. He does not discover in order to know, he knows in order to discover.
The Orginsation of Thought
Chapter II (p. 37)
Greenwood Press Publishers. Westport, Connecticut, USA. 1974

Whitney, Willis Rodney   1868–1958
American chemical and electrical engineer

For the engineer "safety first" is a good slogan, but "safety last" is better for the man of research.
The Stimulation of Research in Pure Science Which Has Resulted from the Needs of Engineers and of Industry
Science, Volume 65, Number 1862, March 25, 1927 (p. 289)

Wordsworth, William   1770–1850
English poet

If the labours of men of Science should ever create any material revolution, direct or indirect, in our condition, and in the impressions which we habitually receive, the Poet will sleep then no more than at present, but he will be ready to follow the steps of the Man of Science, not only in those general indirect effects, but he will be at his side, carrying sensation into the midst of the objects of the Science itself. The remotest discoveries of the Chemist, the Botanist, or Mineralogist will be as proper objects of the Poet's art as any upon which I can be employed, if the time should ever come when these things shall be familiar to us…In R.L. Brett and A.R. Jones (eds.)
Lyrical Ballads
Preface (pp. 259–260)
Methuen & Company Ltd. London, England. 1963

SCIENCE, PROGRESS OF

Abernethy, John   1680–1740
Irish Presbyterian minister, theologian, and dissenter

Although knowledge has at times appeared to exhibit something of uniformity in its advances, yet it can not have escaped the least observant that, as a whole, the Progress of Science has been marked by very variable activity. At once time marvelously rapid; at another, indefinitely slow; now merged in darkness or obscurity, and not blazing forth with meridian splendor.
Memoirs of John Abernethy
Chapter I (p. 1)
Harper & Brothers. New York, New York, USA. 1853

Ardrey, Robert   1908–80
American anthropologist

The contemporary revolution in the natural sciences has proceeded in something more striking than silence. It has proceeded in secret. Like our tiny, furry, squirrel-like, earliest primate ancestors, seventy million years ago, the revolution has found obscurity its best defence and modesty the key to its survival. For it has challenged larger orthodoxies than just those of science, and its enemies exist beyond counting. From seashore and jungle, from ant-heap and travertine cave have been collected the inflammable materials that must some day explode our most precious intellectual movement seeking light under darkest cover.
African Genesis
Chapter I, Section 2 (p. 13)
Athenaeum. New York, New York, USA. 1968

Bernard, Claude   1813–78
French physiologist

The progress of experimental method consists in this, — that the sum of truths grows larger in proportion as the sum of errors grows less. But each one of these particular truths is added to the rest to establish more general truths. In this fusion, the names of promoters of science disappear little by little, and the further science advances, the more it takes an impersonal form and detaches itself from the past.
An Introduction to the Study of Experimental Medicine
Part I, Chapter II, Section iv (p. 42)
Henry Schuman, Inc. New York, New York, USA. 1927

Bronowski, Jacob   1908–74
Polish-born British mathematician and polymath

The progress of science is the discovery at each step of a new order which gives unity to what had long seemed unlike.
Science and Human Value
The Creative Mind (p. 26)
Harper & Row, Publishers. New York, New York, USA. 1965

Cohen, Morris Raphael   1880–1947
American philosopher

…the progress of science always depends upon our questioning the plausible, the respectably accepted, and the seemingly self-evident.
Reason and Nature
Book III
Chapter One, Section II (p. 348)
Free Press. Glencoe, Georgia, USA. 1953

Coles, Abraham   1813–91
American physician, hymnist, and poet

Believing needless ignorance a crime,
You strive to reach the summit of your time;
To old age learning up from early youth
Your life one long apprenticeship to truth.
Wisely suspicious sometimes of the new,
Ye give alert acceptance to the true:
Even though it make old science obsolete,
It with a thousand welcomes still you greet…
Each Year adds something — many things ye know
Your sires knew not a Hundred Years ago.
The Microcosm and Other Poems
The Microcosm
Physician's Character and Aims — Science Progressive
D. Appleton & Company. New York, New York, USA. 1881

Daly, Reginald Aldworth   1871–1957
Canadian-American geologist

Inasmuch as cosmogony and geology are both young sciences, consensus of opinions about the earth's origin and history is still reserved for the future. Meantime these sciences are advancing through the erection and testing of competing hypotheses; in other words, through speculation, controlled by all the available facts. Science progresses through systematic guessing in the good sense of the world.
Our Mobile Earth
Introduction (p. xx)
Charles Scribner's Sons. New York, New York, USA. 1926

Dewar, James   1842–1923
English physicist and chemist

In a legitimate sense all genuine scientific workers feel that they are "inheritors of unfulfilled renown." The battlefields of science are the centers of a perpetual warfare, in which there is no hope of final victory, although partial conquest is ever triumphantly encouraging the continuance of the disciplined and strenuous attack on the seemingly impregnable fortress of nature.
Annual Report of the Board of Regents of the Smithsonian Institution, 1902
History of Cold and the Absolute Zero (p. 240)
Government Printing Office. Washington, D.C. 1903

Dryden, John   1631–1700
English poet, dramatist, and literary critic

Is it not evident, in these last hundred years (when the Study of Philosophy has been the business of all the Virtuosi in Christendom) that almost a new Nature has been reveal'd to us? that more errors of the School have been detected, more useful Experiments in Philosophy have been made, more Noble Secrets in Opticks, Medicine, Anatomy, Astronomy, discover'd, than in all those credulous and doting Ages from Aristotle to us? so true it is that nothing spreads more fast than Science, when rightly and generally cultivated.
Of Dramatick Poesie: An Essay (Volume 1) (p. 12)
Printed for Henry Herringman. London, England. 1684

Duclaux, Pierre Émile   1840–1904
French biochemist

It is because science is sure of nothing that it is always advancing.
In William Osler
Evolution of Modern Medicine
Chapter VI (p. 219)
Yale University Press. New Haven, Connecticut, USA. 1921

A series of judgments, revised without ceasing, goes to make up the incontestable progress of science.
In W. Mansfield Clark
The Determination of Hydrogen Ions
Chapter VIII (p. 177)
The Williams & Wilkins Company. Baltimore, Maryland, USA. 1928

Einstein, Albert   1879–1955
German-born physicist
Infeld, Leopold   1898–1968
Polish physicist

Science forces us to create new ideas, new theories. Their aim is to break down the wall of contradictions which frequently blocks the way of scientific progress. All the essential ideas in science were born in a dramatic conflict between reality and our attempts at understanding.
The Evolution of Physics
Quanta (p. 280)
Simon & Schuster. New York, New York, USA. 1934

Enriques, Federigo   1871–1946
Italian mathematician

[T]he progress of science is dependent upon science itself, it is an extension and not a creation.
Problems of Science
Chapter 3, Section 37 (p. 165)
The Open Court Publishing Company. Chicago, Illinois, USA. 1914

Foster, Sir Michael   1836–1907
English physiologist and educator

The path [of progress in science] may not always be in a straight line; there may be swerving to this side and to that; ideas may seem to return again and again to the same point of the intellectual compass; but it will always

be found that they have reached a higher level.... Moreover, science is not fashioned as is a house, by putting brick to brick, that which is once put remaining as it was put to the end. The growth of science is that of a living being. As is the embryo, phases follows phase, and each member or body puts on in succession different appearances, though all the while the same member, so a scientific conception of one age seems to differ from that of a following age...
Annual Report of the Smithsonian Institution For 1899
The Growth of Science in the Nineteenth Century
Government Printing Office. Washington, D.C. 1900

France, Anatole (Jean Jacques Brousson)   1844–1924
French writer

The progress of science renders useless the very books which have been the greatest aid to that progress. As those works are no longer useful, modern youth is naturally inclined to believe they never had any value; it despises them, and ridicules them if they happen to contain any superannuated opinion whatever.
Translated by Lafcadio Hern
The Crime of Sylvester Bonnard
June 4 (p. 168)
Harper & Brothers. New York, New York, USA. 1890

Free, E. E.
No biographical data available

Like a man on a bicycle science cannot stop; [science] must progress or collapse.
The Electric Brains in the Telephone
The World's Work, Volume LIII, Number 4, February 1927 (p. 429)

Garrod, Archibald   1857–1936
English physician

In these days of rapid scientific progress there is a tendency to accept the facts of nature, as at present known, without glancing back at the slow and difficult stages by which the knowledge of these facts has been arrived at. Yet such a retrospect is by no means unprofitable, since it warns us that hasty generalizations upon insufficient data retard rather than advance the progress of knowledge, and that the theories of the day must not be accepted as necessarily expressing absolute truths.
In Alexander G. Bearn
Archibald Garrod and the Individuality of Man
Chapter 3 (p. 25)
Clarendon Press. Oxford, England. 1993

Greene, Brian   1963–
American physicist

Progress in science proceeds in fits and starts. Some periods are filled with great breakthroughs, in other times researchers experience dry spells. Scientists put forward results, both theoretical and experimental. The results are debated by the community, sometimes they are discarded,

sometimes they are modified, and sometimes they provide inspirational jumping-off points for new and more accurate ways of understanding the physical universe. In other words, science proceeds along a zigzag path toward what we hope will be ultimate truth, a path that began with humanity's earliest attempts to fathom the cosmos and whose end we cannot predict.
The Elegant Universe
Chapter 1 (p. 20)
W.W. Norton & Company, Inc. New York, New York, USA.2003

Heisenberg, Werner Karl   1901–76
German physicist and philosopher

Science progresses not only because it helps to explain newly discovered facts, but also because it teaches us over and over again what the word "understanding" may mean.
Physics and Beyond
Chapter 10 (p. 124)
Harper & Row, Publishers. New York, New York, USA. 1971

Kuhn, Thomas S.   1922–96
American historian of science

…we must explain why science — our surest example of sound knowledge — progresses as it does, and we first must find out how, in fact, it does progress.
In Imre Lakatos and Alan Musgrave (eds.)
Criticism and the Growth of Knowledge
Logic of Discovery or Psychology of Research (p. 20)
Cambridge University Press. Cambridge, England. 1970

Does a field make progress because it is a science, or is it a science because it makes progress?
The Structure of Scientific Revolutions (2nd edition)
Chapter XIII (p. 162)
The University of Chicago Press. Chicago, Illinois, USA. 1970

Lee, Tsung Dao   1926–
Chinese-born American nuclear physicist

The progress of science has always been the result of a close interplay between our concepts of the universe and our observations on nature. The former can only evolve out of the latter and yet the latter is also conditioned greatly by the former. Thus, in our exploration of nature, the interplay between our concepts and our observations may sometimes lead to totally unexpected aspects among already familiar phenomena.
Nobel Lectures, Physics 1942–1962
Nobel lecture for award received in 1957
Weak Interactions and Nonconservation of Parity (p. 417)
Elsevier Publishing Company. Amsterdam, Netherlands. 1964

Lindley, David   1956–
English astrophysicist and author

Progress in science is a matter of jumping to conclusions. The trick is to jump to useful and interesting conclusions. Generalizing from small scraps of evidence may lead one astray, but sticking strictly to what limited evidence one has, and refusing to countenance anything that is not directly provable, leads nowhere at all. The scientist has to generate new ideas and hypotheses, then act upon them.
Where Does the Weirdness Go? Why Quantum Mechanics Is Strange, but Not as Strange as You Think
An Engineer, a Physicist, and a Philosopher… (p. 157)
Basic Books, Inc. New York, New York, USA. 1996

Lowie, Robert H.   1883–1957
Austrian-born American anthropologist

The clarification of concepts…directly gauges scientific progress.
The History of Ethnological Theory
Chapter XIV (p. 281)
Rinehart & Company, Inc. New York, New York, USA. 1937

Mayr, Ernst   1904–2005
German-born American biologist

Any scientific revolution has to accept all sorts of black boxes, for if one had to wait until all black boxes are opened, one would never have any conceptual advances.
One Long Argument: Charles Darwin and the Genesis of Modern Evolutionary Thought
Chapter Ten (p. 146)
Harvard University Press. Cambridge, Massachusetts, USA. 1991

Medawar, Sir Peter Brian   1915–87
Brazilian-born English zoologist

It can be said that Science progresses only by peeling away, one after another, all the covering of apparent stability in the world; disclosing beneath the immobility of the infinitely small, movement of extra rapidity, and beneath the immobility of the Immense, movement of extra slowness.
The Future of Man
Some Reflections on Progress (p. 62)
Methuen & Company Ltd. London, England. 1960

Planck, Max   1858–1947
German physicist

It is a rather zigzag pattern than the curve of scientific progress follows; indeed I might say that the forward movement is of an explosive type, where the rebound is an attendant characteristic of the advance. Every applied hypothesis which succeeds in throwing the searchlight of a new vision across the field of physical science represents a plunge into the darkness; because we cannot at first reduce the vision to a logical statement. Then follows the birth-struggle of a new theory. Once this has seen the light of day it has to go forward willy-nilly until the stamp of its destiny is put on it when the test of the research measurements is applied.
Where Is Science Going?
Nature's Image in Science (pp. 90–91)
W.W. Norton & Company, Inc. New York, New York, USA.1932

An important scientific innovation rarely makes its way by gradually winning over and converting its opponents:

it rarely happens that Saul becomes Paul. What does happen is that its opponents gradually die out and that the growing generation is familiarized with the idea from the beginning...
The Philosophy of Physics
Chapter III (p. 97)
W.W. Norton & Company, Inc. New York, New York, USA.1936

Popper, Karl R.   1902–94
Austrian/British philosopher of science

In science it would be a tremendous loss if we were to say: "We are not making very much progress. Let us sweep away all science and start afresh." The rational procedure is to correct it and to revolutionize it, but not to sweep it away. You may create a new theory, but the new theory is created in order to solve those problems which the old theory did not solve.
Conjectures and Refutations
Chapter 4 (p. 132)
Harper & Row, Publishers. New York, New York, USA. 1963

...in order that a new theory should constitute a discovery or a step forward it should conflict with its predecessor...it should contradict its predecessor; it should overthrow it. In this sense, progress in science — or at least a striking progress — is always revolutionary.
In Rom Harré
Problems of Scientific Revolution
The Rationality of Scientific Revolutions (pp. 82–83)
The Clarendon Press. Oxford, England. 1975

Price, Don K.   1910–1995
American presidential advisor and educator

...most scientists are prepared to work most of the time within the framework of ideas developed by their acknowledged leaders. In that sense...science is ruled by oligarchs who hold influence as long as their concepts and systems are accepted as the most successful strategy.... Once in a great while, a rival system is proposed; then there can usually be no settlement of the issue by majority opinion. The metaphor of "scientific revolution" suggests the way in which the losing party is displaced from authority, discredited and its doctrines eliminated from textbooks.
The Scientific Estate
Chapter 6 (p. 172)
Harvard University Press. Cambridge, Massachusetts, USA. 1965

Priestley, Joseph   1733–1804
English theologian and scientist

If the progress continues the same in another period, of equal length, what a glorious science shall we see unfold, what a fund of entertainment is there in store for us, and what important benefits must derive mankind.
Quoted by John G. McEvoy
Electricity, Knowledge, and the Nature of Progress in Priestley's
Thought *The British Journal for the History of Science*, Volume 12,
Number 40, 1979 (p. 76)

Richet, Charles   1850–1935
French physiologist

One can only progress in the sciences — with the exception of Mathematics — at the price of great pecuniary sacrifice.
Translated by Sir Oliver Lodge
The Natural History of a Savant
Chapter II (p. 21)
J.M. Dent & Sons Ltd. London, England. 1927

All progress in science is progress in civilization, and consequently contributes to the welfare of man.
Translated by Sir Oliver Lodge
The Natural History of a Savant
Chapter XIII (p. 145)
J.M. Dent & Sons Ltd. London, England. 1927

Rindos, David   1947–96
American educator

Progress in science depends not only upon new data but also upon the careful elaboration of new approaches to old data as well as new.
In Michael B. Schiffer (ed.)
Archaeological Method and Theory (Volume 1)
Chapter I (p. 1)
University of Arizona Press. Tucson, Arizona, USA. 1989

Romanoff, Alexis Lawrence   1892–1980
Russian soldier and scientist

Science is the key to the progress of the world.
Encyclopedia of Thoughts
Aphorism 20
Ithaca Heritage Books. Ithaca, New York, USA. 1975

Thomson, Sir George   1892–1975
English physicist

...the progress of science is a little like making a jig-saw puzzle. One makes collections of pieces which certainly fit together, though at first it is not clear where each group should come in the picture as a whole, and if at first one makes a mistake in placing it, this can be corrected later without dismantling the whole group.
The Inspiration of Science
Introduction (pp. 5–6)
Oxford University Press, Inc. London, England. 1961

von Bertalanffy, Ludwig   1901–72
Austrian biologist

The evolution of science is not a movement in an intellectual vacuum; rather it is both an expression and a driving force of the historical process.
Problems of Life: An Evaluation of Modern Biological Thought
Chapter Six (p. 202)
Watts & Company. London, England. 1952

von Goethe, Johann Wolfgang   1749–1832
German poet, novelist, playwright, and natural philosopher

...we might well mention from our perspective, that there are advantages to entering a field of science which is in a state of crisis, and in which we also find an active, extraordinary person. We are young with young methods, our beginnings reach into a new epoch.
In Karl J. Fink
Goethe's History of Science
Chapter 6 (p. 88)
Cambridge University Press. Cambridge, England. 1991

von Liebig, Justus   1803–73
German organic chemist

Thus the progress of science is, like the development of nature's works, gradual and expansive. After the buds and branches spring forth the leaves and blossoms, after that blossoms the fruit.
Familiar Letters on Chemistry
Letter I (p. 10)
Taylor & Walton. London, England. 1843

von Neumann, John   1903–57
Hungarian-American mathematician
Morgenstern, Oskar   1902–77
German-born American economist

The great progress in every science came when, in the study of problems which were modern as compared with ultimate aims, methods were developed which could be extended further and further. The free fall is a very trivial physical phenomenon, but it was the study of this exceedingly simple fact and its comparison with the astronomical material, which brought forth mechanics.
Theory of Games and Economic Behavior
Chapter 1.3.2 (p. 6)
Princeton University Press. Princeton, New Jersey, USA. 1947

Whitehead, Alfred North   1861–1947
English mathematician and philosopher

The progress of Science consists in observing interconnections and in showing with a patient ingenuity that the events of this ever-shifting world are but examples of a few general relations, called laws. To see what is general in what is particular, and what is permanent in what is transitory, is the aim of scientific thought.
An Introduction to Mathematics
Chapter 1 (p. 4)
Oxford University Press, Inc. New York, New York, USA. 1958

[S]cience started its modern career by taking over ideas derived from the weakest side of the philosophies of Aristotle's successors. In some respects it was a happy choice. It enabled the knowledge of the seventeenth century to be formulated so far as physics and chemistry were concerned, with a completeness which lasted to the present time. But the progress of biology and psychology has probably been checked by the uncritical assumption of half-truths.

Science and the Modern World
Chapter I (pp. 16–17)
The Macmillan Company. New York, New York, USA. 1929

In the conditions of modern life the rule is absolute, the race which does not value trained intelligence is doomed. Not all your heroism, not all your social charm, not all your wit, not all your victories on land or at sea, can move back the finger of fate. Today we maintain ourselves. Tomorrow science will have moved forward yet one more step, and there will be no appeal from the judgment which will then be pronounced on the uneducated.
The Organisation of Thought
Chapter I (p. 28)
Greenwood Press Publishers. Westport, Connecticut, USA. 1974

SCIENCE AND ART

Asimov, Isaac   1920–92
American author and biochemist

How often people speak of art and science as though they were two entirely different things, with no interconnection. An artist is emotional, they think, and uses only his intuition; he sees all at once and has no need of reason. A scientist is cold, they think, and uses only his reason; he argues carefully step by step, and needs no imagination. That is all wrong. The true artist is quite rational as well as imaginative and knows what he is doing; if he does not, his art suffers. The true scientist is quite imaginative as well as rational, and sometimes leaps to solutions where reason can follow only slowly; if he does not, his science suffers.
The Roving Mind
Chapter 25 (p. 116)
Prometheus Books. Buffalo, New York, New York, USA. 1983

Author undetermined

Art is personal and science is universal.
In Lecomte du Nouy
The Road to Reason
Chapter 1 (p. 31)
Longmans, Green & Company. London, England. 1949

Blake, William   1757–1827
English poet, painter, and engraver

He who would do good to another must
do it in Minute Particulars:
General Good is the plea of the scoundrel,
hypocrite and flatterer;
For art and science cannot exist but in
minutely organized Particulars.
The Complete Poetry and Prose of William Blake
Jerusalem
The Holiness of Minute Particulars, 3, Section 55 (p. 399)
University of California Press. Berkeley, California, USA. 1982

Brecht, Bertolt   1898–1956
German writer

But science and art meet on this ground, that both are there to make man's life easier, the one setting out to maintain, the other to entertain us. In the age to come art will create entertainment from that new productivity which can so greatly improve our maintenance, and in itself, if only it is left unshackled, may prove to be the greatest pleasure of all.
Translated by John Willett
Brecht on Theatre: The Development of an Aesthetic
A Short Organon for the Theater, 20 (p. 185)
Hill & Wang. New York, New York, USA. 1964

Campbell, Norman R.   1880–1949
English physicist and philosopher

Science is the noblest of the arts and men of science the most artistic of all artists.
Physics: The Elements
Chapter VIII (p. 227–228)
At The University Press. Cambridge, England. 1920

Science, like art, should not be something extraneous, added as a decoration to other activities of existence; it should be part of them, inspiring our most trivial actions as well as our noblest thoughts.
What Is Science?
Chapter VIII (p. 183)
Dover Publications. New York, New York, USA. 1952

Cassidy, Harold Gomes
No biographical data available

If humans understood science and would effectively make their voices heard, they could, with the aid of scientists, control the forces of cultural change in the process of their actual generation, directing them in the ways that lead toward the morally and ethically just ends that arise from the union of art and science. This union, when it is a union of whole science and whole art, supports and illuminates anew a noble image of man.
The Sciences and the Arts: A New Alliance
Chapter 11 (p. 165)
Harper & Brothers. New York, New York, USA. 1962

Cassirer, Ernst   1874–1945
German philosopher

Since art and science move in entirely different planes, they cannot contradict or thwart one another.
An Essay on Man: An Introduction to a Philosophy of Human Culture
Chapter IX (p. 170)
Yale University Press. New Haven, Connecticut, USA. 1944

Cohen, I. Bernard   1914–2003
American physicist and science historian

Great creations whether of science or art — can never be viewed dispassionately.
In the 1952 printing
Optics

Preface (p. ix)
Encyclopædia Britannica, Inc. Chicago, Illinois, USA. 1952

Connolly, Cyril   1903–74
English critic and editor

Today the function of the artist is to bring imagination to science and science to imagination, where they meet, in the myth.
The Unquiet Grave
Part III (p. 86)
Hamish Hamilton. London, England. 1945

Crick, Francis Harry Compton   1916–2004
English biochemist

People with training in the arts still feel that in spite of the alterations made in their life by technology — by the internal combustion engine, by penicillin, by the Bomb — modern science has little to do with what concerns them most deeply. As far as today's science is concerned this is partly true, but tomorrow's science is going to knock their culture right out from under them.
Of Molecules and Men
The Prospect Before Us (p. 95)
University of Washington Press. Seattle, Washington, USA. 1966

De Gourmont, Rémy   1858–1915
French critic and novelist

Art includes everything that stimulates the desire to live; science, everything that sharpens the desire to know. Art, even the most disinterested, the most disembodied, is the auxiliary of life. Born of the sensibility, it sows and creates it in its turn. It is the flower of life and, as seed, it gives back life. Science, or to use a broader term, knowledge, has its end in itself, apart from any idea of life and propagation of the species.
Translated by Glenn S. Burne
Selected Writings
Art and Science (p. 170)
The University of Michigan Press. Ann Arbor, Michigan, USA. 1966

Delbrück, Max   1906–81
German-born American biologist

The books of the great scientists are gathering dust on the shelves of learned libraries. And rightly so. The scientist addresses an infinitesimal audience of fellow composers. His message is not devoid of universality but its universality is disembodied and anonymous. While the artist's communication is linked forever with its original form, that of the scientist is modified, amplified, fused with the ideas and results of others, and melts into the stream of knowledge and ideas which forms our culture.
A Physicist's Renewed Look at Biology: Twenty Years Later
Science, Volume 168, Number 3937, June 12, 1970 (p. 1314)

The scientist has in common with the artist only this: that he can find no better retreat from the world than his work and also no stronger link with the world than his work.

A Physicist's Renewed Look at Biology: Twenty Years Later
Science, Volume 168, Number 3937, June 12, 1970 (p. 1314)

Dubos, René Jules   1901–82
French-born American microbiologist and environmentalist

Directly or indirectly, the various forms of art reflect the strivings, the struggles, and the sufferings of mankind. The state of health and the ills of a society are recorded not only in the writings of its physicians and scholars but also in the themes and moods of its artists and poets.
Mirage of Health
Chapter VII (p. 215)
Harper & Brothers Publishers. New York, New York, USA. 1959

Durant, William James   1885–1981
American historian and essayist

Every science begins as philosophy and ends as art; it arises in hypothesis and flows into achievement.
The Story of Philosophy
Introduction (p. 2)
Simon & Schuster. New York, New York, USA. 1953

Einstein, Albert   1879–1955
German-born physicist

Science exists for Science's sake, like Art for Art's sake, and does not go in for special pleading or for the demonstration of absurdities.
Cosmic Religion, with Other Opinions and Aphorisms
On Science (p. 100)
Covici-Friede. New York, New York, USA. 1931

…one of the strongest motives that lead men to art and science is escape from everyday life with its painful crudity and hopeless dreariness, from the fetters of one's own ever shifting desires. A finely tempered nature longs to escape from personal life into the world of objective perception and thought…. Man tries to make for himself in the fashion that suits him best a simplified and intelligible picture of the world; he then tries to some extent to substitute this cosmos of his for the world of experience, and thus to overcome it. This is what the painter, the poet, the speculative philosopher, and the natural scientist do, each in his own way.
The World as I See It
Principles of Research (pp. 20–21)
Philosophical Library. New York, New York, USA. 1949

Einstein, Albert   1879–1955
German-born physicist

After a certain high level of technical skill is achieved, science and art tend to coalesce in esthetics, plasticity, and form. The greatest scientists are always artists as well.
In Alice Calaprice (ed.)
The Quotable Einstein (p. 171)
Princeton University Press. Princeton, New Jersey, USA. 1996

Escher, M. C.   1898–1972
Dutch graphic artist

…science and art sometimes can touch one another, like two pieces of the jigsaw puzzle which is our human life, and that contact may be made across the borderline between the two respective domains.
In Doris Schattschneider
Visions of Symmetry: Notebooks, Periodic Drawings, and Related Works of M.C. Escher
Chapter 2 (p. 104)
W.H. Freeman & Company. New York, New York, USA. 1990

Feynman, Richard P.

I've always been very one-sided about science and when I was younger I concentrated almost all my effort on it. I didn't have time to learn and I didn't have much patience with what's called the humanities, even though in the university there were humanities that you had to take. I tried my best to avoid somehow learning anything and working at it. It was only afterwards, when I got older, that I got more relaxed, that I've spread out a little bit. I've learned to draw and I read a little bit, but I'm really still a very one-sided person and I don't know a great deal. I have a limited intelligence and I use it in a particular direction.
In Jeffrey Robbins (ed.)
The Pleasure of Finding Things Out: The Best Short Works of Richard P. Feynman
Chapter 1 (p. 2)
Perseus Books. Cambridge, Massachusetts, USA. 1999

Harvey, William   1578–1657
English physician

On the same terms, therefore, as art is attained to, is all knowledge and science acquired; for as art is a habit with reference to things to be done, so is science a habit in respect to things to be known; as that proceeds from the imitation of types or forms so this proceeds from the knowledge of natural things.
In *Great Books of the Western World* (Volume 28)
Anatomical Exercises on the Generation of Animals
Of the Manner and Order of acquiring Knowledge (p. 333)
Encyclopædia Britannica, Inc. Chicago, Illinois, USA. 1952

Heisenberg, Werner Karl   1901–76
German physicist and philosopher

Both science and art form in the course of the centuries a human language by which we can speak about the more remote parts of reality, and the coherent sets of concepts as well as the different styles of art are different words or groups of words in this language.
Physics and Philosophy: The Revolution in Modern Science
Chapter VI (p. 109)
Harper & Row, Publishers. New York, New York, USA. 1958

Huxley, Aldous   1894–1963
English writer and critic

Unlike art, science is genuinely progressive. Achievement in the fields of research and technology is cumulative;

each generation begins at the point where its predecessor left off.
Science, Liberty and Peace
Chapter I (p. 30)
William Morrow & Company, Inc. New York, New York, USA. 1967

Science and art are only too often a superior kind of dope, possessing this advantage over booze and morphia: that they can be indulged in with a good conscience and with the conviction that, in the process of indulging, one is leading the "higher life."
Ends and Means
Chapter XIV (p. 276)
Chatto & Windus. London, England. 1938

Karanikas, Alexander   1916–2006
Greek-American professor of English

...science pierces reality like a dagger in search of fact and truth while art caresses reality looking for pleasure, grace and beauty.
Tillers of a Myth
Science, the False Messiah (p. 127)
The University of Wisconsin Press. Madison, Wisconsin, USA. 1969

Kepes, Gyorgy   1906–2001
Hungarian-born American artist and theorist

The essential vision of reality presents us not with fugitive appearances but with felt patterns of order which have coherence and meaning for the eye and for the mind. Symmetry, balance and rhythmic sequences express characteristics of natural phenomena: the connectedness of nature — the order, the logic, the living process. Here art and science meet on common ground.
The New Landscape
In Art and Science
Chapter I (p. 24)
Paul Theobald & Company Chicago, Illinois, USA. 1956

Mathematicians who build new spaces and physicists who find them in the universe can profit from the study of pictorial and architectural spaces conceived and built by men of art.
The New Landscape
In Art and Science
Chapter I (p. 28)
Paul Theobald & Company Chicago, Illinois, USA. 1956

Klee, Paul   1879–1940
Swiss expressionist painter

...the worst state of affairs is when science begins to concern itself with art.
The Diaries of Paul Klee 1898–1918
Diary III, Number 747 (p. 194)
University of California Press. Berkeley, California, USA. 1964

Knuth, Donald E.   1938–
Creator of TeX

The difference between art and science is that science is what people understand well enough to explain to a computer. All else is art.
In Robert Slater
Portraits in Silicon
Chapter 31 (p. 351)
The MIT Press. Cambridge, Massachusetts, USA. 1987

Koestler, Arthur   1905–83
Hungarian-born English writer

Einstein's space is no closer to reality than Van Gogh's sky. The glory of science is not in a truth more absolute than the truth of Bach or Tolstoy, but in the act of creation itself. The scientist's discoveries impose his own order on chaos, as the composer or painter imposes his; an order that always refers to limited aspects of reality, and is based on the observer's frame of reference, which differs from period to period as a Rembrandt nude differs from a nude by Manet.
The Act of Creation
Book One, Part Two, Chapter X (p. 252)
The Macmillan Company. New York, New York, USA. 1964

Kraus, Karl   1874–1936
Austrian essayist and poet

Science is spectrum analysis. Art is photosynthesis.
In John D. Barrow
The Artful Universe (p. 114)
Clarendon Press. Oxford, England. 1995

Medawar, Sir Peter Brian   1915–87
Brazilian-born English zoologist

There is no spiritual copyright in scientific discoveries, unless they should happen to be quite mistaken. Only in making a blunder does a scientist do something which, conceivably, no one else might ever do again. Artists are not troubled by matters of priority, but Wagner would certainly not have spent twenty years on The Ring if he had thought it at all possible for someone else to nip in ahead of him with Götterdämmerung.
The Act of Creation
New Statesman, 19 June 1964

Melville, Herman   1819–91
American novelist

One can envisage an end of science no more readily than one can envisage an end of imaginative literature or the fine arts.
Advice to a Young Scientist
Chapter 11 (p. 90)
Basic Books, Inc. New York, New York, USA. 1979

Oppenheimer, J. Robert   1904–67
American theoretical physicist

Both the man of science and the man of art live always at the edge of mystery, surrounded by it; both always,

as to the measure of their creation, have had to do with the harmonization of what is new with what is familiar, with the balance between novelty and synthesis, with the struggle to make partial order in total chaos.
Prospects in the Arts and Sciences
Speech, 26 December 1954, Columbia University Bicentennial

The frontiers of science are separated now by long years of study, by specialized vocabularies, arts, techniques, and knowledge from the common heritage even of a most civilized society; and anyone working at the frontier of such science is in that sense a very long way from home, a long way too from the practical arts that were its matrix and origin, as indeed they were of what we today call art.
Prospects in the Arts and Sciences
Speech, 26 December 1954, Columbia University Bicentennial

Reynolds, Osborne   1842–1912
English fluid dynamics engineer

I have to deal with facts, and I shall try to deal with nothing but facts. Many of these facts, or the conclusions to be immediately drawn from them, may appear to bear on the possibilities — or, rather, the impossibilities — of art. But in the Society of Arts I need not point out that art knows no limit; where one way is found to be closed, it is the function of art to find another. Science teaches us the results that will follow from a known condition of things; but there is always the unknown condition, the future effect of which no science can predict.
Papers on Mechanical and Physical Subjects (Volume 2)
Lecture to the Society of Arts
At The University Press. Cambridge, England. 1900=03

Sagan, Carl   1934–96
American astronomer and author

It is sometimes said that scientists are unromantic, that their passion to figure out robs the world of beauty and mystery. But is it not stirring to understand how the world actually works — that white light is made of colors, that color is the way we perceive the wavelengths of light, that transparent air reflects light, that in so doing it discriminates among the waves, and that the sky is blue for the same reason that the sunset is red? It does no harm to the romance of the sunset to know a little bit about it.
Pale Blue Dot: A Vision of the Human Future in Space
Chapter 10 (pp. 159–160)
Random House, Inc. New York, New York, USA. 1994

Santayana, George (Jorge Augustín Nicolás Ruiz de Santillana)   1863–1952
Spanish-born American philosopher

Science is the response to the demand for information. ...
Art is the response to the demand for entertainment.
The Sense of Beauty
Part I, Section 2 (p. 22)
Transaction Publishers. New Brunswick, New Jersey, USA. 2000

Shlain, Leonard
American surgeon and author

Both art and physics are unique forms of language. Each has a specialized lexicon of symbols that is used in a distinctive syntax. Their very different and specific contexts obscure their connection to everyday language as well as to each other. Nevertheless, it is noteworthy just how often the terms of one can be applied to the concepts of the other. "Volume," "space," "mass," "force," "light," "color," "tension," "relationship," and "density" are descriptive words that are heard repeatedly if you trail along with a museum docent. They also appear on the blackboards of freshman college physics lectures.
Quoted by Gerald Nolton
Art and Physics: Parallel Visions in Space, Time and Light
Chapter 1
William Morrow & Company, Inc. New York, New York, USA. 1991

The proponents of these two diverse endeavors [art and physics] wax poetic about elegance, symmetry, beauty, and aesthetics. While physicists demonstrate that A equals B or that X is the same as Y, artists often choose signs, symbols, and allegories to equate a painterly image with a feature of experience. Both of these techniques reveal previously hidden relationships.
Quoted by Gerald Nolton
Art and Physics: Parallel Visions in Space, Time and Light
Chapter 1
William Morrow & Company, Inc. New York, New York, USA. 1991

Silver, Brian L.
Israeli professor of physical chemistry

Whatever the Sun may be, said D. H. Lawrence, it is certainly not a ball of flaming gas. Helios, the sun god, has more sex appeal than a cloud of gas, however hot.
The Ascent of Science
Part IX, Chapter 36 (p. 485)
Solomon Press Book. New York, New York, USA. 1998

[D. H.] Lawrence [saw] science systematically chipping away at the mysterious, but generally benign, unknown and arrogantly replacing it with the dull, prosaic, down-to-earth known. ...The scientist's rainbow is the result of the different refractive indices of the various frequencies of light that make up solar radiation. But man evidently prefers mystery to math, and the intrusion of science into the movements of the planets and the stars, into the living cell and into that final sanctuary of the spirit, the mind, has undoubtedly cast a chill over that warm, blurred garden, the theocentric universe. The scientist, ruthlessly buying up desirable property, appears to many people to be building an automated factory in the middle of the garden.
The Ascent of Science
Part IX, Chapter 36 (p. 485)
Solomon Press Book. New York, New York, USA. 1998

Smyth, H. D.
No biographical data available

We have a paradox in the method of science. The research man may often think and work like an artist, but he has to talk like a bookkeeper in terms of facts, figures and logical sequence of thought.
Quoted by Gerald Nolton
On the Duality and Growth of Physical Science
American Scientist, Volume 41, 1953 (p. 93)

Spencer, Herbert   1820–1903
English social philosopher

…Science is necessary not only for the most successful production, but also for the full appreciation of the fine arts.
Education: Intellectual, Moral, and Physical
Chapter I (p. 70)
A.L. Fowle. New York, New York, USA. 1860

Sullivan, John William Navin   1886–1937
Irish mathematician

The measure in which science falls short of art is the measure in which it is incomplete as science.
The Justification of the Scientific Method
The Athenaeum, May 1919 (p. 275)

Valéry, Paul   1871–1945
French poet and critic

There is a science of simple things, an art of complicated ones. Science is feasible when the variables are few and can be enumerated; when their combinations are distinct and clear. We are tending toward the condition of science and aspiring to it. The artist works out his own formulas; the interest of science lies in the art of making science.
In Jackson Mathews (ed.)
The Collected Works of Paul Valéry (Volume 14)
Moralités
Analects (p. 64)
Princeton University Press. Princeton, New Jersey, USA. 1971

Waddington, Conrad Hal   1905–75
British biologist and paleontologist

The best of modern art is compatible only with true science, and the bogus science requires a fake art to keep it company.
The Scientific Attitude
Science Is Not Neutral (p. 27)
Penguin Books, Middlesex, England. 1941

Whewell, William   1794–1866
English philosopher and historian

Art and Science differ. The object of Science is Knowledge; the objects of Art are Works. In Art, truth is a means to an end; in Science, it is the only end. Hence the Practical Arts are not to be classed among the Sciences.
The Philosophy of the Inductive Sciences Founded Upon Their History (Volume 2)
Aphorisms Concerning Science, Aphorism XXV (p. 471)
John W. Parker. London, England. 1847

Wordsworth, William   1770–1850
English poet

Enough of Science and of Art;
Close up these barren leaves;
Come forth, and bring with you a heart
That watches and receives.
The Complete Poetical Works of William Wordsworth
The Tables Turned, Stanza 8
Crowell. New York, New York, USA. 1888

SCIENCE AND CIVILIZATION

Ackerman, Diane   1948–
American writer

When we think of science, we often picture arcane quests after minutiae, or efforts to explain underlying principles. But it's amazing that in a civilization as complex as ours, we are still engaged in Adam's task, the naming of animals.
The Rarest of the Rare: Vanishing Animals, Timeless Worlds
Insect Love (p. 160)
Vintage Books. New York, New York, USA. 1997

Compton, Arthur H.   1892–1962
American physicist

I verily believe that in the advancement of science lies the hope of our civilization.
Les Prix Nobel. The Nobel Prizes in 1927
Nobel banquet speech for award received in 1927
Nobel Foundation. Stockholm, Sweden. 1928

Huxley, Julian   1887–1975
English biologist, philosopher, and author

Without the impersonal guidance and the efficient control provided by science civilization will either stagnate or collapse, and human nature cannot make progress towards realizing its possible evolutionary destiny.
What Dare I Think?: The Challenge of Modern Science to Human Action and Belief, Including the Henry La Barre Jayne Foundation Lectures
Chapter V (p. 177)
Harper & Brothers. New York, New York, USA. 1931

Science has two main functions in civilization. One is to give man a picture of the world phenomena, the most accurate and complete picture possible. The other is to provide him with the means of controlling his environment and his destiny.
What Dare I Think?: The Challenge of Modern Science to Human Action and Belief, Including the Henry La Barre Jayne Foundation Lectures
Chapter IV (pp. 127–128)
Harper & Brothers. New York, New York, USA. 1931

Modern civilisation rests upon physical science; take away her gifts to our own country, and our position among the leading nations of the world is gone to-morrow; for it is physical science only that makes intelligence and moral energy stronger than brute force.
Collected Essays (Volume 8)
Discourses, Biological and Geological
A Lobster; or, The Study of Zoology (p. 226)
Macmillan & Company Ltd. London, England. 1904

Lovell, Sir Alfred Charles Bernard   1913–
English physicist, radio astronomer, and author

The pursuit of the good and evil are now linked in astronomy as in almost all science…. The fate of human civilization will depend on whether the rockets of the future carry the astronomer's telescope or a hydrogen bomb.
The Individual and the Universe (p. 72)
Oxford University Press. London, England. 1959

Metropolis, Nicholas C.   1915–99
Mathematician

Science is the locomotive that drives our civilization.
In Sigfried S. Hecker and Gian-Carlo Rota (eds.)
Essays on the Future: In Honor of Nick Metropolis
Belated Thoughts (p. xv)
Birkhäuser. Boston, Massachusetts, USA. 2000

Ortega y Gasset, José   1883–1955
Spanish philosopher

…experimental science has progressed thanks in great part to the work of men astoundingly mediocre, and even less than mediocre. That is to say, modern science, the root and symbol of our actual civilisation, finds a place for the intellectually commonplace man and allows him to work therein with success…. A fair amount of the things that have to be done in physics or in biology is mechanical work of the mind which can be done by anyone or almost anyone. For the purpose of innumerable investigations it is possible to divide science into small sections, to enclose oneself in one of these, and leave out of consideration all the rest.
The Revolt of the Masses
Chapter 12 (p. 110, 111)
W.W. Norton & Company, Inc. New York, New York, USA.1960

Parton, H. N.
No biographical data available

A successful blending of the sciences and the humanities is necessary for the health of our civilization.
Science Is Human
Science Is Human (p. 31)
University of Otago Press. Dunedin, New Zealand. 1972

Sagan, Carl   1934–96
American astronomer and science writer

The very method of mathematical reasoning that Isaac Newton introduced to explain the motion of the planets around the Sun has led to most of the technology of our modern world. The Industrial Revolution, for all its shortcomings, is still the global model of how an agricultural nation can emerge from poverty. These debates have bread-and-butter consequences.
Pale Blue Dot: A Vision of the Human Future in Space
Chapter 4 (p. 56)
Random House, Inc. New York, New York, USA. 1994

SCIENCE AND MORALS

Bronowski, Jacob   1908–74
Polish-born British mathematician and polymath

It is not the business of science to inherent the earth, but to inherit the moral imagination; because without that man and beliefs and science will perish together.
The Ascent of Man
Chapter 13 (p. 432)
Little, Brown & Company. Boston, Massachusetts, USA. 1973

Chesterton, G. K. (Gilbert Keith)   1874–1936
English author

…when any part of the general public is drawn into a debate on physical science, we may be certain that it has already become a debate on moral science.
All Is Grist: A Book of Essays
On Gossip about Heredity (p. 96)
Methuen & Company Ltd. London, England.1931

Compton, Karl Taylor   1887–1954
American educator and physicist

I would emphasize the fact that scientific discovery is, per se, neither good nor bad. It simply produces knowledge and with knowledge, opportunity and responsibility. I think it fair to say that the advance of science carries with it powerful demands on morality if the results are to be beneficial rather than harmful.
A Scientist Speaks: Excerpts from Addresses by Karl Taylor Compton During the Years 1930–1949 (p. 5)
Undergraduate Association, MIT. Cambridge, Massachusetts, USA. 1955

Dewey, John   1859–1952
American philosopher and educator

Science through its physical technological consequences is now determining the relations which human beings, severally and in groups, sustain to one another. If it is incapable of developing moral techniques which will also determine these relations, the split in modern culture goes so deep that not only democracy but all civilized values are doomed.
Freedom and Culture
Chapter Six (p. 118)
Prometheus Books. Buffalo, New York, USA. 1989

Diderot, Denis   1713–84
French encyclopedist and philosopher of materialism

The moral universe is so closely linked to the physical universe that it is scarcely likely that they are not one and the same machine.
Eléments de Physiologie (pp. xiii–xiv)
Librairie M. Didier. Paris, France. 1964

Ferré, Nels F. S. 1908–71
Swedish-American theologian

Science can be and is being made into an escapist philosophy — into a dodge of moral disciplines and spiritual responsibilities.
Faith and Reason
Chapter II (p. 83)
Harper & Brothers. New York, New York, USA. 1946

Friedenberg, Edgar Z. 1921–2001
Educator, education critic, and sociologist

…only science can hope to keep technology in some sort of moral order.
The Vanishing Adolescent
The Impact of the School, the Clarification of Experience (p. 50)
Beacon Press. Boston, Massachusetts. 1964

Gould, Stephen Jay 1941–2002
American paleontologist, evolutionary biologist, and historian of science

I do not know when the technical and popular prose of science became separated, although I accept the inevitability of such a division as knowledge became increasingly more precise, detailed, and specialized. We have now reached the point where most technical literature not only falls outside the possibility of public comprehension but also (as we would all admit in honest moments) outside our own competence in scientific disciplines far removed from our personal expertise. I trust that we all regard this situation as saddening, even though we accept its necessity.
Take Another Look
Science, Volume 286, Number 5441, October 29, 1999 (p. 899)

Jefferson, Thomas 1743–1826
3rd president of the United States

…if science produces no better fruits than tyranny, murder, rapine and destitution of national morality, I would rather wish our country to be ignorant, honest and estimable, as our neighboring savages are.
The Writings of Thomas Jefferson (Volume 6)
Letter to John Adams, 1812 (p. 37)
Deby & Jackson. New York, New York, USA. 1859

Kruyt, Hugo Rudolph 1882–1959
Dutch colloid chemist

Clearer than ever we understand that knowledge is not all, that we need morals and brotherhood to avoid science becoming a curse.
In John P. Dickinson

International Council of Scientific Unions
First General Assembly Following the Second World War, Science and Scientific Researchers in Modern Society (p. 165)

Lerner, Max 1902–92
American educator and author

Science itself is a humanist in the sense that it doesn't discriminate between human beings, but it is also morally neutral. It is no better or worse than the ethos with and for which it is used.
Manipulating Life
New York Post, January 24, 1968

Masters, William H. 1915–2001
American gynecologist and researcher

Science by itself has no moral dimension. But it does seek to establish truth. And upon this truth morality can be built.
Two Sex Researchers on the Firing Line
Life, 24 June 1966 (p. 49)

Oppenheimer, J. Robert 1904–67
American theoretical physicist

Scientists aren't responsible for the facts that are in nature. It's their job to find the facts. There's no sin connected with it — no morals. If anyone should have a sense of sin, it's God. He put the facts there.
In Lincoln Barnett
J. Robert Oppenheimer
Life, October 10, 1949 (p. 133)

Poincaré, Henri 1854–1912
French mathematician and theoretical astronomer

There can no more be immoral science than there can be scientific morals.
The Foundations of Science
The Values of Science, Introduction (p. 206)
The Science Press. New York, New York, USA. 1913

Snow, Charles Percy 1905–80
English novelist and scientist

…there is a moral component right in the grain of science itself…
The Two Cultures: And a Second Look
Chapter I (p. 13)
At The University Press. Cambridge, England. 1964

Toynbee, Arnold J. 1852–83
English historian

Our western science is a child of moral virtues; and it must now become the father of further moral virtues if its extraordinary material triumphs in our time are not to bring human history to an abrupt, unpleasant and discreditable end.
A Turning Point in Man's Destiny
The New York Times Magazine, December 26, 1954 (p. 5)

Wallace, Henry A.   1888–1935
33rd vice-president of the United States

I can understand the impulse which prompts scientists to defend science against the attacks of the uninformed. Science has achieved so many miracles for society, saved so many lives, made possible so extraordinary an advance in material living standards for so many millions of people, that it is disquieting to think that all the consequences of science can ever be other than good. Yet I don't see what basis we have for assuming that science can and does have only beneficial consequences. Is the product of man's curiosity inevitably good?…It may be disturbing to realize it, but the truth seems to be that science proceeds without moral obligations; it is neither moral nor immoral, but in essence amoral.
Scientists in an Unscientific Society
Scientific Monthly, Volume 150, 1934 (p. 285)

SCIENCE AND PHILOSOPHY

Bacon, Sir Francis   1561–1626
English lawyer, statesman, and essayist

The great and radical difference of capacities, as to philosophy and the sciences, lies here: that some are stronger and fitter to observe the differences of things; others their correspondencies: for a steady and sharp genius can fix it's contemplations, and dwell and fasten upon all the subtlety of differences, while a sublime and ready genius perceives and compares the smallest and most general agreements of things; but both kinds easily fall into excess, by grasping either at the dividing scale, or shadows of things. The former is so taken up with the particles of things, as almost to neglect their structure, whilst the other views their fabrication with such astonishment, as not to enter into the simplicity of nature.
In George Adams
Lectures on Natural and Experimental Philosophy (Volume 1)
Lecture IV (p. 127)
Printed by R. Hindmarsh. London, England. 1794

Burroughs, John   1837–1921
American naturalist and writer

Science displeases literature when it dehumanizes nature and shows us irrefragable laws when we had looked for humanistic divinities.
The Breath of Life
Chapter X (p. 243)
Houghton Mifflin Company. Boston, Massachusetts, USA. 1915

Chesterton, G. K. (Gilbert Keith)   1874–1936
English author

To mix science up with philosophy is only to produce a philosophy that has lost all its ideal value and a science that has lost all its practical value.
All Things Considered
Science and Religion (p. 187)
John Lane Company. New York, New York, USA. 1908

Compton, Karl Taylor   1887–1954
American educator and physicist

Science is not a technique or a body of knowledge, though it uses both. It is rather an attitude of inquiry, or observation and reasoning, with respect to the world. It can be developed, not by memorizing facts or juggling formulas to get an answer, but only by actual practice of scientific observation and reasoning.
A Scientist Speaks: Excerpts from Addresses by Karl Taylor Compton During the Years 1930–1949 (p. 44)
Undergraduate Association, MIT. Cambridge, Massachusetts, USA. 1955

de Casseres, Benjamin   1873–1945
American journalist and author

My studies in speculative philosophy, metaphysics, and science are all summed up in the image of a mouse called man, running in and out of every hole in the cosmos hunting for the absolute cheese.
Harper's Weekly, Volume 19, Number 3164, June 28, 1976

Durant, William James   1885–1981
American historian and essayist

Philosophy…is the front trench in the siege of truth. Science is the captured territory.
The Story of Philosophy
Introduction (p. 2)
Simon & Schuster. New York, New York, USA. 1953

Science without philosophy, facts without perspective and valuation, cannot save us from havoc and despair. Science gives us knowledge, but only philosophy can give us wisdom.
The Story of Philosophy
Introduction (p. 3)
Simon & Schuster. New York, New York, USA. 1953

Science is analytical description, philosophy is synthetic interpretation. Science wishes to resolve the whole into the known.
The Story of Philosophy
Introduction (p. 2)
Simon & Schuster. New York, New York, USA. 1953

Eddy, Mary Baker   1821–1910
American religious writer

Jesus of Nazareth was the most scientific man that ever trod the globe. He plunged beneath the material surface of things, and found the spiritual cause.
Science and Health with Key to the Scriptures
Chapter X (p. 313)
Joseph Armstrong. Boston, Massachusetts, USA. 1906

Fischer, Martin H.   1879–1962
German-American physician

Not fact-finding, but attainment to philosophy, is the aim of science.
Fischerisms (p. 7)
C.C. Thomas. Springfield, Illinois, USA. 1944

Gornick, Vivian
American essayist

Science — like art, religion, political theory, or psycho-analysis — is work that holds out the promise of philosophic understanding, excites in us the belief that we can "make sense of it all."
Women in Science: Portraits from a World in Transition
Part One (p. 66)
Simon & Schuster. New York, New York, USA. 1983

Huxley, Julian   1887–1975
English biologist, philosopher, and author

The attempt to understand this universe, including the nature of man, is the task of science; and as she makes progress with this task, so will she become more and more an indispensable part of philosophy and religion — imagination's touchstone, thought's background, action's base.
Searching for the Elixir of Life
The Century Illustrated Monthly Magazine, Volume 103, Number 4, February 1922

Jeans, Sir James Hopwood   1877–1946
English physicist and mathematician

The philosophy of any period is always largely interwoven with the science of the period, so that any fundamental change in science must produce reactions in philosophy.
Physics and Philosophy
Chapter I (p. 2)
Dover Publications, Inc. New York, New York, USA. 1981

In whatever ways we define science and philosophy their territories are contiguous; wherever science leaves off — and in many places its boundary is ill-defined — there philosophy begins.
Physics and Philosophy
Chapter I (p. 17)
Dover Publications, Inc. New York, New York, USA. 1981

Jones, Steve   1944–
English genetics professor

…philosophy is to science as pornography is to sex.
In Mary Midgley
Can Science Save Its Soul?
New Scientist, 1 August 1992 (p. 25)

Keats, John   1795–1821
English Romantic lyric poet

Do not all charms fly
At the mere touch of cold philosophy?
There was an awful rainbow once in heaven:

We know her woof, her texture; she is given
In the dull catalogue of common things.
Philosophy will clip an Angel's wings,
Conquer all mysteries by rule and line,
Empty the haunted air, and gnom'ed mine —
Unweave a rainbow…
The Complete Poetical Works and Letters of John Keats
Lamia, Part II, l. 229–237
Houghton Mifflin Company. Boston, Massachusetts, USA. 1890

Mercier, André   1913–99
Swiss physicist

Philosophy does not "solve problems", whereas science does. Philosophy, in its relations to science, gathers up the problems of science, which are no longer problems since they have found solutions, and seeks to order them in such a way that the structure of knowledge does, in fact, appear.
Fifty Years of the Theory of Relativity
Nature, Volume 175, Number 4465, May 28, 1955 (p. 919)

Pope Pius XII   1876–1958
Bishop of Rome

Science descends ever more deeply into the hidden recesses of things, but it must halt at a certain point when questions arise which cannot be settled by means of sense observations. At that point the scientist needs a light which is capable of revealing to him truth which entirely escapes his senses. This light is philosophy.
In Philip G. Fothergill
Life and Its Origin
Pontifical Academy of Science, Meeting 1955 (p. 12)

Popper, Karl R.   1902–94
Austrian/British philosopher of science

All science and all philosophy are enlightened common sense.
Objective Knowledge: An Evolutionary Approach
Chapter 2 (p. 34)
Clarendon Press. Oxford, England. 1972

Renan, Ernest   1823–92
French philosopher and Orientalist

Socrates founded philosophy, and Aristotle science. There was philosophy before Socrates, and science before Aristotle; and since Socrates and since Aristotle, philosophy and science have made immense progress: but all has all been built upon the foundations they laid.
The Life of Jesus
Chapter 28 (p. 383)
Modern Library. New York, New York, USA. 1955

Ritchie, Arthur David   1891–1967
Scottish philosopher and science history writer

Philosophers who write about Science and scientists who write about Philosophy are too often preoccupied

with the scientific theories and discoveries of the moment to the detriment of both their Science and their Philosophy.
Scientific Method: An Inquiry into the Character and Validity of Natural Laws
Preface (p. v)
Kegan Paul, Trench, Trubner & Company Ltd. London, England. 1923

Russell, Bertrand Arthur William   1872–1970
English philosopher, logician, and social reformer

The man who has no tincture of philosophy goes through life imprisoned in the prejudices.... To such a man the world tends to become definite, finite, obvious; common objects rouse no questions, and unfamiliar possibilities are contemptuously rejected.
The Problems of Philosophy
Chapter XV (pp. 156–157)
Oxford University Press, Inc. London, England. 1959

It seems to me that science has a much greater likelihood of being true in the main than any philosophy hitherto advanced (I do not, of course, except my own). In science there are many matters about which people are agreed; in philosophy there are none. Therefore, although each proposition in a science may be false, and it is practically certain that there are some that are false, yet we shall be wise to build our philosophy upon science, because the risk of error in philosophy is pretty sure to be greater than in science. If we could hope for certainty in philosophy, the matter would be otherwise, but so far as I can see such a hope would be a chimerical.
The Philosophy of Logical Atomism
Logical Atomism
University of Minnesota Press. Minneapolis, Minnesota, USA. 1959

Thoreau, Henry David   1817–62
American essayist, poet, and practical philosopher

In our science and philosophy, even, there is commonly no true and absolute account of things. The spirit of sect and bigotry has planted its hoof amid the stars. You have only to discuss the problem, whether the stars are inhabited or not, in order to discover it.
The Writings of Henry David Thoreau (Volume 4)
Life Without Principle (p. 469)
Houghton Mifflin Company. Boston, Massachusetts, USA. 1893

Weyl, Hermann   1885–1955
German mathematician

A scientist who writes on philosophy faces conflicts of conscience from which he will seldom extricate himself whole and unscathed; the open horizon and depth of philosophical thoughts are not easily reconciled with that objective clarity and determinacy for which he has been trained in the school of science.
Philosophy of Mathematics and Natural Science
Preface (p. v)
Princeton University Press. Princeton, New Jersey, USA. 1949

SCIENCE AND POETRY

Beston, Henry   1888–1968
American writer

Poetry is as necessary to comprehension as science. It is as impossible to live without reverence as it is without joy.
The Outermost House
Chapter X (p. 221)
Rinehart & Company. New York, New York, USA. 1928

Clifford, William Kingdon   1845–79
English philosopher and mathematician

It is an open secret to the few who know it, but a mystery and a stumbling-block to the many, that Science and Poetry are own sisters; insomuch that in those branches of scientific inquiry which are most abstract, most formal, and most remote from the grasp of the ordinary sensible imagination, a higher power of imagination akin to the creative insight of the poet is most needed and most fruitful of lasting work.
In Leslie Stephen and Frederick Pollock (eds.)
Lectures and Essays (Volume 1)
Introduction (p. 1)
Macmillan & Company. London, England. 1886

Davis, Joel   1948–
No biographical data available

Poetry and science are closer than most people realize. Many poets and scientists already know this, of course. Most of the rest of us are still trapped in dismal stereotypes about both fields of human endeavor. The deep link between the two is vision.
Alternate Realities
In a Grain of Sand (p. 3)
Plenum Trade. New York, New York, USA. 1997

Day-Lewis, C. (Cecil)   1904–72
Irish-born English author and poet

Science is concerned with finding out and stating the facts: poetry's task is to give you the look, the smell, the taste, the "feel" of those facts.
Poetry for You
Chapter I (p. 10)
Basil Blackwell & Mott Ltd. Oxford, England. 1959

Every good poem, in fact, is a bridge built from the known, familiar side of life over into the unknown. Science, too, is always making expeditions into the unknown. But this does not mean that science can supersede poetry. For poetry enlightens us in a different way from science: it speaks directly to our feelings or imagination. The findings of poetry are no more and no less true than science.
Poetry for You
Chapter VIII (p. 92)
Basil Blackwell & Mott Ltd. Oxford, England. 1959

Holton, Gerald 1922–
Research professor of physics and science history

Poets rush in where scientists fear to tread.
Einstein, History, and Other Passions: The Rebellion Against Science at the End of the Twentieth Century
Part One, Chapter 6 (p. 132)
Addison-Wesley Publishing Company. Reading, Massachusetts, USA. 1996

Jones, Frederick Wood 1879–1954
Physician

Whoever wins to a great scientific truth will find a poet before him in the quest.
Medical Journal of Australia, 29 August 1931

Lysaght, Sidney R. 1860–1941
Irish writer

Science in the first place looks for information, poetry for beauty; and, taking different paths, they meet on the borderland of discovery.
A Reading of Life
Chapter II (p. 35)
Macmillan & Company Ltd. London, England. 1936

Melandri, E.
No biographical data available

The existence of poetics of science is undeniable.... Barring poetics from science is the same as barring use of the hypothesis.
In Fernand Hallyn
The Poetic Structure of the World: Copernicus and Kepler
Introduction (p. 7)
Zone Books. New York, New York, USA. 1990

Miller, Hugh 1802–56
Scottish geologist and theologian

Because science flourishes, must poesy decline? The complaint serves but to betray the weakness of the class who urge it.
Sketch-Book of Popular Geology
Lecture Second (p. 80)
William P. Nimmo & Company. Edinburgh, Scotland. 1880

Spencer, Herbert 1820–1903
English social philosopher

...those who have never entered upon scientific pursuits know not a tithe of the poetry by which they are surrounded.
Education: Intellectual, Moral and Physical
A.L. Fowle. New York, New York, USA. 1860

Thoreau, Henry David 1817–62
American essayist, poet, and practical philosopher

The poet uses the results of science and philosophy, and generalizes their widest deductions.
The Writings of Henry David Thoreau (Volume 1)

A Week on the Concord and Merrimac Rivers
Friday (p. 478)
Houghton Mifflin Company. Boston, Massachusetts, USA. 1893

Tillyard, E. M. W. 1889–1962
English classical scholar
Lewis, C. S. (Clive Staples) 1898–1963
British author, scholar, and popular theologian

Only science can tell you where and when you are likely to meet an elm: only poetry can tell you what meeting an elm is like.
The Personal Heresy: A Controversy
Chapter V (p. 110)
Oxford University Press, Inc. London, England. 1939

von Schlegel, Friedrich 1772–1829
German philosopher, critic, and writer

Strictly speaking, the idea of a scientific poem is probably as nonsensical as that of a poetic science.
Dialogue on Poetry and Literary Aphorisms
Selected Aphorisms from the Lyceum
Aphorism 61 (p. 127)
The Pennsylvania State University Press, University Park. Pennsylvania, USA. 1968

Wheelock, John Hall 1886–1978
American poet

The statements of science are hearsay, reports from a world outside the world we know. What the poet tells us has long been known to us all, and forgotten. His knowledge is of our world, the world we are both doomed and privileged to live in, and it is a knowledge of ourselves, of the human condition, the human predicament.
What Is Poetry?
Chapter 6
Charles Scribner's Sons. New York, New York, USA. 1963

Whitman, Walt 1819–92
American poet, journalist, and essayist

Exact science and its practical movements are no checks on the greatest poet but always his encouragement and support. ...there the arms that lifted him first and brace him best...there he returns after all his goings and comings. The sailor and traveller...the anatomist chemist astronomer geologist phrenologist spiritualist mathematician historian and lexicographer are not poets, but they are the lawgivers of poets and their construction underlies the structure of every perfect poem.
Leaves of Grass
Preface to the 1855 edition of "Leaves of Grass" (p. 304)
Doubleday, Doran & Company, Inc. New York, New York, USA. 1940

If there shall be love and content between the father and the son and if the greatness of the son is the exuding of the greatness of the father, there shall be love between the poet and the man of demonstrable science. In the beauty of poems are the tuft and final applause of science.

Leaves of Grass
Preface to the 1855 edition of "Leaves of Grass" (p.304)
Doubleday, Doran & Company, Inc. New York, New York, USA. 1940

Zee, Anthony
American physicist

In science, one tries to say what no one else has ever said before. In poetry, one tries to say what everyone else has already said, but better. This explains, in essence, why good poetry is as rare as good science.
Fearful Symmetry
Chapter 7 (p. 103)
Macmillan Publishing Company. New York, New York, USA. 1986

SCIENCE AND POLITICS

Born, Max   1882–1970
German-born English physicist

…the subordination of fundamental research to political and military authorities is detrimental. The scientists themselves have learned by now that the period of unrestricted individualism in research has come to an end. They know that even the most abstract and remote ideas may one day become of great practical importance — like Einstein's law of equivalence of mass and energy. They have begun to organize themselves and to discuss the problem of their responsibility to human society. It should be left to these organizations to find a way to harmonize the security of the nations with the freedom of research and publication without which science must stagnate.
The Restless Universe
Postscript (p. 308)
Dover Publications, Inc. New York, New York, USA. 1951

Budworth, D.
No biographical data available

Science policy is essentially about the allocation of scarce resources, and is therefore a part of politics…. The scarce resource with which science policy should concern itself in the short term is not money, but that portion of the scientific population which is capable of initiating and leading significant work. Such people are always in short supply, even when the total population itself is greater than the available jobs.
Science Policy Should Be About People
New Scientist, Volume 69, Number 993, 25 March 1976 (pp. 684–685)

Clarke, Arthur C.   1917–
English science and science fiction writer

The menace of interplanetary imperialism can be overcome only by world-wide technical and political agreements well in advance of the actual event, and these will require continual pressure and guidance from the organizations which have studied the subject.

The Challenge of the Spaceship
The Challenge of the Spaceship (p. 8)
Harper & Brothers. New York, New York, USA. 1959

de Maupertuis, Pierre-Louis Moreau   1698–1759
French mathematician and astronomer

There are sciences over which the will of kings has no immediate influence; it can procure advancement there only in so far as the advantages which it attaches to their study can multiply the number and the efforts of those who apply themselves to them. But there are other sciences which for their progress urgently need the power of sovereigns; they are all those which require greater expenditure than individuals can make or experiments which would not ordinarily be practicable.
Lettres su le progrés des sciences, Oeuvres de Maupertius
Dresden (pp. 6–7)
Publisher undetermined

Johnson, Harry G.   1923–1979
American economist

Basic science, per se, contributes to culture; it contributes to our social well-being, including national defence and public health; to our economic well-being; and it is an essential element of the education not only of scientists but also of the population as a whole. In deciding how much science the society needs, one must decide how the support of science bears on these other, politically defined, goals of the society.
In National Academy of Sciences
Basic Research and National Goals: A Report to the Committee on Science and Astronautics
Federal Support of Basic Research: Some Economic Issues,
Summary (p. 5)
U.S. Government Printing Office. Washington,
D.C. 1965

Koestler, Arthur   1905–83
Hungarian-born English writer

No scientist is admired for failing in the attempt to solve problems that lie beyond his competence. The most he can hope for is the kindly contempt earned by the Utopian politician. If politics is the art of the possible, research is surely the art of the soluble. Both are immensely practical-minded affairs.
The Act of Creation
New Statesman, Volume 19, June 1964

Price, Don K.   1910–1995
American presidential advisor and educator

…all sciences are considered by their professors as equally significant; by the politicians, as equally incomprehensible; and by the military as equally expensive.
The Scientific Estate
Chapter 1 (p. 12)
Harvard University Press. Cambridge, Massachusetts,
USA. 1965

…it has begun to seem evident to a great many administrators and politicians that science had become something very close to an establishment, in the old and proper sense of that word: a set of institutions supported by tax funds but largely on faith and without direct responsibility to political control.

The Scientific Estate
Chapter 1 (p. 12)
Harvard University Press. Cambridge, Massachusetts, USA. 1965

Rabinowitch, Eugene   1901–73
Russian-born American biophysicist

Science has assumed such an important role in determining the parameters of national and international life, that participation in national decisions by people whose world picture has been affected by the study and practice of science (even if this picture has its own bias), is indispensable for many major political decisions — to correct the bias of the more traditional molders of national decisions, such as men with legal training.

Open Season on Scientists
The New Republic, January 1, 1966 (p. 21)

SCIENCE AND RELIGION

Adams, George   1750–95
English instrument maker

The two kingdoms of nature and grace, as two parallel lines, correspond to each other, follow a like course, but can never be made to touch. An adequate understanding of this distinction in all its branches, would be the consummation of human knowledge.

Lectures on Natural and Experimental Philosophy (Volume 1)
Lecture VI (p. 242)
Printed by R. Hindmarsh. London, England. 1794

Adams, Henry Brooks   1838–1918
American man of letters

The preacher then went on to criticise the attitude of religion towards science. "If there is still a feeling of hostility between them…it is no longer the fault of religion. There have been times when the church seemed afraid, but she is so no longer. Analyze, dissect, use your microscope or your spectrum till the last atom of matter is reached; reflect and refine till the last element of thought is made clear; the church now knows with the certainty of science what she once knew only by the certainty of faith, that you will find enthroned behind all thought and matter only one central idea, — that idea which the church has never ceased to embody, — I AM!

Democracy, and Esther: Two Novels by Henry Adams
Esther (p. 212)
Peter Smith. Gloucester, Massachusetts, USA. 1965

Allport, Gordon   1897–1967
American psychologist

A narrowly conceived science can never do business with a narrowly conceived religion.

The Individual and His Religion: A Psychological Interpretation
Preface (p. vi)
The Macmillan Company. New York, New York, USA. 1956

Alpher, Ralph Asher   1921–
American physicist

I…reject the argument put forth by many fundamentalists that science has nothing to do with religion because God is not among the things making up the universe in which we live. Surely if a necessity for a god-concept in the universe ever turns up, that necessity will become evident to the scientist.

Theology of the Big Bang
Religious Humanism, Volume 17, Number 1, Winter 1983 (p. 12)

Appleyard, Bryan   1951–
English author and journalist

Science was the lethally dispassionate search for truth in the world whatever its meaning might be; religion was the passionate search for meaning whatever the truth might be. Science can lay a claim to a meaning in the sense of establishing causality, and religion could claim truth in the sense of a transcendent order. But science's meaning does not answer the question Why? And religion's truth had no scientific relevance.

Understanding the Present: Science and the Soul of Modern Man
Chapter 4 (p. 79)
Doubleday. New York, New York, USA. 1992

Berger, Peter L.   1929–
American sociologist

Protestant theologians have been increasingly engaged in playing a game whose rules have been dictated by their cognitive antagonists.

A Rumor of Angels: Modern Society and the Rediscovery of the Supernatural
Chapter 1 (p. 10)
Doubleday & Company, Inc. Garden City, New York, USA. 1970

Bernal, John Desmond   1901–71
Irish-born physicist and x-ray crystallographer

Now the history of scientific advance has shown us clearly that any appeal to Divine Purpose or any supernatural agency, to explain any phenomenon, is in fact only a concealed confession of ignorance, and a bar to genuine research.

Science and Ethics
A Marxist Critique (p. 116)
George Allen & Unwin Ltd. London, England. 1942

The role of God in the material world has been reduced stage by stage with the advance of science, so much so that He only survives in the vaguest mathematical form in the minds of older physicists and biologists.

In C.H. Waddington (ed.)

Science and Ethics
A Marxist Critique (p. 116)
George Allen & Unwin Ltd. London, England. 1942

Boutroux, Émile   1845–1921
French philosopher

In spite of their relations, science and religion remain, and must remain, distinct. If there were no other way of establishing a rational order between things than that of reducing the many to the one, either by assimilation or by elimination, the destiny of religion would appear doubtful.
Translated by Jonathan Nield
Science and Religion in Contemporary Philosophy
Conclusion (pp. 399–400)
Duckworth & Company. London, England. 1912

Buck, Pearl S.   1892–1973
American author

Science and religion, religion and science, put it as I may they are two sides of the same glass, through which we see darkly until these two, focusing together, reveal the truth.
A Bridge for Passing
Chapter III (p. 255)
John Day Company. New York, New York, USA. 1962

Bultmann, R.
No biographical data available

…the New Testament provides a world picture which belongs entirely to Jewish or Gnostic mythology and is incredible or even meaningless in a scientific age.
In H.J. Paton
The Modern Predicament: A Study in the Philosophy of Religion
Chapter XV, Section 3 (p. 228)
Collier Books. New York, New York, USA. 1962

Burroughs, John   1837–1921
American naturalist and writer

The mysteries of religion are of a different order from those of science; they are parts of an arbitrary system of man's own creation; they contradict our reason and our experience, while the mysteries of science are revealed by our reason, and transcend our experience.
Scientific Faith
The Atlantic Monthly, July 1915 (p. 33)

The miracles of religion are to be discredited, not because we cannot conceive of them, but because they run counter to all the rest of our knowledge; while the mysteries of science, such as chemical affinity, the conservation of energy, the indivisibility of the atom, the change of the non-living into the living…extend the boundaries of our knowledge, though the modus operandi of the changes remains hidden.
Scientific Faith
The Atlantic Monthly, July 1915 (p. 33)

Bush, Vannevar   1890–1974
American electrical engineer and physicist

To pursue science is not to disparage the things of the spirit. In fact, to pursue science rightly is to furnish a framework on which the spirit may rise.
Speech, MIT, October 5, 1953

Bushnell, Horace   1802–76
American Congregational minister

As the science of nature goes toward completion, religion, having all the while been watching for it in close company, will have gotten immense breadth and solidity, from the ideas and facts unfolded in its discoveries, and will be as much enlarged in its confidence and the sentiment of worship, as beholding God's deep system in the world signifies more than looking on its surfaces.
Science and Religion
Putnam's Magazine, Volume 1, 1868 (p. 267)

Butler, Samuel   1612–80
English novelist, essayist, and critic

Science and religion are reconciled in amiable and sensible people but nowhere else.
The Note-Books of Samuel Butler (Volume 1)
1874–1883 (p. 118)
University Press of America, Inc. Lanham, Maryland, USA. 1984

Chadwick, Owen   1916–
English historian and Christian scholar

Science versus Religion — the antithesis conjures two hypostatized entities of the later nineteenth century: Huxley St. George slaying Samuel smoothest of dragons; a mysterious undefined ghost called Science against a mysterious indefinable ghost called Religion; until by 1900 schoolboys decided not to have faith because Science, whatever that was, disproved Religion, whatever that was.
The Secularisation of the European Mind in the Nineteenth Century
Part II, Chapter 7 (p. 161)
Cambridge University Press. Cambridge, England. 1990

Clark, W. C.
No biographical data available
Majone, G.
No biographical data available

The social uses of science have always had something in common with the social uses of religion. And in the two decades following the Second World War, modern science took on a most religious-looking numinous legitimacy as an unquestioned source of authority on all manner of policy problems.
Report of the International Institute of Applied Systems Analysis
The Critical Appraisal of Scientific Inquiries with Policy Implications, Laxenburg, Austria, 1984 (p. 35)

Compton, Karl Taylor   1887–1954
American educator and physicist

Science has contributed to the making of religion into a developing dynamic spiritual force.
A Scientist Speaks: Excerpts from Addresses by Karl Taylor Compton During the Years 1930–1949 (p. 19)
Undergraduate Association, MIT. Cambridge, Massachusetts, USA 1955

Conklin, Edwin Grant   1863–1952
American zoologist

Science cannot solve the great mysteries of our existence, — why we are, whither we are bound, and what it all means. Faith alone assures us that there is definite purpose in all experience. This knowledge makes life worth living and service a privilege.
In Edward H. Cotton
Has Science Discovered God?
A Biologist's Religion (p. 89)
Thomas Y. Crowell Company. New York, New York, USA. 1931

Coulson, Charles Alfred   1910–74
English theoretical chemist

…science is one aspect of God's presence, and scientists therefore part of the company of His heralds.
Science and Christian Belief
Scientific Method (p. 30)
The University of North Carolina Press. Chapel Hill, North Carolina, USA. 1955

Davies, Paul Charles William   1946–
British-born physicist, writer, and broadcaster

If the Church is largely ignored today it is not because science has finally won its age-old battle with religion, but because it has so radically reoriented our society that the biblical perspective of the world now seems largely irrelevant. As one television cynic recently remarked, few of our neighbors possess an ass for us to covet.
God and the New Physics
Chapter 1 (p. 2)
Simon & Schuster. New York, New York, USA. 1983

Those who invoke God as an explanation of cosmic organization usually have in mind a supernatural agency, acting on the world in defiance of natural laws. But it is perfectly possible for much, if not all of what we encounter in the universe to be the product of intelligent manipulation of a purely natural kind: within the laws of physics. For example, our galaxy could have been made by a powerful mind who rearranged the primeval gases using carefully placed gravitating bodies, controlled explosions and all the other paraphernalia of a space age astro-engineer.
God and the New Physics
Chapter 15 (p. 208)
Simon & Schuster. New York, New York, USA. 1983

In spite of the fact that religion looks backward to revealed truth while science looks forward to new vistas and discoveries, both activities produce a sense of awe and a curious mixture of humility and arrogance in their

practitioners. All great scientists are inspired by the subtlety and beauty of the natural world that they are seeking to understand. Each new subatomic particle, every unexpected object, produces delight and wonderment. In constructing their theories, physicists are frequently guided by arcane concepts of elegance in the belief that the universe is intrinsically beautiful.
God and the New Physics
Chapter 17 (p. 220)
Simon & Schuster. New York, New York, USA. 1983

Dembski, William A.   1960–
Mathematician and philosopher

Any view of the sciences that leaves Christ out of the picture must be seen as fundamentally deficient.
Intelligent Design: The Bridge Between Science and Theology
Part 3, Chapter 7, Section 7.6 (p. 206)
InterVarsity Press. Downers Grove, Illinois, USA. 1999

Dobzhansky, Theodosius   1900–75
Russian-American scientist

There are still many people who are happy and comfortable adhering to fundamentalist creeds. This should cause no surprise, since a large majority of these believers are as unfamiliar with scientific findings as were people who lived centuries ago.
The Biology of Ultimate Concern
Chapter 5 (p. 95)
The New American Library, Inc. New York, New York, USA. 1967

Science and religion deal with different aspects of existence…[T]hese are the aspect of facts and the aspect of meaning. But there is one stupendous fact…the meaning of which they have ceaselessly tried to discover. This fact is Man.
The Biology of Ultimate Concern
Chapter 5 (p. 96)
The New American Library, Inc. New York, New York, USA. 1967

…nothing gives more pleasure to a rather common type of religious person than to point out that science cannot explain this or cannot account for that!
The Biology of Ultimate Concern
Chapter 5 (p. 97)
The New American Library, Inc. New York, New York, USA. 1967

Draper, John William   1811–82
American scientist, philosopher, and historian

As to Science, she has never sought to ally herself to civil power. She has never attempted to throw odium or inflict social ruin on any human being. She has never subjected any one to mental torment, physical torture, least of all to death, for the purpose of upholding or promoting her ideas. She presents herself unstained by cruelties and crimes. But in the Vatican — we have only to recall the Inquisition — the hands that are now raised in appeals to the Most Merciful are crimsoned.

History of the Conflict between Religion and Science
Preface (p. xi)
D. Appleton and Company. New York, New York, USA. 1898

du Noüy, Pierre Lecomte   1883–1947
French scientist

Any man who believes in God must realize that no scientific fact, as long as it is true, can contradict God. Otherwise, it would not be true. Therefore, any man who is afraid of science does not possess a strong faith.
Human Destiny
Chapter 16 (p. 243)
Longmans, Green & Company. London, England. 1947

Dubos, René Jules   1901–82
French-born American microbiologist and environmentalist

Religion and science…constitute deep-rooted and ancient efforts to find richer experience and deeper meaning than are found in the ordinary biological and social satisfactions. As pointed out by Whitehead, religion and science have similar origins and are evolving toward similar goals.
A God Within
Chapter 12. On Being Human (p. 255)
Charles Scribner's Sons. New York, New York, USA. 1972

Both [religion and science] started from crude observations and fanciful concepts, meaningful only within a narrow range of conditions for the people who formulated them of their limited tribal experience. But progressively, continuously, and almost simultaneously, religious and scientific concepts are ridding themselves of their coarse and local components, reaching higher and higher levels of abstraction and purity.
A God Within
Chapter 12. On Being Human (p. 255)
Charles Scribner's Sons. New York, New York, USA. 1972

Both the myths of religion and the laws of science, it is now becoming apparent, are not so much descriptions of facts as symbolic expressions of cosmic truths.
A God Within
Chapter 12. On Being Human (p. 255)
Charles Scribner's Sons. New York, New York, USA. 1972

Durant, William James   1885–1981
American historian and essayist

Those of you who specialize in science will find it hard to understand religion, unless you feel, as Voltaire did, that the harmony of the spheres reveals a cosmic mind, and unless you realize, as Rousseau did, that man does not live by intellect alone.
Commencement Address
Webb School of Claremont, California, June 7, 1958

We are such microscopic particles in so immense a universe that none of us is in a position to understand the world, much less to dogmatize about it. Pascal trembled at the thought of man's bewildered minuteness between the immensity of the whole and the complexity of each part; "these infinite spaces," he said, "frighten me!" Let us be careful how we pit our pitiful generalizations against the infinite variety, scope, and subtlety of the world.
Commencement Address
Webb School of Claremont, California, June 7, 1958

Dyson, Freeman J.   1923–
American physicist and educator

Professional scientists today live under a taboo against mixing science and religion.
Disturbing the Universe
Chapter 23 (p. 245)
Basic Books, Inc. New York, New York, USA. 1979

Eddington, Sir Arthur Stanley   1882–1944
English astronomer, physicist, and mathematician

The starting-point of belief in mystical religion is a conviction of significance or, as I have called it earlier, the sanction of a striving in the consciousness. This must be emphasized because appeal to intuitive conviction of this kind has been the foundation of religion through all ages, and I do not wish to give the impression that we have now found something new and more scientific to substitute. I repudiate the idea of proving the distinctive beliefs of religion either from the data of physical science or by the methods of physical science.
The Nature of the Physical World
Chapter XV (p. 333)
The Macmillan Company. New York, New York, USA. 1930

It is curious that the doctrine of the running down of the physical universe is so often looked upon as pessimistic and contrary to the aspirations of religion. Since when has the teaching that "heaven and earth shall pass away" become ecclesiastically unorthodox?
New Pathways in Science
Chapter III, Section III (p. 59)
The Macmillan Company. New York, New York, USA. 1935

It is probably true that the recent changes of scientific thought remove some of the obstacles to a reconciliation of religion with science; but this must be carefully distinguished from any proposal to base religion on scientific discovery. For my own part I am wholly opposed to any such attempt.
Science and the Unseen World
Chapter VII (pp. 72–73)
The Macmillan Company. New York, New York, USA. 1929

Einstein, Albert   1879–1955
German-born physicist

All religions, arts and sciences are branches of the same tree.
Out of My Later Years (p. 7)
Thames & Hudson. London, England. 1950

Everyone who is seriously involved in the pursuit of science becomes convinced that a spirit is manifest in the laws of the Universe — a spirit vastly superior to that of man, and one in the face of which we with our modest powers must feel humble. In this way the pursuit of science leads to a religious feeling of a special sort, which is indeed quite different of the religiosity of someone more naive.

In Helen Dukas and Banesh Hoffman
Albert Einstein: The Human Side — New Glimpses from His Archives
Letter dated 20 December, 1935 (p. 33)
Princeton University Press. Princeton, New Jersey, USA. 1979

Scientific research is based on the idea that everything that takes place is determined by laws of nature, and therefore this holds for the actions of people. For this reason, a research scientist will hardly be inclined to believe that events could be influenced by a prayer, i.e., by a wish addressed to a supernatural Being. However, it must be admitted that our actual knowledge of these laws is only imperfect and fragmentary, so that, actually, the belief in the existence of basic all-embracing laws in Nature also rests on a sort of faith. [Belief in basic laws of Nature] has been largely justified so far by the success of scientific research.

In Helen Dukas and Banesh Hoffman
Albert Einstein: The Human Side: New Glimpses from His Archives
Letter dated 24 January 1936 (pp. 32–34)
Princeton University Press. Princeton, New Jersey, USA. 1979

The basis of all scientific work is the conviction that the world is an ordered and comprehensive entity, which is a religious sentiment. My religious feeling is a humble amazement at the order revealed in the small patch of reality to which our feeble intelligence is equal.

Cosmic Religion, With Other Opinions and Aphorisms
On Science (p. 98)
Covici-Fiede. New York, New York, USA. 1931

[E]very one who is seriously involved in the pursuit of science becomes convinced that a spirit is manifest in the laws of the Universe — a spirit vastly superior to that of man, and one in the face of which we with our modest powers must feel humble. In this way the pursuit of science leads to a religious feeling of a special sort, which is indeed quite different from the religigiosity of someone more naive.

In Helen Dukas and Banesh Hoffman
Albert Einstein: The Human Side: New Glimpses from His Archives
Letter dated 24 January 1936 (pp. 32–34)
Princeton University Press. Princeton, New Jersey, USA. 1979

[T]he cosmic religious experience is the strongest and the noblest driving force behind scientific research. No one who does not appreciate the terrific assertions, and, above all, the devotion without which pioneer creations in scientific thought cannot come into being, can judge the strength of the feeling out of which alone such work, turned away as it is from immediate practical life, can

grow. What a deep faith in the rationality of the structure of the world and what a longing to understand even a small glimpse of the reason revealed in the world there must have been in Kepler and Newton to enable them to unravel the mechanism of the heavens in long years of lonely study.

The New York Times Magazine, 9 November 1930

Certain it is that a conviction, akin to religious feeling, of the rationality or intelligibility of the world lies behind all scientific work of a higher order.... This firm belief, a belief bound up with deep feeling, in a superior mind that reveals itself in the world of experience, represents my conception of God.

Ideas and Opinions
On Scientific Truth (p. 261)
Crown Publishers, Inc. New York, New York, USA. 1954

I have never found a better expression than "religious" for this trust in the rational nature of reality and of its peculiar accessibility to the human mind. Where this trust is lacking science degenerates into an uninspired procedure. Let the devil care if the priests make capital out of this. There is no remedy for that.

Lettres a Maurice Solovine (pp. 102–103)
Gauthier-Villars. Paris, France. 1956

I am of the opinion that all the finer speculations in the realm of science spring from a deep religious feeling, and that without such feeling they would not be fruitful.

Science and God: A Dialog
Forum, Volume 83, June 1930 (p. 373)

Science without religion is lame, religion without science is blind.

Out of My Later Years
Science and Religion, II (p. 26)
Thames & Hudson. London, England. 1950

...science not only purifies the religious impulse of the dross of its anthropomorphism but also contributes to a religious spiritualization of our understanding of life.

Out of My Later Years
Science and Religion, II (p. 29)
Thames & Hudson. London, England. 1950

Eiseley, Loren C.   1907–77
American anthropologist, educator, and author

Century after century, humanity studies itself in the mirror of fashion, and ever the mirror gives back distortions, which for the moment impose themselves upon man's real image. In one period we believe ourselves governed by immutable laws; in the next, by chance. In one period angels hover over our birth; in the following time we are planetary waifs, the product of a meaningless and ever altering chemistry. We exchange halos in one era for fangs in another. Our religious and philosophical conceptions change so rapidly that the theological and moral exhortations of one decade become the

wastepaper of the next epoch. The ideas for which millions yielded up their lives produce only bored yawns in a later generation.
The Unexpected Universe
Chapter Eight, Section 2 (p. 179)
Harcourt, Brace & World, Inc. New York, New York, USA. 1969

Emerson, Ralph Waldo   1803–82
American lecturer, poet, and essayist

The Religion that is afraid of science dishonors God and commits suicide.
The Journals and Miscellaneous Notebooks of Ralph Waldo Emerson
(Volume 2)
1826–1832, 4 March 1831 (p. 239)
Harvard University Press. Cambridge, Massachusetts, USA. 1970

Flaubert, Gustave   1821–90
French novelist

A little science takes your religion from you; a great deal brings you back to it.
Dictionary of Accepted Ideas
M. Reinhardt. London, England. 1954

Fosdick, Harry Emerson   1878–1969
American clergyman and educator

What modern science is doing for multitudes of people, as anybody who watches American life can see, is not to disprove God's theoretical existence, but to make him "progressively less essential."
Adventurous Religion
Will Science Displace God? (p. 136)
Harper & Brothers. New York, New York, USA. 1926

Freud, Sigmund   1856–1939
Austrian neurologist and co-founder of psychoanalysis

The scientific spirit brings about a particular attitude towards worldly matters; before religious matters it pauses for a little, hesitates, and finally there too crosses the threshold. In this process there is no stopping; the greater the number of men to whom the treasures of knowledge become accessible, the more widespread is the falling-away from religious belief…
The Future of an Illusion
Chapter VII (p. 38)
W.W. Norton & Company, Inc. New York, New York, USA. 1961

Froude, James Anthony   1818–94
English historian and biographer

The superstitions of science scoff at the superstitions of faith.
The Lives of the Saints
Eclectic Review, February 1852

Garman, Charles E.   1862–1932
No biographical data available

Science is thinking God's thoughts after Him just as truly as when we read the scriptures.
Letters, Lectures, Addresses of Charles Edward Garman; A Memorial Volume, Prepared with the Cooperation of the Class of 1884, Amherst
Science and Theism (p. 231)
Houghton Mifflin Company. Boston, Massachusetts, USA. 1909

Gilkey, Langdon   1919–2004
Protestant theologian

It is because science is limited to a certain level of explanation that scientific and religious theories can exist side by side without excluding one another, that one person can hold both to the scientific accounts of origins and to a religious account, to the creation of all things by God…
Creationism on Trial: Evolution and God at Little Rock
Chapter 5 (p. 117)
Winston Press. Minneapolis, Minnesota, USA. 1985

Gillispie, Charles Coulston   1918–
French writer and editor of philosophy and history of science

If one be clear about the nature of science as a description of the world, declarative but never normative, may not the choice between science and religion be refused? Is it not simply a false problem, arising from a confusion — an ancient confusion going back to the beginning of science — between objects and persona? Science is about nature, after all, not about duties. It is about objects. Christianity is about persons, the relation of the persons of men to the person of God.
The Edge of Objectivity: An Essay in the History of Scientific Ideas
Chapter VIII (pp. 350–351)
Princeton University Press. Princeton, New Jersey, USA. 1960

Goodspeed, Edgar J.   1871–1962
American scholar

Science needs religion, to prevent it from becoming a curse to mankind instead of a blessing.
The Four Pillars of Democracy
Chapter VI (p. 134)
Harper & Brothers. New York, New York, US. 1940

Science sees meaning in every part; religion sees meaning in the whole.
Four Pillars of Democracy
Chapter V (p. 106)
Harper & Brothers. New York, New York, USA. 1940

…religion needs science, to protect it from religion's greatest danger, superstition.
Four Pillars of Democracy
Chapter V (p. 115)
Harper & Brothers. New York, New York, USA. 1940

Grinnel, Frederick

…modern science constitutes a method for understanding and modifying the world but has no inherent direction, whereas modern religion describes a messianic world view but lacks a useful method to bring about this state of affairs.

Complementarity: An Approach to Understanding the Relationship Between Science and Religion
Perspectives in Biology and Medicine, Volume 29, Number 2, Winter 1986 (p. 293)

Gull, Sir William Withey   1816–90
English physician

Realize, if you can, what a paralyzing influence on all scientific inquiry the ancient belief must have had which attributed the operations of nature to the caprice not of one divinity only, but of many. There still remains vestiges of this in most of our minds, and the more distinct in proportion to our weakness and ignorance.
British Medical Journal, Volume 2, 1874 (p. 425)

Haldane, John Burdon Sanderson   1892–1964
English biologist

The wise man regulates his conduct by the theories both of religion and science. But he regards these theories not as statements of ultimate fact but as art-forms.
Possible Worlds and Other Papers
Chapter XXXI (p. 252)
Harper & Brothers. New York, New York, USA. 1928

Hardin, Garrett   1915–2003
American ecologist and microbiologist

We are terribly clever people, we moderns: we bend Nature to our will in countless ways. We move mountains, we make caves, fly at speeds no other organism can achieve and tap the power of the atom. We are terribly clever. The essentially religious feeling of subserviency to a power greater than ourselves comes hard to us clever people. But by our intelligence we are now beginning to make out the limits of our cleverness, the impotence principles that say what can and cannot be. In an operational sense, we are experiencing a return to a religious orientation toward the world.
Nature and Man's Fate
The Search for Truth
The New American Library. New York, New York, USA. 1961

Heisenberg, Werner Karl   1901–76
German physicist and philosopher

In the history of science, ever since the famous trial of Galileo, it has repeatedly been claimed that scientific truth cannot be reconciled with the religious interpretation of the world. Although I am now convinced that scientific truth is unassailable in its own field, I have never found it possible to dismiss the content of religious thinking as simply part of an outmoded phase in the consciousness of mankind, a part we shall have to give up from now on. Thus in the course of my life I have repeatedly been compelled to ponder on the relationship of these two regions of thought, for I have never been able to doubt the reality of that to which they point.

Across the Frontiers
Chapter XVI (p. 213)
Harper & Row, Publishers. New York, New York, USA. 1974

If we are honest — and scientists have to be — we must admit that religion is a jumble of false assertions, with no basis in reality. The very idea of God is a product of human imagination.
Physics and Beyond: Encounters and Conversations
Chapter 7 (p. 85)
Harper & Row, Publishers. New York, New York, USA. 1972

Hertz, Rabbi Richard
No biographical data available

I find no conflict between science and religion. Science teaches what is. Religion teaches what ought to be. Science describes. Religion prescribes. Science analyzes what we can see. Religion deals with what is unseen. Each can help the other.
The American Jew in Search of Himself
Chapter 4 (p. 42)
Bloch Publishing Company. New York, New York, USA. 1962

Hillis, W. Daniel   1956–
American engineer, inventor, and author

...I remain convinced that neither religion nor science has everything figured out.
The Pattern on the Stone: The Simple Ideas that Make Computers Work (p. 152)
Basic Books, Inc. New York, New York, USA. 1998

Hooykaas, Reijer
Dutch historian of science

Metaphorically speaking, whereas the bodily ingredients of science may have been Greek, its vitamins and hormones were biblical.
Religion and the Rise of Modern Science
Epilogue (p. 162)
William B. Eerdmans Publishing Company. Grand Rapids, Michigan, USA. 1972

Huxley, Julian   1887–1975
English biologist, philosopher, and author

...it is no longer possible to maintain that science and religion must operate in thought-tight compartments or concern separate sectors of life; they are both relevant to the whole of human existence.
In Teilhard de Chardin
The Phenomenon of Man
Introduction (p. 26)
Harper & Row, Publishers. New York, New York, USA. 1959

Like the meridians as they approach the poles, science, philosophy and religion are bound to converge as they draw nearer to the whole. I say "converge" advisedly, but without merging, and without ceasing, to the very end, to assail the real from different angles and on different planes.

In Teilhard de Chardin
The Phenomenon of Man
Introduction (p. 30)
Harper & Row, Publishers. New York, New York, USA. 1959

Huxley, Thomas Henry 1825–95
English biologist

True science and true religion…are twin-sisters, and the separation of either from the other is sure to prove the death of both. Science prospers exactly in proportion as it is religious; and religion flourishes in exact proportion to the scientific depth and firmness of its basis. The great deeds of philosophers have been less the fruit of their intellect than of the direction of that intellect by an imminently religious tone of mind. Truth has yielded herself rather to their patience, their love, their single-heartedness, and their self-denial, than to their logical acumen.
In Herbert Spencer
Education: Intellectual, Moral, and Physical
Chapter I (p. 81)
A.L. Fowle. New York, New York, USA. 1860

…the materialistic position that there is nothing in the world but matter, force, and necessity, is as utterly devoid of justification as the most baseless of theological dogmas.
Collected Essays (Volume 1)
On the Physical Basis of Life (p. 162)
Macmillan & Company Ltd. London, England. 1904

Elijah's great question, "Will you serve God or Baal? Choose ye," is uttered audibly enough in the ears of every one of us as we come to manhood. Let every man who tries to answer it seriously ask himself whether he can be satisfied with the Baal of authority, and with all the good things his worshippers are promised in this world and the next. If he can, let him, if he be so inclined, amuse himself with such scientific implements as authority tells him are safe and will not cut his fingers; but let him not imagine he is, or can be, both a true son of the Church and a loyal soldier of science.
Collected Essays (Volume 2)
Darwiniana
Mr. Darwin's Critics (p. 149)
Macmillan & Company Ltd. London, England. 1904

Inge, William Ralph 1860–1954
English religious leader and author

No scientific discovery is without its religious and moral implications.
Outspoken Essays (Second Series)
Confessio Fidei (p. 56)
Longmans, Green & Company. London, England. 1922

Jastrow, Robert 1925–
American space scientist

For the scientist who has lived by his faith in the power of reason, the story ends like a bad dream. He has scaled the mountains of ignorance; he is about to conquer the highest peak; as he pulls himself over the final rock, he is greeted by a band of theologians who have been sitting there for centuries.
God and the Astronomers
Chapter 6 (p. 116)
W.W. Norton & Company, Inc. New York, New York, USA. 1978

Joint Statement of Religious Leaders

The purpose of science is to develop, without prejudice or preconception of any kind, a knowledge of the facts, the laws, and the processes of nature. The even more important task of religion, on the other hand, is to develop the consciences, the ideals, and the aspirations of mankind.
In Robert Andrews Millikan
Science and Life
A Joint Statement upon the Relations of Science and Religion (p. 86)
The Pilgrim Press. Boston, Massachusetts, USA. 1924

Kaempffert, Waldemar 1877–1956
American science editor and museum director

Religion may preach the brotherhood of man; science practices it.
In Edward R. Murrow
This I Believe
2, Michael Faraday (p. 196)
Simon & Schuster. New York, New York, USA. 1952

King, Jr., Martin Luther 1929–68
American civil rights leader and clergyman

Science investigates; religion interprets. Science gives man knowledge which is power; religion gives man wisdom which is control.
Strength to Love
Chapter I (p. 3)
Harper & Row, Publishers. New York, New York, USA. 1963

Lewis, C. S. (Clive Staples) 1898–1963
British author, scholar, and popular theologian

Keep pressing home on him the ordinariness of things. Above all, do not attempt to use science (I mean, the real sciences) as a defense against Christianity. They will positively encourage him to think about realities he can't touch and see. There have been sad cases among the modern physicists.
The Screwtape Letters: Letters from from a Senior to a Junior Devil (p. 4)
Harper & Row, Publishers. New York, New York, USA. 2001

Lewis, Gilbert Newton 1875–1946
American chemist

…in the struggle of life with the facts of existence, Science is a bringer of aid; in the struggle of the soul with the mystery of existence, Science is the bringer of light.
On the Dread and Dislike of Science
Fortnightly Review, Volume 29, 1878

Lynch, Gary
No biographical data available

What you're really seeking are constraints.... You're seeking things that box you in. That's what separates science from most other human endeavors. Religion is not something where people sit down and say, "Well, if there were a god then".... But science is a constant search for that, for those things that hem you in.
In George Johnson
In the Palaces of Memory: How We Build the Worlds Inside Our Heads
Mucking Around in the Wetware (p. 91)
Alfred A. Knopf. New York, New York, USA. 1991

Marguerite of Valois   1553–1615
Queen of France and Navarre

Science conducts us, step by step, through the whole range of creation, until we arrive, at length, at God.
Memoirs of Marguerite de Valois
Letter XII, 1628 (p. 80)
P.F. Collier & Son, Company. New York, New York, USA. 1910

Mather, Kirtley F.   1888–1978
American geologist

The faith by which a man lives must be in accord with the facts which men know. Only that religion, which is in harmony with the current scientific description of man and the universe, can maintain itself effectively in any age.
In Edward H. Cotton
Has Science Discovered God?
Sermons from Stones (p. 6)
Thomas Y. Crowell Company, Publishers. New York, New York, USA. 1931

McCabe, Joseph   1867–1955
English rationalist writer and ex-Franciscan priest

The theist and the scientist are rival interpreters of nature,
The one retreats as the other advances.
The Existence of God
Chapter V (p. 80)
Watts & Company. London, England. 1933

McKenzie, John L.   d. 1991
American Jesuit theologian and Catholic cardinal

Happily, we have survived into a day when science and theology no longer speak to each other in the language of fishmongers.
The Two-Edged Sword: An Interpretation of the Old Testament
Chapter V. Cosmic Origins (p. 74)
The Bruce Publishing Company. Milwaukee, Minnesota, USA. 1968

Mencken, H. L. (Henry Louis)   1880–1956
American journalist and literary critic

To me the scientific point of view is completely satisfying, and it has been so as long as I can remember. Not once in this life have I ever been inclined to seek a rock and a refuge elsewhere. It leaves a good many dark spots in the universe, to be sure, but not a hundredth time as many as theology. We may trust it, soon or late, to throw light upon many of them, and those that remain dark will be beyond illumination by any other agency. It also fails on occasion to console, but so does theology...
In Charles A. Fecher
Mencken: A Study of His Thought (p. 84)
Alfred A. Knopf. New York, New York, USA. 1978

The notion that science does not concern itself with first causes — that it leaves the field to theology or metaphysics, and confines itself to mere effects — this notion has no support in the plain facts. If it could, science would explain the origin of life on earth at once — and there is every reason to believe that it will do so on some not too remote tomorrow. To argue that gaps in knowledge which will confront the seeker must be filled, not by patient inquiry, but by intuition or revelation, is simply to give ignorance a gratuitous and preposterous dignity.
Treatise on the Gods
Chapter 5 (p. 239)
Vintage Books. New York, New York, USA. 1963

The essence of science is that it is always willing to abandon a given idea, however fundamental it may seem to be, for a better one; the essence of theology is that it holds its truths to be eternal and immutable. To be sure, theology is always yielding a little to the progress of knowledge, and only a Holy Roller in the mountains of Tennessee would dare to preach today what the popes preached in the Thirteenth Century, but this yielding is always done grudgingly, and thus lingers a good while behind the event.
Minority Report: H.L. Mencken's Notebooks
No. 232 (p. 166)
Alfred A. Knopf. New York, New York, USA. 1956

The effort to reconcile science and religion is almost always made, not by theologians, but by scientists unable to shake off altogether the piety absorbed with their mother's milk.
Minority Report: H.L. Mencken's Notebooks
No. 232 (p. 166)
Alfred A. Knopf. New York, New York, USA. 1956

Mernissi, Fatima   1940–
Moroccan writer, feminist, and sociologist

Awareness of the stars and their light pervades the Koran, which reflects the brightness of the heavenly bodies in many verses. The blossoming of mathematics and astronomy was a natural consequence of this awareness. Understanding the cosmos and the movements of the stars means understanding the marvels created by Allah. There would be no persecuted Galileo in Islam, because Islam, unlike Christianity, did not force people to believe in a "fixed" heaven.
Translated by Mary Jo Lakeland

Islam and Democracy: Fear of the Modern World
Chapter 9 (p. 133)
Perseus Publishing. New York, New York, USA. 1992

Miller, Kenneth R.   1948–
American biology professor and author

To a believer, God's great gift was to provide us with a means to understand, to master, and to do good using both the strengths and weaknesses of human nature. Where does science sit with all of this? I would argue that any scientist who believes in God possesses the faith that we are given our unique imaginative powers not only to find God, but also to discover as much of His universe as we could. In other words, to a religious person, science can be a pathway towards God, not away from Him, an additional and sometimes even an amazing grace!
Finding Darwin's God
Chapter 9 (pp. 280–281)
HarperCollins Publishers, Inc. New York, New York, USA. 1999

Millikan, Robert Andrews   1868–1953
American physicist

Modern science, of the real sort, is slowly learning to walk humbly with its God, and in learning that lesson it is contributing something to religion.
Evolution in Science and Religion
Chapter III (pp. 94–95)
Yale University Press. New Haven, Connecticut, USA. 1927

The purpose of science is to develop, without prejudice or preconception of any kind, a knowledge of the facts, the laws and the processes of nature. The even more important task of religion, on the other hand, is to develop the consciences, the ideals and the aspirations of mankind. Each of these two activities represents a deep and vital function of the soul of man, and both are necessary for the life, the progress and the happiness of the human race.
Science Serves God
Time, June 4, 1923

It is a sublime conception of God which is furnished by science, and one wholly consonant with the highest ideals of religion, when it represents Him as revealing Himself through countless ages in the development of the earth as an abode for man and in the age long inbreathing of life into its constituent matter, culminating in man with his spiritual nature and all his Godlike powers.
Science Serves God
Time, June 4, 1923

Moore, Benjamin   1748–1816
Episcopal writer and professor of rhetoric

When new scientific facts are suddenly thrown in amongst old pre-conceived ideas of divinity, there may at first appear discords, and zealous champions of natural science and of religious knowledge fly to arms and indulge in acrimonious polemics; but as time advances and things that are crude and adventitious are thrown away one each side, it is discovered that science has added a new beauty to religion, or rather revealed a beauty that was there all the while, but concealed by misconceptions, or by lack of knowledge.
The Origin and Nature of Life
Chapter I (p. 8)
Henry Holt & Company. New York, New York, USA. No date

Moore, John A.
American writer and professor of genetics and biology

A fundamental difference between religious and scientific thought is that the received beliefs in religion are ultimately based on revelations or pronouncements, usually by some long dead prophet or priest.... Dogma is interpreted by a caste of priests and is accepted by the multitude on faith or under duress. In contrast, the statements of science are derived from the data of observations and experiment, and from the manipulation of these data according to logical and often mathematical procedures.
Science as a Way of Knowing: The Foundations of Modern Biology
Chapter 4 (p. 59)
Harvard University Press. Cambridge, Massachusetts, USA. 1993

Morrow, Lance   1942–
American writer and professor of journalism

Sometime after the Enlightenment, science and religion came to a gentleman's agreement. Science was for the real world: machines, manufactured things, medicines, guns, moon rockets. Religion was for everything else, the immeasurable: morals, sacraments, poetry, insanity, death, and some residual forms of politics and statesmanship. Religion became, in both senses of the word, immaterial.
Fishing in the Tiber: Essays
God and Science (p. 195)
Henry Holt & Company. New York, New York, USA. 1988

Science and religion were apples and oranges. So the pact said: render unto apples the things that are Caesar's, and unto oranges the things that are God's. Just as the Maya kept two calendars, one profane and one priestly, so Western science and religion fell into two different conceptions of the universe, two different vocabularies.
Fishing in the Tiber: Essays
God and Science (p. 195)
Henry Holt & Company. New York, New York, USA. 1988

Nemerov, Howard   1920–91
American poet, novelist, and critic

Religion and science both profess peace (and the sincerity of the professors is not being doubted), but each always turns out to have a dominant part in any war that is going or contemplated.

Figures of Thought: Speculations on the Meaning of Poetry and Other Essays
On the Resemblances Between Science and Religion
David R. Godine. Boston Massachusetts, USA. 1979

Paley, William   1743–1805
English theologian

There cannot be design without a designer; contrivance without a contriver; order without choice; arrangement, without any thing capable of arranging; subserviency and relation to a purpose, without that which could intend a purpose; means suitable to an end, without the end ever having been contemplated, or the means accommodated to it. Arrangement, disposition of parts, subserviency of means to an end, relation of instruments to an use, imply the presence of intelligence and mind.
The Works of William Paley, D.D.
Natural Theology
Chapter II, Section III (p. 22)
Ward, Lock & Company. London, England. No date

Planck, Max   1858–1947
German physicist

There can never be any real opposition between religion and science; for the one is the compliment of the other.
Where Is Science Going?
Chapter V (p. 168)
W.W. Norton & Company, Inc. New York, New York, USA.1932

Religion belongs to that realm that is inviolable before the law of causation and therefore closed to science.
Where Is Science Going?
Chapter V (p. 168)
W.W. Norton & Company, Inc. New York, New York, USA.1932

Religion and natural science…are in agreement, first of all, on the point that there exists a rational world order independent from man, and secondly, on the view that the character of this world order can never be directly known but can only be indirectly recognized or suspected. Religion employs in this connection its own characteristic symbols, while natural science uses measurements founded on sense experiences.
Scientific Autobiography and Other Papers
Religion and Natural Science, Part IV (pp. 182–183)
Philosophical Library. New York, New York, USA. 1949

Religion and natural science are fighting a joint battle in an incessant, never relaxing crusade against skepticism and against dogmatism, against disbelief and against superstition, and the rallying cry in this crusade has always been, and always will be, "On to God."
Scientific Autobiography and Other Papers
Religion and Natural Science, Part IV (p. 187)
Philosophical Library. New York, New York, USA. 1949

Polanyi, Michael   1891–1976
Hungarian-born English scientist, philosopher, and social scientist

Admittedly, religious conversion commits our whole person and changes our whole being in a way that an expansion of natural knowledge does not do. But once the dynamics of knowing are recognized as the dominant principle of knowledge, the difference appears only as one of degree…it establishe[s] a continuous ascent from our less personal knowing of inanimate matter to our convivial knowing of living beings and beyond this to knowing our responsible fellow men. Such I believe is the true transition from the sciences to the humanities and also from our knowing the laws of nature to our knowing the person of God.
Faith and Reason
Journal of Religion, Volume 41, Number 4, October 1961 (p. 244, 245)

Polkinghorne, John   1930–
British physicist, Episcopal priest, and writer

Only in the media, and in the popular and polemical scientific writing, does there persist the myth of the light of pure scientific truth confronting the darkness of obscurantist religious error. Indeed, when one reads writers like Richard Dawkins or Daniel Dennett, one sees that nowadays the danger of a facile triumphalism is very much a problem for the secular academy rather than the Christian Church.
Quarks, Chaos, and Christianity (p. 5)
Abingdon Press. Nashville, Tennessee, USA. 2005

Pope John Paul II   1920–2005
Bishop of Rome

Science can purify religion from error and superstition. Religion can purify science from idolatry and false absolutes.
In James Reston
Galileo, a Life (p. 461)
HarperCollins Publishers, Inc. New York, New York, USA. 1994

Pope Pius XII   1876–1958
Bishop of Rome

The more true science advances, the more it discovers God, almost, as though he were standing, vigilant behind every door which science opens.
Address, November 22, 1951

Popper, Karl R.   1902–94
Austrian/British philosopher of science

Science is most significant as one of the greatest spiritual adventures that man has yet known.
In John Oulton Wisdom
Foundations of Inference in Natural Science (p. v)
Methuen & Company Ltd. London, England. 1952

Raven, Charles E.   1885–1964
English writer of theology and science

To mention Science and Religion in the same sentence is…to affirm an antithesis and suggest a conflict.
Science, Religion and the Future
Chapter I (p. 1)
At The University Press. Cambridge, England. 1943

Raymo, Chet   1936–
American physicist and science writer

Everything we have learned in science since the time of Galileo suggests that the nebulas and galaxies are oblivious to our fates. Everything we have learned suggests that our souls and bodies are inseparable. Everything we have learned suggests that the grave is our destiny. Therefore, if the promise of eternal life is to have maximum drawing power, it is essential for Church and guru to undermine the legitimacy of science.
Skeptics and True Believers: The Exhilarating Connection Between Science and Religion
Chapter Four (pp. 66–67)
Walker & Company. New York, New York, USA. 1998

Reichenbach, Hans   1891–1953
German philosopher of science

The belief in science has replaced in large measure, the belief in God. Even where religion was regarded as compatible with science, it was modified by the mentality of the believer in scientific truth.
The Rise of Scientific Philosophy
Chapter 3 (p. 44)
University of California Press. Berkeley, California, USA. 1951

Rice, Laban Lacy   1870–1973
American educator

Science does not regard its currently established truths as final: religion everywhere and in all centuries has followed the trend toward crystallization of belief.
The Universe: Its Origin, Nature and Destiny
Chapter I (p. 15)
Exposition Press. New York, New York, USA. 1951

Roelofs, Howard Dykema   1893–1974
Professor of philosophy

Religion can produce on occasion what science never does, namely, saints. Today we have science and scientists aplenty. We lack saints.
In Herbert J. Muller
Science and Criticism: The Humanistic Tradition in Contemporary Thought
Chapter III (p. 59)
G. Braziller. New York, New York, USA. 1943

Sagan, Carl   1934–96
American astronomer and author

How is it that hardly any major religion has looked at science and concluded, "This is better than we thought! The Universe is much bigger than our prophets said, grander, more subtle, more elegant?"
Pale Blue Dot: A Vision of the Human Future in Space
Chapter 4 (p. 52)
Random House, Inc. New York, New York, USA. 1994

If you lived two or three millennia ago, there was no shame in holding that the Universe was made for us. It was an appealing thesis consistent with everything we knew; it was what the most learned among us taught without qualification. But we have found out much since then. Defending such a position today amounts to willful disregard of the evidence, and a flight from self-knowledge.
Pale Blue Dot: A Vision of the Human Future in Space
Chapter 4 (p. 52)
Random House, Inc. New York, New York, USA. 1994

Heroes who try to explain the world in terms of matter and energy may have arisen many times in many cultures, only to be obliterated by the priests and philosophers in charge of the conventional wisdom.…
The Demon-Haunted World: Science as a Candle in the Dark (p. 310)
Random House, Inc. New York, New York, USA. 1995

If you want to know when the next eclipse of the Sun will be, you might try magicians or mystics, but you'll do much better with scientists.
The Demon-Haunted World: Science as a Candle in the Dark (p. 30)
Random House, Inc. New York, New York, USA. 1995

Shaw, George Bernard   1856–1950
Irish comic dramatist and literary critic

Let the Churches ask themselves why there is no revolt against the dogmas of mathematics though there is one against the dogmas of religions. It is not that the mathematical dogmas are more comprehensible.… It is not that science is free from legends, witchcraft, miracles, biographic boostings of quacks as heroes and saints, and of barren scoundrels as explorers and discoverers.… But no student of science has yet been taught that specific gravity consists in the belief that Archimedes jumped out of the bath and ran naked through the streets of Syracuse shouting Eureka, Eureka, or that the law of inverse squares must be discarded if anyone can prove that Newton was never in an orchard in his life.
Back to Methuselah
Preface (pp. lxxvii–lxxviii)
Constable & Company Ltd. London, England. 1921

Sperry, Roger Wolcott   1913–94
Neuropsychologist

Probably the widest, deepest rift in contemporary culture and the source of its most profound conflict is that separating the two major opposing views of existence upheld by science and by orthodox religions, respectively. Together they represent two totally different kinds of "truth", the former asking us to accept impersonal

mass-energy accounts of the cosmos, the latter requiring faith in varied spiritual explanations.
The New Mentalist Paradigm and Ultimate Concern
Perspectives in Biology and Medicine, Volume 29, Number 3, Part I, Spring 1986 (p. 415)

Stace, Walter Terence   1886–1967
English philosopher and educator

…no scientific argument — by which I mean an argument drawn from the phenomena of nature — can ever have the slightest tendency either to prove or disprove the existence of God…science is irrelevant to religion.
Religion and the Modern Mind
Chapter 5 (p. 76)
J.B. Lippincott Company. Philadelphia, Pennsylvania, USA. 1952

Stapledon, Olaf   1886–1950
English author

Within the chapel, the great Bible was decorously removed and the windows thrown open, to dispel somewhat the odour of sanctity. For though the early and spiritistic interpretations of relativity and quantum theory had by now accustomed men of science to pay their respects to the religions, many of them were still liable to a certain asphyxia when they were actually within the precincts of sanctity. *Last and First Men*
Chapter II (p. 29)
Jeremy P. Tarcher, Inc. Los Angeles, California, USA. 1988

When the scientists had settled themselves upon the archaic and unyielding benches, the President explained that the chapel authorities had kindly permitted this meeting because they realized that, since men of science had gradually discovered the spiritual foundation of physics, science and religion must henceforth be close allies. Morever the purpose of this meeting was to discuss one of those supreme mysteries which it was the glory of science to discover and religion to transfigure.
Last and First Men
Chapter II (p. 29)
Jeremy P. Tarcher, Inc. Los Angeles, California, USA. 1988

Streeter, B. H. (Burnett Hillman)   1874–1937
English theologian and New Testament scholar

Science is the great cleanser of the human thinking; it makes impossible any religion but the highest.
Reality
Chapter IX (p. 272)
Publisher undetermined

Teilhard de Chardin, Pierre   1881–1955
French Jesuit, paleontologist, and biologist

Religion and science are the two conjugated faces of phases of one and the same act of complete knowledge — the only one which can embrace the past and future of evolution so as to contemplate, measure and fulfill them.

The Phenomenon of Man
Chapter Three, Chapter III, Section 2 (pp. 284–285)
Harper & Brothers. New York, New York, USA. 1959

Temple, Frederick   1821–1902
Anglican prelate and archbishop of Canterbury

Science and Religion seem very often to be the most determined foes to each other that can be found. The scientific man often asserts that he cannot find God in Science; and the religious man often asserts that he cannot find Science in God.
The Relations Between Religion and Science (p. 4)
Macmillan & Company. New York, New York, USA. 1884

Science postulates uniformity; Religion postulates liberty.
The Relations Between Religion and Science (p. 70)
Macmillan & Company. New York, New York, USA. 1884

Thomson, J. Arthur   1861–1933
Scottish biologist

When we are thrilled with the wonder of the world, the heights and depths of things, the beauty of it all, we approach the door of natural religion. And when the Nature-feeling is not superficial but informed with knowledge, with no gain of the hard-won analysis unused, we may reach the threshold. And when we feel that our scientific cosmology leaves Isis still veiled, and when our attempts at philosophical interpretation give us a reasoned conviction of a meaning behind the process, we may perhaps enter in.
The System of Animate Nature (Volume 1)
Lecture I (p. 42)
William & Norgate. London, England. 1920

Religious interpretation and scientific description must not be inconsistent, but they are incommensurable.
The Outline of Science (Volume 4)
Chapter XXXVIII (p. 1177)
G.P. Putnam's Sons. New York, New York, USA. 1937

Tillich, Paul   1886–1965
German-born American theologian

The distinction between the truth of faith and the truth of science leads to a warning, directed to theologians, not to use recent scientific discoveries to confirm the truth of faith. Microphysics have undercut some scientific hypotheses concerning the calculability of the universe. The theory of quantum and the principle of indeterminacy have had this effect. Immediately religious writers use these insights for the confirmation of their own ideas of human freedom, divine creativity, and miracles. But there is no justification for such a procedure at all, neither from the point of view of physics nor from the point of view of religion. The physical theories referred to have no direct relation to the infinitely complex phenomenon of human freedom, and the emission of power in quantums has no direct relation to the meaning of miracles.…

The truth of faith cannot be confirmed by latest physical or biological or psychological discoveries — as it cannot be denied by them.
Dynamics of Faith
Chapter V, Section 2 (p. 85)
Harper & Brothers. New York, New York, USA. 1957

…theology cannot rest on scientific theory. But it must relate its understanding of man to an understanding of universal nature, for man is a part of nature and statements about nature underlie every statement about him.
In Theodosius Dobzhansky
The Biology of Ultimate Concern
Chapter 6 (pp. 109–110)
The New American Library, Inc. New York, New York, USA. 1967

Toynbee, Arnold J.   1852–83
English historian

Theology, not religion, is the antithesis to science.
Toynbee's Industrial Revolution
Notes and Jottings (p. 243)
A.M. Kelley. New York, New York, USA. 1969

Before the close of the seventeenth century our forefathers consciously took their treasure out of religion and reinvested it in natural science…
A Turning Point in Man's Destiny
The New York Times Magazine, December 26, 1954 (p. 5)

Tyndall, John   1820–93
Irish-born English physicist

We claim, and we shall wrest from theology, the entire domain of cosmological theory.
The Belfast Address
The Position of Science

Valéry, Paul   1871–1945
French poet and critic

Without religions the sciences would never have existed. For the human brain would not have trained itself to range beyond the immediate, ever — present "facts" of appearance which, for it, constitute reality.
In Jackson Mathews (ed.)
The Collected Works of Paul Valéry (Volume 14)
Analects, XLVIII (p. 285)
Princeton University Press. Princeton, New Jersey, USA. 1971

Virchow, Rudolf Ludwig Karl   1821–1902
German pathologist and archaeologist

There can be no scientific dispute with respect to faith, for science and faith exclude one another.
Translated by Lelland J. Rather
Disease, Life, and Man, Selected Essays
On Man (p. 68)
Stanford University Press. Stanford, California, USA. 1958

The task of science…is not to attack the objects of faith, but to establish the limits beyond which knowledge cannot go and to found a unified self-consciousness within these limits.
Translated by Lelland J. Rather
Disease, Life, and Man, Selected Essays
On Man (p. 69)
Stanford University Press. Stanford, California, USA. 1958

…belief has no place as far as science reaches, and may be first permitted to take root where science stops.
Translated by Lelland J. Rather
Disease, Life, and Man, Selected Essays
On Man (p. 69)
Stanford University Press. Stanford, California, USA. 1958

Whaling, Thornton   1858–1938
American Presbyterian writer

There can be no real conflict between natural science and true religion because their spheres are entirely distinct and separate.… Conflicts between these two are always the result of misinterpretation and misrepresentation of one or the other or both, and history abounds with illustrations of all these forms of confusing contradictions. Science and religion, while thus separate, have various relationships which make each the servant of the other.
Science and Religion Today (pp. 51–52)
The University of North Carolina Press. Chapel Hill, North Carolina. 1929

Dean Inge [English religious leader] remarks, "We may hope for a time when the science of a religious man will be scientific and religion of a scientific man religious."
Science and Religion Today (pp. 51–52)
The University of North Carolina Press. Chapel Hill, North Carolina. 1929

Whewell, William   1794–1866
English philosopher and historian

All speculations on subjects in which Science and Religion bear upon each other are liable to one of the two opposite charges[:] that the speculator sets Philosophy and Religion at variance; or that he warps Philosophy into a conformity with Religion.
Of the Plurality of Worlds
Preface (p. iv)
John W. Parker & Son. London, England. 1853

White, Andrew Dickson   1832–1918
American author, educator, and diplomat

In all modern history, interference with science in the supposed interests of religion…has resulted in the direst evils both to religion and science; and, on the other hand all untrammeled scientific investigation, no matter how dangerous to religion some of its stages may have seemed…has invariably resulted in the highest good both of religion and of science.
A History of the Warfare of Science with Theology in Christendom
Introduction (p. viii)
Macmillan & Company Ltd. London, England. 1896

Whitehead, Alfred North   1861–1947
English mathematician and philosopher

When we consider what religion is for mankind, and what science is, it is no exaggeration to say that the future course of history depends upon the decision of this generation as to the relations between them.
Science and the Modern World
Chapter XII (p. 181)
The Macmillan Company. New York, New York, USA. 1929

Religion will not gain its old power until it can face change in the same spirit as does science. Its principles may be eternal, but the expression of these principles requires continual development.
Science and the Modern World
Chapter XII (p. 189)
The Macmillan Company. New York, New York, USA. 1929

Science suggests a cosmology; and whatever suggests a cosmology suggests a religion.
Religion in the Making
Truth and Criticism (p. 136)
New American Library. New York, New York, USA. 1960

Wilde, Oscar   1854–1900
Irish wit, poet, and dramatist

Science is the record of dead religions.
Phrases and Philosophies for the Use of the Young
L. Smithers. London, England. 1903
Press. Minneapolis, Minnesota, USA. 1996

SCIENCE AND SOCIETY

Asimov, Isaac   1920–92
American author and biochemist

Science with all its faults has brought education and the arts to more people — a larger percentage — than has ever existed before science. In that respect it is science that is the great humanizer. And, if we are going to solve the problems that science has brought us, it will be done by science and in no other way.
Essay 400 — A Way of Thinking
The Magazine of Fantasy and Science Fiction, December 1994

…science must not be viewed as a mysterious black box out of which came toys and goodies, for that way laymen would view scientists as a kind of lab-coated priesthood — and, eventually, fear and hate them.
Essay 400 — A Way of Thinking
The Magazine of Fantasy and Science Fiction,
December 1994

Barry, Dave   1947–
American humor columnist

…a recent survey, conducted by the National Science Foundation…showed that the average American does not understand basic scientific principles. Naturally, the news media reported this finding as though it was shocking, which is silly. This is, after all, a nation that has produced tournament bass fishing and the Home Shopping Channel; we should be shocked that the average American still knows how to walk erect.
In a World of Scientists, No One Really Knows Much of Anything
Dave's World, July 8, 1996

Bauer, Henry H.   1931–
American chemist

The point is that no amount of knowledge of or about science in itself causes individuals or groups to make good decisions about the many quandaries of life: humans readily subjugate their knowledge to their wishes, believing and doing what they want, all scientific facts and knowledge notwithstanding.
Scientific Literacy and the Myth of the Scientific Method
Chapter 1 (p. 13)
University of Illinois Press. Urbana, Illinois, USA. 1992

What is the social value of science? Why should we support it with taxes? Answer: It can keep people honest. Emperors and popes used to insist that people subscribe to lies about the Earth, about the relationships among different sorts of people, and about a lot of other things. They cannot lie to that extent anymore. Science can put and keep politicians and prophets in their proper place, at least over some things.
Scientific Literacy and the Myth of the Scientific Method
Chapter 7 (p. 146)
University of Illinois Press. Urbana, Illinois, USA. 1992

Scientific research is an investment in the future; trying to make it pay off quickly is as counterproductive as is, in the economic sphere, skimming wealth from corporations through leveraged buy-outs instead of investing for the long haul. Science is part of humanity's cultural heritage. Being educated in science is as important as being educated in philosophy, or psychology, or foreign languages because without it one is ignorant, a primitive savage rather than a civilized human being. And to be scientifically literate is to understand that.
Scientific Literacy and the Myth of the Scientific Method
Chapter 7 (p. 147)
University of Illinois Press. Urbana, Illinois, USA. 1992

Studying science is excellent training for the mind, much better than the classically prescribed study of Latin. When you study science in the right way, you learn about reality therapy; and that is worth applying to other things than science.
Scientific Literacy and the Myth of the Scientific Method
Chapter 7 (p. 147)
University of Illinois Press. Urbana, Illinois, USA. 1992

Bernstein, Jeremy   1929–
American physicist, educator, and writer

We live in a complex, dangerous, and fascinating world. Science has played a role in creating the dangers, and one hopes that it will aid in creating ways of dealing with these dangers. But most of these problems cannot, and will not, be dealt with by scientists alone. We need all the help we can get, and this help has got to come from a scientifically literate general public. Ignorance of science and technology is becoming the ultimate self-indulgent luxury.

Cranks, Quarks, and the Cosmos: Writings on Science
Chapter 16 (p. 202)
Basic Books, Inc. New York, New York, USA. 1993

…the first reason for teaching science to non scientists is that many of these nonscientists have a genuine desire to learn about science, and this, after all, is the best reason for teaching anything to anyone.

Cranks, Quarks, and the Cosmos: Writings on Science
Chapter 16 (p. 196)
Basic Books, Inc. New York, New York, USA. 1993

Brin, David   1950–
American scientist and author

We Americans have refined self-righteousness to a high art, cherishing the romantic image of smart outsiders against the establishment. New Age types see themselves as brave truth-seekers, opposed by a rigid technological priesthood. No matter that this priesthood is dedicated to self-criticism, and to sharing whatever they learn. Science represents this era's "establishment," and is therefore automatically suspect.

Otherness
What to Say to a UFO
Bantam Dell Doubleday Publishing Group. New York, New York, USA. 1994

Calvin, William H.   1939–
Theoretical neurophysiologist

Science doesn't merely empower us, as in seeding better technologies; it also helps prevent trouble in the first place. Knowledge can be like a vaccine, immunizing you against false fears and bad moves.

How Brains Think: Evolving Intelligence, Then and Now
Chapter 3 (p. 41)
Basic Books, Inc. New York, New York, USA. 1996

Clarke, Arthur C.   1917–
English science and science fiction writer

One of the factors, ironically enough, which has contributed to popular willingness to accept the incredible is the success of modern science. Because so many technical marvels have been achieved, the public believes that the scientist is a magician who can make anything happen. It does not know where to draw the line between the possible, the plausible, the improbable and the frankly absurd. Admittedly this is often extremely difficult, and even the experts sometimes fall flat on their faces. But usually, all that is needed is a little common sense.

Voices from the Sky: A Preview of the Coming Space Age
The Lunatic Fringe
Harper & Row, Publishers. New York, New York, USA. 1965

Compton, Karl Taylor   1887–1954
American educator and physicist

There is "something new under the sun" in that modern science has given mankind, for the first time in the history of the human race, a way of securing a more abundant life which does not simply consist in taking it away from someone else. Science really creates wealth and opportunity where they did not exist before.

A Scientist Speaks: Excerpts from Addresses by Karl Taylor Compton During the Years 1930–1949 (p. 2)
Undergraduate Association, MIT. Cambridge, Massachusetts, USA. 1955

Cori, Carl   1896–1984
American biochemist

Art and science can best grow and develop in a society which cherishes freedom and which shows respect for the needs, the happiness and the dignity of human beings.

Les Prix Nobel. The Nobel Prizes in 1947
Nobel banquet speech for award received in 1947
Nobel Foundation. Stockholm, Sweden. 1948

Crick, Francis Harry Compton   1916–2004
English biochemist

The average adult can usually enjoy something only if it relates to what he knows already, and what he knows about science is in many cases pitifully inadequate.

What Mad Pursuit: A Personal View of Scientific Discovery
Chapter 7 (p. 80)
Basic Books, Inc. New York, New York, USA. 1988

Dawkins, Richard   1941–
British ethologist, evolutionary biologist, and popular science writer

People certainly blame science for nuclear weapons and similar horrors. It's been said before but needs to be said again: if you want to do evil, science provides the most powerful weapons to do evil; but equally, if you want to do good, science puts into your hands the most powerful tools to do so. The trick is to want the right things, then science will provide you with the most effective methods of achieving them.

Science, Delusion and the Appetite for Wonder
Richard Dimbleby Lecture, BBC1 Television, November 12th, 1996

It has become almost a cliché to remark that nobody boasts of ignorance of literature, but it is socially acceptable to boast ignorance of science and proudly claim incompetence in mathematics.

Science, Delusion and the Appetite for Wonder
Richard Dimbleby Lecture, BBC1 Television, November 12th, 1996

Science provokes more hostility than ever, sometimes with good reason, often from people who know nothing about it and use their hostility as an excuse not to learn. Depressingly many people still fall for the discredited cliché that scientific explanation corrodes poetic sensibility.

Science and Sensibility
Queen Elizabeth Hall Lecture, London, 24ᵗʰ March 1998

Editorial

The scientific illiteracy of politicians, their simple lack of "feel" for what science is and what it can do, prevents them from exploring the deeper questions, among the most important facing humankind: how can science be conducted so that, on the one hand, the thinkers have the freedom to think, for that is the sine qua non; and how, on the other hand, can the products of unfettered thought be harnessed for the needs of society?

Who Cares About Science?
New Scientist, 17 October 1985 (p. 18)

Eisenhower, Dwight David   1890–1969
34ᵗʰ president of the United States

Science, great as it is, remains always the servant and the handmaiden of freedom. And a free science will ever be one of the most effective tools through which man will eventually bring to realization his age-old aspiration for an abundant life, with peace and justice for all.

In Dael Wolfle (ed.)
Symposium on Basic Research
Casper Auditorium of the Rockefeller Institute, May, 1959
Science: Handmaiden of Freedom (p. 142)
American Association of the Advancement of Science. Washington, D.C. 1959

Feyerabend, Paul K.   1924–94
Austrian-born American philosopher of science

[Because] there is trouble in the third world…[it is argued that] the attempt to judge cosmologies by their content may have to be given up. Such a development, far from being undesirable, changes science from a stern and demanding mistress into an attractive and yielding courtesan who tries to anticipate every wish of her lover. Of course, it is up to us to choose either a dragon or a pussy cat for our company. I do not think I need to explain my own preferences.

Realism, Rationalism and Scientific Method
Consolations for the Specialist (p. 161)
Cambridge University Press. Cambridge, England. 1985

Feynman, Richard P.   1918–88
American theoretical physicist

…the things that appear in the newspaper and that seem to excite the adult imagination are always those things which they cannot possibly understand, because they haven't learned anything at all of the much more interesting well-known [to scientists] things that people have found out before. It's not the case with children, thank goodness, for a while — at least until they become adults.

In Jeffrey Robbins, ed.
The Pleasure of Finding Things Out: The Best Short Works of Richard P. Feynman
Chapter 4 (p. 102)
Perseus Books. Cambridge, Massachusetts, USA. 1999

Another value of science is the fun called intellectual enjoyment which some people get from reading and learning and thinking about it, and which others get from working in it. This is an important point, one which is not considered enough by those who tell us it is our social responsibility to reflect on the impact of science on society. Is this mere personal enjoyment of value to society as a whole? No! But it is also a responsibility to consider the aim of society itself. Is it to arrange matters so that people can enjoy things? If so, then the enjoyment of science is as important as anything else.

In Jeffrey Robbins (ed.)
The Pleasure of Finding Things Out: The Best Short Works of Richard P. Feynman
Chapter 6 (p. 143)
Perseus Books. Cambridge, Massachusetts, USA. 1999

I don't believe in the idea that there are a few peculiar people capable of understanding math, and the rest of the world is normal. Math is a human discovery, and it's no more complicated than humans can understand. I had a calculus book once that said, "What one fool can do, another can." What we've been able to work out about nature may look abstract and threatening to someone who hasn't studied it, but it was fools who did it, and in the next generation, all the fools will understand it.

In Jeffrey Robbins (ed.)
The Pleasure of Finding Things Out: The Best Short Works of Richard P. Feynman
Chapter 9 (p. 144)
Perseus Books. Cambridge, Massachusetts, USA. 1999

When we read about this in the newspaper, it says "Scientists say this discovery may have importance in the search for a cure for cancer." The paper is only interested in the use of the idea, not the idea itself. Hardly anyone can understand the importance of an idea, it is so remarkable. Except that, possibly, some children catch on. And when a child catches on to an idea like that, we have a scientist.

In Jeffrey Robbins (ed.)
The Pleasure of Finding Things Out: The Best Short Works of Richard P. Feynman
Chapter 6 (p. 145)
Perseus Books. Cambridge, Massachusetts, USA. 1999

[I]f a thing is not scientific, if it cannot be subjected to the test of observation, this does not mean that it is dead, or wrong, or stupid. We are not trying to argue that science is somehow good and other things are somehow not good. Scientists take all those things that can be analyzed by observation, and thus the things called science are found out. But there are some things left out, for which the method does not work. This does not mean that those things are unimportant. They are, in fact, in many ways the most important.

The Meaning of It All: Thoughts of a Citizen Scientist
Chapter I (p. 16)
Perseus Books. Reading, Massachusetts, USA. 1998

It is odd, but on the infrequent occasions when I have been called upon in a formal place to play the bongo drums, the introducer never seems to find it necessary to mention that I also do theoretical physics.

The Character of Physical Law
Chapter 1 (p. 13)
British Broadcasting Company. London, England. 1965

Freud, Sigmund   1856–1939
Austrian neurologist and co-founder of psychoanalysis

There are the elements, which seem to mock at all human control: the earth, which quakes and is torn apart and buries all human life and its works; water, which deluges and drowns everything in a turmoil; storms, which blow everything before them; there are diseases, which we have only recently recognized as attacks by other organisms; and finally there is the painful riddle of death, against which no medicine has yet been found, nor probably will be. With these forces nature rises up against us, majestic, cruel and inexorable; she brings…to our mind once more our weakness and helplessness, which we thought to escape through the work of civilization.

The Future of an Illusion
Chapter III (pp. 15–16)
W.W. Norton & Company, Inc. New York, New York, USA.1961

Gleick, James   1954–
American author, journalist, and essayist

Einstein's relativity did not speak to human values. Those were, or were not, relative for reasons unrelated to the physics of objects moving at near-light speed. Borrowing metaphors from the technical sciences could be a dangerous practice.

Genius: The Life and Science of Richard Feynman
Epilogue (p. 430)
Pantheon Books. New York, New York, USA. 1992

Gould, Stephen Jay   1941–2002
American paleontologist and evolutionary biologist

Science, since people must do it, is a socially embedded activity. It progresses by hunch, vision, and intuition. Much of its change through time does not record a closer approach to absolute truth, but the alteration of cultural contexts that influence it so strongly. Facts are not pure and unsullied bits of information; culture also influences what we see and how we see it. Theories, moreover, are not inexorable inductions from facts. The most creative theories are often imaginative visions imposed upon facts; the source of imagination is also strongly cultural.

The Mismeasure of Man
Chapter One (pp. 53–54)
W.W. Norton & Company, Inc. New York, New York, USA.1996

…I believe that science must be understood as a social phenomenon, a gutsy, human enterprise, not the work of robots programmed to collect pure information…. Science, since people must do it, is a socially embedded activity. It progresses by hunch, vision, and intuition. Much of its change through time does not record a closer approach to absolute truth, but the alteration of cultural contexts that influence it so strongly. Facts are not pure and unsullied bits of information; culture also influences what we see and how we see it. Theories, moreover, are not inexorable inductions from facts. The most creative theories are often imaginative visions imposed upon facts; the source of imagination is also strongly cultural.

The Mismeasure of Man
Chapter One (p. 53)
W.W. Norton & Company, Inc. New York, New York, USA.1996

Science is accessible to all thinking people because it applies universal tools of intellect to its distinctive material. The understanding of science — one need hardly repeat the litany — becomes ever more crucial in a world of biotechnology, computers, and bombs.

Time's Arrow, Time's Cycle: Myth and Metaphor in the Discovery of Geological Time
Chapter 1 (p. 7)
Harvard University Press. Cambridge, Massachusetts, USA. 1987

Hardy, G. H. (Godfrey Harold)   1877–1947
English pure mathematician

…a science is said to be useful if its development tends to accentuate the existing inequalities in the distribution of wealth, or more directly promotes the destruction of human life.

A Mathematician's Apology
Chapter 22 (p. 120)
Cambridge University Press. Cambridge, England. 1967

Haskins, C. P.
No biographical data available

Science provides a challenge to effort for the individual youth that is far greater than the challenge of militarism. It provides a unity of thought which is far wider, for it transcends all national boundaries. It provides a wider battleground, for the goal of militarism is the conquering of man, but that of science is the understanding and the subjugation of all the rest of our natural environment. And finally, it is an infinitely broader training than totalitarian training can possibly be, for it requires, in addition to great courage, stamina, and drive, the qualities of intellectualism and gentleness.

Science Philosophy and Religion
Scientific Thought and a Democratic Ideology (p. 235)
The Conference on Science, Philosophy and Religion in Their Relation to a Democratic Way of Life, Inc. New York, New York, USA. 1941

Hoffmann, Roald   1937–
Polish-born American chemist

…the overall effect of science is inexorably democratizing, in the deepest sense of the word — by making available to a wider range of people the necessities and comforts that in a previous age were reserved for a privileged elite.
The Same and Not the Same
Part Eight, Chapter 40 (p. 212)
Columbia University Press. New York, New York, USA. 1995

Holmes, Oliver Wendell   1809–94
American physician, poet, and humorist

I am a little afraid that science is breeding us down too fast into coral-insects. A man like Newton or Leibnitz or Haller used to paint a picture of outward or inward nature with a free hand, and stand back and look at it as a whole and feel like an archangel; but nowadays you have a Society, and they come together and make a great mosaic, each man bringing his little bit and sticking it in its place, but so taken up with his petty fragment that he never thinks of looking at the picture the little bits make when they are put together.
The Poet at the Breakfast-Table
Chapter III (p. 79)
Houghton Mifflin Company. Boston, Massachusetts, USA. 1895

Holton, Gerald   1922–
Research professor of physics and science history

Was not the universe of Dante and Milton so powerful and "gloriously romantic" precisely because it incorporated, and thereby rendered meaningful, the contemporary scientific cosmology alongside the moral and aesthetic conceptions? Leaving aside the question of whether Dante's and Milton's contemporaries by and large were living in a rich and fragrant world of gladness, love, and beauty, it is fair to speculate that if our new cosmos is felt to be cold, inglorious, and unromantic, it is not the new cosmology which may be at fault, but the absence of new Dantes and Miltons.
Einstein, History, and Other Passions: The Rebellion Against Science at the End of the Twentieth Century
Part One, Chapter 2 (p. 53)
Addison-Wesley Publishing Company. Reading, Massachusetts, USA. 1996

But making science again a part of every intelligent person's educational resource is the minimum requirement — not because science is more important than other field, but because it is an integral part of a sound contemporary worldview.

Einstein, History, and Other Passions: The Rebellion Against Science at the End of the Twentieth Century
Part One, Chapter 2 (p. 53)
Addison-Wesley Publishing Company. Reading, Massachusetts, USA. 1996

King, Jr., Martin Luther   1929–68
American civil rights leader and clergyman

The means by which we live have outdistanced the ends for which we live. Our scientific power has outrun our spiritual power. We have guided missiles and misguided men.
Strength to Love
Chapter VII (p. 57)
Harper & Row, Publishers. New York, New York, USA. 1963

Lederman, Leon   1922–
American high-energy physicist

In his "Defense of Poetry," the English romantic poet Percy Bysshe Shelley contended that one of the sacred tasks of the artist is to "absorb the new knowledge of the sciences and assimilate it to human needs, color it with human passions, transform it into the blood and bone of human nature." Not many romantic poets rushed to accept Shelley's challenge, which may explain the present sorry state of our nation and planet. If we had Byron and Keats and Shelley and their French, Italian, and Urdu equivalents explaining science, the science literacy of the general public would be far higher than it is now.
The God Particle: If the Universe Is the Answer, What Is the Question
Chapter 9 (p. 382)
Houghton Mifflin Company. Boston, Massachusetts, USA. 1993

Medawar, Sir Peter Brian   1915–87
Brazilian-born English zoologist

Money can't buy ideas, that's for sure, but lack of it can prevent one having them.
The Cost-Benefit Analysis of Pure Research
Hospital Practice, Sept 1973

I am afraid we shall have to regard the funding of "pure" research as a tax levied upon society that is not dissimilar in kind from that which maintains art galleries and opera houses — a "civilization tax", perhaps.
The Cost-Benefit Analysis of Pure Research
Hospital Practice, Sept 1973

It is the great glory and also the great threat of science that anything which is possible in principle — which does not flout a bedrock law of physics — can be done if the intention to do it is sufficiently resolute and long sustained.
Four Score Years and Ten—And Still Counting
Guardian, December 13, 1984

Paulos, John Allen   1945–
American mathematician

In general, almost any mathematically expressed scientific fact can be transformed into a consumer caveat (or lure) that will terrify (or attract) people.
A Mathematician Reads the Newspaper
Asbestos Removal Closes NYC Schools (p. 142)
Basic Books, Inc. New York, New York, USA. 1995

Russell, Bertrand Arthur William   1872–1970
English philosopher, logician, and social reformer

Some men are so impressed by what science knows that they forget what it does not know; others are so much more interested in what it does not know than in what it does that they belittle its achievements.
Unpopular Essays
Philosophy for Laymen (p. 40)
George Allen & Unwin Ltd. London, England. 1950

Can a society in which thought and technique are scientific persist for a long period, as, for example, ancient Egypt persisted, or does it necessarily contain within itself forces which must bring either decay or explosion?
Lloyd Roberts lecture
Can a Scientific Community be Stable, Royal Society of Medicine, London, November 29, 1949

Sagan, Carl   1934–96
American astronomer and author

Many of the dangers we face indeed arise from science and technology — but, more fundamentally, because we have become powerful without becoming commensurately wise. The world-altering powers that technology has delivered into our hands now require a degree of consideration and foresight that has never before been asked of us.
Pale Blue Dot: A Vision of the Human Future in Space
Chapter 22 (p. 384)
Random House, Inc. New York, New York, USA. 1994

Surely…any powerful tool, those in power will try to use…or even monopolize…. Surely scientists, being people, grow up in a society and reflect the prejudices of that society. How could it be otherwise? Some scientists have been nationalists; some have been racists; some have been sexists. But that doesn't undermine the validity of science. It's just a consequence of being human.
Wonder and Skepticism
Skeptical Inquirer, January/February 1995 (p. 24)

There is a reward structure in science that is very interesting: Our highest honors go to those who disprove the findings of the most revered among us…. [I]t's exactly the opposite [in economics, politics, or religion]: There we reward those who reassure us that what we've been told is right, that we need not concern ourselves about it. This difference, I believe, is at least a basic reason why we've made so much progress in science, and so little in some other areas.
Wonder and Skepticism
Skeptical Inquirer, January/February 1995 (p. 24)

The scientific world view works so well, explains so much and resonates so harmoniously with the most advanced parts of our brains that in time, I think, virtually every culture on the Earth, left to its own devices, would have discovered science.
Cosmos
Chapter VII (p. 176)
Random House, Inc. New York, New York, USA. 1980

Schrödinger, Erwin   1887–1961
Austrian theoretical physicist

…— who are we? …I consider this not only one of the tasks, but the task of science, the only one that really counts.
Science and Humanism: Physics in Our Time
The Alleged Break-Down of the Barrier Between Subject and Object (p. 51)
At The University Press. Cambridge, England. 1952

Silver, Brian L.
Israeli professor of physical chemistry

Scientific ideas have affected the relationship of man to society, his ideas of God, and his image of himself. Science has influenced the way people write poetry and the way they paint pictures. In the hands of bigots, it has provided a theoretical justification for the sterilization of some human beings and the enslavement of others. Science, as a source of ideas, is a major character in the human drama.
The Ascent of Science
Preface (p. xvi)
Solomon Press Book. New York, New York, USA. 1998

Snow, Charles Percy   1905–80
English novelist and scientist

Literary intellectuals at one pole — at the other scientists. — …. Between the two a gulf of mutual incomprehension.
The Two Cultures and the Scientific Revolution
Chapter I (p. 4)
Cambridge University Press. New York, New York, USA. 1961

Stenger, Victor J.   1935–
American physicist

Most humans on this planet use the fruits of science in every phase of their lives. I become very irritated at those who decry science while accepting its every benefit. It is especially ironic how the antiscientists use modern communications to get their messages to the public.
Physics and Psychics: The Search for a World Beyond the Senses
Chapter 14 (p. 297)
Prometheus Books. Buffalo, New York, USA. 1990

Tennyson, Alfred (Lord)   1809–92
English poet

Science grows and Beauty dwindles.
Alfred Tennyson's Poetical Works

Locksley Hall. Sixty Years After, Stanza 123
Oxford University Press, Inc. London, England. 1953

Thomas, Lewis 1913–93
American physician and biologist

The cloning of humans is on most of the lists of things to worry about from Science, along with behavior control, genetic engineering, transplanted heads, computer poetry and the unrestrained growth of plastic flowers.
The Medusa and the Snail: More Notes of a Biology Watcher
On Cloning Human Beings (pp. 51–52)
The Viking Press. New York, New York, USA. 1979

Weinberg, Steven 1933–
American nuclear physicist

It is simply a logical fallacy to go from the observation that science is a social process to the conclusion that the final product, our scientific theories, is what it is because of the social and historical forces acting in this process. A party of mountain climbers may argue over the best path to the peak, and these arguments may be conditioned by the history and social structure of the expedition, but in the end either they find a good path to the peak or they do not, and when they get there they know it. (No one would give a book about mountain climbing the title Constructing Everest.)
Dreams of a Final Theory: The Scientist's Search for the Ultimate Laws of Nature
Chapter VI (p. 165)
Pantheon Books. New York, New York, USA. 1992

It certainly feels to me that we are discovering something real in physics, something that is what it is without any regard to the social and historical conditions that allowed us to discover it.
Dreams of a Final Theory: The Scientist's Search for the Ultimate Laws of Nature
Chapter VI (p. 165)
Pantheon Books. New York, New York, USA. 1992

Weinberg, Steven 1933–
American nuclear physicist

We may need to rely again on the influence of science to preserve a sane world. It is not the certainty of scientific knowledge that fits it for this role, but its uncertainty. Seeing scientists change their minds again and again about matters that can be studied directly in laboratory experiments, how can one take seriously the claims of religious tradition or sacred writings to certain knowledge about matters beyond human experience?
Dreams of a Final Theory: The Scientist's Search for the Ultimate Laws of Nature
Chapter VII (p. 188)
Pantheon Books. New York, New York, USA. 1992

Weisskopf, Victor Frederick 1908–2002
Austrian-American physicist

Science cannot develop unless it is pursued for the sake of pure knowledge and insight. It will not survive unless it is used intensely and wisely for the betterment of humanity and not as an instrument of domination. Human existence depends upon compassion and curiosity. Curiosity without compassion is inhuman; compassion without curiosity is ineffectual.
Science Yesterday, Today, and Tomorrow
Speech, 1993

Wilson, Edward O. 1929–
American biologist and author

The love of complexity without reductionism makes art; the love of complexity with reductionism makes science.
Consilience: The Unity of Knowledge
Chapter 4 (p. 54)
Alfred A. Knopf. New York, New York, USA. 1998

Wolpert, Lewis 1929–
British embryologist

Mary Shelley's *Dr. Frankenstein*, H. G. Wells's *Dr. Moreau* and Aldous Huxley's *Brave New World*…are evidence of a powerfully emotive anti-science movement. Science is dangerous, so the message goes — it dehumanizes; it takes away free will; it is materialistic and arrogant. It removes magic from the world and makes it prosaic. But note where these ideas come from — not from the evidence of history, but from creative artists who have molded science by their own imagination.
The Unnatural Nature of Science
Introduction (p. x)
Harvard University Press. Cambridge, Massachusetts, USA. 1992

Zelinsky, Wilbur 1921–
American cultural geographer

Perhaps the greatest discovery of twentieth century science has to do with its own essential nature: That it is, first and always, a social activity, an organized band of human beings who obey the same fundamental rules of organized behavior as do any other complex group of people, not a disembodied flock of angels soaring unswervingly upward toward the elysian fields of truth.
The Demigod's Dilemma
Annals of the Association of American Geographers, Volume 65, 1975
(p. 133)

SCIENCE AND STATE

Duprée, Hunter 1921–
American historian of science and technology

The mighty edifice of government science dominated the scene in the middle and twentieth century as a Gothic cathedral dominated a thirteenth century landscape. The work of many hands over the years, it universally inspired admiration, wonder and fear.

Science in the Federal Government
Chapter XIX (p. 375)
Harvard University Press. Cambridge, Massachusetts. 1957

Feyerabend, Paul K.   1924–94
Austrian-born American philosopher of science

…the separation of state and church must be complemented by the separation of state and science, that most recent, most aggressive, and most dogmatic religious institution.
Against Method: Outline of an Anarchistic Theory of Knowledge
Chapter 18 (p. 295)
Verso. London, England. 1978

Mellanby, Kenneth   1908–93
English ecologist and entomologist

…the corridors of power have a strong attraction for even the most devoted investigator, and these corridors seldom lead back to the laboratory.
Disorganisation of Scientific Research
New Scientist, Volume 59, Number 86023 August 1973 (p. 436)

Ramón y Cajal, Santiago   1852–1934
Spanish neuropathologist

Today's statesmen undoubtedly have limitations, one of which is not realizing that the greatness and might of nations are the products of science, and that justice, order, and good laws are important but secondary factors in prosperity.
Advice for a Young Investigator
Chapter 6 (p. 91)
The MIT Press. Cambridge, Massachusetts, USA. 1999

Rutherford, Ernest   1871–1937
English physicist

It is essential for men of science to take an interest in the administration of their own affairs or else the professional civil servant will step in — and then the Lord help you.
Bulletin of the Institute of Physics, 1950, 1, Number 1, cover

Wiener, Norbert   1894–1964
American mathematician

Neither the public nor the big administrator [of science] has too good an understanding of the inner continuity of science, but they have both seen its world-shaking consequences, and they are afraid of it. Both of them wish to decerebrate the scientist, as the Byzantine State emasculated its civil servants. Moreover the great administrator who is not sure of his own intellectual level can aggrandize himself only by cutting his scientific employees down to size.
I Am a Mathematician
Epilogue (p. 363)
Doubleday. Garden City, NY 1956

SCIENCE AND SUPERSTITION

Einstein, Albert   1879–1955
German-born physicist

By furthering logical thought and a logical attitude, science can diminish the amount of superstition in the world. There is no doubt that all but the crudest scientific work is based on a firm belief — akin to religious feeling — in the rationality and comprehensibility of the world.
Cosmic Religion, With Other Opinions and Aphorisms
On Science (p. 98)
Covici-Fiede. New York, New York, USA. 1931

Lovecraft, H. P. (Howard Phillips)   1890–1937
American writer of fantasy, horror, and science fiction

We were not…in any sense childishly superstitious, but scientific study and reflection had taught us that the known universe of three dimensions embraces the merest fraction of the whole cosmos of substance and energy.
The Shunned House
Section IV
The Recluse Press. Athol, Massachusetts, USA. 1928

To say that we actually believed in vampires or werewolves would be a carelessly inclusive statement. Rather must it be said that we were not prepared to deny the possibility of certain unfamiliar and unclassified modifications of vital force and attenuated matter; existing very infrequently in three-dimensional space because of its more intimate connection with other spatial units, yet close enough to the boundary of our own to furnish us occasional manifestations which we, for lack of a proper vantage-point, may never hope to understand.
The Shunned House
Section IV
The Recluse Press. Athol, Massachusetts, USA. 1928

Machen, Arthur   1863–1947
Welsh author

I have told you that I was of skeptical habit; but though I understood little or nothing, I began to dread, vainly proposing to myself the iterated dogmas of science that all life is material, and that in the system of things there is no undiscovered land, even beyond the remotest stars, where the supernatural can find a footing. Yet there struck in on this the thought that matter is as really awful and unknown as spirit, that science itself but dallies on the threshold, scarcely gaining more than a glimpse of the wonders of the inner place.
The Novel of the Black Seal (p. 18)
Kessinger Publishing. Whitefish, Montana, USA.

Pagels, Heinz R.   1939–88
American physicist and science writer

I like to browse in occult bookshops if for no other reason than to refresh my commitment to science.
The Dreams of Reason: The Computer and the Rise of the Sciences of Complexity
Chapter 11 (p. 242)
Simon & Schuster. New York, New York, USA. 1988

SCIENCE AND WOMEN

Bolton, Henrietta
No biographical data available

As a general rule the scientific woman must be strong enough to stand alone, able to bear the often unjust sarcasm and dislike of men who are jealous of seeing what they consider their own field invaded.
Women in Science
Popular Science Monthly, Volume 53, 1898 (p. 511)

Buckley, Arabella B.   1840–1929
English author

I have promised to introduce you today to the fairy-land of science, — a somewhat bold promise, seeing that most of you probably look upon science as a bundle of dry facts, while fairy-land is all that is beautiful, and full of poetry and imagination. But I thoroughly believe myself, and hope to prove to you, that science is full of beautiful pictures, of real poetry, and of wonder-working fairies...
The Fairy-Land of Science
Lecture I (p. 7)
D. Appleton & Company. New York, New York, USA. 1892

Cannon, Annie Jump   1863–1941
American astronomer

...a life spent in the routine of science need not destroy the attractive human element of a woman's nature.
Williamina Patton Fleming
Science, Volume 33, Number 861, June 30, 1911 (p. 988)

de Lamennais, Félicité Robert   1782–1854
French nobleman and ecclesiastic scholar

I have never met a woman who was competent to follow a course of reasoning the half of a quarter of an hour — *un demi quart d'heure*. She has qualities which are wanting in us, qualities of a particular, inexpressible charm; but, in the matter of reason, logic, the power to connect ideas, to enchain principles of knowledge and perceive their relationships, woman, even the most highly gifted, rarely attains to the height of a man of mediocre capacity.
In H.J. Mozans
Women in Science
Chapter III (p. 136)
The MIT Press. Cambridge, Massachusetts, USA. 1974

de Pizan, Christine   1364–ca. 1430
Venician Medieval writer and analyst

I'll give you some conclusive examples. I repeat — and don't doubt my word — that if it were the custom to send little girls to school and to teach them all sorts of different subjects there, as one does with little boys, they would grasp and learn the difficulties of all the arts and sciences just as easily as the boys do.
Translated by Rosalind Brown-Grant
The Book of the City of Ladies
Part I, 26 (p. 57)
Penguin Books. London, England. 1999

Gildersleeve, Virginia Crocheron   1877–1965
American educator

If we could produce one or two more Madame Curies, that would accomplish far more for the advancement of women than any amount of agitation, argument and legislation.
Many a Good Crusade: Memoirs of Virginia Crocheron Gildersleeve
Part I. The Advancement of Women, (p. 104)
The Macmillan Company. New York, New York, USA. 1954

Kant, Immanuel   1724–1804
German philosopher

All abstract speculations, all knowledge which is dry, however useful it may be, must be abandoned to the laborious and solid mind of man.... For this reason women will never learn geometry.
In H.J. Mozans
Women in Science
Chapter III (p. 136)
The MIT Press. Cambridge, Massachusetts, USA. 1974

Kass-Simon, G.
American neurobiologist
Farnes, Patricia   1931–85
American health care writer

For women in science to be remembered, not only must their work be thought right, but usually it must have such impact upon scientific thought that exclusion is impossible. If women scientists are wrong, or if they narrowly miss the mark, or if they propound ideas that are ultimately superseded, not only are their ideas quickly forgotten, but as often as not, the women are ostracized by their contemporaries or treated with derision.
Women of Science: Righting the Record
Introduction (p. xiii)
Indiana University Press. Bloomington, Indiana, USA. 1990

Lamy, Étienne   1845–1919
French essayist, politician, and lawyer

Women...group themselves at the center of human knowledge, whereas men disperse themselves toward its outer boundaries. While men are always pushing analysis to its utmost limits, women are seeking a synthesis. While men are becoming more technical, women are becoming more intellectual. They are better placed to observe the correlations of the different sciences, and to subordinate them to the common and unique source of truth form which they all descend. We seem, indeed, to be approaching a time when women will become the conservers of general ideas.

In H.J. Mozans
Women in Science.
Chapter XII (pp. 409–410)
The MIT Press. Cambridge, Massachusetts, USA. 1974

Marcet, Jane Haldimand   1769–1858
English expository author in chemistry, botany, religion, and economics

In writing these pages, the author was more than once checked in her progress by the apprehension that such an attempt might be considered by some, either as unsuited to the ordinary pursuits of her sex, or ill justified by her own recent and imperfect knowledge of the subject. But, on the one hand, she felt encouraged by the establishment of those public institutions, open to both sexes, for the dissemination of philosophical knowledge, which clearly prove that the general opinion no longer excludes women from an acquaintance with the elements of science...

Conversations on Chemistry, in Which the Elements of that Science Are Familiarly Explained and Illustrated by Experiments
Preface (p. iv)
Sidney's Press for Cooke. New Haven, Connecticut, USA. 1809

Merchant, Carolyn   1936? –
American ecofeminist philosopher

While learned ladies had always been present among the educated of nobility, and women had contributed to science and mathematics from earliest times, the "scientific lady" was a product of the Scientific Revolution.

The Death of Nature: Women, Ecology, and the Scientific Revolution
Chapter 11 (p. 269)
Harper & Row, Publishers. San Francisco, California, USA. 1980

Mitchell, Maria   1818–89
American astronomer and educator

Women, more than men, are bound by tradition and authority. What the father, the brother, the doctor, and the minister have said has been received undoubtingly. Until women throw off this reverence for authority, they will not develop. When they do this, when they come to truth through their investigations, when doubt leads them to discovery, the truth which they get will be theirs, and their minds will work on and on unfettered.

In Eve Merrian
Growing Up Female in America
Maria Mitchell (p. 96)
Doubleday & Company, Inc. Garden City, New York, USA. 1971

Mozans, H. J. (John Augustine Zahm)   1851–1921
American priest, professor of physics, and science writer

Whilst men of science will be forced to continue as specialists as long as the love of fame, to consider no other motives of research, continue to be a potent influence in their investigations, it is probable that women will have less love for the long and tedious processes involved in the more difficult kinds of specialization. They will, it seems likely, be more inclined to acquire a general knowledge of the whole circle of the sciences — a knowledge that will enable them to take a comprehensive survey of nature. And it will be fortunate for themselves, as well as for the men who must perforce remain specialists, if they elect to do so. For nothing gives falser views of nature as a whole, nothing more unfits the mind for a proper apprehension of higher and more important truths, nothing more incapacities one for the enjoyment of the masterpieces of literature or the sweeter amenities of life, than the narrow occupation of a specialist who sees nothing in the universe but electrons, microbes and protozoa.

Women in Science
Chapter XII (pp. 408–409)
The MIT Press. Cambridge, Massachusetts, USA. 1974

Myrdal, Sigrid
American scientist and inventor

There's the question of how you react when your data do not turn out the way you want them to. One possibility is to think "Oh no, something went wrong, my experiment failed." or "Did I ask the question wrong?" and put the data in the drawer. I think the feminine approach is to ask "What's this trying to tell me?" and consider that nature may be more interesting and complicated than... expected, but therefore probably a bit more elegant. By actually having to deal with the data, I've gone to totally different interpretations. If something turns out quite screwy, I give it a chance. It's possible that it's more feminine to give something a chance.

In Linda Jean Shepherd
Lifting the Veil: The Feminine Face of Science
Receptivity (p. 86)
Shambhala. Boston, Massachusetts, USA. 1993

Plato   428 BCE–347 BCE
Greek philosopher

Nothing can be more absurd than the practice, which prevails in our country, of men and women not following the same pursuits with all their strength and with one mind, for thus the state, instead of being a whole, is reduced to a half.

In H.J. Mozans
Women and Science (p. 2)
The MIT Press. Cambridge, Massachusetts, USA. 1974

Poullain de la Barre, François   1647–1723
French feminist theorist and philosopher

L'esprot n'a point de sexe.

The mind has no sex.
De l'éducation des dames pour la conduite de l'esprit dans les sciences et dans les moeurs
Paris, France. 1674.

Rich, Adrienne   1929–
American poet

The belief that established science and scholarship — which have so relentlessly excluded women from their making — are "objective" and "value-free" and that feminist studies are "unscholarly," "biased," and "ideological" dies hard. Yet the fact is that all science, and all scholarship, and all art are ideological; there is no neutrality in culture!
Blood, Bread and Poetry
Chapter 1 (p. 3)
W.W. Norton & Company. New York, New York, USA. 1986

Yentsch, Clarice M.
No biographical data available
Sindermann, Carl J.
No biographical data available

Science, as a remarkably conservative human institution despite its relatively brief history, has typically cast women in supporting roles in which they were subservient to male professionals, usually dreadfully underpaid, and totally unrecognized.
The Woman Scientist: Meeting the Challenges for a Successful Career
Chapter 2 (p. 27)
Plenum Press. New York, New York, USA. 1992

SCIENCE CREED

Emerson, Ralph Waldo   1803–82
American lecturer, poet, and essayist

Science corrects the old creeds....
The Complete Works of Ralph Waldo Emerson (Volume 8)
Letters and Social Aims
Chapter VII (p. 228)
Houghton Mifflin Company. Boston, Massachusetts, USA. 1904

Hall, Asaph   1829–1907
American astronomer

The scientific creed is constantly growing and expanding, and we have no fears, but rejoice at its growth. We need no consistory of bishops, no synod of ministers, to tell us what to believe. Everything is open to investigation and criticism.
In Sir Richard Arman Gregory
Discovery; or, The Spirit and Service of Science
Chapter II (pp. 30–31)
Macmillan & Company Ltd. London, England. 1918

Huxley, Thomas Henry   1825–95
English biologist

...science...commits suicide when it adopts a creed.
Collected Essays (Volume 2)
Darwiniana
The Darwin Memorial (p. 252)
Macmillan & Company Ltd. London, England. 1904

LeShan, Lawrence
No biographical data available
Margenau, Henry   1901–97
American physicist

1. We believe that the search for truth is a never-ending quest; yet we pledge ourselves to seek it.
2. We will not recognize or accept any kind of truth that pretends to be ultimate or absolute. We will consider and weigh all claims as provisional conclusions. If examination shows them to be stop signs on the road of inquiry, we will ignore them; if they are signposts, we will note them and move on.
3. We recognize no subjects and no facts that are alleged to be forever closed to inquiry or understanding; for science, every mystery is but a challenge.
4. We believe that new principles of understanding are constantly created through the efforts of man, and that a philosophy which sees the answers to all questions already implied in what is now called science is presumptuous and contrary to the spirit of science.
5. We are confident that scientific illumination can be made to penetrate not only the realms now affirmed as scientific, but also the shadowy regions that surround human consciousness, the essence of the mind, including features that are still obscure or occult and mysterious.

Einstein's Space and Van Gogh's Sky
Chapter 4 (pp. 70–71)
The Macmillan Company. New York, New York, USA. 1982

SCIENCE FICTION

Asimov, Isaac   1920–92
American author and biochemist

Individual science fiction stories may seem as trivial as ever to the blinder critics and philosophers of today — but the core of science fiction, its essence...has become crucial to our salvation if we are to be saved at all.
In Robert Holdstock (ed.)
The Encyclopedia of Science Fiction
Foreword (p. 7)
Octopus Books Ltd. London, England. 1978

Ballard, James Graham   1930–
English writer

Everything is becoming science fiction. From the margins of an almost invisible literature has sprung the intact reality of the 20th century.
Fictions of Every Kind
Books and Bookmen, February 1971

Hawking, Stephen William   1942–
English theoretical physicist

There is a two-way trade between science fiction and science. Science fiction suggests ideas that scientists incorporate into their theories, but sometimes science turns up notions that are stranger than any science fiction.
In Lawrence M. Krauss
The Physics of Star Trek
Forward (p. xii)
Harp Perennial Publishers. New York, New York, USA. 1995

We may not yet be able to boldly go where no man (or woman) has gone before, but at least we can do it in the mind.
In Lawrence M. Krauss
The Physics of Star Trek
Foreword (pp. xi–xii)
Harp Perennial Publishers. New York, New York, USA. 1995

Nevertheless, today's science fiction is often tomorrow's science fact. The physics that underlies Star Trek is surely worth investigating. To confine our attention to the terrestrial matters would be to limit the human spirit.
In Lawrence M. Krauss
The Physics of Star Trek
Forward (p. xiii)
Harp Perennial Publishers. New York, New York, USA. 1995

SCIENCE GEEK

Willis, Connie   1945–
American science fiction writer

The effect, especially with the Coke-bottle glasses, should have been science geek, but it wasn't.... Science geeks wear black shoes and white socks. he wasn't even wearing a pocket protector, though he should have been.
Bellwether (p. 12)
Bantam Spectra. New York, New York, USA. 1997

SCIENTIFIC COMMUNITY

Latour, Bruno   1947–
French sociologist of science
Woolgar, S.
No biographical data available

...a body of practices widely regarded by outsiders as well organized, logical, and coherent, in fact consists of a disordered array of observations with which scientists struggle to produce order.... Despite participants' well-ordered reconstructions and rationalizations, actual scientific practice entails the confrontation and negotiation of utter confusion.
Laboratory Life: The Social Construction of Scientific Facts
Chapter 1 (p. 36)
Sage Publications. Beverly Hills, California, USA. 1979

Oppenheimer, J. Robert   1904–67
American theoretical physicist

In any science there is harmony between practitioners. A man may work as an individual, learning of what his colleagues do through reading or conversation; he may be working as a member of a group on problems whose technical equipment is too massive for individual effort. But whether he is part of a team or solitary in his own study, he, as a professional, is a member of a community. His colleagues in his own branch of science will be grateful to him for the inventive or creative thoughts he has, will welcome his criticism....
The Open Mind
Chapter VIII (pp. 137–138)
Simon & Schuster. New York, New York, USA. 1955

Rabi, Isidor Isaac   1898–1988
Austrian-born American physicist

There isn't a scientific community. It is a culture. It is a very undisciplined organization.
In Daniel S. Greenberg
The Politics of Pure Science
Book One, Chapter I (p. 3)
New American Library. New York, New York, USA. 1967

SCIENTIFIC CRITICISM

Pearson, Karl   1857–1936
English mathematician

In an age like our own, which is essentially an age of scientific inquiry, the prevalence of doubt and criticism ought not to be regarded with despair or as a sign of decadence. It is one of the safeguards of progress; — *la critique est la vie de la science*, I must again repeat. One of the most fatal (and not so impossible) futures for science would be the institution of a scientific hierarchy which would brand as heretical all doubt as to its conclusions, all criticism of its results.
The Grammar of Science
Chapter II, Section 7 (p. 66)
Charles Scribner's Sons. London, England. 1892

Tagore, Rabindranath   1861–1941
Indian poet and philosopher

Our scientific world is our world of reasoning. It has its greatness and uses and attractions. We are ready to pay homage due to it. But when it claims to have discovered the real world for us and laughs at the worlds of all simple-minded men, then we must say it is like a general grown intoxicated with his power, usurping the throne of his king.
Personality
The World of Personality (p. 70)
The Macmillan Company. New York, New York, USA. 1917

Schrödinger, Erwin   1887–1961
Austrian theoretical physicist

Our age is possessed by a strong surge towards the criticism of traditional customs and opinions. A new spirit is arising which is unwilling to accept anything on authority, which does not so much permit as demand independent, rational thought on every subject, and which refrains from hampering any attack based upon such thought, even though it be directed against things which formerly were considered to be as sacrosanct as you please. …Its results can only be advantageous: no scientific structure falls entirely into ruin: what is worth preserving preserves itself and requires no protection.
Science and the Human Temperament
Chapter I (p. 38)
W.W. Norton & Company, Inc. New York, New York, USA.1935

von Goethe, Johann Wolfgang   1749–1832
German poet, novelist, playwright, and natural philosopher

A scientific researcher must always think of himself as a member of a jury. His only concern should be the adequacy of the evidence and the clarity of the proofs which support it. Guided by this, he will form his opinion and cast his vote without regard for whether he shares the author's views.
Scientific Studies (Volume 12)
Chapter VIII (pp. 306–307)
Suhrkamp. New York, New York, USA. 1988

von Mises, Ludwig   1881–1973
Austrian economist

…scientific criticism has no nobler task than to shatter false beliefs.
Socialism: An Economic and Sociological Analysis
Preface to the Second German Edition (p. 19)
Yale University Press. New Haven, Connecticut, USA. 1951

Whitehead, Alfred North   1861–1947
English mathematician and philosopher

If science is not to degenerate into a medley of ad hoc hypotheses, it must become philosophical and must enter upon a thorough criticism of its own foundations.
Science and the Modern World
Chapter I (pp. 16–17)
The Macmillan Company. New York, New York, USA. 1929

Ziman, John M.   1925–2005
British physicist

The community of those who are competent to contribute to, or criticize, scientific knowledge must not be closed; it must be larger, and more open, than the group of those who entirely accept a current consensus or orthodoxy. It is an essential element in the health of Science, or of a science, or of the sciences, that self-confirming, mutually validating circles be unable to close. Yet it is also essential that technical scientific discussion be not smothered in a cloud of ignorant prejudices and cranky speculations.
Public Knowledge: An Essay Concerning the Social Dimension of Science
Chapter 4 (p. 64)
Cambridge University Press. Cambridge, England. 1968

SCIENTIFIC DOUBT

Richet, Charles   1850–1935
French physiologist

Scientific doubt is a first-class quality, but rather eliminates piquancy from controversy.
The Natural History of a Savant
Chapter III (p. 25)
J.M. Dent & Sons Ltd. London, England. 1927

SCIENTIFIC INQUIRY

Dewey, John   1859–1952
American philosopher and educator

The routine of custom tends to deaden even scientific inquiry; it stands in the way of discovery and of the active scientific worker. For discovery and inquiry are synonymous as an occupation. Science is a pursuit, not a coming into possession of the immutable; new theories as points of view are more prized than discoveries that quantitatively increase the store on hand.
Reconstruction in Philosophy
Introduction (p. xvii)
Beacon Press. Boston, Massachusetts, USA. 1920

Heinlein, Robert A.   1907–88
American science fiction writer

There ought not to be anything in the whole universe that man can't poke his nose into — that's the way we're built and I assume there's some reason for it.
Methuselah's Children
Chapter 8 (p. 160)
Aeonian Press. Mattituck, New York, USA. 1976

Herschel, Sir John Frederick William   1792–1871
English astronomer and chemist

A mind which has once imbibed a taste for scientific inquiry, and has learnt the habit of applying its principles readily to the cases which occur, has within itself an inexhaustible source of pure and exciting contemplations…
A Preliminary Discourse on the Study of Natural Philosophy
Part I, Chapter I, Section 11 (pp. 14–15)
Printed for Longman, Rees, Orme, Brown & Green. London, England. 1831

Herwitz, Daniel
No biographical data available

Cosmological inquiry takes place in the space between mathematics, theory, experiment, simulation, observation, and philosophic speculation. It is where science lives.
In Heather Wax and Gerald Shaw
Master of His Universe
Science & Spirit, November–December 2004

Medawar, Sir Peter Brian   1915–87
Brazilian-born English zoologist

If the purpose of scientific methodology is to prescribe or expound a system of enquiry or even a code of practice for scientific behavior, then scientists seem able to get on very well without it.
Pluto's Republic
Induction and Intuition in Scientific Thought (p. 78)
Oxford University Press, Inc. Oxford, England. 1982

The purpose of scientific enquiry is not to compile an inventory of factual information, nor to build up a totalitarian world picture of Natural Laws in which every event that is not compulsory is forbidden. We should think of it rather as a logically articulated structure of justifiable beliefs about nature. It begins as a story about a Possible World — a story which we invent and criticize and modify as we go along, so that it winds by being, as nearly as we can make it, a story about real life.
Pluto's Republic
Mainly About Intuition, Section 4 (pp. 110–111)
Oxford University Press, Inc. Oxford, England. 1982

…it is high time that laymen abandoned the misleading belief that scientific enquiry is a cold dispassionate enterprise, bleached of imaginative qualities, and that a scientist is a man who turns the handle of discovery; for at every level of endeavor scientific research is a passionate undertaking and the Promotion of Natural Knowledge depends above all on a *sortée* into what can be imagined but is not yet known.
Imagination and Hypothesis
The Times Literary Supplement (London), October 25, 1963 (p. 850)

Rothschild, Lord Nathaniel Mayer   1910–90
English banker

It is sometimes said in justification of basic research, that chance observations made during such work, and their subsequent study may be just as important as those made during applied R & D. While there is some truth in this contention, the country's needs are not so trivial as to be left to the mercies of a form of scientific roulette, with many more than the conventional 37 numbers on which the ball may land.
A Framework for Government Research and Development (p. 3)
Her Majesty's Stationery Office. London, England. 1971

Thomson, J. Arthur   1861–1933
Scottish biologist

Scientific inquiry may be likened to fishing in the sea of reality with a particular kind of tackle. The tackle has well-known excellences, but it has also recognized limitations; and there may be much in the sea that the net used will not catch, being of too wide a mesh.
The New World of Science
The Atlantic Monthly, June 1930

SCIENTIFIC INVESTIGATION

Agassiz, Jean Louis Rodolphe   1807–73
Swiss-born American naturalist, geologist, and teacher

Scientific investigation in our day should be inspired by a purpose as animating to the general sympathy, as was the religious zeal which built the Cathedral of Cologne or the Basilica of St. Peter's. The time is passed when men expressed their deepest convictions by these wonderful and beautiful religious edifices; but it is my hope to see, with the progress of intellectual culture, a structure arises among us which may be a temple of the revelations written in the material universe.
Louis Agassiz: His Life and Correspondence (Volume 2)
Dredging Expedition (pp. 670–671)
Houghton Mifflin Company. Boston, Massachusetts, USA. 1885

Bayliss, William Maddock   1860–1925
English physiologist

It is not going too far to say that the greatness of a scientific investigator does not rest on the fact of his having never made a mistake, but rather on his readiness to admit that he has done so, whenever the contrary evidence is cogent enough.
Principles of General Physiology
Preface (pp. xvi–xvii)
Longmans, Green & Company. London, England. 1920

Born, Max   1882–1970
German-born English physicist

The Scientist's urge to investigate, like the faith of the devout or the inspiration of the artist, is an expression of mankind's longing for something fixed, something at rest in the universal whirl: God, Beauty, Truth.
The Restless Universe
Chapter V (p. 278)
Dover Publications, Inc. New York, New York, USA. 1951

Boycott, A. E.
No biographical data available

The difficulty in most scientific work lies in framing the questions rather than in finding the answers.
The Transition from Live to Dead
Nature, Volume 123, January 19, 1929 (p. 93)

Carryl, Charles Edward   1841–1920
American writer

Then we gather as we travel,
Bits of moss and dirty gravel,
And we chip off little specimens of stone;
And we carry home as prizes
Funny bugs, of handy sizes,
Just to give the day a scientific tone.
In Franklin P. Adams
Innocent Merriment: An Anthology of Light Verse
Robinson Crusoe's Story
McGraw-Hill Book Company, Inc. New York, New York, USA. 1942

Chargaff, Erwin   1905–2002
Austrian biochemist

The scientific professions began to develop a momentum of their own, thereby creating a vested interest in always having more science, bigger science, better-endowed science. This is, incidentally, quite in contrast, for instance, to orchestra musicians whose influence on the number of orchestra pieces being written is minimal.
Voices in the Labyrinth
Perspectives in Biology and Medicine, Volume 18, Number 3, Spring 1975 (p. 324)

Crookes, Sir William   1832–1919
English chemist and physicist

To stop short in any research that bids fair to widen the gates of knowledge, to recoil from fear of difficulty or adverse criticism, is to bring reproach on science. There is nothing for the investigator to do but to go straight on; to explore up and down, inch by inch, with the taper of his reason; to follow the light wherever it may lead, even should it at times resemble a will-o'-the-wisp.
Address
British Association for the Advancement of Science, Bristol, England (1898)

Doyle, Sir Arthur Conan   1859–1930
Scottish writer

Holmes is a little too scientific for my tastes — it approaches to cold-bloodedness. I could imagine his giving a friend a little pinch of the latest vegetable alkaloid, not out of malevolence, you understand, but simply out of a spirit of inquiry in order to have an accurate idea of the effects. To do him justice, I think that he would take it himself with the same readiness. He appears to have a passion for definite and exact knowledge.
In William S. Baring-Gould (ed.)
The Annotated Sherlock Holmes (Volume 1)
A Study in Scarlet, Chapter 1 (p. 149)
Wings Books. New York, New York, USA. 1967

Farrington, Benjamin   1891–1974
Irish scholar

Just as all scientific investigation is fruitless which is not pursued in a spirit of truth, so the results of all scientific endeavor are wasted if the continuity of tradition cannot be assured. It is of the very essence of science to be a co-operation and that not only of the men of the same generation, but of the generations successively.
In Lloyd William Taylor
Physics: The Pioneer Science (Volume 2)
Chapter 35 (p. 503)
Houghton Mifflin Company. Boston, Massachusetts, USA. 1941

Hall, Alfred Rupert   1920–
English historian of science

The cumulative growth of science, arising from the employment of methods of investigation and reasoning which have been justified by their fruits and their resistance to the corrosion of criticism, cannot be reduced to any single themes. We cannot say…why some men can perceive the truth, or a technical trick, which has eluded others. From the bewildering variety of experience in its social, economic and psychological aspects it is possible to extract only a few factors, here and there, which have had a bearing on the development of science. At present at least, we can only describe, and begin to analyse, where we should like to understand.
The Scientific Revolution, 1500–1800
Introduction (p. xiv)
Longmans, Green & Company. London, England. 1954

Hertz, Heinrich   1857–94
German physicist

I have never forgotten what I often used to say to myself, that I would rather be a great scientific investigator than a great engineer, but would rather be a second-rate engineer than a second-rate investigator.
Miscellaneous Papers
Introduction (p. x)
Macmillan & Company Ltd. London, England. 1896

Huxley, Thomas Henry   1825–95
English biologist

The method of scientific investigation is nothing but the expression of the necessary mode of working of the human mind. It is simply the mode in which all phenomena are reasoned about, rendered precise and exact.
Collected Essays (Volume 2)
Darwiniana
Six Lectures to Working Men (p. 363)
Macmillan & Company London, England. 1904

Peirce, Charles Sanders   1839–1914
American scientist, logician, and philosopher

The man of action has to believe, the inquirer has to doubt; the scientific investigator is both.
In J.B. Conant
Modern Science and Modern Man
Science and Spiritual Values (p. 103)
Columbia University Press. New York, New York, USA. 1952

von Lommel, Eugen   1837–99
German physicist

The deeds of a man of science are his scientific investigations. Truth once discovered does not remain shut up in the study or the laboratory. When the moment comes, it bursts its narrow bonds and joins the quick pulse of life. That which has been discovered in solitude, in the unselfish struggle for knowledge, in pure love of science, is often fated to be the mighty lever to advance the culture of our race.

In Sir Richard Arman Gregory
Discovery; or, The Spirit and Service of Science
Chapter VII (p. 184)
Macmillan & Company Ltd. London, England. 1918

Wells, H. G. (Herbert George)   1866–1946
English novelist, historian, and sociologist

The popular idea of scientific investigation is a vehement, aimless collection of little facts, collected as a bower of birds collects shells and pebbles, in methodical little rows, and out of this process, in some manner unknown to the popular mind, certain conjuring tricks — the celebrated "wonders of science" — in a sort of accidental way emerge.

The Discovery of the Future (p. 34)
B.W. Huebsch. New York, New York, USA. 1913

Wittgenstein, Ludwig Josef Johann   1889–1951
Austrian-born English philosopher

In the course of a scientific investigation we say all kinds of things; we make many utterances whose role in the investigation we do not understand. For it isn't as though everything we say has a conscious purpose; our tongues just keep going. Our thoughts run in established routines, we pass automatically from one thought to another according to the techniques we have learned. And now comes the time for us to survey what we have said. We have made a whole lot of movements that do not further our purpose, or that even impede it, and now we have to clarify our thought processes philosophically.

Translated by Peter Winch
Culture and Value (p. 64e)
The University of Chicago Press. Chicago, Illinois, USA. 1980

SCIENTIFIC LITERACY

Deason, Hilary J.   d. 1971
No biographical data available

Scientific literacy has become a real and urgent matter for the informed citizen. Many have allowed themselves to lapse into a coma of scientific illiteracy because of the misconception that science and mathematics are beyond their understanding, have no personal appeal, and can be rejected or ignored. For them a tocsin [alarm bell] has been sounded by hundreds of intelligent men and women whose personal crusade is the awakening of people everywhere to scientific awareness. They admonish, "read, mark, learn, and inwardly digest."

A Guide to Science Reading
Foreword to the First Edition (p. ix)
The New American Library. New York, New York, USA. 1966

SCIENTIFIC METHOD

Agre, Peter   1949–
American biologist

The field was essentially stuck, but following the well known scientific approach known as "sheer blind luck," we stumbled upon the protein that is the answer to the question: do water channels exist?

Les Prix Nobel. The Nobel Prizes in 2003
Nobel lecture for award received in 2003
Nobel Foundation. Stockholm, Sweden. 2004

Bauer, Henry H.   1931–
American chemist

There is no good reason to discard the scientific method as an ideal; rather, there is good reason to keep it so. Myths, after all, even if not literally true, are stories that embody moral truths.

Scientific Literacy and the Myth of the Scientific Method
Chapter 2 (p. 39)
University of Illinois Press. Urbana, Illinois, USA. 1992

One of the things wrong with the popular, classical definition of the scientific method is the implication that solitary people can successfully do good science, for example frame hypotheses and test them. In practice, however, the people who put forward the hypotheses are not usually the same people who apply the best test to them.

Scientific Literacy and the Myth of the Scientific Method
Chapter 3 (p. 52)
University of Illinois Press. Urbana, Illinois, USA. 1992

Becker, Carl L.   1873–1945
American historian

It is one of the engaging ironies of modern thought that the scientific method, which it was once fondly hoped would banish mystery from the world, leaves it every day more inexplicable.

The Heavenly City of the Eighteenth Century Philosophers
Chapter I (p. 24)
Yale University Press. New Haven, Connecticut, USA. 1932

Bernard, Claude   1813–78
French physiologist

I believe, in a word, that the true scientific method confines the mind without suffocating it, leaves it as far as possible face to face with itself, and guides it, while respecting the creative originality and the spontaneity which are its most precious qualities.

Translated by Henry Copley Greene

An Introduction to the Study of Experimental Medicine
Part Three, Chapter III, Section iv (p. 226)
Henry Schuman, Inc. New York, New York, USA. 1927

Born, Max   1882–1970
German-born English physicist

There are two objectionable types of believers: those who believe the incredible and those who believe that "belief" must be discarded and replaced by "the scientific method."
Natural Philosophy of Cause and Chance
Appendix One (p. 209)
At The Clarendon Press. Oxford, England. 1949

Bridgman, Percy Williams   1882–1961
American physicist

The scientific method, as far as it is a method, is nothing more than doing one's damnedest with one's mind, no holds barred.
Reflections of a Physicist
Chapter 21 (p. 351)
Philosophical Library. New York, New York, USA. 1950

It seems to me that there is a good deal of ballyhoo about scientific method. I venture to think that the people who talk most about it are the people who do least about it. Scientific method is what working scientists do, not what other people or even they themselves may say about it.… Scientific method is something talked about by people standing on the outside and wondering how the scientist manages to do it.…
Reflections of a Physicist
Chapter 5 (p. 81)
Philosophical Library. New York, New York, USA. 1955

[S]cience is what scientists do, and there are as many scientific methods as there are individual scientists.
Reflections of a Physicist
Chapter 5 (p. 81)
Philosophical Library. New York, New York, USA. 1955

Butler, Nicholas Murray   1862–1947
American educator and university administrator

The making of a few score of admirable specialists, and the annual production of a small army of youths with narrow, if minute, information useful in some particular vocation, is a sorry substitute for reaching the great mass of the population with the influence and ideals of scientific inquiry and the scientific method.
In Bernard Jaffe
New World of Chemistry
Preface (p. vii)
Silver, Burdett & Company. New York, New York, USA. 1935

Campbell, Norman R.   1880–1949
English physicist and philosopher

If the discovery of laws could be reduced to a set of formal rules, anyone who learnt the rules could discover laws. But there is no broad road to progress. Herein lies the most serious objection to much that has been written on the methods of science. There is no method, and it is because there is no method which can be expounded to all the world that science is a delight to those who possess the instincts which make methods unnecessary.
Physics: The Elements
Chapter IV (p. 112)
At The University Press. Cambridge, England. 1920

Conant, James Bryant   1893–1978
American educator and scientist

There is no such thing as the scientific method. If there were, surely an examination of the history of physics, chemistry, and biology would reveal it. For as I have already pointed out, few would deny that it is the progress in physics, chemistry and experimental biology which gives everyone confidence in the procedures of the scientist. Yet, a careful examination of these subjects fails to reveal any one method by means of which the masters in these fields broke new ground.
Science and Common Sense
Chapter Three (p. 45)
Yale University Press. New Haven, Connecticut, USA. 1951

Crick, Francis Harry Compton   1916–2004
English biochemist

What, then, do Jim Watson and I deserve credit for? If we deserve any credit at all, it is for persistence and the willingness to discard ideas when they became untenable. One reviewer thought that we couldn't have been very clever because we went on so many false trails, but that is the way discoveries are usually made. Most attempts fail not because of lack of brains but because the investigator gets stuck in a cul-de-sac or gives up too soon.
What Mad Pursuit: A Personal View of Scientific Discovery
Chapter 6 (p. 74)
Basic Books, Inc. New York, New York, USA. 1988

Davies, Paul Charles William   1946–
British-born physicist, writer, and broadcaster

The success of the scientific method at unlocking the secrets of nature is so dazzling it can blind us to the greatest scientific miracle of all: science works.
The Mind of God: The Scientific Basis for a Rational World
Chapter 1 (p. 20)
Simon & Schuster. New York, New York, USA. 1992

Dubos, René Jules   1901–82
French-born American microbiologist and environmentalist

…like its literary and artistic counterparts, the process of scientific creation is a completely personal experience for which no technique of observation has yet been devised. Moreover, out of false modesty, pride, lack of inclination or psychological insight, very few of the great

discoverers have revealed their own mental processes; at the most they have described methods of work — but rarely their dreams, urges, struggles and visions.
Louis Pasteur: Free Lance of Science
Chapter XIII (p. 369)
Little, Brown & Company. Boston, Massachusetts, USA. 1950

Feyerabend, Paul K. 1924–94
Austrian-born American philosopher of science

The idea that science can, and should, be run according to fixed and universal rules, is both unrealistic and pernicious.
Against Method: Outline of an Anarchistic Theory of Knowledge
Chapter 18 (p. 295)
Verso. London, England. 1978

Feynman, Richard P. 1918–88
American theoretical physicist

After we look for the evidence we have to judge the evidence. There are the usual rules about the judging the evidence; it's not right to pick only what you like, but to take all of the evidence, to try to maintain some objectivity about the thing — enough to keep the thing going — not to ultimately depend upon authority. Authority may be a hint as to what the truth is, but is not the source of information. As long as it's possible, we should disregard authority whenever the observations disagree with it.
In Jeffrey Robbins (ed.)
The Pleasure of Finding Things Out: The Best Short Works of Richard P. Feynman
Chapter 4 (p. 104)
Perseus Books. Cambridge, Massachusetts, USA. 1999

Feynman, Richard P. 1918–88
American theoretical physicist
Leighton, Robert B. 1919–97
American physicist
Sands, Matthew L. 1919–
American physicist

Observation, reason, and experiment make up what we call the scientific method.
The Feynman Lectures on Physics (Volume 1)
Chapter 2–1 (p. 2–1)
Addison-Wesley Publishing Company. Reading, Massachusetts, USA. 1983

Flexner, Abraham 1866–1959
American educator

So long as men strive to transcend their native powers, to rid themselves of prejudice and preconception, to observe phenomena in a dry light, the effort is scientific, whether at the moment it attains mathematical accuracy or not.
Medical Education: A Comparative Study
Chapter I (p. 3)
The Macmillan Company. New York, New York, USA. 1925

France, Anatole (Jean Jacques Brousson) 1844–1924
French writer

…the scientific reasons for preferring one piece of evidence to another are sometimes very strong, but they are never strong enough to outweigh our passions, our prejudices, our interests, or to overcome that levity of mind common to all grave men. If follows that we continually present the facts in a prejudiced or frivolous manner.
Penguin Island
Preface (p. vi)
Dodd, Mead & Company. New York, New York, USA. 1923

Garrod, Archibald 1857–1936
English physician

[Scientific method] acts as a check, as well as a stimulus, sifting the value of the evidence, and rejecting that which is worthless, and restraining too eager flights of the imagination and too hasty conclusions.
In Alexander G. Bearn
Archibald Garrod and the Individuality of Man
Chapter 7 (p. 82)
Clarendon Press. Oxford, England. 1993

Gould, Stephen Jay 1941–2002
American paleontologist, evolutionary biologist, and historian of science

It is important that we, as working scientists, combat these myths of our profession as something superior and apart. The myths may serve us well in the short and narrow as rationale for a lobbying strategy — give us the funding and leave us alone, for we know what we're doing and you don't understand anyway. But science can only be harmed in the long run by its self-proclaimed separation as a priesthood guarding a sacred rite called the scientific method. *Time's Arrow, Time's Cycle: Myth and Metaphor in the Discovery of Geological Time*
Chapter 1 (p. 7)
Harvard University Press. Cambridge, Massachusetts, USA. 1987

The Nobel prizes focus on quantitative nonhistorical, deductively oriented fields with their methodology of perturbation by experiment and establishment of repeatable chains of relatively simple cause and effect. An entire set of disciplines, different though equal in scope and status, but often subjected to ridicule because they do not follow this pathway of "hard" science, is thereby ignored: the historical sciences, treating immensely complex and nonrepeatable events (and therefore eschewing prediction while seeking explanation for what has happened) and using the methods of observation and comparison.
Balzan Prize to Ernst Mayr
Science, Volume 223, Number 4633, January 20, 1984 (p. 255)

If justification required eyewitness testimony, we would have no sciences of deep time — no geology, no ancient human history either. (Should I believe Julius Caesar

ever existed? The hard bony evidence for human evolution...surely exceeds our reliable documentation of Caesar's life.)
Dorothy, It's Really Oz
Time Magazine, August 23, 1999 (p. 59)

[O]ur ways of learning about the world are strongly influenced by the social preconceptions and biased modes of thinking that each scientist must apply to any problem. The stereotype of a fully rational and objective "scientific method," with individual scientists as logical (and interchangeable) robots, is self-serving mythology.
This View of Life. In the Mind of the Beholder
Natural History, Volume 103, Number 2, February 1994 (p. 14)

Hoffman, Paul   1934–
American writer

...sometimes serious scientific problems are solved by a scientific method that can be described only as playful.
Playing for Keeps
Discover, October 1990 (p. 4)

Huxley, Thomas Henry   1825–95
English biologist

I am not afraid of the priests in the long-run. Scientific method is the white ant which will slowly but surely destroy their fortifications. And the importance of scientific methods in modern practical life — always growing and increasing — is the guarantee for the gradual emancipation of the ignorant upper and lower classes, the former of whom especially are the strength of the priests.
Collected Essays (Volume 3)
Science and Education
Life and Letters (p. 330)
Macmillan & Company Ltd. London, England. 1904

Kropotkin, Peter Alekseyevich   1842–1921
Russian revolutionary and geographer

He who has once in his life experienced this joy of scientific creation will never forget it; he will be longing to renew it; and he cannot but feel with pain that this sort of happiness is the lot of so few of us, while so many could also live through it, — on a small or on a grand scale, — if scientific methods and leisure were not limited to a handful of men.
Memoirs of a Revolutionist
Part IV, I (p. 6)
Houghton Mifflin Company. Boston, Massachusetts, USA. 1899

Krutch, Joseph Wood   1893–1970
American naturalist, conservationist, and writer

We are committed to the scientific method, and measurement is the foundation of that method; hence we are prone to assume that whatever is measurable must be significant and that whatever cannot be measured may as well be disregarded.

Human Nature and the Human Condition
Chapter V (p. 78)
Random House, Inc. New York, New York, USA. 1959

Maxwell, James Clerk   1831–79
Scottish physicist

Nature is a journal of science, and one of the several tests of a scientific mind is to discern the limits of the legitimate application of scientific methods.
In W.D. Niven (ed.)
The Scientific Papers of James Clerk Maxwell (Volume 2)
Paradoxical Philosophy (p. 759)
Dover Publications, Inc. New York, New York, USA. 1965

Medawar, Sir Peter Brian   1915–87
Brazilian-born English zoologist

The scientific method, as it is sometimes called, is a potentiation of common sense.
Advice to a Young Scientist
Chapter 11 (p. 93)
Basic Books, Inc., Publishers. New York, New York, USA. 1979

The accusation is sometimes directed against scientists that there is in reality no such thing as the scientific method, i.e., that there is no logically accountable and intellectually rigorous process by which we may proceed directly to the solution of a given problem. Scientific method works only in retrospect. This accusation is perfectly just but it doesn't in practice amount to anything more than saying that there is no set of cut-and-dried rules for writing a poem or passage of music or conducting any other imaginative exercise.
The Cost-Benefit Analysis of Pure Research
Hospital Practice, Sept 1973

Ask a scientist what he conceives the scientific method to be, and he will adopt an expression that is at once solemn and shifty-eyed; solemn because he feels he ought to declare an opinion; shifty-eyed, because he is wondering how to conceal the fact that he has no opinion to declare.
Induction and Intuition in Scientific Thought
Chapter I, Section 2 (p. 11)
American Philosophical Society. Philadelphia, Pennsylvania, USA. 1969

Pearson, Karl   1857–1936
English mathematician

I assert that the encouragement of scientific investigation and the spread of scientific knowledge by largely inculcating scientific habits of mind will lead to more efficient citizenship and so to increased social stability. Minds trained to scientific methods are less likely to be led by mere appeal to the passions or by blind emotional excitement to sanction acts which in the end may lead to social disaster.
The Grammar of Science
Introductory, Section 3 (pp. 10–11)
Charles Scribner's Sons. London, England. 1892

Pirsig, Robert M.   1928–
American writer

Traditional scientific method has always been at the very best, 20–20 hindsight. It's good for seeing where you've been.
Zen and the Art of Motorcycle Maintenance: An Inquiry Into Values
Part III, Chapter xxiv (p. 280)
William Morrow & Company, Inc. New York, New York, USA. 1974

When I think of formal scientific method an image sometimes comes to mind of an enormous juggernaut, a huge bulldozer — slow, tedious, lumbering, laborious, but invincible. It takes twice as long, five times as long, maybe a dozen times as long as informal mechanic's techniques, but you know in the end you're going to get it. There's no fault isolation problem in motorcycle maintenance that can stand up to it. When you've hit a really tough one, tried everything, racked your brain and nothing works, and you know that this time Nature has really decided to be difficult, you say, "Okay, Nature, that's the end of the nice guy," and you crank up the formal scientific method.
Zen and the Art of Motorcycle Maintenance: An Inquiry into Values
Part II, Chapter 9 (p. 107)
William Morrow & Company, Inc. New York, New York, USA. 1974

The real purpose of scientific method is to make sure Nature hasn't misled you into thinking you know something you don't actually know. There's not a mechanic or scientist or technician alive who hasn't suffered from that one so much that he's not instinctively on guard.... If you get careless or go romanticizing scientific information, giving it a flourish here and there, Nature will soon make a complete fool out of you.
Zen and the Art of Motorcycle Maintenance: An Inquiry into Values
Part II, Chapter 9 (p. 108)
William Morrow & Company, Inc. New York, New York, USA. 1974

Polanyi, Michael   1891–1976
Hungarian-born English scientist, philosopher, and social scientist

Scientists who believe that the old, tried, and true is sufficient or who underestimate and fail to understand the need for change may soon be lost in a challenging and exciting period of history. But those who have the vision to see beyond the obvious, the wisdom to search for and recognize the truth, and the ability to apply basic knowledge for the good of mankind will find this period one of great reward and satisfaction.
Challenges to Editors of Scientific Journals
Science, Volume 141, Number 3585, September 13, 1963 (p. 1017)

Raymo, Chet   1936–
American physicist and science writer

How is it that astronomers can tell such stories, stories more fabulous than any myth of gods and nymphs, when the ink of night offers to the eye only pinpricks of light?

The answer is both simple and complex. We look, we invent, we look again. We test our inventions against what we see, and we insist that our inventions be consistent with one another, that our stories of the stars be consistent with our stories of the earth, of life, and of matter and energy.... The story of the falling apple and the story of the stars must resonate together. Only then, when our stories of the world vibrate with a symphonic harmony, are we confident that our inventions partake of reality.
The Virgin and the Mousetrap: Essays in Search of the Soul of Science
Chapter 19 (p. 175)
The Viking Press. New York, New York, USA. 1991

Russell, Bertrand Arthur William   1872–1970
English philosopher, logician, and social reformer

Whatever knowledge is attainable, must be attained by scientific methods; and what science cannot discover, mankind cannot know.
Religion and Science
Science and Ethics (p. 243)
Henry Holt & Company. New York, New York, USA. 1935

Scientific method...consists mainly in eliminating those beliefs which there is reason to think a source of shocks, while retaining those against which no definite argument can be brought.
Human Knowledge: Its Scope and Limits
Part III, Chapter III (p. 185)
Simon & Schuster. New York, New York, USA. 1948

Scientific method...consists in observing such facts as will enable the observer to discover general laws governing facts of the kind in question.
The Scientific Outlook
Chapter I (p. 15)
George Allen & Unwin Ltd. London, England. 1931

In arriving at a scientific law there are three main stages: The first consists in observing the significant facts; the second in arriving at a hypothesis, which, if it is true, would account for these facts; the third is deducing from this hypothesis consequences which can be tested by observation. If the consequences are verified, the hypothesis is provisionally accepted as true, although it will usually require modification later on as a result of the discovery of further facts.
The Scientific Outlook
Chapter II (p. 58)
George Allen & Unwin Ltd. London, England. 1931

Stansfield, William D.   1930–
American biologist

Most scientific theories, however, are ephemeral. Exceptions will likely be found that invalidate a theory in one or more of its tenets. These can then stimulate a new round of research leading either to a more comprehensive theory or perhaps to a more restrictive (i.e., more precisely

defined) theory. Nothing is ever completely finished in science; the search for better theories is endless.
The Science of Evolution
Introduction (p. 8)
Macmillan Company. New York, New York, USA. 1977

The interpretation of a scientific experiment should not be extended beyond the limits of the available data. In the building of theories, however, scientists propose general principles by extrapolation beyond available data. When former theories have been shown to be inadequate, scientists should be prepared to relinquish the old and embrace the new in their never-ending search for better solutions. It is unscientific, therefore, to claim to have "proof of the truth" when all that scientific methodology can provide is evidence in support of a theory.
The Science of Evolution
Introduction (p. 8)
Macmillan Publishing Company. New York, New York, USA. 1977

Skinner, Burrhus Frederick   1904–90
American psychologist

Here was a first principle not formally recognized by scientific methodologists: When you run into something interesting, drop everything else and study it.
A Case History in Scientific Method
The American Psychologist, Volume 11, 1956 (p. 223)

Skolimowski, Henryk   1930–
Polish philosopher

We are the proud inheritors and perpetuators of the scientific tradition. But perhaps also the slaves of certain modes of thinking; subjects to a conceptual tyranny which we glorify, thus being perfect slaves — slaves who enjoy their imprisonment.
In A.J. Ayala (ed.)
Studies in the Philosophy of Biology: Reduction and Related Problems
Problems of Rationality in Biology (p. 213)
Macmillan & Company Ltd. London, England. 1974

Sullivan, John William Navin   1886–1937
Irish mathematician

To judge from the history of science, the scientific method is excellent as a means of obtaining plausible conclusions which are always wrong, but hardly as a means of reaching the truth.
The Justification of the Scientific Method
The Athenaeum, Number 4644, 2 May 1919 (p. 275)

Tate, Allen   1899–1979
American poet, teacher, and novelist

Scientific approaches, because each has its own partial conventions momentarily arrogating to themselves the authority of total explanation, must invariably fail to see all the experience latent in the work.
Critical Responsibility
The New Republic, Volume 51, Number 663, August 17, 1927 (p. 340)

Tennant, F. R.
No biographical data available

Half a century ago, it was taught that the scientific method is the sole means of approach to the whole realm of possible knowledge: that there were no reasonably propounded questions worth discussing to which its method was inapplicable. Such belief is less widely held today. Since many men of science became their own epistemologists, science has been more modest.
Philosophical Theology (Volume 1)
Chapter XIII (p. 333)
At The University Press. Cambridge, England. 1956

Thomson, Sir George   1892–1975
English physicist

The scientific method is not a royal road leading to discoveries in research, as Bacon thought, but rather a collection of pieces of advice, some general, some rather special, which may help to guide the explorer in his passage through the jungle of apparently arbitrary facts.
The Inspiration of Science
Chapter II (p. 7)
Oxford University Press, Inc. London, England. 1961

Weisz, Paul B.   1919–
German-born American chemical engineer and biomedical researcher

All science begins with observation, the first step of the scientific method. At once this delimits the scientific domain; something that cannot be observed cannot be investigated by science.
Elements of Biology (p. 40)
McGraw-Hill Book Company, Inc. New York, New York, USA. 1965

Wells, H. G. (Herbert George)   1866–1946
English novelist, historian, and sociologist

For the true scientific method is this:
To trust no statements without verification,
to test all things as rigorously as possible,
to keep no secrets, to attempt no monopolies,
to give out one's best modestly and plainly,
serving no other end but knowledge.
In William Beebe
Edge of the Jungle
Cover page (p. 1)
Garden City Publishing Company, Inc. Garden City, New York, USA. 1925

Wolpert, Lewis   1929–
British embryologist

Even distinguished philosophers of science... recognize the failure of philosophy to help understand the nature of science. They have not discovered a scientific method that provides a formula or prescriptions for how to make discoveries. But many famous scientists have given advice: try many things; do what makes

your heart leap; think big; dare to explore where there is no light; challenge expectation; cherchez le paradox; be sloppy so that something unexpected happens, but not so sloppy that you can't tell what happened; turn it on its head; never try to solve a problem until you can guess the answer; precision encourages the imagination; seek simplicity; seek beauty.... One could do no better than to try them all.
The Unnatural Nature of Science
Chapter 6 (p. 108)
Harvard University Press. Cambridge, Massachusetts, USA. 1992

No one method, no paradigm, will capture the process of science. There is no such thing as the scientific method.
The Unnatural Nature of Science
Chapter 6 (p. 108)
Harvard University Press. Cambridge, Massachusetts, USA. 1992

Zimmerman, Michael   1946–
American biologist

Having a scientific outlook means being willing to divest yourself of a pet hypothesis, whether it relates to easy self-help improvements, homeopathy, graphology, spontaneous generation, or any other concept, when the data produced by a carefully designed experiment contradict that hypothesis. Retaining a belief in a hypothesis that cannot be supported by data is the hallmark of both the pseudoscientist and the fanatic. Often the more deeply held the hypothesis, the more reactionary is the response to nonsupportive data.
Science, Nonscience, and Nonsense: Approaching Environmental Literacy
Chapter Two (p. 37)
The Johns Hopkins University Press. Baltimore, Maryland, USA. 1995

SCIENTIFIC MIND

Abbey, Edward   1927–89
American environmentalist and nature writer

Any good poet, in our age at least, must begin with the scientific view of the world; and any scientist worth listening to must be something of a poet, must possess the ability to communicate to the rest of us his sense of love and wonder at what his work discovers.
The Journey Home: Some Words in Defense of the American West
Chapter 8 (p. 87)
E.P. Dutton & Company. New York, New York, USA. 1977

Ackerman, Edward A.   1911–73
American geographer

The mind of the scientist, no less than that of the poet or musician, must be structured by thought and experience before it reaches the creative stage.
Where Is a Research Frontier?
Annals of the Association of American Geographers, Volume 30, 1931 (p. 433)

Bauer, Henry H.   1931–
American chemist

Science progresses not because scientists as a whole are passionately open-minded but because different scientists are passionately closed-minded about different things.
Scientific Literacy and the Myth of the Scientific Method
Chapter 4 (p. 76)
University of Illinois Press. Urbana, Illinois, USA. 1992

Beveridge, William Ian Beardmore   1908–
Australian zoologist

It is true that much time and effort is devoted to training and equipping the scientist's mind, but little attention is paid to the techniques of making the best use of it.
The Art of Scientific Investigation
Preface (p. viii)
W.W. Norton & Company, Inc. New York, New York, USA.1957

Compton, Karl Taylor   1887–1954
American educator and physicist

Science requires straight and independent thinking. Every hypothesis or idea is capable of definite proof or disproof. The habit of mind that subjects every idea to rigid test is of utmost value. Much of the loose thinking in social, educational, political, and economic affairs would be avoided if the workers in these fields could be given a real training in accurate scientific thinking.
A Scientist Speaks: Excerpts from Addresses by Karl Taylor Compton During the Years 1930–1949 (p. 39)
Undergraduate Association, MIT. Cambridge, Massachusetts, USA. 1955

Dewey, John   1859–1952
American philosopher and educator

The future of our civilization depends on the widening spread and deepening hold of the scientific habit of mind.
In Lloyd William Taylor
Physics: The Pioneer Science (Volume 1)
Chapter 11 (p. 137)
Houghton Mifflin Company. Boston, Massachusetts, USA. 1941

Doyle, Sir Arthur Conan   1859–1930
Scottish writer

The true scientific mind is not to be tied down by its own conditions of time and space. It builds itself an observatory erected upon the border line of present, which separates the infinite past from the infinite future. From this sure post it makes its sallies even to the beginning and to the end of all things.
The Poison Belt
Chapter Three (p. 84)
The Macmillan Company. New York, New York, USA. 1964

Emerson, Ralph Waldo   1803–82
American lecturer, poet, and essayist

The scientific mind must have a faith which is science.
The Complete Works of Ralph Waldo Emerson (Volume 6)
The Conduct of Life
Chapter VI (p. 240)
Houghton Mifflin Company. Boston, Massachusetts, USA. 1904

Fisher, Sir Ronald Aylmer   1890–1962
English statistician and geneticist

In scientific subjects, the natural remedy for dogmatism
has been found in research. By temperament and training,
the research worker is the antithesis of the pundit. What
he is actively and constantly aware of is his ignorance,
not his knowledge; the insufficiency of his concepts, of
the terms and phrases in which he tries to excogitate his
problems: not their final and exhaustive sufficiency. He
is, therefore, usually only a good teacher for the few who
wish to use their mind as a workshop, rather than to store
it as a warehouse.
Eugenics, Academic and Practical
Eugenics Review, Volume 27, 1935

Foster, Sir Michael   1836–1907
English physiologist and educator

…the mind which has been already sharpened by the
methods of one science takes a keener edge, and the more
quickly, when it is put on the whetstone of another science,
than does a mind which knows nothing of no science.
Annual Report of the Board of Regents of the Smithsonian Institution,
1898
Recent Advances in Science, and Their Bearing on Medicine and
Surgery (p. 340)
Government Printing Office. Washington, D.C. 1899

Kingsley, Charles   1819–75
English clergyman and author

In one word, [scientific] men [have] acquired just the
habit of mind which the study of Natural Science can
give, and must give; for without it there is no use studying Natural Science; and the man who has not got that
habit of mind, if he meddles with science, will merely
become a quack and a charlatan, only fit to get his bread
as a spirit-rapper, or an inventor of infallible pills.
Town Geology
Preface
D. Appleton & Company. New York, New York, USA. 1873

Knickerbocker, William Skinkle   1892–1972
American professor of English and author

…the scientific…mind produces many of the virtues
which in old-fashioned courses of ethics were taught as
objectively as a problem in geometry. Patience, endurance, humility, teachableness, honesty, accuracy — without these it is impossible for a scientist to work.

Classics of Modern Science
Preface
Alfred A Knopf. New York, New York, USA. 1927
Houghton Mifflin Company. Boston, Massachusetts, USA. 1904

Kruger, Otto   1885–1974
American actor

My mind is just as open as it ever was, professor. But
it's a scientific mind, and there's no place in it for
superstitions.
Dracula's Daughter
Film (1936)

Large, E. C.
No biographical data available

There were two age-old tendencies toward stagnation
in scientific thought which those of youthful spirit had
always to resist. One was the human weakness of accepting the uncorroborated say-so of eminent authorities, and
the other was the human stupidity of regarding natural
science as something divisible into watertight compartments.
The Advance of the Fungi
Chapter XXIII (p. 317)
Henry Holt & Company. New York, New York, USA. 1940

Lévi-Strauss, Claude   1908–
French social anthropologist and structuralist

The scientific mind does not so much provide the right
answers as ask the right questions.
Translated by John and Doreen Weightman
The Raw and the Cooked
Overture (p. 7)
Harper & Row, Publishers. New York, New York, USA. 1975

Maxwell, James Clerk   1831–79
Scottish physicist

…one of the severest tests of a scientific mind is to dis-
cern the limits of the legitimate application of the scien-
tific method.
In W.D. Niven (ed.)
The Scientific Papers of James Clerk Maxwell (Volume 2)
Paradoxical Philosophy (p. 759)
At The University Press. Cambridge, England. 1890

Medawar, Sir Peter Brian   1915–87
Brazilian-born English zoologist

There is no such thing as a Scientific Mind. Scientists are
people of very dissimilar temperaments doing different
things in very different ways. Among scientists are collectors, classifiers and compulsive tidiers-up; many are
detectives by temperament and many are explorers; some
are artists and others artisans. There are poet-scientists and
philosopher-scientists and even a few mystics. What sort
of mind or temperament can all these people be supposed
to have in common? Obligative scientists must be very

rare, and most people who are in fact scientists could easily have been something else instead.
The Art of the Soluble
Hypothesis and Imagination (p. 132)
Methuen & Company Ltd. London, England. 1967

Menzel, Donald H.   1901–76
American astronomer and astrophysicist
Boyd, Lyle B.
No biographical data available

The creative scientist, eternally curious, keeps an open mind toward strange phenomena and novel ideas, knowing that we have only begun to understand the universe we live in. He remembers, too, that Biot's discovery that meteorites were "stones from the sky" was at first greeted with disbelief, and he hopes never to be guilty or similar obtuseness. But an open mind does not mean credulity or a suspension of the logical faculties that are man's most valuable asset.
The World of Flying Saucers: A Scientific Examination of a Major Myth of the Space Age
Chapter VIII (p. 289)
Doubleday & Company, Inc. Garden City, New York, USA. 1963

Pagels, Heinz R.   1939–88
American physicist and science writer

I like to browse in occult bookshops if for no other reason than to refresh my commitment to science.
The Dreams of Reason: The Computer and the Rise of the Sciences of Complexity
Chapter 11 (p. 242)
Simon & Schuster. New York, New York, USA. 1988

Russell, Bertrand Arthur William   1872–1970
English philosopher, logician, and social reformer

…the scientific attitude is in some degree unnatural to man; the majority of our opinions are wish-fulfillments, like dreams in the Freudian theory.
The Scientific Outlook
Chapter I (p. 16)
George Allen & Unwin Ltd. London, England. 1931

Thomson, J. Arthur   1861–1933
Scottish biologist

The scientific mood is especially marked by a passion for facts, by cautiousness of statement, by clearness of vision, and by a sense of the inter-relatedness of things.
Introduction to Science
Chapter I (p. 34)
Williams & Norgate Ltd. London, England. 1916

Trotter, Wilfred   1872–1939
British surgeon and sociologist

The truly scientific mind is altogether unafraid of the new, and while having no mercy for ideas which have served their turn or shown their uselessness, it will not grudge to any unfamiliar conception its moment of full and friendly attention, hoping to expand rather than to minimize what small core of usefulness it may happen to contain.
Observation and Experiment and Their Use in the Medical Sciences
British Medical Journal, Volume 2, 1930

Valéry, Paul   1871–1945
French poet and critic

Each mind can regard itself as a laboratory in which processes peculiar to the individual are used for transforming a substance common to all.

The results obtained by certain individuals are a source of wonderment to others. Starting out with ordinary carbon, one man produces a diamond, by means of temperatures and pressures that others never dreamt of. "Why, it's only carbon!" they say, after analyzing it. But they don't know how to do what he did.
In Jackson Mathews (ed.)
The Collected Works of Paul Valéry (Volume 14)
Analects (p. 482)
Princeton University Press. Princeton, New Jersey, USA. 1971

Weidlein, Edward Ray
Chemical engineer

A true scientist never grows old in his way of thinking. His mind is constantly working to improve his surroundings and to better understand the laws of nature. He expects to live in a changing world.
Annual Report of the Board of Regents of the Smithsonian Institution, 1938
A World of Change (p. 199)
Government Printing Office. Washington, D.C. 1939

Weil, Simone   1909–43
French philosopher and mystic

A scientific conception of the world doesn't prevent one from observing what is socially fitting.
Translated by Arthur Wills
The Need for Roots: Prelude to a Declaration of Duties Toward Mankind
Part Three (p. 248)
The Beacon Press. Boston, Massachusetts, USA. 1952

Weisskopf, Victor Frederick   1908–2002
Austrian-American physicist

Some people maintain that scientific insight has eliminated the need for meaning. I do not agree. The scientific worldview established the notion that there is a sense and purpose in the development of the universe when it recognized the evolution from the primal explosion to matter, life, and humanity. In humans, nature begins to recognize itself.
The Joy of Insight: Passions of a Physicist
Chapter Fourteen (pp. 317–318)
Basic Books, Inc. New York, New York, USA. 1991

Whitehead, Alfred North   1861–1947
English mathematician and philosopher

The aim of scientific thought are to see the general in the particular and the eternal in the transitory.
OMNI Magazine
Volume 2, Number 41, November 1979

A man who only knows his own science, as a routine peculiar to that science, does not even know that. He has no fertility of thought, no power of quick seizing the bearing of alien ideas. He will discover nothing, and be stupid in practical applications.
The Organisation of Thought
Chapter II (p. 46)
Greenwood Press Publishers. Westport, Connecticut, USA. 1974

SCIENTIFIC PROGRESS

Bush, Vannevar   1890–1974
American electrical engineer and physicist

Scientific progress on a broad front results from free play of free intellects, working on subjects of their own choice, in the manner dictated by their curiosity for the exploration of the unknown.
Science: The Endless Frontier
Chapter 1 (p. 2)
United States Government Printing Office. Washington, D.C. 1945

Compton, Karl Taylor   1887–1954
American educator and physicist

The whole history of scientific progress illustrates the importance of free communication of ideas, of co-operative work at all levels, of adequate support and facilities, and above all, of high grade research workers and top-notch leadership.
A Scientist Speaks: Excerpts from Addresses by Karl Taylor Compton During the Years 1930–1949 (p. 11)
Undergraduate Association, MIT. Cambridge, Massachusetts, USA. 1955

The geographical pioneer is now supplanted by the scientific pioneer.... Without the scientific pioneer our civilization would stand still and our spirit would stagnate; with him mankind will continue to work toward his higher density. This being so, our problem is to make science as effective an element as possible in our American program for social progress.
A Scientist Speaks: Excerpts from Addresses by Karl Taylor Compton During the Years 1930–1949 (p. 2)
Undergraduate Association, MIT. Cambridge, Massachusetts, USA. 1955

Crick, Francis Harry Compton   1916–2004
English biochemist

It can be confidently stated that our present knowledge of the brain is so primitive — approximately at the stage of the four humours in medicine or of bleeding in therapy (what is psychoanalysis but mental bleeding?) — that when we do have fuller knowledge our whole picture of ourselves is bound to change radically. Much that is now culturally acceptable will then seem to be nonsense. People with training in the arts still feel that in spite of the alterations made in their life by technology — by the internal combustion engine, by penicillin, by the Bomb — modern science has little to do with what concerns them most deeply. As far as today's science is concerned this is partly true, but tomorrow's science is going to knock their culture right out from under them.
Of Molecules and Men
The Prospect Before Us (p. 94)
University of Washington Press. Seattle, Washington, USA. 1966

Gould, Stephen Jay   1941–2002
American paleontologist and evolutionary biologist

The most important scientific revolutions all include, as their only common feature, the dethronement of human arrogance from one pedestal after another of previous convictions about our centrality in the cosmos.
Dinosaur in a Haystack: Reflections in Natural History
Part Three, Chapter 13 (p. 164)
Random House, Inc. New York, New York, USA. 1995

Gregory, Sir Richard Arman   1864–1952
British science writer and journalist

The discovery of a law of Nature is always of great advantage to scientific progress. By the warp and woof of experiment, the man of science weaves a pattern from the threads of evidence, and presents the result to the world for anyone to improve.
Discovery: Or the Spirit and Service of Science
Chapter VII (p. 183)
Macmillan & Company Ltd. London, England. 1918

Jacob, François   1920–
French biologist

Contrary to what I once thought, scientific progress did not consist simply in observing, in accurately formulating experimental facts and drawing up a theory from them. It began with the invention of a possible world, or a fragment thereof, which was then compared by experimentation with the real world. And it was this constant dialogue between imagination and experiment that allowed one to form an increasingly fine-grained conception of what is called reality.
In William Calvin
The Cerebral Symphony: Seashore Reflections on the Structure of Consciousness
Chapter 10 (p. 206)
Bantam Books. New York, New York, USA. 1989

Mather, Kirtley F.   1888–1978
American geologist

The more we know about the world, the more mysterious and marvelous it becomes. The arrogance which characterized so many scientists of preceding generations has given place to a true humility, admirably displayed by most of the leaders in contemporary scientific progress.
In Edward H. Cotton
Has Science Discovered God?
Sermons from Stones (p. 3)
Thomas Y. Crowell Company, Publishers. New York, New York, USA. 1931

Trimble, George S.   d. 1863
No biographical data available

Actually the biggest deterrent to scientific progress is a refusal of some people, including scientists, to believe that things that seem amazing can really happen.
In Charles Berlitz and William Moore
The Philadelphia Experiment: Project Invisibility (p. 8)
Souvenir Press Ltd. London, England. 1979

Virchow, Rudolf Ludwig Karl   1821–1902
German pathologist and archaeologist

…if we would serve science, we must extend her limits, not only as far as our own knowledge is concerned, but in the estimation of others.
Translated by Frank Chase
Cellular Pathology: As Based Upon Physiological and Pathological History
Authors Preface (p. 7)
Dover Publications, Inc. New York, New York, USA. 1971

von Goethe, Johann Wolfgang   1749–1832
German poet, novelist, playwright, and natural philosopher

When science appears to be slowing down and, despite the efforts of many energetic individuals, comes to a dead stop, the fault is often to be found in a certain basic concept that treats the subject too conventionally. Or the fault may lie in a terminology which, once introduced, is unconditionally approved and adopted by the great majority, and which is discarded with reluctance even by independent thinkers, and only as individuals in isolated cases.
Goethe's Botanical Writings
An Attempt to Evolve a General Comparative Theory (p. 81)
University of Hawaii Press. Honolulu, Hawaii, USA. 1952

Wells, H. G. (Herbert George)   1866–1946
English novelist, historian, and sociologist

…Who put up that big City and Guilds place at South Kensington? Enterprising business men! They fancy they'll have a bit of science going on, they want a handy Expert ever and again, and there you are! And what do you get for research when you've done it? Just a bare living and no outlook. They just keep you to make discoveries, and if they fancy they'll use 'em they do.
Tono-Bungay

Book the Second, Chapter the Second, II (p. 156)
Duffield & Company. New York, New York, USA. 1921

SCIENTIFIC PUBLISHING

Agnew, Neil McK.
No biographical data available
Pyke, Sandra W.
No biographical data available

…perhaps the most deceptive myth of all is that the Ph.D. represents the last hurdle in some kind of knowledge race. A student who has just cleared the jump should enjoy this illusion while it lasts. The science game now shifts to a new ground with new rules. Your rating first depends on getting some articles out; then, once you've demonstrated that you can publish, your rating depends on whether you are publishing in respectable journals; then your rating depends on whether you have a good book out; and then…
The Science Game
Chapter 12 (p. 146)
Prentice-Hall, Inc. Englewood Cliffs, New Jersey, USA. 1969

Arber, Agnes Robertson   1879–1960
English botanist

A record of research should not resemble a casual pile of quarried stone; it should seem "not built, but born", as Vasari said in praise of a building.
The Mind and the Eye: A Study of the Biologist's Standpoint
Chapter V (p. 50)
At The University Press. Cambridge, England, USA. 1954

Author undetermined

As this paper contains nothing which deserves the name either of experiment or discovery, and as it is, in fact, destitute of every species of merit, we should have allowed it to pass among the multitude of those articles which must always find their way into the collections of a society which is pledged to publish two or three volumes every year.… We wish to raise our feeble voice against innovations, that can have no other effect than to check the progress of science, and renew all those wild phantoms of the imagination which Bacon and Newton put to flight from her temple.
Review of Dr. Young's Bakerian Lecture
Edinburgh Review, January 1803 (p. 450)

Batchelor, G. K.   1920–2000
English applied mathematics professor and fluid mechanics engineer

Reading a paper is a voluntary and demanding task, and a reader needs to be enticed and helped and stimulated by the author.
Preoccupations of a Journal Editor
Journal of Fluid Mechanics, Volume 106, 1981 (p. 8)

Buckle, Henry Thomas 1821–62
English historian

The publications of our scientific authors overflow with minute and countless details, which perplex the judgment, and which no memory can retain. In vain do we demand that they should be generalized, and reduced into order. Instead of that, the heap continues to swell. We want ideas, and get more facts. We hear constantly what nature is doing, but we rarely hear what man is thinking.
History of Civilization in England (Volume 2)
Chapter VI (p. 396)
D. Appleton & Company. New York, New York, USA. 1891

Casimir, Hendrik B. G. 1909–2000
Dutch physicist

I should like to find a way of discouraging unnecessary publications, but I have not found a solution, save the radical one…that all scientific papers be published anonymously.
In Praise of Smallness — How Can We Return to Small Science
Perspectives in Biology and Medicine, Volume 23, Number 3, Spring 1980 (p. 383)

Chargaff, Erwin 1905–2002
Austrian biochemist

Scientific journals must remain the preserve of articles capable of affecting the consensus of the scientific public. Books are the place for opinions, speculations, and fanciful accounts of ricocheting planets. The publisher has only to convince enough buyers to cover their cost of publication. In a free society with a vigorous press, there is little danger that an important idea will not get a fair hearing.
Uncommon Sense: The Heretical Nature of Science
Chapter 8 (pp. 149–150)
Oxford University Press, Inc. New York, New York, USA. 1993

Comroe, Jr., Julius H. 1911–84
American physician

Almost every scientist working today can get his work published, somewhere, once he decides to "write it up"; maybe it will be in the Bulletin of the Podunk Medical Society rather than in a journal with international prestige or readership, or maybe it will be published only as an abstract. The main determinant of what is or is not published therefore seems to be the scientist, for it is he who decides to become or not to become an author.
Publish and/or Perish
American Review of Respiratory Disease, Volume 113, 1976

Dubos, René Jules 1901–82
French-born American microbiologist and environmentalist

…a scientific paper should never try to make more than one point.
In B.D. Davis

Two Perspectives
Perspectives in Biology and Medicine, Volume 35, Number 1, Autumn 1991 (p. 38)

Dyson, Freeman J. 1923–
American physicist and educator

Most of the papers which are submitted to the Physical Review are rejected, not because it is impossible to understand them, but because it is possible. Those which are impossible to understand are usually published.
Innovation in Physics
Scientific American, Volume 199, Number 3, September 1958

Elder, Joseph
No biographical data available

Publication is the end-product of research. Research without publication is sterile.
Jargon—Good and Bad
Science Volume 119, Number 3095, 23 April 1954 (p. 536)

Gastel, Barbara
American medical science writing educator

Every master's thesis or doctoral dissertation should e accompanied by a lay summary or press release written by the graduate student (with the guidance, if possible, of a science writing instructor or public inromation officer at the student's institution).
Earth and Life Science Editing, Volume 24, 1985 (p.3)

Gelernter, David 1955–
Computer scientist

Scientists nowadays rarely know how to read seriously. They are accustomed to strip-mining a paper to get the facts out and then moving on, not to mollycoddling the thing in search of nuances; there probably aren't any.
In John Brokman and Katinka Matson (eds.)
How Things Are: A Tool Kit For the Mind
Study Talmud (p. 213)
William Morrow & Company. New York, New York, USA. 1995

Glaisher, James Whitbread Lee 1848–1928
English mathematician

In other branches of science, where quick publication seems to be so much desired, there may possibly be some excuse for giving to the world slovenly or ill-digested work, but there is no excuse in mathematics. The form ought to be as perfect as the substance, and the demonstrations as rigorous as those of Euclid. The mathematician has to deal with the most exact facts of Nature, and he should spare no effort to render his interpretation worthy of his subject, and to give to his work its highest degree of perfection.
Presidential Address, British Association for the Advancement of Science, *Nature*, Section A (1890), Volume 42 (p. 467)

Gould, Stephen Jay 1941–2002
American paleontologist, evolutionary biologist, and historian of science

I do not know when the technical and popular prose of science became separated, although I accept the inevitability of such a division as knowledge became increasingly more precise, detailed, and specialized. We have now reached the point where most technical literature not only falls outside the possibility of public comprehension but also (as we would all admit in honest moments) outside our own competence in scientific disciplines far removed from our personal expertise. I trust that we all regard this situation as saddening, even though we accept its necessity.
Take Another Look
Science, Volume 286, Number 5441, October 29, 1999 (p. 899)

Hagstrom, Warren O.
No biographical data available

Manuscripts submitted to scientific periodicals are often called "contributions," and they are, in fact, gifts. Authors do not usually receive royalties or other payments, and their institutions may even be required to aid in the financial support of the periodical. On the other hand, manuscripts for which the scientific authors do receive financial payments, such as textbooks and popularizations, are, if not despised, certainly held in much lower esteem than articles containing original research results.
The Scientific Community
Basic Books, Inc. New York, New York, USA. 1965

Haldane, John Burdon Sanderson 1892–1964
English biologist

Four stages of acceptance: i) this is worthless nonsense; ii) this is an interesting, but perverse, point of view; iii) this is true, but quite unimportant; iv) I always said so.
Journal of Genetics, Volume 58 (p. 464)

Hudson, Jeffrey
No biographical data available

There's this desert prison…with an old prisoner, resigned to his life, and a young one just arrived. The younger one talks constantly of escape, and, after a few months, he makes a break. He's gone a week, and then he's brought back by the guard. He's half dead, crazy with hunger and thirst. He describes how awful it was to the old prisoner. The endless stretches of sand, no oasis, no sign of life anywhere. The old prisoner listens for awhile, then says. "Yep. I know. I tried to escape myself, twenty years ago." The younger prisoner says "You did? Why didn't you tell me, all these months I was planning my escape? Why didn't you let me know it was impossible?" And the old prisoner shrugs, and says, "So who publishes negative results?"

A Case of Need
Tuesday, 11 October
Nine (p. 121)
The World Publishing Company. New York, New York, USA. 1968

Huth, Edward Janavel 1923–
American physician

Why should the investigators confine themselves to one paper when they can slice up data and interpretations into two, three, five, or more papers that will better serve their needs when they face promotion or tenure committees? "Salami science" does not always equal baloney, but such divided publication is often an abuse of scientific publication.
Irresponsible Authorship and Wasteful Publication
Annals of Internal Medicine, Volume 104, 1986

Ingle, Dwight J. 1907–78
Biologist and endocrinologist

Science cannot be equated to measurement, although many contemporary scientists behave as though it can. For example, the editorial policies of many scientific journals support the publication of data and exclude the communication of ideas.
Principles of Research in Biology and Medicine
Chapter 1 (p. 3)
Lippincott. Philadelphia, Pennsylvania, USA. 1958

Kennedy, Donald
No biographical data available

All the thinking, all the textual analysis, all the experiments and the data-gathering aren't anything until we write them up. In the world of scholarship we are what we write.
Academic Duty (p. 186)
Harvard University Press. Cambridge, Massachusetts, USA. 1997

Maslow, A. H. 1908–70
American psychologist

I do not recall seeing in the literature with which I am familiar any paper that criticized another paper for being unimportant, trivial or inconsequential.
Motivation and Personality
Chapter 2 (p. 14)
Harper & Row, Publishers. New York, New York. 1970

Mayo, William J. 1861–1939
American physician

Reading papers is not for the purpose of showing how much we know and what we are doing, but is an opportunity to learn.
The Value of the Weekly General Staff Meeting
Proceedings of Staff Meetings, Mayo Clinic, Volume 10, January 30, 1935

Medawar, Sir Peter Brian 1915–87
Brazilian-born English zoologist

...it is no use looking to scientific "papers," for they not merely conceal but actively misrepresent the reasoning that goes into the work they describe.... Nor is it any use listening to accounts of what scientists say they do, for their opinions vary widely enough to accommodate almost any methodological hypothesis we may care to devise. Only unstudied evidence will do — and that means listening at a keyhole.
The Art of the Soluble
Hypothesis and Imagination (p. 151)
Methuen & Company Ltd. London, England. 1967

Nelkin, Dorothy   1933–2003
American sociologist

...too often science in the press is more a subject for consumption than for public scrutiny, more a source of entertainment than of information. Too often science is presented as an arcane activity outside and above the sphere of normal human understanding, and therefore beyond our control. Too often the coverage is promotional and uncritical, encouraging apathy, a sense of impotence, and the ubiquitous tendency to defer to expertise.
Selling Science: How the Press Covers Science and Technology
Chapter 10 (p. 173)
W.H. Freeman & Company. New York, New York, USA. 1995

Price, Derek John de Solla   1922–83
English science historian and information scientist

...scientists have a strong urge to write papers but only a relatively mild one to read them.
Little Science, Big Science
Chapter 3 (pp. 69–70)
Columbia University Press. New York, New York, USA. 1963

Rowland, Henry Augustus   1848–1901
American physicist

A hermit philosopher we can imagine might make many useful discoveries. Yet, if he keeps them to himself, he can never claim to have benefited the world in any degree. His unpublished results are his private gain, but the world is not better off until he has made them known in language strong enough to call attention to them and to convince the world of their truth.
The Physical Papers of Henry Augustus Rowland
The Highest Aim of the Physicist (p. 669)
Johns Hopkins Press. Baltimore, Maryland, USA. 1902

Schrödinger, Erwin   1887–1961
Australian theoretical physicist

...a typical scientific paper has never pretended to be more than another little piece in a larger jigsaw — not significant in itself but as an element in a grander scheme. This technique, of soliciting many modest contributions to the vast store of human knowledge, has been the secret of Western science since the seventeenth century, for it achieves a corporate, collective power that is far greater than any one individual can exert.
Information, Communication, Knowledge
Nature, Volume 224,Number 5217, October 25, 1969 (p. 324)

Primary scientific papers are not meant to be final statement of indisputable truths; each is merely a tiny tentative step forward, through the jungle of ignorance.
Information, Communication, Knowledge
Nature, Volume 224,Number 5217, October 25, 1969 (p. 324)

Shrady, George
No biographical data available

The time is already past when any man can hope to rise to be an authority in any department of medical science through any royal road of social influence, political manipulations, or even personal charms. Those who are to be the leaders and guides of medical science for the coming generation must earn their position by persistent, original investigation, and by faithfully recording their experience in the permanent literature of the day.
Medical Authorship
Medical Record, Volume 2, 1867

Simpson, Michael A.
No biographical data available

We still consistently overvalue poor research and semi-literate publication; again, partly, because quantity, in number of publications, is easier to measure than quality.
A Mythology of Medical Education
Lancet, Volume 3, 1974

Wilson, Logan
No biographical data available

Results unpublished are little better than those never achieved.... One must write something and get it into print. Situational imperatives dictate a "publish or perish" credo within the ranks.
The Academic Man: A Study in the Sociology of a Profession (p. 197)
Oxford University Press, Inc. London, England. 1942

Woolley, Sir Charles Leonard   1880–1960
English archaeologist

The prime duty of the field archaeologist is to collect and set in order material with not all of which he can himself deal at first hand. In no case will the last word be with him; and just because that is so his publication of the material must be minutely detailed, so that from it others may draw not only corroboration of his views but fresh conclusions and more light.
Digging Up the Past
Chapter V (pp. 133–134)
Charles Scribner's Sons. New York, New York, USA. 1931

Ziman, John M.   1925–2005
British physicist

The moment of truth for many young scientists comes when they first act as a referee for a scientific paper; having striven for years to get their own work published against the criticism of anonymous referees, they find themselves, by psychological role-reversal, on the other side of the fence. Thus do we eventually internalize the "scientific attitude."

Reliable Knowledge
Chapter 6 (fn 13, p. 132)
Cambridge University Press. Cambridge, England. 1978

SCIENTIFIC SPIRIT

Bernard, Claude   1813–78
French physiologist

In my opinion the true scientific spirit is that whose high aspiration fertilize the sciences and draw them on in search of truths which are still beyond them but which must not be suppressed, because they have been attacked by stronger and more delicate philosophic minds.

Translated by Henry Copley Greene
An Introduction to the Study of Experimental Medicine
Part Three, Chapter III, Section iv (p. 223)
Henry Schuman, Inc. New York, New York, USA. 1927

Clifford, William Kingdon   1845–79
English philosopher and mathematician

There is no scientific discoverer, no poet, no painter, no musician, who will not tell you that he found ready made his discovery or poem or picture — that it came to him from outside, and that he did not consciously create it from within.

In Leslie Stephen and Frederick Pollock (eds.)
Lectures and Essays (Volume 1)
Some of the Conditions of Mental Development (p. 99)
Macmillan & Company Ltd. London, England. 1879

The subject of science is the human universe; that is to say, everything that is, or has been, or may be related to man.

In Leslie Stephen and Frederick Pollock (eds.)
Lectures and Essays (Volume 1)
On the Aims and Instruments of Scientific Thought (p. 126)
Macmillan & Company Ltd. London, England. 1879

Compton, Arthur H.   1892–1962
American physicist

The spirit of science knows no national or religious boundaries, and it is thus a powerful force for the peace of the world.

Les Prix Nobel. The Nobel Prizes in 1927
Nobel banquet speech for award received in 1927
Nobel Foundation. Stockholm, Sweden. 1928

Franklin, Benjamin   1706–90
American printer, scientist, and diplomat

Furnished as all Europe now is with Academies of Science, with nice instruments and the spirit of experiment, the progress of human knowledge will be rapid and discoveries made of which we have at present no conception. I begin to be almost sorry I was born so soon, since I cannot have the happiness of knowing what will be known a hundred years hence.

The Writings of Benjamin Franklin
Letter, July 27, 1783, to naturalist Sir Joseph Banks (p. 74)
Macmillan & Company Ltd. London, England. 1906

Garrod, Archibald   1857–1936
English physician

…scientific method is not the same as the scientific spirit. The scientific spirit does not rest content with applying that which is already known, but is a restless spirit, ever pressing forward towards the regions of the unknown.…

Archibald Garrod and the Individuality of Man
Chapter 7 (p. 82)
Clarendon Press. Oxford, England. 1993

Hocking, W. E.   1873–1966
American philosopher

We are scientific people and we want our students to feel the enthusiasm and promise of the scientific method. We want them to feel the moral quality of exact technique, as exact as the subject matter permits. We want them to feel that science is a spiritual experience.

In Lloyd William Taylor
Physics: The Pioneer Science (Volume 1)
Chapter 6 (p. 63
Houghton Mifflin Company. Boston, Massachusetts, USA. 1941

Huxley, Thomas Henry   1825–95
English biologist

…the scientific spirit is of more value than its products, and irrationally held truths may be more harmful than reasoned errors. Now the essence of the scientific spirit is criticism. It tells us that whenever a doctrine claims our assent we should reply, "Take it if you can compel it." The struggle for existence holds as much in the intellectual as in the physical world. A theory is a species of thinking, and its right to exist is coextensive with its power of resisting extinction by its rivals.

Collected Essays (Volume 2)
Darwiniana
The Coming of Age of "The Origin of Species" (p. 229)
Macmillan & Company Ltd. London, England. 1904

Millikan, Robert Andrews   1868–1953
American physicist

The God of science is the Spirit of rational order, and of orderly development. Atheism as I understand it is the denial of the existence of the spirit. Nothing could therefore be more antagonistic to the whole spirit of science.

Evolution in Science and Religion

Chapter III (p. 88)
Yale University Press. New Haven, Connecticut, USA. 1927

Peirce, Charles Sanders   1839–1914
American scientist, logician, and philosopher

...the scientific spirit requires a man to be at all times ready to dump his whole cartload of beliefs, the moment experience is against them.
In Justus Buchler (ed.)
Philosophical Writings of Peirce
Chapter 4 (pp. 46–47)
Dover Publications, Inc. New York, New York, USA. 1955

Rabi, Isidor Isaac   1898–1988
Austrian-born American physicist

The essence of the scientific spirit is to use the past only as a springboard to the future.
In A.A. Warner, Dean Morse, and T.E. Cooney (eds.)
The Environment of Change
The Revolution in Science (p. 47)
Columbia University Press. New York, New York, USA. 1969

Weisskopf, Victor Frederick   1908–2002
Austrian-American physicist

Science, of course, is not the only way of giving sense to our lives. Art does it; so does religion. But when this sense is missing, that's when spiritual pollution is present, when people don't know why they are here. We can have leisure time diversions, of course. But until we learn to fill the vacuum in our minds with content — with meaning, with sense — we will never find solutions to our problems.
The Privilege of Being a Physicist
Chapter 13 (p. 107)
W.H. Freeman & Company. New York, New York, USA. 1989

SCIENTIFIC TRENDS

Cooper, Leon   1930–
American physicist

I like to say sometimes that scientific fashion is like fashion in men's and women's clothes.... One year the ties are wide; the next year they're narrow. One year the skirts are high; the next year they're low. And if everyone is wearing a short skirt, you're just hopelessly out of fashion if you're wearing a long skirt. That's the way it sometimes seems with science. You want to be in the middle of what everyone is talking about; you want to be in the mainstream. And the next year it might be something completely different.
In George Johnson
In the Palaces of Memory: How We Build the Worlds Inside Our Heads
A Model of Memory (p. 149)
Alfred A. Knopf. New York, New York, USA. 1991

SCIENTIFIC TRUTH

Agassiz, Jean Louis Rodolphe   1807–73
Swiss-born American naturalist, geologist, and teacher

...the time has come when scientific truth must cease to be the property of the few, when it must be woven into the common life of the world; for we have reached the point where the results of science touch the very problem of existence, and all men listen for the solving of that mystery.
Methods of Study in Natural History
Chapter IV (p. 42)
Ticknor & Fields. Boston, Massachusetts, USA. 1863

Black, Max   1909–88
Anglo-American philosopher

Scientists can never hope to be in a position to know the truth, nor would they have any means of recognizing it if it came into their possession.
Critical Thinking: An Introduction to Logic and Scientific Method
Chapter 19 (p. 396)
Prentice-Hall. New York, New York, USA. 1952

Broad, William   1951–
Science writer
Wade, Nicholas
British-born scientific writer

Like any other profession, science is ridden with clannishness and clubbiness. This would be in no way surprising, except that scientists deny it to be the case. The pursuit of scientific truth is held to be a universal quest that recognizes neither national boundaries nor the barriers of race, creed or class. In fact, researchers tend to organize themselves into clusters of overlapping clubs.
Betrayers of the Truth
Chapter 9 (p. 180)
Simon & Schuster. New York, New York, USA. 1982

Chargaff, Erwin   1905–2002
Austrian biochemist

The initial incommunicability of truth, scientific or otherwise, shows that we think in groves, and that it is painful for us to be torn away from the womblike security of accepted concepts.
Heraclitean Fire: Sketches from a Life Before Nature
Part II
The Exquisiteness of Minute Differences (p. 86)
Rockefeller University Press. New York, New York, USA. 1978

Crichton, Michael   1942–
American novelist

Scientists have an elaborate line of bullshit about how they are seeking to know the truth about nature.
Jurassic Park
Aviary (p. 284)
Alfred A. Knopf. New York, New York, USA. 1990

Dunne, Finley Peter   1867–1936
American journalist and humorist

There's always wan encouragin' thing about th' sad scientific facts that come out ivry week in th' pa-apers. They're usually not thrue.
Mr. Dooley on Making a Will and Other Necessary Evils
On the Descent of Man (p. 90)
Charles Scribner's Sons. New York, New York, USA. 1919

Eddington, Sir Arthur Stanley   1882–1944
English astronomer, physicist, and mathematician

Amid all our faulty attempts at expression the kernal of scientific truth steadily grows; and of this truth it may be said — The more it changes, the more it remains the same.
The Nature of the Physical World
Conclusion (p. 353)
The Macmillan Company. New York, New York, USA. 1930

Einstein, Albert   1879–1955
German-born physicist

It is difficult even to attach a precise meaning to the term "scientific truth." Thus the meaning of the word "truth" varies according to whether we deal with a fact of experience, a mathematical proposition, or a scientific theory. "Religious truth" conveys nothing clear to me at all.
Ideas and Opinions
On Scientific Truth (p. 261)
Crown Publishers, Inc. New York, New York, USA. 1954

Feynman, Richard P.   1918–88
American theoretical physicist

No government has the right to decide on the truth of scientific principles, nor to prescribe in any way the character of the questions investigated.
The Meaning of It All: Thoughts of a Citizen Scientist
Chapter II (p. 57)
Perseus Books. Reading, Massachusetts, USA. 1998

Geikie, Sir Archibald   1835–1924
English geologist

In scientific as in other mundane questions there may often be two sides, and the truth may ultimately be found not to lie wholly with either.
Annual Report of the Board of Regents of the Smithsonian Institution, 1892
Geological Change and Time (p. 125)
Government Printing Office. Washington, D.C. 1893

Gesenius, Wilhelm   1786–1842
German orientalist and biblical critic

Unwearied personal observation and an impartial examination of the researches of others; the grateful admission and adoption of every real advance and illustration of science; but also a manly foresight and caution, which does not with eager levity adopt every novelty thrown out in haste and from the love of innovation, all these must go hand in hand, wherever scientific truth is to be successfully promoted.
Hebrew Grammar
Preface (p. 7)
Gould, Kendall & Lincoln. Boston, Massachusetts, USA. 1834

Gregory, Sir Richard Arman   1864–1952
British science writer and journalist

It is necessary to believe in the holiness of scientific work in order to preserve to the end; for without the encouragement which such belief gives, many investigators would fall by the wayside.
Discovery; or, The Spirit and Service of Science
Chapter I (p. 12–13)
Macmillan & Company Ltd. London, England. 1918

Scientific truth is not won by prayer and fasting, but by patient observation and persistent inquiry.
Discovery; or, The Spirit and Service of Science
Chapter I (p. 12)
Macmillan & Company Ltd. London, England. 1918

Hawkins, Michael   1942–
British astrophysicist

"Scientific truths" is simply another way of saying "the fittest, most beautiful, and most elegant survivors of scientific debate and testing."
Hunting Down the Universe: The Missing Mass, Primordial Black Holes, and Other Dark Matters
Chapter 1 (p. 6)
Addison-Wesley Publishing Company. Reading, Massachusetts, USA. 1997

Hoagland, Hudson   1899–1982
American physiologist

In a scientific oriented society the quest for truth is the important thing, even though we know that ultimate, final truth with a capital T is not to be found.
Science and the New Humanism
Science, Volume 143, Number 3062, 10 January 1964 (p. 112)

Inscription

The works of those who have stood the test of ages have a claim to that respect and veneration to which no modern can pretend.
On the dome of the National Gallery

Jeffers, Robinson   1887–1962
American poet

…they work alongside the truth
Never touching it; their equations are false
But the things work.
In Tim Hunt (ed.)
The Collected Poetry of Robinson Jeffers (Volume 3)
The Mathematicians and Physics Men (p. 459) Stanford University Press. Stanford, California, USA. 1988

Lee, Oliver Justin   1881–1964
American astronomer

The truth to a scientist is not the vague metaphysical concept about which philosophers talk and write so much and know so little. To him truth is that body of statements and conclusions about any set of features and phenomena in nature which represent most accurately all the best observations he can make and which conform most closely to all findings in adjacent or related phases and fields of investigations. He may and does often wonder what the so-called ultimate truth may be, but he does not worry about it. He knows that a priori pure thinking will never reveal it so far as knowledge of the physical universe is concerned, and that observation and deduction alone in the manner of science will ever do it.
Measuring Our Universe: From the Inner Atom to Outer Space
Chapter 14 (pp. 149–150)
The Ronald Press Company. New York, New York, USA. 1950

Maxwell, James Clerk   1831–79
Scottish physicist

For the sake of these different types, scientific truth should be presented in different forms, and should be regarded as equally scientific, whether it appears in the robust form of vivid colouring of a physical illustration, or in the tenuity and paleness of a symbolical expression.
The Collected Papers of James Clerk Maxwell (Volume 2)
Chapter XLI, Address to the Mathematical and Physical Sections of the British Association, September 15, 1870 (p. 220)
At the University Press. Cambridge, England. No data

Mendeléeff, Maria   ca. 1800–ca. 1850
Siberian factory manager and mother of Russian chemist Dmitri Mendeléeff

Refrain from illusions, insist on work, and not on words, patiently search divine and scientific truth.
In Benjamin Harrow
Eminent Chemists of Our Time
Dmitri Ivanowitch Mendeléeff (p. 22)
D. Van Nostrand Company, Inc. New York, New York, USA. 1927

Poincaré, Henri   1854–1912
French mathematician and theoretical astronomer

For a superficial observer, scientific truth is beyond the possibility of doubt; the logic of science is infallible, and if the scientists are sometimes mistaken, this is only from their mistaking its rules.
The Foundations of Science
Science and Hypothesis, Introduction (p. 27)
The Science Press. New York, New York, USA. 1913

Raymo, Chet   1936–
American physicist and science writer

Science is not a smorgasbord of truths from which we can pick and choose.
The Virgin and the Mousetrap: Essays in Search of the Soul of Science

Chapter 16 (p. 144)
The Viking Press. New York, New York, USA. 1991

Scientific truths are tentative and partial, and subject to continual revision and refinement, but as we tinker with truth in science — amending here, augmenting there — we always keep our ear attuned to the timbre of the web.
The Virgin and the Mousetrap: Essays in Search of the Soul of Science
Chapter 16 (p. 145)
The Viking Press. New York, New York, USA. 1991

Renan, Ernest   1823–92
French philosopher and Orientalist

Orthodox people have as a rule very little scientific honesty. They do not investigate, they try to prove, and this must necessarily be so. The result has been given to them beforehand; this result is true, undoubtedly true. Science has no business with it, science which starts from doubt without knowing whither it is going, and gives itself up bound hand and foot to criticism which leads it wheresoever it lists.
The Future of Science
Chapter III (p. 33)
Roberts Brothers. Boston, Massachusetts, USA. 1893

Serge, Corrado
No biographical data available

Many times a scientific truth is placed as it were on a lofty peak, and to reach it we have at our disposal at first only dark paths along perilous slopes whence it is easy to fall into the abysses where dwells error; only after we have reached the peak by these paths is it possible to lay out safe roads which lead there without peril. Thus it has frequently happened that the first way of obtaining a result has not been quite satisfactory, and that only afterwards did the science succeed in completing the demonstration.
On Some Tendencies in Geometric Investigations
Bulletin of the American Mathematical Society, 2nd Series, Volume 10, June 1904 (pp. 453–454)

Shaw, George Bernard   1856–1950
Irish comic dramatist and literary critic

In the Middle Ages people believed that the earth was flat, for which they at least had the evidence of their senses: we believe it to be round, not because as many as one percent of us could give the physical reasons for so quaint a belief, but because modern science has convinced us that nothing that is obvious is true, and that everything that is magical, improbable, extraordinary, gigantic, microscopic, heartless, or outrageous is scientific.
Man and the Gods: Three Tragedies
Saint Joan, Preface, The Real Joan Is Not Marvellous for Us (p. 132)
Harcourt, Brace & World, Inc. New York, New York, USA. 1964

Spencer, Herbert   1820–1903
English social philosopher

Scientific truths, of whatever order, are reached by eliminating perturbing or conflicting factors, and recognizing only fundamental factors.
The Data of Ethics
Chapter XV, Section 104 (p. 311)
William & Norgate. London, England. 1907

Thomas, Lewis   1913–93
American physician and biologist

The only solid piece of scientific truth about which I feel totally confident is that we are profoundly ignorant about nature.
The Medusa and the Snail: More Notes of a Biology Watcher
The Hazard of Science (p. 73)
The Viking Press. New York, New York, USA. 1979

Thoreau, Henry David   1817–62
American essayist, poet, and practical philosopher

The eye which can appreciate the naked and absolute beauty of a scientific truth is far more rare than that which is attracted by a moral one. Few detect the morality in the former, or the science in the latter.
The Writings of Henry David Thoreau (Volume 1)
A Week on the Concord and Merrimack Rivers
Friday (p. 476)
Houghton Mifflin Company. Boston, Massachusetts, USA. 1893

Weaver, Warren   1894–1978
American mathematician

…[one] finds unresolved and apparently unresolvable disagreement among scientists concerning the relationship of scientific thought to reality — and concerning the nature of reality itself…that the explanations of science have utility, but that they do in sober fact not explain. He finds that the old external appearance of inevitability completely vanished, for he discovers a charming capriciousness in all the individual events. He finds that logic, so generally supposed to be infallible and unassailable, is in fact shaky and incomplete. He finds that the whole concept of objective truth is a will-o-the-wisp.
The Imperfections of Science
American Scientist, Volume 49, 1961 (pp. 99–113)

Wells, H. G. (Herbert George)   1866–1946
English novelist, historian, and sociologist

Scientific truth is the remotest of mistresses; She hides in strange places, she is attained by tortuous and laborious roads, but she is always there! Win to her and she will not fail you; she is yours and mankind's forever. She is reality, the one reality I have found in this strange disorder of existence…
Tono-Bungay
Book the Third, Chapter the Third, I (p. 324)
Duffield & Company. New York, New York, USA. 1921

SCIENTIFIC WORK

Einstein, Albert   1879–1955
German-born physicist

To be sure, it is not the fruits of scientific research that elevate a man and enrich his nature, but the urge to understand, the intellectual work, creative or receptive.
Ideas and Opinions
Good and Evil (p. 12)
Crown Publishers, Inc. New York, New York, USA. 1954

Fisher, Sir Ronald Aylmer   1890–1962
English statistician and geneticist

The concept that the scientific worker can regard himself as an inert item in a vast co-operative concern working according to accepted rules is encouraged by directing attention away from his duty to form correct scientific conclusions, to summarize them and to communicate them to his scientific colleagues, and by stressing his supposed duty mechanically to make a succession of automatic "decisions"…
Statistical Methods and Scientific Inference
Chapter IV (p. 101)
Hafner Publishing Company. New York, New York, USA. 1959

Heisenberg, Werner Karl   1901–76
German physicist and philosopher

Take from your scientific work a serious and incorruptible method of thought, help to spread it, because no understanding is possible without it. Revere those things beyond science which really matter and about which it is so difficult to speak.
Philosophic Problems of Nuclear Science
Chapter 8 (p. 128)
Faber & Faber Ltd. London, England. 1952

Huxley, Thomas Henry   1825–95
English biologist

The only people, scientific or other, who never make mistakes are those who do nothing.
Collected Essays (Volume 5)
Science and Christian Traditions
An Episcopal Trilogy (p. 156)
Macmillan & Company Ltd. London, England. 1904

Lorand, Arnold
American physician and prolongevity advocate

…we have often observed in persons whose lives have been devoted to serious scientific work, which has entirely absorbed them, a total absence of sexual desire for a long time, and even impotence.
Old Age Deferred
Chapter XLIX (p. 399)
F.A. Davis Company, Publishers. Philadelphia, Pennsylvania, USA. 1911

Payne-Gaposchkin, Cecelia   1900–79
British-American astronomer

Do not undertake a scientific career in quest of fame or money. There are easier and better ways to reach them. Undertake it only if nothing else will satisfy you; for nothing else is probably what you will receive. Your reward will be the widening of the horizon as you climb. And if you achieve that reward you will ask no other.
Cecilia Payne-Gaposchkin: An Autobiography and Other Recollections
Chapter 22 (p. 227)
Cambridge University Press. New York, New York, USA. 1984

Rosenthal-Schneider, Ilse   1891–1990
German physicist and author of history and philosophy of science

The deep satisfaction found in scientific work, akin to the delight derived from genuine art, is one of the fundamental human emotions which is highly intensified by personal contact with the creative mind.
In Paul Arthur Schlipp (ed.)
Albert Einstein: Philosopher-Scientist
Presuppositions and Anticipations in Einstein's Physics (p. 145)
The Library of Living Philosophers, Inc. Evanston, Illinois, USA. 1949

Snow, Charles Percy   1905–80
English novelist and scientist

Scientific work…has a value of its own, whether you're liking it or not. It's — there. It's permanent. It's work which is always going to last. It's a real creation.
The Search
Part IV, Chapter II, Section II (p. 326)
The Bobbs-Merrill Company. Indianapolis, Indiana, USA. 1935

Weber, Max   1864–1920
German founder of modern sociology and economic thinker

In science, each of us knows that what he has accomplished will be antiquated in ten, twenty, fifty years. That is the fate to which science is subjected; it is the very meaning of scientific work, to which it is devoted in a quite specific sense, as compared with other spheres of culture…. Every scientific "fulfillment" raises new "questions"; it asks to be surpassed and outdated. Whoever wishes to serve science has to resign himself to this fact…. We cannot work without hoping that others will advance further than we have.
In H.H. Gerth and C. Wright Mills (eds.)
From Max Weber: Essays in Sociology
Science as a Vocation (p. 138)
Oxford University Press, Inc. New York, New York, USA. 1970

SCIENTIST

Adams, Douglas   1952–2001
English author, comic radio dramatist, and musician

You can't possibly be a scientist if you mind people thinking that you're a fool.
The Ultimate Hitchhiker's Guide to the Galaxy
So Long And Thanks For All The Fish
Chapter 31 (p. 587)
The Ballantine Book Company. New York, New York, USA. 2002

Agnew, Ralph Palmer
American mathematician

Scientists, like professional golfers and piano players, should sometimes concentrate upon a task until they can perform it with professional skill.
Differential Equations
Chapter 1, Problem 1.49 (p. 16)
McGraw-Hill Book Company, Inc. New York, New York, USA. 1972

Appleton, Sir Edward   1892–1965
English physicist

It seems to me that we must recognize that the proper use of science is one of the most important challenges of the present day. And here I think the scientist has a twofold mission, not only of extending the frontiers of knowledge but also of interpreting his results to his fellow-men.
Les Prix Nobel. The Nobel Prizes in 1947
Nobel banquet speech for award received in 1947
Nobel Foundation. Stockholm, Sweden. 1948

Appleyard, Bryan   1951–
Author and journalist

Scientists inevitably take on the mantle of the wizards, sorcerers and witch doctors. Their miracle cures are our spells, their experiments our rituals.
Understanding the Present: Science and the Soul of Modern Man
Chapter 1 (p. 9)
Doubleday. New York, New York, USA. 1992

Artuad, Antonin   1896–1948
French poet, actor, and director

But how is one to make a scientist understand that there is something unalterably deranged about differential calculus, quantum theory, or the obscene and so inanely liturgical ordeals of the precession of the equinoxes — ….
In Susan Sontag
Selected Writings
Part 33, Van Gogh, the Man Suicided by Society (p. 497)

Auden, W. H.   1907–72
English-born poet

The true men of action in our time, those who transform the world, are not the politicians and statesmen, but the scientists.
The Dyer's Hand
Part II, The Poet and The City (p. 81)
Random House. New York, New York, USA. 1962

When I find myself in the company of scientists, I feel like a shabby curate who has strayed by mistake into a drawing room full of dukes.
The Dyer's Hand

Part II, The Poet and The City (p. 81)
Random House. New York, New York, USA. 1962

Bacon, Sir Francis   1561–1626
English lawyer, statesman, and essayist

Those who have treated of the sciences have been either empirics or dogmatical. The former like ants only heap up and use their store, the latter like spiders spin out their own webs. The bee, a mean between both, extracts matter from the flowers of the garden and the field, but works and fashions it by its own efforts. The true labor of philosophy resembles hers, for it neither relies entirely nor principally on the powers of the mind, nor yet lays up in the memory of matter afforded by the experiments of natural history and mechanics in its raw state, but changes and works it in the understanding.
In *Great Books of the Western World* (Volume 30)
Novum Organum
First Book, Aphorism 95 (p. 126)
Encyclopædia Britannica, Inc. Chicago, Illinois, USA. 1952

Baker, Russell   1925–
American writer and journalist

Two leading Congressional scientists, Senator Helms and Representative Hyde, have been doing pioneering research on the nature of life. This has produced the Helms–Hyde theory which states that scientific fact can be established by a majority vote of the United States Congress.
The Rescue of Miss Yaskel and Other Pipe Dreams
Congdom & Weed. New York, New York, USA. 1983

Barr, Amelia Edith Huddleston   1831–1919
Anglo-American novelist

Whatever the scientists may say, if we take the supernatural out of life, we leave only the unnatural.
All the Days of My Life
Chapter 26 (p. 477)
Arno Press. New York, New York, USA. 1980

Beveridge, William Ian Beardmore   1908–
Australian zoologist

The scientist who has an independent mind and is able to judge the evidence on its merits rather than in light of prevailing conceptions is the one most likely to be able to realize the potentialities in something really new.
The Art of Scientific Investigation
Chapter Three (p. 35)
W.W. Norton & Company, Inc. New York, New York, USA. 1957

Bohr, Niels Henrik David   1886–1962
Danish physicist

Every sentence I utter must be understood not as an affirmation, but as a question.
New York Times Book Review, October 20, 1957

Born, Max   1882–1970
German-born English physicist

It seems to me that the scientists who led the way to the atomic bomb were extremely skillful and ingenious, but not wise men. They delivered the fruits of their discoveries unconditionally into the hands of politicians and soldiers; thus they lost their moral innocence and their intellectual freedom…
The Restless Universe
Postscript (p. 280)
Dover Publications, Inc. New York, New York, USA. 1951

Brain, Lord Walter Russell   1895–1966
British neurologist

Scientists…meet one another to exchange ideas, to promote their own particular branch of science, or science in general, or because they are aware of its social implications. Nevertheless, such collective activities…play a small part in their lives. Scientists, though they must always be aware of the work of their fellows in their own fields, are essentially individualists; and the body of knowledge to which they are contributing is an impersonal one. Apart from contributing to it, they have no collective consciousness, interest, or aim.
Science and Antiscience
Science, Volume 148, Number 3667, April 1965 (p. 193)

Brewster, Edwin Tenney   1866–1960
Educator

For scientific people are after all precisely like the rest of us, and can no more resist — most of them — the urge to speculate where they can not prove than other men can.
This Puzzling Planet
Chapter XIX (p. 301)
The Bobbs-Merrill Company, Indianapolis, Indiana. 1928

Bronowski, Jacob   1908–74
Polish-born British mathematician and polymath

Dissent is the native activity of the scientist, and it has got him into a good deal of trouble in the last years. But if that is cut off, what is left will not be a scientist. And I doubt whether it will be a man.
Science and Human Values
The Sense of Human Dignity (p. 61)
Harper & Row, Publishers. New York, New York, USA. 1965

The society of scientists must be a democracy. It can keep alive and grow only by a constant tension between dissent and respect; between independence from the view of others, and tolerance from them.
Science and Human Values
The Sense of Human Dignity (pp. 62–63)
Harper & Row, Publishers. New York, New York, USA. 1965

The most remarkable discovery made by scientists is science itself.

A Sense of the Future: Essays in Natural Philosophy
Chapter 2 (p. 6)
The MIT Press. Cambridge, Massachusetts, USA. 1977

It is important that students bring a certain ragamuffin, barefoot irreverence to their studies; they are not here to worship what is known, but to question it.
The Ascent of Man
Chapter 11 (p. 360)
Little, Brown & Company. Boston, Massachusetts, USA. 1973

Buchanan, Scott   1895–1968
American educator and philosopher

The scientist is the contemporary monk copyist, writing over old literature on the palimpsest of experience, triumphantly announcing his faithfulness and accuracy in transferring the copy.
Poetry and Mathematics
Chapter I
The University of Chicago Press. Chicago, Illinois, USA. 1975

Buck, Pearl S.   1892–1973
American author

No one really understood music unless he was a scientist, her father had declared, and not just a scientist, either, oh, no, only the real ones, the theoreticians, whose language mathematics. She had not understood mathematics until he had explained to her that it was the symbolic language of relationships. "And relationships," he had told her, "contained the essential meaning of life."
The Goddess Abides
Part I
John Day Company. New York, New York, USA. 1972

Burroughs, William S.   1914–97
American writer

Too many scientists seem to be ignorant of the most rudimentary spiritual concepts. And they tend to be suspicious, bristly, paranoid-type people with huge egos they push around like some elephantiasis victim with his distended testicles in a wheelbarrow terrified no doubt that some skulking ingrate of a clone student will sneak into his very brain and steal his genius work.
The Adding Machine: Selected Essays
Immortality (p. 132)
Seaver Books. New York, New York, USA. 1986

Calder, Alexander   1898–1976
American sculptor and inventor of the mobile

Scientists leave their discoveries, like foundlings, on the doorstep of society, while the stepparents do not know how to bring them up.
In Alan J. Friedman and Carol C. Donley
Einstein as Myth and Muse
Chapter 1 (p. 7)
Cambridge University Press. Cambridge, England. 1985

Chargaff, Erwin   1905–2002
Austrian biochemist

Great scientists are particularly worth listening to when they speak about something of which they know little; in their own specialty they are usually great and dull.
Heraclitean Fire: Sketches from a Life before Nature
Part II
The Hereditary Code-Script (p. 85)
Rockefeller University Press. New York, New York, USA. 1978

…outside his own ever-narrowing field of specialization, a scientist is a layman. What members of an academy of science have in common is a certain form of semiparasitic living.
Bitter Fruits from the Tree of Knowledge
Perspectives in Biology and Medicine, Section III, Volume 16, Number 4, Summer 1973 (p. 492)

A scientific autobiography belongs to a most awkward literary genre. If the difficulties facing a man trying to record his life are great — and few have overcome them successfully — they are compounded in the case of scientists, of whom many lead monotonous and uneventful lives and who, besides, often do not know how to write…
Book Review of *The Double Helix*
Science, Volume 159, Number 3822, 29 March 1968 (p. 1448)

Chesterton, G. K. (Gilbert Keith)   1874–1936
English author

Apparently a scientist is a man who surveys all the sciences, without any particular study of them, and then gives expression to his own moral principles or prejudices.
All Is Grist: A Book of Essays
On Mr. Mencken and Fundamentalism (p. 50)
Methuen & Company Ltd. London, England. 1931

Clarke, Arthur C.   1917–
English science and science fiction writer

When a distinguished but elderly scientist states that something is possible, he is almost certainly right. When he states that something is impossible, he is very probably wrong.
Profiles of the Future: An Inquiry into the Limits of the Possible
Chapter 2 (p. 14)
Harper & Row, Publishers. New York, New York, USA. 1973

…scientists of over fifty are good for nothing except board meetings and should at all costs be kept out of the laboratory!
Profiles of the Future: An Inquiry into the Limits of the Possible
Chapter 2 (pp. 14–15)
Harper & Row, Publishers. New York, New York, USA. 1973

Conant, James Bryant   1893–1978
American educator and scientist

…scientists today represent the progeny of one line of descent who migrated, so to speak, some centuries ago into certain fields which were ripe for cultivation. Once science had become self-propagating, those who till these fields have had a relatively easy time keeping up the tradition of their forebears.
Science and Common Sense
Chapter One (p. 13)
Yale University Press. New Haven, Connecticut, USA. 1951

Cornforth, John W.   1917–2004
English organic chemist

Scientists do not believe; they check.
Scientists as Citizens
Australian Journal of Chemistry, Volume 46, 1993 (p. 266)

Cousteau, Jacques-Yves   1910–77
French naval officer and ocean explorer

What is a scientist after all? It is a curious man looking through a keyhole, the keyhole of nature, trying to know what's going on.
Christian Science Monitor, 21 July 1971

Cramer, F.
No biographical data available

In the long run it pays the scientist to be honest, not only by not making false statements, but by giving full expression to facts that are opposed to his views. Moral slovenless is visited with far severer penalties in the scientific than in the business world.
In W.I.B. Beveridge
The Art of Scientific Investigation
Chapter Eleven (p. 142)
W.W. Norton & Company, Inc. New York, New York, USA.1957

Crichton, Michael   1942–
American novelist

I sometimes think scientists really don't notice that their colleagues have flaws. But in my experience scientists are very human people: which means that some are troubled, deceitful, petty or vain.
Science Views Media, January 25, 1999

Cronenberg, David   1943–
Canadian film director

…everybody's a mad scientist, and life is their lab. We're all trying to experiment to find a way to live, to solve problems, to fend off madness and chaos.
In Chris Rodley
Cronenberg on Cronenberg
Chapter 1 (p. 7)
Faber & Faber Ltd. London, England. 1992

de Jouvenel, Bertrand   1903–87
French man of letters

No one can become a scientist who is not driven by a primary urge for discovery, who is not the ardent suitor of a hidden beauty. Somewhat romantically, scientists can be likened to a company of knights dispersed in search of Sleeping Princesses, all of whom are more or less distantly related. The spirit of the quest is essential to the making of a scientist, and forms a fundamental bond between scientists.
The Logic of Personal Knowledge
The Republic of Science
The Free Press. Glencoe, Illinois, USA. 1961

de Madariaga, Salvador   1886–1978
Spanish writer and statesman

There are two kinds of scientists: they were once described…as the "why" and the "how." The how-scientist is mainly interested in the way things happen; the why-scientist seeks to find out the cause of things. The first is more of a technician; the second, more of a philosopher. The first is more of a man of talent; the second, more of a man of genius.
Essays with a Purpose
Science and Freedom (p. 43)
Hollis & Carter. London, England. 1954

Devine, Betsy
No biographical data available
Cohen, Joel E.
No biographical data available

Scientists are funny people. Not just the great ones who think they've discovered the secret of life or of the brain or of the common cold. Even ordinary day-to-day scientists are funny, because they all think that the world makes sense! Most people know better.
Absolute Zero Gravity: Science Jokes, Quotes, and Anecdotes
Fireside/Simon & Shuster. New York, New York, USA.

Dr. Kemp
Fictional character

Straightforward scientists have no need for barred doors and drawn blinds.
The Invisible Man
Film (1933)

du Noüy, Pierre Lecomte   1883–1947
French scientist

The scientist with imagination is the pioneer of progress.
The Road to Reason
Chapter 3 (p. 81)
Longmans, Green & Company. London, England. 1949

Dubos, René Jules   1901–82
French-born American microbiologist and environmentalist

Scientists, like artists, unavoidably reflect the characteristics of the civilization and the time in which they live.

In this sense, they are "enchained"…by the inexorable logic of their time and their work.
Louis Pasteur: Free Lance of Science
Introduction (p. xxxviii)
Little, Brown & Company. Boston, Massachusetts, USA. 1950

…like other men, scientists become deaf and blind to any argument or evidence that does not fit into the thought pattern which circumstances have led them to follow.
Louis Pasteur: Free Lance of Science
Chapter VII (p. 197)
Little, Brown & Company. Boston, Massachusetts, USA. 1950

Dyson, Freeman J.   1923–
American physicist and educator

When something ceases to by mysterious it ceases to be of absorbing concern to scientists. Almost all the things scientists think and dream about are mysterious.
Infinite in All Directions
Part One, Chapter Two (p. 14)
HarperCollins Publisher, Inc. New York, New York, USA. 1988

Egler, Frank E.   1911–96
American botanist and ecologist

Scientists are only men, and are subject to all the foibles of their kind. They have the same drives for freedom, security, certainty, image and status as have other men. …the same attraction for the known the familiar and the comfortable, and will cling to old and sterile ideas like a broody hen sitting on boiled eggs. Like those others, there is a lunatic fringe, and a reasonable quota of social misfits, small-pool big-frogs, megalomaniacs, prima donnas, nymphomaniacs, gold diggers, entrepreneurs, prophets and devout disciples.
The Way of Science
The Nature of Science (p. 1)
Hafner Publishing Company. New York, New York, USA. 1970

Einstein, Albert   1879–1955
German-born physicist

The eyes of the scientist are directed upon those phenomena which are accessible to observation, upon their apperception and conceptual formulation.
Concepts of Space: The History of Theories of Space in Physics
Preface (p. xi)
Harvard University Press. Cambridge, Massachusetts, USA. 1954

For the scientist, there is only "being," but no wishing, no valuing, no good, no evil — in short, no goal. As long as we remain within the realm of science proper, we can never encounter a sentence of the type: "Thou shalt not lie."
In Philipp Frank
Relativity — A Richer Truth
The Laws of Science and the Laws of Ethics (p. 9)
Jonathan Cape. London, England. 1951

…the scientist finds his reward in what Henri Poincaré calls the joy of comprehension, and not in the possibilities of application to which any discovery may lead.
In Max Planck
Where Is Science Going?
Epilogue (p. 211)
W.W. Norton & Company, Inc. New York, New York, USA. 1932

Emelyanov, A. S.
No biographical data available

A scientist cannot be a "pure" mathematician, biophysicist or sociologist for he cannot remain indifferent to the fruits of his work, to whether they will be useful or harmful to mankind. An indifferent attitude as to whether people will be better or worse off as a result of scientific achievement is cynicism, if not a crime.
In E.H.S. BurhopIn Maurice Goldsmith and Alan Mackay (eds.)
Society and Science
Scientist and Public Affairs (p. 31)
Simon & Schuster. New York, New York, USA. 1965

England, Terry
No biographical data available

…if three scientists ever agree completely on anything, it's a cult.
Rewind (p. 71)
Avon Books. New York, New York, USA. 1997

Eysenck, Hans Jurgen   1916–97
Founder of theory of personality

Scientists, especially when they leave the particular field in which they have specialized, are just as ordinary, pigheaded and unreasonable as anybody else.
Continuum
OMNI Magazine, Volume 2, December 1979 (p. 49)

Faulkner, William   1897–1962
American novelist and short story writer

Our privacy…has been slowly and steadily and increasingly invaded until now our very dream of civilization is in danger. Who will save us but the scientist and the humanitarian? Yes, the humanitarian in science, and the scientist in the humanity of man.
Quoted in Warren Weaver
Science and People
Science, Volume 122, Number 3183, December 30, 1955 (p. 1259)

Feibleman, James K.   1904–1987
American philosopher

It is not the business of scientists to investigate just what the business of science is.
Pure Science, Applied Science, Technology, Engineering: An Attempt at Definitions
Technology and Culture, Volume II, Number 4, Fall 1961 (p. 305)

Feyerabend, Paul K.   1924–94
Austrian-born American philosopher of science

Scientists are sculptors of reality — but sculptors in a special sense. They do not merely act causally upon the world (though they do that, too, and they have to if they want to "discover" new entities); they also create semantic conditions engendering strong inferences from known effects to novel projections and, conversely, from the projections to testable effects.

Realism and the Historicity of Knowledge
The Journal of Philosophy, Volume LXXXVI, Number 8, 1989
(pp. 404–405)

Feynman, Richard P.   1918–88
American theoretical physicist

It is our responsibility as scientists, knowing the great progress which comes from a satisfactory philosophy of ignorance, the great progress which is the fruit of freedom of thought, to proclaim the value of this freedom; to teach how doubt is not to be feared but welcomed and discussed; and to demand this freedom as our duty to all coming generations.

What Do You Care What Other People Think?
The Value of Science (p. 248)
W.W. Norton & Company, Inc. New York, New York, USA.1988

I would like to point out that people are not honest. Scientists are not honest at all, either. It's useless. Nobody's honest. Scientists are not honest. And people usually believe that they are. That makes it worse. By honest I don't mean that you only tell what's true. But you make clear the entire situation. You make clear all the information that is required for somebody else who is intelligent to make up their mind.

The Meaning of It All: Thoughts of a Citizen Scientist
Chapter III (p. 106)
Perseus Books. Reading, Massachusetts, USA. 1998

Finniston, Sir Monty   1912–91
British industrialist

You mustn't think scientists are stupid.

Saying of the Week
Observer, 16 January 1983

Fitzgerald, Penelope   1916–2000
English novelist and biographer

If they don't depend on true evidence, scientists are no better than gossips.

The Gate of Angels
Chapter 3 (p. 24)
Doubleday & Company. New York, New York, USA. 1992

Foster, Alan Dean   1946–
American science fiction writer

A man of science is helpless by himself, but two of them constitute an entity capable of ignoring starvation, freezing, and prospects of imminent death just by chatting about some item of mutual interest.

Icerigger (p. 116)
Ballantine Books. New York, New York, USA. 1974

Fox, Robin   1934–
English anthropologist, poet, and essayist

The real poet, like any artist, tries all the time to see the general in the particular. In this he is no different from the scientist. They are siblings under the skin.

In Paul R. Gross, Norman Levitt, and Martin W. Lewis (eds.)
The Flight from Science and Reason
State of the Art/Science in Anthropology (p. 343)
New York Academy of Sciences. New York, New York, USA. 1996

Scientists, being children of Adam, can be fools and charlatans or even just blind and biased. Indeed, my own despair at the "academic/scientific enterprise" which raised the initial question is a despair over the inevitability of human frailty, not over the ideals of scientific discovery.

In Paul R. Gross, Norman Levitt, and Martin W. Lewis (eds.)
The Flight from Science and Reason
State of the Art/Science in Anthropology (p. 329)
New York Academy of Sciences. New York, New York, USA. 1996

Freud, Sigmund   1856–1939
Austrian neurologist and co-founder of psychoanalysis

I am not really a man of science, not an observer, not an experimenter, and not a thinker. I am nothing but by temperament a conquistador — an adventurer if you want to translate the word.

Mighty Minds (1900 letter to Fliess)
New Scientist, 4 April 1998 (p. 11)

Fuller, R. Buckminster   1895–1983
American engineer and architect

…public journals, assumedly bespeaking public opinion, [say] scientists "wrest order out of chaos." But the scientists who have made the great discoveries have been trying their best to tell the public that…they have never found chaos to be anything other than the superficial confusion of innately a priori human ignorance at birth — an ignorance that is often burdened by the biases of others to remain gropingly unenlightened throughout its life.

In L.L. Larison Cudmore
The Center of Life: A Natural History of the Cell (p. xi)
New York Times Book Company. New York, New York, USA. 1977

What the scientists have always found by physical experiment was an a priori orderliness of nature, or Universe always operating at an elegance level that made the discovering scientist's working hypotheses seem crude by comparison. The discovered reality made the scientists' exploratory work seem relatively disorderly.

In L.L. Larison Cudmore
The Center of Life: A Natural History of the Cell (p. xi)
New York Times Book Company. New York, New York, USA. 1977

Galston, Arthur William 1920–
American plant biologist

In my view, the only recourse for a scientist concerned about the social consequences of his work is to remain involved with it to the end.
Science and Social Responsibility
Annals of the New York Academy of Science, Volume 196, 1972 (p. 223)

Galton, Sir Francis 1822–1911
English anthropologist, explorer, and statistician

A special taste for science seems frequently to be so ingrained in the constitution of scientific men, that it asserts itself throughout their whole existence.
In Karl Pearson
The Life, Letters and Labours of Francis Galton (Volume 2) (p. 152)
At The University Press. Cambridge, England. 1914–30

Gardner, Martin 1914–
American writer and mathematics games editor

When reputable scientists correct flaws in an experiment that produced fantastic results, then fail to get those results when they repeat the test with flaws corrected, they withdraw their original claims. They do not defend them by arguing irrelevantly that the failed replication was successful in some other way, or by making intemperate attacks on whomever dares to criticize their competence.
Reply to Claims for ESP
The New York Review of Books, February 19, 1981

Gauss, Johann Carl Friedrich 1777–1855
German mathematician, physicist, and astronomer

A taste for the abstract sciences in general and, above all, for the mysteries of numbers, is very rare: this is not surprising, since the charms of this sublime science in all their beauty reveal themselves only to those who have the courage to fathom them. But when a woman, because of her sex, our customs, and prejudices, encounters infinitely more obstacles than men in familiarizing herself with their knotty problems, yet overcomes these fetters and penetrates that which is most hidden, she doubtless has the most noble courage, extraordinary talent, and superior genius.
Letter, Carl Friedrich Gauss to Sophie Germain, 30 April, 1807

Gell-Mann, Murray 1929–
American physicist

But the practitioners of science are, after all, human beings. They are not immune to the normal influences of egotism, economic self-interest, fashion, wishful thinking and laziness. A scientist may try to steal credit, knowingly initiate a worthless project for gain, or take a conventional idea for granted instead of looking for a better explanation. From time to time scientists even fudge their results, breaking one of the most serious taboos of their profession.
The Quark and the Jaguar: Adventures in the Simple and the Complex (p. 80)
W.H. Freeman & Company. New York, New York, USA. 1994

Gibran, Kahlil 1883 1931
Lebanese-American philosophical essayist

A scientist without imagination is a butcher with dull knives and out-worn scales.
Sand and Foam: A Book of Aphorisms (p. 46)
Alfred A. Knopf. New York, New York, USA. 1959

Glashow, Sheldon L. 1932–
American physicist

Many scientists are deeply religious in one way or another, but all of them have a certain rather peculiar faith — they have a faith in the underlying simplicity of nature; a belief that nature is, after all, comprehensible and that one should strive to understand it as much as we can. Now this faith in simplicity, that there are simple rules — a few elementary particles, a few quantum rules to explain the structure of the world — is completely irrational and completely unjustifiable. It is therefore a religion.
The Quantum Universe
Coproduced by WETA-TV and The Smithsonian Institution (1990).

Gleick, James 1954–
American author, journalist, and essayist

Scientists still ask the what if questions. What if Edison had not invented the electric light — how much longer would it have taken? What if Heisenberg had not invented the S matrix? What if Fleming had not discovered penicillin? Or (the king of such questions) what if Einstein had not invented general relativity? "I always find questions like that…odd," Feynman wrote to a correspondent who posed one. Science tends to be created as it is needed. "We are not that much smarter than each other," he said.
Genius: The Life and Science of Richard Feynman
Caltec (p. 329)
Pantheon Books. New York, New York, USA. 1992

Children and scientists share an outlook on life. If I do this, what will happen? is both the motto of the child at play and the defining refrain of the physical scientist…. The unfamiliar and the strange — these are the domain of all children and scientists.
Genius: The Life and Science of Richard Feynman
The Rockaway (p. 19)
Pantheon Books. New York, New York, USA. 1992

Goldenweiser, Alexander 1880–1940
American anthropologist

The scientist, when in his laboratory, is craftsman and inventor in one. He also faces nature as a learner. Like

the craftsman, he is prepared to commit errors and, having learned from them, to revise his procedure. Like the inventor, he is after something new, he plans his experiments deliberately, watches carefully, ever on the alert for a promising lead — a discovery.
Robots or Gods
Chapter IV (p. 44)
Alfred A. Knopf. New York, New York, USA. 1931

A scientist who is no longer capable of framing a hypothesis — or never was — is not a scientist but a methodological fossil.
Robots or Gods
Chapter IV (p. 48)
Alfred A. Knopf. New York, New York, USA. 1931

Goldstein, A.
No biographical data available

Science is always a race…and scientists are competitive people. Because the monetary rewards are minimal, they go for ego rewards.…
In J. Goldberg
Anatomy of a Scientific Discovery
Locks and Keys (p. 25)
Bantam Books. Toronto, Ontario, Canada. 1988

Gornick, Vivian
American critic, essayist, and memoirist

To do science today is to experience a dimension unique in contemporary working lives; the work promises something incomparable: the sense of living both personally and historically. That is why science now draws to itself all kinds of people — charlatans, mediocrities, geniuses — everyone who wants to touch the flame, feel alive in the time.
Women in Science: Portraits from a World in Transition
Part One (p. 26)
Simon & Schuster. New York, New York, USA. 1983

Whatever a scientist is doing — reading, cooking, talking, playing — science thoughts are always there at the edge of the mind. They are the way the world is taken in; all that is seen is filtered through an everpresent scientific musing.
Women in Science: Portraits from a World in Transition
Part One (p. 39)
Simon & Schuster. New York, New York, USA. 1983

Gray, George W.
Freelance science writer

The modern scientist is like a detective who finds clues, but never gets a glimpse of the fugitive he seeks.
New Eyes of the Universe
The Atlantic Monthly, Volume 155, Number 5, May 1935 (p. 607)

Harding, Rosamund E. M.
No biographical data available

If the scientist has, during the whole of his life, observed carefully, trained himself to be on the look-out for analogy and possessed himself of relevant knowledge, then the "instrument of feeling"…will become a powerful divining rod leading the scientist to discover order in the midst of chaos by providing him with a clue, a hint, or an hypothesis upon which to base his experiments.
An Anatomy of Inspiration
Chapter V (p. 86)
W. Heffer & Sons Ltd. Cambridge, England. 1940

Harnwell, G. P. 1903–1982
No biographical data available

The motivations of the pure scientist would appear to many at first thoughts as whimsical and abstract as the immediate results he achieves. The briefest explanation of why he works is curiosity rather than the necessity of earning a livelihood.
Annual Report of the Board of Regents of the Smithsonian Institution, 1939
Our Knowledge of Atomic Nuclei (p. 189)
Government Printing Office. Washington, D.C. 1940

Heinlein, Robert A. 1907–88
American science fiction writer

Most "scientists" are bottle washers and button sorters.
Time Enough for Love
Intermission (p. 257)
G.P. Putnam's Sons. New York, New York, USA. 1973

Highet, Gilbert 1906–78
American classicist

There are naive people all over the world — some of them scientists — who believe that all problems, sooner or later, will be solved by Science. The word Science itself has become a vague reassuring noise, with a very ill-defined meaning and a powerful emotional charge: It is now applied to all sorts of unsuitable subjects and used as a cover for careless and incomplete thinking in dozens of fields. But even taking Science at the most sensible of its definitions, we must acknowledge that it is unperfect as are all activities of the human mind.
Man's Unconquerable Mind
Part Two, Chapter 4 (p. 106)
Cambridge University Press. New York, New York, USA. 1954

Hogan, James P. 1946–
English writer of hard science fiction

Scientists are the easiest to fool…They think in straight, predictable, directable, and therefore misdirectable, lines. The only world they know is the one where everything has a logical explanation and things are what they appear to be. Children and conjurers — they terrify me. Scientists are no problem; against them I feel quite confident.
Code of the Lifemaker
Chapter I (p. 14)
Ballantine Books. New York, New York, USA. 1983

Hubbard, Elbert   1856–1915
American editor, publisher, and author

The scientist who now takes off his shoes knows that the place whereon he stands is holy ground. Science is reverent and speaks with lowered voice, for she has caught glimpses of mysteries undefinable, and to her have come thoughts that are beyond speech. Science cultivates the receptive heart and hospitable mind, and her prayer is for more light, and to that prayer the answer is even now coming.
In Albert Lane
Elbert Hubbard and His Work (p. 100)
The Blanchard Press. Worcester, Massachusetts, USA. 1901

Hubble, Edwin Powell   1889–1953
American astronomer

Scientists in general are not very articulate; they work in comparative seclusion and they do not cultivate the art of persuasion.
The Nature of Science and Other Lectures
Part I, The Nature of Science (p. 3)
The Huntington Library. San Marino, California, USA. 1954

The scientist, in his purely scientific moods, seeks to understand the world — not to reform it, not to control it, but merely to understand it.
The Nature of Science and Other Lectures
Part I, Science and Technology (p. 20)
The Huntington Library. San Marino, California, USA. 1954

Huggins, Charles   1901–1997
Canadian born-American surgeon

…there are two kinds of scientists — The "gee whiz" kind and the "so what" kind. Flies around the urine cause the first type to exclaim: "Gee whiz, what could that mean?" whereas the other says: "so what, let's clean up this mess and get on with a proper experiment."
In Elwood V. Jensen
The Science of Science
Perspectives in Biology and Medicine, Volume 12, Number 2, Winter 1969 (p. 283)

Hull, David L.   1935–
American philosopher of biology

From the beginning of their careers, scientists are presented with a dilemma. They can make their work look as conventional as possible — just one more brick in the edifice of science — or as novel and controversial as possible — declaring a whole new theory or possibly even a whole new science…. From my own reading of the recent history of science, I see no strong correlation between my own estimates of the novelty of an idea and the strategy that an author adopts.
Science as a Process: An Evolutionary Account of the Social and Conceptual Development of Science
Chapter Six (p. 202)
The University of Chicago Press. Chicago, Illinois, USA. 1988

Husserl, Edmund   1859–1938
German philosopher

When it is actually natural science that speaks, we listen gladly and as disciples. But it is not always natural science that speaks when natural scientists are speaking…
Translated by W.R. Boyce Gibson
Ideas: General Introduction to Pure Phenomenology
Second Chapter, Section 20 (p. 86)
George Allen & Unwin Ltd. London, England. 1931

Ian
Fictional character

Your scientists were so preoccupied with whether or not they could, they didn't stop to think if they should.
Jurassic Park
Film (1993)

Imhof, Peter
German computer scientist and social science modeler

…scientists are not a select few intelligent enough to think in terms of "broad sweeping theoretical laws and principles." Instead, scientists are people specifically trained to build models that incorporate theoretical assumptions and empirical evidence. Working with models is essential to the performance of their daily work; it allows them to construct arguments and to collect data.
Tools for Thinking (book review)
Science, Volume 287, 1935–1936

Ingram, Jay   1945–
Canadian author and television host

The caricature of the nerdy scientist in his/her lab coat, complete with pocket protector, uttering incomprehensible jargon is bad enough. But the implied character of the person behind the wardrobe is worse: strait-jacketed by conservatism, too quick to demand hard data, hellbent on reducing life's mysteries to uninteresting sets of numbers and graphs. …The truth is that scientists love a mystery as much as anyone (it's their business to chase mysteries after all) even when…there is almost no chance it will be solved. Why? Because it's intriguing, challenging, and fun.
The Barmaid's Brain and other Strange Tales from Science
The Burning Mirrors of Syracuse

Jensen, Elwood V.
No biographical data available

Research among the less imaginative scientists has been likened to a fox-hunt. A creative investigator shouts "Tally-ho", and the entire troop rides off in the same direction.
The Science of Science
Perspectives in Biology and Medicine, Volume 12, Number 2, Winter 1969 (p. 278)

Katscher, F.
No biographical data available

That great scientists were believing Christians does not prove anything. In this century many free-thinkers have also made great contributions to science, scientific thinking and ethical questions regarding the application of science.
Correspondence
Nature, Volume 363, Number 6428, 3 June 1993 (p. 390)

Killian, Jr., James R.   1904–88
American manager

The scientist, it is repeatedly said, should be on tap but not on top. He thus is considered to be merely one of the hired men who has no business doing anything but what he is told to do in the field of his specialty.... I do not imply that the scientist has any right or unique qualifications to be on top. I am disturbed by the attitude that because a man is a scientist, he is disqualified for public and private administrative responsibility even though he may have the qualifications.
The Shortage Re-Examined
American Scientist, Spring, April 1956 (p. 126)

Kingsley, Charles   1819–75
English clergyman and author

[Scientists] Good men, honest men, accurate men, righteous men, patient men, self-restraining men, fair men, modest men. Men who are aware of their own vast ignorance compared with the vast amount that there is to be learned in such a universe as this. Men who are accustomed to look at both sides of a question; who, instead of making up their minds in haste like bigots and fanatics, wait like wise men, for more facts, and more thought about the facts. *Town Geology*
Preface
D. Appleton & Company. New York, New York, USA. 1873

Koestler, Arthur   1905–83
Hungarian-born English writer

[Scientists are] Peeping Toms at the keyhole of eternity.
The Roots of Coincidence
Chapter 5, Section 9 (p. 140)
Random House. New York, New York, USA. 1972

Kolb, Edward W. (Rocky)   1951–
American cosmologist

No one devotes a lifetime to science in the hope of making small advances. Every scientist secretly or overtly hopes to make great discoveries.
Blind Watchers of the Sky
Chapter Two (p. 42)
Addison-Wesley Publishing Company. Reading, Massachusetts, USA. 1996

Most scientists spend their lives groping around trying to find their way, as if lost in a fog.
Blind Watchers of the Sky
Chapter Eight (p. 203)
Addison-Wesley Publishing Company. Reading, Massachusetts, USA. 1996

Kornberg, Arthur   1918–
American biochemist

A scientist...shouldn't be asked to judge the economic and moral value of his work. All we should ask the scientist to do is to find the truth — and then not keep it from anyone.
San Francisco Examiner, December 19, 1971

Kuhn, Thomas S.   1922–96
American historian of science

Scientists, it should already be clear, never learn concepts, laws, and theories in the abstract and by themselves. Instead, these intellectual tools are from the start encountered in a historically and pedagogically prior unit that displays them with and through their applications.
The Structure of Scientific Revolutions
Chapter V (p. 46)
The University of Chicago Press. Chicago, Illinois, USA. 1970

Though many scientists talk easily and well about the particular individual hypotheses that underlie a concrete piece of current research, they are little better than laypersons at characterizing the established basis of their field, its legitimate problems and methods. If they have learned such abstractions at all they show it mainly through their ability to do successful research.
The Structure of Scientific Revolutions
Chapter V (p. 47)
The University of Chicago Press. Chicago, Illinois, USA. 1970

Larrabee, Eric   1922–90
Historian

Perhaps the time has come for [scientists] to wonder about why they sometimes jar the nerves and try the patience of non-scientists.

...Scientists seem able to go about their business in a state of indifference to, if not ignorance of, anything but the going, currently acceptable doctrine of their several disciplines....
Commentary
Science and the Common Reader, June 1966 (p. 48)

The only thing wrong with scientists is that they don't understand science. They don't know where their own institutions come from, what forces shaped and are still shaping them, and they are wedded to an anti-historical way of thinking which threatens to deter them from ever finding out.
Commentary
Science and the Common Reader, June 1966 (p. 48)

Lebowitz, Fran 1951–
American comedian

Scientists are rarely to be counted among the fun people. Awkward at parties, shy with strangers, deficient in irony they have had no choice but to turn their attention to the close study of everyday objects.

Metropolitan Life
Science (p. 106)
Fawcett Crest. New York, New York, USA. 1978

Lederman, Leon 1922–
American high-energy physicist

Physicists today feel the same emotions that scientists have felt for centuries. The life of a physicist is filled with anxiety, pain, hardship, tension, attacks of hopelessness, depression, and discouragement. But these are punctuated by flashes of exhilaration, laughter, joy, and exultation. These epiphanies come at unpredictable times. Often they are generated simply by the sudden understanding of something new and important, something beautiful, that someone else has revealed.

The God Particle: If the Universe Is the Answer, What Is the Question?
Chapter 1 (p. 7)
Houghton Mifflin Company. Boston, Massachusetts, USA. 1993

[If] you are mortal, like most of the scientists I know, the far sweeter moments come when you yourself discover some new fact about the universe. It's astonishing how often this happens at 3 A.M., when you are alone in the lab and you have learned something profound, and you realize that not one of the other five billion people on earth knows what you now know. Or so you hope. You will, of course, hasten to tell them as soon as possible. This is known as "publishing."

The God Particle: If the Universe Is the Answer, What Is the Question?
Chapter 1 (p. 7)
Houghton Mifflin Company. Boston, Massachusetts, USA. 1993

When I talk about the pain and hardship of a scientist's life, I'm speaking of more than existential angst. Galileo's work was condemned by the Church; Madame Curie paid with her life, a victim of leukemia wrought by radiation poisoning. Too many of us develop cataracts. None of us gets enough sleep. Most of what we know about the universe we know thanks to a lot of guys (and ladies) who stayed up late at night.

The God Particle: If the Universe Is the Answer, What Is the Question
Chapter 1 (p. 16)
Houghton Mifflin Company. Boston, Massachusetts, USA. 1993

Leon, Mark
No biographical data available

"Read Popper on the philosophy of science," Alan said. "I don't wholly agree with him, but he has a point when he says the scientist's job is to disprove rather than prove theories. Humans are passionate believers. The history of philosophy and religion is a grand testament to our will to believe. Scientists just try to inject a little sanity into the whole business, and that often requires a passion not to believe."

The Unified Field (p. 11)
Avon Books. New York, New York, USA. 1996

Leonard, Jonathan Norton 1903–75
No biographical data available

[A scientist's] real work is done in the silent hours of thought, the apparently aimless days of puttering around in the laboratory, and the mighty searching through reference books.

Steinmetz, Jove of Science, Part II
The World's Work, February 1929 (p. 140)

Levi, Primo 1919–87
Italian writer and chemist

A scientist's life, the author says, is indeed conflictual, formed by battles, defeats, and victories: but the adversary is always and only the unknown, the problem to be solved, the mystery to be clarified. It is never a matter of civil war; even though of different opinions, or of different political leanings, scientists dispute each other, they compete, but they do not battle: they are bound together by a strong alliance, by the common faith "in the validity of Maxwell's or Boltzmann's equations," and by the common acceptance of Darwinism and the molecular structure of DNA.

Translated by Raymond Rosenthal
The Mirror Maker: Stories and Essays by Primo Levi
Bacteria Roulette (p. 123)
Shocken Books. New York, New York, USA. 1989

Levinson-Lessing, F. Y. 1861–1939
Russian geologist

A scientist lacking imagination can at best become a splendid walking library and source of information — he absorbs, but does not create.

Compiled by V.V. Vorontsov
Words of the Wise: A Book of Russian Quotations
Translated by Vic Schneierson
Progress Publishers. Moscow, Russia. 1979

Lewis, Gilbert Newton 1875–1946
American chemist

The scientist is a practical man and his are practical aims. He does not seek the ultimate but the proximate. He does not speak of the last analysis but rather of the next approximation. … On the whole, he is satisfied with his work, for while science may never be wholly right it certainly is never wholly wrong; and it seems to be improving from decade to decade.

The Anatomy of Science
Chapter I (pp. 6–7)
Yale University Press. New Haven, Connecticut, USA. 1926

Lewis, Sinclair   1885–1951
American novelist

…the scientist is intensely religious — …he will not accept quarter-truths, because they are an insult to his faith. He wants that everything should be subject to inexorable laws. He is the only real revolutionary, the authentic scientist, because he alone knows how little he knows. He lives in a cold, clear light. Yet he is not cold nor heartless. And he prays for unclouded eyes and freedom from haste, for a quiet and relentless anger against all pretence and all pretentious work and all work left slack and unfinished,…a restlessness whereby he may neither sleep nor accept praise till his observed results equal his calculated results.…
Arrowsmith (p. 278)
Harcourt, Brace & World, Inc. New York, New York, USA. 1952

He had never dined with a duchess, never received a prize, never been interviewed, never produced anything which the public could understand, nor experienced anything since his schoolboy amours which nice people could regard as romantic.

He was, in fact, an authentic scientist.
Arrowsmith
Chapter XII, Section I (p. 128)
Harcourt, Brace & World, Inc. New York, New York, USA. 1952

To be a scientist — it is not just a different job so that a man should choose between being a scientist and being an explorer or a bond-salesman or a physician or a king or a farmer. It is a tangle of very obscure emotions, like mysticism, or wanting to write poetry; it makes its victim all different from the good natural man.
Arrowsmith
Chapter XXVI, Section I (p. 290)
Harcourt, Brace & World, Inc. New York, New York, USA. 1952

The normal man, he does not care much what he does except that he should eat and sleep and make love. But the scientist is intensely religious — he is so religious that he will not accept quarter truths, because they are an insult to his faith.
Arrowsmith
Chapter XXVI, Section I (p. 290)
Harcourt, Brace & World, Inc. New York, New York, USA. 1952

[The scientist] hates the preachers who talk their fables, but he is not too kindly to the anthropologists and historians who can only make guesses, yet they have the nerve to call themselves scientists!
Arrowsmith
Chapter XXVI, Section I (p. 290)
Harcourt, Brace & World, Inc. New York, New York, USA. 1952

Lightman, Alan   1948–
Physicist, novelist, and essayist

Scientists turn reckless and mutter like gamblers who cannot stop betting.

Einstein's Dreams
3 May 1905 (p. 41)
Pantheon Books. New York, New York, USA. 1993

Professor Oliver Lindenbrook
Fictional character

A scientist who cannot prove what he has accomplished, has accomplished nothing.
Journey to the Center of the Earth
Film (1959)

Martel, Yann   1963–
Canadian novelist

Scientists are a friendly, atheistic, hard-working, beer-drinking lot whose minds are preoccupied with sex, chess and baseball when they are not preoccupied with science.
Life of Pi (p.5)
Harcourt Inc. Orlando, Florida, USA. 2001

Maxwell, James Clerk   1831–79
Scottish physicist

[Scientists'] actions and thoughts, being more free from the influence of passion than those of other men, are all the better materials for the study of the calmer parts of human nature.… [B]y aspiring to noble ends… [scientists] have risen above the region of storms into a clearer atmosphere, where there is no misrepresentation of opinion, nor ambiguity of expression, but where one mind comes into closest contact with another at the point where both approach nearest to the truth.
In C.C. Gillispie (ed.)
The Edge of Objectivity: An Essay in the History of Scientific Ideas
Forward (pp. vii, viii)
Princeton University Press. Princeton, New Jersey, USA. 1960

Mayer, Joseph   1904–83
American chemist

What does one have to do to be called a scientist? I decided that anyone who spent on science more than 10% of his waking, thinking time for a period of more than a year would be called a scientist, at least for that year.
In "The Way it Was"
Annual Review of Physical Chemistry, Volume 33, 1982 (pp. 1–2)

Mayo, Charles Horace   1865–1939
American physician

The scientist is not content to stop at the obvious.
Problems in Medical Education
Collected Papers of the Mayo Clinic & Mayo Foundation, Volume 18, 1926

Medawar, Sir Peter Brian   1915–87
Brazilian-born English zoologist

To be creative, scientists need libraries and laboratories and the company of other scientists; certainly a quiet and

untroubled life is a help. A scientist's work is in no way deepened or made more cogent by privation, anxiety, distress, or emotional harassment.
Advice to a Young Scientist
Chapter 5 (p. 40)
Basic Books, Inc. New York, New York, USA. 1979

[T]he private lives of scientists may be strangely and even comically mixed up, but not in ways that have any special bearing on the nature and quality of their work. If a scientist were to cut off an ear, no one would interpret such an action as evidence of an unhappy torment of creativity; nor will a scientist be excused any bizarre [action], however extravagant, on the grounds that he is a scientist, however brilliant.
Advice to a Young Scientist
Chapter 5 (p. 40)
Basic Books, Inc. New York, New York, USA. 1979

Much of a scientist's pride and sense of accomplishment turns…upon being the first to do something — upon being the man who did actually speed up or redirect the flow of thought and the growth of understanding.

The most heinous offense a scientist as a scientist can commit is to declare to be true that which is not so; if a scientist cannot interpret the phenomenon he is studying, it is a binding obligation upon him to make it possible for another to do so.
The Limits of Science
An Essay on Scians [Science] (p. 6)
Harper & Row, Publishers. New York, New York, USA. 1984

[James] Watson's childlike vision makes them seem like the creatures of a Wonderland, all at a strange contentious noisy tea-party which made room for him because for people like him, at this particular kind of party, there is always room.
Lucky Jim
New York Review of Books, 28 March 1968

…Watson had one towering advantage over all of [his classmates in other disciplines]: in addition to being extremely clever he had something important to be clever about. This is an advantage which scientists enjoy over most other people engaged in intellectual pursuits, and they enjoy it at all levels of capability. To be a first-rate scientist it is not necessary (and certainly not sufficient) to be extremely clever, anyhow in a pyrotechnic sense.
Lucky Jim
New York Review of Books, 28 March 1968

One of the great social revolutions brought about by scientific research has been the democratization of learning. Anyone who combines strong common sense with an ordinary degree of imaginativeness can become a creative scientist, and a happy one besides, in so far as happiness depends upon being able to develop to the limit of one's abilities.

Lucky Jim
New York Review of Books, 28 March 1968

People who criticize scientists for wanting to enjoy the satisfaction of intellectual ownership are confusing possessiveness with pride of possession. Meanness, secretiveness and, sharp practice are as much despised by scientists as by other decent people in the world of ordinary everyday affairs; nor, in my experience, is generosity less common among them, or less highly esteemed.
Lucky Jim
New York Review of Books, 28 March 1968

Before a good scientist tries to persuade others that he is on to something good, he must first convince himself.
Florey Story (London Review of Books, 20 December 1979)
Reprinted in "The Strange Case of the Spotted Mice and Other Classic Essays on Science"
Oxford University Press, Inc. New York, New York, USA. 1996

…scientists tend not to ask themselves questions until they can see the rudiments of an answer in their minds. Embarrassing questions tend to remain unasked or, if asked, to be asked rudely.
The Future of Man: The BBC Reith Lectures 1959
Chapter 4 (p. 62)
Methuen & Company Ltd. London, England. 1960

Mencken, H. L. (Henry Louis)   1880–1956
American journalist and literary critic

The scientist who yields anything to theology, however slight, is yielding to ignorance and false pretenses, and as certainly as if he granted that a horse-hair put into a bottle of water will turn into a snake.
Minority Report: H.L. Mencken's Notebooks
No. 45 (p. 33)
Alfred A. Knopf. New York, New York, USA. 1956

Mendeleyev, Dmitry   1834–1907
Russian chemist

Science exists separately from scientists, it lives autonomously, it is the sum of knowledge worked out by the whole mass of scientists, similar to how the acknowledged political order of a country is worked out by the mass of persons who live in it. Science is authoritative, separate scientists are not. A scientist can only and should only use this authority when he is following science, just as in a well-ordered state the authority of power is used only by the person who observes the law.
In Michael D. Gordin
A Well-Ordered Thing: Dmitrii Mendeleev and the Shadow of the Periodic Table
Chapter 4 (p. 103)
Basic Books, Inc. New York, New York, USA. 2004

Mitchell, Maria   1818–89
American astronomer and educator

The true scientist must be self-forgetting. He knows that under the best circumstances he is sowing what others must reap — or rather he is striking the mine which others must open up — for human life at longest has not the measure of a single breath in the long life of science.

In Helen Wright
Sweeper in the Sky
Chapter 9 (p. 168)
Macmillan & Company. New York, New York, USA. 1949

It is the highest joy of the true scientist…that he can reap no lasting harvest — that whatever he may bring into the storehouse today will be surpassed by the gleaners tomorrow — he studies Nature because he loves her and rejoices to "look through Nature up to Nature's God."

In Helen Wright
Sweeper in the Sky
Chapter 9 (p. 168)
Macmillan & Company. New York, New York, USA. 1949

Montessori, Maria   1870–1952
Italian educationist

…what is a scientist?…We give the name scientist to the type of man who has felt experiment to be a means guiding him to search out the deep truth of life, to lift a veil from its fascinating secrets, and who, in this pursuit, has felt arising within him a love for the mysteries of nature, so passionate as to annihilate the thought of himself.

Translated by Anne E. George
The Montessori Method
Chapter I (p. 8)
Frederick A. Stokes Company. New York, New York, USA. 1912

The scientist is not the clever manipulator of instrument, he is the worshipper of nature and he bears the external symbols of his passion as does the follower of some religious order.

Translated by Anne E. George
The Montessori Method
Chapter I (p. 8)
Frederick A. Stokes Company. New York, New York, USA. 1912

Motz, Lloyd   1910–2004
American astronomer
Weaver, Jefferson Hane
American science author

To the nonscientist, science, particularly in its modern dress and as pursued today, is a glittering intellectual jewel, mysterious, forbidding, and even threatening except to a few chosen ones, the scientists, who appear to be superior beings, endowed with an ability to probe and understand nature far beyond that of the average layman.

The Concepts of Science: From Newton to Einstein
Preface (p. vii)
Plenum Press. New York, New York, USA. 1988

Muppets
Fictional characters

German scientist: We are going to perform an electronic cerebractomy.

Doc Hooper: A what?

German Scientist: An electronic cerebractomy! It's something so sensational, you'll have to hold on to your hat.… Look, when a German scientist says hold on to your hat he isn't making casual conversation, he means to hold on to your hat. Hat! Hold!

The Muppet Movie
Film (1979)

National Academy of Sciences

Scientists must be fact-seekers, open-minded, and willing to accept changes indicated by the signposts of evidence.

Science and Creationism — A View from the National Academy of Sciences (p. 5).
National Academy Press. Washington, D.C. 1984

Ninotchka
Fictional character

[Cyd Charrise talking down love to Fred Astaire] He was one of our greatest scientists. He has proved, beyond any question, that physical affection is purely electrochemical.

Silk Stockings
Film (1957)

Pasteur, Louis   1822–95
French chemist

When moving forward toward the discovery of the unknown, the scientist is like a traveler who reaches higher and higher summits from which he sees in the distance new countries to explore.

In Rene Dubos
Louis Pasteur: Free Lance of Science
Chapter III (p. 87)
Little, Brown & Company. Boston, Massachusetts, USA. 1950

Peabody, Francis Weld   1881–1927
American physician

…the popular conception of the scientist as a man who works in a laboratory and who uses instruments of precision is as inaccurate as it is superficial, for a scientist is known, not by his technical processes, but by his intellectual processes; and the essence of the scientific method of thought is that it precedes in an orderly manner toward the establishment of truth.

The Care of the Patient
The Care of the Patient (p. 21)
Harvard University Press. Cambridge, Massachusetts, USA. 1928

Pearse, A. S.   1877–1956
No biographical data available

Science is always right because it seeks only for truth, and truth hurts no one. Unfortunately, scientists are not always right.
Adventure, Romance and Science
Science, Volume 58, Number 1492, 3 August, 1923 (p. 78)

Perelman, S. J. (Sidney Joseph)   1904–79
American comic writer

I guess I'm just an old mad scientist at bottom. Give me an underground laboratory, half a dozen atom smashers, and a beautiful girl in a diaphanous veil waiting to be turned into a chimpanzee, and I care not who writes the nation's laws.
Crazy Like a Fox
Captain Future, Block that Kick (p. 210)
Random House, Inc. New York, New York, USA. 1944

Perry, Ralph Barton   1876- 1957
American philosopher and educator

Every scientist, furthermore, is himself a "self-made man." He owes his strictly scientific attainment to his own efforts and to the endowment with which nature has equipped him. Whatever elevation in life he reaches is not an artificial status created by institutions or traditions, but a measure of solid achievement. The scientist, therefore, respects man for what he is rather than for his class or station.
The Present Conflict of Ideals: A Study of the Philosophical Background of the World War
Chapter IX (pp. 101–102)
Longmans, Green. New York, New York, USA. 1918

Planck, Max   1858–1947
German physicist

Since the real world, in the absolute sense of the word, is independent of individual personalities, and in fact of all human intelligence, every discovery made by an individual acquires a completely universal significance. This gives the inquirer, wrestling with his problem in quiet seclusion, the assurance that every discovery will win the unhesitating recognition of all experts throughout the entire world, and in this feeling of the importance of his work lies his happiness. It compensates him fully for many a sacrifice which he must make in his daily life.
Scientific Autobiography and Other Papers
The Meaning and Limits of Exact Science, Part III (p. 103)
Philosophical Library. New York, New York, USA. 1949

Poincaré, Henri   1854–1912
French mathematician and theoretical astronomer

The scientist should not waste his time on the achievement of practical goals. He will surely reach such goals, but this must be marginal with respect to his principal activity. He should never forget that the specific object he is investigating is part of a whole which is infinitely greater than this object; love for this whole and an interest in it should constitute the only motives of the actions of the scientist. Science has marvelous applications, but a science in which applications were the only aim would no longer be science but only a kitchen.
In Stefan Amsterdamski
Between History and Method
Chapter V. Crisis of the Modern Ideal (p. 94)
Kluwer Academic Publishers. Dordrecht, Netherlands. 1992

Polanyi, Michael   1891–1976
Hungarian-born English scientist, philosopher, and social scientist

There are differences in rank between scientists, but these are of secondary importance: everyone's position is sovereign. The Republic of Science realizes the ideal of Rousseau, of a community in which each is an equal partner in a General Will. But this identification makes the General Will appear in a new light. It is seen to differ from any other will by the fact that it cannot alter its own purpose. It is shared by the whole community because each member of it shares in a joint task.
Science, Faith and Society
Background and Prospect (pp. 16–17)
The University of Chicago Press. Chicago, Illinois. 1964

Popper, Karl R.   1902–94
Austrian/British philosopher of science

…what is to be called "science" and who is to be called a "scientist" must always remain a matter of convention or decision.
The Logic of Scientific Discovery
Part I, Chapter II, Section 10 (p. 52)
Basic Books, Inc. New York, New York, USA. 1959

Price, Derek John de Solla   1922–83
English science historian and information scientist

The ivory tower of the artist can be a one-man cell; that of the scientist must contain many apartments so that he may be housed among his peers.
Little Science, Big Science
Chapter 3 (p. 69)
Columbia University Press. New York, New York, USA. 1963

Primas, Hans   1928–
German spectroscopy scientist

The legendary image of a scientist as a humble searcher for truth is more and more replaced by the image of a scientist as a well-paid brilliant expert, speaking an unintelligible professional jargon, highly competent in a narrowly defined domain but arrogantly extending his competence into fields in which he knows nothing, and neglecting the fact that science is only a small subdivision of human knowledge.

Chemistry, Quantum Mechanics and Reductionism: Perspectives in Theoretical Chemistry
Chapter 1, Section 1.1 (p. 24)
Springer-Verlag, Berlin, West Germany. 1981

Prusiner, Stanley B.   1942–
American neurologist

Being a scientist is a special privilege: for it brings the opportunity to be creative, the passionate quest for answers to nature's most precious secrets, and the warm friendships of many valued colleagues. Collaborations extend far beyond the scientific achievements, no matter how great the accomplishments might be, the rich friendships which have no national borders are treasured even more.
Les Prix Nobel. The Nobel Prizes in 1997
Nobel banquet speech for award received in 1997
Nobel Foundation. Stockholm, Sweden. 1998

Mark O'Brian
Fictional character

I'm a scientist also, Dr. Holden. I know the value of the cold light of reason. But I also know the deep shadows that light can cast. The shadows that can blind men to truth.
The Curse of the Demon
Film (1957)

Ramon y Cajal, Santiago   1852–1934
Spanish neuropathologist

It is certainly true that the scientist's fame is not as great as the playwright or artist's glamour and popularity. People live in a world of sentiment, and it is asking too much of them to provide warmth and support for the heroes of reason.
Advice for a Young Investigator
Chapter 3 (p. 44)
The MIT Press. Cambridge, Massachusetts, USA. 1999

Richards, Ivor Armstrong   1893–1979
English literary critic

We believe a scientist because he can substantiate his remarks, not because he is eloquent and forcible in his enunciation. In fact, we distrust him when he seems to be influencing us by his manner.
Science and Poetry
Chapter II (p. 24)
Kegan Paul, Trench, Trubner & Company Ltd. London, England. 1926

Richet, Charles   1850–1935
French physiologist

Probably, what characterizes all scientists, whatever they may be, archivists, mathematicians, chemists, astronomists, physicists, is that they do not seek to reach a practical conclusion by their work.
The Natural History of a Savant

Chapter I (p. 4)
J.M. Dent & Sons Ltd. London, England. 1927

Roe, Anne
No biographical data available

Science is the creation of scientists and every scientific advance bears somehow the mark of the man who made it.... The creative scientist, whatever his field, is very deeply involved emotionally and personally in his work, and…he himself is his own most essential tool.
The Psychology of the Scientist
Science, Volume 134, Number 3477, August 18, 1961 (p. 456)

Roszak, Theodore   1933–
American social critic

…science rests itself not in the world the scientist beholds at any particular point in time, but in his mode of viewing that world. A man is a scientist not because of what he sees, but because of how he sees it.
The Making of a Counter Culture: Reflections on the Technocratic Society and Its Youthful Opposition
Chapter VII (p. 213)
Doubleday & Company, Inc., Garden City, New York, USA; 1969

Rothman, Milton A.   1919–2001
American nuclear physicist and science writer

It makes no sense to complain about a lack of imagination in scientists when their failure is simply that they cannot make the world be what it is not, and they cannot make the world do what it cannot do.
The Science Gap: Dispelling the Myths and Understanding the Reality of Science
Prometheus Books. Buffalo, New York, USA. 1992

Rothman, Tony   1953–
American cosmologist

The makers of Revenge of the Nerds know, as do millions who have seen it, that all scientists when young are undernourished sociophobics who relate best to a computer terminal through coke bottle eyeglasses after midnight in a basement laboratory.
A Physicist on Madison Avenue
Chapter 1 (p. 3)
Princeton University Press. Princeton, New Jersey, USA. 1991

Ruse, Michael   1940–
English historian and philosopher of science

A scientist should not cheat or falsify data or quote out of context or do any other thing that is intellectually dishonest. Of course, as always, some individuals fail; but science as a whole disapproves of such action. Indeed, when transgressors are detected, they are usually expelled from the community.
Response to the Commentary: Pro Judice
Science, Technology and Human Values, Volume 7, Number 41, Fall 1982 (p. 74)

Rushton, John Phillipe   1941–
British/Canadian experimental psychology professor and writer

Research has suggested that scientists differ from non-scientists by exhibiting a high level of curiosity, especially at an early age, and in demonstrating a relatively low level of sociability. Scientists also tend to be shy, lonely, slow in social development, and indifferent to close personal relationships, group activities and politics. Other attributes include skepticism, preoccupation, reliability, and a facility for precise, critical thinking. Generally they are cognitively complex, independent, non-conformist, assertive, and unlikely to suppress thoughts and impulses; and, like successful entrepreneurs, eminent scientists are also calculated risk-takers.
Journal of Social and Biological Structure, Volume 11, 1980 (p. 140)

Sabin, Albert   1906–93
American medical researcher

No matter how good you are, you cannot be a scientist unless you learn to live with frustration.
I Only Ask for a Place to Work
New Scientist, Volume 57, Number 835, 1 March 1973 (pp. 491–492)

Sagan, Carl   1934–96
American astronomer and author

[Scientists] are capable of self-deception.... All sorts of socially abhorrent doctrines have at one time or another been supported by scientists, well-known scientists, famous brand-name scientists. And, of course, politicians. And respected religious leaders. Slavery, for instance, or the Nazi brand of racism. Scientists make mistakes, theologians make mistakes, everybody makes mistakes...
Contact: A Novel (p. 167)
Simon & Schuster. New York, New York, USA. 1985

Who discovered that CFCs [chlorofluorocarbons] posed a threat to the ozone layer? Was it the principal manufacturer, the DuPont Corporation, exercising corporate responsibility? Was it the Environmental Protection Agency protecting us? Was it the Department of Defense defending us? No, it was two ivory-tower, white-coated university scientists working on something else — Sherwood Rowland and Mario Molina of the University of California, Irvine. Not even an Ivy League university. No one instructed them to look for dangers to the environment. They were pursuing fundamental research. They were scientists following their own interests. Their names should be known to every schoolchild.
Pale Blue Dot: A Vision of the Human Future in Space
Chapter 14 (pp. 221–222)
Random House, Inc. New York, New York, USA. 1994

Seifriz, William   1888–1955
Professor of Botany

It is no matter of chance that the greatest scientists of all time, Copernicus, Newton, Kepler, Linnaeus, Faraday, Darwin, and Maxwell, were men of noble character, modest, straightforward, and full of human sympathy. The great French mathematician, Henri Poincaré, stated that the chief end of life is contemplation, not action.
A New University
Science, Volume 120, Number 3107, 16 July 1954 (pp. 88–89)

Selye, Hans   1907–82
Austrian-American endocrinologist

Scientists are probably the most individualistic bunch of people in the world. All of us are and should be essentially different; there would be no purpose in trying to fit us into a common mold.
From Dream to Discovery: On Being a Scientist
Introduction
McGraw-Hill Book Company, Inc. New York, New York, USA. 1950

Sheckley, Robert   1928–2005
American writer

The scientist, who examines everything, should look at himself. Tentatively I would define him as a discovery-producing animal whose products fall from him as naturally and as thoughtlessly as a hen produces eggs. Like the hen, he is largely indifferent to the use made of his products. Scientists are mostly not in favor of atom bombs, of course, and hens presumably dislike omelets; but both are realists and go along with the conditions they find.
In Damon Knight
The Observers (p. 189)
Tor. New York, New York, USA. 1988

Shelley, Mary   1797–1851
English Romantic writer

The modern masters of chemistry promise very little; they know that metals cannot be transmuted and that the elixir of life is a chimera. But these philosophers, whose hands seem only made to dabble in dirt, and their eyes to pore over the microscope or crucible, have indeed performed miracles. They penetrate into the recesses of nature and show how she works in her hiding places. They ascend into the heavens; they have discovered how the blood circulates, and the nature of the air we breathe. They have acquired new and almost unlimited powers; they can command the thunders of heaven, mime the earthquake, and even mock the invisible world with its own shadows.
Frankenstein
Chapter 3 (p. 39)
Running Press. Philadelphia, Pennsylvania, USA. 1990

Sigurdsson, Haraldur   1939–
Icelandic volcanologist

The scientist or scholar is the keeper of the flame of knowledge and he or she also advances our knowledge in a chosen field of research, but should also be responsible for linking the present with the past and maintaining a record of the history of knowledge in that field.
Melting the Earth
Preface (p. viii)
Oxford University Press, Inc. New York, New York, USA. 1999

Silver, Brian L.
Israeli professor of physical chemistry

Scientists come in many colors, of which the green of jealousy and the purple of rage are fashionable shades.
The Ascent of Science
Preface (p. xiii)
Solomon Press Book. New York, New York, USA. 1998

Sinsheimer, Robert L.   1920–
American molecular biologist

I am a scientist, a member of a most fortunate species. The lives of most people are filled with ephemera. All too soon, much of humanity becomes mired in the tepid tracks of their short lives. But a happy few of us have the privilege to live with and explore the eternal.
The Strands of a Life: The Science of DNA and the Art of Education
The University of California Press. Berkeley, California, USA. 1994

Smith, Homer W.
Renal physiologist

A scientist is one who, when he does not know the answer, is rigorously disciplined to speak up and say so unashamedly; which is the essential feature by which modern science is distinguished from primitive superstition, which knew all the answers except how to say, "I do not know."
From Fish to Philosopher
Chapter XIII (p. 210)
Little, Brown & Company. Boston, Massachusetts, USA. 1953

Snow, Charles Percy   1905–80
English novelist and scientist

I believe the intellectual life of the whole of western society is increasingly being split into two polar groups. … Literary intellectuals at one pole — at the other scientists, and as the most representative, the physical scientists. Between the two a gulf of mutual incomprehension — sometimes (particularly among the young) hostility and dislike, but most of all lack of understanding.
The Two Cultures: And a Second Look
Chapter I (p. 3)
At The University Press. Cambridge, England. 1964

Spallanzani, Lazzaro   1729–99
Italian natural philosopher

If I set out to prove something, I am no real scientist — I have to learn to follow where the facts lead me — I have to learn to whip my prejudices.
In R. Coope
The Quiet Art (p. 4)
E.S. Livingstone Ltd. Edinburgh, Scotland. 1952

Standen, Anthony   1907–??
Anglo-American science writer

When a white-robed scientist, momentarily looking away from his microscope or cyclotron, makes some pronouncement for the general public, he may not be understood but at least he is certain to be believed. Scientists are exalted beings who stand at the very topmost pinnacle of popular prestige, for they have the monopoly of the formula "It has been scientifically proved.", which appears to rule out all possibility of disagreement. Thus the world is divided into Scientists, who practice the art of infallibility, and non-scientists, sometimes contemptuously called "laymen", who are taken in by it.
Science Is a Sacred Cow
Chapter I (p. 13)
E.P. Dutton & Company. New York, New York, USA. 1950

We are having wool pulled over our eyes if we let ourselves be convinced that scientists as a group are anything special in the way of brains. They are very ordinary professional men, and all they know is their own trade, just like all other professional men.
Science Is a Sacred Cow
E.P. Dutton & Company. New York, New York, USA. 1950

Szent-Györgyi, Albert   1893–1986
Hungarian-born American biochemist

The real scientist…is ready to bear privations and, if need be, starvation rather than let anyone dictate to him which direction his work must take.
Science Needs Freedom
World Digest, Volume 55, 1943

In the great struggle between ignorance, distrust and brutality on one side, knowledge, understanding and peace on the other the scientist must stand fearlessly on the side of the latter, strengthening link between man and man and preaching that the only effective weapon of self-defense is good-will to others.
Les Prix Nobel. The Nobel Prizes in 1937
Nobel banquet speech for award received in 1937
Nobel Foundation. Stockholm, Sweden. 1938

Good science is made by good scientists, poor science by poor scientists, and the most brilliant project is worthless in the hands of a poor scientist, while, conversely, a good scientist has a good chance to come up with something valuable whatever he touches, because "*die Welt rundet sich im Tautropfen*" (Goethe), which could be translated by saying that all the great laws of nature are represented in a drop of dew.

Research Grants
Perspectives in Biology and Medicine, Volume 18, Number 1, Autumn 1974 (p. 41)

Tait, Peter Guthrie 1831–1901
Scottish physicist and mathematician

The life of a genuine scientific man is, from the common point of view, almost always uneventful. Engrossed with the paramount claims of inquiries raised high above the domain of mere human passions, he is with difficulty tempted to come forward in political discussions, even when they are of national importance; and he regards with surprise, if not with contempt, the petty municipal squabbles in which local notoriety is so eagerly sought.

In W.J. Miller (ed.)
Scientific Papers: By W.J. Macquorn Rankine
Memoir (p. ix)
Charles Griffin & Company. London, England. 1881

To [the scientific man] the discovery of a new law of nature, or even of a new experimental fact, or the invention of a novel mathematical method, no matter who has been the first to reach it, is an event of an order altogether different from, and higher than, those which are so profusely chronicled in the newspapers.

In W.J. Miller (ed.)
Scientific Papers: By W.J. Macquorn Rankine
Memoir (p. ix)
Charles Griffin & Company. London, England. 1881

Taylor, Alfred Maurice 1903–76
English optics physicist

The three attributes of commitment, imagination, and tenacity seem to be the distinguishing marks of greatness in a scientist. A scientist must be as utterly committed to the pursuit of truth as the most dedicated of mystics; he must be as pertinacious in his struggle to advance into uncharted country as the most indomitable pioneers; his imagination must be as vivid and ingenious as a poet's or a painter's. Like other men, for success he needs ability and some luck; his imagination may be sterile if he has not a flair for asking the right questions, questions to which nature's reply is intelligible and significant.

Imagination and the Growth of Science
Chapter I (p. 5)
Schocken Books. New York, New York, USA. 1970

Thomas, Lewis 1913–93
American physician and biologist

Scientists at work have the look of creatures following genetic instructions; they seem to be under the influence of a deeply placed human instinct. They are, despite their efforts at dignity, rather like young animals engaged in savage play. When they are near to an answer their hair stands on end, they sweat, they are awash in their own adrenaline. To grab the answer, and grab it first, is for them a more powerful drive than feeding or breeding or protecting themselves against the elements.

The Lives of a Cell: Notes of a Biology Watcher
Natural Science (p. 101)
The Viking Press. New York, New York, USA. 1974

Thorne, Kip S. 1940–
American theoretical physicist

I do not aspire to a historian's standards of completeness, accuracy, or impartiality.

Black Holes and Time Warps: Einstein's Outrageous Legacy
Preface (p. 19)
W.W. Norton & Company, Inc. New York, New York, USA. 1994

Ting, Samuel C. C. 1936–
Chinese-American physicist

…scientists must go beyond what is taught in the textbook, and they must think independently. Also, they cannot hesitate to ask questions, even when their view may be unpopular.

In Janet Nomura Morey and Wendy Dunn
Famous Asian Americans
Samuel C.C. Ting (p. 143)

Tolkien, J. R. R. 1892–1973
English philologist, writer, and professor

Merry stared at the lines of marching stones: they were worn and black; some were leaning, some were fallen, some were cracked or broken; they looked like rows of old and hungry teeth. He wondered what they could be…

The Lord of the Rings
The Return of the King, Book Three (p. 795)
HarperCollins Publishers. 2004

Toulmin, Stephen 1922–
English philosopher

No doubt, a scientist isn't necessarily penalized for being a complex, versatile, eccentric individual with lots of extra-scientific interests. But it certainly doesn't help him a bit.

Civilization and Science in Conflict or Collaboration
CIBA Foundation Symposium
Associated Scientific Publishers. Amsterdam, Netherlands. 1972

Twain, Mark (Samuel Langhorne Clemens) 1835–1910
American author and humorist

That is their way, those plagues, those scientists — peg, peg, peg — dig, dig, dig — plod, plod, plod. I wish I could catch a cargo of them for my place; it would be an economy. Yes, for years, you see. They never give up. Patience, hope, faith, perseverance; it is the way of all the breed.

Europe and Elsewhere
Sold to Satan
Harper & Brothers. New York, New York, USA. 1923

That is the way of the scientist. He will spend thirty years in building up a mountain range of facts with the intent to prove a certain theory; then he is so happy in his achievement that as a rule he overlooks the main chief fact of all — that his accumulation proves an entirely different thing.
What Is Man? and Other Essays, 1917 ed
The Bee Essay (p. 283)
Harper & Brothers. New York, New York, USA. 1917

Such is professional jealousy; a scientist will never show any kindness for a theory which he did not start himself. There is no feeling of brotherhood among these people. Indeed, they always resent it when I call them brother. To show how far their ungenerosity can carry them, I will state that I offered to let Prof. H — y publish my great theory as his own discovery; I even begged him to do it; I even proposed to print it myself as his theory. Instead of thanking me, he said that if I tried to fasten that theory on him he would sue me for slander.
A Tramp Abroad
Chapter XLIII (p. 321)
Penguin Books. New York, New York, USA. 1997

Scientists have odious manners, except when you prop up their theory; then you can borrow money of them.
What Is Man? and Other Essays, 1917 ed
The Bee Essay (p. 283)
Harper & Brothers. New York, New York, USA. 1917

University of California, Berekely

Scientists work better when they're all mixed-up.
Advertisement insert
Fortune, April 1986 (p. 814)

Varese, Edgar   1883–1965French American composer

Scientists are the poets of today.
Artspace, Volume 9, Fall 1985 (p. 30)

Vernadskii, Vladimir Ivanovich   1863–1945
Russian mineralogist

Scientists are in fact imaginers and artists; they are not free with their ideas; they can work well and hard only at what their thinking accepts and what their feelings are drawn to. Ideas alternate; impossible and often mad ones appear; they swarm and whirl, fuse and sparkle. Scientists live among these ideas and work for them.
Compiled by V.V. Vorontsov
Words of the Wise: A Book of Russian Quotations
Translated by Vic Schneierson
Progress Publishers. Moscow, Russia. 1979

von Frisch, Karl   1886–1982
Austrian zoologist

No competent scientist *ought* to believe these things on first hearing.
Bees: Their Vision, Chemical Senses, and Language
Foreword (p. vii)
Cornell University Press. Ithaca, New York, USA. 1950

Wald, George   1906–97
American biologist and biochemist

A scientist should be the happiest of men. Not that science isn't serious; but as everyone knows, being serious is one way of being happy…
Les Prix Nobel. The Nobel Prizes in 1967
Nobel banquet speech for award received in 1967
Nobel Foundation. Stockholm, Sweden. 1968

A scientist is in a sense a learned small boy. There is something of the scientist in every small boy. Others must outgrow it. Scientists can stay that way all their lives.
Les Prix Nobel. The Nobel Prizes in 1967
Nobel banquet speech for award received in 1967
Nobel Foundation. Stockholm, Sweden. 1968

Walshe, Sir F. M. R.
No biographical data available

It often is the cloistered scientist who knows least about men who is apt to pontificate most loudly and confidently about Man. Beware of him when he assures you that he knows all the answers about us, for too often his is one of those Peter Pans of science that every generation produces: a clever boy who hasn't grown up.
Canadian Medical Association Journal, Volume 67, 1962 (p. 395)

Weil, Simone   1909–43
French philosopher and mystic

On could count on one's fingers the number of scientists throughout the world with a general idea of the history and development of their particular science: there is none who is really competent as regards sciences other than his own. As science forms an indivisible whole, one may say that there are no longer, strictly speaking, scientists, but only drudges doing scientific work.
Translated by Arthur Wills and John Petrie
Oppression and Liberty
Prospects (p. 13)
Routledge & Kegan Paul Ltd. London, England. 1958

Weinberg, Alvin Martin   1915–2006
American physicist

The traditional working scientists are at the bottom rung — each one knows almost everything about almost nothing; as one progresses toward the top of the ladder, the subject matter becomes more abstract until one finally reaches the philosopher at the top who knows almost nothing about almost everything.
Reflections on Big Science
Chapter II (p. 47)
The MIT Press. Cambridge, Massachusetts, USA. 1967

Weiss, Paul A.   1898–1989
Austrian-born American biologist

Just like the painter, who steps periodically back from his canvas to gain perspective, so the laboratory scientist emerges above ground occasionally from the deep shaft of his specialized preoccupation to survey the cohesive, meaningful fabric developing from innumerable component tributary threads, spun underground much like his own. Only by such shuttling back and forth between the worm's eye view of detail and the bird's eye view of the total scenery of science can the scientist gain and retain a sense of perspective and proportions.
In A. Koestler and J. R. Smithies
Beyond Reductionism: New Perspectives in the Life Sciences
The Living System (p. 3)
Beacon Press. Boston, Massachusetts, USA. 1969

Weisskopf, Victor Frederick   1908–2002
Austrian-American physicist

Science has become adult; I am not sure whether scientists have.
In Anthony R. Michaelis & Hugh Harvey eds.
Scientists in Search of Their Conscience
Conclusion (p. 193)
Springer-Verlag. Berlin, Germany. 1973

Wells, H. G. (Herbert George)   1866–1946
English novelist, historian, and sociologist

What there is great about [scientists] is an annoyance to their fellow scientists and a mystery to the general public, and what is not is evident. There is no doubt about what is not great, no race of men have such obvious littlenesses.... And withal the reef of science that these little "scientists" built and are yet building is so wonderful, so portentous, so full of mysterious half-shapen promises for the mighty future of man! They do not seem to realise the things they are doing.
Seven Famous Novels by H.G. Wells
The Food of the Gods
Chapter I. (p. 533)
Alfred A. Knopf. New York, New York, USA. 1934

No doubt long ago even Mr. Bensington, when...he consecrated his life to the alkaloids and their kindred compounds had some inkling of the vision — more than an inkling. Without some great inspiration, for such glories and positions only as a "scientist" may expect, what young man would have given his life to this work, as young men do? No, they must have seen the glory, they must have had the vision, but so near that it has blinded them, mercifully, so that for the rest of their lives they can hold the light of knowledge in comfort — that we may see.
Seven Famous Novels by H.G. Wells
The Food of the Gods
Chapter I. (p. 533)
Alfred A. Knopf. New York, New York, USA. 1934

Weyl, Hermann   1885–1955
German mathematician

One of the great differences between the scientist and the impatient philosopher is that the scientist bides his time. We must await the further development of science, perhaps for centuries, perhaps for thousands of years, before we can design a true and detailed picture of the interwoven texture of matter, life and soul. But the old classical determinism of Hobbes and LaPlace need not oppress us any longer.
The Open World: Three Lectures in the Metaphysical Implications of Science
Lecture II (p. 55)
Yale University Press. New Haven, Connecticut, USA. 1932

Whewell, William   1794–1866
English philosopher and historian

We need very much a name to describe a cultivator of science in general. I should incline to call him a Scientist. Thus we might say, that as an Artist is a Musician, Painter, or Poet, a Scientist is a Mathematician, Physicist, or Naturalist.
Novum Organum Renovatum
Aphorisms Concerning the Language of Science
John W. Parker & Son. London, England. 1858

Whitehead, Alfred North   1861–1947
English mathematician and philosopher

Many a scientist has patiently designed experiments for the purpose of substantiating his belief that animal operations are motivated by no purpose...Scientists animated by the purpose of proving that they are purposeless constitute an interesting subject for study.
The Function of Reason
Chapter I (p. 12)
Beacon Press. Boston, Massachusetts, USA. 1929

A few generations ago the clergy, or to speak more accurately, large sections of the clergy were the standing examples of obscurantism. Today their place has been taken by scientists.
The Function of Reason
Chapter I (pp. 34–35)
Beacon Press. Boston, Massachusetts, USA. 1929

Whitney, Willis Rodney   1868–1958
American chemical and electrical engineer

We humans want better minds, broader horizons, and greater understanding. Scientists everywhere are at work in their respective fields searching for new truths to improve the process by which our minds, our horizons, our powers, and our outlooks grow.
Annual Report of the Board of Regents of the Smithsonian Institution, 1924
The Vacuum — There's Something in It (p. 193)
Government Printing Office. Washington, D.C. 1925

Wiener, Norbert 1894–1964
American mathematician

...the first industrial revolution,... of the "dark satanic mills," [devalued] the human arm by the competition of machinery.... The modern industrial revolution is similarly bound to devalue the human brain...in its simpler and more routine decisions. Of course, just as the skilled carpenter, the skilled mechanic, the skilled dressmaker have in some degree survived the first industrial revolution, so the skilled scientist and the skilled administrator may survive the second. However,... the average human being of mediocre attainment or less has nothing to sell that is worth anyone's money to buy.
Cybernetics: Or Control and Communication in the Animal and the Machine
Introduction (pp. 27–28)
The MIT Press. Cambridge, Massachusetts, USA. 1961

...the degradation of the position of the scientist as independent worker and thinker to that of a morally irresponsible stooge in a science-factory has proceeded even more rapidly and devastatingly than I had expected.
A Rebellious Scientist after Two Years
Bulletin of the Atomic Scientists, Volume 4, Number 11, November 4, 1948 (p. 338)

[A scientist] must live in a world where science is a career, where he has companions with whom to talk, and in contact with whom he may bring out his own self. It may be true that 95 percent of the really original scientific work is done by less than 5 per cent of the professional scientists, but the greater part of it would not be done at all if the other 95 per cent were not there and did not create a high level of public scientific opinion.
Science, Monkeys, and Mozart
Saturday Review of Literature, November 20, 1956

Wikström, J. E.
Former Swedish Minister of Education and Cultural Affairs

...Scientists are like children playing with fire without heeding the disastrous consequences of their games or, even worse, "they are like incendiaries which completely destroy property."
In Torgny Segerstedt
Ethics for Science Policy: Proceedings of a Nobel Symposium Held at Södergarn, Sweden, 20–25 August 197
Opening Remarks (p. xiii)
Pergamon Press. Oxford, England. 1979

Wilder, Thornton 1897–1975
American playwright and novelist

Then there is technology, the excess of scientists who learn how to make things much faster than we can learn what to do with them.
In Plora Lewis
Thornton Wilder at 65 Looks Ahead – And Back
New York Times Magazine, 15 April 1962 (p. 28)

Wilf, Alexander
No biographical data available

A scientist can not be measured quantitatively by the number of degrees or the accumulation of information. A true scientist should have a measure of courage to correct error and seek truth — no matter how painful. The alternative is more painful. To build error upon error is to drift into dogmas, metaphysics, science fiction, and mythology.
Origin and Destiny of the Moral Species (p. 9)
A.S. Barnes. South Brunswick, New Jersey, USA. 1969

Wilson, Edward O. 1929–
American biologist and author

The ideal scientist thinks like a poet and works like a bookkeeper, and I suppose that if gifted with a full quiver, he also writes like a journalist. As a painter stands before canvas or a novelist recycles old emotion with eyes closed, he searches his imagination for subjects as much as for conclusions, for questions as much as for answers.
Scientists, Scholars, Knaves and Fools
American Scientist, Volume 86, January–February 1998 (p. 7)

Scientists, I believe, are divided into two categories: those who do science in order to be a success in life, and those who become a success in life in order to do science.
Naturalist
The Forms of Things Unknown (p. 210)
Island Press. Washington, D.C. 1994

Scientists do not discover in order to know, they know in order to discover.
Biophilia
The Poetic Species (p. 58)
Harvard University Press. Cambridge, Massachusetts. 1984

Scientists live and die by their ability to depart from the tribe and go out into an unknown terrain and bring back, like a carcass newly speared, some new discovery or fact or theoretical insight and lay it in front of the tribe; and then they all gather and dance around it. Symposia are held in the National Academy of Sciences and prizes are given [The symposia are] fundamentally no [different] from a paleothic camp site celebration.
In Edward Lueders
Writing Natural History: Dialogues with Authors
Dialogue One (p. 25)
University of Utah Press. Salt Lake City, Utah, USA. 1989

The scientist is not a very romantic figure.... [His work] amounts to a sort of puttering: trying to find a good problem, thinking up experiments, mulling over data, arguing in the corridor with colleagues, and making guesses with the aid of coffee and chewed pencils until finally something — usually small — is uncovered. Then comes a flurry of letters and telephone calls, followed by the writing of a short paper in an acceptable jargon. The great majority of scientists are hard-working, pleasant

journeymen, not excessively bright, making their way through a congenial occupation.
Biophilia
The Poetic Species (p. 59)
Harvard University Press. Cambridge, Massachusetts. 1984

Wittgenstein, Ludwig Josef Johann   1889–1951
Austrian-born English philosopher

What a curious attitude scientists have — : "We still don't know that; but it is knowable and it is only a matter of time before we get to know it!" As if that went without saying.
Translated by Peter Winch
Culture and Value (p. 40e)
The University of Chicago Press. Chicago, Illinois, USA. 1980

Wolpert, Lewis   1929–
British embryologist

Both Newton and Darwin were driven by the data and were forced to recognize that they couldn't explain everything. It may be a characteristic of great scientists to know what to accept and what to leave out.
The Unnatural Nature of Science
Chapter 4 (p. 72)
Harvard University Press. Cambridge, Massachusetts, USA. 1992

Young John Zachary   1907–97
English zoologist

One of the characteristics of scientists and their work, curiously enough, is a certain confusion, almost a muddle. This may seem strange if you have come to think of science with a big S as being all clearness and light.
Doubt and Certainty in Science: A Biologist's Reflections on the Brain
First Lecture (p. 1)
Oxford University Press, Inc. Oxford, England. 1960

…in his laboratory he does not spend much of his time thinking about scientific laws at all. He is busy with other things, trying to get some piece of apparatus to work, finding a way of measuring something more exactly.… You may feel that he hardly knows himself what law he is trying to prove. He is continually observing, but his work is a feeling out into the dark, as it were. When pressed to say what he is doing he may present a picture of uncertainty or doubt, even of actual confusion.
Doubt and Certainty in Science: A Biologist's Reflections on the Brain
First Lecture (p. 2)
Oxford University Press, Inc. Oxford, England. 1960

SCRIBBLES

Ulam, Stanislaw   1909–84
Polish-born mathematician

It is still an unending source of surprise for me how a few scribbles on a blackboard or on a piece of paper can change the course of human affairs.

Adventures of a Mathematician
Prologue (p. 5)
Charles Scribner's Sons. New York, New York, USA. 1976

SEA

Beebe, William   1877–1962
American ornithologist

When once it has been seen, it will remain forever the most vivid memory in life, solely because of its cosmic chill and isolation, the eternal and absolute darkness and the indescribable beauty of its inhabitants.
Half Mile Down
Chapter 9 (p. 175)
Harcourt, Brace & Company. New York, New York, USA. 1934

Berger, John   1926–
English art critic, novelist, painter and author

The sun is low in the sky and the sea is calm. Like a mirror as they say. Only it is not like a mirror. The waves which are scarcely waves, for they come and go in many different directions and their rising and falling is barely perceptible, are made up of innumerable tiny surfaces at variegating angles to one another — of these surfaces those which reflect the sunlight straight into one's eyes, sparkle with a white light during the instant before their angle, relative to oneself and the sun, shifts and they merge again into the blackish blue of the rest of the sea.
G
Chapter 10 (p. 310)
The Viking Press. New York, New York, USA. 1972

[A]s the sea recedes towards the sun, the number of sparkling surfaces multiplies until the sea indeed looks somewhat like a silver mirror. But…it is not still. Its granular surface is in continual agitation. The further away the ricocheting grains, of which the mass become silver and the visibly distinct minority a dark leaden colour, the greater is their apparent speed. Uninterruptedly receding towards the sun, the transmission of its reflexions becoming ever faster, the sea neither requires nor recognizes any limit. The horizon is the straight bottom edge of a curtain arbitrarily and suddenly lowered on a performance.
G
Chapter 10 (p. 310)
The Viking Press. New York, New York, USA. 1972

Beston, Henry   1888–1968
American writer

Listen to the surf, really lend it to your ears, and you will hear in it a world of sounds: hollow boomings and heavy roarings, great watery tumblings and tramplings, long hissing seethes, sharp, rifle-shot reports, splashes, whispers, the grinding undertone of stones, and sometimes vocal sounds that might be the half-heard talk of people in the sea.

The Outermost House
Chapter III (p. 43)
Rinehart & Company. New York, New York, USA. 1928

Bradley, Jr., John Hodgdon   1898–1962
American geologist

Like the turbulent crowd at the rim of an arena, the sea surrounds the lands and awaits the issue of combat. Fretfully the colossus chafes at the margins of battlefields it is eager but impotent to enter.
Autobiography of Earth
Chapter VI (p. 168)
Coward-McCann, Inc. New York, New York, USA. 1935

Broch, Hermann   1886–1951
Austrian writer

Those who live by the sea can hardly form a single thought of which the sea would not be part.
The Spell
Forward (p. 3)
North Point Press. San Francisco, California, USA. 1989

Carson, Rachel   1907–64
American marine biologist and author

There we see the parts of the plan fall into place: the water receiving from earth and air the simple materials, storing them up until the gathering energy of the spring sun wakens the sleeping plants to a burst of dynamic activity, hungry swarms of planktonic animals growing and multiplying upon the abundant plants, and themselves falling prey to the shoals of fish; all, in the end, to be redissolved into their component substances when the inexorable laws of the sea demand it.
Undersea
Atlantic Monthly, September 1937 (p. 29)

It is a curious situation that the sea, from which life first arose should now be threatened by the activities of one form of that life. But the sea, though changed in a sinister way, will continue to exist; the threat is rather to life itself.
The Sea Around Us
Preface (p. xiii)
Oxford University Press, Inc. New York, New York, USA. 1989

Cousteau, Jacques-Yves   1910–77
French naval officer and ocean explorer

The sea is not a bargain basement.
The Living Sea
Chapter Seventeen (p. 313)
Harper & Row, Publishers. New York, New York, USA. 1963

From the vast expanses of its surface waters to its beaches and marshes and tidelands and mangrove swamps, from its many thousands of miles of rocky shores to its deepest and darkest abyss, the sea produces life in fantastic abundance.

The Ocean World of Jacques Cousteau: The Adventure of Life
Chapter I (p. 10)
The World Publishing Company. New York, New York, USA. 1973

The sea is the universal sewer.
House Committee on Science and Astronautics
28 January 1971

Cromie, William J.   1930–
American journalist and writer

In the open ocean one sees no green meadows or fertile prairies. Away from the rim of seaweed around the coasts one is aware of only an endless confusion of seemingly barren waves. Yet there are lush pastures in the open sea.
The Living World of the Sea
Chapter 3 (p. 37)
Prentice-Hall, Inc. Englewood Cliffs, New Jersey, USA. 1966

The sea gave birth to life and without its waters no living thing could ever survive on Earth. Now man, rightly or wrongly, is looking to the ocean as his ultimate safety valve, the answer to his problems of food, waste and even space. But the sea has no mind and the sea is not inexhaustible.
The Living World of the Sea
Chapter 15 (p. 332)
Prentice-Hall, Inc. Englewood Cliffs, New Jersey, USA. 1966

Diolé, Philippe
French archaeologist

Between the air and the water a steel blade quivers. What people call the surface is also a ceiling: a mirror from above, watered silk from beneath. Nothing is torn on the way through. Only a few bubbles mark the diver's channel and behind him the frontier soon closes. But once the threshold is crossed, one can turn back slowly and look up: that dazzling screen is the border between two worlds, as clear to one as to the other. Behind the looking glass the sky is made of water.
Translated by Alan Ross
The Undersea Adventure
Chapter 1 (pp. 6–7)
Julian Messner, Inc. New York, New York, USA. 1953

Emerson, Ralph Waldo   1803–82
American lecturer, poet, and essayist

To the geologist the sea is the only firmament…
The Complete Works of Ralph Waldo Emerson (Volume 5)
English Traits (p. 29)
Houghton Mifflin Company. Boston, Massachusetts, USA. 1904

The sea, washing the equator and the poles, offers its perilous aid, and the power and empire that follow it.…
"Beware of me," it says, "but if you can hold me, I am the key to all the lands."
Ralph Waldo Emerson: Essays and Lectures
The Conduct of Life

Wealth (p. 991)
The Library of America. New York, New York, USA. 1983

Empedocles of Acragas   ca. 490 BCE–430 BCE
Greek pre-Socratic philosopher

The sea is the sweat of the earth.
In Arthur Fairbanks
The First Philosophers of Greece
Book I
Fragment 165 (p. 179)

Flecker, James Elroy   1884–15
English poet and playwright

The dragon-green, the luminous, the dark, the serpent-haunted sea.
The Collected Poems of James Elroy Flecker
The Gates of Damascus
West Gate
Doubleday, Page & Company. New York, New York, USA. 1916

Garfield, James A.   1831–81
20th president of the United States

I have seen the sea lashed into fury and tossed into spray, and its grandeur moves the soul of the dullest man; but I remember that it is not the billows, but the calm level of the sea from which all heights and depths are measured.
Proceedings of the Republican National Convention
Chicago, Illinois, June 2–8, 1880 (p. 184)

Hardy, Thomas   1840–1928
English poet and regional novelist

Who can say of a particular sea that it is old? Distilled by the sun, kneaded by the moon, it is renewed in a year, in a day, or in an hour.
The Return of the Native
Book the First, Chapter I (p. 14)
The New American Library. New York, New York, USA. 1959

Hazlitt, William Carew   1834–1913
English bibliographer

I hate to be near the sea, and to hear it roaring and raging like a wild beast in its den. It puts me in mind of the everlasting efforts of the human mind, struggling to be free, and ending just where it began.
Common Places

Houot, Georges   1913–2000
French underwater explorer
Willm, Pierre
No biographical data available

We are entering on the last stage of man's march toward a knowledge of the surface of the globe. The battle that remains to be fought will be long and hard. Despite the progress of science, the sea remains a hostile element, particularly so at the frontier between the water and the atmosphere.
Translated by Michael Bullock
2000 Fathoms Down (p. 182)
E.P. Dutton & Company. New York, New York, USA. 1955

Hugo, Victor   1802–85
French author, lyric poet, and dramatist

When it wishes to be, the sea is gay.
Translated by Isabel F. Hapgood
The Toilers of the Sea
Part II, Book Third, Chapter III (p. 406)
The Heritage Press. New York, New York, USA. 1961

Knight, Norman L.
No biographical data available

O Sea! Thou saline and undulant aqueous solution of halides, carbonates, phosphates, sulfates, and other soluble inorganic compounds! What mysterious colloids are dispersed within thy slightly alkaline bosom? What silent and unseen reactions vibrate in dynamic equilibrium, constantly destroyed and instantly restored, among thy unnumbered oscillating molecules? What uncounted myriads of restless ions migrate perpetually throughout thy tentatively estimated volume? What unguessed phenomena of catalysis, metathesis, and osmosis transpire in thy secret fluid profundities under excessively increased pressure? What cosmic precipitates descend in countless kilograms upon thy argillaceous, gelatinous, siliceous, diatomaceous, and totally unillumined bottom? In short, most magnificent reservoir, what is thy flow-chart and complete analysis?
A Chemist Addresses the Ocean
Industrial and Engineering Chemistry: News Edition, Volume 8, Number 22, September 20, 1930

Ledbetter, B. G.
No biographical data available

The sea often appears tranquil and serene, but its permanent hills and valleys illustrate that in fact it is restless, always in motion, mixing and flowing in ceaseless search for peace.
Sea Level Isn't Level — It's Hilly
Science Digest, Volume 68, Number 1, July 1970 (p. 72)

Lindbergh, Anne Marrow   1906–2001
American aviator and writer

The sea does not reward those who are too anxious, too greedy, or too anxious. To dig for treasures shows not only impatience and greed, but a lack of faith. Patience, patience, patience, is what the sea teaches. Patience and faith. One should lie empty, open, choiceless as a beach-waiting for a gift from the sea.
Gift from the Sea
Part I, The Beach (p. 17)
Pantheon Books, Inc. New York, New York, USA. 1955

Longfellow, Henry Wadsworth   1807–82
American poet

Learn the secret of the sea?
Only those who brave its dangers
Comprehend its mystery.
The Seaside and the Fireside
The Secret of the Sea
Ticknor, Reed & Fields. Boston, Massachusetts, USA. 1850

…my soul is full of longing for the secrets of the sea.
And the heart of the great ocean
Sends a thrilling pulse through me.
The Seaside and the Fireside
The Secret of the Sea
Ticknor, Reed & Fields. Boston, Massachusetts, USA. 1850

Lowell, James Russell   1819–91
American poet, critic, and editor

The sea was meant to be looked at from shore, as mountains are from the plain.
Fireside Travels
At Sea (p. 155)
Ticknor & Fields. Boston, Massachusetts, USA. 1864

There is nothing so desperately monotonous as the sea, and I no longer wonder at the cruelty of pirates.
Fireside Travels
At Sea (p. 157)
Ticknor & Fields. Boston, Massachusetts, USA. 1864

Lubbock, Sir John   1834–1913
English banker, author, and scientist

The Sea is outside time. A thousand, ten thousand, or a million years ago it must have looked just as it does now, and as it will ages hence.
The Beauties of Nature and the Wonders of the World We Live In
Chapter IX (p. 340)
Macmillan & Company New York, New York, USA. 1893

Maury, Matthew Fontaine   1806–73
American hydrographer and naval officer

The sea…has its offices and duties to perform;… its currents, and so, too, its inhabitants; consequently, he who undertakes to study its phenomena must cease to regard it as a waste of waters. He must look upon it as a part of that exquisite machinery by which the harmonies of nature are preserved, and then he will begin to perceive the developments of order and the evidence of design; these make it a most beautiful and interesting subject for contemplation.
In J.A. Colin Nicol
The Biology of Marine Animals (2nd edition)
Chapter I (p. 1)
Sir Isaac Pitman & Sons Ltd. London, England. 1967

Could the waters of the Atlantic be drawn off so as to expose to view this great sea-gash…it would present a scene most rugged, grand, and imposing. The very ribs of the solid earth, with the foundations of the sea would be brought to light, and we should have presented to us at one view, in the empty cradle of the ocean, "a thousand fearful wrecks," with that dreadful array of dead men's skulls, great anchors, heaps of pearl and inestimable stones, which, in the poet's eye, lie scattered in the bottom of the sea, making it hideous with sights of ugly death.
The Physical Geography of the Sea
Chapter XII (p. 208)
Harper & Brothers. New York, New York, USA. 1855

Harmonious in their action, the air and sea are obedient to law and subject to order in all their movements; when we consult them in the performance of their offices, they teach us lessons concerning the wonders of the deep, the mysteries of the sky, the greatness, and the wisdom, and goodness of the Creator.
The Physical Geography of the Sea
Chapter III (p. 96)
Harper & Brothers. New York, New York, USA. 1855

Astronomers had measured the volumes and weighed the masses of the most distant planets, and increased thereby the stock of human knowledge. Was it creditable to the age that the depths of the sea should remain in the category of an unsolved problem?… Indeed, telescopes of huge proportions and of vast space-penetrating powers had been erected here and there by the munificence of individuals, and attempts made with them to gauge the heavens and sound out the regions of space. Could it be more difficult to sound out the sea than to gauge the blue ether and fathom the vault of the sky?
The Physical Geography of the Sea
Chapter XI (pp. 201, 202)
Harper & Brothers. New York, New York, USA. 1855

Melville, Herman   1819–91
American novelist

There is, one knows not what sweet mystery about the sea, whose gently awful stirrings seem to speak of some hidden soul beneath.
In *Great Books of the Western World* (Volume 48)
Moby Dick
Chapter 107 (p. 354)
Encyclopædia Britannica, Inc. Chicago, Illinois, USA. 1952

Miller, Robert C.
No biographical data available

We can no longer think of the sea as a vast, illimitable dumping ground for products that man does not know what to do with on land. It must instead be recognized as our greatest natural resource, and one to be conserved in every possible way. So regarded, and wisely used, it can be a permanent source of raw materials, of food, of life-giving water, and of recreation, enjoyment and adventure.

The Sea
Chapter 15 (p. 311)
Random House, Inc. New York, New York, USA. 1966

Muir, John   1838–1914
American naturalist

…both ocean and sky are already about as rosy as possible; the one with stars, the other with dulse, and foam, and wild light.
Steep Trails
Chapter I (p. 3)
Norman S. Berg, Publisher. Dunwoody, Georgia, USA. 1970

Plattes, Gabriel
No biographical data available

…the Sea never resting, but perpetually winning land in one place and losing in another, doth show what may be done in length of time by a continual operation not subject unto ceasing or intermission.
Discovery of Subterraneal Treasure
Chapter XI (p. 22)
Printed by Robert Bell. Philadelphia, Pennsylvania, USA. 1784

Sandburg, Carl   1878–1967
American poet and biographer

The sea folds away from you like a mystery. You can look and look at it and mystery never leaves it.
Remembrance Rock (p. 75)
Harcourt, Brace & World, Inc. New York, New York, USA. 1948

Sexton, Anne   1928–74
American poet

The sea is mother-death and she is a mighty female, the one who wins, the one who sucks us all up.
In Howard Moss (ed.)
The Poet's Story
A Small Journal (p. 219)
19 November, 1971
The Macmillan Company. New York, New York, USA. 1973

Stevenson, Robert Louis   1850–94
Scottish essayist and poet

"And what is the sea?" asked Will.

"The sea!" cried the miller. "Lord help us all, it is the greatest thing God made! …There are great fish in it five times bigger than a bull, and one old serpent as long as our river and as old as all the world, with whiskers like a man, and a crown of silver on her head.
Strange Case of Dr. Jekyll and Mr. Hyde. The Merry Men and Other Tales and Fables
Will O' the Wisp
The Plain and the Stars (pp. 77–78)
Current Literature Publishing Company New York, New York, USA. 1912

The Bible

Others there are who go to sea in ships, plying their trade on the wide ocean. These have seen what the Lord has done, his marvelous actions in the deep.
The Revised English Bible
Psalms 107:23–24
Oxford University Press, Inc. Oxford, England. 1989

Maury, Matthew Fontaine   1806–73
American hydrographer and naval officer

We never tire of the sea; it is a laboratory in which delightful processes are continually being wrought out for our admiration and use. Its flora and its fauna, its waves and its tides, its salts and its currents, all afford grand and profitable themes of study and thought.
The Physical Geography of the Sea, and Its Meteorology
Chapter XVIII, section 740 (p. 394)
Sampson Low, Son & Marston. London, England. 1868

Young, Louise B.
Science writer

Throughout the planet's history the sea has carved and molded the character of the land. She has scooped out steep escarpments and deep gorges, impressed the rhythm of her movement on the hard rocky shores of the continents. But still she is ever yielding. Beneath her smiling, enigmatic face there are grave depths where silence and darkness dwell always. Here in these hidden places she watches impassively while the earth tears itself violently apart and makes itself anew. Quietly giving way to make room for the growing landmass, she receives and holds this newborn substance in her soft embrace.
The Blue Planet
Chapter 2 (p. 46)
Little, Brown & Company. Boston, Massachusetts, USA. 1983

Verne, Jules   1828–1905
French novelist

Yes; I do love it! The sea is everything. It covers seven-tenths of the terrestrial globe. Its breath is pure and healthy. It is an immense desert, where man is never lonely, for he feels life stirring on all sides. The sea is only the embodiment of a supernatural and wonderful existence. …
Translated by Mercier Lewis
Twenty Thousand Leagues Under the Sea
Part One, Chapter 10 (p. 58)
Nelson Doubleday, Inc. Garden City, New York, USA. 1900

Nature manifests herself in it by her three kingdoms, mineral, vegetable, and animal. The sea is the vast reservoir of Nature. The globe began with sea, so to speak; and who knows if it will not end with it? In it is supreme tranquility. The sea does not belong to despots. Upon its surface men can still exercise unjust laws, fight, tear

one another to pieces, and be carried away with terrestrial horrors. But at thirty feet below its level, their reign ceases, their influence is quenched, and their power disappears.
Translated by Mercier Lewis
Twenty Thousand Leagues Under the Sea
Part One, Chapter 10 (p. 58)
Nelson Doubleday, Inc. Garden City, New York, USA. 1900

SEA SERPENT

19th Century Naval Song

Strange things come up to look at us —
The masters of the deep.
The Modern Traveler
The Return of the Admiral (p. 107)
Printed for T. Lowndes. London, England. 1776–1777

Pontoppidan, Erich   1698–1754
Bishop in Bergen, Norway

Amongst the many great things which are in the ocean, and concealed from our eyes or only presented to our view for a few minutes, is the Kraken. This creature is the largest and most surprising of all the animal creation, and consequently well deserves such an account as the nature of the thing, according to the Creator's wise ordinance, will admit of.
Natural History of Norway

Tennyson, Alfred (Lord)   1809–92
English poet

Below the thunders of the upper deep;
Far, far beneath in the abysmal sea,
His ancient, dreamless, uninvaded sleep
The Kraken sleepeth…
Alfred Tennyson's Poetical Works
The Kraken
Oxford University Press, Inc. London, England. 1953

SEA SICKNESS

Charlie Chan
Fictional character

Mention of food more painful than surgeon's knife without anesthetic.
Charlie Chan on Broadway
Film (1937)

Jerome, Jerome K.   1859–1927
English author

It is a curious fact, but nobody ever is seasick — on land. At sea, you come across plenty of people very bad indeed, whole boat-loads of them; but I never met a man yet, on land, who had ever known at all what it was to be seasick.

Three Men in a Boat, to Say Nothing of the Dog!
Chapter 1 (p. 10)
Time Incorporated. New York, New York, USA. 1964

SEASIDE

Spencer, Herbert   1820–1903
English social philosopher

Whoever at the seaside has not had a microscope and aquarium, has yet to learn what the highest pleasures of the seaside are.
Education: Intellectual, Moral, and Physical
Chapter I (pp. 72–73)
A.L. Fowle. New York, New York, USA. 1860

SEDIMENT

Geikie, Sir Archibald   1835–1924
English geologist

I know no recent observation in physical geography more calculated to impress deeply the imagination than the testimony of this presumably meteoric iron from the most distant abysses of the ocean. To be told that mud gathers on the floor of these abysses at an extremely slow rate conveys but a vague notion of the tardiness of the process. But to learn that it gathers so slowly, that the very star-dust which falls from outer space forms an appreciable part of it, brings home to us, as hardly anything else could do, the idea of undisturbed and excessively slow accumulation.
The Harvard Classics
Scientific Papers: Physiology, Medicine, Surgery, Geology: With Introductions and Notes
Geographical Evolution (p. 347)
P.F. Collier & Son. New York, New York, USA. 1910

Kipling, Rudyard   1865–1936
British writer and poet

There is no sound, no echo of sound, in the deserts of the deep,
Or the great grey level plains of ooze where the shell-burred cables creep.
Here is the womb of the world — here on the tie-ribs of earth
Words, and the words of men, flicker and flutter and beat.
Rudyard Kipling's Verse
The Deep Sea Cables
Hodder & Stroughton. London, England. 1919

SEED

Baker, Henry   1698–1774
English naturalist

Each seed includes a Plant: that Plant, again,

Has other Seeds, which other Plants contain:
Those other Plants have All their Seeds, and Those
More Plants again, successively, inclose.
Thus ev'ry single Berry that we find,
Has, really, in itself whole Forests of its Kind.
The Discovery of a Perfect Plant in Semine
Philosophical Transactions of the Royal Society of London, Number
457, 1740 (p. 451)

A ripe seed falling to the earth is in the condition of the
ovum of an animal getting loose from its ovary and dropping into the uterus, and, to go on with the analogy, the
juices of the earth swell and extend the vessels of the
seed as the juices of the uterus do those of the ovum, till
the seminal leaves unfold and perform the office of a pla-
centa to the infant included plant; which, imbibing suitable and sufficient moisture, gradually extends its parts,
fixes its own root, shoots above the ground, and may be
said to be born.
The Discovery of a Perfect Plant in Semine
Philosophical Transactions of the Royal Society of London, Number
457, 1740 (p. 451)

de la Mare, Walter   1873–1956
English poet and novelist

The seeds I sowed —
For weeks unseen —
Have pushed up pygmy
Shoots of green;
So frail you'd think
The tiniest stone
Would never let
A Glimpse be shown.
Rhymes and Verses: Collected Poems for Children
Seeds
H. Holt & Company. New York, New York, USA. 1947

Muir, John   1838–1914
American naturalist

The dispersal of Juniper seeds is effected by the plum and
cherry plan of living birds at the cost of their board, and
thus obtaining the use of a pair of extra good wings.
Our National Parks
Chapter IV (p. 121)
Houghton Mifflin Company. Boston, Massachusetts, USA. 1901

Sequoia seeds have flat wings and glint and glance in
their flight like a boy's kite.
Our National Parks
Chapter IV (p. 121)
Houghton Mifflin Company. Boston, Massachusetts, USA. 1901

Ruskin, John   1819–1900
English writer, art critic and social reformer

The reason for seeds is that flowers may be; not the rea-
son of flowers that seeds may be.
*The Queen of the Air Being a Study of the Greek Myths of Cloud and
Storm*

II, Section 60, Athena in the Earth (p. 67)
Smith, Elder & Company London, England. 1869

Tabb, John Banister   1845–1909
American poet

Bearing a life unseen,
Thou lingerest between
A flower withdrawn,
And — what thou ne'er shalt see —
A blossom yet to be
When thou art gone.
The Poetry of Father Tabb
Nature — Miscellaneous, The Seed
Dodd, Mead. New York, New York, USA. 1928

Wilde, Oscar   1854–1900
Irish wit, poet, and dramatist

To look at a thing is very different from seeing a thing.
The Decay of Lying (p. 19)
Sunflower Company. New York, New York, USA. 1902

SEISMOGRAPH

Oldham, Richard Dixon   1858–1936
English geologist

…the seismograph, recording the unfelt motion of distant
earthquakes, enables us to see up to a certain point into
the Earth and determine its nature with as great a certainty, as if we could drive a tunnel through it and take
samples of the matter passed through.
Geological Society, The Constitution of the Interior of the Earth as
Revealed by Earthquakes
Quarterly Journal, Volume 62, August 1906 (p. 456)

SEISMOGRAPHER

Karch, Carroll S.
No biographical data available

Seismographer: Shudder bug.
Quote, the Weekly Digest
August 4, 1968 (p. 97)

SEISMOGRAPHY

Richter, Charles   1900–85
American seismologist

Since my first attachment to seismology, I have had the
horror of [earthquake] predictions and of predictors.
Journalists and the general public rush to any suggestion of earthquake prediction like hogs toward a full
trough.
Annals of the New York Academy of Sciences, Ethical and Scientific
Issues Posed by Human Uses of Molecular Genetics, Announcements
Bulletin American Seismological Society, Volume 67, Number 4, August
1977 (p. 1246)

Schrödinger, Erwin   1887–1961
Austrian theoretical physicist

...there are natural sciences which have obviously no practical bearing at all on the life of the human society: astrophysics, cosmology, and some branches of geophysics. Take, for instance, seismology. We know enough about earthquakes to know that there is very little chance of foretelling them, in the way of warning people to leave their houses, as we warn trawlers to return when a storm is drawing near.
Science and Humanism: Physics in Our Time
The Spiritual Bearing of Science on Life (pp. 2–3)
At The University Press. Cambridge, England. 1952

SELF

Pascal, Blaise   1623–62
French mathematician and physicist

When I consider the small span of my life absorbed in the eternity of all time, or the small part of space which I can touch or see engulfed by the infinite immensity of spaces that I know not and that know me not, I am frightened and astonished to see myself here instead of there...now instead of then.
In Rudy Rucker
Infinity and the Mind
Chapter I (p. 2)
Princeton University Press. Princeton, New Jersey, USA. 1995

SELF-AWARENESS

Nuland, Sherwin B.   1930–
American surgeon and teacher of bioethics and medicine

Self-awareness has never been the strong suit of those who choose to become doctors. When so much fuel is readily available for stoking the fires of ego, there is little inclination to apply it in raising the candlepower of the searching light that might illumine the inner man or woman.
The Uncertain Art: The Whole Law of Medicine
The American Scholar, Volume 67, Number 3, Summer, 1998

SELF-DELUSION

Sylvester, James Joseph   1814–97
English mathematician

It is difficult to estimate the lengths to which human self-delusion can be carried.
The Collected Mathematical Papers of James Joseph Sylvester
(Volume 3)
A Lady's Fan on Parallel Motion, and on an Orthogonal Web of Jointed Rods (p. 82)
University Press. Cambridge, England. 1904–1912

SEMINAR

Djerassi, Carl   1929–
Austrian-born American organic chemist and educator

Seminar is not yet officially a transitive verb. Still, most graduate students in any large research-oriented university have at times felt themselves more the helpless objects of a seminar than its active participants. Seminared into numbness describes that feeling of oversaturation.
Cantor's Dilemma 1989
Chapter 14 (p. 122)
Penguin Group Inc. New York, New York, USA.

SENSES

Einstein, Albert   1879–1955
German-born physicist

We can only see the universe by the impressions of our senses reflecting indirectly the things of reality.
Cosmic Religion, with Other Opinions and Aphorisms
On Science (p. 101)
Covici-Fiede. New York, New York, USA. 1931

SERIES

Abel, Niels Henrik   1802–29
Norwegian mathematician

With the exception of the geometric series, there does not exist in all of mathematics a single infinite series whose sum has been determined rigorously.
In Eli Maor
To Infinity and Beyond: A Cultural History of the Infinite (p. 29)
Birkhäuser. Boston, Massachusetts, USA. 1987

The divergent series are the invention of the devil, and it is a shame to base on them any demonstration whatsoever. By using them, one may draw any conclusion he pleases and that is why these series have produced so many fallacies and so many paradoxes...
In Eli Maor
To Infinity and Beyond: A Cultural History of the Infinite (p. 33)
Birkhäuser. Boston, Massachusetts, USA. 1987

SET THEORY

Armstrong, David Malet   1926–
Australian philosopher

Set theory is peculiarly important...because mathematics can be exhibited as involving nothing but set-theoretical propositions about set-theoretical entities.
A Combinatorial Theory of Possibility
Part I, Chapter 1, Section II (p. 10)
Cambridge University Press. Cambridge, England. 1989

Philosophers have not found it easy to sort out sets...

A Combinatorial Theory of Possibility
Part II, Chapter 9, Section iv (p. 133)
Cambridge University Press. Cambridge, England. 1989

Barwise, Jon   1942–2000
American mathematician, philosopher, and logician
Moss, Lawrence
No biographical data available

Set theory has a dual role in mathematics. In pure mathematics, it is the place where questions about infinity are studied. Although this is a fascinating study of permanent interest, it does not account for the importance of set theory in applied areas. There the importance stems from the fact that set theory provides an incredibly versatile toolbox for building mathematical models of various phenomena.
Vicious Circles: On the Mathematics of Non-Wellfounded Phenomena
Chapter 1 (p. 5)
Center for the Study of Language and Information. Stanford, California, US. 1996

Byron, George Gordon, 6ᵗʰ Baron Byron   1788–1824
English Romantic poet and satirist

For a "mixt company" implies, that, save
Yourself and friends, and half a hundred more,
Whom you may bow to without looking grave,
The rest are but a vulgar set.
The Complete Poetical Works of Byron
Beppo: A Venetian Story
Houghton Mifflin. Boston, Massachusetts, USA. 1933

Cleveland, Richard
No biographical data available

We can't be assured of a full set,
Or even a reasonable dull set.
It wouldn't be clear
That there's any set here,
Unless we assume there's a null set.
The Axioms of Set Theory
Mathematics Magazine, Volume 52, Number 4, September 1979
(pp. 256–257)

Poincaré, Henri   1854–1912
French mathematician and theoretical astronomer

Later generations will regard *Mengenlehre* as a disease from which one has recovered.
In Jeremy Gray
Did Poincaré Say "Set Theory Is a Disease"?
The Mathematical Intelligencer, Volume 13, Number 1, Winter 1991
(p. 19)

Quine, Willard Van Orman   1908–2000
American logician and philosopher

To say that mathematics in general has been reduced to logic hints at some new firming up of mathematics at its foundations. This is misleading. Set theory is less settled and more conjectural than the classical mathematical superstructure than can be founded upon it.
Elementary Logic
Chapter IV, Section 48 (p. 125)
Harper & Row, Publishers. New York, New York, USA. 1965

SEX

Heinlein, Robert A.   1907–88
American science fiction writer

Sex is a learned art, as much so as ice skating or tight wire walking or fancy diving; it is not instinct. Oh, two animals couple by instinct, but it takes intelligence and patient willingness to turn copulation into a high and lively art.
Time Enough for Love
Chapter XII (p. 314)
G.P. Putnam's Sons. New York, New York, USA. 1973

Jung, Carl G.   1875–1961
Swiss psychiatrist and founder of analytical psychology

Our civilization bothers us less with food tabus than with sexual restrictions. In modern society these have come to play the role of an injured deity that is getting its own back in every sphere of human activity, including psychology, where it would reduce "spirit" to sexual repression.
Translated by R.F.C. Hull
Flying Saucers: A Modern Myth of Things Seen in the Skies
Chapter Two (p. 45)
Routledge & Kegan Paul. London, England. 1959

Stoppard, Tom   1937–
Czech-born English playwright

Lending one's bicycle is a form of safe sex, possibly the safest there is.
Arcadia
Act I, Scene Four (p. 51)
Faber & Faber Ltd. London, England. 1993

Watts, Alan Wilson   1915–73
American philosopher

Perhaps one of the subordinate reasons why sex is a matter for laughter is that there is something ridiculous in "doing" it with set purpose and deliberation...
Nature, Man, and Woman
Part II, Chapter 8 (p. 201)
Vintage Books. New York, New York, USA. 1970

SEXUALITY

Linnaeus, Carl (von Linné)   1707–78
Swedish botanist and explorer

The organs of generation, which in the animal kingdom are by nature generally removed from sight, in the

vegetable kingdom are exposed to the eyes of all, and that when their nuptials are celebrated, it is wonderful what delight they afford to the spectator by their most beautiful colors and delicious odors.
Oeconomia naturae
Amoenitates Academicae, Volume 2, 1752 (p. 16)

By what mechanisms are the sexuality of the worker naked mole rats suppressed, and how does the queen exert her supremacy? Research at London's Institute of Zoology by Chris Faulkes and others shows surprisingly that the main mechanism are not pheromonal (chemical) as we might immediately suppose. Mysteriously, it is the queenly presence, her behavior, that keeps the rest so firmly switched off; which one of the British researcher has called the "Thatcher effect."
New Scientist, Volume 131, No 1780, 3 August 1991 (p. 43)

SHADOW

Eddington, Sir Arthur Stanley   1882–1944
English astronomer, physicist, and mathematician

In the world of physics we watch a shadowgraph performance of familiar life. The shadow of my elbow rests on the shadow-tables as the shadow-ink flows over the shadow-paper…The frank realisation that physical science is concerned with a world of shadows is one of the most significant of recent advances.
The Nature of the Physical World
Introduction (p. xiv, xv)
The Macmillan Company. New York, New York, USA. 1930

SHAPES

Lovecraft, H. P. (Howard Phillips)   1890–1937
American writer of fantasy, horror, and science fiction

They told him that every figure of space is but the result of the intersection by a plane of some corresponding figure of one or more dimension — as a square is cut from a cube, or a circle from a sphere.
Through the Gates of the Silver Key
Weird Tales, Chapter 5, Volume 24, Number 1, July 1934

SHELL

Glaessner, M. F.   1906–1989
Australian paleontologist and professor of pre-Cambrian life

The naive assumption that shells are acquired because they protect soft bodies seems influenced by anthropocentric thinking: man uses shields for protection from aggressors.
The Dawn of Animal Life: A Biohistorical Study
Chapter 4.6 (p. 174)
Cambridge University Press. Cambridge, England. 1984

Hooke, Robert   1635–1703
English physicist

I…humbly conceive (tho' some possibly may think there is too much notice taken of such a trivial thing as a rotten Shell, yet) that Men do generally too much…pass over without regard these Records of Antiquity which Nature have left as Monuments and Hieroglyphick Characters of preceding Transactions in the like duration or Transactions of the Body of the Earth, which are infinitely more evident and certain tokens than any thing of Antiquity that can be fetched out of Coins or Medals…since the best of those ways may be counterfeited or made by Art and Design…
The Posthumous Works of Robert Hooke
A Discourse on Earthquakes (p. 411)
S. Smith & B. Walford. London, England. 1705

Hsi, Chu (Zhu Xi)   1130–1200
Chinese philosopher

I have seen on high mountains conchs and oyster shells, often embedded in the rocks. These rocks in ancient times were earth or mud, and the conchs and oysters lived in water. Subsequently everything that was at the bottom came to be at the top, and what was originally soft became solid and hard. One should meditate deeply on such matters, for these facts can be verified.
In Joseph Needham
Science and Civilisation in China (Volume 3)
Chapter 23 (p. 598 ff)
The University Press. Cambridge, England. 1954

Tennyson, Alfred (Lord)   1809–92
English poet

See what a lovely shell,
Small and pure as pearl,
Lying close to my foot,
Frail, but a work divine,
Made so fairly well,
With delicate spire and whorl,
How exquisitely minute,
A miracle of design!
Alfred Tennyson's Poetical Works
Maud, Part II, Section II, Stanza I
Oxford University Press, Inc. London, England. 1953

SHORE

Carson, Rachel   1907–64
American marine biologist and author

…the shore has a dual nature, changing with the swing of the tides, belonging now to the land, now to the sea. On the ebb tide it knows the harsh extremes of the land world, being exposed to heat and cold, to wind, to rain and drying sun. On the flood side it is a water world, returning briefly to the relative stability of the open sea.

The Edge of the Sea
Chapter I (p. 1)
Houghton Mifflin Company. Boston, Massachusetts, USA. 1955

The shore is an ancient world, for as long as there has been an earth and sea there has been this place of the meeting of land and water.
The Edge of the Sea
Chapter I (p. 2)
Houghton Mifflin Company. Boston, Massachusetts, USA. 1955

Emerson, Ralph Waldo   1803–82
American lecturer, poet, and essayist

I with my hammer pounding evermore
The rocky coast, smite Andes into dust,
Strewing by beds and, in another age,
Rebuild a continent for better men.
The Complete Works of Ralph Waldo Emerson (Volume 9)
Seashore (p. 243)
Houghton Mifflin Company. Boston, Massachusetts, USA. 1904

SICKNESS

Barnes, Djuna   1892–1982
American author

No man needs curing of his individual sickness; his universal malady is what he should look to.
Nightwood
La Somnambule (p. 41)
Harcourt, Brace & Company. New York, New York, USA. 1937

Burton, Robert   1577–1640
English clergyman and scholar

Sickness is the mother of modesty, putteth us in minde of our mortality; and, when wee are in the full careere of worldly pompe and jollity, she pulleth us by the eare, and maketh us knowe ourselves.
The Anatomy of Melancholy (Volume 1)
Part II, Sect. III, Memb. IV, Subsect. 6 (p. 399)
AMS Press, Inc. New York, New York, USA. 1973

Chrysostom, John   ?–407
Christian bishop and preacher

Princes, Masters, Parents, Magistrates, Judges, Friends, Eniemies, faire or foule meanes cannot containe, us; but a little sicknesse will correct and amend us.
In Robert Burton
The Anatomy of Melancholy (Volume 2)
Part II, Sect. III, Memb. II (p. 157)
AMS Press, Inc. New York, New York, USA. 1973

Donne, John   1572–1631
English poet and divine

And can there be worse sickness, than to know that we are never well, nor can be so?
An Anatomy of the World
The First Anniversary, l. 93–4

Presented for presentation to members of the Roxburghe Club. Cambridge, England. 1951

Dunlap, William   1766–1839
American dramatist and theatrical manager

He seems a little under the weather, somehow; and yet he's not sick.
The Memoirs of a Water Drinker (Volume 1)
Chapter VIII (p. 80)
Saunders & Otley. New York, New York, USA. 1837

Dunne, Finley Peter   1867–1936
American journalist and humorist

…whin a man's sick, he's sick an' nawthin' will cure him or annything will.
Mr. Dooley Says
Drugs (p. 97)
Charles Scribner's Sons. New York, New York, USA. 1910

Emerson, Ralph Waldo   1803–82
American lecturer, poet, and essayist

It is dainty to be sick, if you have leisure and convenience for it.
Journals of Ralph Waldo Emerson 1838–1841
February 7, 1839 (p. 162)
Houghton Mifflin Company. Boston, Massachusetts, USA. 1911

For sickness is a cannibal which eats up all the life and youth it can lay hold of, and absorbs its own sons and daughters.
Ralph Waldo Emerson: Essays and Lectures
The Conduct of Life
Consideration by the Way (p. 1088)
The Library of America. New York, New York, USA. 1983

Fuller, Thomas   1608–61
English clergyman and author

He who was never sick dies the first fit.
In Thomas Fuller
Gnomologia: Adages and Proverbs, Wise Sentences, and Witty Sayings.
Ancient and Modern, Foreign and British
Proverb
Printed for Thomas & Joseph Allman. London, England. 1816

Harris, Joel Chandler   1848–1908
American journalist

You know w'at de jay-bird say ter der squinch-owls!, "I'm sickly but sassy."
The Complete Tales of Uncle Remus
Chapter 50 (p. 311)
Houghton Mifflin Company. New York, New York, USA. 1955

Halsted, Anna Roosevelt   1906–75
Daughter of Franklin and Eleanor Roosevelt

There are so many indignities to being sick and helpless…
In Joseph P. Lash
Eleanor: The Years Alone

To The End, Courage, Letter to David Gray, November 1, 1962 (p. 327)
W.W. Norton & Company, Inc. New York, New York, USA. 1972

Hood, Thomas   1799–1845
English poet and editor

I'm sick of gruel, and the dietics,
I'm sick of pills, and sicker of emetics,
I'm sick of pulse, tardiness or quickness,
I'm sick of blood, its thinness or its thickness, —
In short, within a word, I'm sick of sickness!
The Poetical Works of Thomas Campbell (Volume 2)
Fragment (p. 424)
Wiley & Long. New York, New York, USA. 1836

Jonson, Ben   1573?–1637
English dramatist and poet

Take heed, sickness, what you do,
I shall fear you'll surfeit too.
Live not we, as all they stalls,
Spittles, pest-house, Hospitals,
Scarce will take our present store?
In Robert Bell
The Poems of Robert Greene, Christopher Marlowe, and Ben Jonson
The Forest, viii. To Sickness
Hurst & Company. New York, New York, USA. ca. 1880

Johnson, Samuel   1696–1772
English critic, biographer, and essayist

…what can a sick man say, but that he is sick?
Boswell's "Life of Samuel Johnson"
August, 1784 (p. 1347)
Oxford University Press, Inc. Oxford, England. 1965

Lamb, Charles   1775–1834
English essayist and critic

How sickness enlarges the dimensions of a man's self to himself! he is his own exclusive object. Supreme selfishness is inculcated upon him as his only duty.
Essays of Elia
The Convalescent (p. 330)
Henry Altemus. Philadelphia, Pennsylvania, USA. 1893

If there be a regal solitude, it is a sick-bed. How the patient lords it there; what caprices he acts without control! how kinglike he sways his pillow-tumbling, and tossing, and shifting, and lowering, and thumping, and flatting, and molding it, to the ever-varying requisitions of his throbbing temples.
Essays of Elia
The Convalescent (p. 329)
Henry Altemus. Philadelphia, Pennsylvania, USA. 1893

Luttrell, Henry   1765–1851
English wit and writer

Come, come, for trifles never stick:
Most servants have a failing;
Yours, it is true, are sometimes sick,
But mine are always ale-ing.

In William Davenport Adams
English Epigrams
On Ailing and Ale-ing, dclxxiii
G. Routledge. London, England. 1878

Milton, John   1608–74
English poet

…all maladies
Of ghastly Spasm, or racking torture, qualms
Of heart-sick Agonie, all feverous kinds,
Convulsions, Epilepsies, fierce Catarrhs,
Intestine Stone and Ulcer, Colic pangs,
Dropsies and Asthmas, and Joint-racking Rheums.
In *Great Books of the Western World* (Volume 32)
Paradise Lost
Book XI, l. 480–485
Encyclopædia Britannica, Inc. Chicago, Illinois, USA. 1952

O'Connor, Flannery   1926–64
American author

I have never been anywhere but sick. In a sense sickness is a place, more instructive than a long trip to Europe, and it's always a place where there's no company, where nobody can follow. Sickness before death is a very appropriate thing and I think those who don't have it miss one of God's mercies.
The Habit of Being
Part II (p. 163)
Farrar, Straus & Giroux, Inc. New York, New York, USA. 1988

Roy, Gabrielle   1909–83
Canadian author

The Christian Scientists held that it was not God Who wanted sickness, but man who [put] himself in the way of suffering. If this were the case, though, wouldn't we all die in perfect health?
Translated by Harry Binsse
The Cashier
Chapter 3 (pp. 36–37)
Harcourt, Brace. New York, New York, USA. 1955

Shakespeare, William   1564–1616
English poet, playwright, and actor

This sickness doth infect
The very life-blood of our enterprise.
In *Great Books of the Western World* (Volume 26)
The Plays and Sonnets of William Shakespeare (Volume 1)
The First Part of King Henry the Fourth
Act IV, Scene i, l. 28–29
Encyclopædia Britannica, Inc. Chicago, Illinois, USA. 1952

What, is Brutus sick,
And will he steal out of his wholesome bed,
To dare the vile contagion of the night?
In *Great Books of the Western World* (Volume 26)
The Plays and Sonnets of William Shakespeare (Volume 1)
Julius Caesar
Act II, Scene i, l. 263–265
Encyclopædia Britannica, Inc. Chicago, Illinois, USA. 1952

Sickness is catching.
In *Great Books of the Western World* (Volume 26)
The Plays and Sonnets of William Shakespeare (Volume 1)
A Midsummer-Night's Dream
Act I, Scene i, l. 186
Encyclopædia Britannica, Inc. Chicago, Illinois, USA. 1952

My long sickness
Of health and living now begins to mend,
And nothing brings me all things.
In *Great Books of the Western World* (Volume 27)
The Plays and Sonnets of William Shakespeare (Volume 2)
Timon of Athens
Act V, Scene i, l. 189–191
Encyclopædia Britannica, Inc. Chicago, Illinois, USA. 1952

Sterne, Laurence   1713–68
English novelist and humorist

I am sick as a horse...
The Life and Opinions of Tristram Shandy, Gentleman, and A Sentimental Journey Through France and Italy (Volume 2)
Book VII, Chapter II (p. 66)
Macmillan & Company Ltd. London, England. 1900

Swift, Jonathan   1667–1745
Irish-born English writer

Poor Miss, she's sick as a Cushion...
The Prose Works of Jonathan Swift (Volume the Fourth)
Polite Conversation, Dialogue I (p. 153)
Printed at the Shakespeare Head Press. Oxford, England. 1939–1968

Weingarten, Violet
Writer

Sickness, like sex, demands a private room, or at the very least, a discrete curtain around the ward bed.
Intimations of Mortality (p. 3)
Alfred A. Knopf. New York, New York, USA. 1978

Wolfe, Thomas   1900–38
American novelist

Most of the time we think we're sick it's all in the mind.
Look Homeward, Angel
Part I, Chapter 1 (p. 11)
Simon & Schuster. New York, New York, USA. 1995

SIGHT

Holmes, Oliver Wendell   1809–94
American physician, poet, and humorist

I refer to the use of dioptric media which correct the diminished refracting power of the humors of the eye, — in other words, spectacles. I don't use them. All I ask is a large, fair type, a strong daylight or gas-light, and one yard of focal distance, and my eyes are as good as ever.
The Autocrat of the Breakfast–Table
Chapter VII (p. 173)
Houghton Mifflin Company. Boston, Massachusetts, USA. 1891

Marsh, George Perkins   1801–82
American scholar, author, and statesman

Sight is a faculty; seeing, an art.
The Earth as Modified by Human Action: A New Edition of Man and Nature
Chapter I (p. 12)
Scribner, Armstron & Company. New York, New York, USA. 1874

Plato   428 BCE–347 BCE
Greek philosopher

The sight in my opinion is the source of the greatest benefit to us, for had we never seen the stars, and the sun, and the heaven, none of the words which we have spoken about the universe would ever have been uttered. But now the sight of day and night, and the months and the revolutions of the years, have created number, and have given us a conception of time, and the power of enquiring about the nature of the universe...
In *Great Books of the Western World* (Volume 7)
Timaeus
Section 40 (p. 455)
Encyclopædia Britannica, Inc. Chicago, Illinois, USA. 1952

SIMPLICITY

Bailey, Janet
No biographical data available

It is an article of faith in physics that the world's bewildering mask of complexity hides an ultimate simplicity.
The Good Servant: Making Peace with the Bomb at Los Alamos
Chapter 4 (p. 110)
Simon & Schuster. New York, New York, USA. 1995

Chandrasekhar, Subrahmanyan   1910–95
Indian-born American astrophysicist

The simple is the seal of the true and beauty is the splendor of truth.
Nobel Lectures, Physics 1981–1990
Nobel lecture for award received in 1983
On Stars, Their Evolution and Their Stability (p. 163)
World Scientific Publishing Company. Singapore. 1993

Coleridge, Samuel Taylor   1772–1834
English lyrical poet, critic, and philosopher

We study the complex in the simple; and only from the intuition of the lower can we safely proceed to the intellection of the higher degrees. The only danger lies in the leaping from low to high, with the neglect of the intervening gradations.
Hints Towards the Formation of a More Comprehensive Theory of Life
Physiology of Life (p. 41)
Lea & Blanchard. Philadelphia, Pennsylvania, USA, 1848

Davies, Paul Charles William   1946–
British-born physicist, writer, and broadcaster

Whether mathematical simplicity is God's affair or our, the fact remains that this feature more than any other remains the mainspring of progress in the physical sciences.
The Edge of Infinity: Where the Universe Came from and How It Will End
Chapter 9 (p. 188)
Simon & Schuster. New York, New York, USA. 1981

Davy, Sir Humphry   1778–1829
English chemist

Complexity almost always belongs to the early epochs of every science; and the grandest results are usually obtained by the most simple means.
The Collected Works of Sir Humphry Davy (Volume 4)
Elements of Chemical Philosophy
Part I, Introduction (p. 41)
Smith, Elder & Company. London, England. 1839–1840

The more the phenomena of the universe are studied, the more distinct their connection appears, the more simple their causes, the more magnificent their design, and the more wonderful the wisdom and power of their author.
The Collected Works of Sir Humphry Davy (Volume 4)
Elements of Chemical Philosophy
Part I, Introduction (p. 42)
Smith, Elder & Company. London, England. 1839–1840

Have you ever thought…about whatever man builds, that all of man's industrial efforts, all his calculations and computations, all the nights spent over working draughts and blueprints, invariably culminate in the production of a thing whose sole and guiding principle is the ultimate principle of simplicity?
Wind, Sand and Stars
Chapter 3 (p. 65)
Reynal & Hitchcock. New York, New York, USA. 1939

In any thing at all, perfection is finally attained, not when there is no longer anything to add, but when there is no longer anything to take away.
Wind, Sand and Stars
Chapter 3 (p. 66)
Reynal & Hitchcock. New York, New York, USA. 1939

du Noüy, Pierre Lecomte   1883–1947
French scientist

The complex is not always profound; but the profound is not necessarily simple.
The Road to Reason
Chapter 5 (p. 115)
Longmans, Green & Company. London, England. 1949

Einstein, Albert   1879–1955
German-born physicist

Control by experiment…is, of course, an essential prerequisite of the validity of any theory. But one can't possibly test everything. That is why I am so interested in your remarks about simplicity. Still, I should never claim that I really understood what is meant by the simplicity of natural laws.
In Werner Heisenberg
Physics and Beyond: Encounters and Conversations
Chapter 5 (p. 69)
Harper & Row, Publishers. New York, New York, USA. 1971

In every important advance the physicist finds that the fundamental laws are simplified more and more as experimental research advances. He is astonished to notice how sublime order emerges from what appeared to be chaos. And this cannot be traced back to the workings of his own mind but is due to a quality that is inherent in the world of perception. Leibniz well expressed this quality by calling it a pre-established harmony.
In Max Planck
Where Is Science Going?
Prologue (p. 11)
W.W. Norton & Company, Inc. New York, New York, USA.1932

Emerson, Ralph Waldo   1803–82
American lecturer, poet, and essayist

Nothing is more simple than greatness; indeed, to be simple is to be great.
Ralph Waldo Emerson: Essays and Lectures
Nature: Addresses, and Lectures (p. 100)
The Library of America. New York, New York, USA. 1983

Feynman, Richard P.   1918–88
American theoretical physicist

Perhaps a thing is simple if you can describe it fully in several different ways without immediately knowing that you are describing the same thing.
Nobel Lectures, Physics 1963–1970
Nobel lecture for award received in 1965
The Development of the Space-Time View of Quantum Electrodynamics
Elsevier Publishing Company. Amsterdam, Netherlands. 1972

The answer to all these questions may not be simple. I know there are some scientists who go about preaching that Nature always takes on the simplest solutions. Yet the simplest solution by far would be nothing, that there should be nothing at all in the universe. Nature is far more inventive than that, so I refuse to go along thinking it always has to be simple.
Feynman Lectures on Gravitation
Lecture 13 (p. 186)
Addison-Wesley Publishing Company. Reading, Massachusetts, USA. 1995

Fuller, Thomas   1608–61
English clergyman and author

Generally nature hangs out a sign of simplicity in the face of a fool.
The Holy and Profane State
Book III, Chapter XII, Maxim I
Maxim I (p. 171)
Printed for Thomas Tegg. London, England. 1841

Gibran, Kahlil   1883–1931
Lebanese-American philosophical essayist

The obvious is that which is never seen until someone expresses it simply.
Sand and Foam: A Book of Aphorisms (p. 54)
Alfred A. Knopf. New York, New York, USA. 1959

Goodman, Nelson   1906–98
American philosopher

All scientific activity amounts to the invention of and the choice among systems of hypotheses. One of the primary considerations guiding this process is that of simplicity. Nothing could be much more mistaken than the traditional idea that we first seek a true system and then, for the sake of elegance alone, seek a simple one.
The Test of Simplicity
Science, Volume 128, 1958 (p. 1064)

Gore, George   1826–1909
English electrochemist

Simplicity, whether truthful or not, is often attractive to unphilosophical minds, because it requires less intellectual exertion.
The Art of Scientific Discovery
Chapter IV (p. 29)
Longmans, Green & Company. London, England. 1878

Grimaux, L. E.
No biographical data available
Gerhardt, C.
No biographical data available

The chemist must always compare the results of his experiments with those which precede them; for it is by this comparison alone that little by little we arrive at general laws, and consequently at the simplification of science.
In Russell McCormmach (ed.)
Historical Studies in the Physical Sciences (Volume 6)
In John Hedley Brooke
Laurent, Gerhardt, and the Philosophy of Chemistry (p. 424)
Princeton University Press. Princeton, New Jersey, USA. 1975

Haldane, John Burdon Sanderson   1892–1964
English biologist

In scientific thought we adopt the simplest theory which will explain all the facts under consideration and enable us to predict new facts of the same kind. The catch in this criterion lies in the word "simplest." It is really an aesthetic canon such as we find implicit in our criticisms of poetry or painting. The layman finds such a law as $x/t = k(2x/y^2)$ less simple than "it oozes," of which it is the mathematical statement. The physicist reverses this judgment.
On Being the Right Size and Other Essays
Science and Theology as Art-Forms (pp. 33–34)
Oxford University Press, Inc. Oxford, England. 1985

Heisenberg, Werner Karl   1901–76
German physicist and philosopher

You [to Einstein] must have felt this too: the frightening simplicity and wholeness of the relationships which nature suddenly spreads out before us and for which none of us was in the least prepared.
Physics and Beyond: Encounters and Conversations
Chapter 5 (p. 69)
Harper & Row, Publishers. New York, New York, USA. 1971

You may object that by speaking of simplicity and beauty I am introducing aesthetic criteria of truth, and I frankly admit that I am strongly attracted by the simplicity and beauty of the mathematical schemes which nature presents us. You must have felt this too: the almost frightening simplicity and wholeness of the relationships, which nature suddenly spreads out before us…
Physics and Beyond: Encounters and Conversations
Chapter 5 (pp. 68–69)
Harper & Row, Publishers. New York, New York, USA. 1971

Hoagland, Mahlon   1921–
American biochemist

It is often the scientist's experience that he senses the nearness of truth when such connections are envisioned. A connection is a step toward simplification, unification. Simplicity is indeed often the sign of truth and a criterion of beauty.
Toward the Habit of Truth: A Life in Science
Preface (p. xxiii)
W.W. Norton & Company, Inc. New York, New York, USA.1990

Hoffer, Eric   1902–83
American longshoreman and philosopher

It is not at all simple to understand the simple.
The Passionate State of Mind, and Other Aphorisms
No. 230
Harper & Brothers. New York, New York, USA. 1955

Hoffmann, Hans   1848–1904
German novelist

The ability to simplify means to eliminate the unnecessary so that the necessary may speak.
In Bradley Efron and Robert J. Tibshirani
An Introduction to the Bootstrap
Preface (p. xiv)
Chapman & Hall. New York, New York, USA. 1993

Jevons, William Stanley   1835–82
English economist and logician

Simplicity is naturally agreeable to a mind of limited powers, but to an infinite mind all things are simple.
The Principles of Science: A Treatise on Logic and Scientific Method
Book V, Chapter XXVII (p. 625)
Macmillan & Company. London, England. 1887

Jung, Carl G.   1875–1961
Swiss psychiatrist and founder of analytical psychology

It would be simple enough, if only simplicity were not the most difficult of all things.
Translated by R.F.C. Hull
Alchemical Studies
Modern Psychology Offers a Possibility of Understanding (p. 16)
Princeton University Press. Princeton, New Jersey, USA. 1967

Lavoisier, Antoine Laurent   1743–94
French chemist

In performing experiments, it is a necessary principle, which ought never to be deviated from, that they be simplified as much as possible, and that every circumstance capable of rendering their results complicated be carefully removed.
Elements of Chemistry in a New Systematic Order
Translated by Kerr (p. 103)
W. Creech. Edinburgh, Scotland. 1790

Lindley, Dennis V.   1923–
American statistician

I believe that almost all important, useful ideas are simple. Peter Whittle has recently put it nicely in an autobiographical essay. "If a piece of work is heavy and complicated then it is wrong. ..." Some writers feel that to express their ideas in simple terms is degrading. Some use complexity to disguise the paucity of their material. In fact, simplicity is a virtue and when, as here, it is both original and useful, it can represent a real advance in knowledge.
Simplicity
RSS News, April 1995 (p. 1)

Oppenheimer, J. Robert   1904–67
American theoretical physicist

The point is to simplify and to order knowledge. The profession I'm part of has as its whole function the rendering of the physical world understandable and beautiful. Otherwise, you have only tables and statistics.
With Oppenheimer on an Autumn Day
Look, Volume 30, Number 26, 27 December 1966 (p. 63)

Percy, Walker   1916–90
American writer

It is not merely the truth of science that makes it beautiful, but its simplicity.
Signposts in a Strange Land
From Fact to Fiction (p. 187)
Farrar, Straus & Giroux. New York, New York, USA. 1991

Poincaré, Henri   1854–1912
French mathematician and theoretical astronomer

...it is because simplicity, because grandeur, is beautiful that we preferably seek simple facts, sublime facts, that we delight now to follow the majestic courses of the stars, now to examine with the microscope that prodigious littleness which is also grandeur, now to seek in geologic time the traces of a past which attracts because it is far away.
The Foundations of Science
Science and Method, Book I
Chapter I (p. 367)
The Science Press. New York, New York, USA. 1913

Rainich, G. Y.   1886–1968
Ukrainian mathematical physicist

...the really fundamental things have a way of appearing to be simple once they have been stated by a genius. ...
Analytic Function and Mathematical Physics
Bulletin of the American Mathematical Society, October 1931 (p. 700)

Reid, Thomas   1710–96
Scottish philosopher

Men are often led into errors by the love of simplicity, which disposes us to reduce things to few principles, and to conceive a greater simplicity in nature than there really is.
Essays on the Intellectual Powers of Man
Essay VI, Chapter VIII (p. 656)
Printed for John Bell. London, England. 1785

Schumacher, Ernst Friedrich   1911–77
German-born English economist

...it is rather more difficult to recapture directness and simplicity than to advance in the direction of ever more sophistication and complexity. Any third-rate engineer or researcher can increase complexity; but it takes a certain flair of real insight to make things simple again.
Small Is Beautiful
Part II, Chapter 5 (p. 146)
Harper & Row, Publishers. New York, New York, USA. 1973

Slobodkin, Lawrence B.
American ecologist and evolution scientist

The awkward richness of possibilities seems to shatter any possible coherent theory of simplicity...
Simplicity and Complexity in Games of the Intellect
Chapter 10 (p. 204)
Harvard University Press. Cambridge, Massachusetts, USA. 1992

Smith, George Otis   1871–1944
American geologist

I am convinced that, at its best, science is simple — that the simplest arrangement of facts that sets forth the truth best deserves the term scientific. So the geology I plead for is that which states facts in plain words — in language understood by the many rather than only by the few. Plain geology needs little defining, and I may state my case best by trying to set forth the reasons why we have strayed so far away from the simple type.

Plain Geology
Economic Geology, Volume 17, Number 1, 1922 (p. 34)

Stone, David
No biographical data available

One man's "simple" is another man's "huh?"
OMNI Magazine, May 1979

Teague, Jr., Freeman
No biographical data available

Nothing is so simple it cannot be misunderstood.
OMNI Magazine, May 1979

Teller, Edward   1908–2003
Hungarian-born American nuclear physicist
Sylvester, James Joseph   1814–97
English mathematician

No endeavor that is worthwhile is simple in prospect; if it is right, it will be simple in retrospect.
The Pursuit of Simplicity
Chapter Five (p. 152)
Pepperdine University Press, Malibu. 1981

Thoreau, Henry David   1817–62
American essayist, poet, and practical philosopher

Simplify. Simplify.
The Writings of Henry David Thoreau (Volume 2)
Walden
Chapter II (p. 144)
Houghton Mifflin Company. Boston, Massachusetts, USA. 1893

Twain, Mark (Samuel Langhorne Clemens)   1835–1910
American author and humorist

It's as simple as tit-tat-toe, three-in-a-row, and as easy as playing hooky. I should hope we can find a way that's a little more complicated than that…
The Adventures of Huckleberry Finn
Chapter XXXIV (p. 298)
Grosset & Dunlap Publishers. New York, New York, USA. 1948

Wright, Frank Lloyd   1867–1959
American architect

To know what to leave out and what to put in; just where and just how, ah, that is to have been educated in knowledge of simplicity…
Frank Lloyd Wright: An Autobiography
Simplicity (p. 144)
Duell, Sloan & Pearce. New York, New York, USA. 1943

SIMULTANEITY

Bridgman, Percy Williams   1882–1961
American physicist

Einstein, in thus analyzing what is involved in making a judgment of simultaneity, and in seizing on the act of the observer as the essence of the situation, is actually adopting a new point of view as to what the concepts of physics should be, namely, the operational view.
The Logic of Modern Physics
Chapter I (p. 8)
The Macmillan Company. New York, New York, USA. 1927

Medawar, Sir Peter Brian   1915–87
Brazilian-born English zoologist

Simultaneous discovery is utterly commonplace, and it was only the rarity of scientists, not the inherent improbability of the phenomenon, that made it remarkable in the past. Scientists on the same road may be expected to arrive at the same destination, often not far apart.
The Act of Creation
New Statesman, 19 June 1964

SINGULARITY

Barrow, John D.   1952–
English theoretical physicist

In recent years cosmologists have begun to discuss the spontaneous creation of the Universe as a problem in physics. Those who do this assume that a future synthesis of quantum theory and relativity which reveals how gravity behaves when matter is enormously compressed will evade the predictions of a real singularity of the type required by the singularity theorems. Although the assumptions of the singularity theorems are not expected to hold near the singularity, we do not know whether to expect a singularity or not as yet. But even in the absence of this singularity to denote the beginning of the Universe, it has been speculated that the application of quantum theory to the whole Universe may allow physical content to be given to the concept of "creation of the Universe out of Nothing." The goal of this research is to show that the creation of an expanding universe is inevitable. The reason there is something rather than nothing is that "nothing" is unstable.
The World within The World (p. 230)
Clarendon Press. Oxford, England. 1988

Hawking, Stephen William   1942–
English theoretical physicist

We showed that if general relativity is correct, any reasonable model of the universe must start with a singularity. This would mean that science could predict that the universe must have had a beginning, but that it could not predict how the universe should begin: For that, one would have to appeal to God…. Now, as a result of the singularity theorems, nearly everyone believes that the universe began with a singularity, at which the laws of

physics broke down. However, I now think that although there is a singularity, the laws of physics can still determine how the universe began.
Black Holes and Baby Universes and Other Essays
Chapter Nine (p. 91)
Bantam Books. New York, New York, USA. 1987

SITE

Bagnold, Ralph A.   1896–1990
English officer and engineer

There is an unfailing joy in identifying oneself with the actual sites where great things happened long ago. It appeals to a very human trait in all of us.
Libyan Sands: Travel in a Dead World
Chapter III (p. 74)
Hodder & Stoughton. London, England. 1941

Woolley, Sir Charles Leonard   1880–1960
English archaeologist

If the field archaeologist had his will, every ancient capital would have been overwhelmed by the ashes of a conveniently adjacent volcano. It is with green jealousy that the worker on other sites visits Pompeii and sees the marvelous preservation of its buildings, the houses standing up to the second floor, the frescoes on the walls and all the furniture and household objects still in their places as the owners left them as they fled from the disaster.
Digging Up the Past
Chapter I (p. 19)
Charles Scribner's Sons. New York, New York, USA. 1931

SIZE

Hardy, Thomas   1840–1928
English poet and regional novelist

There is a size ate at which dignity begins; further on there is a size at which grandeur begins; further on there is a size at which solemnity begins; further on, a size at which awfulness begins; further on, a size at which ghastliness begins. That size faintly approaches the size of the stellar universe.
Two on a Tower
Chapter IV (p. 35)
Harper & Brothers. New York, New York, USA. No date

The vastness of the field of astronomy reduces every terrestrial thing to atomic dimensions.
Two on a Tower
Chapter XXXIV (p. 258)
Harper & Brothers. New York, New York, USA. No date

Scott Cary
Fictional character

The unbelievably small and the unbelievably vast eventually meet — like the closing of a gigantic circle. I looked up, as if somehow I would grasp the heavens. The universe, worlds beyond number, God's silver tapestry spread across the night. And in that moment, I knew the answer to the riddle of the infinite. I had thought in terms of man's own limited dimension. I had presumed upon nature. That existence begins and ends in man's conception, not nature's. And I felt my body dwindling, melting, becoming nothing. My fears melted away. And in their place came acceptance. All this vast majesty of creation, it had to mean something. And then I meant something, too. Yes, smaller than the smallest. I meant something, too. To God there is no zero. I still exist!
The Incredible Shrinking Man
Film (1957)

Shapley, Harlow   1885–1972
American astronomer

The atomically small leads directly to the size really immense.
Annual Report of the Board of Regents of the Smithsonian Institution, 1946
On the Astronomical Dating of the Earth's Crust (p. 140)
Government Printing Office. Washington, D.C. 1947

SKELETON

Selzer, Richard   1928–
American physician and essayist

What man does not ponder the whereabouts of his skeleton — the place where it will lie? Say what you will, all sanitary and pragmatic considerations aside, these jaunty saunterers that have held us upright, have stiffened us against the grate and grind of life, are dear to us. What stands closer to a man all his days than his bones?
Mortal Lessons
Bone (pp. 54–55)
Simon & Schuster. New York, New York, USA. 1976

SKEPTICISM

Darwin, Charles Robert   1809–82
English naturalist

I am not very skeptical — a frame of mind which I believe to be injurious to the progress of science. A good deal of skepticism in a scientific man is advisable to avoid too much loss of time, but I have met with not a few men, who, I feel sure, have often thus been deterred from experiment or observations which would have proved directly or indirectly serviceable.
In Francis Darwin (ed.)
The Life and Letters of Charles Darwin (Volume 1)
Chapter II (p. 83)
D. Appleton & Company. New York, New York, USA. 1896

Gould, Stephen Jay   1941–2002
American paleontologist, evolutionary biologist, and historian of science

Skepticism or debunking often receives the bad rap reserved for activities — like garbage disposal — that absolutely must be done for a safe and sane life, but seem either unglamorous or unworthy of overt celebration. Yet the activity has a noble tradition, from the Greek coinage of "skeptic" (a word meaning "thoughtful") to Carl Sagan's…The Demon-Haunted World…. Skepticism is the agent of reason against organized irrationalism — and is therefore one of the keys to human social and civic decency…. Skepticism's bad rap arises from the impression that, however necessary the activity, it can only be regarded as a negative removal of false claims. Not so…. Proper debunking is done in the interest of an alternate model of explanation, not as a nihilistic exercise. The alternate model is rationality itself, tied to moral decency — the most powerful joint instrument for good that our planet has ever known.
In Michael Shermer
Why People Believe Weird Things: Pseudoscience, Superstition, and Other Confusions of Our Time
Foreword (pp. ix–xii)
Henry Holt & Company. New York, New York, USA. 2002

Raymo, Chet   1936–
American physicist and science writer

Skepticism is a critical reluctance to take anything as absolute truth, even one's own most cherished beliefs. Astonishment is the ability to be dazzled by the commonplace. At first glance these two qualities might seem opposed. The Skeptic is often thought to lack passionate commitment. The easily astonished person is sometimes thought of as gullible. In fact reasoned skepticism does not preclude passionate belief, and astonishment is enhanced by knowledge.
Skeptics and True Believers: The Exhilarating Connection Between Science and Religion
Chapter Fourteen (pp. 252–253)
Walker & Company. New York, New York, USA. 1998

The difference between Skeptics and True Believers is not that Skeptics believe what is sensible and obvious, while True Believers accept what is fanciful and far-fetched. Often, it is the other way around.
Skeptics and True Believers: The Exhilarating Connection Between Science and Religion
Chapter Two (p. 27)
Walker & Company. New York, New York, USA. 1998

Sagan, Carl   1934–96
American astronomer and author

…science requires the most vigorous and uncompromising skepticism, because the vast majority of ideas are simply wrong, and the only way to winnow the wheat from the chaff is by critical experiment and analysis.
The Demon-Haunted World: Science as a Candle in the Dark
Chapter 17 (p. 305)
Random House, Inc. New York, New York, USA. 1995

Finding the occasional straw of truth awash in a great ocean of confusion and bamboozle requires vigilance, dedication, and courage. But if we don't practice these tough habits of thought, we cannot hope to solve the truly serious problems that face us — and we risk becoming a nation of suckers, a world of suckers, up for grabs by the next charlatan who saunters along.
The Fine Art of Baloney Detection
Parade, February 1, 1987

Schrödinger, Erwin   1887–1961
Austrian theoretical physicist

Skepticism alone is a cheap and barren affair. Skepticism in a man who has come nearer to the truth than anyone before, and yet clearly recognizes the narrow limits of his own mental construction, is great and fruitful, and does not reduce but doubles the value of the discoveries.
Nature and the Greeks
Chapter II (p. 31)
At The University Press. Cambridge, England. 1954

SKIN

Levi, Primo   1919–87
Italian writer and chemist

I live in my house as I live inside my skin: I know more beautiful, more ample, more sturdy and more picturesque skins: but it would seem to me unnatural to exchange them for mine.
Other People's Trades
My House
Summit Books. New York, New York, USA. 1989

Selzer, Richard   1928–
American physician and essayist

I sing of skin, layered fine as baklava, whose colors shame the dawn, at once the scabbard upon which is writ our only signature, and the instrument by which we are thrilled, protected, and kept constant in our natural place. Here is each man bagged and trussed in perfect amiability.
Mortal Lessons
Skin (p. 105)
Simon & Schuster. New York, New York, USA. 1976

SKY

Astronomy Survey Committee

Nature offers no greater splendor than the starry sky on a clear, dark night. Silent, timeless, jeweled with the constellations of ancient myth and legend, the night sky has inspired wonder throughout the ages.

Astronomy and Astrophysics for the 1980s
Volume 1, Report to the Astronomy Survey Committee (p. 3)

Brahe, Tycho   1546–1601
Danish astronomer

O crassia ingenia, O coecos coeli spectatores.
O thick wits. Oh blind watchers of the sky.
De Nova Stella
Preface

Brandt, John C.
No biographical data available

Chapman, Robert D.
No biographical data available

…each step forward in unraveling the mystery of comets (or any other natural phenomenon) brings great pleasure to all who look to the sky as a source of beauty and intellectual challenge.
Introduction to Comets
Chapter 10 (p. 226)
Cambridge University Press. Cambridge, England. 1981

Browning, Robert   1812–89
English poet

Sky — what a scowl of cloud
Till, near and far,
Ray on ray split the shroud
Splendid, a star!
The Poems and Plays of Robert Browning
The Two Poets of Croisic
The Modern Library. New York, New York, USA. 1934

Bunch, Sterling
American poet and editor

In starry skies, long years ago,
I found my Science. Heart aglow
I watched each night unfold a maze
Of mystic suns and worlds ablaze,
That spoke: "Know us and wiser grow."
In Starry Skies
Popular Astronomy, Volume 34, 1926 (p. 288)

de Saint-Exupéry, Antoine   1900–44
French aviator and writer

A sky as pure as water bathed the stars and brought them out.
Southern Mail
Chapter I (p. 3)
Harcourt, Brace & Company New York, New York, USA. 1971

Emerson, Ralph Waldo   1803–82
American lecturer, poet, and essayist

The sky is the daily bread of the eyes.
In Edward Waldo Emerson (ed.)
Journals of Ralph Waldo Emerson 1841–1844
25 May 1843 (p. 410)
Houghton Mifflin Company. Boston, Massachusetts, USA. 1911

Flammarion, Camille   1842–1925
French astronomer and author

Better than the spectacle of the sea calm or agitated, grander than the spectacle of mountains adorned with forests or crowned with perpetual snow, the spectacle of the sky attracts us, envelops us, speaks to us of the infinite, gives us the dizziness of the abyss; for more than any other, it seizes the contemplative mind and appeals to it, being the truth, the infinite, the eternal, the all.
Popular Astronomy: A General Description of the Heavens
Book VI, Chapter I (p. 554)
Chatto & Windus. London, England. 1894

Friedman, Herbert   1916–2000
American space scientist and astrophysicist

It is impossible for any sensitive person to look at a star-filled sky without being stirred by thoughts of creation and eternity. The mystery of the origin and destiny of the universe haunts us throughout our lives.
The Amazing Universe
Chapter 7 (p. 166)
National Geographic Society. Washington, D.C. 1980

Kreymborg, Alfred   1883–1966
American poet and anthologist

The sky is that beautiful old parchment in which the sun and the moon keep their diary.
In Louis Untermeyer (ed.)
Modern American Poetry
Old Manuscript
Harcourt, Brace & Company. New York, New York, 1936

Lowell, Amy   1874–1925
American poet

A wise man,
Watching the stars pass across the sky,
Remarked:
In the upper air the fireflies move more slowly.
The Complete Poetical Works of Amy Lowell
Meditation
Houghton Mifflin. Boston, Massachusetts, USA. 1955

Manilius, Marcus   fl. 10 AD
Roman poet

It is my delight to traverse the very air and spend my life touring the boundless skies, learning of the constellations and the contrary motions of the planets.
Astronomica
Book I
Publisher undetermined

Maunder, Edward Walter   1851–1928
English astronomer

The oldest picture book in our possession is the Midnight Sky.
The Oldest Picture-Book of All

Nineteenth Century, Volume 48, Number CCLXXXIII, September 1900
(p. 451)

Moulton, Forest Ray   1872–1952
American astronomer

It is doubtful whether there is in the whole range of human experience any more awe-inspiring spectacle than that presented by the sky on a clear and moonless night. Under the vault of the sparkling heavens one is raised, if ever, to an actual realization of the fact that the earth beneath his feet is a relatively tiny mass in comparison with the infinite cosmos spread out above.
Astronomy
Chapter II (p. 14)
The Macmillan Company. New York, New York, USA. 1931

Schaefer, Bradley E.
American professor of physics

The sky is beautiful and vast and harbors many secrets.
Inventory of Cosmic Mysteries
Sky & Telescope, Volume 94, Number 4, October 1994 (p. 68)

Shakespeare, William   1564–1616
English poet, playwright, and actor

My soul is in the sky.
In *Great Books of the Western World* (Volume 26)
The Plays and Sonnets of William Shakespeare (Volume 1)
A Midsummer-Night's Dream
Act V, Scene I
Encyclopædia Britannica, Inc. Chicago, Illinois, USA. 1952

Shore, Jane   1947–
American poet

Each night the sky splits open like a melon
its starry filaments
the astronomer examines with great intensity.
Eye Level
An Astronomer's Journal (p. 31)
The University of Massachusetts Press. Amherst, Massachusetts, USA. 1977

Smoot, George   1945–
American astrophysicist
Davidson, Keay
American science writer

There is something about looking at the night sky that makes a person wonder.
Wrinkles in Time
Chapter 1 (p. 1)
William Morrow & Company, Inc. New York, New York, USA. 1993

Swings, Pol   1906–63
Belgian astrophysicist

The sky belongs to everyone, with stars to spare for all.
In Henry Margenau and David Bergamini (eds.)
The Scientist (p. 116)
Time Inc. New York, New York, 1964

The Bible

He stretches out the sky like a curtain, spreads them out like a tent to live in…
The Revised English Bible
Isaiah 40:22
Oxford University Press, Inc. Oxford, England. 1989

Thomas, Lewis   1913–93
American physician and biologist

Taken all in all, the sky is a miraculous achievement. It works, and for what it is designed to accomplish it is as infallible as anything in nature. I doubt whether any of us could think of a way to improve on it, beyond maybe shifting a local cloud from here to there on occasion.
The Lives of a Cell: Notes of a Biology Watcher
The World's Biggest Membrane (p. 148)
The Viking Press. New York, New York, USA. 1974

Upgren, Arthur
No biographical data available

A dark sky filled with stars has always been one of our most cherished sights. This wonder need not and must not fade into the baleful orange glare above our cities; let the stars continue to twinkle with the fireflies along country lanes. Those stars come from one shared legacy of all people around the world, and it is by the heavens they define that we all ultimately find our way.
Night Has a Thousand Eyes: A Naked-Eye Guide to the Sky, Its Science, and Lore
Afterword (p. 275)
Plenum Trade. New York, New York, USA. 1998

Whitman, Walt   1819–92
American poet, journalist, and essayist

Over all the sky — the sky! Far, far out of reach, studded, breaking out, the eternal stars.
Complete Poetry and Collected Prose
Bivouac on a Mountain Side
The Library of America. New York, New York, USA. 1982

SLEEP

Aristotle   384 BCE–322 BCE
Greek philosopher

The vigorous are no better than the lazy during one half of life, for all men are alike when asleep.
In *Great Books of the Western World* (Volume 8)
Eudemian Ethics
Book II Chapter 1
Encyclopædia Britannica, Inc. Chicago, Illinois, USA. 1952

SNOW

Bentley, Wilson   1865–1931
American photographer of snowflakes

Besides combining her greatest skill and artistry in the production of snowflakes, Nature generously fashions the most beautiful specimens on a very thin plane so that they are specially adapted for photomicrographical study.
Photographing Snowflakes
Popular Mechanics Magazine, Volume 37, 1922 (p. 309)

Longfellow, Henry Wadsworth   1807–82
American poet

Out of the bosom of the Air,
Out of the cloud-folds of her garments shaken,
Over the woodlands brown and bare
Over the harvest-fields forsaken,
Silent and soft and slow
Descends the snow.
The Poetical Works of Henry Wadsworth Longfellow
Snow-Flakes
Houghton Mifflin Company. Boston, Massachusetts, USA. 1883

Muir, John   1838–1914
American naturalist

To lie out alone in the mountains of a still night and be touched by the first of these small silent messengers from the sky is a memorable experience, and the fineness of that touch none will forget.
Steep Trails
Chapter IV (p. 75)
Norman S. Berg, Publisher. Dunwoody, Georgia, USA. 1970

The faint lisp of snowflakes as they alight is one of the smallest sounds mortal can hear.
Our National Parks
Chapter IX (p. 274)
Houghton Mifflin Company. Boston, Massachusetts, USA. 1901

The fertile clouds, descending, glide about and hover in brooding silence, as if thoughtfully examining the forests and streams with reference to the work before them…
Our National Parks
Chapter VIII (p. 249)
Houghton Mifflin Company. Boston, Massachusetts, USA. 1901

Small flakes or single crystals appear, glinting and swirling in zigzags and spirals; and soon the thronging feathery masses fill the sky and make darkness like night, hurrying wandering mountaineers to their winter quarters.
Our National Parks
Chapter VIII (p. 249)
Houghton Mifflin Company. Boston, Massachusetts, USA. 1901

Thoreau, Henry David   1817–62
American essayist, poet, and practical philosopher

How full of creative genius is the air in which these are generated! I should hardly admire them more if real stars fell and lodged on my coat.
In Bradford Torrey and Francis H. Allen (eds.)

The Journal of Henry D. Thoreau (Volume 8)
January 5, 1856 (p. 87)
Houghton Mifflin Company, Boston, Massachusetts, USA. 1949

SOIL

Burroughs, John   1837–1921
American naturalist and writer

The youth of the earth is in the soil and in the trees and verdure that springs from it…
Under the Apple-Trees
Chapter II (p. 40)
Houghton Mifflin Company. Boston, Massachusetts, USA. 1916

Emerson, Ralph Waldo   1803–82
American lecturer, poet, and essayist

Every plant is a manufacturer of soil.
The Complete Works of Ralph Waldo Emerson (Volume 7)
Society and Solitude
Chapter VI (p. 144)
Houghton Mifflin Company. Boston, Massachusetts, USA. 1904

Fuller, Wallace H.   1915–2006
American geologist

A thin rind of loose material covering the continents of the earth is all that stands between life and lifelessness.
Soils of the Desert Southwest
A Word from the Author (p. xiii)
University of Arizona Press. Tucson, Arizona, USA. 1975

Grit and grime, crumbling rock and decaying organic residue-abrading by wind and water — weather into soil — Mother Earth. This soft and yielding earth lives and continually changes under the forces of climate, having formed through the ages as a result of meteorological, geological, and biological action on rock. The soil not only lives, but it continually renews life as well. Animal and plant residues decay into simpler constituents, and nutrient elements again are made available for new life in a perpetual cycle.
Soils of the Desert Southwest
A Word from the Author (p. xiii)
University of Arizona Press. Tucson, Arizona, USA. 1975

Molloy, Les
No biographical data available

…for only rarely have we stood back and celebrated our soils as something beautiful, and perhaps even mysterious. For what other natural body, worldwide in its distribution, has so many interesting secrets to reveal to the patient observer? The great events of long ago — volcanic eruptions, dust storms, floods and Ice Ages — have left their imprints as have the agricultural practices of earlier times.
Soils in the New Zealand Landscape: The Living Mantle
Mallison Rendel Publishers Ltd. Wellington, New Sealand. 1988

The soil can…tell us much about our present day environment. It is the home of millions of living things and a recycling factory for so much of the solar and geochemical energy that sustains life. In its form and properties it expresses the combined influences of local climate, shape of the land, and rocks and organisms that are broken down and incorporated into it.
Soils in the New Zealand Landscape: The Living Mantle
Mallison Rendel Publishers Ltd. Wellington, New Sealand. 1988

Simonson, Roy
American soil scientist

Be it deep or shallow, red or black, sand or clay, the soil is the link between the rock core of the earth and the living things on its surface. It is the foothold for the plants we grow. Therein lies the main reason for our interest in soils.
USDA Yearbook of Agriculture, 1957

SOLAR SYSTEM

Burroughs, John   1837–1921
American naturalist and writer

When I look up at the starry heavens at night and reflect upon what it is that I really see there, I am constrained to say, "There is no God." …I see no lineaments of personality, no human traits, but an energy upon whose currents solar systems are but babbles.
The Light of the Day: Religious Discussions and Criticisms from the Naturalist's Point of View (p. 224)
Houghton Mifflin Company. Boston, Massachusetts, USA. 1900

Carlyle, Thomas   1795–1881
English historian and essayist

Did not the Boy Alexander weep because he had not two Planets to conquer; or a whole Solar System; or after that, a whole universe?
Sartor Resartus
Book II, Chapter VIII (p. 165)
Ginn & Company. Boston, Massachusetts, USA. 1897

Clarke, Arthur C.   1917–
English science and science fiction writer

The Solar System is rather a large place, though whether it will be large enough for so quarrelsome an animal as *Homo sapiens* remains to be seen.
The Challenge of the Spaceship
The Challenge of the Spaceship (p. 8)
Harper & Brothers. New York, New York, USA. 1959

The Solar System, comprising the nine known worlds of our Sun and their numerous satellites, is a relatively compact structure, a snug little celestial oasis in an endless desert.
The Challenge of the Spaceship
The Planets Are Not Enough (p. 54)
Harper & Brothers. New York, New York, USA. 1959

Davies, Paul Charles William   1946–
British-born physicist, writer, and broadcaster

The secret of our success on planet Earth is space. Lots of it. Our solar system is a tiny island of activity in an ocean of emptiness.
The Last Three Minutes: Conjectures About the Ultimate Fate of the Universe
Chapter I (p. 4)
Basic Books, Inc. New York, New York, USA. 1994

Eddington, Sir Arthur Stanley   1882–1944
English astronomer, physicist, and mathematician

The solar system is not the typical product of development of a star; it is not even a common variety of development; it is a freak.
Man's Place in the Universe
Harper's Magazine, October 1928 (p. 574)

Emerson, Ralph Waldo   1803–82
American lecturer, poet, and essayist

The solar system has no anxiety about its reputation…
Ralph Waldo Emerson: Essays and Lectures
The Conduct of Life
Worship (p. 1055)
The Library of America. New York, New York, USA. 1983

Flammarion, Camille   1842–1925
French astronomer and author

Like a shower of stars the worlds whirl, borne along by the winds of heaven, and are carried down through immensity, suns, earths, satellites, comets, shooting stars, humanities, cradles, graves, atoms of the infinite, seconds of eternity, perpetually transform beings and things.
Popular Astronomy: A General Description of the Heavens
American Book Company. New York, New York, USA. 1899

Hey, Nigel S.   1936–
American science writer

Naturally their [men and women] writings deal with spaceflight, the solar system, and the Cosmos…reveal the profoundly human aspects of this great adventure, from the excitement of solving the problems of spacecraft that are millions of miles distant to the self-examination that occurs when considering whether we might someday send robots, and not people, as our ambassadors to distant star systems. Some of the nobility of the human condition, so often obscured, shines through their words.
Solar System
Introduction (p. 8)
Weidenfield & Nicolson. London, England. 2002

Our explorations of the solar system are the first halting steps in a journey that will transform our kind into a species that knowingly lives among the stars, in mind and possibly in body. Our destiny awaits in the planets

and their moons, in this star system and in the galaxy beyond. Without exaggeration this journey is epochal in its significance to the human race.
Why People Need Space
Lecture, National Space Centre, October 2002

Horowitz, Norman H. 1915–2005
American genticist

If the exploration of the solar system in our time bring home to us a realization of the uniqueness of our small planet and thereby increase our resolve to avoid self-destruction, [it] will have contributed more than just science to the human future.
To Utopia and Back: The Search For Life in the Solar System
Chapter Eight (p. 146)
W.H. Freeman & Company. San Francisco, California, USA. 1986

Jeffreys, Sir Harold 1891–1989
English astronomer and geophysicist

Damn the solar system. Bad light; planets too distant; pestered with comets; feeble contrivance; could make a better one myself.
In John D. Barrow
The Artful Universe (p. 34)
Clarendon Press. Oxford, England. 1995

Lambert, Johann Heinrich 1728–77
Swiss-German mathematician and astronomer

Nothing is more simple than the plan of the Solar System…
Translated by James Jacque
The System of the World
Part I, Chapter I (p. 1)
Printed for Vernor & Hood. London, England. 1800

Lowell, Percival 1855–1916
American astronomer

Now when we think that each of these stars is probably the center of a solar system grander than our own, we cannot seriously take ourselves to be the only minds in it all.
Mars
Chapter I (p. 5)
Houghton Mifflin Company. Boston, Massachusetts, USA. 1895

Maclennan, Hugh 1907–99
Canadian author and professor of English

We have just reached the outer fringes of the Solar System. Can any sane man possibly argue that we should stop there?
Scotchman's Return and Other Essays
Remembrance Day, 2010 A.D. (p. 89)
Charles Scribner's Sons. New York, New York, USA. 1960

Patten, W.
No biographical data available

A solar system has attributes and powers that can not be defined or measured in terms of its members, or of its ultimate chemical elements, for a solar system is not merely an aggregate, or the algebraic sum of its various elements and qualities.… It is a system, a new type of individuality, with special creative powers of its own.
The Grand Strategy of Evolution (p. 34)
Richard G. Badger. Boston, Massachusetts, USA. 1920

Peterson, Ivars
Mathematics and computer writer and editor

We can…be thankful that the solar system in which we live has been unreasonably kind throughout the long history of human efforts to understand its dynamics and to extend that knowledge to the rest of the universe. At each step along the way, it has served as a perspicacious teacher, posing questions just difficult enough to prompt new observations and calculations that have led to fresh insights, but not so difficult that any further study becomes mired in a morass of confusing detail.
Newton's Clock: Chaos in the Solar System
Chapter 12 (p. 286)
W.H. Freeman & Company. New York, New York, USA. 1993

Pliny (C. Plinius Secundus) 23–79
Roman savant and author

Most men are not acquainted with a truth known to the founders of the science from their arduous study of the heavens, that what when they fall to earth are termed thunderbolts are the fires of the three upper planets, particularly those of Jupiter, which is in the middle position — possibly because it voids in this way the charge of excessive moisture from the upper circle (of Saturn) and of excessive heat from the circle below (of Mars); and that this is the origin of the myth that thunderbolts are the javelins hurled by Jupiter. Consequently heavenly fire is spit forth by the planet as crackling charcoal flies from a burning log, bringing prophecies with it. And this is accompanied by a very great disturbance of the air, because moisture collected causes an overflow or because it is disturbed by the birth-pangs so to speak of the planet in travail.
Natural History
Volume 1, Book II, sec 84
Harvard University Press. Cambridge, Massachusetts, USA. 1947

Sagan, Carl 1934–96
American astronomer and author

The emerging picture of the early Solar System does not resemble a stately progression of events designed to form the Earth. Instead, it looks as if our planet was made, and survived, by mere lucky chance, amid unbelievable violence. Our world does not seem to have been sculpted by a master craftsman. Here, too, there is no hint of a Universe made for us.

Pale Blue Dot: A Vision of the Human Future in Space
Chapter 17 (p. 295)
Random House, Inc. New York, New York, USA. 1994

Somerville, Mary   1780–1872
English mathematician

Yonder starry sphere
Of planets and of fix'd, in all her wheels,
Resembles nearest mazes intricate,
Eccentric, intervolved, yet regular,
Then most, when most irregular they seem.
The Connexion of the Physical Sciences (9th edition)
Section III (p. 23)
John Murray. London, England. 1858

Tsiolkovsky, Konstantin Eduardovich   1857–1935
Russian research scientist

Should man penetrate the solar system, should he learn
to comport himself there as the mistress in her home —
would the secrets of the world then open for him? Not in
the least. Not anymore that inspecting a pebble or shell
would reveal to him the secrets of the ocean.
Compiled by V.V. Vorontsov
Words of The Wise: A Book of Russian Quotations
Translated by Vic Schneierson
Progress Publishers. Moscow, Russia. 1979

SOLID STATE

Updike, John   1932–
American novelist, short story writer, and poet

Textbooks & Heaven only are Ideal;
Solidity is an imperfect state,
Within the cracked and dislocated Real
Nonstoichiometric crystals dominate.
Stray Atoms sully and precipitate;
Strange holes, excitons, wander loose; because
Of Dangling Bonds, a chemical Substrate
Corrodes and Catalyzes — surface Flaws
Help Expitazial Growth to fix absorptive claws.
Midpoint and Other Poems
The Dance of the Solids
Stanza 9
Fawcett Publications, Inc. Greenwich, Connecticut, USA. 1970

SOLUBILITY

Witt, Otto N.   1853–1915
German chemist

In the strictly scientific sense of the word insolubility
does not exist, and even those substances characterized
by the most obstinate resistance to the solvent action of
water may properly be designated as extraordinarily difficult of solution, not as insoluble.
In Joseph William Mellor
Mellor's Modern Inorganic Chemistry

Chapter 13 (p. 177)
Longmans, Green & Company Ltd. London, England. 1967

SOLUTION

Bohr, Niels Henrik David   1886–1962
Danish physicist

Every great and deep difficulty bears in itself its own
solution. It forces us to change our thinking in order to
find it.
In Brian VanDeMark
Pandora's Keepers
Chapter 1 (p. 29)
Little, Brown & Company. Boston, Massachusetts, USA. 2003

Eddington, Sir Arthur Stanley   1882–1944
English astronomer, physicist, and mathematician

In science we sometimes have convictions as to the right
solution of a problem which we cherish but cannot justify; we are influenced by some innate sense of the fitness
of things.
The Nature of the Physical World
Chapter XV (p. 337)
The Macmillan Company. New York, New York, USA. 1930

Heisenberg, Werner Karl   1901–76
German physicist and philosopher

…the genuine solution of a difficult problem is neither
more nor less than a glimpse of the wider context, a
glimpse that helps us to clear away other difficulties as
well, including many whose existence we do not even
suspect.
Physics and Beyond: Encounters and Conversations
Chapter 8 (p. 102)
Harper & Row, Publishers. New York, New York, USA. 1971

Kosko, Bart   1960–
American electrical engineer

A solution has a way of bubbling up out of your subcon-
scious if you brood about a problem long enough.
Fuzzy Thinking
Chapter 3 (p. 61)
Hyperion. New York, New York, USA. 1993

MacCready, Paul   1925–?
American aeronautical engineer

When you do dome up with a solution, you can always
explain it logically, even though it's the absurd approach
that gave you the solution.
In Kenneth A. Brown
Inventors at Work
Paul MacCready (p. 11)
Microsoft Press. Redmond, Washington, USA. 1988

Mencken, H. L. (Henry Louis)   1880–1956
American journalist and literary critic

...there is always a well-known solution to every human problem — neat, plausible, and wrong.
Prejudices: Second Series
The Divine Afflatus (p. 158)
Alfred A. Knopf. New York, New York, USA. 1922

Szilard, Leo   1898–1964
Hungarian-born American nuclear physicist

Once a man has missed the solution to a problem when he passes by, it is less likely he will find it the next time.
In Editors of International Science and Technology
The Way of the Scientist: Interviews from the World of Science and Technology
Leo Szilard (p. 28)
Simon & Schuster. New York, New York, USA. 1966

SOUL

Russell, Sir Edward John   1872–1965
British agriculturalist and writer

Those young people of today, who will be the leaders of thought and of action tomorrow, are faced with the problem of enduring that, in gaining control over Nature, man does not lose his own soul.
Science and Modern Life (p. 101)
Philosophical Library. New York, New York, USA. 1955

SOUND

Hooke, Robert   1635–1703
English physicist

'Tis not impossible to hear a whisper a furlong's distance, it having been already done; and perhaps the nature of the thing would not make it more impossible, though that furlong should be ten times multiply'd...for that [air] that is not the only medium, I can assure the Reader, that I have, by the help of a distended wire, propagated the sound to a very considerable distance in an instant, or with as seemingly quick a motion as that of light, at least, incomparably swifter then that, which at the same time was propagated through the Air...
Microgarphia
Preface
Printed for Jo. Martyn and Ja. Allestry. London, England. 1665

SPACE

Adams, Douglas   1952–2001
English author, comic radio dramatist, and musician

Space...is big, really big...You may think it's a long way down the street to the chemists' but that's just peanuts to space.
The Ultimate Hitchhiker's Guide to the Galaxy
The Hitchhiker's Guide to the Galaxy
Chapter 8 (p. 53)
The Ballantine Book Company. New York, New York, USA. 2002

Alfven, Hannes   1908–95
Swedish physicist

Having probes in space was like having a cataract removed.
In Eric J. Lerner
The Big Bang Never Happened
Chapter 1 (p. 45)
Random House, Inc. New York, New York, USA. 1991

Bailey, Philip James   1816–1902
English poet

Unimaginable space,
As full of suns as is earth's sun of atoms.
Festus: A Poem
Scene IV (p. 61)
George Routledge & Sons, Ltd. London, England. 1893

Barnes, Bishop
Bishop of San Bernardino

It is fairly certain that our space is finite though unbounded. Infinite space is simply a scandal to human thought.
In Joseph Silk
The Big Bang (p. 81)
W.H. Freeman & Company. San Francisco, California, USA. 1980

Bergaust, Erik   1925–95
American writer and journalist

As far as man on Earth is concerned, space begins at the high border of the Earth's atmosphere and extends to infinity.
Wernher von Braun
Are Flying Saucers Real? (p. 546)
National Space Institute. Washington, D.C. 1976

Bergson, Henri   1859–1941
French philosopher

For it is scarcely possible to give any other definition of space: space is what enables us to distinguish a number of identical and simultaneous sensations from one another; it is thus a principle of differentiation, and consequently it is a reality with no quality.
Translated by F.L. Pogson
Time and Free Will: An Essay on the Immediate Data of Consciousness
Chapter II (p. 95)
George Allen & Unwin Ltd. London, England. 1950

Bradbury, Ray   1920–
American writer

Man does not need escape so much as he needs release into a new spirit, a transcendent knowledge of himself that only Space can give him.
In Ray Bradbury, Arthur C. Clarke, Bruce Murray, Carl Sagan, and Walter Sullivan
Mars and the Mind of Man
Foreword (p. XI)
Harper & Row, Publishers. New York, New York, USA. 1973

Bradley, Jr., John Hodgdon 1898–1962
American geologist

A sea whose shores no eyes have ever seen, whose depth no instrument can fathom, whose waters no scientist can analyze — such is the sea of space. Nothing can be as empty and cold as the gulf wherein our destinies are immersed.
Parade of the Living
Part I, Chapter I (p. 3)
Coward-McCann, Inc. New York, New York, USA. 1930

Bruno, Giordano 1548–1600
Italian philosopher and pantheist

There are countless constellations, suns and planets; we see only the suns because they give light; the planets remain invisible, for they are small and dark. There are also numberless earths circling around their suns, no worse and no less than this globe of ours.
On the Infinite Universe and Worlds
Henry Schuman, Inc. New York, New York, USA. 1950

There is a single general space, a single vast immensity which we may freely call Void: in it are innumerable globes like this on which we live and grow; this space we declare to be infinite, since neither reason, convenience, sense-perception nor nature assign it a limit.
In Joseph Silk
The Big Bang (p. 81)
W.H. Freeman & Company. San Francisco, California, USA. 1980

Captain Kirk
Fictional character

Space, the final frontier…
Opening lines
Star Trek
Television series

Clarke, Arthur C. 1917–
English science and science fiction writer

Go out beneath the stars on a clear winter night, and look up at the Milky Way spanning the heavens like a bridge of glowing mist. Up there, ranged one beyond the other to the end of the Universe, suns without number burn in the loneliness of space. Down to the south hang the brilliant, unwinking lanterns of other worlds — the electric blue of Jupiter, the glowing ember of Mars. Across the zenith, a meteor leaves a trail of fading incandescence, and a tiny voyager of space has come to a flaming end.
In Neil McAleer
Odyssey: The Authorized Biography of Arthur C. Clarke (p. 34)
Victor Gollancz. London, England. 1993

The sea which beats against the coasts of Earth, which seems so endless and so eternal, is as the drop of water on the slide of a microscope compared with the shoreless sea of space.

The Challenge of the Spaceship
Across the Sea of Stars (p. 128)
Harper & Brothers. New York, New York, USA. 1959

In space there are no horizons; the questing eye reaches out forever, in all directions, and finds no fixed point at which to rest.
The Challenge of the Spaceship
Which Way Is Up? (p. 143)
Harper & Brothers. New York, New York, USA. 1959

Collins, Billy 1941–
American poet

Here's to the wind blowing against this lighted house and to the vast, windless spaces between the stars.
The Art of Drowning
Chapter XXXIII, Cheers
University of Pittsburgh Press. Pittsburgh, Pennsylvania, USA. 1995

Davies, Paul Charles William 1946–
British-born physicist, writer, and broadcaster

For those who think of space as emptiness, the assignment of an adjective, especially one as enigmatic as "curved", might be regarded as cryptic.
The Encyclopaedia of Ignorance: Everything You Ever Wanted to Know About the Unknown
Curved Space (p. 78)
Pergamon Press. Oxford, England. 1977

Deudney, Daniel
American political scientist

Space is only 80 miles from every person on earth — far closer than most people are to their own national capitals…
Space: The High Frontier in Perspective
Introduction (p. 6)
Worldwatch Institute. Washington, D.C. 1982

Eddington, Sir Arthur Stanley 1882–1944
English astronomer, physicist, and mathematician

To put the conclusion rather crudely — space is not a lot of points close together; it is a lot of distances interlocked.
The Mathematical Theory of Relativity
Chapter I, Section 1 (p. 10)
At The University Press. Cambridge, England. 1930

Empson, William 1906–84
English literary critic and poet

Space is like earth, rounded, a padded cell;
Plumb the stars' depth, your lead bumps you behind…
The Complete Poems of William Empson
The World's End (p. 13)
University Press of Florida. Gainesville, Florida, USA. 2001

Ferris, Timothy 1944–
American science writer

While walking with Heisenberg, the physicist Felix Bloch, who had just read Weyl's *Space, Time and Matter*,

felt moved to declare that space is simply the field of linear equations. Heisenberg replied, "Nonsense. Space is blue and birds fly through it." "What he meant, Bloch writes, "was that it was dangerous for a physicist to describe Nature in terms of idealized abstractions too far removed from the evidence of actual observation."
The Whole Shebang: A State-of-The Universe's Report
Notes, 3 (p. 320)
Simon & Schuster. New York, New York, USA. 1996

Frost, Robert   1874–1963
American poet

Space ails us moderns: we are sick with space.
Its contemplation makes us out as small
As a brief epidemic of microbes.
Complete Poems of Robert Frost
The Lesson for Today
Henry Holt & Company. New York, New York, USA. 1949

Gauss, Johann Carl Friedrich   1777–1855
German mathematician, physicist, and astronomer

We must confess in all humility that, while number is a product of our mind alone, space has a reality beyond the mind whose rules we cannot completely prescribe.
In Charles W. Misner et al.
Gravitation
Part III, Chapter 8 (p. 195)
W.H. Freeman & Company. San Francisco, California, USA. 1973

Gibran, Kahlil   1883–1931
Lebanese-American philosophical essayist

Space is not space between the earth and the sun to one who looks down from the windows of the Milky Way.
Sand and Foam: A Book of Aphorisms (p. 7)
Alfred A. Knopf. New York, New York, USA. 1959

Glenn, Jr., John   1921–
American astronaut and politician

In space one has the inescapable impression that here is a virgin area of the universe in which civilized man, for the first time, has the opportunity to learn and grow without the influence of ancient pressures. Like the mind of a child, it is yet untainted with acquired fears, hate, greed, or prejudice.
In Kevin W. Kelley
The Home Planet
With Plate 136
Addison-Wesley Publishing Company. Reading, Massachusetts, USA. 1988

Heisenberg, Werner Karl   1901–76
German physicist and philosopher

Space is blue and birds fly through it.
In Harald Fritzsch
Translated by Jean Steinberg
The Creation of Matter: The Universe from Beginning to End
Chapter 1 (pp. 12–13)
Basic Books, Inc. New York, New York, USA. 1984

Hey, Nigel S.   1936–
American science writer

People…regardless of their educational, religious, or economic background, are blessed with the ability to look into the sky and marvel at the greatness of all that is out there. … It transports us beyond ourselves and our artifacts. It is another way in which we are able to emerge from our self-centered psychological neighborhoods, to explore a multidimensional realm where self is of no particular significance. It is one path among many through which individual people may comprehend the close community of all life and all humanity, and, with the accession of humility, the rightness of compassion and peace.
Solar System
Introduction (pp. 7–8)
Weidenfield & Nicolson. London, England. 2002

Hubble, Edwin Powell   1889–1953
American astronomer

The outstanding feature, however, is the possibility that the velocity-distance relation may represent the de Sitter effect, and hence that numerical data may be introduced into discussions of the general curvature of space.
A Relation Between Distance and Radial Velocity Among Extra-Galactic Nebulae
Proceedings of the National Academy of Science, Volume 15, 1929 (p. 168)

Jammer, Max   1915–
Israeli physicist and philosopher

Like all science, the science of space must still be classed as unfinished business.
Concepts of Space: The History of Theories of Space in Physics
Chapter V (p. 190)
Harvard University Press. Cambridge, Massachusetts, USA. 1954

Jeans, Sir James Hopwood   1877–1946
English physicist and mathematician

The immensity of space is paralleled by that of time.
Annual Report of the Board of Regents of the Smithsonian Institution, 1928
The Wider Aspects of Cosmogony (p. 171)
Government Printing Office. Washington, D.C. 1929

…space, regarded as a receptacle for radiant energy, is a bottomless pit.
Supplement to "Nature"
Nature, Volume 122, Number 3079, November 3, 1928 (p. 698)

Kant, Immanuel   1724–1804
German philosopher

Space is not a conception which has been derived from outward experiences. For, in order that certain sensations may relate to something without me (that is, to something which occupies a different part of space from that in which I am); in like manner, in order that I may represent them not merely as without, of, and near to each other, but also in separate places, the representation of space must already exist as a foundation. Consequently, the representation of space cannot be borrowed from the relations of external phenomena through experience; but, on the contrary, this external experience is itself only possible through the said antecedent representation.

In *Great Books of the Western World* (Volume 42)
Critique of Pure Reason
First Part, Of Space, Metaphysical Exposition of this Conception, 1
Encyclopædia Britannica, Inc. Chicago, Illinois, USA. 1952

Leonov, Aleksei   1934–
Soviet cosmonaut

What struck me most was the silence. It was a great silence, unlike any I have encountered on Earth, so vast and deep that I began to hear my own body: my heart beating, my blood vessels pulsing, even the rustle of my muscles moving over each other seemed audible. There were more stars in the sky than I had expected. The sky was deep black, yet at the same time bright with sunlight.

The View from Out There: In Words and Pictures
Life, Volume 11, Number 13, November 1988 (p. 197)

Lewis, Gilbert Newton   1875–1946
American chemist

…when we analyze the highly refined concept of space used by mathematicians we find it to be quite similar to the concept of number.

The Anatomy of Science
Chapter II (p. 29)
Yale University Press. New Haven, Connecticut, USA. 1926

Macvey, John W.
No biographical data available

The land lies sleeping under the enveloping mantle of night. Bright stars gleam like jewels from out the velvet darkness of the moonless sky. Beyond these points of celestial beauty, in depths frightening in their sheer immensity, lies realms powdered in stellar glory.

Whispers from Space
Chapter 1
Macmillan Publishing Company. New York, New York, USA. 1973

Maxwell, James Clerk   1831–79
Scottish physicist

…the aim of the space-crumplers is to make its curvature uniform everywhere, that is over the whole of space whether that whole is more or less than Ñ. The direction of the curvature is not related to one of the $x\ y\ z$ more than another or to $-x\ -y\ -z$ so that as far as I understand we are once more on a pathless sea, starless, windless and poleless…

The Scientific Letters and Papers of James Clerk Maxwell: Volume 2, 1862–1873

Postcard to Peter Guthrie Tait, 11 November, 1874 (p. 137)
Clarendon Press. Oxford, England. 1988

Murray, Bruce   1932–
American professor of planetary science and geology

Space…is a colorful thread intimately woven into the enormous tapestry of human existence and experience.

In Ray Bradbury, Arthur C. Clarke, Bruce Murray, Carl Sagan, and Walter Sullivan
Mars and the Mind of Man
Bruce Murray (p. 47)
Harper & Row, Publishers. New York, New York, USA. 1973

Newton, Sir Isaac   1642–1727
English physicist and mathematician

Absolute space, in its own nature, without relation to anything external, remains always similar and immovable.

Mathematical Principles of Natural Philosophy
Scholium, II
E.P. Dutton & Company. New York, New York, USA. 1922

Ockels, Wubbo   1946–
Dutch astronaut and aerospace engineer

Space is so close: It took only eight minutes to get there and twenty to get back.

The View from Out There: In Words and Pictures
Life, Volume 11, Number 13, November 1988 (p. 198)

Poincaré, Henri   1854–1912
French mathematician and theoretical astronomer

Space is only a word that we have believed a thing.

The Foundations of Science
Author's Preface to Translation (p. 5)
The Science Press. New York, New York, USA. 1913

Siegel, Eli   1902–78
American philosopher, poet, critic and founder of Aesthetic Realism

Space won't keep still, and it won't budge either: so give up trying.

Damned Welcome
Aesthetic Realism, Maxims, Part Two, #396 (p. 153)
Definition Press. New York, New York, USA. 1972

Smith, Logan Pearsall   1865–1946
American author

So gazing up on hot summer nights at the London stars, I cool my thoughts with a vision of the giddy, infinite, meaningless waste of Creation, the blazing Suns, the Planets and frozen Moons, all crashing blindly forever across the void of space.

Trivia
Book II, Mental Vice (p. 97)
Doubleday, Page & Company. Garden City, New York, USA. 1917

I think of Space, and the unimportance in its unmeasured vastness, of our toy solar system; I lose myself in speculations of the lapse of Time, reflecting how at the best

our human life on this minute and perishing planet is as brief as a dream.
Trivia
Book II, Self-Analysis (pp. 121–122)
Doubleday, Page & Company. Garden City, New York, USA. 1917

Sylvester, James Joseph   1814–97
English mathematician

Space is the Grand Continuum from which, as from an inexhaustible reservoir, all the fertilizing ideas of modern analysis are derived…
The Collected Mathematical Papers of James Joseph Sylvester
(Volume 2)
Presidential Address to the British Association
Exeter British Association Report (1869) (p. 659)
University Press. Cambridge, England. 1904–1912

Tennyson, Alfred (Lord)   1809–92
English poet

…The clear galaxy
Shorn of its hoary lustre, wonderful,
Distinct and vivid with sharp point of light,
Blaze within blaze, an unimagin'd depth
And harmony of planet-girded suns
And moon — encircled planets, wheel in wheel,
Arch'd the wan sapphire. Nay, the hum of men.
Or other things talking in unknown tongues,
And notes of busy life in distant worlds
Beat like a far wave on my anxious ear.
Alfred Tennyson's Poetical Works
Timbuctoo, l. 105–113
Oxford University Press, Inc. London, England. 1953

The X-Files

MULDER: Hey, Scully, we send those men up into space to unlock the doors of the universe, and we don't even know what's behind them.
The X-Files
Space
Television program
Season 1, 1993

Thomson, J. Arthur   1861–1933
Scottish biologist

There is grandeur in the spectacle of the star-strewn sky, so apparently crowded, but there are thousands of worlds unseen for every one our unaided eyes can image, and yet the astronomers tell us that the emptiness of space is its most striking characteristic.
The System of Animate Nature (Volume 1)
Lecture I (p. 30)
William & Norgate. London, England. 1920

Tsiolkovsky, Konstantin Eduardovich   1857–1935
Russian research scientist

Man will not always stay on earth; the pursuit of light and space will lead him to penetrate the bounds of the atmosphere, timidly at first, but in the end to conquer the whole of solar space.
In Herbert Friedman
The Amazing Universe
Chapter 1 (p. 28)
National Geographic Society. Washington DC. 1980

vas Dias, Robert
Anglo-American poet and writer

The premise…is that outer space is as much a territory of the mind as it is a physical concept.
Inside Outer Space: New Poems of the Space Age
Introduction (p. xxxix)
Anchor Press. Garden City, New York, USA. 1970

von Bitter Rucker, Rudy   1946–
American mathematician and science fiction writer

What is the shape of space? Is it flat, or is it bent? Is it nicely laid out, or is it warped and shrunken? Is it finite, or is it infinite? Which of the following does space resemble more: (a) a sheet of paper, (b) an endless desert, (c) a soap bubble, (d) a doughnut, (e) an Escher drawing, (f) an ice cream cone, (g) the branches of a tree, or (h) a human body?
The Fourth Dimension: Toward a Geometry of Higher Reality
Chapter 7 (p. 91)
Houghton Mifflin Company. Boston, Massachusetts, USA. 1984

von Braun, Wernher   1912–77
German-American rocket science

…the progress of mankind here on Earth is directly linked to the future than man builds for himself in space.
In Erik Bergaust
Wernher von Braun
A Horse Named Susie (p. 373)
National Space Institute. Washington, D.C. 1976

…don't tell me that man doesn't belong out there. Man belongs wherever he wants to go — and he'll do plenty well when he gets there.
Reach for the Stars
Time, Volume 71, 17 February 1958 (p. 25)

Weyl, Hermann   1885–1955
German mathematician

Nowhere do mathematics, natural sciences, and philosophy permeate one another so intimately as in the problem of space.
Philosophy of Mathematics and Natural Science
Part I, Chapter III (p. 67)
Princeton University Press. Princeton, New Jersey, USA. 1949

Wheeler, John Archibald   1911–
American physicist and educator
Thorne, Kip S.   1940–
American theoretical physicist

Space tells matter how to move…and matter tells space how to curve.
In Charles W. Misner et al
Gravitation
Part I, Chapter 1 (p. 23)
W.H. Freeman & Company. San Francisco, California, USA. 1973

…in essence, the curvature in space created by the electromagnetic field is the electromagnetic field; and this curvature can in principle be detected by purely geometric measurements.
International Science and Technology
The Dynamics of Space-Time, December 1963 (p. 72)

Whitehead, Alfred North   1861–1947
English mathematician and philosopher

All space measurement is from stuff in space to stuff in space.
The Aims of Education and Other Essays
Chapter X (p. 233)
The Macmillan Company. New York, New York, USA. 1959

Whitman, Walt   1819–92
American poet, journalist, and essayist

Now while the great thoughts of space and eternity fill me I will measure myself by them. And now touch'd with the lives of other globes arrived as far long as those of the earth or waiting to arrive, or pass'd on farther than those of the earth, I henceforth no more ignore them than I ignore my own life.
Complete Poetry and Collected Prose
Night on the Prairies
The Library of America. New York, New York, USA. 1982

Every cubic inch of space is a miracle.
Complete Poetry and Collected Prose
Miracles
The Library of America. New York, New York, USA. 1982

Winchell, Alexander   1824–91
American geologist

In the midst of this universe of seething movements is our home. The mind, uplifted in the effort to contemplate them and grasp their method, grows giddy and impotent. How sublime these activities! To what a numerous and lofty companionship does our little planet belong! Hard it seems to be imprisoned here while the realm of the universe tempts us to its exploration. How can a human soul content itself to roll and whirl through space during its mortal days, and eat and sleep and trifle, like rats in a ship at sea, without wondering where we are and whither we are bound.
World-Life or Comparative Geology
Part I, Chapter II, Section 4.7 (p. 142)
S.C. Griggs & Company. Chicago, Illinois, USA. 1883

Zubrin, Robert   1952–
Engineer

Like the philosophy of Greece, the paintings of the Renaissance and the music of the Enlightenment, the explosion of knowledge about our solar system and the surrounding universe will be remembered for thousands of years as the defining brilliance of our age. To destroy [the space] program for the sake of bean counting, or perhaps as part of some obscure political maneuver, is not tolerable. It is not just a mistake, it is a crime — an infamous crime against civilization that is comparable to the burning of the Library of Alexandria.
Space News, September 13, 1999

Americans are proud of our space exploration program, and rightly so. It is a statement that we continue to be a nation of explorers and pioneers. But more than that, it is a statement that we are a truly great nation, great not because of our military might…but because we do great things for all humanity and for all time. Killing our space exploration program amounts to nothing less than pulling some of the stars off our flag. This is a desecration we cannot allow.
Space News, September 13, 1999

SPACE AGE

Clarke, Arthur C.   1917–
English science and science fiction writer

Across the gulf of centuries, the blind smile of Homer is turned upon our age. Along the echoing corridors of time, the roar of the rockets merges now with the creak of the wind-taut rigging. For somewhere in the world today, still unconscious of his destiny walks the boy who will be the first Odysseus of the Age of Space…
The Challenge of the Spaceship
Envoi (p. 213)
Harper & Brothers. New York, New York, USA. 1959

SPACE EXPLORATION

Armstrong, Neil A.   1930–
American astronaut

The *Eagle* has landed.
The Washington Post, July 21, 1969 (p. 1)

Arnold, James R.
No biographical data available

Space is the empty place next to the full place where we live. I believe we will be true to our nature and go there.
The Frontier in Space. Will One Be True to Our Nature and Accept the Challenge of the Next Frontier?
American Scientist, Volume 68, Number 3, May–June 1980 (p. 304)

Asimov, Isaac   1920–92
American author and biochemist

Throughout the history of humanity, we have been extending our range until it is now planet-wide, covering all parts of the Earth's surface and reaching to the bottom of the ocean, to the top of the atmosphere, and beyond it to the Moon. We will flourish only as long as we continue to extend that range, and although the potential range is not infinite, it is incredibly vast even by present standards. We will eventually extend our range to cover the whole of the solar system, and then we will head outward to the stars.

In James Burke, Jules Bergman and Isaac Asimov
The Impact of Science on Society
Our Future in the Cosmos — Space (p. 79)
National Aeronautice and Space Administration. Washington, D.C. 1985

Unless we are willing to settle down into a world that is our prison, we must be ready to move beyond Earth. ...

In James Burke, Jules Bergman and Isaac Asimov
The Impact of Science on Society
Our Future in the Cosmos — Space (p. 80)
National Aeronautice and Space Administration. Washington, D.C. 1985

Bernal, John Desmond   1901–71
Irish-born physicist and x-ray crystallographer

On earth, even if we should use all the solar energy which we receive, we should still be wasting all but one two-billionths of the energy the sun gives out. Consequently, when we have learnt to live on this solar energy and also to emancipate ourselves from the earth's surface, the possibilities of the spread of humanity will be multiplied accordingly.... There will, from desire or necessity, come the idea of building a permanent home for men in space.... At first space navigators, and then scientists whose observations would be best conducted outside the earth, and then finally those who for any reason were dissatisfied with earthly conditions would come to inhabit these bases and found permanent spatial colonies.

The World, the Flesh and the Devil: An Enquiry Into the Future of the Three Enemies of the Rational Soul
Chapter II (pp. 11–12)
Indiana University Press. Bloomington, Indiana, USA. 1969

Blagonravov, Anatoly A.   1894–1975
Russian scientist

The exploration of the cosmos — the moon and the planets — is a noble aim. Our generation has the right to be proud of the fact that it has opened the space era of mankind.

In Mose L. Harvey
The Lunar Landing and the US–Soviet Equation
Bulletin of the Atomic Scientists, Volume 25, Number 7, September 1969 (p. 29)

Bradbury, Ray   1920–
American writer

Get along to Mars and beyond.

The journey is long, the end uncertain, and there is more dark along the way than light, but you can whistle. Come with me by the wall of the great tombyards of all time which lie a billion years ahead. What shall we whistle as we stroll in our rocket, hoping to make it by the vast darkness where shadows wait to seize and keep us?
Follow me.
I know a tune.
Here...listen.

In Ray Bradbury, Arthur C. Clarke, Bruce Murray, Carl Sagan, and Walter Sullivan
Mars and the Mind of Man
Ray Bradbury (p. 143)
Harper & Row, Publishers. New York, New York, USA. 1973

...I would not see our candle blown out in the wind. It is a small thing, this dear gift of life handed us mysteriously out of immensity. I would not have that gift expire. ...What's the use of looking at Mars through a telescope, sitting on panels, writing books, if it isn't to guarantee, not just the survival of mankind, but mankind surviving forever!"

In Ray Bradbury, Arthur C. Clarke, Bruce Murray, Carl Sagan and Walter Sullivan
Mars and the Mind of Man
Ray Bradbury (p. 133)
Harper & Row, Publishers. New York, New York, USA. 1973

Clarke, Arthur C.   1917–
English science and science fiction writer

There are still some scientists who consider that there is no point in sending men into space, even when it becomes technically possible; machines, they argue, can do all that is necessary. Such an outlook is incredibly shortsighted; worse than that, it is stupid, for it completely ignores human nature.

Lecture
St Martin's Technical School on Charing Cross Road, October 5, 1946

Though the specific ideals of astronautics are new, the motives and impulses underlying them are old as the race — and in the ultimate analysis, they owe as much to emotion as to reason. Even if we could learn nothing in space that our instruments would not already tell us, we should go there just the same.

Lecture
St Martin's Technical School on Charing Cross Road, October 5, 1946

To find anything comparable with our forthcoming ventures into space, we must go back far beyond Columbus, far beyond Odysseus — far, indeed, beyond the first ape-man. We must contemplate the moment, now irrevocably lost in the mists of time, when the ancestor of all of us came crawling out of the sea.

Profiles of the Future: An Inquiry into the Limits of the Possible
Chapter 8 (p. 94)
Harper & Row, Publishers. New York, New York, USA. 1973

This [the sea] is where life began, and where most of this planet's life remains to this day, trapped in a meaningless cycle of birth and death. Only the creatures who dared the hostile, alien land were able to develop intelligence; now that intelligence is about to face a still greater challenge. It may even be that this beautiful Earth of ours is no more than a brief resting-place between the sea of salt where we were born, and the sea of stars on which we must now venture forth.

Profiles of the Future: An Inquiry into the Limits of the Possible
Chapter 8 (p. 94)
Harper & Row, Publishers. New York, New York, USA. 1973

With the landing of the first spaceship on Mars and Venus, the Childhood of our race was over and history as we know it began…

The Exploration of Space
Chapter 18 (p. 195)
Harper & Brothers. New York, New York, USA. 1951

The challenge of the great spaces between the worlds is a stupendous one; but if we fail to meet it, the story of our race is drawing to a close. Humanity will have turned its back upon the still untrodden heights and will be descending again the long slope that stretches, across a thousand million years of time, down to the shores of the primeval sea.

Interplanetary Flight: An Introduction to Astronautics
Chapter 10 (p. 127)
Harper & Row, Publishers. New York, New York, USA. 1960

Even if we never reach the stars by our own efforts, in the millions of years that lie ahead it is almost certain that the stars will come to us. Isolationism is neither a practical policy on the national or the cosmic scale. And when the first contact with the outer universe is made, one would like to think that Mankind played an active and not merely a passive role — that we were the discoverers, not the discovered.

The Exploration of Space
Chapter 17 (p. 182)
Harper & Brothers. New York, New York, USA. 1951

Long before the Sun's radiation has shown any measurable increase, Man will have explored all the Solar System and, like a cautious bather testing the temperature of the sea, will be making breathless little forays into the abyss that separates him from the stars.

The Challenge of the Spaceship
The Challenge of the Spaceship (p. 4)
Harper & Brothers. New York, New York, USA. 1959

Interplanetary travel is now the only form of "conquest and empire" compatible with civilization. Without it, the human mind, compelled to circle forever in its planetary goldfish bowl, must eventually stagnate.

The Challenge of the Spaceship
The Challenge of the Spaceship (p. 7)
Harper & Brothers. New York, New York, USA. 1959

…there is no way back into the past; the choice, as Wells once said, is the universe — or nothing. Though men and civilizations may yearn for rest, for the dream of the lotus-eaters, that is a desire that merges imperceptibly into death. The challenge of the great spaces between the worlds is a stupendous one; but if we fail to meet it, the story of our race will be drawing to its close.

Interplanetary Flight: An Introduction to Astronautic
Chapter 10 (p. 127)
Harper & Row, Publishers. New York, New York, USA. 1960

…who can guess what strange roads there may yet be on which we may travel to the stars?

The Promise of Space
To the Stars (p. 299)
Harper & Row, Publishers. New York, New York, USA. 1968

Commoner, Barry   1917–
American biologist, ecologist, and educator

Explorations of space, like the earlier explorations, are great adventures because they are bold, and they are bold because they are hazardous.

Science and Survival
Chapter 4 (p. 56)
The Viking Press. New York, New York, USA. 1966

Cousins, Norman   1912–90
American editor and author

The justification for exploring the cosmos rests not on tangible benefits, but on philosophical grounds and on our instinctive need to evolve.

Rendezvous with Infinity
Cosmic Search Magazine, Volume 1, Number 1, January 1, 1979 (p. 30)

Deudney, Daniel
American political scientist

…for all our looking and probing of the universe, we have yet to find any place as habitable as the remotest, most forbidding parts of this planet. Space exploration has taught us just how rare and precious the earth is.

Space: The High Frontier in Perspective
Toward an Earth-Oriented Space Program (p. 51)
Worldwatch Institute. 1982

Dyson, Freeman J.   1923–
American physicist and educator

When we are a million species spreading through the galaxy, the question "Can man play God and still stay sane?" will lose some of its terrors. We shall be playing God, but only as local deities and not as lords of the universe. There is safety in numbers. Some of us will become insane, and rule over empires as crazy as Doctor Moreau's island. Some of us will shit on the morning star. There will be conflicts and tragedies. But in the long run, the sane will adapt and survive better than the insane.

Disturbing the Universe

Chapter 21 (pp. 236–237)
Basic Books, Inc. New York, New York, USA. 1979

Nature's pruning of the unfit will limit the spread of insanity among species in the galaxy, as it does among individuals on earth. Sanity is, in its essence, nothing more than the ability to live in harmony with nature's laws.
Disturbing the Universe
Chapter 21 (pp. 236–237)
Basic Books, Inc. New York, New York, USA. 1979

Ferris, Timothy   1944–
American science writer

We who came down from out of the forest seek to grow a forest of knowing among the stars.
The Mind's Sky: Human Intelligence in a Cosmic Context
It (p. 222)
Bantam Books. New York, New York, USA. 1992

We don't know whether human music will mean anything to nonhuman intelligences on other planets. But any creature that comes across Voyager and recognizes the record as an artifact can realize that it was dispatched with no hope of return. That gesture may speak more clearly than music. It says: However primitive we seem, however crude this spacecraft, we knew enough to envision ourselves citizens of the cosmos. …However small we were, something in us was large enough to want to reach out to discoverers unknown, in times when we shall have perished or changed beyond recognition. … Whoever and whatever you are, we too once lived in this house of stars, and we thought of you.
Murmurs of Earth: The Voyager Instellar Record
Voyager's Music
Random House, Inc. New York, New York, USA. 1978

Firsoff, Valdemar Axel   1910–82
English astronomer and author

Yet if we go into space, let us do so humbly, in the spirit of cosmic piety. We know very little. We are face to face with the great unknown and have no right to assume that we are alone in the Solar System.
Exploring the Planets
Chapter XV (p. 160)
Sidgwick & Jackson. London, England. 1964

Gregory, Sir Richard Arman   1864–1952
British science writer and journalist

Man as a physical being is but a microscopic part of the universe, yet his mind carries him ever upward, and with spirit bold and unconquerable he seeks to reach the summit of Mount Olympus. Infinite space remains to humble his pride in spite of the knowledge he has obtained of the starry heavens; yet he pursues his inquiries into the unknown, and his children's children will continue to search.

Discovery; or, The Spirit and Service of Science
Chapter I (p. 21)
Macmillan & Company Ltd. London, England. 1918

Hale, George Ellery   1868–1938
American astronomer

Like buried treasures, the outposts of the universe have beckoned to the adventurous from immemorial times. Princes and potentates, political or industrial, equally with men of science, have felt the lure of the uncharted seas of space, and through their provision of instrumental means the sphere of exploration has rapidly widened…
Possibilities of Large Telescopes
Harper's Magazine, April 1928 (p. 639)

Hawking, Stephen William   1942–
English theoretical physicist

I don't think the human race will survive the next thousand years, unless we spread into space. There are too many accidents that can befall life on a single planet. But I'm an optimist. We will reach out to the stars.
By Roger Highfield, Science Editor
Colonies in space may be only hope, says Hawking
Telegraph, Filed: 16/10/2001

Heinlein, Robert A.   1907–88
American science fiction writer

But space travel can't ease the pressure on a planet grown too crowded not even with today's ships and probably not with any future ships — because stupid people won't leave the slopes of their home volcano even when it starts to smoke and rumble. What space travel does do is drain off the best brains: those smart enough to see a catastrophe before it happens and with the guts to pay the price — abandon home, wealth, friends, relatives, everything — and go. That's a tiny fraction of one percent. But that's enough.
Time Enough for Love
Chapter XIV (p. 413)
G.P. Putnam's Sons. New York, New York, USA. 1973

Heppenheimer, T. A.   1947–
Aviation writer

…if humanity persists and endures, in time we will come face to face with the evolution of our sun. In a few billion years its slow brightening will speed up as it swells into a red giant. Earth will then be uninhabitable, as will the inner regions of the Solar System. Yet there will be other more clement stars to which our descendents may wish to migrate. Certainly a society that has developed space flight and space colonization will have the advantage of never thereafter having to stand hostage to fortune.
Toward Distant Suns
Chapter 13 (p. 244)
Stackpole Books. Harrisburg, Pennsylvania, USA. 1979

Hey, Nigel S.   1936–
American science writer

Space scientists and engineers serve the intangible needs of humankind, and share common ground with the poet. It is self-deceptive to suppose that society is wholly bound up in supplying life-or-death needs. Humans are thinkers, explorers, wonderers, and dreamers. If we were not, we would not need space exploration; but then we would also lead a listless and uncreative existence.
How We Will Explore the Outer Planets (p. 142)
G.P. Putnam's Sons. New York, New York, USA. 1973

Most of our knowledge of this marvel-filled universe is due to astronomy, telescopes and to robotic spaceflight. It is impossible to think of any thing that more exquisitely embodies the technical genius of humankind, in so small a package, as the interplanetary spacecraft.
Solar System
Chapter 3 (p. 61)
Weidenfield & Nicolson. London, England. 2002

Hoyle, Sir Fred   1915–2001
English mathematician and astronomer

Space isn't remote at all. It's only an hour's drive away if your car could go straight upwards.
Observer, September 9, 1979

The seemingly insuperable difficulties of deep-space travel suggest an intention to keep us fixed at home in our own solar system, and the physical nature of our part of the Universe, as well as the basic rules of physics and chemistry, have a warning look about them, like barriers designed to isolate intelligent life. This means that for us, unlike the situation for humble microorganisms, deep-space travel is probably a stark impossibility.
The Intelligent Universe
Chapter 6 (p. 156)
Holt, Rinehart & Winston. New York, New York, USA. 1983

Hubble, Edwin Powell   1889–1953
American astronomer

Thus the explorations of space end on a note of uncertainty. And necessarily so. We are, by definition, in the very center of the observable region. We know our immediate neighborhood rather intimately. With increasing distance, our knowledge fades, and fades rapidly. Eventually, we reach the dim boundary — the utmost limits of our telescopes. There, we measure shadows, and we search among ghostly errors of measurement for landmarks that are scarcely more substantial.

The search will continue. Not until the empirical resources are exhausted, need we pass on to the dreamy realms of speculation.
The Realm of the Nebulae
Chapter VIII (p. 202)
Dover Publications, Inc. New York, New York, USA. 1958

The exploration of space has swept outward in successive waves, first, through the system of the planets, then, through the stellar system, and finally, into the realm of the nebulae. Today we study a region of space so vast and so homogeneous that it may well be a fair sample of the universe. At any rate, we are justified in adopting the assumption as a working hypothesis and attempting to infer the nature of the universe from the observed characteristics of the sample.
Annual Report of the Board of Regents of the Smithsonian Institution, 1942
The Problem of the Expanding Universe (p. 119)
Government Printing Office. Washington, D.C. 1943

Johnson, Lyndon B.   1908–73
36th president of the United States

No national sovereignty rules in outer space. Those who venture there go as envoys of the entire human race. Their quest, therefore, must be for all mankind, and what they find should belong to all mankind.
News Conference
Johnson City, Texas, 29 August, 1965

Kennedy, John F.   1917–63
35th president of the United States

We choose to go to the Moon in this decade and do the other things, not because they are easy — but because they are hard!
Speech, Rice University, 12 September, 1962

I believe that this nation should commit itself to achieving the goal, before this decade is out, of landing a man on the Moon and returning him safely to Earth. No single space project in this period will be more impressive to mankind, or more important in the long-range exploration of space; and none will be so difficult or expensive to accomplish.
Announcement to American Congress
25 May, 1961

Kepler, Johannes   1571–1630
German astronomer

Provide ship or sails adapted to the heavenly breezes, and there will be some who will not fear even that void.… So, for those who will come shortly to attempt this journey, let us establish the astronomy: Galileo, you of Jupiter, I of the moon.
In John Lear
Kepler's Dream
Introduction and Interpretation, I (p. 3)
University of California Press. Berkeley, California, USA. 1965

Lewis, John S.
American professor of planetary science

It is in the interests of all the residents of Earth to see exploration continue and to see our realm of competence

expand to fill the Solar System. Like our ancient ancestors at the time of their emergence from the sea onto the
land, we are challenged by events to master this great
new environment, to drink of its knowledge, and to feast
on its boundless resources. Let us not squander this
golden opportunity.
In John S. Lewis
Physics and Chemistry of the Solar System
Chapter XII (p. 517)
Academic Press. San Diego, California, USA. 1995

Lowell, Percival   1855–1916
American astronomer

From time immemorial travel and discovery have called
with strange insistence to him who, wandering on the
world, felt adventure in his veins. The leaving familiar
sights and faces to push forth into the unknown has with
magnetic force drawn the bold to great endeavor and
fired the thought of those who stayed at home.
Mars and Its Canals
Chapter I (p. 3)
Houghton Mifflin Company. Boston, Massachusetts, USA. 1895

Lucretius   ca. 99 BCE–55 BCE
Roman poet

…he passed far beyond the flaming walls of the world
and traversed throughout in mind and spirit the immea-
surable universe…
In *Great Books of the Western World* (Volume 12)
The Nature of the Universe
Book I, 62 (p. 2)
Encyclopædia Britannica, Inc. Chicago, Illinois, USA. 1952

Makarov, Oleg   1933–2003
Soviet cosmonaut

You keep returning to the thought that only very thin
walls separate you from the deathly cold and incomprehensible emptiness of space, which can extinguish life
instantly and piteously.
The View from Out There: In Words and Pictures
Life, Volume 11, Number 13, November 1988 (p. 198)

Martin, Charles-Noël   1923–
French nuclear physicist

As men travel further and further into space they are
bound to meet sights beyond their wildest expectations.
Translated by A.J. Pomerans
The Role of Perception in Science
Chapter 4 (p. 92)
Hutchinson of London. London, England. 1963

Moulton, Forest Ray   1872–1952
American astronomer

…there is not the slightest possibility of [travel to other
worlds]. There is not in sight any source of energy that
would…be necessary to get us beyond the gravitative

control of the earth; there is not theory that would guide
us through interplanetary space to another world even if
we could control our departure from the earth; there is no
means of carrying the large amount of oxygen, water, and
food that would be necessary for such a long journey; and
there is no known way of easing our ether ship down onto
the surface of another world, if we could get there at low
enough speed to avoid destruction.
Consider the Heavens
Chapter VII
Chapter II (p. 107)
Doubleday, Doran & Company, Inc. Garden City, New York, USA. 1935

Nixon, Richard M.   1913–94
37th president of the United States

We must see our space effort, then, not only as an adventure of today but also as an investment in tomorrow.
We did not go to the Moon merely for the sport of it.
To be sure, those undertakings have provided an exciting
adventure for all mankind and we are proud that it was
our nation that met this challenge. But the most important
thing about man's first footsteps on the Moon is what
they promise for the future.
Statement by President Nixon on the Space Program
Released from the Office of the White House
Press Secretary, Key Biscayne, Florida, 7 March 1970

From time immemorial, man has insisted on venturing
into the unknown despite his inability to predict precisely
the value of any given exploration. He has been willing
to take risks, willing to be surprised, willing to adapt to
new experiences. Man has come to feel that such quests
are worthwhile in and of themselves — for they represent
one way in which he expands his vision and expresses the
human spirit. A great nation must always be an exploring
nation if it wishes to remain great.
Statement by President Nixon on the Space Program
Released from the Office of the White House
Press Secretary, Key Biscayne, Florida, 7 March 1970

As we enter a new decade, we are conscious of the fact
that man is also entering a new historic era. For the first
time, he has reached beyond his planet; for the rest of
time, we will think of ourselves as men from the planet
Earth. It is my hope that as we go forward with our space
program, we can plan and work in a way which makes us
proud both of the planet from which we come and of our
ability to travel beyond it.
Statement by President Nixon on the Space Program
Released from the Office of the White House
Press Secretary, Key Biscayne, Florida, 7 March 1970

Oberth, Hermann   1894–1989
German mathematician and physicist

This is the goal:
To make available for life every place where life is
possible.

To make inhabitable all worlds as yet uninhabitable, and all life purposeful.
Translated by G.P.H. de Freville
Man into Space: New Projects for Rocket and Space Travel
Chapter VIII (p. 167)
Harper & Brothers. New York, New York, USA. 1957

O'Neill, Gerard K.   1927–92
American physicist

Clearly our first task is to use the material wealth of space to solve the urgent problems we now face on Earth: to bring the poverty-stricken segments of the world up to a decent living standard, without recourse to war or punitive action against those already in material comfort; to provide for a maturing civilization the basic energy vital to its survival.
The High Frontier
Bantam Dell Doubleday Publishing Group. New York, New York, USA. 1978

Purcell, Edward   1912–97
American physicist

All this stuff about traveling around the universe… belongs back where it came from, on the cereal box.
In A.G.W. Cameron (ed.)
Interstellar Communication; A Collection of Reprints and Original Contributions
Radio Astronomy and Communication Through Space (p. 143)
W.A. Benjamin, Inc. New York, New York, USA. 1963

Reade, Winwood   1838–75
English philosopher and historian

A time will come when science will transform [our bodies] by means which we cannot conjecture…. And then, the earth being small, mankind will migrate into space, and will cross the airless Saharas which separate planet from planet, and sun from sun. The earth will become a Holy Land which will be visited by pilgrims from all quarters of the universe.
The Martyrdom of Man
Chapter IV (pp. 459, 460)
E.P. Dutton & Company. New York, New York, 1926

Roddenberry, Gene   1921–91
American television producer and writer

Let me end with an explanation of why I believe the move into space to be a human imperative. It seems to me obvious in too many ways to need listing that we cannot much longer depend upon our planet's relatively fragile ecosystem to handle the realities of the human tomorrow. Unless we turn human growth and energy toward the challenges and promises of space, our only other choice may be the awful risk, currently demonstrable, of stumbling into a cycle of fratricide and regression which could end all chances of our evolving further or of even surviving.

Hailing Frequencies Open!
Planetary Report, Volume 1, April/May 1981 (p. 3)

Russen, David
No biographical data available

Since Springiness is a cause of forcible motion; and a Spring will, when bended and let loose, extend its self to its length; could a Spring of well-tempered steel be framed, whose basis being fastened to the Earth, and on the other end placed a Frame or Seat, wherein a Man with other necessaries could abide in safety, this Spring being with Cords, Pullies, or other Engines bent, and then let loose by degrees by those who manage the Pullies, the other end…reach the Moon, where the Person who ascended landing, the Spring might again be bent, till the end touching the earth, should discharge the passenger again in safety.
In Noel Deisch
The Navigation of Space in Early Speculation and in Modern Research
Popular Astronomy, Volume 38, Number 2, February 1930 (p. 81)

Sagan, Carl   1934–96
American astronomer and author

Since, in the long run, every planetary civilization will be endangered by impacts from space, every surviving civilization is obliged to become spacefaring — not because of exploratory or romantic zeal, but for the most practical reason imaginable: staying alive…. If our long-term survival is at stake, we have a basic responsibility to our species to venture to other worlds.
Pale Blue Dot: A Vision of the Human Future in Space
Chapter 21 (p. 371)
Random House, Inc. New York, New York, USA. 1994

Centuries hence, when current social and political problems may seem as remote as the problems of the Thirty Years' War are to us, our age may be remembered chiefly for one fact: It was the time when the inhabitants of the earth first made contact with the vast cosmos in which their small planet is embedded.
The Solar System
Scientific American, Volume 233, Number 3, 1975 (p. 30)

Thoreau, Henry David   1817–62
American essayist, poet, and practical philosopher

…perchance, coming generations will not abide the dissolution of the globe, but, availing themselves of future inventions in aerial locomotion, and the navigation of space, the entire race may migrate from the earth, to settle some vacant and more western planet…. It took but little art, a simple application of natural laws, a canoe, a paddle, and a sail of matting, to people the isles of the Pacific, and a little more will people the shining isles of space. Do we not see in the firmament the lights carried along the shore by night, as Columbus did? Let us not despair or mutiny.

The Maine Woods
Paradise (to Be) Regained (p. 58)
Houghton Mifflin Company. New York, New York, USA. 1893

Tipler, Frank   1947–
American physicist

If the human species, or indeed any part of the biosphere, is to continue to survive, it must eventually leave the Earth and colonize space. For the simple fact of the matter is, the planet Earth is doomed.… Let us follow many environmentalists and regard the Earth as Gaia, the mother of all life (which indeed she is). Gaia, like all mothers, is not immortal. She is going to die. But her line of descent might be immortal…Gaia's children might never die out — provided they move into space. The Earth should be regarded as the womb of life — but one cannot remain in the womb forever.
The Physics of Immortality: Modern Cosmology, God and the Resurrection of the Dead
Chapter II (p. 18, 18–19)
Doubleday & Company, Inc. New York, New York, USA. 1994

van der Riet Wooley, Sir Richard   1906–1986
British Astronomer Royal

…the whole procedure [of shooting rockets into space]… presents difficulties of so fundamental a nature, that we are forced to dismiss the notion as essentially impracticable, in spite of the author's insistent appeal to put aside prejudice and to recollect the supposed impossibility of heavier-than-air flight before it was actually accomplished.
x
Rockets in Space
Nature, Supplement, March 14, 1936 (p. 442)

It's utter bilge. I don't think anybody will ever put up enough money to do such a thing…What good would it do us? If we spend the same amount of money on preparing first-class astronomical equipment we would learn much more about the universe…It is all rather rot.
Utter Bilge
Time, January 16, 1956 (p. 42)

Verne, Jules   1828–1905
French novelist

…I repeat that the distance between the earth and her satellite is a mere trifle, and undeserving of serious consideration. I am convinced that before twenty years are over one-half of our earth will have paid a visit to the moon.
From the Earth to the Moon and Round the Moon
From Earth to the Moon, Chapter XIX (p. 99)
A.L. Burt Company. New York, New York, USA. 1890

In spite of the opinions of certain narrow-minded people, who would shut up the human race upon this globe, as within some magic circle which it must never outstep, we shall one day travel to the moon, the planets, and the

stars, with the same facility, rapidity, and certainty as we now make the voyage from Liverpool to New York.
From the Earth to the Moon and Round the Moon
From Earth to the Moon, Chapter XIX (p. 97)
A.L. Burt Company. New York, New York, USA. 1890

von Braun, Wernher   1912–77
German-American rocket scientist

[Space travel] will free man from his remaining chains, the chains of gravity which still tie him to this planet. It will open to him the gates of heaven.
The Jupiter People
Time, February 10, 1958 (p. 18)

Wells, H. G. (Herbert George)   1866–1946
English novelist, historian, and sociologist

All this world is heavy with the promise of greater things, and a day will come, one day in the unending succession of days, when beings, beings who are now latent in our thoughts and hidden in our loins, shall stand upon this earth as one stands upon a footstool and laugh and reach out their hands amidst the stars.
The Discovery of the Future
Nature, Volume 65, Number 1684, February 6, 1902 (pp. 326–331)

Whipple, Fred L.   1906–2004
American comet research pioneer

In conquering space, man will take his greatest single step forward in his ever-expanding struggle against the limitations set by nature. The scientist, by knowing more about the universe, may find paths that will lead to still further conquests of nature. And it may be truly said that man will no longer be limited to seeing "as through a glass darkly." The universe will be spread out clearly before him.
In Cornelius Ryan (ed.)
Across the Space Frontier
The Heavens Open (p. 143)
The Viking Press. New York, New York, USA. 1952

Wilkins, John   1614–72
English writer

We see a great ship swims as well as a small cork, and an eagle flies in the air as well as a little gnat…. 'Tis likely enough that there may be means invented of journeying to the moon; and how happy they shall be that are first successful in this attempt.
A Discourse Concerning a New World and Another Planet
Book 1, Chapter 14
J. Maynard. London, England. 1640

Wren, Sir Christopher   1632–1723
English mathematician and architect

A time would come when Men should be able to stretch out their Eyes…they should see the Planets like our Earth.
Inauguration Speech, Gresham College, 1657

SPACE FLIGHT

Burroughs, Edgar Rice   1875–1950
American writer

I knew that I had ample room in which to wander, since science has calculated the diameter of space to be eighty-four thousand million light years, which, when one reflects that light travels at the rate of one hundred eighty-six thousand miles a second, should satisfy the wanderlust of the most inveterate roamer.
Pirates of Venus
Chapter Two (p. 19)
University of Nebraska Press. Lincoln, Nebraska, USA. 2001

…man is an artifact designed for space travel. He is not designed to remain in his present biologic state any more than a tadpole is designed to remain a tadpole.
The Adding Machine: Selected Essays
Civilian Defense (p. 82)
Seaver Books. New York, New York, USA. 1986

Clarke, Arthur C.   1917–
English science and science fiction writer

It has often been said — and though it is becoming platitudinous it is nonetheless true — that only through space-flight can mankind find a permanent outlet for its aggressive and pioneering instincts. The desire to reach the planets is only an extension of the desire to see what is over the next hill.
The Exploration of Space
Pocket Books. New York, New York, USA. 1979

Cousins, Norman   1912–90
American editor and author

What was most significant about the first lunar voyage was not that men set foot on the moon, but that they set eye on earth.
Rendezvous with Infinity
Cosmic Search Magazine, Volume 1, Number 1, January 1, 1979 (p. 31)

Dyson, Freeman J.   1923–
American physicist and educator

There are three reasons, …apart from scientific considerations, mankind needs to travel in space. The first…is garbage disposal; we need to transfer industrial processes into space so that the earth may remain a green and pleasant place for our grandchildren to live in. The second…to escape material impoverishment: the resources of this planet are finite, and we shall not forego forever the abundance of solar energy and minerals and living space that are spread out all around us. The third…our spiritual need for an open frontier.
Disturbing the Universe
Chapter 10 (p. 116)
Basic Books, Inc. New York, New York, USA. 1979

When will the third romantic age in the history of space-flight begin? The third romantic age will see little model sailboats spreading their wings to the sun in space…
Disturbing the Universe
Chapter 10 (p. 116)
Basic Books, Inc. New York, New York, USA. 1979

Feynman, Richard P.   1918–88
American theoretical physicist

[Regarding space shuttle concept] It appears that there are enormous differences of opinion as to the probability of a failure with loss of vehicle and of human life. The estimates range from roughly 1 in 100 to 1 in 100,000. The higher figures come from the working engineers, and the very low figures from management…. For a successful technology, reality must take precedence over public relations, for nature cannot be fooled.
Roger's Commission Report on the Space Shuttle Challenger Accident
Personal observations on the reliability of the Shuttle, Appendix

Haber, Heinz   1868–1934
German physical chemist

The conquest of space hinges on man's survival in space. And the crews of rocket ships and space stations, while they can never be completely protected against hazards such as meteors, will probably be safer than pedestrians crossing a busy street at a rush hour.
In Cornelius Ryan (ed.)
Across the Space Frontier
Can We Survive in Space? (p. 97)
The Viking Press. New York, New York, USA. 1952

Jung, Carl G.   1875–1961
Swiss psychiatrist and founder of analytical psychology

Space flights are merely an escape, a fleeing away from oneself, because it is easier to go to Mars or to the moon than it is to penetrate one's own being.
In Miguel Serrano
C.G. Jung and Hermann Hesse
The Farewell (p. 102)
Schocken Books. New York, New York, USA. 1966

Kepler, Johannes   1571–1630
German astronomer

There will certainly be no lack of human pioneers when we have mastered the art of flight. Who would have thought that navigation across the vast ocean is less dangerous and quieter than in the narrow, threatening gulfs of the Adriatic, or the Baltic, or the British straits? Let us create vessels and sails adjusted to the heavenly ether, and there will be plenty of people unafraid of the empty wastes. In the meantime, we shall prepare, for the brave sky travelers, maps of the celestial bodies — I shall do it for the moon, you, Galileo, for Jupiter.
In Arthur Koestler
The Watershed — A Biography of Johannes Kepler

Letter from Kepler to Galileo
April 1610 (p. 195)
Doubleday & Company, Inc. Garden City. New York, New York, USA.
1960

Mallove, Eugene F.   1947–2004
American physicist

Starflight is not just very hard, it is very, very, very hard.
Starflight Handbook
Introduction (p. 5)
John Wiley & Sons, Inc. New York, New York, USA. 1989

von Braun, Wernher   1912–77
German-American rocket scientist

With our present knowledge, we can respond to the
challenge of stellar space flight solely with intellectual
concepts and purely hypothetical analysis. Hardware
solutions are still entirely beyond our reach and far, far
away.
Can We Ever Go to the Stars?
Popular Science, Volume 183, Number 1, July 1963 (p. 170)

SPACE SETTLEMENT

Wolfe, Steven
No biographical data available

Remember, the space settlement dream was born in you
so that you would strive for its fulfillment in this generation, not defer it to the next. It was, and is, a call to you
to take some action in this lifetime; and if you are not
meant to see it through to completion, than you must at
least lay a foundation on which those who will follow
can build.
Space Settlement: The Journey Inward
Ad Astra, Jan/Feb/Mar 2004

SPACE-TIME

Barnett, Lincoln   1909–79
Science writer

…the universe is not a rigid and inimitable edifice where
independent matter is housed in independent space and
time; it is an amorphous continuum, without any fixed
architecture, plastic and variable, constantly subject
to change and distortion. Wherever there is matter and
motion, the continuum is disturbed. Just as a fish swimming in the sea agitates the water around it, so a star, a
comet, or a galaxy distorts the geometry of the space-
time through which it moves.
The Universe and Dr. Einstein
Chapter 11 (pp. 81–82)
William Sloane Associates. New York, New York, USA. 1948

Berlinski, David   1942–
American mathematician

Yet everything has a beginning, everything comes to an
end, and if the universe actually began in some dense
explosion, thus creating time and space, so time and
space are themselves destined to disappear, the measure
vanishing with the measured, until with another ripple
running through the primordial quantum field, something
new arises from nothingness once again.
A Tour of the Calculus
Chapter 26 (p. 309)
Pantheon Books. New York, New York, USA. 1995

Bohr, Niels Henrik David   1886–1962
Danish physicist

We must, therefore, be prepared to find that further
advance into this region will require a still more extensive renunciation of features which we are accustomed to
demand of the space time mode of description.
Atomic Theory and the Description of Nature
Introductory Survey (p. 14)
Cambridge University Press. Cambridge, England. 1934

Carlyle, Thomas   1795–1881
English historian and essayist

Deepest of all illusory Appearances, for hiding Wonder,
as for many other ends, are your two grand fundamental
world-enveloping Appearances, Space and Time.
Sartor Resartus
Chapter VIII
Ginn & Company. Boston, Massachusetts, USA. 1897

Clarke, Arthur C.   1917–
English science and science fiction writer

Through all the ages, man has fought against two great
enemies — time and space. Time he may never wholly
conquer, and the sheer immensity of space may also
defeat him when he has ventured more than a few light-
years from the Sun. Yet on this little Earth, at least, he
may one day claim a final victory.
Profiles of the Future: An Inquiry into the Limits of the Possible
Chapter 8 (p. 81)
Harper & Row, Publishers. New York, New York, USA. 1973

Cole, K. C.   1946–
American science writer

…Space and time are us…
*The Hole in the Universe: How Scientists Peered Over the Edge of
Emptiness and Found Everything*
Chapter 5 (p. 109)
Harcourt, Inc. New York, New York, USA. 2001

de Beauregard, Costa
No biographical data available

There can no longer be any objective and essential…
division of space-time between "events which have
already occurred" and "events which have not yet
occurred."…Relativity is a theory in which everything

is "written" and where change is only relative to the perceptual mode of living beings.
In J.T. Fraser
The Voices of Time: A Cooperative Survey of Man's Views of Time as Expressed by the Sciences and by the Humanities
Time in Relativity Theory: Arguments for a Philosophy of Being (p. 429)
G. Braziller. New York, New York, USA. 1966

de Broglie, Louis   1892–1987
French physicist

In space-time, everything which for each of us constitutes the past, the present, and the future is given in block, and the entire collection of events, successive for us, which form the existence of a material particle is represented by a line, the world-line of the particle. Each observer, as his time passes, discovers, so to speak, new slices of space-time which appear to him as successive aspects of the material world, though in reality the ensemble of events constituting space-time exist prior to his knowledge of them.
In Paul Arthur Schlipp (ed.)
Albert Einstein: Philosopher-Scientist
A General Survey of the Scientific Work of Albert Einstein (p. 114)
The Library of Living Philosophers, Inc. Evanston, Illinois, USA. 1949

Dixon, William MacNeile   1866–1946
English author and scholar

Everything lies within space, and everything happens within time.
The Human Situation (p. 328)
Longmans, Green & Company. London, England. 1937

Dyson, Freeman J.   1923–
American physicist and educator

Not only is Space from the point of view of life and humanity empty, but Time is empty also. Life is like a little glow, scarcely kindled yet, in these void immensities.
Infinite in All Directions
Part One, Chapter One (p. 9)
HarperCollins Publisher, Inc. New York, New York, USA. 1988

Eddington, Sir Arthur Stanley   1882–1944
English astronomer, physicist, and mathematician

It [the physical world is a *thing*; not like space, which is a mere negation; nor like time, which is — Heaven knows what!
The Nature of the Physical World
Introduction (p. ix)
The Macmillan Company. New York, New York, USA. 1930

Einstein, Albert   1879–1955
German-born physicist

Time and space are modes by which we think and not conditions in which we live.

In Alyesa Forsee
Albert Einstein, Theoretical Physicist (p. 81)
The Macmillan Company. New York, New York, USA. 1963

Ferris, Timothy   1944–
American science writer

Newton viewed space and time as separate and absolute. As conceived by Einstein they are united in a flexible continuum that responds to the presence of matter. The stars and planets wrap the spacetime continuum around themselves, in a sense, each sitting in the center of a sort of spacetime whirlpool.
The Red Limit: The Search for the Edge of the Universe
Chapter 3 (p. 71)
William Morrow & Company, Inc. New York, New York, USA. 1977

[Einstein explained] the commerce we call gravity occurs because objects follow the easiest, most efficient course over the undulations of the continuum. Earth in its orbit glides along inside the sun's spactime vortex like a roulette ball whirling above the wheel, balancing its velocity against its tendency to slide toward the sun. That tendency is equivalent to gravity, but no "force" of gravity is postulated. Light beams also follow the dips and hills of the continuum. They trace trajectories we call "bent," though that is just three-dimensional parochialism talking; they are going just as straight as the shape of spacetime allows.
The Red Limit: The Search for the Edge of the Universe
Chapter 3 (p. 71)
William Morrow & Company, Inc. New York, New York, USA. 1977

Hawking, Stephen William   1942–
English theoretical physicist

The theory of relativity does, however, force us to change fundamentally our ideas of space and time. We must accept that time is not completely separate from and independent of space, but is combined with it to form an object called space-time.
A Brief History of Time: From the Big Bang to Black Holes
Chapter 2 (p. 123)
Bantam Books. Toronto, Ontario, Canada. 1988

Hoffmann, Banesh   1906–86
Mathematician and educator

What is it that pulls the apple to the ground, bends the circling moon to the earth and makes the planets captive of the sun? ...It is intangible time and space themselves, acting in awesome concert as curved space-time holding sway over all things in the universe.
Relativity and Its Roots
Chapter 6 (pp. 156–157)
W.H. Freeman & Company. New York, New York, USA. 1983

John Shade
Fictional character

Space is a swarming of the eyes, and Time a singing in the ears.
In Vladimir Nabokov
Ada or Ardor: A Family Chronicle
Part Four (p. 542)
McGraw-Hill Book Company, Inc. New York, New York, USA. 1969

Joubert, Joseph   1754–1824
French moralist

There is something divine about the ideas of space and eternity which is wanting in those of pure duration and simple extension.
Translated by H.P. Collins
Pensées and Letters of Joseph Joubert
Chapter XII (p. 90)
Books for Libraries Press. Freeport, New York, USA. 1972

Lamb, Charles   1775–1834
English essayist and critic

Nothing puzzles me more than time and space; and yet nothing troubles me less, as I never think about them.
Quoted by James R. Newman
The World of Mathematics (Volume 1)
Letter to Thomas Manning, January 2, 1806 (p. 552)
Simon & Schuster. New York, New York, USA. 1956

MacLeish, Archibald   1892–1982
American poet and Librarian of Congress

Spacetime has no beginning and no end.
It has no door where anything can enter.
How break and enter what will only bend?
Songs for Eve
Reply to Mr. Wordsworth (p. 39)
Houghton Mifflin Company. Boston, Massachusetts, USA. 1954

Maeterlinck, Maurice   1862–1949
Belgian playwright and poet

To attempt to explain space by time and time by space is to seek to explain the night by darkness and the darkness by the night; it is to revolve hopelessly in the circle of the unknowable.
Translated by Bernard Miall
The Life of Space
The Fourth Dimension, XXIX (pp. 96–97)
Dodd, Mead & Company. New York, New York, USA. 1928

Space is the present made visible. Time is space that is on the move and becoming the future or the past. Space is time extended; it is horizontal time; time is space perpendicular, vertical space. Space is time that endures; time is space that flies.
Translated by Bernard Miall
The Life of Space
The Fourth Dimension, XXIX (p. 97)
Dodd, Mead & Company. New York, New York, USA. 1928

Maxwell, James Clerk   1831–79
Scottish physicist

March on, symbolic host! with step sublime,
Up to the flaming bounds of Space and Time!
There pause, until by Dickenson depicted,
In two dimensions, we the form may trace
Of him whose soul, too large for vulgar space,
In *n* dimensions flourished unrestricted.
In Lewis Campbell and William Garnett
The Life of James Clerk Maxwell with Selections from his Correspondence and Occasional Writings
To the Committee of the Cayley Portrait Fund (p. 637)
Macmillan & Company. London, England. 1882

Minkowski, Hermann   1864–1909
German mathematician

From this hour on, space as such and time as such shall recede to the shadows and only a kind of union of the two retain significance.
In A.P. French
Einstein: A Centenary Volume
Chapter 12 (p. 231)
Harvard University Press. Cambridge, Massachusetts, USA. 1979

The views of space and time which I wish to lay before you have sprung from the soil of experimental physics, and therein lies their strength. Henceforth space by itself and time by itself, are doomed to fade away into mere shadows, and only a kind of union of the two will preserve an independent reality.
Space and Time
80th Assembly of German Natural Scientists and Physicians, September 21, 1908

The objects of our perception invariably include places and times in combination. Nobody has ever noticed a place except at a time, or a time except at a place. But I still respect the dogma that both space and time have independent significance. A point of space at a point of time, that is a system of values x, y, z, t, I will call a world-point.
The Principle of Relativity: A Collection of Original Memoirs on the Special and General Theory of Relativity
Space and Time (p. 76)
Dover Publications, Inc. New York, New York, USA. 1952

Murchie, Guy   1907–97
American biologist

…the key to comprehending space-time is the obvious (to me) fact that space is the relationship between things and other things while time is the relationship between things and themselves.
The Seven Mysteries of Life: An Exploration of Science and Philosophy
Part Three, Chapter 12 (p. 331)
Houghton Mifflin Company. Boston, Massachusetts, USA. 1978

Pope, Alexander   1688–1744
English poet

Ye Gods! annihilate but space and time,
And make two lovers happy.

The Complete Poetical Works
Martinus Scriblerus of The Art of Sinking in Poetry, 11
Houghton Mifflin Company. New York, New York, USA. 1903

Reichenbach, Hans   1891–1953
German philosopher of science

It appears that the solution of the problem of time and space is reserved to philosophers who, like Leibnitz, are mathematicians, or to mathematicians who, like Einstein, are philosophers.
In Paul Arthur Schlipp (ed.)
Albert Einstein: Philosopher-Scientist
The Philosophical Significance of the Theory of Relativity, IV (p. 307)
The Library of Living Philosophers, Inc. Evanston, Illinois, USA. 1949

Stenger, Victor J.   1935–
American physicist

Great physicists from Galileo to Einstein have clarified the meanings of space and time for us, not overthrown their basic conceptions nor declared them obsolete.
Physics and Psychics: The Search for a World Beyond the Senses
Chapter 13 (p. 295)
Prometheus Books. Buffalo, New York, USA. 1990

Synge, John L.   1897–1995
Irish mathematician and physicist

Anyone who studies relativity without understanding how to use simple space-time diagrams is as much inhibited as a student of functions of a complex variable who does not understand the Argads diagram.
Relativity: The Special Theory (p. 63)
North-Holland Publishing Company. Amsterdam, Netherlands. 1965

Taylor, Edwin F.
American physicist
Wheeler, John Archibald   1911–
American physicist and educator

Never make a calculation until you know the answer: Make an estimate before every calculation, try a simple physical argument (symmetry! invariance! conservation!) before every derivation, guess the answer to every puzzle. Courage: no one else needs to know what the guess is. Therefore make it quickly, by instinct. A right guess reinforces this instinct. A wrong guess brings the refreshment of surprise. In either case, life as a spacetime expert, however long, is more fun!
Spacetime Physics
Chapter 1 (p. 60)
W.H. Freeman & Company. San Francisco, California, USA. 1966

Thorne, Kip S.   1940–
American theoretical physicist

...spacetime is like a piece of wood impregnated with water. ...the wood represents space, the water represents time.... [W]ood and water; space and time...are tightly interwoven, unified. The singularity and the laws of quantum gravity that rule it are like a fire into which the water impregnated wood is thrown. The fire boils the water out of the wood, leaving the wood alone and vulnerable; in the singularity, the laws of quantum gravity destroy time, leaving space alone and vulnerable. The fire then converts the wood into a froth of flakes and ashes; the laws of quantum gravity then convert space into a random, probabilistic froth.
Black Holes and Time Warps: Einstein's Outrageous Legacy
Chapter 13 (p. 477)
W.W. Norton & Company, Inc. New York, New York, USA.1994

...Space and time, unified as spacetime, do not merely witness great masses struggling to bend the motion of other masses. Like the gods of ancient Greece, spacetime helps guide the battle and itself participates.... The scope and power of this century's new view of gravity and spacetime is seen nowhere more dramatically than in its prediction of the expansion of the universe. To have predicted...against all expectation, a phenomenon so fantastic is the greatest token yet of our power to understand this strange and beautiful universe.
A Journey into Gravity and Spacetime
Chapter 1 (p. 2)
Scientific American Library. New York, New York, USA. 1990

Valéry, Paul   1871–1945
French poet and critic

Space is an imaginary body, as time is fictive movement.

When we say "in space" or "space is filled with" we are positing a body.
In Jackson Mathews (ed.)
The Collected Works of Paul Valéry (Volume 14)
Analects, CIX (p. 321)
Princeton University Press. Princeton, New Jersey, USA. 1971

SPECIALIZATION

Agassiz, Jean Louis Rodolphe   1807–73
Swiss-born American naturalist, geologist, and teacher

A man cannot be a professor of zoology on one day, and of chemistry on the next, and do good work in both. As in a concert all are musicians — one plays one instrument, and one another, but none all in perfection.
In Charles Frederick Holder
Louis Agassiz: His Life and Work
At Penikese (p. 174)
G.P. Putnam's Sons. New York, New York, USA. 1893

You cannot do without one specialty; you must have some base-line to measure the work and attainments of others.
In Charles Frederick Holder
Louis Agassiz: His Life and Work

At Penikese (p. 174)
G.P. Putnam's Sons. New York, New York, USA. 1893

Asimov, Isaac   1920–92
American author and biochemist

…the orchard of science is a vast globe-encircling monster, without a map, and known to no one man; indeed, to no group of men fewer than the whole international mass of creative scientists. Within it, each observer clings to his own well-known and well-loved clump of trees. If he looks beyond, it is usually with a guilty sigh.
View from a Height
Introduction (p. 7)
Avon Books. New York, New York, USA. 1975

Bell, E. T. (Eric Temple)   1883–1960
Scottish-American mathematician and educator

To interrupt one's own researches in order to follow those of another is a scientific pleasure which most experts delegate to their assistants. Consequently, the confusion of tongues increases as the square of the number of talkers, until only ever more select coteries of narrow specialists really understand the refinements of their esoteric vocabularies.
The Development of Mathematics (p. 510)
McGraw-Hill Book Company, Inc. New York, New York, USA. 1945

Burnet, Frank Macfarlane   1899–1985
Australian immunologist and virologist

We may well find that the men who staff the hospitals of next century will include many who are much more mathematicians and biochemists than physicians as we know them today, but there will still be wide range of surgical and other specialists.… I fancy that those men will still need to be able to apply common sense, courage, and compassion in dealing with all the human difficulties that escape the machines.
Changing Patterns: An Atypical Autobiography (pp. 251–252)
William Heineman. London, England. 1968

Groen, Janny
No biographical data available
Smit, Eefke
No biographical data available

Scientific information is essential, not only for the scientist. The politician, the entrepreneur and the public at large need to know about it too. The people in business find that neither the mass media not the specialized scientific press are providing the information needed. General information is no longer enough, specialist information is only digestible for the learned. Who will bridge the gap?
The Discipline of Curiosity: Science in the World
Introduction (p. 4)
Elsevier Science. Amsterdam, Netherlands. 1990

Heinlein, Robert A.   1907–88
American science fiction writer

A human being should be able to change a diaper, plan an invasion, butcher a hog, conn a ship, design a building, write a sonnet, balance accounts, build a wall, set a bone, comfort the dying, take orders, give orders, cooperate, act alone, solve equations, analyze new problems, pitch manure, program a computer, cook a tasty meal, fight efficiently, die gallantly.

Specialization is for insects.
Time Enough for Love
Intermission (pp. 265–266)
G.P. Putnam's Sons. New York, New York, USA. 1973

Morrow, Prince Albert   1846–1913
American dermatologist and sociologist

The genius of modern medical literature is clearly in the direction of division of labor and associated effort.
A System of Genito-Urinary Diseases, Syphilology, and Dermatology
Preface
D. Appleton & Company. New York, New York, USA. 1893–4

Ortega y Gasset, José   1883–1955
Spanish philosopher

For the purpose of innumerable investigations it is possible to divide science into small sections, to enclose oneself in one of these, and to leave out of consideration all the rest. The solidity and exactitude of the methods allow…thistemporary but…disarticulation of knowledge. The work…done under [such] methods [is] as with a machine, and in order to obtain quite abundant results it is not even necessary to have rigorous notions of their meaning and foundations.
The Revolt of the Masses
Chapter 12 (p. 111)
W.W. Norton & Company, Inc. New York, New York, USA. 1960

[The division of science into specialties means] the majority of scientists help the general advance of science while shut up in the narrow cell of their laboratory, like the bee in the cell of its hive, or the turnspit in its wheel.
The Revolt of the Masses
Chapter 12 (p. 111)
W.W. Norton & Company, Inc. New York, New York, USA. 1960

Osler, Sir William   1849–1919
Canadian physician and professor of medicine

The extraordinary development of modern science may be her undoing. Specialism, now a necessity, has fragmented the specialties themselves in a way that makes the outlook hazardous. The workers lose all sense of proportion in a maze of minutiae.
The Old Humanities and the New Science
Chapter III (p. 49)
Houghton Mifflin Company. Boston, Massachusetts, USA. 1920

Peattie, Donald Culrose   1896–1964
American botanist, naturalist, and author

It is my contention that specialization should be left to those who are not mentally gifted at generalization.
An Almanac for Moderns
September Twenty-First (p. 199)
G.P. Putnam's Sons. New York, New York, USA. 1935

Stevens, Rosemary   1935–
No biographical data available

In the whole process of reassessment…of the medical profession…has come the recognition of medicine as an interdependent, not independent, profession and as one consisting of a complex of specialties rather than one general discipline.
American Medicine and the Public Interest (p. 413)
Yale University Press. New Haven, Connecticut, USA. 1971

Vesalius, Andreas   1514–64
Flemish physician and anatomist

Great harm is caused by too wide a separation of the disciplines which work toward the perfection of each individual art, and much more by the meticulous distribution of the practices of this art to different workers.
The Fabric of the Human Body
Preface
1543

Weiner, Jonathan   1953–
American fiction and non-fiction writer

Specialization has gotten out of hand. There are more branches in the tree of knowledge than there are in the tree of life. A petrologist studies rocks; a pedologist studies soils. The first one sieves the soil and throws away the rocks. The second one picks up the rocks and brushes off the soil. Out in the field, they bump into each other only like Laurel and Hardy, by accident, when they are both backing up.
The Next One Hundred Years: Shaping the Fate of Our Living Earth
Chapter 10 (pp. 198–199)
Bantam Books. New York, New York, USA. 1990

SPECIES

Bessey, Charles E.   1845–1915
American botanist

Nature produces individuals and nothing more
The Taxonomic Aspect of the Species Question
The American Naturalist, Volume 42, Number 496, April 1908 (p. 218)

Blumenbach, Johann Friedrich   1752–1840
German naturalist and anthropologist

What is species? We say that animals belong to one and the same species if they agree so well in form and constitution that those things in which they differ may have arisen from degeneration.… Now we come to the real difficulty, which is to set forth the characters by which in the natural world we may distinguish mere varieties from genuine species.
The Anthropological Treatises of Johann Friedrich Blumenbach
Section II (p. 188)
Longman, Green, Longman, Roberts & Green. London, England. 1865

Darwin, Charles Robert   1809–82
English naturalist

Unless we suppose the same species to have been created in two different countries, we ought not to expect any closer similarity between the organic beings on the opposite sides of the Andes than on shores separated by a broad strait of the sea.
In Francis Darwin (ed.)
The Life and Letters of Charles Darwin (Volume 1)
Chapter X (p. 365)
D. Appleton & Company. New York, New York, USA. 1896

Widely ranging species, abounding in individuals, which have already triumphed over many competitors in their own widely extended homes will have the best chance of seizing on new places, when they spread into new countries.
In *Great Books of the Western World* (Volume 49)
The Origin of Species by Means of Natural Selection
Chapter XII (p. 182)
Encyclopædia Britannica, Inc. Chicago, Illinois, USA. 1952

It is really laughable to see what different ideas are prominent in various naturalists' minds, when they speak of "species:" in some, resemblance seems to go for nothing, and Creation the reigning idea — in some, descent is the key, — in some, sterility an unfailing test, with others it is not worth a farthing. It comes, I believe, from trying to define the undefinable.
In Francis Darwin (ed.)
The Life and Letters of Charles Darwin (Volume 1)
Letter to J.D. Hooker, December 24, 1856 (p. 446)
D. Appleton & Company. New York, New York, USA. 1896

…I look at the term species as one arbitrarily given, for the sake of convenience, to a set of individuals closely resembling each other…
In *Great Books of the Western World* (Volume 49)
The Origin of Species by Means of Natural Selection
Chapter II (p. 29)
Encyclopædia Britannica, Inc. Chicago, Illinois, USA. 1952

Falk, Donald
American ecologist and biologist

We consider species to be like a brick in the foundation of a building. You can probably lose one or two or a dozen bricks and still have a standing house. But by the time you've lost 20 per cent of species, you're going to destabilize the entire structure. That's the way ecosystems work.
Christian Science Monitor, 26 May 1989

Forbes, Edward   1815–54
English naturalist

...every true species presents in its individuals, certain features, specific characters, which distinguish it from every other species; as if the Creator had set an exclusive mark or seal on each type.
The Natural History of the European Seas
Chapter I (p. 8)
John Van Voorst. London, England. 1859

Gregory, Sir Richard Arman   1864–1952
British science writer and journalist

When scientific work is instituted solely with the object of securing commercial gain, its correlative selfishness; when it is confined to the path of narrow specialisation, it leads to arrogance; and when its purpose is materialistic domination, without regard for the spiritual needs of humanity, it is a social danger and may become an excuse for learned barbarity.
Discovery: Or the Spirit and Service of Science
Preface (pp. v–vi)
Macmillan & Company Ltd. London, England. 1918

Lamarck, Jean-Baptiste Pierre Antoine   1744–1829
French biologist

It is not a futile purpose to decide definitely what we mean by the so-called species among living bodies, and to enquire if it is true that species are of absolute constancy, as old as nature, and have all existed from the beginning just as we see them today; or if, as a result of changes in their environment, albeit extremely slow, they have not in course of time changed their characters and shape.
Translated by Hugh Elliot
Zoological Philosophy: An Exposition with Regard to the Natural History of Animals
Chapter III (p. 35)
The University of Chicago Press. Chicago, Illinois, USA. 1984

What a swarm of mollusk shells are furnished by every country and every sea, eluding our means of distinction and draining our resources.
Translated by Hugh Elliot
Zoological Philosophy: An Exposition With Regard to the Natural History of Animals
Chapter III (p. 38)
The University of Chicago Press. Chicago, Illinois, USA. 1984

The idea of bringing together under the name of species a collection of like individuals, which perpetuate themselves unchanged by reproduction and are as old as nature, involved the assumption that the individuals of one species could not unite in reproductive acts with individuals of another species.

Unfortunately, observation has proved and continues every day to prove that this assumption is unwarranted; for the hybrids so common among plants, and the copulations so often noticed between animals of very different species, disclose the fact that the boundaries between these alleged constant species are not so impassable as had been imagined.
Translated by Hugh Elliot
Zoological Philosophy: An Exposition With Regard to the Natural History of Animals
Chapter III (p. 39)
The University of Chicago Press. Chicago, Illinois, USA. 1984

Leakey, Richard Erskine   1944–
Kenyan paleoanthropologist and politician

As a species, we are blessed with a curiosity about the world of nature and our place in it. We want to know — *need* to know — how we came to be as we are, and what our future is.
The Origin of Humankind
Preface (p. xv)
Basic Books, Inc. New York, New York, USA. 1994

Locke, John   1632–1704
English philosopher and political theorist

...the boundaries of the species, whereby men sort them, are made by men.
In *Great Books of the Western World* (Volume 35)
An Essay Concerning Human Understanding
Book III, Chapter VI, Section 37 (p. 279)
Encyclopædia Britannica, Inc. Chicago, Illinois, USA. 1952

Lyell, Sir Charles   1797–1875
English geologist

...species are abstractions, not realities — are like genera. Individuals are the only realities. Nature neither makes nor breaks molds — all is plastic, unfixed, transitional, progressive, or retrograde. There is only one great resource to fall back upon, a reliance that all is for the best, trust in God, a belief that truth is the highest aim, that if it destroys some idols it is better that they should disappear, that the intelligent ruler of the universe has given us this great volume as a privilege, that its interpretation is elevating.
In Leonard G. Wilson (ed.)
Sir Charles Lyell's Scientific Journals on the Species Question
Journal II, July 10, 1856 (p. 121)
Yale University Press. New Haven, Connecticut, USA. 1970

Mayr, Ernst   1904–2005
German-born American biologist

We had an international conference in Rome in 1981 on the mechanisms of speciation. It was attended by many of the leading botanists, zoologists, paleontologists, geneticists, cytologists and biologists. The one thing on which they all agreed was that we still have no idea what happens genetically during speciation. That's a damning statement, but it's the truth.

OMNI Magazine
February, 1983 (p. 78)

It may not be exaggeration if I say that there are probably as many species concepts as there are thinking systematists and students of speciation.
Systematics and the Origin of Species
Chapter V (p. 115)
Harvard University Press. Cambridge, Massachusetts, USA. 1942

Morton, Ron L.
No biographical data available

Species come,
species go;
Some real fast,
some real slow…
Music of the Earth: Volcanoes, Earthquakes and Other Geological Wonders
Chapter 10 (p. 267)
Plenum Press. New York, New York, USA. 1996

Nietzsche, Friedrich   1844–1900
German philosopher

The species does not grow into perfection: the weak again and again get the upper hand of the strong, — their large number, and their greater cunning are the cause of it.
In Alexander Tille (ed.)
The Works of Friedrich Nietzsche
Volume 11, The Twilight of the Idols, Roving Expeditions of an Inopportune Philosopher, Section 14 (p. 174)

Terborgh, John   1936 –

Species are the units of evolution.
Diversity and the Tropical Rain Forest
Chapter 1 (p. 6)
Scientific American Library. New York, New York, USA. 1992

Wallace, Alfred Russel   1823–1913
English humanist, naturalist, and geographer

The rule…I have endeavored to adopt [in determining what is a species and what is a species variety] is, that when the difference between two forms inhabiting separate areas seems quite constant, when it can be defined in words, and when it is not confined to a single peculiarity only, I have considered such forms to be species. When… the individuals of each locality vary among themselves, so as to cause the distinctions between the two forms to become inconsiderable and indefinite, or where the differences, though constant, are confined to one particular only, such as size, tint, or a single point of difference in marking or in outline, I class one of the forms as a variety of the other.
On the Phenomena of Variation and Geographical Distribution as Illustrated by the Papilionidae of the Malayan Region
Transactions of the Linnean Society of London, Volume 25, 1865 (p. 4)

Species are merely those strongly marked races or local forms which, when in contact, do not intermix, and when inhabiting distinct areas are generally regarded to have had a separate origin, and to be incapable of producing a fertile hybrid offspring.
On the Phenomena of Variation and Geographical Distribution as Illustrated by the Papilionidae of the Malayan Region
Transactions of the Linnean Society of London, Volume 25, 1865 (p. 12)

[A]s the test of hybridity cannot be applied [to species identification] in one case in ten thousand, and even if it could be applied, would prove nothing, since it is founded on an assumption of the very question to be decided — and as the test of origin is in every case inapplicable — and as, further, the test of non-intermixture is useless, except in those rare cases where the most closely allied species are found inhabiting the same area, it will be evident that we have no means whatever of distinguishing so-called "true species" from the several modes of variation here pointed out, and into which they so often pass by an insensible gradation.
On the Phenomena of Variation and Geographical Distribution as Illustrated by the Papilionidae of the Malayan Region
Transactions of the Linnean Society of London, Volume 25, 1865 (p. 12)

…The essential character of a species in biology is that it is a group of living organisms, separated from all other such groups by a set of distinctive characters, having relations to the environment not identical with those of any other group of organisms, and having the power of continuously reproducing its like. Genera are merely assemblages of a number of these species which have a closer resemblance to each other in certain important and often prominent characters than they have to any other species…
Fortnightly Review, Volume 57, New Series, 1895 (p. 441)

SPECIFICATION

Alger, John R. M.
American design engineer
Hays, Carl V.
No biographical data available

Once a problem is recognized clearly and all the parties concerned have agreed on its nature, the development of detailed specifications becomes vital.
Creative Synthesis in Design (p. 15)
Prentice-Hall. Englewood Cliffs, New Jersey, USA. 1964

A good engineer is adroit in negotiating changes in specifications or trade-offs…
Creative Synthesis in Design (p. 16)
Prentice-Hall. Englewood Cliffs, New Jersey, USA. 1964

Hoover, Herbert   1874–1964
31st president of the United States

Specifications are the formulated, definite, and complete statements of what the buyer requires of the seller.

National Directory of Commodity Specifications
M 65, Forward (p. 1)

Matthews, J. A.

Good sense is highly desirable in writing specifications and is even more necessary in interpreting them. If they only contained the minimum number of requirements to define the character of material wanted…the matter would be greatly simplified. Rarely do they cover the only material suited to the purpose intended, and more rarely do they cover the best material for the purpose intended. Once written, they become as the laws of the Medes and Persians, which alter not. They acquire a sort of sanctity, like the Ten Commandments or the Constitution before the adoption of the Eighteenth Amendment.
Present Tendencies in Engineering Materials
Mechanical Engineering, Volume 48, Number 8, August 1926 (p. 792)

SPECTROSCOPE

Clerke, Agnes Mary   1842–1907
Irish astronomer

Custom can never blunt the wonder with which we must regard the achievement of compelling rays emanating from a source devoid of sensible magnitude through immeasurable distances, to reveal, by its distinctive qualities, the composition of that source.
A Popular History of Astronomy During the Nineteenth Century
Part II, Chapter XII (p. 372)
A. & C. Black. London, England. 1908

Crookes, Sir William   1832–1919
English chemist and physicist

The spectroscope reveals that the elementary components of the stars and the earth are pretty much the same.
In Frederick Houk Law
Science in Literature
The Romance of the Diamonds (p. 111)
Harper & Brothers. New York, New York, USA. 1929

Draper, John William   1811–82
American scientist, philosopher, and historian

And now, while we have accomplished only a most imperfect examination of objects that we find on earth, see how, on a sudden, through the vista that has been opened by the spectroscope, what a prospect lies beyond us in the heavens! I often look at the bright yellow ray emitted from the chromosphere of the sun, by that unknown element, Helium, as the astronomers have ventured to call it. It seems trembling with excitement to tell its story, and how many unseen companions it has. And if this be the case with the sun, what shall we say of the magnificent hosts of the stars? May not every one of them have special elements of its own? Is not each a chemical laboratory in itself?

Presidential Address
American Chemical Society, November 16, 1876

Huggins, Sir William   1824–1910
English astronomer

One important object of this original spectroscopic investigation of the light of the stars and other celestial bodies, namely to discover whether the same chemical elements as those of our earth are present throughout the universe, was most satisfactorily settled in the affirmative; a common chemistry, it was shown, exists throughout the universe.
The Scientific Papers of Sir William Huggins
Spectra of the Fixed Stars (p. 49)
W. Wesley & Son. London, England. 1909

I looked into the spectroscope. No spectrum such as I expected! A single bright line only!…The riddle of the nebulae was solved. The answer, which had come to us in the light itself, read: Not an aggregation of stars, but a luminous gas. Stars after the order of our own sun, and of the brighter stars, would give a different spectrum; the light of this nebula had clearly been emitted by a luminous gas.
The Scientific Papers of Sir William Huggins
Historical Statement (p. 106)
W. Wesley & Son. London, England. 1909

Maxwell, James Clerk   1831–79
Scottish physicist

The vast interplanetary and interstellar regions will no longer be regarded as waste places in the universe, which the Creator has not seen fit to fill with the symbols of the manifold order of His kingdom. We shall find them to be already full of this wonderful medium; so full, that no human power can remove it from the smallest portion of space, or produce the slightest flaw in its infinite continuity. It extends unbroken, from star to star; and when a molecule of hydrogen vibrates in the dog-star, the medium receives the impulses of these vibrations; and after carrying them in its immense bosom for three years, delivers them in due course, regular order, and full tale into the spectroscope…
In W.D. Niven (ed.)
The Scientific Papers of James Clerk Maxwell (Volume 2)
Action at a Distance (p. 322)
At The University Press. Cambridge, England. 1890

SPECTRUM

Thomson, James   1700–48
Scottish poet

First the flaming red Sprang vivid forth; the tawny orange next, And next delicious yellow; by whose side Fell the kind beams of all-refreshing green. Then the pure blue that swells autumnal skies, Ethereal play'd; and then,

of sadder hue Emerged the deeper indigo (as when The heavy-skirted evening droops with frost), While the last gleamings of refracted light Died in the fainting violet away.
Poetical Works of James Thomson
A Poem Sacred to the Memory of Sir Isaac Newton
Reeves & Turner. London, England. 1895

SPECTRUM ANALYSIS

de la Rue, Warren   1815–89
English astronomer and inventor

…if we were to go to the sun, and to bring away some portions of it and analyze them in our laboratories, we could not examine them more accurately than we can by this new mode of spectrum analysis.…
Chemical News, Volume 4, 1861 (p. 130)

Lockyer, Joseph Norman   1836–1920
English astronomer and physicist

…we believe that each molecular vibration disturbs the ether; that spectra are thus begotten' each wavelength of light resulting from a molecular tremor of corresponding wavelength. The molecule is, in fact, the sender, the ether the wire, and the eye the receiving instrument, in this new telegraphy.
Studies in Spectrum Analysis
Chapter IV (pp. 118–119)
D. Appleton & Company. New York, New York, USA. 1878

White, H. E.
No biographical data available

That photographs are an extremely important feature of any book on atomic spectra may be emphasized by pointing out that, of all the theories and knowledge concerning atoms, the spectrum lines will remain the same for all time.
Introduction to Atomic Spectra (p. vii)
McGraw-Hill Book Company, Inc. New York, New York, USA. 1934

SPECULATION

Comfort, Alex   1920–2000
English gerontologist and author

Rash speculation does not bother the physicists — it has got them where they are today. And it is high time that the life sciences looked critically at the solidity of their tribal idols, including stochastic-genetic evolution, morphogenesis and the "mind-body problem" — while being mindful that, in the present climate, work on some quite unrelated matter may prove, incidentally and quite unwittingly, to have altered the entire face of the problem. Nor will the answers obtained lie within any existing frame of discourse.

On Physics and Biology: Getting Our Act Together
Perspectives in Biology and Medicine, Volume 29, Number 1, Autumn 1985 (p. 9)

Darwin, Charles Robert   1809–82
English naturalist

All young geologists have a great turn for speculation; I have burned my fingers pretty sharply in that way, and am now inclined to cavil at speculation when the direct and immediate effect of a cause in question cannot be shown.
In Francis Darwin (ed.)
More Letters of Charles Darwin (Volume 2)
Letter 487, Darwin to C.H.L. Wood, March 4, 1850 (p. 133)
D. Appleton & Company. New York, New York, USA. 1903

Dr. Seuss (Theodor Seuss Geisel)   1904–1991
American children's book author and illustrator

Some have two feet and some have four. Some have six feet and some have more. Where do they come from? I can't say. But I bet they have come a long long way.
One Fish, Two Fish, Red Fish, Blue Fish (p. 12)
Beginner Books Inc. New York, New York, USA. 1988

Eddington, Sir Arthur Stanley   1882–1944
English astronomer, physicist, and mathematician

If we are not content with the dull accumulation of experimental facts, if we make any deductions or generalizations, if we seek for any theory to guide us, some degree of speculation cannot be avoided. Some will prefer to take the interpretation which seems to be most immediately indicated and at once adopted as an hypothesis; others will rather seek to explore and classify the widest possibilities which are not definitely inconsistent with the facts. Either choice has its dangers: the first may be too narrow a view and lead progress into a cul-de-sac; the second may be so broad that it is useless as a guide and diverge indefinitely from experimental knowledge.
The Internal Constitution of the Stars
Observatory, Volume 43, 1920 (p. 356)

Einstein, Albert   1879–1955
German-born physicist

I think that only daring speculation can lead us further and not accumulation of facts.
In Michele Besso
Correspondence 1903–1955
Letter to M. Besso, October 8, 1952 (p. 487)
Hermann. Paris, France. 1972

Feuerbach, Ludwig   1804–72
German philosopher

Speculation is philosophy intoxicated; let philosophy get sober again; it will then be to the mind what pure spring water is to the body.
In Ludwig Buchner

Force and Matter
Preface to the First Edition (p. xx)
Trubner & Company. London, England. 1864

Jeffreys, Sir Harold   1891–1989
English astronomer and geophysicist

The problem of the origin and development of the solar system suffers from the label "speculative." It is frequently said that as we were not there when the system was formed, we cannot legitimately arrive at any idea as to how it was formed.
In B. Gutenberg (ed.)
Internal Constitution of the Earth
The Origin of the Solar System
Dover Publications. New York, New York, USA. 1951

Ramsay, Sir William   1852–1916
English chemist

Speculation…has a deep fascination for many minds…
Essays Biographical and Chemical
Chemical Essays
What Is an Element? (p. 149)
Archibald Constable & Company Ltd. London, England. 1908

Chemistry and physics are experimental sciences; and those who are engaged in attempting to enlarge the boundaries of science by experiment are generally unwilling to publish speculations; for they have learned, by long experience, that it is unsafe to anticipate events.
Essays Biographical and Chemical
Chemical Essays
Radium and Its Products (p. 179)
Archibald Constable & Company Ltd. London, England. 1908

Twain, Mark (Samuel Langhorne Clemens)   1835–1910
American author and humorist

Spectrum analysis enabled the astronomer to tell when a star was advancing head on, and when it was going the other way. This was regarded as very precious. Why the astronomer wanted to know, is not stated; nor what he could sell out for, when he did know. An astronomer's notions about preciousness were loose. They were not much regarded by practical men, and seldom excited a broker.
Mark Twain's Fables of Man
The Secret History of Eddypus
University of California Press. Berkeley, California, USA. 1972

Whyte, A. Gowans
Scottish writer

The Golden Age of speculation was the Stone Age of knowledge.
The Triumph of Physics
The Rationalist Annual, 1931 (p. 28)

Woodford, F. Peter
American editor

Of course speculation is in order in a Discussion, but it must be reasonable, firmly founded on observation, and subject to test, if it is to get past a responsible editorial board.
Scientific Writing for Graduate Students (p. 29)
Council of Biology Editors. Bethesda, Maryland, USA. 1986

SPIN

Goudsmit, Samuel A.   1902–78
Dutch-born American physicist

It was a little over fifty years ago that George Uhlenbeck and I introduced the concept of spin. It is therefore not surprising that most young physicists do not know that spin had to be introduced. They think that it was revealed in Genesis or perhaps postulated by Sir Isaac Newton, which most young physicists consider to be about simultaneous.
In Anthony French and Edwin Taylor
An Introduction to Quantum Physics (p. 424)
W.W. Norton & Company, Inc. New York, New York, USA. 1978

SPIRAL ARMS

van de Hulst, H. C.   1918–2000
Dutch astronomer

The discovery of spiral arms and — later — of molecular clouds in our Galaxy, combined with a rapidly growing understanding of the birth and decay process of stars, changed interstellar space from a stationary "medium" into an "environment" with great variations in space and time.
In A. Bonetti, J.M. Greenberg and S. Aiello (eds.)
Evolution in Interstellar Dust and Related Topics (p. 5)
North-Holland Publishing Company. Amsterdam, Netherlands. 1989

SPONTANEOUS GENERATION

Urey, Harold Clayton   1893–1981
American chemist

The common assumption is that the earth and its atmosphere have always been as they are now, but if this is assumed it is necessary to account for the present highly oxidized conditions by some processes taking place early in the earth's history. Briefly, the highly oxidized condition is rare in the cosmos.
On the Early Chemical History of the Earth and the Origin of Life
Proceedings of the National Academy of Science USA, Volume 38, 1952

Hacking, Ian   1936–
Canadian-born philosopher of science

STAMP COLLECTING

Alvarez, Luis Walter   1911–88
American experimental physicist

Paleontologists…they're really not very good scientists. They're more like stamp collectors.
New York Times
19 Jan 1988

Birch, Arthur J.   1915–1995
Australian chemist

I have never been emotionally attracted by exactitude of detail, but rather by the broad sweep of ideas collected around philosophically defined examples that can be tested experimentally. I have never been a scientific "stamp collector", although it takes all types to make the world and I have greatly benefited by the "collections" of others…
To See the Obvious
Why Chemistry? (pp. 13–14)
American Chemical Society. Washington, D.C. 1995

Rutherford, Ernest   1871–1937
English physicist

All science is either physics or stamp collecting.
In J.B. Birks
Rutherford at Manchester
Memories of Rutherford (p. 108)
W.A. Benjamin Inc. New York, New York, USA. 1963

Simpson, George Gaylord   1902–84
American paleontologist

Biology starts with biochemistry and goes on to neurophysiology and genetics. All else is stamp-collecting.
This View of Life: The World of an Evolutionist
Chapter Six (p. 108)
Harcourt, Brace & World, Inc. New York, New York, USA. 1964

STANDARD

Woll, Matthew   1880–1956
Luxembourg-born American photo engraver

I know very well that in a great many circles the man who does not enter with a neatly arranged plan, with a set of doctrines, with a rounded and sonorous formula, and with assurance about everything, is set down as something of an old fogey, perhaps reactionary, certainly not one of the elect who are "doing things" and providing guidance for the race. I must assume the risk. I have no formula. [But I shall resist] those who have the formula for so many things and who seek so avidly to force it down the throats of every one else.
Standardization
Annals of the American Academy of Political and Social Science, Volume 137, May 1928 (p. 47)

STAR

Acton, Loren   1936–
American astronaut and solar physicist

When you look out the other way toward the stars you realize it's an awful long way to the next watering hole.
In Kevin W. Kelley
The Home Planet
With Plate 84
Addison-Wesley. Reading, Massachusetts, USA. 1988

Adams, George   1750–95
English instrument maker

New stars offer to the mind a phenomenon more surprising, and less explicable, than almost any other in the science of astronomy.
Lectures on Natural and Experimental Philosophy (Volume 4)
Chapter XLIV (p. 213)
Printed by R. Hindmarsh. London, England. 1794

Aiken, Conrad   1889–1973
American poet, short story writer and novelist

Ice is the silent language of the peak; and fire is the silent language of the star.
Collected Poems
Sonnet 10
Oxford University Press, Inc. New York, New York, USA. 1970

Alighieri, Dante   1265–1321
Italian poet and writer

…and thence we issued forth again to see the stars.
In *Great Books of the Western World* (Volume 21)
The Divine Comedy of Dante Alighieri
Hell, Canto XXXIV, l. 138–139
Encyclopædia Britannica, Inc. Chicago, Illinois, USA. 1952

Andreas, Brian   1956–
American artist, sculpture, and storyteller

We lay there & looked up at the night sky & she told me about stars called blue squares & red swirls & I told her I'd never heard of them. Of course not, she said, the really important stuff they never tell you. You have to imagine it on your own.
Blue Squares

Aratus   271 BCE–213 BCE
Greek statesman

In his fell jaw
Flames a star above all others with searing beams
Fiercely burning, called by mortals Sirius.
In Garrett P. Serviss
Astronomy with the Naked Eye
Chapter III (p. 42)
Harper & Brothers New York, New York, USA. 1908

Aurelius Antoninus, Marcus   121–180
Roman emperor

The Pythagoreans bid us in the morning look to the heavens that we may be reminded of those bodies which continually do the same thing and in the same manner

perform their work, and also be reminded of their purity and nudity. For there is no veil over a star.
In Great Books of the Western World (Volume 12)
The Meditations of Marcus Aurelius
Book XI, # 27 (p. 306)
Encyclopædia Britannica, Inc. Chicago, Illinois, USA. 1952

Look around at the courses of the stars, as if thou wert going along with them; and constantly consider the changes of the elements into one another; for such thoughts purge away the filth of the terrene life.
In Great Books of the Western World (Volume 12)
The Meditations of Marcus Aurelius
Book VII, #47 (p. 282)
Encyclopædia Britannica, Inc. Chicago, Illinois, USA. 1952

Bailey, Philip James   1816–1902
English poet

Surely the stars are images of love.
Festus: A Poem
Scene XXXI (p. 510)
George Routledge & Sons, Ltd. London, England. 1893

…the stars
As dewdrops countless on the aetherial fields
Of the skies…
Festus: A Poem
Scene I (p. 32)
George Routledge & Sons, Ltd. London, England. 1893

Baudelaire, Charles   1821–67
French poet

…those stars whose light speaks a known language…
The Flowers of Evil
Obsession (p. lxxix)
Wesleyan University Press. Middletown, Connecticut, USA. 1979

Benét, William Rose   1886–1950
American poet and editor

One speck within vast star-space lying
Awoke, arose, resumed its clothing,
And crawled another day toward dying.
Animalcule
Stanza 7
George H. Doran Company. New York, New York, USA. 1927

Berry, Richard   1946–
American amateur astronomer and author

It is a pity, in an age of rockets and space telescopes, that so few people have a direct acquaintance with the stars. Learning the stars and following their nightly courses across the sky brings a deep satisfaction, a satisfaction born of familiarity with something both ancient and ageless.
Discover the Stars (p. 2)
Harmony Books. New York, New York, USA. 1987

Blake, William   1757–1827
English poet, painter, and engraver

When the stars threw down their spears,
And water'd heaven with their tears,
Did he smile his work to see?
Did he who made the Lamb make thee?
The Complete Poetry and Prose of William Blake
The Tyger
University of California Press. Berkeley, California, USA. 1982

Borland, Hal   1900–78
American writer

…it is the stars that lure man's mind to the endless immensity of a universe so broad that tangible reality can never span it.
An American Year: Country Life and Landscapes Through the Seasons
June (p. 46)
Simon & Schuster. New York, New York, USA. 1946

Brecht, Bertolt   1898–1956
German writer

SAGREDO [reluctant to go to the telescope]: I feel something not all that remote from fear, Galileo.
GALILEO: I'm about to show you one of the shining milkwhite clouds in the Milky Way. Tell me what it's made up of.
SAGREDO: Those are stars, innumerable stars.
Translated by John Willett
Life of Galileo
Scene 3 (p. 26)
Arcade Publishing. New York, New York, USA. 1994

Brewster, David   1781–1868
English physicist

It is no ways probable that the Almighty, who always acts with infinite wisdom, and does nothing in vain, should create so many glorious suns, fit for so many important purposes, and place them at such distances from one another, without proper objects near enough to be benefited by their influences. Whoever imagines they were created only to give a faint glimmering light to the inhabitants of this globe, must have a very superficial knowledge of astronomy, and a mean opinion of the Divine wisdom: since, by an infinitely less exertion of creating power, the Deity would have given our earth much more light by one single additional moon.
Ferguson's Astronomy, Explained upon Sir Isaac Newton's Principles (Volume 1)
Chapter I (p. 3)
Printed for the author. London, England. 1756

Brood, William J.
No biographical data available

A telescope in the void recently found cosmic "maternity wards" where clouds of interstellar gas and dust appear to be in various stages of giving birth to stars.
"Golden Age" of Astronomy Peers to the Edge of the Universe
New York Times, C1, May 8, 1984

Brown, Fredric   1906–72
American writer of science fiction and mystery

Overhead and in the far distance are the lights in the sky that are stars. The stars they tell us we can never reach because they are too far away. They lie; we'll get there. If rockets won't take us, something will.
The Lights in the Sky Are Stars (p. 20)

Browne, J. Stark
No biographical data available

The stillness of the heavens is, however, apparent only, for commotion of the fiercest kind is raging on all sides. Stars are suns, and the suns are spheres of fire blazing with fury indescribable; scenes of activity so tremendous that no vehemence of tempest or tornado on earth can give the slightest idea of their fearfulness.
The Number and Distances of the Stars
The Rationalist Annual, 1931 (p. 61)

Browning, Robert   1812–89
English poet

All that I know
of a certain star,
Is, it can throw,
(Like the angled spar)
Now a dart of red,
Now a dart of blue.
The Poems and Plays of Robert Browning
Dramatic Lyrics, My Star
The Modern Library. New York, New York, USA. 1934

Bryant, William Cullen   1794–1878
American poet

The sad and solemn night
Hath yet her multitude of cheerful fires;
The glorious host of light
Walk the dark hemisphere till she retires;
All through her silent watches, gliding slow,
Her constellations come and climb the heavens and go.
In Parke Godwin (ed.)
Poems
Hymn to the North Star
D. Appleton & Company. New York, New York, USA. 1874

Meredith, Owen (Edward Robert Bulwer-Lytton, 1st Earl Lytton)   1831–91
English statesman and poet

When stars are in the quiet skies,
Then most I pine for thee;
Bend on me then thy tender eyes,
As stars look on the sea.
The Works of Edward Bulwer-Lytton (Volume 6)
Night and Love (p. 59)
Peter Fenelon Collier. New York, New York, USA. 1892

Bunting, Basil   1900–85
English modernist poet

Furthest, fairest thing, stars, free of our humbug,
each his own, the longer known, the more alone,
wrapt in emphatic fire roaring out to a black flue…
Then is Now. The star you steer by is gone.
Collected Poems
Briggflatts, V (p. 58)
Oxford University Press, Inc. London, England. 1978

Burke, Edmund   1729–97
English statesman and philosopher

The starry heaven, though it occurs so very frequently to our view, never fails to excite an idea of grandeur. This cannot be owing to the stars themselves, separately considered. The number is certainly the cause. The apparent disorder augments the grandeur, for the appearance of care is highly contrary to our ideas of magnificence. Besides, the stars lie in such apparent confusion, as makes it impossible on ordinary occasions to reckon them. This gives them the advantage of a sort of infinity.
A Philosophical Enquiry into the Origin of Our Ideas of the Sublime and Beautiful Magnificence (p. 139)
University of Notre Dame Press. Notre Dame, Indiana, USA. 1968

Burnet, Thomas   1635–1715
English cleric and scientist

They lie carelessly scatter'd, as if they had been sown in the Heaven, like Seed, by handfuls; and not by a skilful hand neither. What a beautiful Hemisphere they would have made, if they had been plac'd in rank and order, if they had been all dispos'd into regular figures, and the little ones set with due regard to the greater. Then all finisht and made up into one fair piece or great Composition, according to the rules of Art and Symmetry.
The Sacred Theory of the Earth (2nd edition)
Book II, Chapter XI (p. 220)
Printed by R. Norton. London, England. 1691

Burritt, Elijah H.   1794–1838
American astronomer

These vast globes of light then, could never have been designed merely to diversify the voids of infinite space, nor to shed a few glimmering rays on our far distant world, for the amusement of a few astronomers, who, but for the most powerful telescopes, had never seen the ten thousandth part of them.
The Geography of the Heavens
Chapter XVI (p. 154)
Huntington & Savage, Mason & Law. New York, New York, USA. 1850

Butler, Samuel   1612–80
English novelist, essayist, and critic

Cry out upon the stars for doing
Ill offices, to cross their wooing.
The Poetical Works of Samuel Butler (Volume 1)
Part III, Canto I, l. 17
Bell & Daldy. London, England. 1835

Campbell, Thomas 1777–1844
Scottish poet

…the sentinel stars set their watch in the sky.
The Complete Poetical Works
The Soldier's Dream
Chadwyck-Healey. Cambridge, England. 1992

Carlyle, Thomas 1795–1881
English historian and essayist

Canopus shining-down over the desert, with its blue diamond brightness (that wild, blue, spirit-like brightness far brighter than we ever witness here), would pierce into the heart of the wild Ishmaelitish man, whom it was guiding through the solitary waste there.
On Heroes and Hero Worship
Lecture I (p. 13)
John B. Alden, Publisher. New York, New York, USA. 1887

…when I gazed into these Stars, have they not looked down on me as if with pity, from their serene spaces; like Eyes glistening with heavenly tears over the little lot of man! Thousands of human generations, all as noisy as our own, have been swallowed up of Time, and there remains no wreck of them any more; and Arcturus and Orion and Sirius and the Pleiades are still shining in their courses, clear and young, as when the Shepherd first noted them in the plain of Shinar.
Sartor Resartus
Book II, Chapter VIII (p. 165)
Ginn & Company. Boston, Massachusetts, USA. 1897

Cernan, Eugene 1934–
American astronaut

I know the stars are my home. I learned about them, needed them for survival in terms of navigation. I know where I am when I look up at the sky. I know where I am when I look up at the Moon; it's not just some abstract romantic idea, it's something very real to me. See, I've expanded my home.
The View from Out There: In Words and Pictures
Life, Volume 11, Number 13, November 1988 (p. 198)

Cicero (Marcus Tullius Cicero) 106 BCE–43 BCE
Roman orator, politician, and philosopher

No one regards things before his feet
But views with care the regions of the sky.
Translated by William Armistead Falconer
Cicero: De Senectute, De Amicitia, De Divinatione
De Divinatione, II, XIII (p. 403)
Harvard University Press. Cambridge, Massachusetts, USA. 1938

Clarke, Arthur C. 1917–
English science and science fiction writer

No man who has lived all his life on the surface of a planet has ever seen the stars, only their feeble ghosts.
The Road to the Sea, Spring
Fiction House Magazine, Volume 1, Number 2, 1951

Overhead, without any fuss, the stars were going out.
The Collected Stories of Arthur C. Clarke
The Nine Billion Names of God (p. 422)
Tom Doherty Associates. New York, New York, USA. 2001

Sooner or later we will come to the edge of the Solar System and will be looking out across the ultimate abyss. Then we must choose whether we reach the stars — or whether we wait until the stars reach us.
The Challenge of the Spaceship
The Planets Are Not Enough (p. 65)
Harper & Brothers. New York, New York, USA. 1959

The thing's hollow — it goes on forever — and — oh my God! — it's full of stars.
2001: A Space Odyssey
V. The Moons of Saturn, Chapter 39 (p. 191)
New American Library, New York, New York, USA. 1968

Clarke, M'Donald (The Mad Poet) 1798–1842
American poet

Whilst twilight's curtain spreading far,
Was pinned with a single star.
Poems of M'Donald Clarke
Death in Disguise, l. 227
J.W. Bell. New York, New York, USA. 1837

Clegg, Johnny 1953–
English musician

…we are the scatterlings of Africa
On a journey to the stars…
Scatterlings of Africa
From the CD Scatterlings of Africa

Cohen, Martin
No biographical data available

To be a star is to know eternal stress. To live as a star is to walk a never ending tightrope, knowing that there can be only one outcome — your fall.
In Byron Preiss (ed.)
The Universe
Star Birth and Maturity (p. 68)
Bantam Books. Toronto, Ontario, Canada. 1987

Cole, Thomas 1627–97
English theologian

How lovely are the portals of the night,
When stars come out to watch the daylight die.
Twilight
Source undetermined

Coleridge, Samuel Taylor 1772–1834
English lyrical poet, critic, and philosopher

…the stars hang bright above her dwelling,
Silent as though they watched the sleeping Earth!
The Complete Poetical Works of Samuel Taylor Coleridge (Volume 1)
Dejection: An Ode, Stanza VIII
The Clarendon Press. Oxford, England. 1912

Coman, Dale Rex   1906–
American research physician and wildlife writer

How dismal a universe it would be without the lights of the stars to probe its infinite blackness.
The Endless Adventure
The Early Days of May (p. 148)
Henry Regnery Company. Chicago, Illinois, USA. 1972

Comte, Auguste   1798–1857
French philosopher

We can imagine the possibility of determining the shapes of stars, their distances, their sizes, and their movements; whereas there is no means by which we will ever be able to examine their chemical composition, their mineralogical structure, or especially, the nature of organisms that live on their surfaces.... Our positive knowledge with respect to the stars is necessarily limited to their observed geometrical and mechanical behavior.
The Positive Philosophy of Auguste Compte (Volume 2) (p. 9)
John Chapman. London, England. 1853

On the subject of stars, all investigations which are not ultimately reducible to simple visual observations are... necessarily denied to us.... We shall never be able by any means to study their chemical composition.
In Neil deGrasse Tyson
Over the Rainbow
Natural History, Volume 110, Number 7, September 2001 (p. 33)

Copernicus, Nicolaus   1473–1543
Polish astronomer

...the first and highest of all is the sphere of the fixed stars, which comprehends itself and all things, and is accordingly immovable.
In *Great Books of the Western World* (Volume 16)
On the Revolutions of the Heavenly Spheres
Book One, Chapter 10 (p. 526)
Encyclopædia Britannica, Inc. Chicago, Illinois, USA. 1952

Crane, Hart   1899–1932
American poet

Stars scribble on our eyes the frosty sagas,
The gleaming cantos of unvanquished space.
In Brom Weber (ed.)
The Complete Poems and Selected Letters and Prose of Hart Crane
Cape Hatteras
Anchor Books. Garden City, New York, USA. 1966

Dawkins, Richard   1941–
British ethologist, evolutionary biologist, and popular science writer

The stars have larger agendas in which the preoccupations of human pettiness do not figure.
Unweaving the Rainbow: Science, Delusion and the Appetite for Wonder
Chapter 6 (p. 117)
Houghton Mifflin Company. Boston, Massachusetts, USA. 1998

The universe we observe has precisely the properties we should expect if there is, at bottom, no design, no purpose, no evil, no good, nothing but blind, pitiless indifference.
River Out of Eden: A Darwinian View of Life
Chapter 4 (p. 133)
Basic Books, Inc. New York, New York, USA. 1995

de la Mare, Walter   1873–1956
English poet and novelist

Wide are the meadows of night
And daisies are shining there,
Tossing their lovely dews,
Lustrous and fair,
And through these sweet fields go,
Wanderers amid the stars —
Venus, Mercury, Uranus, Neptune,
Saturn, Jupiter, Mars.
Peacock Pie: A Book of Rhymes
The Wanderers
A. Constable & Company. London, England. 1913

de Saint-Exupéry, Antoine   1900–44
French aviator and writer

All men have the stars...but they are not the same things for different people. For some, who are travelers, the stars are guides. For others they are no more than little lights in the sky. For others, who are scholars, they are problems. For my businessman they were wealth. But all these stars are silent. You — you alone — will have the stars as no one else has them...
Translated by Katherine Woods
The Little Prince
Chapter XXVI (p. 85)
Harcourt, Brace & Company. New York, New York, USA. 1943

de Tabley, Lord   1835–95
English literary scholar and botanist

The May-fly lives an hour,
The star a million years;
But as a summer flower,
Or as a maiden's fears,
They pass, and heaven is bare
As tho' they never were.
The Collected Poems of Lord de Tabley
Hymn to Astarte
Chapman and Hall. London, England. 1903

Dee, John   1527–1609
English mathematician and occultist

The stars and celestial powers are like seals whose characters are imprinted differently by reason of differences in the elemental matter.
John Dee on Astronomy
XXVI (p. 135)
University of California Press. Berkeley, California, USA. 1978

Dick, Thomas   1600–80
Scottish theologian and philosopher

Come forth, O man! yon azure round survey,
And view those lamps which yield eternal day.
Bring forth thy glasses; clear thy wondering eyes;
Millions beyond the former millions rise;
Look further; — millions more blaze from yonder skies.
The Works of Thomas Dick, LL.D.
The Solar System, Volume 10, Chapter VIII (p. 197)

Dickinson, Emily   1830–86
American lyric poet

"Arcturus" is his other name —
I'd rather call him "star."
It's very mean of Science
To go and interfere!
The Complete Poems of Emily Dickinson
No. 70 (p. 36)
Little, Brown & Company. Boston, Massachusetts, USA. 1960

Disraeli, Benjamin, 1ˢᵗ Earl of Beaconsfield   1804–81
English prime minister, founder of Conservative Party, and novelist

It shows you exactly how a star is formed; nothing can be so pretty! A cluster of vapor, the cream of the milky way, a sort of celestial cheese, churned into light…
Tancred
Book I, Chapter IX
H. Colburn. London, England. 1847

Doyle, Sir Arthur Conan   1859–1930
Scottish writer

Are you conscious of the restful influence which the stars exert? To me they are the most soothing things in Nature. I am proud to say that I don't know the name of one of them. The glamour and romance would pass away from them if they were all classified and ticketed in one's brain. But when a man is hot and flurried, and full of his own little ruffled dignities and infinitesimal misfortunes, then a star bath is the finest thing in the world.
The Stark Munro Letters
Letter VIII (p. 170)
D. Appleton & Company. New York, New York, USA. 1895

Eddington, Sir Arthur Stanley   1882–1944
English astronomer, physicist, and mathematician

We are bits of stellar matter that got cold by accident, bits of a star gone wrong.
New York Times Magazine
October 9, 1932

We can now form some sort of picture of the inside of a star — a hurly burly of atoms, electrons and aether-waves. Disheveled atoms tear along at a hundred miles a second, their normal array of electrons being torn from them in the scrimmage. The lost electrons are speeding a hundred times faster to find new resting places…
Stars and Atoms
Lecture I (p. 26)
Yale University Press. London, England. 1927

Our object in diving into the interior [of a star] is not merely to admire a fantastic world with conditions transcending ordinary experience; it is to get at the inner mechanism which makes stars behave as they do. If we are to understand the surface manifestations, if we are to understand why "one star differeth from another star in glory," we must go below to the engine-room — to trace the beginning of the stream of heat and energy which pours out through the surface.
Stars and Atoms
Lecture I (p. 20)
Yale University Press. London, England. 1927

I am aware that many critics consider the conditions in the stars not sufficiently extreme…the stars are not hot enough. The critics lay themselves open to an obvious retort: we tell them to go and find a hotter place.
Stars and Atoms
Lecture III (p. 102)
Yale University Press. London, England. 1927

…it is reasonable to hope that in a not too distant future we shall be competent to understand so simple a thing as a star.
The Internal Constitution of the Stars
Chapter XIII (p. 393)
At The University Press. Cambridge, England. 1930

Eliot, George (Mary Ann Evans Cross)   1819–80
English novelist

The stars are golden fruit upon a tree
All out of reach.
The Spanish Gypsy
Book II
Blackwood and Sons. Edinburgh, Scotland. 1868

Emerson, Ralph Waldo   1803–82
American lecturer, poet, and essayist

The stars awaken a certain reverence, because though always present, they are inaccessible…
The Complete Works of Ralph Waldo Emerson (Volume 1)
Nature: Addresses and Lectures
Chapter I (p. 7)
Houghton Mifflin Company. Boston, Massachusetts, USA. 1904

But every night comes out the envoys of beauty, and light the universe with their admonishing smile.
Ralph Waldo Emerson: Essays and Lectures
Nature: Addresses, and Lectures
Nature (p. 9)
The Library of America. New York, New York, USA. 1983

If a man would be alone, let him look at the stars. The rays that come from those heavenly worlds, will separate between him and what he touches. One might think the atmosphere was made transparent with this design, to give man, in the heavenly bodies, the perpetual presence of the sublime. Seen in the streets of cities, how great they are!

Ralph Waldo Emerson: Essays and Lectures
Nature: Addresses and Lectures
Nature (p. 9)
The Library of America. New York, New York, USA. 1983

If the stars should appear one night in a thousand years, how would men believe and adore and preserve for many generations the remembrance of the city of God which had been shown.

Ralph Waldo Emerson: Essays and Lectures
Nature: Addresses, and Lectures
Nature (p. 9)
The Library of America. New York, New York, USA. 1983

Flecker, James Elroy   1884–1915
English poet and playwright

West of these out to seas colder than the Hebrides I must go

Where the fleet of stars is anchored and the young Star-captains glow.

The Collected Poems of James Elroy Flecker
The Dying Patriot
Martin Secker. London, England. 1916

Fraunhofer, Joseph von   1787–1826
German optician and physicist

Approximavit sidera
He brought the stars closer
Epitaph on his gravestone

Frost, Robert   1874–1963
American poet

They cannot scare me with their empty spaces

Between stars — on stars where no human race is.

Complete Poems of Robert Frost
Desert Places
Henry Holt & Company. New York, New York, USA. 1949

I could be worse employed
Than as a watcher of the void,
Whose part should be to tell
What star if any fell.
Suppose some seed-pearl sun
Should be the only one;
Yet still I must report
Some cluster one star short.
I should justly hesitate
To frighten church or state
By announcing a star down
From, say, the Cross or Crown.
To make sure what star I missed

I should have to check on my list
Every star in sight.
It might take me all night.

Complete Poems of Robert Frost
On Making Certain Anything Has Happened
Henry Holt & Company. New York, New York, USA. 1949

Gamow, George   1904–68
Russian-born American physicist

Twinkle, twinkle, quasi-star
Biggest puzzle from afar
How unlike the other ones
Brighter than a billion suns
Twinkle, twinkle, quasi-star
How I wonder what you are.

In Louis Berman
Exploring the Cosmos
Chapter 14 (p. 311)
Little, Brown & Company. Boston, Massachusetts, USA. 1973

Whereas all humans have approximately the same life expectancy, the life expectancy of stars varies as much as from that of a butterfly to that of an elephant.

A Star Called the Sun (p. 145)
The Viking Press. New York, New York, USA. 1964

Goddard, Robert H.   1882–1945
American physicist

There can be no thought of finishing, for "aiming at the stars," both literally and figuratively, is a problem to occupy generations, so that no matter how much progress one makes, there is always the thrill of just beginning.

In Eugene Mallove and Gregory Matloff
The Starflight Handbook
Letter to H.G. Wells, April 20, 1932 (p. 1)
John Wiley & Sons, Inc. New York, New York, USA. 1989

Goodenough, Ursula   1943–American biologist

I lie on my back under the stars and the unseen galaxies and I let their enormity wash over me. I assimilate the vastness of the distances, the impermanence, the fact of it all. I go all the way out and then I go all the way down, to the fact of photons without mass and gauge bosons that become massless at high temperatures. I take in the abstractions about forces and symmetries and they caress me, like Gregorian chants because the words are so haunting.

The Sacred Depths of Nature
Chapter I. Reflections (pp. 12–13)
Oxford University Press, Inc. New York, New York, USA. 1998

Greenstein, George   1940–
American astronomer

Overhead, the stars are strewn across a darkness, a blackness so profound that for a moment, for the barest flicker of an instant, I can almost sense their inconceivable

distance. In a sudden; exalting burst of vertigo I fancy what it would be like to fly, to fall up and into that ocean.
The Symbiotic Universe
Prologue (p. 29)
William Morrow & Company, Inc. New York, New York, USA. 1988

Grondal, Florence Armstrong
American astronomer and photographer

If all the diamonds in the world were melted into one huge magical jewel, its sparkling brilliance would pale beside Sirius, the diamond of the heavens.
The Music of the Spheres: A Nature Lover's Astronomy
Chapter VIII (p. 159)
The Macmillan Company. New York, New York, USA. 1926

…if all the wondrous phenomena of visible stars could be seen on but one of the nights of our long ride about the sun, the civilized world would spend its last cent on glasses and sit up until dawn to feast its eyes on the sublimity of the spectacle.
The Music of the Spheres: A Nature Lover's Astronomy
Chapter II (p. 16)
The Macmillan Company. New York, New York, USA. 1926

Guiterman, Arthur   1871–1943
Austrian-American poet

When the bat's on the wing and the bird's in the tree,
Comes the starlighter, whom none may see.
First in the West where the low hills are,
He touches his wand to the Evening Star.
Then swiftly he runs on his rounds on high,
Till he's lit every lamp in the dark blue sky.
Gaily the Troubadour
The Starlighter (p. 190)
E.P. Dutton & Company, Inc. New York, New York, USA. 1936

While poets feign that, passing earthly bars,
We Fireflies shall someday shine as Stars,
Our scientists, more plausibly surmise
That Stars are underdeveloped Fireflies.
Gaily the Troubadour
My Firefly Stars (p. 187)
E.P. Dutton & Company, Inc. New York, New York, USA. 1936

Habington, William   1605–54
English poet

The starres, bright cent'nels of the skies.
The Poems of William Habington
Dialogue between Night and Araphil, l. 3
University Press of Liverpool. Liverpool, England. 1948

Hardy, Thomas   1840–1928
English poet and regional novelist

The sky was clear — remarkably clear — and the twinkling of all the stars seemed to be but throbs of one body, timed by a common pulse.
Far from the Madding Crowd
Chapter 2 (p. 9)
Harper & Row, Publishers. New York, New York, USA. No date

The sovereign brilliance of Sirius pierced the eye with a steely glitter, the star called Capella was yellow, Aldebaran and Betelgueux shone with a fiery red. To persons standing alone on a hill during a clear midnight such as this, the roll of the world eastward is almost a palpable movement.
Far from the Madding Crowd
Chapter 2 (p. 9)
Harper & Row, Publishers. New York, New York, USA. No date

Harjo, Joy   1951–
Native American poet

I can hear the sizzle of newborn stars, and know anything of meaning, of the fierce magic emerging here. I am witness to flexible eternity, the evolving past, and I know we will live forever, as dust or breath in the face of stars, in the shifting pattern of winds.
Secrets from the Center of the World (p. 56)
The University of Arizona Press, Tucson, Arizona. 1989

Hearn, Lafcadio   1850–1904
Greek-born American writer

The infinite gulf of blue above seems a shoreless sea, whose foam is stars, a myriad million lights are throbbing and flickering and palpitating…
In Elizabeth Bisland
The Life and Letters of Lafcadio Hearn (Volume 1)
Letter to H.E. Krehbiel, 1877 (p. 170)
Houghton Mifflin Company. Boston, Massachusetts, USA. 1900

Hegel, Georg Wilhelm Friedrich   1770–1831
German philosopher

The stars are not pulled this way and that by mechanical forces; theirs is a free motion. They go on their way, as the ancients said, like the blessed gods.
Werke
Bd. 7, Abt. I (p. 97)
Publisher undetermined

Hein, Robert
No biographical data available

The stars are luminous dandruff from the deity's beard.
When the god-head combs his hair
A new star appears in the sky;
Yet God is not almost bald…
Quest of the Singing Tree
Stanzas on the Stars
H. Harrison. New York, New York, USA. 1938

Heine, Heinrich   1797–1856
German poet

Perhaps the stars in the sky only appear to us to be so beautiful and pure because we are so far away from them and do not know their intimate lives. Up above there are certainly a few stars that lie and beg; stars that put on airs; stars that are forced to commit all possible transgressions; stars that kiss and betray each other; stars that flatter their enemies and, what is even more painful, their friends, just as we do here below.

The Romantic School and Other Essays
The Romantic School, Book Two, Chapter III (p. 73)
Continuum. New York, New York, USA. 1985

Heppenheimer, T. A.   1947–
American space aviation writer

From the stars has come the matter of our world and of our bodies, and it is to the stars that we will someday return.

Toward Distant Suns
Preface (p. 13)
Stackpole Books. Harrisburg, Pennsylvania, USA. 1979

Herrick, Robert   1591–1674
English poet

The starres of the night
Will lend thee their light
Like Tapers cleare without number.

In J. Max Patrick (ed.)
The Complete Poetry of Robert Herrick
The Night-Piece, to Julia, Stanza 3
W.W. Norton & Company, Inc. New York, New York, USA.1968

Herschel, Friedrich Wilhelm   1738–1822
English astronomer

"We ought perhaps," says [Sir John Frederick William]Herschel, "to look upon certain clusters of stars, and the destruction of a star now and then in some thousands of ages, as the very means by which the whole is preserved and renewed. These clusters may be the laboratories of the universe, wherein the most salutary remedies for the decay of the whole are prepared."

Philosophical Transactions of the Royal Society of London, 1785
(p. 217)

Herschel, Sir John Frederick William   1792–1871
English astronomer and chemist

The stars are the land-marks of the universe…

Essays from the Edinburgh and Quarterly Reviews with Addresses and Other Pieces
An Address
April 11, 1827 (p. 469)
Longman, Brown, Green, Longmans & Roberts. London, England. 1857

If it were not perhaps too hazardous to pursue a former surmise of a renewal in what I figuratively called the Laboratories of the universe, the stars forming these extraordinary nebulæ, by some decay or waste of nature, being no longer fit for their former purposes, and having their projectile forces, if any such they had, retarded in each other's atmosphere, may rush at last together, and either in succession, or by one general tremendous shock, unite into a new body. Perhaps the extraordinary and sudden blaze of a new star in Cassiopeia's chair, in 1572, might possibly be of such a nature.

On the Construction of the Heavens
Philosophical Transactions of the Royal Society of London, Volume 75, 1785

Hodgson, Ralph   1871–1962
English poet

I stood and stared, the sky was lit,
The sky was stars all over it,
I stood, I knew not why,
Without a wish, without a will,
I stood upon that silent hill
And stared into the sky until
My eyes were blind with stars and still
I stared into the sky.

Collected Poems
The Song of Honour
Macmillan & Company Ltd. London, England. 1961

Holmes, Oliver Wendell   1809–94
American physician, poet, and humorist

And Science lifts her still unanswered cry:
"Are all these worlds, that speed their circling flight,
Dumb, vacant, soulless — baubles of the night?"
…Or rolls a sphere in each expanding zone
Crowned with a life as varied as our own?

The Poems of Oliver Wendell Holmes: With Numerous Illustrations
The Secret of The Stars
Houghton Mifflin Company. Boston, Massachusetts, USA. 1887

Hopkins, Gerard Manley   1844–89
English poet and Jesuit priest

Look at the stars! look, look up at the skies!
O look at all the fire-folk sitting in the air!

In Norman H. MacKenzie (ed.)
The Poetical Works of Gerard Manley Hopkins
The Starlight Night
Clarendon Press. Oxford, England. 1990

Horace (Quintus Horatius Flaccus)   65 BCE–8 BCE
Roman philosopher and dramatic critic

With my head exalted I shall touch the stars.

Carmina
I, I, 36
Rupert Hart-Davis. London, England. 1963

Hovey, Richard   1864–1900
American composer, poet, and artist

The dawn is lonely for the sun,
And chill and drear;
The one lone star is pale and wan,
As one in fear.

Along the Trail: A Book of Verse
Chanson de Rosemonde
Small, Maynard & Company. Boston, Massachusetts, 1898

Hoyle, Sir Fred   1915–2001
English mathematician and astronomer

The stars are best seen as a spectacle, not from everyday surroundings where trees and buildings, to say nothing of street lighting, distract the attention too much, but from a steep mountainside on a clear night, or from a ship at sea. Then the vault of heaven appears incredibly large and seems to be covered by an uncountable number of fiery points of light.
The Nature of the Universe
Chapter 3 (p. 51)
The University Press. Cambridge. 1933

It is unlikely that stars die without a spectacular protest.
Frontiers of Astronomy
Chapter Nine (p. 160)
Harper & Row, Publishers. New York, New York, USA. 1955

Hugo, Victor   1802–85
French author, lyric poet, and dramatist

An ant weighs upon the earth; a star can well weigh upon the universe.
Translated by Isabel F. Hapgood
The Toilers of the Sea
Part II, Book Third, Chapter III (p. 409)
The Heritage Press. New York, New York, USA. 1961

Huxley, Julian   1887–1975
English biologist, philosopher, and author

And all about the cosmic sky,
The black that lies beyond our blue,
Dead stars innumerable lie,
And stars of red and angry hue
Not dead but doomed to die.
The Captive Shrew and Other Poems of a Biologist
Cosmic Death
Harper & Brothers. New York, New York, USA. 1933

Huygens, Christiaan   1629–95
Dutch mathematician, astronomer, and physicist

For if 25 years are required for a Bullet out of a Cannon, with its utmost Swiftness, to travel from the Sun to us… such a Bullet would spend almost seven hundred thousand years in its Journey between us and the fix'd Stars. And yet when in a clear night we look upon them, we cannot think them above some few miles over our heads.
The Celestial Worlds Discover'd, or, Conjectures Concerning the Planetary Worlds, Their Inhabitants and Productions (pp. 154–155)
Printed for T. Childe. London, England. 1698

Jacobson, Ethel   1877–1965
New Zealand teacher, newspaper editor, and journalist

Crystal fish
Caught in the seine
Of the trawler, Night.
Stars
Nature Magazine, May 1958 (p. 260)

Jeans, Sir James Hopwood   1877–1946
English physicist and mathematician

The ages of the stars are not the same thing as the age of the universe, nor even are they necessarily comparable with that age. The star may be liken to icebergs coming down from the North and melting as they drift into tropical waters. We can estimate the age of the icebergs within our vision, but we can not say for how long the stream of icebergs has been drifting down from the pole to equator nor for how long new icebergs will continue to form and come down to replace those that pass southward to their doom.
Annual Report of the Board of Regents of the Smithsonian Institution, 1926
The New Outlook in Cosmogony (p. 159)
Government Printing Office. Washington, D.C. 1928

Empty Waterloo Station of everything except six specks of dust and it is still far more crowded with dust than space is with stars.
The Universe Around Us
Chapter I (p. 84)
The Macmillan Company. New York, New York, USA. 1929

Any small bit of the sky does not look very different from what it would if bright and faint stars had been sprinkled out of a celestial pepper pot.
The Universe Around Us
Chapter I (p. 37)
The Macmillan Company. New York, New York, USA. 1929

Jeffers, Robinson   1887–1962
American poet

We know the stars, hotter and more fatal than earth; we have learned lately the fire-wheel galaxies,
Infinite in number or all but infinite, among which our great sun's galaxy's
Flight is as a gnat's, one grain of sand in the Sahara: it is necessary to stretch our minds
To these dimensions…
In Tim Hunt (ed.)
The Collected Poetry of Robinson Jeffers (Volume 4)
Not Solid Earth (p. 538)
Stanford University Press. Stanford, California. USA. 1988

Antares reddens
The great one, the ancient torch, a lord among lost children,
The earth's orbit doubled would not girdle his greatness, one fire
Globed, out of grasp of the mind enormous; but to you O Night
What? Not a spark?…
In Tim Hunt (ed.)
The Collected Poetry of Robinson Jeffers (Volume 1)
Night (p. 115)
Stanford University Press. Stanford, California. USA. 1988

Joyce, James   1882–1941
Irish expatriate writer and poet

…of the parallax or parallactic drift of so called fixed stars, in reality ever moving wanderers from immeasurably

remote eons to infinitely remote futures in comparison with which the years threescore and ten, of allotted life formed a parenthesis of infinitesimal brevity.
Ulysses (p. 683)
Random House, Inc. New York, New York, USA. 1946

Keats, John   1795–1821
English Romantic lyric poet

Bright star, would I were steadfast as thou art —
Not in lone splendor hung aloft the night
And watching, with eternal lids apart,
Like Nature's patient, sleepless Eremite,
The moving waters at their priest-like task
Of pure ablution round earth's human shores.
The Complete Poetical Works and Letters of John Keats
Bright Star
Houghton Mifflin Company. Boston, Massachusetts, USA. 1890

Keill, John   1671–1721
Scottish mathematician and natural philosopher

The fixed stars appear to be of different bignesses, not because they really are so, but because they are not all equally distant from us; those that are nearest will excel in Luster and Bigness; the more remote stars will give a fainter Light, and appear smaller to the Eye.
An Introduction to the True Astronomy
Lecture VI (p. 47)
Printed for Bernard Lintot. London, England. 1721

Krutch, Joseph Wood   1893–1970
American naturalist, conservationist, and writer

The stars are little twinkling rogues who light us home sometimes when we are drunk but care for neither you nor me nor any man.
The Twelve Seasons
June (p. 46)
W. Sloane Associates. New York, New York, USA. 1949

Lambert, Johann Heinrich   1728–77
Swiss-German mathematician and astronomer

While we raise our eyes to the firmament, we see the whole of the stars attached, as it were, to the same vaulted surface; this, however, is an optical illusion; they are, in reality, at very different distances from us, as well as from the Sun, which is the fixed star of our system.
Translated by James Jacque
Kosmologischen Brief
1761

Lee, Nathaniel   1653? –92
English dramatist

The stars, heav'n sentry, wink and seem to die.
Theodosius
Printed for Tho. Chapman. London, England. 1692

Levy, David H.   1948–
Canadian astronomer and science writer

Our fondness for the stars has touched our souls. We all share the feeling of discovery, whether the object we have found is new to all or new only to us. The thrill penetrates our being, as we try to describe…how we have been changed by the universe sharing a secret with us.
David Levy's Guide to the Night Sky
A Miscellany (p. 322)
Cambridge University Press. Cambridge, England. 2000

…observing is an activity that can rapidly become your outlet to relax, your means to commune with the universe, and a vital key to knowing yourself…a voyage on a magic carpet that takes you to other places and other times. Even a casual look at the stars gives you a share in the company of timelessness that they represent. Look through your telescope thoughtfully…for it is more than starlight that the mirror will reflect. Through the vastness of space and time will return also a part of yourself.
The Royal Astronomical Society of Canada
The Joy of Gazing, 1982, Montreal Centre

Longfellow, Henry Wadsworth   1807–82
American poet

Silently one by one, in the infinite meadows of heaven Blossomed the lovely stars, the forget-me-nots of the angels.
The Poetical Works of Henry Wadsworth Longfellow
Evangeline, Part iii
Houghton, Mifflin Company. Boston, Massachusetts, USA. 1883

The stars arise, and the night is holy.
The Poetical Works of Henry Wadsworth Longfellow
Hyperion, Book I, Chapter 1
Houghton, Mifflin. Boston, Massachusetts, USA. 1883

Lowell, Percival   1855–1916
American astronomer

Bright points in the sky or a blow on the head will equally cause one to see stars.
Mars
Chapter IV (p. 159)
Houghton Mifflin Company. Boston, Massachusetts, USA. 1895

Macfie, Ronald Campbell   1867–1931
Scottish poet and physician

One thing is certain, that space is full of millions and millions of shining suns, wherever they came from and however they were evolved, and that there are millions and millions more of dead, dark stars we cannot see. One is apt to forget the dead, dark stars, but they far outnumber those that shine — so much so that Sir Robert Ball says that luminous stars are but the glowworms and fireflies of the universe as compared with the myriads of other animals.
Science, Matter and Immortality
Chapter XI (p. 131)
William & Norgate. London, England. 1909

Mandino, Og 1923–96
American sales guru and author

I will love the light for it shows me the way; yet I will love the darkness for it shows me the stars.
The Greatest Salesman in the World
Chapter Nine (p. 59)
Bantam Trade Edition. New York, New York, USA. 1985

Milton, John 1608–74
English poet

Witness this new-made World, another Heav'n
from Heaven Gate not farr, founded in view
On the clear Hyaline, the Glassie Sea;
Of amplitude almost immense, with starr's
Numerous, and every Starr perhaps a World
Of destined habitation.
In *Great Books of the Western World* (Volume 32)
Paradise Lost
Book VII, l. 617–622
Encyclopædia Britannica, Inc. Chicago, Illinois, USA. 1952

So sinks the day-star in the ocean-bed,
And yet anon repairs his drooping head,
And tricks his beams, and with new-spangled ore
Flames in the forehead of the morning sky.
Lycidas, l. 168–171
Encyclopædia Britannica, Inc. Chicago, Illinois, USA. 1952

…the stars,
That nature hung in heaven, and filled their lamps
with everlasting oil, to give due light
To the misled and lonely traveler.
In *Great Books of the Western World* (Volume 32)
Comus, l. 197–200
Encyclopædia Britannica, Inc. Chicago, Illinois, USA. 1952

Mitchell, Maria 1818–89
American astronomer and educator

When we are chaffed and fretted by small cares, a look at the stars will show us the littleness of our own interests.
In Phebe Mitchell Kendall
Maria Mitchell: Life, Letters, and Journals
Chapter VII (p. 138)
Lee & Shepard. Boston, Massachusetts, USA. 1896

We call the stars garnet and sapphire; but these are, at best, vague terms. Our language has not terms enough to signify the different delicate shades; our factories have not the stuff whose hues might make a chromatic scale for them.
In Phebe Mitchell Kendall
Maria Mitchell: Life, Letters, and Journals
Chapter XI (p. 235)
Lee & Shepard. Boston, Massachusetts, USA. 1896

Moore, Thomas 1779–1852
Irish poet

Thus, when the lamp that lighted
The traveler at first goes out,

He feels awhile benighted,
And looks around in fear and doubt.
But soon, the prospect clearing,
By cloudless starlight on he treads,
And thinks no lamp so cheering
As that light which Heaven sheds.
The Poetical Works of Thomas Moore
I'd Mourn the Hopes
Lee & Shepard. Boston, Massachusetts, USA. 1873

Mullaney, James
Astronomy writer, lecturer, and consultant

The telescope is not just another gadget or material possession, but a magical gift to humankind — a window on creation, a time machine, a spaceship of the mind that enables us to roam the universe in a way that is surely the next best thing to being out there.
Focal Point
Sky & Telescope, March 1990 (p. 244)

…metaphysical aspects of star gazing — its potential as a vehicle for therapeutic relaxation, meditation, and spiritual contact with the awesome creative power manifests in all of nature but is pinnacled in the stars.
Focal Point,
Sky & Telescope, March 1990 (p. 244)

Noyes, Alfred 1880–1958
English poet

Could new stars be born?
Night after night he watched that miracle
Growing and changing colour as it grew…
The Torch Bearers: Watchers of the Sky
Tycho Brahe, IV (p. 57)
Frederick A. Stokes Company. New York, New York, USA. 1922

And all those glimmerings where the abyss of space
Is powdered with a milky dust, each grain
A burning sun, and every sun the lord
Of its own darkling planets…
The Torch Bearers: Watchers of the Sky
Prologue (p. 8)
Frederick A. Stokes Company. New York, New York, USA. 1922

Old Woman

The stars I know and recognize and even call by name. They are my names, of course; I don't know what others call the stars. Perhaps I should ask the priest. Perhaps the stars are God's to name, not ours to treat like pets…
In Robert Coles
The Old Ones of New Mexico
Two Languages, One Soul (p. 10)
University of New Mexico Press. Albuquerque, New Mexico, USA. 1973

Oort, Jan Hendrik 1900–92
Dutch astronomer

Man in the past couple of centuries has been in a position like that of a lookout watching the approach of an armada

of strange objects. The objects came into view first as dim fuzzy forms. As more powerful telescopes brought them closer and closer, they were identified as collections of stars, then distinguished into systems of varied shapes and types; today we can resolve the details of internal structure in many of them.
In Heinz Haber
Stars, Men and Atoms
Chapter 11 (p. 168)
Golden Press. New York, New York, USA. 1962

Pagels, Heinz R.   1939–88
American physicist and science writer

Stars are born, they live and they die. Filling the night sky like beacons in an ocean of darkness, they have guided our thoughts over the millennia to the secure harbor of reason.
Perfect Symmetry: The Search for the Beginning of Time
Part One, Chapter 2 (p. 30)
Simon & Schuster. New York, New York, USA. 1985

Stars are like animals in the wild. We may see the young but never their actual birth, which is a veiled and secret event.
Perfect Symmetry: The Search for the Beginning of Time
Part One, Chapter 2 (p. 44)
Simon & Schuster. New York, New York, USA. 1985

Stars are an image of eternity.
Perfect Symmetry: The Search for the Beginning of Time
Part One, Chapter 3 (p. 54)
Simon & Schuster. New York, New York, USA. 1985

Pasachoff, Jay M.   1943–
American astronomer

Twinkle, twinkle, pulsing star
Newest puzzle from afar.
Beeping on and on you sing —
Are you saying anything?
Twinkle, twinkle more, pulsar,
How I wonder what you are.
Pulsars in Poetry
Physics Today, Volume 22, February 1969 (p. 19)

Peltier, Leslie C.   1900–80
American comet hunter

So clear and sparkling is this autumn night that, with averted vision, I can see quite readily the wraithlike wisps of nebulosity that festoon and enmesh this entire little cluster.
Starlight Nights
Chapter 1 (p. 5)
Harper & Row, Publishers. New York, New York, USA. 1965

The skies were full of stars for me to learn.
Starlight Nights
Chapter 6 (p. 39)
Harper & Row, Publishers. New York, New York, USA. 1965

Peattie, Donald Culrose   1896–1964
American botanist, naturalist, and author

Star gazing is a common name for harmless futility.
An Almanac for Moderns
August Seventeenth (p. 161)
G.P. Putnam's Sons. New York, New York, USA. 1935

Piechowski, Otto Rushe
Astronomy writer

For most of us stargazing remains a soothing balm and intellectual uplift — even if it isn't cutting edge science. It satisfies human needs. Sometimes out of embarrassment, we might shroud these deeper yearnings with scientific talk. But we shouldn't need such "covers." If our romantic encounters with stars reach some psychological, emotional, or spiritual level, so be it.
Sky & Telescope, February 1993

Plato   428 BCE–347 BCE
Greek philosopher

Vain would be the attempt of telling all the figures of them circling as in a dance, and their juxtapositions, and the return of them in their revolutions upon themselves, and their approximations...
In *Great Books of the Western World* (Volume 7)
Timaeus
Section 40 (p. 452)
Encyclopædia Britannica, Inc. Chicago, Illinois, USA. 1952

...he who has not contemplated the mind of nature which is said to exist in the stars, and gone through the previous training, and seen the connection of music with these things, and harmonized them all with laws and institutions, is not able to give a reason of such things as have a reason.
In *Great Books of the Western World* (Volume 7)
Laws
Book XII, 967 (p. 798)
Encyclopædia Britannica, Inc. Chicago, Illinois, USA. 1952

Poe, Edgar Allan   1809–49
American short story writer

Were the succession of stars endless, then the background of the sky would present us a uniform luminosity, like that displayed by the galaxy — since there could be absolutely no point, in all that background, at which would not exist a star. The only mode, therefore, in which, under such a state of affairs, we could comprehend the voids which our telescopes find in innumerable directions, would be by supposing the distance of the invisible background so immense that no ray from it has yet been able to reach us at all.
Eureka
Line 12 (p. 100)
Geo. P. Putnam. New York, New York, USA. 1848

Look down into the abysmal distances! — attempt to force the gaze down the multitudinous vistas of the stars, as we sweep slowly through them thus — and thus — and thus! Even the spiritual vision, is it not all points arrested by the continuous golden walls of the universe? — the walls of the myriads of the shining bodies that mere number has appeared to blend into unity?
In H. Beaver (ed.)
The Science Fiction of Edgar Allan Poe
The Power of Words (p. 171)
Penguin Books. Hammondsworth, England. 1976

Poincaré, Henri   1854–1912
French mathematician and theoretical astronomer

The stars are majestic laboratories, gigantic crucibles, such as no chemist could dream.
The Foundations of Science
The Value of Science, Astronomy (p. 295)
The Science Press. New York, New York, USA. 1913

Pope, Alexander   1688–1744
English poet

The Dog-star rages!
The Complete Poetical Works
Epistle to Dr. Arbuthnot
Houghton Mifflin Company. New York, New York, USA. 1903

Ptolemy   85–165
Greek astronomer

I know that I am mortal and ephemeral. But when I search for the close-knit encompassing convolutions of the stars, my feet no longer touch the earth, but in the presence of Zeus himself I take my fill of ambrosia which the gods produce.
In Johannes Kepler
Mysterium Cosmographicum
Title page

Raymo, Chet   1936–
American physicist and science writer

We are not ruled by stars; we and the stars are one.
The Virgin and the Mousetrap: Essays in Search of the Soul of Science
Chapter 21 (p. 196)
The Viking Press. New York, New York, USA. 1991

I weigh out nebulas. I dam up the Milky Way and use it to grind my grain. I put up summer stars like vegetables in jars for my delectation in winter. I have winter stars folded in boxes in the attic for cloudy summer nights.
Night Brought to Numbers
Sky & Telescope, Volume 71, Number 6, June 1966 (p. 555)

Rhodes, Cecil   1853–1902
British-born South African businessman, mining magnate, and politician

The world is nearly all parceled out, and what there is left of it is being divided up, conquered and colonized. To think of these stars that you see overhead at night, these vast worlds which we can never reach. I would annex the planets if I could; I often think of that. It makes me sad to see them so clear and yet so far.
The Last Will and Testament of Cecil John Rhodes: With Elucidatory Notes to Which Are Added Some Chapters Describing the Political and Religious Ideas of the Testator
Part III (p. 190)
"Review of Reviewers" Office. London, England. 1902

Rilke, Ranier Maria   1875–1926
Czech-born German language poet and novelist

…between stars, what distances…
Sonnets to Orpheus
Second Part, XX
University of California Press. Berkeley, California, USA. 1977

Russell, Peter   1921–2003
British poet and editor

…The fixed stars
Are moving really, and the whole Galaxy turning
Round and round on its own axis agitatedly…
All for the Wolves
Elegiac
Black Swan Books. Redding Ridge, Connecticut, USA. 1971

Sagan, Carl   1934–96
American astronomer and science writer

It will not be we who reach Alpha Centauri and the other nearby stars. It will be a species very much like us, but with more of our strengths and fewer of our weaknesses, a species returned to circumstances more like those for which it was originally evolved, more confident, farseeing, capable, and prudent — the sorts of beings we would want to represent us in a Universe that, for all we know, is filled with species much older, much more powerful, and very different.
Pale Blue Dot: A Vision of the Human Future in Space
Chapter 19 (p. 329)
Random House, Inc. New York, New York, USA. 1994

If we long to believe that the stars rise and set for us, that we are the reason there is a Universe, does science do us a disservice by deflating our conceits?
The Demon-Haunted World: Science as a Candle in the Dark
Chapter 1 (p. 12)
Random House, Inc. New York, New York, USA. 1995

Sagan, Carl   1934–96
American astronomer and science writer
Druyan, Ann   1949–
American author and television producer

Nothing lives forever, in Heaven as it is on Earth. Even the stars grow old, decay, and die. They die, and they are born. There was once a time before the Sun and Earth existed, a time before there was day or night, long, long before there was anyone to record the Beginning for those who might come after.

Shadows of Forgotten Ancestors: A Search for Who We Are
Chapter 1 (p. 11)
Random House, Inc. New York, New York, USA. 1992

Seneca (Lucius Annaeus Seneca)   4 BCE–65 AD
Roman playwright

As long as the ordinary course of heaven runs on, custom robs it of its real size. Such is our constitution that objects of daily occurrence pass us unnoticed even when most worthy of our admiration. On the other hand, the sight even of trifling things is attractive if their appearance is unusual. So this concourse of stars, which paints with beauty the spacious firmament on high, gathers no concourse of the nation. But when there is any change in the wonted order, then all eyes are turned to the sky.
Physical Science in the Time of Nero, Being a Translation of the Quaestiones Naturales of Seneca
Book VII, Chapter I (p. 271)
Macmillan & Company Ltd. London, England. 1910

Non est ad astra mollis e terris via.
There is no easy way to the stars from the earth.
Hercules Furens
Act II, 437
Cornell University Press. Ithaca, New York, USA. 1987

Service, Robert William   1874–1958
Canadian poet and novelist

The waves have a story to tell me,
As I lie on the lonely beach;
Chanting aloft in the pine-tops,
The wind has a lesson to teach;
But the stars sing an anthem of glory
I cannot put into speech.
Collected Poems of Robert Service
The Three Voices
Dodd, Mead & Company New York, New York, USA. 1961

Serviss, Garrett P.   1851–1929
American science fiction writer

Regarded in their broader relations and constraints, the stars as a whole possess a marvelous harmony of effect. It is the true music of the spheres, for who shall say that the universally felt influence of the star-bedight heavens does not arise from our instinctive, but as yet uneducated, perception of a concord which is not of "sweet sounds," but of light and color, whose range of vibrations in the ether infinitely exceeds that of sonant oscillations in the atmosphere?
Astronomy with the Naked Eye
Chapter I (p. 13)
Harper & Brothers. New York, New York, USA. 1908

The stars are the true landmarks which are never changed.
Astronomy with the Naked Eye
Chapter I (p. 1)
Harper & Brothers. New York, New York, USA. 1908

It was the friendly stars that first led men round the globe. As long as those well-known sentinels shone, tranquil and steadfast overhead, they had courage to go on and on. If the stars had deserted, even Columbus would have lost heart.
Astronomy with the Naked Eye
Chapter I (p. 1)
Harper & Brothers. New York, New York, USA. 1908

As long as men have eyes to see and minds to think, it needs but a word, a hint, a glance, to turn them with rapt and ever increasing attention to the wonders overhead.
Astronomy with the Naked Eye
Chapter I (p. 15)
Harper & Brothers. New York, New York, USA. 1908

Shakespeare, William   1564–1616
English poet, playwright, and actor

Look, th' unfolding star calls up the shepherd.
In *Great Books of the Western World* (Volume 27)
The Plays and Sonnets of William Shakespeare (Volume 2)
Measure for Measure
Act IV, Scene iii, l. 218
Encyclopædia Britannica, Inc. Chicago, Illinois, USA. 1952

But I am constant as the northern star,
Of whose true-fix'd and resting quality
There is no fellow in the firmament.
The skies are painted with unnumber'd sparks,
They are all fire and every one doth shine;
But there's but one in all doth hold his place.
In *Great Books of the Western World* (Volume 26)
The Plays and Sonnets of William Shakespeare (Volume 1)
Julius Caesar
Act III, Scene i, l. 60–65
Encyclopædia Britannica, Inc. Chicago, Illinois, USA. 1952

Sherrod, P. Clay
American astronomer and educator

Above us, the sparkling stars of the night skies stretch out like thousands of diamonds suspended on the curtain of space. Unfolding through the beauty and the mysteries of this seemingly endless expanse are patterns and answers familiar to those willing to study them...

There is an affinity for the eternity of space experienced by all mankind, a kind of motherhood in the stars to those who study space.
A Complete Manual of Amateur Astronomy
Preface (p. xii)
Prentice-Hall, Inc., Englewood Cliffs, New Jersey, USA. 1981

Smith, Logan Pearsall   1865–1946
American author

"But what are they really? What do they say they are?" the young lady asked me. We were looking up at the Stars.

Trivia
Book I, The Starry Heaven (p. 51)
Doubleday, Page & Company. Garden City, New York, USA. 1917

Smythe, Daniel   1908–81
American poet

The years of sky are now galactic,
So deep that we have little trace.
Our spectrographs, cool and emphatic,
Betray the depths of stars and space.
What do we seek on dizzying borders
Or groups of systems we have classified?
We cannot search in these huge orders
And find the answers they have passed.
Strange Horizons
Nature Magazine, February 1958 (p. 101)

Spenser, Edmund   1552–99
English poet

He that strives to touch the stars
Oft stumbles at a straw.
The Complete Poetical Works of Edmund Spenser
The Shepherdess Calendar
Houghton Mifflin Company. Boston, Massachusetts, USA. 1908

Stapledon, Olaf   1886–1950
English author

Very soon the heavens presented an extraordinary appearance, for all the stars directly behind me were now deep
red, while those directly ahead were violet. Rubies lay
behind me, amethysts ahead of me. Surrounding the ruby
constellations there spread an area of topaz stars, and
round the amethyst constellations an area of sapphires.
Last and First Men, and Star Maker
Chapter II (p. 262)
Dover Publications, Inc. New York, New York, USA. 1968

Great are the stars, and man is of no account of them.
Last and First Men
Chapter XVI (p. 303)
Jeremy P. Tarcher, Inc. Los Angeles, California, USA. 1988

Tagore, Rabindranath   1861–1941
Indian poet and philosopher

Kind Nature has held before our eyes the smoked glass of
the night and of the distance. And what do we see through
it? We see that the world of stars is still. For we see these
stars in their relation to each other, and they appear to
us like chains of diamonds hanging on the neck of some
god of silence. But Astronomy like a curious child plucks
out an individual star from that chain and then we find it
rolling about.
Personality
The World of Personality (p. 59)
The Macmillan Company. New York, New York, USA. 1917

Taylor, Anne   1782–1866
English poet and children's author

Twinkle, twinkle, little star!
How I wonder what you are.
Up above the world so high,
Like a diamond in the sky…
Rhymes for the Nursery
The Star
Printed and sold by Peter B. Gleason & Company. Hartford,
Connecticut, USA. 1813

Taylor, Bayard   1825–78
American journalist and author

Each separate star
Seems nothing, but a myriad scattered stars
Break up the night, and make it beautiful.
Lars: A Pastoral of Norway
Book III, Conclusion
James R. Osgood & Company. Boston, Massachusetts, USA. 1873

Teasdale, Sara   1884–1933
American writer and poet

Stars over snow,
And in the west a planet
Swinging below a star —
Look for a lovely thing and you will find it
It is not far — It will never be far.
The Collected Poems of Sara Teasdale
Night (p. 197)
Collier Books. New York, New York, USA. 1966

Tennyson, Alfred (Lord)   1809–92
English poet

"The stars," she whispers, "blindly run:
A web is wov'n across the sky;
From our waste places comes a cry,
And murmurs from the dying sun."
Alfred Tennyson's Poetical Works
In Memoriam A.H.H., Section III, Stanza II
Oxford University Press, Inc. London, England. 1953

Many a night I saw the Pleiades, rising
thro' the mellow shade,
Glitter like a swarm of fireflies, tangled in
a silver braid.
Alfred Tennyson's Poetical Works
Locksley Hall, Stanza 5
Oxford University Press, Inc. London, England. 1953

…the fiery Sirius alters hue
And bickers into red and emerald.
Alfred Tennyson's Poetical Works
The Princess, Part Fifth, l. 252–253
Oxford University Press, Inc. London, England. 1953

The Bible

Lift up your eyes to the heavens; consider who created
these, led out their host one by one, and summoned
each by name. Through his great might, his strength and
power, not one is missing.
The Revised English Bible

Isaiah 40:26
Oxford University Press, Inc. Oxford, England. 1989

…the morning stars sang in chorus,
and the sons of God all shouted for joy…
The Revised English Bible
Job 38:7
Oxford University Press, Inc. Oxford, England. 1989

Can you bind the cluster of Pleiades or loose Orion's belt?
The Revised English Bible
Job 38:31
Oxford University Press, Inc. Oxford, England. 1989

The Hon. Mrs. Ward
No biographical data available

Stars — each, perhaps a sun! Far, far away from the earth and its troubles is the mind carried by such thoughts and remembrances.
The Telescope
Preface, Dedication to William Parsons, the Earl of Rosse 1870

Thompson, Francis   1859–1907
English writer

Thou canst not stir a flower
Without troubling of a star.
Complete Poetical Works of Francis Thompson
The Mistress of Vision, Stanza XXII
Boni & Liveright, Inc. New York, New York, USA. 1923

Thomson, James   1700–48
Scottish poet

But who can count the stars of Heaven?
Seasons
Winter, l. 528
Printed by John Mycall. Newburyport, Massachusetts, USA. 1790

Thoreau, Henry David   1817–62
American essayist, poet, and practical philosopher

Truly the stars were given for a consolation to man.
The Writings of Henry David Thoreau (Volume 9)
A Walk to Wachusett (p. 178)
Houghton Mifflin Company. Boston, Massachusetts, USA. 1893

When I consider how, after sunset, the stars come out gradually in troops from behind the hills and woods, I confess that I could not have contrived a more curious and inspiring night.
Journal (Volume 1I)
July 26, 1840 (p. 158)
Princeton University Press. Princeton, New Jersey, USA. 1981

When I look at the stars, nothing which the astronomers have said attaches to them, they are so simple and remote.
In Bradford Torrey and Francis H. Allen (eds.)
The Journal of Henry D. Thoreau (Volume 7)
September 29, 1854, The Red of the Young Oak (p. 60)
Houghton Mifflin Company. Boston, Massachusetts, USA. 1949

The stars are the apexes of what wonderful triangles! What distant and different beings in the various mansions of the universe are contemplating the same one at the same moment!
The Writings of Henry David Thoreau (Volume 2)
Walden
Chapter I (p. 19)
Houghton Mifflin Company. Boston, Massachusetts, USA. 1893

Thorne, Kip S.   1940–
American theoretical physicist

A star is only a glowing pause in the inescapable contraction of a gas cloud to an uncertain, sometimes fantastic end.
The Death of a Star
The Physics Teacher, Volume 9, Number 6, June 1971 (p. 326)

Travers, Pamela Lyndon   1899–1996
Australian-born English writer

[Jane] was watching Mrs. Corry splashing the glue on the sky and Mary Poppins sticking on the star…

"What I want to know," said Jane, "is this: Are the stars gold paper or is the gold paper stars?"

There was no reply to her question and she did not expect one. She knew that only someone very much wiser than Michael could give her the right answer.
Mary Poppins (Revised edition)
Chapter 8 (p. 128)
Harcourt, Inc. San Diego, California, USA. 1981

Trevelyan, George Macaulay   1876–1962
English historian

The stars out there rule the sky more than in England, big and lustrous with the honour of having shone upon the ancients and been named by them.
Clio, A Muse, and Other Essays
Walking
Longmans, Green & Company London, England. 1930

Twain, Mark (Samuel Langhorne Clemens)   1835–1910
American author and humorist

There are too many stars in some places and not enough in others, but that can be remedied presently, no doubt.
Eve's Diary
Sunday (p. 7)
Harper & Brothers. New York, New York, USA. 1906

There's another trouble about theories: there's always a hole in them somewheres, sure, if you look close enough. It's just so with this one of Jim's. Look what billions and billions of stars there is. How does it come that there was just exactly enough star-stuff, and none left over? How does it come there ain't no sand-pile up there?
The Complete Works of Mark Twain (Volume 14)

Tom Sawyer Abroad (pp. 78–79)
Harper & Brothers Publishers. New York, New York, USA. 1899

Stars are good, too. I wish I could get some to put in my hair. But I suppose I never can.
Eve's Diary
Saturday (p. 11)
Harper & Brothers. New York, New York, USA. 1906

The stars ain't so close together as they look to be.
Collected Tales, Sketches, Speeches, & Essays 1891–1910
Extract from Captain Stormfield's Visit to Heaven (pp. 829–830)
The Library of America. New York, New York, USA. 1992

It's lovely to live on a raft. We had the sky, up there, all speckled with stars, and we used to lay on our backs and look up at them, and discuss about whether they was made, or only just happened — Jim he allowed they was made, but I allowed they happened; I judged it would have took too long to make so many. Jim said the moon could have a laid them; well, that looked kind of reasonable, so I didn't say nothing against it because I've seen a frog lay most as many, so of course it could be done. We used to watch the stars that fell, too, and see them streak down. Jim allowed they'd got spoiled and was hove out of the nest.
The Adventures of Huckleberry Finn
Chapter XIX (pp. 153–154)
Grosset & Dunlap Publishers. New York, New York, USA. 1948

Updike, John 1932–
American novelist, short story writer, and poet

Welcome, welcome, little star
I'm delighted that you are
Up in Heaven's vast extent,
No bigger than a continent.
Telephone Poles and Other Poems
White Dwarf (p. 10)
Alfred A. Knopf. New York, New York, USA. 1969

When, on those anvils at the center of stars,
and those even more furious anvils
of the exploding supernovae,
the heavy elements were beaten together
to the atomic number 94…
Facing Nature
Ode to Crystallization
Alfred A. Knopf. New York, New York, USA. 1985

van Gogh, Vincent Willem 1853–90
Dutch painter

To express hope by some star, the eagerness of a soul by a sunset.
The Complete Letters of Vincent Van Gogh with Reproductions of All the Drawings in the Correspondence (Volume 3)
Letter 531 (p. 26)
New York Graphic Society. Greenwich, Connecticut, USA. 1958

One night I went for a walk by the sea along the empty shore. It was not gay, but neither was it sad — it was — beautiful. The deep blue sky was flecked with clouds of a blue deeper than the fundamental blue of intense cobalt, and others of a clearer blue, like the blue whiteness of the Milky Way. In the blue depth the stars were sparkling, greenish, yellow, white, rose, brighter, flashing, more like jewels, than they do at home — even in Paris.
The Complete Letters of Vincent Van Gogh with Reproductions of All the Drawings in the Correspondence (Volume 2)
Letter 499 (p. 589)
New York Graphic Society. Greenwich, Connecticut, USA. 1958

That does not prevent me from having a terrible need of — shall I say the word? — of religion. Then I go out and paint the stars…
The Complete Letters of Vincent Van Gogh with Reproductions of All the Drawings in the Correspondence (Volume 3)
Letter 543 (p. 56)
New York Graphic Society. Greenwich, Connecticut, USA. 1958

Stars, you are unfortunate, I pity you,
Beautiful as you are, shining in your glory…
Night Thoughts
Printed by R. Nobels for R. Edwards. London, England. 1797

Vaughan, Henry 1621–95
English metaphysical poet

The Jewel of the Just,
Shining nowhere but in the dark;
What mysteries do lie beyond thy dust,
Could man outlook that mark!
Poetry and Selected Prose
Accession Hymn
Oxford University Press, Inc. London, England. 1963

Weil, Simone 1909–43
French philosopher and mystic

The stars, those marvelous, brilliant, inaccessible objects, at least as remote as the horizon, which we can neither change nor touch, and which in their turn, touch only our eyes, are what is furthest away from us and closest to us.
On Science, Necessity, and the Love of God
Classified Science and After (p. 40)
Oxford University Press, Inc. London, England. 1968

Wells, H. G. (Herbert George) 1866–1946
English novelist, historian, and sociologist

Looking at these stars suddenly dwarfed my own troubles and all the gravities of terrestrial life. I thought of their unfathomable distance, and the slow inevitable drift of their movements out of the unknown past into the unknown future.
In Robert M. Hutchins and Mortimer J. Adler (eds.)
The Great Ideas Today, 1971
The Time Machine
Chapter Seven (p. 484)
Encyclopædia Britannica, Inc. Chicago, Illinois, USA. 1971

Whitman, Walt 1819–92
American poet, journalist, and essayist

I believe a leaf of grass is no less than the journey-work of the stars.
Complete Poems and Collected Prose
Song of Myself
Section 31
The Library of America. New York, New York, USA. 1982

I was thinking the day most splendid till I saw what the not-day exhibited;

I was thinking of this globe enough till there sprang out so noiseless around me myriads of other globes.
Complete Poetry and Collected Prose
Night on the Prairies
The Library of America. New York, New York, USA. 1982

Wilcox, Ella Wheeler   1850–1919
American poet and journalist

Since Sirius crossed the Milky Way
Full sixty thousand years have gone,
Yet hour by hour and day by day
This tireless star speeds on and on.
In Garrett P. Serviss
Astronomy with the Naked Eye
Chapter III (p. 43)
Harper & Brothers New York, New York, USA. 1908

Wilde, Oscar   1854–1900
Irish wit, poet, and dramatist

LORD DARLINGTON: We are all in the gutter, but some of us are looking at the stars.
The Works of Oscar Wilde (Volume 5)
Lady Windermere's Fan
Act Three
Lamb Publishing Company. New York, New York, USA. 1909

Williams, Sarah   1837–68
American poet

I have loved the stars too fondly to be fearful of the night.
The Best Loved Poems of the American People
The Old Astronomer to His Pupil
Garden City publishing Company. Garden City, New York USA. 1936

Wordsworth, William   1770–1850
English poet

The stars are mansions built by Nature's hand,
And, haply, there the spirits of the blest
Dwell clothed in radiance, their immortal vest…
The Complete Poetical Works of William Wordsworth
Miscellaneous Sonnets, XXV
Crowell. New York, New York, USA. 1888

Look for the stars, you'll say that there are none;
Look up a second time, and, one by one,
You mark them twinkling out with silvery light,
And wonder how they could elude the sight!
The Complete Poetical Works of William Wordsworth
Calm Is the Fragrant Air
Crowell. New York, New York, USA. 1888

Yeats, William Butler   1865–1939
Irish poet and playwright

Under the passing stars, foam of the sky
Live on this lonely face.
The Collected Poems of W.B. Yeats
The Rose of the World (p. 36)
The Macmillan Company. New York, New York, USA. 1956

Young, Edward   1683–1765
English poet and dramatist

How distant some of these nocturnal Suns?
So distant (says the Sage) 'twere not absurd
To doubt, if Beams, set out at Nature's Birth,
Are yet arriv'd at this so foreign World…
Night Thoughts
Night IX, l. 1226–1229
Printed by R. Nobels for R. Edwards. London, England. 1797

STARLIGHT

Hale, George Ellery   1868–1938
American astronomer

Starlight is falling on every square mile of the earth's surface, and the best we can do at present is to gather up and concentrate the rays that strike an area 100 inches in diameter.
The Possibilities of Large Telescopes
Harper's Weekly, April 1928 (p. 640)

Huggins, Sir William   1824–1910
English astronomer

Within this unraveled starlight exists a strange cryptography. Some of the rays may be blotted out, others may be enhanced in brilliancy. Their differences, countless in variety, form a code of signals, in which is conveyed to us, when once we have made out the cipher in which it is written, information of the chemical nature of the celestial gases…. It was the discovery of this code of signals, and of its interpretation, which made possible the rise of the new astronomy.
In D.R. Danielson
The Book of the Cosmos: Imaging the Universe from Heraclitus to Hawking
Chapter 52 (p. 319)
Perseus Publishing. Cambridge, Massachusetts, USA. 2000

Stein, Gertrude   1874–1946
American writer

Star-light, what is star-light, star-light is a little light that is not always mentioned with the sun, it is mentioned with the moon and the sun, it is mixed up with the rest of the time.
Three Lives & Tender Buttons
Rooms (p. 295)
Penguin Putnam, Inc. New York, New York, USA. 2003

STATISTICAL TEST

Anscombe, Francis John   1918–2001
English-born American statistician

Rejection rules are not significance tests.
Rejection of Outliers
Technometrics, Volume 2, 1960 (p. 126)

Chatfield, Christopher
British statistician

The result is that non-statisticians tend to place undue reliance on single "cookbook" techniques, and it has for example become impossible to get results published in some medical, psychological and biological journals without reporting significance values even if of doubtful validity. It is sad that students may actually be more confused and less numerate at the end of a "service course" than they were at the beginning, and more likely to overlook a descriptive approach in favor of some inferential method which may be inappropriate or incorrectly executed.
The Initial Examination of Data
Journal of the Royal Statistical Society, Series A, Volume 148, 1985

Clark, C. A.
No biographical data available

The null hypothesis of no difference has been judged to be no longer a sound or fruitful basis for statistical investigation.... Significance tests do not provide the information that scientists need, and, furthermore, they are not the most effective method for analyzing and summarizing data.
Hypothesis Testing in Relation to Statistical Methodology
Review of Educational Research, Volume 33, 1963

Cochran, William G.   1909–80
Scottish-born American statistician
Cox, Gertrude M.   1900–78
American statistician

A useful property of a test of significance is that it exerts a sobering influence on the type of experimenter who jumps to conclusions on scanty data, and who might otherwise try to make everyone excited about some sensational treatment effect that can well be ascribed to the ordinary variation in his experiment.
Experimental Designs (2nd edition)
Chapter 1 (p. 5)
John Wiley & Sons, Inc. New York, New York, USA. 1992

Cohen, Jacob   1923–
American behavioral psychologist and statistical analyst

After four decades of severe criticism, the ritual of null hypothesis significance testing — mechanical dichotomous decisions around a sacred .05 criterion — still persist. This article reviews the problems with this practice.... What's wrong with [null hypothesis significance testing]? Well, among many other things, it does not tell us what we want to know, and we so much want to know what we want to know that, out of desperation, we nevertheless believe that it does!
The Earth Is Round (p < .05)
American Psychologist, Volume 49, December 1994 (p. 997)

A little thought reveals a fact widely understood among statisticians: The null hypothesis, taken literally (and that's the only way you can take it in formal hypothesis testing), is always false in the real world.... If it is false, even to a tiny degree, it must be the case that a large enough sample will produce a significant result and lead to its rejection. So if the null hypothesis is always false, what's the big deal about rejecting it?
Things I Have Learned (So Far)
American Psychologist, December 1990 (p. 1308)

Cox, Sir David Roxbee   1924–
English statistician

It has been widely felt, probably for thirty years and more, that significance tests are overemphasized and often misused and that more emphasis should be put on estimation and prediction. While such a shift of emphasis does seem to be occurring, for example in medical statistics, the continued very extensive use of significance tests is on the one hand alarming and on the other evidence that they are aimed, even if imperfectly, at some widely felt need.
Some General Aspects of the Theory of Statistics
International Statistical Review, Volume 54, 1986

Deming, William Edwards   1900–93
American statistician, educator, and consultant

Under the usual teaching, the trusting student, to pass the course must forsake all the scientific sense that he has accumulated so far, and learn the book, mistakes and all.
On Probability as Basis for Action
American Statistician, Volume 29, Number 4, November 1975

Pencil and paper for construction of distributions, scatter diagrams, and run-charts to compare small groups and to detect trends are more efficient methods of estimation than statistical inference that depends on variances and standard errors, as the simple techniques preserve the information in the original data.
On Probability as Basis for Action
American Statistician, Volume 29, Number 4, November 1975

Devons, Ely   1913–67
English economist

I cannot oscillate a time series or properly analyze a variance...
Essays in Economics

Chapter 6 (p. 105)
Greenwood Press, Publishers, Westport, Connecticut, USA. 1961

Fisher, Sir Ronald Aylmer   1890–1962
English statistician and geneticist

There is no more pressing need in connection with the examination of experimental results than to test whether a given body of data is or is not in agreement with any suggested hypothesis.
Statistical Methods for Research Workers
Chapter VIII (p. 256)
Oliver & Boyd. Edinburgh, Scotland. 1938

McCloskey, D. N.
Economist
Ziliak, S. T.
No biographical data available

The low and falling cost of calculation, together with a widespread though unarticulated realization that after all the significance test is not crucial to scientific questions, has meant that statistical significance has been valued at its cost. Essentially no one believes a finding of statistical significance or insignificance. This is bad for the temper of the field. My statistical significance is a "finding"; yours is an ornamented prejudice.
The Standard Error of Regressions
Journal of Economic Literature, Volume 34, 1996

Morrison, D. E.
No biographical data available
Henkel, R. E.
No biographical data available

What do we do without the tests, then? What we do without the tests has always in some measure been done in behavioral science and needs only to be done more and better: the application of imagination, common sense, and informed judgment, and the appropriate remaining research methods to achieve the scope, form, process, and purpose of scientific inference.
The Significance Test Controversy — A Reader
Significance Tests in Behavioral Research: Skeptical Conclusions and Beyond
Aldine Publishing Company. Chicago, Illinois, USA. 1970

Parkhurst, D. F.
No biographical data available

Failing to reject a null hypothesis is distinctly different from proving a null hypothesis; the difference in these interpretations is not merely a semantic point. Rather, the two interpretations can lead to quite different biological conclusions.
Interpreting Failure to Reject a Null Hypothesis
Bulletin of the Ecological Society of America, Volume 66, 1985

Rozeboom, W. W.
No biographical data available

The statistical folkways of a more primitive past continue to dominate the local scene.
The Fallacy of the Null-Hypothesis Significance Test
Psychological Bulletin, Volume 57, 1960

Salsburg, David S.
No biographical data available

Most readers of this journal will recognize the limited value of hypothesis testing in the science of statistics. I am not sure that they all realize the extent to which it has become the primary tool in the religion of Statistics. Since the practitioners of that faith seem unable to cure their own folly, it is time we priests of the faith brought them around to realizing that there are more appropriate ways to get useful answers.
The Religion of Statistics as Practiced in Medical Journals
American Statistician, Volume 39, 1985

Schmidt, Frank L.
American industrial and organizational psychology researcher

I believe that…false beliefs are a major cause of the addiction of researchers to significance tests. Many researchers believe that statistical significance testing confers important benefits that are in fact completely imaginary.
Statistical Significance Testing and Cumulative Knowledge in Psychology: Implications for Training of Researchers
Psychological Methods, Volume 1, Number 2, 1996

If the null hypothesis is not rejected, [Sir Ronald] Fisher's position was that nothing could be concluded. But researchers find it hard to go to all the trouble of conducting a study only to conclude that nothing can be concluded.
Statistical Significance Testing and Cumulative Knowledge in Psychology: Implications for Training of Researchers
Psychological Methods, Volume 1, Number 2, 1996

Yates, Frances   1899–1981
English historian

The most commonly occurring weakness in the application of Fisherian methods is, I think, undue emphasis on tests of significance, and failure to recognize that in many types of experimental work, estimates of the treatment effects, together with estimates of the error to which they are subject, are the quantities of primary interest.
Sir Ronald Fisher and the Design of Experiments
Biometrics, Volume 20, 1964

STATISTICIAN

Bailey, W. B.   1873–?
No biographical data available

Cummings, John
No biographical data available

While, therefore, tabulation is a final process, the formulation of the scheme of tabulation should be the initial process, preceding even the formulation of the schedule, which should be determined by the character of the tables to be produced. Failure to observe this fundamental principle in statistical practice, perhaps more than any other characteristic, distinguishes the work of the amateur from that of the expert, the work of the untrained social investigator from that of the experienced scientific statistician.
Statistics (p. 26)
A.C. McClurg & Company. Chicago, Illinois, USA. 1917

Balchin, Nigel   1908–70
English novelist

He divided people into statisticians, people who knew about statistics, and people who didn't. He liked the middle group best. He didn't like the real statisticians much because they argued with him, and he thought people who didn't know any statistics were just animal life.
The Small Back Room (p. 137)
Collins. London, England. 1943

Belloc, Hilaire   1870–1953
French-born poet and historian

The statistician was let loose.
The Silence of the Sea
On Statistics (p. 172)
Sheed & Ward. New York, New York, USA. 1940

Bellow, Saul   1915–
American novelist

An utterly steady, reliable woman, responsible to the point of grimness. Daisy was a statistician for the Gallup Poll.
Herzog (p. 221)
The Viking Press. New York, New York, USA. 1964

Blodgett, James H.
American statistician

The individual statistician must scan closely the authority on which he rests, and guard his statements with all the cautionary words which imperfect knowledge requires, or some mere child will point out the errors in his statements and his conclusions and set people wondering of what value the rest of his work may be.
Obstacles to Accurate Statistics
Journal of the American Statistical Association, New Series Number 41, March 1898 (p. 19)

Bowley, Arthur L.   1869–1957
English statistician and economist

Perhaps statisticians themselves have not always fully recognized the limitations of their work.
Elements of Statistics
Part I, Chapter I (p. 13)
P.S. King & Son Ltd. London, England. 1937

Chernoff, H.   1923–
American mathematician and statistician
Moses, L. E.   1921–
American statistician and social scientist

Years ago a statistician might have claimed that statistics deals with the processing of data…today's statistician will be more likely to say that statistics is concerned with decision making in the face of uncertainty.
Elementary Decision Theory
Introduction (p. 1)
John Wiley & Sons, Inc. New York, New York, USA. 1959

Deming, William Edwards   1900–93
American statistician, educator, and consultant

The minute a statistician steps into the position of the executive who must make decisions and defend them, the statistician ceases to be a statistician.
Sample Design in Business Research (p. 13)
John Wiley & Sons, Inc. New York, New York, USA. 1960

The statistician accepts in any engagement certain responsibilities and obligations to his client and to the people that he works with. In the first place, he is the architect of a survey or experiment. It is his business to fit the various skills together to make them effective. It is important that he clarify the various responsibilities at the outset of the study.
Sample Design in Business Research (p. 10)
John Wiley & Sons, Inc. New York, New York, USA. 1960

The only useful function of a statistician is to make predictions, and thus to provide a basis for action.
In W.A. Wallis
The Statistical Research Group
Journal of the American Statistical Association, 1942–1945, Volume 75, Number 370, June 1980 (p. 321)

It should be emphasized that the statistician is not necessarily abler at handling data than his colleagues trained in economics, sociology, engineering, physics, business, etc. However, because of the high transferability of the statistician's mathematical techniques, and because he acquires a broad knowledge in many fields, he is frequently adept at discovering and measuring errors in data and determining the source of the errors. He avoids drawing wrong conclusions from data whether the data be good or bad.
On the Classification of Statisticians
The American Statistician, Volume 2, Number 2, April 1948 (p. 16)

A statistician's responsibility is not confined to plans: he must also seek assurance of cooperation in field and

office, and maintain constant touch with the work, also with the interpretation of the results.
Some Theory of Sampling (p. 8)
John Wiley & Sons, Inc. New York, New York, USA. 1950

Finney, D. J.
English biometric statistician

Too often, in many fields of science, the statistician is regarded as someone who comes on stage after data have been collected, performs standard calculations, delivers a verdict "Significant" or "Not Significant", and then departs.
The Questioning Statistician
Statistics in Medicine, Volume 1, 1982 (p. 5)

Fisher, Sir Ronald Aylmer   1890–1962
English statistician and geneticist

We have the duty of formulating, of summarizing, and of communicating our conclusions, in intelligible form, in recognition of the right of other free minds to utilize them in making their own decisions.
B, Statistical methods and scientific induction
Journal of the Royal Statistical Society, Volume 17, 1955

The statistician cannot evade the responsibility for understanding the process he applies or recommends.
The Design of Experiments
Introduction (p. 1)
Hafner Publishing Company. New York, New York, USA. 1971

The statistician cannot excuse himself from the duty of getting his head clear on the principles of scientific inference, but equally no other thinking man can avoid a like obligation.
The Design of Experiments
Introduction (p. 2)
Hafner Publishing Company. New York, New York, USA. 1971

Fleiss, Joseph L.
American statistician

There was a statistician from Needham,
Who was so bright, his clients would heed him.
Yet his embarrassed confession
Was that, in linear regression,
He'd never subtract an extra degree of freedom.
Letters to the Editor
The American Statistician, Volume 21, Number 4, October 1967 (p. 49)

There was a statistician from Knossus,
Who had a nonnormal neurosis.
With techniques of newness, He'd measure the skewness,
And also the data's kurtosis.
Letters to the Editor
The American Statistician, Volume 21, Number 4, October 1967 (p. 49)

There was a biometrician named Mabel,
Who'd never look at populations unstable.
Using intricate relations,
She'd find life expectations,
From the lx's of the life table.
Letters to the Editor
The American Statistician, Volume 21, Number 4, October 1967 (p. 49)

Forster, E. M. (Edward Morgan)   1879–1970
English novelist

We are not concerned with the very poor. They are unthinkable, and only to be appreciated by the statistician or the poet.
Howards End
Chapter VI (p. 45)
Vintage Books. New York, New York, USA. 1954

Good, I. J.   1916–
English statistician and cryptographer

The mathematician, the statistician, and the philosopher do different things with a theory of probability. The mathematician develops its formal consequences, the statistician applies the work of the mathematician and the philosopher describes in general terms what this application consists in. The mathematician develops symbolic tools without worrying overmuch what the tools are for; the statistician uses them; the philosopher talks about them. Each does his job better if he knows something about the work of the other two.
Kinds of Probability
Science, Volume 129, Number 3347, February 20, 1959 (p. 443)

Hooke, Robert   1635–1703
English physicist

It is commonly believed that anyone who tabulates numbers is a statistician. This is like believing that anyone who owns a scalpel is a surgeon.
How to Tell the Liars from the Statisticians
Chapter 1 (p. l)
Marcel-Dekker, Inc. New York, New York, USA. 1983

Hopkins, Harry
No biographical data available

Increasingly, we find ourselves caught up in the new contemporary dualism; there is the muddling-on, verbalizing, impressionistic, human old world down there, and there is that Other, Finer, Rational World to which the better statisticians have already been called. Communications between the two can be tenuous.
The Numbers Game: The Bland Totalitarianism
Chapter 6 (p. 134)
Little, Brown & Company. Boston, Massachusetts, USA. 1973

Kerridge, D. F.
No biographical data available

It is not primarily the responsibility of a statistician to make decisions for other people — not in general at any rate.... It is for someone else to say what decisions should

be made with [inferential]…information. In other words, ideally, it is the statisticians job to inform not to decide.
Discussion on Paper by Dr. Marshall and Professor Olkin
Journal of the Royal Statistical Society, Series B, Volume 30, 1968 (p. 440)

Kruskal, William   1919–2005
American mathematician and statistician

An occupational hazard to which we statisticians are exposed occurs in the context of a social occasion, perhaps a dinner party. I am, let us say, seated next to a charming lady whom I have just met, and, as an initial conversational ice-breaking, she turns to me with a winning smile and says: "Now tell me what is it you do?" We must tell the truth, of course, so I reply that I am a statistician. That usually ruins a fine conversation, for in 8.6 cases out of 10 the lady's smile disappears, she turns to my rival on her other side, and I attack the fried chicken in lonely, misunderstood dignity.
Statistics, Moliere, and Henry Adams
American Scientist Magazine, 1967 (p. 416)

Miksch, W. F.   1861–1927
No biographical data available

A couple of government statisticians recently threw dust on the wedding ring business by coming right out with the fact that for every male there are 1.03 females. It's about time they stop shoving the American taxpayer behind decimal points.
The Average Statistician
Collier's, Volume 125 June 17, 1950 (p. 10)

Moroney, M. J.
American statistician

There is more than a germ of truth in the suggestion that, in all society where statisticians thrive, liberty and individuality are likely to be emasculated.
Facts from Figures
Statistics Undesirable (p. 1)
Penguin Books Ltd., Harmondsworth, England. 1951

The statistician's job is to draw general conclusions from fragmentary data. Too often the data supplied to him for analysis are not only fragmentary but positively incoherent, so that he can do next to nothing with them. Even the most kindly statistician swears heartily under his breath whenever this happens.
Facts from Figures
What Happens When We Take a Sample (p. 120)
Penguin Books Ltd., Harmondsworth, England. 1951

Sarton, George   1884–1956
Belgian-born American scholar and writer

I like to think of the constant presence in any sound Republic of two guardian angels: the Statistician and the Historian of Science. The former keeps his finger on the pulse of Humanity…
Sarton on the History of Science
Quetelet (p. 241)
Harvard University Press. Cambridge, Massachusetts, USA. 1962

Seaton, G. L.
No biographical data available

…as the job of finding the truth and explaining it continues to become more complex and more difficult, management again casts a doubtful eye at the statistician, for a different reason. Management's big question is no longer "What can the statistician do for us that we can't do just as well ourselves?"; the question now is, "Do our statisticians have the tools and the capacity and the experience and the persistence and the breadth of vision to seek the truth and to know it when they have found it?
The Statistician and Modern Management
The American Statistician, Volume 2, Number 6, December 1948 (p. 10)

Snedecor, G. W.
Statistician

The characteristic which distinguishes the present-day professional statistician, is his interest and skill in the measurement of the fallibility of conclusions.
On a Unique Feature of Statistics
Presidential Address to the American Statistical Association, December 1948
Journal of the American Statistical Association, Volume 44, Number 245, March 1949

Stamp, Josiah   1880–1941
English economist and financier

Most of you would as soon be told that you are cross-eyed or knock-kneed as that you are destined to be a statistician…
Some Economic Factors in Modern Life
Chapter VIII (p. 253)
P.S. King & Son Ltd. London, England. 1929

I sometimes think that statisticians do not deserve quite all the hard things that are said about them. They are supposed to be cold, unemotional, bloodless and steely-eyed. But, as a matter of fact, we are all statisticians nowadays. We are either forming opinions on other people's statistics, whether we like it or not, or we are providing the raw material of statistics.
Some Economic Factors in Modern Life
Chapter VIII (p. 253)
P.S. King & Son Ltd. London, England. 1929

Thurber, James   1894–1961
American writer and cartoonist

Though statisticians in our time have never kept the score, Man wants a great deal here below and Women even more.
Further Fables of Our Times
The Godfather and His Godchild
Simon & Schuster. New York, New York, USA. 1956

Tukey, John W.   1915–2000
American statistician

Predictions, prophecies, and perhaps even guidance — those who suggested this title to me must have hoped for such — even though occasional indulgences in such actions by statisticians has undoubtedly contributed to the characterization of a statistician as a man who draws straight lines from insufficient data to foregone conclusions!
Where Do We Go From Here?
Journal of the American Statistical Association, Volume 55, Number 289, March 1960 (p. 80)

The most important maxim for data analysis to heed, and one which many statisticians seem to have shunned is this: "Far better an approximate answer to the right question, which is often vague, than an exact answer to the wrong question, which can always be made precise." Data analysis must progress by approximate answers, at best, since its knowledge of what the problem really is will at best be approximate.
The Future of ata Analysis
Annals of Mathematical Statistics, Volume 33, Number 1, March 1962 (pp. 13–14)

(The experimental statistician dare not shrink from the war cry of the analyst "Only a fool would use it, but it's better than we used to use!")
Unsolved Problems of Experimental Statistics
Journal of the American Statistical Association, Volume 49, Number 268, December 1954 (p. 718)

Wang, Chamont   1949–
American statistician

Flip a coin 100 times. Assume that 99 heads are obtained. If you ask a statistician, the response is likely to be: "It is a biased coin." But if you ask a probabilist, he may say: "Wooow, what a rare event."
Sense and Nonsense of Statistical Inference: Controversy, Misuse, and Subtlety
Chapter 6 (p. 154)
Marcel-Dekker, Inc. New York, New York, USA. 1993

Wells, H. G. (Herbert George)   1866–1946
English novelist, historian and sociologist

Behind the adventurer, the speculator, comes that scavenger of adventurers, the statistician.
The Work, Wealth and Happiness of Mankind
Chapter Nine, Part 10 (p. 390)
Doubleday, Doran & Company, Inc. Garden City, New York, USA. 1931

…the movement of the last hundred years is all in favor of the statistician.
The Work, Wealth and Happiness of Mankind
Chapter Nine, Part 10 (p. 391)
Doubleday, Doran & Company, Inc. Garden City, New York, USA. 1931

Wigner, Eugene Paul   1902–95
Hungarian-born American physicist

There is a story about two friends, who were classmates in high school, talking about their jobs. One of them became a statistician and was working on population trends. He showed a reprint to his former classmate. The reprint started, as usual with the Gaussian distribution and the statistician explained to his former classmate the meaning of the symbols for the actual population, for the average population, and so on. His classmate was a bit incredulous and was not quite sure whether the statistician was pulling his leg
The Unreasonable Effectiveness of Mathematics in the Natural Sciences
Communications on Pure and Applied Mathematics, Volume 13, Number 1, February 1960 (p. 1)

Yule, G. U.   1871–1951
Scottish statistician

Since the statistician can seldom or never make experiments for himself, he has to accept the data of daily experiences, and discuss as best he can the relations of a whole group of changes…
On the Theory of Correlation
Journal of the Royal Statistical Society, Volume LX, December 1897 (p. 812)

STATISTICS

Adams, Henry Brooks   1838–1918
American man of letters

History has never regarded itself as a science of statistics. It was the Science of Vital Energy in relation with time; and of late this radiating center of life has been steadily tending, — together with every form of physical and mechanical energy, — toward mathematical expression.
A Letter to American Teachers of History
Chapter I (p. 115)
Press of J.H. Furst Company. Washington, D.C. 1910

Advertisement

…and you thought "impressive" statistics were 36–24–36.
The American Statistician, Volume 33, Number 4, November 1979 (p. 248)

Allen, Roy George Douglas   1906–83
English economist and mathematician

A knowledge of statistical methods is not only essential for those who present statistical arguments it is also needed by those on the receiving end.
Statistics for Economists
Chapter I (p. 9)
Hutchinson's University Library. London, England. 1951

Angell, Roger   1920–
American fiction writer and essayist

Statistics are the food of love.

Late Innings: A Baseball Companion
Chapter 1 (p. 9)
Simon & Schuster. New York, New York, USA. 1982

Baines, J. A.
No biographical data available

Once again, but not I hope, too often, or for the last time, do I dip into the well of Mr. Courtney's sagacity:–

"We may quote to one another with a chuckle the words of the Wise Statesman, lies, damned lies and statistics, still there are some easy figures which the simplest must understand but the astutest cannot wriggle out of."
Parliamentary Representation in England Illustrated by the Elections of 1892 and 1895
Journal of the Royal Statistical Society, Volume 59, 1896 (p. 87)

Balchin, Nigel   1908–70
English novelist

Organic chemist!" said Tilley expressively. "Probably knows no statistics whatever."
The Small Back Room (p. 136)
Collins. London, England. 1943

Bailey, W. B.   1873–?
No biographical data available
Cummings, John
No biographical data available

Statistical tables are essentially specific in their meaning, and they require data that are uniformly specific in the same kind and degree.
Statistics (p. 33)
McClurg. Chicago, Illinois, USA. 1917

Barrett-Browning, Elizabeth   1806–61
English poet

There's too much abstract willing, purposing,
In this poor world. We talk by aggregates,
And think by systems and being used to face
Our evils in statistics, are inclined
To cap them with unreal remedies
Drawn out in haste on the other side.
The Complete Poetical Works of Elizabeth Barrett Browning
Aurora Leigh, Eighth Book, l. 800
Houghton Mifflin Company. Boston, Massachusetts, USA. 1900

Bartlett, Maurice Stevenson   1910–2002
English statistician

[Statistics] is concerned with things we can count. In so far as things, persons, are unique or ill-defined, statistics are meaningless and statisticians silenced; in so far as things are similar and definite — so many workers over 25, so many nuts and bolts made during December — they can be counted and new statistical facts are born.

Essays on Probability and Statistics
Some Remarks on the Theory of Statistics (p. 11)
Methuen & Company Ltd. London, England. 1962

Baudrillard, Jean   1929–
French cultural theorist

Like dreams, statistics are a form of wish fulfillment.
Translated by Chris Turner
Cool Memories
October 1983 (p. 147)
Verso. London, England. 1990

Bell, E. T. (Eric Temple)   1883–1960
Scottish-American mathematician and educator

The statistical method is social mathematics par excellence.
The Development of Mathematics (p. 582)
McGraw-Hill Book Company, Inc. New York, New York, USA. 1945

Mankind in the mass is more despotically governed by the laws of chance than it ever was by the decrees of any tyrant. If our shambling race is ever to get anything but suicidal destruction out of science, it may be a necessary first step that half a dozen human beings in every hundred thousand understand the mass-reactions of creatures who, as individuals, occasionally show that they can stand erect and walk like men. To grasp and analyze mass-reactions, whether of atoms or of human beings, a mastery of the modern statistical method is essential.
The Development of Mathematics (p. 582)
McGraw-Hill Book Company, Inc. New York, New York, USA. 1945

Belloc, Hilaire   1870–1953
French-born poet and historian

It has long been recognized by public men of all kinds… that statistics come under the head of lying, and that no lie is so false or inconclusive as that which is based on statistics.
The Silence of the Sea
On Statistics (p. 170)
Sheed & Ward. New York, New York, USA. 1940

Before the curse of statistics fell upon mankind we lived a happy, innocent life, full of merriment and go, and informed by fairly good judgment.
The Silence of the Sea
On Statistics (p. 171)
Sheed & Ward. New York, New York, USA. 1940

Statistics are the triumph of the quantitative method, and the quantitative method is the victory of sterility and death.
The Silence of the Sea
On Statistics (p. 173)
Sheed & Ward. New York, New York, USA. 1940

Berger, J. O.
No biographical data available

Berry, D. A.
No biographical data available

...to acknowledge the subjectivity inherent in the interpretation of data is to recognize the central role of statistical analysis as a formal mechanism by which new evidence can be integrated with existing knowledge. Such a view of statistics as a dynamic discipline is far from the common perception of a rather dry, automatic technology for processing data.
Statistical Analysis and the Illusion of Objectivity
American Scientist, Volume 76, 1988 (p. 159)

Bernard, Claude   1813–78
French physiologist

Only when a phenomenon includes conditions as yet undefined, can we compile; we must learn, therefore, that we compile statistics only when we cannot possibly help it.
Translated by Henry Copley Greene
An Introduction to the Study of Experimental Medicine
Part Two, Chapter II, Section IX (p. 137)
Henry Schuman, Inc. New York, New York, USA. 1927

I do not understand how we can teach practical and exact science on the basis of statistics.
Translated by Henry Copley Greene
An Introduction to the Study of Experimental Medicine
Part Two, Chapter II, Section ix (p. 138)
Henry Schuman, Inc. New York, New York, USA. 1927

Billings, John Shaw   1838–1913
American surgeon and librarian

Statistics are somewhat like old medical journals, or like revolvers in newly opened mining districts. Most men rarely use them, and find it troublesome to preserve them so as to have them easy of access; but when they do want them, they want them badly.
On Vital and Medical Statistics
Medical Record, Volume 36, 1889

Blalock, Jr., Hubert M.   1927–91
American sociologist and statistical methods researcher

The manipulation of statistical formulas is no substitute for knowing what one is doing.
Social Statistics
Chapter 19 (p. 448)
McGraw-Hill Book Company, Inc. New York, New York, USA. 1960

Bloch, Arthur   1948–
American humorist

If enough data is collected, anything may be proved by statistical methods.
Murphy's Law
William and Holland's Law (p. 47)
Price/Stern/Sloan, Publishers. Los Angeles, California, USA. 1981

Blodgett, James H.
American statistician

In statistical work we should be able to presume upon honesty, fidelity, and diligence.
Obstacles to Accurate Statistics
Journal of the American Statistical Association, New Series Number 41, March 1898 (p. 1)

Boorstin, Daniel J.   1914–2004
American historian

...statistics have tended to make facts into norms.
The Decline of Radicalism
Chapter I (p. 18)
Random House, Inc. New York, New York, USA. 1969

...statistics, which first secured prestige here by a supposedly impartial utterance of stark fact, have enlarged their dominion over the American consciousness by becoming the most powerful statement of the "ought" — displacers of moral imperatives, personal ideals, and unfulfilled objectives.
The Decline of Radicalism
Chapter I (p. 19)
Random House, Inc. New York, New York, USA. 1969

Bowley, Arthur L.   1869–1957
English statistician and economist

Some of the common ways of producing a false statistical argument are to quote figures without their context, omitting the cautions as to their incompleteness, or to apply them to a group of phenomena quite different to that to which they in reality relate; to take these estimates referring to only part of a group as complete; to enumerate the events favorable to an argument, omitting the other side; and to argue hastily from effect to cause, this last error being the one most often fathered on to statistics. For all these elementary mistakes in logic, statistics is held responsible.
Elements of Statistics
Part I, Chapter I (pp. 12–13)
P.S. King & Son Ltd. London, England. 1937

A statistical estimate may be good or bad, accurate or the reverse; but in almost all cases it is likely to be more accurate than a casual observer's impression, and the nature of things can only be disproved by statistical methods.
Elements of Statistics
Part I, Chapter I (p. 9)
P.S. King & Son Ltd. London, England. 1937

Great numbers are not counted correctly to a unit, they are estimated; and we might perhaps point to this as a division between arithmetic and statistics, that whereas arithmetic attains exactness, statistics deals with estimates, sometimes very accurate, and very often sufficiently so for their purpose, but never mathematically exact.

Elements of Statistics
Part I, Chapter I (p. 3)
P.S. King & Son Ltd. London, England. 1937

A knowledge of statistics is like a knowledge of foreign languages or of algebra; it may prove of use at any time under any circumstances.
Elements of Statistics
Part I, Chapter I (p. 4)
P.S. King & Son Ltd. London, England. 1937

Bowman, Scotty   1933–
Canadian Hockey Coach

Bowley, Arthur L.   1869–1957
English statistician and economist

Statistics are for losers.
A Lot More Where They Come From
Sports Illustrated, April 2, 1973

Bowman, W. E.   1912–1985
English amateur hill hiker and satirist

The various estimates of the height of the true summit vary considerably, but by taking an average of these figures it is possible to say confidently that the summit of Rum Doodle is 40,000 1/2 feet above sea level.
The Ascent of Rum Doodle
Chapter 3 (pp. 32–33)
The Vanguard Press. New York, New York, USA. 1956

Buchner, Ludwig   1824–99
German physician and philosopher

The science of statistics, which has only been turned to proper account in modern times, has the great honor of having proved the existence of definite rules in a number of phenomena, which had hitherto been looked upon as merely accidental or as owing their origin to an arbitrary power.
Force and Matter
Free Will (p. 367)
Truth Seeker. New York, New York, USA. 1950

Burgess, Robert W.   1887–1969
American statistician

The fundamental gospel of statistics is to push back the domain of ignorance, prejudice, rule-of-thumb, arbitrary or premature decisions, tradition, and dogmatism and to increase the domain in which decisions are made and principles are formulated on the basis of analyzed quantitative facts.
The Whole Duty of the Statistical Forecaster
Journal of the American Statistical Association, Volume 32, Number 200, December 1937 (p. 636)

Burnan, Tom
No biographical data available

No matter how much reverence is paid to anything purporting to be "statistics," the term has no meaning unless the source, relevance, and truth are all checked.
The Dictionary of Misinformation
Statistics, Use, Misuse, and Abuse Of (p. 271)
Ballantine Books. New York, New York, USA. 1975

…the worship of statistics has had the particularly unfortunate result of making the job of the plain, outright liar that much easier.
The Dictionary of Misinformation
Statistics, Use, Misuse, and Abuse Of (p. 274)
Ballantine Books. New York, New York, USA. 1975

Carlyle, Thomas   1795–1881
English historian and essayist

Statistics is a science which ought to be honorable, the basis of many most important sciences; but it is not to be carried on by steam, this science, any more than others are; a wise head is requisite for carrying it on.
English and Other Critical Essays
Chartism (p. 170)
J.M. Dent & sons Ltd. London, England. 1950

Statistics is a science which ought to be honourable, the basis of many most important sciences; but it is not to be carried on by steam, this science, any more than others are; a wise hand is requisite for carrying it on. Conclusive facts are inseparable from unconclusive except by a head that already understands and knows.
Critical and Miscellaneous Essays
Chartism, II
D. Appleton & Company. New York, New York, USA. 1860

Statistics, one may hope, will improve gradually, and become good for something. Meanwhile, it is to be feared the crabbed satirist was partly right, as things go.
English and Other Critical Essays
Chartism, Chapter II (p. 171)
J.M. Dent & sons Ltd. London, England. 1950

Carroll, Lewis (Charles Dodgson)   1832–98
English writer and mathematician

"And on the dead level our pace is — ?" the younger suggested; for he was weak in statistics, and left all such details to his aged companion.
The Complete Works of Lewis Carroll
A Tangled Tale
Knot I (p. 983)
The Modern Library. New York, New York, USA. 1936

Chatfield, Christopher
British statistician

Thus statistics should generally be taught more as a practical subject with analyses of real data. Of course some theory and an appropriate range of statistical tools need to be learnt, but students should be taught that Statistics is much more than a collection of standard prescriptions.

The Initial Examination of Data
Journal of the Royal Statistical Society, Series A, Volume 148, 1985

Cogswell, Theodore R.   1918–87
American science fiction author

Statistics show that you have nothing to worry about.
In Harry Harrison
Astounding: John W. Campbell Memorial Anthology
Probability Zero (p. 329)
Random House, Inc. New York, New York, USA. 1973

Cohen, Jacob   1923–
American behavioral psychologist and statistical analyst

Since statistical significance is so earnestly sought and devoutly wished for by behavioral scientists, one would think that the a priori probability of its accomplishment would be routinely determined and well understood. Quite surprisingly, this is not the case.
Statistical Power Analysis for the Behavioral Sciences
Chapter 1 (p. 1)
Lawrence Erlbaum Associates. Hillsdale, New Jersey, USA. 1988

I have learned repeatedly, however, that the typical behavioral scientist approaches applied statistics with considerable uncertainty (if not actual nervousness), and requires a verbal-intuitive exposition, rich in redundancy and with many concrete illustrations.
Statistical Power Analysis for the Behavioral Sciences
Preface to the Original Edition (p. xx)
Lawrence Erlbaum Associates, Publishers. Hillsdale, New Jersey, USA. 1988

Cox, Sir David Roxbee   1924–
English statistician
Hinkley, D. V.   1924–
English statistician

Statistical methods of analysis are intended to aid the interpretation of data that are subject to appreciable haphazard variability.
Theoretical Statistics (p. 1)
Introduction (p. 1)
Chapman & Hall. London, England. 1974

Crichton, Michael   1942–
American novelist

Conversation and statistics. Really boring.
Rising Sun
Second Day (p. 254)
Ballantine Books. New York, New York, USA. 1993

Darwin, Charles Robert   1809–82
English naturalist

One has, however, no business to feel so much surprise of one's ignorance, when one knows how impossible it is without statistics to conjecture the duration of life and percentage of deaths to births in mankind.

In Francis Darwin (ed.)
The Life and Letters of Charles Darwin (Volume 1)
C. Darwin to L. Jenyns, [1845?] (p. 394)
D. Appleton & Company. New York, New York, USA. 1896

Davies, John Tasman   1924–?
Chemist

Operational research is the application of methods of the research scientist to various rather complex practical operations…. A paucity of numerical data with which to work is a usual characteristic of the operations to which operational research is applied.
The Scientific Approach
Chapter 7 (p. 86)
Academic Press. London, England. 1965

Davis, Joseph S.
No biographical data available

Statistics are proverbially dry — forgive me if I say they are far better dry than "wet" — but to give them optimum moisture content is simply a matter of mastering fundamentals that no one should hold in contempt.
Statistics and Social Engineering
Journal of the American Statistical Association, Volume 32, Number 197, March 1937 (p. 6)

Dawkins, Richard   1941–
British ethologist, evolutionary biologist, and popular science writer

The essence of life is statistical improbability on a colossal scale.
The Blind Watchmaker
Chapter 11 (p. 317)
W.W. Norton & Company, Inc. New York, New York, USA.1986

de Jonnes, Moreau
No biographical data available

Statistics are like the hieroglyphics of ancient Egypt, where the lessons of history, the precepts of wisdom, and the secrets of the future were concealed in mysterious characters.
Elements de Statistique (p. 5)

de Leeuw, A. L.
No biographical data available

The method used by the scientist to find probable exact truth is what he calls "the method of least squares."
Rambling Through Science
Gambling (p. 88)
Whittlesey House. London, England. 1932

de Madariaga, Salvador   1886–1978
Spanish writer and statesman

Statistics only work well when they dwell on large numbers of absolutely free motions, or what has been described as "perfect disorder." If an element of deliberate direction, of conscious "order", meddles with their

utter "innocence", the facts in question cease to follow statistical laws.
Essays with a Purpose
Freedom and Science (p. 50)
Hollis & Carter. London, England. 1954

Deming, William Edwards   1900–93
American statistician, educator, and consultant

You need not be a mathematical statistician to do good statistical work, but you will need the guidance of a first class mathematical statistician.
Some Principles of the Shewhart Method of Quality Control
Mechanical Engineering, Volume 66, March 1944

A good engineer, or a good economist, or a good chemist, already has a good start, because the statistical method is only good science brought up to date by the recognition that all laws are subject to the variations which occur in nature.
Some Principles of the Shewhart Method of Quality Control
Mechanical Engineering, Volume 66, March 1944

Your study of statistical methods will not displace any other knowledge that you have; rather, it will extend your knowledge of engineering, chemistry, or economics, and make it more useful.
Some Principles of the Shewhart Method of Quality Control
Mechanical Engineering, Volume 66, March 1944

Statistical research is particularly necessary in the government service because of the high level of quality and economy that the public has the right to expect in government statistics.
Some Theory of Sampling (p. viii)
John Wiley & Sons, Inc. New York, New York, USA. 1950

The statistical method is more than an array of techniques. The statistical method is a Mode of Thought; it is Sharpened Thinking; it is Power.
Paper presented at meeting of the International Statistical Institute, September 1953

Unfortunately and inadvertently, intellectual gulfs have grown up between writers in statistics, least squares, and curve fitting. Each of the three groups has gone its own way, rediscovering developments long since discovered by the others, or — what is worse — not rediscovering them.
Statistical Adjustment of Data (p. iv)
John Wiley & Sons, Inc. New York, New York, USA. 1938

Devons, Ely   1913–67
English economist

The experience of falling in love could be adequately described in terms of statistics. A record of heart beats per minute, the stammering and hesitation in speech, the number of calories consumed per day, the heightening of poetic vision, measured by the number of lines of poetry

written to the beloved — I won't go on; no doubt you can think of further measures.
Essays in Economics
Chapter 6 (p. 105)
Greenwood Press, Publishers, Westport, Connecticut, USA. 1961

The two most important characteristics of the language of statistics are first, that it describes things in quantitative terms, and second, that it gives this description an air of accuracy and precision.
Essays in Economics
Chapter 6 (p. 106)
Greenwood Press, Publishers, Westport, Connecticut, USA. 1961

How to use a language which by its very nature implies objectivity, precision and accuracy, in such a way that the subjective element of judgment, imprecision and inaccuracy are fully taken into account? It is because this task is so difficult and so rarely achieved that statistics are frequently referred to as "the hard facts", and yet we talk of three kinds of lies — "lies, damn lies, and statistics."
Essays in Economics
Chapter 6 (p. 111)
Greenwood Press, Publishers, Westport, Connecticut, USA. 1961

…"statistics are only for the statistician", and even then, I might add, only for the good statistician.
Essays in Economics
Chapter 6 (p. 118)
Greenwood Press, Publishers, Westport, Connecticut, USA. 1961

This exaggerated influence of statistics resulting from willingness, indeed eagerness, to be impressed by the "hard facts" provided by the "figures", may play an important role in decision-making.
Essays in Economics
Chapter 7 (p. 134)
Greenwood Press, Publishers, Westport, Connecticut, USA. 1961

Devons, Ely   1913–67
English economist

There are those who are so impressed by the notion that "quantification" is the only form of scientific knowledge, that they see no danger in the distorted, misleading, or simply ineffective picture that a statistical description of events may give. To such people the statistical picture is always to be preferred as the most meaningful and objective. It is indeed because this view is so widespread, that an argument stated in statistical terms has such a powerful influence in policy decision, and induces everyone to try to impress their case on public attention by peppering it with statistics.
Essays in Economics
Chapter 6 (p. 106)
Greenwood Press, Publishers. Westport, Connecticut, USA. 1961

Statistical magic, like its primitive counterpart, is a mystery to the public; and like primitive magic it can never be proved wrong.… The oracle is never wrong; a mistake merely reinforces the belief in magic.

Essays in Economics
Chapter 7 (p. 135)
Greenwood Press, Publishers. Westport, Connecticut, USA. 1961

Dewey, John   1859–1952
American philosopher and educator

Factual science may collect statistics, and make charts. But its predictions are, as has been well said, but past history reversed.
Art as Experience
Chapter XIV (p. 346)
Milton, Balch and Company. New York, New York, USA. 1934

Dickens, Charles   1812–70
English novelist

Mr. Gradgrind sat writing in the room with the deadly statistical clock, proving something no doubt — probably, in the main, that the Good Samaritan was a bad economist.
Hard Times
Book the Second, Chapter XII (p. 192)
J.M. Dent & Sons Ltd. London, England. 1966

Edgeworth, Francis Ysidro   1845–1926
Irish economist and statistician

…Statistics reigns and revels in the very heart of Physics.
On the Use of the Theory of Probabilities in Statistics Relating to Society
Journal of the Royal Statistical Society, January 1913 (p. 167)

Edwards, A. W. F.   1935–
English statistician, geneticist, and evolution biologist

There comes a time in the life of a scientist when he must convince himself either that his subject is so robust from a statistical point of view that the finer points of statistical inference he adopts are irrelevant or that the precise mode of inference he adopts is satisfactory.
Likelihood (p. xi)
Cambridge University Press. Cambridge, England. 1972

Edwards, Tyron   1809–94
American theologian

Statistics is the science of learning from experiences, especially experiences that arrive a little bit at a time.
An Introduction to the Bootstrap
Chapter I (p. 1)
Chapman & Hall. New York, New York, USA. 1993

Efron, Bradley   1938–
American statistician
Tibshirani, Robert J.
No biographical data available

Statistics is a subject of amazingly many users and surprisingly few effective practitioners.

An Introduction to the Bootstrap
Preface (p. xiv)
Chapman & Hall. New York, New York, USA. 1993

Einstein, Albert   1879–1955
German-born physicist
Infeld, Leopold   1898–1968
Polish physicist

By applying the statistical method we cannot foretell the behavior of an individual in a crowd. We can only foretell the chance, the probability, that it will behave in some particular manner.
The Evolution of Physics
Probability Waves (p. 284)
Simon & Schuster. New York, New York, USA. 1961

Eisenhart, Churchill   1913–94
American statistician

The primary function of a statistical consultant in a research organization is to furnish advice and guidance in the collection and use of numerical data to provide quantitative foundations for decisions.
The Role of a Statistical Consultant in a Research Organization
The American Statistician, Volume 2, Number 2, April 1948 (p. 6)

Ellis, Havelock   1859–1939
English sexuality researcher

…the methods of statistics are so variable and uncertain, so apt to be influenced by circumstances, that it is never possible to be sure that one is operating with figures of equal weight.
The Dance of Life
Chapter VII, I (p. 286)
Houghton Mifflin Company. Boston, Massachusetts, USA. 1923

Emerson, Ralph Waldo   1803–82
American lecturer, poet, and essayist

One more fagot of these adamantine bandages is the new science of Statistics.
Ralph Waldo Emerson: Essays and Lectures
The Conduct of Life
Fate (p. 950)
The Library of America. New York, New York, USA. 1983

Farr, William   1807–83
English statistician

You complain that your report would be dry. The dryer the better. Statistics should be the driest of all reading.
Nightingale on Quetelet
Journal of the Royal Statistical Society, Series A, 1981 (p. 144)

Ferguson, Kitty
Science writer

Compared with the adventure of discovering them, the statistics themselves often seem terribly dry.

Measuring the Universe: Our Historic Quest to Chart the Horizons of Space and Time
Prologue (p. 3)
Walker & Company. New York, New York, USA. 1999

Fienberg, Stephen E.   1942–
American statistician
Although advice on how and when to draw graphs is available, we have no theory of statistical graphics…
Graphical Methods in Statistics
The American Statistician, Volume 13, Number 4, November 1979
(p. 165)

Fisher, Sir Ronald Aylmer   1890–1962
English statistician and geneticist

This rather tumultuous overflow of statistical techniques from the quiet backwaters of theoretical methodology… into the working part of going concerns of the largest size, suggest that hidden causes have been at work…preparing men's minds, and shaping the institutions through which they work…
The Expansion of Statistics
American Scientist Magazine, Volume 42, Number 2, April 1954 (p. 277)

Statistical procedure and experimental design are only two different aspects of the same whole, and that whole is the logical requirements of the complete process of adding to natural knowledge by experimentation.
The Design of Experiments
Introduction (p. 3)
Hafner Publishing Company. New York, New York, USA. 1971

In the original sense of the word, "statistics" was the science of Statecraft: to the political arithmetician of the eighteenth century, its function was to be the eyes and ears of the central government.
Presidential Address, First Indian Statistical Conference
Sankhya, 1938, Volume 4, 1938 (p. 14)

Fitzgerald, F. Scott   1896–1940
American novelist and short story writer

I was counting the waves," replied Amory gravely, "I'm going in for statistics."
This Side of Paradise (p. 213)
Ann Arbor Media Group, LLC. Ann Arbor, Michigan, USA. 2006

Freeman, Linton C.
No biographical data available

We are all victims of statistics.
Elementary Applied Statistics
Section A (p. 1)
John Wiley & Sons, Inc. New York, New York, USA. 1965

Gallup, George   1901–84
American journalist and statistician

I could prove God statistically.
OMNI Magazine, Volume 2, Issue 2, November 1979 (p. 42)

Galton, Sir Francis   1822–1911
English anthropologist, explorer, and statistician

The object of statistical science is to discover methods of condensing information concerning large groups of allied facts into brief and compendious expressions suitable for discussion. The possibility of doing this is based on the constancy and continuity with which objects of the same species are found to vary.
Inquiries into Human Faculty and Its Development
Statistical Methods (p. 33)
AMS Press. New York, New York, USA. 1973

Gann, Ernest K.   1910–91
Author, sailor, fisherman, film producer, and airline captain

No, Mother dear, I do not hop into bed with every man I meet, despite your nasty little secret thoughts, but I do very much enjoy a more than occasional roll in the hay, which, if I have my statistics right, is a good deal more often than the average wife enjoys.
Brain 2000 (pp. 27–28)
Doubleday & Company, Inc. New York, New York, USA. 1980

Gissing, George   1857–1903
English novelist

…bits of jokes, bits of statistics, bits of foolery.
New Grub Street
The Sunny Way (p. 492)
The Modern Library. New York, New York, USA. 1926

Green, Celia   1935–
English philosopher and psychologist

When people talk about "the sanctity of the individual" they mean "the sanctity of the statistical norm."
The Decline and Fall of Science
Aphorisms (p. 4)
Hamilton. London, England. 1976

Greenwood, M.
No biographical data available

Sometimes a David felled a Goliath of a statistical difficulty with a smooth stone. It might take a mathematician to prove how truly the stone was aimed.
Discussion, to the paper "Some Aspects of the Teaching of Statistics"
Journal of the Royal Statistical Society, Volume 102, 1939 (p. 522)

Gregory, John   1724–73
Scottish physician and philosopher

The advancement of the sciences…requires only an attention to probabilities…a quick discernment where the greatest probability lies, and habits, of acting in consequence of this with facility and vigor.
Lectures on the Duties and Qualifications of a Physician (p. 164)
W. Strahan. London, England. 1772

Habera, Audrey
Statistician

Runyon, Richard P.
Statistician

When we can't prove our point through the use of sound reasoning, we fall back upon statistical "mumbo jumbo" to confuse and demoralize our opponents.
General Statistics
Chapter 1 (p. 3)
Addison-Wesley Publishing Company. Reading, Massachusetts, USA. 1973

Statistics is the refuge of the uninformed.
General Statistics
Chapter 1 (p. 3)
Addison-Wesley Publishing Company. Reading, Massachusetts, USA. 1973

Statistics is "hocus-pocus" with numbers.
General Statistics
Chapter 1 (p. 3)
Addison-Wesley Publishing Company. Reading, Massachusetts, USA. 1973

Hailey, Arthur   1920–2004
English/Canadian author

Legal proceedings are like statistics. If you manipulate them, you can prove anything.
Airport
Part 3, Chapter 11 (p. 385)
Doubleday & Company, Inc. Garden City, New York, USA. 1968

Hancock, William Keith   1808-?
Australian author

Oratory is dying; a calculating age has stabbed it to the heart with innumerable dagger-thrusts of statistics.
Australia (p. 146)
E. Benn Ltd. London, England. 1945

Hand, D. J.
No biographical data available

Statistics has been likened to a telescope. The latter enables one to see further and to make clear objects which were diminished or obscured by distance. The former enables one to discern structure and relationships which were distorted by other factors or obscured by random variation.
The Role of Statistics in Psychiatry
Psychological Medicine, Volume 15, 1985 (p. 471)

Hawkins, Francis Bisset   1796–1894
No biographical data available

Statistics has become the key to several sciences…and there is reason to believe, that a careful cultivation of it, would materially assist the completion of a philosophy of medicine.… Medical statistics affords the most convincing proofs of the efficacy of medicine.
Elements of Medical Statistics (pp. 2–3)
Longman. London, England. 1829

Hayford, F. Leslie
No biographical data available

In the everyday use of statistics in business, complicated statistical methods rarely are necessary and always are to be avoided if possible. Simplicity of treatment and presentation is a requisite in the making of statistics useful in executive control.
Some Uses of Statistics in Executive Control
Journal of the American Statistical Association, Volume 31, Number 193, March 1936 (p. 36)

…neither statistics nor the statistician can ordinarily give the executive the final answer to his problems.
Some Uses of Statistics in Executive Control
Journal of the American Statistical Association, Volume 31, Number 193, March 1936 (p. 36)

Heinlein, Robert A.   1907–88
American science fiction writer

Oh, the hell with! — it did not change the statistical outcome.
Time Enough for Love
Chapter VI (p. 208)
G.P. Putnam's Sons. New York, New York, USA. 1973

Hoel, P. G.
Statistician

Statistical methods are essentially methods for dealing with data that have been obtained by repetitive operations.
Introduction to Mathematical Statistics
Chapter 1 (p. 1)
John Wiley & Sons, Inc. New York, New York, USA. 1954

Hogben, Lancelot   1895–1975
English zoologist

Acceptability of a statistically significant result of an experiment on animal behavior in contradistinction to a result which the investigator can repeat before a critical audience naturally promotes a high output of publication. Hence, the argument that the techniques work has a tempting appeal to young biologists.
Statistical Theory: The Relationship of Probability, Credibility and Error (p. 27)
George Allen & Unwin Ltd. London, England. 1957

The word statistics has at least six different meanings in current use, four in the context of statistical theory alone.
Science in Authority
The Present Crisis in Statistical Theory (pp. 94–95)
Unwin University Books. London, England, USA. 1963

Holmes, Jr., Oliver Wendell   1841–1935
American jurist

For the rational study of the law the black-letter man may be the man of the present, but the man of the future is the man of statistics and the master of economics.

Path of the Law
The Harvard Law Review, Volume 10, 1897

Hooke, Robert   1635–1703
English physicist

Don't waste time arguing about the merits or demerits of something if you can gather some statistics that will answer the question realistically.
In J.M. Tanur
Statistics: A Guide to the Unknown
Statistics, Sports, and Some Other Things (p. 195)
Wadsworth & Brooks. Pacific Grove, California, ISA. 1989

Do remember that your experiment is merely a hodge-podge of statistics, consisting of those cases that you happen to remember. Because these are necessarily small in number and because your memory may be biased toward one result or another, your experience may be far less dependable than a good set of statistics.
In J.M. Tanur
Statistics: A Guide to the Unknown
Statistics, Sports, and Some Other Things (p. 195)
Wadsworth & Brooks. Pacific Grove, California, ISA. 1989

Hopkins, Harry
No biographical data available

Confidence in the omnicompetence of statistical reasoning grows by what it feeds on.
The Numbers Game: The Bland Totalitarianism
Chapter 6 (p. 132)
Little, Brown & Company. Boston, Massachusetts, USA. 1973

And when, in pursuit of the black cat of definitive truth, more refined techniques of statistical analysis, factor analysis, and so forth, are developed, the researcher is more and more distanced from the subject of his pursuit, and the real human world in which it exists. He raises as by a sort of mathematical levitation, into that other, finer sphere, where black cats are clawless, mewless and abstract...
The Numbers Game: The Bland Totalitarianism
Chapter 7 (p. 141)
Little, Brown & Company. Boston, Massachusetts, USA. 1973

You can't argue with statistics; generally you can't even get at them.
The Numbers Game: The Bland Totalitarianism
Chapter 11 (p. 232)
Little, Brown & Company. Boston, Massachusetts, USA. 1973

Horace (Quintus Horatius Flaccus)   65 BCE–8 BCE
Roman philosopher and dramatic critic

As they put it in Greek, we simply don't COUNT. We consume.
The Satires and Epistles of Horace
Epistle I, to Lollius Maximus

Hotelling, Harold   1895–1973
American mathematical economist

Research in statistical theory and technique is necessarily mathematical, scholarly, and abstract in character, requiring some degree of leisure and detachment, and access to a good mathematical and statistical library.
Memorandum to the Governor of India
24 February 1940

The purely random sample is the only kind that can be examined with entire confidence by means of statistical theory, but there is one thing wrong with it. It is so difficult and expensive to obtain for many uses that sheer cost eliminates it.
How to Lie with Statistics
Chapter 1 (p. 21)
W.W. Norton & Company, Inc. New York, New York, USA.1954

Huff, Darrell   1913–2001
American writer

The secret language of statistics, so appealing in a fact-minded culture, is employed to sensationalize, inflate, confuse, and oversimplify.
How to Lie with Statistics
Introduction (p. 8)
W.W. Norton & Company, Inc. New York, New York, USA.1954

A well-wrapped statistic is better than Hitler's "big lie"; it misleads, yet it cannot be pinned on you.
How to Lie with Statistics
Introduction (p. 9)
W.W. Norton & Company, Inc. New York, New York, USA.1954

Jahoda, Marie   1907–2001
Austrian social psychologist
Deutsch, Morton   1920–
American social psychologist

The use of available statistical records requires, first, that the social scientist be familiar with the better known sources of such data and that he display some ingenuity in discovering less obvious material.
Research Methods in Social Relations
Basic Process, Part I (p. 232)
Dryden Press. New York, New York, USA. 1951

Johnson, Lyndon B.   1908–73
36th president of the United States

The economy was never stronger in your lifetime. But statistics must not be sedatives. Economic power is important only as it is put to human use.
Speech
United Automobile Worker's Convention, Atlantic City, N.J., 23 March, 1964

Johnson, Palmer O.   1891–?
No biographical data available

There was a time when statistics as a tool in experimentation was almost completely ignored by the experimenter;

in fact, it was regarded [as] "introducing unnecessary confusion into otherwise plain issues."
Modern Statistical Science and its Function in Educational and Psychological Research
The Scientific Monthly, June 1951 (p. 385)

Jones, Franklin P.
No biographical data available

[S]tatistics had to be invented...because people were so unstable and irrational, taken one at a time.
The Non-Statistical Man (p. 15)
Belmont Books, New York, New York, USA. 1964

In statistics, you look for the common factor in order to lump otherwise dissimilar items in a single category.
The Non-Statistical Man (p. 17)
Belmont Books, New York, New York, USA. 1964

Statistical laws enable the insurance company to function, and make a profit for its shareholders. But what does statistics do for the policyholder? Not one damn thing!
The Non-Statistical Man (p. 32)
Belmont Books, New York, New York, USA. 1964

Sarah Bascomb was well aware that she didn't live in the same world with her husband, and that made it rather nice, she thought. It would have been exceedingly boring if they both talked of nothing but expectancy tables and statistical probabilities, or the PTA and young Chuck's music lessons.
The Non-Statistical Man (p. 10)
Belmont Books, New York, New York, USA. 1964

Kahn, S. J.
No biographical data available

Statistics in Israel today are like potatoes: they lie in the mud, but they're growing!
Menorah Journal, Volume 42, 1954 (p. 125)

Kaplan, Abraham   1918–93
American philosopher of science, author, and educator

...statistical techniques are tools of thought, and not substitutes for thought.
The Conduct of Inquiry: Methodology for Behavioral Science
Chapter VI, Section 29 (p. 257)
Chandler Publishing Company. San Francisco, California, USA. 1964

Kendall, Maurice G.   1907–83
Statistician
Stuart, A.   1932–
American theoretical and physical chemist

Statistics is the branch of scientific method which deals with the data obtained by counting or measuring the properties of populations of natural phenomena. In this definition "natural phenomena" includes all the happenings of the external world, either human or not.
The Advanced Theory of Statistics (Volume 1)

Chapter 1, Section 1.4 (p. 2)
Charles Griffin & Company Ltd. London, England. 1947

King, Willford   1880–1962
American economist and statistician

Archaeologists unearthed today in Babylon a remarkable set of clay tablets recording the minutes of the 1242 annual meeting of the Babylonical Statistical Association.
Consolidating Our Gains
Journal of the American Statistical Association, Volume 31, Number 193, March 1936 (p. 2)

Koshland, Jr., Daniel E.   1920–
American biochemist

Science. I'm afraid, Dr. Noitall, you do not have any understanding of statistics.
Editorial
Science, Volume 263, Number 5144, 14 January 1994 (p. 155)

Kruskal, William   1919–2005
American mathematician and statistician

It is all too easy to notice the statistical sea that supports our thoughts and actions. If that sea loses its buoyancy, it may take a long time to regain the lost support.
Coordination Today: A Disaster or a Disgrace
The American Statistician, Volume 37, Number 3, 1983 (p. 179)

What is there about the word "statistics" that so often provokes strained silence?
Statistics, Moliere, and Henry Adams
American Scientist Magazine (p. 416)

Statistics is the art of stating in precise terms that which one does not know.
Statistics, Moliere, and Henry Adams
American Scientist Magazine (p. 417)

...each of us has been doing statistics all his life, in the sense that each of us has been busily reaching conclusions based on empirical observations ever since birth.
Statistics, Moliere, and Henry Adams
American Scientist Magazine (p. 417)

LaGuardia, Fiorello   1882–1947
American civil servant, congressman, and New York City mayor

Statistics are like alienists — they will testify for either side.
The Banking Investigation
Liberty, May 13, 1933

Lang, Andrew   1844–1912
Scottish scholar and man of letters

He uses statistics as a drunken man uses lamp-posts — for support rather than illumination.
In Evan Esar
The Dictionary of Humorous Quotations
Doubleday & Company, Inc. Garden City, New York, USA. 1949

Lapin, Lawrence
No biographical data available

Statistics is a body of methods and theory applied to numerical evidence in making decisions in the face of uncertainty.
Statistics for Modern Business Decisions
Chapter I (p. 2)
Harcourt Brace Jovanovich, Inc. New York, New York, USA. 1973

Leacock, Stephen   1869–1944
Canadian humorist

"I've been reading some very interesting statistics," he was saying to the other thinker.
"Ah, statistics!" said the other, "wonderful things, sir, statistics; very fond of them myself."
Literary Lapses
A Force of Statistics (p. 74)
John Lane. London, England. 1911

Lewis, Clarence Irving   1883–1964
American philosopher

…the statistical prediction of the future from the past cannot be generally valid, because whatever is future to any given past, is in turn past for some future. That is, whoever continually revises his judgment of the probability of a statistical generalization by its successively observed verifications and failures, cannot fail to make more successful predictions than if he should disregard the past in his anticipation of the future. This might be called the "Principle of statistical accumulation."
Mind and the World-Order: Outline of a Theory of Knowledge
Chapter XI (p. 386)
Charles Scribner's Sons. New York, New York, USA. 1929

Lippmann, Walter   1889–1974
American journalist and author

The statistical method is of use only to those who have found it out.
A Preface to Politics
The Golden Rule and After (p. 92)
M. Kennerley. New York, New York, USA. 1913

Statistics then is no automatic device for measuring facts.
A Preface to Politics
The Golden Rule and After (p. 92)
M. Kennerley. New York, New York, USA. 1913

Even the most refined statistics are nothing but abstractions.
A Preface to Politics
The Golden Rule and After (pp. 93–94)
M. Kennerley. New York, New York, USA. 1913

…all statistical devices are open to abuse and require constant correction.
A Preface to Politics
The Golden Rule and After (p. 91)
M. Kennerley. New York, New York, USA. 1913

Lloyd George, David, 1st Earl of Dwfor   1863–1945
English prime minister

You can't feed the hungry on statistics.
Advocating Tariff Reform
Speech 1904

Longacre, William A.   1937–
American anthropologist

…statistical techniques are not magical.
Current Thinking in American Archaeology
Bulletin of the American Anthropological Association, Volume 3, Number 3, Part 2, 1970 (p. 132)

Louis, Pierre-Charles-Alexandre   1787–1872
French physician

As to different methods of treatment, it is possible for us to assure ourselves of the superiority of one or other…by enquiring if the greater number of individuals have been cured by one means than another. Here it is necessary to count. And it is, in great part at least, because hitherto this method has not at all, or rarely been employed, that the science of therapeutics is so uncertain.
Translated by P. Martin
Essay on Clinical Instruction (pp. 26, 28)
S. Highley. London, England. 1834

Ludlum, Robert   1927–2001
American writer

"There are three major and perhaps a dozen minor rental agencies, not counting the hotels, which we've covered separately. These are manageable statistics, but, of course, the garages are not."
The Bourne Supremacy
Chapter 18 (p. 260)
Random House, Inc. New York, New York, USA. 1986

Daniel's a statistician. He sees numbers — fractions, equations, totals — and they spell out the odds for him. God knows he's brilliant at it; he's saved the lives of hundreds with those statistics.
The Parsifal Mosaic
Chapter 10 (p. 137)

I don't believe you. Not because you're a poor liar, but because it doesn't conform with the facts. I work with statistics, Mr. Washburn, or Mr. Bourne, or whatever your name is. I respect observable data and I can spot inaccuracies; I'm trained to do that.
The Bourne Identity
Chapter 9 (p. 128)
Richard Marek Publishers. New York, New York, USA. 1980

Death is a statistic for the computers.
The Bourne Identity
Chapter 29 (p. 401)
Richard Marek Publishers. New York, New York, USA. 1980

Marshall, Alfred 1842–1924
English economist

Statistics are the straw out of which I, like every other economist, have to make the bricks.
In Arthur L. Bowley
Elements of Statistics
Part I, Chapter I (p. 8)
P.S. King & Son Ltd. London, England. 1926

Maxwell, James Clerk 1831–79
Scottish physicist

…molecular science teaches us that our experiments can never give us anything more than statistical information, and that no law deduced from them can pretend to absolute precision. But when we pass from the contemplations of our experiment to that of the molecules themselves, we leave the world of chance and change, and enter a region where everything is certain and immutable.
In W.D. Niven (ed.)
The Scientific Papers of James Clerk Maxwell (Volume 2)
Molecules (p. 374)
At The University Press. Cambridge, England. 1890

Meitzen, August 1822–1910
German geographer

No statistical judgment deals with the unit, but strictly and only with the aggregate. The variable elements of persons and things otherwise typical, that are enumerated, are always counted in a specific aggregate and under certain specific circumstances. The qualities of the objects themselves, so far as they are not typical, or the subject of the investigation, are completely unknown.
History, Theory and Techniques of Statistics
American Academy of Political and Social Sciences, May 1898 (p. 168)

No matter what the statistical problem may be, it must proceed according to a plan. It is always a specific question which may be answered in several more or less accurate ways. The end in view and the reasoning which can be drawn upon will indicate in which manner and within which limits the answer is to be given. According to the choice made, it may be very simple or very complicated. But under all circumstances a definite plan providing for all the detail is an absolute prerequisite.
History, Theory and Techniques of Statistics *American Academy of Political and Social Sciences*, May 1898 (p. 168)

Meyers, Jr., G. J.
No biographical data available

Statistical methods serve as land marks which point to further improvement beyond that deemed obtainable by experienced manufacturing men. Hence, after all obvious correctives have been exhausted and all normal logic indicates no further gain is to be made, statistical methods still point toward a reasonable chance for yet further gains; thereby giving the man who is doing trouble shooting sufficient courage of his convictions to cause him to continue to the ultimate gain, in spite of expressed opinion on all sides that no such gain exists.
American Society of Mechanical Engineers, Discussion of E.G. Olds,
On Some of the Essentials of the Control Chart Analysis
Transactions, Volume 64, July 1942

Moroney, M. J.
American statistician

The organized charity, scrimped and iced,
O'Reilly, John Boyle
In the name of a cautious, statistical Christ.
In Bohemia
In Bohemia

A statistical analysis, properly conducted, is a delicate dissection of uncertainties, a surgery of suppositions.
Facts from Figures
Statistics Undesirable (p. 3)
Penguin Books Ltd. Harmondsworth, England. 1951

Historically, Statistics is no more than State Arithmetic, a system of computation by which differences between individuals are eliminated by the taking of an average. It has been used — indeed, still is used — to enable rulers to know just how far they may safely go in picking the pockets of their subjects.
Facts from Figures
Statistics Undesirable (p. 1)
Penguin Books Ltd. Harmondsworth, England. 1951

If you are young, then I say: Learn something about statistics as soon as you can. Don't dismiss it through ignorance or because it calls for thought…. If you are older and already crowned with the laurels of success, see to it that those under your wing who look to you for advice are encouraged to look into this subject. In this way you will show that your arteries are not yet hardened, and you will be able to reap the benefits without doing overmuch work yourself. Whoever you are, if your work calls for the interpretation of data, you may be able to do without statistics, but you won't do as well.
Facts from Figures
Statistics Desirable (p. 463)
Penguin Books Ltd., Harmondsworth, England. 1951

Mr. Gregory
Fictional character

Well, statistics prove that your far safer in a modern plane than in a bathtub.
Charlie Chan at Treasure Island
Film (1939)

Neuman, James R.
No biographical data available

Statistics was founded by John Graunt of London, a "haberdasher of small-wares" in a tiny book called

Natural and Political Observations Made upon the Bills of Mortality.
The World of Mathematics (Volume 3)
Commentary on an Ingenious Army Captain and on a Generous and Many-sided Man (p. 1416)
Simon & Schuster. New York, New York, USA. 1956

O. Henry (William Sydney Porter)   1862–10
American short story writer and journalist

His mathematics carried with it a momentary qualm and a lesson. The thought had not occurred to him that the thought could possibly occur to me not to ride at his side on that red road to revenge and justice. It was the higher calculus. I was booked for the trail. I began to eat more beans.
Tales of O. Henry
A Technical Error (p. 1059)
Doubleday & Company, Inc. Garden City, New York, USA. 1953

"What you've got," says Idaho, "is statistics, the lowest grade of information that exists. They poison your mind…"
Tales of O. Henry
The Handbook of Hymen (p. 113)
Doubleday & Company, Inc. Garden City, New York, USA. 1969

Orwell, George (Eric Arthur Blair)   1903–50
English novelist and essayist

The fabulous statistics continued to pour out of the telescreen. As compared with last year there was more food, more clothes, more houses, more furniture, more cooking pots, more fuel, more ships, more helicopters, more books, more babies — more of everything except disease, crime, and insanity.
Nineteen Eighty-Four
Part One, Chapter V (p. 59)
Buccaneer Books. Cutchogue, New York, USA. 1949

Statistics were just as much a fantasy in their original version as in their rectified version. A great deal of the time you were expected to make them up out of your head. For example, the Ministry of Plenty's forecast had estimated the output of boots for the quarter at a hundred and forty-five millions pairs. The actual output was given as sixty-two millions. Winston, however, in rewriting the forecast, marked the figure down to fifty-seven millions, so as to allow for the usual claim that the quota had been overfilled. In any case, sixty-two millions was no nearer the truth than fifty-seven millions, or a hundred and forty-five millions. Very likely no boots had been produced at all. Likelier still, nobody knew how many had been produced, much less cared.
Nineteen Eighty-Four
Part One, Chapter IV (pp. 41–42)
Buccaneer Books. Cutchogue, New York, USA. 1949

Paulos, John Allen   1945–
American mathematician

It's been estimated that, because of the exponential growth of the world's population, between 10 and 20 percent of all the human beings who have ever lived are alive now. If this is so, does this mean that there isn't enough statistical evidence to conclusively reject the hypothesis of immortality?
Innumeracy
5 Statistics, Trade-offs, and Society (p. 99)
Hill & Wang. New York, New York, USA. 1988

Pearson, E. S.   1895–1980
English statistician
Hartley, H. Q.
No biographical data available

…it is a function of statistical method to emphasize that precise conclusions cannot be drawn from inadequate data.
Biometrika Tables for Statisticians (Volume 1) (p. 83)

Pearson, Karl   1857–1936
English mathematician

There is much value in the idea of the ultimate laws being statistical laws, though why the fluctuations should be attributed to a Lucretian "Chance", I cannot say. It is not an exactly dignified conception of the Deity to suppose him occupied solely with first moments and neglecting second and higher moments!
The History of Statistics in the 17th and 18th Centuries Against the Changing Background of Intellectual, Scientific, and Religious Thought (p. 160)

[Florence Nightingale's] statistics were more than a study, they were indeed her religion. For her Quetelet was the hero as scientist, and the presentation copy of his Physique sociale is annotated by her on every page. Florence Nightingale believed — and in all the actions of her life acted upon the belief — that the administrator could only be successful if he were guided by statistical knowledge. The legislator — to say nothing of the politician — too often failed for want of this knowledge.
Life, Letters and Labours of Francis Galton (Volume 2) (p. 57)
At The University Press. Cambridge, England. 1914–30

[Florence Nightingale]…held that the universe — including human communities — were evolving in accordance with a divine plan; that it was man's business to endeavor to understand this plan and guide his actions in sympathy with it. But to understand God's thoughts, she held we must study statistics, for these are the measure of His purpose. Thus, the study of statistics was for her a religious duty.
Life, Letters and Labours of Francis Galton (Volume 2) (p. 57)
At The University Press. Cambridge, England. 1914–30

Perrin, Jean   1870–1945
French physicist

It is thus that statistics reveals more and more the inconstancy and the irregularity of much social phenomena, when in lieu of applying it to a great nation altogether, one descends to a province, a town, a village.
In Mary Jo Nye
Molecular Reality: A Perspective on the Scientific Work of Jean Perrin (p. 25)
MacDonald. London, England. 1972

Playfair, William 1759–1823
English publicist

No study is less alluring or more dry and tedious than statistics, unless the mind and imagination are set to, or that the person studying is particularly interested in, the subject; which last can seldom be the case with young men in any rank of life.
The Statistical Breviary (p. 16)
J. Wallis. London, England. 1801

Statistical knowledge, though in some degree searched after in the most early ages of the world, has not till within these last 50 years become a regular object of study.
The Statistical Breviary
J. Wallis. London, England. 1801

Price, Derek John de Solla 1922–83
English science historian and information scientist

His passion was to count everything and reduce it to statistics.
Little Science, Big Science
Chapter 2 (p. 33)
Columbia University Press. New York, New York, USA. 1963

Proschan, Frank
American statistician

Pronouncing each word with great deliberateness, Rep. Resent asked, "Are you now, or have you ever been, a member of the American Statistical Association?"

...

Looking Rep. Resent straight in the eye, Minnie defiantly replied, "I refuse to answer on the grounds that it might incriminate me."
Investigation of Latin Squares
Industrial Quality Control, Volume 11, Number 1, July 1954 (p. 31)

Puckett, Andrew
British crime novelist

They were in monthly columns. I added them and then compared the two tables. Well, there was a difference, and a difference on the right side, more blood packs had been separated in the Centre than plasma packs had arrived in CPPL, but it wasn't as large as I would have thought. I stared at the figures for a moment, then I worked out a statistical error rate on them. The difference between

them was sot significant; it could be explained by random error. Statistics don't lie, not in the right hands.
Bloodstains (p. 79)
Doubleday & Company, Inc. Garden City, New York, USA. 1987

For the first five months they were virtually identical, but for the past four, they showed an increasing difference! With shaking fingers, I worked out a Standard Deviation on the sets of totals. There was no doubt: the difference[s] between the Centre's and CPPL's totals were significant. Statistics don't lie...
Bloodstains (p. 80)
Doubleday & Company, Inc. Garden City, New York, USA. 1987

Puzo, Mario 1920–99
American novelist and screenwriter

"You got a ninety percent chance," he said.
Osno said quickly, "How do you get that figure?" He always did that whenever somebody pulled a statistic on him. He hated statisticians.
Fools Die: A Novel
Chapter 24 (p. 270)
Putnam. New York, New York, USA. 1978

Pynchon, Thomas 1937–
American novelist

"I'm sorry. That's the Monte Carlo Fallacy. No matter how many have fallen inside a particular square, the odds remain the same as they always were. Each hit is independent of all the others. Bombs are not dogs. No link. No memory. No conditioning."
Gravity's Rainbow
Part 1 (p. 56)
The Viking Press. New York, New York, USA. 1973

That he must always be lovable, in need of her and never, as now, the hovering statistical cherub who's never quite been to hell but speaks as if he's one of the most fallen.
Gravity's Rainbow
Part 1 (p. 57)
The Viking Press. New York, New York, USA. 1973

Reynolds, H. T.
No biographical data available

...statistics — whatever their mathematical sophistication and elegance — cannot make bad variables into good ones.
Analysis of Nominal Data
Chapter 1 (p. 8)
Sage Publications. Beverly Hills, California, USA. 1977

Rogers, Will 1879–1935
American actor and humorist

Everything is figured out down to a Gnat's tooth according to some kind of statistics.
The Writings of Will Rogers
Volume 4-3 (p. 254)
Oklahoma State University Press. Stillwater, Oklahoma, USA. 1973

Romanoff, Alexis Lawrence   1892–1980
Russian soldier and scientist

Statistics can provide a ready proof
For doubtful facts which ought to stay aloof.
Encyclopedia of Thoughts
Couplets
Ithaca Heritage Books. Ithaca, New York, USA. 1975

Russell, Bertrand Arthur William   1872–1970
English philosopher, logician, and social reformer

Statistics, ideally, are accurate laws about large groups;
they differ from other laws only in being about groups,
not about individuals.
The Analysis of Matter
Chapter XIX (p. 191)
Harcourt, Brace & Company, Inc. New York, New York, USA. 1927

Samuels, Ernest   1903–96
American biographer and lawyer

No honest historian can take part with — or against —
the forces he has to study. To him even the extinction of
the human race should be merely a fact to be grouped
with other vital statistics.
In Ernest Samuels (ed.)
The Education of Henry Adams
Chapter XXX (p. 447)
Houghton Mifflin Company. Boston, Massachusetts, USA. 1974

Taking for granted that the alternative to art was arithmetic, he plunged deep into statistics, fancying that
education would find the surest bottom there; and the
study proved the easiest he had ever approached. Even
the Government volunteered unlimited statistics, endless
columns of figures, bottomless averages merely for the
asking. At the Statistical Bureau, Worthington Ford supplied any material that curiosity could imagine for filling
the vast gaps of ignorance, and methods for applying the
plasters of fact.
In Ernest Samuels (ed.)
The Education of Henry Adams
Chapter XXIII (p. 351)
Houghton Mifflin Company. Boston, Massachusetts, USA. 1974

Schölzer, Ludwig   1735–1809
No biographical data available

History is statistics in a state of progression; statistics is
history at a stand.
Transactions of the Statistical Society of London
Article II, Footnote on page 72
Westminster Review, Volume 1, Part I, April-August 1838

History is for him continuous statistics, statistics
stationary history.
In August Meitzen
History, Theory, and Technique of Statistics (p. 37)
American Academy of Political Science. Philadelphia, Pennsylvania,
USA. 1891

Segal, Erich   1937–
American novelist

The emergency room was a madhouse. The stormy holi-
day roads had yielded more than the statistical expecta-
tion of traffic accidents.
Man, Woman and Child
Chapter 26 (p. 191)
Harper & Row, Publishers. New York, New York, USA. 1980

"How are you, Mrs. Coleman?"
"Not too bad. How's yer statistics?"
Man, Woman and Child
Chapter 1 (p. 8)
Harper & Row, Publishers. New York, New York, USA. 1980

He turned over on his side and picked up the American
Journal of Statistics. Better than a sleeping pill. He idly
leafed through a particularly unoriginal piece on stochastic
processes, and thought, Christ, I've said this stuff a million
times. And then he realized that he himself was the author.
Man, Woman and Child
Chapter 5 (p. 42)
Harper & Row, Publishers. New York, New York, USA. 1980

"I mean, here you are a professor of statistics."
"So?"
"So you have one lousy affair in your whole life.
For a few lousy days. And you get a kid as evidence.
Christ, what are the odds of that happening to any-
body?"
"Oh," said Bob bitterly, "about a billion to one."
Man, Woman and Child
Chapter 13 (p. 109)
Harper & Row, Publishers. New York, New York, USA. 1980

"I am Professor Beckworth," he pronounced in a kind
of soprano-baritone. "Would you like to ask me some
statistics, sir?"
"Yes," replied Bob. "What are the chances of this damn
rain stopping today, Professor?"
"Mmm," said Jean-Claude, pondering earnestly, "You'll
have to see me tomorrow about that."
Man, Woman and Child (p. 178)
Harper & Row, Publishers. New York, New York, USA. 1980

Shapiro, Karl Jay   1913–2000
American poet

We ask for no statistics of the killed,
For nothing political impinges on
This single Casualty, or all those gone,
Missing or healing, sinking or dispersed,
Hundreds of thousands counted, millions lost.
Collected Poems 1940–1978
Elegy for a Dead Soldier
Stanza V (p. 90)
Random House, Inc. New York, New York, USA. 1978

Shaw, George Bernard   1856–1950
Irish comic dramatist and literary critic

Even trained statisticians often fail to appreciate the extent to which statistics are vitiated by the unrecorded assumptions of their interpreters.
The Doctor's Dilemma
Preface on Doctors
Statistical Illusions (p. lxii)
Brentano's. New York, New York, USA. 1920

Smith, Reginald H.   1889–1966
American lawyer and social activist

Lawyers like words and dislike statistics.
A Sequel: The Bar Is Not Overcrowded
American Bar Association Journal, Volume 45, September 1959 (p. 945)

Spengler, Oswald   1880–1936
German philosopher

Statistics belong, like chronology, to the domain of the organic, to fluctuating Life, to Destiny and Incident, and not to the world of laws and timeless causality.
The Decline of the West
Chapter X (p. 218)
Alfred A. Knopf. New York, New York, USA. 1962

Stalin, Joseph   1879–1953
Soviet Russian political leader and general secretary of Communist Party

A single death is a tragedy, a million deaths is a statistic.
Quoted by Anne Fremantle
The New York Times Book Review
Unwritten Pages at the End of the Diary, September 28, 1958 (p. 3)

Stamaty, Mark Alan
American cartoonist and children's book writer

I propose that infinitely refutable statistics be declared the official language of politics.
Washingtoon
Time, September 25, 1995 (p. 21)

Stamp, Josiah   1880–1941
English economist and financier

You cannot escape the statistical method, so you may as well make friend with it. You think it is cold and inhuman and impersonal, but, as a matter of fact, it is fuller of red blood and human nature than half the descriptive literature in the world.
Some Economic Factors in Modern Life
Chapter VIII (p. 256)
P.S. King & Son Ltd. London, England. 1929

Stekel, Wilhelm   1868–1940
Austrian psychoanalyst

Statistics is the art of lying by means of figures.
Marriage at the Crossroads
Chapter II (p. 20)
W. Godwin, Inc. New York, New York, USA. 1931

Sterne, Laurence   1713–68
English novelist and humorist

It was demonstrated however very satisfactorily, that such a ponderous mass of heterogeneous matter could not be congested and conglomerated to the nose, whilst the infant was *in Utero*, without destroying the statistical balance of the foetus, and throwing it plump upon its head nine months before the time.
The Life and Opinions of Tristram Shandy, Gentleman, and A Sentimental Journey Through France and Italy (Volume 1)
Book IV (p. 228)
Macmillan & Company Ltd. London, England. 1900

Stevenson, Robert Louis   1850–94
Scottish essayist and poet

Here he comes, big with statistics…
The Complete Poems of Robert Louis Stevenson
Troubled and sharp about fac's, LXVI
Charles Scribner's Sons. New York, New York, USA. 1923

Stigler, Stephen M.   1941–
American historian and statistician

…elementary statistics texts tell us that the method of least squares was first discovered about 1805. Whether it had one or two or more discoverers can be argued; still the method dates from no later than 1805. We also read that Sir Francis Galton discovered regression about 1885, in studies of heredity. Already we have a puzzle — a modern course in regression analysis is concerned almost entirely with the method of least squares and its variations. How could the core of such a course date from both 1805 and 1885? Is there more than one way a sum of squared deviations can be made small?
The History of Statistics
Introduction (p. 2)
Harvard University Press. Cambridge, Massachusetts, USA. 1986

Stout, Rex   1886–1975
American writer

There are two kinds of statistics, the kind you look up and the kind you make up.
Death of a Doxy (p. 90)
Bantam Book. New York, New York, USA. 1967

Statistics show that seventy-four per cent of wives open letters, with or without a teakettle.
Death of a Doxy (p. 120)
Bantam Book. New York, New York, USA. 1967

Strunsky, Simeon   1879–1948
American essayist

Statistics are the heart of democracy.
Topics of The Times, November 30, 1944

Tchekhov, Anton   1860–1904
Russian writer

Everything is quiet, peaceful and against it all is only the silent protest of statistics…
Tchekhov's Plays and Stories
Gooseberries
J.M. Dent & Sons Limited. London, England. 1958

The Editors

To some people, statistics is "quartered pies, cute little battleships and tapering rows of sturdy soldiers in diversified uniforms." To others, it is columns and columns of numerical facts. Many regard it as a branch of economics. The beginning student of the subject considers it to be largely mathematics.
Statistics, the Physical Sciences and Engineering
The American Statistician, Volume 11, Number 4, August 1948

Thoreau, Henry David   1817–62
American essayist, poet, and practical philosopher

But lo! men have become the tools of their tools.
The Writings of Henry David Thoreau (Volume 2)
Walden
Chapter I (p. 61)
Houghton Mifflin Company. Boston, Massachusetts, USA. 1893

Thorn, John   1947–
Sports historian

While he is racing to the hole, the shortstop is figuring: Based on the speed of the runners and how hard the ball is hit, he probably has no chance of a double play; he may have a little chance of a play at second; and he almost certainly has no play at first. He throws to third because the distance from the hole to the bag is short, his calculation of the various probabilities led him to conclude that this was his "percentage play." Now not so much as a glimmer of any number entered the shortstop's head in this time, yet he was thinking statistically.
The Hidden Game of Baseball: A Revolutionary Approach to Baseball and Its Statistics (p. 5)
Doubleday & Company Inc. Garden City, New York, USA. 1984

Thurstone, Louis Leon   1887–1955
American psychologist and psychometrician

Factor analysis is useful especially in those domains where basic and fruitful concepts are essentially lacking and where crucial experiments have been difficult to conceive. … They enable us to make only the crudest first map of a new domain. But if we have scientific intuition and sufficient ingenuity, the rough factorial map of a new domain will enable us to proceed beyond the factorial stage to the more direct form of psychological exploration in the laboratory.
Current Issues in Factor Analysis
Psychological Bulletin, Volume 37, April 1940 (p. 189)

It is not wise for a statistician who knows factor analysis to attempt problems in a science which he has not himself mastered.

Current Issues in Factor Analysis
Psychological Bulletin, Volume 37, April 1940 (p. 235)

Trollope, Anthony   1815–82
English novelist

We have no statistics to tell us whether there be any such disproportion in class where men do not die early from overwork.
The Eustace Diamond (Volume 1)
Chapter XXIV (p. 223)
Oxford University Press, Inc. London, England. 1973

As one of the legislators of the country I am prepared to state that statistics are always false.
The Eustace Diamond (Volume 1)
Chapter XXIV (p. 223)
Oxford University Press, Inc. London, England. 1973

Tukey, John W.   1915–2000
American statistician

A sort of question that is inevitable is: "someone taught my students exploratory, and now (boo hoo) they want me to tell them how to assess significance or confidence for all these unusual functions of the data. (Oh, what can we do?)" To this there is an easy answer: TEACH them the JACKKNIFE.
The American Statistician, Volume 34, Number 1, February 1980 (p. 25)

Twain, Mark (Samuel Langhorne Clemens)   1835–1910
American author and humorist

Personally, I never care for fiction or story-books. What I like to read about are facts and statistics of any kind.
In Rudyard Kipling
From Sea to Sea
An Interview with Mark Twain
Macmillan & Company Ltd. London, England. 1900

Sometimes, half a dozen figures will reveal, as with a lighting-flash, the importance of a subject which ten thousand labored words with the same purpose in view, had left at last but dim and uncertain.
Life on the Mississippi
Chapter XXVIII (p. 241)
Harper & Row, Publishers. New York, New York, USA. 1951

July 4. Statistics show that we lose more fools on this day than in all the other days of the year put together.
Pudd'nhead Wilson
Chapter XVII (p. 164)
Harper & Brothers Publishers. New York, New York, USA. 1904

I was deducing from the above that I have been slowing down steadily in these thirty-six years, but I perceive that my statistics have a defect: 3,000 words in the spring of 1868, when I was working seven or eight or nine hours at a sitting, has little or no advantage over the sitting of today, covering half the time and producing half the output. Figures often beguile me, particularly when I have the arranging of them myself; in which case the remark attributed to Disraeli would often apply with justice and force:

"There are three kinds of lies: lies, damned lies, and statistics."

In Albert Bigelow Paine (ed.)
Mark Twain's Autobiography (Volume 1)
Chapters Added in Florence (p. 246)
Harper & Brothers Publishers. New York, New York, USA. 1924

van der Post, Laurens   1906–96
Afrikaner author

Thinking has its place...but, only when one is confronted with known facts and statistics. When you're in the unknown and the dark...you surrender your thinking in trust to the feelings that come to you out of the bush.
A Far-Off Place
Chapter 9 (p. 183)
The Hogarth Press. London, England. 1974

von Mises, Richard   1883–1953
Austrian-born American mathematician

The problems of statistical physics are of the greatest in our time, since they lead to a revolutionary change in our whole conception of the universe.
Probability, Statistics, and Truth
Sixth Lecture (p. 219)
Dover Publications, Inc. New York, New York, USA. 1981

Walcott, Derek   1930–
West Indian dramatist and poet

Statistics justify and scholars seize
The salients of colonial policy.
Collected Poems
A Far Cry from Africa, l. 7–8
Farrar, Straus & Giroux. New York, New York, USA. 1986

Walker, Marshall John
American physicist

Statistics provides a quantitative example of the scientific process usually described qualitatively by saying that scientists observe nature, study the measurements, postulate models to predict new measurements, and validate the model by the success of prediction.
The Nature of Scientific Thought
Chapter IV (p. 46)
Prentice-Hall, Inc., Englewood Cliffs, New Jersey, USA. 1963

Mathematical statistics provides an exceptionally clear example of the relationship between mathematics and the external world. The external world provides the experimentally measured distribution curve; mathematics provides the equation (the mathematical model) that corresponds to the empirical curve. The statistician may be guided by a thought experiment in finding the corresponding equation.

The Nature of Scientific Thought
Chapter IV (p. 50)
Prentice-Hall, Inc., Englewood Cliffs, New Jersey, USA. 1963

Wang, Chamont   1949–
American statistician

Statistics has been called a science. It is said to connect its facts by a chain of causation: if it did so, it would be a science, though even then not a distinct and separate science. But the observations of astronomy may be called the science of astronomy as properly as statistics may be denominated a science. No mere record and arrangement of facts can constitute a science.
Transactions of the Statistical Society of London, Art II
Westminster Review, Volume I, Part I, 1838 (p. 69)

As a matter of fact, the whole notion of "statistical inference" often is more of a plague and less of a blessing to research workers.
Sense and Nonsense of Statistical Inference: Controversy, Misuse, and Subtlety
Chapter 2 (p. 29)
Marcel Dekker. New York, New York, USA. 1993

But statistics is not a science, and cannot be one. Studied as the statistical council have decreed statistics shall be studied, no department of human knowledge ever could become a science — a collection of theories — because they have put their veto on theorizing. But statistics is not even a department of human knowledge; it is merely a form of knowledge — a mode of arranging and stating facts which belong to various sciences.
Transactions of the Statistical Society of London, Art II
Westminster Review, Volume I, Part I, 1838 (p. 70)

Waugh, Evelyn   1903–66
English author of satirical novels

O god thou has appointed three score years and ten as man's allotted span but O god statistics go to prove that comparatively few ever attain that age...
In Mark Amory
The Letters of Evelyn Waugh
Letter to Laura Herbert, dated October 1935 (p. 99)
Weidenfeld & Nicolson. London, England. 1980

Wells, H. G. (Herbert George)   1866–1946
English novelist, historian and sociologist

Satan delights equally in statistics and in quoting scripture.
The Undying Fire
Chapter the First Section 3 (p. 9)
The Macmillan Company. New York, New York, USA. 1919

Statistical thinking will one day be as necessary for efficient citizenship as the ability to read and write.
In Warren Weaver
Statistics
Scientific American, Volume 186, Number 1, January 1952 (p. 60)

White, William Frank   1872–1952
No biographical data available

Just as data gathered by an incompetent observer are worthless — or by a biased observer, unless the bias can be measured and eliminated from the result — so also conclusions obtained from even the best data by one unacquainted with the principles of statistics must be of doubtful value.
A Scrap-Book of Elementary Mathematics: Notes, Recreations, Essays
The Mathematical Treatment of Statistics (p. 156)
The Open Court Publishing Company. La Salle, Illinois, USA. 1942

Whitehead, Alfred North   1861–1947
English mathematician and philosopher

There is a curious misconception that somehow the mathematical mysteries of Statistics help Positivism to evade its proper limitation to the observed past. But statistics tell you nothing about the future unless you make the assumption of the permanence of statistical form. For example, in order to use statistics for prediction, assumptions are wanted as to the stability of the mean, the mode, the probable error, and the symmetry or skewness of the statistical expression of functional correlation.
Adventures of Ideas
Chapter VIII (p. 161)
The Macmillan Company. New York, New York, USA. 1956

Wigner, Eugene Paul   1902–95
Hungarian-born American physicist

With classical thermodynamics, one can calculate almost everything crudely; with kinetic theory, one can calculate fewer things, but more accurately; and with statistical mechanics, one can calculate almost nothing exactly.
In Edward B. Stuart, Benjamin Gal-Or, and Alan J. Brainard (eds.)
A Critical Review of Thermodynamics (p. 205)
Publisher undetermined

Wilson, Edwin B.   1879–1964
American statistician

Figures may not lie, but statistics compiled unscientifically and analyzed incompetently are almost sure to be misleading, and when this condition is unnecessarily chronic the so-called statisticians may be called liars.
Bulletin of the American Mathematical Society, Volume 18, 1912

Wolfowitz, J.   1910–
No biographical data available

Except perhaps for a few of the deepest theorems, and perhaps not even these, most of the theorems of statistics would not survive in mathematics if the subject of statistics itself were to die out. In order to survive the subject must be more responsive to the needs of application.
In R.C. Bose and others (eds.)
Essays In Probability and Statistics (p. 748)
University of North Carolina Press. Chapel Hill, North Carolina, USA. 1970

Wonnacott, Ronald J.
No biographical data available

"Those Platonists are a curse." he said,
"God's fire upon the wan,
A diagram hung there instead,
More women born than men."
The Collected Poems of W.B. Yeats
Statistics
The Macmillan Company. New York, New York, USA. 1940

Yates, Frances   1899–1981
English historian

It is very easy to devise different tests which, on the average, have similar properties, …they behave satisfactorily when the null hypothesis is true and have approximately the same power of detecting departures from that hypothesis. Two such tests may, however, give very different results when applied to a given set of data. The situation leads to a good deal of contention amongst statisticians and much discredit of the science of statistics. The appalling position can easily arise in which one can get any answer one wants if only one goes around to a large enough number of statisticians.
Discussion on the Paper by Dr. Box and Dr. Andersen
Journal of the Royal Statistical Society, Series B, Volume 17, 1955 (p. 31)

STATISTICS AND MEDICINE

Bernard, Claude   1813–78
French physiologist

[S]tatistics…are given a great role in medicine, and they therefore raise a medical question which we should examine here. The first requirement in using statistics is that the facts treated shall be reduced to comparable units. Now this is very often not the case in medicine. Everyone familiar with hospitals knows what errors may mark the definitions on which statistics are based. The names of diseases are very often given are haphazard, either because the diagnosis is obscure, or because the cause of death is carelessly recorded by a student who has not seen the patient, or by an employee unfamiliar with medicine. For this reason pathological statistics can be valid only when compiled from data collected by the statistician himself.
Translated by Henry Copley Greene
An Introduction to the Study of Experimental Medicine
Part Two, Chapter II, Section ix (p. 136)
Henry Schuman, Inc. New York, New York, USA. 1927

…the goal of scientific physicians…is to reduce the indeterminate. Statistics therefore apply only to cases in which the cause of the facts observed is still indeterminate.
Translated by Henry Copley Greene
An Introduction to the Study of Experimental Medicine
Part Two, Chapter II, Section IX (p. 139)
Henry Schuman, Inc. New York, New York, USA. 1927

When a physician is called to a patient, he should decide on the diagnosis, then the prognosis, and then the treatment. … Physicians must know the evolution of the disease, its duration and gravity in order to predict its course and outcome. Here statistics intervene to guide physicians, by teaching them the proportion of mortal cases; and if observation has also shown that the successful and unsuccessful cases can be recognized by certain signs, then the prognosis is more certain.
Translated by Henry Copley Greene
An Introduction to the Study of Experimental Medicine
Part Three, Chapter IV, Section III (p. 213)
Henry Schuman, Inc. New York, New York, USA. 1927

Blane, Gilbert Sir   1749–1834
Scottish physician

There is…a great difficulty attending all practical inquiries in medicine; for in order to ascertain truth, in a manner that is satisfactory to a mind habituated to chaste investigation, there must be a series of patient and attentive observations upon a great number of cases, and the different trials must be varied, weighed, and compared, in order to form a proper estimate of the real efficacy of different remedies and modes of treatment.
Observations on the Diseases Incident to Seamen (p. ix)
Joseph Cooper. London, England. 1785

Fenger, Carl Emil   1814–84
Danish physician and politician

The use of the numerical method in medicine is not essentially new. From the time of Hippocrates to our day any doctor would say that this symptom is rare in a particular disease, but that one common; that this cause is more common than that one; that this treatment cures more patients than that one. All these expressions rare, common, more, etc. are indeterminate numerical expressions and presuppose a count, be it methodical or not.
Om den numeriske metode
Ugeskr Laeger, Volume 1, 1839

Salsburg, David S.
No biographical data available

After 17 years of interacting with physicians, I have come to realize that many of them are adherents of a religion they call Statistics. Statistics refers to the seeking out and interpretation of p values. Like any good religion, it involves vague mysteries capable of contradictory and irrational interpretation. It has a priesthood and a class of mendicant friars. And it provides Salvation: Proper invocation of the religious dogmas of Statistics will result in publication in prestigious journals.
The Religion of Statistics as Practiced in Medical Journals
The American Statistician, Volume 39, Number 3, August 1985 (p. 220)

STATISTICS AND SOCIETY

Boorstin, Daniel J.   1914–2004
American historian

The science of statistics is the chief instrumentality through which the progress of civilization is now measured and by which its development hereafter will be largely controlled.
The Decline of Radicalism
Chapter I (p. 8)
Random House, Inc. New York, New York, USA. 1969

Coats, R. H.

Beginning softly, statistics has long been handmaid to these exact sciences, apprenticed in the scullery, but now risen housekeeper, eating with the family.
Science and Society
Journal of the American Statistical Association, Volume 34, Number 205, March 1939 (p. 3)

Devons, Ely   1913–67
English economist

What more tempting facade of rationality than the portrayal of some statistics that seem to point to policy in one direction rather than another?
Essays in Economics
Chapter 7 (p. 134)
Greenwood Press, Publishers, Westport, Connecticut, USA. 1961

No Chancellor of the Exchequer could introduce his proposals for monetary and fiscal policy in the House of Commons by saying "I have looked at all the forecasts, some go one way, some another; so I decided to toss a coin and assume inflationary tendencies if it came down heads and deflationary if it came down tails.…" And statistics, however uncertain, can apparently provide some basis.
Essays in Economics
Chapter 7 (p. 134)
Greenwood Press, Publishers, Westport, Connecticut, USA. 1961

…there seems to be striking similarities between the role of economic statistics in our society and some of the functions which magic and divination play in primitive society.
Essays in Economics
Chapter 7 (p. 135)
Greenwood Press, Publishers, Westport, Connecticut, USA. 1961

Kline, Morris   1908–92
American mathematics professor and writer

The Mathematical Theory of Ignorance: The Statistical Approach to the Study of Man
Mathematics in Western Culture
Title to Chapter XXII
Oxford University Press, Inc. New York, New York, USA. 1953

Lippmann, Walter   1889–1974
American journalist and author

You and I are forever at the mercy of the census-taker and the census maker. That impertinent fellow who goes from house to house is one of the real masters of the statistical situation. The other is the man who organizes the results.
A Preface to Politics
The Golden Rule and After (p. 92)
M. Kennerley. New York, New York, USA. 1913

Ramsey, James B.
No biographical data available

The political practice of citing only agreeable statistics can never settle economic arguments.
Economic Forecasting — Models or Markets?: An Introduction to the Role of Econometrics in Economic Policy (p. 77)
Institute of Economic Affairs. London, England. 1977

Robinson, Lewis Newton
No biographical data available

In this country the statistical side of criminology is very imperfectly developed, and while the same cannot be said with equal force of other English-speaking countries, it yet remains true that the statistical terminology of this social science is characterized, so far as the English language is concerned, by great vagueness and uncertainty.
History and Organization of Criminal Statistics in the United States
Chapter I (p. 1)
Houghton Mifflin Company. Boston, Massachusetts, USA. 1911

Rogers, Will   1879–1935
American actor and humorist

The government keeps statistics on every known thing. But there is yet to be a statistics on how many laws we are living under.
The Writings of Will Rogers
Volume 4-1 (p. 167)
Oklahoma State University Press. Stillwater, Oklahoma, USA. 1973

Smith, Logan Pearsall   1865–1946
American author

For I am one of the unpraised, unrewarded millions without whom Statistics would be a bankrupt science. It is we who are born, who marry, who die, in constant ratios.
Trivia
Book II, Where Do I Come In? (p. 106)
Doubleday, Page & Company. Garden City, New York, USA. 1917

STETHOSCOPE

Byford, W. H.   1817–90
No biographical data available

The flexible stethoscope is a very handy instrument to relieve us from a fatiguing and not very delicate posture.

Advantages of the Prone Position in Examining the Foetal Circulation as a Diagnostic Sign of Pregnancy
Chicago Medical Journal, Volume 15, 1858

Laennec, René-Théophile-Hyacinthe   1781–1826
French physician

I had not imagined it would be necessary to give a name to such a simple device, but others thought differently. If one wants to give it a name, the most suitable would be "stethoscope."
Traité de l'Auscultation Médiate
Volume 1 (p. 11)
J.A. Brosson & J.S. Chaude. Paris, France. 1819

Stokes W.
No biographical data available

The stethoscope is an instrument, not, as some represent it, the bagatelle of a day, the brain-born fancy of some speculative enthusiast, the use of which, like the universal medicine of animal magnetism, will be soon forgotten, or remembered only to be ridiculed. It is one of those rich and splendid gifts which Science now and then bestows upon her most favored votaries, which, while they extended our views and open to us wide and fruitful fields of inquiry, confer in the meantime the richest benefits and blessings on mankind.
An Introduction to the Use of the Stethoscope
Machlachlan & Stewart. Edinburgh, Scotland. 1825

STOMACH

Athenaeus   ca. 200
Greek writer

Every investigation which is guided by principles of Nature fixes its ultimate aim entirely on gratifying the Stomach.
The Deipnosophists
VII

STONE

Linnaeus, Carl (von Linné)   1707–78
Swedish botanist and explorer

My mind reels when, on this height, I look down on the long ages that have flowed by like waves in the sound and have left traces of the ancient world, traces so nearly obscured that they can only whisper now that everything else has been silenced.
In A.G. Nathorst
Annual Report of the Board of Regents of the Smithsonian Institution, 1908
Carl von Linné as a Geologist (p. 738)
Government Printing Office. Washington, D.C. 1909

Parkinson, Cornelia
No biographical data available

Was it a flash of divine insight, or the slower process of observation and deduction, that led human beings to perceive [an] esoteric quality in stones? They saw beauty in the sunrise; but the sun became blinding by midday. There was color in leaves and flowers, until they withered. Water sparkled, but it could not be worn for long. Of all the natural wonders of the earth, only the stones endured. They must indeed be magical; and those who possess magical things can sometimes put to work the magic in them.
Gem Magic: The Wonder of Your Birthstone
Ballentine Books. New York, New York, USA. 1988

STORM

Conrad, Joseph   1857–1924
Polish-born English novelist

An earthquake, a landslide, an avalanche, overtake a man incidentally, as it were — without passion. A furious gale attacks him like a personal enemy, tries to grasp his limbs, fastens upon his mind, seeks to rout his very spirit out of him.
Typhoon
Chapter X (p. 77)
Doubleday, Page & Company. Garden City, New York, USA. 1920

Muir, John   1838–1914
American naturalist

Even the storms are friendly and seem to regard you as a brother, their beauty and tremendous fateful earnestness charming alike.
Our National Parks
Chapter IV (p. 99)
Houghton Mifflin Company. Boston, Massachusetts, USA. 1901

STRATIGRAPHY

Brown, Hugh Auchincloss   1878–1975
Electrical engineer

Each single layer of earth
Tells a story that's all its own;
The sands of the ancient beaches
Have changed into strata of stone.
Cataclysms of the Earth
The Earth Is a Great Stone Book (p. 275)
Twayne Publishers. New York, New York, USA. 1967

Chandler, Mary
No biographical data available

The shatter'd Rocks and Strata seem to say,
"Nature is old, and tends to her Decay":
Yet, lovely in Decay and green in Age,
Her Beauty lasts to her latest Stage.
In Robert Arnold Aubin
Topographical Poetry in XVIII Century England

Chapter IV (p. 164)
The Modern Language Association of America. New York, New York, USA. 1936

Savage, D. E.
American paleontologist

The fossil-mammal worker accepts that many mammals, marine or nonmarine, contributed fossils which are admirable tools for paleontologic stratigraphy and geochronology, and especially for age-magnitude correlations from continent to continent.
In E. Kauffman and J.E. Hazel (eds.)
Concepts and Methods of Biostratigraphy
Aspects of Vertebrate Paleontological Stratigraphy and Geochronology
Dowden, Hutchinson & Ross. Stroudsburg, Pennsylvania, USA. 1977

Shaw, Alan
No biographical data available

Preoccupation with the unattainable is a stultifying approach to any problem. Practical paleontology cannot be concerned with any of the fossils we cannot find. Geologically, we can only be interested in finding the total stratigraphic range through which a species is preserved. While the life and death of millions of unrepresented individuals is of theoretical interest, we cannot gain practically useful information from them.
Time in Stratigraphy
Chapter 17 (p. 103)
McGraw-Hill Book Company. New York, New York, USA. 1964

STREAM

Muir, John   1838–1914
American naturalist

…silvery branches interlacing on a thousand mountains, singing their way home to the sea…
Our National Parks
Chapter VIII (p. 241)
Houghton Mifflin Company. Boston, Massachusetts, USA. 1901

…ungovernable energy, rushing down smooth inclines in wide foamy sheets fold over fold, springing up here and there in magnificent whirls, scattering crisp clashing spray for the sunbeams to iris, bursting with hoarse reverberating roar through ragged gorges and boulder dams, booming in falls, gliding, glancing with cool soothing, murmuring…
Our National Parks
Chapter VIII (p. 242)
Houghton Mifflin Company. Boston, Massachusetts, USA. 1901

STREPTOMYCIN

Waksman, Selman A.   1888–1973
Ukrainian-born American biochemist

The highest scientific award and honor presented to me the day before yesterday gives me the opportunity to summarize briefly the discovery and utilization of strep-tomycin for disease control, notably in the treatment of tuberculosis, the "Great White Plague" of man.
Nobel Lectures, Physiology or Medicine 1942–1962
Nobel lecture for award received in 1952
Streptomycin: Background, Isolation, Properties, and Utilization
(p. 370)
Elsevier Publishing Company. Amsterdam, Netherlands. 1964

STRING THEORY

Amati, Danielle
Italian physicist

String theory is twenty-first century physics that fell acci-dentally into the twentieth century.
Attributed

Ferris, Timothy   1944–
American science writer

The odd thing about string theory was very odd indeed. It required that the universe have at least ten dimensions. As we live in a universe of only four dimensions, the the-ory postulated that the other dimensions…had collapsed into structures so tiny that we do not notice them.
Coming of Age in the Milky Way
Chapter 16 (p. 331)
William Morrow & Company, Inc. New York, New York, USA. 1988

STRUCTURE

Agassiz, Jean Louis Rodolphe   1807–73
Swiss-born American naturalist, geologist, and teacher

It is much more important for a naturalist to understand the structure of a few animals than to command the whole field of scientific nomenclature.
In Burt G. Wilder
Louis Agassiz, Teacher
The Harvard Graduate's Magazine, June, 1907

Clark, R. B.   1923–
Zoologist

It is an indispensable principle that structure must be considered in relation to function; in isolation it is mean-ingless.
Dynamics in Metazoan Evolution: The Origin of the Coelom and Seg-ments
Conclusion (p. 260)
Clarendon Press. Oxford, England. 1964

STUDENT

Kleiner, Israel   1948–
Ukrainian-born author

Can we not at least have a better appreciation of students' difficulties…having witnessed mathe-maticians of the first rank make mistakes, "prove" erroneous theorems, and often come to the right conclusions for insufficient or invalid reasons?
Thinking the Unthinkable: The Story of Complex Numbers (with a Moral)
Mathematics Teacher, October 1988

Osler, Sir William   1849–1919
Canadian physician and professor of medicine

Learn to love the freedom of the student life, only too quickly to pass away; the absence of the coarser cares of after days, the joy in comradeship, the delight in new work, the happiness in knowing that you are making progress. Once only can you enjoy these pleasures.
Aequanimitas, with Other Addresses to Medical Students, Nurses, and Practitioners of Medicine
The Master-Word in Medicine (p. 362)
The Blakiston Company. Philadelphia, Pennsylvania, USA. 1932

STUDY

Agassiz, Jean Louis Rodolphe   1807–73
Swiss-born American naturalist, geologist, and teacher

Lay aside all conceit. Learn to read the book of nature for yourself. Those who have succeeded best have fol-lowed for years some slim thread which has once in a while broadened out and disclosed some treasure worth a life-long search.
In David Stair Jordan
Popular Science Monthly, Volume 40, 1891/92

Bacon, Sir Francis   1561–1626
English lawyer, statesman, and essayist

Studies serve for delight, for ornament, and for ability.
In C.W. Eliot (ed.)
The Harvard Classics, Volume 3
Of Studies
P.F. Collier & Son. New York, New York, USA. 1909–10

Einstein, Albert   1879–1955
German-born physicist

Never regard your study as a duty, but as the enviable opportunity to learn to know the liberating influence of beauty in the realm of the spirit for your own personal joy and to the profit of the community to which your later work belongs.
In Helen Dukas and Banesh Hoffman
Albert Einstein: The Human Side: New Glimpses from His Archives (p. 96)
Princeton University Press. Princeton, New Jersey, USA. 1979

Platt, John R.
No biographical data available

We praise the "lifetime of study," but in dozens of cases, in every field, what was needed was not a lifetime but

rather a few short months or weeks of analytical inductive inference. In any new area we should try, like Roentgen, to see how fast we can pass from the general survey to analytical inductive inference. We should try, like Pasteur, to see whether we can reach strong inferences that encyclopedism could not discern.
Strong Inference
Science, Volume 146, Number 3641, 16 October 1964 (p. 251)

Rowland, Henry Augustus   1848–1901
American physicist

The whole universe is before us to study. The greatest labor of the greatest minds has only given us a few pearls; and yet the limitless ocean, with its hidden depths filled with diamonds and precious stones, is before us. The problem of the universe is yet unsolved, and the mystery involved in one single atom yet eludes us. The field of research only opens wider and wider as we advance, and our minds are lost in wonder and astonishment at the grandeur and beauty unfolded before us.
The Physical Papers of Henry Augustus Rowland
A Plea for Pure Science (p. 613)
Johns Hopkins Press. Baltimore, Maryland, USA. 1902

Skinner, B. F. (Burrhus Frederick)   1904–90
American psychologist

When you run into something interesting, drop everything else and study it.
Cumulative Record
A Case History in Scientific Method (p. 81)
Appleton-Century-Crofts, Inc. New York, New York, USA. 1959

Tsiolkovsky, Konstantin Eduardovich   1857–1935
Russian research scientist

To place one's feet on the soil of asteroids, to lift a stone from the moon with your hand, to construct moving stations in ether space, to organize inhabited rings around Earth, moon and sun, to observe Mars at the distance of several tens of miles, to descend to its satellites or even to its own surface — what could be more insane! However, only at such a time when reactive devices are applied, will a great new era begin in astronomy: the era of more intensive study of the heavens.
In M.K. Tikhonravov (ed.)
Works on Rocket Technology
The Investigation of Universal Space by Means of Reactive Devices (p. 95)
Publisher undetermined

von Goethe, Johann Wolfgang   1749–1832
German poet, novelist, playwright, and natural philosopher

The universe is a harmonious whole, each creature is but a note, a shade of a great harmony, which man must study in its entirety and greatness, lest each detail should remain a dead letter.
Letter to C.L. Knebel, November 17, 1784

STUPIDITY

Eddington, Sir Arthur Stanley   1882–1944
English astronomer, physicist, and mathematician

I am an Evolutionist, not a Multiplicationist. It seems rather stupid to keep doing the same thing over and over again.
The Nature of the Physical World
Chapter IV (p. 86)
The Macmillan Company. New York, New York, USA. 1930

Einstein, Albert   1879–1955
German-American physicist

Everyone has to sacrifice at the altar of stupidity from time to time, to please the Deity and the human race.
In Max Born
The Born–Einstein Letters: Correspondence Between Albert Einstein and Max and Hedwig Born from 1916 to 1955
Letter 219 September, 1920 (p. 35)
Walker & Company. New York, New York, USA. 1971

Levi, Primo   1919–87
Italian writer and chemist

To accuse another of having weak kidneys, lungs, or heart, is not a crime; on the contrary, saying he has a weak brain is a crime. To be considered stupid and to be told so is more painful than being called gluttonous, mendacious, violent, lascivious, lazy, cowardly: every weakness, every vice, has found its defenders, its rhetoric, its ennoblement and exaltation, but stupidity hasn't.
Other People's Trades
The Irritable Chess-Players
Summit Books. New York, New York, USA. 1989

SUBSTANCE

Jeans, Sir James Hopwood   1877–1946
English physicist and mathematician

…we now know that there is, in principle, no permanence in substance; it is mere bottled energy, and possesses no more inherent permanence than bottled beer…
Physics and Philosophy
Chapter II (p. 41)
Dover Publications, Inc. New York, New York, USA. 1981

SUBTERRANEAN

Clarke, John
No biographical data available

Beneath the earth likewise there are laws of nature, less familiar to us, but no less fixed. Be assured that there exists below everything that you see above. There, too, there are antres vast, immense recesses, and vacant spaces, with mountains overhanging on either hand.

There are yawning gulfs stretching down into the abyss, which have buried in their depths their mighty ruins. These retreats are filled with air, for nowhere is there a vacuum in nature; through their ample spaces stretch marshes over which darkness ever broods. Animals also are produced in them, but they are slow-paced and shapeless; the air that conceived them is dark and clammy, the waters are torpid through inaction.
Physical Science in the Time of Nero, Being a Translation of the Quaestiones Naturales of Seneca
Book III, Chapter XVI (pp. 128–129)
Macmillan & Company Ltd. London, England. 1910

Cloos, Hans   1885–1951
German geologist

A subterranean landscape is the ramified labyrinth of crevices and caves, of pores and seams, through which the trolls and earth-sprites climb up and down.
Conversations with the Earth
Chapter IV (p. 48)
Alfred A. Knopf. New York, New York, USA. 1953

Deetz, James   1930–2000
American archaeologist

Archaeology by its formal etymology, is the study of the old; and the old, more often than not, is buried. As a result, archaeologists have traditionally been concerned with the subterranean world. Like Lewis Carroll's Alice, they are confronted with the curious underground world, and attempt to understand and explain it.
In James Deetz (ed.)
Man's Imprint from the Past: Readings in the Methods of Archaeology
Chapter 1 (p. 4)
Little, Brown & Company. Boston, Massachusetts, USA. 1971

SUN

Adams, Douglas   1952–2001
English author, comic radio dramatist, and musician

Several billion trillion tons of super hot exploding hydrogen nuclei rose slowly above the horizon and managed to look small, cold and slightly damp.
The Ultimate Hitchhiker's Guide to the Galaxy
Life, the Universe, and Everything
Chapter 7 (p. 349)
Ballantine Books. New York, New York, USA. 2002

Allen, Woody   1935–
American film director and actor

The sun, which is made of gas, can explode at any moment, sending our entire planetary system hurtling to destruction; students are advised what the average citizen can do in such a case.
Getting Even
Spring Bulletin (p. 58–59)
Random House, Inc. New York, New York, USA. 1971

Bailey, Philip James   1816–1902
English poet

The sun, bright keystone of Heaven's world-built arch…
Festus: A Poem
Scene I (p. 32)
George Routledge & Sons, Ltd. London, England. 1893

Bourdillon, Francis William

The night has a thousand eyes,
And the day but one;
Yet the light of the bright world dies,
With the dying Sun.
Among the Flowers, and Other Poems
The Night Has a Thousand Eyes
Marcus Ward & Company. London, England. 1878

Bradley, Jr., John Hodgdon   1898–1962
American geologist

Since a time so remote that imagination falters in the attempt to conceive it, the sun has mothered her brood of planets.
Autobiography of Earth
Chapter V (p. 133)
Coward-McCann, Inc. New York, New York, USA. 1935

Cook, J. Gordon   1916–
No biographical data available

The darkness of night is dissolving in light that flows steadily across the sky. Over the eastern horizon a curved shoulder of fire appears; our sun has arrived, bringing with it another day of glorious light.
We Live by the Sun
Chapter 2 (p. 18)
The Dial Press. New York, New York, USA. 1957

Copernicus, Nicolaus   1473–1543
Polish astronomer

Since the newness of the hypotheses of this work — which sets the earth in motion and puts an immovable sun at the centre of the universe — has already received a great deal of publicity, I have no doubt that certain of the savants have taken grave offense and think it wrong to raise any disturbance among liberal disciplines which have had the right set-up for a long time now.
In *Great Books of the Western World* (Volume 16)
On the Revolutions of the Heavenly Spheres
Introduction, to the Reader Concerning the Hypothesis of this Work
(p. 505)
Encyclopædia Britannica, Inc. Chicago, Illinois, USA. 1952

In the center of all rests the sun. For who would place this lamp of a very beautiful temple in another or better place than this wherefrom it can illuminate everything at the same time? As a matter of fact, not unhappily do some call it the lantern; others, the mind and still others, the

pilot of the world. Trismegistus calls it a "visible god"; Sophocles' Electra, "that which gazes upon all things." And so the sun, as if resting on a kingly throne, governs the family of stars which wheel around.
In *Great Books of the Western World* (Volume 16)
On the Revolutions of the Heavenly Spheres
Book One, Chapter 10 (pp. 526–528)
Encyclopædia Britannica, Inc. Chicago, Illinois, USA. 1952

Crane, Stephen   1871–1900
American writer

The sun was pasted in the sky like a wafer.
The Red Badge of Courage
IX (p. 115)
Random House, Inc. New York, New York, USA. 1925

Crosby, Harry   1898–1929
American financial heir, bon vivant, and poet

The Sun! The Sun! a fish in the aquarium of sky or golden net to snare the butterfly of soul or else the hole through which stars have disappeared...
Sun Rhapsody
Six Poems 1928
Publisher undetermined

de Fontenelle, Bernard le Bovier   1657–1757
French author

Our sun enlightens the planets that belong to him; why may not every fixed star also have planets to which they give light?
Conversations on the Plurality of Worlds
The Fifth Evening (p. 151)
Printed for Peter Wilson. Dublin, Ireland. 1761

Deutsch, Armin J.   1918–1969
American astronomer and science fiction writer

The face of the sun is not without expression, but it tells us precious little of what is in its heart.
The Sun
Scientific American, Volume 179, Number 5, November 1948 (p. 38)

Dryden, John   1631–1700
English poet, dramatist, and literary critic

The glorious lamp of heaven, the radiant sun,
Is Nature's eye...
The Poetical Works of Dryden
The Fable of Acis, Polyphemus, and Galatea from the Thirteenth Book of Ovid's Metamorphoses (p. 405)
The Riverside Press. Cambridge, Massachusetts, USA. 1949

And since the vernal equinox, the sun
In Aries twelve degrees, or more, had run.
The Poetical Works of Dryden
The Cock and the Fox, l. 448–449
The Riverside Press. Cambridge, Massachusetts, USA. 1949

Behold him setting in his western skies,
The shadows lengthening as the vapours rise.
The Poetical Works of Dryden

Absalom and Achitophel
Part I, l. 268
The Riverside Press. Cambridge, Massachusetts, USA. 1949

Ehrlich, Gretel   1946–
American travel writer, novelist, and essayist

We forget that our sun is only a star destined to someday burn out. The time scale of its transience so far exceeds our human one that our unconditional dependence on its life-giving properties feels oddly like an indiscretion we'd rather forget.
The Solace of Open Spaces
To Live in Two Worlds (p. 105)
Penguin Books. New York, New York, USA. 1986

Eudoxus of Cnidus   ca. 400 BCE–ca. 350 BCE
Greek astronomer, mathematician, and physician

Willingly would I burn to death like Phaeton, were this the price for reaching the sun and learning its shape, its size, and its substance.
In Carl B. Boyer
A History of Mathematics (p. 91)
John Wiley & Sons, Inc. New York, New York, USA. 1968

Falconer, William   1744–1824
Poet

High in his chariot glow'd the lamp of day.
The Shipwreck
Canto I, III, l. 3

Flammarion, Camille   1842–1925
French astronomer and author

Dazzling source of light and heat, of motion, life, and beauty, the inimitable sun has in all ages received the earnest and grateful homage of mortals. The ignorant admire it because they feel the effects of its power and its value; the savant appreciates it because he has learned its unique importance in the system of the world; the artist salutes it because he sees in its splendor the virtual cause of all harmonies.
Popular Astronomy: A General Description of the Heavens
Book III, Chapter I (p. 207)
Chatto & Windus. London, England. 1894

Galilei, Galileo   1564–1642
Italian physicist and astronomer

The doctrine of the movement of the earth and the fixity of the sun is condemned on the ground that the Scriptures speak in many places of the sun moving and the earth standing still. The Scriptures not being capable of lying or erring, it followeth that the position of those is erroneous and heretical who maintain that the sun is fixed and the earth in motion.
The Authority of Scripture in Philosophical Controversies
Section I
The Defenders of Fallacy

Gilbert, Sir William Schwenck   1836–1911
English playwright and poet

Sullivan, Arthur   1842–1900
English composer

The Sun, whose rays
Are all ablaze
With ever-living glory,
Does not deny
His majesty —
He scorns to tell a story!
The Complete Plays of Gilbert and Sullivan
The Mikado
Act II (p. 322)
W.W. Norton & Company, Inc. New York, New York, USA.1976

Heraclitus   540 BCE–480 BCE
Greek philosopher

The sun…is new each day.
In G.S Kirk and J.E. Raven
The Pre-Socratic Philosophers: A Critical History with a Selection of Texts
Fragment 228 (p. 202)
At the University Press. Cambridge, England. 1963

Kelvin, Lord William Thomson   1824–1907
Scottish engineer, mathematician, and physicist

Now, if the sun is not created a miraculous body, to shine on and give out heat forever, we must suppose it to be a body subject to the laws of matter (I do not say there may not be laws which we have not discovered) but, at all events, not violating any laws we have discovered or believe we have discovered. We should deal with the sun as we should with any large mass of molten iron, or silicon, or sodium.
On Geological Time (p. 18)
Address
Geological Society of Glasgoe
February 27, 1868

It seems, therefore, on the whole most probable that the sun has not illuminated the earth for 100,000,000 years, and almost certain that he has not done so for 500,000,000 years. As for the future, we may say, with equal certainty, that inhabitants of the earth can not continue to enjoy the light and heat essential to their life for many million years longer unless sources now unknown to us are prepared in the great storehouse of creation.
The Age of the Sun's Heat
Macmillan's Magazine, March 5, 1862 (p. 293)

Langley, Samuel Pierpoint   1834–1906
American astronomer and aviation pioneer

As the thought of man is widened with the process of the suns, let us hope that we shall one day know more.
The New Astronomy
The Century Illustrated Monthly Magazine, Volume 28, New Series
Volume 6 (p. 936)

…the fields glitter with snow-crystals in the winter noon, and the eye is dazzled with a reflection of the splendor which the sun pours so fully into every nook that by it alone we appear to see everything.
The New Astronomy
The Century Illustrated Monthly Magazine, Volume 28, New Series, Volume 6 (p. 922)

Longfellow, Henry Wadsworth   1807–82
American poet

Down sank the great red sun, and in golden glimmering vapours
Veiled the light of his face, like the Prophet descending from Sinai.
The Poetical Works of Henry Wadsworth Longfellow
Evangeline, Part I, Section 4
Houghton Mifflin Company. Boston, Massachusetts, USA. 1883

Macpherson, James   1736–1796
Scottish poet

Whence are thy beams, O sun! thy everlasting light! Thou comest forth in thy awful beauty; the stars hide themselves in the sky; the moon, cold and pale, sinks in the western wave; but thou thyself movest alone.
The Poems of Ossian
Carthon (p. 233)
Printed by Dewick & Clarke. London, England. 1806

Mann, Thomas   1875–1955
German-born American novelist

"He does seem rather weird," was Hans Castorp's view. "Some of the things he said were very queer: it sounded as if he meant to say that the sun revolves round the earth."
The Magic Mountain
Chapter VI
Of City of God, and Deliverance by Evil (p. 407)
Alfred A. Knopf. New York, New York, USA. 1966

Mayer, Robert
No biographical data availabale

The Sun…is an inexhaustible source of physical force — that continuously wound-up spring which sustains in motion the mechanism of all the activities on Earth.
In L.I. Ponomarev
The Quantum Dice (p. 228)
Institute of Physics Publishing. Bristol, England. 1993

Melville, Herman   1819–91
American novelist

Life or death, weal or woe, the sun stays not his course. Oh: over battlefield and bower; over tower, and town, he speeds, — peers in at births, and death-beds; lights up cathedral, mosque, and pagan shrine; — laughing over all; — a very Democritus in the sky; and in one brief day sees more than any pilgrim in a century's round.

Typee, Omoo, Mardi
Mardi
Chapter 184 (p. 1277)
The Library of America. New York, New York, USA. 1982

Moulton, Forest Ray   1872–1952
American astronomer

If the sun were created expressly to light and heat the earth, what a waste of energy!
In H.H. Newman (ed.)
The Nature of the World and of Man
Astronomy (p. 17)
The University of Chicago Press. Chicago, Illinois, USA. 1927

Muir, John   1838–1914
American naturalist

The sun, looking down on the tranquil landscape, seems conscious of the presence of every living thing on which he is pouring his blessings, while they in turn, with perhaps the exception of man, seem conscious of the presence of the sun as a benevolent father and stand hushed and waiting.
Steep Trails
Chapter XVII (p. 226)
Norman S. Berg, Publisher. Dunwoody, Georgia, USA. 1970

Parker, E. N.   1927–
No biographical data available

The riddles the sun presents are signposts to new horizons.
The Sun
Scientific American, Volume 233, Number 3, September 1975 (p. 50)

Pascal, Blaise   1623–62
French mathematician and physicist

Let man then contemplate the whole of nature in her full and grand majesty, and turn his vision from the low objects which surround him. Let him gaze on that brilliant light, set like an eternal lamp to illumine the universe.
In *Great Books of the Western World* (Volume 33)
Pensées
Section II, 72
Encyclopædia Britannica, Inc. Chicago, Illinois, USA. 1952

Plato   428 BCE–347 BCE
Greek philosopher

That there might be some visible measure of their relative swiftness and slowness as they proceeded in their eight courses, God lighted a fire, which we now call the sun, in the second from the earth of these orbits…
In *Great Books of the Western World* (Volume 7)
Timaeus
Section 39 (p. 451)
Encyclopædia Britannica, Inc. Chicago, Illinois, USA. 1952

Raymo, Chet   1936–
American physicist and science writer

For 5 billion years the sun has exhaled a faint breath as it burns, bathing the Earth in the flux of its exhalations, a wind of atoms and subatomic particles that feeds the Earth's atmosphere and ignites auroras.
The Soul of the Night
Chapter 8 (p. 81)
Prentice-Hall, Inc. Englewood Cliffs, New Jersey, USA. 1985

Riley, James Whitcomb   1849–1916
American poet

And the sun had on a crown
Wrought of gilded thistledown,
And a scarf of velvet vapor
And a raveled rainbow gown;
And his tinsel-tangled hair
Tossed and lost upon the air
Was glossier and flossier
Than any anywhere.
The Complete Works of James Whitcomb Riley in Ten Volumes (Volume 4)
The South Wind and the Sun
Harper & Brothers Publishers. New York, New York, USA. 1916

Rutherford, Mark (William Hale White)   1831–1913
English writer

The sun, we say, is the cause of heat, but the heat is the sun, hence on this window-ledge.
More Pages from a Journal
Notes (p. 120)
H. Frowde. London, England. 1910

Sagan, Carl   1934–96
American astronomer and science writer
Druyan, Ann   1949–
American author and television producer

The immense, overpowering blackness is relieved here and there by a faint point of light — which, upon closer approach, is revealed to be a mighty sun, blazing with thermonuclear fire and warming a small surrounding volume of space.
Shadows of Forgotten Ancestors: A Search for Who We Are
Prologue (p. 3)
Random House, Inc. New York, New York, USA. 1992

Shakespeare, William   1564–1616
English poet, playwright, and actor

…the glorious sun,
Stays in his course and plays the alchemist,
Turning with splendor of his precious eye
The meagre cloddy earth to glittering gold.
In *Great Books of the Western World* (Volume 27)
The Plays and Sonnets of William Shakespeare (Volume 2)
The Twelfth Night
Act V, Scene i, l. 77–80
Encyclopædia Britannica, Inc. Chicago, Illinois, USA. 1952

Smart, Christopher   1722–71
English poet

Glorious the sun in mid-career;
Glorious th' assembled fires appear.
Collected Poems
A Song to David, LXXXIV
Routledge & Kegan Paul. London, England. 1949

Starr, Victor P.
No biographical data available

Gilman, Peter A.
No biographical data available

It has always been easier to record and describe solar events than to provide theoretical explanations for them.
The Circulation of the Sun's Atmosphere
Scientific American, Volume 218, Number 1, January 1968 (p. 100)

Stoppard, Tom   1937–
Czech-born English playwright

Meeting a friend in a corridor, Wittgenstein said: "Tell me, why do people always say it was natural for men to assume that the sun went round the earth, rather than that the earth was rotating?" His friend said, "Well, obviously, because it looks as if the sun is going round the earth." To which the philosopher replied, "Well, what would it have looked like if it had looked as if the earth was rotating?"
Jumpers
Act Two (p. 65)
Grove Press, Inc. New York, New York, USA. 1972

Swift, Jonathan   1667–1745
Irish-born English writer

These people [Laputians] are under continual disquietudes, never enjoying a minute's peace of mind; and their disturbances proceed from causes which very little affect the rest of mortals. Their apprehension arises from several changes they dread in the celestial bodies. For instance…that the sun, daily spending its rays without any nutriment to supply them, will at last be wholly consumed and annihilated; which must be attended with the destruction of this earth, and all the planets that receive their light from it.
In *Great Books of the Western World* (Volume 36)
Gulliver's Travels
Part III, Chapter II (p. 98)
Encyclopædia Britannica, Inc. Chicago, Illinois, USA. 1952

Tennyson, Alfred (Lord)   1809–92
English poet

There sinks the nebulous star we call the sun.
Alfred Tennyson's Poetical Works
The Princess, Part IV, l. 19
Oxford University Press, Inc. London, England. 1953

Thoreau, Henry David   1817–62
American essayist, poet, and practical philosopher

It is true, I never assisted the sun materially in his rising; but doubt not, it was of the last importance only to be present at it.
The Writings of Henry David Thoreau (Volume 2)
Walden
Chapter I (p. 30)
Houghton Mifflin Company. Boston, Massachusetts, USA. 1893

Updike, John   1932–
American novelist, short story writer, and poet

The zeros stared back, every one a wound leaking the word "poison." "That's the weight of the Sun," Caldwell said.
The Centaur
Chapter I (p. 37)
Alfred A. Knopf. New York, New York, USA. 1995

Wells, H. G. (Herbert George)   1866–1946
English novelist, historian, and sociologist

…the sun, red and very large, halted motionless upon the horizon, a vast dome glowing with a dull heat, and now and then suffering a momentary extinction…[it] grew larger and duller in the westward sky, and the life of the old earth ebbed away. At last, more than thirty million years hence, the huge red-hot dome of the sun had come to obscure nearly a tenth part of the darkling heavens.
In Robert M. Hutchins and Mortimer J. Adler (eds.)
The Great Ideas Today, 1971
The Time Machine, Chapter 11
Encyclopædia Britannica, Inc. Chicago, Illinois, USA. 1971

Whitman, Walt   1819–92
American poet, journalist, and essayist

Give me a splendid silent sun, with all his beams full-dazzling.
Complete Poetry and Collected Prose
Give Me a Splendid Sun
The Library of America. New York, New York, USA. 1982

Xenophanes   ca. 575 BCE–ca. 478 BCE
Greek philosopher

The sun comes into being each day from little pieces of fire that are collected…
In G.S Kirk and J.E. Raven
The Pre-Socratic Philosophers: A Critical History with a Selection of Texts
Fragment 178 (p. 172)
At the University Press. Cambridge, England. 1963

SUNSPOT

Birrell, Augustine   1850–1933
English author and politician

The sun is not all spots.
Obiter Dicta
Second Series
John Milton
Elliot Stock. London, England. 1884

Galilei, Galileo   1564–1642
Italian physicist and astronomer

Neither the satellites of Jupiter nor any other stars are spots or shadows, nor are the sunspots stars. It is indeed true that I am quibbling over names, while I know that anyone may impose them to suit himself. So long as a man does not think that by names he can confer inherent and essential properties on things, it would make little difference whether he calls these "stars."
Translated by Stillman Drake
Discoveries and Opinions of Galileo
Letters on Sunspots, Third Letter on Sunspots, From Galileo Galilei to Mark Welser (p. 139)
Doubleday & Company, Inc. New York, New York, USA. 1957

Harris, John   1667?–1719
No biographical data available

Spots! Said she, What, are there Spots in the Sun, which sometimes appear there, and sometimes not; for God's sake what are those Spots?
Astronomical Dialogues Between a Gentleman and a Lady (p. 74)
Printed by T. Wood for Benj. Cowse. London, England. 1719

Zirin, Harold
Astrophysicist

Just like the green fields and virgin forests, the granules, the sunspots, the elegant prominences reflect the pure beauty of nature. They offer aesthetic pleasure, as well as scientific challenge, to those who study them.
Astrophysics of the Sun
Preface (p. ix)
Cambridge University Press. Cambridge, England. 1988

SUPERNOVA

Crowley, Abraham   1618–67
English poet

So when by various Turns of the Celestial Dance,
In many thousand years,
A Star, so long unknown, appears,
Though Heaven it self more beauteous by it grow,
It troubles and alarms the World below,
Does to the Wise a Star, to Fools a Meteor show.
In Thomas Sprat
The History of the Royal-Society of London for Improving of Natural Knowledge
To the Royal Society
Printed for A. Millar. London, England. 1756–1757

Schaaf, Fred
No biographical data available

...a star gone to seed — a star spectacularly sowing space with heavy elements and the promise of new stars, worlds, life, and eyes.
The Starry Room: Naked Eye Astronomy in the Intimate Universe

Chapter 11 (p. 194)
John Wiley & Sons, Inc. New York, New York, USA. 1988

Woosley, Stan
No biographical data available
Weaver, Tom
No biographical data available

The collapse and explosion of a massive star is one of nature's grandest spectacles. For sheer power nothing can match it. During the supernova's first 10 seconds...it radiates as much energy from a central region 20 miles across as all the other stars and galaxies in the rest of the visible universe combined.... It is a feat that stretches even the well-stretched minds of astronomers.
The Great Supernova of 1987
Scientific American, Volume 261, Number 2, August 1989 (p. 32)

SUPERSTITION

Gell-Mann, Murray   1929–
American physicist

Unscientific approaches to the construction of models of the world around us have characterized much of human thinking since the time immemorial, and they are still widespread. Take, for example, the version of sympathetic magic based on the idea that similar things must be connected. It seems natural to many people around the world that, when in need of rain, they should perform a ceremony in which water is procured and poured on the ground.
The Quark and the Jaguar: Adventures In the Simple and the Complex (p. 89)
W.H. Freeman & Company. New York, New York, USA. 1994

Laplace, Pierre Simon   1749–1827
French mathematician, astronomer, and physicist

...the beauty of the universe and the order of celestial things force us to recognize some superior nature which ought to be remarked and admired by the human race. But as far as it is proper to propagate religion, which is joined to the knowledge of nature, so far it is necessary to work toward the extirpation of superstition, for it torments one, importunes one, and pursues one continually and in all places.
A Philosophical Essays on Probabilities
Chapter XVI (p. 174)
Dover Publications, Inc. New York, New York, USA. 1951

Van Sloan, Edward   1882–1964
American actor

Superstition? Who can define the boundary line between the superstition of yesterday and the scientific fact of tomorrow?
Dracula's Daughter
Film (1936)

SUPERSTRING

Dyson, Freeman J.   1923–
American physicist and educator

Superstrings and butterflies are examples illustrating two different aspects of the universe and two different notions of beauty. Superstrings come at the beginning and butterflies at the end because they are extreme examples. Butterflies are at the extreme of concreteness, superstrings at the extreme of abstraction. They mark the extreme limits of the territory over which science claims jurisdiction. Both are, in their different ways, beautiful. Both are, from a scientific point of view, poorly understood. Scientifically speaking, a butterfly is at least as mysterious as a superstring....
Infinite in All Directions
Part One, Chapter Two (p. 14)
HarperCollins Publisher, Inc. New York, New York, USA. 1988

SUPPOSITION

Stoney, George Johnstone   1826–1911
Irish physicist

A theory is a supposition which we hope to be true, a hypothesis is a supposition which we expect to be useful; fictions belong to the realm of art; if made to intrude elsewhere, they become either makebelieves or mistakes.
In Sir William Ramsay
Essays Biographical and Chemical
Chemical Essays
Radium and Its Products (p. 179)
Archibald Constable & Company Ltd. London, England. 1908

SURFACE TENSION

Roth, V. Louise
American zoologist

You may have inner tranquility, but you can't escape surface tension.
In Steven Vogel
Life's Devices: The Physical World of Animal and Plants
Chapter 5 (p. 82)
Princeton University Press. Princeton, New Jersey, USA. 1988

SURGEON

Author undetermined

A good surgeon must have an eagle's eye, a lion's heart, and a lady's hand.
In John Timbs
Doctors and Patients, or, Anecdotes of the Medical World and Curiosities of Medicine (Volume 2) (p. 155)
Richard Bentley & Son. London, England. 1873

Aylett, Robert   1583–1655?
Religious poet

For Mercy doth like skilfull Surgeon deal,
That hath for ev'ry sore a remedy:
If gentle drawing plaisters cannot heal
The wound, because it festreth inwardly,
He sharper corrasives must then apply,
And as he oft cuts off some member dead,
Or rotten, lest the rest should putrifie,
So Mercy wicked Members off doth shred,
Lest they should noysome prove to body and the head.
Peace with Her Foure Gardners
The Brides Ornaments, Meditation III, l. 307–315

Caldwell, George W.
No biographical data available

Who is the man in sterile white
Delving deep at the point of light,
With nurses, trained, at left and right?
Poet Physician: An Anthology of Medical Poetry Written by Physicians
The Surgeon (p. 136)
C.C. Thomas. Springfield, Illinois, USA. 1945

Carnochan, John Murray   1817–87
American surgeon

While respect for life will dictate to the surgeon the greatest prudence — will counsel him to attempt no operation which he would not be willing to perform on his own child — it will also teach him, that if the extremes of boldness are to be shunned, pusillanimity is not the necessary alternative.
Contributions to Operative Surgery and Surgical Pathology
Preface
Lindsay & Blakiston. Philadelphia, Pennsylvania, USA. 1857

Celsus, Aulus Cornelius   fl. 14 AD
Roman medical writer

A surgeon ought to be young, or at any rate, not very old; his hand should be firm and steady, and never shake; he should be able to use his left hand with as much dexterity as his right; his eye-sight should be acute and clear; his mind intrepid, and so far subject to pity as to make him desirous of the recovery of his patient, but not so far as to suffer himself to be moved by his cries; he should neither hurry the operation more than the case requires, nor cut less than is necessary, but do everything just as if the other's screams made no impression upon him.
In Samuel Evans Massengill
A Sketch of Medicine and Pharmacy and a View of Its Progress by the Massengill Family from the Fifteenth to the Twentieth Century (p. 30)
The S.E. Massengill Company. Bristol, Tennessee, USA. 1943

Crichton-Browne, Sir James   1840–1938
English physician

"Oh, sir, we made a terrible mistake in the case of that man yesterday! We amputated the wrong leg!"

"Ah well," the surgeon replied, complacently, "it's of no consequence, for I have just been looking at the other leg, and it's going to get better."
The Doctor's After Thoughts (p. 15)
E. Benn Ltd. London, England. 1932

Every great surgeon, it used to be said, shakes, swears or sweats when he operates.
The Doctor Remembers
Bret Harte (p. 170)
Duckworth & Company. London, England. 1938

Croll, Oswald   1560–1609
German chemist and physician

…it is necessary that every Surgeon should be a Physitian, and every Physitian a Chyrugion, that there may be a sound Bridegroom for a sound Bride…
Philosophy Reformed and Improved in Four Profound Tractates (p. 151)
Printed by M.S. for Lodowick Lloyd. London, England. 1657

Cvikota, Raymond J.

Surgeon: Fee lancer.
Quote, the Weekly Digest, June 9, 1968 (p. 457)

da Costa, J. Chalmers   1863–1933
American physician

A vain surgeon is like a milking stool; of no use except when sat upon.
The Trials and Triumphs of the Surgeon (p. 17)
Dorrance & Company. Philadelphia, Pennsylvania, USA. 1944

A surgeon is like a postage stamp. He is useless when stuck on himself.
The Trials and Triumphs of the Surgeon (p. 17)
Dorrance & Company. Philadelphia, Pennsylvania, USA. 1944

Davies, Robertson   1913–95
Canadian novelist

A man mentioned casually to me this afternoon that his brother was in a hospital, having his appendix removed. This operation is now undertaken without qualm; surgeons regard it as a pastime, something to keep the hands busy, like knitting or eating salted nuts.
The Table Talk of Samuel Marchbanks (p. 176)
Clarke, Irwin. Toronto, Ontario, Canada. 1949

de Chauliac, Guy
No biographical data available

The surgeon should be learned, skilled, ingenious, and of good morals. Be bold in things that are sure, cautious in dangers; avoid evil cures and practices; be gracious to the sick, obliging to his colleagues, wise in his predictions. Be chaste, sober, pitiful, and merciful; not covetous nor extortionate of money, but let the recompense be moderate, according to the work, the means of the sick, the character of the issue or event, and its dignity.

In Samuel Evans Massengill
A Sketch of Medicine and Pharmacy and a View of Its Progress by the Massengill Family from the Fifteenth to the Twentieth Century (p. 262)
The S. E. Massengill Company. Bristol, Tennessee, USA. 1943

de Mondeville, Henri   1260–1320
French pioneer surgeon

The surgeon should be fairly audacious [yet] he should operate with prudence and sagacity; he should never commence perilous operations unless he has provided everything in order to avoid danger; …he should not sing his own praises; he should not cover his colleagues with blame; he should not cause envy among them; he should work always with the idea of acquiring a reputation of probity; he should be reassuring to his patients by kind words and acquiesce to their requests when nothing harmful will result from them as to their cure.
In C.G. Cumston
Henry de Mondeville, the Man and His Writings, with Translations of Several Chapters of His Works
Buffalo Medical Journal, Volume 42, 1903

A Surgeon ought to be fairly bold. He ought not to quarrel before the laity, and although he should operate wisely and prudently, he should never undertake any dangerous operation unless he is sure that it is the only way to avoid a greater danger. His limbs, and especially his hands, should be well shaped with long, delicate and supple fingers which must not be tremulous.
In John Arderne
Treatise of Fistula in Ano (pp. xx)

Dickens, Charles   1812–70
English novelist

What! don't you know what a Sawbones is, Sir? inquired Mr. Weller. I thought everybody know'd as a Sawbones was a Surgeon.
The Posthumous Papers of the Pickwick Club
Chapter XXX (p. 348)
Dodd, Mead & Company. New York, New York, USA. 1944

Dickinson, Emily   1830–86
American lyric poet

Surgeons must be very careful
When they take the knife!
Underneath their fine incisions
Stirs the Culprit, — Life!
The Complete Poems of Emily Dickinson
No. 108 (p. 52)
Little, Brown & Company. Boston, Massachusetts, USA. 1960

Dimmick, Edgar L.
No biographical data available

The public views his *status regas*,
In the profession he's "The Eagle."
With super-supple fingers slim
(Pus, blood, and guts don't bother him),

Up to his elbows, filled with glee,
With snick and slice sadistically,
Into a jar, up on a shelf
He puts a fragment of yourself.
For him no diagnostic doubt —
He'll operate, and so find out.
The In-Side of Two
Journal of the American Medical Association, Volume 199, Number 6,
1967 (p. 274)

Dunlop, William 1766–1839
American dramatist and theatrical manager

There is hardly on the face of the earth a less enviable
situation than that of an Army Surgeon after a battle…
Recollection of the American War, 1812–1814
Chapter III (p. 54)
Historical Publishing Company. Toronto, Ontario, Canada. 1906

Eliot, T. S. (Thomas Stearns) 1888–1965
American expatriate poet and playwright

The wounded surgeon plies the steel
That questions the distempered part;
Beneath the bleeding hands we feel
The sharp compassion of the healer's art
Resolving the enigma of the fever chart.
The Collected Poems and Plays 1909–1950
East Coker, Part IV (p. 127)
Harcourt, Brace & World, Inc. New York, New York, USA. 1952

Findley, Thomas
Physician

The surgeon…is a man of action. He lives in an exhilarating world of knives, blood, and groans. His tempo is of
necessity rapid. He is inclined to look at his less kinetic
colleague with an air of puzzled condescension but may,
in a relaxed moment, admit that the medical man is occasionally able to assist uncomfortable dowagers in the
selection of a cathartic.
The Obligations of an Internist to a General Surgeon
Surgery, Volume 16, 1944 (p. 557)

Gilbertus, Anglicus
No biographical data available

Why in God's name is there such a great difference
between a physician and a surgeon?
Surgery
Time, May 3, 1963 (p. 44)

Gogarty, Oliver St. John 1878–1957
Irish author

Let Surgeon MacCardle confirm you in Hope.
A jockey fell off and his neck it was broke.
He lifted him up like a fine, honest man;
And he said "He is dead; but I'll do all I can."
The Collected Poems of Oliver St. John Gogarty
The Three (p. 109)
Constable. London, England. 1951

Gross, S. D. 1805–84
American emeritus professor of surgery

It is impossible for any man to be a great surgeon if he
is destitute…of the finer feelings of our nature.… I do
not think that it is possible for a criminal to feel much
worse the night before his execution than a surgeon when
he knows that upon his skill and attention must depend
the fate of a valuable citizen, husband, father, mother or
child. Surgery under such circumstances is a terrible task
master, feeding like a vulture upon a man's vitals.
Autobiography of Samuel D. Gross, MD. (p. 172)
George Barrie. Philadelphia, Pennsylvania, USA. 1887

Hazlitt, William Carew 1834–1913
English bibliographer

An ignorant drunken Surgeon that kil'd all men that came
under his hands, boasted himself a better man than the
Parson; for, said he, your Cure maintains but yourself,
but my Cures maintaine all the Sextons in the Towne.
Shakespeare Jest Books (Volume 3)
Conceit, Clichés, Flashes and Whimzies, Number 163
Willis & Sotheran. London, England. 1864

Helmuth, William Tod 1833–1902
American physician

…doctors are the Devil's progeny,
While surgeons come directly down from God!
Scratches of a Surgeon
Surgery vs. Medicine (p. 66)
W.A. Chatterton & Company. Chicago, Illinois, USA. 1879

Jones J.
No biographical data available

It might…be of singular advantage to young surgeons,
particularly before they begin an operation to go through
every part of it attentively in their own minds to consider every possible accident which may happen and to
have the proper remedies at hand in case they should;
and in all operations of delicacy and difficulty to act with
deliberation…
*Plain, concise, practical remarks on treatment of wounds and fractures;
to which is added, a short appendix on camp and military hospitals;
principally designed for the use of young military surgeons in North
America* (pp. 39–40)
John Holt. New York, New York, USA. 1775

Kafka, Franz 1883–1924
German-language novelist

That is what people are like in my district. Always
expecting the impossible from the doctor. They have lost
their ancient beliefs; the parson sits at home and unravels
his vestments, one for another; but the doctor is supposed
to be omnipotent with his merciful surgeon's hands.
Well, as it pleases them; I have not thrust my services
on them…
The Complete Stories

A Country Doctor (p. 224)
Schocken Books. New York, New York, USA. 1971

Massinger, Philip   1583–1640
English dramatic poet

WELL.: Thou wert my surgeon; you must tell no tales;
Those days are done.
I will pay you in private.
A New Way to Pay Old Debts
Act IV, Scene II (p. 123)

Ogilvie, Sir Heneage   1887–1971
English physician

A surgeon conducting a difficult case is like the skipper of an ocean-going racing yacht. He knows the port he must make, but he cannot foresee the course of the journey.
A Surgeon's Life
The Lancet, Volume 255, July 3, 1948 (p. 1)

Paretsky, Sara   1947–
American author

Heart surgeons do not have the world's smallest egos: when you ask them to name the world's three leading practitioners, they never can remember the names of the other two.
In Marilyn Wallace (ed.)
Sisters in Crime (Volume 1)
The Case of the Pietro Andromache II (p. 116)
Berkley Books. New York, New York, USA. 1989

Selzer, Richard   1928–
American physician and essayist

In the operating room the patient must be anaesthetized in order that he or she feel no pain. The surgeon too must be anaesthetized, insulated against the emotional heat of the event so that he can perform this act of laying open the body of a fellow human being, which, take away the purpose for which it is being done, is no more than an act of assault and battery. A barbaric act. So the surgeon dons a carapace which keeps him from feeling. It is what gives many surgeons the appearance of insensitivity.
Speech
Humanities Symposium, Dalhousie University, 1991

Shadwell, Thomas   1642?–92
English dramatist and poet

Oh this Surgeon! this damn'd Surgeon, will this Villainous Quack never come to me? Oh this Plaster on my Neck! It gnaws more than Aqua-Fortis: this abominable Rascle has mistaken sure, and given me the same Caustick he appli'd to my Shins, when they were open'd last.
The Complete Works of Thomas Shadwell (Volume 1)

The Humorists, The First Act (p. 193)
The Fortune Press. London, England. 1927

Twain, Mark (Samuel Langhorne Clemens)   1835–1910
American author and humorist

It is a gratification to me to know that I am ignorant of art, and ignorant also of surgery. Because people who understand art find nothing in pictures but blemishes, and surgeons and anatomists see no beautiful women in all their lives, but only a ghastly stack of bones with Latin names to them, and a network of nerves and muscles and tissues.
Mark Twain's Travels with Mr. Brown
Academy of Design (p. 238)
Alfred A. Knopf. New York, New York, USA. 1946

SURGERY

Clendening, Logan   1884–1945
No biographical data available

Surgery does the ideal thing — it separates the patient from his disease.
Modern Methods of Treatment
Part I, Chapter I (p. 17)
The C.V. Mosby Company. St. Louis, Missouri, USA. 1924

Dennis, F. S.
No biographical data available

There is no science that calls for greater fearlessness, courage, and nerve than that of surgery, none that demands more of self-reliance, principle, independence and the determination in the man. These were the characteristics which were chiefly conspicuous in the early settlers of this country. And it is these old-time Puritan qualities, which descending to them in succeeding generations, have passed into surgeons of America, giving them boldness in their art, and enabling them to win that success in surgery, which now commands the admiration of the civilized world.
Address
The History and Development of Surgery during the Past Century,
International Congress of Arts and Sciences
St. Louis, Missouri, September 1904

Fenger, Carl Emil   1814–84
Danish physician and politician

We must naturally ask ourselves: Does suffering humanity gain anything by this operation? or, in other words, Does the operation enable us to save, or only to prolong, life, and is it worth while for patients having uterine cancer to undergo this severe operation?
The Total Extirpation of the Uterus Through the Vagina
American Journal of Medical Science, Volume 83, January 1883 (p. 45)

Giles, Roscoe C.
No biographical data available

It cannot be too often emphasized, however, that the post-operative treatment is as essential as the operation, and the surgeon is as much responsible for the post-operative treatment as for the operation.
Rickets, the Surgical Treatment of the Chronic Deformities of, with Emphasis on Bow-Legs and Knock-Knees
Journal of the National Medical Association, Volume 14, 1922

Helmuth, William Tod   1833–1902
American physician

There is not one man in a hundred outside of the medical profession, and scarcely one man in ten in it, who understands and appreciates the marvels of modern surgery.
Scratches of a Surgeon
Some of the Wonders of Modern Surgery (p. 42)
W.A. Chatterton & Company. Chicago, Illinois, USA. 1879

Hippocrates   460 BCE–377 BCE
Greek physician

The prime object of the physician in the whole art of medicine should be to cure that which is diseased; and if this can be accomplished in various ways, the least troublesome should be selected; for this is more becoming a good man, and one well skilled in the art, who does not covet popular coin of base alloy.
In *Great Books of the Western World* (Volume 10)
Hippocratic Writings
On the Articulations, 78 (p. 119)
Encyclopædia Britannica, Inc. Chicago, Illinois, USA. 1952

The things relating to surgery, are — the patient; the operator; the assistants; the instruments; the light, where and how; how many things, and how; where the body, and the instruments; the time; the manner; the place.
In *Great Books of the Western World* (Volume 10)
Hippocratic Writings
On the Surgery, 2 (p. 70)
Encyclopædia Britannica, Inc. Chicago, Illinois, USA. 1952

Hubbard, Elbert   1856–1915
American editor, publisher, and author

SURGERY: An adjunct, more or less valuable to the diagnostician.
The Roycroft Dictionary Concocted by Ali Baba and the Bunch on Rainy Days (p. 143)
The Roycrofters. East Aurora, New York, USA. 1914

Johnson, Ernest
No biographical data available

[Back fusions are] like killing a fly on a windowpane with a sledgehammer. The fly is dead, but you've also broken the glass.
That Aching Back
Time, 14 July 1980 (p. 34)

Jones J.
No biographical data available

The exterior of this science, has nothing pleasing or attractive in it, but is rather disgusting to nice, timid, and delicate persons; Its objects too, except in time of war, lying chiefly among the poor and lower classes of mankind, do not excite the industry of the ambitious or avaricious, who find their best account among the rich and great.
Plain, concise, practical remarks on treatment of wounds and fractures; to which Is added, a short appendix on camp and military hospitals; principally designed for the use of young military surgeons in North America (pp. ii–iii)
John Holt. New York, New York, USA. 1775

Kirklin, John   1917–2004
American cardiovascular surgeon

Surgery…is always second best. If you can do something else, it's better. Surgery is limited. It is operating on someone who has no place else to go.
Surgery
Time, May 3, 1963 (p. 60)

Mayo, William J.   1861–1939
American physician

Surgery is more a matter of mental grasp than it is of handicraftsmanship.
Master Surgeons of America; Frederic S. Dennis
Surgery, Gynecology and Obstetrics, Volume 67, October 1938

Ogilvie, Sir Heneage   1887–1971
English physician

Surgery thus attracts the man whose interest in medicine is humanitarian rather than scientific, who loves his fellow men, who wishes to help them and to see that his help is effective. It appeals to the craftsman who enjoys the use of his hands, to the artist whose mind works on visual images, to the romantic who enjoys the drama of life, particularly when it affords him the opportunity to play a decisive role, to the extrovert.
A Surgeon's Life
The Lancet, Volume 255, July 3, 1948 (p. 1)

O'Malley, Austin   1858–1932
American physician and humorist

Surgery: by far the worst snob among the handicrafts.
In Herbert V. Prochnow and Herbert V. Prochnow, Jr.
A Treasury of Humorous Quotations: For Speakers, Writers, and Home Reference
#5724 (p. 322)
Harper & Row, Publishers. New York, New York, USA. 1969

Ovid   43 BCE–17 AD
Roman poet

But that which is incurable must be cut away with the knife, lest the untainted part also draw infection.
Translated by Frank Justus Miller

Metamorphoses (Volume 1)
Book I, l. 190–191 (p. 15)
William Heinemann. London, England. 1916

Selzer, Richard   1928–
American physician and essayist

One enters the body in surgery, as in love, as though one were an exile returning at last to his hearth, daring uncharted darkness in order to reach home.
Mortal Lessons
The Surgeon as Priest (p. 25)
Simon & Schuster. New York, New York, USA. 1976

…surgery is the red flower that blooms among the leaves and thorns that are the rest of Medicine.
Letters to a Young Doctor
Letter II (p. 510)
Simon & Schuster, Inc. New York, New York, USA. 1982

Shaw, George Bernard   1856–1950
Irish comic dramatist and literary critic

We do not go to the operating table as we go to the theatre, to the picture gallery, to the concert room, to be entertained and delighted; we go to be tormented and maimed, lest a worse thing should befall us.… The experts on whose assurance we face this horror and suffer this mutilation should have no interests but our own to think of; should judge our cases scientifically; and should feel about them kindly.
The Doctor's Dilemma
Preface on Doctors
Psychology of Self Respect in Surgeons (p. xxii)
Brentano's. New York, New York, USA. 1920

The notion that therapeutics or hygiene or surgery is any more or less scientific than making or cleaning boots is entertained only by people to whom a man of science is still a magician who can cure diseases, transmute metals, and enable us to live forever.
The Doctor's Dilemma
Preface on Doctors
The Technical Problem (p. lxxxii)
Brentano's. New York, New York, USA. 1920

Sigerist, Henry E.   1891–1957
German-born medical historian

Ignorance is more immediately fatal in surgery than in medicine, or rather, mistakes are more easily apparent to the layman.
A History of Medicine (Volume 2)
Chapter III, Section 1 (p. 203)
Oxford University Press, Inc. New York, New York, USA. 1961

Yeo, R.
No biographical data available

The work was in a moment done.
If possible, without a groan:
So swift thy hand, I could not feel

The progress of the cutting steel.…
For quicker e'en than sense, or thought,
The latent ill view was brought;
And I beheld with ravish'd eyes,
The cause of all my agonies.
And above all the race of men,
I'll bless my GOD for Cheselden.
The Grateful Patient
Gentlemen's Magazine, Volume 2, 1732 (p. 769)

SURPRISE

Faraday, Michael   1791–1867
English physicist and chemist

Let us now consider, for a little while, how wonderfully we stand upon this world. Here it is we are born, bred, and live, and yet we view these things with an almost entire absence of wonder to ourselves respecting the way in which all this happens. So small, indeed, is our wonder, that we are never taken by surprise.
On the Various Forces of Nature and Their Relations to Each Other: A Course of Lectures Delivered Before a Juvenile Audience at the Royal Institution
Lecture I (p. 14)
George Routledge & Sons. New York, USA. 1874

Planck, Max   1858–1947
German physicist

…compared with immeasurably rich, ever young Nature, advanced as man may be in scientific knowledge and insight, he must forever remain the wondering child and must constantly be prepared for new surprises.
Scientific Autobiography and Other Papers
The Meaning and Limits of Exact Science, Part IV (p. 117)
Philosophical Library. New York, New York, USA. 1949

Thomas, Lewis   1913–93
American physician and biologist

The safest and most prudent of bets to lay money on is surprise. There is a very high probability that whatever astonishes us in biology today will turn out to be useable, and useful, tomorrow.
The Medusa and the Snail: More Notes of a Biology Watcher
Medical Lessons from History (p. 172)
The Viking Press. New York, New York, USA. 1979

Watson, Sir William   1858–1935
English author of lyrical and political verse

Strange the world about me lies,
Never yet familiar grown — Still disturbs me with surprise,
Haunts me like a face half known.
The Poems of William Watson
World Stangeness
The Macmillan Company. New York, New York, USA. 1893

SURVEY

Deming, William Edwards   1900–93
American statistician, educator, and consultant

A perfect survey is a myth.
Some Theory of Sampling (p. 24)
John Wiley & Sons, Inc. New York, New York, USA. 1950

A questionnaire is never perfect: some are simply better than others.
Some Theory of Sampling (p. 31)
John Wiley & Sons, Inc. New York, New York, USA. 1950

The only excuse for taking a survey is to enable a rational decision to be made on some problem that has arisen and on which decision, right or wrong, will be made.
Some Theory of Sampling (p. 545)
John Wiley & Sons, Inc. New York, New York, USA. 1950

Deutscher, I.
No biographical data available

…neither the interviewer nor the instrument should act in any way upon the situation. The question, ideally, should be so put and so worded as to be unaffected by contextual contaminations. The interviewer must be an inert agent who exerts no influence or response by tone, expression, stance, or statement. The question must be unloaded in that it does not hint in any way that one response is more desirable or more correct than any other response. It must be placed in the sequence of the instrument in such a way that the subject's response is not affected by previous queries or by his own previous responses.
In S.Z. Nagi and R.G. Corwin
The Social Contexts of Research
Public and Private Opinions: Social Situations and Multiple Realities (p. 325)
John Wiley & Sons, Inc. New York, New York, USA. 1972

Fisher, Sir Ronald Aylmer   1890–1962
English statistician and geneticist

No aphorism is more frequently repeated with field trial, than that we must ask Nature few questions or, ideally, one question at a time. The writer is convinced that this view is wholly mistaken. Nature, he suggests, will best respond to a logical and carefully thought out question-naire; indeed, if we ask her a single question, she will often refuse to answer until some other topic has been discussed.
Journal of the Ministry of Agriculture of Great Britain, Volume 33 (p. 511)

Heinlein, Robert A.   1907–88
American science fiction writer

But what is the purpose of your survey? he asked.
"Does it need a purpose? I tell you, I just made it up."

"But your numbers are too few to be significant. You can't fair a curve with so little data. Besides, your conditions are uncontrolled. Your results don't mean anything."
Beyond this Horizon
Chapter One (p. 2)
Gregg Press. Boston, Massachusetts, USA. 1981

Norton, John K.
No biographical data available

The time of busy people is sometimes wasted by time-consuming questionnaires dealing with incon-sequential topics, worded so as to lead to worthless replies, and circulated by untrained and inexperienced individuals, lacking in facilities for summarizing and disseminating any worthwhile information which they may obtain.
In Douglas R. Berdie and John F. Anderson
Questionnaires: Design and Use (p. ix)
The Scarecrow Press, Inc. Metuchen, New Jersey, USA. 1974

Oppenheim, Abraham Naftali   1924–
No biographical data available

A questionnaire is not just a list of questions or a form to be filled out. It is essentially a scientific instrument for measurement and for collection of particular kinds of data. Like all such instruments, it has to be specifi-cally designed according to particular specifications and with specific aims in mind, and the data it yields are sub-ject to error. We cannot judge a questionnaire as good or bad, efficient or inefficient, unless we know what job it was meant to do. This means that we have to think not merely about the wording of particular questions, but first and foremost, about the design of the investigation as a whole.
Questionnaire Design and Attitude Measurement
Chapter 1 (pp. 2–3)
Basic Books, Inc. New York, New York, USA. 1966

Perelman, Sidney Joseph   1904–79
American comic writer

There is nothing like a good, painstaking survey full of decimal points and guarded generalizations to put a glaze like a Sung vase on your eyeballs.
Keep it Crisp
Sleepy-Time Extra (p. 173)
Random House, Inc. New York, New York, USA. 1946

Strong, Lydia
No biographical data available

Your sales last year just paralleled the sales of rum cokes in Rio de Janeiro, as modified by the sum of the last digits of all new telephone numbers in Toronto. So, why bother with surveys of your own market? Just send away for the data from Canada and Brazil.
Sales Forecasting: Problems and Prospects
Management Review, September 1956 (p. 803)

SURVIVAL

Arnold, Sir Edwin   1832–1904
English poet

How lizard fed on ant, and snake on him,
And kite on both; and how the fish-hawk robbed
The fish-tiger of that which it had seized;
The shrike chasing the bulbul, which did chase
The jeweled butterflies; till everywhere
Each slew a slayer and in turn was slain,
Life living upon death.
Edwin Arnold's Poetical Works (Volume 1)
The Light of Asia, Book the First (p. 21)
Roberts Brothers. Boston, Massachusetts, USA. 1889

Darwin, Charles Robert   1809–82
English naturalist

What a trifling difference must often determine which
shall survive, and which shall perish.
In Francis Darwin (ed.)
The Life and Letters of Charles Darwin (Volume 1)
Darwin to Asa Gray, September 5, 1857 (p. 480)
D. Appleton & Company. New York, New York, USA. 1896

I should premise that I use this term [Struggle for Exis-
tence] in a large and metaphorical sense including dependence of one being on another, and including (which is
more important) not only the life of the individual, but
success in leaving progeny. ...
In *Great Books of the Western World* (Volume 49)
The Origin of Species by Means of Natural Selection
Chapter III (p. 33)
Encyclopædia Britannica, Inc. Chicago, Illinois, USA. 1952

Sagan, Carl   1934–96
American astronomer and science writer

If we have a profound respect for other human beings
as co-equal recipients of this precious patrimony of 4.5
billion years of evolution, why should the identification
not apply also to all the other organisms on Earth which
are equally the product of 4.5 billion years of evolution?
We care for a small fraction of the organisms on Earth
— dogs, cats, and cows, for example — because they are
useful or because they flatter us. But spiders and sala-
manders, salmon and sunflowers are equally our brothers
and sisters.
The Cosmic Connection: An Extraterrestrial Perspective
Chapter 1 (p. 7)
Dell Publishing, Inc. New York, New York, USA. 1975

Thomson, J. Arthur   1861–1933
Scottish biologist

The shore is almost noisy with the conjugation of the
verb to eat in its many tenses.
The Outline of Science (Volume 1)
Chapter III (p. 117)
G.P. Putnam's Sons. New York, New York, USA. 1937

Walker, Marshall
American physicist

The survival technique of the tyrannosaurus was feroc-
ity; it is extinct. The survival technique of the dodo was
passive resistance; it is extinct. The survival technique of
man is science...
The Nature of Scientific Thought
Chapter XV (p. 179)
Prentice-Hall, Inc., Englewood Cliffs, New Jersey, USA. 1963

SYMBIONTICISM

Wallin, Ivan E.   1883–1969
American anatomist

Their universal presence in the cell, coupled with the
known properties of bacteria, appear to indicate that
mitochondria represent the end adjustment of a fundamental biologic process. The establishment of intimate
microsymbiotic complexes has been designated "symbionticism" by the author. ...
Symbionticism and the Origin of Species
Chapter I (p. 8)
Williams & Wilkins Company. Baltimore, Maryland, USA. 1927

Symbionticism, then is proposed as the fundamental
factor or the cardinal principle involved in the origin of
species.
Symbionticism and the Origin of Species
Chapter I (p. 8)
Williams & Wilkins Company. Baltimore, Maryland, USA. 1927

SYMBIOSIS

de Bary, Anton   1831–88
German scieintist and physician

Parasitism, mutualism, lichenism, etc., are each special
cases of that one general association for which the term
symbiosis is proposed as the collective name.
Vortrag auf der Versammlung der Naturforscher und Ärtze zu Cassel
Die Erscheinung der Symbiose (p. 21)
Publisher undetermined

Merezhkovskii, Konstantine   1855–1921
Russian biologist

Above all, a plant, an oak for example, is an animal. An
enormous animal in which live parasites or rather sym-
bionts, an infinite multitude of small microscopic green
organisms, of the species of unicellular "algae," cyano-
phyceae.
In Jan Sapp
Evolution by Association: A History of Symbiosis
Chapter 4 (p. 47)
Oxford University Press, Inc. New York, New York, USA. 1994

Pound, Roscoe   1870–1964
American jurist

Ethically, there is nothing in the phenomena of symbiosis to justify the sentimentalism they have excited in certain writings. Practically, in some instances, symbiosis seems to result in mutual advantage. In all cases it results advantageously to one of the parties, and we can never be sure that the other would not have been nearly as well off, if left to itself.
Symbiosis and Mutualism
The American Naturalist, Volume 27, Number 318, June 1893 (p. 520)

Sapp, Jan
No biographical data available

We have located studies of symbiosis peering through the cracks and creeping across the boundaries which separated ecology from evolution, plants from animals, health from disease, nurture from nature, and the individual from the community. In doing so, we have uncovered layers of oppositions, doctrines and disciplines, and diverse phenomena that have led to disparate interpretations of symbiosis, its scope and significance. In summarizing them here, we see that this history is not a matter of peeling off obstacles to come closer to some hidden core of naked truth. Symbiosis is as much like an onion today as it was a century ago.
Evolution by Association: A History of Symbiosis
Concluding Remarks (p. 205)
Oxford University Press, Inc. New York, New York, USA. 1994

SYMBIOTE

Portier, Paul   1866–1922
French biologist

All Living beings, all animals from Amoeba to Man, all plants from Cryptogams to Dicotyledons are constituted by an association, the "*emboîtement*" of two different beings.
Each living cell contains in its protoplasm formations which histologists designate by the name of "mitochondria." These organelles are, for me, nothing other than symbiotic bacteria, which I call "symbiotes."
Les Symbiotes (p. vii)
Masson. Paris, France. 1918

SYMBOL

Brodie, Sir Benjamin Collins   1817–80
English chemist

A symbol, however, should be something more than a convenient and compendious expression of facts. It is, in the strictest sense, an instrument for the discovery of facts, and is of value mainly with reference to this end, by its adaptation to which it is to be judged.
The Calculus of Chemical Observations
Philosophical Transactions of the Royal Society of London, Volume 156, 1866 (p. 857)

Buchanan, Scott   1895–1968
American educator and philosopher

Symbols, formulae and proofs have another hypnotic effect. Because they are not immediately understood, they, like certain jokes, are suspected of holding in some sort of magic embrace the secret of the universe, or at least some of its more hidden parts.
Poetry and Mathematics
Chapter 1 (p. 37)
The University of Chicago Press. Chicago, Illinois, USA. 1975

Each symbol used in mathematics, whether it be a diagram, a numeral, a letter, a sign, or a conventional hieroglyph, may be understood as a vehicle which someone has used on a journey of discovery.
Poetry and Mathematics
Chapter 2 (p. 47)
The University of Chicago Press. Chicago, Illinois, USA. 1975

Butler, James Newton
No biographical data available
Bobrow, Daniel Gureasko
No biographical data available

"When I use a symbol, it means just what I choose it to mean — neither more nor less."
"The question is, whether you can make symbols mean so many different things?"
"The question is, which is to be master — that's all."
The Calculus of Chemistry
Chapter 2 (p. 7)
W.A. Benjamin, Inc. New York, New York, USA. 1965

Eddington, Sir Arthur Stanley   1882–1944
English astronomer, physicist, and mathematician

The symbol A is not the counterpart of anything in familiar life. To the child the letter A would seem horribly abstract; so we give him a familiar conception along with it.
The Nature of the Physical World
Introduction (p. xiv)
The Macmillan Company. New York, New York, USA. 1930

Goldstein, Herbert   1922–2005
American Physicist and author

It has been remarked in a jocular vein that if H stands for the Hamiltonian, K must stand for the Kamiltonian!
Classical Mechanics (2nd edition)
Classical Transformations (p. 380)
Addison-Wesley Publishing Company. Reading, Massachusetts, USA. 1980

Hilbert, David   1862–1943
German mathematician

Arithmetical symbols are written diagrams and geometrical figures are graphic formulas.
Bulletin of the American Mathematical Society
Mathematical Problems, Volume 8, July 1902 (p. 443)

In the beginning there was the symbol.
In Walter R. Fuchs
Mathematics for the Modern Mind
Chapter 7, Section 7.1 (p. 164)
The Macmillan Company. New York, New York, USA. 1967

Huxley, Aldous   1894–1963
English writer and critic

...some of the greatest advances in mathematics have been due to the invention of symbols, which it afterwards became necessary to explain; from the minus sign proceeded the whole theory of negative quantities.
Jesting Pilate
India and Burma (p. 108)
Chatto & Windus. London, England. 1926

Jung, Carl G.   1875–1961
Swiss psychiatrist and founder of analytical psychology

Thus a word or an image is symbolic when it implies something more than its obvious and immediate meaning. It has a wider "unconscious" aspect that is never precisely defined or fully explained.... As the mind explores the symbols it is led to ideas that lie beyond the grasp of reason.
Man and His Symbols
Part I. The Importance of Dreams (p. 20)
Doubleday & Company, Inc. Garden City, New York, USA. 1964

The Sign is always less than the concept it represents, while a symbol always stands for something more than its obvious and immediate meaning. Symbols, moreover, are natural and spontaneous products.
Man and His Symbols
Part I. The Analysis of Dreams (p. 55)
Doubleday & Company, Inc. Garden City, New York, USA. 1964

Nicholas of Cusa   1401–64
German philosopher, mathematician, and physician

If we approach the Divine through symbols, then it is most suitable that we use mathematical symbols, these have an indestructible certainty.
In Stanley Gudder
A Mathematical Journey (p. 349)
McGraw-Hill Book Company, Inc. New York, New York, USA. 1976

Tillyard, E. M. W.   1889–1962
English classical scholar
Lewis, C. S. (Clive Staples)   1898–1963
English author

Two kinds of symbol must surely be distinguished. The algebraic symbol comes naked into the world of mathematics and is clothed with value by its masters. A poetic symbol — like the Rose, for Love, in Guillaume de Lorris — comes trailing clouds of glory from the real world, clouds whose shape and colour largely determine and explain its poetic use. In an equation, x and y will do as well as a and b; but the Romance of the Rose could not,

without loss, be re-written as the Romance of the Onion, and if a man did not see why, we could only send him back to the real world to study roses, onions, and love, all of them still untouched by poetry, still raw.
The Personal Heresy: A Controversy
Chapter V (p. 97)
Oxford University Press, Inc. London, England. 1939

Truesdell, Clifford   1919–2000
American mathematician, mathematics historian, and natural philospher

There is nothing that can be said by mathematical symbols and relations which cannot also be said by words. The converse, however, is false. Much that can be and is said by words cannot successfully be put into equations, because it is nonsense.
Six Lectures on Modern Natural Philosophy
III, Thermodynamics of Visco-Elasticity (p. 35)
Springer-Verlag. Berlin, West Germany. 1966

Whitehead, Alfred North   1861–1947
English mathematician and philosopher

There is an old epigram which assigns the empire of the sea to the English, of the land to the French, and of the clouds to the Germans. Surely it was from the clouds that the Germans fetched + and − ; the ideas which these symbols have generated are much too important to the welfare of humanity to have come from the sea or from the land.
An Introduction to Mathematics
Chapter 6 (p. 60)
Oxford University Press, Inc. New York, New York, USA. 1958

Mathematics is often considered a difficult and mysterious science, because of the numerous symbols which it employs. Of course, nothing is more incomprehensible than a symbolism which we do not understand. Also a symbolism, which we only partially understand and are unaccustomed to use is difficult to follow. In exactly the same way the technical terms of any profession or trade are incomprehensible to those who have never been trained to use them. But this is not because they are difficult in themselves. On the contrary they have invariably been introduced to make things easy.
An Introduction to Mathematics
Chapter 5 (p. 40)
Oxford University Press, Inc. New York, New York, USA. 1958

...in mathematics, granted that we are giving any serious attention to mathematical ideas, the symbolism is invariably an immense simplification.
An Introduction to Mathematics
Chapter 5 (p. 40)
Oxford University Press, Inc. New York, New York, USA. 1958

SYMBOLIC LOGIC

Whitehead, Alfred North   1861–1947
English mathematician and philosopher

Symbolic Logic has been disowned by many logicians on the plea that its interest is mathematical, and by many mathematicians on the plea that its interest is logical.
Universal Algebra
Preface (p. 6)
Cambridge University Press. Cambridge, England. 1898

SYMMETRY

Aristotle   384 BCE–322 BCE
Greek philosopher

A rose which varies from the ideal of straightness to a hook or snub may still be of good shape and agreeable to the eye.
Politics
Book V, Chapter 9, 1309b [20]
Encyclopædia Britannica, Inc. Chicago, Illinois, USA. 1952

Blake, William   1757–1827
English poet, painter, and engraver

Tyger, Tyger, burning bright
In the forest of the night,
What immortal hand or eye
Could frame thy fearful symmetry?
The Complete Poetry and Prose of William Blake
The Tyger
University of California Press. Berkeley, California, USA. 1982

Borges, Jorge Luis   1899–1986
Argentine writer

…reality favors symmetry.
In Richard Burgin
Conversations with Jorge Luis Borges
Chapter VI (p. 109)
Holt, Rinehart & Winston. New York, New York, USA. 1969

Bulwer, John   1606–56
English physician and writer

True and native beauty consists in the just composure and symmetry of the parts of the body.
Anthropometamorphosis

Carroll, Lewis (Charles Dodgson)   1832–98
English writer and mathematician

You boil it in saw dust: you salt it in glue:
You condense it with locust and tape:
Still keeping one principle object in view —
To preserve its symmetrical shape.
The Complete Works of Lewis Carroll
The Hunting of the Snark
Fit the Fifth (p. 772)
The Modern Library. New York, New York, USA. 1936

Perhaps Looking-glass milk isn't good to drink…
The Complete Works of Lewis Carroll
Through the Looking-Glass
Chapter I (p. 147)
The Modern Library. New York, New York, USA. 1936

Chesterton, G. K. (Gilbert Keith)   1874–1936
English author

Suppose some mathematical creature from the moon were to reckon up the human body; he would at once see that the essential thing about it was that it was duplicate. A man is two men, he on the right exactly resembling him on the left. Having noted that there was an arm on the right and one on the left, a leg on the right and one on the left, he might go further and still find on each side the same number of fingers, the same number of toes, twin eyes, twin ears, twin nostrils, and even twin lobes of the brain. At last he would take it as a law; and then, where he found a heart on one side, would deduce that there was another heart on the other. And just then, where he most felt he was right, he would be wrong.
Orthodoxy
Chapter VI (p. 149)
John Lane Company. New York, New York, USA. 1918

Descartes, René   1596–1650
French philosopher, scientist, and mathematician

Anyone who, upon looking down at his bare feet, doesn't laugh, has either no sense of symmetry or no sense of humor.
In Abdus Salam
The Role of Chirality in the Origin of Life
Journal of Molecular Evolution, Volume 33, Number 2, August 1991 (p. 105)

Ferris, Timothy   1944–
American science writer

…let us pause to slake our thirst one last time at symmetry's bubbling spring.
Coming of Age in the Milky Way
Chapter 20 (p. 385)
William Morrow & Company, Inc. New York, New York, USA. 1988

Frankland, A.
No biographical data available

When the formulae of inorganic compounds are considered even a superficial observer is struck with the general symmetry of their construction; the compounds of nitrogen, phosphorous, antimony and arsenic especially exhibit the tendency of these elements to form compounds containing three or five equivalents of other elements, and it is in these proportions that their affinities are best satisfied.
Philosophical Transactions of the Royal Society of London, Volume 67, 1852 (p. 417)

Herbert, George   1593–1633
English clergyman and metaphysical poet

My body is all symmetry,
Full of proportions, one limb to another,
And all to all the world besides:

Each part may call the farthest, brother:
For head with foot hath private smity,
And both with moon and tides.
The Works of George Herbert
The Temple, Man
Thomas Y. Crowell & Company. Birmingham, England. No date

Joyce, James   1882–1941
Irish expatriate writer and poet

ZOE: Come and I'll peel off.
BLOOM: (Feeling his occiput dubiously with the unpar-
alleled embarrassment of a harassed peddler gauging
the symmetry of her peeled pears.) Somebody would be
dreadfully jealous if she knew.
Ulysses (p. 490)
Random House, Inc. New York, New York,
USA. 1946

Kaku, Michio   1947–
Japanese-American theoretical physicist
Thompson, Jennifer Trainer
American author

…nature, at the fundamental level, does not just prefer
symmetry in a physical theory; nature demands it.
Beyond Einstein: The Cosmic Quest for the Theory of the Universe
Chapter 6 (p. 108)
Bantam Books. Toronto, Ontario, Canada. 1987

Mackay, Charles   1814–89
English poet and journalist

Truth…and if mine eyes
Can bear its blaze, and trace its symmetries,
Measure its distance, and its advent wait,
I am no prophet — I but calculate.
The Poetical Works of Charles Mackay
The Prospects of the Future
G. Routledge & Sons. London, England. 1857

Mao Zedong   1893–1976
Chinese political and military leader and Communist Party chairman

Tell me why should symmetry be of importance?
In T.D. Lee
Symmetries, Asymmetries, and the World of Particles
30 May, 1974 (p. xi)
Washington University Press. Seattle, Washington, USA, and London,
England. 1988

Newman, James Roy   1907–66
Mathematician and mathematical historian

Symmetry establishes a ridiculous and wonderful cous-
inship between objects, phenomena, and theories outwardly unrelated: terrestrial magnetism, women's veils,
polarized light, natural selection, the theory of groups,
invariants and transformations, the work habits of bees
in the hive, the structure of space, vase designs, quantum physics, scarabs, flower petals, X–ray interference

patterns, cell division in sea urchins, equilibrium posi-
tions in crystals, Romanesque cathedrals, snowflakes,
music, the theory of relativity.
The World of Mathematics (Volume 1)
Commentary On Symmetry (p. 670)
Simon & Schuster. New York, New York, USA. 1956

Pascal, Blaise   1623–62
French mathematician and physicist

Those who make antitheses by forcing words are like
those who make false windows for symmetry.
In *Great Books of the Western World* (Volume 33)
Pensées
Section I, 27
Encyclopædia Britannica, Inc. Chicago, Illinois, USA. 1952

Symmetry is what we see at a glance…
In *Great Books of the Western World* (Volume 33)
Pensées
Section I, 28
Encyclopædia Britannica, Inc. Chicago, Illinois, USA. 1952

Valéry, Paul   1871–1945
French poet and critic

The universe is built on a plan the profound symmetry of
which is somehow present in the inner structure of our
intellect.
In Jefferson Hane Weaver
The World of Physics (Volume 2)
O.2 (p. 521)
Simon & Schuster. New York, New York, USA. 1987

Warner, Sylvia Townsend   1893–1978
English novelist and poet

An umbrella, Lueli, when in use resembles the — the
shell that would be formed by rotating an arc of curve
about its axis of symmetry, attached to a cylinder of small
radius whose axis is the same as the axis of symmetry of
the generating curve of the shell. When not in use it is
properly an elongated cone, but it is more usually heli-
codial in form.
Lueli made no answer. He lay down again, this time face
downward.
Mr. Fortune's Maggot
Mr. Fortune's Maggot (p. 115)
New York Review of Books. New York, New York, USA. 1927

Weyl, Hermann   1885–1955
German mathematician

Symmetry, as wide or as narrow as you may define its
meaning, is one idea by which man through the ages
has tried to comprehend and create order, beauty, and
perfection.
Symmetry
Bilateral Symmetry (p. 5)
Princeton University Press. Princeton, New Jersey, USA. 1960

As far as I can see, all a priori statements in physics have their origin in symmetry.
Symmetry
Crystals. The General Mathematical Idea of Symmetry (p. 126)
Princeton University Press. Princeton, New Jersey, USA. 1952

Symmetry is a vast subject, significant in art and nature. Mathematics lies at its root, and it would be hard to find a better one on which to demonstrate the working of the mathematical intellect.
Symmetry
Crystals: The General Mathematical Idea of Symmetry (p. 145)
Princeton University Press. Princeton, New Jersey, USA. 1960

Wickham, Anna (Edith Alice Mary Harper)   1884–1947
English poet

God, Thou great symmetry,
Who put a biting lust in me
From whence my sorrows spring
For all the frittered days
That I have spent in shapeless ways
Give me one perfect thing.
The Contemplative Quarry, and The Man with a Hammer
Envoi
Harcourt Brace. New York, USA. 1921

Yang, Chen Ning   1922–
Chinese-born American theoretical physicist

Nature seems to take advantage of the simple mathematical representations of the symmetry laws. When one pauses to consider the elegance and the beautiful perfection of the mathematical reasoning involved and contrast it with the complex and far-reaching physical consequences, a deep sense of respect for the power of the symmetry laws never fails to develop.
Nobel Lectures, Physics 1942–1962
Nobel lecture for award received in 1957
The Law of Parity Conservation and Other Symmetry Laws of Physics (p. 394)
Elsevier Publishing Company. Amsterdam, Netherlands. 1964

Zee, Anthony
Chinese-American physicist and author

Pick your favorite group: write down the Yang-Mills theory with your groups as its local symmetry group; assign quark fields, lepton fields, and Higgs fields to suitable representations; let the symmetry be broken spontaneously. Now watch to see what the symmetry breaks down to. ...that, essentially, is all there is to it. Anyone can play. To win, one merely has to hit on the choice used by the Greatest Player of all time. The prize? Fame and glory, plus a trip to Stockholm.
Fearful Symmetry
Chapter 14 (pp. 253–254)
Macmillan Publishing Company. New York, New York, USA. 1986

SYMPTOM

Hare, Hobart Amory   1862–1931
American physician

A clear understanding by the physician of the value of the symptoms of disease which he sees and of those described by the patient is of vital importance for the purpose of diagnosis and treatment, and one of the advantages of older physicians over their younger brethren is the ability which they have gained through long training to grasp the essential details of a case almost at their first glance at the patient.
Practical Diagnosis
Introduction (p. 17)
Lea Brothers & Company. Philadelphia, Pennsylvania, USA. 1902

Latham, Peter Mere   1789–1875
English physician

It is by symptoms, and by symptoms only, that we can learn the existence, and seat, and nature, of diseases in the living body, or can direct and methodize their treatment.
In William B. Bean
Aphorisms from Latham (p. 59)
Prairie Press. Iowa City, Iowa, USA. 1962

SYNTHESIS

Berthelot, Marcellin   1827–1907
French chemist

The domain in which chemical synthesis exercises its creative power is vaster than that of nature herself.
In Philip Ball
Designing the Molecular World: Chemistry at the Frontier (p. 13)
Princeton University Press. Princeton, New Jersey, USA. 1994

Dalton, John   1766–1844
English chemist and physicist

Chemical analysis and synthesis go no farther than the separation of particles one from another, and their reunion. No new creation or destruction of matter is within the reach of chemical agency. We might as well attempt to introduce a new planet into the solar system, or to annihilate one already in existence, as to create or destroy a particle of hydrogen.
A New System of Chemical Philosophy (Volume 1)
Part I, Chapter III (p. 212)
R. Bickerstaff. London, England. 1810

Mayr, Ernst   1904–2005
German-born American biologist

What is still lacking is a critical analysis of the writings of the architects of the synthesis.
The Growth of Biological Thought: Diversity, Evolution, Inheritance
Chapter 12 (p. 568)
Harvard University Press. Cambridge, Massachusetts, USA. 1982

We didn't sit down together and forge a synthesis. We all knew each other's writings; all spoke with each other. We all had the same goal, which was simply to understand fully the evolutionary process.... By combining our knowledge, we managed to straighten out all the conflicts and disagreements so that finally a united picture of evolution emerged.
In Pamela Weintraub (ed.)
The Omni Interviews
Darwin Flights (p. 47)
Ticknor & Fields. New York, New York, USA. 1984

The term "evolutionary synthesis" was introduced by Julian Huxley in Evolution: The Modern Synthesis to designate the general acceptance of two conclusions: gradual evolution can be explained in terms of small genetic changes ("mutations") and recombination, and the ordering of this variation by natural selection; and the observed evolutionary phenomena, particularly macroevolutionary processes and speciation, can be explained in a manner that is consistent with the known genetic mechanisms.
The Evolutionary Synthesis
Prologue: Some Thoughts on the History of Evolutionary Synthesis (p. 1)
Harvard University Press. Cambridge, Massachusetts, USA. 1980

Seebach, D.
No biographical data available

No matter what the narrow goal of any particular project, whether the work involved is groundbreaking or of a more routine nature, synthesis and analysis are crucial to every chemist's activities.
Organic Synthesis — Where Now?
Angewandet Chemie International Edition in English, Volume 29, 1990
(p. 1321)

Vivilov, N. I.
No biographical data available

We are now entering an epoch of differential ecological, physiological and genetic classification. It is an immense work. The ocean of knowledge is practically untouched by biologists. It requires the joint labors of many different specialists — physiologists, cytologists, geneticists, systematists, and biochemists. It requires international spirit, the cooperative work of investigators throughout the whole world...it will bring us logically to the next step: integration and synthesis.
In Julian Huxley
The New Systematics
The New Systematics of Cultivated Plants (p. 565)
University Press. Oxford, England. 1940

von Goethe, Johann Wolfgang   1749–1832
German poet, novelist, playwright, and natural philosopher

"Affinities begin really to interest only when they bring about separations."

"What...is that miserable word, which unhappily we hear so often now-a-days — in the world; is that to be found in naure's lessons too?"
"Most certainly," answered Edward; "the title with which chemists were supposed to be most honorably distinguished was, artists of separation."
"It is not so any more," replied Charlotte; "and it is well that it is not. It is a higher art, and it is a higher merit, to unite. An artist of union is what we should welcome in every province of the universe."
Elective Affinities
Chapter IV (p. 38)
Frederick Unger Publishing Company. New York, New York, USA. 1962

Woodward, Robert Burns   1917–79
American chemist

The structure known, but not yet accessible by synthesis, is to the chemist what the unclimbed mountain, the uncharted sea, the untilled field, the unreached planet, are to other men.... The unique challenge which chemical synthesis provides for the creative imagination and the skilled hands ensures that it will endure as long as men write books, paint pictures, and fashion things which are beautiful, or practical, or both.
In William H. Brock
The Norton History of Chemistry
Chapter 16 (p. 633)
W.W. Norton & Company, Inc. New York, New York, USA.1993

[S]ynthetic objectives are seldom if ever taken by chance, nor will the most painstaking, or inspired, purely observational activities suffice. Synthesis must always be carried out by plan, and the synthetic frontier can be defined only in terms of the degree to which realistic planning is possible, utilizing all of the intellectual and physical tools available. It can scarcely be gainsaid that the successful outcome of a synthesis of more than thirty stages provides a test of unparalleled rigor of the predictive capacity of the science, and of the degree of its understanding of its portion of the environment.
In A.R. Todd (ed.)
Perspectives in Organic Chemistry
Synthesis (p. 155)
Interscience, Inc. New York, New York, USA. 1956

SYSTEM

Coates, Robert M.
No biographical data available

He has so clearly laid open and set before our eyes the most beautiful frame of the System of the World, that if King Alphonse were now alive he would not complain for want of the graces of simplicity or of harmony in it.
In Robert H. March
Physics for Poets

Preface to the Principia (p. 35)
McGraw-Hill, Inc. New York, New York, USA. 1996

Lavoisier, Antoine Laurent   1743–94
French chemist

Systems in physical science…are no more than appropriate instruments to aid the weakness of our organs: they are, properly speaking, approximate methods which put us on the path to the solution of the problem; these are the hypotheses which, successively modified, corrected, and changed in proportion as they are found false, should lead us infallibly one day, by a process of exclusion, to the knowledge of the true laws of nature.
Mémoires de l'Académie Royale des Sciences 1777
Memoir on Combustion in General (p. 592)

Facts, observations, experiments — these are the materials of a great edifice, but in assembling them we must combine them into classes, distinguish which belongs to which order and to which part of the whole each pertains.
Mémoires de l'Académie Royale des Sciences 1777
Memoir on Combustion in General (p. 592)

As dangerous as is the desire to systematize in the physical sciences, it is, nevertheless, to be feared that in storing without order a great multiplicity of experiments we obscure the science rather than clarify it, render it difficult of access to those desirous of entering upon it, and finally, obtain at the price of long and tiresome work only disorder and confusion.
Mémoires de l'Académie Royale des Sciences 1777
Memoir on Combustion in General (p. 592)

Peacock, Thomas Love   1785–1866
English writer

All philosophers who find
Some favorite system to their mind,
In every point to make it fit
Will force all nature to submit.
Headlong Hall (p. 44)
J.M. Dent & Company London, England. No date

Thomas, Lewis   1913–93
American physician and biologist

You cannot meddle with one part of a complex system from the outside without the almost certain risk of setting off disastrous events that you hadn't counted on in other, remote parts.
The Medusa and the Snail: More Notes of a Biology Watcher
On Meddling (p. 110)
The Viking Press. New York, New York, USA. 1979

Thompson, W. R.   1887–?
Canadian entomologist

The good systematist develops what the medieval philosophers called a habitus, which is more than a habit and is better designated by its other name of *secunda natura*. Perhaps, like a tennis player or a musician, he works best when he does not get too introspective about what he is doing.
The Philosophical Foundation of Systematics
Canadian Entomology, Volume 84, 1952 (p. 5)

von Goethe, Johann Wolfgang   1749–1832
German poet, novelist, playwright, and natural philosopher

Natural system — a contradiction in terms. Nature has no system; she has, she is life and its progress from an unknown center toward an unknowable goal. Scientific research is therefore endless, whether one proceed analytically into minutiae or follow the trail as a whole, in all its breadth and height.
Goethe's Botanical Writings
Problems (p. 116)
University of Hawaii Press. Honolulu, Hawaii, USA. 1952

SYSTEMATICS

Darwin, Charles Robert   1809–82
English naturalist

Systematize and study affinities.
The Autobiography of Charles Darwin, 1809–1882: With Original Omissions Restored
Appendix, Quotations (p. 160)
Harcourt, Brace. New York, New York, USA. 1959

de Queiroz, K.
Phylogeneticist
Donoghue, M. J.
Phylogeneticist

If the goal of systematics is to depict relationships accurately, then any tradition that interferes with this goal should be abandoned.
Phylogenetic Systematics of Nelson's Version of Cladistics
Cladistics, Volume 4, Number 4, December 1988 (p. 332)

Elton, Charles S.   1900–91
English biologist

The extent to which progress in ecology depends upon accurate identification and upon the existence of sound systematic groundwork for all groups of animals, cannot be too much impressed upon the beginner in ecology. This is the essential basis of the whole thing; without it the ecologist is helpless, and the whole of his work may be rendered useless.
Animal Ecology
Chapter XI (p. 165)
Sidgwick & Jackson, Ltd. London, England. 1927

Hennig, W.
No biographical data available

In order to be able to judge correctly the position of systematics in the field of biology and the role that it is called upon to play in the solution of the basic problems of this science, one must first make clear that there is a systematics not only in biology, but that it is rather an integrating part of any science whatever. It is surprising and peculiar to see to what degree the original significance of this concept has been forgotten in biology in the course of the fundamentally inadmissible but now general limitation of the concept of systematics to a particular subdivision of the science as a whole.
In George Gaylord Simpson
Principles of Animal Taxonomy
Systematics, Taxonomy, Classification, Nomenclature (p. 6)
Columbia University Press. New York, New York, USA. 1961

Hutton, James   1726–97
Scottish geologist, chemist, and naturalist

…the system of this earth has either been intentionally made imperfect, or has not been the work of infinite power and wisdom.
The Theory of the Earth (Volume 1)
Part I, Chapter I, Section I (p. 17)
Messrs Cadwell, Junior, and Davies. London, England. 1795

Mayr, Ernst   1904–2005
German-born American biologist

The amount of diversity in the living world is staggering. About 1 million species of animals and half a million species of plants have already been described, and estimates on the number of still undescribed species range from 3 to 10 million.… It would therefore be impossible to deal with this enormous diversity if it were not ordered and classified. Systematic zoology endeavors to order the rich diversity of the animal world and to develop methods and principles to make this task possible.
Principles of Systematic Zoology
Chapter 1 (p. 1)
McGraw-Hill Book Company, Inc. New York, New York, USA. 1969

The systematist who studies the factors of evolution wants to find out how species originate, how they are related, and what this relationship means. He studies species not only as they are, but also their origin and changes. He tries to find his answers by observing the variability of natural populations under different external conditions and he attempts to find out which factors promote and which inhibit evolution. He is helped in this endeavor by his knowledge of the habits and the ecology of the studied species.
Systematics and the Origin of Species
Chapter I (p. 11)
Harvard University Press. Cambridge, Massachusetts, USA. 1942

Simpson, George Gaylord   1902–84
American paleontologist

Systematics is the scientific study of the kinds and diversity of organisms and of any and all relationships among them.
Principles of Animal Taxonomy
Systematics, Taxonomy, Classification, Nomenclature (p. 7)
Columbia University Press. New York, New York, USA. 1961

T

TABLE

Carlyle, Thomas 1795–1881
English historian and essayist

Tables are like cobwebs, like the sieve of the Danaides; beautifully reticulated, orderly to look upon, but which will hold no conclusion. Tables are abstractions.... There are innumerable circumstances; and one circumstance left out may be the vital one on which all turned.... Conclusive facts are inseparable from inconclusive except by a head that already understands and knows.
English and Other Critical Essays
Chartism, Chapter II (p. 170)
J.M. Dent & Sons Ltd. London, England. 1950

Devons, Ely 1913–67
English economist

The way statistics are presented, their arrangement in a particular way in tables, the juxtaposition of sets of figures, in itself reflects the judgment of the author about what is significant and what is trivial in the situation which the statistics portray.
Essays in Economics
Chapter 6 (p. 109)
Greenwood Press, Publishers. Westport, Connecticut, USA. 1961

Fisher, Sir Ronald Aylmer 1890–1962
English statistician and geneticist

[Referring to] ...witty comments made by A.L. Bowley] (a) The terms used in the headings and margins of the table are all employed in a technical sense, known only to the officers who compiled it, and which they are unable for official reasons to divulge. (b) The sub-divisions of the table and the region to which it refers have been changed since the last return was published. (c) Before tabulation the data have been subjected to numerous adjustments, allowances and other corrections, of a kind to vitiate any tests of significance which the reader may be tempted to apply to them.
Presidential Address, First Indian Statistical Conference, 1938
Sankhya, Volume 4, 1938 (p. 15)

Playfair, William 1759–1823
English publicist

Information that is imperfectly acquired, is generally as imperfectly retained; and a man who has carefully investigated a printed table, finds, when done, that he has only a very faint and partial idea of what he has read; and that like a figure imprinted on sand, is soon totally erased and defaced.
The Commercial and Political Atlas (p. 3)
Printed for J. Debrett. London, England. 1786

TACHYON

Herbert, Nick
American physicist

Although most physicists today place the probability of the existence of tachyons only slightly higher than the existence of unicorns, research into the properties of these hypothetical FTL [faster than light] particles has not been entirely fruitless.
Faster Than Light: Superluminal Loopholes in Physics
Chapter 7 (p. 137)
New American Library. New York, New York, USA. 1988

Nahin, Paul J.
American electrical engineering professor and author

...if tachyons are one day discovered...the day before the momentous occasion a notice from the discoverers should appear in newspapers announcing "tachyons have been discovered tomorrow."
Time Machines: Time Travel in Physics, Metaphysics, and Science Fiction
Notes and References, Note 36 (p. 408)
Springer-Verlag. New York, New York, USA. 1993

TAXONOMIST

Moss, W. W.
No biographical data available

Taxonomists have always had the reputation of being difficult. Intransigence may be rooted in the necessity of defending prolonged self-immersion in a taxon that others find a total bore; it is frustrating to have one's work greeted with a yawn.
In J. Felsenstein (ed.)
Numerical Taxonomy
Taxa, Taxonomists, and Taxonomy (p. 73)
Springer-Verlag. New York, New York, USA. 1983

Numerical taxonomists have proved to be just as prickly as conventional taxonomists, possibly more so because some of the brightest people in systematics are involved in the current taxonomic battles. The political maneuvering and character assassination that characterize certain taxonomists today may not be atypical for science; they certainly provide a fine example of its seamier side. If Feyerabend is correct, it may even be a requirement of human nature that scientific progress occur in this manner.
In J. Felsenstein (ed.)
Numerical Taxonomy
Taxa, Taxonomists, and Taxonomy (p. 73)
Springer-Verlag. New York, New York, USA. 1983

TAXONOMY

Abbott, Donald Putnam   1920–86
American marine biologist and professor

You'll be tempted to grouse about the instability of taxonomy: but stability occurs only where people stop thinking and stop working.
In Galen Howard Hilgard (ed.)
Observing Marine Invertebrates: Drawings from the Laboratory
Author's Preface (p. xvi)
Stanford University Press. Stanford, California, USA. 1987

Blackwelder, R. E.
No biographical data available

Much has been written in recent decades about subspecies and their use in taxonomy. There are strong feelings that they are usable, useful and desirable. There are also strong feelings that they are not really relevant to taxonomy and are an unnecessary encumbrance to classification and nomenclature.
Taxonomy: A Text and Reference Book (p. 171)
John Wiley & Sons, Inc. New York, New York, USA. 1967

Brew, John O.   1906–88
American archaeologist

...classificatory systems are merely tools, tools of analysis, manufactured and employed by students, just as shovels, trowels and whisk brooms are tools of excavation.
Papers of the Peabody Museum of American Archaeology and Ethnology
The Archaeology of Alkali Ridge, Southeastern Utah
The Use and Abuse of Taxonomy, Volume 21 (p. 46)
Peabody Museum. Cambridge, Massachusetts, USA. 1946

Burtt, B. L.
No biographical data available

Numerical taxonomy uses statistical methods to form groups whereas traditional taxonomy only uses them to discriminate more precisely between groups already perceived. If it becomes increasingly apparent that there is a fundamental divergence here, let us remember Whitehead's dictum, that a clash of doctrines is not a disaster — it is an opportunity.
Andanson and Modern Taxonomy
Edinburgh Royal Botanic Gardens, Notes, Volume 26, 1966

Cain, A. J.
Taxonomist

It is not extraordinary that young taxonomists are trained like performing monkeys, almost wholly by imitation, and that in only the rarest cases are they given any instruction in taxonomic theory.
In George Gaylord Simpson
Principles of Animal Taxonomy

Preface (p. vii)
Columbia University Press. New York, New York, USA. 1961

Constance, L.
No biographical data available

Plant taxonomy has not outlived its usefulness: it is just getting under way on an attractively infinite task.
Plant Taxonomy in an Age of Experiment
American Journal of Botany, Volume 44, Number 1, January 1957
(p. 92)

Gould, Stephen Jay   1941–2002
American paleontologist and evolutionary biologist

Parochial taxonomies are a curse of intellectual life.
Time's Arrow, Time's Cycle: Myth and Metaphor in the Discovery of Geological Time
Chapter 1 (p. 3)
Harvard University Press. Cambridge, Massachusetts, USA. 1987

Taxonomy (the science of classification) is often undervalued as a glorified form of filing — with each species in its folder, like a stamp in its prescribed place in an album; but taxonomy is a fundamental and dynamic science, dedicated to exploring the causes of relationships and similarities among organisms. Classifications are theories about the basis of natural order, not dull catalogues compiled only to avoid chaos.
Wonderful Life: The Burgess Shale and the Nature of History
Chapter III (p. 98)
W.W. Norton & Company, Inc. New York, New York, USA. 1989

Heywood, V. H.   1947–
No biographical data available

In these days when Molecular Biology is beginning to be seen as a restricted science, narrowing our vision by concentrating on the basic uniformity of organisms at the macromolecular level, the need for taxonomists to draw attention to the enormous diversity and variation of this earth's biota becomes more and more pressing.
In Tod. F. Stuessy
Plant Taxonomy: The Systematic Evaluation of Comparative Data
Plant Taxonomy (p. xvii)
Columbia University Press. New York, New York, USA. 1990

Jeffrey, C.   1866–1952
Canadian-American botanist

The bringing to light of overlooked names in the old literature is perhaps nearing completion...it is hoped that this will lead to name-changes for nomenclature reasons becoming ever fewer and fewer until eventually they cease to trouble us. Unfortunately, the same cannot be said of name-changes that become necessary for taxonomic reasons. These arise from taxonomic research itself and are inevitable accompaniments of our systems of classifications.
Biological Nomenclature (3rd edition) (p. 31)
Edward Arnold. London, England. 1989

Kevan, D. Keith McE.   1920–1991
British ethnoentomologist

Bad taxonomy, of which there has been plenty, persists. Unlike bad chemistry or bad physiology, of which there has probably been equally as much, it cannot be ignored; it must be undone and redone. Poor taxonomy is not only an ill unto itself; it is contagious, often with a very long incubation period.… One assumes that when [experimental biologists] state that they used 5 ml ethanol, they were not using 6 ml of methanol; and yet, if the experimental animal is wrongly identified, what are the grounds for such an assumption?
The Place of Classical Taxonomy in Modern Systematic Entomology
Canadian Entomology, Volume 105, 1973 (p. 1212)

Rollins, R. C.   1911–1998
No biographical data available

In other words, the field of taxonomy in a way epitomizes the work of all other branches of biology centered on the organism itself, and brings the varied factual information from them to bear on the problems of interrelationship, classification and evolution. Thus taxonomy, as has been aptly remarked, is at once the alpha and omega of biology.
Taxonomy of the Higher Plants
American Journal of Botany, Volume 44, Number 1, January 1957
(p. 188)

Simpson, George Gaylord   1902–84
American paleontologist

Taxonomy is a science, but its application to classification involves a great deal of human contrivance and ingenuity, in short, of art. In this art there is leeway for personal taste, even foibles, but there are also canons that help to make some classifications better, more meaningful, more useful than others.
Principles of Animal Taxonomy
From Taxonomy to Classification (p. 107)
Columbia University Press. New York, New York, USA. 1961

Stace, C.   1938–
English botanist and author

The species is the lowest rank which it is essential to recognize for general taxonomic purposes. It is not, however, the lowest rank which it is desirable and useful to recognize.
Plant Taxonomy and Biosystematics (2nd edition) (p. 192)
Edward Arnold Publishers Ltd. London, England. 1989

Stuessy, Tod F.   1943–
No biographical data available

We as taxonomists celebrate diversity. We celebrate the wildness of the planet. We celebrate the numerous human attempts to understand this wilderness, and we mourn its loss through human miscalculation. We sense the aesthetic of life and much of our efforts are aimed at reflecting this composition. Above all we celebrate the challenges of being alive and dealing with the living world. There is no greater responsibility, privilege, nor satisfaction.
Plant Taxonomy: The Systematic evaluation of Comparative Data
Epilogue (p. 406)
Columbia University Press. New York, New York, USA. 1990

The rise beyond the generic level in classification is to enter a world of much greater uncertainty.… Taxa at higher levels will be well-defined or ill-defined depending on the group in question.
Plant Taxonomy: The Systematic Evaluation of Comparative Data
(p. 207)
Columbia University Press. New York, New York, USA. 1990

Wald, George   1906–97
American biologist and biochemist

The most important thing about a name, after all, is that it remain attached to the thing it designates. One wishes that once a name had come into common use for an organism, it could be stabilized for the use of busy persons who want nothing but that each animal have a name.
In E.S. Guzman Barron (ed.)
Modern Trends in Physiology and Biochemistry
Biochemical Evolution (fn p. 339)
Academic Press, Inc. New York, New York, USA. 1952

TEACHER

Alexander, Burton F.
American mathematics teacher

The responsibility of developing and improving the general and technical vocabularies that are associated with elementary mathematics lies in the hand of the subject teacher.
Language Development in Mathematics Through Vocabularies
The Mathematics Teacher, Volume 40, Number 8, December 1947
(p. 389)

Nunn, T. F.
No biographical data available

It is not sufficient that the teacher should have a competent knowledge of the subject which he professes…he must (in addition) have considered his science from the point of view at which it appears as a human acquisition.
In Lloyd William Taylor
Physics: The Pioneer Science (Volume 1)
Chapter 8 (p. 111)
Houghton Mifflin Company. Boston, Massachusetts, USA. 1941

TEACHING

Acton, F. S.
No biographical data available

When an engineer apologetically approaches a statistician, graph in hand, and asks how he should fit a straight line to these points, the situation is not unlike the moment when one's daughter inquires where babies come from. There is a need for tact, there is a need for delicacy, but here is opportunity for enlightenment and it must not be discarded casually — or destroyed with the glib answer.
National Bureau of Standards Report 12–10–51 (p. 1)
U.S. Government Printing Office, Washington, D.C. 1951

Author undetermined

It is not needful in the present day to discourage thinkers, they are not too numerous.
Exclusion of Opinion
Westminster Review, Volume 29, 1838 (p. 49)

Barzun, Jacques   1907–
French-born American educator, historian, and author

To begin with, no school subject should be treated like a bitter pill that will go down only if sugar-coated. The merest hint of this confirms the pupil's belief that he faces something dreadful and is a victim.
Begin Here: The Forgotten Conditions of Teaching and Learning (p. 81)
University of Chicago Press. Chicago, Illinois, USA. 1992

Berrill, Norman John   1903–96
English-born American biologist

A great teacher is not simply one who imparts knowledge to his students but is one who awakens their interest in the subject and makes them eager to pursue it for themselves. An outstanding teacher is a spark plug, not a fuel line.
In B.W. Rossiter
Journal of Chemical Education, Volume 49, 1972 (p. 388)

Blake, William   1757–1827
English poet, painter, and engraver

To teach doubt and Experiment
Certainly was not what Christ meant.
The Complete Writings of William Blake
The Everlasting Gospel, d, l. 49
Houghton Mifflin Company. Boston, Massachusetts, USA. 1904–1917

Boerhaave, Herman   1668–1738
Dutch Chemist, physician, and botanist

As you have put your selves under my care, to instruct you in the knowledge of Chemistry, I shall think it my duty to endeavor as much as possible to answer your expectations.
Elements of Chemistry (Volume 1)
The Design (p. 1)
Printed for J. & J. Pemberton. London, England. 1735

Bruner, Jerome Seymour   1915–
American psychologist

The teaching of probabilistic reasoning, so very common and important a feature of modern science, is hardly developed in our educational system before college.
The Process of Education
Chapter 3 (p. 45)
Harvard University Press. Cambridge, Massachusetts, USA. 1961

Darwin, Charles Robert   1809–82
English naturalist

I am inclined to give up the attempt as hopeless. Those who do not understand, it seems cannot be made to understand.
In Francis Darwin (ed.)
The Life and Letters of Charles Darwin (Volume 2)
Darwin to Hooker, June 5th, 1860 (p. 110)
D. Appleton & Company. New York, New York, USA. 1896

Ehrensvärd, Gosta Carl Henrik   1910–1980
Swedish biochemist

...consciousness will always be one dimension above comprehensibility.
Translated by Lennart Rodén
Man on Another World
Chapter X (p. 151)
The University of Chicago Press. Chicago, Illinois, USA. 1961

Eldridge, Paul   1888–1982
American educator

What difference does it make how often we lower and raise the bucket into the well if the bucket has no bottom?
Maxims for a Modern Man
484
T. Yoseloff. New York, New York, USA. 1965

Faraday, Michael   1791–1867
English physicist and chemist

The most prominent requisite to a lecturer, though perhaps not really the most important, is a good delivery; for though to all true philosophers science and nature will have charms innumerably in every dress, yet I am sorry to say that the generality of mankind cannot accompany us one short hour unless the path is strewed with flowers.
In J.M. Thomas
Michael Faraday — and the Royal Institution
Chapter 5 (p. 97)
Adam Hilger. Bristol, England. 1991

...a truly popular lecture cannot teach, and a lecture that truly teaches cannot be popular.
In J.M. Thomas
Michael Faraday — and the Royal Institution
Chapter 8 (p. 192)
Adam Hilger. Bristol, England. 1991

A lecturer should appear easy and collected, undaunted and unconcerned, his thoughts about him and his mind clear for the contemplation and description of his subject. His action should be slow, easy and natural consisting principally in changes of posture of the body, in order to avoid the air of stiffness or sameness that would otherwise be unavoidable.

In J.M. Thomas
Michael Faraday — and the Royal Institution
Chapter 3 (p. 18)
Adam Hilger. Bristol, England. 1991

Feyerabend, Paul K.   1924–94
Austrian-born American philosopher of science

A good teacher will not just make people accept a form of life, he will also provide them with means of seeing it in perspective and perhaps of even rejecting it.

In Gerard Radnitzky and Gunnar Andersson
The Structure and Development of Science
Dialogue on Method (p. 86)
D. Ridel Publishing Company. Dordrecht, Germany. 1979

Feynman, Richard P.   1918–88
American theoretical physicist

What I am going to tell you about is what we teach our physics students in the third or fourth year of graduate school.… It is my task to convince you not to turn away because you don't understand it. You see my physics students don't understand it.… That is because I don't understand it. Nobody does.

QED: The Strange Theory of Light and Matter
Chapter 1 (p. 9)
Princeton University Press. Princeton, New Jersey, USA. 1985

Gauss, Johann Carl Friedrich   1777–1855
German mathematician, physicist, and astronomer

I am giving this winter two courses of lectures to three students, of which one is only moderately prepared, the other less than moderately, and the third lacks both preparation and ability. Such are the onera of a mathematical profession.

Briefwechsel zwischen Gauss und Bessel
Letter 4, Letter to Bessel, January 7, 1810 (p. 107)

Heiss, E. D.   1899–?
No biographical data available
Osbourn, E. S.   1897–?
No biographical data available

Science teaching has long concerned itself chiefly with the mastery of laws, facts, and principles to the neglect of certain of the less tangible, but non the less desirable outcomes, such as attitude of mind.

Modern Methods and Materials for Teaching Science
Chapter 2 (p. 15)
The Macmillan Company. New York, New York, USA. 1940

Herbert, George   1593–1633
English metaphysical poet

Teach me, my God and King,
In all things thee to see,
And what I do in any thing,
To do it as for thee…
The Elixir

Hutchison, Sir Robert Grieve   1871–1960
English radiologist

Those of us who have the duty of training the raising generation of doctors…must not inseminate the virgin minds of the young with the tares of our own fads. It is for this reason that it is easily possible for teaching to be top "up to date." It is always well, before handing the cup of knowledge to the young, to wait until the froth has settled.

British Medical Journal, Volume 1, 1925 (p. 995)

Huxley, Aldous   1894–1963
English writer and critic

Ram it in, ram it in!
Children's heads are hollow.
Ram it in, ram it in!
Still there's more to follow.
Proper Studies
Education (p. 111)
Chatto & Windus. London, England. 1957

Huxley, Thomas Henry   1825–95
English biologist

Therefore, the great business of the scientific teacher is, to imprint the fundamental, irrefragable facts of his science, not only by words upon the mind, but by sensible impressions upon the eye, and ear, and touch of the student, in so complete a manner, that every term used, or law enunciated, should afterwards call up vivid images of the particular structural, or other, facts which furnished the demonstration of the law, or the illustration of the term.

Lay Sermons, Addresses and Reviews
On the Study of Zoology (p. 112)
New York, New York, USA. 1872

Jaffe, Bernard   1896–1968
Freelance science writer

An effective way to teach the methods of science is to show how our great scientists reached their goals and how their minds worked in the process.

Journal of Chemical Education, Volume 15, 1938 (p. 383)

Milne, A. A. (Alan Alexander)   1882–1956
English poet, children's writer, and playwright

He learns.
He becomes educated.… He instigates knowledge.
The Complete Tales and Poems of Winnie-the-Pooh
The House at Pooh Corner (p. 254)
Dutton Children's Books. New York, New York, USA. 2001

Moroney, M. J.
American statistician

Statistics is not the easiest subject to teach, and there are those to whom anything savoring of mathematics is regarded as for ever anathema.
Facts from Figures
Statistics Desirable (p. 458)
Penguin Books Ltd., Harmondsworth, England. 1951

Noble, D. F.
No biographical data available

You will please keep in mind that this is a college and not a technical school. The students who come here are not to be trained as chemists, or geologists or physicists. They are to be taught the great fundamental truths of all sciences. The object aimed at is culture, not practical knowledge.
In George E. Peterson
The New England College in the Age of the University (pp. 4–7)
Amherst College Press. Amherst, Massachusetts, USA. 1964

Olds, Edwin G.
No biographical data available

It is hard to understand why he failed to appreciate the pedagogical value of designing an experiment to illustrate a point of theory, predicting the result, running the experiment, and then taking the consequences if it turned out wrong.
Teaching Statistical Quality Control for Town and Gown
Journal of the American Statistical Association, Volume 44, 1949
(pp. 223–224)

Regnault, Nöel   1702–62
Jesuit mathematician

Will you discover to me...those Secrets which Nature has imparted to you?
Philosophical Conversations (Volume 1)
Conversation XII (p. 154)
Printed for W. Innys, C. Davis, and N. Prevost. London, England. 1731

Rohrlich, Fritz
No biographical data available

Come, join me on an adventure of the mind. The going will be a little demanding at times but there will be rich rewards. The new vistas are spectacular. Come with me, I am an experienced guide!
From Paradox to Reality: Our Basic Concepts of the Physical World
Preface (p. vii)
Cambridge University Press. Cambridge, England. 1987

Romanoff, Alexis Lawrence   1892–1980
Russian soldier and scientist

True teaching is a grave profession — Demands most thoughtful, wise discretion.
Encyclopedia of Thoughts
Couplets
Ithaca Heritage Books. Ithaca, New York, USA. 1975

Schufle, J. A.
No biographical data available

The teaching of chemistry has for too long been a process in which facts are transmitted from the notebook of the professor into the notebook of the student without going through the heads of either.
The Use of Case Histories in the Teaching of History of Science
The Texas Journal of Science, Volume 21, Number 1, October 1969
(p. 101)

Stoppard, Tom   1937–
Czech-born English playwright

THOMASINA: If you do not teach me the true meaning of things, who will?

SEPTIMUS: Ah. Yes, I am ashamed. Carnal embrace is sexual congress, with the insertion of the male genital organ into the female genital organ for purposes of procreation and pleasure. Fermat's last theorem, by contrast, asserts that when x, y, and z are whole numbers each raised to power of n, the sum of the first two can never equal the third when n is greater than 2. (Pause.)
THOMASINA: Eurghhh!
SEPTIMUS: Nevertheless, that is the theorem.
THOMASINA: It is disgusting and incomprehensible. Now when I am grown to practice it myself I shall never do so without thinking of you.
Arcadia
Act I, Scene One (p. 3)
Faber & Faber Ltd. London, England. 1993

Sylvester, James Joseph   1814–97
English mathematician

May the time never come when the two offices of teaching and researching shall be sundered in this University [Johns Hopkins]! So long as man remains a gregarious and sociable being, he cannot cut himself off from the gratification of the instinct of imparting what he is learning, of propagating through others the ideas and impressions seething in his own brain, without stunting and atrophying his moral nature and drying up the surest sources of his future intellectual replenishment.
The Collected Mathematical Papers of James Joseph Sylvester
(Volume 3)
Proof of the Fundamental Theorem of Invariants (1878) (p. 77)
University Press. Cambridge, England. 1904–1912

...the two functions of teaching and working in science should never be divorced.
The Collected Mathematical Papers of James Joseph Sylvester
(Volume 3)
An Inquiry into Newton's Rule for the Discovery of Imaginary Roots
(p. 75)
University Press. Cambridge, England. 1904–1912

Truesdell, Clifford   1919–2000
American mathematician, natural philosopher, historian of mathematics

Formerly, the beginner was taught to crawl through the underbrush, never lifting his eyes to the trees; today he is often made to focus on the curvature of the universe, missing even the earth.
Six Lectures on Modern Natural Philosophy
Lecture I (p. 22)
Springer-Verlag. New York, New York, USA. 1966

Tukey, John W.   1915–2000
American statistician

Teaching data analysis is not easy, and the time allowed is always far from sufficient.
The Future of Data Analysis
Annals of Mathematical Statistics, Volume 33, Number 1, March 1962 (p. 11)

Weinberg, Alvin Martin   1915–
American physicist

Very typically a field that was once fashionable eventually ceases to command the interest of the scientists in that field and becomes the concerns of scientists in another field. Nuclear chemistry is a good example of this trend: it began as nuclear physics, was taken over by the chemists, and now, insofar as nuclear properties of radionuclide are important for technology, parts of nuclear chemistry are being taken over by engineers. This tendency for fashions in science to came and go greatly complicates the teaching of science. For, as science proliferates, the discrepancy tends to widen between the older, consolidated body of scientific knowledge and the parts of science that excite the active researcher.
Reflections on Big Science
Chapter II (p. 46)
The MIT Press. Cambridge, Massachusetts, USA. 1967

Wells, H. G. (Herbert George)   1866–1946
English novelist, historian, and sociologist

No man can be a good teacher when his subject becomes inexplicable.
Experiment in Autobiography
Chapter 5, Section 2 (p. 176)
The Macmillan Company. New York, New York, USA. 1934

Wilde, Oscar   1854–1900
Irish wit, poet, and dramatist

Education is an admirable thing, but it is well to remember from time to time that nothing that is worth knowing can be taught.
Epigrams: Phrases and Philosophies for the Use of the Young
Sebastian Melmoth
A.R. Keller. London, England. 1907

Wright, Charles R. A.
No biographical data available

...in teaching the science of chemistry it is preferable, first, to enumerate the facts in language independent of any hypothesis, and then to enunciate the various hypotheses that have been and are held, showing how far each is in accordance or contradiction with the observed facts; rather than to mix up from the outset one particular hypothesis with the facts, so as finally to impress on the mind the manifestly erroneous conclusion that the facts have no evidence apart from the hypothesis that more or less clearly explains them.
Atoms
The Athenaeum, Number 2398, 11 October 1873 (p. 468)

TECHNOLOGY

Abbey, Edward   1927–89
American environmentalist and nature writer

High technology has done us one great service: It has retaught us the delight of performing simple and primordial tasks — chopping wood, building a fire, drawing water from a spring...
A Voice Crying in the Wilderness: Notes from a Secret Journal
Chapter 10 (p. 92)
St. Martin's Press. New York, New York, USA. 1989

Adler, Alfred   1870–1937
Austrian psychiatrist

The confusion of science with technology is understandable. Certainly the two often appear to be aspects of a single larger process, as when science proposes new laws of physics, which inspire the development of a technology for their exploration, which in turn exposes inaccuracies in the laws and forces science to seek a more profound level of theory. But in fact their divergence is great. It is in the divergence of engagement from fulfillment, of means from ends....
Atlantic Monthly, Volume 279, Number 2, February 1997 (p. 16)

If truth is a path, then science explores it, and the brief stops along the way are where technologies begin (they build towns and pave a highway). Technology is results, science is process; though the two fuse and separate and then fuse once more, as ends and means must, their opposition is profound.
Atlantic Monthly, Volume 279, Number 2, February 1997 (p. 16)

Allen, Charles M.
American academic

If the human race wants to go to hell in a basket, technology can help it get there by jet. It won't change the desire or the direction, but it can greatly speed the passage.
Speech
Wake Forest University, Winston–Salem, North Carolina, April 25, 1967

Ashby, Sir Eric   1904–82
British botanist and educator

The habit of apprehending a technology in its completeness: this is the essence of technological humanism, and this is what we should expect education in the higher technology to achieve. I believe it could be achieved by making specialists' studies the core around which are grouped liberal studies which are relevant to these specialist studies. But they must be relevant; the path to culture should be through a man's specialism, not by-passing it.
Technology and the Academics: An Essay on Universities and the Scientific Revolution
Chapter 4 (p. 84)
St. Martin's Press. New York, New York, USA. 1959

A student who can weave his technology into the fabric of society can claim to have a liberal education; a student who cannot weave his technology into the fabric of society cannot claim even to be a good technologist.
Technology and the Academics: An Essay on Universities and the Scientific Revolution
Chapter 4 (p. 85)
St. Martin's Press. New York, New York, USA. 1959

Association of American Colleges

...we have become a people unable to comprehend the technology we invent...
Integrity in the College Curriculum, February 1985 (p. 2)

Ballard, James Graham   1930–
English writer

Science and technology multiply around us. To an increasing extent they dictate the languages in which we speak and think. Either we use those languages, or we remain mute.
Crash
Introduction (p. 7)
Flamingo. London, England. 1993

Barzun, Jacques   1907–
French-born American educator, historian, and author

...something pervasive that makes the difference, not between civilized man and the savage, not between man and the animals, but between man and the robot, grows numb, ossifies and falls away like black mortified flesh when techne assails the senses and science dominates the mind.
In Theodosius Dobzhansky
The Biology of Ultimate Concern
Chapter 5 (p. 103)
The New American Library, Inc. New York, New York, USA. 1967

Beer, Stafford   1926–2002
English theorist academic and consultant

If it works, it's out of date.
Brain of the Firm: A Development in Management Cybernetics
Dedication (p. v)
Herder & Herder. New York, New York, USA. 1972

Bronowski, Jacob   1908–74
Polish-born British mathematician and polymath

Every civilization has been grounded on technology: what makes ours unique is that for the first time we believe that every man is entitled to all its benefits.
Technology and Culture in Evolution
The American Scholar, Volume 41, Number 2, Spring 1972 (p. 207)

Bunge, Mario   1919–
Argentine philosopher and physicist

...whereas science elicits changes in order to know, technology knows in order to elicit changes.
In G. Bugliarello and D. B. Doner (eds.)
The History and Philosophy of Technology
Chapter 15 (p. 264)
University of Illinois Press. Urbana, Illinois, USA. 1979

Byrom, Gletcher L.
No biographical data available

Technology is like having a spouse who helps you with the problems that you wouldn't have had if you hadn't gotten married in the first place.
Technology Is...and Where It's Taking Us
Science Digest, Volume 83, Number 2, February 1978 (p. 25)

Clarke, Arthur C.   1917–
English science and science fiction writer

Any sufficiently advanced technology is indistinguishable from magic.
The Lost Worlds of 2001
Chapter 34 (p. 189)
New American Library. New York, New York, USA. 1972

...we know now that comet and asteroid impacts have changed history, and soon we will have the technology to avert that disaster. ...[T]he dinosaurs became extinct because they didn't have a space program.
Interview, *Discover Magazine*, July 1999

Commoner, Barry   1917–
American biologist, ecologist, and educator

Despite the dazzling successes of modern technology and the unprecedented power of modern military systems, they suffer from a common and catastrophic fault. While providing us with a bountiful supply of food, with great industrial plants, with high-speed transportation, and with military weapons of unprecedented power, they threaten our very survival.
Science and Survival
Chapter 7 (p. 126)
The Viking Press. New York, New York, USA. 1966

de Bono, Edward   1933–
Maltese psychologist and writer

Those who assert that technology has done more harm than good are thinking of a romantic dream world in which a select elite lived a short life of ease and intellectual sophistication surrounded by a population living an even shorter life of poverty, starvation and disease.
Technology Today, 1971

de Saint-Exupéry, Antoine   1900–44
French aviator and writer

…the machine does not isolate man from the great problems of nature but plunges him more deeply into them.
Wind, Sand, and Stars
Chapter 3 (p. 67)
Reynal & Hitchcock. New York, New York, USA. 1939

DeSimone, Daniel V.
No biographical data available
Cross, Hardy   1885–1959
American engineer

Technological invention and innovation are the business of engineering. They are embodied in engineering change.
Education for Innovation
Introduction (p. 4)
Pergamon Press. New York, New York, USA. 1968

Drexler, K. Eric   1955–
American nanotechnology engineer and researcher, and futurist

The promise of technology lures us onward, and the pressure of competition makes stopping virtually impossible. As the technology race quickens, new developments sweep toward us faster, and a fatal mistake grows more likely. We need to strike a better balance between our foresight and our rate of advance. We cannot do much to slow the growth of our technology, but we can speed growth of foresight. And with better foresight, we will have a better chance to steer the technology race in safe directions.
Engines of Creation
Finding the Facts (p. 203)
Anchor Press/Doubleday. Garden City, New York, USA. 1986

Dubos, René Jules   1901–82
French-born American microbiologist and environmentalist

The idealistic and the demonic forces in nationalism are as powerful today as they were in the past but their expressions are changing, because human history is moving from its hallowed parochial traditions to the era of global technology…. The one credo of technology which has been accepted practically all over the world is that nature is to be regarded as a source of raw materials to be exploited for human ends rather than as an entity to be appreciated for its own value.
A God Within
Chapter 10 A Demon Within (p. 204)
Charles Scribner's Sons. New York, New York, USA. 1972

Dyson, Freeman J.   1923–
American physicist and educator

If we had a reliable way to label our toys good and bad, it would be easy to regulate technology wisely. But we can rarely see far enough ahead to know which road leads to damnation. Whoever concerns himself with big technology, either to push it forward or to stop it, is gambling in human lives.
Disturbing the Universe
Chapter 1 (p. 7)
Basic Books, Inc. New York, New York, USA. 1979

Editorial

Technology, when misused, poisons air, soil, water and lives. But a world without technology would be prey to something worse: the impersonal ruthlessness of the natural order, in which the health of a species depends on relentless sacrifice of the weak.
New York Times
29 August 1986

Embree, Alice   1945–
American peace and women's rights activist

America's technology has turned in upon itself; its corporate form makes it the servant of profits, not the servant of human needs.
In Robin Morgan
Sisterhood Is Powerful: An Anthology of Writings from the Women's Liberation Movement
Media Images 1 (p. 211)
Random House, Inc. New York, New York, USA. 1970

Feynman, Richard P.   1918–88
American theoretical physicist

For a successful technology, reality must take precedence over public relations, for Nature cannot be fooled.
What Do You Care What Other People Think?
Appendix F (p. 237)
W.W. Norton & Company, Inc. New York, New York, USA. 1988

Frisch, Max   1911–
Novelist, playwright and diarist

We live technologically, with man as the master of nature, man as the engineer, and let anyone who raises his voice against it stop using bridges not built by nature…. No electric light bulbs, no engines, no atomic energy, no calculating machines, no anesthetics — back to the jungle.
Translated by Michael Bullock
Homo Faber: A Report (p. 103)
Harcourt Brace Jovanovich. San Diego, California, USA. 1959

Technology is the knack of so arranging the world that we don't have to experience it.
In Rollo May
The Cry for Myth

Chapter Three (p. 57)
W.W. Norton & Company, Inc. New York, New York, USA.1991

Gábor, Dennis 1900–79
Hungarian-English physicist

The most important and urgent problems of the technology of today are no longer the satisfactions of the primary needs or of archetypal wishes, but the reparation of the evils and damages by technology of yesterday.
Innovations: Scientific, Technological and Social
Introduction (p. 9)
Oxford University Press, Inc. Oxford, England. 1970

Galbraith, John Kenneth 1908–2006
Canadian-American economist

The imperatives of technology and organization, not the images of ideology, are what determine the shape of economics.
The New Industrial State
Chapter I, Section 3 (p. 7)
Houghton Mifflin Company. Boston, Massachusetts, USA. 1967

It is a commonplace of modern technology that there is a high measure of certainty that problems have solutions before there is knowledge of how they are to be solved.
The New Industrial State
Chapter II, Section 4 (p. 19)
Houghton Mifflin Company. Boston, Massachusetts, USA. 1967

Gore, Al 1948–
45th vice-president of the United Sates and environmentalist

Mistakes in our dealings with Mother Nature can now have much larger uninednded consequences, because many of our new technologies confer upon us new power without automatically giving us new wisdom.
An Inconvenient Truth: The Planetary Emergency of Global Warming and What We Can Do About It (p. 247)
Rodale. New York, New York, USA. 2006

Hanham, H. J.
No biographical data available

Great technological advances are always around the corner.
Clio's Weapons
Daedalus, Spring 1971 (p. 509)

Heidegger, Martin 1889–1976
German philosopher

…the essence of technology…is nothing technological.
Basic Writings
The Question Concerning Technology (p. 285)
Harper & Row. New York, New York, USA. 1977

Hoffer, Eric 1902–83
American longshoreman and philosopher

Where there is the necessary technical skill to move mountains, there is no need for the faith that moves mountains.

The Passionate State of Mind, and Other Aphorisms
No. 12
Harper & Brothers. New York, New York, USA. 1955

Hoyle, Sir Fred 1915–2001
English mathematician and astronomer

Science is unique to human activities in that it possesses vast areas of certain knowledge. The collective opinion of scientists in these areas about any problem covered by them will almost always be correct. It is unlikely that much in these areas will be changed in the future, even in a thousand years. And because technology rests almost exclusively on these areas the products of technology work as they are intended to do.
The Origin of the Universe and the Origin of Religion (p. 17)
Moyer Bell. Wakefield, Rhode Island, USA. 1993

Huxley, Aldous 1894–1963
English writer and critic

Advances in technology do not abolish the institution of war; they merely modify its manifestations.
Science, Liberty and Peace
Chapter II (p. 47)
William Morrow & Company, Inc. New York, New York, USA. 1967

Jones, Barry
No biographical data available

The reality is that many of the changes in science and technology are complex because of the complexity of them.
Sayings of the Wee k
Sydney Morning Herald, 12 July 1986

Kaysen, Carl 1920–
American economist

…the advance of technology, like the growth of population and industry, has also been proceeding exponentially.
Limits to Growth
Foreign Affairs, Volume 50, Number 4, July 1972 (p. 664)

Krutch, Joseph Wood 1893–1970
American naturalist, conservationist, and writer

Technology made large populations possible; large populations now make technology indispensable.
Human Nature and the Human Condition
Chapter VIII (p. 145)
Random House, Inc. New York, New York, USA. 1959

Lerner, Max 1902–92
American educator and author

…a world technology means either a world government or world suicide.
Actions and Passions: Notes on the Multiple Revolution of Our Time
The Imagination of H.G. Wells (p. 17)
Simon & Schuster. New York, New York, USA. 1949

Lilienthal, David E.   1899–1981
American businessman and Tennessee Valley Authority administrator

The machine that frees man's back of drudgery does not thereby make his spirit free. Technology has made us more productive, but it does not necessarily enrich our lives. Engineers can build us great dams, but only great people make a valley great. There is no technology of goodness. Men must make themselves spiritually free.
TVA: Democracy on the March (p. 218)
Harper & Brothers. New York, New York, USA. 1944

Lovins, Amory B.   1947–
American physicist and industry consultant

Any demanding high technology tends to develop influential and dedicated constituencies of those who link its commercial success with both the public welfare and their own. Such sincerely held beliefs, peer pressures, and the harsh demands that the work itself places on time and energy all tend to discourage such people from acquiring a similarly thorough knowledge of alternative policies and the need to discuss them.
Energy Strategy
Foreign Affairs, Volume 55, Number 1, October 1976 (p. 93)

Meadows, Donella H.   1941–2001
American scientist, sustainability advocate, and writer
Meadows, Dennis L.
American professor of systems science

Technology can relieve the symptoms of a problem without affecting the underlying causes. Faith in technology as the ultimate solution to all problems can thus divert our attention from the most fundamental problem — the problem of growth in a finite system — and prevent us from taking effective action to solve it.
The Limits to Growth: The 30 Year Update
Chapter IV (p. 154)
Chelsea Green Publishing. White River Junction, Vermont, USA. 2004

Oppenheimer, J. Robert   1904–67
American theoretical physicist

The open society, the unrestricted access to knowledge, the unplanned and uninhibited association of men for its furtherance — these are what may make a vast, complex, ever growing, ever changing, ever more specialized and expert technological world, nevertheless a world of human community.
Science and the Common Understanding
Chapter 6 (p. 95)
Simon & Schuster. New York, New York, USA. 1953

In fact, most people — when they speak of Science as a good thing — have in mind such Technology as has altered the condition of their life.
The Sacred Beetle and Other Great Essays in Science

Physics in the Contemporary World (p. 194)
Prometheus Books. Buffalo, New York, USA. 1984

Organisation for Economic Co-Operation and Development

Science and technology…have a number of distinguishing characteristics [that] cause special problems or complications. One…is ubiquity: they are everywhere. They are at the forefront of social change. They not only serve as agents of change, but provide the tools for analysing social change. They pose, therefore, special challenges to any society seeking to shape its own future and not just to react to change or to the sometimes undesired effects of change.
Technology on Trial: Public Participation in Decision-Making Related to Science and Technology
Chapter I, Section B (p. 16)
Organization for Economic Co-operation and Development. Paris, France. 1979

Pirsig, Robert M.   1928–
American writer

The way to solve the conflict between human values and technological needs is not to run away from technology, that's impossible. The way to resolve the conflict is to break down the barriers of dualistic thought that prevent a real understanding of what technology is — not an exploitation of nature, but the fusion of nature and the human spirit into a new kind of creation that transcends both.
Zen and the Art of Motorcycle Maintenance: An Inquiry into Values
Part III, Chapter 25 (p. 291)
William Morrow & Company, Inc. New York, New York, USA. 1974

Pope Pius XII   1876–1958
Bishop of Rome

The Church welcomes technological progress and receives it with love, for it is an indubitable fact that technological progress comes from God and, therefore, can and must lead to Him.
Christmas message, 1953

Reich, Charles A.
No biographical data available

Technology and production can be great benefactors of man, but they are mindless instruments, and if undirected they career along with a momentum of their own. In our country, they pulverize everything in their path: the landscape, the natural environment, history and tradition, the amenities and civilities, the privacy and spaciousness of life, beauty, and the fragile, slow-growing social structures which bind us together.
The Greening of America
Chapter 1 (pp. 5–6)
Bantam Books. New York, New York, USA. 1971

Rickover, Hyman G.   1900–86
American naval nuclear engineer

...technology can have no legitimacy unless it inflicts no harm.
A Humanistic Technology
Mechanical Engineering, November 1982 (p. 45)

Sagan, Carl   1934–96
American astronomer and science writer

Many visionary leaders have imagined a time when the allegiance of an individual human being is not to his particular nation-state, religion, race, or economic group, but to mankind as a whole; when the benefit to a human being of another sex, race, religion, or political persuasion ten thousand miles away is as precious to us as to our neighbor or our brother. The trend is in this direction, but is agonizingly slow. There is a serious question whether such a global self-identification of mankind can be achieved before we destroy ourselves with the technological forces our intelligence has unleashed.
The Cosmic Connection: An Extraterrestrial Perspective
Chapter 1 (p. 6)
Dell Publishing, Inc. New York, New York, USA. 1975

The so-called generation gap is a consequence of the rate of social and technological change.

Even within a human lifetime, the change is so great that many people are alienated from their own society. Margaret Mead had described older people as involuntary immigrants from the past to the present.
The Cosmic Connection: An Extraterrestrial Perspective
Chapter 5 (p. 36)
Dell Publishing, Inc. New York, New York, USA. 1975

Schumacher, Ernst Friedrich   1911–77
German-born British economist

...the system of nature, of which man is a part, tends to be self-balancing, self-adjusting, self-cleansing. Not so with technology.... The technology of mass production is inherently violent, ecologically damaging, self-defeating in terms of non-renewable resources, and stultifying for the human person.
Small Is Beautiful
Part II, Chapter V (p. 145)
Harper & Row, Publishers. New York, New York, USA. 1973

Snow, Charles Percy   1905–80
English novelist and scientist

Technology...is a queer thing. It brings you great gifts with one hand, and it stabs you in the back with the other.
New York Times, 15 March 1971

Sophocles   496 BCE–406 BCE
Greek playwright

Wonders are many, and none is more wonderful than man; the power that crosses the white sea, driven by the stormy south-wind, making a path under surges that threaten to engulf him...turning the soil with the offspring of horses, as the ploughs go to and fro from year to year.... And speech, and windswift thought, and all the moods that mould a state, hath he taught himself; and how to flee the arrows of the frost, when 'tis hard lodging under the clear sky, and the arrows of the rushing rain; yea, he hath resource for all....
In *Great Books of the Western World* (Volume 5)
The Plays of Sophocles
Antigone, l. 333–340, 349–354
Encyclopædia Britannica, Inc. Chicago, Illinois, USA. 1952

Soulé, Michael E.
American conservation biologist

Since we have no choice but to be swept along by this vast technological surge, we might as well learn to surf.
In David Western and Mary C. Pearl
Conservation for the Twenty-First Century
Conservation Biology in the Twenty-First Century: Summary and Outlook (p. 303)
Oxford University Press. New York, New York, USA. 1989

Stevenson, Adlai E.   1900–65
American political leader and diplomat

Technology, while adding daily to our physical ease, throws daily another loop of fine wire around our souls. It contributes hugely to our mobility, which we must not confuse with freedom. The extensions of our senses, which we find so fascinating, are not adding to the discrimination of our minds, since we need increasingly to take the reading of a needle on a dial to discover whether we think something is good or bad, or right or wrong.
My Faith in Democratic Capitalism
Fortune Magazine, October, 1955 (p. 156)

Thoreau, Henry David   1817–62
American essayist, poet, and practical philosopher

Our inventions are wont to be pretty toys, which distract our attention from serious things. They are but improved means to an unimproved end.
The Writings of Henry David Thoreau (Volume 2)
Walden
Chapter I (p. 84)
Houghton Mifflin Company. Boston, Massachusetts, USA. 1893

Toffler, Alvin   1928–
American writer and futurist

...[a] great, growling engine of change — technology.
Future Shock
Chapter 2 (p. 25)
Random House, Inc. New York, New York, USA. 1979

...technology feeds on itself. Technology makes more technology possible...

Future Shock
Chapter 2 (p. 27)
Random House, Inc. New York, New York, USA. 1979

If technology…is to be regarded as a great engine, a mighty accelerator, then knowledge must be regarded as its fuel. And we thus come to the crux of the accelerative process in society, for the engine is being fed a richer and richer fuel every day.
Future Shock
Chapter 2 (pp. 29–30)
Random House, Inc. New York, New York, USA. 1979

Wilkins, Maurice   1916–2004
New Zealand-born English molecular biologist

Science, with technology, is the only way we have to avoid starvation, disease, and premature death. The misapplication of science and technology is due to the fact that the politics are wrong. Now my own view is that the politics are indeed wrong; but politics and science are so closely interrelated that they can hardly be separated.
In Horace Freeland Judson
The Eighth Day of Creation: Makers of the Revolution in Biology
DNA, You Know, Is Midas' Gold (p. 97)
Simon & Schuster. New York, New York, USA. 1979

TEETH

Alison, Richard   fl. 1606
English poet

Those cherries fairly do enclose
Of orient pearl a double row,
Which, when her lovely laughter shows,
They look like rosebuds fill'd with snow.
An Howre's Recreation in Musike

Berry, James H.
No biographical data available

Brush them and floss them and take them to the dentist,
Care for them and they will stay with you. Ignore them, and they'll go away.
Special Advertising Section
Time, February 11, 1985 (p. 21)

Christie, Agatha   1890–1976
English author

Beastly things, teeth…. Give us trouble from the cradle to the grave.
At Bertram's Hotel
Chapter X (p. 93)
Dodd, Mead & Company. New York, New York, USA. 1966

de la Salle, St. Jean Baptiste   1651–1719
French educational reformer and priest

It is necessary to clean the teeth frequently, more especially after meals, but not on any account with a pin, or the point of a penknife, and it must never be done at table.
The Rules of Christian Manners and Civility
Chapter I

Editor of the Louisville Journal

Probably the reason why women's teeth decay sooner than men's is not the perpetual friction of their tongues upon the pearl, but rather the intense sweetness of their lips.
In George Denison Prentice
Prenticeana (p. 35)
Derby & Jackson. New York, New York, USA. 1859

Franklin, Benjamin   1706–90
American printer, scientist, and diplomat

Hot things, sharp things, sweet things, old things, all rot the teeth.
Poor Richard's Almanac
1734

Herrick, Robert   1591–1674
English poet

Some ask'd how pearls did grow, and where,
Then spoke I to my girle,
To part her lips, and showed them there
The quarelets of pearl.
The Works of Robert Herrick
The Rock of Rubies and the Quarry of Pearls
Reprinted for W. & C. Tait. Edinburgh, Scotland. 1823

Hood, Thomas   1582–98
English poet and editor

The best of friends fall out, and so
His teeth had done some years ago.
The Complete Poetical Works of Thomas Hood
A True Story, Stanza 2
Greenwood Press, Publishers. Westport, Connecticut, USA. 1980

Lamb, Charles   1775–1834
English essayist and critic

The fine lady, or fine gentleman, who show me their teeth, show me bones.
The Complete Works and Letters of Charles Lamb
The Praise of Chimney-Sweeps (p. 99)
Modern Library. New York, New York, USA. 1935

Martial (Marcus Valerius Martialis)   ca. 40–ca.103
Latin poet and epigrammatist

Thais has black, Laecania snowy teeth. What is the reason? One has those she purchased, the other her own.
Translated by Walter C.A. Ker
Epigrams (Volume 1)
Book V, Epigram XLIII (p. 327)
William Heinemann. London, England. 1930

O'Donoghue, Michael   1940–94
American writer and performer

Tough teeth make tough soldiers.
National Lampoon Tenth Anniversary Anthology
Frontline Dentists (p. 111)
National Lampoon. New York, New York, USA. 1979

Perelman, Sidney Joseph   1904–79
American comic writer

I'll dispose of my teeth as I see fit, and after they've gone, I'll get along. I started off living on gruel, and by God, I can always go back to it again.
Crazy Like a Fox
Nothing But the Tooth (p. 72)
Random House, Inc. New York, New York, USA. 1944

Shakespeare, William   1564–1616
English poet, playwright, and actor

Last scene of all,
That ends this strange eventful history,
Is second childishness and mere oblivion,
Sans teeth, sans eyes, sans taste, sans everything.
In *Great Books of the Western World* (Volume 26)
The Plays and Sonnets of William Shakespeare (Volume 1)
As You Like It
Act II, Scene vii, l. 163–166
Encyclopædia Britannica, Inc. Chicago, Illinois, USA. 1952

Bid them wash their faces,
And keep their teeth clean.
In *Great Books of the Western World* (Volume 27)
The Plays and Sonnets of William Shakespeare (Volume 2)
Coriolanus
Act II, Scene iii, l. 65–66
Encyclopædia Britannica, Inc. Chicago, Illinois, USA. 1952

Swift, Jonathan   1667–1745
Irish-born English writer

…sweet Things are bad for the Teeth.
The Prose Works of Jonathan Swift (Volume the Fourth)
Polite Conversation, Dialogue II (p. 181)
Printed at the Shakespeare Head Press. Oxford, England. 1939–1968

Twain, Mark (Samuel Langhorne Clemens)   1835–1910
American author and humorist

Adam and Eve had many advantages, but the principal one was that they escaped teething.
The Tragedy of Pudd'nhead Wilson
Chapter IV (p. 39)
New American Library. New York, New York, USA. 1980

Wheeler, Hugh   1912–87
English-born playwright

To lose a lover or even a husband or two during the course of one's life can be vexing. But to lose one's teeth is a catastrophe.
Four by Sondheim
A Little Night Music
Act II, Scene I (p. 269)
Applause. New York, New York, USA. 2000

TEKTITES

Faul, Henry   1920–81
Czech-American geochronologist

To anyone who has worked with them, tektites are probably the most frustrating stones ever found on Earth.
Tektites are Terrestrial
Science, Volume 152, Number 3727, 3 June 1966 (p. 1341)

TELEOLOGY

Ayala, Francisco J.   1934–
Spanish-American biologist and philosopher

Biological evolution can however be explained without recourse to a Creator or a planning agent external to the organisms themselves. The evidence of the fossil record is against any directing force, external or immanent, leading the evolutionary process toward specified goals. Teleology in the stated sense is, then, appropriately rejected in biology as a category of explanation.
Biology as an Autonomous Science
American Scientist, Volume 56, Number 3, Autumn 1968 (p. 213)

Butler, Samuel   1612–80
English novelist, essayist, and critic

Of all the questions now engaging the attention of those whose destiny has commanded them to take more or less exercise of mind, I know of none more interesting than that which deals with what is called teleology — that is to say, with design or purpose, as evidenced by the different parts of animals and plants.
Evolution, Old and New
Chapter I (p. 1)
Hardwicke & Bogue. London, England. 1879

Reichenbach, Hans   1891–1953
German philosopher of science

Teleology is analogism, is pseudo explanation; it belongs in speculative philosophy, but has no place in scientific philosophy.
The Rise of Scientific Philosophy
Chapter 12 (p. 195)
University of California Press. Berkeley, California, USA. 1951

von Brücke, Ernst   1819–92
German-born physicist and physiologist

Teleology is a lady without whom no biologist can live. Yet he is ashamed to show himself with her in public.
In W.I.B. Beveridge
The Art of Scientific Investigation
Chapter Five (p. 61)
W.W. Norton & Company, Inc. New York, New York, USA. 1957

TELESCOPE

Bierce, Ambrose 1842–1914
American newspaperman, wit, and satirist

Telescope (n): A device having a relation to the eye similar to that of a telephone to the ear, enabling distant objects to plague us with a multitude of needless details.
The Devil's Dictionary
Doubleday & Company, Inc. Garden City, New York, USA. 1967

Brecht, Bertolt 1898–1956
German writer

LUDOVICO: …look at that queer tube thing they're selling in Amsterdam. I gave it a good looking-over. A green leather casing and a couple of lenses, one this way [he indicates a concave lens].… One of them's supposed to magnify and the other reduces. Anyone in his right mind would expect them to cancel out. They don't. The thing makes everything appear five times the size. That's science for you.
Translated by John Willett
Life of Galileo
Scene 1 (pp. 12–13)
Arcade Publishing. New York, New York, USA. 1994

Butler, Samuel 1612–80
English novelist, essayist, and critic

And now the lofty telescope, the scale
By which they venture Heav'n itself t' assail,
Was raised, and planted full against the Moon…
The Poetical Works of Samuel Butler (Volume 2)
Elephant in the Moon
Bell & Daldy. London, England. 1835

Carroll, Lewis (Charles Dodgson) 1832–98
English writer and mathematician

All this time the Guard was looking at her, first through a telescope, then through a microscope, and then through an opera-glass.
The Complete Works of Lewis Carroll
Through the Looking-Glass
Chapter III (pp. 169–170)
The Modern Library. New York, New York, USA. 1936

Cedering, Siv 1939–
Swedish-American poet, painter, and sculptor

I have helped him polish the mirrors and lenses of our new telescope. It is the largest in existence. Can you imagine the thrill of turning it to some new corner of the heavens to see something never before seen from the earth?
Science 84
Letter from Caroline Herschel (1750–1840)

Copeland, Leland S. 1886–?
American poet and writer

Then outward turn an optic tube

From some high, lonely hill,
That we may glance at cosmic nooks
And marvels rich, until
The morning glow conceals those realms
Where precious things distill,
Far-forth beyond the utmost reach
Of human hope and will.
All Night with the Stars
Sky & Telescope, November 1949

Many of us find that to leave bright room and cozy chair for the dark world outside is contrary to nature, like a moth flying from the light. As a bather plunges into cold water, so the sky hunter must immerse himself in darkness before he will find it comfortable in the night.… So rich is this nocturnal wonderland that even for smallest telescopes numerous objects await observation. A larger lens or mirror is not an assured benefit. Devotion and patience are as important as light grasp.
All Night with the Stars
Sky & Telescope, November 1949

de Grasse Tyson, Neil 1958–
American astrophysicist and writer

The most famous telescope in modern times is, of course, the Hubble, known to the public primarily through the beautiful, full-color, high-resolution images it has produced of objects in the universe. The problem here is that after viewing such exhibits, you wax poetic about the beauty of the universe yet are no closer than before to understanding how it all works.…
Over the Rainbow
Natural History, Volume 110, Number 7, September 2001 (p. 34)

While much good science has come from the Hubble telescope (including the most reliable measure to date for the expansion rate of the universe), you would never know from media accounts that the foundation of our cosmic knowledge continues to flow primarily from the analysis of spectra and not from looking at pretty pictures.
Over the Rainbow
Natural History, Volume 110, Number 7, September 2001 (p. 34)

Dressler, Alan
American astronomer

To look into space is to look back into time, and telescopes are time machines we can ride nearly all the way back to creation itself.
Observing Galaxies Through Time
Sky and Telescope, Volume 82, Number 2, August 1991 (p. 126)

Emerson, Ralph Waldo 1803–82
American lecturer, poet, and essayist

The sight of a planet through a telescope is worth all the course on astronomy…
Ralph Waldo Emerson: Essays and Lectures

Essays: Second Series
New England Reformers (p. 594)
The Library of America. New York, New York, USA. 1983

Everett, Edward   1794–1865
American statesman, educator, and orator

The telescope may be likened to a wondrous cyclopean eye, endued with superhuman power, by which the astronomer extends the reach of his vision to the further heavens, and surveys galaxies and universes compared with which the solar system is but an atom floating in the air.
An Oration
The Uses of Astronomy, Versility of Genius, Albany, New York, 28 July 1856 (p. 22)
Ross & Tousey. New York, New York, USA. 1856

Ferris, Timothy   1944–
American science writer

Telescopes are like musical instruments, in that you get out of them what you put into them.
Seeing in the Dark
Appendix A (p. 299)
Simon & Schuster. New York, New York, USA. 2002

Frost, Robert   1874–1963
American poet

Man acts more like the poor bear in a cage
That all day fights a nervous inward rage…
The toenail click at the shuffle of his feet,
The telescope at one end of his beat,
And at the other end the microscope,
Two instruments of nearly equal hope…
Complete Poems of Robert Frost
The Bear
Henry Holt & Company. New York, New York, USA. 1949

He burned his house down for the fire insurance
And spent the proceeds on a telescope
To satisfy a lifelong curiosity
About our place among the infinities.
Complete Poems of Robert Frost
The Star Splitter
Henry Holt & Company. New York, New York, USA. 1949

Galilei, Galileo   1564–1642
Italian physicist and astronomer

O telescope, instrument of much knowledge, more precious than any scepter! Is not he who holds thee in his hand made king and lord of the works of God?
In William H. Jefferys and R. Robert Robbins
Discovering Astronomy (p. 174)
John Wiley & Sons, Inc. New York, New York, USA. 1995

Gamow, George   1904–68
Russian-born American physicist

My telescope
Has dashed your hope;

Your tenets are refuted.
Let me be terse:
Our universe
Grows daily more diluted!
Mr. Tompkins in Paperback
Chapter 6 (p. 63)
At The University Press. Cambridge, England. 1965

Hankins, Arthur Preston   1880–1932
No biographical data available

Midnight — with Cole of Spyglass Mountain seated high up on his ladder, his far-seeing blue-gray eye glued to the powerful five-hundred-diameter eye-piece of his telescope. Unnoticeably the refractor followed the planet in its endless flight. The driving clock purred softly, the only sound on Spyglass Mountain — cold and still and fraught with uncanny tensity.
Cole of Spyglass Mountain (p. 302)
Dodd, Mead & Company. New York, New York, USA. 1923

Hastings, C. S.
No biographical data available

Let us therefore congratulate the possessors of this noble instrument, wish them God speed in their search after knowledge, while we remind them that although no astronomer can ever make another discovery that will rival that made by the insignificant tube first directed toward the heavens by the Paduan philosopher, yet no mind can weigh the importance of any truth, however trivial in appearance, which may be added to that store which we call "science."
Annual Report of the Board of Regents of the Smithsonian Institution, 1893
The History of the Telescope (p. 109)
Government Printing Office. Washington, D.C. 1894

Herschel, Friedrich Wilhelm   1738–1822
English astronomer

In the old Telescope's tube we sit,

And the shades of the past around us flit…
Herschel at the Cape
Introduction (p. xix)
University of Texas Press. Austin, Texas, USA. 1969

I will make such telescopes and see such things!
In Grant Allen
Biographies of Working Men
Chapter IV

Holmes, Oliver Wendell   1809–94
American physician, poet, and humorist

I was riding on the outside of a stage-coach from London to Windsor, when all at once a picture familiar to me from my New England village childhood came upon me like a reminiscence rather than a revelation. It was a mighty bewilderment of slanted masts and spars and ladders and

ropes, from the midst of which a vast tube, looking as if it might be a piece of ordnance such as the revolted angels battered the walls of Heaven with, according to Milton, lifted its muzzle defiantly towards the sky.
The Poet at the Breakfast-Table
Chapter VIII (p. 219)
Houghton Mifflin Company. Boston, Massachusetts, USA. 1895

I love all sights of earth and skies,
From flowers that grow to stars that shine;
The comet and the penny show,
All curious things above, below…
But most I love the tube that spies
The orbs celestial in their march;
That shows the comet as it whisks
Its tail across the planet's disk,
Or wheels so close against the sun
We tremble at the thought of risks
Our little spinning ball may run.
The Complete Poetical Works of Oliver Wendell Holmes
The Flaneur
Houghton Mifflin Company. Boston, Massachusetts, USA. 1899

Hubble, Edwin Powell   1889–1953
American astronomer

The exploration of space is an achievement of great telescopes.
Annual Report of the Board of Regents of the Smithsonian Institution, 1938
The Nature of the Nebulae (p. 137)
Government Printing Office. Washington, D.C. 1939

Kepler, Johannes   1571–1630
German astronomer

What now, dear reader, shall we make out of our telescope? Shall we make a Mercury's magic-wand to cross the liquid ether with, and like Lucian lead a colony to the uninhabited evening star, allured by the sweetness of the place? Or shall we make it a Cupid's arrow which, entering by our eyes, has pierced our inmost mind, and fired us with love of Venus?…
Dioptrice
Preface (p. 86)

O telescope, instrument of much knowledge, more precious than any scepter! Is not he who holds thee in his hand made king and lord of the works of God?
Dioptrice
Preface (p. 103)

Kitchiner, William   1775–1827
English doctor/optician and telescope inventor

Immense telescopes are only about as useful as the enormous spectacles which are suspended over the doors of opticians!
In William Sheehan
Planets and Perception: Telescopic Views and Interpretations, 1609–1909

Chapter 9 (p. 113)
The University of Arizona Press. Tucson, Arizona, USA. 1988

Longair, Malcolm   1941–
Scottish astronomer

Twas brillig and the slithy toves
Brought plans of telescopes fair to see.
The Jabberwock, he clapped his hands
And said, "That's just for me."
Alice and the Space Telescope
Chapter 2 (p. 7)
The John Hopkins University Press, Baltimore, Maryland, USA. 1989

Lovell, Sir Alfred Charles Bernard   1913–
English astrophysicist

Astronomy has marched forward with the growth in size of its telescopes.
Radio Stars
Scientific American, Volume 188, Number 1, January 1953 (p. 21)

Maunder, Edward Walter   1851–1928
English astronomer

We have no right to assume, and yet we do habitually assume, that our telescopes reveal to us the ultimate structure of the planet.
The Canals of Mars
Knowledge, Number 1, 1894 (p. 251)

Milton, John   1608–74
English poet

…a spot like which perhaps
Astronomer in the Sun's lucent Orbe
Through his glaz'd Optic Tube yet never saw.
In *Great Books of the Western World* (Volume 32)
Paradise Lost
Book III, l. 588–590
Encyclopædia Britannica, Inc. Chicago, Illinois, USA. 1952

Mitchell, Maria   1818–89
American astronomer and educator

The tube of Newton's first telescope…was made from the cover of an old book-a little glass at one end of the tube and a large brain at the other…
In Helen Wright
Sweeper in the Sky
Chapter 9 (p. 168)
Macmillan & Company. New York, New York, USA. 1949

Moulton, Forest Ray   1872–1952
American astronomer

The eye of the fabled Cyclops was not even prophetic of the great telescope at Mt. Wilson, the pupil of whose eye, so to speak, is 100 inches in diameter.
In H.H. Newman (ed.)
The Nature of the World and of Man
Astronomy (p. 1)
The University of Chicago Press. Chicago, Illinois, USA. 1927

Mullaney, James
American astronomy writer, lecturer, and consultant

The telescope in particular need to be regarded as not just another gadget or material possession but a wonderful, magical gift to humankind — a window on creation, a time machine, a spaceship of the mind that enables us to roam the universe in a way that is surely the next best thing to being out there.
Focal Point
Sky & Telescope, March 1990 (p. 244)

Newton, Sir Isaac   1642–1727
English physicist and mathematician

If the theory of making telescopes could at length be fully brought into practice, yet there would be certain bounds beyond which telescopes could not perform. For the air through which we look upon the stars is in a perpetual tremor; as may be seen by the tremulous motion of shadows cast from high towers, and by the twinkling of the fixed stars.... The only remedy is a most serene and quiet air, such as may perhaps be found on the tops of the highest mountains above the gossamer clouds.
In *Great Books of the Western World* (Volume 34)
Optics
Book 1, Part 1, Proposition viii, problem 2 (p. 423)
Encyclopædia Britannica, Inc. Chicago, Illinois, USA. 1952

Noyes, Alfred   1880–1958
English poet

My periwig's askew, my ruffle stained
With grease from my new telescope!
The Torch Bearers: Watchers of the Sky
William Herschel Conducts (p. 231)
Frederick A. Stokes Company. New York, New York, USA. 1922

Panek, Richard
No biographical data available

The relationship between the telescope and our understanding of the dimensions of the universe is in many ways the story of modernity. It's the story of how the development of one piece of technology has changed the way we see ourselves and of how the way we see ourselves has changed this piece of technology, each set of changes reinforcing the other over the course of centuries until, in time, we've been able to look back and say with some certainty that the pivotal division between the world we inhabit today and the world of our ancestors was the invention of this instrument.
Seeing and Believing: How the Telescope Opened Our Eyes and Minds to the Heavens
Prologue (p. 4)
The Viking Press. New York, New York, USA. 1998

Peltier, Leslie C.   1900–80
American comet hunter

Old telescopes never die, they are just laid away.
Starlight Nights
Chapter 28 (p. 232)
Harper & Row, Publishers. New York, New York, USA. 1965

Rowan-Robinson, Michael   1942–
English astronomer, astrophysicist, and professor

Once it was the navigators crossing the oceans to find new continents and new creatures, the globe opening up before their eyes, and at the same time the unknown areas, white on the map, shrinking.

Now it is the astronomers' telescopes penetrating the void to find new worlds, voyages of discoveries made with giant metal eyes, seeing light we cannot see.
Our Universe: An Armchair Guide
Cosmic Landscape
After Preface
W.H. Freeman & Company. New York, New York, USA. 1990

Ryder-Smith, Roland
No biographical data available

All night he watches roving worlds go by
Through tempered glass, his window on the sky
Feels in his own beat
Of some far mightier heart, and hears
The mystic concert of the spheres.
Astronomer
The Scientific Monthly, Volume 67, Number 4, October 1948 (p. 253)

Toogood, Hector B.
No biographical data available

The telescope, an instrument which, if held the right way up, enables us to examine the stars and constellations at close quarters. If held the wrong way up, however, the telescope is of little or no use.
The Outline of Everything with a Critical Survey of the World's Knowledge
Chapter VIII (p. 96)
Little, Brown & Company. Boston, Massachusetts, USA. 1923

Tsiolkovsky, Konstantin Eduardovich   1857–1935
Russian research scientist

All that which is marvelous, and which we anticipate with such thrill, already exists but we cannot see it because of the remote distances and the limited power of our telescopes...
In Adam Starchild (ed.)
The Science Fiction of Konstantin Tsiolkovsky
Dreams of the Earth and Sky (p. 154)
University Press of the Pacific, Inc. Seattle, Washington, USA. 1979

Vehrenberg, Hans   1910–91
German astronomer

It is a fundamental human instinct to collect, whether berries and roots in the prehistoric past or knowledge of the universe today. For several decades, my favorite pastime

has been to collect celestial objects in photographs. I will never forget the many thousands of hours I have spent with my instruments, working peacefully in my telescope shelter as I listened to good music and dreamed about the infinity of the universe.
Atlas of Deep Sky Splendors
Preface
Treugesell-Verlag. Düsseldorf, Germany. 1978

Vezzoli, Dante
No biographical data available

Cyclopean eye that sweeps the sky,
Whose silvered iris gathers light
From galaxies that unseen pierce
The silent blanket of the night.
Eye of Palomar
The Sky, January 1940 (p. 8)

Wittgenstein, Ludwig Josef Johann   1889–1951
Austrian-born English philosopher

A curious analogy could be based on the fact that even the hugest telescope has to have an eye-piece no larger than the human eye.
Translated by Peter Winch
Culture and Value (p. 17e)
The University of Chicago Press. Chicago, Illinois, USA. 1980

Wordsworth, William   1770–1850
English poet

WHAT crowd is this? what have we here! we must not pass it by;
A Telescope upon its frame, and pointed to the sky:
Long is it as a barber's pole, or mast of little boat,
Some little pleasure-skiff, that doth on Thames's waters float.
The Complete Poetical Works of William Wordsworth
Star-Gazers
Crowell. New York, New York, USA. 1888

Zwicky, Fritz   1898–1974
Swiss astronomer and physicist

Only Galileo and I really knew how to use a small telescope.
In Richard Preston
First Light
Part 2 (p. 119)
Random House, Inc. New York, New York, USA. 1996

TEMPERATURE

Feynman, Richard P.   1918–88
American theoretical physicist

John and his father go out to look at the stars. John sees two blue stars and a red star. His father sees a green star, a violet star, and two yellow stars. What is the total temperature of the stars seen by John and his father?

Surely You're Joking, Mr. Feynman!: Adventures of a Curious Character
Judging Books by Their Covers (p. 293)
W.W. Norton & Company, Inc. New York, New York, USA.1985

TEMPLE OF SCIENCE

Einstein, Albert   1879–1955
German-born physicist

Many kinds of men devote themselves to Science, and not all for the sake of Science herself. There are some who come into her temple because it offers them the opportunity to display their particular talents. To this class of men science is a kind of sport in the practice of which they exult, just as an athlete exults in the exercise of his muscular prowess. There is another class of men who come into the temple to make an offering of their brain pulp in the hope of securing a profitable return. The men are scientists only by the chance of some circumstance which offered itself when making a choice of career…it is clear that if the men who have devoted themselves to science consisted only of the two categories I have mentioned, the edifice could never have grown to its present proud dimensions. …
In Max Planck
Where Is Science Going?
Prologue (p. 7)
W.W. Norton & Company, Inc. New York, New York, USA.1932

I am inclined to agree with Schopenhauer in thinking that one of the strongest motives that lead people to give their lives to art and science is the urge to flee from everyday life, with its drab and deadly dullness, and thus to unshackle the chains of one's own transient desires, which supplant one another in interminable succession so long as the mind is fixed on the horizon of daily environment.
In Max Planck
Where Is Science Going?
Prologue (p. 7)
W.W. Norton & Company, Inc. New York, New York, USA.1932

In the temple of science are many mansions, and various indeed are they that dwell therein and the motives that have led them thither. Many take to science out of a joyful sense of superior intellectual power; science is their own special sport to which they look for vivid experience and the satisfaction of ambition; many others are to be found in the temple who have offered the products of their brains on this altar for purely utilitarian purposes. Were an angel of the Lord to come and drive all the people belonging to these two categories out of the temple, the assemblage would be seriously depleted, but there would still be some men, of both present and past times, left inside.
Ideas and Opinions
Principles of Research (p. 224)
Crown Publishers, Inc. New York, New York, USA. 1954

Gregory, Sir Richard Arman   1864–1952
British science writer and journalist

To qualify for admission into the temple of science it is necessary to offer sacrifices at the alter of knowledge; and only those with sincere regard for truth will find their gifts acceptable.
Discovery; or, The Spirit and Service of Science
Chapter III (p. 54)
Macmillan & Company Ltd. London, England. 1918

Pasteur, Louis   1822–95
French chemist

Preconceived ideas are like searchlights which illuminate the path of the experimenter and serve him as a guide to interrogate nature. They become a danger only if he transforms them into fixed ideas — this is why I should like to see these profound words inscribed on the threshold of all the temples of science: "The greatest derangement of the mind is to believe in something because one wishes it to be so"…
In Réne Dubos
Louis Pasteur: Free Lance of Science
Speech to the French Academy of Medicine, July 18, 1876, Chapter XIII (p. 376)
Little, Brown & Company. Boston, Massachusetts, USA. 1950

Planck, Max   1858–1947
German physicist

Anybody who has been seriously engaged in scientific work of any kind realizes that over the entrance to the gates of the temple of science are written the words: *Ye must have faith*. It is a quality which the scientist cannot dispense with.
Where Is Science Going?
Epilogue (p. 214)
W.W. Norton & Company, Inc. New York, New York, USA.1932

TENSOR

Bell, E. T. (Eric Temple)   1883–1960
Scottish-American mathematician and educator

…the tensor calculus that cost Einstein an effort to master is now a regular part of an undergraduate course in the better technical schools. The subject has been so thoroughly emulsified that even an eighteen-year-old can swallow it without regurgitating. But this does not prove that wither his brain or his stomach is stronger than Einstein's was.
Mathematics: Queen and Servant of Science
A Metrical Universe (p. 211)
McGraw-Hill Book Company, Inc. New York, New York, USA. 1951

Benford, Gregory   1941–
American physicist and science fiction novelist

There was a blithe certainty that came from first comprehending the full Einstein field equations, arabesques of Greek letters clinging tenuously to the page, a gossamer web. They seemed insubstantial when you first saw them, a string of squiggles. Yet to follow the delicate tensors as they contracted, as the superscripts paired with subscripts, collapsing mathematically into concrete classical entities — potential; mass; forces vectoring in a curved geometry — that was a sublime experience. The iron fist of the real, inside the velvet glove of airy mathematics.
Timescape
Chapter 15 (pp. 175–176)
Simon & Schuster. New York, New York, USA. 1980

Bester, Alfred   1913–87
American science fiction writer

Tenser, said the Tensor
Tenser, said the Tensor
Tension, apprehension,
And dissension have begun.
The Demolished Man
Chapter iii (p. 48)
Shasta Publishers. Chicago, Illinois, USA. 1953

Einstein, Albert   1879–1955
German-born physicist

…the energy tensor can be regarded only as a provisional means of representing matter. In reality, matter consists of electrically charged particles…
The Meaning of Relativity (p. 82)
Princeton University Press. Princeton, New Jersey, USA. 1945

…in the case of the equations of gravitation it is the four-dimensionality and the symmetric tensor as expression for the structure of space that, together with the invariance with respect to the continuous transformation group, determine the equations all but completely.
Translated by Paul Arthur Schlipp
Albert Einstein: Autobiographical Notes (p. 85)
The Open Court Publishing Company. La Salle, Illinois, USA. 1979

Russell, Bertrand Arthur William   1872–1970
English philosopher, logician, and social reformer

We want to express physical laws in such a way that it shall be obvious when we are expressing the same law by reference to two different systems of co-ordinates, so that we shall not be misled into supposing we have different laws when we have one law in different words. This is accomplished by the method of tensors.
The ABC of Relativity
Chapter XII (p. 110)
George Allen & Unwin Ltd. London, England. 1958

van Dine, S. S.   1888–1939
American art critic and author

The tensor is known to all advanced Mathematicians. It is one of the technical expressions used in non-Euclidean geometry; and though it was discovered by Riemann in connection with a concrete problem in physics, it has now become of widespread importance in the mathematics of relativity. It's highly scientific in the abstract sense, and can have no direct bearing on Sprigg's murder.
The Bishop Murder Case
Chapter 9
Charles Scribner's Sons. New York, New York, USA. 1929

Weyl, Hermann   1885–1955
German mathematician

The conception of tensors is possible owing to the circumstance that the transition from one co-ordinate system to another expresses itself as a linear transformation in the differentials. One here uses the exceedingly fruitful mathematical device of making a problem "linear" by reverting to infinitely small quantities.
Translated by Henry L. Brose
Space — Time — Matter
Chapter II, Section 13 (p. 104)
Dover Publications, Inc. New York, New York, USA. 1922

Whitehead, Alfred North   1861–1947
English mathematician and philosopher

The idea that physicists would in future have to study the theory of tensors created real panic amongst them following the first announcement that Einstein's predictions had been verified.
In Jean-Pierre Luminet
Black Holes (p. 47)
Cambridge University Press. New York, New York, USA. 1992

TESTING

Colton, Charles Caleb   1780–1832
English sportsman and writer

Examinations are formidable, even to the best prepared, for the greatest fool may ask more than the wisest man can answer.
Lacon; or Many Things in a Few Words (p. 170)
William Gowans. New York, New York, USA. 1849

da Vinci, Leonardo   1452–1519
Italian High Renaissance painter and inventor

It is by testing that we discern fine gold.
Leonardo da Vinci's Note Books (p. 60)
Duckworth & Company. London, England. 1906

Raleigh, Sir Walter   1552–1618
Renaissance English courtier and poet

No instrument smaller than the World is fit to measure men and women: Examinations measure Examinees.
Laughter from a Cloud

Some Thoughts on Examinations (p. 120)
Constable. London, England. 1923

Wilde, Oscar   1854–1900
Irish wit, poet, and dramatist

Examinations are pure humbug from beginning to end.
Epigrams: Phrases and Philosophies for the Use of the Young
Oscariana
A.R. Keller. London, England. 1907

In examinations the foolish ask questions that the wise cannot answer.
Epigrams: Phrases and Philosophies for the Use of the Young
Phrases and Philosophies
A.R. Keller. London, England. 1907

THEOREM

Feynman, Richard P.   1918–88
American theoretical physicist

We decided that "trivial" means "proved." So we joked with the mathematicians: "We have a new theorem — that mathematicians can prove only trivial theorems, because every theorem that's proved is trivial."
Surely You're Joking, Mr. Feynman!: Adventures of a Curious Character
A Map of a Cat? (p. 70)
W.W. Norton & Company, Inc. New York, New York, USA. 1985

Hofstadter, Douglas R.   1945–
American academic

All the limitative Theorems of metamathematics and the theory of computation suggest that once the ability to represent your own structure has reached a certain critical point, that is the kiss of death: it guarantees that you can never represent yourself totally. Gödel's Incompleteness Theorem, Church's Undecidability Theorem, Turing's Halting Theorem, Tarski's Truth Theorem — all have the flavor of some ancient fairy tale which warns you that "To seek self-knowledge is to embark on a journey which…will always be incomplete, cannot be charted on any map, will never halt, cannot be described."
Gödel, Escher, Bach: An Eternal Golden Braid
Part II, Chapter XX (p. 697)
Basic Books, Inc. New York, New York, USA. 1979

Howe, Roger   1945–
American mathematician

Everybody knows that mathematics is about Miracles, only mathematicians have a name for them: Theorems.
MAA address, Baltimore, January 9, 1998

Huxley, Aldous   1894–1963
English writer and critic

Too much theorizing, as we all know, is fatal to the soul…
Tomorrow and Tomorrow and Tomorrow and Other Essays

The Education of an Amphibian (p. 7)
Harper & Brothers. New York, New York, USA. 1956

Papert, Seymour   1928–
South African mathematician

For what is important when we give children a theorem to use is not that they should memorize it. What matters most is that by growing up with a few very powerful theorems one comes to appreciate how certain ideas can be used as tools to think with over a lifetime. One learns to enjoy and to respect the power of powerful ideas. One learns that the most powerful idea of all is the idea of powerful ideas.
Mindstorms: Children, Computers and Powerful Ideas
Chapter 3 (p. 76)
Basic Books, Inc. New York, New York, USA. 1980

Planck, Max   1858–1947
German physicist

…Leibniz's theorem…sets forth fundamentally that of all the worlds that may be created, the actual world is that which contains, besides the unavoidable evil, the maximum good.
Translated by R. Jones and D.H. Williams
A Survey of Physics: A Collection of Lectures and Essays
The Principle of Least Action (p. 71)
Methuen & Company Ltd. London, England. 1925

Poincaré, Henri   1854–1912
French mathematician and theoretical astronomer

I beg your pardon; I am about to use some technical expressions, but they need not frighten you for you are not obliged to understand them. I shall say, for example, that I have found the demonstration of such a theorem under such circumstances. This theorem will have a barbarous name unfamiliar to many, but that is unimportant; what is of interest for the psychologist is not the theorem but the circumstances…
The Foundations of Science
Science and Method. Book I
Chapter III (p. 387)
The Science Press. New York, New York, USA. 1913

Sylvester, James Joseph   1814–97
English mathematician

No mathematician now-a-days sets any store on the discovery of isolated theorems, except as affording hints of an unsuspected new sphere of thought, like meteorites detached from some undiscovered planetary orb of speculation.
The Collected Mathematical Papers of James Joseph Sylvester (Volume 2)
Notes to the Exeter Association Address (p. 717)
University Press. Cambridge, England. 1904–1912

Truesdell, Clifford   1919–2000
American mathematician, natural philosopher, historian of mathematics

Noll, Walter
Mathematician

Pedantry and sectarianism aside, the aim of theoretical physics is to construct mathematical models such as to enable us, from the use of knowledge gathered in a few observations, to predict by logical processes the outcomes in many other circumstances. Any logically sound theory satisfying this condition is a good theory, whether or not it be derived from "ultimate" or "fundamental" truth. It is as ridiculous to deride continuum physics because it is not obtained from nuclear physics as it would be to reproach it with lack of foundation in the Bible.
The Non-Linear Field Theories of Mechanics (2nd edition)
Introduction, Section 2 (p. 4)
Springer-Verlag. Berlin, Germany. 1992

Veblen, Oswald   1880–1960
American mathematician

The abstract mathematical theory has an independent, if lonely existence of its own. But when a sufficient number of its terms are given physical definitions it becomes a part of a vital organism concerning itself at every instant with matters full of human significance. Every theorem can be given the form "if you do so and so, such and such will happen."
Remarks on the Foundation of Geometry
Bulleting of the American Mathematical Association, Volume 31, 1925 (p. 135)

THEORIST

Boltzmann, Ludwig Edward   1844–1906
Austrian physicist

A friend of mine has defined the practical man as one who understands nothing of theory and the theoretician as an enthusiast who understands nothing at all.
Theoretical Physics and Philosophical Problems. Selected Writings
On the Significance of Theories (p. 33)
Reidel Publishing Company. Boston, Massachusetts, USA. 1974

Cardozo, Benjamin N.   1870–1938
American jurist

The theorist has a hard time to make his way in an ungrateful world. He is supposed to be indifferent to realities; yet his life is spent in the exposure of realities, which, till illuminated by his searchlight, were hidden and unknown.
The Growth of the Law
Chapter II (p. 21)
Yale University Press. New Haven, Connecticut, USA. 1924

Crick, Francis Harry Compton   1916–2004
English biochemist

Theorists almost always become too fond of their own ideas, often simply by living with them too long. It is difficult to believe that one's cherished theory, which really works rather nicely in some respects, may be completely false.
What Mad Pursuit: A Personal View of Scientific Discovery
Chapter 13 (p. 141)
Basic Books, Inc. New York, New York, USA. 1988

Theorists will often complain that experimentalists ignore their work. Let a theorist produce just one theory of the type sketched above (i.e., one that makes non-obvious verified predictions) and the world will jump to the conclusion (not always true) that he has a special insight into difficult problems.
What Mad Pursuit: A Personal View of Scientific Discovery
Chapter 13 (p. 142)
Basic Books, Inc. New York, New York, USA. 1988

[I]t is virtually impossible for a theorist, by thought alone, to arrive at the correct solution to a set of biological problems.... The best a theorist can hope to do is to point an experimentalist in the right direction....
What Mad Pursuit: A Personal View of Scientific Discovery
Chapter 10 (pp. 109–110)
Basic Books, Inc. New York, New York, USA. 1988

Lederman, Leon   1922–
American high-energy physicist

The sequence of theorist, experimenter, and discovery has occasionally been compared to the sequence of farmer, pig, truffle. The farmer leads the pig to an area where there might be truffles. The pig searches diligently for the truffles. Finally, he locates one, and just as he is about to devour it, the farmer snatches it away.
The God Particle: If the Universe Is the Answer, What Is the Question?
Chapter 1 (p. 16)
Houghton Mifflin Company. Boston, Massachusetts, USA. 1993

Marshall, Alfred   1842–1924
English economist

...the most reckless and treacherous of all theorists is he who professes to let facts and figures speak for themselves.
In A.C. Pigou (ed.)
Memorials of Alfred Marshall
Chapter VI (p. 168)
Macmillan & Company Ltd. London, England. 1925

Spencer-Brown, George   1923–
English mathematician and polymath

A theorem is no more proved by logic and computation than a sonnet is written by grammar and rhetoric, or than a sonata is composed by harmony and counterpoint, or a picture painted by balance and perspective. ...[T]hese forms are, in the final analysis, parasitic on — they have no existence apart from — the creativity of the work

itself. Thus the relation of logic to mathematics is seen to be that of an applied science to its pure ground, and all applied science is seen as drawing sustenance from a process of creation with which it can combine to give structure, but which it cannot appropriate.
Laws of Form
Chapter 12 (p. 102)
George Allen & Unwin Ltd. London, England. 1969

Truesdell, Clifford   1919–2000
American mathematician, natural philosopher, historian of mathematics
Noll, Walter
Mathematician

The task of the theorist is to bring order into the chaos of the phenomena of nature, to invent a language by which a class of these phenomena can be described efficiently and simply.
The Non-Linear Field Theories of Mechanics (2nd edition)
Introduction, Section 2 (p. 4)
Berlin: Springer-Verlag. 1992

THEORY

Adams, Douglas   1952–2001
English author, comic radio dramatist, and musician

There is a theory which states that if ever anyone discovers exactly what the Universe is for and why it is here, it will instantly disappear and be replaced by something even more bizarre and inexplicable.

There is another which states that this has already happened.
The Ultimate Hitchhiker's Guide to the Galaxy
The Restaurant at the End of the Universe
Chapter 1 (p. 148)
Ballantine Books. New York, New York, USA. 2002

Asimov, Isaac   1920–92s
American writer and biochemist

Theories are not so much wrong as incomplete.
The Relativity of Wrong
The Relativity of Wrong (p. 222)
Doubleday & Company, Inc. New York, New York, USA. 1988

Once we learn to expect theories to collapse and to be supplanted by more useful generalizations, the collapsing theory becomes not the gray remnant of a broken today, but the herald of a new and brighter tomorrow.
In Timothy Ferris (ed.)
The World Treasury of Physics, Astronomy, and Mathematics
The Nature of Science (p. 783)
Little, Brown & Company. Boston, Massachusetts, USA. 1991

Scientific theories have a tendency to fit the intellectual fashions of the time.
Asimov on Chemistry
The Weighting Game (p. 3)
Anchor Press/Doubleday. Garden City, New York, USA. 1974

Ayer, Alfred Jules   1910–89
English philosopher

There never comes a point where a theory can be said to be true. The most that one can claim for any theory is that it has shared the successes of all its rivals and that it has passed at least one test which they have failed.
Philosophy in the Twentieth Century
Chapter IV (p. 133)
Random House, Inc. New York, New York, USA. 1982

Berkeley, Edmund C.   1909–88
American computer theoretician

The World is more complicated than most of our theories make it out to be.
Right Answers — A Short Guide for Obtaining Them
Computers and Automation, Volume 18, Number 10, September 1969
(p. 20)

Bernard, Claude   1813–78
French physiologist

A theory is merely a scientific idea controlled by experiment.
Translated by Henry Copley Greene
An Introduction to the Study of Experimental Medicine
Part One, Chapter I, Section vi (p. 26)
Henry Schuman, Inc. New York, New York, USA. 1927

Bernstein, Jeremy   1929–
American physicist, educator, and writer

I would insist that any proposal for a radically new theory in physics or in any other science, contain a clear explanation of why the precedent science worked. What new domain of experience is being explored by the new science, and how does it meld with the old?
Cranks, Quarks, and the Cosmos: Writings on Science
Chapter 1 (p. 18)
Basic Books, Inc. New York, New York, USA. 1993

Berzelius, Jöns Jacob   1779–1848
Swedish chemist

All our theory is but a means of consistently conceptualizing the inward processes of phenomena, and it is presumable and adequate when all scientifically known facts can be deduced from it.
In Edward O. Wilson
The Diversity of Life
Chapter One (p. 8)
W.W. Norton & Company, Inc. New York, New York, USA.1992

This mode of conceptualization [theorizing] can...well be false and...is so frequently. Even though, at a certain period in the development of science, it may match the purpose just as well as a true theory. Experience is augmented, facts appear which do not agree with it, and one is forced to go in search of a new mode of conceptualization within which these facts can also be accommodated;

and in this manner, no doubt, modes of conceptualization will be altered from age to age, as experience is broadened, and the complete truth may ever be attained.
In Edward O. Wilson
The Diversity of Life
Chapter One (pp. 8–9)
W.W. Norton & Company, Inc. New York, New York, USA. 1992

Bethe, Hans   1906–2005
American physicist

Scientific theories are not overthrown; they are expanded, refined, and generalized.
In Victor F. Weisskopf
Physics in the Twentieth Century: Selected Essays
Forward (p. x)
The MIT Press. Cambridge, Massachusetts, USA. 1972

Bigelow, S. Lawrence   1870–1947
American physical chemistry professor

Our laws summarize our knowledge, our theories summarize our beliefs.... Laws codify established facts, they are history; theories contain the possibilities of the future. Theories may be considered as knowledge in the state of flux.
Theoretical and Physical Chemistry
Section I, Chapter II (pp. 15–16)
The Century Company. New York, New York, USA. 1917

Bohr, Niels Henrik David   1886–1962
Danish physicist

...your theory is crazy. The question which divides us is whether it is crazy enough to have a chance of being correct.
In Martin Gardner
The Ambidextrous Universe (p. 280)

Bondi, Sir Hermann   1919–2005
English mathematician and cosmologist

No theory, however attractive, merits scientific consideration unless it sticks out its neck sufficiently to be disproved by experiment or observation.
In Robert M. Hutchins and Mortimer J. Adler (eds.)
The Great Ideas Today, 1966
Astronomy and the Physical Sciences (p. 263)
Encyclopædia Britannica, Inc. Chicago, Illinois, USA. 1966

Born, Max   1882–1970
German-born English physicist

The human mind is conservative, and the scientist makes no exception from this rule. He will accept a new theory only if it stands the trial of many experimental tests.
Les Prix Nobel. The Nobel Prizes in 1954
Nobel banquet speech for award received in 1954
Nobel Foundation. Stockholm, Sweden. 1955

A theory, to be of any real use to us, must satisfy two tests. In the first place, it must not make use of any ideas which

are not confirmed by experiment. Special assumptions must not be dragged in merely to meet some particular difficulty. In the second place, the theory must not only explain all the facts known already, but must also enable us to foresee other facts which were not known before and can be tested by further experiment.
The Restless Universe
Chapter I (pp. 5–6)
Dover Publications, Inc. New York, New York, USA. 1951

Bradbury, Ray   1920–
American writer

Theories are invigorating and tonic. Give me an ounce of fact and I will produce you a ton of theory by tea this afternoon.
In Ray Bradbury, Arthur C. Clarke, Bruce Murray, Carl Sagan, and Walter Sullivan
Mars and the Mind of Man
Foreword (p. x)
Harper & Row, Publishers. New York, New York, USA. 1973

Bradley, Jr., John Hodgdon   1898–1962
American geologist

But alas for theories that germinate in the minds of men! They wilt in the bright light of advancing knowledge.
Parade of the Living
Part I, Chapter I (pp. 6–7)
Coward-McCann, Inc. New York, New York, USA. 1930

Bridgman, Percy Williams   1882–1961
American physicist

Every new theory as it arises believes in the flush of youth that it has the long-sought goal; it sees no limits to its applicability, and believes that at long last it is the fortunate to achieve the "right" answer.
The Nature of Physical Theory
Chapter X (p. 136)
Princeton University Press. Princeton, New Jersey, USA. 1936

British Association for the Advancement of Science

Great physical theories with their trains of practical consequences, are pre-eminently national objects, whether for glory or utility.
In John Frederick William Herschel
Essays from the Edinburgh and Quarterly Reviews with Addresses and Other Pieces
Meeting, Newcastle, 1838 (p. 109)
Longman, Brown, Green, Longmans & Roberts. London, England. 1857

Buckland, Frank
No biographical data available

Your theory is most excellent, and I shall endeavor to collect facts for you with a view to its elucidation.
In Karl Pearson
The Life, Letters, and Labours of Francis Galton (Volume 2) (p. 87)
At The University Press. Cambridge, England. No date

Burroughs, Edgar Rice   1875–1950
American writer

…even theories must have foundations.
Pirates of Venus
Chapter Four (p. 44)
University of Nebraska Press. Lincoln, Nebraska, USA. 2001

Campbell, Norman R.   1880–1949
English physicist and philosopher

To those who have not the power to think, theory will always be dangerous.
Physics: The Elements
Chapter VI (p. 121)
At The University Press. Cambridge, England. 1920

Space and time are the conceptions of theory, not of laws. They are neither necessary nor useful in the statement of the results of any experiment.
Theory and Experiment in Relativity
Nature, Volume 106, Number 2677, February 17, 1921 (p. 804)

Cantor, Georg   1845–1918
German mathematician

My theory stands as firm as a rock; every arrow directed against it will return quickly to its archer. How do I know this? Because I have studied it from all sides for many years; because I have examined all objections which have ever been made against the infinite numbers; and above all because I have followed its roots, so to speak, to the first infallible cause of all created things.
In Joseph Dauben
Georg Cantor: His Mathematics and Philosophy of the Infinite
Chapter 12 (p. 298)
Princeton University Press. Princeton, New Jersey, USA. 1990

Chamberlin, T. C.   1843–1928
American geologist

The moment one has offered an original explanation for a phenomenon which seems satisfactory, that moment affection for his intellectual child springs into existence, and as the explanation grows into a definite theory his parental affections cluster about his offspring and it grows more and more dear to him…. There springs up also unwittingly a pressing of the theory to make it fit the facts and a pressing of the facts to make them fit the theory…
Journal of Geology, Volume 5, 1897 (p. 837)

The mind lingers with pleasure upon the facts that fall happily into the embrace of the theory, and feels a natural coldness toward those that seem refractory….
The Method of Multiple Working Hypotheses
Science, Volume 148, Number 3671, 7 May 1965 (p. 755)

Charlie Chan
Fictional character

Theory like mist on eyeglasses. Obscure facts.

Charlie Chan in Egypt
Film (1935)

Charcot, Jean-Martin   1825–93
French neurologist

Theory is good, but it doesn't prevent things from existing.
The Complete Psychological Works of Sigmund Freud (Volume 3)
Charcot (p. 13)
The Hogarth Press. London, England. 1962

Chesterton, G. K. (Gilbert Keith)   1874–1936
English author

A man warmly concerned with any large theories has always a relish for applying them to any triviality.
The Wisdom of Father Brown
The Absence of Mr. Glass (p. 16)
Dodd, Mead & Company. New York, New York, USA. 1927

Crichton, Michael   1942–
American novelist

Most areas of intellectual life have discovered the virtues of speculation, and have embraced them wildly. In academia, speculation is usually dignified as theory.
Why Speculate
Talk, International Leadership Forum, La Jolla, California, April 26, 2002

Clarke, Arthur C.   1917–
English science and science fiction writer

"I'd be glad to settle without the theory," remarked Kimball, "if I could even understand what this thing is — or what it's supposed to do."
The Lost Worlds of 2001
Chapter 31 (p. 179)
New American Library. New York, New York, USA. 1972

Colton, Charles Caleb   1780–1832
English sportsman and writer

Theory is worth little, unless it can explain its own phenomena, and it must effect this with out contradicting itself; therefore, the facts are sometimes assimilated to the theory, rather than the theory to the facts.
Lacon; or Many Things in a Few Words (p. 77)
William Gowans. New York, New York, USA. 1849

Professors in every branch of the sciences prefer their own theories to truth: the reason is, that their theories are private property, but truth is common stock.
Lacon; or Many Things in a Few Words (p. 189)
William Gowans. New York, New York, USA. 1849

Cooper, Leon   1930–
American physicist

A theory is a well-defined structure that we hope is in correspondence with what we observe. It's an architecture, a cathedral.

In George Johnson
In the Palaces of Memory: How We Build the Worlds Inside Our Heads
The Memory Machine (pp. 114–115)
Alfred A. Knopf. New York, New York, USA. 1991

Couderc, Paul
No biographical data available

The only conceptions that succumb are those that pretend to fix the image of a profound reality: true relations among things survive, united to the true new relations in the burgeoning theory. Let us then rejoice at the massacre of old theories because this is the criterion of progress. There is, I think, no ground for fear that nature will undernourish the seekers. Nothing should diminish our enthusiasm for the experimental victories, decisive and definitive, of the past thirty years.
In Lucienne Felix
The Modern Aspect of Mathematics (p. 31)
Basic Books, Inc. New York, New York, USA. 1960

Couper, Archibald Scott   1831–92
Scottish scientist

The end of chemistry is its theory. The guide in chemical research is a theory. It is therefore of the greatest importance to ascertain whether the theories at present adopted by chemists are adequate to the explanation of chemical phenomena, or are at least based upon the true principles which ought to regulate scientific research.
The London, Edinburgh and Dublin Philosophical Magazine and Journal of Science, Volume 16, Number 4, 1858

Crease, Robert P.
American science historian
Mann, Charles C.
American journalist and science writer

[P.A.M.] Dirac said often he was beset by fears when he came up with his new theory…. [H]e soon discovered that his fear was justified. Quantum electrodynamics was indeed a great step forward, but it came at a great price. Dirac had set down the beginnings of the modern theory of electromagnetism — the first solid piece of the standard model — but he had also unwittingly let loose an onslaught of conceptual demons that would change our views of space and matter.
In T. Ferris (ed.)
World Treasury of Physics, Astronomy, and Mathematics
Uncertainty and Complementarity (p. 69)
Little, Brown & Company. Boston, Massachusetts, USA. 1991

Crick, Francis Harry Compton   1916–2004
English biochemist

…a theory will always command more attention if it is supported by unexpected evidence, particularly evidence of a different kind.
What Mad Pursuit: A Personal View of Scientific Discovery

Chapter 10 (p. 115)
Basic Books, Inc. New York, New York, USA. 1988

Cromer, Alan   1935–
American physicist and educator

The word theory, as used in the natural sciences, doesn't mean an idea tentatively held for purposes of argument — that we call a hypothesis. Rather, a theory is a set of logically consistent abstract principles that explain a body of concrete facts. It is the logical connections among the principles and the facts that characterize a theory as truth. No one element of a theory…can be changed without creating a logical contradiction that invalidates the entire system. Thus, although it may not be possible to substantiate directly a particular principle in the theory, the principle is validated by the consistency of the entire logical structure.
Uncommon Sense: The Heretical Nature of Science
Chapter 7 (p. 137)
Oxford University Press, Inc. New York, New York, USA. 1993

d'Abro, Abraham
No biographical data available

…a theory of mathematical physics is not one of pure mathematics. Its aim and its raison d'être are not solely to construct the rational scheme of some possible world, but to construct that particular rational scheme of the particular real world in which we live and breathe. It is for this reason that a theory of mathematical physics, in contradistinction to one of pure mathematics, is constantly subjected to the control of experiment.
The Evolution of Scientific Thought
Chapter XXI (p. 215)
Dover Publications, Inc., New York, New York, USA. 1950

Dark, K. R.
No biographical data available

It would be easy to imagine that archaeological theory is daunting, or irrelevant, or both. Theorists often use jargon-laden sentences, quote obscure works, discuss periods and areas distant from those of one's own interest, and are keen to promote their own views.
Theoretical Archaeology
Introduction (p. 1)
Duckworth & Company. London, England. 1995

Darwin, Charles Robert   1809–82
English naturalist

Let theory guide your observations, but till your reputation is well established, be sparing in publishing theory. It makes persons doubt your observations.
In Francis Darwin (ed.)
More Letters of Charles Darwin (Volume 2)
Letter 646, Darwin to Scott, June 6, 1863 (p. 323)
D. Appleton & Company. New York, New York, USA. 1903

In October 1838, that is, fifteen months after I had begun my systematic inquiry, I happened to read for amusement Malthus on Population, and being well prepared to appreciate the struggle for existence which everywhere goes on from long-continued observation of the habits of animals and plants, it at once struck me that under these circumstances favorable variations would tend to be preserved, and unfavorable ones to be destroyed. The results of this would be the formation of a new species. Here, then I had at last got a theory by which to work.
Autobiography

…for without the making of theories I am convinced there would be no observation.
In Francis Darwin (ed.)
The Life and Letters of Charles Darwin (Volume 2)
C. Darwin to C. Lyell, June 1st [1860] (p. 108)
D. Appleton & Company. New York, New York, USA. 1896

Darwin, G. H.   1809–82
No biographical data available

A theory is, then, a necessity for the advance of science, and we may regard it as the branch of a living tree, of which facts are the nourishment.
Address to British Association, Section A
Nature, Volume 34, Number 879, September 2, 1886 (p. 420)

Davies, John Tasman   1924–
Chemist

Theories are generalizations and unifications, and as such they cannot logically follow only from our experiences of a few particular events. Indeed we often generalize from a single event, just as a dog does who, having once seen a cat in a certain driveway, looks eagerly around whenever he passes that place in future. Evidently this latter activity is equivalent to testing the theory…that "there is always a cat in that driveway."
The Scientific Approach
Chapter 1 (p. 11)
Academic Press. London, England. 1965

…a theory arises from a leap of the imagination…
The Scientific Approach
Chapter 1 (p. 11)
Academic Press. London, England. 1965

Davies, Paul Charles William   1946–
British-born physicist, writer, and broadcaster

The basis of this theory is that in nature there is an inherent uncertainty or unpredictability that manifests itself only on an atomic scale. For example, the position of a subatomic particle such as an electron may not be a well-defined concept at all; it should be envisaged as jiggling around in a random sort of a way. Energy, too, becomes a slightly nebulous concept, subject to capricious and unpredictable changes.
The Edge of Infinity: Where the Universe Came from and How It Will

End
Chapter 4 (p. 90)
Simon & Schuster. New York, New York, USA. 1981

Davis, Philip J.   1923–
American mathematician
Hersh, Reuben   1927–
American mathematician

If the number of theorems is larger than one can possibly survey, who can be trusted to judge what is "important?" One cannot have survival of the fittest if there is no interaction.
The Mathematical Experience
Ulam's Dilemma (p. 21)
Birkhäuser. Boston, Massachusetts, USA. 1981

Davy, Sir Humphry   1778–1829
English chemist

Theories ought to be made for time, and be considered capable of improvement.
The Collected Works of Sir Humphry Davy (Volume 1)
Memories of the Life of Sir Humphry Davy
Chapter II (p. 70)
London, England. 1839–1840

de Fermat, Pierre   1601–65
French mathematician

We have found a beautiful and most general proposition, namely, that every integer is either a square, or the sum of two, three or at most four squares. This theorem depends on some of the most recondite mysteries of number, and is not possible to present its proof on the margin of this page.
In Tobias Dantzig
Number: The Language of Science (4th edition) (p. 269)
The Macmillan Company. New York, New York, USA. 1954

I have found a very great number of exceedingly beautiful theorems.
In E.T. Bell
Men of Mathematics (p. 56)
Simon & Schuster. New York, New York, USA. 1937

de Grasse Tyson, Neil   1958–
American astrophysicist and writer

Scientific evidence in support of a theory sometimes takes you places where your senses have never been. Twentieth-century science has largely been built upon data collected with all manner of tools that enable us to see the universe in decidedly uncommon ways. As a consequence, while we have always required that a theory make mathematical sense, we no longer require that a theory make common sense. We simply demand that it be consistent with the results of observations and experiments....
In Defense of the Big Bang
Natural History, Volume 105, Number 12, December 1996 (p. 76)

A well-constructed theory must explain some of what is not understood, predict previously unknown phenomena, and, to be successful, have its predictions consistently confirmed. Furthermore, skeptics should not hesitate to question every possible assumption, no matter how basic.
In Defense of the Big Bang
Natural History, Volume 105, Number 12, December 1996 (p. 76)

Dingle, Herbert   1890–1978
English astrophysicist

Success in scientific theory is won, not by rigid adherence to the rules of logic, but by bold speculation which dares even to break those rules if by that means new regions of interest may be opened up.
Through Science to Philosophy
Part II, Chapter XV (p. 346)
At The Clarendon Press. Oxford, England. 1937

Doyle, Sir Arthur Conan   1859–1930
Scottish writer

One forms provisional theories and waits for time or fuller knowledge to explode them. A bad habit, Mr. Ferguson, but human nature is weak.
In William S. Baring-Gould (ed.)
The Annotated Sherlock Holmes (Volume 2)
The Adventure of the Sussex Vampire (pp. 467–468)
Wings Books. New York, New York, USA. 1967

I don't mean to deny that the evidence is in some ways very strong in favor of your theory, I only wish to point out that there are other theories possible.
In William S. Baring-Gould (ed.)
The Annotated Sherlock Holmes (Volume 2)
Adventures of the Norwood Builder (p. 421)
Wings Books. New York, New York, USA. 1967

Duhem, Pierre-Maurice-Marie   1861–1916
French physicist and mathematician

A physical theory is not an explanation. It is a system of mathematical propositions, deduced from a small number of principles, which aim...to represent as simply, as completely, and as exactly as possible a group of experimental laws.
The Aim and Structure of Physical Theory
Part I, Chapter II (p. 19)
Princeton University Press. Princeton, New Jersey, USA. 1954

The sole purpose of physical theory is to provide a representation and classification of experimental laws; the only test permitting us to judge a physical theory and pronounce it good or bad is the comparison between the consequences of this theory and the experimental laws it has to represent and classify.
The Aim and Structure of Physical Theory
Part II, Chapter VI (p. 180)
Princeton University Press. Princeton, New Jersey, USA. 1954

Unlike the reduction to absurdity employed by geometers, experimental contradiction does not have the power to transform a physical hypothesis into an indisputable truth; in order to confer this power on it, it would be necessary to enumerate completely the various hypotheses which may cover a determinate group of phenomena; but the physicist is never sure he has exhausted all the imaginable assumptions. The truth of a physical theory is not decided by heads or tails.
The Aim and Structure of Physical Theory
Part II, Chapter VI (p. 190)
Princeton University Press. Princeton, New Jersey, USA. 1954

When several taps of the beak break the shell of an egg from which the chick escapes, a child may imagine that this rigid and immobile mass, similar to the white shells he picks up on the edge of a stream, had suddenly taken life and produced the bird who runs away with a chirp; but just where his childish imagination sees a sudden creation, the naturalist recognizes the last stage of a long development; he thinks back to the first fusion of two microscopic nuclei in order to review next the series of divisions, differentiations, and reabsorptions which, cell by cell, have built up the body of the chick. The ordinary layman judges the birth of [scientific] theories as the child the appearance of the chick.
The Aim and Structure of Physical Theory
Part II, Chapter VII (p. 221)
Princeton University Press. Princeton, New Jersey, USA. 1954

Contemplation of a set of experimental laws does not, therefore, suffice to suggest to the physicist what hypotheses he should choose in order to give a theoretical representation of these laws; it is also necessary that the thoughts habitual with those among whom he lives and the tendencies impressed on his own mind by his previous studies come and guide him, and restrict the excessively great latitude left to this day a merely empirical form until circumstances prepare the genius of a physicist to conceive the hypothesis which will organize them into a theory!
The Aim and Structure of Physical Theory
Part II, Chapter VII (p. 255)
Princeton University Press. Princeton, New Jersey, USA. 1954

Duschl, Richard Alan   1951–
American science education professor and researcher

In order to say we have developed a knowledge of science, we must be able to say we have an understanding of the function, structure, and generation of scientific theories.
Restructuring Science Education: The Importance of Theories and Their Development
Chapter 6 (p. 96)
Teachers College Press. New York, New York, USA. 1990

Eddington, Sir Arthur Stanley   1882–1944
English astronomer, physicist, and mathematician

…a reasoned theory is preferable to blind extrapolation.
The Expanding Universe
Chapter I, Section IV (p. 18)
The University Press. Cambridge. 1933

There was just one place where the theory did not seem to work properly, and that was — infinity. I think Einstein showed his greatness in the simple and drastic way in which he disposed of difficulties at infinity. He abolished infinity. He slightly altered his equations so as to make space at great distances bend round until it closed up. So that, if in Einstein's space you kept going right on in one direction, you do not get to infinity; you find yourself back at your starting-point again. Since there was no longer any infinity, there could be no difficulties at infinity. Q. E. D.
The Expanding Universe
Chapter I, Section V (pp. 21–22)
The University Press. Cambridge. 1933

The relativity theory of physics reduces everything to relations; that is to say, it is structure, not material, which counts. The structure cannot be built up without material; but the nature of the material is of no importance.
Space, Time and Gravitation: An Outline of the General Relativity Theory
Chapter XII (p. 197)
At The University Press. Cambridge, England. 1953

We have found a strange footprint on the shores of the unknown. We have devised profound theories, one after another to account for its origin. At last, we have succeeded in reconstructing the creature that made the footprint. And lo! It is our own.
Space, Time and Gravitation: An Outline of the General Relativity Theory
Chapter XII (p. 201)
At The University Press. Cambridge, England. 1921

Einstein, Albert   1879–1955
German-born physicist

These fundamental concepts and postulates, which cannot be further reduced logically, form the essential part of a theory, which reason cannot touch. It is the grand object of all theory to make these irreducible elements as simple and as few in number as possible, without having to renounce the adequate representation of any empirical content whatever.
Translated by Alan Harris
Essays in Science
On the Method of Theoretical Physics (p. 15)
Philosophical Library. New York, New York, USA. 1934

No fairer destiny could be allotted to any physical theory, than that it should of itself point out the way to the introduction of a more comprehensive theory, in which it lives on as a limiting case.
Translated by Robert W. Lawson
Relativity: The Special and General Theory

Part II, Chapter 22 (pp. 98–99)
Pi Press. New York, New York, USA. 2005

Of the general theory of relativity you will be convinced, once you have studied it. Therefore I am not going to defend it with a single word.
In Paul Arthur Schlipp (ed.)
Albert Einstein: Philosopher-Scientist
Letter to A. Sommerfeld, 8 February 1916 (p. 101)
The Library of Living Philosophers, Inc. Evanston, Illinois, USA. 1949

Physical theory has two ardent desires, to gather up as far as possible all pertinent phenomena and their connections, and to help us not only to know how Nature is and how her transactions are carried through, but also to reach as far as possible the perhaps utopian and seemingly arrogant aim of knowing why Nature is thus and not otherwise. Here lies the highest satisfaction of a scientific person.
In G. Holton
Thematic Origins of Scientific Thought
Chapter 8 (p. 242)
Harvard University Press. Cambridge, Massachusetts, USA. 1973

Theories are evolved and are expressed in short compass as statements of a large number of individual observations in the form of empirical laws, from which the general laws can be ascertained by comparison. Regarded in this way, the development of a science bears some resemblance to the compilation of a classified catalogue. It is, as it were, a purely empirical enterprise.
Translated by Robert W. Lawson
Relativity: The Special and General Theory
Appendix III, The Experimental Confirmation of the General Theory of Relativity
Pi Press. New York, New York, USA. 2005

For the creation of a theory the mere collection of recorded phenomena never suffices — there must always be added a free invention of the human mind that attacks the heart of the matter. And: the physicist must not be content with the purely phenomenological considerations that pertain to the phenomenon. Indeed, he should press on to the speculative method, which looks for the underlying pattern.
In Helen Dukas and Banesh Hoffman
Albert Einstein: The Human Side: New Glimpses from His Archives
Lecture at the Berlin Planetarium, 4 October 1931 (pp. 29–30)
Princeton University Press. Princeton, New Jersey, USA. 1979

The scientific theorist is not to be envied. For Nature, or more precisely experiment, is an inexorable and not very friendly judge of his work. It never says "Yes" to a theory. In the most favorable cases it say "Maybe," and in the great majority of cases simply "No." If an experiment agrees with a theory it means for the latter "Maybe," and if it does not agree it means "No." Probably every theory will some day experience its "No" — most theories [do], soon after conception.
In Helen Dukas and Banesh Hoffman

Albert Einstein: The Human Side: New Glimpses from His Archives
Note dated 11 November 1922 (p. 18)
Princeton University Press. Princeton, New Jersey, USA. 1979

If my theory of relativity is proven successful, Germany will claim me as a German and France will declare that I am a citizen of the world. Should my theory prove untrue, France will say that I am a German, and Germany will declare that I am a Jew.
The Great Quotations
Address at the Sorbonne (p. 226)
New York Times, 16 February 1930

A theory is the more impressive the greater the simplicity of its premises, the more different kinds of things it relates, and the more extended its area of applicability.
Translated by Paul Arthur Schlipp
Albert Einstein: Autobiographical Notes (p. 31)
Open Court. La Salle, Illinois, USA. 1979

I have learned something else from the theory of gravitation: no collection of empirical facts however comprehensive can ever lead to the setting up of such complicated equations. A theory can be tested by experience, but there is no way from experience to the construction of a theory. Equations of such complexity as are the equations of the gravitational field can be found only through the discovery of a logically simple mathematical condition that determines the equations completely or almost completely. Once one has obtained those sufficiently strong formal conditions, one requires only little knowledge of facts for the construction of the theory…
Translated by Paul Arthur Schlipp
Albert Einstein: Autobiographical Notes (p. 31)
Open Court. La Salle, Illinois, USA. 1979

Einstein, Albert   1879–1955
German-born physicist
Infeld, Leopold   1898–1968
Polish physicist

There are no eternal theories in science. It always happens that some of the facts predicted by a theory are disproved by experiment. Every theory has its period of gradual development and triumph, after which it may experience a rapid decline.
The Evolution of Physics
The Two Electric Fluids (p. 77)
Simon & Schuster. New York, New York, USA. 1961

Creating a new theory is not like destroying an old barn and erecting a skyscraper in its place. It is rather like climbing a mountain, gaining new and wider views, discovering unexpected connections between our starting point and its rich environment. But the point from which we started out still exists and can be seen, although it appears smaller and forms a tiny part of our broad view gained by the mastery of the obstacles on our adventurous way up.

The Evolution of Physics
The Mechanical Scaffold (pp. 158–159)
Simon & Schuster. New York, New York, USA. 1961

Eliot, George (Mary Ann Evans Cross) 1819–80
English novelist

The possession of an original theory which has not yet been assailed must certainly sweetens the temper of a man who is not beforehand ill-natured.
Impressions of Theophrastus Such
How We Encourage Research (p. 26)
William Blackwood. London, England. 1879

Faraday, Michael 1791–1867
English physicist and chemist

The world little knows how many of the thoughts and theories which have passed through the mind of a scientific investigator have been crushed in silence and secrecy; that in the most successful instances not a tenth of the suggestions, the hopes, the wishes, the preliminary conclusions have been realized.
In W.I.B. Beveridge
The Art of Scientific Investigation
Chapter Five (p. 58)
W.W. Norton & Company, Inc. New York, New York, USA.1957

Feynman, Richard P. 1918–88
American theoretical physicist

Another thing I must point out is that you cannot prove a vague theory wrong. If the guess that you make is poorly expressed and rather vague, and the method that you use for figuring out the consequences is a little vague — you are not sure, and you say, "I think everything's right because it's all due to so and so,…and I can sort of explain how this works"…then you see that this theory is good, because it cannot be proved wrong! Also if the process of computing the consequences is indefinite, then with a little skill any experimental results can be made to look like the expected consequences.
The Character of Physical Law
Chapter 7 (pp. 158–159)
BBC. London, England. 1965

If someone were to propose that the planets go around the sun because all planet matter has a kind of tendency for movement, a kind of motility, let us call it an "oomph," this theory could explain a number of other phenomena as well. So this is a good theory, is it not? It is nowhere near as good as the proposition that the planets move around the sun under the influence of a central force which varies exactly inversely as the square of the distance from the center. The second theory is better because it is so specific; it is so obviously unlikely to be the result of chance. It is so definite that the barest error in the movement can show that it is wrong; but the planets could wobble all over the place, and, according

to the first theory, you could say, "Well, that is the funny behavior of the oomph."
The Meaning of It All: Thoughts of a Citizen Scientist
Chapter I (pp. 19–20)
Perseus Books. Reading, Massachusetts, USA. 1998

This is the key of modern science and…the beginning of the true understanding of Nature — this idea to look at the thing, to record the details, and to hope that in the information thus obtained might lie a clue to one or another theoretical interpretation.
The Character of Physical Law
Chapter 1 (p. 15)
BBC. London, England. 1965

Fischer, Martin H. 1879–1962
German-American physician

Don't confuse hypothesis and theory. The former is a possible explanation; the latter, the correct one. The establishment of theory is the very purpose of science.
In Howard Fabing and Ray Marr
Fischerisms
C.C. Thomas. Springfield, Illinois, USA. 1944

Frost, Robert 1874–1963
American poet

A theory if you hold it hard enough
And long enough gets rated as a creed.…
Complete Poems of Robert Frost
Etherealizing
Henry Holt & Company. New York, New York, USA. 1949

Gay-Lussac, Joseph Louis 1778–1850
French chemist and physicist

In order to draw any conclusion…it is prudent to wait until more numerous and exact observations have provided a solid foundation on which we may build a rigorous theory.
In Maurice Crosland
Gay-Lussac: Scientist and Bourgeois
Chapter 4 (p. 71)
Cambridge University Press. Cambridge, England. 1978

George, William H.
No biographical data available

Theories come into fashion and theories go out of fashion, but the facts connected with them stay.
The Scientist in Action: A Scientific Study of His Methods
Chapter XII (p. 218)
Williams & Norgate Ltd. London, England. 1936

Gilbert, Sir William Schwenck 1836–1911
English playwright and poet
Sullivan, Arthur 1842–1900
English composer

About binomial theorem I'm teeming with a lot o' news
—With many cheerful facts about the square of the
hypotenuse.
The Complete Plays of Gilbert and Sullivan
The Pirates of Penzance
Act I (p. 133)
W.W. Norton & Company, Inc. New York, New York, USA.1976

Glashow, Sheldon L. 1932–
American physicist

No matter how compelling or elegant it is, a theory of
physics must be subjected to experimental verification or
it differs little from medieval theology.
*Interactions: A Journey Through the Mind of a Particle Physicist and
the Matter of This World*
Chapter 4 (p. 77)
Warner Books. New York, New York, USA. 1988

Goddard, Robert H. 1882–1945
American physicist

In this present age of science, when no problem seems too
baffling for the inventor, and no mysterious phenomenon
too much in the dark for elucidation by the discoverer, it
is surprising that so few fundamental scientific theories
can be proved.
The Papers of Robert H. Goddard (Volume 1)
On Taking Things for Granted (p. 63)
McGraw-Hill Book Company, Inc. New York, New York, USA. 1970

Goldhaber, Maurice 1911–
Austrian-American physicist

Antaeus was the strongest person alive, invincible as long
as he was in contact with his mother, the earth. Once he
lost contact with the earth, he grew weak and was van-
quished. Theories in physics are like that. They have to
touch the ground for their strength.
In Robert P. Crease and Charles C. Mann
How the Universe Works
The Atlantic Monthly, August 1984 (p. 91)

Goodman, Nicholas P.
No biographical data available

There are no deep theorems — only theorems that we
have not understood very well.
Reflections on Bishops Philosophy of Mathematics
The Mathematical Intelligence, Volume 5, Number 3, 1983 (p. 63)

Gould, Stephen Jay 1941–2002
American paleontologist and evolutionary biologist

As the new Darwinian orthodoxy swept through Europe,
its most brilliant opponent, the aging embryologist Karl
Ernst von Baer, remarked with bitter irony that every
triumphant theory passes through three stages: first it
is dismissed as untrue; then it is rejected as contrary to
religion; finally, it is accepted as dogma and each scientist
claims that he had long appreciated its truth.

Ever Since Darwin: Reflections in Natural History
Chapter 20. The Validation of Continental Drift (pp. 161–162)
W.W. Norton & Company, Inc. New York, New York, USA.1977

I would say usually, theories act as straitjackets to
channel observations toward their support, and to fore-
stall data that might refute them. Such theories cannot be
rejected from within, for we will not conceptualize the
disapproving observations.
Dinosaur in a Haystack: Reflections in Natural History
Part Three, Chapter 12 (p. 151)
Random House, Inc. New York, New York, USA. 1995

All great theories are expansive, and all notions so rich
in scope and implication are underpinned by visions
about the nature of things. You may call these visions
"philosophy," or "metaphor," or "organizing principle,"
but one thing they are surely not — they are not simple
inductions from observed facts of the natural world.
*Time's Arrow, Time's Cycle: Myth and Metaphor in the Discovery of
Geological Time*
Chapter 1 (p. 9)
Harvard University Press. Cambridge, Massachusetts, USA. 1987

Gratzer, Walter Bruno 1932–
English science writer

It is also of course difficult to renounce a cherished
theory, the product of costly intellectual and emotional
investment, and to accept the cost to ambition, reputation
and pride of a humiliating retraction. As the economist
J. K. Galbraith put it, "faced with the choice between
changing one's mind and proving that there is no need to
do so, almost everyone gets busy with the proof." And so
a fatuous optimism triumphs over the caution that must
guide all scientists through most of their working lives.
*The Undergrowth of Science: Delusion, Self-Deception and Human
Frailty*
Chapter 3 (p. 81)
Oxford University Press, Inc. Oxford, England. 2000

Gribbin, John 1946–
British science writer and astronomer

Good theories are the ones that get those predictions
right; the best theories enable us to "get right" the calculation of how the Universe came into being and then
exploded into its present form. But that doesn't mean that
they convey ultimate truth, or that there "really are" little
hard particles rattling around against each other inside
the atom. Such truth as there is in any of this work lies in
the mathematics; the particle concept is simply a crutch
ordinary mortals can use to help them towards an understanding of the mathematical laws.
The Search of Superstrings, Symmetry, and the Theory of Everything
(pp. 51–52)
Little, Brown & Company. Boston, Massachusetts, USA. 1998

Haldane, John Burdon Sanderson 1892–1964
English biologist

"The results so far obtained are consistent with the view that…" has taken the place of "Thus saith the Lord…" as an introduction to a new theory.
Possible Worlds and Other Papers
Chapter XXX (p. 223)
Harper & Brothers. New York, New York, USA. 1928

Hamilton, Edith   1868–1963
German-born classicist and educator

Theories that go counter to the facts of human nature are foredoomed.
The Roman Way
Comedy's Mirror
W.W. Norton & Company, Inc. New York, New York, USA.1932

Hawking, Stephen William   1942–
English theoretical physicist

…a good theory is characterized by the fact that it makes a number of predictions that could in principle be disproved or falsified by observation.
A Brief History of Time: The Updated and Expanded Edition
Chapter 1 (p. 10)
Bantam Books. Toronto, Ontario, Canada. 1988

Heaviside, Oliver   1850–1925
English electrical engineer, mathematician, and physicist

Theory always tends to become more abstract as it emerges successfully from the chaos of facts by processes of differentiation and elimination, whereby the essentials and their connections become recognized, while minor effects are seen to be secondary or unessential, and are ignored temporarily, to be explained by additional means.
In J.W. Mellor
Higher Mathematics for Students of Chemistry and Physics (p. 370)
Dover Publications. New York, New York, USA. 1955

Heinlein, Robert A.   1907–88
American science fiction writer

Permit me to say, speaking from experience, all theories are empty.
Time Enough for Love
Chapter XIII (p. 390)
G.P. Putnam's Sons. New York, New York, USA. 1973

Modern theory did not arise from revolutionary ideas which have been, so to speak, introduced into the exact sciences from without. On the contrary, they have forced their way into research which was attempting consistently to carry out the program of classical physics — they arise out of its very nature.
In Heinz R. Pagels
The Cosmic Code: Quantum Physics as the Language of Nature
Part I, Chapter 3 (p. 67)
Simon & Schuster. New York, New York, USA. 1982

Heitler, W.   1904–1981
German theoretical physicist

It is usually the fate of good physical theory that after its initial success, difficulties or limitations of its applicability become apparent. Eventually it is superseded by a better theory in which some of the difficulties are removed or which has a wider field of application, as the case may be.
The Quantum Theory of Radiation
Introduction (p. xi)
At The Clarendon Press. Oxford, England. 1954

Herschel, Sir John Frederick William   1792–1871
English astronomer and chemist

A philosophical theory does not shoot up like the tall and spiry pine in graceful and unencumbered natural growth, but, like a column built by men, ascends amid extraneous apparatus and shapeless masses of materials…
Essays from the Edinburgh and Quarterly Reviews with Addresses and Other Pieces
Terrestrial Magnetism (p. 67)
Longman, Brown, Green, Longmans & Roberts. London, England. 1857

Hilton, James   1900–1954
English-born novelist

And I believe that the Binomial Theorem and a Bach Fugue are, in the long run, more important than all the battles of history.
This Week Magazine, 1937

Holt, Michael
No biographical data available

Nobody knows why, but [the only] scientific theories that really work are the mathematical ones.
Mathematics in Art
Chapter 2 (p. 33)
Studio Vista. London, England. 1971

Horton, Robin
British anthropologist and philosopher

To say of the traditional African thinker that he is interested in supernatural rather than natural causes makes little more sense…than to say of the physicist that he is interested in nuclear rather than natural causes. In fact, both are making the same use of theory to transcend the limited vision of natural causes provided by common sense.
African Traditional Thought and Western Science
Africa, Volume 37, Number 1, 1967 (p. 57)

Hoyle, Sir Fred   1915–2001
English mathematician and astronomer

It is true that we must not accept a theory on the basis of emotional preference but it is not an emotional preference to attempt to establish a theory that would place us in a position to obtain a complete understanding of the Universe. The stakes are high, and win or lose, are worth playing for.
Frontiers of Astronomy

Epilogue (pp. 354–355)
Harper & Row, Publishers. New York, New York, USA. 1955

Hoyle, Sir Fred   1915–2001
English mathematician and astronomer
Hoyle, Geoffrey   1942–
English science fiction writer

Be suspicious of a theory if more and more hypotheses are needed to support it as new facts become available, or as new considerations are brought to bear.
Evolution from Space (p. 135)
Simon & Schuster. New York, New York, USA. 1982

Hubble, Edwin Powell   1889–1953
American astronomer

Many theories are formulated but relatively few endure the tests. The survivors, in general, must be occasionally revised to conform with the growing body of knowledge. The ability to theorize is highly personal; it involves art, imagination, logic, and something more. An outstanding genius may invent a successful new type of theory; first-rate men may follow the lead and develop other theories on the same pattern; less competent minds are embarrassed by the custom of testing predictions.
The Realm of the Nebulae
Introduction (p. 5)
Dover Publications, Inc. New York, New York, USA. 1958

No theory is sacred. When a theory fails to meet the test of verified predictions, it is modified to include the larger field, or, vary rarely, it may be abandoned completely.
The Nature of Science and Other Lectures
Part I, Experiment and Experience (p. 41)
The Huntington Library. San Marino, California, USA. 1954

Huggins, Sir William   1824–1910
English astronomer

A theory which sweeps the astronomical horizon of so many mysteries must not only arouse our profound interest, but claim the respectful consideration of men of science.
Annual Report of the Board of Regents of the Smithsonian Institution, 1902
Stellar Evolution in the Light of Recent Research (p. 192)
Government Printing Office. Washington, D.C. 1903

Hutton, James   1726–97
Scottish geologist, chemist, and naturalist

But when, in framing a theory of the earth, a geologist shall indulge his fancy in framing, without evidence, that which had preceded the present order of things, he then either misleads himself, or writes a fable for the amusement of his reader.
The Theory of the Earth (Volume 1)
Part I, Chapter III (pp. 280–281)
Messrs. Cadwell, Junior & Davies. London, England. 1795

Huxley, Thomas Henry   1825–95
English biologist

The struggle for existence holds as much in the intellectual as in the physical world. A theory is a species of thinking, and its right to exist is coextensive with its power of resisting extinction by its rivals.
Collected Essays (Volume 2)
Darwiniana
The Coming of Age of "The Origin of Species" (p. 229)
Macmillan & Company Ltd. London, England. 1904

James, William   1842–1910
American philosopher and psychologist

Theories thus become instruments, not answers to enigmas, in which we can rest. We don't lie back upon them, we move forward, and, on occasion, make nature over again by their aid.
Pragmatism: A New Name for Some Old Ways of Thinking
Lecture II (p. 53)
Longmans, Green & Company London, England. 1914

…the classic stages of a theory's career. First, you know, a new theory is attacked as absurd; then it is admitted to be true, but obvious and insignificant; finally it is seen to be so important that its adversaries claim that they themselves discovered it.
Pragmatism: A New Name for Some Old Ways of Thinking
Lecture VI (p. 198)
Longmans, Green & Company London, England. 1914

Jastrow, Joseph   1863–1944
Polish-born psychologist

Theories without facts or based on uncritically selected facts are vain, and facts without theoretical interpretation, blind.
In Joseph Jastrow (ed.)
The Story of Human Error
Introduction (p. 33)
D. Appleton-Century Company, Inc. New York, New York, USA. 1936

Joos, Georg   1894–1959
German physicist

While it is true that theory often sets difficult, if not impossible tasks for the experiment, it does, on the other hand, often lighten the work of the experimenter by disclosing cogent relationships which make possible the indirect determination of inaccessible quantities and thus render difficult measurements unnecessary.
Theoretical Physics
Introduction (p. 1)
Blackie & Son Ltd. London, England. 1968

Kelvin, Lord William Thomson   1824–1907
Scottish engineer, mathematician, and physicist

A scientific theory is a tool and not a creed.
In Richard Willstätter
From My Life: The Memoirs of Richard Willstatter

Chapter 12 (p. 388)
W.A. Benjamin. New York, New York, USA. 1965

Keyser, Cassius Jackson   1862–1947
American mathematician

For scientific theories are, each and all of them, and they will continue to be, built upon and about notions which, however sublimated, are nevertheless derived from common sense.
Mathematics (pp. 5–6)
Columbia University Press. New York, New York, USA. 1907

Kitaigorodski, Aleksandr Isaakovich   1914–
No biographical data available

A first-rate theory predicts; a second-rate theory forbids and a third-rate theory explains after the event.
Lecture, ICU, Amsterdam, August 1975

Koestler, Arthur   1905–83
Hungarian-born English writer

The history of cosmic theories…may without exaggeration be called a history of collective obsessions and controlled schizophrenias.
The Sleepwalkers
Preface (p. 15)
The Macmillan Company. New York, New York, USA. 1966

Kuhn, Thomas S.   1922–96
American historian of science

The scientist must…be concerned to understand the world and to extend the precision and scope with which it has been ordered. That commitment must, in turn, lead them to scrutinize, either for themselves or through colleagues, some aspect of nature in great empirical detail. And, if that scrutiny displays pockets of apparent disorder, then these must challenge the scientist to a new refinement of their observational techniques or to a further articulation of their theories.
The Structure of Scientific Revolutions
Chapter IV (p. 42)
The University of Chicago Press. Chicago, Illinois, USA. 1970

…no theory ever solves all the puzzles with which it is confronted at a given time; nor are the solutions already achieved often perfect.
The Structure of Scientific Revolutions
Chapter XII (p. 146)
The University of Chicago Press. Chicago, Illinois, USA. 1970

One often hears that successive theories grow ever closer to, or approximate more and more closely to, the truth. Apparently generalizations like that refer not to the puzzle-solutions and the concrete predictions derived from a theory but rather to its ontology, to the match, that is, between the entities with which the theory populates nature and what is "really there."
The Structure of Scientific Revolutions

Postscript–1969 (p. 206)
The University of Chicago Press. Chicago, Illinois, USA. 1970

Lakatos, Imre   1922–74
Hungarian-born philosopher

No theory forbids some state of affairs specifiable in advance; it is not that we propose a theory and Nature may shout No. Rather, we propose a maze of theories, and Nature may shout INCONSISTENT.
Criticism and the Methodology of Scientific Research Programmes
Proceedings of the Aristotelian Society, Volume 69, 1968–1969 (p. 162)

Scientists want to make their theories respectable, deserving of the title "science", that is genuine knowledge. Now the most relevant knowledge in the seventeenth century, when science was born, concerned God, the Devil, Heaven and Hell. If one got one's conjectures about matters of divinity wrong, the consequences of one's mistake was eternal damnation.
The Methodology of Scientific Research Programmes (p. 2)
Cambridge University Press. Cambridge, England. 1978

Laszlo, E.
No biographical data available

Ours is a complex world. But human knowledge is finite and circumscribed. "Nature does not come as clean as you can think it," warned Alfred North Whitehead, and went on to propound an extremely clean and elegant cosmology. Since theories, like window panes, are clear only when they are clean, and the world does not come as cleanly as all that, we must know where we perform a clean-up operation.
The Systems View of the World: The Natural Philosophy of the New Developments in the Sciences
Chapter 1, Section 2 (p. 13)
George Braziller. New York, New York, USA. 1972

Scientific theories, while simpler than reality, must nevertheless reflect its essential structure. Science then must beware of rejecting the structure for the sake of simplicity; that would be to throw out the baby with the bath water.
The Systems View of the World: The Natural Philosophy of the New Developments in the Sciences
Chapter 1, Section 2 (p. 13)
George Braziller. New York, New York, USA. 1972

Lauden, Larry   1945–
American philosopher of science

…the rationale for accepting or rejecting any theory is thus fundamentally based on the idea of problem-*solving progress*. If one research tradition has solved more important problems than its rivals, then accepting that tradition is rational precisely to the degree that we are aiming to "progress," i.e., to maximize the scope of solved problems. In other words, the choice of one tradition over its rivals is a progressive (and thus a rational) choice

precisely to the extent that the chosen tradition is a better problem solver than its rivals.
Progress and Its Problems: Toward a Theory of Scientific Growth
Chapter Three (p. 109)
University of California Press. Berkeley, California, USA. 1977

Lederman, Leon   1922–
American high-energy physicist

...a comment on the word "theory," which lends itself to popular misconceptions. "That's your theory" is a popular sneer. Or "That's only a theory." Our fault for sloppy use. The quantum theory and the Newtonian theory are well-established, well-verified components of our world view. They are not in doubt. It's a matter of derivation. Once upon a time it was Newton's (as yet unverified) "theory." Then it was verified, but the name stuck. "Newton's theory" it will always be. On the other hand, superstrings and GUTs are speculative efforts to extend current understanding, building on what we know.
The God Particle: If the Universe Is the Answer, What Is the Question?
Chapter 9 (p. 389)
Houghton Mifflin Company. Boston, Massachusetts, USA. 1993

The better theories are verifiable. Once upon a time that was the sine qua non of any theory. Nowadays, addressing events at the Big Bang, we face, perhaps for the first time, a situation in which a theory may never be experimentally tested.
The God Particle: If the Universe Is the Answer, What Is the Question?
Chapter 9 (p. 389)
Houghton Mifflin Company. Boston, Massachusetts, USA. 1993

Lewontin, Richard C.   1929–
American evolutionary geneticist and philosopher of science

Theory generally should not be an attempt to say how the world is. Rather, it is an attempt to construct the logical relations that arise from various assumptions about the world.
In E. Mayr and W.B. Provine (eds.)
The Evolutionary Synthesis
Part One, Chapter 1
Theoretical Population Genetics in the Evolutionary Synthesis (p. 65)
Harvard University Press. Cambridge, Massachusetts, USA. 1980

It is not always appreciated that the problem of theory building is a constant interaction between constructing laws and finding an appropriate set of descriptive state variables such that laws can be constructed.
The Genetic Basis of Evolutionary Change
Chapter 1 (p. 8)
Columbia University Press. New York, New York, USA. 1974

...we cannot go out and describe the world in any old way we please and then sit back and demand that an explanatory and predictive theory be built on that description.
The Genetic Basis of Evolutionary Change
Chapter 1 (p. 8)
Columbia University Press. New York, New York, USA. 1974

Libes, Antoine   1752–1832
French physicist

Let us add a word in favor of theories, which certain physicists still dare to present as invincible obstacles to the discovery of truth. It is incontestable that experience and observation ought to serve as the basis of our physical knowledge. But without the help of theory the most well-certified experiments, the most numerous observations will be only isolated facts in the hands of the physicist, isolated facts which cannot serve for the advancement of physics. The man of genius must seize upon these scattered links and bring them together skillfully to form a continuous chain. This continuity constitutes the theory, which alone can give us a glimpse of the relations which bind the facts to one another and of their dependence on the causes which have produced them.
In Russell McCormmach (ed.)
Historical Studies in the Physical Sciences (Volume 4)
In Robert H. Silliman
Fresnel and the Emergence of Physics as a Discipline (p. 143)
Princeton University Press. Princeton, New Jersey, USA. 1974

Lichtenberg, Georg Christoph   1742–99
German physicist and satirical writer

This whole theory is good for nothing except disputing about.
Lichtenberg: Aphorisms & Letters
Aphorisms (p. 57)
Jonathan Cape. London, England. 1969

Lindley, David   1956–
English astrophysicist and author

Theoretical physicists can invent theories in which the speed of light is not an absolute limit, but those theories do not correspond to the world we inhabit. The speed of light does not have to be finite, but in our world, as distinct from all the imaginary worlds a mathematician might invent, it is. Some things in the end can be determined only empirically, by looking at the world and figuring out how it works.
The End of Physics: The Myth of a Unified Theory
Prologue (p. 6)
Basic Books, Inc. New York, New York, USA. 1993

Ultimately, theories must be useful and accurate, and in that sense it does not matter whether they were arrived at through piercing insight or blind luck.
The End of Physics: The Myth of a Unified Theory
Prologue (p. 10)
Basic Books, Inc. New York, New York, USA. 1993

Lodge, Sir Oliver   1851–1940
English physicist

A physical theory cannot take the whole universe into account; but if it is to be complete enough to be

satisfactory, and to make trustworthy predictions, it must take all relevant factors into account.
Contributions to a British Association Discussion on the Evolution of the Universe
Nature, Supplement, October 24, 1931 (p. 722)

Mach, Ernst 1838–1916
Austrian physicist and philosopher

The object of natural science is the connection of phenomena; but the theories are like dry leaves which fall away when they have long ceased to be the lungs of the tree of science.
History and Root of the Principle of the Conservation of Energy
Chapter IV (p. 74)
The Open Court Publishing Company. Chicago, Illinois, USA. 1911

Maimonides, Moses 1135–1204
Spanish-born philosopher, jurist, and physician

Man knows only these poor mathematical theories about the heavens, and only God knows the real motions of the heavens and their causes.
In Phillip Frank
Modern Science and its Philosophy
Chapter 13 (p. 222)
Harvard University Press, Cambridge, England. 1952

Malthus, Thomas Robert 1766–1834
English economist

Each pursues his own theory, little solicitous to correct or improve it by an attention to what is advanced by his opponents.
In Robert M. Hutchins and Mortimer J. Adler (eds.)
The Great Ideas Today, 1961
Essays on the Principle of Population (p. 473)
Encyclopædia Britannica, Inc. Chicago, Illinois, USA. 1961

Matthew, William Diller 1871–1930
Canadian-American paleontologist

Many a false theory gets crystallized by time and absorbed into the body of scientific doctrine through lack of adequate criticism when it is formulated.
Supplementary Note
Climate and Evolution (Volume 1) (p. 159)
Special Publications of the New York Academy of Sciences. December 1950

Mayes, Jr., Harlan

Theory, glamorous mother of the drudge experiment.
In Eric M. Rogers
Physics for the Inquiring Mind
Chapter 40 (p. 648)
Princeton University Press. Princeton, New Jersey, USA. 1960

Mayo, John
No biographical data available

This theory of the learned author is certainly very ingenious, but I am not sure that it is in the same degree in accordance with truth.
Medico-Physical Works
Fourth Treatise
Chapter I (p. 231)
The Alembic Club. Edinburgh, Scotland. 1907

Medawar, Sir Peter Brian 1915–87
Brazilian-born English zoologist

Scientific theories…begin as imaginative constructions. They begin, if you like, as stories, and the purpose of the critical or rectifying episode in scientific reasoning is precisely to find out whether or not these stories are about real life.
Pluto's Republic
Science and Literature, Section 4 (p. 53)
Oxford University Press, Inc. Oxford, England. 1982

Mencken, H. L. (Henry Louis) 1880–1956
American journalist and literary critic

…for a professor must have a theory, as a dog must have fleas.
Prejudices: First Series
Criticism of Criticism of Criticism (p. 12)
Alfred A. Knopf. New York, New York, USA. 1923

Merezhkovskii, Konstantine 1855–1921
Russian biologist

The Germans compare German science to a lighthouse. I would as well, but then to a lighthouse without sacred fire to light up the world. The Germans carry stones to construct a solid base without which there would be no lighthouse, and in this task no other nation surpasses them. But, it is left for others to arrive and light the fire. Now, without fire, there is no lighthouse.
In Jan Sapp
Evolution by Association: A History of Symbiosis
Chapter 4 (p. 56)
Oxford University Press, Inc. New York, New York, USA. 1994

Neumann, John von 1903–57
Hungarian-American mathematician

It must be emphasized that it is not a question of accepting the correct theory and rejecting the false one. It is a matter of accepting that theory which shows greater formal adaptability for a correct extension. This is a formalistic esthetic criterion, with a highly opportunistic flavor.
Collected Works (Volume 6)
Method in the Physical Sciences (p. 498)
Pergamon Press. New York, New York, USA. 1961–1963

Newell, A.
No biographical data available

Working with theories is not like skeet shooting, where theories are lofted up and bang, they are shot down with

a falsification bullet, and that's the end of that theory. Theories are more like graduate students—once admitted you try hard to avoid flunking them out, it being much better for them and for the world if they can become long-term contributors to society.
Unified Theories of Cognition
Introduction (p. 14)
Harvard University Press. Cambridge, Massachusetts, USA. 1990

Nietzsche, Friedrich   1844–1900
German philosopher

It is certainly not the least charm of a theory that it is refutable…
Beyond Good and Evil
Chapter I, 18 (pp. 18–19)
The Modern Library. New York, New York, USA. 1917

Nizer, Louis   1902–94
English-born American lawyer

The argument seemed sound enough, but when a theory collides with a fact, the result is a tragedy.
My Life in Court (p. 433)
Doubleday & Company, Inc. New York, New York, USA. 1961

Novalis (Friederich von Hardenberg)   1772–1801
German poet

Theories are like fishing: it is only by casting into unknown waters that you may catch something.
In Jean-Pierre Luminet
Black Holes (p. 1)
Cambridge University Press. New York, New York, USA. 1992

Oman, John   1860–1939
English Presbyterian theologian

To refuse to consider any possibility is merely the old habit of making theory the measure of reality.
The Natural and the Supernatural
Chapter XV (p. 269)
The Macmillan Company. New York, New York, USA. 1931

Oppenheimer, J. Robert   1904–67
American theoretical physicist

The theory of our modern technical [era] shows that nothing is as practical as theory.
Reflex, July 1977

Papoulis, Athanasios   1921–2002
Greek-American engineer and applied mathematician

Scientific theories deal with concepts, never with reality. All theoretical results are derived from certain axioms by deductive logic. In physical sciences the theories are so formulated as to correspond in some useful sense to the real world, whatever that may mean. However, this correspondence is approximate, and the physical justification of all theoretical conclusions is based on some form of inductive reasoning.
Probability, Random Variables, and Stochastic Processes
Chapter 1 (p. 3)
McGraw-Hill Book Company, Inc. New York, New York, USA. 1965

Parsons, Talcott   1902–79
American sociologist

Theory not only formulates what we know but also tells us what we want to know, that is, the questions to which an answer is needed.
The Structure of Social Action
Part I, Chapter I (p. 9)
The Free Press. Glencoe, Illinois, USA. 1949

Pasteur, Louis   1822–95
French chemist

Without theory, practice is but routine born of habit. Theory alone can bring forth and develop the spirit of invention.
In René Dubos
Louis Pasteur: Free Lance of Science
Chapter I (p. 11)
Little, Brown & Company. Boston, Massachusetts, USA. 1950

Petit, Jean-Pierre   1937–
French astrophysicist

Sir, please believe me, it's the first time this has ever happened. Have another try, don't get upset. You know our Theorems are GUARANTEED.
Euclid Rules OK? (p. 11)
John Murray Ltd. London, England. 1982

Playfair, John   1748–1819
Scottish geologist, physicist, and mathematician

A theory is never more unfairly dealt with, than when those parts are separated which were meant to support one another, and each left to stand or fall by itself.
Illustrations of the Huttonian Theory of the Earth
Section 303 (p. 340)
Dover Publications, Inc. New York, New York, USA. 1964

The want of theory, then, does not secure the candor of an observer, and it may very much diminish his skill. The discipline that seems best calculated to promote both is a thorough knowledge of the methods of inductive investigation; an acquaintance with the history of physical discovery; and the careful study of those sciences in which the rules of philosophizing have been most successfully applied.
Illustrations of the Huttonian Theory of the Earth
Note XXVI, 459 (p. 528)
Dover Publications, Inc. New York, New York, USA. 1964

Poincaré, Henri   1854–1912
French mathematician and theoretical astronomer

It is not sufficient for a theory to affirm no false relations; it must not hide true relations.
The Foundations of Science

Science and Hypothesis, Part IV
Chapter X (p. 145)
The Science Press. New York, New York, USA. 1913

At the first blush it seems to us that theories last only a day and that ruins upon ruins accumulate.... But if we look more closely, we see that what thus succumb are the theories properly so called, those which pretend to teach us what things are. But there is in them something which usually survives. If one of them taught us a true relation, this relation is definitively acquired, and it will be found again under a new disguise in the other theories which will successively come to reign in place of the old.
The Foundations of Science
The Value of Science, Science and Reality (p. 351)
The Science Press. New York, New York, USA. 1913

Popper, Karl R.   1902–94
Austrian/British philosopher of science

...scientific theories, if they are not falsified, for ever remain hypotheses or conjectures.
Unended Quest: An Intellectual Autobiography
Chapter 16 (p. 79)
Open Court Publishing Company. La Salle, Illinois, USA. 1976

The initial stage, the act of conceiving or inventing a theory, seems to me neither to call for logical analysis nor to be susceptible of it. The question how it happens that a new idea occurs to a man — whether it is a musical theme, or a dramatic conflict, or a scientific theory — may be of great interest to empirical psychology; but it is irrelevant to the logical analysis of scientific knowledge.
The Logic of Scientific Discovery
Part I, Chapter I, Section 2 (p. 31)
Basic Books, Inc. New York, New York, USA. 1959

Theories are nets cast to catch what we call "the world": to rationalize, to explain, and to master it. We endeavor to make the mesh ever finer and finer.
The Logic of Scientific Discovery
Part II, Chapter III (p. 59)
Basic Books, Inc. New York, New York, USA. 1959

Never let yourself be goaded into taking seriously problems about words and their meanings. What must be taken seriously are questions of fact, and assertions about facts: theories and hypotheses; the problems they solve; and the problems they raise.
Unended Quest: An Intellectual Autobiography
Chapter 7 (p. 19)
Open Court Publishing Company. La Salle, Illinois, USA. 1976

Scientific theories are not the digest of observations, but they are inventions — conjectures boldly put forward for trial, to be eliminated if they clashed with observations; with observations which were rarely accidental, but as a rule undertaken with the definite intention of testing a theory by obtaining, if possible, a decisive refutation.

Conjectures and Refutations: The Growth of Scientific Knowledge
Chapter 1, Section IV (p. 46)
Harper & Row, Publishers. New York, New York, USA. 1963

As with our children, so with our theories, and ultimately with all the work we do: our products become largely independent of their makers. We may gain more knowledge from our children or from our theories than we ever imparted to them.
Unended Quest: An Intellectual Autobiography
Chapter 40 (p. 196)
Open Court Publishing Company. La Salle, Illinois, USA. 1976

Every "good" scientific theory is one which forbids certain things to happen; the more a theory forbids, the better it is.
In C.A. Mace (ed.)
British Philosophy in the Mid-Century
Philosophy of Science: A Personal Report I (p. 159)
George Allen & Unwin Ltd. London, England. 1957

...our critical examinations of our theories lead us to attempts to test and to overthrow them; and these lead us further to experiments and observations of a kind which nobody would ever have dreamed of without the stimulus and guidance both of our theories and of our criticisms of them. For indeed, the most interesting experiments and observations were carefully designed in order to test our theories, especially our new theories.
Conjectures and Refutations: The Growth of Scientific Knowledge
Chapter 10, Section I (pp. 215–216)
Harper & Row, Publishers. New York, New York, USA. 1963

We have no reason to regard the new theory as better than the old theory — to believe that it is nearer to the truth — until we have derived from the new theory new predictions which were unobtainable from the old theory (the phases of Venus...) and until we have found that these new predictions were successful.
Conjectures and Refutations: The Growth of Scientific Knowledge
Chapter 10 (p. 246)
Harper & Row, Publishers. New York, New York, USA. 1963

...a high probability cannot be one of the aims of science. For the scientist is most interested in theories with a high content. He does not care for highly probable trivialities but for bold and severely testable (and severely tested) hypotheses. If (as Carnap tells us) a high degree of confirmation is one of the things we aim at in science, then degree of confirmation cannot be identified with probability. ...if high probability were an aim of science, then scientists should say as little as possible, and preferably utter tautologies only. But their aim is to "advance" science, that is to add to its content. Yet this means lowering its probability. And in view of the high content of universal laws, it is [not] surprising to find that their probability is zero...
Conjectures and Refutations: The Growth of Scientific Knowledge
Chapter 11, Section VI (p. 286)
Harper & Row, Publishers. New York, New York, USA. 1963

The dogmatic attitude of sticking to a theory as long as possible is of considerable significance. Without it we could never find out what is in a theory — we should give the theory up before we had real opportunity of finding out its strength; and in consequence no theory would ever be able to play its role of bringing order into the world, of preparing us for future events, of drawing our attention to events we should otherwise never observe.

Conjectures and Refutations: The Growth of Scientific Knowledge
Chapter 15, fn 1 (p. 312)
Harper & Row, Publishers. New York, New York, USA. 1963

Pratchett, Terry   1948–
English author

...theories, diverse as they are, have two things in common. They explain the observed facts, and they are completely and utterly wrong.

The Light Fantastic (p. 165)
Colin Smythe. Gerrards Cross, England. 1986

Priestley, J. B.
No biographical data available

A first encounter with any grand fantastic theory, not political or economic, delights me.

Delight
Fantastic Theories (p. 52)
William Heinemann. London, England. 1949

Raup, David Malcolm   1933–
American geophysicist and paleontologist

[A] new theory is guilty until proven innocent, and the pre-existing theory is innocent until proven guilty.... Continental drift was guilty until proven innocent.

The Nemesis Affair: A Story of the Death of the Dinosaurs and the Ways of Science
Chapter 12 (pp. 195, 205)
W.W. Norton & Company, Inc. New York, New York, USA.1986

Reiser, Anton   1628–86
German Lutherischer theologian

Our experience and observations alone never lead to finalities. Theory, however, creates reliable roads over which we may pursue our journeys through the world of observations.

In Bernard Jaffe
New World of Chemistry
Chapter 11 (p. 133)
Silver, Burdett & Company. New York, New York, USA. 1935

Richards, Dickinson W.   1895–1973
American physician and physiologist

The problems are the ones that we have always known. The little gods are still with us, under different names. There is conformity: of technique, leading to repetition; of language, encouraging if not imposing conformity of thought. There is popularity: it is so easy to ride along on an already surging tide; to plant more seed in an already well-ploughed field; so hard to drive a new furrow into stony ground. There is laxness: the disregard of small errors, of deviations, of the unexpected response; the easy worship of the smooth curve. There is also fear: the fear of speculation; the overprotective fear of being wrong. We are forgetful of the curious and wayward dialectic of science, whereby a well-constructed theory even if it is wrong, can bring a signal advance.

Transactions of the Association of American Physicians, Volume 75, 1962 (p. I)

Richet, Charles   1850–1935
French physiologist

I often recall to my students the history of Don Quixote, who, having constructed a helmet of cardboard and wood, wished to prove its solidity. Alas, the poor helmet flew to bits when his own good sword struck it. Then the knight, no whit discouraged, made a new and stronger helmet. He raised his sword.

The Natural History of a Savant
Chapter II (p. 17)
J.M. Dent & Sons Ltd. London, England. 1927

Romanoff, Alexis Lawrence   1892–1980
Russian soldier and scientist

A theory is worthless without good supporting data.

Encyclopedia of Thoughts
Aphorisms 2410
Ithaca Heritage Books. Ithaca, New York, USA. 1975

Rota, Gian-Carlo   1932–99
Italian-born American mathematician
Pringsheim, Alfred   1850–1941
German mathematician

Theorems are not to mathematics what successful courses are to a meal. The nutritional analogy is misleading.

In Philip J. Davis and Reuben Hersh
The Mathematical Experience
Introduction (pp. xviii–xix)
Birkhäuser. Boston, Massachusetts, USA. 1981

Rothman, Tony   1953–
American cosmologist
Sudarshan, George   1931–
Indian physicist

The everyday usage of "theory" is for an idea whose outcome is as yet undetermined, a conjecture, or for an idea contrary to evidence. But scientists use the word in exactly the opposite sense. [In science] "theory"...refers only to a collection of hypotheses and predictions that is amenable to experimental test, preferably one that has been successfully tested. It has everything to do with the facts.

Doubt and Certainty: The Celebrated Academy: Debates on Science, Mysticism, Reality, in General on the Knowable and Unknowable
First Debate (p. 2)
Perseus Books. Reading, Massachusetts. USA. 1998

It is not difficult to calculate that if one inflated the world to keep up with the current rate of population growth, then after 2598 years the earth would be expanding at the speed of light. The growth of science is proceeding even faster. Several years ago, in physics at least, we crossed the point at which the expected lifetime of a theory became less than the lead time for publication in the average scientific journal. Consequently, most theories are born dead on arrival and journals have become useless, except as historical documents.
A Physicist on Madison Avenue
Chapter 8 (p. 118)
Princeton University Press. Princeton, New Jersey, USA. 1991

A theory is accepted only when the last of its opponents dies off. The Copernican Revolution was a great shift in mankind's thinking, but did not take place overnight.
Instant Physics: From Aristotle to Einstein, and Beyond
Chapter 1 (p. 15)
Ballantine Books. New York, New York, USA. 1995

Russell, Bertrand Arthur William   1872–1970
British philosopher, logician, and social reformer

...it is only theory that makes men completely incautious.
Unpopular Essays
Ideas that Have Harmed Mankind (p. 210)
George Allen & Unwin Ltd. London, England. 1950

Santayana, George (Jorge Augustín Nicolás Ruiz de Santillana)   1863–1952
Spanish-born American philosopher

Theory helps us to bear our ignorance of facts.
The Sense of Beauty
Part III, Section 30 (p. 125)
Transaction Publishers. New Brunswick, New Jersey, USA. 2000

Sayers, Dorothy L.   1893–1957
English novelist and essayist

Very dangerous things, theories.
The Unpleasantness at the Bellona Club
Chapter 4 (p. 27)
HarperPaperback. New York, New York, USA. 1995

Schegel, Richard
No biographical data available

We must accept, I think, that there is an inherent limitation in the structure of science that prevents a scientific theory from ever giving us an adequate total explanation of the universe. Always, there is a base in nature (or, correspondingly, a set of assumptions in theory) which cannot be explained by reference to some yet more fundamental property. This feature of science has been commented on by many writers in the philosophy of science; and, certainly the limitation is a point of difference between science and those religious or metaphysical systems in which there is an attempt to present a doctrine that gives answers for all ultimate questions.
Completeness in Science
Chapter 14, Section 2 (p. 252)
Appleton-Century-Crofts. New York, New York, USA. 1967

Schiller, Ferdinand Canning Scott   1864–1937
English philosopher

It is the business of theories to forecast "facts", and of facts to form points of departure for theories, which again, when verified by the new facts to which they have successfully led, will extend the borders of knowledge.
In Charles Singer (ed.)
Studies in the History and Method of Science (Volume 1)
Scientific Discovery and Logical Proof (p. 275)
At The Clarendon Press. Oxford, England. 1917

Schön, Donald A.   1930–97
American philosopher of practice and learning theory

...there is a high, hard ground where practitioners can make effective use of research-based theory and technique, and there is a swampy lowland where situations are confusing "messes" incapable of technical solution. ...[I]n the swamp are the problems of greatest human concern.
The Reflective Practitioner: How Professionals Think in Action (p. 42)
Aldershot Press. Avebury, England. 1983

Schramm, David N.   1945–97
American astrophysicist
McKee, Christopher F.   1942–
American astrophysicist

When theory runs too far ahead of what can be measured, a field becomes more philosophy than science.
Astronomy in the Mind and the Lab
Sky and Telescope, Volume 82, Number 4, October 1991 (p. 352)

Seeger, Raymond J.

It is noteworthy that the etymological root of the word theatre is the same as that of the word theory, namely a view. A theory offers us a better view.
Journal of the Washington Academy of Sciences, Volume 36, 1946 (p. 286)

Shaw, George Bernard   1856–1950
Irish comic dramatist and literary critic

The weakness of the man who, when his theory works out into a flagrant contradiction of the facts, concludes "so much the worse for the facts: let them be altered," instead of "so much the worse for my theory."
Liberty

New York
A Degenerate's View of Nordau, July 27, 1895

Silver, Brian L.
Israeli professor of physical chemistry

Some see the fragility of scientific theory as an indication of a basic inability of science to explain the universe. But scientific change is almost always accompanied by an increase in our ability to rationalize and predict the course of nature. Newton could explain far more than Aristotle, Einstein far more than Newton. Science frequently stumbles, but it gets up and carries on. The road is long.
The Ascent of Science
Preface (p. xiii)
Solomon Press Book. New York, New York, USA. 1998

Facts may be regarded as indisputable; theories are not.
The Ascent of Science
Part I, Chapter 2 (p. 19)
Solomon Press Book. New York, New York, USA. 1998

Skolimowski, Henryk   1930–
Polish philosopher

Theories, like old soldiers, fade away rather than being killed on the scientific battlefield.
In A.J. Ayala (ed.)
Studies in the Philosophy of Biology: Reduction and Related Problems
Problems of Rationality in Biology (p. 217)
Macmillan & Company Ltd. London, England. 1974

Slater, John C.   1900–76
American physicist and theoretical chemist

A theoretical physicist in these days asks just one thing of his theories: if he uses them to calculate the outcome of an experiment, the theoretical prediction must agree, within limits, with the result of the experiment. He does not ordinarily argue about philosophical implications of his theory. Almost his only recent contribution to philosophy has been the operational idea, which is essentially only a different way of phrasing the statement I have just made, that the one and only thing to be done with a theory is to predict the outcome of an experiment. As a physicist, I find myself very well satisfied with this attitude.
Electrodynamics of Ponderable Bodies
Journal of the Franklin Institute, Volume 225, Number 3, March 1938
(pp. 277–287)

Questions about a theory which do not affect its ability to predict experimental results correctly seem to me quibbles about words, rather than anything more substantial, and I am quite content to leave such questions to those who derive some satisfaction from them.
Electrodynamics of Ponderable Bodies
Journal of the Franklin Institute, Volume 225, Number 3, March 1938
(pp. 277–287)

Slosson, Edwin E.   1865–1929
American chemist and journalist

The scientist does not abandon a theory because it has inconsistencies any more than he divorces his wife because she has inconsistencies.
Easy Lesson in Einstein: A Discussion of the More Intelligible Features of the Theory of Relativity
Scientific Versus Legal Laws (p. 106)
Harcourt, Brace & Company. New York, New York, USA. 1920

Stenger, Victor J.   1935–
American physicist

The fact that a theory may eventually test wrong does not detract from its original merit as a worthy try. On the other hand, if an idea is poorly formulated, often because the terms used are not clearly defined, then how can we even test it?…We cannot determine that gibberish is anything but gibberish.
Physics and Psychics: The Search for a World Beyond the Senses
Chapter 3 (p. 58)
Prometheus Books. Buffalo, New York, USA. 1990

Stevenson, Robert Louis   1850–94
Scottish essayist and poet

It is better to emit a scream in the shape of a theory than to be entirely insensible to the jars and incongruities of life and take everything as it comes in a forlorn stupidity.
Virginibus Puerisque and Familiar Studies of Men and Books
Crabbed Age and Youth (p. 42)
J.M. Dent & Sons Ltd. London, England. No date

Sussmann, Hector   1946–
Argentinean-American mathematician

A mark of a good theory is that it proves even the most trivial results.
Gelfand Workshop
Rutgers University, February 11, 2002

Synge, John L.   1897–1995
Irish mathematician and physicist

A well built theory has three merits: (i) it has an aesthetic appeal, (ii) it is comparatively easy to understand, and (iii), if its postulates are clearly stated, it may be taken out of its original physical context and applied in another.
The Hamiltonian Method and Its Application to Water Waves
Proceedings of the Irish Academy, Volume 63, Section A, Number 1, May 1962 (p. 1)

Teall, Sir J. J. Harris   1849–1924
British geologist

It is only when a theory has proved its usefulness as a coordinator of fact that it becomes worthy of the dignity of publication. It may be true or false, most likely the latter; but if it coordinates more facts than any other it is at any rate useful and may be conveniently retained until replaced by a better.
Annual Report of the Board of Regents of the Smithsonian Institution, 1902

The Evolution of Petrological Ideas (p. 289)
Government Printing Office. Washington, D.C. 1903

Toulmin, Stephen   1922–
English philosopher

It is part of the art of the sciences, which has to be picked up in the course of the scientist's training, to recognize the situations in which any particular theory or principle can be applied to, and when it will cease to hold.
The Philosophy of Science: An Introduction
Chapter III (pp. 92–93)
Harper & Row, Publishers. New York, New York, USA. 1960

Turner, Michael S.
American astrophysicist

If all you have are observations, that's botany. If all you have is theory, that's philosophy.
In John Hogan
Universal Truths
Scientific American, Volume 264, Number 4, October 1990 (p. 117)

Twain, Mark (Samuel Langhorne Clemens)   1835–1910
American author and humorist

…the trouble about arguments is, they ain't nothing but theories, after all, and theories don't prove nothing, they only give you a place to rest on, a spell, when you are tuckered out butting around and around trying to find out something there ain't no way to find out. … There's another trouble about theories: there's always a hole in them somewheres, sure, if you look close enough.
Tom Sawyer Abroad; Tom Sawyer, Detective and Other Stories, etc., etc.
Tom Sawyer Abroad
Chapter IX (p. 70)
Harper & Brothers. New York, New York, USA. 1902

van Fraassen, Bas C.   1941–
Dutch-born philosopher

Science aims to give us, in its theories, a literally true story of what the world is like; and acceptance of a scientific theory involves the belief that it is true.
The Scientific Image
Chapter 2, section 1.1 (p. 8)
Clarendon Press. Oxford, England. 1990

…the success of current scientific theories is no miracle. It is not even surprising to the scientific (Darwinist) mind. For any scientific theory is born into a life of fierce competition, a jungle red in tooth and claw. Only the successful theories survive — the ones which in fact latched on to the actual regularities in nature.
The Scientific Image
Chapter 2, Section 7 (p. 40)
Clarendon Press. Oxford, England. 1980

von Goethe, Johann Wolfgang   1749–1832
German poet, novelist, playwright and natural philosopher

No phenomenon can be explained in and of itself; only many comprehended together, methodically arranged, in the end yield something that could be regarded as theory.
In Karl J. Fink
Goethe's History of Science
Chapter 2 (p. 26)
Cambridge University Press. Cambridge, England. 1991

Dear friend, all theory is grey
And green the golden tree of life.
In *Great Books of the Western World* (Volume 47)
Faust
The First Part
Faust's Study (2), l. 2038–2039
Encyclopædia Britannica, Inc. Chicago, Illinois, USA. 1952

Waddington, Conrad Hal   1905–75
British biologist and paleontologist

A scientific theory cannot remain a mere structure within the world of logic, but must have implications for action and that in two different ways. In the first place, it must involve the consequence that if you do so and so, such and such results will follow. That is to say it must give, or at least offer the possibility of controlling the process; and secondly — and this is a point not so often mentioned by those who discuss the nature of scientific theories — its value is quite dependent on its power of suggesting the next step in scientific advance.
The Nature of Life
Chapter I (pp. 11–12)
Harper & Row, Publishers. New York, New York, USA. 1960

Weinberg, Steven   1933–
American nuclear physicist

This is often the way it is in physics — our mistake is not that we take our theories too seriously, but that we do not take them seriously enough.
The First Three Minutes
Chapter VI (p. 131)
Basic Books, Inc. New York, New York, USA. 1988

Our theories are very esoteric — necessarily so, because we are forced to develop these theories using a language, the language of mathematics, that has not become part of the general equipment of the educated public. Physicists generally do not like the fact that our theories arc so esoteric. On the other hand, I have occasionally heard artists talk proudly about their work being accessible only to a band of cognoscenti and justify this attitude by quoting the example of physical theories like general relativity that also can be understood only by initiates. Artists like physicists may not always be able to make themselves understood by the general public, but esotericism for its own sake is just silly.
Dreams of a Final Theory: The Scientist's Search for the Ultimate Laws of Nature
Chapter VI (p. 150)
Pantheon Books. New York, New York, USA. 1992

Wells, H. G. (Herbert George)   1866–1946
English novelist, historian, and sociologist

Very simple was my explanation, and plausible enough — as most wrong theories are!
In Robert M. Hutchins and Mortimer J. Adler (eds.)
The Great Ideas Today, 1971
The Time Machine
Chapter Four (p. 468)
Encyclopædia Britannica, Inc. Chicago, Illinois, USA. 1971

Whewell, William   1794–1866
English philosopher and historian

It is a test of true theories not only to account for, but to predict phenomena.
The Philosophy of the Inductive Sciences Founded upon Their History
(Volume 2)
Aphorisms, Aphorisms Concerning Science, XII (p. 468)
John W. Parker. London, England. 1847

…there is a mask of theory over the whole face of nature…
The Philosophy of the Inductive Sciences Founded upon Their History
(Volume 1)
Part I, Book I, Chapter II, sect. 10 (p. 42)
John W. Parker. London, England. 1847

Winchell, Alexander   1824–91
American geologist

When a great theory has grown into existence, and the general assent of competent judges has converted a sublime conception from the state of a provisional hypothesis to the position of a strengthening doctrine, there is unusual interest in glancing over the progress of science and noting the actual steps by which the guess became theory, and the theory, doctrine.
World-Life or Comparative Geology
Part IV (p. 550)
S.C. Griggs & Company. Chicago, Illinois, USA. 1883

White, Henry S.
No biographical data available

The accepted truths of today, even the commonplace truths of any science, were the doubtful or the novel theories of yesterday.
Bulletin of the American Mathematical Society, Volume 15, 1909
(p. 325)

Whitehead, Alfred North   1861–1947
English mathematician and philosopher

On the absolute theory, bare space and bare time are such very odd existences, half something and half nothing.
The Idealistic Interpretations of Einstein's Theory
Proceedings of the Aristotelian Society, N.S. Volume 22, Part III,
(p. 131)

…to come very near a true theory and to grasp its precise application are two very different things.

The Organization of Thought
Chapter VI (p. 127)
Greenwood Press Publishers. Westport, Connecticut, USA. 1974

Wilson, Edward O.   1929–
American biologist and author

Nothing in science — nothing in life, for that matter, makes sense without theory. It is our nature to put all knowledge into context in order to tell a story, and to re-create the world by this means. …We are enchanted by the beauty of the natural world. Our eye is caught by the dazzling visual patterns of polar star trails, for example, and the choreography of chromosomes in dividing root tip cells of plant. Both disclose processes that are also vital to our lives. In unprocessed form, however, without theoretical frameworks of heliocentric astronomy and Mendelian heredity, they are no more than beautiful patterns of light.
Consilience: The Unity of Knowledge (p. 58)
Alfred A. Knopf. New York, New York, USA. 1998

Wisdom, John O.
No biographical data available

Sometimes [the word theory] is used for a hypothesis, sometimes for a confirmed hypothesis; sometimes for a train of thought; sometimes for a wild guess at some fact, or for a reasoned claim of what some fact is — or even for a philosophical speculation.
Foundations of Inference in Natural Sciences
Chapter III (p. 33)
Methuen & Company Ltd. London, England. 1952

Woodbridge, Frederick James Eugene   1867–1940
American philosopher

It is a theory of nature, that system of things which allows a plant to grow, an animal to graze, and a man to think, fully as much as it allows the sun to be eclipsed or bodies to be in motion or at rest.
Aristotle's Vision of Nature
Chapter III (p. 49)
Columbia University Press. New York, New York, USA. 1965

Woodger, Joseph Henry   1894–1981
English biologist

Theoretical statements, it is clear, cannot be verified because we can never know whether they are true. All we can do is to go on testing their consequences until an observation record turns up which contradicts them. Then we have the choice of two courses: we can say that the theoretical statement is false and reject it; or we can assume that we have been mistaken in our observation and retain the theoretical statements.
Biology and Language
Lecture II (p. 57)
At The University Press. Cambridge, England. 1952

Wurtz, Charles Adolphe   1817–84
French organic chemist

The triumph of a theory is to embrace the greatest number and the greatest variety of facts.
A History of Chemical Theory from the Age of Lavoisier to the Present Time
Lavoisier
I (p. 7)
Macmillan & Company Ltd. London, England. 1869

Wyndham, John   1903–69
English science fiction writer

…I do refuse to accept a bad theory simply on the grounds that there is not a better [one]…
The Midwich Cuckoos
Chapter Twenty (p. 221)
M. Joseph. London, England. 1977

Ziman, John M.   1925–2005
British physicist

The verb "to theorize" is now conjugated as follows: "I built a model; you formulated a hypothesis; he made a conjecture."
Reliable Knowledge
Chapter 2 (fn 20, p. 22)
Cambridge University Press. Cambridge, England. 1978

…a significant fraction of the ordinary scientific literature in any field is concerned with essentially irrational theories put forward by a few well-established scholars who have lost touch with reality.
Some Pathologies of the Scientific Life
Nature, Volume 227, 5 September 1970

…the sooner we all face up to the fact that theory and practice are indissoluble, and that there is no contradiction between the qualities of usefulness and beauty, the better.
Growth and Spread of Science
Nature, Volume 221, Number 5180, February 8, 1969 (p. 521)

THEORY OF FUNCTIONS

Keyser, Cassius Jackson   1862–1947
American mathematician

The Modern Theory of Functions — that stateliest of all the pure creations of the human intellect.
In Columbia University
Lectures on Science, Philosophy and Art 1907–1908 (p. 16)
New York, New York, USA. 1908

Volterra, Vito   1860–1940
Italian mathematician

The theory that has had the greatest development in recent times is without any doubt the theory of functions.
In Stanley Gudder

A Mathematical Journey (p. 32)
McGraw-Hill Book Company, Inc. New York, New York, USA. 1976

THERMODYNAMICS

Allen, Woody   1935–
American film director and actor

It's the Second Law of Thermodynamics — sooner or later everything turns to shit. That's my phrasing, not the Encyclopedia Britannica's.
Husbands and Wives
Film (1992)

Atkins, Peter William   1940–
English physical chemist and writer

Everything is driven by motiveless, purposeless decay.
The Creation
Chapter 2 (p. 23)
W.H. Freeman. San Francisco, California, USA. 1981

Barnett, Lincoln   1909–79
American science writer

Although it is true that the amount of matter in the universe is perpetually changing, the change appears to be mainly in one direction — toward dissolution. The sun is slowly but surely burning out, the stars are dying embers, and everywhere the cosmos heart is turning to cold; matter is dissolving into radiation, and energy is being dissipated into empty space. "The universe is thus progressing toward an ultimate 'heat death' or, as it is technically defined, a condition of maximum entropy" [quoting Einstein].… And there is no way of avoiding this destiny.

[T]he fateful principle known as the Second Law of Thermodynamics, which stands today as the principal pillar of classical physics left intact by the march of science, proclaims that the fundamental processes of nature are irreversible. Nature moves only one way.
The Universe and Dr. Einstein
Chapter 14 (p. 99)
William Sloane Associates. New York, New York, USA. 1948

Blum, Harold   1899–?
No biographical data available

No matter how carefully we examine the energetics of living systems we find no evidence of defeat of thermodynamic principles.
Time's Arrow and Evolution
Princeton University Press. Princeton, New Jersey, USA. 1951

Bohr, Niels Henrik David   1886–1962
Danish physicist

The old thermodynamics…is to statistical thermodynamics what classical mechanics is to quantum mechanics.

In Werner Heisenberg
Physics and Beyond: Encounters and Conversations
Chapter 9 (p. 107)
Harper & Row, Publishers. New York, New York, USA. 1971

Boltzmann, Ludwig Edward 1844–1906
Austrian physicist

General thermodynamics proceeds from the fact that, as far as we can tell from our experience up to now, all natural processes are irreversible. Hence according to the principles of phenomenology, the general thermodynamics of the second law is formulated in such a way that the unconditional irreversibility of all natural processes is asserted as a so-called axiom.
Translated by Stephen G. Brush
Lectures on Gas Theory (pp. 444–445)
University of California Press. Berkeley, California, USA. 1964

The Second Law can never be proved mathematically by means of the equations of dynamics alone.
On Certain Questions of the Theory of Gases
Nature, Volume 51, 1895 (p. 413)

Bridgman, Percy Williams 1882–1961
American physicist

...the laws of thermodynamics have a different feel from most of the other laws of the physicist...they smell more of their human origin.
The Nature of Thermodynamics
Chapter I (p. 3)
Harvard University Press. Cambridge, Massachusetts, USA. 1941

Cardenal, Ernesto 1925–
Nicaraguan poet and Roman Catholic priest

The second law of thermodynamics!:
energy is indestructible in quantity
but continually changes in form.
And it always runs down like water.
Translated by John Lyons
Cosmic Canticle
Cantiga 3, Autumn Fugue (p. 29)
Curbstone Press. Willimantic, Connecticut, USA. 1993

Dickerson, Richard E.
American molecular biologist

It is possible to know thermodynamics without understanding it...
Molecular Thermodynamics
Chapter 7 (p. 387)
W.A. Benjamin. New York, New York, USA. 1969

Eddington, Sir Arthur Stanley 1882–1944
English astronomer, physicist, and mathematician

The law of entropy always increases — the Second Law of Thermodynamics — holds, I think, the supreme position among the laws of Nature. If someone points out to you that your pet theory of the universe is in disagreement with Maxwell's equation — then so much the worse for Maxwell's equations. If it is found to be contradicted by observation — well, these experimentalists do bungle things sometimes. But if your theory is found to be against the Second Law of Thermodynamics I can give you no hope; there is nothing for it but to collapse in deepest humiliation.
The Nature of the Physical World
Chapter IV (p. 74)
The Macmillan Company. New York, New York, USA. 1930

Epstein, P. S. 1883–1966
German-born physicist

Thermodynamics deals with systems which, in addition to mechanical and electromagnetic parameters, are described by a specifically thermal one, namely, the temperature or some equivalent of it. Thermodynamics is essentially a science about the conditions of equilibrium of systems and about the processes which can go on in states little different from the state of equilibrium.
Textbook of Thermodynamics
Chapter I (p. 2)
John Wiley & Sons, Inc. New York, New York, USA. 1937

Feynman, Richard P. 1918–88
American theoretical physicist
Leighton, Robert B. 1919–97
American physicist
Sands, Matthew L. 1919–
American physicist

In doing a problem involving a given mass of some substance, the condition of the substance at any moment can be described by telling what its temperature is and what its volume is. If we know the temperature and volume of a substance, and that the pressure is some function of the temperature and volume, then we know the internal energy. One could say, "I do not want to do it that way. Tell me the temperature and the pressure and I will tell you the volume. I can think of the volume as a function of temperature and pressure, and so on." That is why thermodynamics is hard, because everyone uses a different approach. If we could only sit down once and decide on our variables, and stick to them, it would be fairly easy.
The Feynman Lectures on Physics (Volume 1)
Chapter 44–5 (p. 44–9)
Addison-Wesley Publishing Company. Reading, Massachusetts, USA. 1983

Hoffmann, Roald 1937–
Polish-born American chemist

My second law, your second law, ordains that local order, structure in space and time, be crafted in ever-so-losing contention with proximal disorder in this neat but getting messier universe.

The Metamict State
The Devil Teaches Thermodynamics (p. 3)
University of Central Florida Press. Orlando, Florida, USA. 1987

Hogan, Graig J.
No biographical data available

Everything that happens in the universe consists of the same basic stuff, "mass-energy," transfigured in space and time from one form into another.
The Little Book of the Big Bang: A Cosmic Primer (p. 25)
Copernicus. New York, New York, USA. 1998

Kelvin, Lord William Thomson   1824–1907
Scottish engineer, mathematician, and physicist

It is impossible, by means of inanimate material agency, to derive mechanical effect from any portion of matter by cooling it below the temperature of the coldest of the surrounding objects.
On the Dynamical Theory of Heat, with Numerical Results Deduced from Mr. Joule's Equivalent of a Thermal Unit, and M. Regnault's Observations on Steam
Transactions of the Royal Society of Edinburgh, March 1851

Lewis, Gilbert Newton   1875–1946
American chemist
Randall, Merle   1888–1950
American chemist

The fascination of a growing science lies in the work of the pioneers at the very borderland of the unknown, but to reach this frontier one must pass over well traveled roads; of these one of the safest and surest is the broad highway of thermodynamics.
Revised by Kenneth S. Pitzer and Leo Brewer
Thermodynamics (p. x)
McGraw-Hill Book Company, Inc. New York, New York, USA. 1961

The second law of thermodynamics not only is a principle of wide reaching scope and application, but also is one which has never failed to satisfy the severest test of experiment. The numerous quantitative relations derived from this law have been subjected to more and more accurate experimental investigation without the detection of the slightest inaccuracy.
Thermodynamics (p. 87)
McGraw-Hill Book Company, Inc. New York, New York, USA. 1961

Maxwell, James Clerk   1831–79
Scottish physicist

I do not think…that the perfect identity which we observe between different portions [of molecules] of the same kind of matter can be explained on the statistical principle of the stability of the averages of large numbers of quantities each of which may differ from the mean. For if of the molecules of some substance such as hydrogen, some were of slightly greater mass than others, we have the means of producing a separation between molecules of different masses, and in this way we should be able to produce two kinds of hydrogen, one of which would be somewhat denser than the other. As this cannot be done, we must admit that the equality which we assert to exist between the molecules of hydrogen applies to each individual molecule, and not merely to the average of groups of millions of molecules.
Theory of Heat
Limitation of the Second Law of Thermodynamics (p. 309)
Longmans, Green & Company. London, England. 1871

Meixner, J.   1908–94
German theoretical physicist

A careful study of the thermodynamics of electrical networks has given considerable insight into these problems and also produced a very interesting result: the non-existence of a unique entropy value in a state which is obtained during an irreversible process…. I would say, I have done away with entropy. The next step might be to let us also do away with temperature.
In Edward B. Stuart, Benjamin Gal-Or and Alan J. Brainard (eds.)
A Critical Review of Thermodynamics: The Proceedings of the International Symposium
University of Pittsburgh. Pittsburgh, Pennsylvania, April 7–8, 1969

Morowitz, Harold J.   1927–
American biophysicist

The use of thermodynamics in biology has a long history rich in confusion…
Beginnings of Cellular Life: Metabolism Recapitulates Biogenesis
Chapter 6 (p. 69)
Yale University Press. New Haven, Connecticut, USA. 1992

Pippard, A. B.   1920–
English physicist

It may be objected by some that I have concentrated too much on the dry bones [of thermodynamic theory], and too little on the flesh which clothes them, but I would ask such critics to concede at least that the bones have an austere beauty of their own.
Classical Thermodynamics
Preface (p. vii)
Cambridge University Press. Cambridge, England. 1966

Reiss, H.
No biographical data available

Almost all books on thermodynamics contain some errors which are not purely typographical.
Methods of Thermodynamics
Preface (p. ix)
Blaisdell Publishing Company. New York, New York, USA. 1965

…the almost certain truth [is] that nobody (authors included) understands thermodynamics completely. The writing of a book therefore becomes a kind of catharsis in which the author exorcises his own demon of incomprehension and prevents it from occupying the soul of another.

Methods of Thermodynamics
Preface (p. vii)
Blaisdell Publishing Company. New York, New York, USA. 1965

Ross, John   1920–
American physical chemist

I have written that there are no known violations of the second law of thermodynamics (*Chemical and Engineering News*, July 27, 1980). Unfortunately I have been intentionally misinterpreted by creationists who say that this quote proves that evolution is impossible. This is nonsense: evolution is in no way a violation of the second law.
Letter to Carl Gaither
19 June 2007

Seifert, H. S.   1911–77
American aeronautics and astronautics scientist

The first and second laws of thermodynamics are of course known to us as well as the Ten Commandments, and probably obeyed more consistently.
Can We Decrease Our Entropy?
American Scientist, Summer, June 1961 (p. 124A)

Sommerfield, Arnold   1868–1951
German physicist

The science of thermodynamics introduces a new concept, that of temperature.
Thermodynamics and Statistical Mechanics, Lectures on Theoretical Physics (Volume 1)
Translated by J. Kestin (p. 1)
Academic Press. New York, New York, USA. 1956

Stenger, Victor J.   1935–
American physicist

Scientists speak of the Law of Inertia or the Second Law of Thermodynamics as if some great legislature in the sky once met and set down rules to govern the universe.
Not by Design
Chapter 1 (p. 14)
Prometheus Books. Buffalo, New York, USA. 1988

Truesdell, Clifford   1919–2000
American mathematician, natural philosopher, historian of mathematics

Every physicist knows exactly what the first and the second law mean, but…no two physicists agree about them.
In Mario Bunge (ed.)
Delaware Seminar in the Foundations of Physics
Foundations of Continuum Mechanics (p. 37)

…thermodynamics is the kingdom of deltas.
The Tragicomical History of Thermodynamics
Chapter 1 (p. 1)
Springer-Verlag. New York, New York, USA. 1980

THERMOMETER

Fahrenheit, Daniel Gabriel   1686–1736
German physicist

It then came into my mind what that most careful observer of natural phenomena had written about the correction of the barometer; for he had observed that the height of the column of mercury in the barometer was a little (though sensibly enough) altered by the varying temperature of the mercury. From this I gathered that a thermometer might perhaps be constructed with mercury, which would not be so hard to construct, and by the use of which it might be possible to carry out the experiment which I so greatly desired to try.
Experimenta Circa Gradum Caloris Liquorum Nonnullorum Ebullientium Instituta
Philosophical Transactions of the Royal Society of London, Volume 33, Number 1, 1724

THINKING

Asimov, Isaac   1920–92
American author and biochemist

Many adults, whether consciously or unconsciously, find it beneath their adult dignity to do anything as childish as read a book, think a thought, or get an idea. Adults are rarely embarrassed at having forgotten what little algebra or geography they once learned.
The Roving Mind
His Own Particular Drummer
Prometheus Books. Buffalo, New York, New York, USA. 1983

Crick, Francis Harry Compton   1916–2004
English biochemist

Some scientists work so hard there is no time left for serious thinking.
What Mad Pursuit: A Personal View of Scientific Discovery
Basic Books, Inc. New York, New York, USA. 1988

Cromer, Alan   1935–
American physicist and educator

Scientific thinking, which is analytic and objective, goes against the grain of traditional human thinking, which is associative and subjective.
Uncommon Sense: The Heretical Nature of Science
Chapter 1 (pp. 1–2)
Oxford University Press, Inc. New York, New York, USA. 1993

Dewey, John   1859–1952
American philosopher and educator

The first distinguishing characteristic of thinking is facing the facts — inquiry, minute and extensive scrutinizing, observation.
Reconstruction in Philosophy
Chapter VI (p. 140)
Beacon Press. Boston, Massachusetts, USA. 1920

Intelligent thinking means an increment of freedom in action — an emancipation from chance and fatality. "Thought" represents the suggestion of a way of response that is different from that which would have been followed if intelligent observation had not effected an inference as to the future.
Reconstruction in Philosophy
Chapter VI (p. 144)
Beacon Press. Boston, Massachusetts, USA. 1920

Edison, Thomas 1847–1931
American inventor

I am going to have a sign put up all over my plant, reading "There is no expedient to which a man will not resort to avoid the real labor of thinking."
In Dogbert D. Runes (ed.)
The Diary and Sundry Observations of Thomas Alva Edison
Chapter XXIX (p. 167)
Philosophical Library. New York, New York, USA. 1948

Einstein, Albert 1879–1955
German-born physicist

Thinking for its own sake, as in music! When I have no special problem to occupy my mind, I love to reconstruct proofs of mathematical and physical theorems that have long been known to me. There is no goal in this, merely an opportunity to indulge in the pleasant occupation of thinking.
In Helen Dukas and Banesh Hoffman
Albert Einstein: The Human Side: New Glimpses from His Archives
Letter from Spring 1918 (p. 17)
Princeton University Press. Princeton, New Jersey, USA. 1979

Harrington, John W. 1918–
American naturalist

We cannot see without thinking; we cannot think without seeing.
Dance of the Continents
Epilogue (p. 232)
J.P. Tarcher. Los Angeles, California, USA. 1983

Heisenberg, Werner Karl 1901–76
German physicist and philosopher

...one extreme is the idea of an objective world, pursuing its regular course in space and time, independently of any kind of observing subject; this has been the guiding image from modern science. At the other extreme is the idea of a subject, mystically experiencing the unity of the world and no longer confronted by an object or by any objective world; this has been the guiding image of Asian mysticism. Our thinking moves somewhere in the middle, between these two limiting conceptions; we should maintain the tension resulting from these opposites.
Across the Frontiers
Chapter XVI (p. 227)
Harper & Row, Publishers. New York, New York, USA. 1974

Rothman, Tony 1953–
American cosmologist

Sudarshan, George 1931–
Indian physicist

...one result of unimaginative, mechanistic thinking was that societies eventually ceased to burn people at the stake for witchcraft.
Doubt and Certainty: The Celebrated Academy: Debates on Science, Mysticism, Reality, in General on the Knowable and Unknowable
Fourth Debates (p. 74)
Perseus Books. Reading, Massachusetts. USA. 1998

Sagan, Carl 1934–96
American astronomer and science writer

The scientific way of thinking is at once imaginative and disciplined. This is central to its success. Science invites us to let the facts in, even when they don't conform to our preconceptions. It counsels us to carry alternative hypotheses in our heads and see which best fit the facts. It urges on us a delicate balance between no-holds-barred openness to new ideas, however heretical, and the most rigorous skeptical scrutiny of everything — new ideas and established wisdom. This kind of thinking is also an essential tool for a democracy in an age of change.
The Demon-Haunted World: Science as a Candle in the Dark
Chapter 2 (p. 27)
Random House, Inc. New York, New York, USA. 1995

Steiner, Rudolf 1861–1925
Austrian philosopher and scientist

In thinking, we have that element given us which welds our separate individuality into one whole with the cosmos.
The Philosophy of Freedom: The Basis for a Modern World Conception
Chapter 5 (p. 70)
Rudolf Steiner Press. London, England. 1999

THOUGHT

Baldwin, J. Mark 1861–1934
American philosopher and psychologist

We do not scatter our thoughts as widely as possible in order to increase the chances of getting a true one; on the contrary, we call the man who produces the most thought-variations a "scatter-brain," and expect nothing inventive from him...we succeed in thinking well by thinking hard; we get the valuable thought-variations by concentrating attention upon the body of related [data] which we already have; we discover new relations among the data of experience by running over and over the links and couplings of the apperceptive systems with which our minds are already filled.
On Selective Thinking
The Psychological Review, Volume 5, Number 1, 1889 (p. 4)

Byron, George Gordon, 6ᵗʰ Baron Byron   1788–1824
English Romantic poet and satirist

The power of Thought; — the magic of the Mind!
The Complete Poetical Works of Byron
The Corsair
Canto I, Stanza 8
Houghton Mifflin. Boston, Massachusetts, USA. 1933

Carlyle, Thomas   1795–1881
English historian and essayist

Thought once awakened does not again slumber.
On Heroes and Hero Worship
Lecture I (p. 24)
John B. Alden, Publisher. New York, New York, USA. 1887

And what is that Science, which the scientific head alone, were it screwed off, and (like the Doctor's in the Arabian Tale) set in a basin to keep it alive, could prosecute without a heart, but one other of the mechanical and menial handicrafts, for which the Scientific Head (having a Soul in it) is too noble an organ? I mean that Thought without Reverence is barren, perhaps poisonous; at best, dies like cookery with the day that called it forth; does not live, like sowing, in successive tilts and wider-spreading harvests, bringing food and plenteous increase to all Time.
Sartor Resartus
Sartor Resartus

Coman, Dale Rex   1906–
American research physician and wildlife writer

Thoughts are timid things. They are frightened away by noise and they make none themselves. They flutter as silently as do owls on soft-edged wings.
The Endless Adventure
Sanctuaries (p. 176)
Henry Regnery Company. Chicago, Illinois, USA. 1972

Davy, Sir Humphry   1778–1829
English chemist

My real, my working existence is among the objects of scientific research. Common amusements and enjoyments are necessary to me only as dreams to interrupt the flow of thoughts too nearly analogous to enlighten and vivify.
In Sir William Ramsay
Essays Biographical and Chemical
The Great London Chemists
Section II (pp. 47–48)
Archibald Constable & Company Ltd. London, England. 1908

Douglas, A. Vibert   1894–1988
Canadian astronomer

Guided by some of the great thinkers of today, our thoughts have traversed aeons of time, contemplating some of the changes taking place with majestic deliberation throughout the vastness of space. "Time rolls his ceaseless course." A million million years suffice for the birth of a star and its early development; a few hundred thousand years will tell the tale of the life of mankind upon this planet; and as for man, an individual man, the years of his life are three score years and ten, and yet such is the power of a great mind that, despite the brevity of its allotted span, it can wrestle with the problems of nature and learn something at least of the immensities of space and time.
Annual Report of the Board of Regents of the Smithsonian Institution, 1925
Time and Space (p. 155)
Government Printing Office. Washington, D.C. 1926

Einstein, Albert   1879–1955
German-born physicist

Scientific thought is a development of pre-scientific thought.
Translated by Alan Harris
Essays in Science
The Problem of Space, Ether, and the Field in Physics (p. 61)
Philosophical Library. New York, New York, USA. 1934

Emerson, Ralph Waldo   1803–82
American lecturer, poet, and essayist

Look sharply after your thoughts. They come unlooked for, like a new bird seen on your trees, and, if you turn to your usual task, disappear; and you shall never find that perception again; never, I say — but perhaps years, ages, and I know not what events and worlds may lie between you and its return!
Journals of Ralph Waldo Emerson 1864–1876
October 1872 (p. 365)
Houghton Mifflin Company. Boston, Massachusetts, USA. 1911

The crystal sphere of thought is as concentrical as the geological structure of the globe. As our soils and rocks lie in strata, concentric strata, so do all men's thinkings run laterally, never vertically.
Ralph Waldo Emerson: Essays and Lectures
The Method of Nature (p. 117)
The Library of America. New York, New York, USA. 1983

Haldane, John Scott   1860–1936
Scottish physiologist

…scientific thought does not involve physical realism…
The Philosophical Basis of Biology: Donnellan Lectures, University of Dublin, 1930
Lecture III, The Deeper Meaning of Berkeley's Reasoning (p. 120)
Doubleday, Doran & Company, Inc. Garden City, New York, USA. 1931

Heisenberg, Werner Karl   1901–76
German physicist and philosopher

…when new groups of phenomena compel changes in the pattern of thought…even the most eminent of physicists find immense difficulties. For the demand for change in the thought pattern may engender the feeling that the ground is to be pulled from under one's feet. … I believe that the difficulties at this point can hardly be overestimated. Once one has experienced the desperation with which clever and conciliatory men of science react to the demand for a change in the thought pattern, one can only be amazed that such revolutions in science have actually been possible at all.

In Robert M. Augros & George N. Stanciu
The New Story of Science: Mind and the Universe
Chapter III (p. 45)
Bantam Books, Inc. New York, New York, USA. December 1986

Hobbes, Thomas   1588–1679
English philosopher and political theorist

From desire ariseth the thought of some means we have seen produce the like of that which we aim at; and from the thought of that, the thought of means to that mean; and so continually till we come to some beginning within our own power.

In *Great Books of the Western World* (Volume 23)
Leviathan
Chapter III (p. 53)
Encyclopædia Britannica, Inc. Chicago, Illinois, USA. 1952

Holmes, Oliver Wendell   1809–94
American physician, poet, and humorist

A thought is often original, though you have uttered it a hundred times.

The Autocrat of the Breakfast-Table
Chapter I (p. 7)
Houghton Mifflin Company. Boston, Massachusetts, USA. 1891

…little-minded people's thoughts move in such small circles that five minutes' conversation gives you an arc long enough to determine their whole curve. An arc in the movement of a large intellect does not sensibly differ from a straight line.

The Autocrat of the Breakfast-Table
Chapter I (p. 10)
Houghton Mifflin Company. Boston, Massachusetts, USA. 1891

Jespen, G. L.
No biographical data available

Habits of thought in the tradition of a science are not readily changed; it is not easy to deviate from the customary channels of accumulated experience in the conventionalized subjects.

In G.L. Jespen, E. Mayr, and G.G. Simpson (eds.)
Genetics, Paleontology and Evolution
Foreword (p. v)
Princeton University Press. Princeton, New Jersey, USA. 1949

Keynes, John Maynard   1883–1946
British economist

Anyone who has ever attempted pure scientific or philosophical thought knows how one can hold a problem momentarily in one's mind and apply all one's powers of concentration to piercing through it, and how it will dissolve and escape and you find what you are surveying is a blank.

Essays in Biography
Newton the Man (p. 312)
Horizon Press, Inc. New York, New York, USA. 1951

Ludmerer, Kenneth M.   1947–
American physician and professor or medicine and history

Critical thinking, in short, offers the way to keep science and technology in harness.

Learning to Heal: The Development of American Medical Education
Chapter 14 (p. 280)
Basic Books, Inc. New York, New York, USA. 1985

Martin, Charles-Noël   1923–
French physicist

It is always extremely difficult to express thoughts. Words and phrases are so many fretters by which our spirit is bound. Words are mere symbols of reality, and the written word is not more than a one-dimensional flow across the two-dimensional page of a three-dimensional book.

Translated by A.J. Pomerans
The Role of Perception in Science
Chapter 1 (p. 15)
Hutchinson of London. London, England. 1963

Moore, H. P.
No biographical data

…today, there are many guardians of culture who are more shocked at a misspelled word (even in our quite unsystematic English spelling) than at a hazily expressed thought.

Engineering Culture
Science, Volume 73, Number 1881, January 16, 1931 (p. 51)

Nash, John F.   1928-
American mathematician

Rationality of thought imposes a limit on a person's concept of his relation to the cosmos.

Les Prix Nobel. The Nobel Prizes in 1994
Autobiography
Nobel Foundation. Stockholm, Sweden. 1995

Planck, Max   1858–1947
German physicist

Is there something in the nature of man, some inner realm, that science cannot touch? …Or to speak more concretely, is there a point at which the causal line of thought ceases and beyond which science cannot go?

Where Is Science Going?
Chapter V (p. 160)
W.W. Norton & Company, Inc. New York, New York, USA.1932

Shakespeare, William   1564–1616
English poet, playwright, and actor

Flout 'em and scout 'em
And scout 'em and flout 'em;
Thought is free.
In *Great Books of the Western World* (Volume 27)
The Plays and Sonnets of William Shakespeare (Volume 2)
The Tempest
Act III, Scene ii, l. 130–133
Encyclopædia Britannica, Inc. Chicago, Illinois, USA. 1952

Whitehead, Alfred North   1861–1947
English mathematician and philosopher

…the first man who noticed the analogy between a group
of seven fishes and a group of seven days made a notable
advance in the history of thought.
Science and the Modern World
Chapter II (p. 30)
The Macmillan Company. New York, New York, USA. 1929

Wittgenstein, Ludwig Josef Johann   1889–1951
Austrian-born English philosopher

Thoughts rise to the surface slowly, like bubbles.
(Sometimes it's as though you could see a thought,
an idea, as an indistinct point far away on the hori-
zon; and then it often approaches with astonishing
swiftness.)
Translated by Peter Winch
Culture and Value (p. 63e)
The University of Chicago Press. Chicago, Illinois, USA. 1980

THUNDERBOLT

Lewis, Edwin Herbert   1866–1938
American rhetorician, novelist, and poet
The thunderbolts were imprisoned in crucibled crystalline
ore,
And locked in the laughing ocean, and shut in the shining
shore,
And lulled in the light of evening, and hushed in gentle
grain
And unimperiled lilies impearled with quiet rain.
White Lightning
Cover page
Covici-McGee. Chicago, Illinois, USA. 1923

TIDAL BORE

Adam, John A.
No biographical data

What is a bore? The answer will vary depending on
whether one is at a cocktail party…or the Bay of Fundy
in Nova Scotia.
Mathematics in Nature: Modeling Patterns in the Natural World
Chapter Nine (p. 194)
Princeton University Press. Princeton, New Jersey, USA. 2003

TIDE

Coman, Dale Rex   1906–
American research physician and wildlife writer

It does not take long to realize that, instead of clocks,
the tides beat out the measure of the marsh and shore,
and that all you see, plant and animal, must adapt to the
periodic changes of water level.
The Endless Adventure
The Sand Dunes and Salt Marshes in November (p. 62)
Henry Regnery Company. Chicago, Illinois, USA. 1972

Defant, Albert   1884–1974
Austrian meteorologist and oceanographer

The tides are the heartbeat of the ocean, a pulse that can
be felt all over the world.
Ebb and Flow: The Tides of Earth, Air, and Water
Chapter I (p. 9)
The University of Michigan Press. Ann Arbor, Michigan, USA. 1958

Harrington, Thomas
No biographical data available

The benefit which God designed for man by the Tides in
giving a perpetual motion to the Waters was to prevent
their corrupting, and thereby breeding any infection that
might arise from too long a stagnation of them.
Science Improved; or the Theory of the Universe
Section III (p. 18)
Printed for the Author. London, England. 1774

Kelvin, Lord William Thomson   1824–1907
Scottish engineer, mathematician, and physicist

The subject on which I have to speak this evening is
the tides, and at the outset I feel in a curiously difficult position. If I were asked to tell what I mean by
the Tides I should feel it exceedingly difficult to answer the question. The tides have something to do with
motion of the sea. Rise and fall of the sea is sometimes called a tide; but I see, in the Admiralty Chart
of the Firth of Clyde, the whole space between Ailsa
Craig and the Ayrshire coast marked "very little tide
here." Now, we find there a good ten feet rise and fall,
and yet we are authoritatively told there is very little
tide. The truth is, the word "tide" as used by sailors at
sea means horizontal motion of the water; but when
used by landsmen or sailors in port, it means vertical
motion of the water.
Lecture
The British Association at the Southampton Meeting, Friday, August
25, 1882

TIME

Adams, George   1750–95
English instrument maker

Nothing can be more shocking to reason than eternal time; infinite divisibility is not less absurd.
Lectures on Natural and Experimental Philosophy (Volume 3)
Chapter XXIV (p. 12)
Printed by R. Hindmarsh. London, England. 1794

Aristotle   384 BCE–322 BCE
Greek philosopher

...time is "number of movement in respect to the before and after", and is continuous since it is an attribute of what is continuous.
In *Great Books of the Western World* (Volume 8)
Physics
Book IV, Chapter 11 (p. 300)
Encyclopædia Britannica, Inc. Chicago, Illinois, USA. 1952

Aurelius Antoninus, Marcus   121–180
Roman emperor

Time is like a river made up of the events which happen, and a violent stream; for as soon as a thing has been seen, it is carried away, and another comes in its place, and this will be carried away too.
In *Great Books of the Western World* (Volume 12)
The Meditations of Marcus Aurelius
Book IV, #43 (p. 267)
Encyclopædia Britannica, Inc. Chicago, Illinois, USA. 1952

Bacon, Sir Francis   1561–1626
English lawyer, statesman, and essayist

...time is the greatest innovator...
In Fred Allison Howe (ed.)
The Essays or Counsels
Civil and Moral, XXIV, Of Innovations (p. 75)

...time, which is the author of authors, be not deprived of his due, which is, further and further to discover truth.
In *Great Books of the Western World* (Volume 30)
Advancement of Learning
First Book, Chapter IV, Section 12
Section IV, 12 (p. 15)
Encyclopædia Britannica, Inc. Chicago, Illinois, USA. 1952

Barnett, Lincoln   1909–79
American science writer

Time itself will come to an end. For entropy points the direction of time. Entropy is the measure of randomness. When all system and order in the universe have vanished, when randomness is at its maximum, and entropy cannot be increased, when there is no longer any sequence of cause and effect, in short when the universe has run down, there will be no direction to time — there will be no time.
The Universe and Dr. Einstein
Chapter 14 (p. 100)
William Sloane Associates. New York, New York, USA. 1948

Barrow, Isaac   1630–77
English clergyman and mathematician

Because Mathematicians frequently make use of Time, they ought to have a distinct idea of the meaning of that Word, otherwise they are Quacks...
In Paul Davies
About Time: Einstein's Unfinished Revolution
Header (p. 183)
Simon & Schuster. New York, New York, USA. 1995

Bergson, Henri   1859–1941
French philosopher

Whereever anything lives, there is, open somewhere, a register in which time is being inscribed.
Translated by Arthur Mitchell
Creative Evolution
Chapter I (p. 17)
The Modern Library. New York, New York, USA. 1944

Time is an invention or it is nothing at all. But of time-invention physics can take no account.... Modern physics...rests altogether on a substitution of time-length for time-invention.
Translated by Arthur Mitchell
Creative Evolution
Chapter IV (p. 361)
The Modern Library. New York, New York, USA. 1944

Blake, William   1757–1827
English poet, painter, and engraver

I see the Past, Present, and Future existing all at once before me.
The Complete Poetry and Prose of William Blake
Jerusalem, l. 15
University of California Press. Berkeley, California, USA. 1982

Bohm, David   1917–92
American physicist

Eternity can be affected by what happens in time.
Quoted by Renée Weber
Dialogues with Scientists and Sages: The Search for Unity (p. 91)
Routledge & Kegan Paul. London, England. 1986

But the puzzle is, what happened before time began?
Quoted by Renée Weber
Dialogues with Scientists and Sages: The Search for Unity (p. 199)
Routledge & Kegan Paul. London, England. 1986

Bondi, Sir Hermann   1919–2005
English mathematician and cosmologist

Time must never be thought of as pre-existing in any sense; it is a manufactured quantity.
In Paul Davies
About Time: Einstein's Unfinished Revolution
Header (p. 21)
Simon & Schuster. New York, New York, USA. 1995

Borges, Jorge Luis   1899–1986
Argentine writer

Time is a river which sweeps me along, but I am the river; it is a tiger which mangles me, but I am the tiger; it

is a fire which consumes me, but I am the fire. The world, unfortunately, is real; I, unfortunately, am Borges.
Translated by Anthony Kerrigan, Alastair Reid et al.
A Personal Anthology
A New Refutation of Time (p. 64)
Grove Press. New York, New York, USA. 1967

Our destiny...is not frightful because it is unreal; it is frightful because it is irreversible and ironbound. Time is the substance of which I am made. Time is a river which sweeps me along, but I am the river; it is a tiger which mangles me, but I am the tiger; it is a fire which consumes me, but I am the fire. The world, unfortunately, is real; I unfortunately, am Borges.
Translated by Anthony Kerrigan, Alastair Reid and others
A Personal Anthology
A New Refutation of Time (p. 64)
Grove Press. New York, New York, USA. 1967

...he believed in an infinite series of times, a growing, dizzying web of divergent, convergent, and parallel times. That fabric of times that approach one another, fork, are snipped off, or are simply unknown for centuries, contains all possibilities. In most of those times, we do not exist; in some, you exist but I do not; in others, I do and you do not; in others still, we both do. In this one, which the favouring hand of chance has dealt me, you have come to my home; in another, when you come through my garden you find me dead; in another, I say these same words, but I am an error, a ghost.
In Donald A. Yates & James E. Irby (eds.)
Labyrinths: Short Stories & Other Writings
The Garden of Forking Paths (p. 28)
A New Direction Book. New York, New York, USA. 1964

Borland, Hal   1900–78
American writer

Forget that second-ticking clock.
Time is the seed
Waiting to fly from the milkweed pod.
Time is the speed
Of a dragonfly.
Time is the rabbit's desperate scut.
Time's dimensions are hidden in rocks,
In wind and rain, but never in clocks.
Borland Country
Foreword (p. 5)
J.B. Lippincott Company. Philadelphia, Pennsylvania, USA. 1971

Bradbury, Ray   1920–
American writer

There was a smell of Time in the air tonight. He smiled and turned the fancy in his mind. There was a thought. What did Time smell like? Like dust and clocks and people. And if you wondered what Time sounded like it sounded like water running in a dark cave and voices crying and dirt dropping down upon hollow box lids, and rain. And, going further, what it looked like a silent film

in an ancient theater, one hundred billion faces falling like those New Year balloons, down and down into nothing. That was how Time smelled and looked and sounded. And tonight...tonight you could almost touch Time.
The Martian Chronicles

Bridgman, Percy Williams   1882–1961
American physicist

But in no case is there any question of time flowing backward, and in fact the concept of backward flow of time seems absolutely meaningless. ... If it were found that the entropy of the universe were decreasing, would one say that time was flowing backward, or would one say that it was a law of nature that entropy decreases with time?
Reflections of a Physicist
Chapter 8 (p. 165)
Philosophical Library. New York, New York, USA. 1950

It seems to me that in any operational view of the meaning of natural concepts the notion of time must be used as a primitive concept, which cannot be analysed but must be accepted, so that it is meaningless to speak of a reversal of the direction of time.
Reflections of a Physicist
Chapter 8 (p. 165)
Philosophical Library. New York, New York, USA. 1950

Brillouin, Léon   1889–1969
French physicist

...one of the most important features about time is its irreversibility. Time flows on and never comes back. When the physicist is confronted with this fact he is greatly disturbed. All the laws of physics in their elementary forms are reversible.
In Walter Buckley and Anatol Rapaport (eds.)
Modern Systems Research for the Behavioral Scientist: A Sourcebook
Life, Thermodynamics, and Cybernetics (p. 150)
Aldine Publishing Company. Chicago, Illinois, USA. 1968

Browne, Sir Thomas   1605–82
English author and physician

Time which antiquates antiquities, and hat an art to make dust of all things.
Hydriotophia
Chapter V (p. 69)
Printed for Hen. Brome. London, England. 1658

Carlyle, Thomas   1795–1881
English historian and essayist

That great mystery of Time, were there no other; the illimitable, silent, never-resting thing called Time, rolling, rushing on, swift, silent, like an all-embracing ocean-tide, on which we and all the Universe swim like exhalations, like apparitions which are, and then are not. ...
On Heroes and Hero Worship
Lecture I (p. 12)
John B. Alden, Publisher. New York, New York, USA. 1887

...no hammer in the Horologe of Time peals through the universe when there is a change from Era to Era. Men understand not what is among their hands...
On Heroes and Hero Worship
John B. Alden, Publisher. New York, New York, USA. 1887

Carroll, Lewis (Charles Dodgson)   1832–98
English writer and mathematician

Alice sighed wearily. "I think you might do something better with the time," she said, "than wasting it in asking riddles with no answers."

"If you knew Time as well as I do," said the Hatter, "you wouldn't talk about wasting it."
The Complete Works of Lewis Carroll
Alice's Adventures in Wonderland
Chapter VII (p. 78)
The Modern Library. New York, New York, USA. 1936

Chaucer, Geoffrey   1343–1400
English poet

The tyme, that may not sojourne
But goth, and never may retourne,
As water that down runneth ay,
But never drope retourne may;
Ther may no-thing as tyme endure,
Metal, nor erthely creature;
For alle thing it fret, and shal:
The tyme eek, that chaungeth al,
And all doth waxe and fostred be,
And alle thing destroyeth he.
The Romaunt of the Rose

Christianson, Gale E.
No biographical data available

Historical time is a tricky thing; it flows at an ever accelerating speed, like a river approaching a great waterfall. We must soon learn to cope with the awesome power given to us by the heirs of Copernicus or our species, and all others on the planet, are doomed to a painful and purposeless extinction. Were this to happen, only the stars would remain; and what are the stars, after all, without the eyes of man to gaze upon them or the human mind to contemplate the vastness of their wonders?
This Wild Abyss: The Story of the Men Who Made Modern Astronomy
Chapter 9 (p. 434)
The Free Press. New York, New York, USA. 1978

Clemence, G. M.
No biographical data available

The measurement of time is essentially a process of counting.
Time and Its Measurement
American Scientist, Volume 40, Number 2, April 1952 (p. 261)

Cleugh, Mary F.
Psychologist

It cannot be too often emphasized that physics is concerned with the measurement of time, rather than with the essentially metaphysical question as to its nature. ... We must not believe that physical theories can ultimately solve the metaphysical problems that time raises.
Time and Its Importance in Modern Thought
Chapter II (p. 51)
Methuen & Company Ltd. London, England. 1937

Cummings, Ray   1887–1957
American science fiction writer

This same Space; the spread of this lawn...what would it be in another hundred years? Or a thousand? This little Space, from the Beginning to the End so crowded with events and only Time to hold them apart.
The Shadow Girl
Gerald G. Swan. London, England. 1946

"Time," said George, "why I can give you a definition of time. It's what keeps everything from happening at once."
The Man Who Mastered Time
Chapter I (p. 1)
A.L. Burt Company. New York, New York, USA. 1929

Davies, Paul Charles William   1946–
British-born physicist, writer, and broadcaster

Relativity physics has shifted the moving present out from the superstructure of the universe, into the minds of human beings, where it belongs.
The Physics of Time Asymmetry (p. 2)
University of California Press. Berkeley, California, USA. 1976

Davis, Philip J.   1923–
American mathematician
Hersh, Reuben   1927–
American mathematician

Time, that mysterious something, that flow, that relation, that mediator, that arena for event, envelops us and confounds us all.
Descartes' Dream: The World According to Mathematics
Chapter IV
Of Time and Mathematics (p. 189)
Harcourt Brace Jovanovich. San Diego, California, USA. 1986

...we still cannot say what time is; we cannot agree whether there is one time or many times, cannot even agree whether time is an essential ingredient of the universe or whether it is the grand illusion of the human intellect.
Descartes' Dream: The World According to Mathematics
Chapter IV
Of Time and Mathematics (p. 189)
Harcourt Brace Jovanovich. San Diego, California, USA. 1986

Dewey, John   1859–1952
American philosopher and educator

Time and memory are true artists; they remold reality nearer to the heart's desire.
Reconstruction in Philosophy
Chapter V (p. 104)
Beacon Press. Boston, Massachusetts, USA. 1920

Dillard, Annie   1945–
American poet, essayist, novelist, and writing teacher

Time is the continuous loop, the snakeskin with scales endlessly overlapping without beginning or end, or time is an ascending spiral if you will, like a child's toy Slinky.
Pilgrim at Tinker Creek
Chapter 5 (p. 76)
Harper's Magazine Press. New York, New York, USA. 1974

Eddington, Sir Arthur Stanley   1882–1944
English astronomer, physicist, and mathematician

Whatever may be time *de jure*, the Astronomer Royal's time is time *de facto*. His time permeates every corner of physics.
The Nature of the Physical World
Chapter III (p. 36)
The Macmillan Company. New York, New York, USA. 1930

In any attempt to bridge the domains of experience belonging to the spiritual and physical sides of our nature, time occupies the key position.
The Nature of the Physical World
Chapter V (p. 91)
The Macmillan Company. New York, New York, USA. 1930

The great thing about time is that it goes on.
In Paul Davies
About Time: Einstein's Unfinished Revolution
Header (p. 25)
Simon & Schuster. New York, New York, USA. 1995

The philosopher discusses the significance of time; the astronomer measures time. The astronomer goes confidently about his business and does not think of asking the philosopher what exactly is this thing he is supposed to be measuring; nor does the philosopher always stop to consider whether time in his speculations is identical with the time which the world humbly accepts from the astronomer. In these circumstances it is not surprising that some confusion should have arisen.
The Relativity of Time
Nature, Volume CVI, Number 2677, February 17, 1921 (p. 802)

Einstein, Albert   1879–1955
German-born physicist

Till now it was believed that time and space existed by themselves, even if there was nothing — no Sun, no Earth, no stars — while now we know that time and space are not the vessel for the Universe, but could not exist at all if there were no contents, namely, no Sun, no Earth, and other celestial bodies.
New York Times, April 4, 1921

The distinction between past, present and future is only an illusion, even if a stubborn one.
In Paul Davies
About Time: Einstein's Unfinished Revolution
Header (p. 70)
Simon & Schuster. New York, New York, USA. 1995

Michele [Besso] has left this strange world just before me. This is of no importance. For us convinced physicists the distinction between past, present and future is an illusion, although a persistent one.
In Eric J. Lerner
The Big Bang Never Happened
Chapter 7 (p. 283)
Random House, Inc. New York, New York, USA. 1991

Eliot, George (Mary Ann Evans Cross)   1819–80
English novelist

Men can do nothing without the make-believe of a beginning. Even Science, the strict measurer, is obliged to start with a make-believe unit, and must fix on a point in the stars' unceasing journey when his sidereal clock shall pretend that time is Nought. His less accurate grandmother Poetry has always been understood to start in the middle; but on reflection it appears that her proceeding is not very different from his; since Science, too, reckons backward as well as forward, divides his unit into billions, and with his clock-finger at Nought really sets off in medias res. No retrospect will take us to the true beginning; and whether our prologue be in heaven or on earth, it is but a fraction of that all-presupposing fact with which our story sets out.
Daniel Deronda
Book I, Chapter I (p. 5)
A.L. Burt Company. New York, New York, USA. 18??

Eliot, T. S. (Thomas Stearns)   1888–1965
American expatriate poet and playwright

Time present and time past
Are both perhaps present in time future,
And time future contained in time past.
The Collected Poems and Plays 1909–1950
Burnt Norton (p. 117)
Harcourt, Brace & World, Inc. New York, New York, USA. 1952

Fraser, Julius Thomas
No biographical data available

The resulting dichotomy between time felt and time understood is a hallmark of scientific-industrial civilization, a sort of collective schizophrenia.
In Eric J. Lerner
The Big Bang Never Happened
Chapter 7 (p. 283)
Random House, Inc. New York, New York, USA. 1991

Froude, James Anthony   1818–94
English historian and biographer

Time has no relation to Being, conceived mathematically; it would be absurd to speak of circles or triangles as any older today than they were at the beginning of the world.
Short Studies on Great Subjects (Volume 1)
Spinoza (p. 359)
Longmans, Green & Company. London, England. 1879

Gale, Richard M.
No biographical data available

…"time" is indefinable…due to the fact that temporal notions are implicitly involved in all of the basic concepts by means of which we think and talk about the world.
The Language of Time
Part One, Chapter I (p. 5)
Humanities Press. New York, New York, USA. 1968

Galton, Sir Francis   1822–1911
English anthropologist, explorer, and statistician

It is difficult to withstand a suspicion that the three dimensions of space and the fourth dimension of time may be four independent variables of a system that is neither space nor time, but something else wholly unconceived by us. Our present enigma as to how a First Cause could itself have been brought into existence — how the tortoise of the fable, that bears the elephant that bears the world, is itself supported, — may be wholly due to our necessary mistranslation of the four or more variables of the universe, limited by inherent conditions, into the three unlimited variables of Space and the one of Time.
Inquiries into Human Faculty and Its Development
The Observed Order of Events (p. 196)
AMS Press. New York, New York, USA. 1973

Gardner, Earl Stanley   1889–1970
American author

Time is really nothing but a huge circle. You divide a circle of three hundred and sixty degrees into twenty-four hours, and you get fifteen degrees of arc that is the equivalent of each hour.
The Case of the Buried Clock (p. 82)
Grosset & Dunlap. New York, New York, USA. 1943

Haughton, Samuel   1821–97
Irish scientific writer

The infinite time of the geologists is in the past; and most of their speculations regarding this subject seem to imply the absolute infinity of time, as if the human imagination was unable to grasp the period of time requisite for the formation of a few inches of sand or feet of mud, and its subsequent consolidation into rock.
Manual of Geology
Lecture IV (p. 80)
Longmans, Green, Reader, and Dyer. London, England. 1866

Hawking, Stephen William   1942–
English theoretical physicist

Imaginary time is another direction of time, one that is at right angles to ordinary, real time. We could get away from this one-dimensional, linelike behavior of time.… Ordinary time would be a derived concept we invent for psychological reasons. We invent ordinary time so that we can describe the universe as a succession of events in time, rather than as a static picture, like a surface map of the earth.… Time is just like another direction in space.
Playboy, Interview, April 1990

Heraclitus   ca. 540 BCE–ca. 475 BCE
Greek philosopher

Time is like a river flowing endlessly through the universe.
In Franzo H. Crawford
Introduction to the Science of Physics
Chapter 10 (p. 160)
Harcourt, Brace & World, Inc. New York, New York, USA. 1968

Høeg, Peter   1957–
Danish author

Time refuses to be simplified or reduced. You cannot say that it is only found in the mind or only in the universe, that it runs only in one direction, or in every one imaginable. That it exists only in biological substructure, or is only a social convention. It is all of these things.
Borderliners
Chapter Seven (p. 259)
Farrar, Straus & Giroux. New York, New York, USA. 1994

Housman, A. E. (Alfred Edward)   1859–1936
English poet, scholar, and satirist

Three minutes' thought would suffice to find this out; but thought is irksome and three minutes is a long time.
Selected Prose
Chapter II, Section 3 (p. 56)
At The University Press. Cambridge, England. 1961

Hoyle, Sir Fred   1915–2001
English mathematician and astronomer

If there is one thing we can be sure enough of in physics it is that all time exists with equal reality.
October the First Is Too Late
Chapter Six (p. 75)
Harper & Row, Publishers. New York, New York, USA. 1966

Hurley, Patrick M.   1912–2000
British geophysicist

How majestic are those broad reaches of time! Looking into an abyss, one senses the gigantic form of the void only in comparison to one's own minute stature. It is almost incomprehensible that only a few billion years ago our galaxy was born in a gigantic bomb-flash of nuclear energy. What an inspiring picture of the process of creation! But awesome and inspiring as it is to contemplate this mighty spectacle, the true reward is not to be found

in whether our calculations are correct, give or take a few million years; it lies in the discoveries, in the advancement of human knowledge and philosophy that are the inevitable products of scientific search for law in nature.
How Old Is the Earth?
Chapter VI (p. 152)
Greenwood Press, Publishers. Westport, Connecticut, USA. 1959

Hutton, James   1726–97
Scottish geologist, chemist, and naturalist

Time, which measures every thing in our idea, and is often deficient to our schemes, is to nature endless and as nothing.
Theory of the Earth
Transactions of the Royal Society of Edinburgh, Volume 1, 1788 (p. 215)

Huxley, Thomas Henry   1825–95
English biologist

Biology takes her time from geology. The only reason we have for believing in the slow rate of the change in living forms is the fact that they persist through a series of deposits which, geology informs us, have taken a long while to make. If the geological clock is wrong, all the naturalist will have to do is to modify his notions of the rapidity of change accordingly.
Quarterly Journal of the Geological Society London, Volume 25 (p. xxxviii)

Jeans, Sir James Hopwood   1877–1946
English physicist and mathematician

…time figures as the mortar which binds the bricks of matter together…
The Mysterious Universe
Chapter V (p. 144)
The Macmillan Company. New York, New York, USA. 1932

…time leaves its mark, its wrinkles and its grey hairs, on the stars, so that we can guess their ages tolerable well, and the evidence is all in favor of stellar lives, not of thousands of millions, but of millions of millions, of years.
The Universe Around Us
Chapter I (p. 78)
The Macmillan Company. New York, New York, USA. 1929

Kant, Immanuel   1724–1804
German philosopher

Time is not an empirical conception. For neither coexistence nor succession would be perceived by us, if the representation of time did not exist as a foundation a priori. Without this presupposition we could not represent to ourselves that things exist together at one and the same time, or at different times, that is, contemporaneously, or in succession.
In *Great Books of the Western World* (Volume 42)
Critique of Pure Reason

First Part, Of Time, Metaphysical Exposition of this Conception, 5
Encyclopædia Britannica, Inc. Chicago, Illinois, USA. 1952

Krauss, Lawrence M.   1954–
American theoretical physicist

The possibilities of space travel beckon us every time we gaze up at the stars, yet we seem to be permanent captives in the present. The question that motivates not only dramatic license but a surprising amount of modern theoretical physics research can be simply put: Are we or are we not prisoners on a cosmic temporal freight train that cannot jump the tracks?
The Physics of Star Trek
Chapter Two (p. 13)
HarperPerennial. New York, New York, USA. 1995

Lamarck, Jean-Baptiste Pierre Antoine   1744–1829
French biologist

Time is insignificant and never a difficulty for Nature. It is always at her disposal and represents an unlimited power with which she accomplishes her greatest and smallest tasks.
Translated by Albert V. Carozzi
Hydrogeology
Chapter 3 (p. 61)
University of Illinois Press. Urbana, Illinois, USA. 1964

Oh, how very ancient the earth is!
Translated by Albert V. Carozzi
Hydrogeology
Chapter 3 (p. 75)
University of Illinois Press. Urbana, Illinois, USA. 1964

Lapworth, Charles   1842–1920
English geologist

Far be it from me to suggest that geologists should be reckless in their drafts upon the bank of Time; but nothing whatever is gained, and very much is lost, by persistent niggardliness in this direction.
Proceedings of the Geological Society of London, Volume 59, 1903 (p. lxxii)

Lightman, Alan   1948–
American physicist, novelist, and essayist

There is a place where time stands still…illuminated by only the most feeble red light, for light is diminished to almost nothing at the center of time, its vibrations slowed to echoes in vast canyons, its intensity reduced to the faint glow of fireflies.
Einstein's Dreams
14 May 1905 (p. 70, 72–73)
Pantheon Books. New York, New York, USA. 1993

Lyell, Sir Charles   1797–1875
English geologist

Such views of the immensity of past time, like those unfolded by the Newtonian philosophy in regard to space,

were too vast to awaken ideas of sublimity unmixed with a painful sense of our incapacity to conceive a plan of such infinite extent. Worlds are seen beyond worlds immeasurably distant from each other, and beyond them all innumerable other systems are faintly traced on the confines of the visible universe.
Principles of Geology (Volume 1)
Chapter IV (p. 63)
John Murray. London, England. 1830

...until we habituate ourselves to contemplate the possibility of an indefinite lapse of time having been comprised within each of the modern periods of earth's history, we shall be in danger of forming most erroneous and partial views of geology.
Principles of Geology (Volume 3)
Chapter VIII (p. 97)
John Murray. London, England. 1830

In vein do we aspire to assign limits to the works of creation in space, whether we examine the starry heavens, or that world of minute animalcules which is revealed to us by the microscope. We are prepared, therefore, to find that in time also the confines of the universe lie beyond the reach of mortal ken.
Principles of Geology (Volume 3)
Concluding Remarks (p. 384)
John Murray. London, England. 1830

Mann, Thomas   1875–1955
German-born American novelist

Time has no division to mark its passage, there is never a thunder-storm or blare of trumpets to announce the beginning of a new month or year. Even when a new century begins it is only we mortals who ring bells and fire off pistols.
The Magic Mountain
Chapter V
Whims of Mercurius (p. 225)
Alfred A. Knopf. New York, New York, USA. 1966

Mason, Rick
No biographical data available

With a bit of a mind slip
You're in for a time slip
And nothing can ever be the same.
Time Warp
Dance song
The Rocky Horror Picture Show
Film (1975)

Mehlberg, Henry   1904–79
Polish-American philosopher of science

It seems to me that it would be either a miracle or an unbelievable coincidence if all the major scientific theories...somehow managed to co-operate with each other so as to conceal time's arrow from us. There would be neither a miracle nor an unbelievable coincidence in the

concealment of time's arrow from us only if there were nothing to conceal — that is, if time had no arrows.
In Robert S. Cohen (ed.)
Time, Causality, and the Quantum Theory (Volume 1) (p. 207)
Reidel. Dordrecht, Netherland. 1980

Milton, John   1608–74
English poet

Fly envious Time, till thou run out thy race...
The Complete Poetical Works of John Milton
On Time, l. 1
Houghton Mifflin Company. Boston, Massachusetts, USA. 1924

Misner, Charles W.
American physicist
Thorne, Kip S.   1940–
American theoretical physicist

Time is defined so that motion looks simple.
In Charles W. Misner et al.
Gravitation
Part I, Chapter 1 (p. 23)
W.H. Freeman & Company. San Francisco, California, USA. 1973

Morris, Richard   1939–2003
American physicist and science writer

Though science has not yet probed all the depths of the subject of time, it at least knows what we should be asking about the subject. Knowing what to ask is often the most significant step on the road to understanding.
Time's Arrow: Scientific Attitudes Toward Time
Chapter 12 (p. 218)
Simon & Schuster. New York, New York, USA. 1985

Nabokov, Vladimir   1899–1977
Russian-American writer

Pure Time, Perceptual Time, Tangible Time, Time free of content, context, and running commentary — this is my time and theme. All the rest is numerical symbol or some aspect of Space. The texture of Space is not that of Time, and the piebald four-dimensional sport bred by relativists is a quadruped with one leg replaced by the ghost of a leg. My time is also Motionless Time (we shall presently dispose of "flowing" time, water-clock time, water-closet time).
Ada or Ardor: A Family Chronicle
Part Four (p. 539)
McGraw-Hill Book Company, Inc. New York, New York, USA. 1969

Newton, Sir Isaac   1642–1727
English physicist and mathematician

Absolute, true, and mathematical time, of itself, and from its own nature, flows equably without relation to anything external.
Mathematical Principles of Natural Philosophy
Definitions, Scholium I
E.P. Dutton & Company. New York, New York, USA. 1922

Penrose, Roger   1931–
English mathematical physicist

…our present picture of physical reality, particularly in relation to the nature of time, is due for a grand shake-up — even greater, perhaps, than that which has already been provided by present-day relativity and quantum mechanics.
The Emperor's New Mind: Concerning Computers, Minds, and the Laws of Physics
Chapter 8 (p. 371)
Oxford University Press, Inc. Oxford, England. 1989

Pirsig, Robert M.   1928–
American writer

The past cannot remember the past. The future can't generate the future. The cutting edge of this instant right here and now is always nothing less than the totality of every thing there is.
Zen and the Art of Motorcycle Maintenance: An Inquiry into Values
Part III, Chapter 24 (p. 283)
William Morrow & Company, Inc. New York, New York, USA. 1974

Plato   428 BCE–347 BCE
Greek philosopher

Time, then, and the heaven came into being at the same instant in order that, having been created together, if ever there was to be a dissolution of them, they might be dissolved together.
In *Great Books of the Western World* (Volume 7)
Timaeus
Section 38 (p. 451)
Encyclopædia Britannica, Inc. Chicago, Illinois, USA. 1952

…The creator…sought to make the universe eternal, so far as might be. Now the nature of the ideal being was everlasting, but to bestow this attribute in its fullness upon a creature was impossible. Wherefore he resolved to have a moving image of eternity, and when he set in order the heaven, he made this image eternal but moving according to number, while eternity itself rests in unity, and this image we call time.
In *Great Books of the Western World* (Volume 7)
Timaeus
Section 37 (p. 450)
Encyclopædia Britannica, Inc. Chicago, Illinois, USA. 1952

Playfair, John   1748–1819
Scottish geologist, physicist, and mathematician

The mind seemed to grow giddy by looking so far into the abyss of time.
Biographical Account of the Late James Hutton, F.R.S.
Transactions of the Royal Society of Edinburgh, Volume V, Part III, 1805 (p. 73)

Plotinus   ca. 205–270
Egyptian-Roman philosopher

Time at first — in reality before that "first" was produced by desire of succession — Time lay, though not yet as Time, in the Authentic Existent together with the Cosmos itself; the Cosmos also was merged in the Authentic and motionless within it.
In *Great Books of the Western World* (Volume 17)
The Six Enneads
Third Ennead VII.11 (p. 126)
Encyclopædia Britannica, Inc. Chicago, Illinois, USA. 1952

The origin of Time, clearly, is to be traced to the first stir of the Soul's tendency towards the production of the sensible Universe with the consecutive act ensuing. This is how "Time" — as we read — "came into Being simultaneously with" this All: the Soul begot at once the Universe and Time; in that activity of the Soul this Universe sprang into being; the activity is Time, the Universe is the content of Time.
In *Great Books of the Western World* (Volume 17)
The Six Enneads
Third Ennead VII.11 (p. 127)
Encyclopædia Britannica, Inc. Chicago, Illinois, USA. 1952

Poinsot, Louis   1777–1859
French mathematician and physicist

If anyone asked me to define time, I should reply: "Do you know what it is that you speak of?" If he said "Yes," I should answer, "Very well, let us talk about it." If he said "No," I should answer, "Very well, let us talk about something else."
In William Maddock Bayliss
Principles of General Physiology (3rd edition)
Preface (p. xvii)
Longmans, Green & Company. London, England. 1920

Prigogine, Ilya   1917–2003
Russian-born Belgian physical chemist

Time is creation. The future is just not there.
In Eric J. Lerner
The Big Bang Never Happened
Chapter 7 (p. 321)
Random House, Inc. New York, New York, USA. 1991

The irreversibility [of time] is the mechanism that brings order out of chaos.
In Eric J. Lerner
The Big Bang Never Happened
Chapter 7 (p. 283)
Random House, Inc. New York, New York, USA. 1991

Putnam, H.
No biographical data available

I do not believe that there are any longer any philosophical problems about Time; there is only the physical problem of determining the exact physical geometry of the four-dimensional continuum that we inhabit.
Time and Physical Geometry
Journal of Philosophy, Volume 64, April 1967

Reichenbach, Hans   1891–1953
German philosopher of science

There is no other way to solve the problem of time than the way through physics. More than any other science, physics has been concerned with the nature of time. If time is objective the physicist must have discovered that fact.
The Direction of Time (p. 16)
University of California Press. Berkeley, California, USA. 1956

There is no other way to solve the problem of time than the way through physics. ... If time is objective the physicist must have discovered the fact. If there is Becoming, the physicist must know it. ... If there is a solution to the philosophical problem of time, it is written down in the equations of mathematical physics.
The Direction of Time (p. 16)
University of California Press. Berkeley, California, USA. 1956

Russell, Bertrand Arthur William   1872–1970
English philosopher, logician, and social reformer

...to realise the unimportance of time is the gate of wisdom.
Our Knowledge of the External World
Lecture VI (p. 167)
Houghton Mifflin Company. Boston, Massachusetts, USA. 1891

...there is some sense — easier to feel than to state — in which time is an unimportant and superficial characteristic of reality. Past and future must be acknowledged to be as real as the present, and a certain emancipation from slavery to time is essential to philosophic thought.
Mysticism and Logic and Other Essays
Chapter I, Section III (p. 21)
Longmans, Green & Company London, England. 1925

Saint Augustine of Hippo   354–430
Theologian and doctor of the Church

Time is like a river made up of events which happen, and its current is strong; no sooner does anything appear than it is swept away.
In Paul Davies
Other Worlds: A Portrait of Nature in Rebellion, Space, Superspace, and the Quantum Universe
Chapter 10 (p. 186)
Simon & Schuster. New York, New York, USA. 1980

...all time past to be driven away by time to come; and all time to come, to follow upon the past; and that all both past and to come, is made up, and flows out of that which is always present? Who now shall so hold fast this heart of man, that it may stand, and see, how that eternity ever still standing, gives the word of command to the times past or to come, itself being neither past nor to come?
St. Augustine's Confessions (Volume 2)
Book XI, XI (p. 233)
William Heinemann. London, England. 1912

Clear now it is and plain, that neither things to come, nor things past, are. Nor do we properly say, there be three times, past, present, and to come; but perchance it might be properly said, there be three times: a present time of past things; a present time of present things; and a present time of future things. ... The present time of past things is our memory; the present time of present things is our sight; the present time of future things our expectation.
St. Augustine's Confessions (Volume 2)
Book XI, XX (p. 253)
William Heinemann. London, England. 1912

For what is time? Who is able easily and briefly to explain that? Who is so much as in thought to comprehend it, so as to express himself concerning it? And yet what in our usual discourse do we more familiarly and knowingly make mention of than time? And surely we understand it well enough, when we speak of it: we understand it also, when in speaking with another we hear it named.
St. Augustine's Confessions (Volume 2)
Book XI, XIV (p. 237, 239)
William Heinemann. London, England. 1912

What is time then? If nobody asks me, I know: but if I were desirous to explain it to one that should ask me, plainly I know not.
St. Augustine's Confessions (Volume 2)
Book XI, XIV (p. 237, 239)
William Heinemann. London, England. 1912

...if nothing were coming, there should be no time to come: and if nothing were, there should now be no present time. These two times therefore, past and to come, in what sort are they, seeing the past is now no longer, and that to come is not yet? As for the present, should it always be present and never pass into times past, verily it should not be time but eternity.
St. Augustine's Confessions (Volume 2)
Book XI, XIV (p. 239)
William Heinemann. London, England. 1912

Santayana, George (Jorge Augustín Nicolás Ruiz de Santillana)   1863–1952
Spanish-born American philosopher

The essence of nowness runs like a fire along the fuse of time.
Realms of Being
Chapter IX (p. 491)
Cooper-Square Publishers, Inc. New York, New York, USA. 1972

Shakespeare, William   1564–1616
English poet, playwright, and actor

Unless hours were cups of sack, and minutes capons, and clocks the tongues of bawds, and dials the signs of leaping-houses, and the blessed sun himself a fair hot wench in flame-colour'd taffeta, I see no reason why thou shouldst be so superfluous to demand the time of the day.

In *Great Books of the Western World* (Volume 26)
The Plays and Sonnets of William Shakespeare (Volume 1)
The First Part of King Henry the Fourth
Act I, Scene ii, l. 7–10
Encyclopædia Britannica, Inc. Chicago, Illinois, USA. 1952

There are many events in the womb of time which will be delivered.
In *Great Books of the Western World* (Volume 27)
The Plays and Sonnets of William Shakespeare (Volume 2)
Othello, The Moor of Venice
Act I, Scene iii, l. 376
Encyclopædia Britannica, Inc. Chicago, Illinois, USA. 1952

What seekest thou else
In the dark backward and abysm of time?
In *Great Books of the Western World* (Volume 27)
The Plays and Sonnets of William Shakespeare (Volume 2)
The Tempest
Act VI, Scene ii, l. 50
Encyclopædia Britannica, Inc. Chicago, Illinois, USA. 1952

Silesius, Angelus   1624–77
German poet

Do not compute eternity
as light-year after year
One step across
that line called Time
Eternity is here.
The Book of Angelus Silesius
Of Time and Eternity (p. 42)
Alfred A. Knopf. New York, New York, USA. 1976

Stevenson, Robert Louis   1850–94
Scottish essayist and poet

She is settling fast? said the First Lieutenant as he returned from shaving.

"Fast, Mr. Spoker?" asked the Captain. "The expression is a strange one, for Time (if you will think of it) is only relative."
Fables
The Sinking Ship
Charles Scribner's Sons. New York, New York, USA. 1923

Suess, Eduard   1831–1914
Austrian geologist

The astronomer, in order to render conceivable the immensity of celestial space, points to the parallelism of the stellar rays or to the white clouds of the Milky Way. There is no such means of comparison by which we can illustrate directly the great length of cosmic periods, and we do not even possess a unit with which such periods might be measured. The distance in space of many stars from the earth has been determined; for the distance in time of the latest strand-line on Capri or the last shell-bed on Tromsö, we cannot suggest an estimate even in approximate figures.
The Face of the Earth (Volume 2)

Part III, Chapter XIV (p. 556)
At The Clarendon Press. Oxford, England. 1906

We hold the organic remains of the remote past in our hand and consider their physical structure, but we know not what interval of time separates their epoch from our own; they are like those celestial bodies without parallax, which inform us of their physical constitution by their spectrum, but furnish no clue to their distance. As Rama looks out upon the Ocean, its limits mingling and uniting with heaven on the horizon, and as he ponders whether a path might not be built into the Immeasurable, so we look over the Ocean of time, but nowhere do we see signs of a shore.
The Face of the Earth (Volume 2)
Part III, Chapter XIV (p. 556)
At The Clarendon Press. Oxford, England. 1906

Swinburne, Richard   1943–
English philosopher

It would be an error to suppose that if the universe is infinitely old, and each state of the universe at each instant of time has a complete explanation which is a scientific explanation in terms of a previous state of the universe and natural laws (and so God is not invoked), that the existence of the universe throughout infinite time has a complete explanation, or even a full explanation. It has not. It has neither. It is totally inexplicable.
The Existence of God
Chapter 7 (p. 122)
At the Clarendon Press. Oxford, England. 1979

The X-Files

SCULLY: No, wait a minute. You're saying that, that time disappeared. Time can't just disappear, it's, it's, it's a universal invariant!

MULDER: Not in this zip code.
Pilot
Television program
Season 1 (1993)

Thoreau, Henry David   1817–62
American essayist, poet, and practical philosopher

Time is but the stream I go a-fishin in.
The Writings of Henry David Thoreau (Volume 2)
Walden
Chapter II (p. 155)
Houghton Mifflin Company. Boston, Massachusetts, USA. 1893

Tuttle, Hudson   1836–1910
American spiritualist

Thousands of years are, in the chronometer of nature, one stroke of the pendulum — a moment.
In Ludwig Buchner
Force and Matter
Chapter IX (p. 56)
Trubner & Company. London, England. 1864

Urey, Harold Clayton   1893–1981
American chemist

However, the evolution from inanimate systems of bio-chemical compounds, e.g., the proteins, carbohydrates, enzymes and many others, of the intricate systems of reactions characteristic of living organisms, and the truly remarkable ability of molecules to reproduce themselves seems to those most expert in the field to be almost impossible. Thus a time from the beginning of photosynthesis of two billion years may help to accept the hypothesis of the spontaneous generation of life.
On the Early Chemical History of the Earth and the Origin of Life
Proceedings of the National Academy of Science, Volume 38, 1952
(p. 362)

Virgil   70 BCE– 19 BCE
Roman epic, didactic, and idyllic poet

Time is flying — flying never to return.
In Paul Davies
Other Worlds: A Portrait of Nature in Rebellion, Space, Superspace, and the Quantum Universe
Chapter 10 (p. 186)
Simon & Schuster. New York, New York, USA. 1980

von Helmholtz, Hermann   1821–94
German scientist and philosopher

Physico-Mechanical laws are, as it were, the telescopes of our spiritual eyes which can penetrate into the deepest night of time, past and to come.
In Alexander Winchell
World-Life or Comparative Geology
Part II, Chapter IV (p. 451)
S.C. Griggs & Company. Chicago, Illinois, USA. 1883

Vyasa   ca. 3100 BCE
Vedic and Puranic scribe

Time does not sleep when all things sleep,
Only Time stands straight when all things fall.
Is, was, and shall be are Time's Children.
Is, was, and shall be are Time's Children.
The Mahabharata of Vyasa
The Beginning (p. 65)
Publisher undetermined

Weil, Simone   1909–43
French philosopher and mystic

Time is an image of eternity, but it is also a substitute for eternity.
Gravity and Grace
Renunciation of Time (p. 18)
Routledge & Kegan Paul. London, England. 1952

Wells, H. G. (Herbert George)   1866–1946
English novelist, historian, and sociologist

"Can an instantaneous cube exist?"
"Don't follow you," said Filby.

"Can a cube that does not last for any time at all, have a real existence?"
Filby became pensive. "Clearly," the Time Traveler proceeded, "any real body must have extension in four directions: it must have Length, Breadth, Thickness, and — Duration.... There are really four dimensions, three which we call the three planes of Space and a fourth, Time. There is, however, a tendency to draw an unreal distinction between the former three dimensions and the latter, because it happens that our consciousness moves intermittently in one direction along the latter from the beginning to the end of our lives."
In Robert M. Hutchins and Mortimer J. Adler (eds.)
The Great Ideas Today, 1971
The Time Machine, Chapter 1
Encyclopædia Britannica, Inc. Chicago, Illinois, USA. 1971

Wheeler, John Archibald   1911–
American theoretical physicist and educator

Should we be prepared to see some day a new structure for the foundations of physics that does away with time?
In Paul Davies
About Time: Einstein's Unfinished Revolution
Header (p. 178)
Simon & Schuster. New York, New York, USA. 1995

Time is nature's way to keep everything from happening at once.
In Paul Davies
About Time: Einstein's Unfinished Revolution
Header (p. 236)
Simon & Schuster. New York, New York, USA. 1995

Time ends. That is the lesson of the "big bang." It is also the lesson of the black hole.
The Lesson of the Black Hole
Proceedings of the American Philosophical Society, Volume 125, 25

Of all the obstacles to a thoroughly penetrating account of existence, none looms up more dismayingly than "time." Explain time? Not without explaining existence. Explain existence? Not without explaining time. To uncover the deep and hidden connection between time and existence, to close on itself our quartet of questions, is a task for the future.
Hermann Weyl and the Unity of Knowledge
American Scientist, Volume 74, July–August 1986 (p. 374)

White, Henry Kirke   1785–1806
English poet

...it is fearful then
To steer the mind in deadly solitude
Up the vague stream of probability
To wind the mighty secrets of the past
And turn the key of time.
Poetical Works
Time
Bell & Daldy. London, England. 1870

Whitehead, Alfred North   1861–1947
English mathematician and philosopher

Time and space express the universe as including the essence of transition and the success of achievement. The transition is real, and the achievement is real. The difficulty is for language to express one of them without explaining away the other.
Modes of Thought
Chapter II, Lecture Five (pp. 139–140)
The Macmillan Company. New York, New York, USA. 1938

Apart from time there is no meaning for purpose, hope, fear, energy. If there be no historic process, then everything is what it is, namely, a mere fact. Life and motion are lost.
Modes of Thought
Chapter II, Lecture Five (p. 139)
The Macmillan Company. New York, New York, USA. 1938

It is impossible to meditate on time and the mystery of the creative process of nature without an overwhelming emotion at the limitations of human intelligence.
The Concept of Nature
Chapter III (p. 73)
At The University Press. Cambridge, England. 1920

Whitrow, G. J.   1912–2000
English mathematician, cosmologist, and science historian

The basic objection to attempts to deduce the unidirectional nature of time from concepts such as entropy is that they are attempts to reduce a more fundamental concept to a less fundamental one.
The Natural Philosophy of Time (2nd ed.)
Chapter 7 (p. 338)
Clarendon Press. Oxford, England. 1980

…the history of natural philosophy is characterized by the interplay of two rival philosophies of time — one aiming at its "elimination" and the other based on the belief that it is fundamental and irreducible.
The Natural Philosophy of Time (2nd edition)
Chapter 7 (p. 370)
Clarendon Press. Oxford, England. 1980

Wittgenstein, Ludwig Josef Johann   1889–1951
Austrian-born English philosopher

What Eddington says about "the direction of time" and the law of entropy comes to this: time would change its direction if men should start walking backwards one day. Of course you can call it that if you like; but then you should be clear in your mind that you have said no more than that people have changed the direction they walk in.
Translated by Peter Winch
Culture and Value (p. 18e)
The University of Chicago Press. Chicago, Illinois, USA. 1980

Yeats, William Butler   1865–1939
Irish poet and playwright

Time drops in decay,
Like a candle burnt out.
The Collected Poems of W.B. Yeats
The Moods (p. 54)
The Macmillan Company. New York, New York, USA. 1956

Zebrowski, George   1945–
Polish-American science fiction writer

Time is a relationship that we have with the rest of the universe; or more accurately, we are one of the clocks, measuring one kind of time. Animals and aliens may measure it differently. We may even be able to change our way of marking time one day, and open up new realms of experience, in which a day today will be a million years.
OMNI Magazine, 1994

TIME TRAVEL

Allen, Elizabeth Akers   1832–1911
Journalist and poet

Backward, turn backward, O Time, in your flight.
Make me a child again just for to-night.
Rock Me to Sleep, Mother
Rock Me to Sleep, Mother (p. 11)
Estes & Lauriat. Boston, Massachusetts, USA. 1883

Bester, Alfred   1913–87
American science fiction writer

"We're like millions of strands of spaghetti in the same pot. No time traveler can ever meet another time traveler in the past or the future. Each of us must travel up or down his own strand alone."

"But we're meeting each other now."

"We're no longer time travelers, Henry. We've become the spaghetti sauce."
Starlight: The Great Short Fiction of Alfred Bester
The Man Who Murdered Mohammed (p. 100)
Doubleday & Company, Inc. Garden City, New York, USA. 1976

Clarke, Arthur C.   1917–
English science and science fiction writer

The most convincing argument against time travel is the remarkable scarcity of time travelers. However unpleasant our age may appear to the future, surely one would expect scholars and students to visit us, if such a thing were possible at all. Though they might try to disguise themselves, accidents would be bound to happen — just as they would if we went back to Imperial Rome with cameras and tape-recorders concealed under our nylon togas.
Profiles of the Future: An Inquiry into the Limits of the Possible
Chapter 11 (p. 132)
Harper & Row, Publishers. New York, New York, USA. 1973

Kaku, Michio   1947–
Japanese-American theoretical physicist

…imagine the chaos that would arise if time machines were as common as automobiles, with tens of millions of them commercially available. Havoc would soon break loose, tearing at the fabric of our universe. Millions of people would go back in time to meddle with their own past and the past of others, rewriting history in the process.… It would be impossible to take a simple census to see how many people there were at any given time.
Hyperspace : A Scientific Odyssey Through Parallel Universes, Time Warps, and the 10th Dimension
Chapter 11 (p. 234)
Oxford University Press, Inc. New York, New York, USA. 1995

Krauss, Lawrence M.   1954–
American theoretical physicist

While every one of us is a time traveler, the cosmic pathos that elevates human history to the level of tragedy arises precisely because we seem doomed to travel in only one direction — into the future.
The Physics of Star Trek
Chapter Two (p. 13)
Harp Perennial Publishers. New York, New York, USA. 1995

Schickel, Richard
American film critic

Time travel is the thinking person's UFO, an improbability that nevertheless resonates with mysterious and sometimes marvelous possibilities.
Review
Back to the Future, Part II
Time, December 4, 1989

Wells, H. G. (Herbert George)   1866–1946
English novelist, historian, and sociologist

Man…can go up against gravitation in a balloon, and why should he not hope that ultimately he may be able to stop or accelerate his drift along the Time-Dimension, or even turn about and travel the other way.
The Great Ideas Today, 1971
The Time Machine
Chapter One (p. 451)
Encyclopædia Britannica, Inc. Chicago, Illinois, USA. 1971

I'm afraid I cannot convey the peculiar sensations of time traveling. They are excessively unpleasant.
In Robert M. Hutchins and Mortimer J. Adler (eds.)
The Great Ideas Today, 1971
The Time Machine
Chapter Three (p. 458)
Encyclopædia Britannica, Inc. Chicago, Illinois, USA. 1971

TOOL

Beecher, Henry Ward   1813–87
American Congregational preacher and orator

A tool is but the extension of a man's hand, and a machine is but a complex tool. He that invents a machine augments the power of a man and the well-being of mankind.
In Lenox R. Lohr
Centennial of Engineering: History and Proceedings of Symposia: 1852–1952
Historical Background (p. 340)
Centennial of Engineering. Chicago, Illinois. 1952

Bergson, Henri   1859–1941
French philosopher

Science has equipped man in less than fifty years with more tools than he had made during the thousands of years he had lived on earth. Each new machine being for man a new organ — an artificial organ — his body became suddenly and prodigiously increased in size, without his soul being at the same time able to dilate to the dimensions of his body.
In Lenox R. Lohr
Centennial of Engineering: History and Proceedings of Symposia: 1852–1952
Historical Background (p. 343)
Centennial of Engineering. Chicago, Illinois. 1952

Carlyle, Thomas   1795–1881
English historian and essayist

Man is a Tool-using Animal. Weak in himself, and of small stature, he stands on a basis, at most for the flattest-soled, of some half square foot, insecurely enough; Has to straddle out his legs, lest the very wind supplant him. Feeblest of bipeds! Three quintals are a crushing load for him; the steer of the meadow tosses him aloft like a waste rag. Nevertheless he can use Tools: with these the granite mountain melts into light dust before him, seas are his smooth highway, winds and fire his unwearying steeds. Nowhere do you find him without Tools; without Tools he is nothing, with Tools he is all.
Sartor Resartus
Book I, Chapter V (pp. 35–36)
Ginn & Company. Boston, Massachusetts, USA. 1897

TOOTH

Baxter, Richard
No biographical data available

An aching tooth is better out than in.
To lose a rotten member is a gain.
Hypocrisy

Fischer, Martin H.   1879–1962
German-American scientist

I find that most men would rather have their bellies opened for five hundred dollars than have a tooth pulled for five.
In Howard Fabing and Ray Marr
Fischerisms
C.C. Thomas. Springfield, Illinois, USA. 1944

Hazlitt, William Carew   1834–1913
English bibliographer

One said a tooth drawer was a kind of unconscionable trade, because his trade was nothing else but to take away those things whereby every man gets his living.
Shakespeare Jest Books (Volume 3)
Conceit, Clichés, Flashes and Whimzies, Number 84
Willis & Sotheran. London, England. 1864

Mayo, Charles Horace   1865–1939
American physician

A crowned tooth is not a "crown of glory" and may cover a multitude of germs.
Problems of Infection
Minnesota Medicine, Volume 1, 1918

TOOTHACHE

Burns, Robert   1759–96
English author

My curse upon your venom'd stang,
That shoots my tortur'd gooms alang,
An' thro' my lug gies monie a twang
Wi' gnawing vengeance,
Tearing my nerves wi' bitter pang,
Like racking engines!
The Poems and Songs of Robert Burns (Volume 2)
Address to the Toothache, Stanza I
Clarendon Press. Oxford, England. 1968

Busch, Wilhelm   1832–1908
German cartoonist, painter, and poet

A toothache, not to be perverse,
Is an unmitigated curse…
German Satirical Writings
The Poet Thwarted (p. 161)
Continuum. New York, New York, USA. 1984

Carroll, Lewis (Charles Dodgson)   1832–98
English writer and mathematician

"Do I look very pale?" said Tweedledum, coming up to have his helmet tied on. (He called it a helmet, though it certainly looked much more like a saucepan.)

"Well — yes — a little", Alice replied gently.

"I'm very brave generally", he went on in a low voice: "only to-day I happen to have a headache."
"And I've got a toothache!" said Tweedledee, who had overheard the remark. I'm far worse than you!"
The Complete Works of Lewis Carroll
Through the Looking-Glass
Chapter IV (p. 193)
The Modern Library. New York, New York, USA. 1936

Collins, John   1742–1808
No biographical data available

Maria one Morning was smitten full sore,
With the Tooth-ache's unmerciful Pang;
And she vow'd, if she liv'd to the Age of Five-score,
That she still should remember the Fang.
Scripscrapologia
Excuse for Oblivion, l. 1–4
Published by the author. Birmingham, England. 1804

Fuller, Thomas   1608–61
English clergyman and author

The tongue is ever turning to the aching tooth.
Gnomologia: Adages and Proverbs, Wise Sentences, and Witty Sayings. Ancient and Modern, Foreign and British
No. 4796
Printed for Thomas & Joseph Allman. London, England. 1816

Gilbert, Sir William Schwenck   1836–1911
English playwright and poet
Sullivan, Arthur   1842–1900
English composer

Roll on, thou ball, roll on!
Through pathless realms of Space
Roll on!
What though I'm in a sorry case?
What though I cannot meet my bills?
What though I suffer toothache's ills?
What though I swallow countless pills?
In Helen and Lewis Melville
An Anthology of Humorous Verse
To the Terrestrial Globe
Dodd, Mead & Company New York, New York, USA. 1924

Heath-Stubbs, John   1918–2006
English critic, anthologist, translator, and poet

Venerable Mother Toothache
Climb down from the white battlements,
Stop twisting in your yellow fingers
The fourfold rope of nerves.
Collected Poems 1943–1987
A Charm Against the Toothache (p. 312)
Carcanet Press Ltd. Manchester, England. 1988

Hood, Thomas   1582–98
English poet and editor

One tooth he had with many fangs,
That shot at once as man pangs,
It had an universal sting;
One tough of that ecstatic stump
Could jerk his limbs, and make him jump,
Just like a puppet on a string.
The Complete Poetical Works of Thomas Hood
A True Story
Greenwood Press, Publishers. Westport, Connecticut, USA. 1980

James, Henry   1843–1916
American novelist

He might have been a fine young man with a bad tooth-ache; with the first even of his life. What ailed him above all, she felt, was that trouble was new to him…
The Spoils of Poynton
Chapter 8 (p. 102)
New Directions Houghton Mifflin Company. Norfolk, Connecticut, USA, 1924

Josselyn, John   1630–75
English gentleman

…for the Toothache I have found the following medicine very available, Brimstone and Gunpowder compounded with butter; rub the mandible with it, the outside being first warm'd.
Two Voyages to New-England
The Second Voyage (pp. 128–129)

Mather, Cotton   1663–1728
American minister and religious writer

O Man, Since the Hardest and Strongest Things thou hast about thee, are so fast Consuming; Do not imagine that the rest of thy Body will remain Long Unconsumed, or that any Bones of thy Body shall not soon Moulder into Dust.
The Angel of Bethesda
Capsila XI (p. 63)
American Antiquarian Society and Barre Publishers. Barre, Massachu-setts, USA. 1972

a Thigh-bone of a Toad, applied unto an aking Tooth, rarely fails of easing the Pain.
The Angel of Bethesda
Capsila XI (p. 64)
American Antiquarian Society and Barre Publishers. Barre, Massachu-setts, USA. 1972

Melville, Herman   1819–91
American novelist

Another [sailor] has the toothache: the carpenter out pin-cers, and clapping one hand upon his bench bids him be seated there; but the poor fellow unmanageably winces under the unconcluded operation; whirling round the handle of his wooden vice, the carpenter signs him to clap his jaw in that, if he would have him draw the tooth.
In *Great Books of the Western World* (Volume 48)
Moby Dick
Chapter 107 (p. 344)
Encyclopædia Britannica, Inc. Chicago, Illinois, USA. 1952

Proverb

The tooth-ache is more ease,
than to deale with ill people.
In George Herbert
Outlandish Proverbs
#558
Printed by T. Maxey for T. Garthwait. London, England. 1651

Ray, John   1627–1705
English naturalist

Who hath aching teeth hath ill tenants.
A Complete Collection of English Proverbs (p. 26)
Printed for G. Cowie. London, England. 1813

Shakespeare, William   1564–1616
English poet, playwright, and actor

For there was never yet philosopher
That could endure the toothache patiently.
In *Great Books of the Western World* (Volume 26)
The Plays and Sonnets of William Shakespeare (Volume 1)
Much Ado About Nothing
Act V, Scene i, l. 35–36
Encyclopædia Britannica, Inc. Chicago, Illinois, USA. 1952

He that sleeps feels not the toothache.
In *Great Books of the Western World* (Volume 27)
The Plays and Sonnets of William Shakespeare (Volume 2)
Cymbeline
Act V, Scene iv, l. 177
Encyclopædia Britannica, Inc. Chicago, Illinois, USA. 1952

Shaw, George Bernard   1856–1950
Irish comic dramatist and literary critic

The man with toothache thinks everyone happy whose teeth are sound.
The Revolutionist's Handbook & Pocket Companion
Maxims for Revolutionists, Greatness (p. 56)
Publisher undetermined. USA. 1962

TRACK

Defoe, Daniel   1660–1731
English pamphleteer, journalist, and novelist

…there was exactly the print of a foot, toes, heel and every part of a foot; how it came thither, I knew not, nor could I in the least imagine…
Robinson Crusoe (p. 113)
Dodd, Mead & Company. New York, New York, USA. 1946

Doyle, Sir Arthur Conan   1859–1930
Scottish writer

"Not a bird my dear Roxton — not a bird?"
"A beast?"
"No; a reptile — a dinosaur. Nothing else could have left such a track."
The Lost World
Chapter X (p. 168)
The Colonial Press, Clinton, Massachusetts, USA. 1959

Melville, Herman   1819–91
American novelist

That turnpike earth! — that common highway all over dented with the marks of…heels and hoofs.
In *Great Books of the Western World* (Volume 48)
Moby Dick
Chapter 13 (p. 44)
Encyclopædia Britannica, Inc. Chicago, Illinois, USA. 1952

Mills, Enos A.   1870–1922
Naturalist, writer and nature guide

The tracks and records in the snow which I read in passing made something of a daily newspaper for me. They told much of the news of the wilds.
In William H. Carr
The Stir of Nature
Chapter Three (p. 37)
Oxford University Press, Inc. New York, New York, USA. 1930

TRACKING

Milne, A. A. (Alan Alexander)   1882–1956
English poet, children's writer, and playwright

"Hallo" said Piglet, "What are you doing?"
"Tracking something" said Winnie-the Pooh very mysteriously.
"Tracking what?" said Piglet, coming closer.
"That's just what I ask myself. I ask myself, What?"
The Complete Tales & Poems of Winnie-the-Pooh
Winnie-the-Pooh, Winnie-the-Pooh, Pooh and Piglet Go Hunting and Nearly Catch a Woozle (p. 34)
Dutton Children's Books. New York, New York, USA. 2001

TRADITION

von Goethe, Johann Wolfgang   1749–1832
German poet, novelist, playwright, and natural philosopher

It would have been good, if Bacon had not poured the child out with the bath water, if he had seen the value of existing tradition and had advanced this point of view, if he would have known how to value and to make use of existing experiences, rather than in his style to refer to that which is indeterminable and infinite. He knew of Gilbert's work on magnetism, for example, but seemed to have no idea of the monumental worth which already existed in this discovery.
In Karl J. Fink
Goethe's History of Science
Chapter 6 (p. 75)
Cambridge University Press. Cambridge, England. 1991

TRANSISTOR

Landauer, Rolf   1927–99
German-American physicist

An ordinary transistor circuit is like a door.... You slam it open, you slam it shut. You don't have to have a delicate regard for the amount of force you use when you push it one way or the other. These quantum systems are not like that. Quantum computers don't use just an open door or a shut door. Both the open door and the shut door are present simultaneously. The problems all relate to the fact that the process is not perfect. It doesn't do exactly what you want it to do.
In Tim Folger
The Best Computer in All Possible Worlds
Discover Magazine, October 1995

TREE

Abbey, Edward   1927–89
American environmentalist and nature writer

...unless the need were urgent, I could no more sink the blade of an ax into the tissues of a living tree than I could drive it into the flesh of a fellow human.
The Journey Home: Some Words in Defense of the American West
Chapter 19 (p. 208)
E.P. Dutton. New York, New York, USA. 1977

Arnold, Sir Edwin   1832–1904
English poet

Almond blossom, sent to teach us
That the spring days soon will reach us.
Poems
Almond Blossoms
Roberts Brothers. Boston, Massachusetts, USA. 1880

Author undetermined

At that awful hour of the Passion, when the Savior of the world felt deserted in His agony, when —

The sympathizing sun, his light withdrew, and wonder'd how the stars their dying Lord could view — when earth shaking with horror, rung the passing bell for Deity, and universal nature groaned, then from the loftiest tree to the lowliest flower all felt a sudden thrill, and trembling, bowed their heads, all save the proud and obdurate aspen, which said, "Why should we weep and tremble, we trees, and plants, and flowers are pure and never sinned!"

Ere it ceased to speak, an involuntary trembling seized its every leaf, and the word went forth that it should never rest, but tremble on until the day of judgment.
Legend [of the quaking aspen tree]
Notes and Queries, First Series, Volume 6, Number 161

Bailey, William Whitman   1843–1914
American botanist

Nature is especially fond of tassels. With them she clothes many of her noblest trees.
Willows: "Pussy" and Other
The American Botanist, Volume VI, Number 2, February 1904 (p. 23)

Bailey, Liberty Hyde   1858–1954
American horticulturalist and botanist

The heavier palms are the big game of the plant world.
Gentes Herbarium
Palms, and Their Characteristics, 3, Fasc. 1
L.H. Bailey Hortorium of the New York State College of Agriculture and Life Sciences. Ithaca, New York, USA.

Barrett-Browning, Elizabeth 1806–61
English poet

A great acacia with its slender trunk
And overpoise of multitudinous leaves
(In which a hundred fields might spill their dew
And intense verdure, yet find room enough)
Stood reconciling all the place with green.
The Complete Poetical Works of Elizabeth Barrett Browning
Aurora Leigh, Book VI, l. 536–541
Houghton Mifflin Company. Boston, Massachusetts, USA. 1900

Borland, Hal 1900–78
American writer

Trees are the oldest living things we know. Rooted in the earth and reaching for the stars, they partake of immortality.
Our Natural World
The Woodlands (p. 4)
J.B. Lippincott Company. Philadelphia, Pennsylvania, USA. 1969

Only the unobservant sees nothing but trees in a forest. Any woodland is a complex community of plants and animal life with its own laws of growth and survival. But if you would know strength and majesty and patience, welcome the company of trees.
Beyond Your Doorstep: A Handbook to the Country
Chapter 4 (p. 75)
Alfred A. Knopf. New York, New York, USA. 1962

Bronte, Emily 1818–48
English novelist

My love for Linton is like the foliage in the woods. Time will change it, I'm well aware, as winter changes the trees — my love for Heathcliff resembles the eternal rocks beneath — a source of little visible delight, but necessary.
Wuthering Heights
Chapter IX (p. 88)
J.M. Dent & Sons Ltd. London, England. 1907

Bryant, Alice Franklin 1900–77
No biographical data available

Like a cathedral in some old world town
Rising above all mundane buildings, rears
The banyan tree, a growth of long slow years,
Towering above the palms. Its verdant crown
Fashions a far-spread roof, from which falls down
A diamond and tinted light with jeweled spears
Of sunbeam piercing through. The whole appears
An ornate Gothic pile of world renown.
The Banyan Tree
Nature Magazine, Volume 50, Number 5, May 1957 (p. 265)

Bryant, William Cullen 1794–1878
American poet

The tulip-tree, high up,
Opened, in airs of June, her multitude
Of golden chalices to humming-birds

And silken-winged insects of the sky.
Poems
The Fountain, Stanza 3
D. Appleton & Company. New York, New York, USA. 1874

Meredith, Owen (Edward Robert Bulwer-Lytton, 1st Earl Lytton) 1831–91
English statesman and poet

Trees that, like the poplar, lift upwards all their boughs, give no shade and no shelter, whatever their height. Trees that most lovingly shelter and shade us, when, like the willow, the higher soar their summits, the lowlier droop their boughs.
What Will He Do with It? (Volume 2)
Book XI, Chapter X, Introductory lines (p. 359)
P.F. Collier & Son. New York, New York, USA. 1902

Burns, Robert 1759–96
English author

Green, slender, leaf-clad holly-boughs
Were twisted graceful', round her brows;
I took her for some Scottish Muse,
By that same token;
And come to stop those reckless vows,
Would soon be broken.
The Complete Poetical Works of Robert Burns
The Vision, Duan First, Stanza 9
Houghton Mifflin Company. Boston, Massachusetts, USA. 1897

Byron, George Gordon, 6th Baron Byron 1788–1824
English Romantic poet and satirist

Dark tree — still sad when others' grief is fled,

The only constant mourner o'er the dead!
The Complete Poetical Works of Byron
The Giaour, l. 286
Houghton Mifflin Company. Boston, Massachusetts, USA. 1933

Campbell, Thomas 1777–1844
Scottish poet

Oh, leave this barren spot to me!
Spare, woodman, spare the beechen tree!
The Complete Poetical Works
The Beech-Tree's Petition, Stanza I
Chadwyck-Healey. Cambridge, England. 1992

Comstock, Anna Botsford 1854–1930
American illustrator, writer, and educator

The mortal who has never enjoyed a speaking acquaintance with some individual tree is to be pitied; for such an acquaintance, once established, naturally ripens into a friendliness that brings serene comfort to the human heart, whatever the heart of the tree may or may not experience. To those who know them, the trees, like other friends, seem to have their periods of reaching out for sympathetic understanding. How often this outreaching is met with repulse will never be told; for tree friends

never reproach us — but wait with calm patience for us to grow into comprehension.
Trees at Leisure
Comstock Publishing Company. Ithaca, New York, USA. 1916

Dampier-Whetham, William   1867–1952
English scientific writer

There was a young man who said, "God
To you it must seem very odd
That a tree as a tree simply ceases to be
When there's no one about in the Quad."…
Young man, your astonishment's odd,
I am always about in the Quad
And that's why the tree continues to be
As observed by, Yours faithfully, God.
In Joseph Needham and Walter Pagel (eds.)
Background to Modern Science
From Aristotle to Galileo (pp. 40–41)
The Macmillan Company. New York, New York, USA. 1938

de Saint-Exupéry, Antoine   1900–44
French aviator and writer

Now there were some terrible seeds on the planet that was the home of the little prince; and these were the seeds of the baobab. The soil of that planet was infested with them. A baobab is something you will never, never be able to get rid of if you attend to it too late. It spreads over the entire planet. It bores clear through it with its roots. And if the planet is too small, and the baobabs are too many, they split it to pieces…
Translated by Katherine Woods
The Little Prince
Chapter V (p. 21)
Harcourt, Brace & Company. New York, New York, USA. 1943

Dickens, Charles   1812–70
English novelist

They whirled past the dark trees, as feathers would be swept before a hurricane. Houses, gates, churches, hay-stacks, objects of every kind they shot by, with a velocity and noise like roaring waters suddenly let loose. Still the noise of pursuit grew louder, and still my uncle could hear the young lady wildly screaming, "Faster! Faster!"
The Posthumous Papers of the Pickwick Club
Chapter XLIX (p. 597)
Dodd, Mead & Company. New York, New York, USA. 1944

The earth covered with a sable pall as for the burial of yes-terday; the clumps of dark trees, its giant plumes of funeral feathers, waving sadly to and fro: all hushed, all noiseless, and in deep repose, save the swift clouds that skim across the moon, and the cautious wind, as, creeping after them upon the ground, it stops to listen, and goes rustling on, and stops again, and follows, like a savage on the trail.
Martin Chuzzlewit
Chapter XV (p. 232)
Dodd, Mead & Company. New York, New York, USA. 1944

Douglas, Andrew Ellicott   1867–1962
American astronomer

By translating the story told by tree rings, we have pushed back the horizons of history in the United States for nearly eight centuries before Columbus reached the shores of the New World…
The Secret of the Southwest Solved by Talkative Tree Rings
National Geographic, Volume 56, Number 6, 1929 (p. 737)

Dryden, John   1631–1700
English poet, dramatist, and literary critic

The monarch oak, the patriarch of the trees,
Shoots rising up, and spreads by slow degrees.
Three centuries he grows, and three he stays
Supreme in state; and in three more decays.
The Poetical Works of Dryden
Tales from Chaucer, Palamon and Arcite, Book III, l. 1058
The Riverside Press. Cambridge, Massachusetts, USA. 1949

English, Thomas Dunn   1819–1902
American lawyer, physician, and poet

That was a day of delight and of wonder,
While lying the shade of the maple-trees under —
He felt the soft breeze at its frolicsome play;
He smelled the sweet odor of newly mown hay…
The Select Poems of Dr. Thomas Dunn English
Under the Trees
Published by private subscription. Newark, New Jersey, USA. 1894

Forster, E. M. (Edward Morgan)   1879–1970
English novelist

What is the good of your stars and trees, your sunrise and the wind, if they do not enter into our daily lives?
Howards End
Chapter XVI (p. 143)
Vintage Books. New York, New York, USA. 1954

The tree rustled. It had made music before they were born, and would continue after their deaths, but its song was of the moment.
Howards End
Chapter XL (p. 315)
Vintage Books. New York, New York, USA. 1954

Hardy, Thomas   1840–1928
English poet and regional novelist

To dwellers in a wood almost every species of tree has its voice as well as its feature. At the passing of the breeze the fir-trees sob and moan no less distinctly than they rock; the holly whistles as it battles with itself; the ash hisses amid its quiverings; the beech rustles while its flat boughs rise and fall.
Under the Greenwood Tree; or The Mellstock Quire
Part the First, Chapter I (p. 3)
Harper & Brothers. New York, New York, USA. 1939

The instinctive act of humankind was to stand and listen, and learn how the trees on the right and the trees on he left wailed or chaunted to each other in the regular antiphonies of a cathedral choir; how hedges and other shapes to leeward then caught the note, lowering it to the tenderest sob; and how the hurrying gust then plunged into the south, to be heard no more.

Far from the Madding Crowd
Chapter 2 (p. 9)
Harper & Row, Publishers. New York, New York, USA. No date

Hawthorne, Nathaniel 1804–64
American novelist and short story writer

And what is more melancholy than the old apple-trees that linger about the spot where once stood a homestead, but where there is now only a ruined chimney rising out of a grassy and weed-grown cellar? They offer their fruit to every wayfarer — apples that are bitter-sweet with the moral of time's vicissitude.

Mosses from an Old Manse: The Procession of Life
The Old Manse (p. 8)
A.L. Burt Company, Publishers. New York, New York, USA. No date

Hay, John
No biographical data available

They [trees] hang on from a past no theory can recover. They will survive us. The air makes their music. Otherwise they live in savage silence, though mites and nematodes and spiders teem at their roots, and though the energy with which they feed on the sun and are able to draw water sometimes hundreds of feet up their trunks and into their twigs and branches calls for a deafening volume of sound.

The Undiscovered Country
Living with Trees (p. 110)
W.W. Norton & Company, Inc. New York, New York, USA. 1981

Hayne, Paul H. 1830–1886
American poet

Where drooping lotos-flowers, distilling balm,
Dream by the drowsy streamlets sleep hath crown'd,
While Care forgets to sigh, and Peace hath balsamed
Pain.

Sonnets, and Other Poems
Pent in this Common Sphere
Harper & Calvo. Charleston, South Carolina, USA. 1857

Heine, Heinrich 1797–1856
German poet

If thou lookest on the lime-leaf,
Thou a heart's form wilt discover;
Therefore are the lindens ever
Chosen seats of each fond lover.

The Book of Songs
New Spring, Number 23, Stanza 3 (p. 110)
The Roycrofters. East Aurora, New York, USA. 1903

A pine tree standeth lonely
On a far norland height:

It slumbereth, while round it
The snow falls thick and white.

The Book of Songs
Lyrical Interlude, Number 34 (pp. 63–64)
The Roycrofters. East Aurora, New York, USA. 1903

Hemans, Felicia D. 1793–1835
English poet

I have looked on the hills of the stormy North,
And the larch has hung all his tassels forth…

The Poetical Works of Mrs. Felicia Hemans
The Voice of Spring, Stanza 3
Crosby, Nichols, Lee & Company. Boston, Massachusetts, USA. 1860

Herbert, George 1593–1633
English metaphysical poet

Great trees are good for nothing but shade.

Outlandish Proverbs
Printed by T. Maxey for T. Garthwait. London, England. 1651

Ingemann, Bernhard S. 1789–1862
Danish poet and novelist

What whispers so strange at the hour of midnight,
From the aspen leaves trembling so wildly?
Why in the lone wood sings it sad, when the bright
Full moon beams upon it so mildly?

In George Barrow
The Songs of Scandinavia and Other Poems and Ballads (Volume 2)
The Aspen
Constable & Company Ltd. London, England. 1923

Ingelow, Jean 1820–97
English poet and novelist

And when I see the chestnut letting
All her lovely blossoms falter down, I think
"Alas the day!"

Poems
The Warbling of Blackbirds
Longmans, Green, Reader & Dyer. London, England. 1867

Irving, Washington 1783–1859
American essayist and short story writer

It was…a fine autumnal day; the sky was clear and serene, and nature wore that rich and golden livery which we always associate with the idea of abundance. The forests had put on their sober brown and yellow, while some trees of the tenderer kind had been nipped by the frosts into brilliant dyes of orange, purple, and scarlet.

Essays from the Sketch Book
The Legend of Sleepy Hollow (p. 55)
Houghton Mifflin Company. Boston, Massachusetts, USA. 1891

Leyden, John 1775–1811
Scottish poet

Beneath a shivering canopy reclined,
Of aspen leaves that wave without a wind,
I love to lie, when lulling breezes stir
The spiry cones that tremble on the fir.

The Poetical Works of Dr. John Leyden
Scenes of Infancy
W.P. Nimmo. London, England. 1875

Longfellow, Henry Wadsworth   1807–82
American poet

Sweet is the air with the budding haws, and the valley stretching for miles below

Is white with blossoming cherry-trees, as if just covered with lightest snow.
The Complete Writings of Henry Wadsworth Longfellow (Volume 5)
Christus, Golden Legend
Part IV (p. 265)
Houghton Mifflin Company. Boston, Massachusetts, USA. 1904–1917

O hemlock-tree!
O hemlock-tree!
how faithful
are thy branches!
Green not alone in summer time,
But in the winter's frost and rime!
O hemlock-tree! O hemlock-tree! how faithful
are thy branches!
The Complete Writings of Henry Wadsworth Longfellow (Volume 6)
The Hemlock Tree
Stanza 1
Houghton Mifflin Company. Boston, Massachusetts, USA. 1904–1917

Lowell, James Russell   1819–91
American poet, critic, and editor

The ash her purple drops forgivingly
And sadly, breaking not the general hush;
The maple's swamps glow like a sunset sea,
Each leaf a ripple with its separate flush;
All round the wood's edge creeps the skirting blaze,
Of bushes low, as when, on cloudy days,
Ere the rain falls, the cautious farmer burns his brush.
The Poetical Works of James Russell Lowell
An Indian-Summer Reverie, 11
Houghton Mifflin Company. Boston, Massachusetts, USA. 1890

The pine is the mother of legends.
The Poetical Works of James Russell Lowell
The Growth of a Legend
Houghton Mifflin Company. Boston, Massachusetts, USA. 1890

Rippling through thy branches goes the sunshine,
Among thy leaves that palpitate forever,
And in the sea, a pining nymph had prisoned
The soul, once of some tremulous inland river,
Quivering to tell her woe, but ah! dumb, dumb forever.
The Poetical Works of James Russell Lowell
The Birch Tree
Houghton, Mifflin and Company. Boston, Massachusetts, USA. 1890

Melville, Herman   1819–91
American novelist

For, as when the red-cheeked, dancing girls, April and May, trip home to the wintry, misanthropic woods; even

the barest, ruggedest, most thunder-cloven old oak will at least send forth some few green sprouts, to welcome such glad-hearted visitants…
In *Great Books of the Western World* (Volume 48)
Moby Dick
Chapter 28 (p. 91)
Encyclopædia Britannica, Inc. Chicago, Illinois, USA. 1952

Milton, John   1608–74
English poet

Awake, the morning shines, and the fresh field
Call us; we lose the prime, to mark how spring
Our tended Plants, how blows the Citron Grove,
What drops the Myrrhe, & what the balmie Reed,
How Nature paints her colours, how the Bee
Sits on the Bloom, extracting liquid sweet.
In *Great Books of the Western World* (Volume 32)
Paradise Lost
Book V, l. 20–25
Encyclopædia Britannica, Inc. Chicago, Illinois, USA. 1952

Moore, Thomas   1779–1852
Irish poet

And the wind, full of wantonness, woos like a lover
The young aspen-trees till they tremble all over.
The Poetical Works of Thomas Moore
Lalla Rookh, Light of the Harem
Lee & Shepard. Boston, Massachusetts, USA. 1873

Morris, George P.
No biographical data available

Woodman, spare that tree!
Touch not a single bough!
In youth it sheltered me,
And I'll protect it now.
Poems
Woodman, Spare that Tree
Charles Scribner's Sons. New York, New York, USA. 1853

Muir, John   1838–1914
American naturalist

When a man plants a tree he plants himself.
Steep Trails
Chapter X (p. 141)
Norman S. Berg, Publisher. Dunwoody, Georgia, USA. 1970

Few are altogether deaf to the preaching of pine trees. Their sermons on the mountains go to our hearts; and if people in general could be got into the woods, even for once, to hear the trees speak for themselves, all difficulties in the way of forest preservation would vanish.
The National Parks and Forest Reservations
Sierra Club Bulletin, Volume 1, Number 7, January 1896

I have seen oaks of many species in many kinds of exposure and soil, but those of Kentucky excel in grandeur all I had ever before beheld. They are broad and dense and green. In the leafy bowers and caves of their ong branches dwell magnificent avenues of shade, and every tree seems to be blessed with a double portion of strong exulting life.

A Thousand Mile Walk to the Gulf
Chapter I (p. 2)
Houghton Mifflin Company. Boston Massachusetts, USA. 1916

We all travel the milky way together, trees and men; but it never occurred to me until this stormday, while swinging in the wind, that trees are travelers, in the ordinary sense. They make many journeys, not extensive ones, it is true; but our own little journeys, away and back again, are only little more than tree wavings, many of them not so much.

Mountains of California
Chapter X (p. 256)
The Century Company. New York, New York, USA. 1911

There is something wonderfully attractive in this king tree, even when beheld from afar, that draws us to it with indescribable enthusiasm; its superior height and massive smoothly rounded outlines proclaiming its character in any company; and when one of the oldest attains full stature on some commanding ridge it seems the very god of the woods.

Our National Parks
Chapter IX (p. 287)
Houghton Mifflin Company. Boston, Massachusetts, USA. 1901

The Big Tree (*Sequoia gigantea*) is Nature's forest masterpiece, and, so far as I know, the greatest of living things.

Our National Parks
Chapter IX (p. 268)
Houghton Mifflin Company. Boston, Massachusetts, USA. 1901

Resolute, consummate, determined in form, always beheld with wondering admiration, the Big Tree always seems unfamiliar, standing alone, unrelated, with peculiar physiognomy, awfully solemn and earnest.

Our National Parks
Chapter IX (p. 272)
Houghton Mifflin Company. Boston, Massachusetts, USA. 1901

…Sequoias, kings of their race, growing close together like grass in a meadow, poised their brave domes and spires in the sky, three hundred feet above the ferns and lilies that enameled the ground; towering serene through the long centuries, preaching God's forestry fresh from heaven.

Our National Parks
Chapter IX (p. 334)
Houghton Mifflin Company. Boston, Massachusetts, USA. 1901

…they never lose their god-like composure, never toss their arms or bow or wave like the pines, but only slowly, solemnly nod and sway, standing erect, making no sign of strife, none of rest, neither in alliance nor at war with the winds, too calmly unconsciously noble and strong to strive with or bid defiance to anything.

Our National Parks
Chapter IX (pp. 283–284)
Houghton Mifflin Company. Boston, Massachusetts, USA. 1901

[The Sugar Pine is] the largest, noblest, and most beautiful of all the seventy or eighty species of pine trees in the world. …

Our National Parks
Chapter IV (p. 109)
Houghton Mifflin Company. Boston, Massachusetts, USA. 1901

The mighty trees getting their food are seen to be wide awake, every needle thrilling in the welcome nourishing storms, chanting and bowing low in glorious harmony, while every raindrop and snowflake is seen as a beneficent messenger from the sky.

Our National Parks
Chapter I (p. 26)
Houghton Mifflin Company. Boston, Massachusetts, USA. 1901

I never saw a discontented tree. They grip the ground as though they liked it; and though fast rooted, they travel about as far as we do.

In Linnie Marsh Wolfe (ed.)
John of the Mountains
Chapter VII, Section 2, June-July 1890 (p. 313)
Houghton Mifflin Company. Boston, Massachusetts, USA. 1938

…many of nature's five hundred kinds of wild trees had to make way for orchards and cornfields.

Our National Parks
Chapter IX (p. 335)
Houghton Mifflin Company. Boston, Massachusetts, USA. 1901

As far as man is concerned [trees] are the same yesterday, today, and forever, emblems of permanence.

Our National Parks
Chapter IX (p. 269)
Houghton Mifflin Company. Boston, Massachusetts, USA. 1901

…God has cared for these trees, saved them from drought, disease, avalanches, and a thousand straining, leveling tempests and floods; but he cannot save them from fools, only Uncle Sam can do that.

Our National Parks
Chapter X (p. 365)
Houghton Mifflin Company. Boston, Massachusetts, USA. 1901

Peattie, Donald Culrose 1896–1964
American botanist, naturalist and author

A Tree in its old age is like a bent but mellowed and wise old man; it inspires our respect and tender admiration; it is too noble to need our pity.

An Almanac for Moderns
October Twenty-Seventh (p. 241)
G.P. Putnam's Sons. New York, New York, USA. 1935

Pope, Alexander 1688–1744
English poet

A spring there is, whose silver waters show
Clear as a glass the shining sands below:

A flowering lotos spreads its arms above,
Shades all the banks, and seems itself a grove.
The Complete Poetical Works
Sappho to Phaon, l. 177
Houghton Mifflin Company. New York, New York, USA. 1903

Pownall, Thomas   1722–1805
English statesman and soldier

The individual Trees of those Woods grow up, have their
Youth, their old Age, and a Period to their Life, and die
as we Men do. You will see many a Sapling growing up,
many an old Tree tottering to its Fall, and many fallen
and rotting away, while they are succeeded by others of
their Kind, just as the Race of Man is: By this Succession of Vegetation this Wilderness is kept cloathed with
Woods just as the human Species keeps the Earth peopled
by its continuing Succession of Generations.
A Topographical Description of the Dominions of the United States
Section I, On the Face of the Country (p. 24)
University of Pittsburgh Press. Pittsburgh, Pennsylvania, USA. 1949

Proust, Marcel   1871–1922
French novelist

We have nothing to fear and a great deal to learn from
trees, that vigorous and pacific tribe which without stint
produces strengthening essences for us, soothing balms,
and in whose gracious company we spend so many cool,
silent and intimate hours.
Translated by Louise Varèse
Pleasures and Regrets
Regrets, Reveries, Changing Skies, Chapter XXVI (p. 165)
Crown Publishers. New York, New York, USA. 1948

Shelley, Mary   1797–1851
English Romantic writer

But I am a blasted tree; the bolt has entered my soul; and
I felt then that I should survive to exhibit what I shall
soon cease to be — a miserable spectacle of wrecked hu-
manity, pitiable to others and intolerable to myself.
Frankenstein
Chapter 19 (p. 114)
Running Press. Philadelphia, Pennsylvania, USA. 1990

Spenser, Edmund   1552–99
English poet

Like to an almond tree mounted hye
On top of greene Selinis all alone,
With blossoms brave bedecked daintily;
Whose tender locks do tremble every one,
At everie little breath, that under heaven is blowne.
The Complete Poetical Works of Edmund Spenser
The Faerie Queene
Book I, Canto VII, Stanza 32
Houghton Mifflin Company. Boston, Massachusetts, USA. 1908

St. Bernard of Clairvaux   1091–1153
French monk

Believe me who have tried. Thou wilt find something
more in woods than in books. Trees and rocks will teach
what thou canst not hear from a master.
Epistle 106
Source undetermined

Steinbeck, John   1902–68
American novelist

The redwoods once seen, leave a mark or create a vision
that stays with you always…. It's not only their unbelievable stature, nor the color which seems to shift and
vary under your eyes, no, they are not just like any trees
we know, they are ambassadors from another time.
Travels with Charley: In Search of America
Part Three (p. 168)
The Viking Press. New York, New York, USA. 1962

Sterling, John   1808–44
Irish-born writer and clergyman

The Spice Tree lives in the garden green,
Beside it the fountain flows;
And a fair Bird sits the boughs between,
And sings his melodious woes.
Poems
The Spice Tree, Stanza 1
Edward Moxon. London, England. 1839

Taylor, Bayard   1825–78
American journalist and author

Ancient Pines,
Ye bear no record of the years of man.
Spring is your sole historian…
The Poetical Works of Bayard Taylor
The Pine Forest of Monterey, Stanza 4
Houghton, Osgood. Boston, Massachusetts, USA. 1880

Tennyson, Alfred (Lord)   1809–92
English poet

In crystal vapour everywhere
Blue isles of heaven laugh'd between,
And far, in forest-deeps unseen,
The topmost elm-tree gather'd green
From draughts of balmy air.
Alfred Tennyson's Poetical Works
Sir Lancelot and Queen Guinevere, Stanza I
Oxford University Press, Inc. London, England. 1953

Thackeray, William Makepeace   1811–63
English writer

Know ye the willow-tree,
Whose grey leaves quiver,
Whispering gloomily
To yon pale river?
The Complete Poems of W.M. Thackeray
The Willow-Tree
White, Stokes & Allen. New York, New York, USA. 1884

Christmas is here;

Winds whistle shrill,
Icy and chill,
Little care we;
Little we fear
Weather without,
Sheltered about
The Mahogany-Tree.
The Complete Poems of W.M. Thackeray
The Mahogany-Tree
White, Stokes & Allen. New York, New York, USA. 1884

The Bible

I shall plant cedar in the wilderness, acacias, myrtle, and wild olives; I shall grow pines on the barren heath side by side with fir the box tree…
The Revised English Bible
Isaiah 41:19
Oxford University Press, Inc. Oxford, England. 1989

Thoreau, Henry David   1817–62
American essayist, poet, and practical philosopher

It is remarkable how closely the history of the apple tree is connected with that of man.
The Writings of Henry David Thoreau (Volume 9)
Wild Apples (p. 356)
Houghton Mifflin Company. Boston, Massachusetts, USA. 1893

Twain, Mark (Samuel Langhorne Clemens)   1835–1910
American author and humorist

I once heard a grouty Northern invalid say that a coconut tree might be poetical, possibly it was; but it looked like a feather-duster struck by lightning.
Roughing It (Volume 2)
Chapter XVIII (p. 215)
Harper & Brothers. New York, New York, USA. 1899

Wordsworth, William   1770–1850
English poet

Of vast circumference and gloom profound
This solitary Tree! a living thing
Produced too slowly ever to decay;
Of form and aspect too magnificent
To be destroyed.
The Complete Poetical Works of William Wordsworth
Yew-Trees
Crowell. New York, New York, USA. 1888

TREE OF LIFE

Mason, Frances
No biographical data available

Evolution does not move in a straight course, symbolized by the links in a chain; the tree is the symbol of nature's plan of creation. The trunk represents the main course of life through the ages; the branches are the great groups of plants and animals that have appeared during the growth of the tree; the plants and animals now living are the green twigs at the tips of the branches. In the evolution of forms there are no offshoots leading from one branch to another; the branches start from below and diverge as they grow, each branch maintaining its own course.
In Frances Mason
Creation by Evolution
Frontispiece (p. ii)
The Macmillan Company. New York, New York, USA. 1928

TREE RINGS

Burroughs, John   1837–1921
American naturalist and writer

An old tree, unlike an old person, as long as it lives at all, always has a young streak, or rather ring, in it. It wears a girdle of perpetual youth.
Studies in Nature and Literature
Bird Life in an Old Apple Tree (p. 38)
Houghton Mifflin Company. Boston, Massachusetts, USA. 1908

Leopold, Aldo   1886–1948
American naturalist

We sensed that these two piles of sawdust were something more than wood: that they were the integrated transect of a century; that our saw was biting its way, stroke by stroke, decade by decade, into the chronology of a lifetime, written in concentric annual rings of oak.
A Sand County Almanac, with Essays on Conservation from Round River
Part I, February (p. 10)
Sierra Club. San Francisco, California, USA. 1970

TRIAL AND ERROR

Born, Max   1882–1970
German-born English physicist

…I believe that there is no philosophical highroad in science, with epistemological signposts. No, we are in a jungle and find our way by trial and error, building our road behind us as we proceed. We do not find signposts at crossroads, but our own scouts erect them, to help the rest.
Experiment and Theory in Physics (p. 44)
Cambridge University Press. Cambridge, England. 1944

TRIANGLE

Beckett, Samuel   1906–89
Irish playwright

…do not despair. Remember there is no triangle, however obtuse, but the circumference of some circle passes through its wretched vertices.

The Collected Works of Samuel Beckett
Murphy
Chapter 10 (p. 213)
Grover Press, Inc., New York, New York, USA. 1970

Creele, August   1780–1856

German civil engineer and mathematician

It is indeed wonderful that so simple a figure as the triangle is so inexhaustible in properties. How many as yet unknown properties of other figures may there not be?
School Science and Mathematics (p. 672)
School Science and Mathematical Association, 1905

TRIGONOMETRY

Chesterton, G. K. (Gilbert Keith)   1874–1936
English author

A straight liner is straight
And a square mile is flat:

But you learn in trigonometrics a trick worth two of that.
The Collected Poems of G.K. Chesterton
Songs of Education, V, The Higher Mathematics (p. 97)

Howell, Scott   1959–
American conservative political consultant

As long as schools continue to teach trigonometry and algebra, there will always be a moment of silence, and indeed prayer, in our public schools.
On why he sees no need to formalize a moment of silence in Utah schools

Philips, J. D.
No biographical data available

The notion that anyone other than a scientist will ever use even the most elementary trigonometry or algebra is laughable. Imagine the absurdity of being in a car or on a plane when suddenly the need arises to solve a quadratic equation or to graph a trigonometric function. But this is precisely the scenario that the traditional defense has coerced us into accepting as realistic. Clearly this is absurd. And so is our complicity.
Mathematics as an Aesthetic Discipline
Humanistic Mathematics Network Journal, Number 12, October 1995

TRILOBITE

Conrad, Timothy   1803–77
American geologist and malacologist

The race of man shall perish, but the eyes
Of trilobites eternal be in stone,
And seem to stare about in mild surprise
At changes greater than they have yet known.

A Geological Vision and Other Poems
Murphy & Bechtel. Trenton, New Jersey, USA. 1871

Howell, G. K.
No biographical data available

Thou man of hammer and the disreputable trilobite I have some what to say unto thee. The hammer is an honest instrument that advertises what it does when it smashes — but for the trilobite ah what shall I say? I say that an animal that used 20 000 eyes must have been essentially a sneak! — not the one to meet a foe squarely but one that would be peeking around in all directions out of some if its headlights to be ready to run at the first sign of an adversary.
In Ellis L. Yochelson
Charles Doolittle Walcott, Paleontologist
Letter to Walcott, October 31, 1879 (p. 118)
The Kent State University Press. Kent, Ohio, USA. 1998

And then you never know how to class [a trilobite] — he wasn't a mollusk or a fish, and he wasn't a bird nor again an honest square reptile like the gay alligator. And he wasn't an Englishman — well perhaps you don't have Pinafore out among the Utes and the prairie dogs.
In Ellis L. Yochelson
Charles Doolittle Walcott, Paleontologist
Letter to Walcott, October 31, 1879 (p. 118)
The Kent State University Press. Kent, Ohio, USA. 1998

Levi-Setti, Riccardo
No biographical data available

Trilobites tell me of ancient marine shores teeming with budding life, when silence was only broken by the wind, the breaking of the waves, or by the thunder of storms and volcanoes. The struggle for survival already had its toll in the seas, but only natural laws and events determined the fate of evolving life forms. No footprints were to be found on those shores, as life had not yet conquered land. Genocide had not been invented as yet, and the threat to life on Earth resided only with the comets and asteroids.
Trilobites
Preface (p. vii)
The University of Chicago Press. Chicago, Illinois, USA. 1993

All fossils are, in a way, time capsules that can transport our imagination to unseen shores, lost in the sea of eons that preceded us. The time of trilobites is unimaginably far away, and yet, with relatively little effort, we can dig out these messengers of our past and hold them in our hand. And, if we learn the language, we can read the message.
Trilobites
Preface (p. vii)
The University of Chicago Press. Chicago, Illinois, USA. 1993

Newman, Joseph S.   1892–1960
American poet

A million years ago, or six...perhaps as much as seven,
When rhizopods were spewing forth the chalky cliffs of Devon,
Upon a cool and mossy rock, beneath a bed of sedum,
A trilobite named Annie lived in trilobitish freedom.
Poems for Penguins and Other Lyrical Lapses
The Trilobite
Greenburg. New York, New York, USA. 1941

TRUTH

Abbey, Edward   1927–89
American environmentalist and nature writer

...I am sometimes forced to the conclusion that the whole truth is not always represented in certain of the orthodox attitudes. The intuitions of a lover are not always to be trusted; but neither are those of the loveless.
In Joseph Wood Krutch
The Great Chain of Life
Prologue (p. xi)
Houghton Mifflin Company. Boston, Massachusetts, USA. 1957

Adams, George   1750–95
English instrument maker

Truth, though destined to be the guide of man, is not bestowed with an unconditional profusion; but is hidden in darkness, and involved in difficulties; intended, like all the other gifts of heaven, to be fought and cultivated by all the different powers and exertions of human reason.
Lectures on Natural and Experimental Philosophy (Volume 1)
Lecture II (p. 62)
Printed by R. Hindmarsh. London, England. 1794

You should, therefore, set out in the search of truth as of a stranger, not in search of arguments to support as of a stranger, not in search of arguments to support your own opinions, and endeavor to maintain your mind in a state of equilibrium, an indifference for everything but known and well attested truth, totally regardless of the place from whence it comes, or that to which it tends...
Lectures on Natural and Experimental Philosophy (Volume 1)
Lecture II (p. 28)
Printed by R. Hindmarsh. London, England. 1794

Aristotle   384 BCE–322 BCE
Greek philosopher

The investigation of the truth is in one way hard, in another easy. An indication of this is found in the fact that no one is able to attain the truth adequately, while, on the other hand, we do not collectively fail, but every one says something true about the nature of things, and while individually we contribute little or nothing to the truth, by the union of all a considerable amount is amassed. Therefore, since the truth seems to be like the proverbial door, which no one can fail to hit, in this respect it must be easy, but the fact that we can have a whole truth and not the particular part we aim at shows the difficulty of it.
In *Great Books of the Western World* (Volume 8)
Metaphysics
Book II, Chapter I (p. 511)
Encyclopædia Britannica, Inc. Chicago, Illinois, USA. 1952

Aronowitz, Stanley   1933–
American sociologist, labor/union advocate, and writer

The power of science consists, in the first place, in its conflation of knowledge and truth. Devising a method of proving the validity of propositions about objects taken as external to the knower has become identical with what we mean by truth.
Science as Power: Discourse and Ideology in Modern Society
Preface (p. vii)
University of Minnesota Press. Minneapolis, Minnesota, USA. 1988

Avedon, Richard   1923–2004
American photographer

The moment an emotion or fact is transformed into a photograph it is no longer fact but an opinion. There is no such thing as inaccuracy in a photograph. All photographs are accurate. None of them is the truth.
The Chronicle of Higher Education, July 10, 1991, (p. B2)

Bacon, Sir Francis   1561–1626
English lawyer, statesman, and essayist

The human understanding resembles not a dry light, but admits a tincture of the will and passions, which generate their own system accordingly; for man always believes more readily that which he prefers.
In *Great Books of the Western World* (Volume 30)
Novum Organum
First Book, Aphorism 49 (p. 111)
Encyclopædia Britannica, Inc. Chicago, Illinois, USA. 1952

Balfour, Arthur James   1848–1930
British prime minister

It is not by mere accumulation of material, nor even by a plant-like development, that our beliefs grow less inadequate to the truths which they strive to represent. Rather we are like one who is perpetually engaged in altering some ancient dwelling in order to satisfy new-born needs. The ground-plan of it is being perpetually modified. We build here; we pull down there. One part is kept in repair, another part is suffered to decay. And even those portions of the structure which may in themselves appear quite unchanged, stand in such new relations to the rest, and are put to such different uses, that they would scarce be recognized by their original designer.
The Foundations of Belief
Appendix, Section I (p. 350)
Longmans, Green & Company. London, England. 1912

Barfield, Owen   1898–1997
British philosopher, critic, and anthroposophist

It was not simply a new theory of the nature of the celestial movements that was feared, but a new theory of the nature of theory; namely, that, if a hypothesis saves all the appearances, it is identical with truth.

Saving the Appearances: A Study in Idolatry
Chapter VII (pp. 50–51)
Faber & Faber Ltd. London, England. 1957

Beaumont, William   1785–1853
American army surgeon

Truth, like beauty, when "unadorned, is adorned the most"; and in prosecuting these experiments and inquiries, I believe I have been guided by its light.

In William Osler
Aequanimitas, with Other Addresses to Medical Students, Nurses, and Practitioners of Medicine
The Army Surgeon (p. 113)
The Blakiston Company. Philadelphia, Pennsylvania, USA. 1932

Becker, Ernest   1925–74
Canadian anthropologist

The man of knowledge in our time is bowed under a burden he never imagined he would ever have: the overproduction of truth that cannot be consumed. For centuries man lived in the belief that truth was slim and elusive and that once he found it the troubles of mankind would be over. And here we are in the closing of the 20th century, choking on truth.

The Denial of Death
Preface (p. x)
The Free Press. New York, New York, USA. 1973

Bernard, Claude   1813–78
French physiologist

It seems, indeed, a necessary weakness of our mind to be able to reach truth only across a multitude of errors and obstacles.

Translated by Henry Copley Greene
An Introduction to the Study of Experimental Medicine
Part Three, Chapter I, Section ii (p. 170)
Henry Schuman, Inc. New York, New York, USA. 1927

Men of science, then, do not seek for the pleasure of seeking; they seek the truth to possess it, and they possess it already within the limits in the present state of the sciences. But men of science must not halt on the road; they must climb ever higher and strive toward perfection; they must always seek, as long as they see anything to be found.

Translated by Henry Copley Greene
An Introduction to the Study of Experimental Medicine
Part Three, Chapter III, Section iv (p. 222)
Henry Schuman, Inc. New York, New York, USA. 1927

Bohr, Niels Henrik David   1886–1962
Danish physicist

Truth lies in the abyss.

In Gerald Holton
Scientific Optimism and Societal Concerns
A Note of the Psychology of Scientists (p. 83)
Publisher undetermined

The opposite of a correct statement is a false statement. But the opposite of a profound truth may well be another profound truth.

In Werner Heisenberg
Physics and Beyond: Encounters and Conversations
Chapter 8 (p. 102)
Harper & Row, Publishers. New York, New York, USA. 1971

Born, Max   1882–1970
German-born English physicist

Truth is what the scientist aims at. He finds nothing at rest, nothing enduring, in the universe. Not everything is knowable, still less is predictable. But the mind of man is capable of grasping and understanding at least a part of Creation; amid the flight of phenomena stands the immutable pole of law.

The Restless Universe
Chapter V (p. 278)
Dover Publications, Inc. New York, New York, USA. 1951

My optimistic enthusiasm about the disinterested search for truth has been severely shaken. ...

The Restless Universe
Postscript (p. 279)
Dover Publications, Inc. New York, New York, USA. 1951

Bronowski, Jacob   1908–74
Polish-born British mathematician and polymath

Truth in science is like Everest, an ordering of the facts.

Science and Human Values
The Sense of Human Dignity (p. 52)
Harper & Row, Publishers. New York, New York, USA. 1965

We cannot define truth in science until we move from fact to law. And within the body of laws in turn, what impresses us as truth is the orderly coherence of the pieces. They fit together like the characters of a great novel, or like the words of a poem. Indeed, we should keep that last analogy by us always, for science is a language, and like a language it defines its parts by the way they make up a meaning. Every word in a sentence has some uncertainty of definition, and yet the sentence defines its own meaning and that of its words conclusively. It is the internal unity and coherence of science which gives it truth, and which makes it a better system of prediction than any less orderly language.

The Common Sense of Science
Chapter VIII, Section 5 (p. 131)
Harvard University Press. Cambridge, Massachusetts, USA. 1953

Brown, John   1810–82
Scottish physician and author

You may come to the chest of knowledge. It is shut, it is bolted, but…you have the key; put it in steadily and home. But what is the key? It is the love of truth; neither more or less; no other key opens it; no false one, however cunning can pick that lock; no assault of hammer, however stout, can force it open; but with its own key, a little child may open it; often does open it.

In Sir Richard Arman Gregory
Discovery; or, The Spirit and Service of Science
Chapter II (p. 28)
Macmillan & Company Ltd. London, England. 1918

Bush, Vannevar 1890–1974
American electrical engineer and physicist

It is a great truth of science that every ending is a beginning, that each question answered leads to new problems to solve, that each opportunity grasped and utilized engenders fresh and greater opportunities.

In Helen Wright
Palomar: The World's Largest Telescope
Dedication of the Hale Telescope (p. 183)
The Macmillan Company. New York, New York, USA. 1952

Calvin, Melvin 1911–97
American biochemist

The true student will seek evidence to establish fact rather than confirm his own concept of truth, for truth exists whether it is discovered or not.

Chemical Evolution
Chapter 11 (p. 252)
Oregon State System of Higher Education. Eugene, Oregon, USA. 1961

Carmichael, Robert Daniel 1879–1967
American mathematician

He who discovers a fact or makes known a new law or adds a novel beauty to truth in any way makes every one of us his debtor. How beautiful upon the highway are the feet of him who comes bringing in his hands the gift of a new truth to mankind.

The Logic of Discovery
Chapter IX (p. 273)
The Open Court Publishing. Chicago, Illinois, USA. 1930

Chandler, Raymond Thornton 1888–1959
American novelist

There are two kinds of truth: the truth that lights the way and the truth that warms the heart. The first of these is science, and the second is art.… With art science would be as useless as a pair of high forceps in the hands of a plumber. Without science art would become a crude mess of folklore and emotional quackery.

The Notebooks of Raymond Chandler
Great Thought (p. 7)
Ecco Press. New York, New York, USA. 1976

Chargaff, Erwin 1905–2002
Austrian biochemist

…I prefer the search for the truth to its possession.

Serious Questions
Knowledge Industry (p. 111)
Birkhäuser. Boston, Massachusetts, USA. 1986

Charlie Chan
Fictional character

Truth, like football, receive many kicks before reaching goal.

Charlie Chan at the Olympics
Film (1937)

Clarke, Arthur C. 1917–
English science and science fiction writer

Faiths come and go, but Truth abides. Out there among the stars lie such truths as we may understand, whether we learn them by our own efforts, or from the strange teachers who are waiting for us along the infinite road on which our feet are now irrevocably set.

The Challenge of the Spaceship
Of Space and the Spirit (p. 212)
Harper & Brothers. New York, New York, USA. 1959

Cole, William 1530–1600
English man of letters

Whoever attempts to erect a building, should take care that the foundation be securely laid; so also in our inquiries after truth, all our proceedings should be founded upon just and incontrovertible grounds.

Philosophical Remarks on the Theory of Comets, a Dissertation on the Nature and Properties of Light
Introduction (p. xi)
B.J. Holdsworth. London, England. 1823

Compton, Arthur H. 1892–1962
American physicist

The truths that science teaches are of common interest the world over. The language of science is universal, and is a powerful force in bringing the peoples of the world closer together. We are all acquainted with the sharp divisions which religions draw between men. In science there are no such divisions: all peoples worship at the shrine of truth.

Les Prix Nobel. The Nobel Prizes in 1927
Nobel banquet speech for award received in 1927
Nobel Foundation. Stockholm, Sweden. 1928

Cornforth, John W. 1917–2004
English organic chemist

…truth is so seldom the sudden light that shows new order and beauty; more often, truth is the uncharted rock that sinks [a] ship in the dark.

Nobel Banquet Chemistry 1975
Speech

D'Alembert, Jean Le Rond 1717–83
French mathematician

Geometrical truths are in a way asymptotes to physical truths, that is to say, the latter approach the former indefinitely near without ever reaching them exactly.
In Alphonse Rebiére
Mathematiques et Mathematiciens: Pensées et Curiosites (p. 10)

Darwin, Charles Robert   1809–82
English naturalist

The truth will not penetrate a preoccupied mind.
In Francis Darwin (ed.)
More Letters of Charles Darwin (Volume 1)
Letter 222, Darwin to Hooker, July 28, 1868 (p. 305)
D. Appleton & Company. New York, New York, USA. 1903

Davy, Sir Humphry   1778–1829
English chemist

To explain nature and the laws instituted by the Author of nature and to apply the phenomena presented in the external world to useful purposes are the great ends of physical investigation, and these ends can only be obtained by the exertion of all the faculties of the mind. And the imagination, the memory, and the reason are perhaps equally essential to the development of great and important truths.
Humphry Davy on Geology: The 1805 Lectures for the General Audience
Lecture Four (p. 58)
The University of Wisconsin Press. Madison, Wisconsin, USA. 1980

de Fontenelle, Bernard le Bovier   1657–1757
French author

Truth enters so naturally into the mind, that when we learn any thing for the first time, it appears as if we only remembered the thing learned, or exerted the faculty of our memory.
Conversations on the Plurality of Worlds
The Second Evening (p. 70)
Printed for Peter Wilson. Dublin, Ireland. 1761

Descartes, René   1596–1650
French philosopher, scientist, and mathematician

…we must believe that all the sciences are so interconnected, that it is much easier to study them all together than to isolate one from all the others. If, therefore, anyone wishes to search out the truth of things in serious earnest, he ought not to select one special science; for all the sciences are cojoined with each other and interdependent.…
In *Great Books of the Western World* (Volume 31)
Rules for the Direction of the Mind
Rule 1 (p. 1)
Encyclopædia Britannica, Inc. Chicago, Illinois, USA. 1952

Dewey, John   1859–1952
American philosopher and educator

There is but one sure road of access to truth — the road of cooperative inquiry operating by means of observation, experiment, record, and controlled reflection.
Common Faith
Chapter II (p. 32)
Yale University Press. New Haven, Connecticut, USA. 1934

Drake, Daniel   1785–1852
American physician

The love of pleasure and the love of science may coexist, but cannot be indulged at the same time; though in fact they are seldom united. A student should draw his pleasures from the discovery of truth, and find his amusements in the beauties and wonders of nature. He should seek for recreation not debauchery
Physician to the West (p. 298)
University Press of Kentucky. Lexington, Kentucky, USA. 1970

Dumas, Jean Baptiste-Andre   1800–84
French biochemist

Truth is so beautiful that it deserves every effort a man can bestow to attain it; it is so fruitful that it carries along with it its own recompense. By keeping the end in view, without occupying ourselves with particulars, we find the ordinary details of prosperity and riches fall into their proper places.
In Faraday Lectures
Lectures Delivered Before the Chemical Society
The First Faraday Lecture (p. 3)
The Chemical Society. London, England. 1928

Eddington, Sir Arthur Stanley   1882–1944
English astronomer, physicist, and mathematician

An addition to knowledge is won at the expense of an addition to ignorance. It is hard to empty the well of Truth with a leaky bucket.
The Nature of the Physical World
Chapter X (p. 229)
The Macmillan Company. New York, New York, USA. 1930

In science as in religion the truth shines ahead as a beacon showing us the path; we do not ask to attain it; it is better far that we be permitted to seek.
Science and the Unseen World
Chapter II (p. 23)
The Macmillan Company. New York, New York, USA. 1929

Accidental truth of a conclusion is no compensation for erroneous deduction.
Space, Time and Gravitation: An Outline of the General Relativity Theory
Chapter I (p. 29)
At The University Press. Cambridge, England. 1921

Einstein, Albert   1879–1955
German-born physicist

Truth is what stands the test of experience.
In Philipp Frank
Relativity — A Richer Truth
The Laws of Science and the Laws of Ethics (p. 10)
Jonathan Cape. London, England. 1951

The search for truth is more precious than its possession.
The American Mathematical Monthly, Volume 100, Number 3, March 1993 (p. 254)

As for the search for truth, I know from my own painful searching, with its many blind alleys, how hard it is to take a reliable step, be it ever so small, towards the understanding of that which is truly significant.
In Helen Dukas and Banesh Hoffman
Albert Einstein: The Human Side: New Glimpses from His Archives
Letter dated 13 February 1934 (p. 18)
Princeton University Press. Princeton, New Jersey, USA. 1979

It is the most beautiful reward for one who has striven his whole life to grasp some little bit of truth if he sees that other men have real understanding of and pleasure with his work.
In Helen Dukas and Banesh Hoffman
Albert Einstein: The Human Side: New Glimpses from His Archives
Letter dated 9 December, 1952 (p. 29)
Princeton University Press. Princeton, New Jersey, USA. 1979

But the years of searching in the dark for a truth that one feels, but cannot express; the intense desire and the alternations of confidence and misgiving, until one breaks through to clarity and understanding, are only known to him who has himself experienced them.
In Ronald W. Clark
Einstein: The Life and Times
Part Five, Chapter 21 (p. 590)
The World Publishing Company. New York, New York, USA. 1971

I want to know how God created this world. I am not interested in this or that phenomenon, in the spectrum of this or that element. I want to know His thoughts, the rest are details.
In Ronald W. Clark
Einstein: The Life and Times
The World Publishing Company. New York, New York, USA. 1971

Eliot, George (Mary Ann Evans Cross)   1819–80
English novelist

Approximate truth is the only truth attainable, but at least one must strive for that, and not wade off into arbitrary falsehood.
The George Eliot Letters (Volume 4) (p. 43)
Yale University Press. New Haven, Connecticut, USA. 1954–1978

Errera, Leo   1858–1905
Belgian botanist

Truth is on a curve whose asymptote our spirit follows eternally.
In J.A. Thomson
Introduction to Science
Chapter V (p. 125)
Williams & Norgate Ltd. London, England. 1916

Esquivel, Laura   1951?–
Mexican novelist

Anything could be true or false, depending on whether one believed it.
Like Water for Chocolate
July (p. 127)
Doubleday & Company, Inc. New York, New York, USA. 1989

Everett, Edward   1794–1865
American statesman, educator, and orator

In the pure mathematics we contemplate absolute truths, which existed in the Divine Mind before the morning stars sang together, and which will continue to exist there, when the last of their radiant host shall have fallen from heaven.
In E.T. Bell
Mathematics: Queen and Servant of Science
Mathematical Truth (p. 21)
McGraw-Hill Book Company, Inc. New York, New York, USA. 1951

Feynman, Richard P.   1918–88
American theoretical physicist

We've learned from experience that the truth will come out. Other experimenters will repeat your experiment and find out whether you were wrong or right. Nature's phenomena will agree or they'll disagree with your theory. And, although you may gain some temporary fame and excitement, you will not gain a good reputation as a scientist if you haven't tried to be very careful in this kind of work. And it's this type of integrity, this kind of care not to fool yourself, that is missing to a large extent in much of the research in cargo cult science.
Surely You're Joking, Mr. Feynman!: Adventures of a Curious Character
(Caltech commencement address, 1974) Cargo Cult Science (p. 342)
W.W. Norton & Company, Inc. New York, New York, USA.1985

It is possible to know when you are right way ahead of checking all the consequences. You can recognize truth by its beauty and simplicity.
The Character of Physical Law
Chapter 7 (p. 171)
BBC. London, England. 1965

Fourcroy, Antoine-François   1755–1809
French chemist

The general truths in any science are continually multiplied, as its perfection advances, and its means of investigation are improved. Such has been the fortune of chemistry.
Translated by R. Heron
Elements of Chemistry and Natural History (Volume 1)
Advertisement (p. 1)
Printed for G. Mudie & Son. Edinburgh, Scotland. 1796

Frederick the Great   1712–86
German king

The greatest and noblest pleasure which men can have in this world is to discover new truths; and the next is to shake off old prejudices.

In Lloyd William Taylor
Physics: The Pioneer Science (Volume 2)
Chapter 47 (p. 729)
Houghton Mifflin Company. Boston, Massachusetts, USA. 1941

Galilei, Galileo   1564–1642
Italian physicist and astronomer

Two truths cannot contradict one another.
Translated by Stillman Drake
Discoveries and Opinions of Galileo
Letter to Madame Christina of Lorraine (p. 186)
Doubleday & Company, Inc. New York, New York, USA. 1957

Galilei, Vincenzio   1520–1591
Father of Galileo Galilei

It appears to me that they who in proof of any assertion rely simply on the weight of authority, without adducing any argument in support of it, act very absurdly. I, on the contrary, wish to be allowed freely to question and freely to answer you without any sort of adulation, as well becomes those who are in search of truth.
In John Joseph Fahie
Galileo
Chapter I (p. 3)
John Murray. London, England. 1903

Gauss, Johann Carl Friedrich   1777–1855
German mathematician, physicist, and astronomer

In the Theory of Numbers it happens rather frequently that, by some unexpected luck, the most elegant new truths spring up by induction.
In G. Polya
Induction and Analogy in Mathematics (Volume 1)
Chapter IV (p. 59)
Princeton University Press. Princeton, New Jersey, USA. 1954

Gore, George   1826–1909
English electrochemist

The deepest truths require still deeper truths to explain them.
The Art of Scientific Discovery
Chapter III (p. 26)
Longmans, Green & Company. London, England. 1878

Gray, George W.
American free lance science writer

No truth is sacrosanct. No belief is too generally accepted, too well established by experiment, to escape the challenge of doubt. And no doubt is too radical to receive a hearing if it is seriously proposed.
The Riddle of Our Reddening Skies
Harper's Monthly Magazine, July 1937 (p. 169)

Gregory, Sir Richard Arman   1864–1952
British science writer and journalist

In the pursuit of truth the man of science spends his days; and for the defense of truth he is prepared to stand against the world.
Discovery; or, The Spirit and Service of Science
Chapter II (p. 24)
Macmillan & Company Ltd. London, England. 1918

A truthful mind is necessary for the discovery of truth in Nature.
Discovery; or, The Spirit and Service of Science
Chapter II (p. 25)
Macmillan & Company Ltd. London, England. 1918

Halmos, Paul R.   1916–2006
Hungarian-born American mathematician

The joy of suddenly learning a former secret and the joy of suddenly discovering a hitherto unknown truth are the same to me — both have the flash of enlightenment, the almost incredibly enhanced vision, and the ecstasy and euphoria of released tension.
I Want to Be a Mathematician
Chapter 1 (p. 3)
Springer-Verlag. New York, New York, USA. 1985

Heaviside, Oliver   1850–1925
English electrical engineer, mathematician, and physicist

We do not dwell in the Palace of Truth. But, as was mentioned to me not long since, "There is a time coming when all things shall be found out." I am not so sanguine myself, believing that the well in which Truth is said to reside is really a bottomless pit.
Electromagnetic Theory
Chapter I, Volume 1 (p. 1)
"The Electrician" printing & publishing company. London, England. 1894–1912

Heinlein, Robert A.   1907–88
American science fiction writer

The hardest part about gaining any new idea is sweeping out the false idea occupying that niche. As long as that niche is occupied, evidence and proof and logical demonstration get nowhere. But once the niche is emptied of the wrong idea that has been filling it — once you can honestly say, "I don't know," then it becomes possible to get at the truth.
The Cat Who Walks Through Walls: A Comedy of Manners
Chapter XVIII (p. 230)
G.P. Putnam's Sons. New York, New York, USA.1985

Herschel, Sir John Frederick William   1792–1871
English astronomer and chemist

It is only when we are wandering and lost in the mazes of particulars, or entangled in fruitless attempts to work our way downwards in the thorny paths of applications, to which our reasoning powers are incompetent, that nature appears complicated: — the moment we contemplate it as

it is, and attain a position from which we can take a commanding view, though but of a small part of its plan, we never fail to recognise that sublime simplicity on which the mind rests satisfied that it has attained the truth.
The Cabinet of Natural Philosophy
Part III, Chapter VI, Section 393 (pp. 360–361)
Longman, Rees, Orme, Brown & Green. London, England. 1831

Holmes, Oliver Wendell 1809–94
American physician, poet, and humorist

Every probability — and most of our common, working beliefs are probabilities — is provided with buffers at both ends, which break the force of opposite opinions clashing against it; but scientific certainty has no spring in it, no courtesy, no possibility of yielding. All this must react on the minds which handle these forms of truth.
The Autocrat of the Breakfast-Table
Chapter III (p. 56)
Houghton Mifflin Company. Boston, Massachusetts, USA. 1891

Huxley, Aldous 1894–1963
English writer and critic

Science is the only way we have of shoving truth down the reluctant throat.
Literature and Science
Chapter 27 (p. 79)
Harper & Row, Publishers. New York, New York, USA. 1963

Huxley, Thomas Henry 1825–95
English biologist

Ecclesiasticism in science is only unfaithfulness to truth.
Collected Essays (Volume 2)
Darwiniana
Mr. Darwin's Critics (p. 149)
Macmillan & Company Ltd. London, England. 1904

The scientific spirit is of more value than its products, and irrationally held truths may be more harmful than reasoned errors.
Collected Essays (Volume 2)
Darwiniana
The Coming of Age of "The Origin of Species" (p. 229)
Macmillan & Company Ltd. London, England. 1904

History warns us, however, that it is the customary fate of new truths to begin as heresies and to end as superstitions. . . .
Collected Essays (Volume 2)
Darwiniana
The Coming of Age of "The Origin of Species" (p. 229)
Macmillan & Company Ltd. London, England. 1904

Science has fulfilled her function when she has ascertained and enunciated truth. . . .
Collected Essays (Volume 7)
Man's Place in Nature, On the Relations of Man to the Lower Animals (p. 151)
Macmillan & Company Ltd. London, England. 1904

Magna est veritas et praevalebit! Truth is great, certainly, but, considering her greatness, it is curious what a long time she is apt to take about prevailing.
Man's Place in Nature and Other Anthropological Essays
Preface (pp. ix–x)
D. Appleton & Company. New York, New York, USA. 1896

But to those whose life is spent, to use Newton's noble words, in picking up here a pebble and there a pebble on the shores of the great ocean of truth — who watch, day by day, the slow but sure advance of that mighty tide, bearing on its bosom the thousand treasures wherewith man ennobles and beautifies life: — it would be laughable, if it were not so sad, to see the little Canutes of the hour enthroned in solemn state, bidding that grteat wave to stay, and threatening to check it beneficent progress.
In Francis Darwin (ed.)
The Life and Letters of Charles Darwin (Volume 2)
Chapter II (p. 77)
D. Appleton & Company. New York, New York, USA. 1896

Inge, WIlliam Ralph 1860–1954
English religious leader and writer

Every truth is a shadow, except the last; but every truth is a substance in its own place, though it be but a shadow in another place; and the shadow is a true shadow, as the substance is a true substance.
Proceedings of the Aristotelian Society, 1918–1919 (p. 272)

Jeffers, Robinson 1887–1962
American poet

The mathematicians and physics men
Have their mythology; they work alongside the truth,
Never touching it; their equations are false
But the things work. Or, when gross error appears,
They invent new ones; they drop the theory of waves
In universal ether and imagine curved space.
The Beginning and the End and Other Poems
The Great Wound (p. 11)
Random House, Inc. New York, New York, USA. 1963

Jones, Raymond F. 1915–94
American writer

. . .in the statistical world you can multiply ignorance by a constant and get truth.
The Non-Statistical Man (p. 58)
Belmont Books, New York, New York, USA. 1964

Jonson, Ben 1573?–1637
English dramatist and poet

If in some things I dissent from others, whose wit, industry, diligence, and judgment I look up at and admire, let me not therefore hear presently of ingratitude and rashness. For I thank those that have taught me, and ever will; but yet dare not think the scope of their labour and inquiry

was to envy their posterity what they also could add and find out.... If I err, pardon me....
Timber; or Discoveries Made upon Man and Matter
Explorata; or, Discoveries (p. 7)
Ginn & Company. Boston, Massachusetts, USA. 1892

Kepler, Johannes   1571–1630
German astronomer

The very truth, and the nature of things, though repudiated and ordered into exile, sneaked in again through the back door, to be received by me under an unwonted guise.
Translated by William H. Donahue
New Astronomy
Part IV, 58 (p. 575)
At The University Press. Cambridge, England. 1992

Lamarck, Jean-Baptiste Pierre Antoine   1744–1829
French biologist

Man is condemned to exhaust all possible errors when he examines any set of facts before he recognises the truth.
Translated by Hugh Elliot
Zoological Philosophy: An Exposition with Regard to the Natural History of Animals
Chapter V (p. 57)
The University of Chicago Press. Chicago, Illinois, USA. 1984

...both individual and public reason, when they find themselves exposed to any alteration, usually set up so great an obstacle to it, that it is often harder to secure the recognition of a truth than it is to discover it.
Translated by Hugh Elliot
Zoological Philosophy: An Exposition with Regard to the Natural History of Animals
Chapter VIII (p. 404)
The University of Chicago Press. Chicago, Illinois, USA. 1984

Laplace, Pierre Simon   1749–1827
French mathematician, astronomer, and physicist

Induction, analogy, hypotheses founded upon facts and rectified continually by new observations, a happy tact given by nature and strengthened by numerous comparisons of its indications with experience, such are the principal means for arriving at truth.
A Philosophical Essay on Probabilities
Chapter XVII (p. 176)
Dover Publications, Inc. New York, New York, USA. 1951

Lawson, Alfred William   1869–1954
American baseball player, popular philosopher and economist

Education is the science of knowing TRUTH.
Miseducation is the art of absorbing FALSITY.
TRUTH is that which is, not that which ain't.
FALSITY is that which ain't, not that which is.
In Martin Gardner
Fads and Fallacies in the Name of Science
Chapter 6 (p. 76)
Dover Publications, Inc., New York, New York, USA. 1957

Le Bon, Gustave   1841–1931
French social psychologist, author, and ameteur physicist

Science has promised us truth — an understanding of such relationships as our minds can grasp; it has never promised us either peace or happiness.
La Psychologie des Foules
Introduction

Levy, Hyman   1889–1975
British mathematician and social activist

Truth is a dangerous word to incorporate within the vocabulary of science. It drags with it, in its train, ideas of permanence and immutability that are foreign to the spirit of a study that is essentially an historically changing movement, and that relies so much on practical examination within restricted circumstances....
The Universe of Science
Chapter V (p. 206)
The Century Company. New York, New York, USA. 1933

Truth is an absolute notion that science, which is not concerned with any such permanency, had better leave alone.
The Universe of Science
Chapter V (p. 207)
The Century Company. New York, New York, USA. 1933

Lewis, Gilbert Newton   1875–1946
American chemist

The theory that there is an ultimate truth, although very generally held by mankind, does not seem useful to science except in the sense of a horizon toward which we may proceed, rather than a point which may be reached.
The Anatomy of Science
Chapter I (p. 7)
Yale University Press. New Haven, Connecticut, USA. 1926

Lodge, Sir Oliver   1851–1940
English physicist

The direct aim of Science is Truth, and the temptation of its devotees is to concentrate too narrowly on this one aim and lose sight of the wealth of existence which gives all the meaning and value to bare fact, thus gaining but a purblind view of the universe, in spite of a large accumulation of knowledge which is accurate as far as it goes, but so incomplete as regards the totality of things as to be liable to mislead.
In J. Arthur Thomson
The Outline of Science (Volume 4)
Chapter XXIV (p. 1077)
G.P. Putnam's Sons. New York, New York, USA. 1937

Mach, Ernst   1838–1916
Austrian physicist and philosopher

Truth suffers herself to be won. She flirts at times disgracefully. Above all, she is determined to be merited,

and has naught but contempt for the man who will win her too quickly.
Popular Scientific Lectures
On the Causes of Harmony (p. 45)
The Open Court Publishing Company. Chicago, Illinois, USA. 1898

The inquirer seeks the truth. I do not know if the truth seeks the inquirer. But were that so, then the history of science would vividly remind us of that classical rendez-vous, so often immortalized by painters and poets. A high garden wall. At the right a youth, at the left a maiden. The youth sighs, the maiden sighs! Both wait. Neither dreams how near the other is.
Popular Scientific Lectures
On the Causes of Harmony (p. 45)
The Open Court Publishing Company. Chicago, Illinois, USA. 1898

Only when Truth is in exceptionally good spirits does she bestow upon her wooer a glance of encouragement. For, thinks Truth, if I do not do something, in the end the fellow will not seek me at all.
Popular Scientific Lectures
On the Causes of Harmony (pp. 45–46)
The Open Court Publishing Company. Chicago, Illinois, USA. 1898

Medawar, Sir Peter Brian   1915–87
Brazilian-born English zoologist

The truth is not in nature waiting to declare itself, and we cannot know a priori which observations are relevant and which are not; every discovery, every enlargement of the understanding begins as an imaginative preconception of what the truth might be. This imaginative preconception — a "hypothesis" — arises by a process as easy or as difficult to understand as any other creative act of mind; it is a brainwave, an inspired guess, the product of a blaze of insight. It comes, anyway, from within and cannot be arrived at by the exercise of any known calculus of discovery.
Advice to a Young Scientist
Chapter 11 (p. 84)
Basic Books, Inc. New York, New York, USA. 1979

Millikan, Robert Andrews   1868–1953
American physicist

…in science, truth once discovered always remains truth.
Science and the New Civilization
Chapter III (p. 76)
Charles Scribner's Sons. New York, New York, USA. 1930

Moulton, Forest Ray   1872–1952
American astronomer

Many a chemist, physicist, biologist, psychologist, and historian, as well as monk, has had as his first and only love The Truth, and as it his greatest reward the approval of his own conscience.
In H.H. Newman (ed.)

The Nature of the World and of Man
Astronomy (p. 2)
The University of Chicago Press. Chicago, Illinois, USA. 1927

Science does not bow down before precedent nor custom nor dogma; it exalts the truth and honestly seeks it.
In H.H. Newman (ed.)
The Nature of the World and of Man
Astronomy (p. 4)
The University of Chicago Press. Chicago, Illinois, USA. 1927

Newton, Sir Isaac   1642–1727
English physicist and mathematician

Truth is the offspring of silence and unbroken meditation.
Attributed to Newton
Source unknown

I do not know what I may appear to the world, but to myself I seem to have been only like a boy playing on the sea-shore, and diverting myself in now and then finding a smoother pebble or a prettier shell than ordinary, whilst the great ocean of truth lay all undiscovered before me.
In David Brewster
Memoirs of the Life, Writings and Discoveries of Sir Isaac Newton
(Volume 2)
Chapter 27 (p. 407)
Hamilton, Adams & Company. London, England. 1855

Orlans, Harold   1912–
American education researcher

A profession which seeks the truth must consider whether silence about motives and restraint in expression serve, on balance, to enhance or suppress it.
Neutrality and Advocacy in Policy Research
Policy Sciences, Volume 6, 1975

Osler, Sir William   1849–1919
Canadian physician and professor of medicine

Truth has been well called the daughter of Time, and even in anatomy, which is a science in a state of fact, the point of view changes with successive generations.
Aequanimitas, with Other Addresses to Medical Students, Nurses, and Practitioners of Medicine
The Leaven of Science (p. 84)
The Blakiston Company. Philadelphia, Pennsylvania, USA. 1932

The truth is the best you can get with your best endeavor, the best that the best men accept — with this you must learn to be satisfied, retaining at the same time with due humility an earnest desire for an ever larger portion.
Selected Writings of Sir William Osler
Chapter 11 (p. 172)
Oress. London, England. 1951

Pagels, Heinz R.   1939–88
American physicist and science writer

The only touchstone for empirical truth is experiment and observation.

Perfect Symmetry: The Search for the Beginning of Time
Part Four, Chapter 1 (p. 355)
Simon & Schuster. New York, New York, USA. 1985

Pascal, Blaise   1623–62
French mathematician and physicist

We may have three main objects in the study of truth: first, to find it when we are seeking it; second, to demonstrate it after we have found it; third, to distinguish it from error by examining it.
In *Great Books of the Western World* (Volume 33)
Scientific Treatises
On Geometrical Demonstration (p. 430)
Encyclopædia Britannica, Inc. Chicago, Illinois, USA. 1952

Pasteur, Louis   1822–95
French chemist

Truth, Sir, is a great coquette. She will not be won by too much passion. Indifference is often more successful with her. She escapes when apparently caught, but she yields readily if patiently waited for. She reveals herself when one is about to abandon the hope of possessing her; but she is inexorable when one affirms her, that is when loves her with too much fervor.
In Rene Dubos
Louis Pasteur: Free Lance of Science
Chapter XIV (p. 389)
Little, Brown & Company. Boston, Massachusetts, USA. 1950

Peirce, Charles Sanders   1839–1914
American scientist, logician, and philosopher

…truths, on the average, have a greater tendency to get believed than falsities have. Were it otherwise, considering that there are myriads of false hypotheses to account for any given phenomenon, against one sole true one (or if you will have it so, against every true one), the first step towards genuine knowledge must have been next door to a miracle.
The Collected Works of Charles Sanders Peirce (Volume 5)
Pragmatism and Pragmaticism (p. 431)

Penrose, Roger   1931–
English mathematical physicist

Scientists do not invent truth — they discover it.
In John Horgan
Quantum Consciousness
Scientific American, Volume 261, Number 5, November 1989 (p. 32)

Planck, Max   1858–1947
German physicist

Conscientiousness and truth are as necessary in research in pure science as in practical life.
A Survey of Physical Theory
Dynamical Laws and Statistical Laws
Methuen & Company Ltd. London, England. 1925

If we seek a foundation for the edifice of exact science which is capable of withstanding every criticism, we must first of all tone down our demands considerably. We must not expect to succeed at a stroke, by one single lucky idea, in hitting on an axiom of universal validity, to permit us to develop, with exact methods, a complete scientific structure. We must be satisfied initially to discover some form of truth which no skepticism can attack. In other words, we must set our sights not on what we would like to know, but first on what we do not know with certainty.
Scientific Autobiography and Other Papers
The Meaning and Limits of Exact Science, Part I (p. 84)
Philosophical Library. New York, New York, USA. 1949

It is not the possession of truth, but the success which attends the seeking after it, that enriches the seeker and brings happiness to him.
In H.A. Ross
Elihu Root Lectures of Carnegie Institution of Washington on the Influence of Science and Research on Current Thought
The Nature of Progress in Science (p. 14)
Washington, D.C. 1945

…"to believe" means "to recognize as a truth," and the knowledge of nature, continually advancing on incontestably safe tracks, has made it utterly impossible for a person possessing some training in natural science to recognize as founded on truth the many reports of extraordinary occurrences contradicting the laws of nature, of miracles which are still commonly regarded as essential supports and confirmations of religious doctrines, and which formerly used to be accepted as facts pure and simple, without doubt or criticism.
Scientific Autobiography and Other Papers
Religion and Natural Science, Part I (p. 154)
Philosophical Library. New York, New York, USA. 1949

…the whole strenuous intellectual work of an industrious research worker would appear, after all, in vain and hopeless, if he were not occasionally through some striking facts to find that he had, at the end of his all criss-cross journeys, at last accomplished at least one step which was conclusively nearer the truth.
Nobel Lectures, Physics 1901–1921
Nobel lecture for award received in 1918 (p. 407)
Elsevier Publishing Company. Amsterdam, Netherlands. 1967

Poe, Edgar Allan   1809–49
American short story writer

Truth is not always in a well. In fact, as regards the more important knowledge, I do believe that she is invariably superficial. The depth lies in the valleys where we seek her, and not upon the mountain-tops where she is found.
Complete Tales and Poems of Edgar Allan Poe
The Murders in the Rue Morgue (p. 153)
The Modern Library. New York, New York, USA. 1965

Priestley, Joseph   1733–1804
English theologian and scientist

When I…compare my last discoveries relating to the constitution of the atmosphere with the first, I see the closest and easiest connexion in the world between them, so as to wonder that I should not have been led immediately from the one to the other. That this was not the case, I attribute to the force of prejudice, which unknown to ourselves, biases not only our judgments, properly so called, but even the perception of our senses: for we may take a maxim so strongly for granted, that the plainest evidence of sense will not entirely change, and often hardly modify, our persuasions; and the more ingenious a man is, the more effectually he is entangled in his errors; …his ingenuity only helping him to deceive himself, by evading the force of truth.
In F.W. Gibbs
Joseph Priestley: Adventurer in Science and Champion of Truth
Chapter 9 (p. 119)
Thomas Nelson & Sons Ltd. London, England. 1965

Reichenbach, Hans   1891–1953
German philosopher of science

He who searches for truth must not appease his urge by giving himself up to the narcotic of belief.
In Ruth Renya
The Philosophy of Matter in the Atomic Era: A New Approach to the Philosophy of Science (p. 16)
Asia Publishing House. Bombay, India. 1962

Renan, Ernest   1823–92
French philosopher and Orientalist

Science has no enemies save those who consider truth as useless and making no difference, and those who granting to truth its priceless value profess to get at it by other roads than those of criticism and rational investigation.
The Future of Science
Chapter IV (p. 68)
Roberts Brothers. Boston, Massachusetts, USA. 1893

The simplest schoolboy is now familiar with truths for which Archimedes would have sacrificed his life.
In L.I. Ponomarev
The Quantum Dice (p. 34)
Institute of Physics Publishing. Bristol, England. 1993

Richet, Charles   1850–1935
French physiologist

Truth, the goddess, the sovereign, the all-powerful, who will freeze with terror those who jeer at her!
Translated by Sir Oliver Lodge
The Natural History of a Savant
Chapter II (p. 25)
J.M. Dent & Sons Ltd. London, England. 1927

…if you would discover a new truth, do not seek to know what use will be made of it.

The Natural History of a Savant
Chapter XII (p. 133)
J.M. Dent & Sons Ltd. London, England. 1927

Romanoff, Alexis Lawrence   1892–1980
Russian soldier and scientist

Science speaks the language of universal truth.
Encyclopedia of Thoughts
Aphorisms 961
Ithaca Heritage Books. Ithaca, New York, USA. 1975

Russell, Bertrand Arthur William   1872–1970
English philosopher, logician, and social reformer

When a man tells you that he knows the exact truth about anything, you are safe in inferring that he is an inexact man.
The Scientific Outlook
Characteristics of Scientific Method (p. 65)
George Allen & Unwin Ltd. London, England. 1931

Science thus encourages abandonment of the search for absolute truth, and the substitution of what may be called "technical" truth, which belongs to any theory that can be successfully employed in inventions or in predicting the future. "Technical" truth is a matter of degree: a theory from which more successful inventions and predictions spring is truer than one which gives rise to fewer.
Religion and Science
Grounds of Conflict (p. 15)
Henry Holt & Company. New York, New York, USA. 1935

Sagan, Carl   1934–96
American astronomer and science writer

The truth may be puzzling. It may take some work to grapple with. It may be counterintuitive. It may contradict deeply held prejudices. It may not be consonant with what we desperately want to be true. But our preferences do not determine what's true.
Wonder and Skepticism
Skeptical Inquirer, Volume 19, Issue 1, January-February 1995

We have a method, and that method helps us to reach not absolute truth, only asymptotic approaches to the truth — never there, just closer and closer, always finding vast new oceans of undiscovered possibilities.
Wonder and Skepticism
Skeptical Inquirer, Volume 19, Issue 1, January–February 1995

Sattler, R.
No biographical data available

Modern philosophy of science has gone far beyond the naive belief that science reveals the truth. Even if it could, we would have no means of proving it. Certainty seems unattainable. All scientific statements remain open to doubt…. We cannot reach the absolute at least as far as science is concerned; we have to content ourselves with the relative.

Biophilosophy
Chapter 1 (p. 41)
Springer-Verlag. Berlin, Germany. 1986

Sendivogius, Michael   1566–1636
Polish alchemist and inventor

There is abundance of knowledge, yet but little Truth known. The generality of our knowledge is as Castles in the Air, or groundless Fancies.
A New Light of Alchymy
To the Reader
Printed by A. Clark. London, England. 1674

Shaw, George Bernard   1856–1950
Irish comic dramatist and literary critic

RIDGEON: The buried truth germinates and breaks through to the light.
The Doctor's Dilemma
Act V (p. 114)
Brentano's. New York, New York, USA. 1920

Shepherd, Linda Jean
American biochemist

…the "truth" has many faces, depending upon the perspective of the observer.
Lifting the Veil: The Feminine Face of Science
Chapter 6 (p. 153)
Shambhala. Boston, Massachusetts, USA. 1993

Smuts, Jan Christian   1870–1950
South African statesman, military leader, and holistic philosopher

Truth is a whole, and the truth of physics will be found to link on and to be but part of that larger truth which is the nature and the character of the universe.
Contributions to a British Association Discussion on the Evolution of the Universe
Nature, Supplement, October 24, 1931 (p. 718)

Spencer-Brown, George   1923–
English mathematician and polymath

To arrive at the simplest truth, as Newton knew and practiced, requires years of contemplation. Not activity. Not reasoning. Not calculating. Not busy behavior of any kind. Not reading. Not talking. Not making an effort. Not thinking. Simply bearing in mind what it is one needs to know. And yet those with the courage to tread this path to real discovery are not only offered practically no guidance on how to do so, they are actively discouraged and have to set about it in secret, pretending meanwhile to be diligently engaged in the frantic diversions and to conform with the deadening personal opinions which are continually being thrust upon them.
Laws of Form
Appendix I (p. 110)
George Allen & Unwin Ltd. London, England. 1969

Teilhard de Chardin, Pierre   1881–1955
French Jesuit, paleontologist, and biologist

We are given to boasting of our age being an age of science…. Yet though we may exalt research and derive enormous benefits from it, with what pettiness of spirit, poverty of means and general haphazardness do we pursue truth in the world today! [W]e leave it to grow as best it can, hardly tending it, like those wild plants whose fruits are plucked by primitive peoples in their forests.
The Phenomenon of Man
Book Four, Chapter III, Section 2 (p. 278, 278, 279)
Harper & Brothers. New York, New York, USA. 1959

Thomson, Sir George   1892–1975
English physicist

Science is essentially a search for truth.
The Inspiration of Science
Introduction (p. 1)
Oxford University Press, Inc. London, England. 1961

Tolstoy, Leo   1828–1910
Russian writer

Some mathematician, I believe, has said that true pleasure lies not in the discovery of truth, but in the search for it.
Anna Karenina
Part II, Chapter XIV (p. 192)
Barnes & Noble Books. New York, New York, USA. 2003

Toynbee, Arnold J.   1852–83
English historian

The Truth apprehended by the Subconscious Psyche finds natural expression in Poetry; The Truth apprehended by the Intellect finds natural expression in science….
In Theodosius Dobzhansky
The Biology of Ultimate Concern
Chapter 6 (p. 115)
The New American Library, Inc. New York, New York, USA. 1967

Trollope, Anthony   1815–82
English novelist

There are certain statements which, though they are false as hell, must be treated as though they were true gospel.
The Eustace Diamond (Volume 2)
Chapter LXXVIII (p. 353)
Oxford University Press, Inc. London, England. 1973

Uzor
Fictional character

Truth will flourish in fantasy only to wither and die in what you call reality.
The Mummy's Curse
Film (1944)

Vaihinger, Hans   1852–1933
German philosopher

We have repeatedly insisted…that the boundary between truth and error is not a rigid one, and we were able ultimately to demonstrate that what we generally call truth, namely a conceptual world coinciding with the external world, is merely the most expedient error.
The Philosophy of "As If"
Part I, Chapter XXIV (p. 108)
Harcourt, Brace & Company, Inc. New York, New York, USA. 1925

van Leeuwenhoek, Antony   1632–1723
Dutch biology researcher and microscope developer

As I aim at nothing but Truth, and so far as in me lieth, to point out Mistakes that may have crept into certain Matters; I hope that in so doing those I chance to censure will not take it ill: and if they would expose any Errors in my own Discoveries, I'd esteem it an Encouragement toward the Attaining of a nicer Accuracy.
Antony van Leeuwenhoek and His "Little Animals"
Envoy: Leeuwenhoeck's Place in Protozoology and Bacteriology (p. 387)
John Bale, Sons & Danielsson Ltd. London, England. 1932

Wallace, Alfred Russel   1823–1913
English humanist, naturalist, and geographer

Truth is born into this world only with pangs and tribulations, and every fresh truth is received unwillingly. To expect the world to receive a new truth, or even an old truth, without challenging it, is to look for one of those miracles which do not occur.
In an interview/obituary by W.B. Northrop
The Outlook (New York), Volume 105, 1913 (p. 622)

Wegener, Alfred   1880–1930
German climatologist and geophysicist

Scientists still do not appear to understand sufficiently that all earth sciences must contribute evidence toward unveiling the state of our planet in earlier times, and that the truth of the matter can only be reached by combing all this evidence.… It is only by combing the information furnished by all the earth sciences that we can hope to determine "truth" here, that is to say, to find the picture that sets out all the known facts in the best arrangement and that therefore has the highest degree of probability.
Translated by John Biram
The Origin of Continents and Oceans (4th edition)
Foreword (p. vii)
Dover Publications, Inc. New York, New York, USA. 1966

Weil, Simone   1909–43
French philosopher and mystic

Truth is a radiant manifestation of reality.
Translated by Arthur Wills
The Need for Roots: Prelude to a Declaration of Duties Toward Mankind
Part Three (p. 253)
The Beacon Press. Boston, Massachusetts, USA. 1952

Weinberg, Steven   1933–
American nuclear physicist

We search for universal truths about nature and when we find them, we show that they can be deduced from deeper truths.
Dreams of a Final Theory: The Scientist's Search for the Ultimate Laws of Nature
Prologue (p. 6)
Pantheon Books. New York, New York, USA. 1992

Weyl, Hermann   1885–1955
German mathematician

We are not very pleased when we are forced to accept a mathematical truth by virtue of a complicated chain of formal conclusions and computations, which we traverse blindly, link by link, feeling our way by touch. We want first an overview of the aim and of the road; we want to understand the idea of the proof, the deeper context.
In Abe Shenitzer
Part II. Topology and Abstract Algebra as Two Roads of Mathematical Comprehension
The American Mathematical Monthly, Volume 102, Number 7, August–September 1995 (p. 646)

Whewell, William   1794–1866
English philosopher and historian

Experience must always consist of a limited number of observations; and however numerous these may be, they can show nothing with regard to the infinite number of cases in which the experiment has not been made.…

[T]ruths can only be known to be general, not universal, if they depend upon experience alone. Experience cannot bestow that universality which she herself cannot have, nor that necessity of which she has no comprehension.
The Philosophy of the Inductive Sciences Founded upon Their History (Volume 1)
Part I, Book I, Chapter V, Article 1, Article 2 (pp. 63, 64)
John W. Parker. London, England. 1847

Whipple, George H.   1878–1976
American pathologist

Any investigator is indeed fortunate who can contribute a tiny stone to the great edifice which we call scientific truth.
Les Prix Nobel. The Nobel Prizes in 1934
Nobel banquet speech for award received in 1934
Nobel Foundation. Stockholm, Sweden. 1935

Wilde, Oscar   1854–1900
Irish wit, poet, and dramatist

JACK: …That, my dear Algy, is the whole truth, pure and simple.
ALGERNON — The truth is rarely pure and never simple.
The Importance of Being Earnest
Act I (p. 13)
Walter H. Baker Company. Boston, Massachusetts, USA. 19 —

It is a terrible thing for a man to find out suddenly that all his life he has been speaking nothing but the truth.
In John D. Barrow
The World Within the World (p. 260)
Clarendon Press. Oxford, England. 1988

Wilkins, John   1614–72
English writer

That the strangeness of this opinion is no sufficient reason why it should be rejected, because other certain truths have been formerly esteemed ridiculous, and great absurdities entertained by common consent.
The Discovery of a World in the Moone (p. 1)

Wilson, Edward O.   1929–
American biologist and author

…if history and science have taught us anything, it is that passion and desire are not the same as truth. The human mind evolved to believe in the gods. It did not evolve to believe in biology. Acceptance of the supernatural conveyed a great advantage throughout prehistory, when the brain was evolving. Thus it is in sharp contrast to biology, which was developed as a product of the modern age and is not underwritten by genetic algorithms. The uncomfortable truth is that the two beliefs are not factually compatible. As a result those who hunger for both intellectual and religious truth will never acquire both in full measure.
Consilience: The Unity of Knowledge
Chapter 11 (p. 262)
Alfred A. Knopf. New York, New York, USA. 1998

Wright, Chauncey   1830–75
American philosopher of science

We receive the truths of science by compulsion. Nothing but ignorance is able to resist them.
In Edward H. Madden (ed.)
The Philosophical Writings of Chauncey Wright
The Philosophy of Herbert Spencer (p. 23)
The Liberal Arts Press. New York, New York, USA. 1958

TUNNELING

Drinker, Henry   1850–1937

A barbarous people may, perhaps, develop a high degree of perfection in the mere art of open-air building, where stone can be piled on stone, and rafter fitted to rafter, in the light of day; but it takes the energy, knowledge, experience, and skill of an educated and trained class of men to cope with the unknown dangers of the dark depths that are to be invaded by the tunnel-man.
In Henry Drinker
Tunneling, Explosive Compounds, and Rock Drilling (p. 32)
John Wiley & Sons, Inc. New York, New York, USA. 1878

TURBULENCE

Feynman, Richard P.   1918–88
American theoretical physicist
Leighton, Robert B.   1919–97
American physicist
Sands, Matthew L.   1919–
American physicist

The next great era of awakening of human intellect may well produce a method of understanding the qualitative content of equations. Today we cannot. Today we cannot see that the water flow equations contain such things as the barber pole structure of turbulence that one sees between rotating cylinders. Today we cannot see whether Schrödinger's equation contains frogs, musical composers, or morality — or whether it does not.
The Feynman Lectures on Physics (Volume 2)
Chapter 41 (p. 41–12)
Addison-Wesley Publishing Company. Reading, Massachusetts, USA. 1983

Lamb, Sir Horace   1848–1934
English applied mathematician

It remains to call attention to the chief outstanding difficulty of our subject [turbulent motion].
Hydrodynamics
Chapter VI, section 365 (p. 663)
Dover Publications, Inc. Mineola, New York, USA. 1945

Saffman, P. G.
No biographical data available

…we should not altogether neglect the possibility that there is no such thing as "turbulence." That is to say, it is not meaningful to talk of the properties of a turbulent flow independently of the physical situation in which it arises. In searching for a theory of turbulence, we are perhaps looking for a chimera.
In H. Fiedler (ed.)
Structure and Mechanisms of Turbulence (Volume 2)
Problems and Progress in the Theory of Turbulence (p. 276)
Springer-Verlag. Berlin, Germany. 1978

TYPHUS

Butler, Samuel   1612–80
English novelist, essayist, and critic

They made a clean sweep of all machinery that had not been in use for more than two hundred and seventy-one years (which period was arrived at after a series of compromises), and strictly forbade all further improvements and inventions under pain of being considered in the eye of the law to be labouring under typhus fever, which they regard as one of the worst of all crimes.

Erewhon and Erewhon Revisited
Chapter IX (pp. 81–82)
The Modern Library. New York, New York, USA. 1955

Nicolle, Charles 1866–1936
French bacteriologist

And this is the ultimate lesson that our knowledge of the mode of transmission of typhus has taught us: Man carries on his skin a parasite, the louse. Civilization rids him of it. Should man regress, should he allow himself to resemble a primitive beast, the louse begins to multiply again and treats man as he deserves, as a brute beast.
Nobel Lectures, Physiology or Medicine 1922–1941
Nobel lecture for award received in 1928
Investigations on Typhus (p. 187)
Elsevier Publishing Company. Amsterdam, Netherlands. 1965

TYPOLOGY

Bordes, Francois 1919–81
French scientist, geologist, and archaeologist

One has to see a great number of implements, classify them, see them again several times, before one acquires a "typological eye."
On Old and New Concepts of Typology
Current Anthropology, Volume 13, Number 1 (p. 141)

Brögger, A. W.
No biographical data available

The proud edifice of chronology built on a foundation of typology is a dangerous mirage.
Kulturgeschichte des Norwegischen Altertums (p. 14)

Hawksworth, D. L.
No biographical data available

The Purpose of typification is to fix permanently the application of names of all ranks governed by the Code so as to preclude the possibility of the same name being used in different senses; i.e., for different plants.
Mycologist's Handbook: An Introduction to the Principles of Taxonomy and Nomenclature in the Fungi and Lichens (p. 127)
Commonwealth Mycological Institute. Kew, England. 1974

Krieger, A.

In speaking of "types" did the author follow any philosophy of typology, or — as is so common — did he merely devise still another "typology" for his own convenience.
Epistemology and Archaeological Theory, Comment on Lowthern
Current Anthropology, Volume 3, 1963 (p. 506)

Malmer, Mats P.
No biographical data available

Archaeology is directed at the general, it aims to depict the important features of existence for groups of people in prehistoric times. It is clear, therefore, that typology is the central method in archaeology: the study of types and their associations. This central archaeological concept is the type. If typology were not the central and unifying factor, all other methods and subsidiary sciences would fall hopelessly apart, and archaeology as a science would cease to exist…. There can be no typology without types, no archaeology without typology.
Acta Archaeologia Lundensis
Jungneolithische Studien, Number 2 (pp. 880–881)

Reed, T. D.
No biographical data available

Towards the end of the last century those strange new gods Typology and Chronology, Athanasian in their relationship, arose and the archaeologists bowed down and worshipped them…. The younger archaeologist of today is a sad, wise, disillusioned, and almost human being.
The Battle for Britain in the 5th Century: An Essay in Dark Age History (pp. 5–6)

Taylor, Walter W. 1913–97
American archaeologist

It is possible to type automobiles on the basis of the length of the scratches in their paint, to classify sand tempered potsherds on the number of sand grains in each, or to group together all chipped stone points which have side notches. It would be possible, but the pertinent question is "So what?"
A Study of Archeology
Part II, Chapter 5 (p. 127)
Southern Illinois University Press. Carbondale, Illinois, USA. 1967

U

UFO

Bramley, William
American author

An in-depth study of the UFO phenomenon reveals that it does not offer a happy little romp through the titillating unknown. The UFO appears more and more to be one of the grimmest realities ever confronted by the human race.
The Gods of Eden
Avon Books. New York, New York, USA. 1989

Sagan, Carl   1934–96
American astronomer and science writer

UFOs: The reliable cases are uninteresting and the interesting cases are unreliable.
Other Worlds (p. 114)
Bantom Books. New York, New York, USA. 1975

After I give lectures — on almost any subject — I am often asked, "Do you believe in UFOs?." I'm always struck by how the question is phrased, the suggestion that this is a matter of belief and not evidence. I'm almost never asked, "How good is the evidence that UFOs are alien spaceships?"
The Demon-Haunted World: Science as a Candle in the Dark
Chapter 3 (p. 82)
Random House, Inc. New York, New York, USA. 1995

UNCERTAINTY

Buffalo Springfield   1966–67
American folk rock group

There's something happening here,
What it is ain't exactly clear.
The Best of Buffalo Springfield
For What It's Worth
Electra CD. 1969

Heisenberg, Werner Karl   1901–76
German physicist and philosopher

In fact, our ordinary description of nature, and the idea of exact laws, rests on the assumption that it is possible to observe the phenomena without appreciably influencing them.
The Physical Principles of the Quantum Theory
Translated by Carl Ekhart and Frank C. Hoyt. (p. 62)
The University of Chicago Press. Chicago, Illinois, USA. 1930

Hoyle, Sir Fred   1915–2001
English mathematician and astronomer

If matters still seem very uncertain it must always be remembered that clearly sign-posted roads are not to be expected at a pioneering frontier.
Frontiers of Astronomy
Chapter Nineteen (p. 341)
Harper & Row, Publishers. New York, New York, USA. 1955

Pagels, Heinz R.   1939–88
American physicist and science writer

Space looks empty only because this great creation and destruction of all the quanta takes place over such short times and distances.
The Cosmic Code: Quantum Physics as the Language of Nature
Part II, Chapter 8 (p. 274)
Simon & Schuster. New York, New York, USA. 1982

Professor Hubert J. Farnsworth
Fictional character

Announcer [on loudspeaker]: And it's a dead heat! They're checking the electron microscope. And the winner is…[A man holds up a "3" in a window.]…number 3, in a quantum finish.

Farnsworth: No fair! You changed the outcome by measuring it.
Futurama
Luck of the Fryrish
Aired 11 March 2001

Vincenti, Walter G.   1917–
American aeronautical engineer

In the end, decreasing uncertainty in the growth of knowledge in a technology comes, I suggest, mainly from the increase in scope and precision (that is, the decrease in unsureness) in the vicarious means of selection. Just as expanding scope tends, as we saw, to widen the field that can be overtly searched, so also the increase in both scope and precision sharpens the ability to weed out variations that won't work in the real environment. Blindness in the variations may by the same token even increase — engineers have freedom to be increasingly blind in their trial variations as their means of vicarious selection become more reliable. One sees engineers today, for example, using computer models to explore a much wider field of possibilities than they were able to select from just a decade ago.
What Engineers Know and How They Know It: Analytical Studies from Aeronautical History
Chapter 8 (p. 250)
The Johns Hopkins University Press. Baltimore, Maryland, USA. 1990

Ziman, John M.   1925–
British physicist

Many philosophers have now sadly come to the conclusion that there is no ultimate procedure which will wring the last drops of uncertainty from what scientists call their knowledge.

Public Knowledge: An Essay Concerning the Social Dimension of Science
Chapter 1 (p. 5)
Cambridge University Press. Cambridge, England. 1968

UNCERTAINTY PRINCIPLE

Stoppard, Tom   1937–
Czech-born English playwright

An electron can be here or there at the same moment. You can choose. It can go from here to there without going in between; it can pass through two doors at the same time, or from one door to another by a path which is there for all to see until someone looks, and then the act of looking has made it take a different path.

[An electron's] movements cannot be anticipated because it has no reasons. It defeats surveillance because when you know what it's doing you can't be certain where it is, and when you know where it is you can't be certain what it's doing: Heisenberg's uncertainty principle; and this is not because you're not looking carefully enough, it is because there is no such thing as an electron with a definite position and a definite momentum; you fix one, you lose the other, and it's all done without tricks, it's the real world, it is awake.
Tom Stoppard: Plays
Hapgood, Act I, Scene 5 (p. 544)
Faber & Faber. London, England. 1999

UNDERSTANDING

Arnott, Neil   1788–1874
Scottish physician

…no man can understand a subject of which he does not carry a distinct outline in his mind…
Elements of Physics, or, Natural Philosophy, General and Medical
Synopsis (p. 4)
Printed for Thomas & George Underwood. London, England. 1827

Atiyah, Sir Michael   1929–
English mathematician

…it is hard to communicate understanding because that is something you get by living with a problem for a long time. You study it, perhaps for years, you get the feel of it and it is in your bones. You can't convey that to anybody else. Having studied the problem for five years you may be able to present it in such a way that it would take somebody else less time to get to that point than it took you, but if they haven't struggled with the problem and seen all the pitfalls, then they haven't really understood it.
An Interview with Michael Atiya
The Mathematical Intelligencer, Volume 6, Number 1, 1984 (p. 17)

Author undetermined

The rabbi spoke three times. The first talk was brilliant; clear and simple. I understood every word. The second was even better; deep and subtle. I didn't understand much, but the rabbi understood all of it. The third was by far the finest; a great and unforgettable experience. I understood nothing, and the rabbi himself didn't understand much either.
In Aage Petersen
The Philosophy of Niels Bohr
Bulletin of the Atomic Scientists, Volume 19, Number 7, September 1963 (p. 8)

I understand the material. I just can't do the problems.
The Physics Teacher, Volume 6, Number 9, December 1968

Bacon, Sir Francis   1561–1626
English lawyer, statesman, and essayist

The eye of the understanding is like the eye of the sense: for as you may see great objects through small crannies or levels; so you may see great axioms of nature through small and contemptible instances.
The Works of Francis Bacon (Volume 1)
Sylva Sylvarum
Century I, 91 (p. 278)
Printed for C. & J. Rivington. London, England. 1826

The human understanding is like a false mirror, which, receiving rays irregularly, distorts and discolors the nature of things by mingling its own nature with it.
In *Great Books of the Western World* (Volume 30)
Novum Organum
First Book, Aphorism 48 (p. 110)
Encyclopædia Britannica, Inc. Chicago, Illinois, USA. 1952

Barbour, Julian   1937–
English physicist

…the higher we climb, the more comprehensive the view. Each new vantage point yields a better understanding of the interconnection of things. What is more, gradual accumulation of understanding is punctuated by sudden and startling enlargements of the horizon, as when we reach the brow of a hill and see things never conceived of in the ascent. Once we have found our bearings in the new landscape, our path to the most recently attained summit is laid bare and takes its honourable place in the new world.
The End of Time: The Next Revolution in Physics
Part 1, Chapter 1 (p. 13)
Weidenfeld & Nicolson. London, England. 1999

Becker, Carl L.   1873–1945
American historian

We really haven't time to stand amazed, either at the starry firmament above or the Freudian complexes within us. The multiplicity of things to manipulate and make use of so fully engage our attention that we have neither the leisure nor the inclination to seek a rational explanation of the force that makes them function so efficiently.

The Heavenly City of the Eighteenth Century Philosophers
Chapter I (pp. 23–24)
Yale University Press. New Haven, Connecticut, USA. 1932

Bergaust, Erik
No biographical data available

Some day in the very, very distant future earthlings may learn to understand the universe…

Wernher von Braun
Are Flying Saucers Real? (p. 547)
National Space Institute. Washington, D.C. 1976

Bergman, Torbern Olaf   1735–84
Swedish chemist and naturalist

A scientist strives to understand the work of Nature. But with our insufficient talents as scientists, we do not hit upon the truth all at once. We must content ourselves with tracking it down, enveloped in considerable darkness, which leads us to make new mistakes and errors. By diligent examination, we may at length little by little peel off the thickest layers, but we seldom get the core quite free, so that finally we have to be satisfied with a little incomplete knowledge.

In J.A. Schufle
Chymia
Torbern Bergman, Earth Scientist, Volume 12, 1967
Lecture to the Royal Swedish Academy of Science
May 23, 1764 (p. 78)
University of Pennsylvania Press. Philadelphia, Pennsylvania, USA.
1948–1967

Berkeley, George   1685–1753
Irish prelate and metaphysical philosopher

PHILONOUS: I am not for imposing any sense on your words: you are at liberty to explain them as you please. Only, I beseech you, make me understand something by them.

Three Dialogues Between Hylas and Philonous
First Dialogue (p. 40)
The Bobbs-Merrill Company, Inc. Indianapolis, Indiana, USA. 1954

Bush, Vannevar   1890–1974
American electrical engineer and physicist

The enthusiasm, the exuberance, that properly accompanies the great achievements of science, the thrill of at last beginning to understand nature and the universe about us, in all their awesome magnificence, continues to lead many men all over the world, especially young men, on to this new materialism.

Science Is Not Enough
Chapter II (pp. 19–20)
William Morrow & Company, Inc. New York, New York, USA. 1967

Chu, Steven   1948–
American physicist

You want to try to put something that you learn in your own language, so that it's no longer something that's merely memorized but something you transfer from your head to your gut — you simplify it and put it in your own language to the point where it seems almost obvious and intuitive. It's only when you understand your science in this very obvious, intuitive way that you have a chance of thinking of something new.

Interview
American Scientist, Volume 86, January–February 1998 (p. 25)

Cole, K. C.   1946–
American science writer

…the need to go to the moon or smash atoms is on a par with the need to have natural history museums: Science provides a handle on who we are and how we fit into the scheme of things. Understanding our place in the sun requires an understanding of the sun's place in the solar system, the cycles of the sky, the nature of the elements, and the improbabilities of life. If what we learn leaves us a little stunned by our limitations and potentials, so be it. Science gives us a sense of scale and a sense of limits, an appreciation for perspective and a tolerance for ambiguity.

First You Build a Cloud and Other Reflections on Physics as a Way of Life
Introduction (p. 11)
Harcourt Brace & Company. New York, New York, USA. 1999

Conrad, Joseph   1857–1924
Polish-born English novelist

Things and men have always a certain sense, a certain side by which they must be got hold of if one wants to obtain a solid grasp and a perfect command.

Under Western Eyes
Section 10 (p. 304)
Harper & Brothers Publishers. New York, New York, USA. 1911

Cortázar, Julio   1914–84
Argentinian novelist and short story writer

It had been some time since Gregorovius had given up the illusion of understanding things, but at any rate, he still wanted misunderstanding to have some sort of order, some reason about them.

Translated by Gregory Rabassa
Hopscotch
Chapter 31 (p. 179)
Pantheon Books. New York, New York, USA. 1966

Dahlberg, Edward   1900–77
American novelist and essayist

It takes a long time to understand nothing.

Reasons of the Heart

On Wisdom and Folly
Horizon Press, Inc. New York, New York, USA. 1965

Eco, Umberto 1932–
Italian novelist, essayist, and scholar

ADSO: "But how does it happen," I said with admiration, "that you were able to solve the mystery of the library looking at it from the outside, and you were unable to solve it when you were inside?"

WILLIAM OF BASKERVILLE: "Thus God knows the world, because He conceived it in His mind, as if from the outside, before it was created, and we do not know its rule, because we live inside it, having found it already made."
Translated by William Weaver
The Name of the Rose
Vespers (p. 218)
Harcourt Brace Jovonovich. San Diego, California, USA. 1983

Einstein, Albert 1879–1955
German-born physicist

The hardest thing to understand is why we understand anything at all.
In Morton Wagman
Cognitive Science and Concepts of Mind (p. 103)
Praeger. New York, New York, USA. 1991

Einstein, Albert 1879–1955
German-born physicist
Infeld, Leopold 1898–1968
Polish physicist

With the help of physical theories we try to find our way through the maze of observed facts, to order and understand the world of our sense impressions.
The Evolution of Physics
Physics and Reality (p. 296)
Simon & Schuster. New York, New York, USA. 1961

The scientist reading the book of nature…must find the solution for himself, for he cannot, as impatient readers of other stories often do, turn to the end of the book. In our case the reader is also the investigator, seeking to explain, at least in part, the relation of events to their rich context. To obtain even a partial solution the scientist must collect the unordered facts available and make them coherent and understandable by creative thought.
The Evolution of Physics
The Great Mystery (pp. 4–5)
Simon & Schuster. New York, New York, USA. 1961

Ferguson, Marilyn 1938–
American writer

Real progress in understanding nature is rarely incremental. All important advances are sudden intuitions, new principles, new ways of seeing. We have not fully recognized this process of leaping ahead, however, in part because textbooks tend to tame revolutions, whether cultural or scientific. They describe the advances as if they had been logical in their day, not at all shocking.
The Aquarian Conspiracy: Personal and Social Transformation in the 1980s
Chapter 1 (p. 28)
J.P. Tarcher, Inc. Los Angeles, California, USA. 1980

Ferris, Timothy 1944–
American science writer

We might eventually obtain some sort of bedrock understanding of cosmic structure, but we will never understand the universe in detail; it is just too big and varied for that. If we possessed an atlas of our galaxy that devoted but a single page to each star system in the Milky Way (so that the sun and all its planets were crammed in on one page), that atlas would run to more than ten million volumes of ten thousand pages each. It would take a library the size of Harvard's to house the atlas, and merely to flip through it, at the rate of a page per second, would require over ten thousand years.
Coming of Age in the Milky Way
Chapter 20 (p. 383)
William Morrow & Company, Inc. New York, New York, USA. 1988

Feynman, Richard P. 1918–88
American theoretical physicist

What I cannot create I do not understand.
In James Gleick
Genius: The Life and Science of Richard Feynman
Epilogue (p. 437)
Pantheon Books. New York, New York, USA. 1992

I would like to be rather more special, and I would like to be understood in an honest way rather than in a vague way.
The Character of Physical Law
Chapter 1 (p. 13)
BBC. London, England. 1965

One does not, by knowing all the physical laws as we know them today, immediately obtain an understanding of anything much.
The Character of Physical Law
Chapter 5 (p. 122)
BBC. London, England. 1965

Galilei, Galileo 1564–1642
Italian physicist and astronomer

We proceed in step-by-step discussion from inference to inference, whereas He conceives through mere intuition. Thus in order to gain insight into some of the properties of the circle, of which it possesses infinitely many, we begin with one of the simplest; we take it for a definition and proceed from it by means of inference to a second property, from this to a third, and hence a fourth, and

so on. The divine intellect, on the other hand, grasps the essence of a circle *senza temporaneo discorso* and thus apprehends the infinite array of the properties.
Opere (VII)
Dialogo (p. 129)

SAGREDO: My brain already reels. My mind, like a cloud momentarily illuminated by a lightening-flash, is for an instant filled with an unusual light, which now beckons to me and which now suddenly mingles and obscures strange, crude ideas.
In *Great Books of the Western World* (Volume 28)
Dialogues Concerning the Two New Sciences
First Day (p. 132)
Encyclopædia Britannica, Inc. Chicago, Illinois, USA. 1952

The vain presumption of understanding everything can have no other basis than never understanding anything. For anyone who had experienced just once the perfect understanding of one single thing, and had truly tasted how knowledge is accomplished, would recognize that infinity of other truths of which he understands nothing.
Translated by Stillman Drake
The Two Chief World Systems
First Day (p. 101)
University of California Press. Berkeley, California, USA. 1953

Hawking, Stephen William   1942–
English theoretical physicist

…there may be no ultimate theory, and even if there is, we may not find it. But it is surely better to strive for a complete understanding than to despair of the human mind.
Black Holes and Baby Universes and Other Essays
Preface (p. ix)
Bantam Books. New York, New York, USA. 1993

Hazlitt, William Carew   1834–1913
English bibliographer

…in what we really understand, we reason but little.
The Collected Works of William Hazlitt
On the Conduct of Life (p. 430)
McClure, Phillips & Company. New York, New York, USA. 1904

Heisenberg, Werner Karl   1901–76
German physicist and philosopher

Whenever we proceed from the known into the unknown we may hope to understand, but we may have to learn at the same time a new meaning of the word "understanding."
Physics and Philosophy
Chapter XI (p. 201)
Harper & Row, Publishers. New York, New York, USA. 1962

The exact sciences also start from the assumption that in the end it will always be possible to understand nature, even in every new field of experience, but that we may make no a priori assumptions about the meaning of the word understand.

In Heinrich O. Proskauer
The Rediscovery of Color: Goethe Versus Newton Today
Preface (p. ix)
Anthroposophic Press. Spring Valley, New York, USA. 1986

Even for a physicist the description in plain language will be a criterion of the degree of understanding that has been reached.
Physics and Philosophy: The Revolution in Modern Science
Chapter X (p. 168)
Harper & Row, Publishers. New York, New York, USA. 1958

…as facts and knowledge accumulate, the claim of the scientist to an understanding of the world in a certain sense diminishes.
Wandlungen in der Grundlagen der Naturwissenschaft
Zur Geschichte der physikalischen Naturerklrung (p. 28)

Hoffman, Roald   1937–
Polish-born applied theoretical chemist and writer

In principle one could go ahead and calculate each molecule.… [H]owever…even if the results were in excellent agreement with experiment, the resultant predictability would not necessarily imply understanding. True understanding implies a knowledge of the various physical factors, the mix of different physical mechanisms, that go into making an observable.
Interaction of Orbitals Through Space and Through Bonds
Accounts of Chemical Research, Volume 4, Number 1, 1971 (p. 1)

Holmes, Oliver Wendell   1809–94
American physician, poet, and humorist

A moment's insight is sometimes worth a life's experience.
The Professor at the Breakfast-Table
Chapter X (p. 301)
Ticknor & Fields. Boston, Massachusetts, USA. 1860

Hoyle, Sir Fred   1915–2001
English mathematician and astronomer

The man who voyages strange seas must of necessity be a little unsure of himself. It is the man with the flashy air of knowing everything, who is always on the ball, always with it, that we should beware of.
Of Men and Galaxies
Motives and Aims of the Scientist (pp. 24–25)
University of Washington Press. Seattle, Washington, USA. 1964

Hugo, Victor   1802–85
French author, lyric poet, and dramatist

…a man may be a fine genius, and yet understand nothing of an art which he has not studied.
Notre-Dame de Paris
Book III, Chapter 2 (p. 126)
J.M. Dent & Sons Ltd. London, England. 1910

Huygens, Christiaan   1629–95
Dutch mathematician, astronomer, and physicist

If any one is resolved to find fault with it, let him first be sure he understands it.
The Celestial Worlds Discover'd, or, Conjectures Concerning the Planetary Worlds, Their Inhabitants and Productions
Book the First, Arguments for the Truth of It (p. 13)
Printed for T. Childe. London, England. 1698

Juster, Norton   1929–
American architect and author

Milo tried very hard to understand all the things he'd been told, and all the things he'd seen, and, as he spoke, one curious thing still bothered him. "Why is it, " he said quietly, "that quite often even the things which are correct just don't seem to be right?"
The Phantom Tollbooth
Chapter 16 (p. 198)
Alfred A. Knopf. New York, New York, USA. 1989

Kaufmann, William J., III   1942–94
American astronomer

We shall speak of things we cannot understand. We shall discuss concepts we cannot grasp. We shall examine processes we cannot comprehend.
Stars and Nebulas
Chapter I (p. 4)
W.H. Freeman & Company. San Francisco, California, USA. 1978

Kelvin, Lord William Thomson   1824–1907
Scottish engineer, mathematician, and physicist

Some people say they cannot understand a million million. Those people cannot understand that twice two makes four. That is the way I put it to people who talk to me about the incomprehensibility of such large numbers. I say finitude is incomprehensible, the infinite in the universe is comprehensible.
Wave Theory of Light
Journal of the Franklin Institute, Volume 118, November 1884

Le Guin, Ursula K.   1929–
American writer of science fiction and fantasy

If the human creatures will not understand Relativity, very well; but they must understand Relatedness.
The Wind's Twelve Quarters
Direction of the Road (p. 223)
Harper & Rowe, Publishers, New York, New York, USA; 1975

Lec, Stanislaw   1909–66
Polish poet and aphorist

Some like to understand what they believe in. Others like to believe in what they understand.
Unkempt Thoughts (p. 159)
St. Martin's Press. New York, New York, USA. 1962

Oppenheimer, Frank   1912–85
American physicist

Understanding is a lot like sex. It's got a practical purpose, but that's not why people do it normally.
In K.C. Cole
The Universe and the Teacup: The Mathematics of Truth and Beauty
Chapter 1 (p. 5)
Harcourt Brace & Company. New York, New York, USA. 1998

Ortega y Gasset, José   1883–1955
Spanish philosopher

To be surprised, to wonder, is to begin to understand.
The Revolt of the Masses
Chapter I (p. 12)
W.W. Norton & Company, Inc. New York, New York, USA.1960

Osler, Sir William   1849–1919
Canadian physician and professor of medicine

To understand the old writers one must see as they saw. Feel as they felt, believe as they believed — and this is hard, indeed impossible! We may get near them by asking the Spirit of the Age in which they lived to enter in and dwell with us, but it does not always come.... Each generation has its own problems to face, look at truth from a special focus, and does not see quite the same outlines as any other.
The Evolution of Medicine
Chapter VI (p. 218)
Yale University Press. New Haven, Connecticut, USA. 1922

Pagels, Heinz R.   1939–88
American physicist and science writer

The attempt to understand the origin of the universe is the greatest challenge confronting the physical sciences. Armed with the new concepts, scientists are rising to meet that challenge, although they know that success may be far away. Yet when the origin of the universe is understood, it will open a new vision that is beautiful, wonderful and filled with the mystery of existence. It will be our intellectual gift to our progeny and our tribute to the scientific heroes who began this great adventure of the human mind, never to see it completed.
Perfect Symmetry: The Search for the Beginning of Time
Part One, Chapter 7 (p. 156)
Simon & Schuster. New York, New York, USA. 1985

Palade, George E.   1912–?
Russian-born American cell biologist

For a scientist, it is a unique experience to live through a period in which his field of endeavor comes to bloom — to be witness to those rare moments when the dawn of understanding finally descends upon what appeared to be confusion only a while ago — to listen to the sound of darkness crumbling.
Les Prix Nobel. The Nobel Prizes in 1974
Nobel banquet speech for award received in 1974
Nobel Foundation. Stockholm, Sweden. 1975

Polanyi, Michael   1891–1976
Hungarian-born English scientist, philosopher, and social scientist

We have a solid tangible object before us.... But we do not know what it is. Then let a team of physicists and chemists inspect the object. Let them be equipped with all the physics and chemistry ever to be known, but let their technological outlook be that of the stone age. Or, if we cannot disregard the practical incompatibility of these two assumptions, let us agree that in their investigations they shall not refer to any operational principles. They will describe the clock precisely in every particular, and in addition, they will predict all its possible future configurations. Yet they will never be able to tell us that it is a clock. The complete knowledge of a machine as an object tells us nothing about it as a machine.
Personal Knowledge
Chapter 11, Section 2 (p. 330)
Harper & Row, Publishers. New York, New York, USA. 1962

Popper, Karl R.   1902–94
Austrian/British philosopher of science

Only a man who understands science (that is scientific problems) can understand its history.... [O]nly a man who has some real understanding of its history (the history of its problem situations) can understand science.
Objective Knowledge
On the Theory of the Objective Mind
Oxford University Press, Inc. Oxford, England. 1972

Bohr...thought of understanding in terms of pictures and models — in terms of a kind of visualization. This was too narrow, I felt; and in time I developed an entirely different view. According to this view what matters is the understanding not of pictures but of the logical force of a theory: its explanatory power, its relation to the relevant problems and to other theories.
Unended Quest: An Intellectual Autobiography
Chapter 18 (p. 93)
Open Court Publishing Company. La Salle, Illinois, USA. 1976

...the activity of understanding is, essentially, the same as that of all problem solving.
Objective Knowledge: An Evolutionary Approach
Chapter 4 (p. 166)
Clarendon Press. Oxford, England. 1972

Rabi, Isidor Isaac   1898–1988
Austrian-born American physicist

Scientific understanding...is an essential step to our finding a home for ourselves in the universe. Through understanding the universe, we become at home in it. In a certain sense we have made this universe out of human concepts and human discoveries. It ceases to be a lonely place, because we can to some extent actually navigate in it.

In A.A. Warner, Dean Morse and T.E. Cooney (eds.)
The Environment of Change
The Revolution in Science (p. 49)
Columbia University Press. New York, New York, USA. 1969

Ramsay, Sir William   1852–1916
English chemist

I trust we have not wearied you in giving some account of our attempts to see the invisible, touch the intangible, and weigh the imponderable.
Annual Report of the Board of Regents of the Smithsonian Institution, 1912
Measurements of Infinitesimal Quantities of Substances (p. 229)
Government Printing Office. Washington, D.C. 1913

Recorde, Robert   1510?–58
English mathematician and writer

I see in the heaven marvelous motions; and in the reste of the worlde straunge transmutations, and therfore desire muche to know what the worlde is, and what are the principall partes of it, and also how all these strange sightes doo come.
The Castle of Knowledge
The First Treatise (p. 3)
Imprinted by R. Wolfe. London, England. 1556

Sagan, Carl   1934–96
American astronomer and science writer

We go about our daily lives understanding almost nothing of the world. We give little thought to the machinery that generates the sunlight that makes life possible, to the gravity that glues us to an Earth that would otherwise send us spinning off into space, or to the atoms of which we are made and on whose stability we fundamentally depend. Except for children (who don't know enough not to ask the important questions), few of us spend much time wondering why nature is the way it is; where the cosmos came from, or whether it was always here; if time will one day flow backward and effects precede causes; or whether there are ultimate limits to what humans can know.
In Stephen W. Hawking
A Brief History of Time: From the Big Bang to Black Holes
Introduction (p. ix)
Bantam Books. Toronto, Ontario, Canada. 1988

If you know something only qualitatively, you know it no more than vaguely. If you know it quantitatively — grasping some numerical measure that distinguishes it from an infinite number of other possibilities — you are beginning to know it deeply.
Billions & Billions: Thoughts on Life and Death at the Brink of the Millennium
Chapter 2 (p. 21)
Random House, Inc. New York, New York, USA. 1997

Stewart, Ian   1945–
English mathematician and science writer

A person who insists on understanding every tiny step before going on to the next is liable to concentrate so much on looking at his feet that he fails to realize he is walking in the wrong direction.
Concepts of Modern Mathematics
Chapter 20 (p. 286)
Dover Publications, Inc. New York, New York, USA. 1995

Swift, Jonathan   1667–1745
Irish-born English writer

…where I am not understood, it shall be concluded, that something very useful and profound is couched underneath…
Gulliver's Travels, the Tale of a Tub, Battle of the Books, Etc.
Tale of a Tub, the Preface (p. 403)
Oxford University Press, Inc. London, England. 1929

Teller, Edward   1908–2003
Hungarian-born American nuclear physicist

What is called understanding is often no more than a state where one has become familiar with what one does not understand.
Better a Shield than a Sword: Perspectives in Defense and Technology
Chapter 30 (p. 218)
The Free Press. New York, New York, USA. 1987

Walker, Kenneth   1882–1966
Physician

It may be said that all understanding of the universe comes from the combined action of two faculties in us, the power to register impressions and the capacity to reason and reflect on them.
Meaning and Purpose
Chapter II (p. 18)
Jonathan Cape. London, England. 1944

Welch, Lew   1926–71?
American Beat poet

Step out onto the Planet. Draw a circle a hundred feet round. Inside the circle are 300 things nobody understands, and maybe nobody's ever really seen. How many can you find?
Hermit Poems
Step Out Onto the Planet
Four Seasons Foundation. San Francisco, California, USA. 1965

Wittgenstein, Ludwig Josef Johann   1889–1951
Austrian-born English philosopher

Telling someone something he does not understand is pointless, even if you add that he will not be able to understand it.
Translated by Peter Winch
Culture and Value (p. 7e)
The University of Chicago Press. Chicago, Illinois, USA. 1980

Woodbridge, Frederick James Eugene   1867–1940
American philosopher

We understand a thing when we have discovered what it can do in relation to other things. In different relations it acts differently, but in every case with a definiteness in accord with its property. Its operation in specific cases is a specific operation which nonetheless illustrates its proper action.
Nature and Mind: Selected Essays of Frederick J.E. Woodbridge (p. 257)
Columbia University Press. New York, New York, USA. 1937

UNEXPECTED

Heraclitus   540 BCE–480 BCE
Greek philosopher

If one does not expect the unexpected one will not find it out, since it is not to be searched out, and difficult to compass.
In G.S. Kirk and J.E. Raven
The Pre-Socratic Philosophers: A Critical History with a Selection of Texts
Fragment 213 (p. 195)
At the University Press. Cambridge, England. 1963

Raymo, Chet   1936–
American physicist and science writer

Delight in the unexpected is part of the lifeblood of science. Almost alone among belief systems, science welcomes the disturbingly new.
The Virgin and the Mousetrap: Essays in Search of the Soul of Science
Chapter 15 (p. 138)
The Viking Press. New York, New York, USA. 1991

Selye, Hans   1907–82
Austrian endocrinologist

…"peripheral vision": the ability not only to look straight at what you want to see, but also to watch continually, through the corner of your eye, for the unexpected. I believe this to be one of the greatest gifts a scientist can have. Usually we concentrate so much upon what we intend to examine that other things cannot reach our consciousness, even if they are far more important. This is particularly true of things so different from the commonplace that they seem improbable. Yet, only the improbable is really worthy of attention! If the unexpected is nevertheless found to be true, the observation usually represents a great step forward.
From Dream to Discovery: On Being a Scientist
McGraw-Hill Book Company, Inc. New York, New York, USA. 1950

UNIFIED FIELD THEORY

Davies, Paul Charles William   1946–
British-born physicist, writer, and broadcaster

Ranged against GUTs, however, is the fact that there is no unique theory, and the unification scale is so remote there is no prospect whatever that it will become accessible to direct experimentation. How, then, are we to discriminate between rival theories? If the GUTs describe a world so small and so energetic that we can never observe it, has not physics degenerated to pure philosophy? Are we not in the same position as Democritus and the other Greek philosophers who mused endlessly about the shapes and properties of atoms without any hope of ever observing them?

Superforce: The Search for a Grand Unified Theory of Nature
Chapter 8 (p. 135)
Simon & Schuster. New York, New York, USA. 1984

Hawking, Stephen William   1942–
English theoretical physicist

The discovery of a complete unified theory, therefore, may not aid the survival of our species. It may not even affect our life-style. But ever since the dawn of civilization, people have not been content to see events as unconnected and inexplicable. They have craved an understanding of the underlying order in the world. Today we still yearn to know why we are here and where we came from. Humanity's deepest desire for knowledge is justification enough for our continuing quest. And our goal is nothing less than a complete description of the universe we live in.

A Brief History of Time: From The Big Bang to Black Holes
Chapter 2 (p. 13)
Bantam Books. Toronto, Ontario, Canada. 1988

Kaku, Michio   1947–
Japanese-American theoretical physicist
Thompson, Jennifer
American author

To a physicist, finally discovering the unified field theory is like being a child left in the middle of a toy store. Far from the end, it is only a beginning.

Beyond Einstein : The Cosmic Quest for the Theory of the Universe
Chapter 11 (p. 204)
Bantam Books. Toronto, Ontario, Canada. 1987

UNIFORMITARIANISM

Gould, Stephen Jay   1941–2002
American paleontologist and evolutionary biologist

Is uniformitarianism necessary?

Is Uniformitarianism Necessary?
Journal of Science, Volume 263, 1965 (p. 223)

Lapworth, Charles   1842–1920
English geologist

Uniformity and Evolution are one.

Report of the British Association for the Advancement of Science (1892)
Presidential Address to the Geology Section (p. 707)

Russell, Bertrand Arthur William   1872–1970
English philosopher, logician, and social reformer

…what is surprising in physics is not the existence of general laws, but their extreme simplicity. It is not the uniformity of nature that should surprise us, for, by sufficient analytic ingenuity any conceivable course of nature might be shown to exhibit uniformity. What should surprise us is the fact that the uniformity is simple enough for us to be able to discover it. But it is just this characteristic of simplicity which it would be fallacious to generalize, for it is obvious that simplicity has been a part of cause of their discovery, and can, therefore, give not ground for the supposition that other undiscovered laws are equally simple.

Scientific Method in Philosophy
Section I (p. 8)
At The Clarendon Press. Oxford, England. 1914

UNIQUENESS

Doyle, Sir Arthur Conan   1859–1930
Scottish writer

As a rule, when I have heard some slight indication of the course of events, I am able to guide myself by the thousands of other similar cases which occur to my memory. In the present instance I am forced to admit that the facts are, to the best of my belief, unique.

In William S. Baring-Gould (ed.)
The Annotated Sherlock Holmes (Volume 1)
The Red-Headed League (p. 419)
Wings Books. New York, New York, USA. 1967

Simon, Herbert Alexander   1916–2001
American social scientist

The definition of man's uniqueness has always formed the kernel of his cosmological and ethical systems. With Copernicus and Galileo, he ceased to be the species located at the centre of the universe, attended by sun and stars. With Darwin, he ceased to be the species created and specially endowed by God with soul and reason. With Freud he ceased to be the species whose behavior was — potentially — governable by rational mind. As we begin to produce mechanisms that think and learn, he has ceased to be the species uniquely capable of complex, intelligent manipulation of his environment.

What Computers Mean for Man and Society
Science, Volume 195, Number 4283, March 18, 1977 (p. 190)

UNITS

Lodge, Sir Oliver   1851–1940
English physicist

Changing the units does not affect the velocity of light. Whether you say light travels at 186,000 miles a second

or whether you say it is so many inches an hour makes no difference to the velocity. An algebraic symbol ought to represent the thing itself, not a mere number of units. Altering the numerical specifications — which is what you do by altering units — means no difference to the thing itself.
Royal Astronomical Society
Monthly Notices, Volume 80, 1919 (p. 107)

UNIVERSE

Adams, Douglas   1952–2001
English author, comic radio dramatist, and musician

For a long period of time there was much speculation and controversy about where the so-called "missing matter" of the Universe had got to. All over the Galaxy the science departments of all the major universities were acquiring more and elaborate equipment to probe and search the hearts of distant galaxies, and then the very center and the very edges of the whole Universe, but when eventually it was tracked down it turned out in fact to be all the stuff which the equipment had been packed in.
The Ultimate Hitchhiker's Guide to the Galaxy
Mostly Harmless
Chapter 17 (p. 756)
Ballantine Books. New York, New York, USA. 2002

The Universe, as has been observed before, is an unsettlingly big place, a fact which for the sake of a quiet life most people tend to ignore.
The Ultimate Hitchhiker's Guide to the Galaxy
The Restaurant at the End of the Universe
Chapter 10 (p. 194)
Ballantine Books. New York, New York, USA. 2002

If the Universe came to an end every time there was some uncertainty about what happened in it, it would never have got beyond the first picosecond. And many of course don't. It's like the human body, you see. A few cuts and bruises here and there don't hurt it. Not even major surgery if it's done properly. Paradoxes are just the scar tissue. Time and space heal themselves up around them and people remember a version of events which makes as much sense as they require it to make.
Dirk Gently's Holistic Detective Agency
Chapter 32 (p. 283)
Simon & Schuster. New York, New York, USA. 1988

Alfven, Hannes   1908–95
Swedish physicist

I have never thought that you can get the extremely clumpy, heterogeneous universe we have today from a smooth and homogenous one dominated by gravitation.
In Eric J. Lerner
The Big Bang Never Happened
Chapter 1 (p. 42)
Random House, Inc. New York, New York, USA. 1991

Amaldi, Ginestra Giovene
Italian physicist

Our imagination has roamed far and wide through distant reaches of the Universe. Understandably, we may have become dazed by the immense dimensions of space and the enormous sizes of some of its occupants.
Our World and the Universe Around Us (Volume 1)
First Steps into Space (p. 124)
Abradale Press. New York, New York, USA. 1966

Apfel, Necia H.   1930–
Hynek, J. Allen   1910–?
No biographical data available

It is hard for us today to assimilate all the new ideas that are being suggested in response to the new information we have. We must remember that our picture of the universe is based not only on our scientific knowledge but also on our culture and our philosophy. What new discoveries lie ahead no one can say. There may well be civilizations in other parts of our galaxy or in other galaxies that have already accomplished much of what lies ahead for mankind. Others may just be beginning. The universe clearly presents an unending challenge.
Architecture of the Universe
Chapter 21 (p. 453)
The Benjamin/Cummings Publishing Company, Inc. Menlo Park, California, USA. 1979

Atkins, Peter William   1940–
English physical chemist and writer

My aim is to argue that the universe can come into existence without intervention, and that there is no need to invoke the idea of a Supreme Being in one of its numerous manifestations.
The Creation
Preface
W.H. Freeman. San Francisco, California, USA. 1981

Aurelius Antoninus, Marcus   121–180
Roman emperor

Either it is a well-arranged universe or a chaos huddled together, but still a universe. But can a certain order subsist in thee, and disorder in the All?
In *Great Books of the Western World* (Volume 12)
The Meditations of Marcus Aurelius
Book IV, #27 (p. 266)
Encyclopædia Britannica, Inc. Chicago, Illinois, USA. 1952

Bacon, Leonard   1887–1954
American poet and critic

Eddington's universe goes phut.
Richard Tolman's can open and shut.
Eddington's bursts without grace or tact,
But Tolman's swells and perhaps may contract.
Rhyme and Punishment

Richard Tolman's Universe
Farrar & Rinehart, Inc. New York, New York, USA. 1936

Bacon, Sir Francis   1561–1626
English lawyer, statesman, and essayist

For the fabric of this universe is like a labyrinth to the contemplative mind, where doubtful paths, deceitful imitations of things and their signs, winding and intricate folds and knots of nature everywhere present themselves, and a way must constantly be made through the forests of experience and particular natures, with the aid of the uncertain light of the senses, shining and disappearing by fits.
In Basil Montague
The Works of Francis Bacon (Volume 3)
The Great Instauration, Preface (p. 336)
Parry & McMillan. Philadelphia, Pennsylvania, USA. 1859

Bagehot, Walter   1826–77
English journalist

We are startled to find a universe we did not expect.
Literary Studies (Volume 2) (p. 403)
J.M. Dent & Sons Limited. London, England. 1951

Taken as a whole, the universe is absurd. There seems an unalterable contradiction between the human mind and its employments.
Literary Studies (Volume 1) (p. 36)
J.M. Dent & Sons Limited. London, England. 1951

Barbellion, Wilhelm Nero Pilate   1889–1919
English author

This great bully of a universe overwhelms me. The stars make me cower. I am intimidated by the immensity surrounding my own littleness.
The Journal of a Disappointed Man
March 2, 1917 (p. 283)
George H. Doran Company. New York, New York, USA. 1919

Barth, John   1930–
American writer

All the scientists hope to do is describe the universe mathematically, predict it, and maybe control it. The philosopher, by contrast, seems unbecomingly ambitious: He wants to understand the universe; to get behind phenomena and operation and solve the logically prior riddles of being, knowledge, and value. But the artist, and in particular the novelist, in his essence wishes neither to explain nor to control nor to understand the universe. He wants to make one of his own, and may even aspire to make it more orderly, meaningful, beautiful, and interesting than the one God turned out. What's more, in the opinion of many readers of literature, he sometimes succeeds.
The Friday Book: Essays and Other Nonfiction
How to Make a Universe (p. 17)
G.P. Putnam's Sons. New York, New York, USA. 1984

Bergson, Henri   1859–1941
French philosopher

The universe is not made, but is being made continually. It is growing, perhaps indefinitely....
Translated by Arthur Mitchell
Creative Evolution
Chapter III (p. 255)
The Modern Library. New York, New York, USA. 1944

Bloch, Arthur   1948–
American humorist

The universe is simmering down, like a giant stew left to cook for four billion years. Sooner or later we won't be able to tell the carrots from the onions.
In John D. Barrow
The World Within the World (p. 221)
Clarendon Press. Oxford, England. 1988

Blount, Sir Thomas Pope   1649–97
English author

Whoever surveys the curious fabric of the universe can never imagine, that so noble a structure should be fram'd for no other use, than barely for mankind to live and breathe in. It was certainly the design of the great Architect, that his creatures should afford not only necessaries and accommodations to our animal part, but also instructions to our intellectual.
A Natural History
Preface
Printed for R. Bentley. London, England. 1693

Born, Max   1882–1970
German-born English physicist

We have sought for firm ground and found none. The deeper we penetrate, the more restless becomes the universe; all is rushing about and vibrating in a wild dance.
The Restless Universe
Chapter V (p. 277)
Dover Publications, Inc. New York, New York, USA. 1951

Bove, Ben
No biographical data available

The universe lies before us. What we know about it today is merely the steppingstone to a greater, deeper understanding.
The Milky Way Galaxy
Chapter 10 (p. 201)
Holt, Rinehart & Winston. New York, New York, USA. 1961

Bradley, Jr., John Hodgdon   1898–1962
American geologist

...man is the only animal who can face with a thought, a dream, and a smile the mystery and the madness and the terrible beauty of the universe.

Autobiography of Earth
Chapter XII, Section III (p. 347)
Coward-McCann, Inc. New York, New York, USA. 1935

Browne, J. Stark
No biographical data available

And we, listening to this wonderful music of the spheres, are filled with emotions of the deepest humility and awe, but at the same time with a great pride in the achievements of the mind of man in wrestling from the dark universe about us some of its long-hidden secrets.
The Rationalist Annual
The Number and Distances of the Stars, 1931 (p. 66)

Bruno, Giordano   1548–1600
Italian philosopher and pantheist

The center of the universe is everywhere, and the circumference nowhere.
In Joseph Silk
The Big Bang (p. 84)
W.H. Freeman & Company. San Francisco, California, USA. 1980

The universe is then one, infinite, immobile.… It is not capable of comprehension and therefore is endless and limitless, and to that extent infinite and indeterminable, and consequently immobile.
Translated by Jack Lindsay
Cause, Principle, and Unity
Fifth Dialogue (p. 135)
International Publishers. New York, New York, USA. 1962

Burritt, Elijah H.   1794–1838
American astronomer

Beyond these are other suns, giving light and life to other systems, not a thousand, or two thousand merely, but multiplied without end, and ranged all around us, at immense distances from each other, attended by ten thousand times ten thousand worlds, all in rapid motion; yet calm, regular and harmonious — all space seems to be illuminated, and every particle of light a world.… And yet all this vast assemblages of suns and worlds may bear no greater proportion to what lies beyond the utmost boundaries of human vision, than a drop of water to the ocean.
The Geography of the Heavens
Chapter XVI (p. 153)
Huntington & Savage, Mason & Law. New York, New York, USA. 1850

Burroughs, William S.   1914–97
American writer

This is a war universe. War all the time. That is its nature. There may be other universes based on all sorts of other principles, but ours seems to be based on war and games.
The War Universe
Grand Street 37, Volume Ten, Number 1, 1991 (p. 95)

Bush, Vannevar   1890–1974
American electrical engineer and physicist

In no other discipline…do men confront mystery and challenge of the order of that which looms down on the astronomers in the long watches of the night. The astronomer knows at first hand…how slight is our earth, how slight and fleeting are mankind.… But more than that, he senses…the majesty which resides in the mind of man because that mind seeks in all its slightness to see, to learn, to understand at least some part of the mysterious majesty of the universe.
In James Mullaney
Some Noted Dreamers Tell of the Skies' Spell
Science Digest, June 1978 (p. 41)

Calder, Alexander   1898–1976
American kinetic sculptor

The universe is real but you can't see it. You have to imagine it.
In Katharine Kuh
The Artist's Voice: Talks with Seventeen Artists
Josef Albers (p. 14)
Da Capo Press Edition. Cambridge, Massachusetts, USA. 2000

Camus, Albert   1913–60
French novelist, essayist, and playwright

…I laid my heart open to the benign indifference of the universe.
The Outsider
Part II, Chapter V (p. 127)
H. Hamilton. London, England. 1946

Card, Orson Scott   1951–
Science fiction writer

Give the universe a push, and you don't know which dominoes will fall. There are always a few you never thought were connected.
Ender's Shadow
Chapter 22 (p. 342)
TOR. New York, New York, USA. 1999

Carlyle, Thomas   1795–1881
English historian and essayist

Margaret Fuller: I accept the Universe.
Thomas Carlyle: Gad! she'd better!
In D.A. Wilson
Carlyle on Cromwell and Others
Looking Round, Margaret Fuller Has to Listen (pp. 349–350)
Kegan Paul, Trench, Trubner & Company Ltd. London, England. 1925

I don't pretend to understand the Universe — it's a great deal bigger than I am.
In D.A. Wilson and D.W. MacArthur
Carlyle in Old Age (1865–1881) (p. 177)
Kegan Paul, Trench, Trubner & Company Ltd. London, England. 1934

Chekhov, Anton Pavlovich   1860–1904
Russian author and playwright

But perhaps the universe is suspended on the tooth of some monster.
Note-Book of Anton Chekhov (p. 20)
B.W. Huebsch, Inc. New York, New York, USA. 1921

Clarke, Arthur C.   1917–
English science and science fiction writer

There is no reason to assume that the universe has the slightest interest in intelligence — or even in life. Both may be random accidental by-products of its operations like the beautiful patterns on a butterfly's wings. The insect would fly just as well without them....
The Lost Worlds of 2001
Chapter 16 (p. 109)
New American Library. New York, New York, USA. 1972

Every thoughtful man has often asked himself: Is our race the only intelligence in the universe, or are there other, perhaps far higher, forms of life elsewhere? There can be few questions more important than this, for upon its outcome may depend all philosophy — yes, and all religion, too.
Lecture
St Martin's Technical School on Charing Cross Road, October 5, 1946

The universe: a device contrived for the perpetual astonishment of astronomers.
In Clifford A. Pickover
Keys to Infinity
Chapter 5 (p. 41)
John Wiley & Sons, Inc. New York, New York, USA. 1995

Many and strange are the universes that drift like bubbles in the foam of the river of time.
The Collected Stories of Arthur C. Clarke
The Wall of Darkness (p. 104)
Tom Doherty Associates. New York, New York, USA. 2001

...the universe has no purpose and no plan...
The Collected Stories of Arthur C. Clarke
The Star (p. 521)
Tom Doherty Associates. New York, New York, USA. 2001

...the universes...drift like bubbles in the foam upon the River of Time.
The Wall of Darkness
Super Science Stories, July 1949

Coleridge, Samuel Taylor   1772–1834
English lyrical poet, critic, and philosopher

It surely is not impossible that to some infinitely superior being the whole universe may be as one plain, the distance between planet and planet being only as the pores in a grain of sand, and the spaces between system and system no greater than the intervals between one grain and the grain adjacent.
The Table Talk and Omniana of Samuel Taylor Coleridge
Omniana
The Universe (p. 415)
George Bell & Sons. London, England. 1884

Conger, George Perrigo   1884–1960
American philosopher

The universe as revealed in modern days and ways is so overwhelming that mind needs some other title than that of self-appointed legislator for it. Mind must register before it can regulate.
A World of Epitomizations: A Study in the Philosophy of the Sciences
Introduction to Division Two (pp. 345–346)
Princeton University Press. Princeton, New Jersey, USA. 1931

Cook, Peter   1937–95
English comedian

I am very interested in the Universe — I am specializing in the Universe and all that surrounds it.
Beyond the Fringe
Disc 2, Sitting on the Bench
EMI International. 1996

Copernicus, Nicolaus   1473–1543
Polish astronomer

But they say that beyond the heavens there isn't any body or place or void or anything at all; and accordingly it is not possible for the heavens to move outward: in that case it is rather surprising that something can be held together by nothing. But if the heavens were infinite and were finite only with respect to a hollow space inside, then it will be said with more truth that there is nothing of heaven, since anything which occupied any space would be in them, but the heavens will remain immobile. For movement is the most powerful reason wherewith they try to conclude that the universe is finite.
In *Great Books of the Western World* (Volume 16)
On the Revolutions of the Heavenly Spheres
Book One, Chapter 8 (p. 519)
Encyclopædia Britannica, Inc. Chicago, Illinois, USA. 1952

Crane, Stephen   1871–1900
American writer

A man said to the universe:
"Sir I exist!"
"However," replied the universe,
"The fact has not created in me
A sense of obligation."
The Collected Poems of Stephen Crane
War Is Kind (p. 101)
Alfred A. Knopf. New York, New York, USA. 1965

Croswell, Ken
American astronomer and author

Is it mere coincidence that the universe happens to possess just those properties which allow part of it to be alive? Some people say yes; it was simply good luck that the universe was born with the particular characteristics that it has. Others say no; our universe is only one of many universes.... Still others, of a more spiritual persuasion,

see the universe's remarkable offspring as a sign that an intelligent creator wrote a tremendous symphony whose melodies the stars, galaxies, and planets now play with beauty and precision…. Whatever the case, and vast and complex though the universe is, its most astonishing features are two of the simplest: it exists, and so do we.
Planet Quest: The Epic Discovery of Alien Solar Systems
Chapter 12 (p. 247)
Oxford University Press, Inc. Oxford, England. 1997

Crowley, Aleister   1875–1947
British occultist and writer

It sometimes strikes me that the whole of science is a piece of impudence; that nature can afford to ignore our impertinent interference. If our monkey mischief should ever reach the point of blowing up the earth by decomposing an atom, and even annihilated the sun himself, I cannot really suppose that the universe would turn a hair.
The Confessions of Aleister Crowley: An Autohagiography
Part One, Chapter 14 (p. 128)
Arkana. London, England. 1989

I have never grown out of the infantile belief that the universe was made for me to suck.
The Confessions of Aleister Crowley: An Autobiography
Part Three, Chapter 54 (p. 460)
Arkana. London, England. 1989

D'Alembert, Jean Le Rond   1717–83
French mathematician

To some one who could grasp the universe from a unified standpoint, the entire creation would appear as a unique truth and necessity.
In Charles W. Misner et al.
Gravitation
Part X, Chapter 44 (p. 1218)
W.H. Freeman & Company. San Francisco, California, USA. 1973

Darling, David   1953–
British astronomer and science writer

In giving birth to us, the universe has performed its most astonishing creative act. Out of a hot, dense melee of subatomic particles…it has fashioned intelligence and consciousness…. Somehow the anarchy of genesis has given way to exquisite, intricate order, so that now there are portions of the universe that can reflect upon themselves….
Equations of Eternity: Speculations on Consciousness, Meaning, and the Mathematical Rules that Orchestrate the Cosmos
Introduction (p. xiii)
Hyperion. New York, New York, USA. 1993

Davies, Paul Charles William   1946–
British-born physicist, writer, and broadcaster

Mathematics and beauty are the foundation stones of the universe. No one who has studied the forces of nature can doubt that the world about us is a manifestation of something very, very clever indeed.
The Forces of Nature (2nd edition)
Conclusion (p. 167)
Cambridge University Press. Cambridge, England. 1983

If there is a purpose to the universe, and it achieves that purpose, then the universe must end, for its continued existence would be gratuitous and pointless. Conversely, if the universe endures forever, it is hard to imagine that there is any ultimate purpose to the universe at all.
The Last Three Minutes: Conjectures About the Ultimate Fate of the Universe
Chapter 11 (p. 155)
Basic Books, Inc. New York, New York, USA. 1994

Dawkins, Richard   1941–
British ethologist, evolutionary biologist, and popular science writer

I believe that an orderly universe, one indifferent to human preoccupations, in which everything has an explanation even if we still have a long way to go before we find it, is a more beautiful, more wonderful place than a universe tricked out with capricious, ad hoc magic.
Unweaving the Rainbow: Science, Delusion and The Appetite for Wonder
Preface (p. xi)
Houghton Mifflin Company. Boston, Massachusetts, USA. 1998

Day, Clarence   1874–1935
American writer

Is it possible that our race may be an accident, in a meaningless universe, living its brief life uncared for, on this dark, cooling star: but so — and all the more — what marvelous creatures we are! What fairy story, what tale from the Arabian Nights of the Jinns, is a hundredth part as wonderful as this story of simians! It is so much more heartening, too, than the tales we invent. A universe capable of giving birth to so many accidents is — blind or not — a good world to live in, a promising universe.
This Simian World
Chapter XIX (p. 91)
Alfred A. Knopf. New York, New York, USA. 1941

de Fontenelle, Bernard le Bovier   1657–1757
French author

How vast then! And beyond all reckoning, and beyond all mensuration must the spaces of the universe be!
Conversations on the Plurality of Worlds
The Fifth Evening (p. 151, fn)
Printed for Peter Wilson. Dublin, Ireland. 1761

…I see the universe so large…that I know not where I am, or what will become of me.
Conversations on the Plurality of Worlds
The Fifth Evening (p. 151)
Printed for Peter Wilson. Dublin, Ireland. 1761

...the universe is but a watch on a larger scale; all its motions depending on determined laws and mutual relation of its parts.

Conversations on the Plurality of Worlds
The First Evening (p. 10)
Printed for Peter Wilson. Dublin, Ireland. 1761

...when the heavens appeared to me as a little blue vault, stuck with stars, methought the universe was too straight and close, I was almost stifled for want of air; but now, it is enlarged in height and breadth, and a thousand and a thousand vortexes taken in, I begin to breathe with more freedom, and think the universe to be incomparably more magnificent than it was before.

Conversations on the Plurality of Worlds
The Fifth Evening (pp. 151–152)
Printed for Peter Wilson. Dublin, Ireland. 1761

de Sitter, Willem   1872–1934
Dutch mathematician, physicist and astronomer

Our conception of the structure of the Universe bears all the marks of a transitory structure. Our theories are decidedly in a state of continuous and just now very rapid evolution.

In J.H.F. Umbgrove
The Pulse of the Earth
Chapter I (p. 1)
Martinus Nijhoff. The Hague, Netherlands. 1947

de Vries, Peter   1910–93
American editor and novelist

The universe is like a safe to which there is a combination but the combination is locked up in the safe.

Let Me Count the Ways
Chapter Twenty-Two (p. 307)
Little, Brown & Company. Boston, Massachusetts, USA. 1965

Dee, John   1527–1609
English mathematician and occultist

The entire universe is like a lyre tuned by some excellent artificer, whose strings are separate species of the universal whole.

Translated by Wayne Schumaker
John Dee on Astronomy
XI (p. 127)
University of California Press. Berkeley, California, USA. 1978

DeLillo, Don   1936–
American novelist

It's the size of things that worries people. No reason for the universe to be so large.

Ratner's Star
Vintage Contemporaries. Toronto, Ontario, Canada. 1976

Deutsch, Karl W.   1912–92
Czech-born American international political scientist

Any universe uneven enough to sustain the life of a flatworm should perhaps be uneven enough to be eventually known by man.

Mechanism, Organism, and Society: Some Models in Natural and Social Science
Philosophy of Science, Volume 18, Number 3, July 1951 (p. 231)

Dillard, Annie   1945–
American poet, essayist, novelist, and writing teacher

The universe was not made in jest but in solemn incomprehensible earnest.

Pilgrim at Tinker Creek
Chapter 15 (p. 270)
Harper's Magazine Press. New York, New York, USA. 1974

du Prel, Karl   1839–99
German hypnosis researcher

The universe as a totality is without cause, without origin, without end.

In Ludwig Buchner
Force and Matter (p. 11)
Truth Seeker. New York, New York, USA. 1950

Dyson, Freeman J.   1923–
American physicist and educator

The hypothesis is that the universe is constructed according to a principle of maximum diversity. The principle of maximum diversity operates both at the physical and at the mental level. It says that the laws of nature and the initial conditions are such as to make the universe as interesting as possible. As a result, life is possible but not too easy. Always when things are dull, something new turns up to challenge us and to stop us from settling into a rut. Examples of things which make life difficult are all around us: comet impacts, ice ages, weapons, plagues, nuclear fission, computers, sex, sin and death. Not all challenges can be overcome, and so we have tragedy. Maximum diversity often leads to maximum stress. In the end we survive, but only by the skin of our teeth.

Infinite in All Directions
Part Two, Chapter Seventeen (p. 298)
Harper Collins Publisher, Inc. New York, New York, USA. 1988

As a working hypothesis to explain the riddle of our existence, I propose that our universe is the most interesting of all possible universes, and our fate as human beings is to make it so.

Infinite in All Directions
Preface (p. vii)
Harper Collins Publisher, Inc. New York, New York, USA. 1988

As we look out into the Universe and identify the many accidents of physics and astronomy that have worked together to our benefit, it almost seems as if the Universe must in some sense have known that we were coming.

In John D. Barrow and Frank J. Tipler

The Anthropic Cosmological Principle
Chapter 5.5 (p. 318)
Clarendon Press. Oxford, England. 1986

...have found a universe growing without limit in richness and complexity, a universe of life surviving forever and making itself known to its neighbors across unimaginable gulfs of space and time. Whether the details of my calculations turn out to be correct or not, there are good scientific reasons for taking seriously the possibility that life and intelligence can succeed in molding this universe of ours to their own purposes.
Infinite in All Directions
Part One, Chapter Six (p. 117)
Harper Collins Publisher, Inc. New York, New York, USA. 1988

I do not feel like an alien in this universe. The more I examine the universe and study the details of its architecture, the more evidence I find that the universe in some sense must have known that we were coming.
Disturbing the Universe
Chapter 23 (p. 250)
Basic Books, Inc. New York, New York, USA. 1979

Eddington, Sir Arthur Stanley   1882–1944
English astronomer, physicist, and mathematician

The unanimity with which the galaxies are running away looks almost as though they had a pointed aversion to us. We wonder why we should be shunned as though our system were a plague spot in the universe.
The Expanding Universe
Chapter I, Section III (p. 12)
The University Press. Cambridge. 1933

Meanwhile the knowledge that has been attained shows only the more plainly how much there is to learn. The perplexities of today foreshadow the discoveries of the future. If we have still to leave the stellar universe a region of hidden mystery, yet it seems as though, in our exploration, we have been able to glimpse the outline of some vast combination which unites even the farthest stars into an organised system.
Stellar Movements and the Structure of the Universe
Chapter XII (p. 261)
Macmillan & Company Ltd. London, England. 1914

I would feel more content that the universe should accomplish some great scheme of evolution and, having achieved whatever may be achieved, lapse back into chaotic changelessness, than its purpose should be banalised by continual repetition. I am an Evolutionist, not a Multiplicationist. It seems rather stupid to keep doing the same thing over and over again.
The Nature of the Physical World
Chapter IV (p. 86)
The Macmillan Company. New York, New York, USA. 1930

Editorial

It is obvious that we must regard the universe as extending infinitely, forever, in every direction; or that we must regard it as not so extending. Both possibilities go beyond us.
Einstein's Finite Universe
Scientific American, Volume 124, Number 11, March 12, 1921 (p. 202)

Ehrmann, Max   1872–1945
American lawyer and writer

You are a child of the universe, no less than the trees and the stars; you have a right to be here. And whether or not it is clear to you, no doubt the universe is unfolding as it should.
Desiderata
Published by author. 1927

Eliot, T. S. (Thomas Stearns)   1888–1965
American expatriate poet and playwright

Do I dare
Disturb the universe?
In a minute there is time
For decisions and revisions which a minute will reverse.
The Collected Poems and Plays 1909–1950
The Love Song of J. Alfred Prufrock (p. 5)
Harcourt, Brace & World, Inc. New York, New York, USA. 1952

Elliot, Hugh   1752–1830
British diplomat and adventurer

No sign of purpose can be detected in any part of the vast universe disclosed by our most powerful telescopes.
Modern Science and Materialism (p. 39)
Longmans, Green & Company. 1919

Emerson, Ralph Waldo   1803–82
American lecturer, poet, and essayist

We are taught by great actions that the universe is the property of every individual in it.
Ralph Waldo Emerson: Essays and Lectures
Nature: Addresses, and Lectures
Beauty (p. 16)
The Library of America. New York, New York, USA. 1983

Philosophically considered, the universe is composed of Nature and the Soul.
Ralph Waldo Emerson: Essays and Lectures
Nature: Addresses, and Lectures
Introduction (p. 8)
The Library of America. New York, New York, USA. 1983

Everything in the universe goes by indirection. There are no straight lines.
The Complete Works of Ralph Waldo Emerson (Volume 7)
Society and Solitude
Works and Days (p. 181)
Houghton Mifflin Company. Boston, Massachusetts, USA. 1904

...the universe does not jest with us, but is in earnest...
Letters and Social Aims
Poetry and Imagination (p. 3)
James R. Osgood & Company. Boston, Massachusetts, USA. 1876

Engard, Charles J.
American botanist

We accept the universe as far as we know it, but we do not attempt to explain why it exists. It is difficult enough to understand how!
In Johann Wolfgang von Goethe, Bertha Mueller, and Charles J. Engard
Goethe's Botanical Writings
Introduction (p. 14)
University of Hawaii Press. Honolulu, Hawaii, USA. 1952

Estling, Ralph
No biographical data available

There is no question about there being design in the Universe. The question is whether this design is imposed from the Outside or whether it is inherent in the physical laws governing the Universe. The next question is, of course, who or what made these physical laws.
The Skeptical Inquirer, Spring 1993

I do not know what, if anything, the Universe has in its mind, but I am quite, quite sure that, whatever it has in its mind, it is not at all like what we have in ours. And, considering what most of us have in ours, it is just as well.
The Skeptical Inquirer, Spring 1993

Euler, Leonhard   1707–83
Swiss mathematician and physicist

For since the fabric of the universe is most perfect and the work of a most wise Creator, nothing at all takes place in the universe in which some rule of maximum or minimum does not appear.
In Morris Kline
Mathematical Thought from Ancient to Modern Times (p. 573)
Oxford University Press, Inc. New York, New York, USA. 1972

Ferris, Timothy   1944–
American science writer

There could be more life out there than we've ever imagined — for if the universe has taught us anything, it is that reality is richer and more resourceful than our wildest dreams.
Life Beyond Earth
The Ice Zone (p. 116)
Simon & Schuster. New York, New York, USA. 2000

We live in a changing universe, and few things are changing faster than our conception of it.
The Whole Shebang: A State-of-the-Universe's Report
Preface (p. 11)
Simon & Schuster. New York, New York, USA. 1996

Feynman, Richard P.   1918–88
American theoretical physicist

This universe has been described by many, but it just goes on, with its edge as unknown as the bottom of the bottomless sea of the other ideas — just as mysterious, just as awe-inspiring, and just as incomplete as the poetic pictures that came before.
The Meaning of It All: Thoughts of a Citizen Scientist
Chapter I (p. 10)
Perseus Books. Reading, Massachusetts, USA. 1998

Is no one inspired by our present picture of the universe? The value of science remains unsung by singers: you are reduced to sharing not a song or poem, but an evening lecture about it. This is not yet a scientific age.
What Do You Care What Other People Think?
The Value of Science (p. 244)
W.W. Norton & Company, Inc. New York, New York, USA.1988

Feynman, Richard P.   1918–88
American theoretical physicist
Leighton, Robert B.   1919–97
American physicist
Sands, Matthew L.   1919–
American physicist

A poet once said, "The whole universe is in a glass of wine." We will probably never know in what sense he meant that, for poets do not write to be understood.... How vivid is the claret, pressing its existence into the consciousness that watches it! If our small minds, for some convenience, divide this glass of wine, this universe, into parts — physics, biology, geology, astronomy, psychology, and so on — remember that nature does not know it! So let us put it all back together, not forgetting ultimately what it is for. Let it give us one more final pleasure: drink it and forget it all!
The Feynman Lectures on Physics (Volume 1)
Chapter 3–7 (p. 3–10)
Addison-Wesley Publishing Company. Reading, Massachusetts, USA. 1983

Field, Edward   1924–
American poet

Look, friend, at this universe
With its spiral clusters of stars
Flying out all over space
Like bedsprings suddenly bursting free;
New and Selected Poems from the Book of My Life
From Stand Up, Friend, with Me (1963) Prologue

Flammarion, Camille   1842–1925
French astronomer and author

When the last human eyelid closes here below, and our globe — after having been for so long the abode of life with its passions, its labour, its pleasures and its pains, its loves and its hatred, its religious and political expectations and all its vain finalities — is enshrouded in the winding-sheet of a profound night, when the extinct sun wakes no more; well, then — then, as to-day, the universe will be as complete, the stars will continue to shine in the sky, other suns will illuminate other worlds, other springs will bring round the bloom of flowers and the illusions of youth, other mornings and other evenings will follow in succession, and the universe will move on as at present; for creation is developed in infinity and eternity.

Popular Astronomy: A General Description of the Heavens
Book II, Chapter VI (p. 164)
Chatto & Windus. London, England. 1894

May we conclude, then, that in these successive endings the universe will one day become an immense and dark tomb. No: otherwise it would already have become so during a past eternity. There is in nature something else besides blind matter; an intellectual law of progress governs the whole creation; the forces which rule the universe cannot remain inactive. The stars will rise from their ashes. The collision of ancient wrecks causes new flames to burst forth, and the transformation of motion into heat creates nebulae and worlds. Universal death shall never reign.

Popular Astronomy: A General Description of the Heavens
Book I, Chapter VII (p. 80)
Chatto & Windus. London, England. 1894

France, Anatole (Jean Jacques Brousson) 1844–1924
French writer

The universe which science reveals to us is a dispiriting monotony. All the suns are drops of fire and all the planets are drops of mud.

In Stanley L. Jacki
Creator
Chapter Two (p. 26)
Scottish Academic Press. Edinburgh, Scotland. 1980

If desire lends a grace to whatsoever be the object of it, then the desire of the unknown makes beautiful the Universe.

My Friend's Book
Chapter XI (p. 159)
Dodd, Mead & Company. New York, New York, USA. 1924

Frayn, Michael 1933–
English dramatist

The complexity of the universe is beyond expression in any possible notation. Lift up your eyes. Not even what you see before you can ever be fully expressed. Close your eyes. Not even what you see now.

Constructions
No. 1
Wildwood House. London, England. 1974

Fritzsch, Harald 1943–
German theoretical physicist

The universe is more than merely an accretion of electrons, quarks, and galaxies, more than space and time. The complex, interrelated world of the earth which created us is part of it. It is our duty not only to ourselves to preserve this world. The universe itself imposes this task on us.

The Creation of Matter: The Universe from Beginning to End
Chapter 17 (p. 282)
Basic Books, Inc. New York, New York, USA. 1984

Frost, Robert 1874–1963
American poet

The Universe is but the Thing of Things
The things but balls all going round in rings
Some of them mighty huge, some mighty tiny
All of them radiant and might shiny.

Complete Poems of Robert Frost
Accidentally on Purpose
Henry Holt & Company. New York, New York, USA. 1949

…all reasoning is in a circle. At least that's why the universe is round.

Complete Poems of Robert Frost
Build the Soil
Henry Holt & Company. New York, New York, USA. 1949

Galilei, Galileo 1564–1642
Italian physicist and astronomer

No one will be able to read the great book of the Universe if he does not understand its language which is that of mathematics.

In A. Zee
Fearful Symmetry
Chapter 9 (p. 122)
Macmillan Publishing Company. New York, New York, USA. 1986

Philosophy is written in this grand book, the universe, which stands continually open to our gaze. But the book cannot be understood unless one first learns to comprehend the language and read the letters in which it is composed. It is written in the language of mathematics, and its characters are triangles, circles, and other geometric figures without which it is humanly impossible to understand a single word of it.

Translated by Stillman Drake
Discoveries and Opinions of Galileo
The Assayer (pp. 237–238)
Doubleday & Company, Inc. New York, New York, USA. 1957

I should think that anyone who considered it more reasonable for the whole universe to move in order to let the Earth remain fixed would be more irrational than one who should climb to the top of a cupola just to get a view of the city and its environs, and then demand that the whole countryside should revolve around him so that he would not have to take the trouble to turn his head.

Dialogues Concerning the Two Chief World Systems

Giraudoux, Jean   1882–1944
French novelist, playwright, and essayist

COUNTESS: I know perfectly well that at this moment the whole universe is listening to us — and that every word we say echoes to the remotest star.
English adaptation by Maurice Valency
The Madwoman of Chaillot
Act Two (p. 94)
Random House, Inc. New York, New York, USA. 1947

Gleiser, Marcello   1959–
Brazilian-born physicist and astronomer

Our planet is not in a special place in the solar system, our Sun is not in a special place in our galaxy and our galaxy is not in a special place in the Universe.
The Dancing Universe
Chapter 9 (p. 274)
The Penguin Group. New York, New York, USA. 1997

Guth, Alan   1947–
American physicist

The universe may be the ultimate free lunch.
In P.C.W. Davies (ed.)
The New Physics
The Inflationary Universe (p. 54)
Cambridge University Press. Cambridge, England. 1989

Guth, Alan   1947–
American physicist
Steinhardt, Paul
American physicist

The inflationary model of the universe provides a possible mechanism by which the observed universe could have evolved from an infinitesimal region. It is then tempting to go one step further and speculate that the entire universe evolved from literally nothing.
The Inflationary Universe
Scientific American, Volume 250, Number 5, May 1984 (p. 128)

Halacy, Jr., D. S.
No biographical data available

Our universe operates not at the whims of those who live in it, but by inexorable natural laws.
They Gave Their Names to Science
Prologue (p. 9)
G.P. Putnam's Sons. New York, New York, USA. 1967

Haldane, John Burdon Sanderson   1892–1964
English biologist

…the universe is not only queerer than we suppose but it is queerer than we can suppose.
Possible Worlds and Other Papers
Chapter XXXIV (p. 298)
Harper & Brothers Publishers. New York, New York, USA. 1928

Hale, George Ellery   1868–1938
American astronomer

It is a far cry from the facile imaginings of the philosopher to the rigorous demonstrations of exact science, and the true structure of the universe is not yet known.
Beyond the Milky Way
Beyond the Milky Way (p. 100)
Charles Scribner's Sons. New York, New York, USA. 1926

Hardy, G. H. (Godfrey Harold)   1877–1947
English pure mathematician

"Imaginary" universes are so much more beautiful than this stupidly constructed "real" one; and most of the finest products of an applied mathematician's fancy must be rejected, as soon as they have been created, for the brutal but sufficient reason that they do not fit the facts.
A Mathematician's Apology
Section 26 (p. 135)
Cambridge University Press. Cambridge, England. 1967

Harrison, Edward Robert
Cosmologist

What determines the design of a universe; is it the Universe, God, fortuity, or the human mind?
Masks of the Universe
Chapter 1 (p. 5)
Macmillan Publishing Company. New York, New York, USA. 1985

We cannot doubt the existence of an ultimate reality. It is the Universe forever masked. We are part of an aspect of it, and the masks figured by us are the Universe observing and understanding itself from a human point of view. When we doubt the Universe we doubt ourselves. The Universe thinks, therefore it is.
Masks of the Universe
Chapter I (p. 14)
Macmillan Publishing Company. New York, New York, USA. 1985

We do not know what sets limits to the Great Chain of hierarchical structures, nor do we know what unifies it. We are clueless as to why atoms exist and why the Universe is structured the way it is. Of course, if the Universe were structured in any other way, we would not be here asking these pertinent questions; or so we are told. But I am a heretic and inclined to think the other way: without us this Universe would not be here.
A Twinkle in the Eye of the Universe
Quarterly Journal of the Royal Astronomical Society, Volume 25, Number 4, December 1984 (p. 428)

From the outset we must decide whether to use Universe or universe. This in not so trivial a matter as it might seem. We know of only one planet called Earth; similarly, we know of only one Universe. Surely then the proper word is Universe?
Cosmology: The Science of the Universe
Chapter 1 (p. 10)
Cambridge University Press. Cambridge, England. 1981

The universe consists only of atoms and the void: all else is opinion and illusion.
Masks of the Universe
Chapter 4 (p. 55)
Macmillan Publishing Company. New York, New York, USA. 1985

Haught, James A.
No biographical data available

The universe is a vast, amazing, seething dynamo which has no discernable purpose except to keep on churning. From quarks to quasars, it's alive with incredible power. But it seems utterly indifferent to any moral laws. It destroys as blindly as it nurtures.
2000 Years of Disbelief: Famous People with The Courage to Doubt
Afterthought (p. 324)
Prometheus Books. Amherst, New York, USA. 1996

Hawking, Stephen William   1942–
English theoretical physicist

There ought to be something very special about the boundary conditions of the universe and what can be more special than the condition that there is no boundary.
In John D. Barrow and Frank J. Tipler
The Anthropic Cosmological Principle
Chapter 6.15 (p. 444)
Clarendon Press. Oxford, England. 1986

We see the universe the way it is because we exist.
A Brief History of Time: The Updated and Expanded Edition
Chapter 8 (p. 128)
Bantam Books. Toronto, Ontario, Canada. 1988

Even if there is only one possible unified theory, it is just a set of rules and equations. What is it that breathes fire into the equations and makes a universe for them to describe? The usual approach of science of constructing a mathematical model cannot answer the questions of why there should be a universe for the model to describe. Why does the universe go to all the bother of existing?
A Brief History of Time: From the Big Bang to Black Holes
Chapter 11 (p. 174)
Bantam Books. Toronto, Ontario, Canada. 1988

I do not agree with the view that the universe is a mystery, something that one can have intuition about but never fully analyze or comprehend.
Black Holes and Baby Universes and Other Essays
Preface (p. viii)
Bantam Books. New York, New York, USA. 1993

I think the universe is completely self-contained. It doesn't have any beginning or end, it doesn't have any creation or destruction.
In Renée Weber
Dialogues with Scientists and Sages: The Search for Unity (p. 89)
Routledge & Kegan Paul. London, England. 1986

…we do not know what is happening at the moment farther away in the universe: the light that we see from distant galaxies left them millions of years ago and in the case of the most distant object that we have seen, the light left some eight thousand million years ago. Thus, when we look at the universe, we are seeing it as it was in the past.
A Brief History of Time: From the Big Bang to Black Holes
Chapter 2 (p. 28)
Bantam Books. Toronto, Ontario, Canada. 1988

Hawkins, Michael   1942–
British astrophysicist

The stars and galaxies that fill our view as we survey the depths of the Universe are really just a froth delineating the massive, dark unseen structures beneath.
Hunting Down the Universe: The Missing Mass, Primordial Black Holes, and Other Dark Matters
Chapter 13 (p. 183)
Addison-Wesley Publishing Company. Reading, Massachusetts, USA. 1997

Hayflick, Leonard   1928–
American microbiologist

Everything in the universe ages, including the universe.
In Nancy Shute
U.S. News and World Report, 18/25 August 1997 (p. 57)

Haynes, Margaret
No biographical data available

The universe is just a bowl of spaghetti.
In Eric J. Lerner
The Big Bang Never Happened
Chapter 1 (p. 49)
Random House, Inc. New York, New York, USA. 1991

Heinlein, Robert A.   1907–88
American science fiction writer

Tomorrow I will seven eagles see, a great comet will appear, and voices will speak from whirlwinds foretelling monstrous and fearful things — This Universe never did make sense; I suspect that it was built on government contract.
The Number of the Beast
Chapter II (p. 19)
Fawcett Columbine Books. New York, New York, USA. 1980

No storyteller has ever been able to dream up anything as fantastically unlikely as what really does happen in this mad universe.
Time Enough for Love
Prelude, Chapter II (p. 51)
G.P. Putnam's Sons. New York, New York, USA. 1973

A zygote is a gamete's way of producing more gametes. This may be the purpose of the universe.
Time Enough for Love
Intermission (p. 262)
G.P. Putnam's Sons. New York, New York, USA. 1973

Henderson, Archibald   1877–1933
American mathematician and writer

We are doomed to dwell within a finite universe a thousand million times greater than the region now accessible to astronomical observation. Our glances are confined for ever within this giant — this all too minute — monad.
The Size of the Universe
Science, Volume 58, Number 1497, 7 September, 1923 (p. 172)

Hinshelwood, Sir Cyril   1897–1967
English chemist

To some men knowledge of the universe has been an end possessing in itself a value that is absolute: to others it has seemed a means of useful application.
The Structure of Physical Chemistry
Chapter I (p. 2)
Clarendon Press. Oxford, England. 1951

Hogan, John
No biographical data

…cosmologists — and the rest of us — may have to forego attempts at understanding the universe and simply marvel at its infinite complexity and strangeness.
Universal Truths
Scientific American, Volume 263, Number 4, October 1990 (p. 117)

Holmes, John Haynes   1879–1964
American clergyman

The universe is not hostile, nor yet is it friendly. It is simply indifferent.
A Sensible Man's View of Religion
Is the Universe Friendly? (p. 39)

Hoyle, Sir Fred   1915–2001
English mathematician and astronomer

There is a coherent plan in the universe, though I don't know what it's a plan for.
Wired, 2/98 (p. 174)

Perhaps the most majestic feature of our whole existence is that while our intelligences are powerful enough to penetrate deeply into the evaluation of this quite incredible Universe, we still have not the smallest clue to our own fate.
The Nature of the Universe
Chapter 7 (p. 142)
The University Press. Cambridge. 1933

The Universe is everything; both living and inanimate things; both atoms and galaxies; and if the spiritual exists as well as the material, of spiritual things also; and if there is a Heaven and a Hell, of Heaven and Hell too; for by its very nature the Universe is the totality of all things.
Frontiers of Astronomy
Chapter Eighteen (p. 304)
Harper & Row, Publishers. New York, New York, USA. 1955

…if there is one important result that comes out of our inquiry into the nature of the Universe it is this: when by patient inquiry we learn the answer to any problem, we always find, both as a whole and in detail, that the answer thus revealed is finer in concept and design than anything we could ever have arrived at by a random guess.
The Nature of the Universe
Chapter 7 (p. 140)
The University Press. Cambridge. 1933

Hubble, Edwin Powell   1889–1953
American astronomer

We find them smaller and fainter, in constantly increasing numbers, and we know that we are reaching out into space, further and ever further, until, with the faintest nebulae that can be detected with the greatest telescope, we arrive at the frontiers of the known universe.
In Joseph Silk
The Big Bang (p. 26)
W.H. Freeman & Company. San Francisco, California, USA. 1980

…equipped with his five senses, man explores the universe around him and calls the adventure science.
The Nature of Science and Other Lectures
Part I, The Nature of Science (p. 6)
The Huntington Library. San Marino, California, USA. 1954

Huygens, Christiaan   1629–95
Dutch mathematician, astronomer, and physicist

What a wonderful and amazing Scheme have we here of the magnificent Vastness of the Universe! So many Suns, so many Earths…!
The Celestial Worlds Discover'd; or, Conjectures Concerning the Planetary Worlds, Their Inhabitants and Productions
Kosmotheoros (p. 222)
Printed for T. Childe. London, England. 1698

Inge, William Ralph   1860–1954
English religious leader and author

If the universe is running down like a clock, the clock must have been wound up at a date which we could name if we knew it. The world, if it is to have an end in time, must have had a beginning in time.
God and the Astronomers
Chapter 3 (p. 48)
W.W. Norton & Company, Inc. New York, New York, USA.1978

Ionesco, Eugene   1912–94
French playwright

…the universe seems to me infinitely strange and foreign. At such a moment I gaze upon it with a mixture of anguish and euphoria; separate from the universe, as though placed at a certain distance outside it; I look and I see pictures, creatures that move in a kind of timeless time and spaceless space, emitting sounds that are a kind of language I no longer understand or ever register.

Notes and Counter Notes: Writing on the Theatre
Part II, Interviews, Brief Notes for Radio (p. 136)

James, William   1842–1910
American philosopher and psychologist

Whatever universe a professor believes in must at any rate be a universe that lends itself to lengthy discourse. A universe definable in two sentences is something for which the professorial intellect has no use. No faith in anything of that cheap kind!
Pragmatism: A New Name for Some Old Ways of Thinking
Lecture I (p. 4)
Longmans, Green & Company. London, England. 1914

Jastrow, Robert   1925–
American space scientist

Theologians generally are delighted with the proof that the Universe had a beginning, but astronomers are curiously upset. Their reactions provide an interesting demonstration of the response of the scientific mind — supposedly a very objective mind — when evidence uncovered by science itself leads to a conflict with the articles of faith in our profession. It turns out that the scientist behaves the way the rest of us do when our beliefs are in conflict with the evidence. We become irritated, we pretend the conflict does not exist, or we paper it over with meaningless phrases.
God and the Astronomers
Epilogue (p. 117)
W.W. Norton & Company, Inc. New York, New York, USA.1978

Thus, the facts indicate that the Universe will expand forever. We still come across pieces of mass here and there in the Universe, and someday we may find the missing matter, but the consensus at the moment is that it will not be found. According to the available evidence, the end will come in darkness.
God and the Astronomers
Epilogue (p. 123)
W.W. Norton & Company, Inc. New York, New York, USA.1978

Jeans, Sir James Hopwood   1877–1946
English physicist and mathematician

The universe can not go on forever as it now is, and neither can it have existed in its present condition from all eternity.
Annual Report of the Board of Regents of the Smithsonian Institution, 1926
The New Outlook in Cosmogony (p. 155)
Government Printing Office. Washington, D.C. 1928

The Universe can be best pictured, although still very imperfectly and inadequately, as consisting of pure thought, the thought of what, for want of a wider word, we must describe as a mathematical thinker.
The Mysterious Universe
Chapter V (p. 136)
The Macmillan Company. New York, New York, USA. 1932

The universe begins to look more like a great thought than a machine.
In Jefferson Hane Weaver
The World of Physics (Volume 2)
Q.3 (p. 632)
Simon & Schuster. New York, New York, USA. 1987

The physical universe never has any choice — it must inevitably move along a single road to a predestined end.
Contributions to a British Association Discussion on the Evolution of the Universe
Nature, Supplement, October 24, 1931 (p. 701)

The universe consists in the main not of stars but of desolate emptiness — inconceivably vast stretches of desert space in which the presence of a star is a rare and exceptional event. … The stars move blindly through space, and the players in the stellar blind-man's-buff are so few and far between that the chance of encountering another star is almost negligible.
The Universe Around Us
Chapter I (pp. 84–85)
The Macmillan Company. New York, New York, USA. 1929

Jeffers, Robinson   1887–1962
American poet

The learned astronomer
Analyzing the light of most remote star-swirls
Has found them — or a trick of distance deludes his prism — All at incredible speeds fleeing outward from ours.
I thought, no doubt they are fleeing the contagion
Of consciousness that infects this corner of space.
In Tim Hunt (ed.)
The Collected Poetry of Robinson Jeffers (Volume 2)
Margrave (p. 161)
Stanford University Press. Stanford, California. USA. 1988

The universe expands and contracts like a great heart.
It is expanding, the farthest nebulae
Rush with the speed of light into empty space.
The Beginning and the End and Other Poems
The Great Explosion (p. 3)
Random House, Inc. New York, New York, USA. 1963

It seemed to Barclay the cloud broke and he saw the stars,
Those of this swarm were many, but beyond them universe past universe
Flared to infinity, no end conceivable. Alien, alien, alien universes.
In Tim Hunt (ed.)
The Collected Poetry of Robinson Jeffers (Volume 1)
The Women at Point Sur (p. 312)
Stanford University Press. Stanford, California. USA. 1988

Jennings, Herbert Spencer   1868–1947
American zoologist

…the universe is not "a mere clockwork mechanical wonder swinging in a vast vacuum," but is a system that, in the course of time, comes to life.

The Universe and Life
Chapter II (p. 33)
Yale University Press. New Haven, Connecticut, USA. 1941

Joad, Cyril Edwin Mitchinson   1891–1953
English philosopher and broadcasting personality

When the scientist leaves his laboratory and speculates about the universe as a whole, the resultant conclusions are apt to tell us more about the scientist than about the universe.
Philosophical Aspects of Modern Science
Chapter XI (p. 339)
George Allen & Unwin Ltd. London, England. 1939

Kaku, Michio   1947–
Japanese-American theoretical physicist

The fact that our universe, like Appleworld, is curved in an unseen dimension beyond our spatial comprehension has been experimentally verified. These experiments, performed on the path of light beams, shows that starlight is bent as it moves across the universe.
Hyperspace : A Scientific Odyssey Through Parallel Universes, Time Warps, and the 10th Dimension
Chapter 1 (p. 17)
Oxford University Press, Inc. New York, New York, USA. 1995

Kepler, Johannes   1571–1630
German astronomer

This very cogitation carries with it I don't know what secret, hidden horror; indeed one finds oneself wandering in this immensity to which are denied limits and centre and therefore also all determinate places.
In Alexander Koyre
From the Closed World to the Infinite Universe
Chapter III (p. 61)
The Johns Hopkins Press. Baltimore, Maryland, USA. 1968

The diversity of the phenomena of Nature is so great, and the treasures hidden in the heavens so rich, precisely in order that the human mind shall never be lacking in fresh nourishment.
Mysterium Cosmographicum
Original Dedication (p. 55)

Keyser, Cassius Jackson   1862–1947
American mathematician

Depend upon it, the universe will never really be understood unless it may be sometime resolved into an ordered multiplicity and made to own itself an everlasting drama of the calculus.
The Human Worth of Rigorous Thinking: Essays and Addresses
Mathematical Emancipations: Dimensionality and Hyperspace (p. 101)
Columbia University Press. New York, New York, USA. 1925

Kirshner, Robert P.
American astronomer

Although the Universe is under no obligation to make sense, students in pursuit of the Ph.D. are.
Exploding Stars and the Expanding Universe
Quarterly Journal of the Royal Astronomical Society, Volume 32, Number 3, September 1991 (p. 240)

If Copernicus taught us the lesson that we are not at the center of things, our present picture of the universe rubs it in.
In John Noble Wilford
From Distant Galaxies, News of a "Stop-and-Go Universe"
New York Times, June 3, 2003

Koestler, Arthur   1905–83
Hungarian-born English writer

There are no longer any absolute directions in space. The universe has lost its core. It no longer has a heart, but a thousand hearts.
The Sleepwalkers
Part Three, Chapter II, Section 6 (p. 217)
The Macmillan Company. New York, New York, USA. 1966

In my youth I regarded the universe as an open book, printed in the language of physical equations, whereas now it appears to me as a text written in invisible ink, of which in our rare moments of grace we are able to decipher a small fragment.
Bricks to Babel
Epilogue (pp. 682–683)
Random House, Inc. New York, New York, USA. 1980

Kolb, Edward W. (Rocky)   1951–
American cosmologist

There are many beautiful and wondrous things to see in the universe, and to discover them we simply have to gaze into the dark night sky.
Blind Watchers of the Sky
Chapter One (p. 1)
Addison-Wesley Publishing Company. Reading, Massachusetts, USA. 1996

Whether our cosmological view of the universe is right or wrong, or just incomplete, we were brave enough to confront our ignorance and look. We looked with all our might, and with boldness and imagination managed to see a little bit farther than our predecessors. We were not proud of our blindness, but neither were we ashamed of it or intimidated by it, for we chose to look for the light of truth fully cognizant of our blindness.
Blind Watchers of the Sky
Chapter Ten (p. 282)
Addison-Wesley Publishing Company. Reading, Massachusetts, USA. 1996

Krauss, Lawrence M.   1954–
American theoretical physicist

There is a maxim about the universe which I always tell my students: That which is not explicitly forbidden is

guaranteed to occur. Or, as Data said in the episode "Parallels," referring to the laws of quantum mechanics, "All things which can occur, do occur."
The Physics of Star Trek
Chapter Two (p. 16)
Harp Perennial Publishers. New York, New York, USA. 1995

Kunitz, Stanley   1905–2006
American poet

I see lines of your spectrum shifting red,
The Universe expanding, thinning out,
Our worlds flying, oh flying, fast apart.
The Collected Poems
The Science of the Night (p. 88)
W.W. Norton & Company, Inc. New York, New York, USA.2000

Kunz, F. L.
No biographical data available

The whole universe is one mathematical and harmonic expression, made up of finite representations of the infinite.
In Renée Weber
Dialogues with Scientists and Sages: The Search for Unity (p. 139)
Routledge & Kegan Paul. London, England. 1986

Lambert, Johann Heinrich   1728–77
Swiss-German mathematician and astronomer

…the firmament must be more enduring than things on our earth and the empires of the world.
Translated by Stanley Jaki
Cosmological Letters on the Arrangement of the World-Edifice
Twentieth Letter (p. 193)
Science History Publications. New York, New York, USA. 1976

Lao Tzu   fl. 6th century BCE
Chinese philosopher and father of Taoism

Something mysteriously formed,
Born before heaven and earth.
In the silence and the void,
Standing alone and unchanging,
Ever present and in motion.
Translated by Gia-Fu Feng and Jane English
Tao Te Ching
Twenty-five
Alfred A. Knopf. New York, New York, USA. 1974

Lederman, Leon   1922–
American high-energy physicist

We hope to explain the entire universe in a single, simple formula that you can wear on your T-shirt.
In Richard Wolkomir article
Quark City
OMNI Magazine, February 1984 (p. 41)

Lemaître, Abbé Georges   1894–1966
Belgian astronomer and cosmologist

Our universe bears the marks of youth and we can hope to reconstruct its story. The documents at our disposal are not buried in the piles of bricks carved by the Babylonians; our library does not risk being destroyed by fire; it is in space, admirably empty, where light waves are preserved better than sound is conserved on the wax of phonograph discs.
The Primeval Atom
Chapter II (p. 75)
D. Van Nostrand Company, Inc. New York, New York, USA. 1950

Lloyd, Seth   1950–
American engineer

I wanted to get a handle on why the universe is so complex…. Or at any rate why there seems to be so much information processing going on. You can look at life and almost all the things — well, all the things — that we see going on around us in terms of information processing. You could say that life is an example of information being processed in the service of getting a free lunch out of your environment. A typical event in evolution, for example, is some organism suddenly, by a mutation, being able to produce an enzyme that allows it to digest something it couldn't get at before. The free lunch is there, but in order to get it, you've got to be able to process information.
In Tim Folger
Discover Magazine
The Best Computer in All Possible Worlds, October 1995

Longfellow, Henry Wadsworth   1807–82
American poet

The Universe as an immeasurable wheel
Turning for evermore
In the rapid and rushing river of Time.
The Poetical Works of Henry Wadsworth Longfellow
Rain in Summer
Houghton Mifflin Company. Boston, Massachusetts, USA. 1883

Lucretius   ca. 99 BCE–55 BCE
Roman poet

…the existing universe is bounded in none of its dimensions; for then it must have an outside.
In *Great Books of the Western World* (Volume 12)
Lucretius: on the Nature of Things
Book One, l. 958 (p. 12)
Encyclopædia Britannica, Inc. Chicago, Illinois, USA. 1952

Mach, Ernst   1838–1916
Austrian physicist and philosopher

The universe is like a machine in which the motion of certain parts is determined by that of others, only nothing is determined about the motion of the whole machine.
History and Root of the Principle of the Conservation of Energy
Chapter IV (p. 62)
The Open Court Publishing Company. Chicago, Illinois, USA. 1911

Marquis, Don   1878–1937
American newspaperman, poet, and playwright

you write so many things
about me that are not true
complained the universe
there are so many things
about you which you seem to be
unconscious of yourself said archy
contain a number of things
which i am trying to forget
rejoined the universe
such as what asked archy
such as cockroaches and poets
replied the universe
the lives and times of archy & mehitabel
poets (p. 289)
Doubleday, Doran & Company, Inc. Garden City, New York, USA. 1933

do not tell me said warty bliggens that there is not a
purpose in the universe
the thought is blasphemy
the lives and times of archy & mehitabel
warty bliggens, the toad (p. 56)
Doubleday, Doran & Company, Inc. Garden City, New York, USA. 1933

the men of science are talking about the size and shape
of the universe
again i thought i had settled that for them years ago
it is as big as you think it is and it is spherical in shape
the lives and times of archy & mehitabel
why the earth is round (p. 284)
Doubleday, Doran & Company, Inc. Garden City, New York, USA. 1933

Masson, David
No biographical data available

Erst, space was nebulous.
It whirled and in the whirl, the nebulous milk
Broke into rifts and curdled into orbs —
Whirled and still curdled, till the azure rifts
Severed and shored vast systems, all of orbs.
In Alexander Winchell
World-Life or Comparative Geology
Part II, Chapter I (p. 145)
S.C. Griggs & Company. Chicago, Illinois, USA. 1883

McAleer, Neil   1942–
American science writer

If science ever knows for certain the fate of the Universe,
what will this tell us? It will tell us the ultimate fate
of the atoms that now make up our living bodies and
brains, the same atoms that allow us to exist and struggle to give meaning to the Universe and to our brief
lives.
The Mind-Boggling Universe
Chapter 6 (p. 239)
Doubleday & Company, Inc. Garden City, New York, USA. 1957

Melville, Herman   1819–91
American novelist

It's too late to make any improvements now. The universe
is finished; the copestone is set on, and the chips were
carted off a million years ago.
In *Great Books of the Western World* (Volume 48)
Moby Dick
Chapter 2 (p. 7)
Encyclopædia Britannica, Inc. Chicago, Illinois, USA. 1952

Minto, Walter
No biographical data available

This immense, beautiful, and varied universe is a book
written by the finger of Omnipotence and raises the
admiration of every attentive beholder.
In John Archibald Wheeler
At Home in the Universe
The Spirit of Colleagueship at Princeton (p. 89)
The American Institute of Physics. Woodbury, New York, USA. 1994

Morris, Richard   1939–2003
American physicist and science writer

How is it that common elements such as carbon,
nitrogen, and oxygen happened to have just the right
kind of atomic structure that they needed to combine
to make the molecules upon which life depends? It is
almost as though the universe had been consciously
designed.
The Fate of the Universe
Chapter 8 (pp. 154–155)
Playboy Press. New York, New York, USA. 1982

Muir, John   1838–1914
American naturalist

When we try to pick out anything by itself, we find it
hitched to everything else in the universe.
My First Summer in the Sierra
July 27 (p. 211)
Houghton Mifflin Company. Boston, Massachusetts, USA. 1911

The clearest way into the Universe is through a forest
wilderness.
The Wilderness World of John Muir
The Philosophy of John Muir (p. 312)
Houghton, Mifflin & Company. Boston, Massachusetts, USA. 2001

How hard to realize that every camp of men or beast has
this glorious starry firmament for a roof! In such places
standing alone on the mountaintop it is easy to realize that
whatever special nests we make of leaves and moss like
marmots and birds, or tents or piled stone we all dwell in
a house of one room the world with the firmament for its
roof and are sailing the celestial spaces without leaving
any track.
The Wilderness World of John Muir
The Philosophy of John Muir (p. 312)
Houghton, Mifflin & Company. Boston, Massachusetts, USA. 2001

Newton, Sir Isaac   1642–1727
English physicist and mathematician

The most beautiful system of the sun, planets, and comets, could only proceed from the counsel and dominion of an intelligent and powerful Being.

A Heavenly Master governs all the world as Sovereign of the universe. We are astonished at Him by reason of His perfection, we honor Him and fall down before Him because of His unlimited power. From blind physical necessity, which is always and everywhere the same, no variety adhering to time and place could evolve, and all variety of created objects which represent order and life in the universe could happen only by the willful reasoning of its original Creator, Whom I call the Lord God.
Principia
General Scholium

…since Space is divisible in infinitum, and Matter is not necessarily in all places, it may be also allow'd that God is able to create Particles of Matter of several Sizes and Figures, and in several Proportions to Space, and perhaps of different Densities and Forces, and thereby to vary the Laws of Nature, and make Worlds of several sort in several Parts of the Universe.
In *Great Books of the Western World* (Volume 34)
Optics
Query 31
Encyclopædia Britannica, Inc. Chicago, Illinois, USA. 1952

Nietzsche, Friedrich   1844–1900
German philosopher

If the universe may be conceived as a definite quantity of energy, as a definite number of centers of energy — and every other concept remains indefinite and therefore useless — it follows therefore that the universe must go through a calculable number of combinations in the great game of chance which constitutes its existence. In infinity, at some moment or other, every possible combination must once have been realized; not only this, but it must have been realized an infinite number of times.
Complete Works
Volume IX (p. 430)
Foulis. Edinburgh, Scotland. 1913

Noyes, Alfred   1880–1958
English poet

This universe exists, and by that one impossible fact
Declares itself a miracle.
The Torch Bearers: Watchers of The Sky
Newton, VII (p. 226)
Frederick A. Stokes Company Publishers. New York, New York, USA. 1922

Oates, Joyce Carol   1938–
American writer

Nothing is accidental in the universe — this is one of my Laws of Physics — except the entire universe itself, which is Pure Accident, pure divinity.

Do What You Will
The Summing Up: Meredith Dawe
Random House, Inc. New York, New York, USA. 1982

Pagels, Heinz R.   1939–88
American physicist and science writer

…the universe contains the record of its past the way that sedimentary layers of rock contain the geological record of the earth's past.
Perfect Symmetry: The Search for the Beginning of Time
Part One, Chapter 2 (p. 24)
Simon & Schuster. New York, New York, USA. 1985

Parker, Barry
Canadian science writer

Looking into the dark night sky we feel a tingle of excitement as we are overcome by its grandeur and beauty. Each point of light we see is the image of a star, an image of light that may have left the star long before we were born. The universe is vast beyond imagination — almost terrifying in its intensity and complexity.
Einstein's Dream: The Search for a Unified Theory of the Universe
Chapter 1 (p. 1)
Plenum Press. New York, New York, USA. 1986

Pascal, Blaise   1623–62
French mathematician and physicist

The spaces of the universe enfold me and swallow me up like a speck; but I, by the power of thought, may comprehend the universe.
In *Great Books of the Western World* (Volume 33)
Pensées
Section VI, 348
Encyclopædia Britannica, Inc. Chicago, Illinois, USA. 1952

[The Universe] is an infinite sphere, the centre of which is everywhere, the circumference nowhere.
In *Great Books of the Western World* (Volume 33)
Pensées
Section II, 72
Encyclopædia Britannica, Inc. Chicago, Illinois, USA. 1952

Pasteur, Louis   1822–95
French chemist

The universe is asymmetrical; for, if the whole of the bodies which compose the solar system moving with their individual movements were placed before a glass, the image in the glass could not be superimposed upon the reality.
In René Dubos
Louis Pasteur: Free Lance of Science
Chapter IV (p. 111)
Little, Brown & Company. Boston, Massachusetts, USA. 1950

Peattie, Donald Culrose   1896–1964
American botanist, naturalist, and author

There is no certainty vouchsafed us in the vast testimony of Nature that the universe was designed for man, nor yet for any purpose, even the bleak purpose of symmetry. The courageous thinker must look the inimical aspects of his environment in the face, and accept the stern fact that the universe is hostile and deadly to him save for a very narrow zone where it permits him, for a few eons, to exist.
An Almanac for Moderns
March Twenty-Ninth (p. 11)
G.P. Putnam's Sons. New York, New York, USA. 1935

Penzias, Arno   1933–
German-American mathematical physicist

Either we've seen the birth of the universe, or we've seen a pile of pigeon shit.
In Roylston Roberts
Serendipity: Accidental Discoveries in Science
John Wiley & Sons, Inc. New York, New York, USA. 1989

Peirce, Charles Sanders   1839–1914
American scientist, logician, and philosopher

The universe ought to be presumed too vast to have any character.
Chance, Love, and Logic: Philosophical Essays (p. 127)
University of Nebraska Press. Lincoln, Nebraska, USA. 1998

Plato   428 BCE–347 BCE
Greek philosopher

Time and the heavens came into being at the same instant, in order that, if they were ever to dissolve, they might be dissolved together. Such was the mind and thought of God in the creation of time.
In James Jeans
The Mysterious Universe
Chapter V (p. 182)
The Macmillan Company. New York, New York, USA. 1932

…had we never seen the stars, and the sun, and the heaven, none of the words which we have spoken about the universe would ever have been uttered.
In *Great Books of the Western World* (Volume 7)
Timaeus
Section 47 (p. 455)
Encyclopædia Britannica, Inc. Chicago, Illinois, USA. 1952

Poe, Edgar Allan   1809–49
American short story writer

Telescopic observations, guided by the laws of perspective, enable us to understand that the perceptible Universe exists as a roughly spherical cluster of clusters irregularly disposed.
Eureka
Line 16 (p. 96)
Geo. P. Putnam. New York, New York, USA. 1848

The Universe is a plot of God.
Eureka

Line 7 (p. 120)
Geo. P. Putnam. New York, New York, USA. 1848

I design to speak of the Physical, Metaphysical and Mathematical — of the Material and Spiritual Universe: — of its Essence, its Origin, its Creation, its Present Condition and its Destiny.
Eureka
Line 9 (p. 7)
Geo. P. Putnam. New York, New York, USA. 1848

Polanyi, Michael   1891–1976
Hungarian-born English scientist, philosopher, and social scientist

The universe is still dead, but it already has the capacity of coming to life.
In Freeman Dyson
Personal Knowledge
Chapter 13, Section 7 (p. 404)
Harper & Row, Publishers. New York, New York, USA. 1962

Pope, Alexander   1688–1744
English poet

Order is Heav'n's first law.
The Complete Poetical Works (Volume 2)
An Essay on Man, Epistle IV, l. 49
Houghton Mifflin Company. New York, New York, USA. 1903

Prigogine, Ilya   1917–2003
Russian-born Belgian physical chemist

I certainly think we are only living in the prehistory of the understanding of our universe.
In Renée Weber
Dialogues with Scientists and Sages: The Search for Unity (p. 199)
Routledge & Kegan Paul. London, England. 1986

Rabi, Isidor Isaac   1898–1988
Austrian-born American physicist

The scientist does not defy the universe. He accepts it. It is his dish to savor, his realm to explore; it is his adventure and never-ending delight. It is complaisant and elusive but never dull. It is wonderful both in the small and in the large. In short, its exploration is the highest occupation for a gentleman.
In Leon Lederman
The God Particle: If the Universe Is the Answer, What Is the Question?
Chapter 4 (p. 104)
Houghton Mifflin Company. Boston, Massachusetts, USA. 1993

Ramón y Cajal, Santiago   1852–1934
Spanish histologist

As long as the brain is a mystery, the universe will also be a mystery.
In Victor Cohn
Charting the Soul's Frail Dwelling-House
The Washington Post, September 5, 1982, Final Edition (p. A1)

Rand, Ayn
Russian-born American novelist and philosopher

know not if this earth on which I stand is the core of the universe or if it is but a speck of dust lost in eternity.
Anthem
Chapter XI (p. 95)
E.P. Dutton & Company. New York, New York, USA. 1995

Raymo, Chet   1936–
American physicist and science writer

Give me the ninety-two elements and I'll give you a universe. Ubiquitous hydrogen. Standoffish helium, Spooky boron. No-nonsense carbon. Promiscuous oxygen. Faithful iron. Mysterious phosphorous. Exotic xenon. Brash tin. Slippery mercury. Heavy-footed lead.
The Soul of The Night
Chapter 7 (p. 65)
Prentice-Hall, Inc. Englewood Cliffs, New Jersey, USA. 1985

Reade, Winwood   1838–75
English philosopher and historian

The universe is anonymous; it is published under secondary laws; these at least we are able to investigate, and in these perhaps we may find a partial solution of the great problem.
The Martyrdom of Man
Chapter IV (p. 465)
E.P. Dutton & Company. New York, New York, 1926

Reed, Ishmael   1938–
American poet, essayist, and writer

The universe is a spiraling Big Band in a polka-dotted speak-easy, effectively generating new lights every one-night stand.
In A. Zee
An Old Man's Toy: Gravity at Work and Play in Einstein's Universe
Chapter 8 (p. 123)
The Macmillan Company. New York, New York, USA. 1989

Reichenbach, Hans   1891–1953
German philosopher of science

Instead of asking for a cause of the universe, the scientist can ask only for the cause of the present state of the universe; and his task will consist in pushing farther and farther back the date from which he is able to account for the universe in terms of laws of nature.
The Rise of Scientific Philosophy
Chapter 12 (p. 208)
University of California Press. Berkeley, California, USA. 1951

Remsen, Ira   1846–1927
American chemist

The universe is inexhaustible, and its mysteries are inexplicable. We may and must strive to learn all we can, but we can not hope to learn all. We are finite; the mysteries we are dealing with are infinite.
The Age of Science
Science, New Series, Volume 20, Number 407, July 15, 1904 (p. 73)

Renard, Maurice   1875–1939
French writer

Man, peeping at the Universe through only a few tiny windows — his senses — catches mere glimpses of the world around him. He would do well to brace himself against unexpected surprises from the vast unknown; from that immeasurable sector of reality that has remained a closed book.
In Charles Noël Martin
The Role of Perception in Science (p. 7)
Hutchinson of London. London, England. 1963

Richards, Theodore William   1868–1928
American chemist

The mystery that enshrouds the ultimate nature of the physical universe has always stimulated the curiosity to the thinking man.
Annual Report of the Board of Regents of the Smithsonian Institution, 1911 (Faraday Lecture) The Fundamental Properties of the Elements (p. 100)
Government Printing Office. Washington, D.C. 1912

Rindler, Wolfgang
Physicist and author

Modern scientific man has largely lost his sense of awe in the Universe. He is confident that given sufficient intelligence, perseverance, time, and money, he can understand all there is beyond the stars.
In M. Taube
Evolution of Matter and Energy
Chapter 2 (p. 18)
Springer-Verlag. New York, New York, USA. 1985

Rothman, Tony   1953–
American cosmologist

When confronted with the order and beauty of the universe and the strange coincidences of nature, it's very tempting to take the leap of faith from science into religion. I am sure many physicists want to. I only wish they would admit it.
In J.L. Casti
Paradigms Lost: Images of Man in the Mirror of Science
Chapter 7 (pp. 482–483)
William Morrow & Company, Inc., New York, New York, USA. 1989

Rubin, Vera   1928–
American astronomer

The joy and fun of understanding the universe we bequeath to our grandchildren — and to their grandchildren. With over 90% of the matter in the universe still to play with, even the sky will not be the limit.
In Marcia Bartusiak
The Woman Who Spins the Stars
Discover, October 1990 (p. 94)

Ruderman, M. A.
No biographical data available

Rosenfeld, A. H.
No biographical data available

We are peeling an onion layer by layer, each layer uncovering in a sense another universe, unexpected, complicated, and — as we understand more — strangely beautiful.
An Explanatory Statement on Elementary Particle Physics
American Scientist, Volume 48, June 1960, Number 2 (p. 210)

Russell, Bertrand Arthur William   1872–1970
English philosopher, logician, and social reformer

So far as scientific evidence goes, the universe has crawled by slow stages to a somewhat pitiful result on this earth, and is going to crawl by still more pitiful stages to a condition of universal death.
Why I Am Not a Christian: And Other Essays on Religion and Related Subjects
Has Religion Made Useful Contributions to Civilization (p. 32)
Watts. London, England. 1927

The Universe may have a purpose, but nothing that we know suggests that, if so, this purpose has any similarity to ours.
Why I Am Not a Christian: And Other Essays on Religion and Related Subjects
Do We Survive Death? (p. 92)
Watts. London, England. 1927

All the labors of the ages, all the devotion, all the inspiration, all the noonday brightness of human genius, are destined to extinction in the vast death of the solar system; and the whole temple of Man's achievement must inevitably be buried beneath the debris of a universe in ruins.
In George Smoot
Wrinkles in Time
Chapter 4 (p. 69)
William Morrow & Company, Inc. New York, New York, USA. 1993

Sagan, Carl   1934–96
American astronomer and science writer

We might have lived in a Universe in which nothing could be understood by a few simple laws, in which Nature was complex beyond our abilities to understand, in which laws that apply on Earth are invalid on Mars, or in a distant quasar. But the evidence — not the preconceptions, the evidence — proves otherwise. Luckily for us, we live in a Universe in which much can be "reduced" to a small number of comparatively simple laws of Nature. Otherwise we might have lacked the intellectual capacity and grasp to comprehend the world.
The Demon-Haunted World: Science As a Candle in the Dark
Chapter 15 (pp. 273–274)
Random House, Inc. New York, New York, USA. 1995

The universe is not required to be in perfect harmony with human ambition.

But the fact that some geniuses were laughed at does not imply that all who are laughed at are geniuses. They laughed at Columbus, they laughed at Fulton, they laughed at the Wright brothers. But they also laughed at Bozo the Clown.
Cosmos
Chapter II (p. 31)
Random House, Inc. New York, New York, USA. 1980

A universe in which everything is known would be static and dull, as boring as the heaven of some weak-minded theologians. A Universe that is unknowable is no fit place for a thinking being.
Broca's Brain: Reflections on the Romance of Science
Part I, Chapter 2 (p. 18)
Random House, Inc. New York, New York, USA. 1979

A religion old or new, that stressed the magnificence of the universe as revealed by modern science, might be able to draw forth reserves of reverence and awe hardly tapped by the conventional faiths. Sooner or later, such a religion will emerge.
Pale Blue Dot: A Vision of the Human Future in Space
Chapter 4 (p. 52)
Random House, Inc. New York, New York, USA. 1994

Sagan, Carl   1934–96
American astronomer and author
Druyan, Ann   1949–
American author and television producer

The Universe is lavish beyond imagining.
Shadows of Forgotten Ancestors: A Search for Who We Are
Chapter 1 (p. 13)
Random House, Inc. New York, New York, USA. 1992

Sandage, Allan   1926–
American astronomer

The present universe is something like the old professor nearing retirement with his brilliant future behind him.
In G. Borner
The Early Universe
Chapter 3 (p. 90)
Springer-Verlag, Berlin, Germany. 1988

Santayana, George (Jorge Augustín Nicolás Ruiz de Santillana)   1863–1952
Spanish-born American philosopher

The universe, as far as we can observe it, is a wonderful and immense engine; its extent, its order, its beauty, its cruelty, make it alike impressive. If we dramatize its life and conceive its spirit, we are filled with wonder, terror, and amusement, so magnificent is that spirit, so prolific, inexorable, grammatical and dull.
In Logan Pearsall Smith
Little Essays Drawn from the Writings of George Santayana
Piety (p. 85)
Books for Libraries Press, Freeport, New York, USA. 1967

Schiller, Ferdinand Canning Scott 1759–1805
German poet, dramatist and philosopher

The universe is one of God's thoughts.
Essays: Aesthetical and Philosophical
Letter 4: Theosophy of Julius

Shelley, Percy Bysshe 1792–1822
English poet

The curtain of the Universe is rent and shattered,
The splendor-winged worlds disperse like wild doves
scattered.
Hellas, Leaves of Grass. A Lyrical Drama

Its easier to suppose that the universe has existed from
all eternity than to conceive a Being beyond its limits
capable of creating it.
The Complete Poetical Works of Percy Bysshe Shelley
Queen Mab
Houghton Mifflin Company. Boston, Massachusetts, USA. 1901

Below lay stretched the boundless universe!
The Complete Poetical Works of Percy Bysshe Shelley
The Daemon of the World Part I, The Daemon and the Spirit
l 241
Houghton Mifflin Company. Boston, Massachusetts, USA. 1901

Siegel, Eli 1902–78
American philosopher, poet, critic, and founder of Aesthetic Realism

The universe is Why, How, and What, in any order, and
all at once.
Damned Welcome
Aesthetic Realism, Maxims, Part One, #69 (p. 28)
Definition Press. New York, New York, USA. 1972

The universe, being clever, has given scientists trouble.
Damned Welcome
Aesthetic Realism, Maxims, Part One, #71 (p. 28)
Definition Press. New York, New York, USA. 1972

The weight of the universe is at one with all its space.
Damned Welcome
Aesthetic Realism, Maxims, Part One, #70 (p. 28)
Definition Press. New York, New York, USA. 1972

Silk, Joseph 1942–
American astronomer and physicist

The development of human awareness of the Universe
evolved from the geocentric cosmology of the ancient
world via the heliocentric cosmology of the Renaissance
and the egocentric cosmology of the nineteenth century,
to the ultimate destination of the Big-Bang theory of the
expanding Universe.
Cosmic Enigmas
Cosmologists and Their Myths (p. 3)
AIP Press. Woodbury, New York, USA. 1994

Smith, Logan Pearsall 1865–1946
American author

I woke this morning…into the well-known, often-dis-
cussed, but, to my mind, as yet unexplained Universe.
Trivia
Book I, To-Day (p. 4)
Doubleday, Page & Company. Garden City, New York, USA. 1917

Smuts, Jan Christiaan 1870–1950
South African statesman, military leader, and holistic philosopher

Truth, beauty, goodness, and love are as much structures
of the evolutionary universe as the sun and the earth and
the moon.
Contributions to a British Association Discussion on the Evolution of
the Universe
Nature, Supplement, October 24, 1931 (p. 718)

Spenser, Edmund 1552–99
English poet

Why then should witless man so much misween,
That nothing is, but that which he hath seene?
What if in the Moones faire shining spheare?
What if in every other starre unseene,
Of other worldes he happily should heare?
That nothing is, but that which he hath seene?
The Complete Poetical Works of Edmund Spenser
The Faerie Queene
Book the Second, Introduction
Houghton Mifflin Company. Boston, Massachusetts, USA. 1908

Stern, S. Alan
American planetary scientist an author

The place we call our Universe is, for the most part, cold
and dark and all but endless. It is the emptiest of emp-
ties. It is old, and yet very young. It contains much that
is dead, and yet much that is alive, forever reinventing
itself, and sometimes inventing something wholly new.
In S. Alan Stern (ed.)
*Our Universe: The Thrill of Extragalactic Exploration as Told by
Leading Experts*
The Frontier Universe: At the Edge of the Night (p. 1)
Cambridge University Press. Cambridge, England. 2001

Sullivan, Walter 1918–96
American science journalist

We do know enough already, however, to believe that no
myth or legend could be as rich in beauty, wonder, and
awe as the full reality of the universe that is our home.
In Ray Bradbury, Arthur C. Clarke, Bruce Murray, Carl Sagan, and
Walter Sullivan
Mars and the Mind of Man
Walter Sullivan (p. 127)
Harper & Row, Publishers. New York, New York, USA. 1973

Swann, William Francis Gray 1884–1962
Anglo-American physicist

There is one great work of art; it is the universe. Ye
men of letters find the imprints of its majesty in your
sense of the beauty of words. Ye men of song find it in

the harmony of sweet sounds. Ye painters feel it in the design of beauteous forms, and in the blending of rich soft colors do your souls mount on high to bask in the brilliance of nature's sunshine. Ye lovers are conscious of its beauties in forms ye can but ill define. Ye men of science find it in the rich harmonies of nature's mathematical design.
The Architecture of the Universe
Chapter XII (p. 424)
The Macmillan Company. New York, New York, USA. 1934

Swimme, Brian   1950–
American mathematical cosmologist

I am convinced that the story of the universe that has come out of three centuries of modern scientific work will be recognized as a supreme human achievement, the scientific enterprise's central gift to humanity, a revelation having a status equal to that of the great religious revelations of the past.
In Connie Barlow (ed.)
Evolution Extended: Biological Debates on the Meaning of Life
The MIT Press. Cambridge, Massachusetts, USA. 1994

Talbot, Michael   1953–92
American physicist

…we have to begin to view the universe as ultimately constituted not of matter and energy, but of pure information.
Beyond the Quantum
Chapter 6 (p. 155)
The Macmillan Company. New York, New York, USA. 1986

Teller, Woolsey   1890–1954
Essayist

…the picture of the universe presented by astronomy is one of dismal stretches of time and space and unparalled desolation. In the eternal abyss of space — bleak, cold, and dark — there are no signs of a Cosmic Consciousness.
The Atheism of Astronomy
Chapter VI (p. 120)
Arno Press & The New York Times. New York, New York, USA. 1972

Tennyson, Alfred (Lord)   1809–92
English poet

This truth within thy mind rehearse,
That in a boundless universe
Is boundless better, boundless worse.
Alfred Tennyson's Poetical Works
The Two Voices, Stanza 9
Oxford University Press, Inc. London, England. 1953

Thompson, Francis   1859–1907
English writer

The universe is his box of toys. He dabbles his fingers in the day-fall. He is gold-dusty with tumbling amidst the stars. He makes bright mischief with the moon. The meteors nuzzle their noses in his hand.
The Works of Francis Thompson
Shelley (p. 18)
Burns & Oats. London, England. 1913

Thoreau, Henry David   1817–62
American essayist, poet, and practical philosopher

The universe is wider than our views of it.
The Writings of Henry David Thoreau (Volume 2)
Walden
Chapter XVIII (p. 493)
Houghton Mifflin Company. Boston, Massachusetts, USA. 1893

…the universe is not rough-hewn, but perfect in its details.
The Writings of Henry David Thoreau (Volume 9)
Natural History of Massachusetts (p. 132)
Houghton Mifflin Company. Boston, Massachusetts, USA. 1893

Toynbee, Arnold J.   1852–83
English historian

Huddled together in our little earth we gaze with frightened eyes into the dark universe.
Toynbee's Industrial Revolution
Notes and Jottings (p. 256)
A.M. Kelley. New York, New York, USA. 1969

Trimble, V.
No biographical data available

Those of us who are not directly involved in the fray can only suppose that the universe is open (W<1) on Wednesday, Friday, and Sunday and closed (W>1) on Thursday, Saturday, and Monday. (Tuesday is choir practice.)
Dark Matter in the Universe: Where, What, and Why?
Contemporary Physics, Volume 29, 1988 (p. 389)

Turner, Michael S.
American astrophysicist

The progress made in our understanding of the universe during the twentieth century is nothing short of stunning.
A Sober Assessment of Cosmology at the New Millennium
Publications of the Astronomical Society of the Pacific, Volume 113, 2001 (p. 653)

Tyron, E. P.
No biographical data available

If it is true that our Universe has a zero net value for all conserved quantities, then it may simply be a fluctuation of the vacuum of some larger space in which our Universe is imbedded. In answer to the question of why it happened, I offer the modest proposal that our Universe is simply one of those things which happen from time to time.
Is the Universe a Vacuum Fluctuation?
Nature, Volume 246, Number 5433, December 14, 1973 (p. 397)

von Goethe, Johann Wolfgang 1749–1832
German poet, novelist, playwright, and natural philosopher

Man is not born to solve the problems of the universe, but to find out where the problems begin, and then to take his stand within the limits of the intelligible.
In Louis Berman
Exploring the Cosmos
Chapter 16 (p. 351)
Little, Brown & Company. Boston, Massachusetts, USA. 1973

Weinberg, Steven 1933–
American nuclear physicist

The effort to understand the universe is one of the very few things that lifts human life a little above the level of farce, and gives it some of the grace of tragedy.
The First Three Minutes
Epilogue (p. 155)
Basic Books, Inc. New York, New York, USA. 1988

The more the universe seems comprehensible, the more it also seems pointless.
The First Three Minutes
Epilogue (p. 154)
Basic Books, Inc. New York, New York, USA. 1988

It is very hard to realize that this all is just a tiny part of an overwhelmingly hostile universe. It is even harder to realize that this present universe has evolved from an unspeakably unfamiliar early condition, and faces a further extinction of endless cold or intolerable heat. The more the universe seems comprehensible, the more it also seems pointless.
The First Three Minutes
Epilogue (p. 154)
Basic Books, Inc. New York, New York, USA. 1988

…the urge to trace the history of the universe back to its beginning is irresistible.
The First Three Minutes
Chapter I (p. 4)
Basic Books, Inc. New York, New York, USA. 1988

Wharton, Edith 1862–1937
American novelist

…she had never been able to understand the laws of a universe which was so ready to leave her out of its calculations.
The House of Mirth
Book I, Chapter III (p. 42)
Charles Scribner's Sons. New York, New York, USA. 1919

Wheeler, John Archibald 1911–
American theoretical physicist and educator

We will first understand how simple the universe is when we recognize how strange it is.
From the Big Bang to the Big Crunch
Cosmic Search Magazine, Volume 1, Number 4, Fall 1979

The Universe is a self-excited circuit.
In Freeman Dyson
Infinite in All Direction
Part I, Chapter Three (p. 53)
Harper Collins Publisher, Inc. New York, New York, USA. 1988

…this is our universe, our museum of wonder and beauty, our cathedral…
A Journey into Gravity and Spacetime
Opening
Scientific American Library. New York, New York, USA. 1990

Wheeler, John Archibald 1911–
American theoretical physicist and educator
Thorne, Kip S. 1940–
American theoretical physicist

A model universe that is closed, that obeys Einstein's geomethermodynamic law, and that contains a nowhere negative density of mass-energy, inevitably develops a singularity. No one sees any escape from the density of mass-energy rising without limit. A computing machine calculating ahead step by step the dynamical evolution of the geometry comes to the point where it can not go on. Smoke, figuratively speaking, starts to pour out of the computer. Yet physics, surely continues to go on if for no other reason than this: Physics is by definition that which does go on its eternal way despite all the shadowy changes in the surface of reality.
Gravitation
Part X, Chapter 44 (p. 1196)
W.H. Freeman & Company. San Francisco, California, USA. 1973

Whitman, Walt 1819–92
American poet, journalist, and essayist

Praised be the fathomless universe

For life and joy and for objects and knowledge curious…
Complete Poetry and Collected Prose
Memories of President Lincoln, Stanza 14
The Library of America. New York, New York, USA. 1982

The whole theory of the universe is directed unerringly to one single individual — namely to you.
Complete Poetry and Collected Prose
Leaves of Grass
By Blue Ontario's Shore, Stanza 15
The Library of America. New York, New York, USA. 1982

The world, the race, the soul —
Space and time, the universes
All bound as is befitting each — all
Surely going somewhere.
Complete Poems and Collected Prose
Going Somewhere
The Library of America. New York, New York, USA. 1982

Wiechert, Emil 1861–1928
Prussian geophysicist

The universe is infinite in all directions.
In Freeman Dyson
Infinite in All Directions
Part One, Chapter Three (p. 53)
HarperCollins Publisher, Inc. New York, New York, USA. 1988

Young, Louise B.
American science writer

The universe is unfinished, not just in the limited sense of an incompletely realized plan but in the much deeper sense of a creation that is a living reality of the present. A masterpiece of artistic unity and integrated Form, infused with meaning, is taking shape as time goes by. But its ultimate nature cannot be visualized, its total significance grasped, until the final lines are written.
The Unfinished Universe
Conclusion (p. 205, 208)
Simon & Schuster. New York, New York, USA. 1986

Zebrowski, George   1945–
Polish-American science fiction writer

The rationality of our universe is best suggested by the fact that we can discover more about it from any starting point, as if it were a fabric that will unravel from any thread.
Is Science Rational?
OMNI Magazine, June 1994 (p. 50)

In a perfectly rational universe, infinities turn back on themselves…
Is Science Rational?
OMNI Magazine, June 1994 (p. 50)

UNIVERSE AND COSMOGENESIS

Ackerman, Diane   1948–
American writer

Fifteen billion years ago, when the Universe
let rip and, in disciplined panic,
Creation spewed mazy star-treacle and resin,
shrinking balls of debut fire smoldered
and glitched.
The Planets: A Cosmic Pastoral
Neptune, IV (p. 129)
William Morrow & Company, Inc. New York, New York, USA. 1976

Adams, Douglas   1952–2001
English author, comic radio dramatist, and musician

In the beginning the universe was created. This made a lot of people very angry and has been widely regarded as a bad move. Many races believe that it was created by some sort of god, though the Jatravartid people of Viltvodle VI believe that the entire Universe was in fact sneezed out of the nose of a being called the Great Green Arkleseizure.
The Ultimate Hitchhiker's Guide to the Galaxy

The Restaurant at the End of the Universe
Chapter 1 (p. 149)
Ballantine Books. New York, New York, USA. 2002

Barrow, John D.   1952–
English theoretical physicist

One day we may be able to say something about the origins of our own cosmic neighborhood. But we can never know the origins of the universe. The deepest secrets are the ones that keep themselves.
The Origin of the Universe
Chapter 8 (p. 137)
Basic Books, Inc. New York, New York, USA. 1994

Bowyer, Stuart   1934–
American astrophysicist

Ultimately, the origin of the universe is, and always will be, a mystery.
In Henry Margenau and Roy Abraham Varghese (eds.)
Cosmos, Bios, Theos
Chapter 2 (p. 32)
Open Court. La Salle, Illinois, USA. 1992

Cardenal, Ernesto   1925–
Nicaraguan poet and Roman Catholic priest

In the beginning there was nothing
neither space
nor time.
The entire universe concentrated
in the space of the nucleus of an atom,
and before that even less,
much less than a proton,
and even less still,
an infinitely dense mathematical point.
Translated by John Lyons
Cosmic Canticle
Cantigua 1, Big Bang (p. 11)
Curbstone Press. Willimantic, Connecticut, USA. 1993

Egyptian Myth   ca. 2500 B.C.

In the beginning, only the ocean existed, upon which there appeared an egg. Out of the egg came the sun-god and from himself he begat four children: Shu and Tefnut, Keb and Nut. All these, with their father, lay upon the ocean of chaos. Then Shu and Tefnut thrust themselves between Keb and Nut. They planted their feet upon Keb and raised Nut on high so that Keb became the earth and Nut the heavens.
In Eric J. Lerner
The Big Bang Never Happened
Chapter 2 (p. 58)
Random House, Inc. New York, New York, USA. 1991

Flaubert, Gustave   1821–90
French novelist

SMARH: How vast creation is! I see the planets rise, I see the fiery stars driven along…. Space opens out as

I rise, worlds revolve around me, and I am the center of this bustling creation.
Early Writings
Smarh (p. 216)
University of Nebraska Press. Lincoln, Nebraska, USA. 1991

Gamow, George   1904–68
Russian-born American physicist

Before we can discuss the basic problem of the origin of our universe, we must ask ourselves whether such a discussion is necessary.
The Creation of the Universe
Chapter I (p. 6)
The Viking Press. New York, New York, USA. 1952

In the beginning God created radiation and ylem. And ylem was without shape or number, and the nucleons were rushing madly over the face of the deep.
My World Line: An Informal Autobiography
Chapter 6 (p. 127)
The Viking Press. New York, New York, USA. 1979

Hawkins, Gerald S.   1928–2003
English archaeoastronomer

In the beginning.... A scientist cannot continue this sentence with absolute certainty. It would be like asking a child to give an account of his birth or a description of his conception.
In *Reader's Digest*
Marvels and Mysteries of the World Around Us
Earth's Ancient Drama (p. 10)
The Reader's Digest Association. Pleasentville, New York, USA. 1972

Hein, Robert
No biographical data available

The first world the cosmic colossus created with a word:
One lightning word from the golden lips of Truth
And electric earth condensed on a creamy cloud,
Adorned with a necklace of blue-gold stars and a chain
Of peppermint planets in the amphitheater of space.
Quest of the Singing Tree
The Larger Creation, Creation of the Earth
H. Harrison. New York, New York, USA. 1938

Hoyle, Sir Fred   1915–2001
English mathematician and astronomer

Without continuous creation, the Universe must evolve toward a dead state in which all the matter is condensed into a vast number of dead stars.
The Nature of the Universe
Chapter 6 (pp. 131–132)
The University Press. Cambridge. 1933

Kipling, Rudyard   1865–1936
British writer and poet

Before the High and Far-Off Times, O my Best Beloved, came the Time of the Very Beginnings; and that was in the days when the Eldest Magician was getting Things ready. First he got the Earth ready; then he got the Sun ready; and then he told all the Animals that they could come out and play.
Just So Stories
The Crab that Played with the Sea (p. 171)
Doubleday & Company, Inc. Garden City, New York, USA. 1952

Marquesas Islanders

In the beginning there was nothing. There arose a swelling, a ferment, a black fire, a spinning vortices, a bubbling, a swallowing — there arose a whole series of pairs of props, posts, or piles, large and small, long and short, crooked and bent, decayed and rotten. Similarly there arose pairs of roots, large and small, long and short, and so forth; there arose countless and infinitely many supports. Above all, there now arose the ground, the foundation, the hard rock, there arose the space for light, there arose rocks of different sorts.
In John A. Wood
Meteorites and the Origin of Planets
Creation Myth (p. v)
McGraw-Hill Book Company, Inc. New York, New York, USA. 1968

Reeves, Hubert   1932–
Canadian astrophysicist

In the beginning was the absolute rule of the flame: The universe was in limbo. Then after countless eras, the fires slowly abated like the sea at the outgoing tide. Matter awoke and organized itself; the flame gave way to music.
Atoms of Silence
Introduction (p. 5)
The MIT Press. Cambridge, Massachusetts. USA. 1984

Sagan, Carl   1934–96
American astronomer and science writer

If the general picture of an expanding universe and a Big Bang is correct, we must then confront still more difficult questions. What were conditions like at the time of the Big Bang? What happened before that? Was there a tiny universe, devoid of all matter, and then the matter suddenly created from nothing? How does that happen? In many cultures it is customary to answer that God created the universe out of nothing. But this is mere temporizing. If we wish courageously to pursue the question, we must of course ask next where God comes from. And if we decide this to be unanswerable, why not save a step and decide that the origin of the universe is an unanswerable question. Or, if we say that God has always existed, why not save a step and conclude that the universe has always existed?
Cosmos
Chapter X (p. 257)
Random House, Inc. New York, New York, USA. 1980

Singh, Jagjit   1919–2002
Indian mathematician and science writer

In the beginning there was neither heaven nor earth,
And there was neither space nor time.
And the Earth, the Sun, the Stars, the Galaxies and the
whole universe were
confined within a small volume like the bottled genie of
the Arabian Nights.
And then God said, "Go!"
And straight way the Galaxies rushed out of their
prison, scattering in all
directions, and they have continued to run away from
one another ever since,
afraid lest some cosmic Hand should gather them again
and put them back in
the bottle (which is not bigger than a pin-point).
And they shall continue to scatter thus till they fade
from each other's ken —
and thus, for each other, cease to exist at all.
Mathematical Ideas: Their Nature and Use
Space and Time (pp. 209–210)
Hutchinson & Company Ltd. London, England. 1972

Smoot, George   1945–
American experimental astrophysicist

The question of "the beginning" is as inescapable for
cosmologists as it is for theologians.
Wrinkles in Time
Chapter 9 (p. 189)
William Morrow & Company, Inc. New York, New York, USA. 1993

Spenser, Edmund   1552–99
English poet

Through knowledge we behold the world's creation,
How in his cradle first he fostered was;
And judge of Natures cunning operation,
How things she formed of a formless mass.
The Complete Poetical Works of Edmund Spenser
The Tears of the Muses, l. 499–502
Houghton Mifflin Company. Boston, Massachusetts, USA. 1908

Sturluson, Snorri   1179–1241
Icelandic writer

Erst was the age when nothing was:
Nor sand nor sea, nor chilling stream-waves;
Earth was not found, nor Ether — Heaven,
— A Yawning Gap, but grass was none.
The Prose Edda
Here Begins the Beguiling of Gylfi (p. 16)
The American-Scandinavian Foundation. New York, New York, USA.
1916

Sufi Creation Myth

I was a hidden treasure and desired to be known: therefore
I created the creation in order to be known.
In George Smoot

Wrinkles in Time
Chapter 1 (p. 1)
William Morrow & Company, Inc. New York, New York, USA. 1993

Tagore, Rabindranath   1861–1941
Indian poet and philosopher

It seems to me that, perhaps, creation is not fettered by
rules,

That all the hubbub, meeting and mingling are blind hap-
penings of fate.
Translated by Indu Dutt
Our Universe (p. 75)
Jaico Publishing House. Bombay, India. 1969

The Bible

In the beginning God created the heaven and the earth.
The Revised English Bible
Genesis 1:1
Oxford University Press, Inc. Oxford, England. 1989

Townes, Charles H.   1915–
American inventor of the laser

I do not understand how the scientific approach alone,
as separated from a religious approach, can explain an
origin of all things. It is true that physicists hope to look
behind the "big bang," and possibly to explain the origin
of our universe as, for example, a type of fluctuation. But
then, of what is it a fluctuation and how did this in turn
begin to exist? In my view the question of origin seems
always left unanswered if we explore from a scientific
view alone.
In Henry Margenau and Roy Abraham Varghese (eds.)
Cosmos, Bios, Theos
Chapter 25 (p. 123)
Open Court. La Salle, Illinois, USA. 1992

Updike, John   1932–
American novelist, short story writer and poet

By computation, they all must have begun at one place
about five billion years ago; all the billions and trillions
and quadrillions squared and squared again of tons of
matter in the universe were compressed into a ball at the
maximum possible density, the density within the nucleus of the atom; one cubic centimeter of this primeval egg
weighed two hundred and fifty tons.
The Centaur
Chapter I (p. 38)
Alfred A. Knopf. New York, New York, USA. 1995

Weinberg, Steven   1933–
American nuclear physicist

…the urge to trace the history of the universe back to the
beginnings is irresistible. From the start of modern science in the sixteenth and seventeenth centuries, physicists
and astronomers have returned again and again to the
problem of the origin of the universe.

The First Three Minutes
Chapter I (p. 4)
Basic Books, Inc. New York, New York, USA. 1988

Wilmot, John (2nd Earl of Rochester)   1647–80
English libertine and satirical and bawdy poet

E'er time and place were, time and place were not,
When Primitive Nothing something straight begot,
Then all proceeded from the great united — What.
Collected Works of John Wilmot Earl of Rochester
Upon Nothing
The Nonesuch Press. London, England. 1926

Zuni Creation Myth

In the beginning of things Awonawilona was alone. There was nothing beside him in the whole of time. Everywhere there was black darkness and void. Then Awonawilona conceived in himself the thought, and the thought took shape and got out into space and through this stepped out into the void, into outer space, and from them came nebulae of growths and mists, full of power and growth.
In Raymond Van Over
Sun Songs
Zuni Creation Myths
Cosmic Creation (p. 23)
New American Library. New York, New York, USA. 1980

UNIVERSE, DEATH OF

Balfour, Arthur James   1848–1930
English prime minister

…the energies of our system will decay, the glory of the sun will be dimmed, and the earth, tideless and inert, will no longer tolerate the race which has for a moment disturbed its solitude. Man will go down into the pit, and all his thoughts will perish.
The Foundations of Belief
Part I, Chapter I, Section III (p. 33)
Longmans, Green & Company. London, England. 1912

Byron, George Gordon, 6th Baron Byron   1788–1824
English Romantic poet and satirist

I had a dream, which was not all a dream.
The bright sun was extinguish'd, and the stars
Did wander darkling in the eternal space,
Rayless, and pathless, and the icy earth
Swung blind and blackening in the moonless air.
The Complete Poetical Works of Byron
Miscellaneous Poems, Darkness
Houghton Mifflin Company. Boston, Massachusetts, USA. 1933

Davies, Paul Charles William   1946–
British-born physicist, writer, and broadcaster

Many billions of years will elapse before the smallest, youngest stars complete their nuclear burning and shrink into white dwarfs. But with slow, agonizing finality perpetual night will surely fall.
The Last Three Minutes: Conjectures About the Ultimate Fate of the Universe
Chapter 5 (p. 50)
Basic Books, Inc. New York, New York, USA. 1994

A universe that came from nothing in the big bang will disappear into nothing at the big crunch. Its glorious few zillion years of existence not even a memory.
The Last Three Minutes: Conjectures About the Ultimate Fate of the Universe
Chapter 9 (p. 123)
Basic Books, Inc. New York, New York, USA. 1994

Dyson, Freeman J.   1923–
American physicist and educator

Since the universe is on a one-way slide toward a state of final death in which energy is maximally degraded, how does it manage, like King Charles, to take such an unconsciously long time a-dying.
Energy in the Universe
Scientific American, Volume 224, Number 3, 1971 (p. 52)

Eddington, Sir Arthur Stanley   1882–1944
English astronomer, physicist, and mathematician

…the universe will finally become a ball of radiation, becoming more and more rarified and passing into longer and longer wave-lengths. The longest waves of radiation are Hertzian waves of the kind used in broadcasting. About every 1500 million years this ball of radio waves will double in diameter; and it will go on expanding in geometrical progression for ever. Perhaps then I may describe the end of the physical world as — one stupendous broadcast.
New Pathways in Science
Chapter III, Section VI (p. 71)
The Macmillan Company. New York, New York, USA. 1935

Eliot, T. S. (Thomas Stearns)   1888–1965
American expatriate poet and playwright

This is the way the world ends
Not with a bang but a whimper.
The Collected Poems and Plays 1909–1950
The Hollow Men (p. 59)
Harcourt, Brace & World, Inc. New York, New York, USA. 1952

Frost, Robert   1874–1963
American poet

Some say the world will end in fire,
Some say in ice.
From what I've tasted of desire
I hold with those who favor fire.
But if it had to perish twice,
I think I know enough of hate
To say that for destruction ice
Is also great

And would suffice.
Complete Poems of Robert Frost
Fire and Ice
Henry Holt & Company. New York, New York, USA. 1949

Gamow, George   1904–68
Russian-born American physicist

Galaxies are ever spinnik,
Stars will burn to final sparrk,
Till ourr universe is thinnink
And is lifeless, cold and darrk.
Mr. Tompkins in Paperback
Chapter 6 (p. 60)
At The University Press. Cambridge, England. 1965

Gribbin, John   1946–
British science writer and astronomer

"Big Crunch" is…an ugly term which hardly seems appropriate for so important an event as the end of the universe. But there is no convention as yet for a label of the moment of destruction at the end of time, and I am free to borrow the term "omega point."
The Omega Point: The Search for the Missing Mass and the Ultimate Fate of the Universe
Introduction (p. 2)
Bantam Books. Toronto, Ontario, Canada. 1988

Harrison, Edward Robert   1919–2007
English-born American cosmologist

The stars begin to fade like guttering candles and are snuffed out one by one. Out of the depths of space the great celestial cities, the galaxies, cluttered with the memorabilia of ages, are gradually dying. Tens of billions of years pass in the growing darkness. Occasional flickers of light pierce the fall of cosmic night, and spurts of activity delay the sentence of a universe condemned to become a galactic graveyard.
Cosmology: The Science of the Universe
Chapter 18 (p. 360)
Cambridge University Press. Cambridge, England. 1981

Hawking, Stephen William   1942–
English theoretical physicist

The present evidence therefore suggests that the universe will probably expand forever, but all we can really be sure of is that even if the universe is going to recollapse, it won't do so for at least another ten thousand million years, since it has already been expanding for at least that long. This should not unduly worry us: by that time, unless we have colonized beyond the Solar System, mankind would long since have died out, extinguished along with our sun!
A Brief History of Time: From the Big Bang to Black Holes
Chapter 3 (p. 46)
Bantam Books. Toronto, Ontario, Canada. 1988

James, William   1842–1910
American philosopher and psychologist

Though the ultimate state of the universe may be its vital and psychical extinction, there is nothing in physics to interfere with the hypothesis that the penultimate state might be the millennium — in other words a state in which a minimum of difference of energy-level might have its exchanges so skillfully canalisés that a maximum of happy and virtuous consciousness would be the only result. In short, the last expiring pulsation of the universe's life might be, "I am so happy and perfect that I can stand it no longer."
The Atlantic Monthly
Letter to Henry Adams dated June 17, 1910
September 1920 (p. 316)

Jastrow, Robert   1925–
American space scientist

Within the isolated galaxies, the old stars burn out one by one, and fewer and fewer new stars are formed to replace them. Stars are the source of the energy by which all beings live. When the light of the last star is extinguished, the Universe fades into darkness, and all life comes to an end.
God and the Astronomers
Epilogue (p. 117)
W.W. Norton & Company, Inc. New York, New York, USA.1978

Jeans, Sir James Hopwood   1877–1946
English physicist and mathematician

Everything points with overwhelming force to a definite event, or series of events, of creation at some time or times, not infinitely remote. The universe cannot have originated by chance out of its present ingredients, and neither can it have been always the same as now. For in either of these events no atoms would be left save such as are incapable of dissolving into radiation; there would be neither sunlight nor starlight but only a cool glow of radiation uniformly diffused through space. This is, indeed, so far as present-day science can see, the final end towards which all creation moves, and at which it must at last arrive.
Eos or the Wider Aspects of Cosmogony (p. 55)
Kegan Paul, Trench, Trubner & Company, Ltd. London, England. 1931

Jeffers, Robinson   1887–1962
American poet

Time will come no doubt
When the sun too shall die; the planets will freeze, and the air on them; frozen gases, white flakes of air
Will be the dust: which no wind will ever stir: this very dust in dim starlight glistening
Is dead wind, the white corpse of wind.
Also the galaxy will die; the glitter of the Milky Way,
our universe, all the stars that have names are dead.

In Tim Hunt (ed.)
The Collected Poetry of Robinson Jeffers (Volume 3)
The Double Axe: The Inhumanist (p. 261)
Stanford University Press. Stanford, California. USA. 1988

For man will be blotted out, the blithe earth die, the
brave sun
Die blind and blacken to the heart…
In Tim Hunt (ed.)
The Collected Poetry of Robinson Jeffers (Volume 1)
To the Stone-Cutters (p. 5)
Stanford University Press. Stanford, California. USA. 1988

It will contract, the immense navies of stars and
galaxies,
Dust-clouds and nebulae
Are recalled home, they crush against each other in one
harbor, they stick in one lump.
The Beginning and the End and Other Poems
The Great Explosion (p. 3)
Random House, Inc. New York, New York, USA. 1963

I seem to have stood a long time and watched the stars
pass.
They also shall perish I believe.
Here to-day, gone to-morrow, desperate wee galaxies
Scattering themselves and shining their substance away
Like a passionate thought. It is very well ordered.
In Tim Hunt (ed.)
The Collected Poetry of Robinson Jeffers (Volume 2)
Margrave (p. 171)
Stanford University Press. Stanford, California. USA. 1988

Joyce, James   1882–1941
Irish expatriate writer and poet

Gasballs spinning about, crossing each other, passing.
Same old dingdong always. Gas, then solid, then world,
then cold, then dead shell drifting around, frozen rock
like that pineapple rock. The moon.
Ulysses (p. 164)
Random House, Inc. New York, New York, USA. 1946

Lucretius   ca. 99 BCE–55 BCE
Roman poet

And so some day,
The mighty ramparts of the mighty universe
Ringed round with hostile force,
Will yield and face decay and come crumbling to ruin.
In Paul Davies
About Time: Einstein's Unfinished Revolution
Header (p. 33)
Simon & Schuster. New York, New York, USA. 1995

MacLeish, Archibald   1892–1982
American poet and Librarian of Congress

And there, there overhead, there, there hung over
Those thousands of white faces, those dazed eyes,
There in the starless dark the poise, the hover,
There with vast wings across the canceled skies,
There in the sudden blackness the black pall

Of nothing, nothing, nothing — nothing at all.
Collected Poems 1917–1952
The End of the World
Houghton Mifflin Company. Boston, Massachusetts, USA. 1952

Nicholson, Norman   1914–87
English poet

And if the universe
Reversed and showed
The colour of its money;
If now observable light
Flowed inward, and the skies snowed
A blizzard of galaxies,
The lens of night would burn
Brighter than the focused sun,
And man turn blinded
With white-hot darkness in his eyes.
The Pot Geranium
The Expanding Universe (p. 212)
Faber & Faber Ltd. London, England. 1994

Russell, Bertrand Arthur William   1872–1970
English philosopher, logician, and social reformer

In the vast death of the solar system, and the whole temple of Man's achievements must inevitable be buried beneath all the labours of the ages, all the devotion, all the
inspiration, all the noonday brightness of human genius,
are destined to extinction debris of a universe in ruins
— all these things, if not quite beyond dispute, are yet so
nearly certain that no philosophy which rejects them can
hope to stand. Only within the scaffolding of these truths,
only on the firm foundation of unyielding despair, can the
soul's habitation henceforth be safely built.
*Why I Am Not a Christian: And Other Essays on Religion and Related
Subjects* (p. 107)
Watts. London, England. 1927

Wells, H. G. (Herbert George)   1866–1946
English novelist, historian, and sociologist

There will be a time when the day will be as long as a
year is now, and the cooling sun, shorn of its beams, will
hang motionless in the heavens.
The Outline of History (Volume 1)
Book I, Chapter I, Section 3 (p. 15)
Garden City Books. Garden City, New York, USA. 1961

…a steady twilight brooded over the Earth…. All traces of the moon had vanished. The circling of the stars,
growing slower and slower, had given place to creeping
points of light…the sun, red and very large, halted motionless upon the horizon, a vast dome glowing with a
dull heat…. The rocks about me were of a harsh reddish
colour, and all the traces of life that I could see at first
was the intensely green vegetation…the same rich green
that one sees on forest moss or on the lichen in caves:
plants which like these grow in a perpetual twilight….

I cannot convey the sense of abominable desolation that hung over the world.
The Great Ideas Today, 1971
The Time Machine
Chapter Eleven (p. 497)
Encyclopædia Britannica, Inc. Chicago, Illinois, USA. 1971

Yeats, William Butler   1865–1939
Irish poet and playwright

When shall the stars be blown about the sky,
Like the sparks blown out of a smithy, and die?
The Collected Poems of W.B. Yeats
The Secret Rose (p. 67)
The Macmillan Company. New York, New York, USA. 1956

UNKNOWN

Asimov, Isaac   1920–92
American author and biochemist

If it is exciting to probe the unknown and shed light on what was dark before, then more and more excitement surely lies ahead of us.
The Universe
Chapter 19 (p. 294)
Walker & Company. New York, New York, USA. 1966

Auvaiyaar   ca. 9th century
Tamil sage and poetess

What is known is a handful; the unknown is as vast as the universe.
Attributed
The Physics Teacher, Volume 15, Number 9, December 1977 (p. 544)

Carlson, A. J.   1875–1956
Swedish-American physiologist

We recognize the unknown but not the unknowable.
Science and the Supernatural
Science, Volume 73, Number 1887, February 27, 1931 (p. 221)

When we know that we don't know, that is itself an achievement, for then the field is cleared of the confusing and obstructing rubbish of tradition, and we are free to use all our ingenuity and imagination in contriving methods to find out.
Science and the Supernatural
Science, Volume 73, Number 1887, February 27, 1931 (p. 221)

Chargaff, Erwin   1905–2002
Austrian biochemist

I have always oscillated between the brightness of reality and the darkness of the unknowable.
Heraclitean Fire: Sketches from a Life before Nature
Part I
The Silence of the Heavens (p. 55)
Rockefeller University Press. New York, New York, USA. 1978

Charles, John
American planetary geologist

We want to research what we call the "known unknowns".… This will reduce total risk in the face of the unknown unknowns, the true surprises.…
In James Olberg
Red Planet Blues
Popular Science, July 2003 (p. 64)

Eddington, Sir Arthur Stanley   1882–1944
English astronomer, physicist, and mathematician

Something unknown is doing we don't know what…
The Nature of the Physical World
Chapter XIII (p. 291)
The Macmillan Company. New York, New York, USA. 1930

Farmer, Philip José   1918–
American science fiction and fantasy writer

Some of you have asked why we should set out for a goal that lies we know not how far away or that might not even exist. I will tell you that we are setting sail because the Unknown exists and we would make it the Known. That's all!
To Your Scattered Bodies Go
Chapter 13 (p. 98)
Berkley Publishing Corporation. New York, New York, USA. 1971

Hinshelwood, Sir Cyril   1897–1967
English chemist

…as the chart of the unknown becomes filled in, judgment of the most profitable course to follow changes. Mysterious inlets may prove dead ends or may open into vast seas.
Science and Scientists
Supplement to Nature, Volume 207, Number 5001, 4 September 1965 (p. 1057)

Huxley, Aldous   1894–1963
English writer and critic

Cheerfully…let us advance together, men of letters and men of science, further and further into the ever-expanding regions of the unknown.
Literature and Science
Chapter 38 (p. 118)
Harper & Row, Publishers. New York, New York, USA. 1963

Huxley, Thomas Henry   1825–95
English biologist

The known is finite, the unknown infinite; intellectually we stand on an islet in the midst of an illimitable ocean of inexplicability. Our business in every generation is to reclaim a little more land.
In Francis Darwin (ed.)
The Life and Letters of Charles Darwin (Volume 1)
Chapter XIV (p. 557)
D. Appleton & Company. New York, New York, USA. 1896

Leiber, Jr., Fritz   1910–92
American writer of science fiction and horror

Science has only increased the area of the unknown. And if there is a god, her name is Mystery.
Our Lady of Darkness (p. 44)
Berkley Publishing Corp. New York, New York, USA. 1977

Lindbergh, Charles H.   1902–74
American aviator

Whether outwardly or inwardly, whether in space or in time, the farther we penetrate the unknown, the vaster and more marvelous it becomes.
Autobiography of Values
Chapter Fifteen (p. 402)
Hartcourt Brace Jovanovich. New York, New York, USA. 1967

Locke, John   1632–1704
English philosopher and political theorist

For the understanding, like the eye, judging of objects only by its own sight, cannot but be pleased with what it discovers, having less regret for what has escaped it, because it is unknown.
In *Great Books of the Western World* (Volume 35)
Concerning Human Understanding
Epistle to the Reader (p. 87)
Encyclopædia Britannica, Inc. Chicago, Illinois, USA. 1952

Melville, Herman   1819–91
American novelist

…I am tormented with an everlasting itch for things remote. I love to sail the forbidden seas…
In *Great Books of the Western World* (Volume 48)
Moby Dick
Chapter 1 (p. 4)
Encyclopædia Britannica, Inc. Chicago, Illinois, USA. 1952

Mencken, H. L. (Henry Louis)   1880–1956
American journalist and literary critic

Penetrating so many secrets, we cease to believe in the unknowable. But there it sits nevertheless, calmly licking its chops.
Minority Report: H.L. Mencken's Notebooks
No. 364 (p. 241)
Alfred A. Knopf. New York, New York, USA. 1956

Nicholson, Norman   1914–87
English poet

No man has seen it; nor the lensed eye
That pin-points week by week the same patch of sky
Records even a blur across its pupil. Only
The errantry of Saturn, the wry
Retarding of Uranus, speak
Of the pull beyond the pattern:
The unknown is shown
Only by a bend in the known.
In Neil Curry (ed.)

Norman Nicholson Collected Works
The Undiscovered Planet (p. 211)
Faber & Faber Ltd. London, England. 1994

Nietzsche, Friedrich   1844–1900
German philosopher

To trace something unknown back to something known is alleviating, soothing, gratifying and gives moreover a feeling of power, Danger, disquiet, anxiety attend the unknown — the first instinct is to eliminate these distressing states. First principle: any explanation is better than none.…
In Alexander Tille (ed.), Thomas Common (trans.)
The Works of Friedrich Nietzsche, Volume 11
Twilight of the Idols, The Four Great Errors, Section 5 (p. 138)
Henry & Company. London, England. 1896

Oppenheimer, J. Robert   1904–67
American theoretical physicist

The problem of doing justice to the implicit, the imponderable and the unknown is always with us in science, it is with us in the most trivial of personal affairs, and it is one of the great problems of all forms of art.
In Lincoln Barnett
Writing on Life: Sixteen Close-Ups
Physicist Oppenheimer (p. 358)
William Sloane Associates, Publishers. New York, New York, USA. 1951

Pinter, Harold   1930–
English absurdist playwright

In other words, apart from the known and the unknown, what else is there?
The Homecoming
Act Two (p. 52)
Methuen & Company Ltd. London, England. 1966

Rabi, Isidor Isaac   1898–1988
Austrian-born American physicist

With the beginning of direct exploration of the solar system and promise; in fact science derives it sustenance from the unknown; all the good things have come from that inexhaustible realm.
Faith in Science
The Atlantic Monthly, Volume 187, Number 1, January 1951 (p. 28)

Rumsfeld, Donald   1932–
American businessman, politician, and secretary of state

As we know, There are known knowns. There are things we know we know. We also know there are known unknowns. That is to say we know there are some things We do not know. But there are also unknown unknowns, The ones we don't know we don't know.
Department of Defense News Briefing
February 12, 2002

Service, Robert William   1874–1958
Canadian poet and novelist

Let us probe the silent places,
let us seek what luck betides us;
Let us journey to a lonely land I know.
There's a whisper on the night-wind,
there's a star agleam to guide us,
And the Wild is calling, calling...let us go.
The Complete Poems of Robert Frost
The Call of the Wild
Stanza 5
Dodd, Mead & Company. New York, New York, USA. 1940

Singer, June   1920–2004
American Gnostic

As knowledge proceeds with spiraling movement to
penetrate the vast universe of black mystery, one is
continually astonished to discover that at the outer limit of
awareness where science interfaces with the unknown, there
is nothing but a growing edge, where knowledge and ignorance meet. The more one learns, the more one discovers the
increasing magnitude of the unknown, as anyone who has
tried to do "exhaustive" research knows very well!
Androgyne: Toward a New Theory of Sexuality (p. 59)
Doubleday & Company, Inc. Garden City, New York, USA. 1967

Vernon, A. G.
No biographical data available

It is the successful, or even the unsuccessful, pursuit of
truth which gives happiness to each generation of scientific men, and not the value of the truth itself — the
energy, the doing, not the thing done. If a time could
arrive when all was known, when there could not be a
new investigation or experiment, our keenest pleasure
would be at an end. We may therefore feel happy in the
thought of how much is still unknown.
In Sir Richard Arman Gregory
Discovery; or, The Spirit and Service of Science
Chapter II (p. 28)
Macmillan & Company Ltd. London, England. 1918

Whitney, Willis Rodney   1868–1958
American chemical and electrical engineer

Scientists know that research merely discloses parts of
the infinite unknown. Paradoxically, the enticing, helpful
"unknown" increases as men continue to subtract from
it. Progress in every line of experimental science follows the same law. The apparently narrow path gradually
expands into unlimited, unexplored territory.
Annual Report of the Board of Regents of the Smithsonian Institution, 1924
The Vacuum — There's Something in It (p. 194)
Government Printing Office. Washington, D.C. 1925

UREA

Wöhler, Friedrich   1800–82
German chemist

The fact that in the union of these substances they appear
to change their nature, and give rise to a new body, drew
my attention anew to the subject, and research gave the
unexpected result that by the combination of cyanic acid
with ammonia, urea is formed, a fact that is more noteworthy inasmuch as it furnishes an example of the artificial production of an organic, indeed a so-called animal
substance, from inorganic material.
In Henry M. Leicester and Herbert S. Klickstein
A Source Book in Chemistry: 1400–1900
Friedrich Wöhler (p. 310)
McGraw-Hill Book Company, Inc. New York, New York, USA. 1952

URIC ACID

Wöhler, Friedrich   1800–82
German chemist

No other substance in organic chemistry attracts the
attention of the physiologist and chemist to a higher
degree than uric acid.
In Rolf Huisgen
Adolf von Baeyer's Scientific Achievements — A Legacy
Angewandet Chemie International Edition in English, Volume 25,
Number 4, April 1986 (p. 302)

URINANALYSIS

Addis, Thomas   1881–1949
English-American physician

When the patient dies the kidneys may go to the pathologist, but while he lives the urine is ours. It can provide
us day by day, month by month, and year by year, with
a serial story of the major events going on within the
kidney.
Glomerular Nephritis: Diagnosis and Treatment (p. 2)
The Macmillan Company. New York, New York, USA. 1948

Harington, John   1561–1612
English inventor of flush toilet

He called for his urinal and having made water in it, he
cast it, & viewed it (as Physicians do) a prettie while; at
last he swore soberly, he saw nothing in that man's water,
but that he might live.
The Metamorphosis of Aiax: A New Discourse of a Stale Subject
1596

Shakespeare, William   1564–1616
English poet, playwright, and actor

FALSTAFF: What says the Doctor to my water?
PAGE: He said, Sir, the water itself was good healthy
water; but, for the party that owned it, he might have
more diseases than he knows for.
In *Great Books of the Western World* (Volume 26)
The Plays and Sonnets of William Shakespeare (Volume 1)
The Second Part of Kink Henry the Fourth
Act I, Scene ii, l. 1–4
Encyclopædia Britannica, Inc. Chicago, Illinois, USA. 1952

V

VACCINATION

Jeffers, Robinson   1887–1962
American poet

They take horses
And give them sicknesses through hollow needles, their
blood saves babies: I am here on the mountain making
Antitoxin for all the happy towns and farms…
In Tim Hunt (ed.)
The Collected Poetry of Robinson Jeffers (Volume 1)
A Redeemer (p. 407)
Stanford University Press. Stanford, California. USA. 1988

VACUUM

Bacon, Roger   1214–92
English philosopher, scientist, and friar

For vacuum rightly conceived of is merely a mathematical
quantity extended in the three dimensions, existing per se
without heat and cold, soft and hard, rare and dense, and
without any natural quality, merely occupying space, as
the philosophers maintained before Aristotle, not only
within the heavens, but beyond.
Opus Majus (Volume 2)
Part Five, Ninth Distinction, Chapter II (p. 485)

Huygens, Christiaan   1629–95
Dutch mathematician, astronomer, and physicist

…but what God has bin pleas'd to place beyond the
Region of the Stars, is as much above our Knowledge, as
it is our Habitation.

Or what if beyond such a determinate space he has left
an infinite Vacuum; to show, how inconsiderable is all
that he has made is, to what his Power could, had he so
pleas'd, have produc'd?
*The Celestial Worlds Discover'd; or, Conjectures Concerning the Plan-
etary Worlds, Their Inhabitants and Productions*
Book the Second (p. 156)
Printed for T. Childe. London, England. 1698

Marquis, Don   1878–1937
American newspaperman, poet, and playwright

he i said is afraid of a vacuum
what is there in a vacuum to make one afraid said the
flea there is nothing in it i said
and that is what makes one afraid to contemplate it
a person can t think of a place with nothing at all in it
without going nutty
and if he tries to think that nothing is something after all
he gets nuttier
the lives and times of archy & mehitabel

the merry flea (p. 45)
Doubleday, Doran & Company, Inc. Garden City, New York, USA. 1933

Morris, Richard   1939–2003
American physicist and science writer

In modern physics, there is no such thing as "nothing."
Even in a perfect vacuum, pairs of virtual particles are
constantly being created and destroyed. The existence of
these particles is no mathematical fiction. Though they
cannot be directly observed, the effects they create are
quite real. The assumption that they exist leads to predictions that have been confirmed by experiment to a high
degree of accuracy.
The Edges of Science
Chapter II (p. 25)
Houghton Mifflin Company. Boston, Massachusetts, USA. 1955

Pagels, Heinz R.   1939–88
American physicist and science writer

Once our minds accept the mutability of matter and the
new idea of the vacuum, we can speculate on the origin
of the biggest thing we know — the universe. Maybe the
universe itself sprang into existence out of nothingness
— a gigantic vacuum fluctuation which we know today
as the big bang. Remarkably, the laws of modern physics
allow for this possibility.
The Cosmic Code: Quantum Physics as the Language of Nature
Part II, Chapter 8 (p. 278)
Simon & Schuster. New York, New York, USA. 1982

Pascal, Blaise   1623–62
French mathematician and physicist

Because…you have believed from childhood that a box
was empty when you saw nothing in it, you have believed
in the possibility of a vacuum.
In *Great Books of the Western World* (Volume 33)
Pensées
Section II, 82
Encyclopædia Britannica, Inc. Chicago, Illinois, USA. 1952

Whitehead, Alfred North   1861–1947
English mathematician and philosopher

You cannot have first space and then things to put into
it, any more than you can have first a grin and then a
Cheshire cat to fit on to it.
In Sir Arthur Stanley Eddington
New Pathways in Science
Chapter II, Section VI (p. 48)
The Macmillan Company. New York, New York, USA. 1935

Williams, Tennessee   1911–83
American playwright

…a vacuum is a hell of a lot better than some of the stuff
that nature replaces it with.
Cat on a Hot Tin Roof

Act 2
A New Directions Book. New York, New York, USA. 1975

VALUE

Bronowski, Jacob   1908–74
Polish-born British mathematician and polymath

The values by which we are to survive are not rules for just and unjust conduct, but are those illuminations in whose light justice and injustice, good and evil, means and ends are seen in fearful sharpness of outline.
Science and Human Values
The Sense of Human Dignity (p. 73)
Harper & Row, Publishers. New York, New York, USA. 1965

Poincaré, Henri   1854–1912
French mathematician and theoretical astronomer

If a new result has value it is when, by binding together long-known elements, until now scattered and appearing unrelated to each other, it suddenly brings order where there reigned apparent disorder.
Annual Report of the Board of Regents of the Smithsonian Institution, 1909
The Future of Mathematics (p. 126)
Government Printing Office. Washington, D.C. 1910

VARIANCE

Boring, Edwin Garrigues   1886–1968
American psychologist

McDougall's freedom was my variance. McDougall hoped that variance would always be found in specifying the laws of behavior, for there freedom might still persist. I hoped then — less wise than I think I am now (it was 31 years ago) — that science would keep pressing variance towards zero as a limit. At any rate this general fact emerges from this example: freedom, when you believe it is operating, always resides in an area of ignorance. If there is a known law, you do not have freedom.
When Is Human Behavior Predetermined
The Scientific Monthly, Volume 84, 1957 (p. 190)

Cooley, Charles Horton   1864–1929
American sociologist

It is clear that one who attempts to study precisely things that are changing must have a great deal to do with measures of change.
Observations on the Measure of Change
Journal of the American Statistical Association, New Series, Number 21, March 1893

Crichton, Michael   1942–
American novelist

The computer informed her that three spaces accounted for eighty-one percent of variance.

The Terminal Man
Chapter 6 (p. 47)
Alfred A. Knopf. New York, New York, USA. 1972

VARIATION

Darwin, Charles Robert   1809–82
English naturalist

...individuals of the same species often present, as is known to every one, great differences of structure, independently of variation, as in the two sexes of various animals, in the two or three castes of sterile females or workers amongst insects, and in the immature and larval states of many of the lower animals.
In *Great Books of the Western World* (Volume 49)
The Origin of Species by Means of Natural Selection
Chapter II (p. 25)
Encyclopædia Britannica, Inc. Chicago, Illinois, USA. 1952

Many laws regulate variation, some few of which can be dimly seen.... I will here only allude to what may be called correlated variation. Important changes in the embryo or larva will probably entail changes in the mature animal.... Breeders believe that long limbs are almost always accompanied by an elongated head...cats which are entirely white and have blue eyes are generally deaf.... [I]t appears that white sheep and pigs are injured by certain plants whilst dark-colored individuals escape....
In *Great Books of the Western World* (Volume 49)
The Origin of Species by Means of Natural Selection
Chapter I (p. 11)
Encyclopædia Britannica, Inc. Chicago, Illinois, USA. 1952

But at present, after drawing up a rough copy on this subject, my conclusion is that external conditions do extremely little, except in causing mere variability. This mere variability (causing the child not closely to resemble the parent) I look at as very different from the formation of a marked variety or new species.
In Francis Darwin (ed.)
The Life and Letters of Charles Darwin (Volume 1)
C. Darwin to J.D. Hooker, November 23rd [1856] (p. 445)
D. Appleton & Company. New York, New York, USA. 1896

...the number of intermediate varieties, which must have formerly existed, [must] be truly enormous. Why then is not every geological formation and every stratum full of such intermediate links? Geology assuredly does not reveal any such finely graduated organic chain; and this, perhaps, is the most obvious and gravest objection which can be urged against the theory.
In *Great Books of the Western World* (Volume 49)
The Origin of Species by Means of Natural Selection
Chapter X (p. 152)
Encyclopædia Britannica, Inc. Chicago, Illinois, USA. 1952

Fieller, E. C.
American statistician

Before the inherent variability of the test-animals was appreciated, assays were sometimes carried out on as few as three rabbits: as one pharmacologist put it, those were the happy days.
The Biological Standardization of Insulin
Supplement to the Journal of the Royal Statistical Society, Volume 7, Number 1, 1940–41 (p. 3)

Harvey, William   1578–1657
English physician

...to me the form of the egg has never appeared to have aught to do with the engenderment of the chick, but to be a mere accident; and to this conclusion I come the rather when I see the diversities in the shapes of the eggs of different hens.
In *Great Books of the Western World* (Volume 28)
Anatomical Exercises on the Generation of Animals
Exercise 59 (p. 462)
Encyclopædia Britannica, Inc. Chicago, Illinois, USA. 1952

Hume, David   1711–76
Scottish philosopher and historian

Nothing so like as eggs; yet no one, on account of this appearing similarity, expects the same taste and relish in all of them.
In *Great Books of the Western World* (Volume 35)
An Enquiry Concerning Human Understanding
Section IV, Part II (p. 462)
Encyclopædia Britannica, Inc. Chicago, Illinois, USA. 1952

Huxley, Thomas Henry   1825–95
English biologist

The student of anatomy is perfectly well aware that there is not a single organ of the human body the structure of which does not vary, to a greater or less extent, in different individuals.
Man's Place in Nature and Other Anthropological Essays
Chapter III (p. 185)
D. Appleton & Company. New York, New York, USA. 1896

Leibniz, Gottfried Wilhelm   1646–1716
German philosopher and mathematician

...there are never in nature two beings which are exactly alike...
Philosophical Papers and Letters (Volume 2)
Monadology, 9 (p. 1044)
The University of Chicago Press. Chicago, Illinois, USA. 1956

Pallister, William Hales   1877–1946
Canadian physician

What shall we say of a plot of ground
Planted in similar seed,
Where thousands of similar plants are found
But one is a new type indeed;
When dissimilar comes from similar,
And freedom has its hour,

When the scion is not as ancestors are,
What is this latent power?
Poems of Science
De Ipsa Natura, Variation (p. 213)
Playford Press. New York, New York, USA. 1931

Peirce, Charles Sanders   1839–1914
American scientist, logician, and philosopher

The endless variety in the world has not been created by law. It is not the nature of uniformity to originate variation, nor of law to beget circumstance.
Collected Papers (Volume 6)
Chapter 6, Section 2 (p. 373)
Harvard University Press. Cambridge, Massachusetts, USA. 1960

Tippett, L. C.
English statistician

Variation is, of course, an important characteristic of populations that individuals cannot have.... A thousand exactly similar steel bearing balls (if such were possible) would be no more than one ball multiplied one thousand times. It is the quality of variation that makes it difficult at first to carry in mind a population in its complexity.
The World of Mathematics (Volume 3)
Sampling and Standard Error (p. 1480)
Simon & Schuster. New York, New York, USA. 1956

Waddington, Conrad Hal   1905–75
British biologist and paleontologist

To suppose that the evolution of the wonderfully adapted biological mechanisms has depended only on a selection out of a haphazard set of variations, each produced by blind chance, is like suggesting that if we went on throwing bricks together into heaps, we should eventually be able to choose ourselves the most desirable house.
The Listener, 13 February 1952

Wheeler, William Morton   1865–1937
American entomologist

Since no two events are identical, every atom, molecule, organism, personality, and society is an emergent and, at least to some extent, a novelty.
Emergent Evolution of the Social
Proceedings of the Sixth International Congress of Philosophers, Cambridge, Massachusetts, USA, 1926.

Ricklefs, R.
No biographical data available

Variation in the environment is a fact of life for all plants and animals, except perhaps for inhabitants of the abyssal depths of the sea.
Ecology (2nd edition) (p. 159)
Chiron Press. New York, New York. USA
American humorist and critic

VARIETY

Cowper, William   1731–1800
English poet

Variety's the very spice of life,
That gives it all its flavor.
The Poetical Works of William Cowper
The Task, Book II (The Timepiece), l. 606
John W. Lovell Company. New York, New York, USA. No date

Feyerabend, Paul K.   1924–94
Austrian-born American philosopher of science

Unanimity of opinion may be fitting for a church, for the frightened or greedy victims of some (ancient, or modern) myth, or for the weak and willing followers of some tyrant. Variety of opinion is necessary for objective knowledge. And a method that encourages variety is also the only method that is comparable with a humanitarian outlook.
Against Method: Outline of an Anarchistic Theory of Knowledge
Chapter 3 (p. 46)
Verso. London, England. 1978

VECTOR

Kelvin, Lord William Thomson   1824–1907
Scottish engineer, mathematician, and physicist

Quaternions came from Hamilton…and have been an unmixed evil to those who have touched them in any way. Vector is a useless survival…and has never been of the slightest use to any creature.
In Jerrold E. Marsden and Anthony J. Tromba
Vector Calculus
Chapter 1 (p. 1)
W.H. Freeman & Company. New York, New York, USA. 2003

Warren, Robert Penn   1905–89
American writer and critic

What if angry vectors veer
Round your sleeping head, and form.
There's never need to fear
Violence of the poor world's abstract storm.
Poems
Lullaby: Smile in Sleep
Louisiana State University Press. Baton Rouge, Louisiana, USA. 1998

VECTOR ANALYSIS

Gibbs, J. Willard   1839–1903
American mathematician

If I wished to attract the student of any of these sciences to an algebra for vectors, I should tell him that the fundamental notions of this algebra were exactly those with which he was daily conversant.… In fact, I should tell him that the notions which we use in vector analysis are those which he who reads between the lines will meet on every page of the great masters of analysis, or of those who have probed the deepest secrets of nature.
Quaternions and the Algebra of Vectors
Nature, Volume 47, Number 1220, 16 March, 1893 (p. 464)

The numerical description of a vector requires three numbers, but nothing prevents us from using a single number for its symbolical designation. An algebra or analytical method in which a single letter or other expression is used to specify a vector may be called a vector algebra or vector analysis.
Elements of Vector Analysis Arranged for the Use of Students in Physics, 1881

VEGETARIAN

Hutchison, Sir Robert Grieve   1871–1960
English radiologist

Don't scrape your insides with much roughage as it is more likely to do harm than good. Vegetarianism is harmless enough though it is apt to fill a man with wind and self-righteousness.
Address
British Medical Association, Winnipeg, Canada, 1930

VEGETATION

Carson, Rachel   1907–64
American marine biologist and author

The earth's vegetation is a part of the web of life in which there are intimate and essential relations between plants and the earth, between plants and other plants, between plants and animals, and we must learn to respect that fine and fragile web if there is to be anything left for the next generation.
Silent Spring
Chapter 6 (p. 64)
Houghton Mifflin Company. Boston, Massachusetts, USA. 1961

White, Gilbert   1720–93
English naturalist and cleric

Vegetation is highly worthy of our attention; and in itself is of the utmost consequence to mankind, and productive of many of the greatest comforts and elegancies of life.
The Natural History of Selborne
Letter XL (p. 192)
Robert M. McBride & Company. New York, New York, USA. 1925

VENUS, TRANSIT OF

Harkness, William   1837–1903
Scottish-American astronomer and surgeon

We are now on the eve of the second transit of a pair [of two transits of Venus as it passes directly between the arth and the Sun], after which there will be no other till he twenty-first century of our era has dawned upon the arth, and the June flowers are blooming in 2004. When he last transit season occurred the intellectual world was awakening from the slumber of ages, and that wondrous cientific activity which has led to our present advanced knowledge was just beginning. What will be the state of cience when the next transit season arrives God only knows. Not even our children's children will live to take art in the astronomy of that day. As for ourselves, we have to do with the present…

Address by William Harkness
Proceedings of the AAAS 31st Meeting (Salem, 1883), August, 1882 (p. 77)

Proctor, Richard A.   1837–88
English astronomer

I think the astronomers of the first years of the twenty first century, looking back over the long transit-less period which will then have passed, will understand the anxiety of astronomers in our own time to utilize to the full whatever opportunities the coming transits may afford…; and I venture to hope…they will not be disposed to judge over harshly what some in our own day may have regarded as an excess of zeal.

Transits of Venus, a Popular Account (p. 231)
Longmans, Green & Company. London, England. 1882
American humorist and critic

VERNAL EQUINOX

Cuppy, Will   1884–1929
American humorist and critic

Among things you might be thinking about today is the vernal equinox — it's March 21, you know. The vernal equinox is the point at which the sun apparently crosses the celestial equator toward the north, or you can say it is the moment at which this occurs, or you can simply say: "Hooray! Spring is here!" Exactly why the sun does this on March 21 is a long story.

How to Get from January to December
March 21 (p. 61)
Holt. New York, New York, USA. 1951

VERNIER

Langley, Samuel Pierpoint   1834–1906
American astronomer and aviation pioneer

That little Vernier, on whose slender lines
The midnight taper trembles as it shines,
Tells through the mist where dazzled Mercury burns,
And marks the point where Uranus returns.

The New Astronomy
Chapter I (p. 3)
Houghton Mifflin Company. Boston, Massachusetts, USA. 1889

VERTEBRATE

Lamarck, Jean-Baptiste Pierre Antoine   1744–1829
French biologist

…if the vertebrates differ markedly from one another in their organisation, it is because nature only started to carry out her plan in their respect with the fishes; that she made further advances with the reptiles; that she carried it still nearer perfection with the birds, and that finally she only attained the end with the most perfect mammals.

Translated by Hugh Elliot
Zoological Philosophy: An Exposition with Regard to the Natural History of Animals
Chapter VI (p. 81)
The University of Chicago Press. Chicago, Illinois, USA. 1984

VIBRATION

Tuttle, Hudson   1836–1910
American spiritualist

There is no breath of air so gentle, no wave breaking on the sands, but the vibrations of these movements run through all space.

In Ludwig Buchner
Force and Matter
Chapter III (p. 16)
Trubner & Company. London, England. 1864

VIEW

Dewey, John   1859–1952
American philosopher and educator

It is not truly realistic or scientific to take short views, to sacrifice the future to immediate pressure, to ignore facts and forces that are disagreeable and to magnify the enduring quality of whatever falls in with immediate desire. It is false that the evils of the situation arise from absence of ideals; they spring from wrong ideals.

Reconstruction in Philosophy
Chapter V (p. 130)
Beacon Press. Boston, Massachusetts, USA. 1920

VIRUS

Cudmore, Lorraine Lee
American cell biologist

All living things need their instruction manual (even nonliving things like viruses) and that is all they need, carried in one very small suitcase.

The Center of Life: A Natural History of the Cell
The Universal Cell (p. 8)
New York Times Book Company. New York, New York, USA. 1977

Newman, Michael
No biographical data available

Observe this virus: think how small
Its arsenal, and yet how loud its call;
It took my cell, now takes your cell,
And when it leaves will take our genes as well.
The Sciences
Cloned Poem, 1982

Thomas, Lewis   1913–93
American physician and biologist

We live in a dancing matrix of viruses; they dart, rather like bees, from organism to organism, from plant to insect to mammal to me and back again, and into the sea, tugging along pieces of this genome, strings of genes from that, transplanting grafts of DNA, passing around heredity as though at a great party. They may be a mechanism for keeping new, mutant kinds of DNA in the widest circulation among us. If this is true, the odd virus disease on which we must focus so much of our attention in medicine, may be looked on as an accident, something dropped.
The Lives of a Cell: Notes of a Biology Watcher
The Lives of a Cell (p. 5)
The Viking Press. New York, New York, USA. 1974

VITALITY

Burroughs, John   1837–1921
American naturalist and writer

Biological science has hunted the secret of vitality like a detective, and it has done some famous work; but it has not yet unraveld the mystery.
The Breath of Life
Chapter IV (p. 76)
Houghton Mifflin Company. Boston, Massachusetts, USA. 1915

VITAMIN

Drummond, Jack Cecil   1891–1952
English biochemist

The suggestion is now advanced that the final "–e" [of Funk's "vitamine"] be dropped, so that the resulting word Vitamin is acceptable under the standard scheme of nomenclature adopted by the Chemical Society.... It is recommended that the somewhat cumbrous nomenclature introduced by McCollum (Fat-soluble A, Water-soluble B), be dropped, and that the substances be spoken of as Vitamin A, B, C, etc.
The Nomenclature of the So-Called Accessory Food Factors (Vitamins)
Biochemical Journal, Volume 14, 1920

VOID

Blake, William   1757–1827
English poet, painter, and engraver

For the Chaotic Voids outside of the Stars are measured
by The Stars...
The Complete Poetry and Prose of William Blake
Milton
Book the Second
University of California Press. Berkeley, California, USA. 1982

Pagels, Heinz R.   1939–88
American physicist and science writer

The nothingness "before" the creation of the universe is the most complete void that we can imagine — no space, time, or matter existed. It is a world without place, without duration or eternity, without number — it is what mathematicians call "the empty set." Yet this unthinkable void converts itself into the plenum of existence — a necessary consequence of physical laws. Where are these laws written into that void? What "tells" the void that is pregnant with a possible universe? It would seem that even the void is subject to law, a logic that exists prior to space and time.
Perfect Symmetry: The Search for the Beginning of Time
Part Three, Chapter 5 (p. 347)
Simon & Schuster. New York, New York, USA. 1985

Thomson, James   1700–48
Scottish poet

With what an awful, world-revolving power,
Were first the unwieldy planets launched along
The illimitable void! There
to remain
Amidst the flux of many
thousand years,
That oft has swept the toiling race of men,
And all their labored monuments, away.
In Eli Maor
To Infinity and Beyond: A Cultural History of the Infinite (p. 206)
Birkhäuser. Boston, Massachusetts, USA. 1987

VOLATILITY

Bowen, Norman L.   1887–1956
Canadian geologist

To many petrologists a volatile component is exactly like a Maxwell demon; it does just what one may wish it to do.
The Evolution of the Igneous Rocks
Chapter XVI (p. 282)
Dover Publication, Inc., New York, New York, USA. 1956

VOLCANO

Anderson, Tempest   1846–1913
British ophthalmic surgeon

Very few branches of science still remain available for the amateur of limited leisure. Electricity, Chemistry,

Bacteriology, most branches of Geology and Mineralogy, have all led to results of highest economic value, and they are cultivated by a large body of professional men subsidized by Colleges or by the Government. They are in a position to give their whole time to their work, and their results are so voluminous that to keep abreast of the literature of any single branch would occupy more than the entire leisure of most men, yet this is a necessary preliminary to any attempt at original work. I was consequently led to seek some branch of Science which gave no prospect of pecuniary return, and I determined on Vulcanology, which had the additional advantage of offering exercise in the open air, and in districts often remote and picturesque.
Volcanic Studies in Many Lands
Preface (p. ix)
John Murray. London, England. 1917

Anyidoho, Kofi   1947–
Ghanaian poet

Our Earth survives recurring furies
of her stomach pains and quakes
From the bleeding anger of her wounds
volcanic ash becomes the hope
that gives rebirth to abundance of seedtimes.
Earthchild, with Brain Surgery: Poems
The Homing Call of Earth
Woeli Publishing Services. 1985

Author undetermined

Among the many wonderful works of God, none exhibits so much of awful grandeur as an active volcano.
Wonders of Creation: A Descriptive Account of Volcanoes and Their Phenomena
Preface
T. Nelson. London, England. 1890

For the clouds that overhang an active volcano during an eruption of its vapours are, in reality, thunderclouds highly charged with electricity. They accordingly produce what Baron Humboldt calls the volcanic storm. It includes all the most terrible of atmospheric phenomena — lightnings of extraordinary vividness; thunders that peal and reverberate as if they would rend the echoes asunder; torrents of rain that pour down upon the mountain and its neighborhood, hissing like thousands of serpents when they fall on the glowing lava-torrent; and whirlwinds that sweep the volcanic ashes round and round in vast eddies, and before whose violence no man of mortal mould is able for a moment to stand.
Wonders of Creation: A Descriptive Account of Volcanoes and Their Phenomena
Chapter I
T. Nelson. London, England. 1890

A volcano is a mountain that is busted and squirts out stuff.

In C. Judson Herrick
The Evolution of Human Nature
Chapter Five (p. 57)
University of Texas Press. Austin, Texas, USA. 1956

Bradley, Jr., John Hodgdon   1898–1962
American geologist

Volcanoes are personalities that resist classification.
Autobiography of Earth
Chapter VII (p. 211)
Coward-McCann, Inc. New York, New York, USA. 1935

Burnet, Thomas   1635–1715
English cleric and scientist

There is nothing certainly more terrible in all Nature than Fiery Mountains, to those that live within the view or noise of them; but it is not easier for us, who never see them nor heard them, to represent to our selves with such just and lively imaginations as shall excite us in the same passions, and the same horror as they would excite, if present to our senses.
The Sacred Theory of the Earth (2nd edition)
Book III, Chapter VII (p. 272)
Printed by R. Norton. London. 1691

Cassius Dio   150–235
Roman senator and historian

Thus day was turned into night and light into darkness. Some thought that the Giants were rising again in revolt (for at this time also many of their forms could be discerned in the smoke and, moreover, a sound as of trumpets was heard), while others believed that the whole universe was being resolved into chaos or fire.
Dio's Roman History
Epitome of Book LXVI (p. 211)
Heinemann. London, England. 1914–27

de Saint-Exupéry, Antoine   1900–44
French aviator and writer

This day, as I fly, the lava world is calm. There is something surprising in the tranquility of this deserted landscape where once a thousand volcanoes boomed to each other in their great subterranean organs and spat forth their fire. I fly over a world mute and abandoned, strewn with black glaciers.
Wind, Sand and Stars
Chapter 5, Section I (pp. 99–100)
Reynal & Hitchcock. New York, New York, USA. 1939

He carefully cleaned out his active volcanoes. He possessed two active volcanoes; and they were very convenient for heating his breakfast in the morning. He also had one volcano that was extinct. But, as he said, "One never knows!" So he cleaned out the extinct volcano, too. If they are well cleaned out, volcanoes burn slowly and steadily, without any eruptions. Volcanic eruptions are like fires in a chimney.

Translated by Katherine Woods
The Little Prince
Chapter IX (p. 32)
Harcourt, Brace & Company. New York, New York, USA. 1943

Decker, Robert 1927–2005
American volcanologist
Decker, Barbara
American science writer

Volcanoes are nature's forges and stills where the elements of the Earth, both rare and common, are moved and sorted.
Volcanoes (3rd edition)
Chapter 13 (p. 168)
W.H. Freeman & Company. San Francisco, California, USA. 1981

Volcanoes assail the senses. They are beautiful in repose and awesome in eruption; they hiss and roar, they smell of brimstone. Their heat warms, their fires consume; they are the homes of gods and goddesses.

Volcanoes are described in words and pictures, but they must be experienced to be known. Their roots reach deep inside the Earth; their products are scattered in the sky. Understanding volcanoes is an unconquered challenge.
Volcanoes (3rd edition)
Preface (p. vii)
W.H. Freeman & Company. San Francisco, California, USA. 1981

Francis, Peter 1944–99
English volcanologist

If mountains can have personalities, then volcanoes are schizophrenic — they have split personalities. For most of their life, they are dormant, and one tends to think of them as graceful unsweeping cones, delicately capped with snow, dreaming serenely over the cherry-blossom-draped landscapes of calendars and travel posters. Sometimes, perhaps not very often during their lifetimes, volcanoes erupt and present a wholly different character. Convolute eruption clouds tower above them, raining hot ashes on the helpless humans who live on their flanks, and glowing tongues of liquid rock ooze inexorably downwards, engulfing the flimsy structures which stand in their way.
Volcanoes: A Planetary Perspective
Chapter 1 (p. 13)
Penguin Books Ltd. Middlesex, England. 1976

Guterson, David 1956–
American writer

"Everything up here is crumbling," he said. "Erosion city or something."
Basalt lava," Christine said.
"Old volcanoes."
"Very old volcanoes."
"Like fifty million years."

"Even older."
East of the Mountains
Chapter Two (pp. 37–38)
Harcourt Brace & Company. New York, New York, USA. 1999

Hutton, James 1726–97
Scottish geologist, chemist, and naturalist

A volcano is not made on purpose to frighten superstitious people into fits of piety and devotion, nor to overwhelm devoted cities with destruction; a volcano should be considered as a spiracle to the subterranean furnace, in order to prevent the unnecessary elevation of land and fatal effects of earthquakes.
The Theory of the Earth (Volume 1)
Part I, Section III (p. 146)
Messrs. Cadwell, Junior & Davies. London, England. 1795

Krafft, Katia 1942–91
French volcanologist and photographer

I would always like to be near craters, drunk with fire, gas, my face burned by the heat…. It's not that I flirt with my death, but at this point I don't care about it, because there is the pleasure of approaching the beast and not knowing if he is going to catch you.
In Stanley Williams and Fen Montaigne
Surviving Galeras
Chapter 6 (p. 101)
Houghton Mifflin Company. Boston, Massachusetts, USA. 2001

Lowry, Malcolm 1909–57
English novelist

…it was in eruption, yet no, it wasn't the volcano, the world itself was bursting, bursting into black spouts of villages catapulted into space with himself falling through it all, through the blazing of ten million bodies, falling…
Under the Volcano
Chapter XII (p. 375)
Penguin Books, USA. New York, New York, USA. 1971

McBirney, Alexander R. 1924–
American geologist and founder of Center for Volcanology

The progress that volcanology has made since ancient scholars explained the lavas of Vesuvius and Etna as products of combustion and subterranean storms is more apparent than real. The sad fact is that we desperately need a coherent and demonstrable theory on volcanism. Why do volcanoes erupt? The only honest answer is that we do not have the vaguest idea.
In Katia Krafft
Volcanoes: Earths Awakening (p. 4)
Hamond World Atlas Corporation. 1916

Michener, James A. 1907?–97
American novelist

For nearly forty million years the first island struggled in the bosom of the sea, endeavoring to be born as

observable land. For nearly forty million submerged years its subterranean volcano hissed and…spewed forth rock, but it remained nevertheless hidden beneath the dark waters of the restless sea…a small climbing pretentious thing of no consequence.
Hawaii
Chapter I (p. 5)
Random House, Inc. New York, New York, USA. 1959

Miller, Hugh   1802–56
Scottish geologist and theologian

The billows fall back in boiling eddies; the solid strata are upheaved into a flat dome, crusted with corals and shells; it cracks, it severs, a dark gulf yawns suddenly in the midst; a dense strongly variegated cloud of mingled smoke and steam arises black as midnight in its central volumes, but checquered, where the boiling waves hiss at its edge, with wreaths of white.…
Sketch-Book of Popular Geology
Lecture Third (p. 109)
William P. Nimmo & Company. Edinburgh, Scotland. 1880

And over the roar of waves or the rush of tides we may hear the growling of a subterranean thunder, that now dies away in low deep mutterings, and no, ere some fresh earthquake-shock tern — pests the sea, bellows wildly from the abyss.
Sketch-Book of Popular Geology
Lecture Third (pp. 109)
William P. Nimmo & Company. Edinburgh, Scotland. 1880

Muir, John   1838–1914
American naturalist

Like gigantic geysers, spouting hot stone instead of hot water, they work and sleep, and we have no sure means of knowing whether they are only sleeping or dead.
Steep Trails
Chapter III (p. 56)
Norman S. Berg, Publisher. Dunwoody, Georgia, USA. 1970

Ovid   43 BCE–17 AD
Roman poet

Near Troezen, ruled by Pittheus, there is a hill, high and treeless, which once was a perfectly level plane, but now a hill; for (horrible to relate) the wild forces of the winds, shut up in dark regions underground, seeking an outlet for their flowing and striving vainly to obtain a freer space, since there was no chink in all their prison through which their breath could go, puffed out and stretched the ground, just as when one inflates a bladder with his breath, or the skin of a horned goat. That swelling in the ground remained, has still the appearance of a high hill, and has hardened as the years went by.
Translated by Frank Justus Miller
Metamorphoses (Volume 2)
Chapter XV (pp. 385–387)
William Heinemann. London, England. 1916

Perry, Lilla Cabot   1848–1933
American poet

Forgive me not! Hate me and I shall know
Some of love's fire still burns in your breast!
Forgiveness finds its home in hearts at rest,
On dead volcanoes only lies the snow.
Ode to Volcanoes and the Living Earth

Scrope, George Poulett   1797–1876
English geologist and political economist

The action of a Volcano, in its simplest and most general form, may be described as the rise of earthly substances in a liquefied state and at a high temperature, from beneath the outer crust of the earth; accompanied by prodigious volumes of elastic fluids, which, appearing to be evolved from the interior of the mass, burst upwards with violent successive detonations, scattering into the air, to a considerable height, numerous fragments, still in a liquid state, of the lava, through which they tear their way, together with shattered blocks of the solid pre-existing rocks, which obstructed their expansion.
Considerations on Volcanoes
Chapter I (pp. 1–2)
W. Phillips & George Yarp. London, England. 1825

Shelley, Percy Bysshe   1792–1822
English poet

Nature's most secret steps
He like her shadow has pursued, where'er
The red volcano overcanopies
Its fields of snow and pinnacles of ice
With burning smoke.
The Complete Poetical Works of Percy Bysshe Shelley
Alastor
Houghton Mifflin Company. Boston, Massachusetts, USA. 1901

Shindler, Tom
Musician

Those lovely white snow peaks, those rulers of mountains,
Who knows what secrets they keep?
But when they look like forever, just stop and remember
The Giants are only asleep.
The Giants Are Only Asleep
Source undetermined

Tazieff, Haroun   1914–98
Polish-born French volcanologist

In all ages volcanoes have frightened, fascinated and attracted man, because what they hold is at once terrifying, splendid, and mysterious.
Craters of Fire
Chapter XVIII (p. 209)
Hamish Hamilton. London, England. 1952

Tennyson, Alfred (Lord)   1809–92
English poet

Fires that shook me once, but now to silent ashes fallen away.

Cold upon the dead volcano sleeps the gleam of dying day.
Alfred Tennyson's Poetical Works
Locksley Hall, Sixty Years After, Stanza 21
Oxford University Press, Inc. London, England. 1953

Had the fierce ashes of some fiery Peak
Been hurled so high they ranged round the World,
For day by day through many a blood-red eve
The wrathful sunset glared.
Alfred Tennyson's Poetical Works
St. Telemachus
Oxford University Press, Inc. London, England. 1953

Thompson, Dick
American science journalist

Volcanoes are magnificent primordial beasts. They are geology's living dinosaurs.
Volcano Cowboys: The Rocky Evolution of a Dangerous Science
Introduction (p. 1)
St. Martin's Press. New York, New York, USA. 2000

Twain, Mark (Samuel Langhorne Clemens)   1835–1910
American writer and humorist

Here was a yawning pit upon whose floor the armies of Russia could camp, and have room to spare…over a mile square of it was ringed and streaked and striped with a thousand branching streams of liquid and gorgeously brilliant fire! Occasionally the molten lava flowing under the superincumbent crust broke through — split a dazzling streak, from five hundred to a thousand feet long, like a sudden flash of lightning, and then acre after acre of the cold lava parted into fragments, turned up edgewise like cakes of ice when a great river breaks up, plunged downwards, and were swallowed in the crimson cauldron.
Roughing It (Volume 2)
Chapter XXXIII (pp. 296, 298, 299)
Harper & Brothers Publishers. New York, New York, USA. 1899

Virgil   70 BCE–19 BCE
Roman epic, didactic, and idyllic poet

There is an isle hard by Sicania's coast.
That rises, and Aeolian Lipare,
With smoking rocks precipitous, where beneath
Thunders a cave, and Aetna's vaults, scooped out
By Cyclopean forges; the strong strokes
Of anvils to the ear bring echoing groans;
Hisses the steel ore through its hollow depths,
And from its furnaces pants fire — the home
of Vulcan, and Vulcania the land's name.
In *Great Books of the Western World* (Volume 13)
The Aeneid
Book VIII, l. 416–424 (p. 270)
Encyclopædia Britannica, Inc. Chicago, Illinois, USA. 1952

VOLCANOLOGIST

Tazieff, Haroun   1914–98
Polish-born French volcanologist

Studying dormant volcanoes is of no more profit to the volcanologist who is attempting to make forecasts than is the study of healthy people for the practicing physician.
In Stanley Williams and Fen Montaigne
Surviving Galeras
Chapter 6 (p. 101)
Houghton Mifflin Company. Boston, Massachusetts, USA. 2001

Twain, Mark (Samuel Langhorne Clemens)   1835–1910
American writer and humorist

I found the reddest-faced set of men I almost ever saw. In the strong light every countenance glowed like red-hot iron, every shoulder was suffused with crimson and shaded rearward into dingy, shapeless obscurity! The place below looked like the infernal regions and these men like half-cooked devils just come up on furlough. The smell of sulphur is strong, but not unpleasant to a sinner.
Mark Twain's Letters from Hawaii
Volcano House, June 3rd — Midnight (p. 294)
The University Press of Hawaii. Honolulu, Hawaii, USA. 1975

VOLUME

Avogadro, Amedeo   1776–1856
Italian chemist

It must then be admitted that very simple relations also exist between the volumes of gaseous substances and the numbers of simple or compound molecules which form them. The first hypothesis to present itself in this connection, and apparently even the only admissible one, is the supposition that the number of integral molecules in any gases is always the same for equal volumes, or always proportional to the volumes.
Essay on a Manner of Determining the Relative Masses of the Elementary Molecules of Bodies, and the Proportions in Which They Enter into These Compounds
Journal de Physique de Chimie d'historire Naturelle et des Artes,
Volume 73, 1811

Shenstone, W. A.   1850–1908
American science teacher and silica glass-blowing inventor

Avogadro's hypothesis affords a bridge by which we can pass from large volumes of gases, which we can handle, to the minuter molecules, which individually are invisible and intangible.
In Joseph William Mellor
Mellor's Modern Inorganic Chemistry
Chapter 5 (p. 73)
Longmans. London, England. 1967

W

WARNING

Churchill, Winston Spencer   1882–1965
British prime minister, statesmen, soldier, and author

The era of procrastination, of half-measures, of sooth-ing and baffling expedients, of delays, is coming to its close. In its place we are entering a period of consequences.
In Al Gore
An Inconvenient Truth: The Planetary Emergency of Global Warming and What We Can Do About It (pp. 101–102)
From a 1936 speech to citizens of Great Britain
Rodale. New York, New York, USA. 2006

WATER

Abbey, Edward   1927–89
American environmentalist and nature writer

Water, water, water.... There is no shortage of water in the desert but [in] exactly the right amount, a perfect ratio of water to rock, of water to sand, insuring that wide, free, open, generous spacing among plants and animals, homes and towns and cities, which makes the arid West so different from any other part of the nation. There is no lack of water here, unless you try to establish a city where no city should be.
Desert Solitaire
Water (pp. 144–145)
Ballantine Books. New York, New York, USA. 1968

Bangs, Richard   1950–
Adventure writer

...of all our planet's activities — geological movements, the reproduction and decay of biota, and even the dis-ruptive propensities of certain species (elephants and humans come to mind) — no force is greater than the hydrologic cycle.
Rivergods: Exploring the World's Great Wild Rivers
Introduction (p. xiii)
Sierra Club Books. San Francisco, California, USA. 1958

Bradley, Jr., John Hodgdon   1898–1962
American geologist

The history of the land has been written very largely in water.
Autobiography of Earth
Chapter III (p. 72)
Coward-McCann, Inc. New York, New York, USA. 1935

Buckley, Arabella B.   1840–1929
English naturalist and science writer

We are going to spend an hour today in following a drop of water on its travels. If I dip my finger in this basin of water and lift it up again, I bring with it a small glistening drop out of the water below and hold it before you. Tell me, have you any idea where this drop has been? What changes it has undergone, and what work it has been doing during all the long ages water has lain on the face of the earth?
The Fairy-Land of Science
Lecture IV (pp. 95–96)
D. Appleton & Company. New York, New York, USA. 1899

Burke, Edmund   1729–97
English statesman and philosopher

Water, when simple is insipid, inodorous, colorless, and smooth; it is found, when not cold, to be a great resolver of spasms, and lubricator of the fibers; this power it probably owes to its smoothness.
On the Sublime and the Beautiful
Part IV, Section XXI (p. 166)
Cassell & Company Ltd. London, England. 1887

Butler, Samuel   1612–80
English novelist, essayist, and critic

Water is frozen steam, and ice frozen water.
The Note-Books of Samuel Butler (Volume 1)
1874–1883 (p. 74)
University Press of America, Inc. Lanham, Maryland, USA. 1984

Coleridge, Samuel Taylor   1772–1834
English lyrical poet, critic, and philosopher

Water, water, everywhere,
And all the boards did shrink;
Water, water, everywhere,
Nor any drop to drink.
The Rime of the Ancient Mariner and Other Poems
Rime of the Ancient Mariner, Part II, l. 114–118
Little Leather Library Corporation. New York, New York, USA. 1915

da Vinci, Leonardo   1452–1519
Italian High Renaissance painter and inventor

The water wears away the mountains and fills up the valleys, and if it had the power it would reduce the earth to a perfect sphere.
Translated by Edward MacCurdy
The Notebooks of Leonardo da Vinci (Volume 1)
Physical Geography (p. 317)
George Braziller. New York, New York, USA. 1958

If a drop of water falls into the sea when it is calm, it must of necessity be that the whole surface of the sea is raised imperceptibly, seeing that water cannot be compressed within itself, like air.
Leonardo da Vinci's Note Books (p. 101)
Duckworth & Company. London, England. 1906

Davy, Sir Humphry   1778–1829
English chemist

...I have come to [the] conclusion...that water is the basis of all the gases, and that oxygen, hydrogen, nitrogen, ammonia, nitrous acid, &c., are merely electrical forms of water....
Fragmentary Remains
Chapter IV
Letter to T.A. Knight (p. 129)
John Churchill. London, England. 1858

de Saint-Exupéry, Antoine   1900–44
French aviator and writer

Water, thou hast no taste, no color, no odor; canst not be defined, art relished while ever mysterious. Not necessary to life, but rather life itself, thou fillest us with a gratification that exceeds the delight of the senses.
Wind, Sand and Stars
Chapter 8 (p. 234)
Reynal & Hitchcock. New York, New York, USA. 1939

Earle, Sylvia Alice   1935–
American oceanographer and education advocate

It doesn't matter where on Earth you live, everyone is utterly dependent on the existence of that lovely, living saltwater soup. There's plenty of water in the universe without life, but nowhere is there life without water.
Sea Change: A Message of the Oceans
Introduction (p. xii)
G.P. Putnam's Sons. New York, New York, USA. 1995

Eiseley, Loren C.   1907–77
American anthropologist, educator, and author

If there is magic on this planet, it is contained in water.
The Immense Journey
The Flow of the River (p. 15)
Vintage Books. New York, New York, USA. 1957

Emerson, Ralph Waldo   1803–82
American lecturer, poet, and essayist

Water...transports vast boulders of rock in its iceberg a thousand miles. But its far greater power depends on its talent of becoming little, and entering the smallest holes and pores. By this agency, carrying in solution elements needful to every plant, the vegetable world exists.
The Complete Works of Ralph Waldo Emerson (Volume 7)
Society and Solitude
Chapter VI (p. 146)
Houghton Mifflin Company. Boston, Massachusetts, USA. 1904

Esar, Evan   1899–1995
American humorist

A beautiful blonde is chemically three-fourths water, but what lovely surface tension.
20,000 Quips and Quotes (p. 127)
Doubleday & Company, Inc. Garden City, New York, USA. 1968

Franks, Felix
English chemist and water researcher

Of all known liquids, water is probably the most studied and least understood....
Water: A Comprehensive Treatise
Introduction — Water, the Unique Chemical (p. 18)
Plenum Press. New York, New York, USA. 1972–82

Herbert, Sir Alan   1890–1971
English novelist, playwright, poet, and politician

The rain is plenteous but, by God's decree,
Only a third is meant for you and me;
Two-thirds are taken by the growing things
Or vanish Heavenward on vapour's wings:
Nor does it mathematically fall
With social equity on one and all.
The population's habit is to grow
In every region where the water's low:
Nature is blamed for failings that are Man's,
And well-run rivers have to change their plans.
Water
Source undetermined

Hugo, Victor   1802–85
French author, lyric poet, and dramatist

The water is supple because it is incompressible. It glides away from under the effort. Borne down on one side, it escapes on the other. It is thus that the water becomes a wave. The wave is its liberty.
Translated by Isabel F. Hapgood
The Toilers of the Sea
Part II, Book Third, Chapter II (p. 402)
The Heritage Press. New York, New York, USA. 1961

Huxley, Thomas Henry   1825–95
English biologist

...we live in the hope and the faith that, by the advance of molecular physics, we shall by and by be able to see our way as clearly from the constituents of water to the properties of water, as we are now able to deduce the operations of a watch from the form of its parts and the manner in which they are put together.
Collected Essays (Volume 1)
Method and Result
On the Physical Basis of Life (p. 152)
Macmillan & Company Ltd. London, England. 1904

Lawrence, D. H. (David Herbert)   1885–1930
English writer

Water is H_2O, hydrogen two parts, oxygen one, but there is also a third thing, that makes water, and nobody knows what that is.
Pansies
The Third Thing
Martin Secker. London, England. 1930

Le Févre, Nicholas   1615–79
French chemist

That insipid liquor which commonly is called Water, hath by the Chymists the name of Phlegm given unto it, when it is separated from all other Mixture…
A Complete Body of Chymistry
Part I
Chapter III, Section II (p. 22)
Printed for O. Pullyn. London, England. 1640

McKay, Christopher
American planetary scientist

If some alien called me up [and said]…, "Hello, this is Alpha, and we want to know what kind of life you have," — I'd say, waterbased.… Earth organisms figure out how to make do without almost anything else. The single non-negotiable thing life requires is water.
Interview
OMNI Magazine, July 1992 (p. 66)

Norse, Elliot A.
American marine conservation biologist

In every glass of water we drink, some of the water has already passed through fishes, trees, bacteria, worms in the soil, and many other organisms, including people.… Living systems cleanse water and make it fit, among other things, for human consumption.
In R.J. Hoage (ed.)
Animal Extinctions: What Everyone Should Know
The Value of Animal and Plant Species for Agriculture, Medicine, and Industry (p. 62)
Smithsonian Institution Press. Washington, D.C. 1985

Overstreet, Harry Allen   1875–1970
American social psychologist and civic awareness advocate

Water, however, is not simply the sum of hydrogen and oxygen. It is something qualitatively new, something that cannot be found by the most searching examination of the gas, hydrogen, nor of the gas, oxygen. No amount of previous knowledge of the atomic structure of hydrogen and oxygen could, apparently, give a knowledge of this peculiar fluid that results from combining the two gasses.
The Enduring Quest
Chapter IV (p. 59)
W.W. Norton & Company, Inc. New York, New York, USA.1931

Russell, Bertrand Arthur William   1872–1970
English philosopher, logician, and social reformer

A drop of water is not immortal; it can be resolved into oxygen and hydrogen. If, therefore, a drop of water were to maintain that it had a quality of aqueousness which would survive its dissolution we should be inclined to be skeptical.
What I Believe
Chapter I (p. 6)
E.P. Dutton & Company. New York, New York, USA. 1925

Snicket, Lemony (Daniel Handler)   1970–
American writer

After a great deal of time examining oceans, investigating rainstorms, and staring very hard at several drinking fountains, the scientists of the world developed a theory regarding how water is distributed around our planet, which they have named the "water cycle." The water cycle consists of three key phenomena — evaporation, precipitation, and collection — and all of them are equally boring.
A Series of Unfortunate Events. Book the Eleventh: The Grim Grotto
HarperCollins Publishers. New York, New York, USA. 2004

Strauss, Maurice B.

In the beginning the abundance of the sea
Led to profligacy.
The ascent through the brackish waters of the estuary
To the salt–poor lakes and ponds
Made immense demands
Upon the glands.
Salt must be saved, water is free.
In the never-ending struggle for security,
Man's chiefest enemy.
According to the bard of Stratford on the Avon,
The banks were climbed and life established on dry land
Making the incredible demand
Upon another gland
That water, too, be saved.
Body Water in Man: The Acquisition and Maintenance of the Body Fluids
Salt and Water
Chapter XII (p. 238)
Little, Brown & Company. Boston, Massachusetts, USA. 1957

Maury, Matthew Fontaine   1806–73
American hydrographer and naval officer

The tooth of running water is very sharp.
The Physical Geography of the Sea
Chapter XIV (p. 321)
Harper & Brothers. New York, New York, USA. 1855

van Helmont, Jean-Baptista   1579–1644
Flemish chemist

That all plants immediately and substantially stem from the element water alone I have learnt from the following experiment. I took an earthen vessel in which I placed two hundred pounds of earth dried in an oven, and watered with rain water. I planted in it the stem of a willow tree weighing five pounds. Five years later it had developed a tree weighing one hundred and sixty-nine pounds and three ounces. Nothing but rain (or distilled water) had been added. The large vessel was placed in earth and covered by an iron lid with a tin-surface that was pierced with many holes. I have not weighed the leaves that came off in the four autumn seasons. Finally I dried the earth in the vessel again and found the same two hundred pounds of it diminished by about two ounces. Hence one hundred and sixty-four pounds of wood, bark and roots had come up from water alone.

In William H. Brock
The Norton History of Chemistry
Introduction (p. xxi)
W.W. Norton & Company, Inc. New York, New York, USA.1933

Walton, Izaak   1593–1683
English writer

And an ingenious Spaniard says that rivers and the inhabitants of the watery element were made for wise men to contemplate, and fools to pass by without consideration…for you may note, that the waters are Nature's storehouse, in which she locks up her wonders.
The Complete Angler
First Day, Chapter I (p. 31, 34)
T.N. Foulis. London, England. 1913

WAVE

Crew, Henry
American physicist

To the mathematician the problems of wave-motion offer a field for his highest power of analysis; to the physicist they suggest experiments demanding all the skill at his disposal; to the engineer and to those who go down to the sea in ships these problems are matters of life and death, while to the poet and the artist they are "the sea dancing to its own music."
In Lloyd William Taylor
Physics: The Pioneer Science (Volume 1)
Chapter 24 (p. 327)
Houghton Mifflin Company. Boston, Massachusetts, USA. 1941

Thomson, Sir George Paget   1892–1975
British physicist

The wind catches the filaments and the spider is carried where the filaments take it. In much the same way the point which represents the energy of the electron is guided by the waves which surround it, and extend possibly to an indefinite distance in all directions. If the waves pass over an obstacle like an atom their direction is modified and the modification is transmitted back to the electron and enables it to guide its path in accordance with the distribution of matter which it finds around it.
The Atom
Chapter VII (p. 110)
Oxford University Press, Inc. London, England. 1956

WAVE MECHANICS

Eddington, Sir Arthur Stanley   1882–1944
English astronomer, physicist, and mathematician

Schrödinger's wave-mechanics is not a physical theory, but a dodge — and a very good dodge too.
The Nature of the Physical World

Chapter X (p. 219)
The Macmillan Company. New York, New York, USA. 1930

Gamow, George   1904–68
Russian-born American physicist

In wave mechanics there are no impenetrable barriers, … as the British physicist R.H. Fowler [also] put it after my lecture on that subject at the Royal Society of London…
My World Line: An Informal Autobiography
Chapter 3 (p. 60)
The Viking Press. New York, New York, USA. 1979

WAVE-PARTICLE DUALITY

Glashow, Sheldon L.   1932–
American physicist

One quantum notion that mystifies the novice is the wave-particle duality. Does light consist of a beam of particles or is it a wave phenomenon? the question is hundreds of years old. Newton thought light was probably a stream of particles. Maxwell seemed to answer the problem decisively by showing light to be an electromagnetic wave. Yet Einstein in 1905, demonstrated that under some circumstances light behaves as if it were a beam of discrete particles, which are now called photons.
Interactions: A Journey Through the Mind of a Particle Physicist and the Matter of This World
Chapter 3 (p. 51)
Warner Books. New York, New York, USA. 1988

WEAPON

Amis, Martin   1949–
English writer

Nuclear weapons…are remarkable artifacts. They derive their power from an equation: when a pound of uranium-235 is fissioned, the "liberated mass" within its 1,132,0 00,000,000,000,000,000,000 atoms is multiplied by the speed of light squared — with the explosive force, that is to say, of 186,000 miles per second times 186,000 miles per second. Their size, their power, has no theoretical limit. They are biblical in their anger.
Einstein's Monsters
Introduction: Thinkability (p. 8)
Jonathan Cape Ltd. London, England. 1987

WEATHER

Hopfield, John   1933–
American physicist and neural scientist

You might understand how a few gas molecules interact with one another, but you wouldn't imagine that putting millions of them together would get you weather.
In Ti Sanders

Weather: A User's Guide to the Atmosphere
Chapter 1 (p. 1)
Icarus Press. South Bend, Indiana, USA. 1985

Mitchell, Margaret   1900–49
American author

You can always tell the weather by the sunsets.
Gone With the Wind
Part One, Chapter I (p. 7)
The Macmillan Company. New York, New York, USA. 1936

Thoreau, Henry David   1817–62
American essayist, poet, and practical philosopher

Who watched the forms of the clouds over this part of the earth a thousand years ago? Who watches them to-day?
The Journal of Henry David Thoreau (Volume 13)
December 13. P.M. (p. 23)
Houghton Mifflin Company. Boston, Massachusetts, USA. 1906

WEED

Author undetermined

Weeds are unuseful flowers.
In John Burroughs
Birds and Bees Essays
Introduction (p. 6)
Houghton Mifflin Company. Boston, Massachusetts, USA. 1914

Bailey, William Whitman   1843–1914
American botanist

Weeds are active enemies, not to be despised so much as hated. They are cut down or uprooted whenever found. So great a pest are they that man has taken them for a type of rank, rapid and useless growth. Yet, when curiosity leads us to observe them, we find beauty even in the meanest.
The American Botanist
Volume I, Number 4, October 1901 (p. 50)

Emerson, Ralph Waldo   1803–82
American lecturer, poet, and essayist

What is a weed? A plant whose virtues have not yet been discovered…
The Complete Works of Ralph Waldo Emerson (Volume 11)
Miscellanies
Chapter XXX (p. 512)
Houghton Mifflin Company. Boston, Massachusetts, USA. 1904

Larcom, Lucy   1824–93
American writer

I like these plants that you call weeds
Sedge, hardhack, mullein, yarrow, —
That knit their roots and sow their seeds
Where any grassy wheel-track leads
Through country by-ways narrow.
I Like These Plants that You Call Weeds: Historicizing American

Women's Nature Writing
Nineteenth Century, Volume 58, June 2003

Shakespeare, William   1564–1616
English poet, playwright, and actor

Now 'tis the spring and weeds are shallow rooted;
Suffer them now and they'll outgrow the garden
And choke the herbs for lack of husbandry.
In *Great Books of the Western World* (Volume 26)
The Plays and Sonnets of William Shakespeare (Volume 1)
The Second Part of King Henry the Sixth
Act III, Scene I
Encyclopædia Britannica, Inc. Chicago, Illinois, USA. 1952

WEIGHT

da Vinci, Leonardo   1452–1519
Italian High Renaissance painter and inventor

Weight, pressure, and accidental movement together with resistance are the four accidental powers in which all the visible works of mortals have their existence and their end.
Leonardo da Vinci's Note Books (p. 55)
Duckworth & Company. London, England. 1906

WEIGHTLESSNESS

Verne, Jules   1828–1905
French novelist

Fancy has depicted men without reflection, others without shadow. But here reality, by the neutralizations of attractive forces, produced men in whom nothing had any weight, and who weighed nothing themselves.
From the Earth to the Moon, and Round the Moon
Chapter VIII (p. 228)
A.L. Burt Company. New York, New York, USA. 1890

WETLANDS

Beebe, William   1877–1962
American ornithologist

The marsh, to him who enters it in a receptive mood, holds, besides mosquitoes and stagnation, — melody, the mystery of unknown waters, and the sweetness of Nature undisturbed by man.
The Log of the Sun
Night Music of the Swamp (p. 172)
Henry Holt and Company. New York, New York, USA. 1906

Lanier, Sidney   1842–81
American writer and musician

Ye marshes, how candid and simple and nothing-with-holding and free
Ye publish yourselves to the sky and offer yourselves to the sea!

In Goodridge Bliss Roberts
Younger American Poets, 1830–1890
The Marshes of Glynn
Griffith, Farran, Okeden & Welsh. London, England.1891

WHIRLPOOL

Poe, Edgar Allan   1809–49
American short story writer

I became possessed with the keenest curiosity about the whirl itself. I positively felt a wish to explore its depths, even at the sacrifice I was going to make; and my principal grief was that I should never be able to tell my old companions on shore about the mysteries I should see.
In H. Beaver (ed.)
The Science Fiction of Edgar Allan Poe
A Descent into the Maelstrom (p. 83)
Penguin Books. Hammondsworth, England. 1976

WHITE DWARF

Updike, John   1932–
American novelist, short story writer, and poet

You offer cheer to tiny Man
'Mid galaxies Gargantuan
A little pill in endless night,
An antidote to cosmic fright.
White Dwarf
Source undetermined

WILDERNESS

Abbey, Edward   1927–89
American environmentalist and nature writer

We would guard and defend and save it [wilderness] as a place for all who wish to discover the nearly lost pleasures of adventure, adventure not only in the physical sense, but also mental, spiritual, moral, aesthetic, and intellectual adventure. A place for the free.
The Journey Home: Some Words in Defense of the American West
Chapter 8 (p. 88)
E.P. Dutton & Company. New York, New York, USA. 1977

The idea of wilderness needs no defense. It only needs more defenders.
The Journey Home: Some Words in Defense of the American West
Chapter 21 (p. 223)
E.P. Dutton & Company. New York, New York, USA. 1977

On this great river [the Colorado River] one could glide forever — and here we discover the definition of bliss, salvation, Heaven, all the old Mediterranean dreams: a journey from wonder to wonder, drifting through eternity into ever deeper, always changing grandeur, through beauty continually surpassing itself: the ultimate Homeric voyage.

The Journey Home: Some Words in Defense of the American West
Chapter 17 (p. 201)
E.P. Dutton & Company. New York, New York, USA. 1977

Now I can do no more than offer one final prayer to the young, to the bold, to the angry, to the questioning, to the lost. Beyond the wall of the unreal city, beyond the security fences topped with barbed wire and razor wire, beyond the asphalt belting of the superhighways, beyond the cemented banksides of our temporarily stopped and mutilated rivers, beyond the rage of lies that poisons the air, there is another world waiting for you. It is the old true world of the deserts, the mountains, the forests, the islands, the shores, the open plains. Go there. Be there. Walk gently and quietly deep within it.
Beyond the Wall
Author's Introduction (p. xvi)
Ballantine Books. New York, New York, USA. 1968

Come on in. The earth, like the sun, like the air, belongs to everyone — and to no one.
The Journey Home: Some Words in Defense of the American West
Chapter 8 (p. 88)
E.P. Dutton & Company. New York, New York, USA. 1977

But the love of wilderness is more than a hunger for what is always beyond reach; it is also an expression of loyalty to the earth, the earth which bore us and sustains us, the only home we shall ever know, the only paradise we ever need — if only we had the eyes to see.
Desert Solitaire
Down the River (p. 190)
Ballantine Books. New York, New York, USA. 1968

How difficult to imagine this place without a human presence; how necessary. I am almost prepared to believe that this sweet virginal primitive land will be grateful for my departure and the absence of the tourists, will breathe metaphorically a collective sigh of relief — like a whisper of wind — when we are all and finally gone and the place and its creations can return to their ancient procedures unobserved and undisturbed by the busy, anxious, brooding consciousness of man. Grateful for our departure? One more expression of human vanity. The finest quality of this stone, these plants and animals, this desert landscape is the indifference manifest to our presence, our absence, our coming, our staying or our going. Whether we live or die is a matter of absolutely no concern whatsoever to the desert.
Desert Solitaire
Bedrock and Paradox (p. 300)
Ballantine Books. New York, New York, USA. 1968

It is my fear that if we allow the freedom of the hills, and the last of the wilderness to be taken from us, then the very idea of freedom may die with it.
Down the River
Part II, Chapter 8 (p. 121)
E.P. Dutton & Company. New York, New York, USA. 1982

Berry, Wendell   1934–
American essayist, poet, critic, and farmer

There does exist a possibility that we can live more or less in harmony with our native wilderness; I am betting my life that such a harmony is possible. But I do not believe that it can be achieved simply or easily or that it can ever be perfect, and I am certain that it can never be made, once and for all, but it is the forever unfinished lifework of our species.
The Land of Harmony
Preserving Wilderness (pp. 138–139)
Five Seasons Press. Hereford, England. 1987

This wilderness, the universe, is somewhat hospitable to us, but it is also absolutely dangerous to us (it is going to kill us, sooner or later), and we are absolutely dependent upon it.
The Land of Harmony
Preserving Wilderness (pp. 138–139)
Five Seasons Press. Hereford, England. 1987

We live in a wilderness, in which we and our works occupy a tiny space and play a tiny part. We exist under its dispensation and by its tolerance.
The Land of Harmony
Preserving Wilderness (pp. 138–139)
Five Seasons Press. Hereford, England. 1987

Brower, David   1912–2000
American environmentalist

My feeling is we need to save wilderness for its own sake, for the mysterious and complex knowledge it has within it. Thoreau was right when he said, "In wilderness is the preservation of the world."
In Jonathan White
Talking on the Water
Sierra Club Books. San Francisco, California, USA. 1994

…we need boundaries around cities, not around wildness.
Let the Mountains Talk, Let the Rivers Run
Chapter 5 (p. 43)
HarperCollins Publishers. New York, New York, USA. 1995

Carr, William H.   1902–85
American desert environmentalist and writer

I am thankful for the wild spaces that are yet untouched. May they not decrease in size and number!
The Stir of Nature
Chapter Eight (p. 116)
Oxford University Press, Inc. New York, New York, USA. 1930

Gerould, Katherine Fullerton   1879–1944
American writer

The wilderness is a good place to cry in; the echoes are magnificent.
Modes and Morals

The Extirpation of Culture (p. 67)
Charles Scribner's Sons. New York, New York, USA. 1920

Gould, Stephen Jay   1941–2002
American paleontologist and evolutionary biologist

I can easily understand why, for most naturalists, the highest form of beauty, inspiration, and moral value might be imputed to increasingly rare patches of true wilderness — that is, to parcels of nature devoid of any human presence, either in current person or by previous incursion.
Leonardo's Mountain of Clams and the Diet of Worms
Introduction (p. 1)
Harmon Brown. New York, New York, USA. 1998

Hopkins, Frederick Gowland   1844–89
English biochemist

What would the world be, once bereft
Of wet and of wilderness? Let them be left,
O let them be left, wilderness and wet;
Long live the weeds and the wilderness yet.
In W.H. Gardner and N.H. MacKenzie (eds.)
The Poems of Gerard Manley Hopkins
Inversnaid, Stanza 4
Oxford University Press, Inc. London, England. 1930

Krutch, Joseph Wood   1893–1970
American naturalist, conservationist, and writer

The Wilderness and the idea of wilderness is one of the permanent homes of the human spirit.
Grand Canyon: Today and All Its Yesterday.
William Sloane Associates, Publishers. New York, New York, USA. 1958

Lindbergh, Charles A.   1902–74
American aviator

In wilderness I sense the miracle of life, and behind it our scientific accomplishments fade to trivia.
The Wisdom of Wilderness
Life, Volume 63, Number 25, December 22, 1967 (p. 10)

Thoreau, Henry David   1817–62
American essayist, poet, and practical philosopher

In wildness is the preservation of the world.
Walking
The Atlantic Monthly, Volume 9, Number 56, June 1862 (p. 665)

WILDLIFE

Borland, Hal   1900–78
American writer

The newcomer to the country will find the first signs of "wild life" in his own house. Even before he explores the dooryard he can sharpen his eyes indoors. He may be surprised at the outsiders who want to share that house with him.

Beyond Your Doorstep: A Handbook to the Country
Chapter 1 (p. 1)
Alfred A. Knopf. New York, New York, USA. 1962

Hornaday, William Temple   1854–1937
American naturalist

And yet the game of North America does not belong wholly and exclusively to the men who kill! The other ninety-seven per cent of the People have vested rights in it.… Posterity has claims upon it that no man can ignore.… A continent without wild life is like a forest with no leaves on the trees.
Our Vanishing Wild Life
Preface (p. ix)
C. Scribner's Sons. New York, New York, USA. 1913

Myers, Norman   1934–
British environmentalist

Without knowing it, we utilize hundreds of products each day that owe their origin to wild animals and plants. Indeed our welfare is intimately tied up with the welfare of wildlife. Well may conservationists proclaim that by saving the lives of wild species, we may be saving our own.
A Wealth of Wild Species: Storehouse for Human Welfare
Wild Species (p. 3)
Westview Press. Boulder, Colorado.1983

Prince Philip (Philip Mountbatten), Duke of Edinburgh   1921–
British naturalist

Miners used to take a canary around the coal mines to warn them when the air was so foul that the canary died. This is the importance of wildlife to us; because if wildlife dies it is our turn next. If any part of the life of this planet is threatened, all is threatened. If you say "not interested" to wildlife conservation then you are signing your own death warrant.
The Times (London), May 17, 1988

WIND

Conrad, Joseph   1857–1924
Polish-born English novelist

It was something formidable and swift, like the sudden smashing of a Vial of Wrath. It seemed to explode all around the ship with an overpowering concussion and a rush of great waters, as if an immense dam had been blown up to windward. It destroyed at once the organised life of the ship by its shattering effect. In an instant the men lost touch of each other. This is the disintegrating power of a great wind. It isolates one from one's kind. An earthquake, a landslip, an avalanche, overtake a man incidentally, as it were — without passion. A furious gale attacks him like a personal enemy, tries to grasp his limbs, fastens upon his mind, seeks to rout the very spirit out of him.
Typhoon
Chapter X (p. 77)
Doubleday, Page & Company. Garden City, New York, USA. 1920

Longfellow, Henry Wadsworth   1807–82
American poet

Through woods and mountain passes
The winds, like anthems, roll.
The Poetical Works of Henry Wadsworth Longfellow
Midnight Mass for the Dying Year
Houghton Mifflin Company. Boston, Massachusetts, USA. 1883

Maury, Matthew Fontaine   1806–73
American hydrographer and naval officer

Properly to appreciate the various offices which the winds and the waves perform, we must regard nature as a whole, for all the departments thereof are intimately connected. If we attempt to study in one of them, we often find ourselves tracing clews which lead us off insensibly into others, and, before we are aware, we discover ourselves exploring the chambers of some other department.
The Physical Geography of the Sea
Chapter X (p. 181)
Harper & Brothers. New York, New York, USA. 1855

Peattie, Donald Culrose   1896–1964
American botanist, naturalist, and author

The oldest voice in the world is the wind. When you see it fitfully turning the blades of a mill lazily to draw water, you think of it as an unreliableservant of man. But in truth it is one of ourmasters, obedient only to the lord sun and thewhirling of the great globe itself.
Weather: A National Journal (p. 24)
Weldon Owen Pty Ltd. Sydney, Australia. 1996

WISDOM

Abelard, Peter   1079–1142
French scholastic philosopher

Assiduous and frequent questioning is indeed the first key to wisdom.
Sic et Non (Yes and No)

Aquinas, St. Thomas   1227?–74
Dominican philosopher and theologian

Wisdom is a kind of science in so far as it has that which is common to all the sciences, namely, to demonstrate conclusions from principles. But since it has something proper to itself above the other sciences, in so far, that is, as it judges of them all, not only as to their conclusions, but also as to their first principles, therefore it is a more perfect virtue than science.

Summma Theologica
1–II, 57, 2

Aristotle   384 BCE–322 BCE
Greek philosopher

Again, we do not regard any of the senses as wisdom; yet surely these give us the most authoritative knowledge of particulars. But they do not tell us the "why" of anything — e.g., why fire is hot; they only say that it is hot.
In *Great Books of the Western World* (Volume 8)
Metaphysics
Book I, Chapter 1 (pp. 499–500)
Encyclopædia Britannica, Inc. Chicago, Illinois, USA. 1952

Bell, E. T. (Eric Temple)   1883–1960
Scottish-American mathematician and educator

Wisdom was not born with us, nor will it perish when we descend into the shadows with a regretful backward glance that other eyes than ours are already lit by the dawn of a new and truer mathematics.
The Queen of the Sciences
Chapter X (p. 138)
The Williams & Wilkins Company. Baltimore, Maryland, USA. 1931

Berrill, Norman John   1903–96
English-born American biologist

Wisdom, the highest product of the human mind, comes late; the young are rarely wise and are not expected to be.
Man's Emerging Mind
Chapter I (p. 14)
Dodd, Mead & Company. New York, New York, USA. 1955

Bush, Vannevar   1890–1974
American electrical engineer and physicist

There are also the old men, whose days of vigorous building are done, whose eyes are too dim to see the details of the arch or the needed form of its keystone, but who have built a wall here and there, and lived long in the edifice; who have learned to love it and who have even grasped a suggestion of its ultimate meaning; and who sit in the shade and encourage the young men.
Endless Horizons
Chapter 17 (p. 181)
Public Affairs Press. Washington, D.C. 1946

Campbell, Donald T.   1916–96
American evolutionary philosopher and social scientist

In going beyond what is already known, one cannot but go blindly. If one can go wisely, this indicates already achieved wisdom of some general sort.
In Paul Arthur Schlipp (ed.)
The Philosophy of Karl R. Popper
Evolutionary Epistemology (p. 422)
Open Court. La Salle, Illinois, USA. 1974

Collingwood, Robin George   1889–1943
English historian and philosopher

A man ceases to be a beginner in any given science and becomes a master in that science when he has learned that…he is going to be a beginner all his life.
The New Leviathan; or, Man, Society, Civilization and Barbarism
Part I, Chapter I, aphorism I.46 (p. 3)
At The Clarendon Press. Oxford, England. 1942

Davies, Robertson   1913–95
Canadian novelist

Knowledge may enable you to memorize the whole of Gray's Anatomy and Osler's Principles and Practice of Medicine, but only wisdom can teach you what to do with what you have learned.
The Merry Heart
Chapter 5 (p. 105)
McClelland & Stewart. Toronto, Ontario, Canada. 1996

Knowledge and Wisdom and they are not the same,- because Knowledge is what you are taught, but Wisdom is what you bring to it.
The Cunning Man (p. 167)
McClelland & Steward. Toronto, Ontario, Canada. 1994

Descartes, René   1596–1650
French philosopher, scientist, and mathematician

…human wisdom…always remains one and the same, however applied to different subjects, and suffers no more differentiation proceeding from them than the light of the sun experiences from the variety of things which it illumines.…
In *Great Books of the Western World* (Volume 31)
Rules for the Direction of the Mind
Rule 1 (p. 1)
Encyclopædia Britannica, Inc. Chicago, Illinois, USA. 1952

Eisenschiml, Otto   1880–1963
Austrian-American chemist and historian

Wisdom, if inarticulate, is as impotent as loud-mouthed stupidity.
The Art of Worldly Wisdom: Three Hundred Precepts for Success Based on the Original Work of Baltasar Gracian
Part Nine (p. 112)
Duell, Sloan & Pearce. New York, New York, USA. 1947

Learn as much as you can, but remember that wisdom is more than a mere accumulation of facts. No one hires a man because he knows the encyclopedia by heart.
The Art of Worldly Wisdom: Three Hundred Precepts for Success Based on the Original Work of Baltasar Gracian
Part Nine (p. 108)
Duell, Sloan & Pearce. New York, New York, USA. 1947

Gregory, Sir Richard Arman   1864–1952
British science writer and journalist

The love of truth is the beginning and end of wisdom.
Discovery; or, The Spirit and Service of Science
Chapter II (p. 27)
Macmillan & Company Ltd. London, England. 1918

Kingsley, Charles   1819–75
English clergyman and author

Wise men know that their business is to examine what is, and not to settle what is not.
The Water-Babies
Chapter II (p. 62)
Dodd, Mead & Company. New York, New York, USA. 1910

Milton, John   1608–74
English poet

…to know
That which before us lies in daily life
Is the prime Wisdom…
In *Great Books of the Western World* (Volume 32)
Paradise Lost
Book VIII, l. 192–194
Encyclopædia Britannica, Inc. Chicago, Illinois, USA. 1952

Narby, Jeremy
Swiss-born anthropologist and author

…wisdom requires not only the investigation of many things, but the contemplation of the mystery.
The Cosmic Serpent: DNA and the Origins of Knowledge
Chapter 11 (p. 162)
Tarcher/Putnam. New York, New York, USA. 1998

Procter, Bryan Waller   1787–1874
English poet

…he who can draw a joy
From rocks, or woods, or weeds, or things
That seem all mute, and does it — is wise.
The Poetical Works of Barry Cornwall
A Haunted Stream
Henry Colburn & Company London, England. 1822

The Bible

…I applied my mind to study and explore by means of wisdom all that is done under heaven. It is a worthless task that God has given to mortals to keep them occupied.
The Revised English Bible
Ecclesiastes 1:13
Oxford University Press, Inc. Oxford, England. 1989

For in much wisdom is much vexation; the more knowledge, the more suffering.
The Revised English Bible
Ecclesiastes 1:18
Oxford University Press, Inc. Oxford, England. 1989

WONDER

Chesterton, G. K. (Gilbert Keith)   1874–1936
English author

The world will never starve for want of wonders; but only for want of wonder.
Tremendous Trifles
Tremendous Trifles, I
Dodd, Mead & Company New York, New York, USA. 1909

Cole, William   1530–1600
English man of letters

When we contemplate the works of creation in the construction of the planetary system, and the stupendous parts of the universe, we are struck with admiration at their magnitude; so, when we descend the scale of nature, and contemplate the minutiae of material objects, we discover an equal harmony and beauty in their disposition, and are equally lost in wonder.
Philosophical Remarks on the Theory of Comets, a Dissertation on the Nature and Properties of Light
Conclusion (p. 93)
B.J. Holdsworth. London, England. 1823

Dawkins, Richard   1941–
British ethologist, evolutionary biologist, and popular science writer

We have an appetite for wonder, a poetic appetite, which real science ought to be feeding but which is being hijacked, often for monetary gain, by purveyors of superstition, the paranormal and astrology.
Unweaving the Rainbow: Science, Delusion and the Appetite for Wonder
Chapter 6 (p. 114)
Houghton Mifflin Company. Boston, Massachusetts, USA. 1998

I believe that astrologers, for instance, are playing on — misusing, abusing — our sense of wonder. I mean when they hijack the constellations, and employ sub-poetic language like the moon moving into the fifth house of Aquarius. Real astronomy is the rightful proprietor of the stars and their wonder. Astrology gets in the way, even subverts and debauches the wonder.
Science, Delusion and the Appetite for Wonder
Richard Dimbleby Lecture, BBC1 Television, November 12th, 1996

The fact that we slowly apprehend our world, rather than suddenly discover it, should not subtract from its wonder.
Unweaving the Rainbow: Science, Delusion and the Appetite for Wonder
Chapter 1 (p. 5)
Houghton Mifflin Company. Boston, Massachusetts, USA. 1998

The feeling of awed wonder that science can give us is one of the highest experiences of which the human psyche is capable. It is a deep aesthetic passion to rank with the finest that music and poetry can deliver. It is truly one of the things that makes life worth living and it does so, if anything, more effectively if it convinces us that the time we have for living it is finite.
Unweaving the Rainbow: Science, Delusion and the Appetite for Wonder
Preface (p. x)
Houghton Mifflin Company. Boston, Massachusetts, USA. 1998

There is an appetite for wonder, and isn't true science well qualified to feed it?
Science, Delusion and the Appetite for Wonder
Richard Dimbleby Lecture, BBC1 Television, November 12th, 1996

The mystic is content to bask in the wonder and revel in a mystery that we were not "meant" to understand. The scientist feels the same wonder but is restless, not content; recognizes the mystery as profound, then adds, "But we're working on it."
Unweaving the Rainbow: Science, Delusion and the Appetite for Wonder
Chapter 2 (p. 17)
Houghton Mifflin Company. Boston, Massachusetts, USA. 1998

Gardner, Martin 1914–
American writer and mathematics games editor

…almost all scientists believe that as their knowledge increases, their sense of wonder also grows. The scientist sees a flower, said physicist John Tyndall, "with a wonder superadded."
Science vs. Beauty?
Skeptical Inquirer, March 1, 1995

Guiducci, Mario 1585–1646
Italian follower of and ghost writer for Galileo Galilei

When new or rarely seen things awaken in our minds more wonder than those which are common and ordinary, our desire to learn their causes should be aroused accordingly, and with it our wish to put to test those things reported to us by others or supplied by our own minds.
Discourse on the Comets
Discourse on the Comet (p. 2)
Published by the author. Florence, Italy. 1619

Hardy, Thomas 1840–1928
English poet and regional novelist

Until a person has thought out the stars and their interspaces, he has hardly learnt that there are things more terrible than monsters of shape, namely, monsters of magnitude without known shape. Such monsters are the voids and waste places of the sky.
Two on a Tower
Chapter IV (p. 34)
Harper & Brothers, Publishers. New York, New York, USA. No date

Herschel, Sir John Frederick William 1792–1871
English astronomer and chemist

Accustomed to trace the operation of general causes, and the exemplification of general laws, in circumstances where the uninformed and unenquiring eye perceives neither novelty nor beauty, [the scientist and natural philosopher] walks in the midst of wonders.
A Preliminary Discourse on the Study of Natural Philosophy
Part I, Chapter I, Section 10 (p. 15)
Printed for Longman, Rees, Orme, Brown & Green. London, England. 1831

Krauss, Lawrence M. 1954–
American theoretical physicist

There is plenty of wonder left in the universe even after we have examined all the clues nature has thrown our way. I really believe that our imaginations have not even begun to exhaust the possibilities of existence. To proclaim the slogan "The Truth IS Out There" is perhaps too trite. I prefer "You ain't seen nothin' yet!"
Beyond Star Trek: Physics from Alien Invasions to the End of Time
Epilogue (p. 175)
Basic Books, Inc. New York, New York, USA. 1997

Poe, Edgar Allan 1809–49
American short story writer

…wonders and wild fancies…strangely rife among mankind.
In H. Beaver (ed.)
The Science Fiction of Edgar Allan Poe
The Conversation of Eiros and Charmion (p. 67)
Penguin Books. Hammondsworth, England. 1976

Pope, Alexander 1688–1744
English poet

Pretty! in amber to observe the forms
Of hairs, of straws, or dirt, or grubs, or worms!
The things, we know, are neither rich nor rare,
But wonder how the devil they got there.
The Complete Poetical Works
Prologue to the Satires, l. 169
Houghton Mifflin Company. New York, New York, USA. 1903

Sagan, Carl 1934–96
American astronomer and science writer

Claims that cannot be tested, assertions immune to disproof are veridically worthless, whatever value they may have in inspiring us or in exciting our sense of wonder.
The Demon-Haunted World: Science as a Candle in the Dark
Chapter 10 (p. 171)
Random House, Inc. New York, New York, USA. 1995

Steef, Duncan
No biographical data available

There is philosophy, and hope, in this great enterprise [of planetary exploration]. What a tribute to the human mind it is! We are replete with ambitions and egotisms, and with our love of the temporal — and yet we project our consciousness to the moons of Jupiter and light-years beyond, to the edges of the universe. This fascination, which has endured in some form since our ancestors first contemplated the stars, has nothing to do with the material. It exists because we already possess the greatest treasure of all, wonder. wonder. For such a species, there is hope indeed.
In Nigel S. Hey
Solar System
Ices in the Solar System (p. 257)
Weidenfield & Nicolson. London, England. 2002

Verhoeven, Cornelis 1928–2001
Dutch philosopher and essayist

Wonder is a certainty which has only just been established and has not yet lost the expectation of seeing its opposite appear. This does not exclude the knowledge of that which is incited by wonder. On the contrary: the more we know about something the more we realize that this knowledge is never exhaustive. Knowledge may nourish wonder since it can postulate the possibility that things may be different than they are.... Wonder that a thing is so is motivated by the possibility that it might be different. This movement is endless since this "difference" remains completely undefined.

Translated by Mary Foran
The Philosophy of Wonder
Two (p. 27)
The Macmillan Company. New York, New York, USA. 1972

WORD

Becker, Carl L.   1873–1945
American historian

If we would discover the little backstairs door that for any age serves as the secret entranceway to knowledge, we will do well to look for certain unobtrusive words with uncertain meanings that are permitted to slip off the tongue or the pen without fear and without research; words which, having from constant repetition lost their metaphorical significance, are unconsciously mistaken for objective realities.

The Heavenly City of the Eighteenth Century Philosophers
Chapter II (p. 47)
Yale University Press. New Haven, Connecticut, USA. 1932

Heisenberg, Werner Karl   1901–76
German physicist and philosopher

Words have no well-defined meaning. We can sometimes by axioms give a precise meaning to words, but still we never know how these precise words correspond to reality, whether they fit reality or not.

In Paul Buckley and F. David Peat (eds.)
Glimpsing Reality: Ideas in Physics and the Link to Biology
Werner Heisenberg (p. 7)
University of Toronto Press. Toronto, Ontario, Canada. 1996

Horace (Quintus Horatius Flaccus)   65 BCE–8 BCE
Roman philosopher and dramatic critic

If so be there are abstruse things which absolutely require new terms to make them clear, it will be in your power to frame words which never sounded in the ears of a cinctured Cethegus, and free pardon will be granted if the license be used modestly.... Each generation has been allowed, and will be allowed still to issue words that bear the mint-mark of the day.

In James Boswell
Boswell's "Life of Samuel Johnson"
Ars Poetica 48 (Wickham) (fn p. 158, year 1750)
Oxford University Press, Inc. Oxford, England. 1965

Lewis, Gilbert Newton   1875–1946
American chemist

There is always the danger is scientific work that some word or phrase will be used by different authors to express so many ideas and surmises that, unless redefined, it loses all real significance.

Valence and Tautomerism
Journal of the American Chemical Society, Volume 35, 191 (p. 144S)

Maxwell, James Clerk   1831–79
Scottish physicist

When a physical phenomenon can be completely described as a change in the configuration and motion of a material system, the dynamical explanation of that phenomenon is said to be complete. We cannot conceive any further explanation to be either necessary, desirable, or possible, for as soon as we know what is meant by the words configuration, mass and force, we see that the ideas which they represent are so elementary that they cannot be explained by means of anything else.

Scientific Papers
II, On the Dynamical Evidence of the Molecular Constitution of Bodies (p. 419)

Popper, Karl R.   1902–94
Austrian/British philosopher of science

The relationship between a theory (or a statement) and the words used in its formulation is in several ways analogous to that between written words and the letters used in writing them down.

In Paul Arthur Schlipp (ed.)
The Philosophy of Karl Popper (Volume 1)
Book I, Part I
Autobiography of Karl Popper
Section 7 (p. 15)
The Open Court Publishing Company. LaSalle, Illinois, USA. 1974

Schuster, Sir Arthur   1851–1934
English physicist

Scientific controversies constantly resolve themselves into differences about the meaning of words.

In C. K. Ogden and I. A. Richards
The Meaning of Meaning
Introductory Quotations (p. xxiv)
Harcourt, Brace & Company. New York, New York, USA. 1949

Whewell, William   1794–1866
English philosopher and historian

I am always glad to hear of the progress of your researches, and never the less so because they require the fabrication of a new word or two. Such a coinage has always taken place at the great epochs of discovery; like the medals that are struck at the beginning of a new reign — or rather like the change of currency produced by the

accession of a new sovereign; for their value and influ-ence consists in their coming into common circulation.
In Silvanius P. Thompson
Michael Faraday: His Life and Work
Chapter IV
Letter from Whewell to Faraday
14 October 1837 (p. 163)
Cassell & Company Ltd. London, England. 1901

Wolfenden, John Frederick   1906–85
English education leader

The everyday difficulty is to use words "pure and sim-ple," without getting entangled in their emotional lives.... [T]he scientist is to a large extent freed from this tempta-tion. He knows very well the danger of using words.
The Gap —The Bridge
Essay on institutional dichotomy

WORK

Darwin, Charles Robert   1809–82
English naturalist

I forget whether I ever told you what the object of my present work is, — it is to view all facts that I can master (eheu, eheu, how ignorant I find I am) in Natural History (as on geographical distribution, palaeontology, classifi-cation, hybridism, domestic animals and plants, &c., &c., &c.) to see how far they favor or are opposed to the no-tion that wild species are mutable or immutable: I mean with my utmost power to give all arguments and facts on both sides.
In Francis Darwin (ed.)
The Life and Letters of Charles Darwin (Volume 1)
Letter to Fox, March 27, 1855 (p. 409)
D. Appleton & Company. New York, New York, USA. 1896

Davy, Sir Humphry   1778–1829
English chemist

The most important part of the history of a man of sci-ence is necessarily recorded in his work.
The Collected Works of Sir Humphry Davy (Volume 1)
Memories of the Life of Sir Humphry Davy
Chapter I (p. 1)
London, England. 1839–1840

Editor

No man is truly equal to his work until he is superior to it.
Editor's Outlook
Journal of Chemical Education, Volume 10, Number 2, February 1933 (p. 66)

Hardy, G. H. (Godfrey Harold)   1877–1947
English pure mathematician

Good work is not done by "humble" men.
A Mathematician's Apology

Chapter 2 (p. 66)
Cambridge University Press. Cambridge, England. 1967

Littlewood, John E.   1885–1977
British mathematician

Most of the best work starts in hopeless muddle and floundering, sustained on the "smell" that something is there.
In Béla Bollabás (ed.)
Littlewood's Miscellany
Academic Life (p. 144)
Cambridge University Press. New York, New York, USA. 1986

Lowell, Percival   1855–1916
American astronomer

Gauge your work by its truth to nature, not by the plau-dits it receives from man. In the end the truth will prevail and though you may never live to see it, your work will be recognized after you are gone.
In William Graves Hoyt
Lowell and Mars
Chapter 15 (p. 300)
University of Arizona Press. Tucson, Arizona, USA. 1976

WORLD

Agassiz, Jean Louis Rodolphe   1807–73
Swiss-born American naturalist, geologist, and teacher

The world is the geologists great puzzle box.
In Loren Eiseley
The Firmament of Time
Chapter II (p. 31)
Athenaeum. New York, New York, USA. 1960

Bennett, Arnold   1867–1931
English novelist and playwright

Well, my deliberate opinion is — it's a jolly strange world.
In A.E. Trueman
This Strange World
Chapter I (p. 15)
The Scientific Book Club. London, England

Bridgman, F. W.
No biographical data available

The world is not a world of reason, understandable by the intellect of man.... It is probable that new methods of education will have to be painfully developed and applied to very young children in order to inculcate the instinctive and successful use of habits of thought so contrary to those which have been naturally acquired.
Harper's Magazine, March 1919

Bronk, William   1918–99
American poet and author

Whether what we sense of this world is the what of this world only, or the what of which of several possible worlds — which what?
The World, the Worldless
Metonymy as an Approach to a Real World
New Directions. New York, New York, USA. 1964

Bruno, Giordano 1548–1600
Italian philosopher and pantheist

God is infinite.… He is glorified not in one, but in countless suns; not in a single earth, a single world, but in a thousand thousand, I say in an infinity of worlds.
On the Infinite Universe and Worlds

Davies, Sir John 1569–1626
English poet

Behold the world how it is whirled round,
And for it is so whirl'd, is named so;
In whose large volume many rules are found
Of this new Art, which it doth fairly show:
For your quick eyes in wandering too and fro
From East to West, on no one thing can glance,
But if you make it well, it seemes to daunce.
Orchestra
Stanza 34
1596

Emerson, Ralph Waldo 1803–82
American lecturer, poet, and essayist

The world is an immense picture-book of every passage in human life.
Letters and Social Aims
Poetry and Imagination (pp. 8–9)
James R. Osgood & Company. Boston, Massachusetts, USA. 1876

Hazlitt, William Carew 1834–1913
English bibliographer

If the world were good for nothing else, it is a fine subject for speculation.
In W. Carew Hazlitt (ed.)
The Round Table; Northcotes Conversations; Characteristics
Characteristics, CCCII (p. 505)
George Bell & Sons. London, England. 1884

Lec, Stanislaw 1909–66
Polish poet and aphorist

Who created the world? So far only God admits to it.
Translated by Jacek Galazka
More Unkempt Thoughts (p. 52)
Funk & Wagnalls. New York, New York, USA. 1968

How did they get a permit to create the world?
Translated by Jacek Galazka
More Unkempt Thoughts (p. 57)
Funk & Wagnalls. New York, New York, USA. 1968

Mach, Ernst 1838–1916
Austrian physicist and philosopher

Properly speaking the world is not composed of "things" as its elements, but colors, tones, pressures, spaces, times, in short what we ordinarily call individual sensations.
The Science of Mechanics (5th edition)
Chapter IV, Part IV, Section 2 (p. 580)
Open Court. La Salle, Illinois, USA. 1942

Mumford, David 1937–
English-born mathematician

The world is a very complicated place, as babies know.
International Congress of Mathematics 2002
Beijing
August 21, 2002

Pascal, Blaise 1623–62
French mathematician and physicist

The whole visible world is only an imperceptible atom in the ample bosom of nature. No idea approaches it. We may enlarge our conceptions beyond all imaginable space; we only produce atoms in comparison with the reality of things.
In *Great Books of the Western World* (Volume 33)
Pensées
Section II, 72
Encyclopædia Britannica, Inc. Chicago, Illinois, USA. 1952

Regnault, Nöel 1702–62
Jesuit mathematician

Till you have discovered to me the Mysteries of the Loadstone, I shall be no more at Quiet than a Loadstone itself, which is not in its natural Situation, and which is seeking out the Poles of the Earth.
Philosophical Conversations (Volume 1)
Conversation XV (p. 196)
Printed for W. Innys, C. Davis & N. Prevost. London, England. 1731

We have not…Eyes piercing enough to penetrate so far as the Surface of the World; we don't see the external Figure of it: But if we judge it by the common Persuasion, and by what is offered to our Senses, when the Weather is serene, and the Heavens sparkle with Stars, the World is round: It is a Sphere.
Philosophical Conversations (Volume 1)
Conversation XIII (p. 158)
Printed for W. Innys, C. Davis & N. Prevost. London, England. 1731

Seneca (Lucius Annaeus Seneca) 4 BCE–65 AD
Roman playwright

The world is a poor affair if it does not contain matter for investigation for the whole world in every age. Nature does not reveal all her secrets at once. We imagine we are initiated in her mysteries: we are as yet, but hanging around her outer courts.
De Aurmentis Scientiarum
De Cometis

Shelley, Percy Bysshe 1792–1822
English poet

Worlds on worlds are rolling ever
From creation to decay
Like the bubbles on a river
Sparkling, bursting, borne away.
The Complete Poetical Works of Percy Bysshe Shelley
Worlds on Worlds
Houghton Mifflin Company. Boston, Massachusetts, USA. 1901

Stevenson, Robert Louis 1850–94
Scottish essayist and poet

The world is so full of a number of things,
I'm sure we should all be as happy as kings.
A Child's Garden of Verses
Happy Thought
Delacorte Press. New York, New York, USA. 1985

Tennyson, Alfred (Lord) 1809–92
English poet

Come, my friends,
'T s not too late to seek a newer world.
…
To sail beyond the sunset, and the baths
Of all the western stars.
Alfred Tennyson's Poetical Works
Ulysses, l. 56–57, 60–61
Oxford University Press, Inc. London, England. 1953

Thoreau, Henry David 1817–62
American essayist, poet, and practical philosopher

This curious world which we inhabit is more wonderful
than it is convenient, more beautiful than it is useful; it is
more to be admired than to be used.
Commencement address
Harvard University, 1837

Toulmin, Stephen 1922–
English philosopher
Goodfield, June
Science writer, screenwriter, and historian

The picture of the natural world we all take for granted today has one remarkable feature, which cannot be
ignored in any study of the ancestry of science: it is a
historical picture. Not content with achieving intellectual
command over the world of their own times, men have
been anxious to go further, and discover how the present
state of things came to be as it is. Having mapped the
existing topography of the heavens and grasped the principles now governing the world of matter, they have also
reached back into the darkness of past time, to a period
which earlier generations would have found inconceivably remote.
The Discovery of Time
Introduction (p. 17)
Harper & Row, Publishers. New York, New York,
USA. 1965

Twain, Mark (Samuel Langhorne Clemens) 1835–1910
American writer and humorist

It takes a long time to prepare a world for man, and
such a thing is not done in a day. Some of the great scientists, carefully ciphering the evidence furnished by
geology, have arrived at the conviction that our world
is prodigiously old, and they may be right, but Lord
Kelvin is not of their opinion. He takes the cautious,
conservative view, in order to be on the safe side, and
feels sure it is not so old as they think. As Lord Kelvin
is the highest authority in science now living, I think
we must yield to him and accept his view. He does not
concede that the world is more than a hundred million
years old.
In Bernard Devoto (ed.)
Letters from the Earth
The Damned Human Race
Chapter I (pp. 211–212)
Harper & Row, Publishers. New York, New York, USA. 1959

Weyl, Hermann 1885–1955
German mathematician

The world does not happen, it simply is.
Symmetry
Bilateral Symmetry (p. 5)
Princeton University Press. Princeton, New Jersey, USA. 1960

Wordsworth, William 1770–1850
English poet

…worlds unthought of till the searching mind
Of Science laid them open to mankind.
The Complete Poetical Works of William Wordsworth
To the Moon, Rydal, l. 40
Crowell. New York, New York, USA. 1888

WRITING

Cohen, I. Bernard 1914–2003
American physicist and science historian

The writings of all great men stand as a perpetual
challenge to each succeeding generation which attempts
to make an interpretation suitable to its own age.
Franklin and Newton
Chapter One (p. 3)
Harvard University Press. Cambridge, Massachusetts, USA. 1966

Einstein, Albert 1879–1955
German-born physicist

Your exposition is of matchless clarity and perspicuity.
You did not dodge any problems but took the bull by
the horns, said all that is essential, and omitted all that
is inessential.
The Collected Papers of Albert Einstein (Volume 8)
Letter 297. Letter to Moritz Schlick, 6 February 1917 (p. 284)

Emerson, William   1701–82
English mathematician

I am very sensible how difficult a thing it is to write well upon the science of Astronomy; by reason the subject is so comprehensive, and consists of so many parts, and is connected with so many other sciences, which it requires the perfect knowledge of; and besides, is a work of so much time, that a man had need have the life of Mathusalem, to go thro' the whole of it.
A System of Astronomy: Containing the Investigation and Demonstration of the Elements of that Science
The Preface (p. iii)
Printed for J. Nourse. London, England. 1769

Kepler, Johannes   1571–1630
German astronomer

…prolixity of phrases has it own obscurity, no less than terse brevity. The latter evades the mind's eye while the former distracts it; the one lacks the light while the other overwhelms with superfluous glitter; the latter does not arouse the sight while the former quite dazzles it.
New Astronomy
Author's introduction (p. 47)
At The University Press. Cambridge, England. 1992

Locke, John   1632–1704
English philosopher and political theorist

I have put into thy hands what has been the diversion of some of my idle and heavy hours. If it has the good luck to prove so of any of thine, and thou hast but half so much pleasure in reading as I had in writing it, thou wilt as little think thy money, as I do my pains, ill bestowed.
In *Great Books of the Western World* (Volume 35)
An Essay Concerning Human Understanding
Epistle to the Reader (p. 87)
Encyclopædia Britannica, Inc. Chicago, Illinois, USA. 1952

Lowell, Percival   1855–1916
American astronomer

I believe that all writing should be a collection of precious stones of truth which is beauty. Only the arrangement differs with the character of the book. You string them into a necklace for the world at large, pigeon hole them in drawers for the scientist. In the necklace you have the cutting of your thought, i.e., the expressing of it and the arrangement of the thoughts among themselves.
In Ferris Greenslet
The Lowells and Their Seven Worlds
Book VII, Chapter I (p. 355)
Houghton Mifflin Company. Boston, Massachusetts, USA. 1946

Moog, Florence   1915–87
American biologist

Good writing, after all, is just clear thinking. Anyone who can think well enough to make advances in any learned field ought to be able to write about his work. I am, of course, aware than many research papers submitted to scientific journals are, from a literary standpoint, putrid; but usually such essays are scientifically not very fragrant either.
Can Scientists Write for the General Public
Science, Volume 119, Number 3095, 23 April, 1954 (p. 567)

Oersted, Hans Christian   1777–1851
Danish physicist and chemist

So much is certain: that nothing is better adapted to form a mind which is capable of great development, than living and participating in great scientific revolutions. I would therefore counsel all those whom the period they live in has not naturally presented with this advantage, to procure it artificially for themselves, by reading the writings of those periods in which the sciences have suffered great changes. To pursue the writings of the most opposite systems, and to extract their hidden truth, to answer questions raised by these opposite systems, to transfer the chief theories of the one system into the other, is an exercise which cannot be sufficiently recommended to the student. He would certainly be rewarded for this labour, by becoming as independent as possible of the narrow opinions of his age.
The Soul in Nature with Supplementary Contributions
Observations on the History of Chemistry: A Lecture 1805–1807 (p. 322)
Henry G. Bohn. London, England. 1852

Rossi, Hugo   1935–
American mathematician

It is extremely hard for mathematicians to do expository writing. It is not in our nature. In fact, the very nature of mathematical meaning and grammar militates against it. However, this puts us at a distinct disadvantage relative to other sciences.… Good exposition should be valued, not only for the success in communication but also as evidence of real mathematical insight. It is no accident that among our greatest mathematicians are our greatest teachers and expositors.
From the Editor
Notices of the American Mathematical Society, Volume 42, Number 1, January 1995 (p. 4)

von Braun, Wernher   1912–77
German-born rocket scientist

When a good scientific paper earns a student as much glory as we shower upon the halfback who scored the winning touchdown, we shall have restored the balance that is largely missing from our schools.
Text of Address by von Braun Before the Publishers' Group Meeting Here
New York Times, 29 April 1960, L 20, column 5

WRONG

Kingsley, Charles   1819–75
English clergyman and author

You must not say that this cannot be, or that that is contrary to nature. You do not know what Nature is, or what she can do; and nobody knows, not even Sir Roderick Murchison, or Professor Owen, or Professor Sedgwick, or Professor Huxley, or Mr. Darwin, or Professor Faraday, or Mr. Grove.... They are very wise men; and you must listen respectfully to all they say: but even if they should say, which I am sure they never would, "That cannot exist. That is contrary to nature, you must wait a little and see; for perhaps even they may be wrong."

The Water-Babies
Chapter II (p. 58)
Dodd, Mead & Company. New York, New York, USA. 1910

X

X-RAY

Author undetermined

The Roentgen Rays, The Roentgen Rays
What is this craze,
The town's ablaze,
With the new phase
of X-rays ways
I'm full of daze,
Shock and amaze,
For nowadays,
I hear they'll gaze,
Thro' cloak and gown — and even stays,
These naughty, naughty Roentgen Rays.
In John G. Taylor
The New Physics
Chapter 2 (p. 46)
Basic Books, Inc. New York, New York, USA. 1972

Bacon, Roger   1214–92
English philosopher, scientist, and friar

No substance is so dense as altogether to prevent rays from passing. Matter is common to all things, and thus there is no substance on which the action involved in the passage of a ray may not produce a change. Thus it is that rays of heat and sound penetrate through the walls of a vessel of gold or brass. It is said by Boethius that a lynx's eye will pierce through thick walls. In this case the wall would be permeable to visual rays. In any case there are many dense bodies which altogether interfere with the visual and other senses of man, so that rays cannot pass with such energy as to produce an effect on human sense, and yet nevertheless rays do really pass, though without our being aware of it.
In Victor Robinson
The Story of Medicine
Chapter VII (pp. 208–209)
The New York Home Library. New York, New York, USA. 1943

Jauncey, G. E. M.   1888–1947
Australian physicist

O Roentgen, then the news is true
And not a trick of idle rumor
That bids us each beware of you
And of your grim and graveyard humor.
Scientific American, February 22, 1896

Lewis, Edwin Herbert   1866–1938
American rhetorician, novelist, and poet

…left alone with the X-ray man, Marvin plied him with questions. He so fascinated the radiographer that presently he was rewarded with a mystery even greater than that of the subtle unseen light. He was taken into a dark closet and permitted to peer into a small instrument containing salts of radium.

He saw a flight of stars, a sheaf of rays, a faint fierce sparkling! The heavy metallic radium atom was exploding! It was bombarding a small black screen with cannon flashes!

Instantly the boy inquired why somebody did not capture the power of that explosion and set it to work. He was told that any such achievement was impossible. The show was not affected by heat or cold, and would continue for a thousand years or more till the radium was all used up.

What were those flashes? How could he learn more about them? He must wait till he had enough physics to follow the writings of a man named Rutherford.
White Lightning
Chapter 2 (p. 8)
Covici-McGee. Chicago, Illinois, USA. 1923

Polanyi, Michael   1891–1976
Hungarian-born English scientist, philosopher, and social scientist

One of the greatest and most surprising discoveries of our own age, that of the diffraction of X-rays by crystals (in 1912) was made by a mathematician, Max von Laue, by the sheer power of believing more concretely than anyone else in the accepted theory of crystals and X-rays.
Personal Knowledge
Chapter 9, Section 5 (p. 277)
Harper & Row, Publishers. New York, New York, USA. 1962

Röntgen, Wilhelm Conrad   1845–1923
German physicist

For brevity's sake I shall use the expression "rays," and to distinguish them from others of this name I shall call them "x-rays."
In Otto Glasser
Dr. W.C. Röntgen
Chapter IV (p. 42)
Charles C. Thomas, Publisher. Springfield, Illinois, USA. 1945

Russell, Bertrand Arthur William   1872–1970
English philosopher, logician, and social reformer

Everybody knows something about X-rays, because of their use in medicine. Everybody knows that they can take a photograph of the skeleton of a living person, and show the exact position of a bullet lodged in the brain. But not everybody knows why this is so. The reason is that the capacity of ordinary matter for stopping the rays varies approximately as the fourth power of the atomic number of the elements concerned.…
The ABC of Atoms (p. 97)
E.P. Dutton & Company. New York, New York, USA. 1923

Russell, L. K.
No biographical data available

She is so tall, so slender, and her bones —
Those frail phosphates, those carbonates of lime —
Are well produced by cathode rays sublime,
By oscillations, amperes and by ohms.
Her dorsal vertebrae are not concealed
By epidermis, but are well revealed.
Line on an X-Ray Portrait of a Lady
Life, March 12, 1896

Shakespeare, William 1564–1616
English poet, playwright, and actor

Come, come, and sit you down, you shall not budge;
You go not till I set you up a glass
Where you may see the innermost part of you.
In *Great Books of the Western World* (Volume 27)
The Plays and Sonnets of William Shakespeare (Volume 2)
Hamlet, Prince of Denmark
Act III, Scene iv, l. 18–20
Encyclopædia Britannica, Inc. Chicago, Illinois, USA. 1952

Y

YELLOW FEVER

Theiler, Max   1899–1972
South African-born American microbiologist

By the intelligent application of antimosquito measures combined with vaccination, public-health officials have now the means available to render what was once a prevalent epidemic disease to one which is now a comparatively rare infection of man.
Nobel Lectures, Physiology or Medicine 1942–1962
Nobel lecture for award received in 1951
The Development of Vaccines Against Yellow Fever (p. 359)
Elsevier Publishing Company. Amsterdam, Netherlands. 1964

Z

ZEEMAN EFFECT

Zeeman, Pieter   1865–1943
Dutch physicist

In consequence of my measurements of Kerr's magneto-optical phenomena, the thought occurred to me whether the period of the light emitted by a flame might be altered when the flame was acted upon by magnetic force. It has turned out that such an action really occurs.
The Effect of Magnetisation on the Nature of Light Emitted by a Substance
Nature, Volume 55, Number 1424, 11 February 1897 (p. 347)

ZERO

Bôcher, Maxime   1867–1918
American mathematician

…there is what may perhaps be called the method of optimism which leads us either willfully or instinctively to shut our eyes to the possibility of evil. Thus the optimist who treats a problem in algebra or analytic geometry will say, if he stops to reflect on what he is doing: "I know that I have no right to divide by zero; but there are so many other values which the expression by which I am dividing might have that I will assume that the Evil One has not thrown a zero in my denominator this time."
The Fundamental Conceptions and Methods in Mathematics
Bulletin of the American Mathematical Society, 2nd Series, Volume 11, 1904 (pp. 134–135)

Dunham, William
American mathematician

Dividing by zero is the closest thing there is to arithmetic blasphemy.
The Mathematical Universe: An Alphabetical Journey Through the Great Proofs, Problems, and Personalities
Quotient (p. 203)
John Wiley & Sons, Inc. New York, New York, USA. 1994

Hugo, Victor   1802–85
French author, lyric poet, and dramatist

One microscopic glittering point; then another; and another, and still another; they are scarcely perceptible, yet they are enormous. This light is a focus; this focus, a star; this star, a sun; this sun, a universe; this universe, nothing. Every number is zero in the presence of the infinite.
Translated by Isabel F. Hapgood
The Toilers of the Sea
Part II, Book Second, Chapter V (p. 370)
The Heritage Press. New York, New York, USA. 1961

Oken, Lorenz   1779–1851
German naturalist

The whole science of mathematics depends upon zero. Zero alone determines the value in mathematics.

Zero is in itself nothing. Mathematics is based upon nothing, and, consequently, arises out of nothing.
Elements of Physiophilosophy
Part I (p. 5)
The Ray Society. London, England. 1847

Zamyatin, Yevgeny   1884–1937
Russian novelist, playwright, and satirist

The circles are at times golden, sometimes they are bloody, but all have 360 degrees. They go from 0 degrees to 10 degrees, 20 degrees, 200 degrees, 360 degrees — and then again 0 degrees. Yes we have returned to zero. But for a mathematically working mind it is obvious that this zero is different.
Translated by Gregory Zilboorg
We
Record Twenty (p. 110)
E.P. Dutton & Company. New York, New York, USA. 1952

ZETA

Conrey, B.
No biographical data available

It's a whole beautiful subject and the Riemann zeta function is just the first one of these, but it's just the tip of the iceberg. They are just the most amazing objects, these L-functions — the fact that they exist, and have these incredible properties are tied up with all these arithmetical things — and it's just a beautiful subject. Discovering these things is like discovering a gemstone or something. You're amazed that this thing exists, has these properties and can do this.
In K. Sabbagh
The Riemann Hypothesis: The Greatest Unsolved Problem in Mathematics
Chapter 12 (p. 196)
Farrar, Straus & Giroux. New York, New York, USA. 2002

Gutzwiller, M. C.
Swiss-born American physicist

The zeta function is probably the most challenging and mysterious object of modern mathematics, in spite of its utter simplicity.… The main interest comes from trying to improve the Prime Number Theorem, i.e., getting better estimates for the distribution of the prime numbers. The secret to the success is assumed to lie in proving a conjecture which Riemann stated in 1859 without much fare, and whose proof has since then become the single most desirable achievement for a mathematician.
Chaos in Classical and Quantum Mechanics

Chapter 17.9 (p. 308)
Springer-Verlag. New York, New York, USA. 1990

Sabbagh, K.
Writer and television producer

In 1859, a German mathematician called Bernhard Riemann, a "timid diffident soul with a horror of attracting attention to himself," published a paper that drew more attention to him than to almost any other mathematician in the 19th century. In it he made an important statement: the non-trivial zeros of the Riemann zeta function all have real part equal to 1/2. That is the Riemann Hypothesis: 15 words encapsulating a mystery at the heart of our number system.
In Ernst Peter Fischer
Beautiful Mathematics
Prospect, January 2002

ZETA FUNCTION

Bombieri, Enrico   1940–
Italian mathematician

I am firmly convinced that the most important unsolved problem in mathematics today is the truth or falsity of a conjecture about the zeros of the zeta function, which was first made by Riemann himself.... Even a single exception to Riemann's conjecture would have enormously strange consequences for the distribution of prime numbers.... If the Riemann hypothesis turns out to be false, there will be huge oscillations in the distribution of primes. In an orchestra, that would be like one loud instrument that drowns out the others — an aesthetically distasteful situation.
Prime Territory: Exploring the Infinite Landscape at the Base of the Number System
The Sciences, Sept/Oct 1992

Borwein, J.   1951–
Scottish mathematician
Bradley, D.
No biographical data available

It is intriguing that any of the various new expansions and associated observations relevant to the critical zeros arise from the field of quantum theory, feeding back, as it were, into the study of the Riemann zeta function. But the feedback of which we speak can move in the other direction, as techniques attendant on the Riemann zeta function apply to quantum studies.
Computational Strategies for the Riemann Zeta Function
Journal of Computational and Applied Mathematics, Volume 121, 2000

Sarnak, P.   1953–
South African-born American mathematician

[It has been] said that the zeros [of the Riemann zeta function] weren't real, nobody measured them. They are as real as anything you will measure in a laboratory — this has to be the way we look at the world.
1999 Mathematical Science Research Institute lecture
Random Matrix Theory and Zeroes of Zeta Functions – A Survey

ZOO

Diolé, Philippe   1908–77
French biologist

The world of animals in captivity offers us at once a prophetic glimpse and a caricature of the world in which modern man lives out his life. The animal suffers psychologically and his suffering is not unlike that of man himself, since its world is characterized by deterioration of its environment and by its own degradation. The causes are the same in both cases: the increase in the number of individual animals in a zoo — baboons, for example — suffer from, and are deformed by, lack of sufficient space for them to lead a harmonious social existence. When captivity has done its work and an animal has become truly dangerous, it then becomes necessary to isolate it in a cage of its own.
Translated by J.F. Bernard
The Errant Ark: Man's Relationship with Animals
The Cruelty of Paradise
Putnam. New York, New York, USA. 1974

Hediger, Heini   1908–92
Swiss zoologist

One of the most frequent misconceptions which is constantly met in the zoo is the business of regarding the animals as prisoners. This is as false and old-fashioned as if in these days everybody still thought that radio and television sets contained little men who talked, sang and danced inside the sets.
Translated by Gwynne Vevers and Winwood Reade
Man and Animal in the Zoo
Chapter 3 (p. 99)
Delacorte Press. New York, New York, USA. 1969

Queneau, Raymond   1903–76
French poet, novelist, and publisher

In the dog days while I was in a bird cage at feeding time I noticed a young puppy with a neck like a giraffe who, like the toad, ugly and venomous, wore yet a precious beaver upon his head. This queer fish obviously had a bee in his bonnet and was quite bats; he started yak-yakking at a wolf in sheep's clothing claiming that he was treading on his dogs with his beetle-crushers, but the sucker got a flea in his ear; that foxed him, and quiet as a mouse he ran like a hare for a perch.

I saw him again later in front of the Zoo with a young buck who was telling him to bear in mind a certain drill about his fevers.

Exercises in Style
Zoological (p. 179)
New Direction Publishing Corporation. New York, New York, USA.
1981

Wynne, Annette
American poet

Excuse us, Animals in the Zoo,
I'm sure we're very rude to you;
Into your private house we stare
And never ask you if you care;
And never ask you if you mind.
Perhaps we really are not kind:
I think it must be hard to stay
And have folks looking in all day,
I wouldn't like my house that way.
All Through the Year
Excuse Us, Animals in the Zoo
Frederick A. Stokes. New York, New York, USA. 1932

ZOOLOGIST

Agassiz, Jean Louis Rodolphe   1807–73
Swiss-born American naturalist, geologist, and teacher

Lay aside all conceit. Learn to read the book of nature
for yourself. Those who have succeeded best have followed for years some slim thread which has once in a
while broadened out and disclosed some treasure worth
a life-long search.
In David Stair Jordan
Popular Science Monthly, Volume 40, 1891

Wheeler, William Morton   1865–1937
American entomologist

…I shall strenuously endeavor to be modern, I can only
beg you, if I fail to come within hailing distance of the
advance guard of present-day zoologists, to remember
that the range of adaptability in all organisms, even zoologists, is very limited.
Essays in Physiological Biology
Essay I (p. 4)
Harvard University Press. Cambridge, Massachusetts, USA. 1939

ZOOLOGY

Bierce, Ambrose   1842–1914
American newspaperman, wit, and satirist

HIPPOGRIFF, n. An animal (now extinct) which was
half horse and half griffin. The griffin was a compound
creature, half lion and half eagle. The hippogriff was,
therefore, only one quarter eagle, which is $2.50 in gold.
Zoology is full of surprises.
The Devil's Dictionary
Doubleday & Company, Inc. Garden City, New York, USA. 1967

Bock, W. J.
No biographical data available

Communication — information exchange — among
zoologists is the core of zoological nomenclature;
everything else pales in the light of the importance of
communication.
History and Nomenclature of Avian Family Group Names
Bulletin of the American Museum of Natural History, Volume 221, 1994
(p. 8)

Doyle, Sir Arthur Conan   1859–1930
Scottish writer

Living, as I do, in an educated and scientific atmosphere,
I could not have conceived that the first principles of
zoology were so little known. Is it possible that you do
not know the elementary fact in comparative anatomy,
that the wing of a bird is really the forearm, while the
wing of a bat consists of three elongated fingers with
membranes between?
The Lost World
Chapter IV (p. 55)
The Colonial Press. Clinton, Massachusetts, USA. 1959

Elton, Charles S.   1900–91
English biologist

…the discoveries of Darwin, himself a magnificent field
naturalist, had the remarkable effect of sending the whole
zoological world flocking indoors, where they remained
hard at work for fifty years or more, and whence they are
now beginning to put forth cautious heads again into the
open air.
Animal Ecology
Chapter I (p. 3)
Sidgwick & Jackson, Ltd. London, England. 1927

Feynman, Richard P.   1918–88
American theoretical physicist

I began to read the paper. It kept talking about extensors
and flexors, the gastrocnemius muscle, and so on. This
and that muscle were named, but I had not the foggiest
idea of where they were located in relation to the nerves
or to the cat. So I went to the librarian in the zoology
section and asked her if she could find me a map of the
cat.

"A map of the cat, sir?" she asked horrified. "You mean
a zoological chart!"
Surely You're Joking, Mr. Feynman!: Adventures of a Curious Character
A Map of a Cat? (p. 72)
W.W. Norton & Company, Inc. New York, New York, USA.1985

Hugo, Victor   1802–85
French author, lyric poet, and dramatist

Zoology is as limitless as cosmography.
Translated by Isabel F. Hapgood
The Toilers of the Sea

Part II, Book Third, Chapter III (p. 418)
The Heritage Press. New York, New York, USA. 1961

Kovalevskii, V. O.
Russian paleontologist

And so, the task of modern zoology consists in this; it should acquaint us with the entire variety of animal forms which populate our world, not in terms of a disorganized multitude from which this or that form happens to catch our attention, but as a structured whole, in which each form occupies a designated place, so one can instantly note and critically analyze all the particularities of each separate member; it should show us the inner structure of these groups and of their individual members, and in what relationship they stand to members of other groups; it should present the history of each member, beginning with its [first] appearance...it should open the ancient tombs of the earth and demonstrate to us the endless series of ancestors and relatives which proceed those animals which we now see.
In William Coleman and Camille Limoges (eds.)
Studies in History of Biology (Volume 2)
Kovalevskii and Paleontology (pp. 112–113)
The Johns Hopkins University Press. Baltimore, Maryland, USA. 1977–84

Polanyi, Michael   1891–1976
Hungarian-born English scientist, philosopher, and social scientist

The existence of animals was not discovered by zoologists, nor that of plants by botanists, and the scientific value of zoology and botany is but an extension of man's pre-scientific interests in animals and plants.
Personal Knowledge
Chapter 6, Section 2 (p. 139)
Harper & Row, Publishers. New York, New York, USA. 1962

Index

Printed in the United States of America